ANNUAL REVIEW OF MICROBIOLOGY

Production Editor: CLEO X. RAY
Bibliographic Quality Control: MARY A. GLASS
Color Graphics Coordinator: EMÉ O. AKPABIO
Electronic Content Coordinator: SUZANNE K. MOSES
Subject Indexer: KYRA KITTS

Annual Review of Microbiology

VOLUME 56, 2002

L. NICHOLAS ORNSTON, *Editor*
Yale University

ALBERT BALOWS, *Associate Editor*
Centers for Disease Control, Atlanta
Emory University Medical School, Emeritus

SUSAN GOTTESMAN, *Associate Editor*
National Cancer Institute, Bethesda

www.annualreviews.org science@annualreviews.org 650-493-4400

ANNUAL REVIEWS
4139 El Camino Way • P.O. BOX 10139 • Palo Alto, California 94303-0139

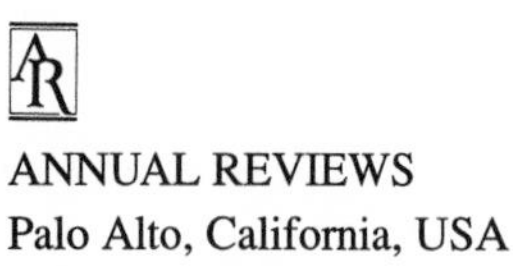

ANNUAL REVIEWS
Palo Alto, California, USA

International Standard Serial Number: 0066-4227
International Standard Book Number: 0-8243-1156-6
Library of Congress Catalog Card Number: 49-432

∞ The paper used in this publication meets the minimum requirements of American National Standards for Information Sciences—Permanence of Paper for Printed Library Materials, ANSI Z39.48-1992.

TYPESET BY TECHBOOKS, FAIRFAX, VA
PRINTED AND BOUND IN THE UNITED STATES OF AMERICA

PREFACE

Who started the rumor that it is difficult to grow spirochetes? As surely as the swallows return to Capistrano, ticks invade my New England garden in the spring and, shortly thereafter, I find myself growing *Borrelia burgdorferi*. Despite relentless ticking off, springtime can become Lyme time, and I count the sensitivity of *Borrelia* to doxycycline among my blessings. How long such blessings are to continue is uncertain, as the threat of antibiotic-resistant microorganisms moves closer to reality. All this illustrates that our perceptions of the smallest of creatures often are indirect, yet we neglect their activities at our peril. It is the charge of the Editorial Committee of the *Annual Review of Microbiology* to cast fresh light into the shadows of microbial activity and to elicit reports that illuminate microbial interactions in their complexity and their pressing significance. The torch is then passed to authors who do the heavy work in the production of the volume. Their load is lightened by the efforts of our Production Editor, Cleo X. Ray. A. C. Matin and John Taylor assisted the Editorial Committee with the selection of topics for this volume. We note with sadness the departure of Julian E. Davies and David G. Russell from the editorial board; their contributions are highly valued. With publication of this volume, we welcome Stephen J. Giovannoni and Karla Kirkegaard to the Editorial Committee; their advent heralds stimulating discussions to come.

Nick Ornston
Editor

Annual Review of Microbiology
Volume 56, 2002

CONTENTS

ERRATA

An online log of corrections to *Annual Review of Microbiology* chapters (if any, 1997 to the present) may be found at http://micro.annualreviews.org/

Related Articles

From the ***Annual Review of Biochemistry***, Volume 71 (2002)

Discoveries of Vitamin B_{12} and Selenium Enzymes, Thressa Campbell Stadtman

Error-Prone Repair DNA Polymerases in Prokaryotes and Eukaryotes, Myron F. Goodman

The Bacterial RecA Protein and the Recombinational DNA Repair of Stalled Replication Forks, Shelley L. Lusetti and Michael M. Cox

Biological Roles of Proteases in Parasitic Protozoa, Michael Klemba and Daniel E. Goldberg

DNA Replication in Eukaryotic Cells, Stephen P. Bell and Anindya Dutta

The La Protein, Sandra L. Wolin and Tommy Cedervall

Lipopolysaccharide Endotoxins, Christian R. H. Raetz and Chris Whitfield

Catalytic Strategies of the Hepatitis Delta Virus Ribozymes, I-hung Shih and Michael D. Been

From the ***Annual Review of Genetics***, Volume 36 (2002)

Chromosome Rearrangements and Transposable Elements, Wolf-Ekkehard Loennig and H. Saedler

Maintaining the Genome: DNA Damage and Replication Checkpoints, Kara Nyberg, Rhett J. Michelson, Charles W. Putnam, and Ted A. Weinert

DNA Topology-Mediated Control of Global Gene Expression in Escherichia coli, Craig J. Benham and G. Wesley Hatfield

Genetics of Cryptococcus neoformans, Christina Hull and Joseph Heitman

Genetics of Influenza Viruses, David A. Steinhauer and John J. Skehel

Genetics of Motility and Chemotaxis of a Fascinating Group of Bacteria: The Spirochetes, Nyles W. Charon and Stuart F. Goldstein

Origins of Spontaneous Mutations: Specificity and Directionality of Base-Substitution, Frameshift, and Sequence-Substitution Mutageneses, Hisaji Maki

From the ***Annual Review of Immunology***, Volume 20 (2002)

Prospects for Vaccine Protection Against HIV-1 Infection and AIDS, Norman L. Letvin, Dan H. Barouch, David C. Montefiori, and Amanda G. Fisher

Genetic Dissection of Immunity to Mycobacteria: The Human Model, Jean-Laurent Casanova and Laurent Abel

CpG Motifs in Bacterial DNA and Their Immune Effects, Arthur M. Krieg

From the ***Annual Review of Plant Biology***, Volume 53 (2002)

Revealing the Molecular Secrets of Marine Diatoms, Angela Falciatore and Chris Bowler

ANNUAL REVIEWS is a nonprofit scientific publisher established to promote the advancement of the sciences. Beginning in 1932 with the *Annual Review of Biochemistry*, the Company has pursued as its principal function the publication of high-quality, reasonably priced *Annual Review* volumes. The volumes are organized by Editors and Editorial Committees who invite qualified authors to contribute critical articles reviewing significant developments within each major discipline. The Editor-in-Chief invites those interested in serving as future Editorial Committee members to communicate directly with him. Annual Reviews is administered by a Board of Directors, whose members serve without compensation.

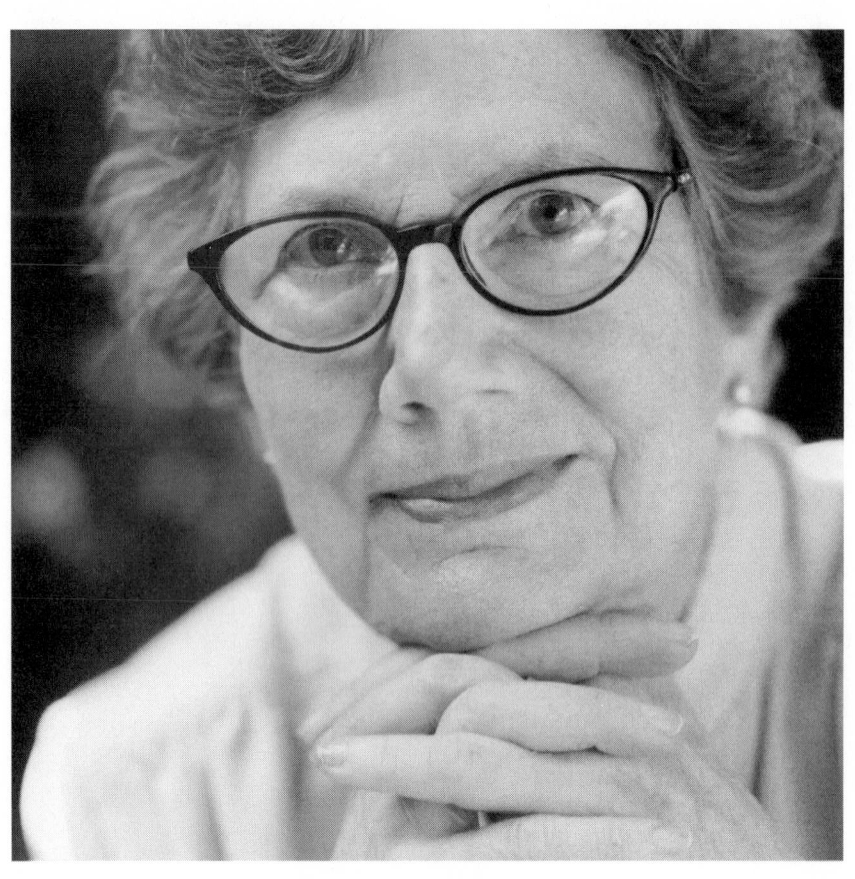

Evelyn Witkin

Annu. Rev. Microbiol. 2002. 56:1–15
doi: 10.1146/annurev.micro.56.012302.161130

First published online as a Review in Advance on April 2, 2002

CHANCES AND CHOICES: Cold Spring Harbor 1944–1955

Evelyn M. Witkin*
Waksman Institute, Rutgers, the State University of New Jersey, New Brunswick, New Jersey 08855-0759; e-mail: ewitkin@aol.com

Key Words *Escherichia coli*, *Microbial Genetics Bulletin*, ultraviolet mutagenesis, DNA repair, SOS response

■ **Abstract** The author describes the circumstances and events that led her to begin graduate work in genetics at Columbia University in 1941 and to spend the summer of 1944 and the years 1945–1955 at the Department of Genetics of the Carnegie Institution of Washington at Cold Spring Harbor. She then recalls incidents and recounts anecdotes meant to convey something of the atmosphere of that place during those memorable years.

CONTENTS

INTRODUCTION

Midway in my graduate studies at Columbia University, in the summer of 1944, I found myself transposed from the Zoology Department laboratories on

*Correspondence should be sent to: One Firestone Court, Princeton, NJ 08540.

Morningside Heights in Manhattan to those of the Carnegie Institution of Washington's Department of Genetics at Cold Spring Harbor, on the north shore of Long Island. In that place, at just that time, the revolution in biology was stirring, and for a young person in love with the mystery of the gene there could surely have been no better place to spend the next eleven years. I am reminded of a line from Wordsworth, who was thrilled in his youth by the glorious promise of the French Revolution and wrote "T'was bliss, in that dawn, to be alive, but to be young was very Heaven!"

I will begin by describing the series of unlikely happenings that led me to Cold Spring Harbor at that felicitous time. Then I'll free-associate about what it was like to be there during the 1940s and 1950s.

WHY GENETICS?

There were no scientists in my family, and I never met any working scientists while I was growing up. My notions of what a life in science might be like came almost entirely from books and to some extent from films like *Pasteur* and *Madame Curie*. A gift on my fifteenth birthday was a wonderful book titled *Outposts of Science*, by Bernard Jaffe. Each chapter described the leading edge of research in a different field, and I found it fascinating. At around the same time, I read Sinclair Lewis' *Arrowsmith* and Paul de Kruif's *Microbe Hunters*, books that made the struggle for truth as a scientist seem eminently romantic and rewarding. I must have noticed that the scientist/heroes in these books were all men, but I don't think it occurred to me that such a life might not be a realistic goal for a girl.

At Washington Irving High School, a public high school for girls in Manhattan, I developed a dual passion, which persists to this day, for biology and literature. My science teachers encouraged me to consider a career in science. This probably would not have happened had I attended a coeducational high school, where boys were usually assumed to be better at science and math than girls. I was strongly attracted to science, but I felt an equally compelling pull toward a literary life.

As a junior at New York University on Washington Square I had a hard time choosing a major subject. I decided on biology, reasoning that I could study literature, but not science, on my own, promising myself that some day I would devote myself to the Victorian poets I loved. I have kept that promise since I retired from active research in genetics in 1991.

Genetics began to take center stage in my thinking when a friend at Harvard introduced me to English translations of some of the writings of the Russian plant breeder Trofim Lysenko. Lysenko believed that the gene, as conceived by Mendel and Morgan, did not exist. He claimed that hereditary traits could be altered in desired directions by appropriate manipulation of the environment. His data (later shown to be largely fraudulent) looked convincing, and I was intrigued. I decided to remain at New York University after graduation to pursue graduate work in genetics, planning to put some of Lysenko's ideas to experimental test. But that was not to be.

THE ROCKY ROAD TO COLD SPRING HARBOR

Early in my senior year at NYU, in the fall of 1940, something happened that changed the course of my life. Some of us learned that our star football player, Leonard Bates, the only black member of the team, would not be allowed to accompany his teammates to the University of Missouri in October, where the first football game of the season was scheduled to take place. NYU was party to a "gentlemen's agreement" with many Southern colleges whereby our black athletes would be kept at home whenever an athletic event was to take place in the South. This capitulation to racism struck me and several of my friends as unacceptable, and we started circulating a petition headed "Bates Must Play" to protest. We obtained thousands of signatures within a few days, but Bates did not play. The same situation arose later when NYU's basketball and track teams traveled to compete in the South. In each case, a black athlete was benched. The protest gathered momentum as the year progressed, and by March 1941, it had become intolerable to the University administrators. They decided to end the protest by suspending its seven student leaders, including me, for three months.

As a result of the suspension, I could not graduate with my class in May and I lost the offer of a graduate assistantship in the NYU Biology Department, where I had intended to start graduate work in the fall. Over the summer, I made up the credits needed to graduate, received an October degree from NYU, and was accepted as a graduate student in Columbia University's Department of Zoology. And that has made all the difference.

By a couple of months into my graduate course in genetics at Columbia, I realized that Lysenko was either an ignoramus or a charlatan, or both, and that my own ignorance had led me to take his false claims seriously. By that time, I had read Hermann Muller's papers on the nature of the gene, and I had become terminally excited about its remarkable and mysterious properties. Reading Muller convinced me that a way to approach gene structure and function was through the study of mutations. I decided, with the approval of my advisor, Professor Theodosius Dobzhansky, that I would attempt to induce mutations with chemical agents in *Drosophila* as a dissertation problem.

Sometime in 1943, Professor Dobzhansky gave me a preprint of the classic paper by Luria & Delbruck published later that year (5). I was to read it and report on it in class. This paper established, in effect, that *Escherichia coli* had genes like other organisms. I was thrilled by the implication that bacteria were ideal material for the study of genes and mutations. My enthusiasm led Dobzhansky to make a startling proposal. He suggested that I do my dissertation research with bacteria, instead of with *Drosophila*, and that I spend the summer of 1944 at Cold Spring Harbor, where Luria and Delbruck would be working and where I could learn to handle microorganisms. My husband (I had been married a few months earlier) was scheduled to spend that summer in Texas doing war research for the Air Force, so I accepted without hesitation. Professor Dobzhansky was a close friend of the director of the laboratories at Cold Spring Harbor, Dr. Milislav Demerec, and my visit to Cold Spring Harbor that summer was easily arranged.

Everything good that ever happened to me professionally came of my association with Cold Spring Harbor. Here, indeed, was serendipity. Had I not been taken in by Lysenko's flawed and dishonest science I would probably not have chosen to go into genetics. Had I not been suspended from NYU for what could be called premature civil rights activism, I would not have gone to Columbia for graduate work and I would not have had Professor Dobzhansky as my sponsor. Had he assigned the article by Luria & Delbruck to another student, I would probably never have set foot in Cold Spring Harbor. [Perhaps I should admit here that I asked Professor Dobzhansky to be my sponsor, while under Lysenko's spell, not because he was a great geneticist, (which I did not yet know) but because he was Russian and would be able to read Lysenko's papers in the original! At a party for the Professor's fiftieth birthday, after I had safely acquired my PhD, I confessed this to him, and I never heard him laugh more heartily.]

BEGINNER'S LUCK

On my first day in Cold Spring Harbor, early in June of 1944, I reported to Dr. Demerec, trying to appear calm, but I was barely able to conceal my excitement. The fantasy born of *Arrowsmith* was about to come true. I would soon be doing research, starting to explore the deepest mystery in biology: the nature of the gene. To be allowed to do that, in such a place with such people, was a privilege beyond anything I could have dreamed. I knew that I would enter the lab every morning as if it were a cathedral.

Dr. Demerec led me to a table in a small room just outside his office. He handed me a culture of *E. coli* B, pointed to an ultraviolet (UV) lamp, and told me to induce mutations. I had never had a course in bacteriology, and I had no idea how to proceed. One of Dr. Demerec's assistants took me in hand, gave me some pointers on sterile technique, and showed me where the supplies were kept. I started a culture for use the next day and planned my first experiment. There were no published UV survival curves for *E. coli*, so I started by trying to determine an appropriate radiation dose range. In my first run, I guessed wrong and used doses that were far too high. Nothing grew on any of the irradiated plates, each of which I had seeded with about a million bacteria, except one, which contained four colonies the next day. I lowered the dose range by increasing the distance bewteen the UV lamp and the plates to be irradiated and repeated the experiment. This time, there was a reasonable range of survivals. I would soon be able to establish a reliable survival curve.

The four bacteria that had survived the first high-radiation exposure piqued my curiosity. There was much talk in the lab of mutants resistant to bacteriophage or to penicillin, and I wondered if I might have selected mutants resistant to UV radiation. I cultured the four survivors and compared their UV resistance to that of the parent strain, streaking them side by side on the same plates. Holding my breath as I opened the doors of the incubator on the next day, I tried to prolong the moment of suspense, as I was to do hundreds of times in the future. I finally looked at the plates for my first (and therefore perhaps sweetest) Eureka! experience. Sure enough, the four survivors were mutants a hundred times more resistant to UV radiation than the wild-type B strain. One of the four became strain B/r, and before long, the investigation of the genetic basis of radiation resistance became the topic of my dissertation research.

During my first week at Cold Spring Harbor I had stumbled on a new mutant that seemed to create something of a stir among the scientists there. Dr. Demerec reacted with considerable excitement when I showed him my results. Salvador Luria, Max Delbruck, Alexander Hollaender, and Barbara McClintock were genuinely interested in my radiation resistant mutant, and so was my advisor Theodosius Dobzhansky, who was there for the summer. Having tasted so early the incomparable joy of discovery, I soon became addicted to *E. coli* and to the exhilaration of digging for its secrets. I spent the next 47 years doing just that.

Why do I call it beginner's luck? Because in many later attempts, I never again succeeded in isolating UV-resistant mutants as the last survivors of a single exposure to a high dose of UV. A dose high enough to kill a large population of sensitive bacteria also eliminates the few relatively (but not absolutely) resistant mutants. It took some time to develop a procedure that allowed the quantitative isolation of the resistant mutants (10).

MORE ABOUT THE SUMMER OF 1944

From my first encounter with Cold Spring Harbor, I was captivated by the serene beauty of the setting and overwhelmed by the people I met. There were two neighboring institutions there: the Department of Genetics of the Carnegie Institution of Washington and the Long Island Biological Association, which was supported by local philanthropists. Dr. Demerec was the director of both. The laboratories were nestled along the shore of an inlet of Long Island Sound, opposite the old whaling village of Cold Spring Harbor. The year-round population at the laboratories was quite small, but summer brought an influx of transient investigators, including many famous scientists. That first summer I met Ernst Mayr, Alfred Mirsky, Bentley Glass, Curt Stern, Tracy Sonneborn, and Alexander Hollaender, among others, all concerned with understanding the nature of the gene, which had become the unifying theme of the laboratories under Demerec's directorship.

Although there was a pervasive sense that phage and bacterial genetics were initiating something important that was bound to revolutionize genetics, of all the older geneticists who had made reputations earlier doing "classical" work with *Drosophila* or plants, only Milislav Demerec had actually joined the early ranks of microbial geneticists, and by 1944 he was not only providing space and encouragement for Luria, Delbruck, and others pioneering in this new direction, but was himself already carrying on several projects with microorganisms. He was my immediate supervisor, and I owe much to his unfailing support and generosity. He went on, of course, to make many important contributions to bacterial genetics, including the first evidence that genes in a related metabolic pathway are often clustered on the *E. coli* chromosome map.

Two of the scientists I met that summer became close and enduring friends. One was Barbara McClintock, a year-round Carnegie staff member, who was already starting the work that led to her discovery of transposable elements and the unsuspected dynamism of the genome. Beginning that summer, and for years thereafter, I had the great good fortune to spend hours in her laboratory every week, studying her spotted maize kernals and listening, fascinated, as she deconstructed their amazing significance. The other was Salvador (Salva) Luria, who took an interest in my work and became my prime mentor. I learned much about rigorous thinking from his kind but unsparing criticism. I admired his brilliance, his integrity, his breadth of culture, and his wicked sense of humor.

Salva Luria was particularly helpful in the early stages of my work with the radiation resistant mutant: It was he who suggested that I look at the irradiated bacteria under the microsocope. When I did, the reason for the wild-type B strain's extreme UV sensitivity was immediately evident. The irradiated cells failed to divide. Instead, they grew for up to several hours to form long filaments, and then most of them died. The B/r mutation restored the ability of the irradiated bacteria to divide, after only a brief delay, and that accounted for the mutants' UV resistance.

I went back to Columbia in the fall of 1944, but the zoology department was poorly equipped for work with microorganisms, and it was a struggle to keep my experiments moving forward at a reasonable pace. Professor Dobzhansky, recognizing the problem, proposed that I return to Cold Spring Harbor and stay as long as necessary to finish my dissertation research. I completed my course requirements by the summer, and, as my husband had entered the Army by then, I gladly agreed to go back.

COLD SPRING HARBOR AGAIN

On August 6, 1945, when I returned to Cold Spring Harbor to continue work on the radiation resistance of B/r, I found the place transformed by the course in bacteriophage genetics that had been offered there for the first time. Max Delbruck, Salva Luria, Herman Kalckar, Rollin Hotchkiss, and the others comprising the initial "phage group" now dominated the summer population. High spirits and humor prevailed in this group, side by side with their scientific seriousness. In fact, with the war ending, the overall atmosphere at the laboratory was festive and sometimes almost euphoric, with sailing, swimming, and square dancing on the lawn, and the occasional hilarious practical joke, providing the background for much hard work and intense conversation. Throughout, one could not help but feel what Delbruck described as "the first principle" of the phage group: "openness . . . that you tell each other what you are doing and thinking, and that you don't care who has priority." The future crackled with promise.

An important feature of Cold Spring Harbor in the mid-1940s was the presence of a good many physicists who, primarily through Delbruck or Erwin Schrodinger's book *What is Life*?, had become interested in genetics and especially in phage genetics. The kind of talk that went on at the beach was heavily influenced by the likes of Leo Szilard, Aaron Novick, and Philip Morrison, who took the phage course in 1947. Gunther Stent and Seymour Benzer were among those who took the course in 1948. The synergistic effect of the diversity of scientific styles represented here at that time has become a matter of legend that I need not belabor. Needless to say, it was heady stuff for a young graduate student. Leo Szilard, for reasons not too hard to imagine, was interested in my mutant B/r, in which a single mutation increased resistance to radiation a hundredfold. More than once he telephoned, at three o'clock in the morning or thereabouts, to ask me a specific question about my experiments. He did not identify himself, nor did he apologize for waking me, but as soon as I picked up the phone, his question hit me without warning. He hung up just as abruptly as soon as I managed a sleepy and probably confused answer.

The impact of the war on the Cold Spring Harbor community had been significant. Several war-related research projects were conducted under contracts with government agencies, including Demerec's effort to develop strains of Penicillium yielding increased amounts of penicillin and Vernon Bryson's collaboration with Dr. E.V. Grace on the use of aerosols to deliver antibiotics in clinical use. Shortages of gas and rationing of food had resulted in contraction of the summer population and cancellation of the Cold Spring Harbor Symposium during three wartime summers from 1943 to 1945.

THE 1946 COLD SPRING HARBOR SYMPOSIUM

When the war ended, planning for the 1946 Symposium could begin. The topic was to be *Heredity and Variation in Microorganisms*, and it would be the first formal conference on this new direction in genetics. One of those charged with organizing the program was Salva Luria. He and his new wife Zella spent the whole year 1945–1946 at Cold Spring Harbor, on leave from the University of Indiana, where Salva was on the faculty and Zella was a graduate student in psychology. (She had been a student of my husband's, and when she left for Indiana in the fall of 1944 to start graduate work, we had suggested that she look up Tracy Sonneborn and Salva Luria—she did, in that order, and Salva quipped in his autobiography that had Sonneborn not already been married, she might have married him instead.)

Salva was aware of a paper published in 1938 by a woman named Mary Bunting. It dealt with color variation in *Serratia marcescens*. He considered it a first-rate piece of work and thought it was perhaps the only study prior to his paper with Delbruck that could be considered real bacterial genetics. Salva declared that Mary Bunting must speak at the 1946 Symposium. She had published nothing since that paper, and nobody seemed to know where she might be. Salva insisted that she be found and that wherever she was and whatever she might be doing, she had to be on the program. I remember him saying that "we simply can't hold the Symposium without her."

Mostly through his efforts, Bunting was located in New Haven, where her husband was on the faculty of the Yale School of Medicine. Bunting herself was busy with three infants, with another to come soon after, and since obtaining her degree had not been able to carry on active research. When Salva approached her about being on the program of the 1946 CSH Symposium, she protested that she could not possibly do it—she was terrified at the prospect. It was only Salva's persistence that gradually wore her down, and she did indeed speak at the meeting, with consequences for her future that can hardly be overestimated. The immediate result was an arrangement at Yale, made with Edward Tatum's help, for Bunting to have laboratory space and some technical assistance, allowing her to pick up some work on a flexible schedule, to attend seminars and to catch up with the fast-moving field, while her children were small.

The rest is on the record: Mary Bunting did indeed resume her work in bacterial genetics with distinction, working at Yale until 1955, then, after her husband's death, becoming Professor of Bacteriology and Dean of Douglass College at Rutgers, and in 1960 accepting the Presidency of Radcliffe College. At Radcliffe, she established the Mary Bunting Institute, which still continues to facilitate the return to professional work of women whose careers have been interrupted by family pressures. After her retirement from Radcliffe, she spent some time at Princeton University, advising women students during the transition from an all-male to a coeducational institution. Salva's attitude and actions in this story were absolutely typical of him—of his scientific and personal generosity, his simple assumption that men and women are equal human beings, his willingness to exert effort to do what he considered right.

The 1946 Symposium brought together an international group of scientists working on the genetics of various microorganisms including *Neurospora*, yeast, *Aspergillus*, bacteria, and viruses. It was here that Joshua Lederberg, then a medical student at Yale, announced that *E. coli* K-12 could undergo genetic recombination, a pivotal discovery for molecular genetics. I remember a group of participants walking to the beach and debating possible names for the new field represented at the Symposium. The consensus was to call it microbial genetics.

AN UNUSUAL ARRANGEMENT

Late in 1946, my husband was discharged from the Army and joined me in Cold Spring Harbor. He was entranced by the beauty of the place and was willing to commute two hours each way to his work in Brooklyn so that I could complete my research and write my thesis without interruption. As it turned out, we stayed on for nearly ten years. After I received my PhD degree from Columbia in 1947, I remained as Dr. Demerec's postdoctoral fellow and then, until 1955, as a member of the scientific staff.

Dr. Vannevar Bush was President of the Carnegie Institution of Washington during my time at Cold Spring Harbor. He visited the department of genetics frequently, got to know everyone, and was kept informed about each person's research. On one of his visits, early in 1949, I asked him if I might take a three-month leave after the birth of my first child, which was expected in March. His reaction astounded me. He made a brief, impassioned speech about the unfairness of the difficulties facing women who want to have children and careers in science and declared that "institutions have to change to make it work." He then asked me what I would need to make it work for me. Scarcely able to believe what I was hearing, I told him that, ideally, I would like to come back to work on a part-time basis for a number of years. "Done!" he said. And that's the way it was until I left Cold Spring Harbor in 1955.

It was possible in that situation to be fully productive while spending limited hours in the laboratory, especially with the help of a full-time assistant. Planning of experiments, analysis of data, and writing could be done at home. There were no committees, no graduate students, no teaching responsibilities (until the three-week summer course in bacterial genetics started in 1950), and no grant proposals to write. The department budget covered salaries, the costs of equipment, materials, and technical help for the staff members. Outside grants were not permitted. Time in the laboratory was time spent at the bench.

MICROBIAL GENETICS BULLETIN

One day early in 1949, physicist-turned-biologist Leo Szilard and microbiologist Bernard Davis knocked on my door. They had an idea. They believed that investigators in the new field of microbial genetics should have an informal newsletter—something like the *Drosophila Information Service* or the *Phage Information Service*—to promote communication and disseminate information. Reminding me that I would have time during my maternity leave, they asked if I would consider organizing and editing such a publication. After some hesitation, I agreed to do so.

We talked about applying for a grant to cover the expenses of typing, mimeographing, and mailing the newsletter, but as it turned out later this was not necessary. Milislav Demerec generously agreed to absorb these costs in the department of genetics budget. We decided to call the newsletter *Microbial Genetics Bulletin*, a title chosen so as to include eukaryotic as well as prokaryotic microorganisms.

Word was passed by mail or telephone, or by direct contact with colleagues, to inform a core group of the plan and to solicit items for the first issue, which appeared in January 1950. It was circulated to an international mailing list of 74 individuals, a list that included everyone we could think of who could be considered a microbial geneticist, including most of the speakers at the 1946 Symposium. Fifty-six of the initial *MGB* recipients worked in the United States, fifteen in Europe (England, France, Germany, Italy, or Denmark) and three elsewhere (Australia, Nigeria, or Canada.) Thirteen future Nobel laureates were included: O.T. Avery, G.W. Beadle, F.M. Burnet, M. Delbruck, R. Dulbecco, A. Hershey, J. Lederberg, S.E. Luria, A. Lwoff, C.M. MacLeod, M. McCarty, J. Monod, and E. Tatum.

The proposed scope and purpose of the *Microbial Genetics Bulletin* was set forth in a Foreword, which stated that "the growth of microbial genetics as a cohesive field suggests the need for an informal medium of contact among investigators in this area of work." All material appearing in the *MGB* was to have the status of personal communications, to be cited or quoted only with the permission of the author. Contributions in the following categories were included in the first issue and were solicited for subsequent issues: (*a*) brief notes—work in progress at the prepublication level, critical and theoretical comments, minor observations, and techniques not scheduled for publication that might be of interest to other workers; (*b*) suggestions on nomenclature, with a view to unification; (*c*) listings of available mutant stocks; (*d*) requests for stocks and information; (*e*) announcements of meetings, talks, and visits; (*f*) personal items (gossip); and (*g*) directory of participants.

Forty-seven issues of the *Microbial Genetics Bulletin* were published, from 1950 to 1979. I held the editorship until 1964, when Howard Adler, at Oak Ridge National Laboratory, took it over with *MGB* #20, and finally, in 1974, with *MGB* #36, James C. Copeland, at Ohio State University, became the editor. *MGB* #36 was mailed to 1184 scientists and libraries in Africa, North, South, and Central America, Asia, Australia, East and West Europe, India, and New Zealand. That issue listed as corporate sponsors Cetus Scientific Laboratories, The Eli Lilly Company, Pfizer Incorporated, G.D. Searle and Company, and the Upjohn Company.

The yellowing pages of the first issues of the *Bulletin* make for an interesting read, fifty years and a whole revolution later. Below, a sampling of items in some of these early issues is offered as a poignant reminder of where we were then.

MGB #1

The first issue contains a Brief Note from Esther M. Lederberg, at the University of Wisconsin, entitled "Lysogenicity of *Escherichia coli* K12." Here she describes a latent bacteriophage, which she calls lambda, carried by most derivatives of *E.coli* K-12 and demonstrable only with the help of a sensitive indicator strain. Bernard Davis and Milislav Demerec submitted items in the Nomenclature category that are also of historical interest. Davis proposes using the terms "auxotrophic" and the corresponding noun "auxotroph" to refer to mutants with increased nutritional requirements, a suggestion that was quickly and universally adopted. Demerec suggests a system of nomenclature and symbolism for bacterial genetics based on the general principles of genetic nomenclature used by *Drosophila* workers. Although the system ultimately adopted for bacterial genetics differed considerably from Demerec's initial proposal, this article stimulated broad discussion (some of it in subsequent issues of *MGB*) and led by 1966 to the standardization of the genetic nomenclature for bacteria in current use (2).

MGB #2

The second issue appeared in April 1950. The Foreword reports enthusiastic reception of the first issue and informs readers that there is no subscription fee, as the costs of preparation and distribution are being carried by the Department of Genetics of the Carnegie Institution of Washington. Forty-two names were added to the mailing list, including S. Benzer, E. Chargaff, R. Dubos, O. Maaloe, R. Sager, C.B. van Niel, and J.D. Watson (then a graduate student in Indiana). By October 1951, the mailing list had grown to 250 names.

Among the Brief Notes in *MGB* #2 is one by F.M. Burnet and P.E. Lind, describing recombination of characters in influenza virus. Another, by Bernard Davis, rigorously rules out a filterable transforming substance or gamete as responsible for the K-12 recombination system of Lederberg, using a U-tube with its two arms separated by a fritted glass disc. A note by A.D. Hershey distinguishes between allelic and nonallelic host range mutants in bacteriophage T2; Raymond Latarjet, at the Institut Curie in Paris, presents the first evidence for induction of mutations in a bacteriophage.

The Nomenclature section of *MGB* #2 contains two critical comments on Demerec's proposal for bacterial genetics nomenclature, which appeared in *MGB* #1. G. Bertani and D. Catcheside make suggestions that come somewhat closer to the system ultimately adopted.

Announcements of two courses to be offered during the summer of 1950 at the Biological Laboratory at Cold Spring Harbor are included in *MGB* #2. A course in bacterial genetics, to be taught for the first time by M. Demerec, V. Bryson, and E. Witkin, was to take place immediately following the course in bacterial viruses (the phage course), which would be offered for the sixth time.

There are several contributions to the Apparatus and Technique section of *MGB* #2. Mark Adams describes an inexpensive substitute for copper pipette containers, made from cardboard mailing cylinders. Anonymous (initials J.L.) recommends the new aluminum culture caps to replace cotton plugs for general use, and especially, with holes drilled through the center to admit glass tubing, to avoid rolling aerating tubes in cotton. M. Demerec warns against the use of Baltimore Biological Laboratory's discs of heavy cardboard as inserts into Petri dish covers to reduce evaporation. When dishes carrying these discs as inserts are oven-sterilized, the discs release a toxic substance that adheres to the glass. Culture medium poured into such dishes becomes toxic to *E. coli*. G. Maccacaro tells, in detail, how to construct an "automatic" colony counter.

MGB #4

In issue #4, dated February 1951, several Brief Notes by M. Demerec and coworkers describe some of the earliest quantitative studies of UV mutagenesis in *E. coli*. A. Hershey provides evidence for heterozygous bacteriophage, and J. Lederberg reports screening large numbers of *E. coli* strains for cross-fertility with K-12. Of 100 isolations from human sources, 8 were found to cross with about the same fertility as K-12, and an equal number appeared to cross less well. Lederberg concludes that recombination is not a unique feature of one or a few isolates of *E. coli* and points out the wealth of material that such an extended survey can provide for studying the evolutionary genetics of bacteria.

MGB #5

There is a story behind two of the items under Methods and Techniques in *MGB* #5, published in October 1951. One is J. Lederberg's first description of the replica plating technique, using velvet discs to "print" bacterial colonies on a variety of selective media. In a postscript initialed J.L. and E.L., repeated replica platings are described as a way to prove that most bacterial mutants occur in clones, and for isolating resistant mutants as pure clones. The authors report that this work is in press.

The second article, by N. Visconti, describes another version of replica plating, similar in principle to Lederberg's, but different in detail and considerably less versatile. Visconti was a brilliant Italian aristocrat who took the phage course in 1950 and stayed in Cold Spring Harbor, working closely with Delbruck to develop a complete theory of phage recombination, until 1953. Sometimes he worked at a bench next to mine. One day he accidentally dropped a Petri dish containing lactose-positive colonies on eosine-methylene blue (EMB) agar, colony side down, on the floor. After picking it up, he stared for several minutes at the neat imprint of the red colonies on the light-colored floor, after which I could almost see a light bulb flash over his head, and he immediately began to experiment with replica plating, using filter papers for the transfer. For a few days, he was full of ideas for improving and applying the technique, until someone showed him a preprint of the Lederbergs' article on replica plating with velvet discs. Sometimes the prepared mind is not enough.

THE EARLY FIFTIES

The unprepared mind, in this case my own, is illustrated in an incident that took place in 1950. That summer, Dr. Demerec, Vernon Bryson, and I were teaching the first bacterial genetics course. One of the experiments was a straightforward repetition of Lederberg's initial mating between two K-12 strains of *E. coli* (4). He had shown that cell-to-cell contact between the bacteria of the two strains was required for the recombination. The mechanism of the genetic exchange was not yet known. As there were eight pairs of students performing the same cross, I asked each pair to vary the procedure by agitating the mating mixture gently on a home-made shaker (an off-centered fan motor clamped to a ring stand) for a specific time, ranging from 0 to 16 minutes, before plating to select recombinants. I expected that the number of recombinants would rise regularly in direct proportion to the time of agitation as the probability of cell-to-cell contact increased. Instead, the results showed the opposite. The yield of recombinants was maximal with no shaking before plating and decreased in perfect inverse proportion to the time of agitation. Puzzled by this strange result, we discussed it at length without arriving at a reasonable explanation. Six years later, our perplexing result suddenly made sense. Francois Jacob, at the 1956 Cold Spring Harbor Symposium, described the "blender experiment" that had led him and Wollman to discover the uses of interrupted mating in bacterial recombination (14). After Jacob's talk, I asked him if the drastic agitation of the blender was really necessary to break apart the mating pairs. He answered that no, the blender was "just for show"—mating could easily be interrupted by nothing more than a good shake!

In the 1951 Cold Spring Harbor Symposium, entitled *Genes and Mutations*, Barbara McClintock gave her first major talk on mobile genetic elements (6). In her own words, "the response was puzzlement, and in some instances hostility." Certainly, as I remember it, there was baffled silence after her talk and little or no discussion of her densely documented evidence and argument for transposable elements and their effects on gene expression. The audience seemed embarrassed by the lack of response, understandably, because McClintock was respected and admired as a great geneticist. Here, her conclusions were too radically in conflict with the entrenched genetic concept of a stable genome, and her data too complex, to allow for rapid or easy acceptance, although a small number of geneticists who had come to know her work well believed it to be profoundly important. Only after transposons were discovered in bacteria, 20 years later, was McClintock's magnificent series of paradigm-shattering discoveries generally appreciated.

The 1951 Symposium included several sessions on bacterial and viral genetics. In one of these sessions, Lederberg reported Zinder's discovery of bacterial transduction. In another, I described experiments I had done to determine whether the staining bodies then called "nuclei" in *E. coli* cells actually carry the genes. In my work on radiation resistance, I had noticed that bacteria at different stages of growth contained different numbers of these staining bodies. Stationary phase cells usually contained only one or two, cells in various stages of the lag phase mainly one to four, and in logarithmically growing cells there were typically four to eight nucleoids. No one had yet demonstrated that bacterial genes were actually contained in these bodies. My idea was to irradiate populations having different average numbers of nucleoids per cell with a mutagenic dose of UV, plate the survivors on EMB-indicator medium, and identify induced lactose-negative (Lac−) mutants, without selection, as white colonies. If these bodies indeed contain the genes, an induced Lac− mutant arising in a cell containing multiple nucleoids should be detected in a mixed clone, as a white sector in an otherwise red lactose-positive (Lac+) colony, owing to segregation of the mutated nucleoid from the unmutated ones during the early divisions of the microcolony.

Two clear predictions followed from the hypothesis that the nucleoids carry the genes: (*a*) The proportion of mixed Lac−/Lac+ clones to whole, unsectored Lac− colonies should increase with the average number of nucleoids per cell, and (*b*) the larger the average number of nucleoids per cell, the smaller should be the size of Lac− sectors in the mixed clones.

The results obtained with six UV-irradiated populations of strains B and B/r showed a clear correlation between the ratio of uninucleate to multinucleate cells and the ratio of intact to sectored Lac-colonies, in good agreement with the prediction (11). Furthermore, large Lac-sectors, occupying three fourths or one half of the colony, predominated in populations having mainly two nucleoids per cell, whereas smaller sectors (one fourth of the colony or smaller) were more frequent in populations having primarily four or more nucleoids per cell. The results were consistent with the assumption that these bodies carry genes and that the chromosome theory of heredity applies to *E. coli*.

It is impossible to think of the early 1950s in Cold Spring Harbor without recalling the powerful dual impact of the 1952 experiment by Hershey & Chase

(3), which suddenly convinced everyone that genes are made of DNA after all, and the dramatic solution of DNA structure by Watson & Crick a year later. Although Avery et al. (1) had demonstrated in 1944 that the genetic material is DNA, the prevailing attitude at Cold Spring Harbor had been respectful skepticism. Some suggested that the transforming DNA in their experiments had activated genetic information already present or had somehow caused a directed mutation. Others believed that the miniscule trace of protein still contaminating the DNA was the active agent. Delbruck declared DNA to be a "stupid" molecule, incapable of carrying genetic information. Hershey & Chase settled that question forever.

In an unscheduled talk at the 1953 Cold Spring Harbor Symposium, Watson electrified the audience when he described the structure of DNA and passed around a three-dimensional model of the double helix. It is no exaggeration to say that it was as if we were handed the secret of life. Biology has never been the same.

FULL CIRCLE

By 1950, having considered filamentous growth and radiation resistance a distraction, albeit an absorbing one, I had begun the work on UV mutagenesis that occupied me for the rest of my working life. I did not know then that there was a deep, hidden connection between these two seemingly unrelated research areas, a connection that would not be revealed until about twenty-five years later. Elsewhere (12, 13), I have described in detail the ideas, experiments, results, and interpretations that comprise my contribution to our understanding of UV mutagenesis, DNA repair, and the DNA damage-inducible SOS response. It was in the SOS response, the focus of my last research efforts, that the connection between UV mutagenesis and filamentous growth emerged. When *E. coli* is exposed to UV, about 30 unlinked but coordinately regulated SOS genes are activated (9). Their diverse products enhance DNA repair and survival and increase genetic variability. One of the SOS-inducible proteins is DNA polymerase V, the error-prone enzyme that effects UV mutagenesis via translesion synthesis (8). Another is the cell division inhibitor, inactivated by the mutation in B/r, that is responsible for filamentous growth in *E.coli* B.[1] "In our beginning is our end," wrote T.S. Eliot in *Four Quartets*. Yes.

[1]In most wild-type *E. coli* strains, the SOS-induced cell division inhibitor (product of the *sulA* gene) has a short half-life. Once DNA repair is completed and SOS expression is rerepressed, the inhibitor is degraded by the Lon protease (7). This permits cell division to resume, thereby maintaining coordination bewteen DNA replication and septation. The wild-type B strain carries an inactivating mutation in *lon*, so the induced inhibitor persists long after DNA repair is completed and DNA replication restarts, preventing septation and causing filamentous growth. The *sulA* mutation in B/r inactivates the inhibitor.

The *Annual Review of Microbiology* is online at http://micro.annualreviews.org

LITERATURE CITED

1. Avery OT, MacLeod CM, McCarty M. 1944. Studies on the chemical nature of the substance inducing transformation of Pneumoccocal types. I. Induction of transformation by a desoxynucleic acid fraction isolated from *Pneumococcus* Type III. *J. Exp. Med.* 79:137–58
2. Demerec M, Adelberg EA, Clark AJ, Hartman PE. 1966. A proposal for a uniform nomenclature in bacterial genetics. *Genetics* 54:61–76
3. Hershey AD, Chase M. 1952. Independent functions of viral protein and nucleic acid in growth of bacteriophage. *J. Gen. Physiol.* 36:39–56
4. Lederberg J. 1947. Gene recombination and linked segregations in *Escherichia coli. Genetics* 32:505–25
5. Luria SE, Delbruck M. 1943. Mutations of bacteria from virus sensitivity to virus resistance. *Genetics* 28:491–511
6. McClintock B. 1951. Chromosome organization and genic expression. *Cold Spring Harbor Symp. Quant. Biol.* 16:13–47
7. Mizusawa S, Gottesman S. 1983. Protein degradation in *Escherichia coli*: The *lon* gene controls the stability of *sulA* protein. *Proc. Natl. Acad. Sci. USA* 80:358–62
8. Tang M, Shen X, Frank EG, O'Donnell M, Woodgate R, et al. 1999. UmuD'$_2$C is an error-prone DNA polymerase, *Escherichia coli* pol V. *Proc. Natl. Acad. Sci. USA* 96:8919–24
9. Walker GC. 1987. The SOS response of *Escherichia coli.* In *Escherichia coli* and *Salmonella typhimurium,* ed. FC Neidhardt, JL Ingraham, KB Low, B Magasanik, M Schaechter, HE Umbarger, pp. 1346–57. Washington, DC: ASM
10. Witkin EM. 1947. Genetics of resistance to radiation in *Escherichia coli. Genetics* 32:221–48
11. Witkin EM. 1951. Nuclear segregation and the delayed appearance of induced mutants in *Escherichia coli. Cold Spring Harbor Symp. Quant. Biol.* 16:357–72
12. Witkin EM. 1989. Ultraviolet mutagenesis and the SOS response in *Escherichia coli*: a personal perspective. *Environ. Mol. Mutagen.* 14(Suppl. 16):30–34
13. Witkin EM. 1994. Mutation frequency decline revisited. *BioEssays* 16:437–44
14. Wollman EL, Jacob F, Hayes W. 1956. Conjugation and genetic recombination in *Escherichia coli* K-12. *Cold Spring Harbor Symp. Quant. Biol.* 21:141–62

Annu. Rev. Microbiol. 2002. 56:17–38
doi: 10.1146/annurev.micro.56.012302.160310

First published online as a Review in Advance on April 26, 2002

FUNCTION OF *PSEUDOMONAS* PORINS IN UPTAKE AND EFFLUX

Robert E. W. Hancock[1] and Fiona S. L. Brinkman[2]

[1]Department of Microbiology and Immunology, University of British Columbia, Vancouver, BC, Canada V6T 1Z3; e-mail: bob@cmdr.ubc.ca

[2]Department of Molecular Biology and Biochemistry, Simon Fraser University, 8888 University Dr., Burnaby, BC, Canada V5A 1S6; e-mail: brinkman@sfu.ca

Key Words antibiotic resistance, *Pseudomonas aeruginosa*, OprM, OprF, OprD

■ **Abstract** Porins are proteins that form water-filled channels across the outer membranes of Gram-negative bacteria and thus make this membrane semipermeable. There are four types of porins: general/nonspecific porins, substrate-specific porins, gated porins, and efflux porins (also called channel-tunnels). The recent publication of the genomic sequence of *Pseudomonas aeruginosa* PAO1 has dramatically increased our understanding of the porins of this organism. In particular this organism has 3 large families of porins: the OprD family of specific porins (19 members), the OprM family of efflux porins (18 members), and the TonB-interacting family of gated porins (35 members). These familial relationships underlie functional similarities such that well-studied members of these families become prototypes for other members. We summarize here the latest information on these porins.

CONTENTS

INTRODUCTION

Pseudomonas aeruginosa is an exceptionally versatile organism that can adopt many ecological niches (33). It is known to exist in the environment, including in soil and attached to rocks in streams, and can opportunistically infect diverse organisms from grasshoppers to humans. It has become especially notorious as a human pathogen, being the third-most-common hospital pathogen, causing approximately 10% of the 2 million hospital infections in the United States annually. It is also the major cause of chronic lung infections in individuals with the genetic disease cystic fibrosis, and it is thought to be associated with progressive deterioration of lung function and eventual death in such individuals. A major reason for its prominence as a hospital pathogen is its high intrinsic resistance to antimicrobials, including antibiotics and disinfectants (16). Antibiotic resistance makes *P. aeruginosa* one of the most difficult organisms to treat. Similarly resistance to disinfection makes it difficult to remove from a hospital environment.

The average hospital strain can be susceptible to several antibiotics such as the recently introduced β-lactam antibiotics, aminoglycosides such as gentamicin, tobramycin, and amikacin, and fluoroquinolones such as ciprofloxacin (16). However, the organism is naturally less susceptible than most bacteria to such agents and naturally resistant to many others, a phenomenon termed "intrinsic resistance." Table 1 compares the minimal inhibitory concentrations for several antimicrobials of typical strains of *Pseudomonas* and *Escherichia coli*. The major impact of high intrinsic resistance is that, even for agents to which *P. aeruginosa* is initially susceptible, a mutation that causes a moderate increase in resistance can make this organism clinically untreatable.

The recent publication of the genomic sequence of *P. aeruginosa* (49) has provided a quantum increase in our knowledge of this organism. Overall, the genome of strain PA01 encodes 5570 genes (on 6.3 mega base pairs of DNA). The nature of the genes that have been annotated is consistent with the versatility of this organism, including four motility systems, a large number of systems for metabolism of carbon sources, the highest percentage of regulators (nearly 1 gene in 9) of any genome to date, and a plethora of transport systems (49). At the time of writing this review, the partial sequence of *P. putida* had been released as an unfinished genome (http://tigrblast.tigr.org/ufmg/index.cgi), and examination of this revealed similar trends for this bacterium. Indeed although none of the other type 1 fluorescent pseudomonads are significant hospital pathogens, they share similarities in having large genomes and substantial metabolic diversity. Being less well studied, especially with respect to outer membrane proteins (17), they will be covered only in passing here. Information supplemental to this review is posted at www.cmdr.ubc.ca/bobh/omps/, and we largely reference articles subsequent to our last major review of this topic (17).

PSEUDOMONAS OUTER MEMBRANES

More than two decades ago we demonstrated (35) that *P. aeruginosa* had low outer membrane permeability (approximately 8% that of *E. coli*) but a large exclusion limit (permitting passage of compounds of around 3000 molecular weight compared to an exclusion limit of around 500 molecular weight for *E. coli*). This at first seemed contradictory, and indeed was challenged in the literature, but was subsequently confirmed (4, 36). In particular, investing *P. aeruginosa* with a raffinose metabolic operon permitted relatively rapid growth on tri- and tetra-saccharides (compared to *E. coli*), confirming the large exclusion limit (4). Furthermore, such studies indicated that the major porin contributing to this large exclusion limit was OprF (see below). Similarly cloning highly permeable porins, either deletion mutants of OprD (19) or *E. coli* OmpF (41), into *P. aeruginosa* led to substantial decreases in minimal inhibitory concentration (MIC) for multiple antibiotics, indicating that outer membrane permeability was limiting for antibiotic susceptibility.

However, low outer membrane permeability is insufficient by itself to explain high intrinsic antibiotic resistance (15). Even a poorly permeable outer membrane, like that of *P. aeruginosa*, will permit antibiotics to diffuse and equilibrate their concentrations across the membrane in 1–100 sec (compared to a doubling time of 2000 sec or more). Other secondary resistance mechanisms have to exist that will take advantage of the relatively low rate of permeation of antibiotics across the outer membrane; two have been described (15, 42). *P. aeruginosa* contains a periplasmic β-lactamase that is inducible (by some β-lactams, notably imipenem). Because this enzyme works catalytically, it will hydrolyze β-lactams at a steady rate that will benefit from the slow passage of β-lactam into the periplasm (15). Indeed it has been demonstrated for β-lactams such as imipenem and panipenem that β-lactamase is the major secondary determinant of intrinsic resistance (34), and for many but not all β-lactams, knockout of efflux has no apparent effect in β-lactamase derepressed mutants.

Conversely, some β-lactams and many other antibiotics are far more influenced by multidrug efflux systems (34, 41). The major system of note for *P. aeruginosa* intrinsic antibiotic resistance is the MexAB-OprM, RND efflux system, which benefits from relatively slow uptake of antibiotics and actively effluxes them from the cell (Table 1).

The consequence of a poorly permeable outer membrane is that many substrates have to utilize specialized pathways to cross the outer membrane at a rate sufficient to support growth. Indeed *P. aeruginosa* utilizes a diversity of outer membrane permeation pathways to support growth (17). As mentioned above, OprF is the major channel for larger substrates and can be considered a general or nonspecific porin (a porin is defined as a trans-outer-membrane protein that encloses a water-filled channel—general porins lack substrate specificity). Other proteins that also

TABLE 1 Susceptibility of *P. aeruginosa* and *E. coli* strains to antibiotics

	MIC (μg/ml)[a]					
Strains	**CTX**	**CB**	**CIP**	**NAL**	**TC**	**CAM**
P. aeruginosa WT	4	32	2	25	8	6.4
P. aeruginosa/oprDΔ*L5*[b]	1	—	0.02	—	0.8	0.8
P. aeruginosa oprM::Ω[c]	1	0.4	0.1	2	0.5	0.2
E. coli WT	0.13	—	0.03	4	2	8

[a]Abbreviations: CTX, cefotaxime β-lactam; CB, carbenicillin β-lactam; CIP, ciprofloxacin fluoroquinolone; NAL, nalidixic acid quinolone; TC, tetracycline; CAM, chloramphenicol; PXB, polymyxin B; GM, gentamicin; WT, wild type. The symbol "—" signifies no data available.

[b]OprDΔL5 has a large channel such that when cloned into *P. aeruginosa* it increases outer-membrane permeability (19, 20).

[c]*oprM*::Ω mutations delete the major efflux system such that this *P. aeruginosa* strain lacks the major intrinsic efflux pathway of *P. aeruginosa* (22).

function as substrate-specific porins can serve as general porins for small substrates. For example, OprD acts as a rate-limiting porin for gluconate (18) and possibly some other low-molecular-weight substrates (S. Tamber & R.E.W. Hancock, unpublished observations), whereas OprB acts as a porin for monosaccharide passage (59). These proteins probably account for the nonspecific passage of most substrates across the outer membrane because *P. aeruginosa* utilizes few substrates larger than ~200 Da, the size of a monosaccharide.

OUTER MEMBRANE PROTEINS

Knowledge of the *P. aeruginosa* genome (49) permits us to predict all of the outer membrane proteins in this organism, by homology to known *P. aeruginosa* outer membrane proteins or those from other organisms, and/or possession of certain motifs (particularly signal sequences plus a propensity for predicted β-strands and/or a conserved C-terminal β-strand). Table 2 describes all of the known outer membrane proteins of *P. aeruginosa* together with their *Pseudomonas* ID number, which will let the reader search for information on these proteins at www.pseudomonas.com. In Table 3 a list of possible additional porins is provided (see also www.cmdr.ubc.ca/bobh/omps/ for additional information about all known and predicted outer membrane proteins). Investigation of porins from other species has revealed that they contain transmembrane anti-parallel β-strands that wrap into a barrel (22) (Figure 1, see color insert). Between 8 and 22 β-strands (from 1 to 3 subunits) make up this β-barrel embedded in the outer membrane bilayer. Generally speaking, these β-strands are interconnected by short turn sequences on the periplasmic side and larger loop sequences on the external face of the membrane, although a somewhat different picture is observed for the efflux channel-tunnels

TABLE 2 Known outer-membrane proteins

Gene	PAID[a]	Protein function and name	Position in genome nucleotides		No. of AAs	Medline Ref. ID	Known porin (class)[b]
			From	To			
algE	PA3544	Alginate production protein AlgE	3968448	3969920	490	92077417	P
aprF	PA1248	Alkaline protease secretion protein AprF	1353827	1355272	481	93051361	EP, PI
fliF	PA1101	Flagella M-ring protein	1192405	1194201	598	96239027	
fptA	PA4221	Fe(III)-pyochelin receptor	4726800	4724638	720	94117363	GP
fpvA	PA2398	Ferripyoverdine receptor	2655187	2657634	815	93328663	GP
hasR	PA3408	Heme uptake receptor HasR	3817335	3814660	891		GP
icmP	PA4370	Insulin-cleaving metalloproteinase; ICMP	4898192	4899532	446	10452958	
lppL	PA5276	Lipopeptide LppL	5941335	5941475	46	90279511	
omlA	PA4765	Lipoprotein OmlA	5352176	5352706	176	9973334	
oprB	PA3186	Glucose/carbohydrate porin OprB; protein D1	3577275	3575911	454	95286479	SP
oprC	PA3790	Putative copper transport porin OprC	4249873	4247702	723	96349120	GP?
oprD	PA0958	Basic amino acid, basic peptide and imipenem porin OprD; also named Porin D, Protein D2	1045314	1043983	443	90368779	SP, P
oprE	PA0291	Anaerobically induced porin OprE; Porin E1	327284	328666	460	93360827	SP
oprF	PA1777	Major porin and structural porin OprF; Porin F	1921174	1922226	350	88086862	P
oprG	PA4067	Outer-membrane protein OprG	4544606	4545304	232	99277900	
oprH	PA1178	PhoP/Q and low Mg^{2+} -inducible outer-membrane protein H1	1277006	1277608	200	89255086	GP?
oprI	PA2853	Outer-membrane lipoprotein OprI	3206914	3207165	83	89327122	
oprJ	PA4597	Multidrug efflux protein OprJ	5151071	5149632	479	97032139	EP
oprL	PA0973	Peptidoglycan-associated lipoprotein OprL	1057400	1057906	168	97312009	
oprM	PA0427	Major intrinsic multiple antibiotic resistance efflux protein OprM	476333 476333	477790 477790	485 485	97312458 97312458	EP EP
oprO	PA3280	Pyrophosphate-specific porin OprO	3674323	3673007	438	93023860	SP
oprP	PA3279	Phosphate-specific porin OprP; protein P	3672548	3671226	440	86296709	SP
pfeA	PA2688	Ferric enterobactin receptor PfeA	3040241	3042481	746	93123148	GP

(*Continued*)

TABLE 2 (*Continued*)

Gene	PAID[a]	Protein function and name	Position in genome nucleotides From	To	No. of AAs	Medline Ref. ID	Known porin (class)[b]
phuR	PA4710	Heme/Hemoglobin uptake receptor PhuR	5289216	5291510	764		GP
pilQ	PA5040	Type 4 fimbrial biogenesis protein PilQ	5677857	5675713	714	94049125	
popD	PA1709	Translocator protein PopD; PepD	1854849	1855736	295	98449523	
popN	PA1698	Type III secretion protein PopN	1847227	1848093	288	98037517	
pscC	PA1716	Type III secretion protein PscC	1859493	1861295	600	97126825	PIII
xcpQ	PA3105	General secretion pathway protein D	3484353	3486329	658	95020542	PII
xcpU	PA3100	General secretion pathway protein H; PilD-dependent protein PddB	3480238	3479720	172	92269572	
xqhA	PA1868	Secretion protein XqhA	2028968	2031298	776	98343806	PII

[a]PAID; *Pseudomonas aeruginosa* gene identity number (see www.pseudomonas.com).

[b]P, general porin; SP, specific porin; GP, putative gated porin; EP, OprM family member of efflux and protein secretion porins; PI, putative type I secretion subfamily; PII, type II secretion channel; PIII, type III secretion channel.

like OprM (Figure 1*B*). The central area of the β-barrels of the general and specific porins contains stretches of amino acids from one of the interconnecting regions, often loop 3, that folds back into the channel region and gives this region many of its important characteristics.

A major finding from the genome sequence is that of the 163 known or predicted *P. aeruginosa* outer membrane proteins, 64 are found as part of 3 families of porins, the OprD-specific porin family, the TonB-dependent gated porin family, and the OprM efflux/secretion family.

GENERAL PORINS

OprF

OprF is a major outer membrane protein in *P. aeruginosa* that has been studied extensively due to its proposed utility as a vaccine component, role in antimicrobial drug resistance, and porin function (17). OprF has been described as a multifunctional protein, as gene disruption and gene deletion analysis has indicated that it is required for cell growth in low-osmolarity medium and for the maintenance of cell shape (44). In addition it appears to have a nonspecific porin function and binds to the underlying peptidoglycan (4, 17). Additional functions of OprF have been identified in other *Pseudomonas* sp., for example, in *P. fluorescens* OprF is a root adhesion.

Many studies of OprF suggest that it resembles *E. coli* OmpA in both function and structure, and it is a structural member of the OmpA family of proteins (17). Through epitope-mapping experiments, and linker-insertion mutagenesis, a 16-β-stranded membrane topology model for *P. aeruginosa* OprF was originally proposed (17). However, subsequent deletion studies and secondary structure predictions indicated that there are 3 domains to this protein: (*a*) a N-terminal domain (first ~160 aa) containing 8 anti-parallel sheets proposed to form a β-barrel structure (7, 46), (*b*) a loop or hinge region (161 to 209 aa) containing a poly-proline-alanine repeat region and two disulfide bonds, and (*c*) a C-terminal domain (210 to 326 aa) highly conserved with the corresponding domains of other OmpA family proteins (44, 50). This latter C-terminal region has also been shown to be the domain that forms the noncovalent linkage with peptidoglycan in the periplasm (44, 50). It is linked to the N-terminal domain by a proline-rich hinge and a loop region that contains two disulfide bonds (note that these disulfide bonds are not found in all *Pseudomonas* species OprF proteins) (44). This three-domain structural model for OprF has been further supported by circular dichroism spectroscopy analysis (7, 50) and three-dimensional modeling of the N terminus of OprF to the crystal structure of the proposed orthologous sequences in *E. coli* OmpA (Figure 1*D*).

Clinical isolates of *P. aeruginosa* that are multiply antibiotic resistant and deficient in the major outer membrane protein OprF have been obtained (17, 43). Sequencing of the *oprF* gene in such a clinical isolate has shown that the *oprF* gene and promoter are intact, indicating that a regulatory mutation may be involved (8). This regulatory mutation has not yet been revealed, although recent analysis of the promoter region of OprF has indicated that it is not just constitutively expressed from a sigma 70 promoter, as originally proposed (8). In addition, there is an extracellular factor (ECF) sigma factor promoter upstream of the gene that appears to be affected by disruption of an upstream ECF sigma factor gene named sigX. This sigma factor gene is not mutated in the clinical isolates that are multiply antibiotic resistant, so the mechanism for such resistance remains unknown.

The porin function of OprF has been extensively studied through liposome swelling experiments and planar lipid bilayer analysis of both the full-length protein and the N-terminal β-barrel domain (17). The size of the channels has been controversial (17). However, OprF channels have been shown to be nonspecific in nature, with weak cation selectivity. Both small (0.36 nS) channels and, rarely, large (2–5 nS) channels appear to form in planar lipid bilayer experiments, with only the small channels forming when the N-terminal domain of the protein is examined for porin function (7, 46). Because the full length of the protein is required for large channel formation and there is evidence that the C terminus of the protein contains both surface-exposed and peptidoglycan-binding regions, it seems possible that OprF forms more than one conformation varying in both structure and channel formation. Evidence supporting this concept of more than one structure and channel size has also been presented for *P. fluorescens* OprF (12).

OprF is also noted for its antigenicity, and vaccine candidates containing portions of OprF have been constructed (e.g., 14, 26). Due to its antigenicity, multifunctional nature, and apparently complex structure, OprF continues to be a fascinating protein that merits further study.

SPECIFIC PORINS

The best-characterized specific porin is *E. coli* LamB, which contains within its channel a substrate-binding site for maltose and maltodextrins (48). The crystal structure of LamB indicates that this porin is rather analogous to the nonspecific porins with quite modest differences (Figure 1). LamB is an 18 (compared to 16)-stranded β-barrel, which contains about 30% extra (~100) amino acids compared to the nonspecific porin OmpF (22). These extra residues are largely found in the surface loops that fold over to constrict the entrance of the channel (loops 4, 6, and 9) or reach over to the adjacent monomer in the LamB trimer (loop 2) (2). The substrate-binding site includes parts of loops 4, 5, and 6 (which are slightly longer than in OmpF) and the barrel wall, and it involves several hydrophobic residues that collectively are termed the "greasy slide." *Pseudomonas* has at least three well-characterized specific porins, OprB, OprP, and OprD, each of which contains one or more less-characterized homologs.

OprP/O

OprP (protein P) is a protein of 48,000 molecular weight that is induced under conditions of low phosphate (<0.15mM) (17). It is involved in the high-affinity, phosphate-starvation inducible transport system (PTS), as studied using an *oprP*::Tn*501* mutant. Purification of OprP and studying it using the planar lipid bilayer model membrane system indicated that the OprP channel contains a binding site for phosphate with a Kd of approximately 0.15 μM. While the channel is permeable to small anions, it is blocked by the binding of phosphate to its binding site (51). Molecular modeling and insertion mutagenesis have led to the proposal that OprP is a 16-stranded β-barrel (17). Systematic site-directed mutagenesis of all the lysine residues in the N-terminal half of OprP to glutamate and glycine revealed that lys-121 in the proposed loop 3 region was part of the phosphate-binding site (51). Two other basic lysine residues, lys-74 and lys-126, when changed to the acidic residue glutamate, but not when changed to the neutral residue glycine, affected the movement of anions through the OprP channel, which indicates that these other lysine residues probably represent secondary (non-rate-limiting) phosphate-binding sites. Thus the OprP channel probably acts as an electrical wire transmitting negatively charged phosphate residues from one positively charged binding site to the next of higher affinity. The lys-121-binding site is the highest-affinity binding site in OprP, but it still is of lower affinity than the periplasmic phosphate-binding protein, and thus, the phosphate will flow along the concentration gradient toward the periplasm. The *oprO* gene resides immediately upstream of *oprP* and shares 76% identical amino acids (17). It is induced under conditions of phosphate starvation in the stationary phase of growth and like *oprP* has upstream pho-box sequences for binding of the regulator PhoB to its promoter. The OprO channel prefers pyrophosphate to phosphate (for OprP the situation is reversed).

OprD Family

OprD was first identified as a protein that was lost when *P. aeruginosa* clinical isolates became resistant to the broad-spectrum β-lactam imipenem (30). This β-lactam strongly resembles a dipeptide containing a positively charged residue. Consistent with this finding, Trias & Nikaido demonstrated that OprD is a specific porin that binds basic amino acids, dipeptides containing a basic residue and imipenem and related zwitterionic carbapenems (including meropenem) (54). This was confirmed in part by planar lipid bilayer analysis (19).

OprD is the closest *P. aeruginosa* homolog of the *E. coli* nonspecific porin OmpF, a fact that assisted in building a model for this porin as a 16-stranded β-barrel (20). This model was tested with reasonable success by PCR-directed site-specific (4–8 amino acid) deletion mutagenesis. Investigation of OprD mutants with deletions in specific loops demonstrated that both loop 2 and loop 3 deletions lose the ability to bind imipenem and mediate imipenem susceptibility (19, 37). Thus, OprD differs from other specific porins (22) in that loop 2 has a role in substrate binding to the channel. Also loops 5, 7, or 8 deletion variants of OprD have increased susceptibility of *P. aeruginosa* to multiple antibiotics, and they correspondingly produce larger channels (that still bind imipenem), which indicates that these loops constrict the channel entrance to limit nonspecific movement of molecules through OprD channels (20).

OprD is found as a moderately expressed outer membrane protein but is regulated by multiple systems. It is repressed by MexT (which also induces the MexEF-OprN efflux system), salicylate, and catabolite repression (23, 39), and it is activated by arginine/ArgR and a variety of other amino acids as carbon and nitrogen sources (38).

The genome sequence (49) revealed that OprD is part of a 19-member family of outer membrane proteins in *P. aeruginosa* which are 46%–57% similar to OprD at the amino acid level. Phylogenetic analysis has revealed two subfamilies, the OprD group and the OpdK group (F.S.L. Brinkman, S. Tamber & R.E.W. Hancock, unpublished data). Eight homologs are more closely related to OprD, and those studied have roles in amino acid or peptide transport. Eleven homologs are more similar to the PhaK porin of *P. putida* that is required for growth on phenyl acetic acid (including the previously studied anaerobically induced porin OprE) (17), and those studied have roles in transport of organic carbon sources (S. Tamber & R.E.W. Hancock, unpublished data). However, study of mutants in each gene indicate that only OprD is involved in antibiotic uptake, in contrast to earlier conclusions made regarding certain OprD homologs (17). Microarray analysis has indicated that only OprD, OprQ, OpdP, OpdQ, and OprE are even moderately produced in minimal medium with succinate as a carbon source (M. Brazas & R.E.W. Hancock, unpublished data).

OprB

The closest homolog in *P. aeruginosa* of the crystallized specific porin LamB (Figure 1*C*) is OprB (17). It is induced by growth on minimal medium supplemented with glucose as the sole carbon source, and catabolite repressed by succinate. In contrast, *E. coli* LamB is induced by maltose, which is not a growth substrate for *P. aeruginosa*. Nevertheless, studies of both OprB and LamB indicate that they form rather similar channels, with small single-channel conductance for KCl that can be blocked by maltodextrins of four sugars more effectively than by glucose. Structural predictions based on regions of homology with LamB indicate that OprB has a cluster of five tryptophan and seven phenylalanine residues that resemble the so-called "greasy slide," which is proposed to guide the diffusion of sugars through the LamB channel.

Interposon mutants lacking OprB were deficient in passage across the outer membrane of a variety of sugars inducing mannitol, fructose, and glycerol (59). Similarly the liposome swelling experiments of Trias et al. indicated that OprB was selective for glucose and xylose (55). Thus, OprB is a general carbohydrate-selective porin. The equivalent porin of *P. putida* was highly similar in many properties (17) and is 80% identical (*P. putida* unfinished genome sequence). Interestingly, *P. aeruginosa* contains one other close OprB homolog, PA2291, which demonstrates 96% identity to OprB, and another, PA4099, which has 24% identical and 12% similar amino acids.

GATED PORINS

Iron is a requirement for virtually all microorganisms and it is of particular importance in aerobic metabolism (56). Thus bacteria have evolved a series of elegant strategies for acquiring iron, including the production and secretion of powerful iron-binding compounds called siderophores, and the direct acquisition of iron from heme or hemoglobin, *Pseudomonas* sp., being aerobes, employ a wide variety of uptake systems for acquisition of iron in conjunction with siderophores (both known and unknown) and heme/hemoglobin. The initial step in uptake involves association with an outer membrane receptor protein. Prototypes of these receptors have been crystallized (22) and form 22-stranded β-barrels, into the center of which folds a 4-stranded β-sheet domain (visualized as a gate) (Figure 1*A*). Engagement of the receptor by the ferric-iron-loaded compound and energy input through the auspices of a periplasm-spanning inner-membrane protein called TonB (in conjunction with ExbB and ExbD) leads to a conformational change that opens the gate and lets the iron-loaded compound through the outer membrane (21). Such receptor proteins are termed TonB-dependent receptors and/or iron-regulated outer membrane proteins (IROMP) and function as gated porins. A major surprise arising from knowledge of the genome sequence was the large number [35] of such gated porin homologs (Tables 2 and 3). Here we describe only those members of this family that have been functionally characterized.

FpvA

P. aeruginosa pyoverdine is a 6,7-dihydroxyyquinolone-containing fluorescent compound joined to a partly cyclic octapeptide. It has high affinity for Fe^{3+}. It is probably the predominant siderophore for iron acquisition from transferrin or serum in vivo (56). Although many *Pseudomonas* sp. produce pyoverdine siderophores, there is chemical heterogeneity and considerable specificity in that each bacterium tends to utilize its own siderophore and few others (13). This specificity is mediated at the level of the outer membrane receptors/gated porins. The receptor for the pyoverdine of *P. aeruginosa* PA01 is FpvA (13).

It was demonstrated that FpvA copurified with iron-free pyoverdine, but this did not lead to productive transport (47). Ferric-pyoverdine displaces this iron-free pyoverdine with rapid kinetics to form FpvA-pyoverdine-Fe^{3+} complexes in a reaction that is dependent on TonB. (In fact, *P. aeruginosa* has two TonB homologs, and one, TonB1, is preferred over TonB2 for this displacement). This then presumably leads to ferric-pyoverdine translocation across the membrane. Insertion mutagenesis of *fpvA* has identified two sites, Y359 and Y402 (13), where incorporation of an 18–amino acid–encoding sequence compromised ferric pyoverdine binding and uptake. These residues are presumed to be extramembranous, and it was hypothesized that they are in a region that is involved in ligand binding.

PupA, PupB

The characterized ferric-pyoverdine receptors of *P. putida* are called PupA and PupB, although pyoverdines are also called pseudobactins in this species (6, 25). These proteins are homologous to FpvA and other putative TonB receptors of *P. aeruginosa* (Table 3). These receptors have different specificities in that PupA is a specific receptor for ferric pseudobactin 358 while PupB facilitates transport via two siderophores, pseudobactin BN7 and BN8, as well as being inducible by a variety of heterologous siderophores. Interestingly, Bitter et al. (6) constructed hybrid siderophores with the *E. coli* ferric-coprogen receptor FhuA, and these hybrids were active and helped define domains of these proteins. For example, it was concluded that the ligand-binding domains were located in different regions of these proteins. Both the genome sequence and specific PCR experiments indicate that *P. putida* contains multiple ferric-pseudobactin (pyoverdine) receptors. This is certainly also true for *P. aeruginosa* PA01 (Table 3).

FptA/PfeA/Heme

Another *P. aeruginosa* siderophore named pyochelin is structurally distinct in possessing neither hydroxamate nor catecholate-chelating groups. When loaded

TABLE 3 Probable outer-membrane porins

PAID	Gene name	Range from	Range to	Similarity	No. of AAs	Probable porin class[a]
PA2760	oprQ	3120072	3121349	59% similar to OprD, named OprE3 in Genbank	425	SP
PA2291	opbA	2522616	2521258	62% similar to OprB of *P. aerguinosa*	452	SP
PA2700	opdB	3053843	3055150	57% similar to OprD	435	SP
PA0162	opdC	184594	185928	58% similar to OprD	444	SP
PA1025	opdD	1110947	1112197	62% similar to PhaK of *P. putida*; OprD family	416	SP
PA0240	opdF	271838	270573	53% similar to OprE; OprD family	421	SP
PA2213	opdG	2432312	2433562	60% similar to PhaK of *P. putida*; OprD family	416	SP
PA0755	opdH	824198	822915	58% similar to OprE; OprD family	427	SP
PA0189	opdI	216908	215550	55% similar to OprD	452	SP
PA2420	opdJ	2702925	2704343	51% similar to OprD	472	SP
PA4898	opdK	5495712	5494459	56% similar to PhaK of *P. putida*; OprD family	417	SP
PA4137	opdL	4626661	4627917	69% similar to PhaK of *P. putida*; OprD family	418	SP
PA4179	opdN	4674943	4676238	58% similar to PhaK of *P. putida*; OprD family	431	SP
PA2113	opdO	2324783	2323554	62% similar to PhaK of *P. putida*; OprD family	409	SP
PA4501	opdP	5038900	5040354	52% similar to OprD	484	SP
PA3038	opdQ	3400683	3401948	65% similar to PhaK of *P. putida*; OprD family	421	SP
PA3588	opdR	4021918	4020668	56% similar to OprE; OprD family	416	SP
PA2505	opdT	2823919	2822573	57% similar to OprD, named OprD3 in Genbank	448	SP
PA1288	fadL	1400505	1399231	47% similar to fatty acid transport protein FadL of *E. coli*	424	SP
PA4589		5140440	5139049	41% similar to fatty acid transport protein FadL of *E. coli*	463	SP
PA1764		1906842	1908440	40% similar to fatty acid transport protein FadL of *E. coli*	532	SP
PA4099		4581392	4582696	36% similar to glucose porin OprB	434	SP
PA0165		189120	189956	46% similar to region of OMP Tsx of *S. typhimurium*	278	SP?
PA2522	czcC	2843304	2842018	59% similar to cation efflux protein CzcC of *R. eutropha*; OprM family	428	EP

TABLE 3 (*Continued*)

PAID	Gene name	Range from	Range to	Similarity	No. of AAs	Probable porin class[a]
PA2837	opmA	3190210	3191649	53% similar to OprN	479	EP
PA2525	opmB	2847778	2846282	50% similar to OprM	498	EP
PA4208	opmD	4710620	4712083	56% similar to OprN	487	EP
PA3521	opmE	3939494	3938019	52% similar to OprN	491	EP
PA4592	opmF	5144533	5143052	40% similar to type I secretion protein CyaE of *B. pertussis*; OprM family	493	EP, PI?
PA5158	opmG	5805679	5807157	53% similar to putative aromatic efflux pump OMP of *S. aromaticivorans*; OprM family	492	EP
PA4974	opmH	5584100	5585548	54% similar to efflux porin TolC of *E. coli*	482	EP, PI?
PA3894	opmI	4362983	4361493	51% similar to putative aromatic efflux pump OMP of *S. aromaticivorans*; OprM family	496	EP
PA1238	opmJ	1340527	1339079	51% similar to OprN	482	EP
PA4144	opmK	4636297	4637712	49% similar to Type I secretion protein CyaE of *B. pertussis*; OprM family.	471	EP, PI?
PA1875	opmL	2043847	2045124	41% similar to AprF; OprM family	425	EP, PI?
PA3404	opmM	3810612	3809257	68% similar to AprF; OprM family	451	EP, PI
PA2391	opmQ	2645303	2646727	48% similar to OprM	474	EP
PA0931	pirA	1018230	1020458	72% similar to ferric enterobactin receptor PfeA	742	GP
PA4514	piuA	5055876	5053615	49% similar to putative iron transport receptor of *E. coli*	753	GP
PA1910	ufrA	2084267	2081853	99% similar to undefined iron transport receptor UfrA of *P. aeruginosa*	804	GP
PA1322	pfuA	1433166	1435364	44% similar to ferrichrome-iron receptor of *S. paratyphi*	732	GP
PA0674	pigC	734159	734875	53% similar to FpvA	238	GP
PA1922	cirA	2097491	2099452	56% similar to iron-regulated colicin I receptor of *E. coli*.	653	GP
PA3901	fecA	4368836	4371190	75% similar to ferric citrate receptor FecA of *E. coli*	784	GP
PA0470	fiuA	532437	530029	98% similar to ferrioxamine receptor of *P. aeruginosa*	802	GP
PA1302	hxuC	1411585	1414140	57% similar to Ton-dependent heme receptor TdhA of *H. ducreyi*	851	GP
PA4675	optH	5243177	5245405	62% similar to ferric aerobactin receptor IutA *E. coli*	742	GP
PA4897	optI	5491345	5494314	52% similar to OM hemin receptor of *P. aeruginosa*	989	GP

(*Continued*)

TABLE 3 (*Continued*)

PAID	Gene name	Range from	Range to	Similarity	No. of AAs	Probable porin class[a]
PA2335	optO	2577150	2579519	37% similar to pesticin receptor of *Y. pestis*	789	GP
PA2466	optS	2785225	2782763	63% similar to ferrioxamine receptor FoxA of *Y. enterocolitica*	820	GP
PA4837		5429841	5427715	45% similar to ferrichrome iron receptor FhuA of *E. agglomerans*	708	GP
PA0151		171047	173434	43% similar to ferric-pseudobactin receptor PupB of *P. putida*	795	GP
PA0192		219172	221544	39% similar to pesticin receptor FyuA of *Y. enterocolitica*	790	GP
PA0434		484964	487156	43% similar to ferric-pseudobactin receptor PupB of *P. putida*	730	GP
PA0781		851319	849256	37% similar to PhuR	687	GP
PA0982		1065103	1064555	46% similar to 27-kDa OMP of *Coxiella burnetii*; probable TonB-dependent receptor	182	GP
PA1271		1381804	1383654	46% similar to BtuB, OM receptor for transport of vitamin B12 of *E. coli*	616	GP
PA1365		1476384	1478825	68% similar to the ferric alcaligin receptor AleB of *R. eutropha*	813	GP
PA1613		1758597	1756489	37% similar to OM receptor for colicin I CirA of *E. coli*	702	GP
PA2057		2251275	2253815	43% similar to ferric-pseudobactin receptor PupB of *P. putida*	846	GP
PA2089		2298012	2300663	40% similar to ferric enterobactin receptor of *B. pertussis*	883	GP
PA2289		2518561	2516429	56% similar to putative OM receptor for iron transport in *E. coli*	710	GP
PA2911		3265847	3268003	42% similar to putative hydroxamate-type ferrisiderophore receptor of *P. aeruginosa*	718	GP
PA3268		3658150	3655985	61% similar to *E. coli* ferric citrate receptor FecA	721	GP
PA4156		4652457	4650373	48% similar to ferric vibriobactin receptor ViuA of *V. cholerae*	694	GP
PA4168		4663853	4666261	54% similar to ferripyoverdine receptor FpvA	802	GP
PA2590		2933461	2930807	50% similar to ferric enterobactin receptor of *Xylella fastidiosa*	884	GP
PA0685	hxcQ	741925	744336	49% similarity to type II secretion protein XcpQ	803	PII
PA1382		1498813	1501092	49% similar to S-protein secretion D of *Aeromonas hydrophila*	759	PII
PA4304		4829628	4828378	47% similar to type II secretion protein of *Mesorhizobium loti*	416	PII

[a]P, general porin; SP, specific porin; GP, putative gated porin; EP, OprM family member of efflux and protein secretion porins; PI, putative type I secretion subfamily; PII, type II secretion channel; PIII, type III secretion channel.

with iron it is taken up by the FptA receptor (3). Interestingly, FptA is a known virulence determinant (56).

The *E. coli* siderophore enterobactin can utilize the PfeA receptor in iron uptake (11). Indeed PfeA shows more than 60% homology to its *E. coli* counterpart, the crystallized gated porin FepA, with especially high homology in the ligand-binding regions. Consistent with this, the cloned *pfeA* gene complemented an *E. coli fepA* mutant to permit enterobactin-dependent iron uptake. It has been postulated that a second, lower-affinity ferric-enterobactin uptake system exists in *P. aeruginosa* (56), and a protein named PirA, which has 72% similarity to PfeA, is a candidate for being responsible for this uptake system.

Another characterized uptake system in *P. aeruginosa* is the heme iron uptake system. This involves two outer membrane receptors, HasR and PhuR (40). Both systems mediate growth on hemin or hemoglobin as its sole iron source, but it requires a double knockout to eliminate growth on hemin and hemoglobin. Another ORF in *P. aeruginosa* termed OptI is 52% similar to HasR (Table 3) but has not been characterized. A variety of other TonB dependent iron uptake receptors exist but only one, Fiu, has been defined. This apparently acts as the receptor for the uptake of ferrioxamine B (56).

OprC

OprC was first described as a nonselective porin that formed slightly anion-selective, small diffusion pores (17). However, later work disproved a role in antibiotic uptake (61, 62). Nakae and collaborators demonstrated that OprC is 65% homologous with *P. stuzeri* NosA (27, 61), an outer membrane porin required for production of the Cu^{2+}-containing nitrate reductase. OprC is only made anaerobically and is repressed by high medium Cu^{2+} concentrations (61). It is interesting that it shows substantial homology to PfeA (25% identity, 40% similarity over 504 amino acids from the N and C termini) and thus appears to be a member of the large TonB-dependent family of proteins, most of which are involved in uptake of complexed iron. It seems likely that the substrate for OprC is actually Cu^{2+}. Another unusual member of the TonB family of outer membrane receptors is BtuB, the *E. coli* receptor for vitamin B12 (10). *P. aeruginosa* contains a gene PA1271 that is 46% similar to BtuB.

OprH

OprH is an outer membrane protein that is upregulated upon Mg^{2+} starvation by the PhoPQ two-component regulatory system, with which it forms the *oprH phoP phoQ* operon (31). Insertion and deletion mutagenesis have demonstrated that it forms an eight-stranded β-barrel (45). While devoid of porin activity in its native form, it forms channels when surface loop 4 is deleted (B. Rehm & R.E.W. Hancock, unpublished observations). Thus, it is possible that OprH is a gated porin for divalent cations.

EFFLUX PORINS

As described above, active efflux is a major contributor to intrinsic multiple antibiotic resistance in *P. aeruginosa.* In addition overexpression of any of at least three efflux operons leads to even higher resistance to a wide range of clinically useful antibiotics (42). The most important efflux systems in *P. aeruginosa* are members of the resistance-nodulation-division (RND) family. This series of efflux systems involves a three-component efflux pathway, which includes a cytoplasmic-membrane pump protein, a peripheral cytoplasmic-membrane linker (sometimes called a membrane-fusion protein), and an elaborate outer membrane/periplasmic channel protein. Each of these proteins is so highly conserved (at around the 20% or greater identity level) that sequence homology searching can easily identify them. We largely concern ourselves here with the outer membrane channel proteins (termed here efflux porins). The best studied of these is the *E. coli* TolC channel-tunnel, which was recently crystallized (24) and is discussed below. This protein has a dual function in multiple antibiotic efflux and as a component of the type I secretion system for hemolysin.

P. aeruginosa has 18 outer membrane proteins with putative functions in efflux (49). Eleven of these, including OprM, OprN, and OprJ, fall into one phylogenetic subclass (www.cmdr.ubc.ca/bobh/omps/phylogenetic.htm) and are presumed to be parts of specialized multiple antibiotic efflux systems. Of the other seven, one is a homolog of CzcC that is involved in cation efflux (as a detoxification mechanism), one is AprF, which is involved in the type I secretion of alkaline protease, and a third OpmH is the closest *P. aeruginosa* homolog of *E. coli* TolC (54% similar). The four others, OpmF, OmpK, OpmL, and OpmM, are similar to CyaE of *Burkholderia pertussis* or to AprF (Table 3), and thus are likely to also be components of type I protein secretion pathways for as-yet-unknown substrates. As is clear for TolC, a single efflux outer membrane protein can serve more than one secretion/efflux system (24) in part because specificity is determined by the pump component in combination with the linker (45).

OprM

OprM is the major outer membrane efflux porin involved in intrinsic multiple antibiotic resistance in *P. aeruginosa.* Deletion of OprM leads to 10–1000-fold increases in susceptibility to many antibiotics from different classes (29), and the cloned OprM gene can complement such deletions. Conversely, mutations in the *nalB* (*mexR*) gene can lead to overexpression of OprM and its neighboring linker and pump proteins, MexA and MexB, and cause resistance to a broad range of antibiotics (42).

It has also been reported that OprM collaborates with the MexX-MexY system to mediate aminoglycoside resistance (1), although certain results (57) and our own unpublished studies are not entirely consistent with this conclusion. OprM shares only 21% identity with TolC, but it can be structurally modeled based on the TolC crystal structure (58) (Figure 1*B*) using a procedure called threading. Studies involving insertions and deletions in two laboratories (28, 58) indicate that the resultant model is reasonably accurate. Thus we can define OprM function by reference to the TolC-like model (24). OprM is assumed to be a trimer of three subunits that comprises a single channel-tunnel spanning the outer membrane and periplasm. The trimer forms a 12-stranded β-barrel (4 β-strands per monomer) that lodges in the outer membrane and sits atop a coiled 12-helix α-helical barrel that spans the periplasm and is presumed to contact the MexB-pump/MexA-linker complex in the cytoplasmic membrane. The α-helical barrel twists into a constricted point at the base proximal to the cytoplasmic membrane and is proposed to open like an iris diaphragm upon contact with the pump/linker complex, energy input, and possibly substrate engagement. Indeed freshly purified OprM formed nice large channels in planar lipid bilayer experiments, but over time of storage the channel conductance became much smaller (58), in fact similar to the conductance of *E. coli* TolC (5), a result that is consistent with open and closed states of the OprM channel.

Mutagenesis of OprM has contributed substantially to the overall picture of how this protein operates (28, 58). Insertions in the surface loop regions of the outer membrane barrel do not influence function, whereas insertions or deletions in most locations within the α-helical barrel are nonpermissive. As shown for TolC, there is a putative girdle around the periplasmic α-helical barrel segment, but deletions or insertions in this region seem to be well tolerated. Similarly, deletions and insertions at both the N and C termini, including removal of the putative N-terminal acylation site (such that OprM cannot become a lipoprotein), are tolerated and largely without functional consequences.

Other Multidrug Efflux Porins

P. aeruginosa has at least two other efflux porins, OprJ and OprN, that are normally silent but can be highly expressed due to mutation, as part of the MexCD-OprJ and MexEF-OprN operons, leading to multidrug resistance (17, 42). Most overexpressing mutants are in the *nfxB* repressor and *mexT* (*nfxC*) activator genes. The latter system is very interesting since *mexT* mutations lead to coordinate upregulation of the MexEF-OprN efflux system and downregulation of OprD (23, 39).

P. putida has several homologous systems that have been largely studied because of their ability to efflux aromatic hydrocarbons (42). However, systems involving the efflux porins ArpC, MepC, TtgC, and TtgI all influence antibiotic susceptibility when overexpressed. We have also gathered preliminary evidence linking OpmG, OpmH, and OpmI to aminoglycoside efflux in *P. aeruginosa* (J. Jo & R.E.W. Hancock, unpublished data).

AprF and Protein Secretion

P. aeruginosa secretes many proteins involved in virulence, utilizing largely type II secretion systems (52, 53). However, alkaline protease, the product of the *aprA* gene, is secreted by a three-component type 1 secretion system, AprDEF, where

AprF is the outer membrane component (53). Based on homology modeling and the known dual function of TolC (24), it seems possible to conclude that the outer membrane efflux component AprF functions similarly to TolC, although clearly folding of AprA must be avoided during secretion, suggesting a potential chaperone-like function. As mentioned above, there are five other homologs that could be engaged in type I secretion.

In addition to type I secretion, there are both type II and type III protein secretion pathways in *P. aeruginosa* (49). Both utilize outer membrane proteins that form ring-like structures with multiple subunits (9, 53). These presumably form the channels for secretion of proteins; however, because a large channel would compromise the low outer membrane permeability of *P. aeruginosa*, it is presumed that these channels are gated, probably by engagement of the specific secreted protein with other components of the secretion apparatus. In *P. aeruginosa*, the major outer membrane channel for the type I general secretion pathway is XcpQ (9). There is one other XcpQ homolog named XqhA (32). For type III secretion, the XcpQ homolog PscC acts as a channel (60).

CONCLUDING REMARKS

Publication of the genome sequence of *P. aeruginosa* (49) has dramatically expanded the extent of our understanding of the *P. aeruginosa* outer membrane and its porins. With the genomic sequencing of *P. putida* finished, and that of *P. syringae* and *P. fluorescens* underway, we are rapidly moving to where we can better understand how the outer membrane contributes to the extraordinary versatility of this group of organisms. The biggest surprise in the genomic sequence of *P. aeruginosa* was the finding of three large families of outer membrane proteins with 18–35 individual members. These families arose from distant gene duplication events followed by evolutionary divergence, almost as if *Pseudomonas* was operating on a rather simple blueprint. Thus these families likely reflect the versatility of this organism as each protein diverged to have slightly differing functions and was selected to be maintained in the organism. A major topic of research will be to attempt to understand how the functions of these proteins relate to their diverged sequences and how the regulatory network permits them to be expressed when they are needed.

ACKNOWLEDGMENTS

We gratefully acknowledge and thank Jennifer L. Gardy (Simon Fraser University) for her assistance with analysis of proposed *P. aeruginosa* outer membrane proteins and her leadership regarding website design. The work of the authors was supported by grants from the Canadian Institutes of Health Research, the Canadian Cystic Fibrosis Foundation (to R.E.W. Hancock) and the Natural Sciences and Engineering Research Council of Canada (to F.S.L. Brinkman). R.E.W. Hancock holds a Canada Research Chair and F.S.L. Brinkman is a Michael Smith Foundation for Health Research Scholar.

LITERATURE CITED

1. Aires JR, Köhler T, Nikaido H, Plesiat P. 1999. Involvement of an active efflux system in the natural resistance of *Pseudomonas aeruginosa* to aminoglycosides. *Antimicrob. Agents Chemother.* 43:2624–28
2. Andersen C, Bachmeyer C, Tauber H, Benz R, Wang J, et al. 1999. In vivo and in vitro studies of major surface loop deletion mutants of the *Escherichia coli* K-12 maltoporin: contribution to maltose and maltooligosaccharide transport and binding. *Mol. Microbiol.* 32:851–67
3. Ankenbauer RG, Quan HN. 1994. FptA, the Fe(III)-pyochelin receptor of *Pseudomonas aeruginosa*: a phenolate siderophore receptor homologous to hydroxamate siderophore receptors. *J. Bacteriol.* 176:307–19
4. Bellido F, Martin NL, Siehnel RJ, Hancock REW. 1992. Reevaluation, using intact cells, of the exclusion limit and role of porin OprF in *Pseudomonas aeruginosa* outer membrane permeability. *J. Bacteriol.* 174:5196–203
5. Benz R, Maier E, Gentschev I. 1993. TolC of *Escherichia coli* functions as an outer membrane channel. *Zentralbl. Bakteriol.* 278:187–96
6. Bitter W, van Leeuwen IS, de Boer J, Zomer HW, Koster MC, et al. 1994. Localization of functional domains in the *Escherichia coli* coprogen receptor FhuE and the *Pseudomonas putida* ferric-pseudobactin 358 receptor PupA. *Mol. Gen. Genet.* 245:694–703
7. Brinkman FS, Bains M, Hancock REW. 2000. The amino terminus of *Pseudomonas aeruginosa* outer membrane protein OprF forms channels in lipid bilayer membranes: correlation with a three-dimensional model. *J. Bacteriol.* 182:5251–55
8. Brinkman FS, Schoofs G, Hancock RE, De Mot R. 1999. Influence of a putative ECF sigma factor on expression of the major outer membrane protein, OprF, in *Pseudomonas aeruginosa* and *Pseudomonas fluorescens*. *J. Bacteriol.* 181:4746–54
9. Brok R, Van Gelder P, Winterhalter M, Ziese U, Koster AJ, et al. 1999. The C-terminal domain of the *Pseudomonas* secretin *XcpQ* forms oligomeric rings with pore activity. *J. Mol. Biol.* 294:1169–79
10. Cadieux N, Bradbeer C, Kadner RJ. 2000. Sequence changes in the ton box region of BtuB affect its transport activities and interaction with TonB protein. *J. Bacteriol.* 182:5954–61
11. Dean CR, Poole K. 1993. Cloning and characterization of the ferric enterobactin receptor gene (pfeA) of *Pseudomonas aeruginosa*. *J. Bacteriol.* 175:317–24
12. El Hamel C, Freulet MA, Jaquinod M, De E, Molle G, Orange N. 2000. Involvement of the C-terminal part of *Pseudomonas fluorescens* OprF in the modulation of its pore-forming properties. *Biochim. Biophys. Acta* 1509:237–44
13. Folschweiller N, Schalk IJ, Celia H, Keiffer B, Abdallah MA, et al. 2000. The pyoverdin receptor FpvA, a TonB receptor involved in iron uptake by *Pseudomonas aeruginosa*. *Mol. Membr. Biol.* 17:123–33
14. Gilleland HE, Gilleland LB, Staczek J, Harty RN, Garcia-Sastre A, et al. 2000. Chimeric animal and plant viruses expressing epitopes of outer membrane protein F as a combined vaccine against *Pseudomonas aeruginosa* lung infection. *FEMS Immunol. Med. Microbiol.* 27:291–97
15. Hancock REW. 1997. The bacterial outer membrane as a drug barrier. *Trends Microbiol.* 5:37–42
16. Hancock REW, Speert DP. 2000. Antibiotic resistance in *Pseudomonas aeruginosa*. Mechanisms and impact on treatment. *Drug Resist. Updat.* 3:247–55
17. Hancock REW, Worobec EA. 1998. Outer

membrane proteins. See Ref. 33, pp. 139–67

18. Huang H, Hancock REW. 1993. Genetic definition of the substrate selectivity of outer membrane porin protein OprD of *Pseudomonas aeruginosa*. *J. Bacteriol.* 175:7793–800
19. Huang H, Hancock REW. 1996. The role of specific surface loop regions in determining the function of the imipenem-specific pore protein OprD of *Pseudomonas aeruginosa*. *J. Bacteriol.* 178:3085–90
20. Huang H, Jeanteur D, Pattus F, Hancock REW. 1995. Membrane topology and site-specific mutagenesis of *Pseudomonas aeruginosa* porin OprD. *Mol. Microbiol.* 16:931–41
21. Klebba PE, Newton SM. 1998. Mechanisms of solute transport through outer membrane porins: burning down the house. *Curr. Opin. Microbiol.* 1:238–47
22. Koebnik R, Locher KP, Van Gelder P. 2000. Structure and function of bacterial outer membrane proteins: barrels in a nutshell. *Mol. Microbiol.* 37:239–53
23. Kohler T, Michea-Hamzehpour M, Henze U, Gotoh N, Curty LK, Pechere J-C. 1997. Characterization of MexE-MexF-OprN, a positively regulated multidrug efflux system of *Pseudomonas aeruginosa*. *Mol. Microbiol.* 23:345–54
24. Koronakis V, Li J, Koronakis E, Stauffer K. 1997. Structure of TolC, the outer membrane component of the bacterial type I efflux system, derived from two-dimensional crystals. *Mol. Microbiol.* 23:617–26
25. Koster M, van de Vossenberg J, Leong J, Weisbeek PJ. 1993. Identification and characterization of the pupB gene encoding an inducible ferric-pseudobactin receptor of *Pseudomonas putida* WCS358. *Mol. Microbiol.* 8:591–601
26. Larbig M, Mansouri E, Freihorst J, Tummler B, Kohler G, et al. 2001. Safety and immunogenicity of an intranasal *Pseudomonas aeruginosa* hybrid outer membrane protein F-I vaccine in human volunteers. *Vaccine* 19:2291–97
27. Lee HS, Hancock REW, Ingraham JL. 1989. Properties of a *Pseudomonas stutzeri* outer membrane channel-forming protein (NosA) required for production of copper-containing N_2O reductase. *J. Bacteriol.* 171:2096–100
28. Li XZ, Poole K. 2001. Mutational analysis of the OprM outer membrane efflux component of the MexA-MexB-OprM multidrug efflux system of *Pseudomonas aeruginosa*. *J. Bacteriol.* 183:12–27
29. Li XZ, Poole K, Nikaido H. 1995. Role of MexA-MexB-OprM in antibiotic efflux in *Pseudomonas aeruginosa*. *Antimicrob. Agents Chemother.* 39:1948–53
30. Lynch MJ, Drusano GL, Mobley HLT. 1987. Emergence of resistance to imipenem in *Pseudomonas aeruginosa*. *Antimicrob. Agents Chemother.* 31:1892–96
31. Macfarlane ELA, Kwasnicka A, Ochs MM, Hancock REW. 1999. PhoP-PhoQ homologues in *Pseudomonas aeruginosa* regulate expression of the outer-membrane protein OprH and Polymyxin B resistance. *Mol. Microbiol.* 34:305–16
32. Martinez A, Ostrovsky P, Nunn DN. 1998. Identification of an additional member of the secretin superfamily of proteins in *Pseudomonas aeruginosa* that is able to function in type II protein secretion. *Mol. Microbiol.* 28:1235–46
33. Montie T, ed. 1998. *Pseudomonas. Biotechnology Handbooks*, Vol. 10. London: Plenum. 334 pp.
34. Nakae T, Nakajima A, Ono T, Saito K, Yoneyama H. 1999. Resistance to β-lactam antibiotics in *Pseudomonas aeruginosa* due to interplay between the MexAB-OprM efflux pump and β-lactamase. *Antimicrob. Agents Chemother.* 43:1301–3
35. Nikaido H, Hancock REW. 1986. Outer membrane permeability of *Pseudomonas aeruginosa*. In *The Bacteria: A Treatise on Structure and Function*, ed. JR Sokatch, pp. 145–93. London: Academic
36. Nikaido H, Nikaido T, Harayama S. 1991. Identification and characterization of

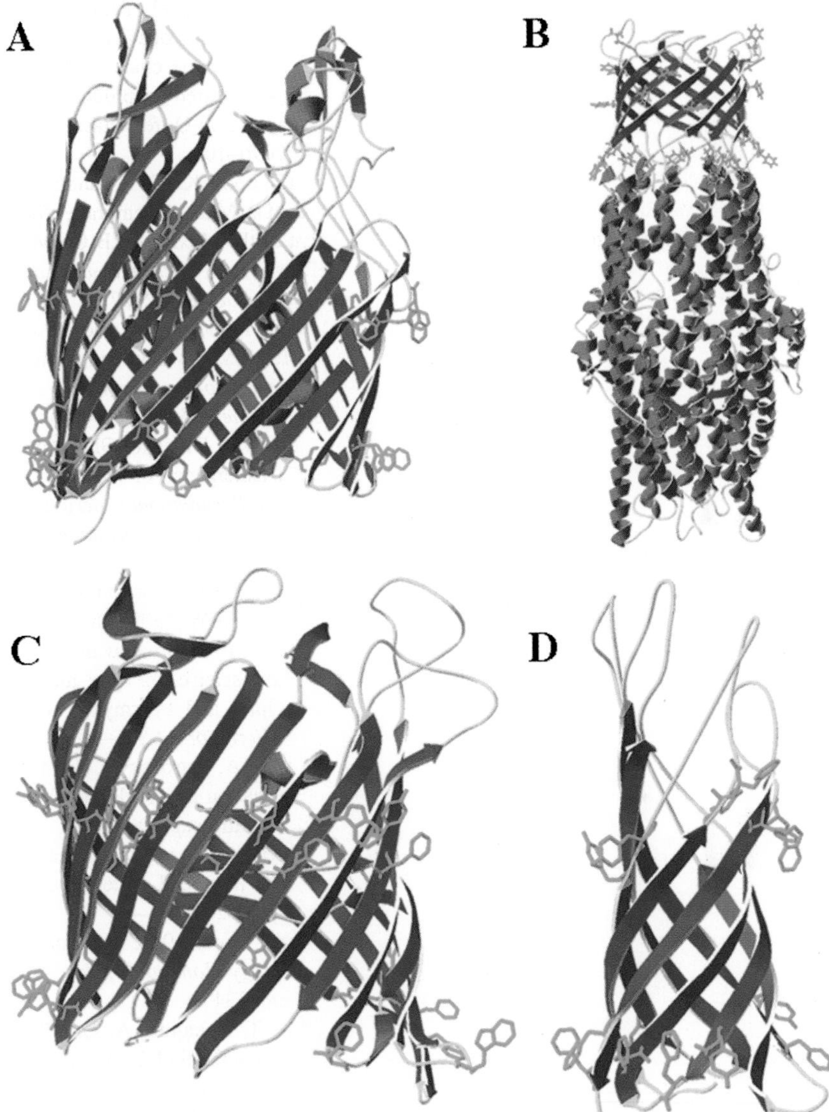

Figure 1 Representative models of the four classes of porins, based on crystal structures of *E. coli* gated porin FepA (*A*) and specific porin LamB (*C*) (22) and homology models of *P. aeruginosa* efflux porin OprM (*B*) and the N-terminal domain of the nonspecific porin OprF (*D*). The homology models were developed by threading to orthologous *E. coli* proteins as previously described (7, 58). Structures are colored to aid visualization of β-strands (*blue*), α-helices (*red*), and loop regions (*yellow*) with aromatic residues that form "rings" around the β-barrels illustrated in *green*. Such rings are proposed to stabilize the barrel in the membrane, being situated at the lipid-solvent interface.

porins in *Pseudomonas aeruginosa*. *J. Biol. Chem.* 266:770–79

37. Ochs MM, Bains M, Hancock REW. 2000. Role of putative loops 2 and 3 in imipenem passage through the specific porin OprD of *Pseudomonas aeruginosa*. *Antimicrob. Agents Chemother.* 44:1983–85
38. Ochs MM, Lu CD, Hancock REW, Abdelal AT. 1999. Amino acid–mediated induction of the basic amino acid–specific outer membrane porin OprD from *Pseudomonas aeruginosa*. *J. Bacteriol.* 181:5426–32
39. Ochs MM, McCusker MP, Bains M, Hancock REW. 1999. Negative regulation of the *Pseudomonas aeruginosa* outer membrane porin OprD selective for imipenem and basic amino acids. *Antimicrob. Agents Chemother.* 43:1085–90
40. Ochsner UA, Johnson Z, Vasil ML. 2000. Genetics and regulation of two distinct haem-uptake systems, *phu* and *has*, in *Pseudomonas aeruginosa*. *Microbiology* 146:185–98
41. Okamoto K, Gotoh N, Nishino T. 2001. *Pseudomonas aeruginosa* reveals high intrinsic resistance to penem antibiotics. Penem resistance mechanisms and their interplay. *Antimicrob. Agents Chemother.* 45:1964–71
42. Poole K. 2001. Multidrug resistance in Gram-negative bacteria. *Curr. Opin. Microbiol.* 4:500–8
43. Pumbwe L, Everett MJ, Hancock REW, Piddock LJ. 1996. Role of gyrA mutation and loss of OprF in the multiple antibiotic resistance phenotype of *Pseudomonas aeruginosa* G49. *FEMS Microbiol. Lett.* 143:25–28
44. Rawling EG, Brinkman FS, Hancock REW. 1998. Roles of the carboxy-terminal half of *Pseudomonas aeruginosa* major outer membrane protein OprF in cell shape, growth in low-osmolarity medium, and peptidoglycan association. *J. Bacteriol.* 180:3556–62
45. Rehm BHA, Hancock REW. 1996. Membrane topology of the outer membrane protein OprH from *Pseudomonas aeruginosa*: PCR-mediated site-directed insertion and deletion mutagenesis. *J. Bacteriol.* 178:3346–49
46. Saint N, El Hamel C, De E, Molle G. 2000. Ion channel formation by N-terminal domain: a common feature of OprFs of *Pseudomonas* and OmpA of *Escherichia coli*. *FEMS Microbiol. Lett.* 190:261–65
47. Schalk IJ, Hennard C, Dugave C, Poole K, Abdallah MA, et al. 2001. Iron-free pyoverdin binds to its outer membrane receptor FpvA in *Pseudomonas aeruginosa*: a new mechanism for membrane iron transport. *Mol. Microbiol.* 39:351–60
48. Schirmer T, Keller TA, Wang YF, Rosenbusch JP. 1995. Structural basis for sugar translocation through maltoporin channels at 3.1 Å resolution. *Science* 267:512–14
49. Stover KC, Pham XQ, Erwin AL, Mizoguchi SD, Warrener P, et al. 2000. Complete genome sequence of *Pseudomonas aeruginosa*: an opportunistic pathogen. *Nature* 406:959–64
50. Sugawara E, Steiert M, Rouhani S, Nikaido H. 1996. Secondary structure of the outer membrane proteins OmpA of *Escherichia coli* and OprF of *Pseudomonas aeruginosa*. *J. Bacteriol.* 178:6067–69
51. Sukhan A, Hancock REW. 1996. The role of specific lysine residues in the passage of anions through the *Pseudomonas aeruginosa* porin OprP. *J. Biol. Chem.* 271: 21239–42
52. Thanassi DG, Hultgren SJ. 2000. Multiple pathways allow protein secretion across the bacterial outer membrane. *Curr. Opin. Cell Biol.* 12:420–30
53. Tommassen J, Filloux A, Bally M, Murgier M, Lazdunski A. 1992. Protein secretion in *Pseudomonas aeruginosa*. *FEMS Microbiol. Rev.* 9:73–90
54. Trias J, Nikaido H. 1990. Protein D2 channel of the *Pseudomonas aeruginosa* outer membrane has a binding site for basic amino acids and peptides. *J. Biol. Chem.* 265:15680–84
55. Trias J, Rosenberg EY, Nikaido H. 1988. Specificity of the glucose channel formed

by protein D1 of *Pseudomonas aeruginosa. Biochim. Biophys. Acta.* 938:493–96

56. Vasil M, Ochsner UA. 1999. The response of *Pseudomonas aeruginosa* to iron: genetics, biochemistry, and virulence. *Mol. Microbiol.* 34:399–413
57. Westbrock-Wadman S, Sherman DR, Hickey MJ, Coulter SN, Zhu YQ, Warrener P. 1999. Characterization of a *Pseudomonas aeruginosa* efflux pump contributing to aminoglycoside impermeability. *Antimicrob. Agents Chemother.* 43:2975–83
58. Wong KKY, Brinkman FSL, Benz RS, Hancock REW. 2001. Evaluation of a structural model of *Pseudomonas aeruginosa* outer membrane protein OprM, an efflux component involved in intrinsic antibiotic resistance. *J. Bacteriol.* 183:367–74
59. Wylie JL, Worobec EA. 1995. The OprB porin plays a central role in carbohydrate uptake in *Pseudomonas aeruginosa. J. Bacteriol.* 177:3021–26
60. Yahr TL, Goranson J, Frank DW. 1996. Exoenzyme S of *Pseudomonas aeruginosa* is secreted by a type III pathway. *Mol. Microbiol.* 22:991–1003
61. Yoneyama H, Nakae T. 1996. Protein C (OprC) of the outer membrane of *Pseudomonas aeruginosa* is a copper-regulated channel protein. *Microbiology* 142:2137–44
62. Yoneyama H, Yamano Y, Nakae T. 1995. Role of porins in the antibiotic susceptibility of *Pseudomonas aeruginosa*: construction of mutants with deletions in the multiple porin genes. *Biochem. Biophys. Res. Commun.* 213:88–95
63. Zgurskaya HI, Nikaido H. 2000. Multidrug resistance mechanisms: drug efflux across two membranes. *Mol. Microbiol.* 37:219–25

Annu. Rev. Microbiol. 2002. 56:39–64
doi: 10.1146/annurev.micro.56.012302.160959

First published online as a Review in Advance on July 15, 2002

The Bittersweet Interface of Parasite and Host: Lectin-Carbohydrate Interactions During Human Invasion by the Parasite *Entamoeba histolytica*

William A. Petri, Jr.,[1] Rashidul Haque,[2] and Barbara J. Mann[1]

[1]Division of Infectious Diseases, University of Virginia, MR4 Bldg Room 2115, Lane Road, Charlottesville Virginia 22908-1340; e-mail: wap3g@virginia.edu; bjm2r@virginia.edu
[2]International Centre for Diarrhoeal Disease Research, G.P.O. Box 128, Dhaka 1000, Bangladesh; e-mail: rhaque@icddrb.org

Key Words amebiasis, adherence, cyst, cytolysis, complement, colon, immunity

■ **Abstract** *Entamoeba histolytica*, as its name suggests, is an enteric parasite with a remarkable ability to lyse host tissues. However, the interaction of the parasite with the host is more complex than solely destruction and invasion. It is at the host-parasite interface that cell-signaling events commit the parasite to (*a*) commensal, noninvasive infection, (*b*) developmental change from trophozoite to cyst, or (*c*) invasion and potential death of the human host. The molecule central to these processes is an amebic cell surface protein that recognizes the sugars galactose (Gal) and N-acetylgalactosamine (GalNAc) on the surface of host cells. Engagement of the Gal/GalNAc lectin to the host results in cytoskeletal reorganization in the parasite. The parasite cytoskeleton regulates the extracellular adhesive activity of the lectin and recruits to the host-parasite interface factors required for parasite survival within its host. If the parasite lectin attaches to the host mucin glycoproteins lining the intestine, the result is commensal infection. In contrast, attachment of the lectin to a host cell surface glycoprotein leads to lectin-induced host cell calcium transients, caspase activation, and destruction via apoptosis. Finally, trophozoite quorum sensing via the lectin initiates the developmental pathway resulting in encystment. The structure and function of the lectin that controls these divergent cell biologic processes are the subject of this review.

CONTENTS

INTRODUCTION

Amebiasis is caused by *Entamoeba histolytica*, an enteric protozoan parasite exclusively of humans with a remarkable ability to kill cells in a contact-dependent manner. The two major clinical syndromes of amebiasis are amebic colitis and amebic liver abscess. Together they are thought to result in 50 million cases of colitis and liver abscess and 100,000 deaths worldwide each year (47, 81). There is no zoonotic reservoir, and no insect vectors required for transmission. The life cycle is therefore simple, with an infectious cyst and invasive trophozoite. Infection occurs when the cyst is ingested via contaminated water and food. This has been seen most recently in epidemic form with an outbreak in Tblissi, Republic of Georgia, due to a fecally contaminated municipal water supply (5), but occurs endemically every day among the poor of developing countries (30).

In developing countries, colonization with *E. histolytica* has been observed in 5% or more of poor children. Less than 10% of colonized individuals develop colitis, with the rest clearing the infection within months (24, 30). Patients with amebic colitis typically present with a several-week history of gradual onset of abdominal pain and tenderness, diarrhea, and bloody stools (dysentery). The pathological lesions in the colon include ulceration of the intestinal epithelium and invasion into the lamina propria by trophozoites (52, 80). Inflammation, with infiltrating neutrophils and mononuclear lymphocytes, is pronounced, but inflammatory cells near the amebae appear lysed with pyknotic nuclei. Liver abscess may present acutely with fever, right-upper abdominal tenderness and pain, or subacutely with prominent weight loss and abdominal pain. Most frequently patients with liver abscess will not have concurrent colitis, although a history of dysentery in the preceding year can sometimes be obtained (47).

The mysteries of the disease include:

1. the parasite, host, and environmental factors that determine if an amebic infection will be noninvasive or result in amebic colitis or amebic liver abscess;
2. the mechanism of parasite persistence in the host that leads to the sometime months-long latent period between infection and disease;
3. the adult male predominance of amebic liver abscess in the face of a lack of any sex or age predisposition to colonization and colitis; and
4. the biological fitness for the parasite that is derived from invasion and potential killing of the host by the noninfectious stage of the parasite.

THE COLONIC ENVIRONMENT AND DETERMINANTS OF INVASION

In the intestine, the ingested cyst excysts into eight trophozoites. It is in the intestine that the determinants of invasive versus commensal infection must come into play (Figure 1). Different strains of the parasite may intrinsically differ in their invasiveness. This is most remarkably the case with "nonpathogenic zymodemes," which never cause disease and are now genetically reclassified as a separate species *E. dispar* (70, 77). However, most *E. histolytica* (sensu stricto) infections also do

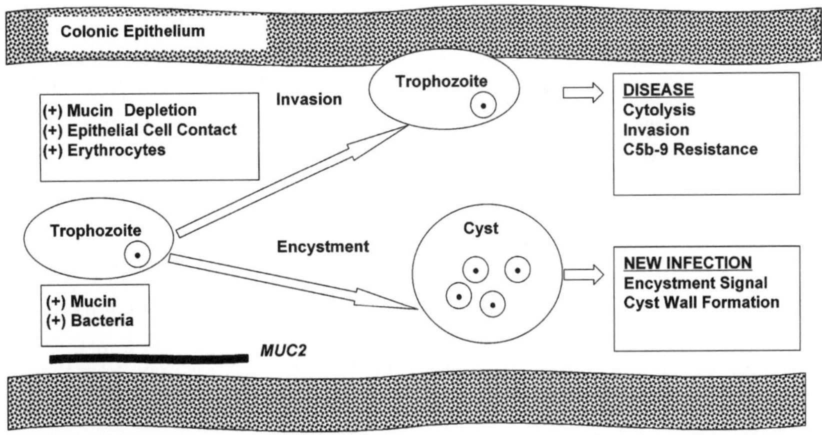

Figure 1 Environmental factors recognized by the Gal/GalNAc lectin that may signal encystment or invasion. Colonic mucin glycoproteins and/or bacteria may initiate a Gal/GalNAc lectin–induced encystment pathway. During encystment the Gal/GalNAc lectin may also be involved in cyst wall formation. Depletion of mucins allows parasite contact with the epithelial cell surface and ingestion of exudated erythrocytes, initiating an invasion pathway. Upon invasion of the host, lectin-mediated killing of immune effector cells and evasion of complement C5b-9 lysis may allow survival of the parasite.

not lead to invasion (30). It is likely that *E. histolytica* strains vary in invasiveness and that host innate and immune responses contribute to whether an infection leads to disease. Subtle interactions between the amebae, the human intestinal mucin glycoproteins, and the resident bacterial flora of the colon are also likely to be important.

One of the factors influencing invasiveness may be the bacterial flora of the intestine. Bacteria appear to be more than just a major food source for the amebae. Amebic trophozoites in the large intestine reside among bacteria in the mucin blanket that coats the intestinal epithelium. Bacteria influence the balance between production of trophozoites and cyst, with only certain mixtures of bacteria capable of supporting encystation in vitro (22). Because of the inability of the cyst to invade, conditions that promote encystment should inhibit invasiveness. Certain strains of bacteria in vitro have also been shown to alter the adherence and cytotoxicity of the trophozoite (7) and to downregulate the levels of the adherence lectin light subunit (see below).

Invasiveness is also controlled by the mucin glycoprotein layer on the intestinal epithelium. The predominant secretory mucin of the intestine is encoded by the gene *MUC2*. The MUC2 protein is extensively modified in the Golgi by the addition of O-linked GalNAc residues (78). The extremely high content of O-linked carbohydrate (80% by mass) of MUC2 makes it a high-affinity ligand for the *E. histolytica* Gal/GalNAc lectin ($K_D = 8.2 \times 10^{-11}$ M) (12, 61). The *MUC2* gene is highly polymorphic, with at least 12 alleles identified in Caucasian populations (78). It is interesting to consider whether polymorphisms in *MUC2* could explain differing sensitivities of individuals to *E. histolytica* infection and disease. As amebae enter the mucin coat, their ability to invade may be prevented by mucins binding to and inhibiting the activity of the lectin and by the physical barrier between parasite and host epithelium, which the mucin coat produces. For example cell lines that produce mucins are more resistant to in vitro killing by *E. histolytica*; this resistance is blocked by O-glycosylation inhibitors, which prevent normal glycosylation of the mucin glycoproteins. The interaction is also modified by the amebae: Amebae degrade mucins at sites of invasion in humans (51, 52), and amebae have a heat-stable mucus secretagogue activity, which results in goblet cell mucus secretion from both glandular and interglandular regions. The nonpathogenic ameba *E. moshkovskii* does not have this secretagogue activity (14). One of the most important roles of mucin glycoproteins may be in the regulation of encystation: In the related parasite *E. invadens* mucins induce trophozoites to transform into cysts (21, 22).

Finally the interaction of ameba with red blood cells likely has a role in invasion. Amebae containing ingested red blood cells are commonly seen in the stool and colonic ulcers of patients with amebic colitis (47). The red cells are endocytosed via the parasite lectin. The noninvasive parasite *E. dispar* is much less likely to ingest red blood cells. It is interesting to speculate that ingestion of erythrocytes induces a virulence program in the amebae leading to colitis or liver abscess, whereas interaction with mucin glycoproteins has the opposite effect by inducing encystation. Virtually nothing is known about the effects of red cell ingestion on in vitro or in vivo pathogenicity of *E. histolytica*.

PARASITE AND HOST GENETIC DETERMINANTS OF INVASIVENESS

Little is understood about genetic differences in *E. histolytica* or humans that control the outcome of infection, with two exceptions: (*a*) Amebic liver abscess is predominantly a disease of adult males, and (*b*) *E. histolytica* has been reclassified into noninvasive (*E. dispar*) and potentially invasive (*E. histolytica*) species. The varied organ tropisms and clinical presentations of infection by *E. histolytica* sensu stricto have stimulated interest in the role of *E. histolytica* genetic diversity in virulence. Polymorphisms in genes encoding repetitive sequences have allowed typing of *E. histolytica* (18). Using these techniques genetic differences between *E. histolytica* from different parts of the world, and even within a single refugee camp, have been demonstrated. For example, 25 distinct polymorphisms were observed among 42 stool isolates, and an additional 9 distinct patterns among the 12 liver abscess isolates in Dhaka, Bangladesh (4). The great number of different strains typed by DNA polymorphisms has to date, however, prevented correlation of a particular strain with invasiveness. From the standpoint of the host, the understanding of genetic determinants of invasiveness has received a major boost from the study of infection in preschool children in Dhaka. In a prospective study 7% of all new *E. histolytica* infections in the population were attributable to a trait associated with susceptibility to infection (R. Haque & W.A. Petri, unpublished data). Delineation of the human genes influencing susceptibility to amebiasis promises to be generally enlightening for understanding parasite interaction with the innate and acquired immune systems.

PARASITE LECTIN-MEDIATED ADHERENCE TO THE HOST

Carbohydrate-protein interactions are responsible for the contact-dependent cytotoxicity for which *E. histolytica* was named. A cell surface lectin of *E. histolytica* trophozoites binds to host galactose (Gal) and N-acetyl-D-galactosamine (GalNAc) (48, 60). Trophozoite adherence to human colonic mucin glycoproteins is inhibited by 90% by Gal or GalNAc (14), as is adherence to human neutrophils and erythrocytes (28), certain bacteria (7), as well as a variety of cell culture lines (36, 37). Studies of human neutrophils have confirmed that the parasite Gal/GalNAc lectin (and not fibronectin, vitronectin, CD11/CD18 integrins, complement or mannose-binding protein) is the only defined adhesin participating in the adherence event (10).

Perhaps the most significant observation is that blockade of lectin activity with Gal or GalNAc prevents the contact-dependent cytotoxicity for which the organism is named (60). Additionally, Chinese hamster ovary (CHO) cell glycosylation-deficient mutants lacking terminal Gal/GalNAc residues on N- and O-linked sugars are nearly totally resistant to amebic adherence and cytolytic activity (36, 37, 64). The first receptor encountered by the lectin may be the human colonic mucin layer of the large intestine (13). Chadee et al. demonstrated that binding of the lectin to colonic mucins is Gal/GalNAc-inhibitable and of high affinity (dissociation constant of $8.2 \times 10^{-11}\ M^{-1}$) (14). Interaction of trophozoites with colonic mucins appears to be a dynamic process, with trophozoites both inducing the secretion of and degrading colonic mucins (12). The mucin layer may protect the host from contact-dependent cytotoxicity by the parasite by binding to and neutralizing the lectin while serving as a site of attachment for the parasite to colonize the large bowel.

Identification of the Gal/GalNAc Lectin

The lectin was originally purified by carbohydrate affinity chromatography and with adherence-inhibitory monoclonal antibodies (mAb) (46, 48), an approach later repeated by Tannich and colleagues (76). The Gal/GalNAc lectin is composed of a 260-kDa heterodimer of disulfide-linked heavy (170 kDa) and light (35/31 kDa) subunits, which is noncovalently associated with an intermediate subunit of 150 kDa (Figure 2). The purified 260-kDa heterodimer retains GalNAc-binding activity to cells (49) and neoglycoproteins (1).

The heavy (170 kDa) subunit gene (*hgl*) sequence contains an amino-terminal 15–amino acid hydrophobic signal sequence, an extracellular cysteine-rich domain

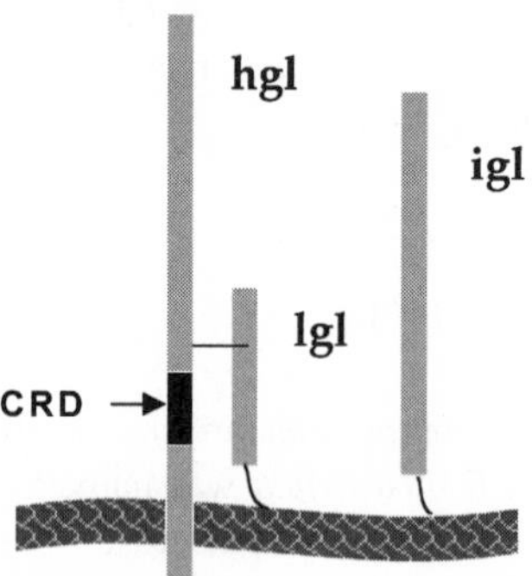

Figure 2 Gal/GalNAc adherence lectin of *Entamoeba histolytica*. The Gal/GalNAc lectin mediates parasite adherence to, and killing of, host cells. It is present on the plasma membrane of the ameba and is composed of three subunits. The integral membrane heavy subunit (Hgl) has a short cytoplasmic tail implicated in intracellular signaling. The carbohydrate recognition domain (CRD) is located within hgl. Hgl is disulfide bonded to a lipid-anchored light subunit (Lgl). Finally the lipid-anchored intermediate subunit (Igl) is noncovalently associated with the hgl-lgl heterodimer. The functions of Lgl and Igl in adherence and killing are unknown.

of 1209 amino acids containing sites for N-linked glycosylation, and transmembrane and cytoplasmic domains of 26 and 41 amino acids, respectively (39, 76). Disulfide bonds apparently fold the lectin into a protease-resistant structure, as the lectin is only sensitive to proteolytic attack upon reduction and alkylation of cysteine residues (39). Approximately 6% of the mass of the heavy subunit is carbohydrate; tunicamycin treatment of amebae prevents N-linked glycosylation of the lectin and results in amebae with dramatically reduced ability to adhere. The heavy subunit gene family consists of five unlinked genes in *E. histolytica* (57). The *hgl* gene products are 89%–95% identical at the amino acid level, with the most variable region being the cysteine-free domain comprising amino acids 188 to 378. The *hgl1* gene product has been sequenced from four different *E. histolytica* isolates and was shown to be remarkably conserved (M. Tanyuksel & W.A. Petri, unpublished data).

The light subunit is encoded by multiple genes encoding isoforms with different posttranslational modifications. The major 35- and 31-kDa isoforms have nearly identical amino acid compositions and CNBr-fragmented peptide patterns (41, 42). The 35-kDa isoform is more highly glycosylated and lacks the acyl-glycosyl-phosphatidylinositol (GPI) anchor present on the 31-kDa isoform. The light subunit gene family has 6–7 members; the light subunit genes that have been sequenced encode proteins with an amino-terminal signal sequence of 13 amino acids and an unusually short 7–amino acid carboxy-terminal GPI-anchor addition motif. There are greater sequence differences between *lgl* genes than between *hgl* genes, with the *lgl* genes sequenced only sharing 79%–85% identity. Four light subunit–specific mAb have been produced, which recognize distinct epitopes on five different light subunit isoforms. Immunoblots with these mAb demonstrate co-migration of light and heavy subunits when nonreduced trophozoite proteins were analyzed by SDS-PAGE, indicating that the subunits do not exist free of the heterodimer in significant quantities. Whereas anti-heavy subunits inhibit adherence, light subunit–specific antibodies do not, consistent with the heavy subunit containing the carbohydrate recognition domain (43).

There is some indirect evidence for a role of Lgl in virulence despite the fact that it apparently does not directly mediate adherence. Co-cultivation of *E. histolytica* with *Escherichia coli* serotype 055 has long been known to lead to a reduction in cytolytic activity of the amebae. Recently it has been recognized that this is associated with a reduction in expression of the lower-molecular-weight isoforms of Lgl (45). Similarly, expression of Lgl mRNA and the Lgl lower-molecular-weight isoforms are decreased in the Rahman strain of *E. histolytica*, which is defective in cytolysis and is avirulent in animal models. Overexpression of Lgl in Rhaman, however, did not restore cytolysis, indicating that the cytolysis defect is not solely due to decreased Lgl expression (45). In strain HM1:IMSS, antisense RNA was used to decrease Lgl protein expression. This had no effect on Hgl protein but surprisingly resulted in a substantial decrease in the native lectin heterodimer (2). It is unclear from this work where the Hgl protein uncomplexed with the heterodimeric lectin is located in the cell. However, the decrease in lectin heterodimer was associated with a 50% decrease in monolayer destruction and cell killing and with an inability to form liver abscess in hamsters. There was no difference in adherence or in erythrocyte endocytosis, both processes mediated by the lectin. This suggests that the levels of lectin on the cell surface of the antisense-expressing cells were high enough to mediate some, but not all, functions of the lectin. Together these experiments are consistent with a role for Lgl in lectin-mediated cytolysis and virulence.

The Hgl-Lgl heterodimer has a unique mechanism of membrane association: Hgl is transmembrane but Lgl is GPI-anchored. The role of the GPI modification in membrane anchoring, assembly, and function of the lectin heterodimer has been tested. Epitope-tagged Lgl with or without the GPI-anchor addition signal was expressed in *E. histolytica* trophozoites. Full-length FLAG-tagged Lgl protein, detected with FLAG-specific mAb, was precipitated in association with Hgl. In contrast the FLAG-tagged Lgl protein lacking the GPI-anchor addition signal did not assemble with Hgl into a lectin heterodimer (56).

The lack of heterodimerization of the GPI-deficient Lgl construct with Hgl enabled determination of the ability of Lgl to bind carbohydrate in the absence of Hgl. The only identified CRD on the lectin is in Hgl. Previously it had not been possible to test the light subunit independently for Gal/GalNAc-binding activity because reduction of the disulfide bonds needed to separate it from Hgl-destroyed lectin carbohydrate-binding activity. Lysates of cells expressing mutant Lgl were incubated with Gal-agarose beads or with control beads [GalNAc-agarose conjugated via the 3-OH position to destroy recognition by the amebic lectin (82)]. The full-length Lgl protein was detected in the eluate from the Gal beads by virtue of its association with Hgl. However, the GPI-deficient Lgl protein, which was not heterodimerized with Hgl, was not bound to the Gal beads. It was concluded that Lgl protein, when not part of the heterodimeric lectin, lacks detectable carbohydrate-binding activity.

Interference with the overall production of Lgl via antisense production results in a reduction in heterodimeric lectin and a decrease in cytotoxicity (2). Inducible or constitutive expression of the Lgl construct lacking the GPI signal sequence had no detectable effect on adherence or cytotoxicity (data not shown) likely because the mutant Lgl could not heterodimerize with Hgl. Additionally, expression of the GPI-Lgl protein did not interfere with assembly of the native endogenous Lgl into functional lectin. A picture is beginning to emerge on the roles of Hgl and Lgl in the amebic lectin: Carbohydrate-binding and cell-signaling activities appear to be limited to Hgl, whereas the GPI-anchor addition sequence of Lgl enables heterodimer formation and may endow the lectin with yet-to-be discovered biologic properties.

The intermediate subunit (Igl), also known as the "150-kDa lectin" orginally described by Tachibana and colleagues, is noncovalently associated with the 260-kDa lectin heterodimer (16). Initially identified as a target of mAbs that block trophozoite adherence, this lectin has been shown to be intimately associated with the Gal/GalNAc lectin in several different ways. Igl is present in small amounts in preparations of the Gal/GalNAc lectin that have been purified by either Gal-affinity chromatography or anti-Hgl mAb-affinity chromatography. Similarly, Hgl is present in small quantities in the Igl purified with mAb directed against Igl. In native gel electrophoresis, Igl and the Hgl-Lgl heterodimer co-migrate at an estimated molecular mass of 380 kDa.

Two different gene copies of Igl have been cloned from *E. histolytica* (15). They encode proteins with 84% amino acid identity. The *igl* genes encode novel proteins of 1101 amino acids with hydrophobic amino- and carboxy-terminal signal sequences consistent with a GPI-anchored plasma membrane protein. Igl1 and Igl2 have calculated molecular masses of 119,987 and 119,513 and predicted isoelectric points (pI) of 5.03 and 5.61, respectively. In contrast, the estimated molecular mass and pI of the native Igl is 150 kDa and 6.9, respectively, which suggests the existence of posttranslational modifications in the native protein. The most abundant amino acid residues are cysteines (12.3%), lysines (9.5%), and threonines and serines (both 8.9%). The amino acid sequences predict 12 potential N-glycosylation sites and 3 O-glycosylation sites. The Igl proteins lack a carbohydrate-recognition motif but have limited sequence identity with the variant surface glycoproteins (VSPs) of *Giardia lamblia* (for example, BLAST e value of $2e^{-42}$; 22% identity of amino acids 32-1036 with amino acids 51-1126 of pir T42017). The sequence identity includes some of the "CXXC" motifs of the VSPs implicated in protein-protein interactions.

Carbohydrate Recognition by the Gal/GalNAc Lectin

The interaction of the lectin with host glycoconjugates is multivalent and of high specificity. This was demonstrated experimentally by the 100,000-fold-greater affinity of Gal/GalNAc-containing polyvalent neoglycoproteins than the monosaccharides Gal or GalNAc. The stereospecificity of the multivalent binding of Gal/GalNAc by amebic and hepatic (also known as the asialoglycoprotein receptor) Gal/GalNAc lectins differs. Small synthetic oligosaccharides (such as NAc-YD(G-GalNAcAH)2), which bind with high affinity to the hepatic Gal/GalNAc lectin, do not bind with high affinity to the *E. histolytica* lectin. The optimal spacing of polyvalent GalNAc residues has not been defined, but work with neoglycoprotins differing in the concentration of GalNAc has indicated that wider GalNAc spacing is tolerated for the parasite lectin than for the hepatic lectin. The amebic lectin may recognize maxiclusters, which are multivalent structures spaced at relatively greater distances on a polypeptide backbone, more than the "miniclusters" formed by branching tri- or tetraantennary termini of N-linked glycoproteins recognized by the hepatic lectin (1). This may reflect the amebic lectin's engagement of mucin glycoproteins, which have terminal Gal/GalNAc O-linked oligosaccharides widely spaced on elongated mucinpolypeptide backbones.

GalNAc is the preferred carbohydrate determinant over Gal, with approximately sevenfold-higher affinity for the monosaccharide and 1000-fold-higher affinity for GalNAc- than Gal-containing oligosaccharides. Lactose (Gal β1-4 glucose) and

N-acetyl-lactosamine (GalNAc β1-4 glucose) are not of higher affinity than GalNAc, indicating that the lectin binds a single saccharide residue. In low-ionic-strength buffers, lectin binding to GalNAc requires calcium, and the presence of a hydrophobic aglycon on p-nitrophenyl β-N-acetylgalactosaminide increased affinity eightfold, which suggests that the carbohydrate-binding domain has a high-affinity site for calcium ions and an aromatic side chain. The substructural specificity of monosaccharide recognition is also different for the parasite lectins than for the hepatic lectins. Both hepatic and amebic lectins require 3-OH and 4-OH groups of GalNAc, whereas only the amebic lectin requires the 6-OH group (82). In contrast, the amebic lectin had greater tolerance for changes at the N-acyl position. For example, replacement of the N-acetyl group with the bulkier N-benzoyl group increased amebic lectin affinity twofold and decreased hepatic lectin affinity eightfold. This suggests that the N-acyl-binding area of the amebic lectin is more spacious than that of the hepatic lectin. Consistent with this, the free amino sugar galactosamine was only eightfold less potent for the amebic lectin but 1000-fold less potent for the hepatic lectin. Overall a picture emerges of a unique multivalent lectin-glycoconjugate interaction that should be amenable to specific (and potentially therapeutic) inhibition with GalNAc-containing oligosaccharides (1).

The Carbohydrate Recognition Domain

The carbohydrate recognition domain (CRD) of the lectin is a potential target for colonization-blocking vaccines or drugs. Adherence-inhibitory mAb epitopes are contained within the extracellular cysteine-rich domain of the heavy subunit, making this a likely location for the CRD (38). Direct assignment of carbohydrate-binding activity to either the 170-kDa or 35-kDa subunits of the native lectin was not possible, as the reduction in disulfide bonds required to separate the subunits resulted in loss of carbohydrate-binding activity (W.A. Petri, unpublished results). Both lectin subunits lack sequence identity to well-characterized lectin carbohydrate-binding domains [including the mammalian C type lectins (mannose-binding protein and E-selectin), mammalian galectins, the legume lectins, wheat germ agglutinin, ricin, *E. coli* heat-labile enterotoxin, cholera toxin, or the influenza virus hemagglutinin] (68). Other investigators have remarked on an area of the heavy subunit with limited sequence identity to wheat germ agglutinin (76); however, none of the wheat germ agglutinin active-site residues and few of the critical cysteines required for intrasubunit disulfide bond formation are present in the corresponding lectin sequence.

It is not surprising that sequence analysis has failed to identify the carbohydrate-binding domain, as even carbohydrate-binding domains that interact with the same monosaccharide can have diverse structures. For example, only the plane formed by the C3-C6 carbons of the Gal residues and the aromatic side chain (tryptophan, tyrosine, or phenylalanine) in the binding site is well conserved for several Gal-binding lectins, while the rest of the binding site residues differ (68). A second difficulty with interpretation of linear sequence data is that the carbohydrate-binding domains are formed not by linear stretches of amino acids but by amino acid residues that are scattered over the primary sequence of one or more subunits and brought together by the tertiary folding of the lectins. Nonetheless, sequence analysis has not aided the identification of the carbohydrate-binding domain.

The first hint that the heavy subunit contained the CRD was the observation that mAb directed against it affected Gal/GalNAc-binding activity. In contrast, mAb and polyclonal antibodies directed against the light subunit isoforms have no significant effect on adherence or cytotoxicity (40). Neutralizing mAb epitopes on the Hgl protein all mapped to the cysteine-rich domain (amino acids 482-1138). Antibodies that blocked or augmented Gal/GalNAc-binding activity were mapped between lectin residues 482-818, suggesting that the carbohydrate-binding domain is located in this part of the cysteine-rich domain (38).

A likely site for the CRD was amino acids 895-998 of the cysteine-rich region where the adherence-inhibitory mAb 8C12 mapped. The sequence of the 104–amino acid fragment is completely conserved among other members of the *E. histolytica* lectin gene family and 89% conserved in the homologous protein expressed by the closely related noninvasive parasite *E. dispar*. The CRD and the unrelated amino-terminal cystein (C) and tryptophan (W)-rich domain of the lectin were expressed in *E. coli* as recombinant polyhistidine fusion proteins. As expected, mAb 8C12 recognized the purified CRD fusion protein. Two adherence-enhancing mAbs (3F4 and 8A3) that mapped to the CRD by deletion analysis did not recognize the polyhistidine CRD fusion protein on a Western blot. This suggests that these mAbs recognized conformational epitopes or epitopes located elsewhere on the protein, which depended on the presence of the CRD for their structures. Two other adherence-inhibitory mAbs recognized epitopes on either side of the 104–amino acid fragment, each about 100 amino acids away. Therefore epitope mapping pointed to, but did not unambiguously identify, the region from amino acids 895-998 as the CRD of the *E. histolytica* lectin. A GalNAc-containing neoglycoprotein (a synthetic glycoprotein) was used to test the hypothesis that the 104–amino acid CRD fragment contained a carbohydrate recognition domain. *E. histolytica* membranes bind to the polyvalent neoglycoprotein, $GalNAc_{20}BSA$, with 500,000-fold-higher affinity than to GalNAc monosaccharide. The purified CRD polyhistidine fusion protein bound ^{125}I-labeled $GalNAc_{20}BSA$, whereas the CW protein did not. Binding of $GalNAc_{20}BSA$ by the CRD fusion protein was calcium-dependent and specifically inhibited by the Gal/GalNAc-terminal glycoprotein asialofetuin. Immunization with CRD resulted in an antisera that completely blocked adherence and passively transferred partial immunity to challenge with *E. histolytica* in the liver of gerbils (20).

Kain and colleagues in vitro transcribed and translated intact and fragmented Hgl2 protein to map regions with carbohydrate-binding activity. The full-length and cysteine-rich fragment (amino acids 356-1143) of Hgl, but not the amino-terminal fragment (amino acids 1-480), exhibited binding to CHO cells that were 60% inhibited with Gal. The fragment from amino acids 480-900 bound to

GalNAc-BSA-coated microtiter wells but not to BSA-coated wells, consistent with this region containing a Gal-binding domain. Kain and colleagues concluded that the CRD is located within the cysteine-rich region, with residues 356-480 and/or 900-1143 required for high-affinity binding (50). One conclusion from the studies of these two groups is that carbohydrate-binding activity is contained within the cysteine-rich domain of Hgl. Somewhat unresolved is whether there is more than one CRD in this domain, as the two groups demonstrated carbohydrate recognition activity in fragments with only minimal overlap [amino acids 895-998 by (20) and amino acids 480-900 by (50)].

Conformational Control of Carbohydrate Binding

A surprise from mAb mapping of the lectin heavy subunit was the discovery that mAb against epitopes 1 and 2, but not epitopes 3–6, dramatically enhanced trophozoite adherence to CHO cells and human colonic mucins (49). Antibody-mediated enhancement of adherence occurred at 4°C and was Gal/GalNAc-specific. Antibody-mediated clustering of the lectin was not the explanation for increased binding, as Fab fragments of epitope 1 mAb also increased binding. Increased Gal/GalNAc-binding activity of the lectin induced by epitope 1 mAb was directly demonstrated by showing enhanced binding of purified lectin to CHO cells. Together the data were consistent with mAb binding inducing a change to a more active conformation of the Gal/GalNAc lectin. The significance of this observation is that the lectin's carbohydrate-binding activity appears to be conformationally controlled, similar to other eukaryotic adhesins including CR3 of human neutrophils and LFA-1 of T lymphocytes. The ability to control lectin activity may provide trophozoites with a mechanism to detach from mucins and epithelial cells as they invade the host.

We hypothesized that this regulation could occur via inside-out signaling mediated by the lectin cytoplasmic domain. This is a particularly attractive hypothesis partly because the lectin cytoplasmic tail contains regions of identity with the cytoplasmic tail of $\beta 2$ integrin-like domains. These domains are important in integrin inside-out signaling. Inducible expression of a fusion protein containing the lectin transmembrane and cytoplasmic tail (using the tetO-inducible expression system), but not a fusion protein containing the transmembrane but not the cytoplasmic tail, resulted in a 50% decrease in adherence. The induced fusion protein containing the cytoplasmic tail was intracellular, did not covalently associate with the light subunit to form heterodimers, and did not affect the level of cell surface wild-type lectin. We therefore concluded that the decreased adherence was a dominant-negative effect of the fusion protein on regulation of lectin activity. Mutation (to alanines) of the $T_{1253}\,I_{1254}\,T_{1255}\ldots Y_{1261}$ β-integrin motif in the lectin cytoplasmic tail motif resulted in an abrogation of the dominant-negative phenotype. These observations raise the possibility of similar mechanisms of inside-out signaling in integrins and lectin (79).

Interference with inside-out signaling caused an 84% decrease in the size of amebic liver abscesses in an animal model, supporting the importance of the lectin, and of adhesive regulation, in virulence. Erythrophagocytosis, serum resistance, and cytolysis of adherent amebae were not affected by the dominant-negative effect on inside-out signaling, indicating that the mechanism by which the lectin mediates these phenotypes is not dependent on inside-out signaling.

Sequence Similarity of CRD to Hepatocyte Growth Factor

Sequence analysis of the CRD revealed that it had limited sequence identity to the receptor-binding domain of hepatocyte growth factor (HGF). Specifically, the region from amino acids 913-939 of CRD had 52% sequence identity with amino acids 59-85 of HGF, which forms part of the receptor-binding domain sufficient for high-affinity HGF binding. HGF is produced by mesenchymal cells and acts as a growth stimulator of epithelial cells. It also stimulates the motility and invasiveness of epithelial and endothelial cells and is also known as a "scatter factor" for that property. It paradoxically has cytotoxic and growth inhibitory properties in some situations. Its high-affinity receptor is the c-*met* proto-oncogene product. HGF binding to c-*met* results in autophosphorylation of c-*met* intracellular tyrosine kinase domain (Y1235), which is associated with activation of its catalytic activity. The increased cell motility and decreased cell-cell adhesions induced by HGF may be mediated in part by tyrosine phosphorylation of β-catenin and focal adhesion kinase, and activation of the small GTP-binding protein rho. It also stimulates the activation or induction of proteases that degrade the extracellular matrix. Therefore, a potential interaction of the lectin (via its CRD domain) with the c-*met* receptor was interesting for the potential promotion of amebic invasiveness via disruption of epithelial barriers and/or dissolution of the extracellular matrix. We tested the ability of the CRD to compete with HGF for binding to its receptor c-*met* using a microtiter plate assay. HGF binding to c-*met* was competed with excess CRD or native lectin but not with the unrelated CW fusion protein. The competition of HGF binding to c-Met was not due to the carbohydrate-binding activity of CRD: Inhibition of HGF binding to c-*met* by 2 μM his-CRD was 54 $\pm$ 5.1% with, and 56 $\pm$ 2.6% without, 50 mM GalNAc (20). The ability of the lectin to compete with HGF for binding to c-*met* is consistent with recognition of c-*met* by the amebae. The contribution of such an interaction to the liver tropism of *E. histolytica* remains to be delineated.

Gal/GalNAc Lectin-Associated Proteins

Evidence is accumulating of proteins that are noncovalently associated with the lectin, either with its extracellular, membrane, or cytoplasmic domains. Hughes and Mann (unpublished data) have used the yeast two-hybrid genetic screen to identify proteins interacting with the cytoplasmic tail of the lectin. The 29-kDa alkylhydroperoxide reductase (peroxidase) was identified with the yeast two-hybrid using the cytoplasmic tail as bait. Interaction with the amebic cytoplasmic tail was confirmed in vivo by confocal microscopy and immunoprecipitation. The peroxidase activity of the protein is potentially important in the detoxification of reactive oxygen and sulfur species that the amebae are exposed to during interaction with the host polymorphonuclear lymphocytes. The biologic importance of the membrane localization of the peroxidase achieved via interaction with the lectin cytoplasmic tail is yet to be determined. Analysis of two-dimensional gels of affinity-purified Gal/GalNAc lectin show a complex pattern of proteins, which suggests that the lectin serves as a nucleation site for a number of different proteins potentially involved and required for interaction with the host.

KILLING OF HOST CELLS

Entamoeba histolytica was named by Schaudinn in 1903 for its ability to lyse tissues. Destruction of host cells is contact-dependent (via the Gal/GalNAc lectin) and extracellular. *E. histolytica* trophozoites in vitro kill a wide variety of tissue culture cell lines as well as human neutrophils, T lymphocytes, and macrophages. Contact-dependent killing of CHO cells and other target cells is nearly completely inhibited by Gal and GalNAc, and CHO cell glycosylation–deficient mutants lacking terminal Gal residues on N- and O-linked sugars are nearly totally resistant to amebic adherence and cytolytic activity (Figure 3).

The mechanism of contact-dependent killing by *E. histolytica* has been the subject of intensive investigation. Killing requires an intact parasite: Amebic filtrates or sonicates are not cytotoxic. The important role of the parasite cytoskeleton in killing has been demonstrated by C3 intoxication of amebic Rho (27), by cytochalasin disruption of the cytoskeleton (60), and by expression of dominant-negative myosin II (3). Intracellular calcium in target cells rises approximately 20-fold within seconds of direct contact by an amebic trophozoite and is associated with membrane blebbing (62). Cell death occurs 5–15 min after the lethal hit is delivered. Extracellular EDTA and treatment of the target cells with the slow sodium-calcium channel blockers verapamil and bepridil (63) significantly reduce amebic killing of target cells in suspension.

Isolation of amebic pore-forming proteins similar in function to pore-forming proteins of the immune system has been reported by a number of investigators. A purified 5-kDa amebapore and a synthetic peptide based on the sequence of its third amphipathic α-helix has cytolytic activity for nucleated cells at high concentrations (10–100 μM) (33, 34). Amebapore has a pH optimum of 5.3 and is inactive at pH 7, which may be of some significance in light of the inhibition of cytotoxicity with weak base treatment of amebae. Interestingly, no DNA degradation was observed in cells lysed in vitro by the purified amebapore, suggesting a different mechanism of cell killing by the purified amebapore than by the intact parasite (6). Constitutive expression of antisense RNA against the amebapore caused an approximately 50% drop in its expression, yet almost completely blocked cytolysis as assessed by trypan blue release (8), and dramatically decreased liver abscess formation. The potential problem with these experiments is that the high levels of drug selection required to achieve partial antisense inactivation of the amebapore mRNA may have artifactually inhibited amebic cytotoxicity. This could be

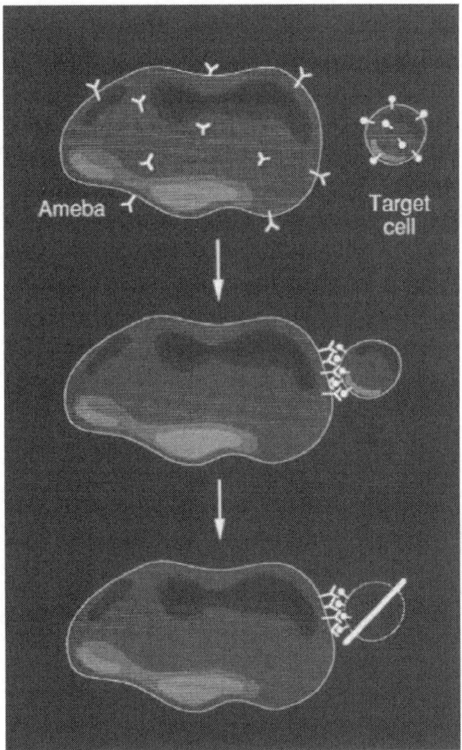

Figure 3 Contact-dependent killing of host cells by *Entamoeba histolytica* trophozoites. Amebae adhere to the target cell via a parasite lectin specific for Gal and GalNAc residues on the host cell. Only after adherence via the lectin does contact-dependent killing of the host occur [reprinted with permission from (40)].

resolved if the experiments were repeated using inducible expression of antisense RNA. This is possible with inducible promoter systems developed for *E. histolytica* based on the tetracycline operator-repressor system (29, 58, 79).

Cells killed by the parasite undergo nuclear chromatin condensation, membrane blebbing, and internucleosomal DNA fragmentation (31, 55). Nuclei of hepatocytes in a mouse model of amebic liver abscess were labeled by terminal deoxynucleotidyl-transferase-mediated dUTP-biotin nick end labeling (TUNEL), consistent with an apoptotic death. Several lines of evidence infer a nonclassical mechanism of apoptotic killing by *E. histolytica*. Overexpression of the Bcl-2 protein that inhibits apoptosis due to a variety of cellular stresses (e.g., serum starvation and UV radiation) did not prevent murine cell DNA fragmentation following exposure to *E. histolytica* (55). Furthermore, *E. histolytica* caused hepatocyte apoptosis in mice deficient in the Fas/Fas ligand and TNF-RI signaling pathways (73). Taken together, these data suggest that *E. histolytica* initiates host cell apoptosis by directly activating the host cell's distal apoptotic machinery.

Huston et al. have demonstrated that killing of the Jurkat human T lymphocyte cell line occurred via apoptosis, as judged by DNA fragmentation and caspase 3 activation. This killing was inhibited by Gal. Classical upstream caspases seemed not to be involved, as caspase 8–deficient cells, resistant to killing by fasL, were readily killed by Eh. Caspase 8–deficient cells treated with a caspase 9 inhibitor (Ac-LEHD-fmk) (at a level sufficient to inhibit apoptosis via etoposide) were readily killed as well. In contrast, the caspase 3 inhibitor Ac-DEVD-CHO at 100 μM (sufficient to block killing via actinomycin D) blocked *E. histolytica* killing, as measured both by DNA fragmentation and C^{r51} release, indicating that it was necessary for both apoptotic death phenotype and necrosis to occur (31). Blockade of caspases blocks amebic liver abscess formation in mice (32). The upstream trigger that activates caspase 3 remains unknown (31).

The importance of apoptosis in disease was indicated by the extensive apoptotic death of host cells at the site of invasion of amebic trophozoites during intestinal infection, the most prevalent form of amebiasis. The parasite's ability to form liver abscesses is impaired in amebae rendered defective in cytotoxicity either by inducible expression of a dominant-negative mutant of the Gal/GalNAc lectin or by expression of antisense amebapore RNA (2, 79). *E. histolytica* cytotoxicity therefore appears crucial to the ability of the parasite to cause disease.

The Role of Lectin in Cytotoxicity

Apposition of amebic and target cell plasma membranes, as can be achieved by centrifugation of target cells and amebae into a pellet, will not lead to cytolysis if the amebic lectin is inhibited with Gal/GalNAc (59) or if the target cell lacks Gal and GalNAc on its surface (37, 64). This is consistent with the lectin mediating not only adherence but also participating in the cytolytic event. The identification of anti-lectin mAb directed against epitope 1 of the heavy subunit, which blocked cytotoxicity but not adherence, directly implicated the lectin in the cytotoxic event (69). The same anti-lectin mAb that blocked cytotoxicity also caused a conformational change in the lectin that increased carbohydrate-binding capacity. One could speculate that the mAb blocks cytotoxicity by preventing a second conformational change in the lectin required for cell killing.

Killing occurs after the lectin engages GalNAc on O-linked target cell surface oligosaccharides: The Lec1 CHO mutant, which lacks lactosamine-containing N-linked carbohydrates but contains normal O-linked carbohydrates, was killed nearly as efficiently as wild-type cells. In contrast, the ldlD/Lec1 CHO cell mutant that lacks both N- and O-linked structures was almost completely resistant to amebic cytotoxicity (36, 37, 64). Lectin-mediated capping of the O-linked structures in Lec1 [which are unaltered compared to wild-type CHO cells and should be Sialic acid-Gal-(Sialic acid)GalNAc] could be deduced to mediate killing.

Transfer of the Lectin to Host Cells

Interaction of the ameba with the host cell results in transfer of the lectin to the host; this is an active process that requires an intact viable parasite (35). Localization of the lectin to the lateral surface of epithelial cells is seen within five minutes of interaction with *E. histolytica* trophozoites, before any destruction of the cells. Transfer of the lectin to epithelial cells is not observed when amebic sonicates or filtrates are applied to the monolayers. Transfer of the lectin is decreased by 86% upon addition of Gal, GalNAc, or lactose, which indicates a requirement for parasite adherence to epithelium for transfer to occur. Transfer of 29-kDa and 96-kDa surface antigens is not observed, indicating some specificity to the transfer event. Transfer of the lectin precedes death of the host cells, raising the question if the transfer is involved in the process of cell death. Transfer is specific to epithelial-derived cell lines and is not seen in fibroblast lines. Because amebae kill both epithelial and fibroblast cell lines, if the lectin transfer event is involved in epithelial cell death, then a different mechanism of killing must be involved in fibroblast death. The significance and mechanism of transfer of the lectin therefore remains intriguing and mysterious.

LECTIN-MEDIATED RESISTANCE TO COMPLEMENT LYSIS

The complement system is one of the first barriers to infection in nonimmune individuals; circumvention of this defense is central to the pathogenesis of amebiasis. Invasion of the colon and hematogenous spread to the liver result in the continuous exposure of the extracellular trophozoite to the human complement system. Trophozoites activate the classical and alternative complement pathways in the absence of anti-amebic antibodies. Incubation of trophozoites in normal human sera results in depletion of human complement as measured by CH50 and C5b-9 hemolytic assays and C3 and C4 depletion (65). The amebic 56-kDa cysteine proteinase cleaves C3 at a site one amino acid distal to that of the human C3 convertase and may be the route by which complement is activated (67). Depletion of complement in hamsters by cobra venom factor treatment increases both the frequency and severity of amebic liver abscess, providing evidence of the protective role of the complement system in amebiasis (11).

E. histolytica freshly isolated from patients with invasive amebiasis, and laboratory strains passed through animals, activates the alternative complement pathway but is resistant to C5b-9 complexes deposited on the membrane surface (65–67). On the other hand, amebae cultured from the stool of asymptomatically infected individuals or virulent amebae attenuated by axenic (in the absence of associated bacteria) culture activate the alternative complement pathway and are killed by C5b-9 (65, 66).

Killing of amebae is mediated by the terminal complement components, and the direct lysis of sensitive but not resistant *E. histolytica* has been demonstrated with purified complement components C5b-9 (66). Resistance to terminal complement attack in *E. histolytica* could be due to an amebic cell surface protein with C5b-9-inhibitory activity or by endocytosis or shedding of the C5b-9 complex. Rapid membrane repair via shedding or endocytosis of the membrane-inserted C5b-9 complex has been postulated to confer C5b-9 resistance to several different cells, including nucleated mammalian cells and the metacyclic (infective promastigote) stage of *Leishmania major*. However, shedding or release of C9 from the membrane is not the explanation for C5b-9 resistance in *E. histolytica*, as C9 binding is higher in resistant than in sensitive amebae.

Braga et al. produced mAb against serum-resistant amebae and identified an antibody that increased *E. histolytica* lysis by human sera and by purified human complement components C5b-9. It was a surprise that the antigen recognized by the antibody was the 170-kDa lectin subunit (9). Inhibition of complement resistance by anti-170-kDa mAb was shown to be specific to mAb-recognizing epitopes 6 and 7. Examination of the sequence of the 170-kDa subunit showed limited identity with CD59, a human inhibitor of C5b-9 assembly, and the purified lectin was recognized by anti-CD59 antibodies. The lectin bound to purified human C8 and C9 and blocked assembly in the amebic membrane of the complement membrane attack complex at the steps of C8 and C9 insertion.

The lectin gene family therefore appears to participate not only in adherence and host cell killing but also in evasion of the complement system of defense via a remarkable mimicry of human CD59. Gal/GalNAc inhibition of lectin activity had only a minor effect on C5b-9 resistance of trophozoites, which suggests that the lectin and complement regulatory domains of the lectin are distinct.

REGULATION OF LECTIN GENE EXPRESSION

The control of transcription of the *hgl5* lectin heavy-subunit gene is beginning to be understood now that gene transfection is possible in *E. histolytica* (44, 53). Primer extension and nuclear run-on analyses are consistent with monocistronic transcription of the gene with a lack of 5′ processing of *hgl5* mRNA (75). Several unique features are present in the core promoters of protein-encoding genes including *hgl5* in *E. histolytica* (54, 75) (Figure 4). An initiator element (Inr) [AAAAATTCA] surrounds the site of transcription initiation and is conserved in 37 *E. histolytica* protein-encoding genes. The sequence is divergent from that of other eukaryotic initiators. Inr controls the site of transcription initiation in *hgl5* if both GAAC and TATA elements are mutated (75). The GAAC element [GAACT] is present between TATA and Inr in 31/37 promoters examined. Mutation of GAAC decreases the rate of transcription and results in new transcription start sites. Positional analysis demonstrates that GAAC directs a new transcription start site 2–7 base pairs downstream of itself (75). An *E. histolytica* nuclear protein of an estimated 36-kDa mass exhibits sequence-specific binding to GAAC (74). Finally a TATA element [GTATTTAAA(G/C)] is present at −30 bases upstream of transcriptional start site.

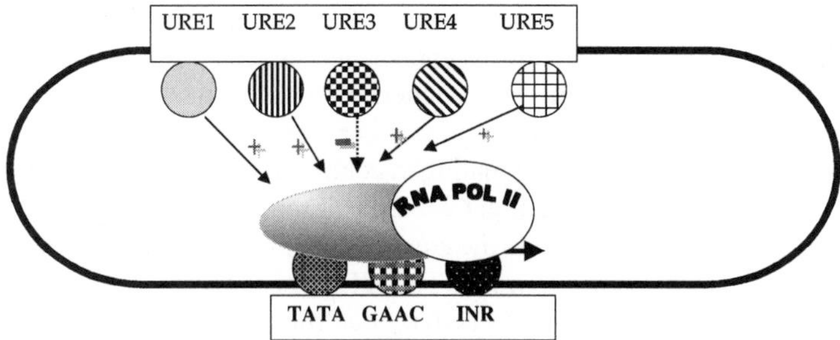

Figure 4 Cartoon depiction of the promoter for the *hgl5* virulence gene of *E. histolytica*. Transcription factors bound to upstream regulatory elements (URE) interact with general transcription factors bound to the core promoter elements (TATA, GAAC, and INR) to promote or inhibit transcription by RNA polymerase II.

Positional analysis demonstrates that TATA directs a new transcription site 30–31 base pairs downstream of itself (75).

The requirement for three core promoter elements is fascinating: Why would transcription of protein-encoding genes in *E. histolytica* be dependent on a third core regulatory region? Perhaps the *E. histolytica* TFIID complex has multiple DNA-binding regions composed of TBP-, GBP-, and Inr-binding proteins. A variety of preassembled TFIID complexes could exist, containing some or all of the core promoter–binding proteins. These different TFIID complexes could differentially regulate a variety of core promoters containing all three or only one or two of these regulatory regions. A second model is based on the fact that TBP in vitro binds to multiple AT-rich sequences. The specificity of TBP binding to the TATA region is conferred in large part by its proximity to other regulatory regions. In an AT-rich organism such as *E. histolytica* a mechanism may have developed in which a factor such as GBP localizes TBP to the promoter. Thus a model can be hypothesized in which transcription in *E. histolytica hgl5* genes may be dependent on protein-protein interactions where GBP functions to tether or localize TBP/TFIID to the core promoter.

The upstream elements of the promoter for *hgl5* were mapped by deletions and scanning mutagenesis to the 272 bases upstream of the start of transcription (54). Five UREs were identified by mutational analysis. The two elements having the most pronounced effect on gene expression have been further studied. URE3 (TATTCTATT at –80 base pairs) differs from the other UREs in that it is a negative regulatory element in *hgl5*. The URE3 sequence is also present in the 5′ upstream regions of at least seven other *E. histolytica* genes (25,26). URE3 is a positive regulatory element in the ferredoxin (*fdx*) gene. Its activity is induced in stationary phase growth (74). Yeast one-hybrid traps were used to identify an amebic protein (URE3-BP) with sequence-specific binding to URE3. Analysis of the protein sequence did not reveal a canonical DNA-binding motif but did demonstrate two calcium-binding EF hand motifs. The recombinant URE3-BP demonstrated calcium-dependent and sequence-specific recognition of the URE3 double-stranded DNA sequence. With only one other calcium-regulated transcription factor identified in a eukaryote, it will be of wide interest to determine the role of calcium signaling in control of amebic virulence gene expression.

Mutation of URE4 (a direct 9–base pair repeat of the sequence AAAAATGAA at –170 base pairs) results in the greatest drop in reporter gene expression controlled by the *hgl5* promoter (54, 72). The URE4 sequence is also present in the *hgl3* 5′ flanking region. Promoter activity controlled by the URE4 element is induced upon serum starvation of trophozoite cultures (74). DNA sequence-specific affinity chromatography identified two URE4 enhancer-binding proteins, EhEBP1 and 2 (71). Recombinant EhEBP1 and 2 bound double-stranded DNA containing the URE4 sequence in a sequence-specific manner, and overexpression of EhEBP1 in trophozoites led to a sevenfold drop in reporter gene activity controlled by URE4, confirming its ability to bind to the URE4 motif in vivo. Interestingly both EhEBP1 and 2 contained the RNA-binding motif RRM (71). Two previously described DNA-binding proteins that contain RRM motifs share with EhEBP1 and 2 recognition of AT-rich DNA sequences. In the context of the AT-rich *E. histolytica* genome, it will be important to determine if transcription factors containing RRM motifs are a common feature.

With the *E. histolytica* genome project nearing completion, and DNA microarrays for gene expression analysis coming online, an exciting and rapidly moving period of discovery of gene expression profiles is in store. The detailed analysis of the *hgl5* promoter should provide a framework for future understanding of virulence gene regulation.

ROLE OF THE LECTIN IN ENCYSTATION

The transformation of trophozoite to cyst in the large intestine completes the life cycle of *E. histolytica*. Without formation of the cyst, the parasite would be unable to infect another individual. The pioneering work of Eichinger and colleagues has shed the first light on the biological mechanisms of encystation (17, 19, 21, 22). Using the reptilian parasite *E. invadens*, which encysts readily in vitro, as a model system, sequential steps in the process have been delineated. Following initiation of encystation by nutrient deprivation and hypo-osmotic stress, individual trophozoites aggregate into large multicellular groups. The aggregation stage is catalyzed by the presence of precise amounts of polyvalent Gal-terminated molecules (serum or mucin glycoproteins or *Crithidia fasciculata*). Encystment medium lacking serum glycoproteins is less efficient at induction of encystation (from 90% in the presence of glycoproteins to 25% encystment in their absence).

Mucin may be one physiologic ligand that induces the aggregation stage of encystment as the mucin blanket is where the amebic trophozoites reside in the large intestine. Mucin has a narrow range of in vitro activity to induce aggregation and subsequent encystment, suggesting that the control over cyst production by the parasite is equally precise. The importance of terminal Gal residues on the mucins was demonstrated by increased potency of the mucins when treated with neuraminidase, and abrogation of activity if the mucins were first treated with β-galactosidase. The lectin may also have a role in cyst wall formation at a later step of the process. A chitin-binding lectin potentially involved in cell wall synthesis may be anchored to the surface of encysting trophozoites by the Gal lectin (23).

The addition of free monomeric Gal blocks the encystment process at the step of aggregation and blocks encystment-specific transcripts ("gene 122," chitinase, and ubiquitin) (19). An attractive hypothesis from this work is that engagement of Gal-rich mucin glycoproteins by the amebic lectin stimulates trophozoite aggregation and/or cell-to-cell signaling to initiate the process of encystation.

ACQUIRED IMMUNE RESPONSE TO THE LECTIN IN CHILDREN

The importance of the Gal/GalNAc lectin for parasite adherence, invasion, and encystation suggested that an immune response against it could mediate protection from amebiasis. Haque and colleagues (30) investigated this issue in an observational study of *E. histolytica* infection in children in Dhaka, Bangladesh. Specifically they asked whether protection from intestinal infection correlated with mucosal or systemic antibody responses to the *E. histolytica* Gal/GalNAc adherence lectin. The presence of stool anti-lectin IgA was associated with the absence of *E. histolytica* colonization: 0% (0/64) of IgA (+) and 13.4% (33/246) of IgA (−) children were infected as determined by antigen detection ($p = 0.001$). This finding was followed up with a one-year prospective study of 289 children ages 2–5 in Dhaka. During the 12 months, 39% (105/269) of the children had at least one new *E. histolytica* infection. Children with stool IgA lectin–specific antibodies at the beginning of the study had 64% fewer new *E. histolytica* infections by 5 months [3/42 IgA (+) versus 47/227 IgA (−); $p = 0.03$]. The conclusion from these studies was that a mucosal IgA anti-lectin antibody response is associated with immune protection against *E. histolytica* colonization. Recent studies demonstrating that immunization with the lectin provides protection from amebic colitis in a mouse model (E. Houpt & W.A. Petri, unpublished data) lends credence to the conclusion that the human immune response against the lectin leads to acquired immunity.

SUMMARY AND CONCLUSIONS

It is at the plasma membrane of the parasite that interactions with the host and environment lead to crucial decisions to invade or to encyst. It is increasingly clear from the past two decades of research that a single molecule, the Gal/GalNAc lectin, has the decisive role in these processes. Interaction of the lectin with precise quantities of mucin glycoproteins leads to formation of the cyst, whereas adherence of the trophozoite to the intestinal epithelium via the lectin leads to contact-dependent cytolysis and death of the host. Our knowledge of the contribution of the lectin to host invasion and to formation of the infectious cyst is not an end but a beginning. What lies ahead is to understand how the cellular processes initiated by the lectin result in infection and disease. The challenge is to comprehend the complex single cell and multicellular behavior of a parasite that remains one of mankind's scourges.

ACKNOWLEDGMENTS

Work from the authors' laboratories was supported by NIH grants AI-32615 (B.J. Mann), AI-26649, AI-37941, and AI-43596 (W.A. Petri), the Burroughs Wellcome Fund, and the Lucille P. Markey Charitable Trust. Dr. Haque is an International Research Scholar of the Howard Hughes Medical Institute, and Dr. Petri is a Burroughs Wellcome Scholar in molecular parasitology.

The *Annual Review of Microbiology* is online at http://micro.annualreviews.org

LITERATURE CITED

1. Adler P, Wood SJ, Lee YC, Lee RT, Petri WA Jr, Schnaar RL. 1995. High affinity binding of the *E. histolytica* lectin to polyvalent N-acetylgalactosaminides. *J. Biol. Chem.* 270:5164–71
2. Ankri S, Padilla-Vaca F, Stolarsky T, Koole L, Katz U, Mirelman D. 1999. Antisense inhibition of expression of the light subunit (35 kDa) of the Gal/GalNac lectin complex inhibits *Entamoeba histolytica* virulence. *Mol. Microbiol.* 33:327–37
3. Arhets P, Olivo JC, Gounon P, Sansonetti P, Guillén N. 1998. Virulence and functions of myosin II are inhibited by overexpression of light meromyosin in *Entamoeba histolytica. Mol. Biol. Cell* 6:1537–47
4. Ayeh-Kumi P, Ali IKM, Lockhart L, Petri WA Jr, Haque R. 2001. *Entamoeba histolytica*: genetic diversity of clinical isolates from Bangladesh as demonstrated by polymorphisms in the serine-rich gene. *Exp. Parasitol.* 99:80–88
5. Barwick RS, Uzicanin A, Lareau S, Malakmadze N, Imnadze P, et al. 1999. *Outbreak of amebiasis in Tbilisi, Republic of Georgia, 1998.* Presented at Am. Soc. Trop. Med. Hyg. Annu. Meet., Nov. 29–Dec. 2, 1999, Washington, DC
6. Berninghausen O, Leippe M. 1997. Necrosis versus apoptosis as the mechanism of target cell death induced by *Entamoeba histolytica. Infect. Immun.* 65:3615–21
7. Bracha R, Mirelman D. 1983. Adherence and ingestion of *Escherichia coli* serotype 055 by trophozoites of *Entamoeba histolytica. Infect. Immun.* 40:882–87
8. Bracha R, Nuchamowitz Y, Leippe M, Mirelman D. 1999. Antisense inhibition of amoebapore expression in *Entamoeba histolytica* causes a decrease in amoebic virulence. *Mol. Microbiol.* 34:463–72
9. Braga LL, Ninomiya H, McCoy JJ, Eacker S, Wiedmer T, et al. 1992. Inhibition of the complement membrane attack complex by the galactose-specific adhesin of *Entamoeba histolytica. J. Clin. Invest.* 90:1131–37
10. Burchard GD, Bilke R. 1992. Adherence of pathogenic and nonpathogenic *Entamoeba histolytica* strains to neutrophils. *Parasitol. Res.* 78:146–53
11. Capin R, Capin NR, Carmona M, Ortiz-Ortiz L. 1980. Effect of complement depletion on the induction of amebic liver abscess in the hamster. *Arch. Invest. Med.* 11(Suppl. 1):173–80

12. Chadee K, Johnson ML, Orozco E, Petri WA, Ravdin JI. 1988. Binding and internalization of rat colonic mucins by the Gal/GalNAc adherence lectin of *Entamoeba histolytica*. *J. Infect. Dis.* 158:398–406
13. Chadee K, Meerovitch E. 1985. *Entamoeba histolytica:* early progressive pathology in the cecum of the gerbil (*Meriones unguiculatus*). *Am. J. Trop. Med. Hyg.* 34:283–91
14. Chadee K, Petri WA Jr, Innes DJ, Ravdin JI. 1987. Rat and human colonic mucins bind to and inhibit the adherence lectin of *Entamoeba histolytica*. *J. Clin. Invest.* 80:1245–54
15. Cheng XJ, Hughes MA, Huston CD, Loftus B, Gilchrist CA, et al. 2001. The 150 kDa Gal/GalNAc lectin co-receptor of *Entamoeba histolytica* is a member of a gene family containing multiple CXXC sequence motifs. *Infect. Immun.* 69:5892–98
16. Cheng XJ, Tsukamoto H, Kaneda Y, Tachibana H. 1998. Identification of the 150 kDa surface antigen of *Entamoeba histolytica* as a galactose- and N-acetyl-D-galactosamine-inhibitable lectin. *Parasitol. Res.* 84:632–39
17. Cho J, Eichinger D. 1998. *Crithidia fasciculata* induces encystation of *Entamoeba invadens* in a galactose-dependent manner. *J. Parasitol.* 84:705–10
18. Clark CG, Diamond LS. 1993. *Entamoeba histolytica*: a method for isolate identification. *Exp. Parasitol.* 77:450–55
19. Coppi A, Eichinger D. 1999. Regulation of *Entamoeba invadens* encystation and gene expression with galactose and N-acetylglucosamine. *Mol. Biochem. Parasitol.* 102:67–77
20. Dodson JM, Lenkowski PW Jr, Eubanks AC, Jackson TFHG, Napodano J, et al. 1999. Role of the *Entamoeba histolytica* adhesin carbohydrate recognition domain in infection and immunity. *J. Infect. Dis.* 179:460–66
21. Eichinger D. 2001. A role for a galactose lectin and its ligands during encystment of *Entamoeba*. *J. Eukaryot. Microbiol.* 48:17–21
22. Eichinger D. 2001. Encystation in parasitic protozoa. *Curr. Opin. Microbiol.* 4:421–26
23. Frisardi M, Ghosh SK, Field F, van Dellen K, Rogers R, et al. 2000. The most abundant glycoprotein of amebic cyst walls (Jacob) is a lectin with five Cys-rich, chitin-binding domains. *Infect. Immun.* 68:4217–24
24. Gathiram V, Jackson TFHG. 1985. Frequency distribution of *Entamoeba histolytica* zymodeme in a rural South African population. *Lancet* 1:719–721
25. Gilchrist CA, Holm CF, Hughes MA, Schaenman J, Mann BJ, Petri WA Jr. 2001. Identification and characterization of an *Entamoeba histolytica* URE3 sequence-specific DNA binding protein containing EF-hand motifs. *J. Biol. Chem.* 276:11838–41
26. Gilchrist CA, Mann BJ, Petri WA Jr. 1998. Control of ferredoxin and Gal/GalNAc lectin gene expression in *Entamoeba histolytica* by a *cis*-acting serum-responsive element. *Infect. Immun.* 66:2383–86
27. Godbold GD, Mann BJ. 2000. Cell killing by the human parasite *Entamoeba histolytica* is inhibited by the Rho-inactivating C3 exoenzyme. *Mol. Biochem. Parasitol.* 108:147–51
28. Guerrant RL, Brush J, Ravdin JI, Sullivan JA, Mandell GL. 1981. Interaction between *Entamoeba histolytica* and human polymorphonuclear neutrophils. *J. Infect. Dis.* 143:83–93
29. Hamann L, Buß H, Tannich E. 1997. Tetracycline-controlled gene expression in *Entamoeba histolytica*. *Mol. Biochem. Parasitol.* 84:83–91
30. Haque R, Ali IKM, Sack RB, Ramakrishnan G, Farr BM, Petri WA Jr. 2001. Amebiasis and mucosal IgA antibody against the *Entamoeba histolytica* adherence lectin in Bangladeshi children. *J. Infect. Dis.* 183:1787–93
31. Huston CD, Houpt ER, Mann BJ, Hahn CS, Petri WA Jr. 2000. Caspase 3 dependent killing of human cells by the parasite *Entamoeba histolytica*. *Cell. Microbiol.* 2:617–25

32. Le Yan L, Stanley SL Jr. 2001. Blockade of caspases inhibits amebic liver abscess formation in a mouse model of disease. *Infect. Immun.* 69:7911–14
33. Leippe M, Ebel S, Schoenberger OL, Horstmann RD, Muller-Eberhard HJ. 1991. Pore-forming protein of pathogenic *Entamoeba histolytica. Proc. Natl. Acad. Sci. USA* 88:7659–63
34. Leippe M, Tannich E, Nickel R, van der Goot G, Pattus F, et al. 1992. Primary and secondary structure of the pore-forming peptide of pathogenic *Entamoeba histolytica. EMBO J.* 3501–6
35. Leroy A, De Bruyne G, Mareel M, Nokkaew C, Bailey G, Nelis H. 1995. Contact-dependent transfer of the galactose-specific lectin of *Entamoeba histolytica* to the lateral surface of enterocytes in culture. *Infect. Immun.* 63:4253–60
36. Li E, Becker A, Stanley SL. 1988. Use of Chinese hamster ovary cells with altered glycosylation patterns to define the carbohydrate specificity of *Entamoeba histolytica* adhesion. *J. Exp. Med.* 167:1725–30
37. Li E, Becker A, Stanley SL. 1989. Chinese hamster ovary cells deficient in N-acetylglucosaminyltransferase I activity are resistant to *Entamoeba histolytica*-mediated cytotoxicity. *Infect. Immun.* 57:8–12
38. Mann BJ, Chung CY, Dodson JM, Ashley LS, Braga LL, Snodgrass TL. 1993. Neutralizing monoclonal antibody epitopes of the *Entamoeba histolytica* galactose adhesin map to the cysteine-rich extracellular domain of the 170-kDa subunit. *Infect. Immun.* 61:1772–78
39. Mann BJ, Vedvick T, Torian B, Petri WA Jr. 1991. Sequence of a cysteine-rich galactose-specific lectin of *Entamoeba histolytica. Proc. Natl. Acad. Sci. USA* 88:3248–52
40. McCoy JJ, Mann BJ, Petri WA Jr. 1994. Adherence and cytotoxicity of *Entamoeba histolytica*, or how lectins let parasites stick around. *Infect. Immun.* 62:3045–50
41. McCoy JJ, Mann BJ, Vedvick T, Pak Y, Heimark DB, Petri WA Jr. 1993. Structural analysis of the light subunit of the *Entamoeba histolytica* adherence lectin. *J. Biol. Chem.* 268:24223–31
42. McCoy JJ, Mann BJ, Vedvick T, Petri WA Jr. 1993. Sequence analysis of genes encoding the *Entamoeba histolytica* galactose-specific adhesin light subunit. *Mol. Biochem. Parasitol.* 61:325–28
43. McCoy JJ, Weaver AM, Petri WA Jr. 1994. Use of monoclonal anti-light subunit antibodies to study the structure and function of the *Entamoeba histolytica* Gal/GalNAc adherence lectin. *Glycoconj. J.* 11:432–36
44. Nickel R, Tannich E. 1994. Transfection and transient expression of chloramphenicol acetyltransferase gene in the protozoan parasite *Entamoeba histolytica. Proc. Natl. Acad. Sci. USA* 91:7095–98
45. Padilla-Vaca F, Ankri S, Bracha R, Koole LA, Mirelman D. 1999. Downregulation of *Entamoeba histolytica* virulence by monoxenic cultivation with *Escherichia coli* O55 is related to a decrease in expression of the light (35-kilodalton) subunit of the Gal/GalNAc lectin. *Infect. Immun.* 67:2096–92
46. Petri WA Jr, Chapman MD, Snodgrass T, Mann BJ, Broman J, Ravdin JI. 1989. Subunit structure of the galactose and N-acetyl-D-galactosamine-inhibitable adherence lectin of *Entamoeba histolytica. J. Biol. Chem.* 264:3007–12
47. Petri WA Jr, Singh U. 1999. State of the art: diagnosis and management of amebiasis. *Clin. Infect. Dis.* 29:1117–25
48. Petri WA Jr, Smith RD, Schlesinger PH, Murphy CF, Ravdin JI. 1987. Isolation of the galactose binding adherence lectin of *Entamoeba histolytica. J. Clin. Invest.* 80: 1238–44
49. Petri WA Jr, Snodgrass TL, Jackson TFHG, Gathiram V, Simjee AE, et al. 1990. Monoclonal antibodies directed against the galactose-binding lectin of *Entamoeba histolytica* enhance adherence. *J. Immunol.* 144:4803–9
50. Pillai DR, Wan PSK, Yau YCW, Ravdin JI, Kain KC. 1999. The cysteine-rich region of the *Entamoeba histolytica* adherence lectin (170-kDa subunit) is sufficient

for high affinity Gal/GalNAc-specific binding in vitro. *Infect. Immun.* 67:3836–41

51. Pittman FE, El-Hashimi WK, Pittman JC. 1973. Studies of human amebiasis. II. Light and electron microscopic observations of colonic mucosal and exudate in acute amebic coolitis. *Gastroenterology* 65:588–603
52. Prathap K, Gilman R. 1970. The histopathology of acute intestinal amebiasis. *Am. J. Pathol.* 60:229–45
53. Purdy JE, Mann BJ, Pho LT, Petri WA Jr. 1994. Transient transfection of the enteric parasite *Entamoeba histolytica* and expression of firefly luciferase. *Proc. Natl. Acad. Sci. USA* 91:7099–103
54. Purdy JE, Pho LT, Mann BJ, Petri WA Jr. 1996. Upstream regulatory elements controlling expression of the *Entamoeba histolytica* lectin. *Mol. Biochem. Parasitol.* 78: 91–103
55. Ragland BD, Ashley LS, Vaux DL, Petri WA Jr. 1994. *Entamoeba histolytica*: target cells killed by trophozoites undergo apoptosis which is not blocked by bcl-2. *Exp. Parasitol.* 79:460–67
56. Ramakrishnan G, Lee S, Mann BJ, Petri WA Jr. 2000. *Entamoeba histolytica*: deletion of the GPI anchor signal sequence on the Gal/GalNAc lectin light subunit prevents its assembly into the lectin heterodimer. *Exp. Parasitol.* 96:57–60
57. Ramakrishnan G, Ragland BD, Purdy JE, Mann BJ. 1996. Physical mapping and expression of gene families encoding the N-acetyl D-galactosamine adherence lectin of *Entamoeba histolytica*. *Mol. Microbiol.* 19:91–100
58. Ramakrishnan G, Vines RR, Mann BJ, Petri WA Jr. 1997. A tetracycline-inducible gene expression system in *Entamoeba histolytica*. *Mol. Biochem. Parasitol.* 84:93–100
59. Ravdin JI, Croft BY, Guerrant RL. 1980. Cytopathogenic mechanisms of *Entamoeba histolytica*. *J. Exp. Med.* 152:377–90
60. Ravdin JI, Guerrant RL. 1981. Role of adherence in cytopathogenic mechanisms of *Entamoeba histolytica*. Study with mammalian tissue culture cells and human erythrocytes. *J. Clin. Invest.* 68:1305–13
61. Ravdin JI, John JE, Johnston LI, Innes DJ, Guerrant RL. 1985. Adherence of *Entamoeba histolytica* to rat and human colonic mucosa. *Infect. Immun.* 48:292–97
62. Ravdin JI, Moreau F, Sullivan JA, Petri WA Jr, Mandell GL. 1988. Relationship of free intracellular calcium to the cytolytic activity of *Entamoeba histolytica*. *Infect. Immun.* 56:1505–12
63. Ravdin JI, Sperelakis N, Guerrant RL. 1982. Effect of ion channel inhibitors on the cytopthogenicity of *Entamoeba histolytica*. *J. Infect. Dis.* 146:335–40
64. Ravdin JI, Stanley P, Murphy CF, Petri WA Jr. 1989. Characterization of cell surface carbohydrate receptors for *Entamoeba histolytica* adherence lectin. *Infect. Immun.* 57:2179–86
65. Reed SL, Curd JG, Gigli I, Gillin FD, Braude AI. 1986. Activation of complement by pathogenic and nonpathogenic *Entamoeba histolytica*. *J. Immunol.* 136: 2265–70
66. Reed SL, Gigli I. 1990. Lysis of complement-sensitive *Entamoeba histolytica* by activated terminal complement components. *J. Clin. Invest.* 86:1815–22
67. Reed SL, Keene WE, McKerrow JH, Gigli I. 1989. Cleavage of C3 by a neutral cysteine proteinase of *Entamoeba histolytica*. *J. Immunol.* 143:189–95
68. Rini J. 1995. Lectin structure. *Annu. Rev. Biophys. Biomol. Struct.* 24:551–77
69. Saffer LD, Petri WA Jr. 1991. Role of the galactose lectin of *Entamoeba histolytica* in adherence-dependent killing of mammalian cells. *Infect. Immun.* 59:4681–83
70. Sargeaunt PG, Williams JE, Grene JD. 1978. The differentiation of invasive and non-invasive *Entamoeba histolytica* by isoenzyme electrophoresis. *Trans. R. Soc. Trop. Med. Hyg.* 72:519–21
71. Schaenman JM, Gilchrist CA, Mann BJ, Petri WA Jr. 2001. Identification of two *Entamoeba histolytica* sequence-specific *hgl5* enhancer-binding proteins with homology

to the RNA-binding motif RRM. *J. Biol. Chem.* 276:1602–9

72. Schaenman JM, Mann BJ, Petri WA Jr. 1998. An upstream regulatory element containing two nine basepair repeats regulates expression of the *Entamoeba histolytica hgl5* lectin gene. *Mol. Biochem. Parasitol.* 94:309–13
73. Seydel KB, Stanley SL Jr. 1998. *Entamoeba histolytica* induces host cell death in amebic liver abscess by a non-Fas-dependent, non-tumor necrosis factor alpha-dependent pathway of apoptosis. *Infect. Immun.* 66:2980–83
74. Singh U, Gilchrist CA, Schaenman JM, Rogers JB, Hockensmith JW, et al. 2002. Context-dependent roles of the *Entamoeba histolytica* core promoter element GAAC in transcriptional activation and protein complex assembly. *Mol. Biochem. Parasitol.* In press
75. Singh U, Rogers JB, Mann BJ, Petri WA Jr. 1997. Transcription initiation is controlled by three core promoter elements in the protozoan parasite *Entamoeba histolytica. Proc. Natl. Acad. Sci. USA* 94:8812–17
76. Tannich E, Ebert F, Horstmann RD. 1991. Primary structure of the 170-kDa surface lectin of pathogenic *Entamoeba histolytica. Proc. Natl. Acad. Sci. USA* 88:1849–53
77. Tannich E, Horstmann RD, Knobloch J, Arnold HH. 1989. Genomic DNA differences between pathogenic and nonpathogenic *Entamoeba histolytica. Proc. Natl. Acad. Sci. USA* 86:5118–22
78. Vinall LE, Hill AS, Piguy P, Pratt WS, Toribara N, et al. 1998. Variable number tandem repeat polymorphism of the mucin genes located in the complex on 11p15. 5. *Hum. Genet.* 102:357–66
79. Vines RR, Ramakrishnan G, Rogers J, Lockhart L, Mann BJ, Petri WA Jr. 1998. Regulation of adherence and virulence by the *Entamoeba histolytica* lectin cytoplasmic domain, which contains an $\beta 2$ integrin motif. *Mol. Biol. Cell* 9:2069–79
80. Wanke C, Butler T, Islam M. 1988. Epidemiologic and clinical features of invasive amebiasis in Bangladesh: a case-control comparison with other diarrheal diseases and postmortem findings. *Am. J. Trop. Med. Hyg.* 38:335–41
81. WHO/PAHO/UNESCO. 1997. Report of a consultation of experts on amoebiasis. *Wkly. Epidemiol. Rep. WHO* 72(14):97–99
82. Yi D, Lee RT, Longo P, Boger ET, Lee YC, et al. 1998. Substructural specificity and polyvalent carbohydrate recognition by the *Entamoeba histolytica* and rat hepatic N-acetylgalactosamine/galactose lectins. *Glycobiology* 8:1037–43

Annu. Rev. Microbiol. 2002. 56:65–91
doi: 10.1146/annurev.micro.56.012302.161052

First published online as a Review in Advance on May 3, 2002

HEAVY METAL MINING USING MICROBES[1]

Douglas E. Rawlings
Department of Microbiology, University of Stellenbosch, Private Bag X1, Stellenbosch 7602, South Africa; e-mail: der@sun.ac.za

Key Words biomining, bioleaching, mineral biooxidation, chemolithotrophic bacteria

■ **Abstract** The use of acidiphilic, chemolithotrophic iron- and sulfur-oxidizing microbes in processes to recover metals from certain types of copper, uranium, and gold-bearing minerals or mineral concentrates is now well established. During these processes insoluble metal sulfides are oxidized to soluble metal sulfates. Mineral decomposition is believed to be mostly due to chemical attack by ferric iron, with the main role of the microorganisms being to reoxidize the resultant ferrous iron back to ferric iron. Currently operating industrial biomining processes have used bacteria that grow optimally from ambient to 50°C, but thermophilic microbes have been isolated that have the potential to enable mineral biooxidation to be carried out at temperatures of 80°C or higher. The development of higher-temperature processes will extend the variety of minerals that can be commercially processed.

CONTENTS

---

[1]Definitions: The term bioleaching refers to the conversion of an insoluble metal (usually a metal sulfide, e.g., CuS, NiS, ZnS) into a soluble form (usually the metal sulfate, e.g., $CuSO_4$, $NiSO_4$, $ZnSO_4$). When this happens, the metal is extracted into water; this process is called bioleaching (48, 89). Because these processes are oxidations, this process may also be termed biooxidation. However, the term biooxidation is usually used to refer to processes in which the recovery of a metal is enhanced by microbial decomposition of the mineral, but the metal being recovered is not solubilized. An example is the recovery of gold from arsenopyrite ores where the gold remains in the mineral after biooxidation and is extracted by cyanide in a subsequent step. The term bioleaching is clearly inappropriate when referring to gold recovery (although arsenic, iron, and sulfur are bioleached from the mineral). Biomining is a general term that may be used to refer to both processes.

INTRODUCTION

The use of microbes to extract metals from ores is simply the harnessing of a natural process for commercial purposes. Microbes have participated in the deposition and solubilization of heavy metals in the earth's crust since geologically ancient times. Most of this activity is linked to the iron and sulfur cycles. Anaerobic sulfate-reducing bacteria generate sulfides that can react with a variety of metals to form insoluble metal sulfides. These sulfide precipitates serve as a source of mineral deposits that may become incorporated into rock formations. The metal sulfides in turn can serve as electron donors for the usually aerobic sulfur-oxidizing microbes, which results in the formation of metal sulfates. Because many of these metal sulfates are soluble, this process serves as a means of leaching the metal from the mineral deposit. This ability of certain microbes to solubilize metals from ores has given rise to a growing biomining industry (72).

The use of microbes in ore processing has some distinct advantages over the traditional physicochemical methods. Almost without exception, microbial extraction procedures are more environmentally friendly. They do not require the high amounts of energy used during roasting or smelting and do not produce sulfur dioxide or other environmentally harmful gaseous emissions. Furthermore, mine tailings and wastes produced from physicochemical processes when exposed to rain and air may be biologically leached, producing unwanted acid and metal pollution. Tailings from biomining operations are less chemically active, and the biological activity they can support is reduced by at least the extent to which they have already been bioleached. Biomining also has a clear advantage in the extraction of metals from certain low-grade ores. For example, copper can be recovered from low-grade ores and dumps left behind from previous mining operations by using the biological activity that takes place during controlled irrigation of the dump (5, 6). Many of these metals are not economically recoverable by nonbiological methods. Where ore-type and geological features permit, metal recovery using leaching solutions produced and regenerated by microbes can be carried out in situ with obvious cost advantages and minimal disturbance to the surrounding environment.

HISTORY

Biomining has a long history, although the early miners did not know that microbes were involved. The use of microorganisms to extract copper has its roots deep in antiquity. It is not clear to what extent microbial activity was used to extract copper by eighteenth-century Welsh miners at Anglesey (North Wales), but it is certain that microbes played a role in even earlier activities at the Rio Tinto mine. Pre-Romans recovered silver and the Romans recovered copper from a deposit located in the Seville province in the south of Spain, which later became known as the Rio Tinto mine. The Rio Tinto (Red River) obtained its name from the red color imparted to the water by the high concentration of ferric iron (Figure 1, see color insert). This dissolved ferric iron (and the less easily seen dissolved copper) is due to natural microbial activity. From earliest records the Rio Tinto has been known as a river devoid of fish and with water that is undrinkable.

The site of the immense workings of the Rio Tinto mines was rediscovered in 1556 when Francisco de Mendoza, under instruction of Philip II of Spain, went searching for assets that could help fill the crown's empty coffers (1a). Diego Delgardo, the priest left behind to investigate the mines, was fascinated by the inhospitable river. Villagers from the area explained that if iron was placed in the river it would disappear, a property confirmed by the priest. The water had another property, which the priest would not commit to paper but on which he wished to report to the king in person. Speculation is that he had discovered the electrochemical phenomenon whereby copper is precipitated from solution as iron dissolves. This gives the appearance of iron being converted into copper. This process is known as cementation and is one of the methods by which dissolved copper is recovered from aqueous solutions. What may have made the priest secretive about this phenomenon is that a method for converting one metal to another had long been an objective of alchemists, and if iron could be converted to copper, he may have believed that they were on the threshold of discovering how to convert iron to gold! Following the visits and report of Delgardo and Mendoza, several attempts were made to reopen mining operations at Rio Tinto, but it was not until about 1750 that sustained mining at the site occurred.

BIOMINING MICROBES: GENERAL CHARACTERISTICS

The primary biomining organisms have several physiological features in common. They are all chemolithoautotrophic and are able to use ferrous iron or reduced inorganic sulfur sources (or both) as electron donors. Because the by-product of sulfur-oxidation is sulfuric acid, these organisms are acidophilic and most will grow within the pH range 1.5–2.0. This extreme acidophily applies even to those biomining organisms that can oxidize only iron. Although biomining bacteria may be able to use electron acceptors other than oxygen (e.g., ferric iron), they generally grow best in highly aerated solutions. All primary biomining organisms fix CO_2, although there is considerable variation in the efficiency with which this is done.

The less-efficient CO_2-fixing species require either elevated levels of CO_2 or a small amount of yeast extract to grow rapidly. As may be expected, biomining bacteria are generally resistant to a range of metal ions, with some variation in metal tolerance between species and isolates within a species. These common properties explain why biomining organisms are ideally suited to growing in the inorganic environment created by the active aeration of a suspension of a suitable iron- or sulfur-containing mineral in water or during the passive aeration that takes place when a heap of the mineral is irrigated with water. Air provides the carbon source (CO_2) and the preferred electron acceptor (O_2), the mineral ore supplies the electron donor (ferrous iron and/or reduced inorganic sulfur), and water is the medium for growth. Some biomining organisms can also fix nitrogen from the air, although they may not be able to do so in a highly aerated environment. Trace elements are provided by the mineral and water. In commercial processes small quantities of inexpensive, fertilizer-grade, ammonium sulfate and potassium phosphate may be added to ensure that nutrient limitation does not occur.

BIODIVERSITY OF BIOMINING MICROBES: ISOLATION AND DETECTION

The variety of microbes identified as being capable of growth in situations that simulate biomining commercial processes is rapidly growing. This is partly because of an increase in the number of environments being screened for such organisms, partly because of an increase in the variety of minerals being tested, and most importantly because of new techniques available to screen for the presence of organisms. Investigations into the presence of the obligately acidophilic chemolithoautotrophs have presented several difficulties. These bacteria are difficult to cultivate on solid media, as they are very sensitive to organic matter including the small quantities of sugar present as impurities in polysaccharide-based gelling agents such as agar or agarose (91). Attempts to use highly purified agars have not been very successful, probably because some of the sugar molecules in the gelling agent are released owing to acid-hydrolysis at low pH, and the released sugars inhibit cell growth. A number of alternative gelling agents have met with partial success, but most of these are difficult to work with.

The most successful approach to using laboratory media has been the development of a double-layer plate technique, whereby freshly grown acidophilic heterotrophic bacteria (which are frequently found growing in close association with the autotrophic iron and sulfur oxidizers) are mixed into an inorganic pour plate medium (43, 47). After the first layer has set, a second layer of inorganic medium without the heterotrophs is poured on top. The starving heterotrophs absorb any free sugars and metabolic waste products, allowing for good growth of the obligate chemoautolithotrophs. However, even a good isolation medium does not solve all the problems. Many of the bacteria in mineral environments grow in biofilms that adhere strongly to the surface of the particulate matter or grow deep within the pores that form during mineral decomposition. Furthermore, there is the possibility that some of these fastidious bacteria may be unculturable on any laboratory media.

A breakthrough in investigating the ecology of biomining processes was achieved by the application of the now widely used techniques of polymerase chain reaction (PCR) amplification of 16S rRNA genes from total DNA extracted from environmental samples (29). PCR-based methods of detecting microorganisms have their drawbacks in that not all microbes may lyse equally well, and attached microoganisms may be difficult to free from solid material. The technique is also not a quantitative estimate of microbial numbers in samples that contain several types of unknown organisms. Nevertheless, PCR-based methods have allowed an analysis of the composition of the microorganisms in heaps and stirred tank reactors in a manner that does not require growth in a laboratory.

Unfortunately, many studies of the biodiversity of biomining organisms have used rather unreliable and selective methods that require being able to cultivate the microbes on laboratory media that do not take into account the vicissitudes of the microbes. Furthermore, in many laboratory leaching experiments, minerals were inoculated with a pure culture or a defined mixture of cultures, and it is uncertain whether such organisms would be competitive in the nonsterile conditions of commercial biomining processes.

Mesophiles, Moderate Thermophiles, and Thermophiles

Mineral decomposition can take place at a variety of temperatures. In the case of heap and dump leaching, no minimum temperature has been established, but most biooxidation probably takes place in the 20°–35°C range. In biooxidation tanks the temperature is properly controlled, and commercially proven processes have been developed that operate at either 40° or 50°C (17, 62, 93). Processes that operate at 75°–78°C are being developed. As may be expected, these operating temperatures are sufficiently different for the types of microbes present that are also different. Nevertheless, it appears that combinations of iron- and sulfur-oxidizing organisms can be selected that are able to facilitate the efficient decomposition of ores at each of these temperatures. Good recent reviews of the biodiversity of microorganisms from acid environments are available (33, 44, 46, 63). The description of biomining organisms given below is not meant to be a complete survey of all possible organisms that could be encountered; rather it concentrates on those that have been reported from current commercial biomining processes and those under development. This is difficult because although several studies on the bacteria present in processes that operate from ambient to 40°C have been reported, there is only a single published study on a process that operates at 50°C (62). Processes that operate at temperatures as high as 75°–80°C are under development, and information on the organisms being used is not yet in the public domain.

Microorganisms Considered Important in Commercial Mineral Biooxidation Processes

In the following survey of important biomining organisms, an attempt has been made to concentrate on organisms found in nonsterile commercial or pilot-scale processes, where the most competitive organisms would have been selected.

ACIDITHIOBACILLUS Biomining bacteria belonging to this genus were previously included in the genus *Thiobacillus*. As a result of 16S rRNA sequence analysis, it became clear that the genus *Thiobacillus* (as described prior to 2000) included sulfur-oxidizing bacteria that belonged to α-, β-, and γ-divisions of the *Proteobacteria*. To solve this anomaly, the genus *Thiobacillus* was subdivided (49) and a new genus, *Acidithiobacillus*, was created to accommodate the highly acidophilic members of the former genus. These members include *Acidithiobacillus ferrooxidans* (previously *Thiobacillus ferrooxidans*), *At. thiooxidans* (previously *T. thiooxidans*), and *At. caldus* (previously *T. caldus*). Phylogenetically, the genus *Acidithiobacillus* is situated very close to the branch point between the β- and γ-subdivisions of the *Proteobacteria* (53, 73). These bacteria appear to be ubiquitous and have been isolated from sites that provide a suitable environment for their growth (such as sulfur springs and acid mine drainage) from many regions throughout the world.

AT. FERROOXIDANS This was the first bacterium discovered that was capable of oxidizing minerals (12). Although typical *At. ferrooxidans* isolates have a genomic G + C mole ratio of 57%–59%, they have been found to belong to at least four DNA-DNA hybridization similarity subgroups (35). The percentage DNA-DNA similarity between some of these subgroups is sufficiently low (10%–50%) for them to be considered separate species. In addition, a strain called *T. ferrooxidans* m1 has a G + C of 65%, is a β-proteobacterium, and belongs to a different, as yet unnamed, genus. Nutritionally, typical *At. ferrooxidans* isolates are considered obligate autotrophs. They are also able to grow on formic acid, provided it is added in small quantities as it is consumed (70). Because formic acid is a C1 compound, this property is not inconsistent with autotrophy. *At. ferrooxidans* is able to use either ferrous iron or a variety of reduced inorganic sulfur compounds as an electron donor. It is preferentially aerobic but also able to grow using ferric iron as an electron acceptor, provided that it has a reduced inorganic sulfur compound as an electron donor (69).

For many years *At. ferrooxidans* was considered to be the most important microorganism in biomining processes that operate at 40°C or less. However, in some of the older studies techniques were used that were less sophisticated than those available since the early to mid-1990s. It is currently understood that *At. ferrooxidans* is not favored in situations in which the ferric iron content is much higher than the ferrous iron (high-redox potential), such as is found in continuously operating stirred tank reactors operating under steady-state conditions (76). Nevertheless, *At. ferrooxidans* may be the dominant bacterium in some dump and heap leaching environments including those involved in uranium and copper oxide/sulfide leaching, especially if the ferrous iron concentration in solution is high (≥5 g/liter) (68). *At. ferrooxidans* is capable of rapid growth relative to many biomining bacteria and is generally favored within the temperature range 20°–35°C and pH 1.8–2.0, providing the ratio of soluble ferrous iron to ferric iron is high. It is also possible to adapt *At. ferrooxidans* so that they are able to grow at pH values outside their optimal range (96). As may be expected from bacteria growing in a mineral-rich environment, these bacteria are remarkably tolerant to a wide range of soluble metal ions [reviewed in (56)].

At. thiooxidans is nutritionally very much like *At. ferrooxidans* except that it is unable to oxidize ferrous iron and is therefore restricted to using reduced sulfur compounds as an electron donor. Like *At. ferrooxidans*, it is mesophilic and probably has a limit for growth of about 35°C, but it is even more acid-tolerant (pH 0.5–5.5). An investigation of copper leaching columns using simulated recycled leach liquor with a high sulfate content and pH 0.7 indicated that *At. thiooxidans* was the dominant sulfur oxidizer (94). The bacteria are different in other respects, with *At. thiooxidans* having a genomic G + C ratio of 53%. The DNA-DNA similarity between *At. thiooxidans* and *At. ferrooxidans* is about 20% or less (35).

At. caldus is similar to *At. thiooxidans* in regard to its ability to oxidize reduced sulfur compounds, but not ferric iron, and its ability to grow at very low pH (32). It is distinguished from *At. thiooxidans* in that it is moderately thermophilic and has an optimum growth temperature of about 45°C. It has a 16S rRNA sequence closely related to *At. thiooxidans*, and for several years there was confusion as to whether new isolates of sulfur-oxidizing bacteria from mineral environments were *At. thiooxidans* or *At. caldus*. We have found *At. caldus* to be the dominant sulfur-oxidizing bacterium in processes operating within the temperature range 35°–50°C and consider it to be the "weed" of biomining bacteria under these conditions (77). Unlike *At. thiooxidans*, some strains (if not all) of *At. caldus* are able to grow mixotrophically using yeast extract or glucose.

LEPTOSPIRILLUM Bacteria of this genus are similar to acidithiobacilli in that they are also highly acid-tolerant (optimum pH ~1.5–1.8), gram-negative, chemolithoautrotrophic bacteria. However, in most other respects they are very different (45). Based on 16S rRNA sequence data, they are not members of the *Proteobacteria* but belong to the division *Nitrospira*. An unusual and unifying characteristic of the leptospirilli is that they are capable of using only ferrous iron as an electron donor. As a result, the leptospirilli have a high affinity for ferrous iron and unlike *At. ferrooxidans*, their ability to oxidize ferrous iron is not inhibited by ferric iron. Because the redox potential of the Fe^{2+}/Fe^{3+} couple is very positive (+770 mV at pH 2), they are forced to use the O_2/H_2O redox couple (+820 mV) as their electron acceptor, and therefore the leptospirilli are obligately aerobic organisms. These bacteria are also acid-tolerant, and in a study of copper leaching at pH 0.7 (94), *Leptospirillum ferrooxidans* was reported to be the dominant iron oxidizer. Leptospirilli have been widely reported in biooxidation processes, usually in combination with a sulfur-oxidizing bacterium such as *At. caldus* or *At. thiooxidans*. In a study carried out in Australia a PCR-based technique was used to analyze the microbial population in a silver-catalyzed column for the leaching of chalcopyrite ore at 37°C (19), *L. ferrooxidans* was readily detected, and no 16S rRNA corresponding to *At. ferrooxidans* or *At. thiooxidans* was apparent. Two species of leptospirilli have been recognized: *L. ferrooxidans* and *L. thermoferrooxidans* (36). Unfortunately, the *L. thermoferrooxidans* reported to grow at 45°C has been lost so it is difficult to compare it with available isolates.

However, it is likely that at least four species of *Leptospirillum* exist. Bacteria previously referred to as *L. ferrooxidans* fall into two G + C groups, 49%–51% and 55%–56%. Genome DNA-DNA hybridization between the groups is 11% or less (12a). As the *L. ferrooxidans*–type strain belongs to the 49%–51% G + C group, it has been proposed that the higher G + C group should be recognized as a third species, named *L. ferriphilum* (12a). The existence of a fourth species is suggested by 16S rRNA sequence data from total DNA extracted from abandoned pyrite mine samples, but no members of this proposed fourth species have been isolated in laboratory culture (3). Interestingly, several (but not all) isolates of *L. ferriphilum*, but no isolates of *L. ferrooxidans*, tested were capable of growth at 45°C. In addition, all strains from commercial biooxidation plants in South Africa were *L. ferriphilum* (12a), but this could have been because the plants were operated at 40°, 45°, or 50°C, and this temperature favored *L. ferriphilum* above *L. ferrooxidans*.

ACIDIPHILIUM With the exception of one species, bacteria belonging to the genus *Acidiphilium* are acid-tolerant, gram-negative heterotrophs rather than iron- or sulfur-oxidizing autotrophs. As such, they are not primarily involved in mineral decomposition (33). They are included because they have been detected in a batch bioreactor (28) and are frequently found growing near bacteria such as *At. ferrooxidans*, where they are believed to feed on the organic waste products produced by the iron and sulfur oxidizers. Indeed strains of *Acidiphilium* are sometimes so closely associated with *At. ferrooxidans* that they have been difficult to separate, and many of the early studies of *At. ferrooxidans* are misleading because they were carried out with mixed cultures. The ability of *Acidiphilium* isolates to detoxify laboratory growth media is the underlying principle of the overlay plating technique, in which *Acidiphilium* strains such as SJH are used in a bottom layer to remove presumably organic toxins from the upper layer, allowing the fastidious, organic matter–intolerant, iron- and sulfur-oxidizing autotrophs to grow (43, 47). Another possible effect of members of the genus *Acidiphilium* is that during heterotrophic growth at low-oxygen tension, they are able to use ferric iron as an electron acceptor, thereby regenerating ferrous iron, which serves as an electron acceptor for the iron oxidizers. An interesting observation is that members of the genus *Acidiphilium* are able to produce bacteriochlorophyll-*a* but are not able to grow using light as their sole energy source (50). One species, *Acidiphilium acidophilum* (previously *Thiobacillus acidophilus*), is unique among members of the genus in that it is able to grow autotrophically using reduced inorganic sulfur, heterotrophically using a variety of carbon sources, or mixotrophically using both organic and inorganic carbon (33).

Potentially Important Bacteria in Mineral Biooxidation Processes

As is described below, mineral biooxidation is a combination of chemistry and biology. The chemistry of biooxidation of minerals such as chalcopyrite only proceeds at an economically viable rate at temperatures of 70°C and higher. Therefore, organisms that are able to regenerate the leaching solutions also need to grow at these high temperatures. Several new commercial processes for the biooxidation of previously untested minerals, which operate at a variety of different temperatures, are being (or are likely to be) developed. As a result, new types of microorganisms that catalyze such processes will almost certainly be discovered. Some organisms that contribute or are likely to contribute to the functioning of these new processes are described below.

SULFOBACILLUS AND RELATIVES Sulfobacilli are moderately thermophilic (40°–60°C), endospore-forming, gram-positive bacteria that have been isolated from heaps of mineral waste and biomining operations. These bacteria are able to grow autotrophically or heterotrophically. When growing autotrophically they use ferrous iron, reduced inorganic sulfur compounds, or sulfide minerals as electron donors (30, 65). However, their ability to fix CO_2 appears to be poor. To grow strongly, they require elevated levels of CO_2 in the atmosphere, small quantities of yeast extract, or close association with heterotrophic iron-oxidizing bacteria such as *Acidimicrobium ferrooxidans* (11). Clark & Norris (11) reported that *Am. ferrooxidans* can outnumber the *Sulfobacillus* species in laboratory reactors. Glucose can serve as a carbon and energy source for sulfobacilli when growing heterotrophically. Furthermore, they are able to grow in the absence of oxygen, using ferric iron as an electron acceptor and either organic or inorganic sulfur compounds as electron donors (4). Although several species of sulfobacilli exist, only two have been named, *Sulfobacillus thermosulfidooxidans* and *Sb. acidophilus*. *Sb. thermosulfidooxidans* has been reported to be more active in the oxidation of iron and sulfide minerals, whereas *Sb. acidophilus* more readily oxidizes sulfur, especially in the absence of organic nutrients (65). It is highly likely that sulfobacilli (or closely related bacteria) form part of the consortium of microorganisms that are used commercially to oxidize a variety of sulfide minerals in the temperature range 45°–55°C (possibly as high as 60°C).

FERROPLASMA AND RELATIVES These organisms are archaea rather than bacteria, are pleomorphic in shape, and lack cell walls. *Ferroplasma acidiphilum* was isolated from a pilot plant bioreactor treating arsenopyrite/pyrite in Kazakstan (31). It oxidizes ferrous iron but not sulfur and appears to be obligately aerobic. The archeaon is mesophilic, growing optimally at 33°C with an upper limit of 45°C. It has an optimum pH for growth of 1.7 and a lower limit of about 1.3. A closely related mixotrophic archeaon, *Ferroplasma acidarmanus*, was isolated from acid mine drainage (20). Similar archaea have been isolated from commercial bioreactors also treating an arsenopyrite/pyrite concentrate operating at 40°C at the Fairview mine in Barberton, South Africa. These organisms appeared in large numbers when pH control of the reactor was removed and the pH was allowed to fall in an unrestrained manner to about pH 0.5 (D.E. Rawlings, unpublished data). More recently, a related archaeon was found in copper ore column/heap-type plants operating at 30°C at high salt concentration (95). Although different from each other, 16S rRNA sequence data indicated that the two archaea isolates were most closely related to *Picrophilus oshimae* and *Thermoplasma acidophilum*. Both of these archaea could be grown on laboratory media; however, they grew in close association with chemolithotrophic bacteria, and attempts to isolate either archaeon as a pure culture were unsuccessful.

SULFOLOBUS The metabolic capability of microorganisms designated as *Sulfolobus* isolates is confusing because many early studies were carried out with impure cultures (63). Many of the most important mineral biooxidation studies have been carried out in laboratories using a strain known as *Sulfolobus* strain BC, now known to be an isolate of *Sulfolobus metallicus* (40). These obligately autotrophic archaea grow by oxidizing ferrous iron, reduced inorganic sulfur compounds, or sulfide ores. *S. metallicus* is thermophilic, with strain BC growing at an optimum of 68°C and preferably within a pH range of about 1.3–1.7. *S. metallicus* is very capable of oxidizing minerals such as arsenopyrite and chalcopyrite, especially if the air is enriched with 1% v/v CO_2 (63). Recently, some *Sulfolobus*-like organisms that are capable of rapidly oxidizing mineral sulfides at 80°–85°C have been discovered (64). The ability of members of the genus *Sulfolobus* to tolerate the abrasion that occurs during the vigorous mixing of high concentrations of mineral has been questioned. Some isolates of *Sulfolobus*-like organisms are more resilient than others, and it is possible that with suitable screening, abrasion-resistant strains will be found.

METALLOSPHAERA These archaea are aerobic iron- and sulfur-oxidizing chemolithotrophs that are also able to grow on complex organics such as yeast extract or casamino acids but not sugars. The species most frequently described in the context of mineral sulfide oxidation is *Metallosphaera sedula* (39). *M. sedula* has been reported to grow at pH 1.0–4.5 and is able to oxidize a variety of minerals at temperatures of 80°–85°C. *Metallosphaera*-like organisms have been reported to be potentially the most efficient at high-temperature bioleaching of recalcitrant chalcopyrite ores (63). However, based on 16S rRNA sequence comparisons, there is some disagreement as to which of these isolates is more *Sulfolobus*- or *Metallosphaera*-like (64).

ACIDIANUS There are several species of this group of archaea that oxidize minerals, although the industrial potential of this group is thought to be less promising than that of *Sulfolobus* and *Metallosphaera*. *Acidianus brierleyi* can grow autotrophically by oxidizing ferrous iron or sulfur, or grow heterotrophically on complex organic substrates. The optimum temperature is 70°C and the optimum pH is 1.5–2.0 (87). *Acidianus infernus* and *Ad. ambivalens* are obligate chemolithotrophs that can grow either aerobically or anaerobically by the oxidation or reduction of inorganic sulfur compounds. *Ad. infernus* has an optimum temperature of 90°C and an optimum pH of 2.0.

MOLECULAR BIOLOGY AND GENOMICS

The biomining organism that has been studied to the greatest extent by far is *At. ferrooxidans*. The 2.7-Mb genome of the ATCC23270-type strain has been almost completely sequenced by two organizations: Integrated Genomics, Chicago, and The Institute for Genomic Research (http://www.tigr.org/), and some of the information has been analyzed (37, 88). Many chromosomal genes and plasmids from related strains of *At. ferrooxidans* have been cloned and shown to be expressed and to function in laboratory strains of *Escherichia coli*. Most of this work has been recently reviewed (73, 74). In addition, DNA has been transformed into several *At. ferrooxidans* isolates using conjugation (67, 58) and by electroporation into a single isolate, although many were tested (51). Furthermore, a *recA* mutant has been constructed by gene replacement, and the tools required to carry out molecular genetic studies as well as to modify the bacterium by genetic engineering are in place (58). However, up until now genetic modification of commercial isolates has been restrained partly by technical difficulties associated with growing and manipulating these bacteria and partly because of public sensitivity to the use of genetically modified organisms. Although biomining microbes are not pathogenic, current processes for the biooxidation of ores and concentrates are not carried out under sterile conditions, so microorganism containment will be impossible without substantial process modifications.

Studies of the molecular genetics of other biomining microorganisms are still very patchy, with one or two genes from several bacteria having been cloned (73), a few plasmids isolated, and a conjugation system for *At. thiooxidans* demonstrated (41). Genetic work on the important genus *Leptospirillum* has been ignored, with the exception of a putative chemotaxis-related gene from *L. ferrooxidans* (14) and a single plasmid from the same species (N.J. Coram & D.E. Rawlings, unpublished data). The genome sequences of the 1.56-Mb *Tp. acidophilum* (81) and the 2.99-Mb *Sulfolobus solfataricus* (http://www.cbr.nrc.ca) have been completed. Although neither of these archaea has been identified in commercial biomining environments, the *Thermoplasma* is related to *Ferroplasma*, and other *Sulfolobus* species are likely to be present in high-temperature biomining processes that may become commercial in the near future.

CURRENT BIOMINING PROCESSES

There are two main types of processes for commercial-scale microbially assisted metal recovery. These are irrigation-type and stirred tank–type processes. Irrigation processes involve the percolation of leaching solutions through crushed ore or concentrates that have been stacked in columns, heaps, or dumps (5, 86). There are also several examples of the irrigation of an ore body in situ, that is, without bringing the ore to the surface. Stirred tank–type processes employ continuously operating, highly aerated stirred tank reactors (17, 75). One feature of both types of processes is that, unlike most other commercial fermentation processes, neither is sterile, and no attempt is made to maintain the sterility of the inoculum. There is no need for sterility because the highly acidophilic chemoautolithotrophic microorganisms create an environment that is not suitable for the growth of other organisms. Second, because the aim of the process is to maximize the decomposition of the mineral, the microorganisms that do this most efficiently will out-compete those that are less efficient. There is a continual selection for those organisms that are most effective at decomposing the mineral, as the mineral provides the only energy source available.

Irrigation-Type Processes

The metal recovered in by far the greatest quantities using bioleaching is copper. Recent estimates of the total quantity of copper ore treated by microbially assisted processes are difficult to obtain. However, in 1999 it was reported that the copper-heap bioleaching plants built since 1991 and remaining in operation processed in excess of 30×10^6 tonnes of ore per annum (7, 9). All current commercial processes for copper recovery are of the dump, heap, or in situ irrigation type, although stirred tank processes for the extraction of copper from chalcopyrite are in an advanced phase of development.

Dump leaching was initiated in the late 1960s, and one of the best-known dump leaching operations is located at the Kennecott Copper mine in Bingham Canyon, Utah (5, 6). The largest of the dumps at this site consists of four billion tonnes of low-grade copper ore waste. An example of a smaller, more recent, and deliberately constructed dump operation is the Bala Ley plant of the Chuquicamata division of Codelco in Chile (86). Dumps consist of run-of-mine ore, which may be piled 350 m high. The dumps are subjected to cycles of preconditioning, irrigation, rest, conditioning, and washing, each of which may extend for over a year. Irrigation is carried out with raffinate, an iron- and sulfate-rich recycled wastewater from which the copper has been removed. Microorganisms growing in the dump catalyze the chemical reactions (see below) that result in insoluble copper sulfides being converted to soluble copper sulfate. The copper sulfate–containing pregnant leach solution is removed from the bottom of the dump, and the modern method for copper metal recovery is by a process of solvent extraction and electrowinning (86).

Heap leaching of copper (Figure 2) is similar to dump leaching except that the process is designed to be more efficient. Ore is crushed, acidified with sulfuric acid, and agglomerated in rolling drums to bind fine particles to coarse particles (86). The agglomerated ore is stacked 2–10 m high on irrigation pads lined with high-density polyethylene to avoid the loss of solution. Aeration pipes may be included during construction to permit forced aeration and speed up the bioleaching process. Although inoculation of heaps with bacteria has been considered, bioleaching bacteria are ubiquitous and it is not clear to what extent inoculation speeds up the process. Inorganic nutrients such as $(NH_4)_2SO_4$ and KH_2PO_4 are frequently added to the raffinate prior to irrigation through drip lines placed on the surface of the heap. The increased efficiency that results from the careful construction

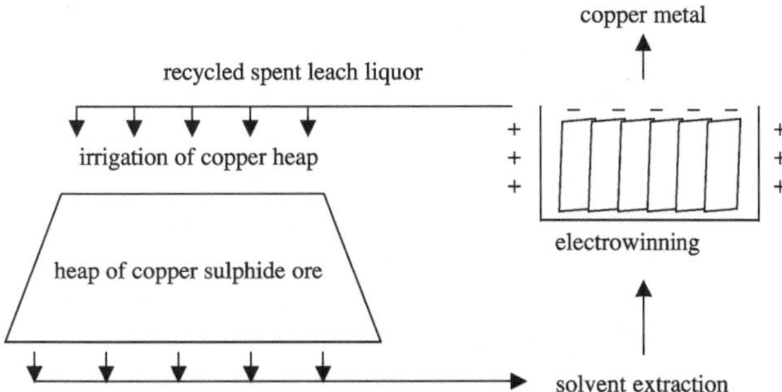

Figure 2 Heap leaching of copper-containing ore. The ore is crushed, agglomerated, and stacked onto plastic-lined pads and irrigated with recycled leach liquor. The bacteria growing on the ore oxidize ferrous iron to ferric iron, which solubilizes the copper, and the pregnant copper-containing solution is recovered from the heap. The copper is concentrated through a solvent extraction process, followed by recovery though electrowinning, and the spent leach liquor is recycled to the heap.

and operation of a heap reactor results in the heap bioleaching processes being completed in a period of months rather than years. An example of a large, modern, heap bioleaching operation is the plant commissioned in 1994 at Quabrada Blanca in northern Chile that produces 75,000 tonnes of copper per annum from chalcocite ore containing 1.3% copper (86).

Heap reactors have also been used for refractory gold-bearing ores. In this case the heap of agglomerated ore is initially irrigated with an acid–ferric iron solution containing bacteria and thereafter with recycled heap reactor effluent (8). The gold remains in the heap, and after a sufficient amount of ore decomposition has taken place the heap is washed to remove acid and excess cyanide-consuming compounds. The heap is taken up, reagglomerated with lime rather than acid, restacked on lined pads, and the gold extracted chemically with a dilute solution of cyanide. Newmont Gold Company has built a demonstration plant using this technology in Carlin, Nevada, and the cost of heap biooxidation and gold extraction is in the range of US$4–6 per tonne of ore processed (8). This process allows low-grade ore with as little as 1 g Au per tonne to be processed, which would otherwise have been considered as waste. A variation of this process has been developed by GeoBiotics Inc., in which a thin layer of gold-bearing sulfide ore concentrate is coated onto a support rock and stacked in heaps (97). The thin coating facilitates rapid biooxidation of the ore concentrate, and the heap is used as a cheaper form of reactor than the stirred tanks in which this type of concentrate is usually treated.

There are several examples of bioleaching operations that use irrigation-type processes and extract metal in situ. One example is the copper recovery process at San Manuel near Tuscon, Arizona (86), and another, Gunpowder's Mammoth mine, Queensland, Australia (7, 9). At San Manuel an array of injection wells is used to introduce acidified leaching solution into the mineral deposit. The leach solution gravitates through the ore and collects in abandoned mine workings or in a centrally placed production well from which it is pumped to the surface. Suitable geology is required for in situ processes to work, and fluid loss at San Manuel is about 13.5% (86). Although no longer in operation, in situ bioleaching for the recovery of uranium took place in several mines within the Elliot Lake district of northern Ontario (60, 61). Typically, underground ore was fractured and a bulkhead built across the opening of a stope. The ore behind the bulkhead was flooded with acid mine drainage liquor and aerated by passing compressed air through perforated pipes. After a period of up to 3 weeks, the liquor was drained and pumped to the surface, where the uranium was extracted. This cycle was repeated until the quantity of uranium fell to a level that was no longer economically viable. In 1988 ~300 tonnes of uranium were recovered at the Denison mine using in situ bioleaching, and it was estimated that this process made an additional 4×10^6 tonnes of ore available that could not have been processed by conventional technologies.

MICROBIOLOGY OF IRRIGATION-TYPE PROCESSES Microbes present in copper heap-leach operations have been studied more extensively than most other irrigation-type processes. Espejo and coworkers (22, 68, 94) have used the species-specific sizes of PCR-amplified 16S-23S rRNA intergenic spacer regions to investigate the microbes in heaps or columns in which Chilean chalcocite/covellite ores are leached. *L. ferrooxidans* and *At. thiooxidans* were found to dominate the process, and no *At. ferrooxidans* was detected. If ferrous iron solution (5 g/liter) was added to the leach solution, 16S–23S rRNA spacer regions corresponding to *At. ferrooxidans* dominated the population, but recovery of copper was enhanced by no more than 5% (68). In a related study agglomerated copper sulfide ore was inoculated and leached by irrigation with recycled leaching solution that contained high concentrations of sulfate ions (120 g/liter). The bacteria were identified by comparing the sizes of 16S–23S 16SrRNA spacer regions and heteroduplex formation with known species. A homogenous population of *At. thiooxidans* and hetergenous populations of *At. ferrooxidans* and *L. ferrooxidans* were found (22). In a similar study organisms capable of leaching copper sulfide ores in conditions of high acidity (pH 0.7) had 16S–23S rRNA spacers of a size consistent with those of *L. ferrooxidans* and *At. thiooxidans*, but no spacers of a size corresponding to *At. ferrooxidans* were detected (94).

Stirred Tank Processes

The use of highly aerated, stirred tank bioreactors provides a step up in the rate and efficiency of mineral biooxidation processes (Figure 3). Because these reactors are expensive to construct and operate, their use is restricted to high-value ores and concentrates. The bioreactors are typically arranged in series and are operated in continuous-flow mode, with feed being added to the first tank and overflowing from tank to tank until biooxidation of the mineral concentrate is sufficiently complete

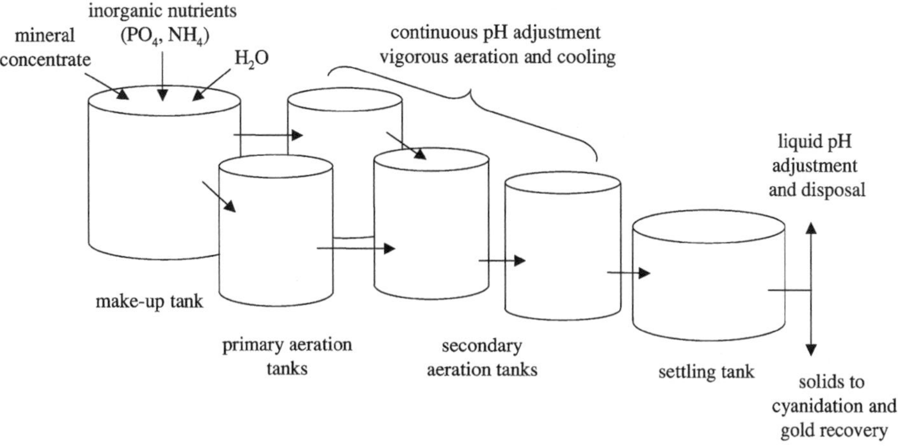

Figure 3 Flow diagram of a typical continuous-flow biooxidation facility for the pretreatment of gold-bearing arsenopyrite concentrate. Finely milled flotation concentrate is added to inorganic nutrients and water in a make-up tank. The mixture is passed through a series of vigorously aerated biooxidation tanks. The microorganisms decompose the arsenopyrite, making the gold accessible to cyanide. Duplication (or doubling in volume) of the number of primary aeration tanks is to increase the hydraulic retention time above that of the cell doubling time to ensure that microbial cell wash-out does not occur during the mineral treatment process. The concentrate, which after pretreatment still contains the gold, is recovered in a thickening tank and sent to the gold recovery process.

(55, 93). The bioreactor tanks in the first stage are usually arranged in parallel to provide sufficient retention time for the microbial cell numbers to reach high steady-state levels without being washed out (16). The feed is mineral concentrate suspended in water to which small quantities of fertilizer-grade $(NH_4)_2SO_4$ and KH_2PO_4 have been added. Mineral biooxidation is an exothermic process, and the bioreactors have to be cooled to remove excess heat. Large volumes of air are blown through each bioreactor, and a large agitator ensures that the solids remain in even suspension and are carried over into the next tank.

Most of the commercial operations that use stirred tank bioreactors are pretreatment processes for the recovery of gold from recalcitrant arsenopyrite concentrates. Recalcitrant ores are those in which the gold is finely divided in a mixture of pyrite/arsenopyrite and cannot easily be solubilized by the usual process of cyanidation. Treatment of the ores to decompose the arsenopyrite is required to allow the cyanide to make contact with the gold. Because the gold is contained in a relatively minor fraction, the ore is crushed and a gold-containing concentrate is prepared by flotation. The finely ground gold-bearing mineral makes up 18%–20% of the feed volume and is fed at a rate that permits a total residence time of about four days in the series of stirred tanks (17). The microbes that oxidize the mineral release ferric iron and sulfate, which acidifies the environment, and the solution is typically maintained at pH 1.5–1.6.

The potential for using microbes for the treatment of these minerals was recognized in the 1980s by the late Eric Livesey-Goldblatt (57). Most of the biooxidation plants were built by Gencor and operate at 40°C (16, 17), but the plant at the Youanmi mine in Western Australia was designed by BacTech and operated at 50°C (62). As may be expected, the consortia of microbes present in the two processes differ substantially, but both mixtures of organisms have proved to be robust and have successfully treated gold-bearing concentrate for extended periods of time. Once the required amount of mineral biooxidation has taken place, the recovered mineral is washed, separated from the liquid fraction in thickeners, neutralized, and the gold extracted with cyanide. The first commercial operation to use stirred tank biooxidation of gold-containing concentrates was commissioned in 1986 at the Fairview mine in Barberton, South Africa. Since then at least six similar plants have been built in other countries, including Australia, Brazil, Ghana, and Peru (7) (Figure 4, see color insert). The plant at Sansu, Ghana, was commissioned in 1994, expanded in 1995, and is probably the largest fermentation process in the world. It consists of 24 tanks of 1,000,000 liters each, processes 1000 tonnes of gold concentrate per day, and earns about half of Ghana's foreign exchange.

Stirred tank processes for other ores and concentrates have also been developed. These include the full-scale cobalt bioleaching plant initiated by BRGM of France at Kasese, Uganda (10), and the pilot-scale BioNIC process for leaching of nickel by Billiton Process Research in South Africa (18). New stirred tank processes to recover copper from chalcopyrite ore that operate at temperatures in excess of 75°C are undergoing trials. These high temperatures are required because chalcopyrite is more stable than ores like chalcocite, and decomposition of chalcopyrite is too slow at 40° or 50°C.

MICROBIOLOGY OF STIRRED TANK PROCESSES Organisms present in continuous-flow, high-rate, stirred biooxidation tanks used to pretreat gold-bearing arsenopyrite concentrates have been studied in most detail. In a study of the bacteria in the biooxidation plant at the Fairview mine, restriction enzyme patterns of 16S rDNA that were amplified from known cultures of *At. ferrooxidans*, *At. thiooxidans*, and *Leptospirillum* were compared with those from total DNA isolated from the biooxidation tanks. There was no restriction pattern corresponding to *At. ferrooxidans*, and the population was reported to be dominated by *Leptospirillum* and *At. thiooxidans* (71). Subsequent studies have shown that the restriction enzyme patterns of *At. thiooxidans* and *At. caldus* are similar and that the bacteria were almost certainly *At. caldus* (26).

An examination of the bacteria in commercial biooxidation tanks operating at 40°C, using an immunofluorescent-antibody microscope count detection technique, differed from the PCR-based work in that *At. ferrooxidans* cells were detected in most samples (54). *At. ferrooxidans* cells were, however, in the minority. The proportions of bacterial types in continuous-flow biooxidation tanks from

São Bento (Brazil) and Fairview were 48%–57% *L. ferrooxidans*, 26%–34% *At. thiooxidans*, and 10%–17% *At. ferrooxidans*. The proportions varied within the ranges given, depending on whether the sample had been treated with TritonX to release attached bacteria and from which tank in the series of biooxidation tanks the sample was derived (17). Using the immunofluorescent technique, a slightly different distribution of bacteria was found in pilot-scale bioreactors treating a nickel pentlandite-pyrrhotite ore. In this case numbers of *At. ferrooxidans* were higher, although they never exceeded about 33% of the total population (54).

Published information on the dominant bacteria in processes operating at 50°C is limited. The fatty acid methyl ester composition was used in an attempt to identify the attached and unattached microbes in a commercial continuous-flow, stirred tank process used to treat an arsenopyrite concentrate operating at 50°C (25). Although the species of bacteria could not be identified, bacteria with a fatty acid profile related to *Thiobacillus* species were found to dominate the culture. A small quantity of fatty acids with a profile related to *Sb. thermosulfidooxidans* was identified in a batch reactor inoculated with the commercial culture. No evidence of *Leptospirillum ferrooxidans* was found using fatty acid composition or by visualization using a microscope.

MECHANISM OF BIOOXIDATION

In 1970 it was reported that *At. ferrooxidans* increased the rate of ferrous iron oxidation by half a million to a million times, compared with the abiotic chemical oxidation of ferrous iron by dissolved oxygen (52). The ferric iron produced as a result was able to chemically oxidize sulfide minerals. There has been a long-standing debate concerning whether the microbially assisted biooxidation of minerals is by a so-called direct or indirect mechanism (59). Some of the disagreement appears to have been caused by a lack of clarity as to what is meant by direct and indirect. There is general agreement as to what is meant by the indirect mechanism, namely the chemical attack by ferric iron or protons on a mineral sulfide that results in the dissolution of the mineral and the formation of ferrous iron and various forms of sulfur. Iron-oxidizing microbes use the ferrous iron as an electron donor, reoxidizing it to ferric iron, thereby regenerating the reactant. If the role of the microbes is nothing more than their ability to regenerate ferric iron and protons (as shown in Figure 5), then the efficiency of biooxidation should be independent of whether the microbes are in contact with the mineral or not. What is understood by the direct mechanism is not always unequivocal. In a loose sense it has meant that attachment of the microbes to the ore enhances the rate of mineral dissolution. In a more strict sense, direct attack is viewed as a process by which components within the bacterial membrane interact directly with the metal and sulfide moieties of the mineral by using an enzymatic type of mechanism (82).

A number of scientists from a variety of disciplines have addressed this problem in recent years. It is not possible to cover fully the points made by each contributor

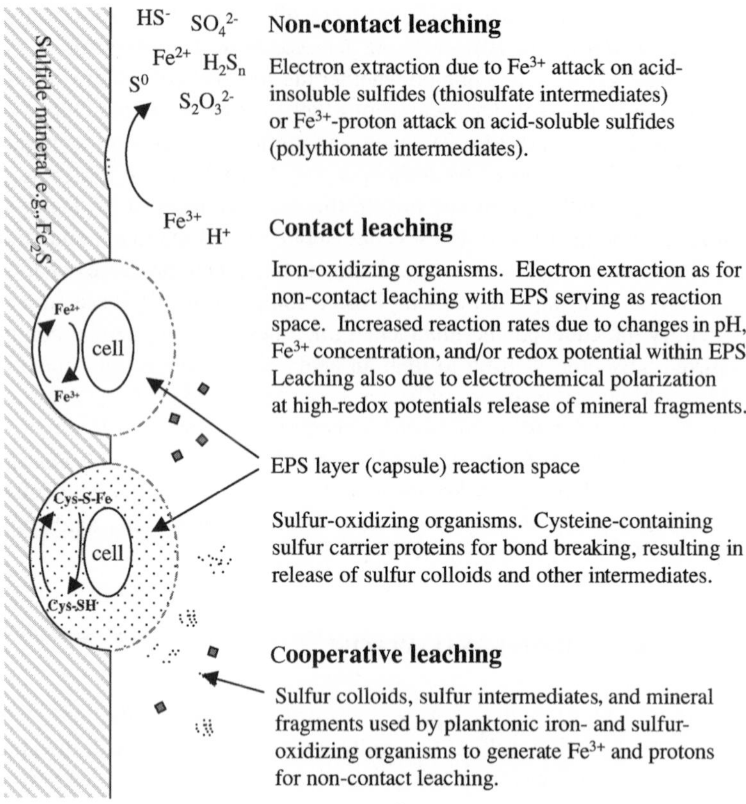

Sulfide mineral e.g., Fe_2S
HS^- SO_4^{2-}
Fe^{2+} H_2S_n
S^0
$S_2O_3^{2-}$
Fe^{3+} H^+
Non-contact leaching
Electron extraction due to Fe^{3+} attack on acid-insoluble sulfides (thiosulfate intermediates) or Fe^{3+}-proton attack on acid-soluble sulfides (polythionate intermediates).
Contact leaching
Iron-oxidizing organisms. Electron extraction as for non-contact leaching with EPS serving as reaction space. Increased reaction rates due to changes in pH, Fe^{3+} concentration, and/or redox potential within EPS. Leaching also due to electrochemical polarization at high-redox potentials release of mineral fragments.
Fe^{2+}
cell
Fe^{3+}
EPS layer (capsule) reaction space
Cys-S-Fe
cell
Cys-SH
Sulfur-oxidizing organisms. Cysteine-containing sulfur carrier proteins for bond breaking, resulting in release of sulfur colloids and other intermediates.
Cooperative leaching
Sulfur colloids, sulfur intermediates, and mineral fragments used by planktonic iron- and sulfur-oxidizing organisms to generate Fe^{3+} and protons for non-contact leaching.

Figure 5 Schematic diagram illustrating the proposed mechanisms of pyrite biooxidation.

to this debate in the space available. However, there is a growing consensus view on many aspects of the mechanism of bioleaching. An understanding of metal sulfide dissolution is complicated because different metal sulfides have different types of crystal structure, and it has been observed that the oxidation of different metal sulfides proceeds via different intermediates (83, 85). In addition, as is discussed below, different iron-oxidizing bacteria may use different bond-breaking strategies (90). For example, the ability of *At. ferrooxidans* to oxidize iron is subject to ferric iron inhibition and is inhibited at high-redox potentials, whereas the ability of *L. ferrooxidans* to oxidize iron is far less inhibited by ferric iron and continues at high-redox potentials (76).

Among the most helpful contributions to an understanding of the mechanism of mineral biooxidation are those from the laboratories of Wolfgang Sand, Frank Crundwell, and Helmut Tributsch. Bacteria like *At. ferrooxidans* (15, 66) or *L. ferrooxidans* (34) have a strong affinity for mineral surfaces like pyrite, to which they rapidly attach. In the case of *At. ferrooxidans* aporusticyanin appears to play an important role, especially during the early stages of attachment (2). Sand and Tributsch and their coworkers have stressed the role played by the exopolysaccharide (EPS) layer produced by *At. ferrooxidans* (27, 78, 79) and *L. ferrooxidans* (90) when attached to a mineral. The EPS layer attaches the bacteria to a mineral surface and forms the matrix in which the cells divide and eventually form a biofilm. Iron species within the EPS layer have been estimated to be ~53 g/liter, a concentration that can be maintained only by the formation of complexes, possibly with glucoronic acid (85). The iron-impregnated EPS layer serves as a reaction space in which the high concentration of ferric iron mounts chemical attack on the valence bonds of the mineral (85). In the process the ferric iron is reduced to ferrous, which is reoxidized to ferric by the iron-oxidizing microbes:

$$14\,Fe^{2+} + 3.5\,O_2 + 14\,H^+ \rightarrow 14\,Fe^{3+} + 7\,H_2O.$$

This microbially mediated reoxidation of ferrous iron can result in a localized increase in pH within the EPS near the surface of the mineral. Researchers from the Crundwell laboratory have evidence to suggest that it is this local rise in pH that assists in mineral dissolution (23, 24, 38).

Sand and coworkers (84, 85) have observed that the oxidation of different metal sulfides proceeds via different intermediates. Therefore, the dissolution reaction is not identical for all metal sulfides. They have proposed a thiosulfate mechanism for the oxidation of acid-insoluble metal sulfides such as pyrite (FeS_2), molybdenite (MoS_2), and tungstenite (WS_2), and a polysulfide mechanism for acid-soluble metal sulfides such as sphalerite (ZnS), chalcopyrite ($CuFeS_2$), and galena (PbS).

In the thiosulfate mechanism, solubilization is through a ferric iron attack on the acid-insoluble metal sulfides, with thiosulfate being the main intermediate and sulfate the main end-product. Using pyrite as an example, the reactions proposed by Schippers & Sand (85) are

$$FeS_2 + 6\,Fe^{3+} + 3\,H_2O \rightarrow S_2O_3^{2-} + 7\,Fe^{2+} + 6\,H^+$$

$$S_2O_3^{2-} + 8\,Fe^{3+} + 5\,H_2O \rightarrow 2\,SO_4^{2-} + 8\,Fe^{2+} + 10\,H^+.$$

In the case of the polysulfide mechanism, solubilization of the acid-soluble metal sulfide is through a combination of ferric ions and protons, with elemental sulfur as the main intermediate. This elemental sulfur is relatively stable but may be oxidized to sulfate by sulfur-oxidizing bacteria. ZnS is an example of an acid soluble mineral:

$$MS + Fe^{3+} + H^+ \rightarrow M^{2+} + 0.5\,H_2S_n + Fe^{2+} (n \geq 2)$$

$$0.5\,H_2S_n + Fe^{3+} \rightarrow 0.125\,S_8 + Fe^{2+} + H^+$$

$$0.125\,S_8 + 1.5\,O_2 + H_2O \rightarrow SO_4^{2-} + 2\,H^+.$$

As Schippers & Sand (85) point out, this explains why strictly sulfur-oxidizing bacteria such as *A. thiooxidans* are able to leach some metal sulfides but not others.

The role of the microorganisms in the solubilization of metal sulfides is, therefore, to provide sulfuric acid for a proton hydrolysis attack and to keep the iron in the oxidized ferric state for an oxidative attack on the mineral.

Tributsch (90) has argued that the break-up of the pyrite crystal structure can be the result of one of five mechanisms: (*a*) reaction of protons with a sulfide to give SH^- ions; (*b*) extraction of electrons from the sulfide valence band resulting in the release of metal ions and sulfur compounds (in the case of *At. ferrooxidans*); (*c*) broken chemical bonds already present in the sulfide (p-type conduction), which leads to higher interfacial dissolution; (*d*) reaction with a polysulfide or metal complex-forming agent; or (*e*) electrochemical dissolution resulting from multiple electron extraction and depolarization of the pyrite, which occurs at high concentrations of ferric ions (in the case of iron-oxidizing bacteria like *L. ferrooxidans*). Mechanisms (*a*), (*b*), and (*c*) are clearly equivalent to the indirect mechanism. However, Tributsch has argued that mechanisms in which a sulfur carrier may be involved (*d*) or those for which a high concentration of ferric iron is necessary (*e*) require contact (or close proximity) between the cell and the mineral surface. These mechanisms would not be as effective if they relied on carriers or chemicals in the bulk liquid. Holmes et al. (38) have argued that the idea that an increase in ferric iron concentration at the surface of a mineral results in an increase in the rate of pyrite dissolution is not consistent with their observations of changes in the mixed potential of bacterial and chemically leached pyrite.

Tributsch and coworkers (79, 90) have shown that during the oxidation of pyrite the EPS layer of *At. ferrooxidans* becomes loaded with colloidal sulfur and that much of this is released into the environment. The sulfur particles and globules released in this apparently wasteful process feed other sulfur-oxidizing bacteria in a cooperative leaching interaction. Unidentified carrier molecules that contain reactive thiol groups provided by the amino acid cysteine greatly assist in pyrite dissolution. Indeed, these workers have shown that on its own cysteine rapidly oxidizes pyrite in the absence of oxygen or bacteria (80). Free-SH groups from the pyrite react with the sulfhydryl group of cysteine. This thiol-disulfide reaction results in cysteine being consumed by the pyrite with the release of iron-sulfur species. The thiol-dependent mechanism alone is probably insufficient to oxidize pyrite at the rates observed. In the case of a strictly iron-oxidizing bacterium like *L. ferrooxidans*, electrochemical dissolution of pyrite owing to high concentrations of ferric iron results in small fragments of pyrite becoming visibly entrapped in the EPS layer. However, there appears to be much less, if any, colloidal sulfur present (80). The observation that the oxidation of ferrous iron is not inhibited by ferric iron in the case of *L. ferrooxidans*, as it is with *At. ferrooxidans*, is consistent with this proposed mechanism. Because what many understood by "direct leaching" implied direct contact of cell membrane components with the mineral, as opposed to only close proximity (in this model), the term "contact leaching" is preferred to direct leaching.

Taking into account the views of most of the important contributors to the direct versus indirect leaching debate, regarding reaction kinetics, stoichiometry,

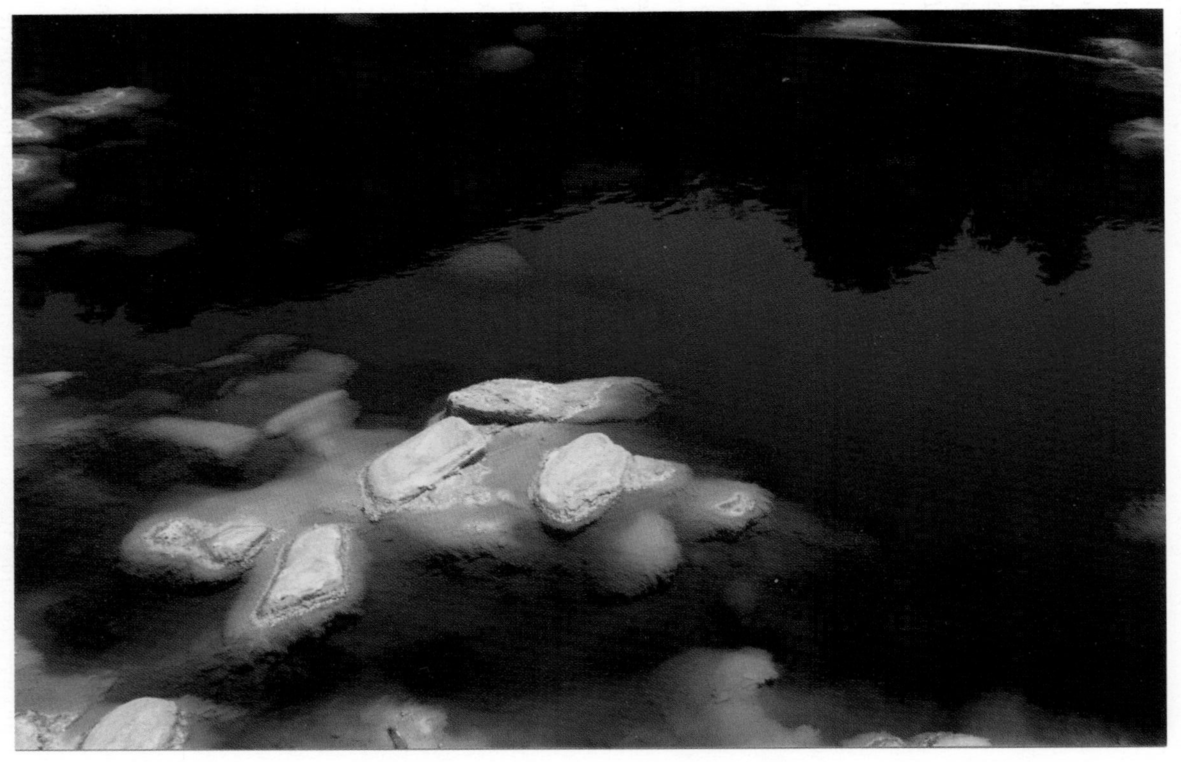

Figure 1 The Rio Tinto in the Seville province of Spain. This acidic river (± pH 2.3) is red owing to high concentrations of dissolved ferric iron resulting from natural microbially mediated mineral decomposition. Early miners discovered that copper ores irrigated with this water resulted in the leaching of copper from the ore.

Figure 4 The biooxidation plant at the Tamboraque mine 90 km east of Lima, Peru. The gold-bearing arsenopyrite mineral concentrate is pretreated by vigorous aeration in a series of stirred tanks to partially decompose the concentrate prior to extraction of the gold using cyanide. (Photo courtesy Martha Ly, Tamboraque mine)

and several other considerations (13) and including a recent scanning electron microscope study of mineral surface leaching (21), a consensus view is that the mechanism of mineral solubilization is *strictu sensu* indirect (physicochemical). Tributsch and colleagues (90) have presented a strong case that a close proximity between the cell and the mineral surface is required for certain types of metal solubilization reactions to proceed efficiently. Bacterial attachment may speed up the rate of leaching owing to the proximity to the mineral surface of a high concentration of ferric iron complexed in the EPS, a localized rise in pH (38), or the proximity of thiol groups of cysteine-containing molecules produced by sulfur-oxidizing bacteria. However, in none of these processes is direct contact between the cell surface (membrane) and the mineral essential. In this sense they are not strictly direct mechanisms. In the absence of firm evidence for the need for direct mineral–cell membrane contact, the term contact leaching should be used to replace direct leaching because of the confusion around this latter term (Figure 5).

FUTURE DIRECTIONS

The use of microbes to extract metals from ores is now well established. Biooxidation plants with tank sizes of 1300 m^3 (with 2000 m^3 tanks under design) and one plant that earns half of Ghana's foreign exchange attest to this. Heap-leach and stirred tank processes that operate at ambient or temperatures of 40°–50°C or less will continue and will be applied to new types of minerals. Exploration of the molecular biology of these organisms has begun with *At. ferrooxidans*, but studies of other, probably more important, players such as members of the leptospirilli and *At. caldus* are in their infancy.

Much of the future of biomining is likely to be hot. Development of processes that operate at 70°C or above will make the recovery of metals from a large number of minerals economically attractive where otherwise the reaction rates are too slow. Many of these will be stirred tank–type processes, although high-temperature irrigation-type processes in which support material is coated with a mineral concentrate may also prove to be economical for copper recovery from chalcopyrite (42). At increased temperatures a variety of largely unstudied archaea will become important (64). Unlike mesophiles, where iron- and sulfur-oxidizing organisms are ubiquitously associated with mineral deposits and can be readily enriched, research on thermophiles will require screening of microbes from high-temperature environments in which suitable organisms are likely to be found. Isolation of organisms with the right combination of iron- and sulfur-oxidizing ability together with the ability to withstand abrasion (stirred tank processes) and resistance to high concentrations of metal ions is a challenge. Nevertheless, candidate organisms have already been found and are largely unexplored territory. These high-temperature organisms present a number of challenges in regard to standard microbiology techniques such as plating, incubation, and preparation of pure cultures. Challenges such as the isolation of mutants and the selection of transformants also apply to studies on the molecular biology and genetics of these high-temperature organisms.

New engineering, and especially new molecular biology tools (genomics and proteomics), make this an exciting field to investigate.

ACKNOWLEDGMENTS

The author thanks colleagues in the international biohydrometallurgy community who are too numerous to name; his early career mentor, David Woods; his long-term postdoctoral associate, Shelly Deane; and numerous talented postgraduate students for many stimulating hours of discussions in the field of biomining. In addition, he thanks BHP-Billiton Minerals Technology, Randburg (previously Gencor, then Billiton Process Research Laboratory), the National Research Foundation, Pretoria, and the Universities of Cape Town and Stellenbosch for many years of financial support.

The *Annual Review of Microbiology* is online at http://micro.annualreviews.org

LITERATURE CITED

1. Amils R, Ballester A, eds. 1999. *Biohydrometallurgy and the Environment Towards the 21st Century. Part A*. Amsterdam: Elsevier
1a. Avery D. 1974. *Not on Queen Victoria's Birthday; The Story of the Rio Tinto Mines*. London: Collins
2. Blake RC, Sasaki K, Ohmura N. 2001. Does aporusticyanin mediate the adhesion of *Thiobacillus ferrooxidans* to pyrite? *Hydrometallurgy* 59:357–72
3. Bond PL, Smriga SP, Banfield JF. 2000. Phylogeny of microorganisms populating a thick, subaerial predominantly lithotrophic biofilm at an extreme acid mine drainage site. *Appl. Environ. Microbiol.* 66: 3842–49
4. Bridge TAM, Johnson DB. 2000. Reductive dissolution of ferric iron minerals by *Acidiphilium* SJH. *Geomicrobiol. J.* 17: 193–206
5. Brierley CL. 1978. Bacterial leaching. *Crit. Rev. Microbiol.* 6:207–62
6. Brierley CL. 1982. Microbiological mining. *Sci. Am.* 247(2):42–51
7. Brierley CL. 1997. Mining biotechnology: research into commercial development and beyond. See Ref. 72, pp. 3–17
8. Brierley JA. 1997. Heap leaching of gold-bearing deposits: theory and operational description. See Ref. 72, pp. 103–15
9. Brierley JA, Brierley CL. 1999. Present and future commercial applications of biohydrometallurgy. See Ref. 1, pp. 81–89
10. Briggs AP, Millard M. 1997. Cobalt recovery using bacterial leaching at the Kasese project, Uganda. In *International Biohydrometallurgy Symposium, IBS97*, pp. M2.4.1–2.4.12. Glenside, South Aust.: Aust. Mineral Found.
10a. Ciminelli VST, Garcia O Jr, eds. 2001. *Biohydrometallurgy: Fundamentals, Technology and Sustainable Development, Part A*. Amsterdam: Elsevier
11. Clark DA, Norris PR. 1996. *Acidimicrobium ferrooxidans* gen. nov., sp. nov.: mixed–culture ferrous iron oxidation with *Sulfobacillus* species. *Microbiology* 142: 785–90
12. Colmer AR, Hinkel ME. 1947. The role of microorganisms in acid mine drainage: a preliminary report. *Science* 106:253–56
12a. Coram NJ, Rawlings DE. 2002. Molecular relationship between two groups of *Leptospirillum* and the finding that the world and the *Leptospirillum ferriphilum* sp. nov. dominates South African commercial biooxidation tanks that operate at

40°C. *Appl. Environ. Microbiol.* 68:838–45

13. Crundwell FK. 2001. How do bacteria interact with minerals? See Ref. 10a, pp. 149–58
14. Delgardo M, Toledo H, Jerez CA. 1998. Molecular cloning, sequencing and expression of a chemoreceptor gene from *Leptospirillum ferrooxidans*. *Appl. Environ. Microbiol.* 64:2380–85
15. Devasia P, Natarajan KA, Sathyanarayana DN, Ramananda Rao G. 1993. Surface chemistry of *Thiobacillus ferrooxidans* relevant to adhesion on mineral surfaces. *Appl. Environ. Microbiol.* 59:4051–55
16. Dew DW. 1995. Comparison of performance for continuous bio-oxidation of refractory gold ore flotation concentrates. In *Biohydrometallurgical Processing*, ed. T Vargas, CA Jerez, JV Wiertz, H Toledo, 1:239–51. Santiago: Univ. Chile Press
17. Dew DW, Lawson EN, Broadhurst JL. 1997. The BIOX® process for biooxidation of gold-bearing ores or concentrates. See Ref. 72, pp. 45–80
18. Dew DW, Miller DM. 1997. The BioNIC process, bioleaching of mineral sulphide concentrates for the recovery of nickel. In *International Biohydrometallurgy Symposium, IBS97*, pp. M7.1.1–7.1.9. Glenside, South Aust.: Aust. Mineral Found.
19. De Wulf-Durand P, Bryant LJ, Sly LI. 1997. PCR-mediated detection of acidophilic, bioleaching-associated bacteria. *Appl. Environ. Microbiol.* 63:2944–48
20. Edwards KJ, Bond PL, Gihring TM, Banfield JF. 2000. An Archaeal iron-oxidizing extreme acidophile important in acid mine drainage. *Science* 279:1796–99
21. Edwards KJ, Hu B, Hamers RJ, Banfield JF. 2001. A new look at microbial leaching patterns on sulfide minerals. *FEMS Microbiol. Ecol.* 34:197–206
22. Espejo RT, Romero J. 1997. Bacterial community in copper sulfide ores inoculated and leached with solution from a commercial-scale copper leaching plant. *Appl. Environ. Microbiol.* 63:1344–48
23. Fowler TA, Holmes PR, Crundwell FK. 1999. Mechanism of pyrite dissolution in the presence of *Thiobacillus ferrooxidans*. *Appl. Environ. Microbiol.* 65:2987–93
24. Fowler TA, Holmes PR, Crundwell FK. 2001. On the kinetics and mechanism of the dissolution of pyrite in the presence of *Thiobacillus ferrooxidans*. *Hydrometallurgy* 59:257–70
25. Franzmann PD, Williams TL. 1997. Biomass and the microbial community composition in an arsenopyrite oxidation bioreactor determined by phospholipid-derived fatty acid methyl ester composition and content. In *International Biohydrometallurgy Symposium, IBS97*, pp. QP1.1–1.10. Glenside, South Aust.: Aust. Mineral Found.
26. Gardner MN, Rawlings DE. 2000. Production of rhodanese by bacteria present in bio-oxidation plants used to recover gold from arsenopyrite ores. *J. Appl. Microbiol.* 89:185–90
27. Gehrke T, Telegdi J, Thierry D, Sand W. 1998. Importance of extracellular polymeric substances from *Thiobacillus ferrooxidans* for bioleaching. *Appl. Environ. Microbiol.* 64:2743–47
28. Goebel BM, Stackebrandt E. 1994. Cultural and phylogenetic analysis of mixed microbial populations found in natural and commercial bioleaching environments. *Appl. Environ. Microbiol.* 60:1614–21
29. Goodfellow M, O'Donnell AG. 1993. Roots of bacterial systematics. In *Handbook of New Bacterial Systematics*, ed. M Goodfellow, AG O'Donnell, pp. 3–54. London: Academic
30. Golovacheva RS, Karavaiko GI. 1979. Sulfobacillus—a new genus of spore-forming thermophilic bacteria. *Microbiology (Mikrobiologiya)* 48:658–65
31. Golyshina OV, Pivovarova TA, Karavaiko G, Kondrat'eva TF, Moore ERB, et al. 2000. *Ferroplasma acidiphilum* gen. nov., sp. nov., an acidophilic, autotrophic, ferrous iron-oxidizing, cell-wall-lacking,

mesophilic member of the *Ferroplasmacaea* fam. nov., comprising a distinct lineage of the Archaea. *Int. J. Syst. Evol. Microbiol.* 50:997–1006

32. Hallberg KB, Lindström EB. 1994. Characterization of *Thiobacillus caldus* sp. nov., a moderately thermophilic acidophile. *Microbiology* 140:3451–56
33. Hallberg KB, Johnson DB. 2001. Biodiversity of acidophilic prokaryotes. *Adv. Appl. Microbiol.* 49:37–84
34. Hallmann R, Friedrich A, Koops H-P, Pommerening-Röser A, Rohde K, et al. 1992. Physiological characteristics of *Thiobacillus ferrooxidans* and *Leptospirillum ferrooxidans* and physicochemical factors influence microbial metal leaching. *Geomicrobiol. J.* 10:193–206
35. Harrison AP. 1982. Genomic and physiological diversity amongst strains of *Thiobacillus ferrooxidans*, and a genomic comparison with *Thiobacillus thiooxidans*. *Arch. Microbiol.* 131:68–76
36. Hippe H. 2000. *Leptospirillum* gen. nov. (ex Markosyan 1972), nom. rev., including *Leptospirillum ferrooxidans* sp. nov. (ex Markosyan 1972) nom. rev. and *Leptospirillum thermoferrooxidans* sp. nov. (Golovacheva et al. 1992). *Int. J. Syst. Evol. Microbiol.* 50:501–3
37. Holmes DS, Barreto M, Valdes J, Dominguez C, Nayibe M, et al. 2001. Whole genome sequence of *Acidithiobacillus ferrooxidans*: metabolic reconstruction, heavy metal resistance and other characteristics. See Ref. 10a, pp. 237–51
38. Holmes PR, Fowler TA, Crundwell FC. 1999. The mechanism of bacterial action in the leaching of pyrite by *Thiobacillus ferrooxidans*. *J. Electrochem. Soc.* 146: 2906–12
39. Huber G, Spinnler C, Gambacorta A, Stetter KO. 1989. *Metallosphaera sedula* gen. and sp. nov. represents a new genus of aerobic, metal-mobilizing, thermoacidophilic archaebacteria. *Syst. Appl. Microbiol.* 12:38–47
40. Huber G, Stetter KO. 1991. *Sulfolobus metallicus*, new species, a novel strictly chemolithotrophic thermophilic archaeal species of metal-mobilizers. *Syst. Appl. Microbiol.* 14:372–78
41. Jin SM, Yan WM, Wang ZN. 1992. Transfer of IncP plasmids to extremely acidophilic *Thiobacillus thiooxidans*. *Appl. Environ. Microbiol.* 58:429–30
42. Johansson C, Schrader V, Suissa J, Adutwum K, Kohr W. 1999. Use of the GEOCOATTM process for the recovery of copper from chalcopyrite. See Ref. 1, pp. 569–76
43. Johnson DB. 1995. Selective solid media for isolating and enumerating acidophilic bacteria. *J. Microbiol. Methods* 23:205–18
44. Johnson DB. 1998. Biodiversity and ecology of acidophilic microorganisms. *FEMS Microbiol. Ecol.* 27:307–17
45. Johnson DB. 2001. Genus II *Leptospirillum* Hippe 2000 (ex Markosyan 1972, 26). In *Bergey's Manual of Comparative Bacteriology*, ed. G Garrity, 1:443–47. Berlin: Springer
46. Johnson DB, Roberto FF. 1997. Heterotrophic acidophiles and their roles in the bioleaching of sulfide minerals. See Ref. 72, pp. 259–79
47. Johnson DB, McGinness. 1991. An efficient and universal solid medium for growing mesophilic and moderately thermophilic, iron-oxidizing, acidophilic bacteria. *J. Microbiol. Method* 13:113–22
48. Kelly DP, Norris PR, Brierley CL. 1979. Microbiological methods for the extraction and recovery of metals. In *Microbial Technology: Current State and Future Prospects*, ed. AT Bull, DG Ellwood, C Ratledge, pp. 263–308. Cambridge: Cambridge Univ. Press
49. Kelly DP, Wood AP. 2000. Re-classification of some species of *Thiobacillus* to the newly designated genera *Acidithiobacillus* gen. nov., *Halothiobacillus* gen. nov. and *Thermithiobacillus* gen. nov. *Int. J. Syst. Evol. Microbiol.* 50:511–16
50. Kishimoto N, Fukaya F, Inagaki K, Sugio

T, Tanaka H, Tano T. 1995. Distribution of bacteriochlorophyll-*a* among aerobic and acidiphilic bacteria and light enhanced CO_2-incorporation in *Acidiphilium rubrum*. *FEMS Microbiol. Ecol.* 16: 291–96

51. Kusano T, Sugawara K, Inoue C, Takeshima T, Numata M, Shiratori T. 1992. Electrotransformation of *Thiobacillus ferrooxidans* with plasmids containing a mer determinant as the selective marker by electroporation. *J. Bacteriol.* 174:6617–23
52. Lacey DT, Lawson F. 1970. Kinetics of the liquid phase oxidation of acid ferrous sulphate by the bacterium *Thiobacillus ferrooxidans*. *Biotechnol. Bioeng.* 12:29–50
53. Lane DJ, Harrison AP, Stahl D, Pace B, Giovannoni SJ, et al. 1992. Evolutionary relationships among sulfur- and iron-oxidizing bacteria. *J. Bacteriol.* 174:269–78
54. Lawson EN. 1997. The composition of mixed populations of leaching bacteria active in gold and nickel recovery from sulphide ores. In *International Biohydrometallurgy Symposium, IBS97*, pp. QP4.1–4.10. Glenside, South Aust.: Aust. Mineral Found.
55. Lindström EB, Gunneriusson E, Tuovinen OH. 1992. Bacterial oxidation of refractory ores for gold recovery. *Crit. Rev. Biotechnol.* 12:133–55
56. Leduc LG, Ferroni GD. 1994. The chemolithotrophic bacterium *Thiobacillus ferrooxidans*. *FEMS Microbiol. Rev.* 14:103–20
57. Livesey-Goldblatt E, Norman P, Livesey-Goldblatt DR. 1983. Gold recovery from arsenopyrite/pyrite ore by bacterial leaching and cyanidation. In *Recent Progress in Biohydrometallurgy*, ed. G Rossi, AE Torma, pp. 627–41. Iglesias, Italy: Assoc. Mineraria Sarda
58. Liu Z, Guiliani N, Appia-Ayme C, Borne F, Ratouchnaik J, Bonnefoy V. 2000. Construction and characterization of a *recA* mutant of *Thiobacillus ferrooxidans* by marker exchange mutagenesis. *J. Bacteriol.* 182:2269–76
59. Lundgren DG, Silver M. 1980. Ore leaching by bacteria. *Annu. Rev. Microbiol.* 34: 263–83
60. McCready RGL. 1988. Progress in the bacterial leaching of metals in Canada. In *Biohydrometallurgy-1987*, ed. PR Norris, DP Kelly, pp. 177–95. Kew/Surrey, UK: Sci. Technol. Lett.
61. McCready RGL, Gould WD. 1989. Bioleaching of uranium at Denison mines. In *Biohydrometallurgy-1989*, ed. J Salley, RGL McCready, PL Wichlacz, pp. 477–85. Ottawa, Canada: CANMET
62. Miller PC. 1997. The design and operating practice of bacterial oxidation plant using moderate thermophiles (the BacTech process). See Ref. 72, pp. 81–102
63. Norris PR. 1997. Thermophiles and bioleaching. See Ref. 72, pp. 247–58
64. Norris PR, Burton NP, Foulis NAM. 2000. Acidophiles in bioreactor mineral processing. *Extremophiles* 4:71–76
65. Norris PR, Clark DA, Owen JP, Waterhouse S. 1996. Characteristics of *Sulfobacillus acidophilus* sp. nov. and other moderately thermophilic mineral-sulphide-oxidizing bacteria. *Microbiology* 142:775–83
66. Ohmura N, Kitamura K, Saiki H. 1993. Selective adhesion of *Thiobacillus ferrooxidans* to pyrite. *Appl. Environ. Microbiol.* 59:4044–50
67. Peng J-B, Yan W-M, Bao X-Z. 1994. Plasmid and transposon transfer to *Thiobacillus ferrooxidans*. *J. Bacteriol.* 176:2892–97
68. Pizarro J, Jedlicki E, Orellana O, Romero J, Espejo RT. 1996. Bacterial populations in samples of bioleached copper ore as revealed by analysis of DNA obtained before and after cultivation. *Appl. Environ. Microbiol.* 62:1323–28
69. Pronk JT, Liem K, Bos P, Kuenen JG. 1991. Energy transduction by anaerobic ferric iron respiration in *Thiobacillus*

ferrooxidans. *Appl. Environ. Microbiol.* 57:2063–68

70. Pronk JT, Meijer WM, Haseu W, van Dijken JP, Bos P, Kuenen JG. 1991. Growth of *Thiobacillus ferrooxidans* on formic acid. *Appl. Environ. Microbiol.* 57:2057–62
71. Rawlings DE. 1995. Restriction enzyme analysis of 16S rRNA genes for the rapid identification of *Thiobacillus ferrooxidans*, *Thiobacillus thiooxidans* and *Leptospirillum ferrooxidans* strains in leaching environments. In *Biohydrometallurgical Processing*, ed. CA Jerez, T Vargas, H Toledo, JV Wiertz, II:9–17. Santiago: Univ. Chile
72. Rawlings DE, ed. 1997. *Biomining: Theory, Microbes and Industrial Processes*. Berlin: Springer-Verlag
73. Rawlings DE. 2001. The molecular genetics of *Thiobacillus ferrooxidans* and other mesophilic, acidophilic, chemolithotrophic, iron- or sulfur-oxidizing bacteria. *Hydrometallurgy* 59:187–201
74. Rawlings DE, Kusano T. 1994. Molecular genetics of *Thiobacillus ferrooxidans*. *Microbiol. Rev.* 58:39–55
75. Rawlings DE, Silver S. 1995. Mining with microbes. *Bio/Technology* 13:773–78
76. Rawlings DE, Tributsch H, Hansford GS. 1999. Reasons why '*Leptospirillum*'-like species rather than *Thiobacillus ferrooxidans* are the dominant iron-oxidizing bacteria in many commercial processes for the biooxidation of pyrite and related ores. *Microbiology* 145:5–13
77. Rawlings, DE, Coram NJ, Gardner MN, Deane SM. 1999. *Thiobacillus caldus* and *Leptospirillum ferrooxidans* are widely distributed in continuous flow biooxidation tanks used to treat a variety of metal containing ores and concentrates. See Ref. 1, pp. 777–86
78. Rodriguez-Leiva M, Tributsch H. 1988. Morphology of bacterial leaching patterns by *Thiobacillus ferrooxidans* on pyrite. *Arch. Microbiol.* 149:401–5
79. Rojas J, Giersig M, Tributsch H. 1995. Sulfur colloids as temporary energy reservoirs for *Thiobacillus ferrooxidans* during pyrite oxidation. *Arch. Microbiol.* 163: 352–56
80. Rojas-Chapana JA, Tributsch H. 2001. Biochemistry of sulfur extraction in bio-corrosion of pyrite by *Thiobacillus ferrooxidans*. *Hydrometallurgy* 59:291–300
81. Ruepp A, Graml W, Santos-Martinez M-L, Koretke KK, Volker C, et al. 2000. The genome sequence of the thermoacidophilic scavenger *Thermoplasma acidophilum*. *Nature* 407:508–13
82. Sand W, Gehrke T, Hallmann R, Schippers A. 1995. Sulfur chemistry, biofilm, and the (in)direct attack mechanism—critical evaluation of bacterial leaching. *Appl. Microbiol. Biotechnol.* 43:961–66
83. Sand W, Gehrke T, Jozsa P-G, Schippers A. 2001. (Bio)chemistry of bacterial leaching—direct vs. indirect process. *Hydrometallurgy* 59:159–75
84. Schippers A, Jozsa P-G, Sand W. 1996. Sulfur chemistry in bacterial leaching of pyrite. *Appl. Environ. Microbiol.* 62: 3424–31
85. Schippers A, Sand W. 1999. Bacterial leaching of metal sulfides proceeds by two indirect mechanisms via thiosulfate or via polysulfides and sulfur. *Appl. Environ. Microbiol.* 65:319–21
86. Schnell HA. 1997. Bioleaching of copper. See Ref. 72, pp. 21–43
87. Segerer A, Neuner A, Kristjansson JK, Stetter KO. 1986. *Acidianus infernus* gen. nov., sp. nov., and *Acidianus brierleyi* comb. nov.: facultatively aerobic, extremely acidophilic thermophilic sulfur-metabolizing archaebacteria. *Int. J. Syst. Bacteriol.* 36:559–64
88. Selkov E, Overbeek R, Kogan Y, Chu L, Vonstein V, et al. 2000. Functional analysis of gapped genomes: amino acid metabolism of *Thiobacillus ferrooxidans*. *Proc. Natl. Acad. Sci. USA* 97:3509–14
89. Torma AE. 1977. The role of *Thiobacillus ferrooxidans* in hydrometallurgical processes. In *Advances in Biochemical*

Engineering, ed. TK Ghose, A Fretcher, N Blackebrough, 6:1–37. New York: Springer

90. Tributsch H. 2001. Direct vs. indirect bioleaching. *Hydrometallurgy* 59:177–85
91. Tuovinen OH, Niemelä SI, Gyllenberg HG. 1971. Effect of mineral nutrients and organic substances on the development of *Thiobacillus ferrooxidans*. *Biotechnol. Bioeng.* 13:517–27
92. Deleted in proof
93. Van Aswegen PC, Godfrey MW, Miller DM, Haines AK. 1991. Developments and innovations in bacterial oxidation of refractory ores. *Miner. Metallurg. Process.* 8:188–92
94. Vásquez M, Espejo RT. 1997. Chemolithotrophic bacteria in copper ores leached at high sulfuric acid concentration. *Appl. Environ. Microbiol.* 63:332–34
95. Vásquez M, Moore ERB, Espejo RT. 1999. Detection by polymerase chain reaction-amplification sequencing of an archaeon in a commercial-scale copper bioleaching plant. *FEMS Microbiol. Lett.* 173:183–87
96. Vian M, Creo C, Dalmastri C, Gionni A, Palazzolo P, Levi G. 1986. *Thiobacillus ferrooxidans* selection in continuous culture. In *Fundamental and Applied Biohydrometallurgy*, ed. RW Lawrence, RMR Branion, HG Ebner, pp. 395–406. Amsterdam: Elsevier
97. Whitelock JL. 1997. Biooxidation of refractory gold ores (the Geobiotics process). See Ref. 72, pp. 117–27

Annu. Rev. Microbiol. 2002. 56:93–116
doi: 10.1146/annurev.micro.56.012302.160854

First published online as a Review in Advance on April 16, 2002

MICROSPORIDIA: Biology and Evolution of Highly Reduced Intracellular Parasites

Patrick J. Keeling and Naomi M. Fast
Department of Botany, Canadian Institute for Advanced Research, University of British Columbia, Vancouver BC, V6T 1Z4, Canada; e-mail: pkeeling@interchange.ubc.ca

Key Words metabolism, genomics, infection, phylogeny, fungi

■ **Abstract** Microsporidia are a large group of microbial eukaryotes composed exclusively of obligate intracellular parasites of other eukaryotes. Almost 150 years of microsporidian research has led to a basic understanding of many aspects of microsporidian biology, especially their unique and highly specialized mode of infection, where the parasite enters its host through a projectile tube that is expelled at high velocity. Molecular biology and genomic studies on microsporidia have also drawn attention to many other unusual features, including a unique core carbon metabolism and genomes in the size range of bacteria. These seemingly simple parasites were once thought to be the most primitive eukaryotes; however, we now know from molecular phylogeny that they are highly specialized fungi. The fungal nature of microsporidia indicates that microsporidia have undergone severe selective reduction permeating every level of their biology: From cell structures to metabolism, and from genomics to gene structure, microsporidia are reduced.

CONTENTS

INTRODUCTION

The microsporidia (or Microspora) are an unusual group of eukaryotic, obligate intracellular parasites that have attracted the curiosity of biologists for more than 100 years. Like many other intracellular parasites, microsporidia are highly specialized and have evolved an extremely sophisticated and unique infection mechanism along with other adaptations to life inside other cells. These adaptations are primarily characterized by reduction. Compared with other eukaryotes, microsporidia

0066-4227/02/1013-0093$14.00

are highly reduced at every level: from morphology and ultrastructure, to biochemistry and metabolism, and even at the level of their molecular biology, genes and genomes.

The first microsporidian was described in the middle of the nineteenth century when pébrine, or "pepper disease," was ravaging silkworms in southern Europe and threatening to destroy the European silk industry. The pébrine agent was observed to be a microscopic parasite that was named *Nosema bombycis* by Nägeli in 1857 (66). Nägeli considered *Nosema* to be a member of the schizomycete fungi, although classification at that time did not reflect the true diversity of microbial life, and schizomycetes were a grab bag of yeasts and bacteria. After further study, Balbiani accordingly created a new group for *Nosema* in 1882, calling it Microsporidia (2), the name still in use.

Today, the microsporidia are known to be an extremely diverse group of parasites. There are currently approximately 150 described genera of microsporidia with over 1200 individual species (78, 79). By far, most microsporidia infect animals, where they have been characterized in all vertebrate orders as well as most invertebrates, including the parasitic myxosporidia (95). While these animal parasites account for the vast majority of microsporidia, a few species have been shown to infect certain protists, such as ciliates and gregarine apicomplexa (95). It is interesting that these gregarines and some of the ciliates are themselves animal parasites, which suggests that the microsporidian probably once infected the same animal hosts as these protists and later adapted to parasitize its neighbor. Given the diversity and abundance of microsporidia known in animals today, it seems likely that the actual number of microsporidia far exceeds those which have been described and that the number of microsporidian species could perhaps approach the number of species of animals.

Although widespread among animals, microsporidia are apparently most prevalent in arthropods and fish. They are used as biological control agents against insect pests and, under natural circumstances, are found to be destructive to apiculture, fish, and some crustacea important to aquaculture (4, 75). The first microsporidian infection described in mammals was *Encephalitozoon cuniculi*, originally found in rabbits in 1922 (106). This species is now known to frequently infect a broad range of mammalian hosts. In 1959 the first clear case of a human microsporidian infection was recorded (60), but such cases were relatively rare until the mid-1970s, when a dramatic increase in recorded infections accompanied the increased prevalence of immunosuppressed individuals, either resulting from infection with HIV or due to the use of immunosuppressing drugs (98). The most common of these opportunistic human microsporidia is *Enterocytozoon bieneusi*, which was first described as a gastrointestinal parasite causing "wasting syndrome" (a potentially lethal diarrhea) in 1985 (20). *E. bieneusi* is now known to infect a wide range of human tissues in AIDS patients and occasionally infects healthy immunocompetent humans where it results in an acute but self-limiting intestinal disorder (74). Presently, 13 species of microsporidia have been found to infect humans (30, 92, 99) leading to a long list of human diseases, including chronic diarrhea and wasting syndrome, keratoconjunctivitis, pneumonia, bronchitis, nephritis, urethritis, prostatitis, hepatitis, encephalitis, myostitis, and peritonitis (30, 98, 99, 105).

THE MICROSPORIDIAN SPORE

The focal point of the microsporidian infection strategy, life history, and diagnosis is the spore, a single, highly organized cell (Figure 1). Spores are the only easily recognizable stage of microsporidia, they are the stage where species can be differentiated, and they are the only stage of microsporidia that is viable outside of a host cell. Spores range in size from as little as 1 μm in *E. bieneusi* to 40 μm in *Bacillidium filiferum* (92) and can be spherical, ovoid, rod-shaped, or crescent-shaped, although most are ovoid. Within a species, spore morphology tends to be fairly regular, although some species do possess different spore types in different stages of their life cycles (92). The spore is bound by a normal unit membrane and two rigid extracellular walls. The exospore wall is composed of a dense, granulofibrous, proteinaceous matrix (7, 93) and is generally uniform at the surface, although it can be highly ornamented in aquatic microsporidia (92). The endospore wall is composed of alpha-chitin (92) and proteins and is of uniform thickness, except at the apex of the spore where the endospore wall is considerably thinner than elsewhere. Within the spore membrane is the sporoplasm, or the cytoplasm of the spore, which is the infectious material of microsporidia. The sporoplasm contains a single nucleus or two nuclei arranged as a diplokaryon (two closely appressed nuclei), cytoplasm enriched with ribosomes, and is otherwise dominated by structures relating to infection. There are three principal structures related to infection: the polaroplast, the polar filament or polar tube, and the posterior vacuole. The polaroplast is a large organization of membranes occupying the anterior

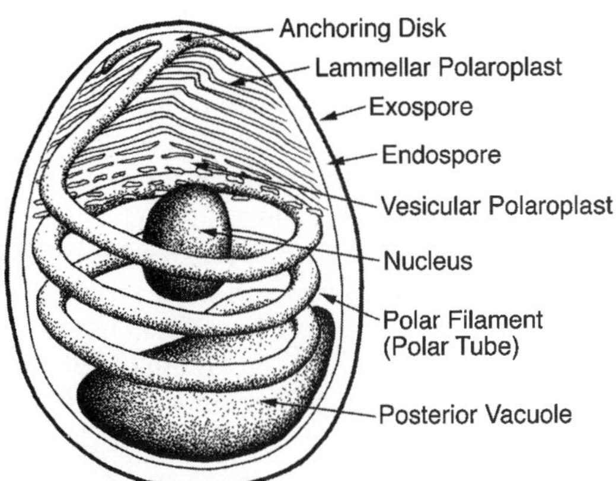

Figure 1 Diagram of a microsporidian spore showing the major structures discussed in the text.

part of the spore. The anterior portion of the polaroplast exists as highly organized, closely stacked membranes called the lammellar polaroplast, whereas the posterior portion is more loosely organized and is called the vesicular polaroplast. The most obvious organelle associated with infection is the polar filament or polar tube. In the sporoplasm this filament is composed of membrane and glycoprotein layers and ranges from 0.1 to 0.2 μm in diameter and 50 to 500 μm in length (51, 92). The polar filament is attached at the apex of the spore via an umbrella-shaped structure called the anchoring disk, from which it extends to the posterior of the spore. For approximately one third to one half the length of the spore the polar filament is straight, and the remainder is helically coiled about the contents of the sporoplasm. The number of coils, their arrangement relative to one another, and even the angle of helical tilt are conserved and diagnostic for a particular species (79, 92). The polar filament terminates at the third major organelle associated with infection, the posterior vacuole. There is apparently some physical association between the end of the polar filament and the posterior vacuole (59, 100), but whether the polar filament actually enters the vacuole or simply contacts it is not known, as the point of contact has never been observed (51, 92, 94).

MICROSPORIDIAN INFECTION AND LIFE CYCLE

Germination of the microsporidian spore is one of the most interesting and dramatic series of subcellular events in biology, involving a build-up and controlled release of tremendous force, a cascade of extremely rapid events in close succession, and a complete alteration of the sporoplasm, which includes some unique restructuring of membrane topology.

Spore germination begins with an environmental trigger that varies for different species depending on their habitat (85) but is largely poorly understood (51). In vitro, spores may be germinated by a number of physical and chemical stimuli including, but not limited to, alterations in pH, dehydration followed by rehydration, hyperosmotic conditions, the presence of anions or cations, or exposure to ultraviolet light or peroxides [for an extensive review of these conditions for various species see (51)]. When a spore is induced to germinate, the first sign is a general swelling of the spore and a specific swelling of the polaroplast and posterior vacuole (59). This is the result of an increase in osmotic pressure in the spore (52, 67, 86), but how this pressure builds up is the subject of some debate. Spores are equipped with aquaporins that specifically transport water across the sporoplasm membrane (31), which explains how the influx of water can take place, but not necessarily why. One intriguing possibility was raised after it was observed that levels of trehalose drop significantly during the germination of *Nosema algerae* (84, 88). Trehalose is a glucose-glucose disaccharide found widely in nature and is the major carbohydrate storage material of microsporidian spores (84, 90) and fungal spores (1, 29). It was suggested that during spore activation in microsporidia, trehalose is degraded to constituent glucose monomers, effectively increasing the number of soluble molecules within the spore and leading to the import of large quantities of water and the concomitant increase in osmotic pressure (84, 86, 88). This model is a fascinating possibility and would represent an ingenious system to generate force in the cell. However, the germination of the spore involves a great number of changing conditions, so the levels of trehalose cannot be tied unequivocally to the rise in internal osmotic pressure. In fungal spores trehalose does not appear to primarily act as an energy reserve but rather as an anti-stress metabolite (1). This also may be the role of trehalose in microsporidia, and trehalose degradation may only be a step in the germination process. Moreoever, the trehalose levels in a related species of *Nosema* do not change during germination (19), suggesting some, or potentially many, other causes for water influx. Indeed, alternative models point to the intracellular concentration of calcium ions as a cause for the influx of water (50) and a possible role for calmodulin in the process (102). Here it is suggested that membrane breakdown during spore activation could release calcium ions from the endomembrane system into the sporoplasm. These ions could induce the influx of water and could also induce the activation of enzymes such as trehalase whose activity could further enhance the hypertonic shift in the sporoplasm (50).

Whatever the proximal cause, the osmotic pressure of the spore builds up and this pressure seems to be the driving force of the subsequent events of germination. The internal pressure of the spore and the breakdown in sporoplasmic membranes culminate in the rupturing of the anchoring disk and the discharge of the polar filament by eversion (Figure 2). Eversion begins at the spore apex, where the discharging polar filament actually breaks through the thinnest region of the spore wall, and is frequently likened to turning the finger of a glove inside-out. As

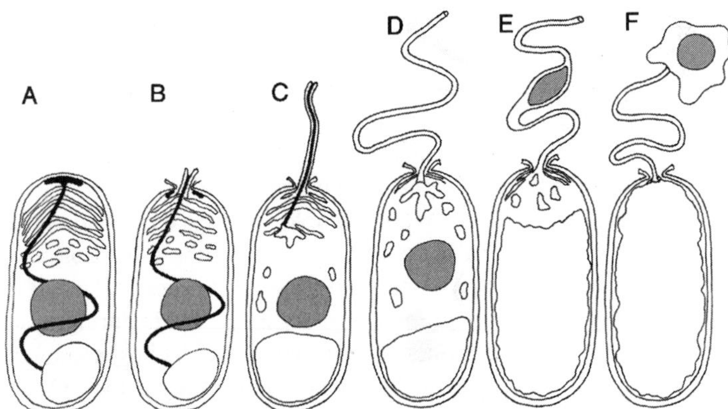

Figure 2 Polar tube eversion during spore germination. (*A*) Dormant spore, showing polar filament (*black*), nucleus (*gray*), polaroplast and posterior vacuole. (*B*) Polaroplast and posterior vacuole swelling, anchoring disk ruptures, and polar filament begins to emerge, everting as it does so. (*C*) Polar filament continues to evert. (*D*) Once the polar tube is fully everted, the sporoplasm is forced into and (*E*) through the polar tube. (*F*) Sporoplasm emerges from the polar tube bound by new membrane.

the filament everts, it becomes a tube and the dense granular, glycoproteinaceous material that filled the filament when inside the sporoplasm is deposited on the outside of the tube (51). The remarkable nature and violent speed of this event cannot be overstated. The discharged polar tube can range in length from 50–500 μm in length (potentially 100 times the length of the spore) and is generated by turning inside out a filament that is coiled around the sporoplasm. Yet, the entire event of germination takes place in fewer than two seconds and the tip of the discharging tube can move through the medium at velocities exceeding 100 μm/s (32).

Clearly, the polar tube is a significant projectile, and if a potential host cell lies nearby, the discharging tube can strike this cell and pierce its membrane. Once the polar tube is fully discharged, the continued pressure within the spore (most likely from the swelling of the posterior vacuole) forces the sporoplasm through the polar tube. Although somewhat elastic, the polar tube is a narrow conduit ranging from just 0.1 to 0.25 μm in diameter, but the sporoplasm is nevertheless forced through the polar tube rapidly, emerging at the tip of the tube in only 15–500 ms (32). If the polar tube has penetrated a host cell, the sporoplasm emerges from the tube directly into the host cytoplasm, thus infecting without the host recognizing the parasite as a foreign invader. This invasion is a remarkable event, since the sporoplasm membrane has been left behind in the now empty spore (87, 101), raising some intriguing questions about membrane topology surrounding spore germination. First, an intracellular membrane-based organelle has broken through the sporoplasm membrane and spore wall while being turned inside out. Yet, the sporoplasm passes out of its bounding membrane and through the polar tube to emerge at the other side with a brand-new membrane. This is thought to be possible because polaroplast membrane is forced into the polar tube at the onset of germination, and it is this membrane that forms the bounding plasma membrane of the intracellular stage of the parasite (101), a remarkable feat of membrane manipulation.

Once inside the host cell, the parasite is referred to as a meront and it begins a stage of growth and division characterized by a high degree of interspecies variation. In general, however, the parasite is found directly within the host cytoplasm and not within a host phagocytotic vacuole, as is the case for many other intracellular parasites. Occasionally the parasite induces the formation of a surrounding membrane known as a parasitophorous vacuole at an early stage of infection (55), but most often they do not (although they commonly do during sporogenesis). At this stage, the parasite is intimately associated with its host and often induces significant changes to the host that are not obviously deleterious. Often the host cell reorganizes around the parasite, so that the parasite can be found surrounded by host organelles such as endoplasmic reticulum (ER), nuclei, or mitochondria (92). In some instances the microsporidian also physically interacts with the host nucleus, resulting in the enlargement of host nuclear pores (57) or even the invasion and infection of the nucleus itself (35). Host cells may also transform in shape and size, often enlarging (103). The archetype of microsporidian-induced host cell transformation is the xenoma, a host cell harnessed and transformed by the parasite to promote its own growth and development. In such cases, the host cell is induced to enlarge and undergo many rounds of nuclear division resulting in an enormous plasmodium filled with parasites in highly ordered strata depending on their stage of development: mature spores at the center with the earlier stages radiating toward the periphery of the xenoma (12, 54, 103).

The onset of sporogony is marked in some species by the separation of diplokaryotic nuclei (11) and in other specific cases by meiosis (34), although synaptonemal complexes have been observed in all life stages of other microsporidia. Morphological features are more consistent indicators of sporogony and include an apparent thickening of the plasma membrane (due to the accumulation of electron-dense material on the membrane) and the increased presence of ER and ribosomes. Both the ER and ribosomes change organization throughout sporogony, with the ER becoming highly ordered, and the ribosomes increasingly forming arrays attached to the ER, known as polyribosomes (92). Although sporogony can occur in direct contact with the host cytoplasm, some species produce a sporophorous vesicle in which the sporonts develop. In most species, this stage of the life cycle is also accompanied by some degree of division, although the number of sporoblasts (presporal cells) produced varies among species from two (bisporous) to many (polysporous). Following division, the extrusion apparatus (including the polar filament, polaroplast, and posterior vacuole) begins to develop. Early observations implicated the Golgi as giving rise to the polar filament (91), and this has been confirmed by histochemical labeling for activity of the Golgi marker, thiamine pyrophosphatase (80). Further, histochemical studies assaying activity of nucleoside diphosphatase, a marker of the ER and sometimes of the outer *cis*-Golgi, also implicate the ER in the formation of the polar filament and associated polaroplast membranes (81). These results suggest that the Golgi and ER (which themselves are homologous membrane systems) give rise to the sophisticated extrusion apparatus of microsporidia. As the extrusion apparatus nears complete formation and the sporoblasts approach maturity, the cells decrease in size and the chitinous endospore layer develops. Once complete, the mature spores are released. Some species produce autoinfective spores that germinate immediately in an attempt to infect the same host, thus spreading the infection quickly though an individual (76). Alternatively, spores are released into the environment (i.e., via host urine, feces, or decomposition) where they can infect other individuals, most often by way of the digestive system.

MICROSPORIDIAN GENOMICS

The genomes of microsporidia have aroused interest since they were first karyotyped and found to be much smaller than expected for eukaryotes. Presently, the sizes of 13 microsporidian genomes are known and they fall between 19.5 Mbp and only 2.3 Mbp [(6, 70); for a summary table see (63)]. Apart from their small size, microsporidian genomes are in all characteristics eukaryotic: They have multiple linear chromosomes, telomeres, and segregate by closed mitosis. However, the reduced size of microsporidian genomes has led to two genome sequence surveys (25, 36), the sequencing of an entire chromosome (69) and now the completely sequenced genome of *E. cuniculi* (44). The general characteristics of the 2.9-Mbp *Encephalitozoon* genome mirror what has been observed in other highly reduced eukaryotic genomes (23): Genes are typically flanked by short intergenic regions (although there are no overlapping genes), there are few repeat sequences, little evidence of selfish elements, and few introns. Introns were predicted to be present in microsporidia based on studies that characterized elements of the spliceosome (21, 27) and one putative intron in a ribosomal protein-coding gene (5). In the 1997 predicted open reading frames in the complete *E. cuniculi* genome, only 11 introns were identified, giving *Encephalitozoon* one of the lowest intron densities among eukaryotes. One of the surprising characteristics of the *Encephalitozoon* genome that demonstrates the extreme degree of reduction is the finding that *Encephalitozoon* genes themselves are actually shorter on average than their homologs from other organisms (44). This suggests that the selection for reduction is very strong indeed, perhaps even overwhelming selection against marginally disadvantageous deletions in protein- and RNA-encoding gene sequences.

ORIGIN AND EVOLUTION OF MICROSPORIDIA

Our conception of the evolutionary history of microsporidia has been radically rewritten on a number of occasions since their original description in the middle of the nineteenth century. When Nägeli considered *Nosema* to be a yeast-like fungus, the concept of a "protist" or a "protozoan" was in its infancy, so it was common to pigeonhole microbial organisms into animals, plants, or sometimes fungi depending on their characteristics. The unique mode of infection seen in *Nosema* eventually led to their separation from fungi, but with no obvious similarity to any other group of eukaryotes, the evolutionary origins of microsporidia were not much clearer. As the real diversity of microbial eukaryotes began to dawn on biologists, more complex hierarchical classification schemes were developed based on certain common features of morphology. One of the cell types to be identified and classified together were spore-forming parasites, collectively called Sporozoa. This group contains what are now known as apicomplexa, myxosporidia, actinomyxidia, haplosporidia, microsporidia, and a handful of individual genera. Within Sporozoa, microsporidia were considered to be most closely related to a variety of other parasites at different times but were most often believed to be akin to myxosporidia and actinomyxidia, which, with microsporidia, were collectively called the Cnidosporidia (53).

For some time, microsporidia were considered to be either Cnidosporidia [although the vast differences between microsporidia on one hand and myxosporidia and actinomyxidia on the other were pointed out (58)] or an independent protist lineage of uncertain affinity (56). However, in 1983, attention was drawn to the possible evolutionary significance of microsporidia in a new way. Cavalier-Smith (13) proposed that the origin of eukaryotes might have preceded the endosymbiotic origin of the mitochondrion by some considerable span of time, implying that there may be protists that evolved before the mitochondrial origin. In other words, there may be primitively amitochondriate eukaryotes, and focusing attention on these protists could unlock some of the secrets surrounding the origin of eukaryotes. Four lineages of amitochondriate protists that could hold this pivotal position were identified, and these were collectively named Archezoa: Archamoebae (e.g., *Entamoeba*), Metamonada (e.g., *Giardia*), Parabasalia (e.g., *Trichomonas*), and Microsporidia (13). Archezoa were also known to have other characteristics that could be considered primitive, for instance the microsporidia and some other Archezoa contain 70S ribosomes with bacterial-sized rRNAs rather than the 80S ribosomes typical of eukaryotes (18, 39). In some articulations of the Archezoa hypothesis, microsporidia were actually singled out as perhaps the most primitive and ancient lineage of all eukaryotes, since in addition to lacking mitochondria they also lack flagella and other 9 + 2 structures and were sometimes thought to lack Golgi as well (68). Shortly after the Archezoa hypothesis was formulated, the tools of molecular phylogenetics began to be applied vigorously to microbial eukaryotes, and the first molecular data from microsporidia lent extraordinary support to the Archezoa hypothesis. The small subunit ribosomal RNA (SSU rRNA) from the microsporidian *Vairimorpha* was shown to be the earliest branch on the eukaryotic tree (96), and of greater interest, the microsporidia were found to be the only eukaryotes to retain the prokaryotic trait of having their 5.8S rRNA fused to the large subunit (LSU) rRNA (97). These two pieces of evidence bolstered the notion that microsporidia were indeed an ancient and primitive lineage, and further evidence seemed to accumulate with the sequencing of every new microsporidian gene: Phylogenies based on elongation factor 1α (43), elongation factor 2 (42), as well as isoleucyl tRNA synthetase (9), all showed the microsporidia branching deeply.

The same apparent early phylogenetic position was also seen with other Archezoa (with the possible exception of *Entamoeba*), and altogether the case of an ancient origin and primitive lack of mitochondria for microsporidia and other Archezoa seemed neatly sewn up. Despite the accumulated evidence in favor of the Archezoa, the highly adapted parasitic lifestyle of the microsporidia was always a source of doubt: Such specialized parasites may appear primitive when in fact they reflect a process of reduction from a more complex ancestor. It was also noted that microsporidian gene sequences are highly divergent and possibly misleading due to an artifact in phylogenetic reconstruction known as "long branch attraction." It was even noted that the fused 5.8S-LSU rRNA in microsporidia could easily be a secondarily derived state since microsporidian rRNAs are extremely strange compared with other eukaryotes, or even when compared with prokaryotes. Microsporidian rRNAs are exceptionally small because they have sustained a number of deletions, sometimes in regions of the sequence that are highly conserved even between prokaryotes and eukaryotes. It was reasoned that a deletion that affected one of the rRNA operon processing sites could generate the fused microsporidian rRNA from an ancestral eukaryotic rRNA operon (14).

These concerns eventually proved to be well founded, as further sampling of microsporidian genes soon revealed that not all evidence supported an ancient origin for the group. The first phylogenies to contradict the evidence for an early origin of microsporidia were those of alpha-tubulin and beta-tubulin (24, 46). Here the microsporidia formed a surprising but extremely well-supported group with fungi. This alternative, fungal origin for microsporidia is in complete disagreement with earlier evidence that microsporidia are ancient or primitive organisms because fungi are not considered ancient or primitive and are now known to be close relatives of animals. How could different genes from the same organisms provide such different phylogenetic trees? One possibility is that the accelerated rates of substitution common to many microsporidian genes led to artifacts in the phylogenetic reconstruction, and it was proposed that the ancient origin of microsporidia was erroneously suggested by the highly derived genes of rRNA and elongation factors. However, tubulin phylogenies are not immune to this problem, and it was noted that fungal and microsporidian tubulins are both highly divergent [perhaps because neither group contains 9 + 2 structures, releasing evolutionary constraints on their tubulin gene sequences (46)], suggesting that these phylogenies may be in error. It has now been demonstrated, however, that both alpha- and beta-tubulins from the flagellated chytrid fungi are highly conserved, and in phylogenies where chytrids are the only fungal representatives, microsporidia still branch with fungi with high statistical support. This shows that the relationship between microsporidia and fungi in tubulin phylogenies is not a long branch artifact (48).

The evidence from tubulin phylogenies led to an immediate re-evaluation of certain characteristics of microsporidia such as an insertion in the EF-1α protein, (43) as well as unusual features of their meiosis that had previously been recognized as resembling homologous processes in fungi (28). In addition, genes encoding TATA-box-binding protein (26), mitochondrial HSP70 (33, 37, 71), glutamyl-tRNA synthetase (10), and the largest subunit of RNA polymerase II (RPB1) (38) were also sequenced from microsporidia, and the phylogenies of each of these genes further supported the fungal origin of the group—strongly in the case of RPB1. The genome of *E. cuniculi* also yielded a number of proteins with strong fungal affinities (44), and a recent analysis of combined molecular data from four genes further supports the fungal origin of microsporidia (3). Also of critical importance is that much of the evidence for the ancient origin of microsporidia has recently been undermined by the re-analysis of genes that had previously supported the early-branching position. Re-analyzing EF-2 and LSU rRNA data using methods that take into account variations in substitution rates at different sites in a sequence showed that these genes not only fail to support the ancient position of microsporidia but actually weakly support the fungal relationship (38, 89). Some examples of these phylogenies are shown in Figure 3, and a summary of molecular

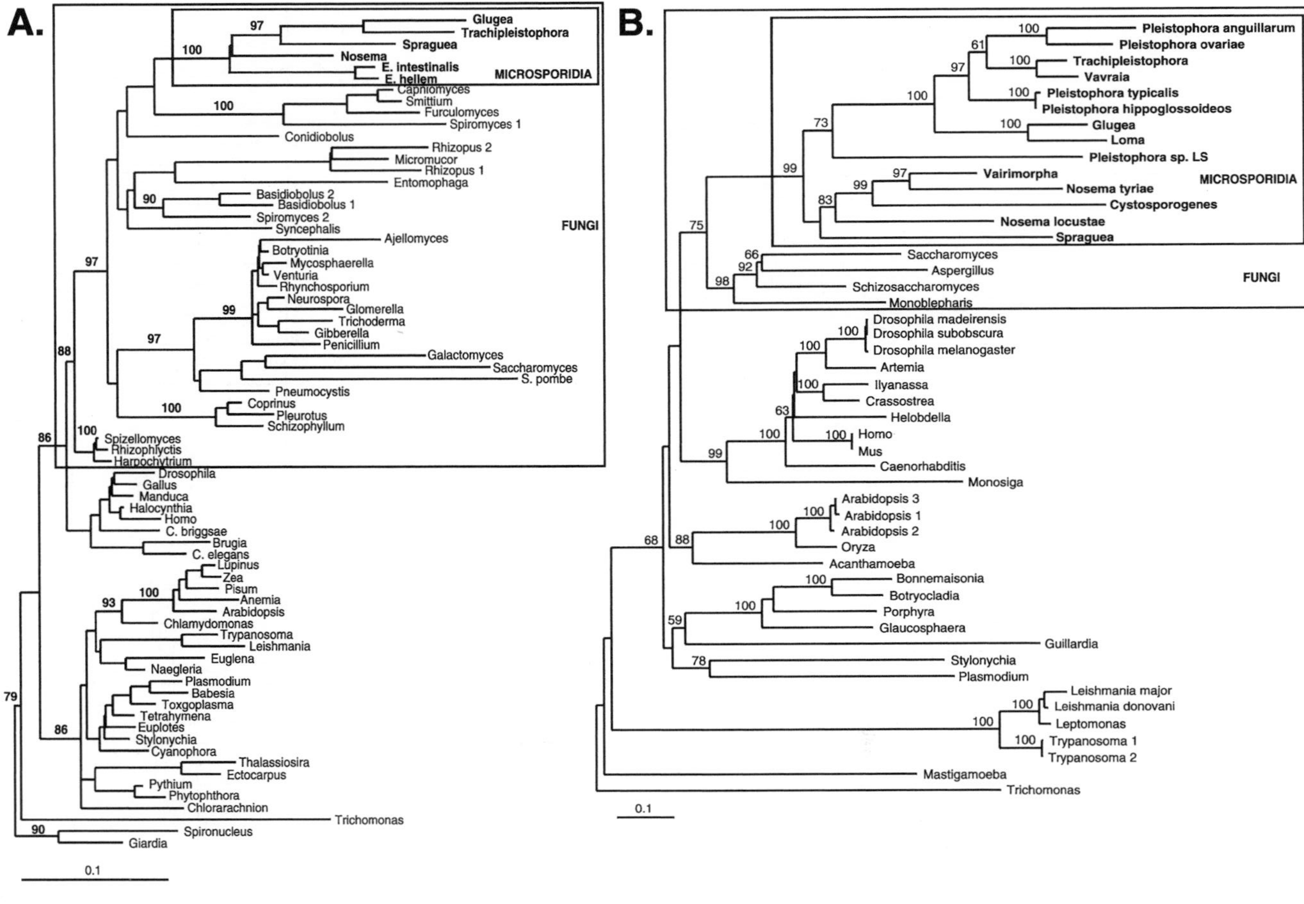

Figure 3 *(Continued on next page)*

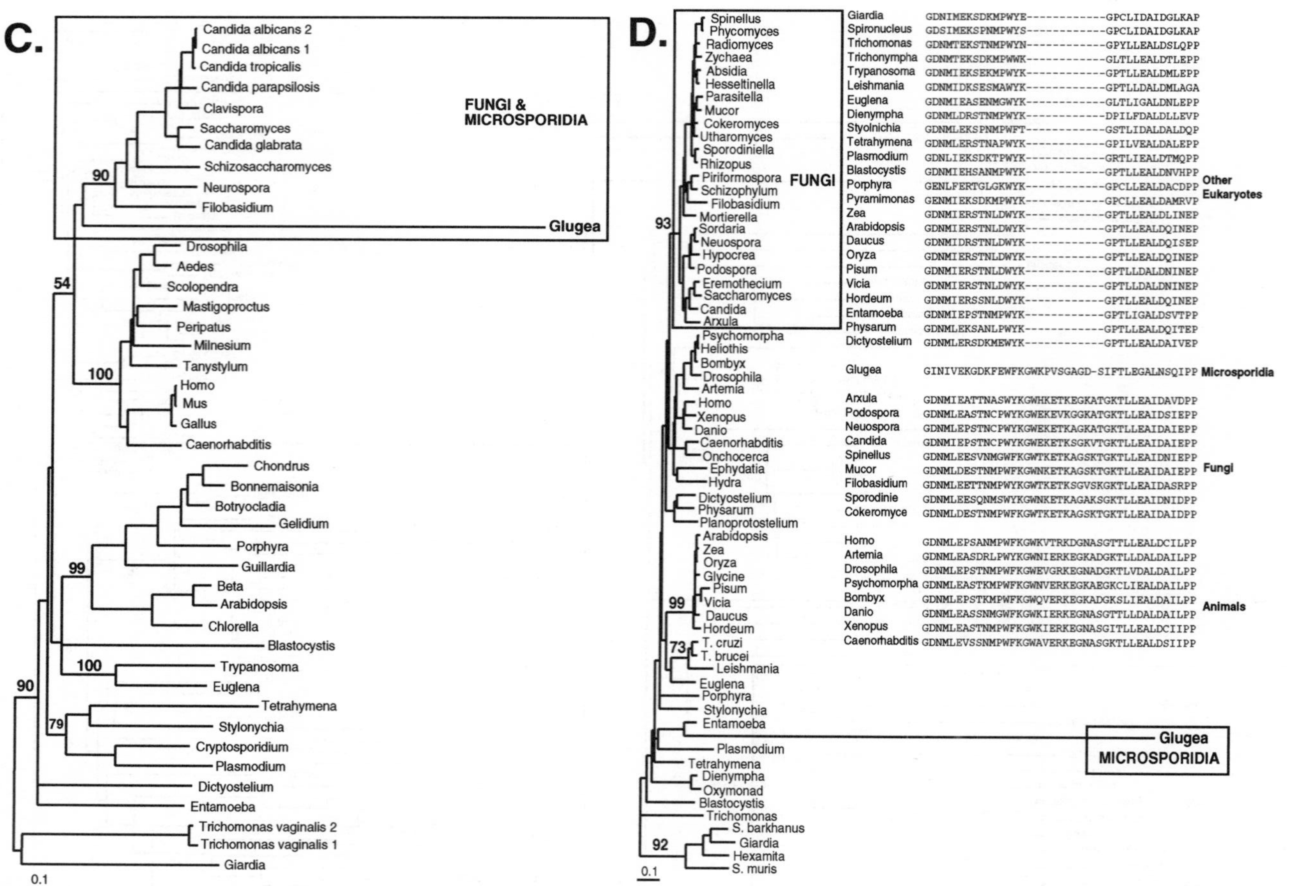

C.
FUNGI &
MICROSPORIDIA
Candida albicans 2
Candida albicans 1
Candida tropicalis
Candida parapsilosis
Clavispora
Saccharomyces
Candida glabrata
Schizosaccharomyces
Neurospora
Filobasidium
Glugea
90
54
Drosophila
Aedes
Scolopendra
Mastigoproctus
Peripatus
Milnesium
Tanystylum
100
Homo
Mus
Gallus
Caenorhabditis
Chondrus
Bonnemaisonia
Botryocladia
Gelidium
Porphyra
Guillardia
99
Beta
Arabidopsis
Chlorella
Blastocystis
100
Trypanosoma
Euglena
90
Tetrahymena
79
Stylonychia
Cryptosporidium
Plasmodium
Dictyostelium
Entamoeba
Trichomonas vaginalis 2
Trichomonas vaginalis 1
Giardia
0.1
D.
FUNGI
Spinellus
Phycomyces
Radiomyces
Zychaea
Absidia
Hesseltinella
Parasitella
Mucor
Cokeromyces
Utharomyces
Sporodiniella
Rhizopus
Piriformospora
Schizophylum
Filobasidium
Mortierella
93
Sordaria
Neuospora
Hypocrea
Podospora
Eremothecium
Saccharomyces
Candida
Arxula
Psychomorpha
Heliothis
Bombyx
Drosophila
Artemia
Homo
Xenopus
Danio
Caenorhabditis
Onchocerca
Ephydatia
Hydra
Dictyostelium
Physarum
Planoprotostelium
Arabidopsis
Zea
Oryza
Glycine
Pisum
99
Vicia
Daucus
Hordeum
73
T. cruzi
T. brucei
Leishmania
Euglena
Porphyra
Stylonychia
Entamoeba
Glugea
MICROSPORIDIA
Plasmodium
Tetrahymena
Dienympha
Oxymonad
Blastocystis
Trichomonas
S. barkhanus
92
Giardia
Hexamita
S. muris
0.1
Giardia GDNIMEKSDKMPWYE------------GPCLIDAIDGLKAP
Spironucleus GDSIMEKSPNMPWYS------------GPCLIDAIDGLKAP
Trichomonas GDNMTEKSTNMPWYN------------GPYLLEALDSLQPP
Trichonympha GDNMTEKSDKMPWWK------------GLTLLEALDTLEPP
Trypanosoma GDNMIEKSEKMPWYK------------GPTLLEALDMLEPP
Leishmania GDNMIDKSESMAWYK------------GPTLLDALDMLAGA
Euglena GDNMIEASENMGWYK------------GLTLIGALDNLEPP
Dienympha GDNMLDRSTNMPWYK------------DPILFDALDLLEVP
Styolnichia GDNMLEKSPNMPWFT------------GSTLIDALDALDQP
Tetrahymena GDNMLERSTNAPWYK------------GPILVEALDALEPP
Plasmodium GDNLIEKSDKTPWYK------------GRTLIEALDTMQPP
Blastocystis GDNMIEHSANMPWYK------------GPTLLEALDNVHPP
Porphyra GENLFERTGLGKWYK------------GPCLLEALDACDPP
Pyramimonas GENMIEKSDKMPWYK------------GPCLLEALDAMRVP
Zea GDNMIERSTNLDWYK------------GPTLLEALDLINEP
Arabidopsis GDNMIERSTNLDWYK------------GPTLLEALDQINEP
Daucus GDNMIDRSTNLDWYK------------GPTLLEALDQISEP
Oryza GDNMIERSTNLDWYK------------GPTLLEALDQINEP
Pisum GDNMIERSTNLDWYK------------GPTLLDALDNINEP
Vicia GDNMIERSTNLDWYK------------GPTLLDALDNINEP
Hordeum GDNMIERSSNLDWYK------------GPTLLEALDQINEP
Entamoeba GDNMIEPSTNMPWYK------------GPTLIGALDSVTPP
Physarum GDNMLEKSANLPWYK------------GPTLLEALDQITEP
Dictyostelium GDNMLERSDKMEWYK------------GPTLLEALDAIVEP
Other
Eukaryotes
Glugea GINIVEKGDKFEWFKGWKPVSGAGD-SIFTLEGALNSQIPP
Microsporidia
Arxula GDNMIEATTNASWYKGWHKETKEGKATGKTLLEAIDAVDPP
Podospora GDNMLEASTNCPWYKGWEKEVKGGKATGKTLLEAIDSIEPP
Neuospora GDNMLEPSTNCPWYKGWEKETKAGKATGKTLLEAIDAIEPP
Candida GDNMIEPSTNCPWYKGWEKETKSGKVTGKTLLEAIDAIEPP
Spinellus GDNMLEESVNMGWFKGWTKETKAGSKTGKTLLEAIDNIEPP
Mucor GDNMLDESTNMPWFKGWNKETKAGSKTGKTLLEAIDAIEPP
Filobasidium GDNMLEETTNMPWYKGWTKETKSGVSKGKTLLEAIDASRPP
Sporodinie GDNMLEESQNMSWYKGWNKETKAGAKSGKTLLEAIDNIDPP
Cokeromyce GDNMLDESTNMPWFKGWTKETKAGSKTGKTLLEAIDAIDPP
Fungi
Homo GDNMLEPSANMPWFKGWKVTRKDGNASGTTLLEALDCILPP
Artemia GDNMLEASDRLPWYKGWNIERKEGKADGKTLLDALDAILPP
Drosophila GDNMLEPSTNMPWFKGWEVGRKEGNADGKTLVDALDAILPP
Psychomorpha GDNMLEASTKMPWFKGWNVERKEGKAEGKCLIEALDAILPP
Bombyx GDNMLEPSTKMPWFKGWQVERKEGKADGKSLIEALDAILPP
Danio GDNMLEASSNMGWFKGWKIERKEGNASGTTLLDALDAILPP
Xenopus GDNMLEASTNMPWFKGWKIERKEGNASGITLLEALDCIIPP
Caenorhabditis GDNMLEVSSNMPWFKGWAVERKEGNASGKTLLEALDSIIPP
Animals

phylogenies including microsporidia is given in Table 1. Furthermore, the notion that microsporidia are primitively amitochondriate has been disproved by two means. First, if they are fungi then we know that microsporidia must be derived from mitochondrial ancestors since fungi and all their close relatives contain mitochondria. Second, genes for proteins derived from mitochondria have been found in microsporidia. Most mitochondria contain a small genome, but the majority of genes for mitochondrial proteins are encoded in the nucleus and their products are post-translationally targeted to the organelle. Therefore, it has been reasoned that an organism that has drastically altered its mitochondrion, or even lost it altogether, could retain relic mitochondrial genes in the nuclear genome (17). One such gene, encoding mitochondrial HSP70, has now been characterized in three species of microsporidia: *N. locustae*, *Vairimorpha necatrix*, and *E. cuniculi*. The presence of this gene in these organisms gives solid confirmation that microsporidia had a mitochondriate ancestry, and in some analyses even supports a fungal origin (33, 37, 71). Now, several more mitochondrion-derived genes have been found in microsporidia (25, 44), the evolutionary implications of which will be considered when the current presence or absence of mitochondria is discussed below.

Altogether, there are now a number of gene phylogenies that provide robust support for some relationship between microsporidia and other fungi, but what exactly is this relationship? Most genes that have been used to test this have only included ascomycetes and occasionally basidiomycetes. With such poor sampling of fungi, it is not clear from these studies if microsporidia actually are fungi, or if they are merely a closely related sister group of fungi. Unfortunately, only two genes have currently been sampled from diverse fungi to better define this relationship, and these are alpha- and beta-tubulins. In the case of beta-tubulin there is strong support for microsporidia actually evolving from within the fungi, but phylogenies fail to distinguish whether microsporidia are specifically related to ascomycetes or zygomycetes (48). Alpha-tubulin also strongly supports microsporidia evolving from within the fungi, but in this case, and in analyses combining both genes, the microsporidia show a specific relationship to zygomycetes (P.J. Keeling, unpublished

←

Figure 3 Phylogenies of four microsporidian proteins showing several types of microsporidian phylogeny. (*A*) Beta-tubulin and (*B*) RPB1 phylogenies show strong support for microsporidia being related to, or derived from, fungi. (*C*) Elongation factor-2 was originally reported to show microsporidia branching early among eukaryotes, but re-analysis taking into account site-to-site rate variation actually shows the microsporidia branching weakly with fungi. (*D*) Elongation factor-1α is an example of a phylogeny that consistently places microsporidia early among eukaryotes; however, it is an extremely divergent gene, calling into question its use as a phylogenetic marker for microsporidia. This is particularly interesting in the case of EF-1α since the microsporidian gene actually contains an insertion otherwise found only in animals and fungi (*inset*). All trees were inferred using gamma-corrected distances and weighted neighbor joining as described in (25).

TABLE 1 Summary of published phylogenetic trees showing position of microsporidia

Gene	"Deep" branching	Unresolved/ intermediate	Fungal
Alpha-tubulin			(24, 46, 48)
Beta-tubulin			(24, 46, 48)
RPB1			(38)
TBP			(26)
Glu-tRNA synthetase			(10)
Ser-tRNA synthetase			(44)
V-ATPase-A			(44)
TF IIB			(44)
GTPase			(44)
mt HSP70	(37, 71)	(71)	(33, 37)
LSU rRNA	(70)	(70)	(89)
EF-2	(42)		(38)
Ile-tRNA synthtase		(9)	
mt PDH-alpha		(25)	
mt PDH-beta		(25)	
EF-1α	(42, 43)	(38)	
SSU rRNA	(96)	(89)	
eIF-2γ	(47)		
Gln-tRNA synthetase	(10)		
Proteosome alpha	(8)		

data). Interestingly, a zygomycete origin for microsporidia was proposed based on the superficial similarities between the polar tube and the apical spore body of harpellalean zygomycetes (15). However, molecular evidence to date does not suggest a specific relationship between microsporidia and harpellalean zygomycetes or any other group of zygomycetes, except in certain analyses where microsporidia are related to entomophthorales (48).

While the exact relationship between microsporidia and fungi remains to be clarified, nearly all current evidence does support one major conclusion: Microsporidia are not ancient eukaryotes, but are instead highly evolved fungi. This conclusion colors nearly all other aspects of microsporidia in a new light: No longer are they primitive in lacking mitochondria, flagella, or peroxisomes—these features result from reductive evolution, probably in response to their growing adaptation to intracellular parasitism (the mitochondrion is a special case discussed below). Similarly, microsporidian biochemistry is not primitive, it is reduced. Even at the molecular level, the tiny genomes of microsporidia evolved from larger genomes by gene loss and compaction, and the unusual genes and gene sequences we find today are highly derived, not ancient (49).

CORE METABOLISM

Much of the metabolism of eukaryotes centers around the mitochondrion, the so-called powerhouse of the cell. However, ultrastructural studies on microsporidia in the 1960s revealed no mitochondrion (91), and no study since then has actually visualized an organelle answering to the description of a typical mitochondrion in any microsporidian (63, 92). We now know that microsporidia evolved from mitochondriate fungi, but two interesting questions remain: What was the fate of the microsporidian mitochondrion, and how has their core metabolism adapted?

Metabolic pathways for energy generation from carbohydrates have more or less been worked out from several amitochondriate protists, in particular the parabasalian *Trichomonas vaginalis*, the diplomonad *Giardia lamblia*, and the entamoebid *Entamoeba histolytica* (data are also beginning to accumulate from the apicomplexan, *Cryptosporidium*) (65, 73). In addition, a number of genes encoding enzymes involved in core metabolic pathways have also been characterized from these organisms. The metabolic picture that has emerged from these studies shows some commonality between these disparate organisms, but also a great deal of variation. In general all these organisms lack electron transfer chains, oxidative phosphorylation, and the tricarboxylic acid (TCA) cycle. All break down glucose using the glycolytic pathway, which is like that of other eukaryotes except that phosphofructokinase is pyrophosphate-dependent rather than ATP-dependent (65). From phosphoenol pyruvate, pyruvate is formed directly or can be formed using a malate bypass not found in typical eukaryotes. Pyruvate metabolism is perhaps the defining difference between these amitochondriates and other eukaryotes. Typically, pyruvate enters the mitochondrion and is oxidatively decarboxylated by the pyruvate dehydrogenase complex (PDHC), but in the amitochondriate parabasalia, diplomonads, and entamoebids, it is decarboxylated by a single enzyme, pyruvate:ferredoxin oxidoreductase (PFOR), an iron-sulphur protein also used by anaerobic bacteria. Electrons are transferred from pyruvate to ferredoxin by PFOR, then to NADH, and ultimately to an organic terminal electron acceptor. In the parabasalia, the terminal electron acceptor is ionic hydrogen, resulting in the production of hydrogen gas, and the reactions following pyruvate production take place within a membrane-bound organelle called the hydrogenosome (64). Conversely, in *Giardia* and *Entamoeba* all the reactions of core carbon metabolism are cytosolic. Altogether, sugar metabolism in these organisms is substantially less efficient than that of mitochondriate eukaryotes, relying as it does on what has been termed "extended glycolysis" and substrate-level phosphorylation (65).

Until recently, microsporidia could not be compared to these other "amitochondriates" since next to nothing was known about their metabolic capacities, largely because their obligate intracellular growth and division presents serious challenges to biochemical assays (83). Nevertheless, some biochemical assays were carried out using purified spores and in vitro–germinated spores maintained for a short time in a cell culture medium (102). Altogether, these studies confirmed the suspected lack of TCA cycle, showed a requirement for ATP in the sustaining media (buttressing suspicions that these parasites probably import ATP from their hosts), determined that the parasites produce lactic acid and pyruvic acid, and demonstrated the presence of several enzymes involved in glycolysis, the pentose-phosphate pathway, as well as trehalose synthesis and degradation (22, 84, 102).

These studies indicate that microsporidia have retained the glycolytic pathway and suggest that they probably use extended glycolysis. However, the first microsporidian gene encoding an enzyme for core carbon metabolism proved that they are very different indeed. The *N. locustae* genome was found to encode both the alpha and beta subunits of pyruvate dehydrogenase (PDH, or PDHC E1), the first enzyme of the pyruvate dehydrogenase complex (25). In mitochondriate eukaryotes, PDH is responsible for the decarboxylation of pyruvate and gives rise to the "active aldehyde" intermediate, 2-alpha-hydroxyethyl-thiamine pyrophosphate (HETPP), which is then converted by PDHC E2 (dihydrolipoamide acetyl transferase) and E3 (dihydrolipoamide dehydrogenase) into acetyl-CoA. Genes encoding PDH subunits have also been found in the genome of *E. cuniculi*, but PDHC E2 and E3 are absent, which suggests that the role of PDH in microsporidia is unique. One possible explanation stems from the observation that PDH and PFOR actually share certain similarities in both structure (16) and biochemical activity; in particular, both use HETPP as an active intermediate (62). Accordingly, microsporidian PDH might be synthesizing HETPP to reduce the iron-sulphur center of a second protein, which then transfers electrons to ferredoxin, in effect mimicking the activity of PFOR but without actually using the enzyme (25). The catalog of genes present in the *Encephalitozoon* genome does indeed confirm that microsporidia use a completely novel form of core metabolism. Most importantly, the *Encephalitozoon* genome encodes genes for PDHC E1 (but not E2 or E3) as well as ferredoxin, ferredoxin:NADPH oxidoreductase, and a number of proteins involved in synthesizing iron-sulphur centers (44). Microsporidia also differ from other amitochondriates in that phosphofructokinase is related to the ATP-dependent enzyme of other fungi rather than being pyrophosphate dependent. A schematic representation of certain pathways inferred to be important in core energy metabolism in microsporidia is shown in Figure 4.

The presence of PDH-based pyruvate metabolism in *Nosema* and *Encephalitozoon* breaks one of the main "rules" of amitochondriate metabolism, namely the use of PFOR or some derivative of PFOR rather than PDH (65). Yet, this difference also epitomizes "amitochondriate" metabolism in that the core carbon metabolic machinery of these protists appears to have been cobbled together piece by piece, probably directed more by what enzymes happened to be available than by which ones might be brought together to make the most efficient system (65). This metabolism provides a fine example of evolution working as a "tinkerer" as proposed by Jacob (40, 41, 45).

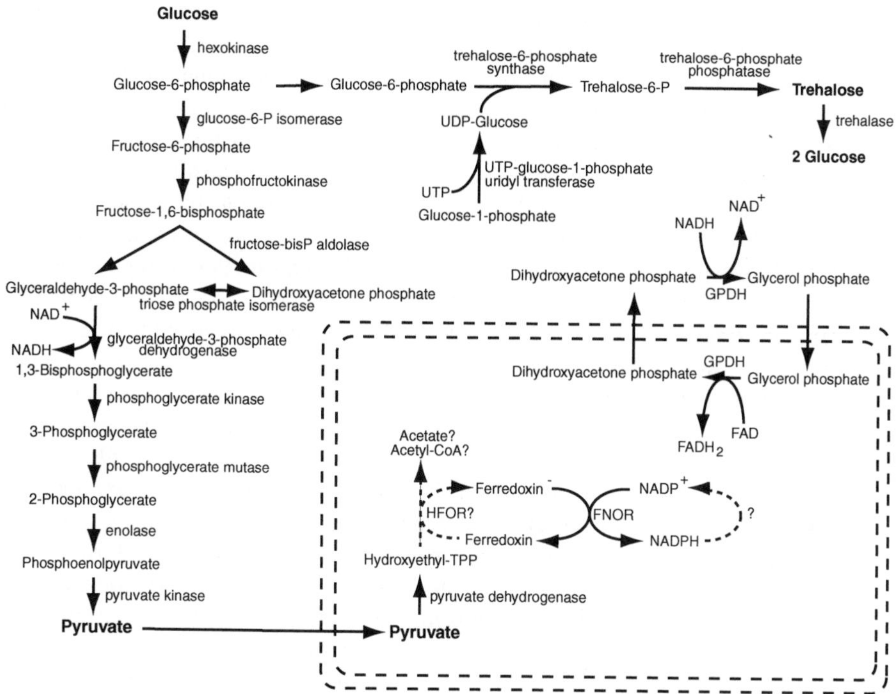

Figure 4 Schematic of microsporidian core carbon metabolism inferred from presence and absence of genes in microsporidia. Only certain pathways focusing on glucose metabolism are shown; others have been left out for clarity and can be seen in (44).

It is abundantly clear from the discovery that microsporidia are related to fungi, their possession of mitochondrion-derived HSP70 and PDH, and from a wealth of data from the *Encephalitozoon* genome that microsporidia evolved from a mitochondrion-containing ancestor. But none of these data actually tells us whether they still contain mitochondria today. The lack of mitochondria was originally proposed based on the absence of identifiable structures in ultrastructural investigations (91), but it has been shown repeatedly that organelles can easily go unnoticed. The apicomplexan plastid was only identified by its tell-tale genome (61, 104), and the mitochondrion of *Entamoeba* was only just identified by localizing one of the handful of proteins predicted to be derived from the mitochondrial endosymbiont; since it lacks a genome its identification is much more difficult (82). Conversely, the mitochondrion-derived Cpn60 protein from the diplomonad *Giardia* has also been localized, and it is apparently not contained in a membrane-bound organelle; so it appears to be possible that mitochondria can be completely lost while the host genome retains genes derived from the endosymbiont (72, 77). In the case of microsporidia, there are conflicting views on whether the mitochondrion may have been lost outright or persists today in a reduced and derived form.

The best evidence to date focuses on the potential amino-terminal leaders of mitochondrion-derived proteins to see if they are potentially mitochondrion-targeting transit peptides. The three initial reports of mitochondrial HSP70 genes all came to different conclusions regarding the amino termini of these genes: It was proposed that they were targeted to a mitochondrion, that they likely were not, or that they could be targeted to peroxisomes instead. A re-analysis of the HSP70 leaders, as well as those of PDH alpha and beta, came to the conclusion that they did not appear to have amino-terminal mitochondrion-targeting sequences (25). However, there is now credible evidence from a variety of proteins encoded in the *Encephalitozoon* genome that several proteins may contain mitochondrial-targeting transit peptides, leading to the proposition that a cryptic mitochondrion, a mitosome, would likely be found in microsporidia (44). If the metabolic profile of microsporidia inferred from the presence and absence of genes in core metabolic pathways is examined, one can also see that some reactions may necessitate the presence of a membrane-bound organelle. One example from the *Encephalitozoon* genome is the glycerol-3-phosphate shuttle, where dihydroxyacetone phosphate (DHAP) is reduced to glycerol-3-phosphate (oxidizing NADH in the process), which is transported to the mitochondrion where it is oxidized to re-form DHAP (reducing FAD in the process). *Encephalitozoon* contains both cytosolic and mitochondrial homologs of the key enzyme in this shuttle, glycerol phosphate dehydrogenase (GPDH) (44). The significance of this pathway in microsporidia is that the shuttle is present in order to move electrons across the mitochondrial envelope, so the presence of this pathway implies that microsporidia do indeed contain a membrane-bound relic mitochondrion and not just a few leftover enzymes. Conclusive evidence for this hypothesis will only come from localizing mitochondrion-derived enzymes to an organelle in microsporidia.

MICROSPORIDIA IN THE NEXT 150 YEARS

In 2001, microsporidian research was transformed by the completion of the *E. cuniculi* genome sequence (44). These data immediately provided insight and compelling answers to a number of questions about microsporidian biology that had previously been matters of hypothesis. Yet, we are really only beginning to scratch the surface of understanding just how these parasites have become so well adapted to their parasitic lifestyle. In terms of microsporidian genome size, *Encephalitozoon* is more of an exception, rather than the rule, having one of the most reduced genomes. Different lineages of microsporidia have reduced their genome sizes at different rates, doubtless losing and retaining pathways differentially. One question that arises from the *Encephalitozoon* genome is, "What makes up the difference between small and large microsporidian genomes?" Are larger genomes less compact, have they undergone less-severe reduction, or have their environments simply allowed for the loss of fewer genes? In addition to questions of biology, our current interpretations of microsporidian origins are still clouded by our lack of specific information about their ancestors. Although we are confident that microsporidia and fungi are related, the specifics of their relationship await further analysis. Once the origin of microsporidia has been unambiguously resolved, it may be possible to reconstruct how they might have evolved to be the highly specialized parasites we see today. This reconstruction process could provide important insights into microsporidian biology, and may also give us a glimpse of the process of adapting to parasitism.

ACKNOWLEDGMENTS

We thank Miklós Müller and William Martin for continuing advice and discussions on anaerobic metabolism, and Jirí Vávra for correspondence on microsporidian biology. We also thank members of the Keeling lab and Jirí Vávra for critical reading of the manuscript. Microsporidian research in the Keeling lab is supported by a New Investigator award in Pathogenic Mycology from the Burroughs-Wellcome Fund and a grant from the Canadian Institutes of Health Research (CIHR). P. J. Keeling is a Scholar of the Canadian Institute for Advanced Research (CIAR) and the Michael Smith Foundation for Health Research (MSFHR). N. M. Fast is supported by postdoctoral fellowships from MSFHR and CIHR.

The *Annual Review of Microbiology* is online at http://micro.annualreviews.org

LITERATURE CITED

1. Arguelles JC. 2000. Physiological roles of trehalose in bacteria and yeasts: a comparative analysis. *Arch. Microbiol.* 174:217–24
2. Balbiani G. 1882. Sur les microsporidies ou sporospermies des articules. *C. R. Acad. Sci.* 95:1168–71
3. Baldauf SL, Roger AJ, Wenk-Siefert I, Doolittle WF. 2000. A kingdom-level phylogeny of eukaryotes based on combined protein data. *Science* 290:972–77
4. Becnel JJ, Andreadis TG. 1999. Microsporidia of insects. See Ref. 105, pp. 447–501
5. Biderre C, Méténier G, Vivarès CP. 1998. A small spliceosomal-type intron occurs in a ribosomal protein gene of the microsporidian *Encephalitozoon cuniculi*. *Mol. Biochem. Parasitol.* 94:283–86
6. Biderre C, Pagès M, Méténier G, David D, Bata J, et al. 1994. On small genomes in eukaryotic organisms: molecular karyotypes of two microsporidian species (Protozoa) parasites of vertebrates. *C. R. Acad. Sci. III* 317:399–404
7. Bigliardi E, Selmi MG, Lupetti P, Corona S, Gatti S, et al. 1996. Microsporidian spore wall: ultrastructural findings on *Encephalitozoon hellem* exospore. *J. Eukaryot. Microbiol.* 43:181–86
8. Bouzat JL, McNeil LK, Robertson HM, Solter LF, Nixon JE, et al. 2000. Phylogenomic analysis of the alpha proteasome gene family from early-diverging eukaryotes. *J. Mol. Evol.* 51:532–43
9. Brown JR, Doolittle WF. 1995. Root of the universal tree of life based on ancient aminoacyl-tRNA synthetase gene duplications. *Proc. Natl. Acad. Sci. USA* 92:2441–45
10. Brown JR, Doolittle WF. 1999. Gene descent, duplication, and horizontal transfer in the evolution of glutamyl- and glutaminyl-tRNA synthetases. *J. Mol. Evol.* 49:485–95

11. Cali A, Takvorian PM. 1999. Developmental morphology and life cycles of the microsporidia. See Ref. 105, pp. 85–128
12. Canning EU, Lom J, Nicholas JP. 1982. Genus *Glugea* Thelohane, 1891 (Phylum Microspora): redescription of the type species *Glugea anomala* (Moniez, 1887) and recognition of its sporogonic development within sporophorous vesicles (pansporoblastic membranes). *Protistologica* 18:193–210
13. Cavalier-Smith T. 1983. A six-kingdom classification and a unified phylogeny. In *Endocytobiology II: Intracellular Space as Oligogenetic*, ed. HEA Schenk, WS Schwemmler, pp. 1027–34. Berlin: de Gruyter
14. Cavalier-Smith T. 1993. Kingdom protozoa and its 18 phyla. *Microbiol. Rev.* 57: 953–94
15. Cavalier-Smith T. 1998. A revised six-kingdom system of life. *Biol. Rev.* 73:203–66
16. Chabriere E, Charon MH, Volbeda A, Pieulle L, Hatchikian EC, Fontecilla-Camps JC. 1999. Crystal structures of the key anaerobic enzyme pyruvate:ferredoxin oxidoreductase, free and in complex with pyruvate. *Nat. Struct. Biol.* 6:182–90
17. Clark CG, Roger AJ. 1995. Direct evidence for secondary loss of mitochondria in *Entamoeba histolytica*. *Proc. Natl. Acad. Sci. USA* 92:6518–21
18. Curgy JJ, Vávra J, Vivarès CP. 1980. Presence of ribosomal RNAs with prokaryotic properties in Microsporidia, eukaryotic organisms. *Biol. Cell.* 38:49–51
19. de Graaf DC, Masschelein G, Vandergeynst F, De Brabander HF, Jacobs FJ. 1993. In vitro germination of *Nosema apis* (Microspora: Nosematidae) spores and its effect on their a-trehalose/d-glucose ratio. *J. Invert. Pathol.* 62:220–25
20. Desportes I, Le Charpentier Y, Galian A, Bernard F, Cochand-Priollet B, et al. 1985. Occurrence of a new microsporidan: *Enterocytozoon bieneusi* n. g., n. sp., in the enterocytes of a human patient with AIDS. *J. Protozool.* 32:250–54
21. DiMaria P, Palic B, Debrunner-Vossbrinck BA, Lapp J, Vossbrinck CR. 1996. Characterization of the highly divergent U2 RNA homolog in the microsporidian *Vairimorpha necatrix*. *Nucleic Acids Res.* 24:515–22
22. Dolgikh VV, Sokolova JJ, Issi IV. 1997. Activities of enzymes of carbohydrate and energy metabolism of the spores of the microsporidian, *Nosema grylli*. *J. Eukaryot. Microbiol.* 44:246–49
23. Douglas S, Zauner S, Fraunholz M, Beaton M, Penny S, et al. 2001. The highly reduced genome of an enslaved algal nucleus. *Nature* 410:1091–96
24. Edlind TD, Li J, Visvesvara GS, Vodkin MH, McLaughlin GL, Katiyar SK. 1996. Phylogenetic analysis of β-tubulin sequences from amitochondrial protozoa. *Mol. Phylogenet. Evol.* 5:359–67
25. Fast NM, Keeling PJ. 2001. Alpha and beta subunits of pyruvate dehydrogenase E1 from the microsporidian *Nosema locustae*: mitochondrion-derived carbon metabolism in microsporidia. *Mol. Biochem. Parasitol.* 117:201–9
26. Fast NM, Logsdon JM Jr, Doolittle WF. 1999. Phylogenetic analysis of the TATA box binding protein (TBP) gene from *Nosema locustae*: evidence for a microsporidia-fungi relationship and spliceosomal intron loss. *Mol. Biol. Evol.* 16:1415–19
27. Fast NM, Roger AJ, Richardson CA, Doolittle WF. 1998. U2 and U6 snRNA genes in the microsporidian *Nosema locustae*: evidence for a functional spliceosome. *Nucleic Acids Res.* 26:3202–7
28. Flegel TW, Pasharawipas T. 1995. A proposal for typical eukaryotic meiosis in microsporidians. *Can. J. Microbiol.* 41:1–11
29. Francois J, Parrou JL. 2001. Reserve carbohydrates metabolism in the yeast *Saccharomyces cerevisiae*. *FEMS Microbiol. Rev.* 25:125–45

30. Franzen C, Muller A. 2001. Microsporidiosis: human diseases and diagnosis. *Microbes Infect.* 3:389–400
31. Frixione E, Ruiz L, Cerbon J, Undeen AH. 1997. Germination of *Nosema algerae* (Microspora) spores: Conditional inhibition by D2O, ethanol and Hg2+ suggests dependence of water influx upon membrane hydration and specific transmembrane pathways. *J. Eukaryot. Microbiol.* 44:109–16
32. Frixione E, Ruiz L, Santillan M, de Vargas LV, Tejero JM, Undeen AH. 1992. Dynamics of polar filament discharge and sporoplasm expulsion by microsporidian spores. *Cell Motil. Cytoskelet.* 22:38–50
33. Germot A, Philippe H, Le Guyader H. 1997. Evidence for loss of mitochondria in Microsporidia from a mitochondrial-type HSP70 in *Nosema locustae*. *Mol. Biochem. Parasitol.* 87:159–68
34. Hazard EI, Brookbank JW. 1984. Karyogamy and meiosis in an *Ambylospora* sp. (Microspora) in the mosquito *Culex salinarius*. *J. Invert. Pathol.* 44:3–11
35. Hendrick RP, Groff JM, Baxa DV. 1991. Experimental infections with *Nucleospora salmonis* n. g., n. s.: an intranuclear microsporidium from chinook salmon (*Oncorhynchus tshawitscha*). *Am. Fish. Soc. Newsl.* 19:5
36. Hinkle G, Morrison HG, Sogin ML. 1997. Genes coding for reverse transcriptase, DNA-directed RNA polymerase, and chitin synthase from the microsporidian *Spraguea lophii*. *Biol. Bull.* 193:250–51
37. Hirt RP, Healy B, Vossbrinck CR, Canning EU, Embley TM. 1997. A mitochondrial Hsp70 orthologue in *Vairimorpha necatrix*: molecular evidence that microsporidia once contained mitochondria. *Curr. Biol.* 7:995–98
38. Hirt RP, Logsdon JM Jr, Healy B, Dorey MW, Doolittle WF, Embley TM. 1999. Microsporidia are related to Fungi: evidence from the largest subunit of RNA polymerase II and other proteins. *Proc. Natl. Acad. Sci. USA* 96:580–85
39. Ishihara R, Hayashi Y. 1968. Some properties of ribosomes from the sporoplasm of *Nosema bombycis*. *J. Invert. Pathol.* 11: 377–85
40. Jacob F. 1977. Evolution and tinkering. *Science* 196:1161–66
41. Jacob F. 2001. Complexity and tinkering. *Ann. NY Acad. Sci.* 929:71–73
42. Kamaishi T, Hashimoto T, Nakamura Y, Masuda Y, Nakamura F, et al. 1996. Complete nucleotide sequences of the genes encoding translation elongation factors 1 alpha and 2 from a microsporidian parasite, *Glugea plecoglossi*: implications for the deepest branching of eukaryotes. *J. Biochem.* 120:1095–103
43. Kamaishi T, Hashimoto T, Nakamura Y, Nakamura F, Murata S, et al. 1996. Protein phylogeny of translation elongation factor EF-1α suggests Microsporidians are extremely ancient eukaryotes. *J. Mol. Evol.* 42:257–63
44. Katinka MD, Duprat S, Cornillot E, Méténier G, Thomarat F, et al. 2001. Genome sequence and gene compaction of the eukaryote parasite *Encephalitozoon cuniculi*. *Nature* 414:450–53
45. Keeling PJ. 2001. Parasites go the Full Monty. *Nature* 414:401–2
46. Keeling PJ, Doolittle WF. 1996. Alpha-tubulin from early-diverging eukaryotic lineages and the evolution of the tubulin family. *Mol. Biol. Evol.* 13:1297–305
47. Keeling PJ, Fast NM, McFadden GI. 1998. Evolutionary relationship between translation initiation factor eIF-2gamma and selenocysteine-specific elongation factor SELB: change of function in translation factors. *J. Mol. Evol.* 47:649–55
48. Keeling PJ, Luker MA, Palmer JD. 2000. Evidence from beta-tubulin phylogeny that microsporidia evolved from within the fungi. *Mol. Biol. Evol.* 17:23–31
49. Keeling PJ, McFadden GI. 1998. Origins of microsporidia. *Trends Microbiol.* 6:19–23

50. Keohane EM, Weiss LM. 1998. Characterization and function of the microsporidian polar tube: a review. *Folia Parasitol.* 45:117–27
51. Keohane EM, Weiss LM. 1999. The structure, function, and composition of the microsporidian polar tube. See Ref. 105, pp. 196–224
52. Kudo RR. 1918. Experiments on the extrusion of polar filaments of cnidosporidian spores. *J. Parasitol.* 4:141–47
53. Kudo RR. 1947. *Protozoology*. Springfield, IL: Thomas. 778 pp.
54. Larsson JIR. 1983. A revisionary study of the taxon *Tuzetia* Maurand, Fize, Fenqick and Michel, 1971, and related forms (Microspora, Tuzetiidae). *Protistologica* 19:323–55
55. Larsson JIR, Ebert D, Vávra J, Voronin VN. 1996. Redescription of *Pleistophora intestinalis* Chatton, 1907, a microsporidian parasite of *Daphnia magna* and *Daphnia pulex*, with establishment of a new genus *Glugoides* (Microspora, Glugeigae). *Eur. J. Protistol.* 32:251–61
56. Levine ND, Corliss JO, Cox FE, Deroux G, Grain J, et al. 1980. A newly revised classification of the protozoa. *J. Protozool.* 27:37–58
57. Liu TP. 1972. Ultrastructural changes in the nuclear envelope of larval fat body cells of *Simulium vittatum* (Diptera) induced by microsporidian infection of *Thelohaniabracteata*. *Tissue Cell* 4:493–502
58. Lom J, Vávra J. 1962. A proposal to the classification within the subphylum Cnidospora. *Syst. Zool.* 11:172–75
59. Lom J, Vávra J. 1963. The mode of sporoplasm extrusion in microsporidian spores. *Acta Protozool.* 1:81–92
60. Matsubayashi H, Koike T, Mikata I, Takei H, Hagiwara S. 1959. A case of *Encephalitozoon*-like body infection in man. *Arch. Pathol.* 67:181–87
61. McFadden GI, Reith ME, Munholland J, Lang-Unnasch N. 1996. Plastid in human parasites. *Nature* 381:482
62. Menon S, Ragsdale SW. 1997. Mechanism of the *Clostridium thermoaceticum* pyruvate:ferredoxin oxidoreductase: evidence for the common catalytic intermediacy of the hydroxyethylthiamine pyropyrosphate radical. *Biochemistry* 36:8484–94
63. Méténier G, Vivarès CP. 2001. Molecular characteristics and physiology of microsporidia. *Microbes Infect.* 3:407–15
64. Müller M. 1993. The hydrogenosome. *J. Gen. Microbiol.* 139:2879–89
65. Müller M. 1998. Enzymes and compartmentataion of core energy metabolism of anaerobic eukaryotes—a special case in eukaryotic evolution? In *Evolutionary Relationships Among Protozoa*, ed. GH Coombs, K Vickerman, MA Sleigh, A Warren, pp. 109–32. London: Chapman & Hall
66. Nägeli K. 1857. Über die neue Krankheit der Seidenraupe und verwandte Organismen. *Bot. Ztg.* 15:760–61
67. Oshima K. 1937. On the function of the polar filament in *Nosema bombycis*. *Parasitology* 29:220–24
68. Patterson DJ. 1994. Protozoa: evolution and systematics. In *Progress in Protozoology*, ed. K Hausmann, N Hülsmann, pp. 1–14. Stuttgart: Fischer-Verlag
69. Peyret P, Katinka MD, Duprat S, Duffieux F, Barbe V, et al. 2001. Sequence and analysis of chromosome I of the amitochondriate intracellular parasite *Encephalitozoon cuniculi* (Microspora). *Genome Res.* 11:198–207
70. Peyretaillade E, Biderre C, Peyret P, Duffieux F, Méténier G, et al. 1998. Microsporidian *Encephalitozoon cuniculi*, a unicellular eukaryote with an unusual chromosomal dispersion of ribosomal genes and a LSU rRNA reduced to the universal core. *Nucleic Acids Res.* 26:3513–20
71. Peyretaillade E, Broussolle V, Peyret P, Méténier G, Gouy M, Vivarès CP. 1998. Microsporidia, amitochondrial protists, possess a 70-kDa heat shock protein gene

of mitochondrial evolutionary origin. *Mol. Biol. Evol.* 15:683–89

72. Roger AJ, Svard SG, Tovar J, Clark CG, Smith MW, et al. 1998. A mitochondrial-like chaperonin 60 gene in *Giardia lamblia*: evidence that diplomonads once harbored an endosymbiont related to the progenitor of mitochondria. *Proc. Natl. Acad. Sci. USA* 95:229–34
73. Rotte C, Stejskal F, Zhu G, Keithly JS, Martin W. 2001. Pyruvate: NADP+ oxidoreductase from the mitochondrion of *Euglena gracilis* and from the apicomplexan *Cryptosporidium parvum*: a biochemical relic linking pyruvate metabolism in mitochondriate and amitochondriate protists. *Mol. Biol. Evol.* 18:710–20
74. Sandfort J, Hannemann A, Gelderblom H, Stark K, Owen RL, Ruf B. 1994. *Enterocytozoon bieneusi* infection in an immunocompetent patient who had acute diarrhea and who was not infected with the human immunodeficiency virus. *Clin. Infect. Dis.* 19:514–16
75. Shaw RW, Kent ML. 1999. Fish microsporidia. See Ref. 105, pp. 418–46
76. Solter LF, Maddox JV. 1998. Timing of an early sporulation sequence of microsporidia in the genus *Vairimorpha* (Microsporidia: Burenellidae). *J. Invertebr. Pathol.* 72:323–29
77. Soltys BJ, Gupta RS. 1994. Presence and cellular distribution of a 60-kDa protein related to mitochondrial Hsp60 in *Giardia lamblia*. *J. Protistol.* 80:580–88
78. Sprague V, Becnel JJ. 1999. Checklist of available generic names for microsporidia with type species and type hosts. See Ref. 105, pp. 517–30
79. Sprague V, Becnel JJ, Hazard EI. 1992. Taxonomy of phylum microspora. *Crit. Rev. Microbiol.* 18:285–395
80. Takvorian PM, Cali A. 1994. Enzyme histochemical identification of the Golgi apparatus in the microsporidian, *Glugea stephani*. *J. Eukaryot. Microbiol.* 41:63S–64S
81. Takvorian PM, Cali A. 1996. Polar tube formation and nucleoside diphosphatase activity in the microsporidian, *Glugea stephani*. *J. Eukaryot. Microbiol.* 43: 102S–3S
82. Tovar J, Fischer A, Clark CG. 1999. The mitosome, a novel organelle related to mitochondria in the amitochondrial parasite *Entamoeba histolytica*. *Mol. Microbiol.* 32:1013–21
83. Trager W. 1974. Some aspects of intracellular parasitism. *Science* 183:269–73
84. Undeen AH, Elgazzar LM, Vander Meer RK, Narang S. 1987. Trehalose levels and trehalase activity in germinated and ungerminated spores of *Nosema algerae* (Microspora: Nosematidae). *J. Invertebr. Pathol.* 50:230–37
85. Undeen AH, Epsky ND. 1990. *In vitro* and *in vivo* germination of *Nosema locustae* (Microsporidia: Nosematidae) spores. *J. Invertebr. Pathol.* 56:371–79
86. Undeen AH, Frixione E. 1990. The role of osmotic pressure in the germination of *Nosema algerae* spores. *J. Protozool.* 37: 561–67
87. Undeen AH, Frixione E. 1991. Structural alteration of the plasma membrane in spores of the microsporidium *Nosema algerae* on germination. *J. Protozool.* 38:511–18
88. Undeen AH, Vander Meer RK. 1994. Conversion of intrasporal trehalose into reducing sugars during germination of *Nosema algerae* (Protista: Microspora) spores: a quantitative study. *J. Eukaryot. Microbiol.* 41:129–32
89. Van de Peer Y, Ben Ali A, Meyer A. 2000. Microsporidia: accumulating molecular evidence that a group of amitochondriate and suspectedly primitive eukaryotes are just curious fungi. *Gene* 246:1–8
90. Van der Meer JW, Gochnauer TA. 1971. Trehalose activity associated with spores of *Nosema apis*. *J. Invertebr. Pathol.* 17: 38–41
91. Vávra J. 1965. Étude au microscope électronique de la morphologie et du

développement de quelques microsporidies. *C. R. Acad. Sci.* 261:3467–70

92. Vávra J, Larsson JIR. 1999. Structure of the microsporidia. See Ref. 105, pp. 7–84
93. Vávra J, Vinckier D, Torpier G, Porchet E, Vivier E. 1986. A freeze-fracture study of the microsporidia (Protozoa: Microsporidia). I. The sporophorous vesicle, the spore wall, the spore plasma membrane. *Protistologica* 22:143–54
94. Vinckier D, Porchet E, Vivier E, Vávra J, Torpier G. 1993. A freeze-fracture study of the microsporidia (Protozoa: Microspora). II. The extrusion apparatus: polar filament polaroplast, posterior vacuole. *Eur. J. Biochem.* 29:370–80
95. Vivier E. 1975. The microsporidia of the protozoa. *Protistology* 11:345–61
96. Vossbrinck CR, Maddox JV, Friedman S, Debrunner-Vossbrinck BA, Woese CR. 1987. Ribosomal RNA sequence suggests microsporidia are extremely ancient eukaryotes. *Nature* 326:411–14
97. Vossbrinck CR, Woese CR. 1986. Eukaryotic ribosomes that lack a 5.8S RNA. *Nature* 320:287–88
98. Weber R, Bryan RT, Schwartz DA, Owen RL. 1994. Human microsporidial infections. *Clin. Microbiol. Rev.* 7:426–61
99. Weber R, Schwartz DA, Deplazes P. 1999. Laboratory diagnosis of microsporidiosis. See Ref. 105, pp. 315–62
100. Weidner E. 1972. Ultrastructural study of microsporidian invasion into cells. *Z. Parasitenkd.* 40:227–42
101. Weidner E, Byrd W, Scarbourough A, Pleshinger J, Sibley D. 1984. Microsporidian spore discharge and the transfer of polaroplast organelle into plasma membrane. *J. Protozool.* 31:195–98
102. Weidner E, Findley AM, Dolgikh V, Sokolova J. 1999. Microsporidian biochemistry and physiology. See Ref. 105, pp. 172–95
103. Weissenberg R. 1976. Microsporidian interactions with host cells. In *Comparative Pathobiology, Vol. 1. Biology of the Microsporidia*, ed. LA Bulla, TC Cheng, pp. 203–38. New York: Plenum
104. Wilson RJ, Denny PW, Preiser PR, Rangachari K, Roberts K, et al. 1996. Complete gene map of the plastid-like DNA of the malaria parasite *Plasmodium falciparum*. *J. Mol. Biol.* 261:155–72
105. Wittner M, Weiss LM. 1999. *The Microsporidia and Microsporidiosis*. Washington, DC: ASM. 553 pp.
106. Wright JH, Craighead EM. 1922. Infectious motor paralysis in young rabbits. *J. Exp. Med.* 36:135–49

Annu. Rev. Microbiol. 2002. 56:117–37
doi: 10.1146/annurev.micro.56.012302.161024

First published online as a Review in Advance on April 8, 2002

BACTERIOCINS: Evolution, Ecology, and Application

Margaret A. Riley and John E. Wertz
Department of Ecology and Evolutionary Biology, Yale University, New Haven, Connecticut 06511; e-mail: margaret.riley@yale.edu; john.wertz@yale.edu

Key Words colicin, resistance, diversity, nisin

■ **Abstract** Microbes produce an extraordinary array of microbial defense systems. These include classical antibiotics, metabolic by-products, lytic agents, numerous types of protein exotoxins, and bacteriocins. The abundance and diversity of this potent arsenal of weapons are clear. Less clear are their evolutionary origins and the role they play in mediating microbial interactions. The goal of this review is to explore what we know about the evolution and ecology of the most abundant and diverse family of microbial defense systems: the bacteriocins. We summarize current knowledge of how such extraordinary protein diversity arose and is maintained in microbial populations and what role these toxins play in mediating microbial population-level and community-level dynamics. In the latter half of this review we focus on the potential role bacteriocins may play in addressing human health concerns and the current role they serve in food preservation.

CONTENTS

INTRODUCTION

What Are Microbial Defense Systems?

Microbes produce an extraordinary array of microbial defense systems. These include broad-spectrum classical antibiotics so critical to human health concerns, metabolic by-products such as the lactic acids produced by lactobacilli, lytic agents such as lysozymes found in many foods, numerous types of protein exotoxins, and bacteriocins, which are loosely defined as biologically active protein moieties with a bacteriocidal mode of action (41, 90). This biological arsenal is striking not only in its diversity but in its natural abundance. For instance lactic acid production is a defining trait of lactic acid bacteria (36). Bacteriocins are found in almost every bacterial species examined to date, and within a species tens or even hundreds of different kinds of bacteriocins are produced (41, 72). Halobacteria universally produce their own version of bacteriocins, the halocins (95). Streptomycetes commonly produce broad-spectrum antibiotics (79). It is clear that microbes invest considerable energy into the production and elaboration of antimicrobial mechanisms. Less clear is how such diversity arose and what roles these biological weapons serve in microbial communities.

One family of microbial defense systems, the bacteriocins, has served as a model for exploring evolutionary and ecological questions. In this review, current knowledge of how the extraordinary range of bacteriocin diversity arose and is maintained in microbial populations is assessed, and the role these toxins play in mediating microbial dynamics is discussed. Fascination with bacteriocins is not restricted to the evolutionary and ecologically minded; in the latter half of this review our attention focuses on the potential application of these toxins to address human health concerns and the current and growing use of bacteriocins to aid in food preservation.

BACTERIOCINS: THE MICROBIAL WEAPON OF CHOICE

Bacteriocins differ from traditional antibiotics in one critical way: They have a relatively narrow killing spectrum and are only toxic to bacteria closely related to the producing strain. These toxins have been found in all major lineages of Bacteria and, more recently, have been described as universally produced by some members of the Archaea (95). According to Klaenhammer, 99% of all bacteria may make at least one bacteriocin and the only reason we haven't isolated more is that few researchers have looked for them (47, 67).

Bacteriocins of Gram-Negative Bacteria

The bacteriocin family includes a diversity of proteins in terms of size, microbial targets, modes of action, and immunity mechanisms. The most extensively studied, the colicins produced by *Escherichia coli*, share certain key characteristics (3, 6, 13, 30, 40, 48, 64). Colicin gene clusters are encoded on plasmids and are composed of a colicin gene, which encodes the toxin; an immunity gene, which encodes a protein conferring specific immunity to the producer cell by binding to and inactivating the toxin protein; and a lysis gene, which encodes a protein involved in colicin release through lysis of the producer cell. Colicin production is mediated by the SOS regulon and is therefore principally produced under times of stress. Toxin production is lethal for the producing cell and any neighboring cells recognized by that colicin. A receptor domain in the colicin protein that binds a specific cell surface receptor determines target recognition. This mode of targeting results in the relatively narrow phylogenetic killing range often cited for bacteriocins. The killing functions range from pore formation in the cell membrane to nuclease activity against DNA, rRNA, and tRNA targets. Colicins, indeed all bacteriocins produced by gram-negative bacteria, are large proteins. Pore-forming colicins range in size from 449 to 629 amino acids. Nuclease bacteriocins have an even broader size range, from 178 to 777 amino acids.

Although colicins are representative of gram-negative bacteriocins, there are intriguing differences found within this subgroup of the bacteriocin family. *E. coli* encodes its colicins exclusively on plasmid replicons (65). The nuclease pyocins of *Pseudomonas aeruginosa*, which show sequence similarity to colicins and other, as yet uncharacterized, bacteriocins are found exclusively on the chromosome (81). Another close relative to the colicin family, the bacteriocins of *Serratia marcesens*, are found on both plasmids and chromosomes (20, 26, 32).

Many bacteriocins isolated from gram-negative bacteria appear to have been created by recombination between existing bacteriocins (6, 51, 68, 76). Such frequent recombination is facilitated by the domain structure of bacteriocin proteins. In colicins, the central domain comprises about 50% of the protein and is involved in the recognition of specific cell surface receptors. The N-terminal domain (~25% of the protein) is responsible for translocation of the protein into the target cell. The remainder of the protein houses the killing domain and the immunity region, which is a short sequence involved in immunity protein binding. Although the pyocins produced by *P. aeruginosa* share a similar domain structure, the order of the translocation and receptor recognition domains are switched (80). As we explore further below, the conserved domain configuration of these toxins is responsible for much of the bacteriocin diversity we find in nature.

Bacteriocins of Gram-Positive Bacteria

Bacteriocins of gram-positive bacteria are as abundant and even more diverse as those found in gram-negative bacteria (39, 90). They differ from gram-negative bacteriocins in two fundamental ways. First, bacteriocin production is not necessarily the lethal event it is for gram-negative bacteria. This critical difference is due to the transport mechanisms gram-positive bacteria encode to release bacteriocin toxin. Some have evolved a bacteriocin-specific transport system, whereas others employ the *sec*-dependent export pathway. In addition, the gram-positive bacteria have evolved bacteriocin-specific regulation, whereas bacteriocins of gram-negative bacteria rely solely on host regulatory networks.

The lactic acid bacteria (LAB) are particularly prolific in bacteriocin production. Klaenhammer distinguishes three classes of LAB bacteriocins (47). Class I bacteriocins are the lantibiotics, so named because they are post-translationally modified to contain amino acids such as lanthionine and B-methyllanthionine, and several dehydrated amino acids (33). Lantibiotics are further divided into two subgroups, A and B, based on structural features and their mode of killing (43). Type A lantibiotics kill the target cell by depolarizing the cytoplasmic membrane (2, 84). They are larger than type B lantibiotics and range in size from 21 to 38 amino acids. Nisin, the archetypal and best-studied gram-positive bacteriocin, is a type A lantibiotic (31). The type B lantibiotics have a more globular secondary structure and are smaller than type A, with none exceeding 19 amino acids in length. Type B lantibiotics function through enzyme inhibition. One example is mersacidin, which interferes with cell wall biosynthesis (7).

Class II LAB bacteriocins are also small, ranging in size from 30 to 60 amino acids, and are heat-stable, nonlanthionine-containing peptides (43). They are organized into subgroups: Class IIa is the largest group and its members are distinguished by a conserved amino-terminal sequence (YGNGVXaaC) and a shared activity against *Listeria*. Like type A lantibiotics, class IIa bacteriocins act through the formation of pores in the cytoplasmic membrane. Examples include pediocin AcH (4), sakacin A (83), and leucocin A (35). Class IIb bacteriocins such as lacticin F (58) and lactococcin G (60) form pores composed of two different proteins in the membrane of their target cells. A third subgroup (IIc) has been proposed, which consists of bacteriocins that are *sec*-dependent (such as acidocin B) (52). Class III bacteriocins are large heat-labile proteins such as helveticins J and V (42, 98) and lactacin B (1). An additional proposed class (VI) requires lipid or carbohydrate moieties for activity. Little is known about the structure and function of this proposed class. Examples include leuconocin S (8) and lactocin 27 (96).

Gram-positive bacteriocins in general and lantibiotics in particular require many more genes for their production than do gram-negative bacteriocins. The nisin gene cluster includes genes for the prepeptide (*nisA*), enzymes for modifying amino acids (*nisB*, *nisC*), cleavage of the leader peptide (*nisP*), secretion (*nisT*), immunity (*nisI*, *nisFEG*), and regulation of expression (*nisR*, *nisK*) (9, 21, 22, 44, 50, 66, 97). These gene clusters are most often encoded on plasmids but are occasionally found on the chromosome. Several gram-positive bacteriocins, including nisin, are located on transposons (18).

The conventional wisdom about the killing range of gram-positive bacteriocins is that they are restricted to killing other gram-positive bacteria. The range of killing can vary significantly, from relatively narrow as in the case of lactococcins A, B, and M, which have been found to kill only *Lactococcus* (77), to extraordinarily broad. For instance, some type A lantibiotics such as nisin A and mutacin B-Ny266 have been shown to kill a wide range of organisms including *Actinomyces*, *Bacillus*, *Clostridium*, *Corynebacterium*, *Enterococcus*, *Gardnerella*, *Lactococcus*, *Listeria*, *Micrococcus*, *Mycobacterium*, *Propionibacterium*, *Streptococcus*, and

Staphylococcus (57). Contrary to conventional wisdom, these particular bacteriocins are also active against a number of medically important gram-negative bacteria including *Campylobacter*, *Haemophilus*, *Helicobacter*, and *Neisseria* (57).

Production of bacteriocins in gram-positive bacteria is generally associated with the shift from log phase to stationary phase. Nisin production begins during mid-log phase and increases to a maximum as the cells enter stationary phase (9). The regulation of expression is not cell cycle–dependent per se, but rather culture density–dependent. It has been demonstrated that nisin A acts as a protein pheromone in regulating its own expression, which is controlled by a two-component signal transduction system typical of many quorum-sensing systems. The genes involved are *nisR* (the response regulator) and *nisK* (the sensor kinase) (16). Nisin transcription can be induced by the addition of nisin to the culture medium with the level of induction directly related to the level of nisin added (49, 50).

Bacteriocins of Archaea

The Archaea produce their own distinct family of bacteriocin-like antimicrobials, known as archaeocins. The only characterized member is the halocin family produced by halobacteria, and few halocins have been described in detail (11, 75, 85). The first halocin discovered, S8, is a short hydrophobic peptide of 36 amino acids, which is processed from a much larger pro-protein of 34 kD (63). Halocin S8 is encoded on a megaplasmid and is extremely hardy; it can be desalted, boiled, subjected to organic solvents, and stored at 4°C for extended periods without losing activity. Expression is growth stage–dependent. Although basal levels are present in low concentrations during exponential growth, there is an explosive ninefold increase in production during the transition to stationary phase (85). The mechanism of halocin action has been established only for halocin H6 (a Na+/H+ antiporter inhibitor), and the immunity mechanism is unknown (94).

Archaeocins are produced as the cells enter stationary phase. When resources are limited, producing cells lyse sensitive cells and enrich the nutrient content of the local environment. As stable proteins, they may remain in the environment long enough to reduce competition during subsequent phases of nutrient flux. The stability of halocins may help explain why there is so little species diversity in the hypersaline environments frequented by halobacteria (85).

As is clear from this brief survey of bacteriocin diversity and distribution, this heterogeneous family of toxins is united only by the shared features of being protein-based toxins that are relatively narrow in killing spectrum and often extremely hardy and stable. What makes these the weapons of choice in the microbial world remains an intriguing question.

EVOLUTION OF BACTERIOCIN DIVERSITY

Colicins as a Model for Evolutionary Studies

The colicins and other enteric bacteriocins, such as klebicins, remain the only bacteriocins for which detailed evolutionary investigations have been undertaken.

Among the colicins, there are two main evolutionary lineages, which also distinguish the two primary modes of killing: pore formation and nuclease activity (70). Studies that include DNA and protein sequence comparisons (6, 68), surveys of DNA sequence polymorphism in natural isolates (62, 74, 92), experimental evolution (28, 91), and mathematical modeling (28) have revealed two primary modes of colicin evolution (93).

The Role of Diversifying Recombination in Colicin Evolution

The more abundant pore-former colicins are generated by domain shuffling, which is mediated by recombination (6, 93). All characterized pore-former colicin proteins share one or more regions with high levels of sequence similarity to other pore-former colicins (Figure 1). This patchwork of shared and divergent sequences suggests frequent recombination. The location of the different patches frequently corresponds to the different functional domains of the proteins. The most recent illustration of the power of diversifying recombination is seen in the first published klebicin sequence (Figure 2), which is a nuclease klebicin that shares sequence similarity with both colicin A–like pore former and pyocin S1–like nuclease sequences (73). Such domain-based shuffling between bacteriocins is responsible for much of the variability observed among gram-negative bacteriocins.

The influence of diversifying recombination is not limited to the closely related bacteriocins of enteric bacteria. As mentioned above, the S pyocins of *P. aeruginosa* are the result of recombination between several pore-former and nuclease colicins with other, as yet uncharacterized, bacteriocins (81, 82). Even altering the domain structure of the protein, as seen for pyocins that have switched the receptor

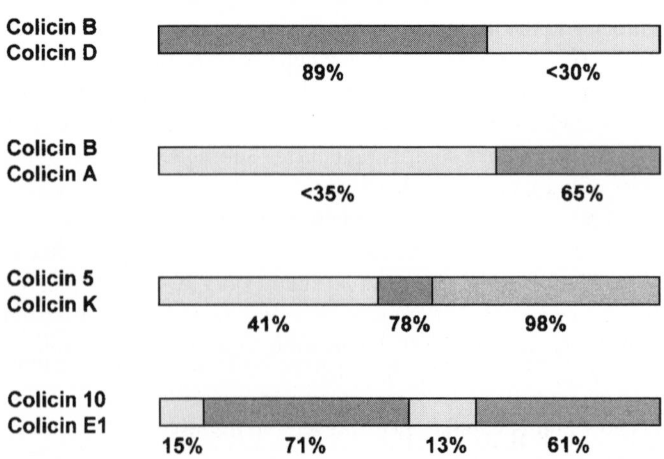

Figure 1 Pairwise comparisons of pore-forming colicin protein sequences. Values below each comparison indicate the percent sequence identity for the region indicated. Colicin proteins are not drawn to scale.

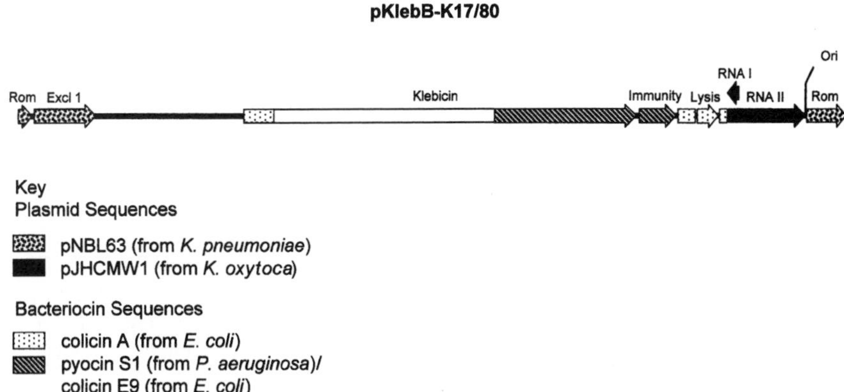

Figure 2 Patterns of sequence similarity in klebicins suggest recombination. The chimeric nature of the pKlebB plasmid sequence is indicated by alternate shadings. The key notes regions of sequence similarity with other bacteriocin gene clusters and plasmids. pKlebB illustrates a pattern typical of other bacteriocin-encoding plasmids where sequences encoding plasmid functions are similar to sequences found in other plasmids segregating in the host species, whereas those sequences composing and flanking the bacteriocin gene cluster show similarity to bacteriocin sequences from other species.

recognition and translocation domains relative to the order found in colicins, has not limited the influence of diversifying recombination.

The Role of Diversifying Selection in Colicin Evolution

An alternative mode of evolution is responsible for the current diversity of nuclease colicins. These colicins, which include both RNase-and DNase-killing functions, share a recent common ancestry. Their DNA sequences are quite similar, ranging from 50% to 97% sequence identity. However, many pairs of nuclease colicins have elevated levels of divergence in the immunity region (Figure 3). To explain this pattern of divergence, Riley and collaborators have proposed a two-step process of mutation and selection (68, 69, 93).

The diversifying selection hypothesis posits the action of strong positive selection acting on mutations that generate novel immunity and killing functions (Figure 4). The first event in this process is the occurrence of a mutation in the immunity gene resulting in a broadened immunity function. The resulting producer cell is now immune to the ancestral version of the colicin as well as having gained immunity to some number of similar colicins. This broadened immunity function increases the fitness of the producer strain in populations where multiple colicins are found, which is the case in all *E. coli* populations sampled to date (29, 72). A second mutation, this time in the colicin gene, is paired with the immunity mutation.

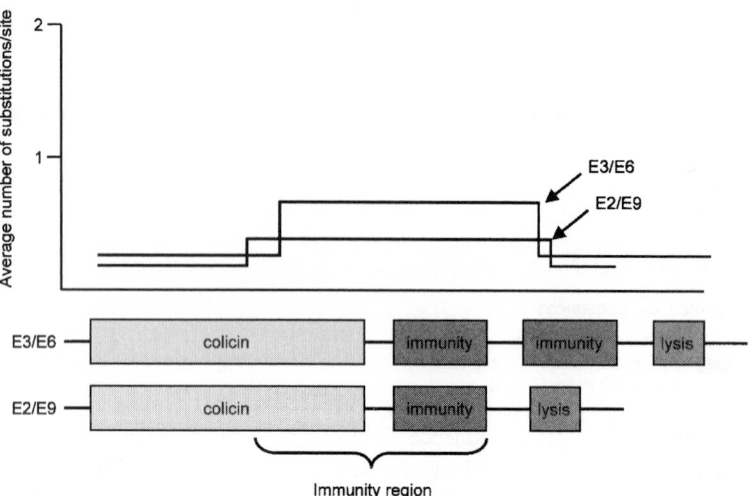

Figure 3 The graph indicates the average number of total nucleotide substitutions between pairs of nuclease-type colicin gene clusters (colicin pairs E2/E9 and E3/E6). Most of the divergence between colicins occurs in the immunity region of the gene cluster (composed of the immunity gene and the immunity-binding region of the colicin gene).

This pair of mutations produces a novel colicin that is no longer recognized by the ancestral immunity protein. Thus, the possessor of the novel colicin will rapidly displace (by killing) the ancestral, formerly abundant bacteriocin-producing strain. This evolved colicin will ultimately be replaced by yet another novel colicin as the cycle repeats itself. This process results in a family of closely related proteins that have diverged most extensively in the region involved in immunity binding and killing function, as seen for nuclease colicins (69).

Recently, the DNA sequence of a new pore-former colicin, Y, was determined (71). Colicin Y is a close relative of colicin U, another pore-former colicin isolated from a different continent over 20 years earlier (86). This pair of colicins has a pattern of DNA substitution identical to that observed among the nuclease colicins with an elevated level of substitution in the immunity region. This observation suggests that the process of diversifying selection is not restricted to nuclease colicins. Further, several E2 colicins isolated from Australia suggest that diversifying recombination is not restricted to pore-former colicins (92). Half of the E2 producers carry the characterized E2 plasmid. The other half carry a recombinant plasmid with sequences derived from colicin E7 and the characterized E2 plasmid. These isolated observations suggest that it is not the case that pore formers diversify only by means of recombination and nuclease colicins by diversifying selection. The evolutionary process is more complex than the proposed simple dichotomy suggests.

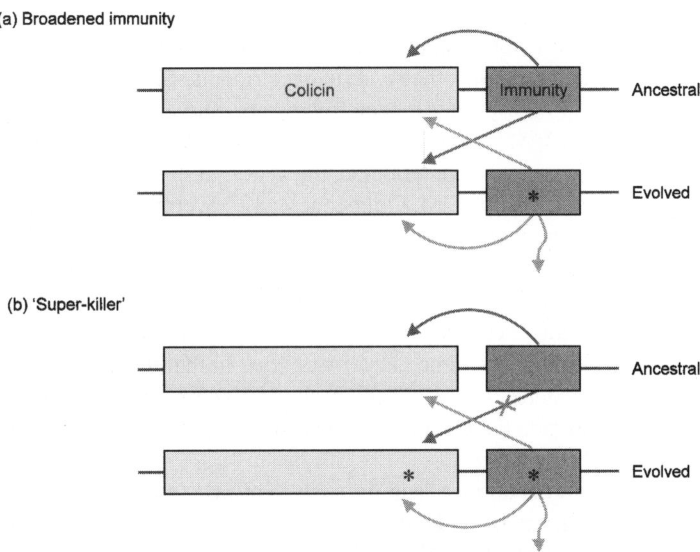

Figure 4 The hypothesis of diversifying selection invokes two steps in the generation of a novel immunity function. (*a*) A point mutation in the immunity gene generates a broadened immunity function (noted with an *asterisk*). The strain with this colicin gene cluster is immune to itself, to its ancestor, and to other closely related colicins (noted with *gray arrows*). The ancestral colicin is immune to itself and to the evolved colicin (noted with *black arrows*). (*b*) A paired mutation occurs in the immunity-binding portion of the evolved colicin gene that generates a "super-killer" (noted with a second *asterisk*). The evolved strain is still immune to itself, its ancestor, and other colicins. However, the ancestral strain is now no longer immune to the evolved strain (noted with an *X*).

A Two-Step Process of Colicin Evolution

Riley has developed a model of colicin diversification that involves two phases (70). When rare, as is currently the case for most nuclease colicins, the occurrence of point mutations that alter immunity function may be the primary mode for generating novel bacteriocin phenotypes. Novel immunity and killing functions are rapidly selected since they allow a cell to avoid being killed by other bacteriocins or allow cells carrying them to displace their ancestors. These novel bacteriocins are then maintained until a new immunity or killing function emerges. When colicins are abundant, as is the case for many pore-former colicins, domain swapping may become a more frequent mode of diversification. This "switch" in evolutionary mechanism is due simply to the requirement for a set of bacteriocins to be abundant enough to serve as templates for recombination. Once abundant, recombination can more rapidly generate additional diversity.

We have only just begun to tap into the diversity of enteric bacteriocins. However, recent work suggests that similar evolutionary mechanisms may play a role in the diversification of other enteric bacteriocins. Sequence comparisons reveal that in several cases, enteric bacteriocins are chimeras of known gram-negative bacteriocins (73; M.A. Riley, C.M. Goldstone & J.E. Wertz, unpublished information.) For other enteric bacteriocins, the action of diversifying selection has been proposed (M.A. Riley, C.M. Goldstone & J.E. Wertz, unpublished information). Finally, some new enteric bacteriocins have no similarity with those characterized previously. A particularly interesting example of this latter observation is the recently described Colicin Js (87). This plasmid-borne bacteriocin has a typical colicin gene cluster composition, with toxin, immunity, and lysis genes. However, the organization of the gene cluster is unique in that the lysis gene is transcribed 5′ to the toxin gene. The genes themselves show no similarity to any known bacteriocin genes, and the encoded toxin is 94 amino acids, which is smaller than any other described colicin.

Bacteriocin-encoding plasmids, such as pColJs (which encodes colicin Js) and pKlebB (which encodes klebicin B), demonstrate another aspect of bacteriocin evolution (73, 87). These bacteriocin plasmids are chimeras with a plasmid "backbone" comprising replication and maintenance sequences typical of plasmids found in the bacteriocins' host species. In the case of pKlebB isolated from *Klebsiella pneumoniae*, the plasmid contains sequences similar to pNBL63 (102) and pJHC-MW1 (17), isolated from *K. oxytoca* and *K. pneumoniae* respectively, encoding plasmid maintenance functions. The sequence surrounding and comprising part of the klebicin B gene cluster shares similarity with colicin A and E9, originally isolated from *E. coli* (73). In the case of pColJs, the plasmid backbone is virtually identical to ColE1, whereas the DNA flanking the colicin Js gene cluster shows high similarity to pPCP1 from *Yersinia pestis* (37). The colicin Js gene cluster itself has a significantly lower G + C content (33.6%) than the rest of the plasmid (52.9%), indicating that it originated from yet a third source (87), perhaps even outside of the *Enterobacteriaceae*. This type of recombination, although not altering the bacteriocin genes proper, results in an increased host range. As we continue to explore bacteriocin diversity, our model of bacteriocin evolution will almost certainly become more elaborate and complex.

ECOLOGICAL ROLE OF BACTERIOCINS

Without question, bacteriocins serve some function in microbial communities. This statement follows from the detection of bacteriocin production in all surveyed lineages of prokaryotes. Equally compelling is the inference of strong positive selection acting on enteric bacteriocins. Such observations argue that these toxins play a critical role in mediating microbial population or community interactions. What remains in question is what, precisely, that role is.

Bacteriocins may serve as anti-competitors enabling the invasion of a strain into an established microbial community. They may also play a defensive role and act to prohibit the invasion of other strains or species into an occupied niche or limit the advance of neighboring cells. An additional role has recently been proposed for gram-positive bacteriocins, in which they mediate quorum sensing (55). It is likely that whatever roles bacteriocins play, these roles change as components of the environment, both biotic and abiotic, change.

Theoretical and Experimental Studies of Bacteriocin Ecology

Early experimental studies on the ecological role of bacteriocins were inconclusive and contradictory (14, 24, 27, 34, 38, 45, 101). More recently a theoretical and empirical base has been established that has defined the conditions that favor maintenance of toxin-producing bacteria in both population and community settings. Almost exclusively, these studies have modeled the action of colicins. Chao & Levin showed that the conditions for invasion of a colicin-producer strain were much broader in a spatially structured environment than in an unstructured one (10). In an unstructured environment with mass-action, a small population of producers cannot invade an established population of sensitive cells. This failure occurs because the producers pay a price for toxin production—the energetic costs of plasmid carriage and lethality of production—but the benefits, the resources made available by killing sensitive organisms, are distributed at random. Moreover, when producers are rare, the reduction in growth rate experienced by the sensitive strain (owing to extra deaths) is smaller than the reduction felt by the producer (owing to its costs), and the producer population therefore becomes extinct. In a physically structured environment, such as on the surface of an agar plate, the strains grow as separate colonies. Toxin diffuses out from a colony of producers, thus killing sensitive neighbors. The resources made available accrue disproportionately to the producing colony owing to its proximity, and therefore killers can increase in frequency even when initially rare.

The Rock-Paper-Scissors Model

Recent modeling efforts have incorporated additional biological reality. Two such efforts introduced a third species, one that is resistant to the toxin but cannot itself produce the toxin (15, 46). Resistance can be conferred through mutations in either the binding site or the translocation machinery required for a bacteriocin to enter the target cell. Acquisition of an immunity gene will also confer resistance to its cognate bacteriocin. The authors in both studies reasonably assume there is a cost to resistance and that this cost is less than the cost of toxin production borne by the killer strain (25). Owing to this third member, pairwise interactions among the strains have the nontransitive structure of the childhood game of rock-paper-scissors (Table 1) (53). The producer strain beats the sensitive strain, owing to the toxin's effects on the latter. The sensitive strain beats the resistant strain because only the latter suffers the cost of resistance. And the resistant strain wins against the producer because the latter bears the higher cost of toxin production and release while the former pays only the cost of resistance. In an unstructured environment,

TABLE 1 Chemical warfare among microbes as a non-transitive, three-way game similar to the "rock-paper-scissors" game

Strain below	Wins against	Loses against
Killer	Sensitive	Resistant
Sensitive	Resistant	Killer
Resistant	Killer	Sensitive

this game allows periodic cycles, in which all three types coexist indefinitely but each with fluctuating abundance. In a structured environment, this game permits a quasi-stable global equilibrium, one in which all three strains can persist with nearly constant global abundance (15).

Further effects of evolution were incorporated into the Czárán et al. model by allowing as many as 14 distinct systems of toxin production, sensitivity, and resistance, along with the genetic processes of mutation and recombination that can alter these traits and their associations (15). The permutations of these systems permit the existence of several million different strains. From this additional complexity emerges two distinct quasi-equilibrium conditions, the "frozen" and "hyper-immunity" states. In the frozen state, all the toxins are maintained globally, but most colonies are single-toxin producers. That is, each colony produces one toxin to which it is also immune. By contrast, in the hyper-immunity state, many colonies produce no toxin, many others make one, still others produce several toxins, but only a few produce most of the available toxins. Resistance shows a different distribution, with all of the colonies being resistant to most or all of the toxins. Which of these two outcomes is obtained depends upon initial conditions. If the evolving system begins with the entire population sensitive to all toxins, then the frozen state results. The hyper-immunity state is reached if the system starts with enough diversity that most colonies already have multiple killer and resistance traits.

Numerous surveys of colicin production in natural populations suggest that populations of *E. coli* may closely match predictions of the Czárán model (29, 72). In *E. coli*, producer strains are found in frequencies ranging from 10% to 50%. Resistant strains are even more abundant and are found at frequencies from 50% to 98%. In fact, most strains are resistant to all co-segregating colicins. Finally, there is a small population of sensitive cells. Figure 5 provides a summary of phenotype distributions in a population of *E. coli* isolated from wild field mice in Australia (29). The Czárán model predicts this distribution of phenotypes results from frequent horizontal transfer of resistance, and the significant cost to colicin production (15). In other words, if a strain can gain resistance and lose production, it will over time—just as was observed in the *E. coli* isolated from the field mouse population over the course of a summer (29).

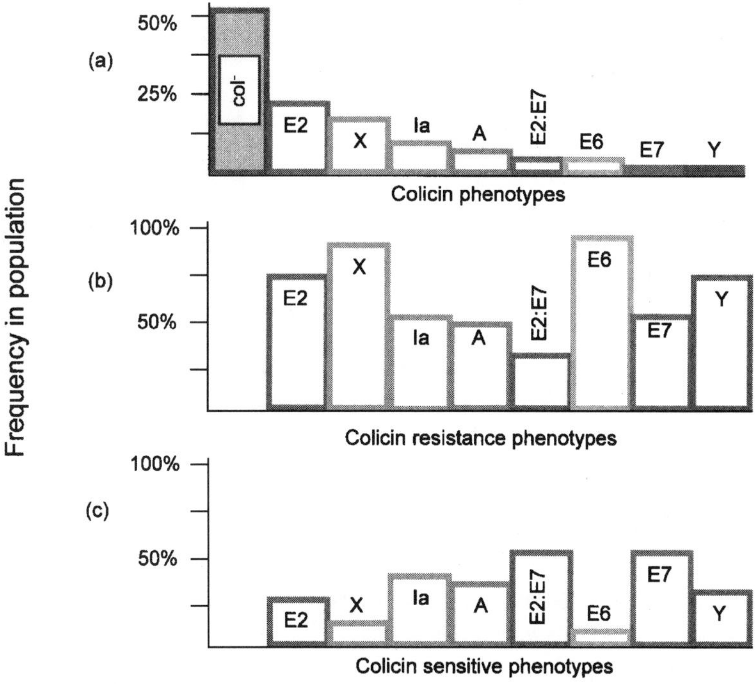

Figure 5 A survey of colicin production and resistance in *E. coli*. Over 400 strains were isolated from two populations of feral mice in Australia over a period of seven months. The isolates were scored for colicin production and resistance. (*a*) Colicin production is abundant with just under 50% of the strains producing eight distinct colicin types. Col^- represents nonproducer strains. (*b*) The majority of isolates are resistant to most co-occurring colicins. (*c*) A small proportion of the population is sensitive to co-occurring colicins.

The Killing Breadth of Bacteriocins

We assume bacteriocins play a role in mediating within-species (or population-level) dynamics. This assumption is based upon the narrow killing range exhibited by most bacteriocins. However, recent work calls this assumption into question. Bacteriocins from natural isolates of several species of enteric bacteria were assayed for their killing effect against a large set of nonproducers isolated from the same sources (M.A. Riley, C.M. Goldstone, J.E. Wertz & D. Gordon, unpublished information). Figure 6 reveals that contrary to expectations killing breadth varies significantly for different bacteriocins. Some are clearly most effective at killing within the producer strains own species. Others kill more broadly or kill quite specifically isolates of a different species. This diversity of killing breadth argues that bacteriocins, contrary to prior suggestions, play an equally compelling role in mediating both population-level and community-level interactions. A more

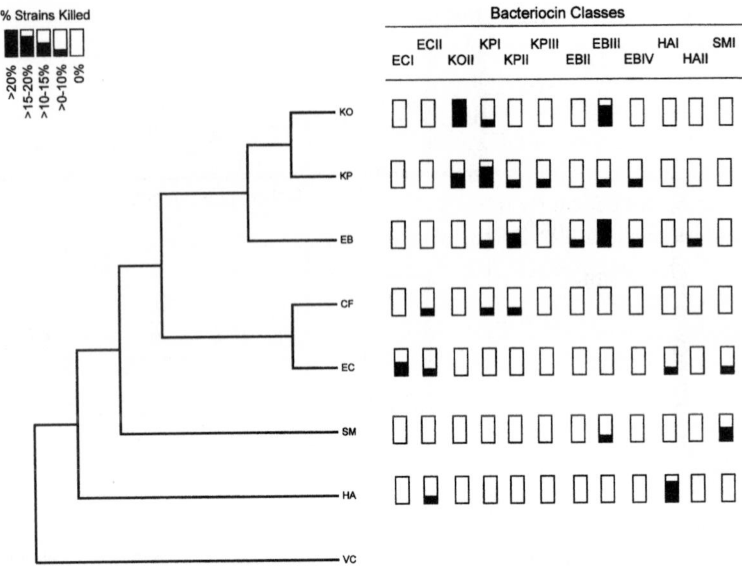

Figure 6 Phylogenetic breadth of bacteriocin killing. The killing spectrum of each class of bacteriocins was cross-referenced with a phylogenetic tree of the enteric species they were screened against. Heights of the black boxes are proportional to the percentage of strains sensitive to each class of bacteriocin. Bacteriocins were screened against 40 natural isolates from each enteric species. The molecular phylogeny of a subset of enteric bacteria is based on a composite of five housekeeping genes (*gapA*, *groEL*, *gyrA*, *ompA*, *pgi*) and 16s ribosomal sequences. The tree is rooted using *Vibrio cholerae* as an outgroup. KO, *Klebsiella oxytoca*; KP, *Klebsiella pneumoniae*; EB, *Enterobacter cloacae*; CF, *Citrobacter freundii*; EC, *E. coli*; SM, *Serratia marcescens*; HA, *Hafnia alvei*; VC, *Vibrio cholerae*.

thorough understanding of how bacteriocins function awaits the development of a more biologically realistic experimental approach. Prior studies have considered how producer, sensitive, and resistant strains within the same species interact. If the goal is to understand the role these toxins play in nature, our experiments must incorporate more complex microbial communities and environments.

BACTERIOCIN APPLICATIONS

Bacteriocins and Human Health

The rapid rise and spread of multi-resistant bacterial pathogens have forced the consideration of alternative methods of combating infection (59, 78). One of the limitations of using broad-spectrum antibiotics is that they kill almost any bacterial species not specifically resistant to the drug. Given such a broad killing spectrum, these antibiotics are used frequently, which results in an intensive selection pressure for the evolution of antibiotic resistance in both pathogen and commensal bacteria (100). Once resistance appears, it is simply a matter of time and the intensity of human-mediated selection before human pathogens will acquire resistance (54).

Current solutions to this dilemma involve developing a more rationale approach to antibiotic use, which involves curtailing the prescription of drugs for anything other than bacterial infections, cycling through different drugs over a shorter time frame, and educating the public about the necessity of taking an entire course of antibiotics (54, 89). Bacteriocins provide an alternative solution. With their relatively narrow spectrum of killing activity, they can be considered "designer drugs," which target specific bacterial pathogens. Given the diversity of bacteriocins produced in nature, it is a relatively simple task to find bacteriocins active against specific human pathogens (R.L. Dorit & M.A. Riley, unpublished information). The development and use of such narrow-spectrum antimicrobials not only increases the number of drugs on the pharmaceutical shelf but, more importantly, extends their shelf life. This latter feature emerges because with a designer drug approach, each antibiotic is used infrequently, which results in a reduction in the intensity of selection for resistance. From an ecological and evolutionary perspective, the use of narrow-spectrum antimicrobials to address the current threat posed by multi-resistance bacterial pathogens makes quite a bit of sense. It leads to a reduction in the collateral killing of nonpathogen species, i.e., commensal species, which in turn leads to a decrease in nosocomial infection levels. It also results in a reduction in the intensity of selection for antibiotic resistance. With so few species of bacteria killed by each designer drug, antibiotic resistance resulting from antibiotic use will evolve and spread more slowly.

Bacteriocins and Food Preservation

The only bacteriocins currently employed in food preservation are those produced by LAB used in the production of fermented foods (56). Because LAB have been used for centuries to ferment foods, they enjoy GRAS (generally regarded as safe) status by the U.S. Food and Drug Administration (FDA). This permits their use in fermented foods without additional regulatory approval (56).

Nisin was the first bacteriocin to be isolated and approved for use in foods, specifically to prevent the outgrowth of *Clostridium botulinum* spores in cheese spreads in England (12). By 1988, the FDA had approved its use as a biopreservative for a narrow range of foods, including pasteurized egg products. Today, nisin is accepted as a safe food preservative by over 45 countries, and it is the most widely used commercial bacteriocin and it remains the only bacteriocin that may be added to U.S. foods.

Over the past decade the recurrence of listeriosis outbreaks, combined with the natural resistance of the causative agent, *Listeria monocytogenes*, to traditional food preservation methods such as its ability to grow at near-freezing temperatures has focused the attention of bacteriocin researchers on this organism (61). This attention has resulted in the isolation of a large number of class IIa bacteriocins, all of which are highly active against *L. monocytogenes* [recently reviewed in (23)].

The next wave of development of bacteriocins as food preservatives is at hand. Bacteriocins have been discovered in cured meats, milk and cheese, spoiled salad dressing, and soybean paste. Luchansky and colleagues have developed a gelatin form of pediocin, a class IIa bacteriocin made by lactic acid–producing bacteria, that protects hot dogs from *Listeria* contamination (67). His team has also added a strain of pediocin-producing bacteria to sausage and found a reduction of *Listeria* numbers to be fewer than one ten-thousandth the original number in untreated sausage. Equally compelling, active pediocin was found in the sausage after two months of refrigeration. At the University of Melbourne in Australia, Barrie Davidson has been targeting *Listeria* with piscicolin, a bacteriocin from yet another lactic acid–producing bacterium (67). Piscicolin has already been patented and it will soon be ready for use in meat products and as a rinse for salad greens or chicken parts (67).

A natural concern about using bacteriocins for the preservation of food is the selection of resistant strains. Studies in LAB have shown that resistance carries a significant fitness cost, with resistant strains having a slower growth rate than their sensitive ancestor (19). Treatment with a combination of bacteriocins, for instance nisin and a class IIa bacteriocin, would theoretically reduce the incidence of resistance (5, 99). There is currently conflicting evidence as to whether resistance to one class of LAB bacteriocins can result in cross-resistance to another class (5, 88).

CONCLUSIONS AND FUTURE DIRECTIONS

Bacteriocins represent one of the best-studied microbial defense systems. Although we are still in the earliest stages of exploring their evolutionary relationships and ecological roles, it is clear from their abundance and diversity that they are the microbial weapons of choice. Sorting out why they are such a successful family of toxins will require a substantial commitment to future research. Answering this question will require a substantial effort to more fully characterize the diversity of bacteriocin proteins, their modes of targeting and killing, the gene clusters that encode them, and the mechanisms of bacteriocin gene regulation. In addition, we require more sophisticated ecological models (both empirical and theoretical) to aid in our growing sense of the diverse roles the toxins play in mediating microbial dynamics and maintaining microbial diversity. The impact of such studies is not solely academic. The potential for bacteriocins to serve as alternatives to classical antibiotics in treating bacterial infections is real, and the application of bacteriocins in food preservation is exploding. The future roles bacteriocins may serve is limited only by our imagination.

ACKNOWLEDGMENTS

We thank Carla Goldstone for her help in preparing this review and acknowledge financial support from the NIH GM 58433 and the Rockefeller Foundation.

The *Annual Review of Microbiology* is online at http://micro.annualreviews.org

LITERATURE CITED

1. Barefoot SF, Klaenhammer TR. 1984. Purification and characterization of the *Lactobacillus acidophilus* bacteriocin lactacin-B. *Antimicrob. Agents Chemother.* 26:328–34
2. Belkum MJ, Hayema BJ, Geis A, Kok J, Venema G. 1989. Cloning of two bacteriocin genes from a lactococcal bacteriocin plasmid. *Appl. Environ. Microbiol.* 55: 1187–91
3. Benedetti H, Geli V. 1996. Colicin transport, channel formation and inhibition. *Hand. Biol. Phys.* 2:665–91
4. Bhunia AK, Johnson MC, Ray B. 1987. Direct detection of an antimicrobial peptide of *Pediococcus acidilactici* in sodium dodecyl sulfate-polyacrylamide gel-electrophoresis. *J. Ind. Microbiol.* 2:319–22
5. Bouttefroy A, Milliere J. 2000. Nisin-curvaticin 13 combinations for avoiding the regrowth of bacteriocin resistant cells of *Listeria monocytogenes* ATCC 15313. *Int. J. Food Microbiol.* 62:65–75
6. Braun V, Pilsl H, Groß P. 1994. Colicins: structures, modes of actions, transfer through membranes, and evolution. *Arch. Microbiol.* 161:199–206
7. Brotz H, Bierbaum G, Markus A, Molitor E, Sahl HG. 1995. Mode of action of the lantibiotic mersacidin—inhibition of peptidoglycan biosynthesis via a novel mechanism. *Antimicrob. Agents Chemother.* 39: 714–19
8. Bruno MEC, Montville TJ. 1993. Common mechanistic action of bacteriocins from lactic-acid bacteria. *Appl. Environ. Microbiol.* 59:3003–10
9. Buchman G, Banerjee S, Hansen J. 1988. Structure, expression and evolution of gene encoding the precursor of nisin, a small protein antibiotic. *J. Biol. Chem.* 263:16260–66
10. Chao L, Levin BR. 1981. Structured habitats and the evolution of anticompetitor toxins in bacteria. *Proc. Natl. Acad. Sci. USA* 78:6324–28
11. Cheung J, Danna K, O'Connor E, Price L, Shand R. 1997. Isolation, sequence, and expression of the gene encoding halocin H4, a bacteriocin from the halophilic archaeon *Haloferax mediterranei* R4. *J. Bacteriol.* 179:548–51
12. Chung K, Dickson J, Crouse J. 1989. Effects of nisin on growth of bacteria attached to meat. *Appl. Environ. Microbiol.* 55:1329–33
13. Cramer WA, Heymann JB, Schendel SL, Deriy BN, Cohen FS, et al. 1995. Structure-function of the channel-forming colicins. *Annu. Rev. Biophys. Biomol. Struct.* 24:611–41
14. Craven JA, Miniats OP, Barnum DA. 1971. Role of colicins in antagonism between strains of *Escherichia coli* in dual-infected gnotobiotic pigs. *Am. J. Vet. Res.* 32:1775–9
15. Czárán TL, Hoekstra RF, Pagie L. 2002. Chemical warfare between microbes promotes biodiversity. *Proc. Natl. Acad. Sci. USA* 99:786–90
16. de Ruyter PGGA, Kuipers OP, deVos WM. 1996. Controlled gene expression systems for *Lactococcus lactis* with the food-grade inducer nisin. *Appl. Environ. Microbiol.* 62:3662–67
17. Dery KJ, Chavideh R, Waters V, Chamorro R, Tolmasky L, Tolmasky M. 1997. Characterization of the replication and mobilization regions of the multiresistance *Klebsiella pneumoniae* plasmid pJHCMW1. *Plasmid* 38:97–105
18. Dodd HM, Horn N, Gasson MJ. 1990. Analysis of the genetic determinant for production of the peptide antibiotic nisin. *J. Gen. Microbiol.* 136:555–66
19. Dykes GA, Hastings JW. 1998. Fitness costs associated with class IIa bacteriocin

resistance in *Listeria monocytogenes* b73. *Lett. Appl. Microbiol.* 26:5–8

20. Enfedaque J, Ferrar S, Guasch JF, Tomas J, Regue M. 1996. Bacteriocin 28b from *Serratia marcescens* N28b: identification of *Escherichia coli* surface components involved in bacterocin binding and translocation. *Can. J. Microbiol.* 42:19–26
21. Engelke G, Gutowski-Eckel Z, Hammelmann M, Entian KD. 1992. Biosynthesis of the lantibiotic nisin: genomic organization and membrane localization of the nisb protein. *Appl. Environ. Microbiol.* 58: 3730–43
22. Engelke G, Gutowskieckel Z, Kiesau P, Siegers K, Hammelman M, Entian KD. 1994. Regulation of nisin biosynthesis and immunity in lactocoocus-lactis 6F3. *Appl. Environ. Microbiol.* 60:814–25
23. Ennahar S, Sashihara T, Sonomoto K, Ishizaki A. 2000. Class IIa bacteriocins: biosynthesis, structure and activity. *FEMS Microbiol. Rev.* 24:85–106
24. Feldgarden M, Golden S, Wilson H, Riley MA. 1995. Can phage defense maintain colicin plasmids in *Escherichia coli*? *J. Microbiol.* 141:2977–84
25. Feldgarden M, Riley MA. 1998. The phenotypic and fitness effects of colicin resistance in *Escherichia coli* K-12. *Evolution* 53:1019–27
26. Ferrer S, Viejo MB, Guasch JF, Enfedaque J, Regue M. 1996. Genetic evidence for an activator required for induction of colicin-like bacteriocin 28b production in *Serratia marcescens* by DNA-damaging agents. *Am. Soc. Microbiol.* 178:951–60
27. Freter R. 1983. Mechanisms that control the microflora in the large intestine. In *Human Intestinal Microflora in Health and Disease*. New York: Academic
28. Gordon D, Riley MA. 1999. A theoretical and empirical investigation of the invasion dynamics of colicinogeny. *Microbiology* 145:655–61
29. Gordon D, Riley MA, Pinou T. 1998. Temporal changes in the frequency of colicinogeny in *Escherichia coli* from house mice. *Microbiology* 144:2233–40
30. Gouaux E. 1997. The long and short of colicin action: the molecular basis for the biological activity of channel-forming colicins. *Structure* 5:313–77
31. Gross E, Morell JL. 1971. Structure of nisin. *J. Am. Chem. Soc.* 93:4634–35
32. Guasch J, Enfedaque J, Ferrer S, Gargallo D, Regue M. 1995. Bacteriocin 28b, a chromosomally encoded bacteriocin produced by most *Serratia marcesens* biotypes. *Res. Microbiol.* 146:477–83
33. Guder A, Wiedemann I, Sahl HG. 2000. Posttranslationally modified bacteriocins—the lantibiotics. *Biopolymers* 55: 62–73
34. Hardy KG. 1975. Colicinogeny and related phenomena. *Bacteriol. Rev.* 39:464–515
35. Hastings JW, Sailer M, Johnson K, Roy KL, Vederas JC, Stiles ME. 1991. Characterization of leucocin A-UAL 187 and cloning of the bacteriocin gene from *Leuconostoc gelidum. J. Bacteriol.* 172: 7491–500
36. Holt JG. 1994. *Bergey's Manual of Determinative Bacteriology*. Baltimore, MD: Williams & Wilkins
37. Hu P, Elliot J, McCready P, Skowronski E, Games J, et al. 1998. Structural organization of virulence-associated plasmids of *Yersinia pestis. J. Bacteriol.* 180:5192–202
38. Ikari NS, Kenta DM, Young VM. 1969. Interaction in the germ-free mouse intestine of colicinogenic and colicin-sensitive microorganisms. *Proc. Soc. Exp. Med.* 130:1280–84
39. Jack RW, Tagg JR, Ray B. 1995. Bacteriocins of gram-positive bacteria. *Microbiol. Rev.* 59:171–200
40. James R, Kleanthous C, Moore GR. 1996. The biology of E colicins: paradigms and paradoxes. *Microbiology* 142:1569–80
41. James R, Lazdunski C, Pattus F, eds. 1991. *Bacteriocins, Microcins and Lantibiotics*,

Vol. 65. New York: Springer-Verlag. 519 pp.

42. Joerger MC, Klaenhammer T. 1986. Characterization and purification of helveticin-J and evidence for a chromosomally determined bacteriocin produced by *Lactobacillus helveticus*-481. *J. Bacteriol.* 167:439–46
43. Jung G, Sahl HG. 1991. Lantibiotics: a survey. In *Nisin and Novel Lantibiotics*, pp. 1–34. Leiden, The Netherlands: ESCOM
44. Kaletta C, Entian K-D. 1989. Nisin, a peptide antibiotic: cloning and sequencing of the *nis* A gene and posttranslational processing of its peptide product. *J. Bacteriol.* 171:1597–601
45. Kelstrup J, Gibbons RJ. 1969. Inactivation of bacteriocins in the intestinal and oral cavity. *J. Bacteriol.* 99:888–90
46. Kerr B, Riley MA, Feldman MW, Bohannan BJM. 2002. Local dispersal promotes biodiversity in a real-life game of rock-paper-scissors. *Nature* 418:171–74
47. Klaenhammer TR. 1988. Bacteriocins of lactic acid bacteria. *Biochimie* 70:337–49
48. Konisky J. 1982. Colicins and other bacteriocins with established modes of action. *Annu. Rev. Microbiol.* 36:125–44
49. Kuipers OP, Beerthuyzen MM, Deruyter PGGA, Luesink EJ, de Vos WM. 1995. Autoregulation of nisin biosynthesis in *Lactococcus lactis* by signal-transduction. *J. Biol. Chem.* 270:27299–304
50. Kuipers OP, Beerthuyzen MM, Siezen RJ, Devos WM. 1993. Characterization of the nisin gene-cluster *nisABTCIPR* of *Lactococcus lactis*—requirement of expression of the *nisA* and *nisI* genes for development of immunity. *Eur. J. Biochem.* 216:281–91
51. Lau PCK, Parsons M, Uchimura T. 1992. Molecular evolution of E colicin plasmids with emphasis on the endonuclease types. See Ref. 41, 65:353–78
52. Leer RJ, Vandervossen J, Vangiezen M, Vannoort JM, Pouwels PH. 1995. Genetic-analysis of acidocin-B, a novel bacteriocin produced by *Lactobacillus acidophilus*. *Microbiology* 141:1629–35
53. Lenski RR, Riley MA. 2002. Chemical warfare from an ecological perspective. *Proc. Natl. Acad. Sci. USA* 99:556–58
54. Lipsitch M, Bergstrom CT, Levin BR. 2000. The epidemiology of antibiotic resistance in hospitals: paradoxes and prescriptions. *Proc. Natl. Acad. Sci. USA* 97: 1938–43
55. Miller M, Bassler B. 2001. Quorum sensing in bacteria. *Annu. Rev. Microbiol.* 55: 165–99
56. Montville TJ, Winkowski K. 1997. Biologically based preservation systems and probiotic bacteria. In *Food Microbiology: Fundamentals and Frontiers*. Washington, DC: ASM
57. Mota-Meira M, Lapointe G, Lacroix C, Lavoie MC. 2000. MICs of mutacin B-Ny266, nisin A, vancomycin, and oxacillin against bacterial pathogens. *Antimicrob. Agents Chemother.* 44:24–29
58. Muriana PM, Klaenhammer TR. 1991. Cloning, phenotypic expression, and DNA sequence of the gene for lactacin F, an antimicrobial peptide produced by *Lactobacillus* spp. *J. Bacteriol.* 173:1779–88
59. Neu HC. 1992. The crisis in antibiotic resistance. *Science* 257:1064–73
60. Nissenmeyer JH, Havarstein LS, Sletten K, Nes IF. 1992. A novel lactococcal bacteriocin whose activity depends on the complementary action of 2 peptides. *J. Bacteriol.* 174:5686–92
61. Palumbo S. 1986. Is refrigeration enough to restrain foodborne pathogens. *J. Food Prot.* 49:1003–9
62. Pinou T, Riley M. 2001. Nucleotide polymorphism in microcin V plasmids. *Plasmid* 46:1–9
63. Price L, Shand R. 2000. Halocin S8: a 36-amino acid microhalocin from the haloarchaeal strain S8a. *J. Bacteriol.* 182:4951–58
64. Pugsley A. 1984. The ins and outs of colicins. *Microbiol. Sci.* 1:168–75, 203–5
65. Pugsley AP, Oudega B. 1987. Methods for

studying colicins and their plasmids. In *Plasmids, a Practical Approach*, ed. KG Hardy, pp. 105–61. Oxford: IRL

66. Ra R, Beerthuyzen M, de Vos WM, Saris PEJ, Kuipers OP. 1999. Effects of gene disruptions in the nisin gene cluster of *Lactococcus lactis* on nisin production and producer immunity. *Microbiology* 145:1227–33
67. Raloff J. 1998. Staging germ warfare in foods. *Sci. News* 153:89–90
68. Riley M. 1993. Molecular mechanisms of colicin evolution. *Mol. Biol. Evol.* 10: 1380–95
69. Riley MA. 1993. Positive selection for colicin diversity in bacteria. *Mol. Biol. Evol.* 10:1048–59
70. Riley MA. 1998. Molecular mechanisms of bacteriocin evolution. *Annu. Rev. Genet.* 32:255–78
71. Riley M, Cadavid L, Collett M, Neely M, Adams M, et al. 2000. The newly characterized colicin Y provides evidence of positive selection in pore-former colicin diversification. *Microbiology* 146:1671–77
72. Riley MA, Gordon DM. 1992. A survey of col plasmids in natural isolates of *Escherichia coli* and an investigation into the stability of col-plasmid lineages. *J. Microbiol.* 138:1345–52
73. Riley MA, Pinou T, Wertz JE, Tan Y, Valletta CM. 2001. Molecular characterization of the klebicin B plasmid of *Klebsiella pneumoniae*. *Plasmid* 45:209–21
74. Riley MA, Tan Y, Wang J. 1994. Nucleotide polymorphism in colicin E1 and Ia plasmids from natural isolates of *Escherichia coli*. *Proc. Natl. Acad. Sci. USA* 91:11276–80
75. Rodriguez-Valera F, Juez G, Kushner DJ. 1982. Halocins: salt dependent bacteriocins produced by extremely halophilic rods. *Can. J. Microbiol.* 28:151–54
76. Roos U, Harkness RE, Braun V. 1989. Assembly of colicin genes from a few DNA fragments. Nucleotide sequence of colicin D. *Mol. Microbiol.* 3:891–902
77. Ross RP, Galvin M, McAuliffe O, Morgan SM, Ryan MP, et al. 1999. Developing applications for lactococcal bacteriocins. *Antonie Van Leeuwenhoek Int. J. Gen. Mol. Microbiol.* 76:337–46
78. Roy PH. 1997. Dissemination of antibiotic resistance. *Med. Sci.* 13:927–33
79. Saadoun I, al-Momani F, Malkawi HI, Mohammad MJ. 1999. Isolation, identification and analysis of antibacterial activity of soil streptomycetes isolates from north Jordan. *Microbios* 100:41–46
80. Sano Y, Kobayashi M, Kageyama M. 1993. Functional domains of S-type pyocins deduced from chimeric molecules. *J. Bacteriol.* 175:6179–85
81. Sano Y, Matsui H, Kobayashi M, Kageyama M. 1990. Pyocins S1 and S2, bacteriocins of *Pseudomonas aeruginosa*. In *Pseudomonas: Biotransformations, Pathogenesis, and Evolving Biotechnology*, ed. S Silver, pp. 352–58. Washington, DC: ASM
82. Sano Y, Matsui H, Kobayashi M, Kageyama M. 1993. Molecular structures and functions of pyocins S1 and S2 in *Pseudomonas aeruginosa*. *J. Bacteriol.* 175: 2907–16
83. Schillinger U, Lucke F-K. 1989. Antibacterial activity of *Lactobacillus sake* isolated from meat. *Appl. Environ. Microbiol.* 55:1901–6
84. Schuller F, Benz R, Sahl HG. 1989. The peptide antibiotic subtilin acts by formation of voltage-dependent multi-state pores in bacterial and artificial membranes. *Eur. J. Biochem.* 182:181–86
85. Shand R, Price L, O'Connor EM. 1998. Halocins: protein antibiotics from hypersaline environments. In *Microbiology and Biogeochemistry of Hypersaline Environments*, ed. A Oren, Chapter 24, pp. 295–306. Boca Raton, FL: CRC
86. Smajs D, Pilsl H, Braun V. 1997. Colicin U, a novel colicin produced by *Shigella boydii*. *J. Bacteriol.* 179:4919–28
87. Smajs D, Weinstock G. 2001. Genetic organization of plasmid colJs, encoding

colicin Js activity, immunity, and release genes. *J. Bacteriol.* 183:3949–57

88. Song HJ, Richard J. 1997. Antilisterial activity of three bacteriocins used at subminimal inhibitory concentrations and cross-resistance of the survivors. *Int. J. Food Microbiol.* 36:155–61
89. Stewart FM, Antia R, Levin BR, Lipsitch M, Mittler JE. 1998. The population genetics of antibiotic resistance. II: analytic theory for sustained populations of bacteria in a community of hosts. *Theoret. Popul. Biol.* 53:152–65
90. Tagg JR, Dajani AS, Wannamaker LW. 1976. Bacteriocins of gram-positive bacteria. *Bact. Rev.* 40:722–56
91. Tan Y, Riley MA. 1996. Rapid invasion by colicinogenic *Escherichia coli* with novel immunity functions. *Microbiology* 142:2175–80
92. Tan Y, Riley MA. 1997. Nucleotide polymorphism in colicin E2 gene clusters: evidence for nonneutral evolution. *Mol. Biol. Evol.* 14:666–73
93. Tan Y, Riley MA. 1997. Positive selection and recombination: major molecular mechanisms in colicin diversification. *Trends Ecol. Evol.* 12:348–51
94. Torreblanca M, Meseguer I, Rodriguez-Valera F. 1989. Halocin H6, a bacteriocin from *Haloferax gibbonsii*. *J. Gen. Microbiol.* 135:2655–61
95. Torreblanca M, Meseguer I, Ventosa A. 1994. Production of halocin is a practically universal feature of archael halophilic rods. *Lett. Appl. Microbiol.* 19: 201–5
96. Upreti GC, Hinsdill RD. 1975. Production and mode of action of lactocin 27—bacteriocin from a homofermentative *Lactobacillus*. *Antimicrob. Agents Chemother.* 7:139–45
97. Vandermeer JR, Polman J, Beerthuyzen MM, Siezen RJ, Kuipers OP, Devos WM. 1993. Characterization of the *Lactococcus-lactis* nisin-a operon genes *nisP*, encoding a subtilisin-like serine protease involved in precursor processing, and *nisR*, encoding a regulatory protein involved in nisin biosynthesis. *J. Bacteriol.* 175:2578–88
98. Vaughan EE, Daly C, Fitzgerald GF. 1992. Identification and characterization of helveticin V-1829, a bacteriocin produced by *Lactobacillus helveticus* 1829. *J. Appl. Bacteriol.* 73:299–308
99. Vignolo G, Palacios J, Farias ME, Sesma F, Schillinger U, et al. 2000. Combined effect of bacteriocins on the survival of various *Listeria* species in broth and meat system. *Curr. Microbiol.* 41:410–16
100. Walker ES, Levy F. 2001. Genetic trends in a population evolving antibiotic resistance. *Evolution* 55:1110–22
101. Wilson KH. 1997. *Biota of the Human Gastrointestinal Tract*. New York: Chapman & Hall. 680 pp.
102. Wu S, Dornbusch K, Kronvall G, Norgren M. 1999. Characterization and nucleotide sequence of a *Klebsiella oxytoca* cryptic plasmid encoding a CMY-type beta-lactamase: confirmation that the plasmid-mediated cephamycinase originated from the *Citrobacter freundii* Amp C beta-lactamase. *Antimicrob. Agents Chemother.* 43:1350–57

Annu. Rev. Microbiol. 2002. 56:139–65
doi: 10.1146/annurev.micro.56.012302.160907

First published online as a Review in Advance on May 16, 2002

EVOLUTION OF DRUG RESISTANCE IN *CANDIDA ALBICANS*

Leah E. Cowen, James B. Anderson, and Linda M. Kohn
Department of Botany, University of Toronto, Mississauga, Ontario, L5L 1C6, Canada; e-mail: lcowen@utm.utoronto.ca; janderso@utm.utoronto.ca; kohn@utm.utoronto.ca

Key Words genomics, experimental evolution, *Saccharomyces cerevisiae*, HIV, adaptation

■ **Abstract** The widespread deployment of antimicrobial agents in medicine and agriculture is nearly always followed by the evolution of resistance to these agents in the pathogen. With the limited availability of antifungal drugs and the increasing incidence of opportunistic fungal infections, the emergence of drug resistance in fungal pathogens poses a serious public health concern. Antifungal drug resistance has been studied most extensively with the yeast *Candida albicans* owing to its importance as an opportunistic pathogen and its experimental tractability relative to other medically important fungal pathogens. The emergence of antifungal drug resistance is an evolutionary process that proceeds on temporal, spatial, and genomic scales. This process can be observed through epidemiological studies of patients and through population-genetic studies of pathogen populations. Population-genetic studies rely on sampling of the pathogen in patient populations, serial isolations of the pathogen from individual patients, or experimental evolution of the pathogen in nutrient media or in animal models. Predicting the evolution of drug resistance is fundamental to prolonging the efficacy of existing drugs and to strategically developing and deploying novel drugs.

CONTENTS

INTRODUCTION

The emergence of drug resistance in all pathogenic microorganisms, including fungi, is an evolutionary process initiated by exposure to antimicrobial agents (65, 118). Resistance evolves because antimicrobial agents are rarely deployed in a way that completely eradicates the pathogen population, with survivors subject to natural selection. Whenever the pathogen population remains large enough over the course of drug treatment, the evolution of resistance is all but inevitable.

The evolution of drug resistance depends on genetic variability, the ultimate source of which is mutation. Once a mutation conferring resistance arises in the pathogen, its fate is determined by key population processes, such as selection, genetic drift, recombination, and migration (including transmission between hosts). The emergence and spread of drug resistance depends on more than the variety of different possible mutations that enable the pathogen to avoid, remove, or inactivate a drug. These resistance mutations interact with the rest of the genome to determine the composite phenotype (48, 131). In the evolutionary process the most important component of a resistance phenotype is reproductive output, or fitness. The relative fitness of sensitive and resistant genotypes determines how quickly resistance will spread in a pathogen population exposed to a drug and whether resistance will persist in the absence of the drug. Although biochemical and genetic mechanisms of antifungal drug resistance are well documented, the evolutionary processes by which these mechanisms spread and persist in pathogen populations await investigation.

Fungi are major pathogens of agricultural plants (86) and important opportunistic pathogens of humans (37, 81, 89), recently ranking as the seventh most common cause of infectious disease–related deaths in the United States (76, 94). Opportunistic fungal pathogens may cause superficial or invasive infections. The majority of invasive mycoses are caused by *Cryptococcus neoformans* and species of *Candida* and *Aspergillus* (94). Despite the increasing importance of opportunistic fungal pathogens, the number of effective antifungal drugs remains limited, with resistance compromising the effectiveness of all but the newest (37, 132).

Antifungal drug resistance has been studied most extensively with the diploid pathogenic yeast *Candida albicans*. *C. albicans* is a ubiquitous commensal, residing on the mucosal surfaces of the mouth, digestive tract, or genitourinary system of 15%–60% of healthy humans, depending on the sample group (82). *C. albicans* is also a good example of an opportunistic pathogen, causing both superficial infections and invasive fungal disease in immunocompromised individuals (32, 51, 103). Species of *Candida* are the fourth most common cause of nosocomial bloodstream infections in the United States, with *C. albicans* the most commonly encountered (91, 98).

C. albicans has been the fungal pathogen of choice for studying drug resistance because it is more easily manipulated and contained than other medically important fungal pathogens, such as *Cryptococcus neoformans*, *Aspergillus fumigatus*, and *Histoplasma capsulatum* (27). *C. albicans* is a wet yeast, classified as a relatively low risk to research personnel. In common with other countries in the American Biological Safety Association (http://www.absa.org/riskgroups/fungi.htm), Health Canada considers *C. albicans* to be a Risk Group 2 agent, unlikely to be a serious hazard to healthy laboratory workers, the community, livestock, or the environment (http://www.hc-sc.gc.ca/hpb/lcdc/biosafty/docs/lbg4_e.html#4.6.2). As a research system, *C. albicans* also offers a range of molecular-genetic tools (27), a complete genome sequence (123), and a sufficiently close phylogenetic relationship to the model yeast system, *Saccharomyces cerevisiae*, that many genes and pathways have highly similar counterparts in both yeasts. Unfortunately, *C. albicans* is, as yet, not amenable to conventional genetic analysis. Although *C. albicans* cells rendered homo- or hemizygous for mating-type-like genes are able to mate, meiosis has not been observed (50, 74). This disadvantage can be overcome with complementary studies using *S. cerevisiae* as a genetic stand-in. It is notable that some of the key studies of drug resistance (18, 52, 56, 57, 99, 109, 115), dimorphism, and virulence (58, 68, 78, 133) have exploited this synergy. *S. cerevisiae* stands on its own as a pathosystem based on recent reports of its clinical isolation from a variety of body sites and patient groups, with subsequent successful infection using these isolates in animal models (44).

This review focuses on the evolution of antifungal drug resistance [for reviews of clinical, molecular, and biochemical aspects see (1, 28, 37, 56, 88, 109, 132)]. The evolution of antifungal drug resistance in the pathogen proceeds on temporal, spatial, and genomic scales. Temporal scales of evolution range from ancient events associated with speciation to contemporary changes occurring within one patient under treatment. Spatial scales range from large-scale global or regional populations of patients to fine-scale niches within the body of one patient. Genomic scales range from the small-scale interactions of primary resistance mutations with one or a few genes to large-scale interactions with many genes. Large-scale interactions result in substantial changes in genome-wide patterns of gene expression in the pathogen. The effects of small- and large-scale interactions are detected as changes in fitness. The introduction and spread in pathogen populations (recruitment) of genetic mechanisms conferring antifungal drug resistance on different temporal, spatial, and genomic scales can be observed through both epidemiological and population-genetic studies. Population-genetic studies are implemented with samples of the pathogen from patient populations, with serial isolations of the pathogen from single patients or, alternatively, through experimental evolution of the pathogen in batch cultures, chemostats, or animal models. In this review we consider the sources and recruitment of antifungal drug resistance in pathogen populations on these temporal, spatial, and genomic scales.

MECHANISMS OF ACTION OF ANTIFUNGAL DRUGS

There are a limited number of antifungal drugs, especially compared with the number of antibacterial drugs (29, 39, 110). Drug targets that distinguish pathogen from host are more difficult to identify in fungi than in bacteria, at least in part because fungi and animals are relatively closely related as crown eukaryotes, whereas bacteria are much more distantly related to their human hosts (8). Because many potential antifungal drug targets have homologs of similar function and susceptibility to inhibition in humans, toxic side effects dramatically reduce the number of antifungal agents that can be used therapeutically. Currently available antifungals have only a small number of targets, including ergosterol and its biosynthesis, nucleic acid synthesis, and cell wall synthesis [for reviews of mechanisms of action of antifungal drugs see (28, 37, 56, 120, 125, 130)]. We summarize the salient features of the main classes of drugs.

Of the five classes of systemic antifungal compounds currently in clinical use, the polyenes, the azoles, and the allylamines all target ergosterol, the major sterol in fungal cell membranes that is not present in animals, whereas the fluoropyrimidines and the echinocandins have other targets (37). The polyenes complex with ergosterol in the fungal cell membrane and compromise the integrity of the cell membrane. Drugs in this class are fungicidal and have the broadest spectrum of activity of any clinically useful antifungal (37). The only systemic polyene in clinical use is amphotericin B, which has both acute and chronic side effects (28, 132). The azoles inhibit the cytochrome P450 enzyme, lanosterol demethylase, in the ergosterol biosynthesis pathway (56, 109). Ergosterol depletion and the accumulation of intermediates in the pathway disrupt the structure of the cell membrane, alter the activity of several membrane-bound enzymes, and may interfere with the hormone-like sparking function in the cell cycle (132). The azoles are fungistatic and have a broad spectrum of activity. Azoles include both the imidazoles (e.g., ketoconazole, as well as miconazole, which is now approved only for topical use) and the triazoles (e.g., fluconazole, itraconazole, and voriconazole), the group of antifungals most intensively under development (37, 110, 120). Systemic azoles are well tolerated and are generally free of serious host toxicities (28). The allylamines inhibit the enzyme squalene epoxidase in the ergosterol biosynthesis pathway. This class of drugs is fungicidal and has a broad spectrum of in vitro activity but has limited clinical efficacy owing to poor pharmacokinetics. Terbinafine is the only systemic antifungal in this class in clinical use (37). The fluoropyrimidines inhibit nucleic acid synthesis. Five-fluorocytosine (5-FC) is the only systemic antifungal drug in this class in clinical use and is fungicidal, with a limited activity spectrum (29). The echinocandins, the newest class, inhibit β-(1,3)-glucan synthesis, resulting in disruption of the fungal cell wall. Caspofungin is the only compound in this class approved for clinical use and is fungicidal with minimal host toxicities (38, 120).

Conventional drug screening procedures have resulted in the discovery and development of few compounds with unique modes of action, with caspofungin being the only one that has been commercially developed (38, 108, 120). Although the identification of novel drug targets has been facilitated by genomics technology (43, 75), resistance to all of the antifungal drugs that have been widely deployed is well documented.

A major public health concern is that the availability of antifungal drugs may not keep pace with the growing need. With the trend of increasing numbers of immunocompromised individuals, especially as a result of HIV infection (79), the epidemiology of opportunistic fungal infections is volatile. While the introduction of treatment therapies for HIV, including protease inhibitors, has reduced the incidence of mucosal candidal infections among treated patients (30), these HIV therapies are associated with high toxicity (31, 49), may have negative interactions with antifungal therapies (29), and are vulnerable to resistance (36). There are two facets to the impact of this resistance on the incidence of fungal infections. Resistance in the virus, with resulting decline in the immune status of the host and associated increase in susceptibility to opportunistic infection is well documented (36). Surprisingly, resistance in the fungus may also be a factor. These same protease inhibitors also reduce the activity and production of *C. albicans*–secreted aspartyl proteases (19, 55), reduce adherence and invasion of *C. albicans* to epithelial cells (10, 55), and reduce growth of *C. albicans* both in vitro and in vivo (19). Still unknown is the evolutionary potential of the fungus to overcome the inhibitory effects of these antiviral drugs. The severity of the AIDS epidemic is compounded by political and economic circumstances restricting access to effective drugs (12). With limited access to both drugs and medical supervision, even those patients receiving treatment may not undergo a complete course of therapy, creating conditions that are favorable for the evolution of resistance. Management of opportunistic fungal infections is a challenging problem, the solution of which will depend on the discovery of new and effective drugs.

WHAT IS ANTIFUNGAL DRUG RESISTANCE?

Drug resistance is a complex manifestation of factors in both host and pathogen. From a clinical perspective, drug resistance may be defined as the persistence or progression of an infection despite appropriate drug therapy. The clinical outcome of treatment depends not only on the susceptibility of the pathogen to a given drug but also on factors including pharmacokinetics, drug interactions, immune status, and patient compliance, as well as several specific conditions such as the occurrence of biofilms on surfaces of catheters and prosthetic valves (132).

Evaluation of how the susceptibility of the pathogen to a drug contributes to clinical outcome of treatment requires that drug susceptibility be measurable and reproducible in vitro. Drug resistance can be measured as the minimum inhibitory concentration (MIC) that curtails the growth of the fungus under standardized in vitro test conditions (104). Before the development of a standardized protocol, MIC determination varied up to 50,000-fold among different laboratories (42). Interlaboratory reproducibility was dramatically improved with the development of the National Committee for Clinical Laboratory Standards (NCCLS) protocol for antifungal susceptibility testing of yeasts (100, 101). Even with a standardized method of susceptibility testing, MICs are only sometimes predictive of clinical outcome. For example, a correlation of in vitro susceptibility to azole drugs with clinical response was observed with mucosal candidal infections in HIV-infected patients (42, 100, 132). In contrast, azole MIC did not correlate with clinical outcome for patients with candidal infections who were not infected with HIV (41). Despite cases of discordance between MIC and clinical outcome, MICs determined according to the NCCLS protocol have been used to determine interpretive breakpoints for fluconazole and itraconazole susceptibility in a classification system for clinicians (100, 102). Given that host variables, especially the immune status of the patient and the site of fungal infection, are likely to affect the efficacy of drug treatment, it is not surprising that the relationship between MIC and clinical outcome is complex. More surprising is the discordance between MIC and pathogen fitness in the presence of a drug in a simple laboratory environment (discussed in "Experimental Evolution," below).

Differences in antifungal drug susceptibilities among fungal individuals, populations, and species reflect different timescales of evolution (note that in yeast populations the individual is the cell). Antifungal drug resistance has been classified as either primary, when a fungus is resistant to a drug prior to any exposure, or secondary, when an initially sensitive fungus becomes resistant after exposure to the drug (102, 132). The major limitation of this classification is that lack of prior drug exposure can never be determined conclusively. Fungal species display intrinsically different susceptibilities to different drugs, with each species showing a distribution of MICs. For example, *Cryptococcus neoformans* and several species of *Aspergillus* are resistant to 5-FC (132). *C. lusitaniae* and *C. guillermondii*, as well as *Fusarium* and *Trichosporon* species, are more resistant to amphotericin B than is *C. albicans* (132). *C. neoformans* and several *Candida* species, including *C. krusei* and *C. glabrata*, are resistant to many of the azole drugs (37). Not surprisingly, species of *Candida* that are intrinsically more resistant to azoles have emerged as important opportunistic pathogens following the widespread deployment of these drugs (20, 80). Differences in intrinsic antifungal susceptibility between species reflect genetic changes that could have arisen before or after speciation, becoming fixed with continued reproductive isolation (7, 117).

On a more recent evolutionary timescale, many fungal species show a broad range of susceptibilities to a given drug, with genotypes that are sensitive and genotypes that are resistant. Populations within the species may have adapted to the drug in question after exposure to the agent at any point in their history. Depending on the extent of the fitness cost of resistance, the resistance phenotype may persist even in the absence of the drug. Resistant genotypes can be transmitted, as reported for fluconazole-resistant *C. albicans* strains among patients in the same hospital, as well as between sexual partners (132). On the most contemporary evolutionary scale is the emergence of drug resistance in an initially sensitive population of yeast cells that adapts to the drug in a patient over the course of treatment (discussed in "Serial Isolations from Individual Patients," below).

Initially sensitive fungal pathogens have become resistant to the clinically important antifungal drugs by a variety of molecular mechanisms summarized here [for reviews see (37, 56, 88, 109, 124, 132)]. Resistance to amphotericin B has emerged in both *Candida* and *Cryptococcus* species after exposure to the drug, but this is relatively rare (37). Fungal resistance to polyenes in general is associated with altered membrane lipids, especially the sterols (37). Based on the number of reports, azole resistance is relatively widespread, especially in *C. albicans* and *C. dublinienesis* (132). Resistance to azoles is associated with mutation in, or overexpression of, the target enzyme *ERG11*, with mutations in other enzymes in the ergosterol biosynthetic pathway, with active efflux of the drug by overexpression of efflux pumps of the ATP-binding cassette transporter or major facilitator families, and with decreased membrane permeability due to alterations in membrane sterols (56, 109). Resistance to 5-FC develops frequently when it is used as the sole antifungal agent. Fungal resistance to 5-FC is associated with impaired cytosine deaminase or mutations in any of the enzymes necessary for 5-FC action, particularly phosphoribosyl transferase (132). The evolution of drug resistance requires the recruitment of such mechanisms in populations.

Studying the Evolution of Drug Resistance

The recruitment of mechanisms conferring antifungal drug resistance in pathogen populations is determined by population-genetic processes including mutation, selection, genetic drift, recombination, and migration (remembering that this includes transmission between hosts). Mutation is the source of genetic variation. The fate of a mutation in a population depends on both the fitness effect of the mutation and on the population size, which determines the relative contribution of selection and genetic drift to evolution. Adaptation may proceed by the accumulation of numerous mutations of small effect or by few mutations of large effect (16, 83). The degree of dominance of a mutation is also important in a diploid background, in which a fully recessive mutation would confer no immediate advantage. Genetic exchange and recombination shuffle mutations that occur in different individuals and combine beneficial mutations in novel combinations in a single individual. Recombination may contribute to efficient adaptation to novel environments (84, 141) or, if the sequence in which beneficial mutations occur is important, may actually retard the rate of adaptation (53). In populations with no genetic exchange among individuals, genotypes that carry different beneficial mutations compete with one another, a process that may interfere with the progression of a mutation to fixation (40). The emergence of resistance mechanisms in individuals and their proliferation in populations are studied in samples of pathogen populations on different temporal and spatial scales.

Sampling Pathogen Populations from Patient Populations

Trends in antimicrobial resistance can be monitored by sampling clinical isolates over long temporal and broad geographic scales. There are both national (91, 93) and international (90) surveillance programs to monitor the frequency of different pathogens encountered in the clinic and their susceptibilities to antifungal drugs. Surveillance data can be especially useful for spotting a trend, such as the pervasiveness of resistance to therapeutic agents, or a correlation, such as an association between treatment regimens (and patient compliance) with emergence of resistance.

The spread of certain strains can be monitored by resolving the genotypes of fungal pathogens recovered from patients (116). Of particular interest is whether strains of restricted origin have a higher degree of genetic similarity than strains of general origin. Restricted samples could include strains from the same anatomical site (72), patient population (137), hospital (92), geographic region (34, 69, 70, 138, 140), or strains with similar antifungal susceptibility profiles (25, 92, 139). For example, if strains isolated from health care workers' hands in an intensive care unit are identical to those isolated from indwelling catheters in patients, then sanitation measures can be specifically targeted to prevent this kind of transmission (107, 128). Alternatively, if isolates from these sources are as dissimilar as isolates randomly chosen from broader samples, then the emphasis remains on more general sanitation practices (126). In both cases, observations of the distribution of pathogen genotypes can improve prophylaxis, treatment, and prevention of pathogen transmission and thereby limit the spread of preexisting drug resistance.

Whether fungal genotypes based on molecular-genetic markers with no known relationship to drug resistance (neutral markers) can be predictive of drug resistance depends on the population structure of the pathogen. In highly clonal populations genomes are reproduced and transmitted intact. In this case neutral markers will predict drug resistance, provided that resistance is maintained as a stable trait in the population (2). Alternatively, in a population with a high frequency of genetic exchange and recombination, alleles that are not tightly linked are shuffled, and any association between neutral markers and drug resistance will decay with time (60). Most fungal species, including those with no known sexual stage in their life cycles, show evidence of genetic exchange and recombination (2, 116).

Inferences of genetic exchange and recombination in diploid genomes are not as clear-cut as for haploid genomes. Because *C. albicans* is diploid, recombination detected with nuclear markers does not necessarily reflect genetic exchange between individuals. The difficulty arises from the ability of a highly heterozygous individual to produce a plethora of genotypes through the intracellular processes of mitotic recombination, which include crossing over and gene conversion. Even in the complete absence of genetic exchange between individuals, mitotic recombination within diploid individuals of *C. albicans* leads to loss of heterozygosity (24) and can produce distributions of genotypes that are indistinguishable from those expected with genetic exchange between individuals (2). From the available evidence based on nuclear markers (45, 95, 119, 127), it is not possible to conclude that genetic exchange between individuals of *C. albicans* is ongoing or has ever been frequent; clonal reproduction clearly predominates in this fungus.

Unlike in diploids, recombination in haploid genomes necessarily reflects genetic exchange between individuals. In haploids recombinant genotypes for nonduplicated loci arise only with genetic exchange between individuals and not with mitotic recombination. Even in diploid fungi such as *C. albicans* the mitochondrial genome is effectively haploid and can be exploited for determining whether genetic exchange between individuals has occurred. Analysis of mitochondrial nucleotide sequence variation in a population of *C. albicans* from HIV-infected patients identified infrequent past events of genetic exchange and recombination followed by clonal proliferation of genotypes (3), again indicating that clonal reproduction predominates. Despite the signature of past recombination in fungal pathogens, little is known about the actual rates or times of recombination. In the host even those species showing evidence for past genetic exchange generally reproduce clonally, with little or no evidence for recombination (59, 116). Taken together, the available evidence suggests that the evolution of resistance in fungal pathogens is more likely to occur by sequential accumulation of mutations within asexual lineages than by recombination of mutations that occurred in different lineages.

Only a few studies have attempted to find an association between drug resistance and molecular-genetic markers in *C. albicans*. Despite the predominantly clonal population structure in *C. albicans*, Pfaller et al. found that DNA fingerprints of fluconazole-resistant bloodstream isolates were no more similar than would be expected from a random set of clinical isolates (92). Cowen et al. showed that multilocus genotypes of oral isolates from HIV-infected patients were not predictive of azole resistance (25). These results suggest that resistance is labile and may be gained or lost too quickly to be predicted in the overall population of *C. albicans* by association with neutral markers. In contrast, Xu et al. identified a small group of genotypically similar fluconazole-resistant strains isolated from patients infected with HIV (139). This suggests that prediction of fluconazole resistance may be possible over a short timescale.

In all studies of pathogen populations based on samples from patient populations, evolutionary processes, including the evolution of drug resistance, cannot be observed directly but must be inferred from indirect evidence. By measuring allele frequencies in the pathogen population, parameters including mutation rates, recombination rates, intensity of selection, population size, and migration rates can be estimated but cannot be known with certainty or controlled. Although this approach provides important information about the distribution of antifungal resistance among genotypes in both pathogen and patient populations, it provides limited insight into the process or the genetic mechanisms underlying the evolution of antifungal drug resistance. Even a comprehensive sample, or census, of the general pathogen population over a long time and a wide geographical scale would be unlikely to provide these insights, given the diversity of the genotypes present and the complex patterns of descent and transmission. Sampling of *C. albicans* populations from individual patients over the course of antifungal drug treatment reduces this complexity and provides the opportunity to identify the emergent molecular mechanisms responsible for the evolution of drug resistance in a clinical setting.

Serial Isolations from Individual Patients

Serial isolations of the pathogen from individual patients over the course of drug treatment can be appropriate for tracking the increase in resistance over a shorter timescale, with the associated genetic alterations. This approach relies on identifying a series of isolates with increasing levels of drug resistance that are determined by high-resolution DNA typing methods to be identical or very similar (132). In this case, in which the isolates are assumed to be clonally related, changes in genotype and phenotype that emerge over the course of treatment can be identified.

Studies based on serial isolations from patients have shown that the emergence of azole resistance in *C. albicans* is often an incremental process that involves the accumulation of multiple mechanisms of resistance (35, 109). For example, in one series of 17 isolates of *C. albicans* from an HIV-infected patient, a gradual increase in resistance was associated with specific molecular alterations: first, overexpression of the *MDR1* multidrug efflux transporter gene; second, overexpression and mutation of the target gene, *ERG11*, and mitotic recombination rendering the mutation homozygous; and third, overexpression of the *CDR* multidrug efflux transporter genes (132). Confirming that the specific genetic changes associated with the emergence of resistance in clinical isolates are causally related to the resistance phenotype relies on in vitro study of both *C. albicans* and *S. cerevisiae* (109, 132).

While providing a biologically realistic view of in vivo changes in drug resistance, there is at least one serious limitation to the use of serial isolations of fungal pathogens from individual patients. The chronological recovery of isolates from patients may not reflect the actual sequence of evolutionary events because evolutionary relationships among the isolates are inferred rather than known with certainty. For example, genotypes may emerge from quiescent reservoirs of the pathogen during relapses of infection. Also, as with studies of populations of the pathogen isolated from populations of the host, parameters including mutation rate, recombination rate, intensity of selection, population size, and transmission of genotypes can only be inferred.

Experimental Evolution

The dynamics and genetic mechanisms of the evolution of antifungal resistance can be more directly observed through experimental evolution of fungal pathogens in the laboratory than in samples of existing pathogen populations or serial isolations from individual patients. The main attraction of this approach is the ability to replicate experiments and to control conditions such as ploidy (73), size of population (129, 141), strength of selection (129, 141), rate of mutation (4, 54, 112, 141, 143), and opportunity for genetic exchange and recombination (141). Experimental evolution subjects populations to natural, as opposed to artificial, selection (105). Any fungal pathogen can be evolved in the presence of any antifungal drug in the laboratory, and both genotypic and phenotypic changes in the evolved populations can be measured relative to the ancestor. An introduction to the features of experimental evolution precedes a discussion of how this approach has been used to study the evolution of drug resistance in *C. albicans*.

Experimental evolution begins with one or more known genotypes and then follows a trajectory of change over time. With experimental populations founded from a single genotype, immigration of genotypes and genetic exchange between individuals may be prevented so that mutation during the experiment is the only source of genetic variation. This allows the specific genetic changes that underlie adaptation and the temporal sequence in which they occur to be identified relative to a known ancestral genotype. Experiments with replicate populations of different sizes address the element of chance in adaptation. Similarly, experiments with replicate populations founded with different ancestral genotypes address the contribution of history to adaptation (122). Whether genetic background constrains the evolution of drug resistance or whether all genotypes show comparable ability to evolve resistance can be determined.

Starting from an ancestral genotype, the evolutionary trajectory of an experimental population depends on the availability of mutations. The two most important parameters affecting availability are mutation rate and population size. Because the number of cell generations is known in experimental populations and the occurrence of mutations can be bracketed in time, mutation rates can be determined (71). The fitness effects of mutations and the relationship of mutation rate to the evolutionary rate can also be measured (4, 113, 136, 141, 142). Given a constant mutation rate, population size becomes the most important determinant of adaptation. In a population of infinite size, evolution should be entirely deterministic, with all possible mutations available all of the time. At any given point, the fittest genotype is favored by natural selection. In reality, populations are not infinitely large. In populations of finite size, the emergence of adaptive mutations and the sequence in which they occur is often stochastic, with only a subset of all possible mutations available at any given time.

The effect of natural selection is more apparent in large populations than in small populations (129, 141). In smaller populations genetic drift is more likely to overwhelm selection, with some genotypes increasing in frequency by chance rather than by dint of their adaptive fitness. Although large experimental population sizes may be more appropriate for studying adaptation, in the real world pathogen populations are often bottlenecked to very small sizes during transmission between hosts; only a small fraction of the population infecting one host is transmitted and successfully colonizes a new host (11). If this population bottleneck is small enough, the best-adapted genotypes are not necessarily transmitted to a new host, especially if these genotypes are at a low frequency in the population (66). In addition, because of the physical association and relatedness of nearby individuals, the effective size of structured populations found within a host will always be smaller than that of unstructured populations with comparable numbers of individuals.

In populations of sufficient size, fitness (or reproductive output over a defined period of time) is expected to improve with adaptation to an environment (61). This expectation is not always met. Evolving populations can actually decrease in fitness relative to their ancestor (85). If genotype B has a fitness advantage over the ancestral genotype A and genotype C has a fitness advantage over genotype B, this does not necessarily imply that genotype C will have a fitness advantage over genotype A. This decrease in fitness could be attributed to negative epistatic interactions, in which observed fitness is less than fitness expected if sequentially accumulated adaptive mutations are additive in their effect in one environment.

Also, adaptation of a population to one environment may result in a decrease in fitness in another environment (a trade-off) (21, 26). It is often assumed that a pathogen that evolved with a drug will be fitter than its ancestor in the presence of the drug; resistant genotypes increase in frequency relative to their sensitive counterparts when the drug is deployed. On the other hand, if drug resistance carries a fitness cost for the pathogen, once the drug is removed, resistant microbes decrease in frequency relative to microbes that are more drug-sensitive but fitter.

Running an evolution experiment requires conditions that are conducive to the evolution of the population and that are appropriate to the experimental question. This requires fine-tuning. Drug concentrations must be adjusted to substantially inhibit the growth of the fungus without resulting in extinction. If the drug under investigation is fungicidal, then the range of possible concentrations is more limited than if the drug in question is fungistatic. Cross-resistance can be observed by testing populations previously adapted to one drug for resistance to another drug, or by evolving populations simultaneously in the presence of different drugs. The stability of resistance can be monitored by subsequent evolution in the absence of any drug.

Setting a timescale for the evolution experiment depends on the question. Evolution experiments of a few hundred generations reflect a timescale approximating that of a pathogen evolving drug resistance in its host during a course of drug treatment. Evolution experiments over thousands of generations test whether the populations ultimately converge on one stable, adaptive optimum (62).

Finally, a choice must be made between an unstructured (Figure 1) and a structured (Figure 2) environment. The host is a spatially structured environment, not

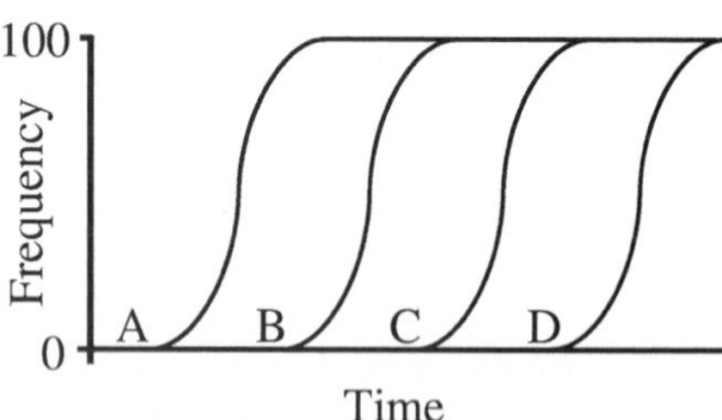

Figure 1 Evolution in an unstructured environment, such as a chemostat or batch culture with constant agitation. Spatial associations between individuals are continually disrupted. In large populations, with strong selection and rare mutation, evolution often proceeds as a series of sweeps in which an adaptive mutation (designated A, B, C, and D) rises to fixation and then serves as the background genotype for another adaptive mutation, which then also rises to fixation [see (134) for an excellent example]. The predominant genotype at the end is ABCD. Because mutations A, B, C, and D all reach fixation quickly and because these mutations accumulate within one lineage, determining their order of occurrence is easily reconstructed by regular sampling of the population over time.

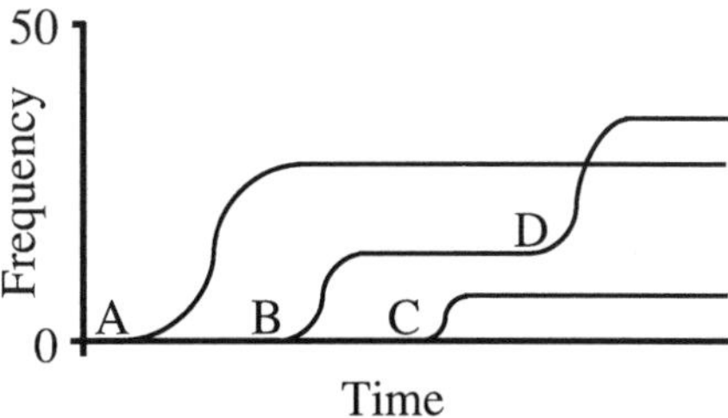

Figure 2 Evolution in a highly structured environment, such as an animal host. Physical and spatial associations between individuals are rarely disrupted. This population may also be spatially subdivided into groups between which migration is restricted. Adaptive mutations (designated A, B, C, and D) therefore tend to remain at low frequency and do not sweep to fixation. Different genotypes may predominate in different spatial sectors, and genotypes may also persist for periods of time in reservoirs without reproducing. Under these conditions an adaptive mutation only rarely (e.g., in the case of mutation B) serves as the background for subsequent mutations. The most frequent genotypes at the end are A, BD, and C. Because mutations A, B, C, and D never approach fixation and because these mutations (except B and D) occur in separate lineages (as a radiation), determining their order of occurrence would not be easily reconstructed by regular sampling of the population over time.

a flask full of liquid. Most experimental evolution studies have evolved microbial populations in liquid media either in a chemostat or in batch cultures that are serially transferred at regular intervals into fresh medium (121, 141). The chemostat affords more uniform conditions, whereas the batch culture affords a higher degree of replication. Despite spatial heterogeneities that may arise from differences between the surface of the test tube or chemostat and the agitating medium, this environment represents our closest approximation of an unstructured system. One would expect a succession of genotypes to increase in frequency in the population. In contrast, a spatially structured environment may favor radiations if different genotypes predominate in different niches (96, 97). Even the surface of a petri dish provides a structured environment in which genotypes are spatially segregated rather than in direct competition. Animal models also provide a structured environment, but with the added factors of the host physiology and immune system.

The biofilm is another important type of naturally occurring, spatial structuring. Fungal biofilms represent a structured environment with highly heterogeneous architecture. *C. albicans* can form biofilms in vivo as well as in vitro. Biofilms are important clinically because their development is associated with increased antifungal resistance (19a, 97a, 97b). Several in vitro models have been developed to study biofilm growth processes, architectural organization, and antifungal resistance. These models use either 96-well microtiter plates or bioprosthetic materials (19a, 97a, 97b). Such models could be used to determine whether the increased drug resistance associated with *C. albicans* biofilm formation is a consequence of the production of extracellular material that physically interacts with antifungal drugs or whether it is due to genetic and biochemical alterations of the fungal cells.

Studies of experimental evolution of fungi [reviewed in (141)], bacteria [reviewed in (121, 129)], and viruses (134, 135) have focused on the evolutionary dynamics of adaptation to general culture conditions. Few studies have addressed the dynamics of adaptation to antimicrobial drugs over a timescale of hundreds of generations. Experimental evolution of drug resistance recapitulates a realistic process in fungi in which mutations conferring resistance arise within a lineage. There is no conclusive evidence of horizontal transfer of antifungal drug resistance between lineages (106). In comparison, given the extensive horizontal transfer of determinants of antibiotic resistance, studies of bacteria have focused on the fate of resistance once it has evolved rather than on the emergence of resistance.

We have studied the dynamics and mechanisms of evolved resistance in both *C. albicans* and *S. cerevisiae*. In *C. albicans* (24) 12 replicate populations were established from a single drug-sensitive cell and were serially transferred each day over 330 generations. There was no genetic exchange among individuals and no migration between populations. Mutation during the experiment was the only source of genetic variability. Six populations were evolved with fluconazole adjusted to twice the minimum inhibitory concentration (MIC) measured at regular intervals, and six were evolved without the drug. All of the populations evolved with fluconazole adapted to the presence of the drug as indicated by an increase in MIC. Their evolutionary trajectories, however, were strikingly different, with each population reaching different levels of drug resistance. The populations showed distinct overexpression patterns of the four genes known to be involved in azole resistance (*CDR1*, *CDR2*, *MDR1*, and *ERG11*). In this experiment chance, in the form of the random occurrence of mutations that confer an adaptive advantage, was a major factor in the evolution of drug resistance. Because the minimum population size with each transfer was large (10^6 cells), genetic drift was less likely a factor than the randomness of mutation in explaining the divergence of trajectories.

We observed a different pattern in haploid experimental populations of *S. cerevisiae* that were 10 times as large (minimum population size was 10^7 cells) as those of *C. albicans*. Three uniquely marked populations were established from a single genotype and were evolved with increasing concentrations of fluconazole. Again, there was no genetic exchange among individuals and no migration between populations. Over 400 generations the three populations followed parallel adaptive trajectories with the same underlying molecular mechanisms of azole resistance, including overexpression of the ATP-binding cassette transporter genes *PDR5* and *SNQ2*. Genetic analysis showed that each population had recruited the same two determinants of fluconazole resistance, one of a large effect occurring first in the evolution experiment and the second of a smaller effect occurring later (J.B. Anderson, unpublished data). The more uniform trajectories in *S. cerevisiae* than in *C. albicans* may have been due to differences in population size, ploidy, or the frequency of mutations contributing to resistance.

Selection under incremental increases in drug concentration and selection under one step at a high drug concentration will likely yield different mechanisms of resistance. Under incremental steps the pathogen adapts, whereas under one-step selection the pathogen has no opportunity for compensatory adaptation. To test this hypothesis, starting with the same genetic background of *S. cerevisiae* as was used in our evolution experiment, we selected fluconazole-resistant mutants by plating cells on medium containing the highest concentration of fluconazole that was reached during the 400 generations of evolution. Each of three resistant mutants recovered from the one-step selection overexpressed the target gene of the azoles, *ERG11*, in contrast to the ones selected over 400 generations of evolution, which overexpressed the ATP-binding cassette transporters *PDR5* and *SNQ2*. Genetic analysis showed that each of the three mutants harbored one mutation and that these three mutations were allelic. Clearly, different mutations are favored in the different selective regimes using the same drug. This means the evolutionary outcome depends on how the drug is deployed.

In another one-step selection experiment Perepnikhatka et al. (87) exposed drug-sensitive clones of *C. albicans* to a high concentration of fluconazole on solid medium and recovered resistant mutants. These mutants were associated with specific chromosomal alterations. Whether these specific chromosomal alterations are directly related to drug resistance is not yet known. Barchiesi et al. (9) conducted a serial transfer experiment in which one replicate of an initially sensitive clone of *C. tropicalis* was evolved in each of three different concentrations of fluconazole; resistant mutants were recovered that showed different degrees of resistance depending on the concentration of the drug used for selection. Fluconazole resistance was associated with overexpression of the multidrug efflux transporters *MDR1* and *CDR1*.

The changes in candidate genes and their expression are not sufficient to explain the molecular basis for adaptation (111). We used DNA microarrays to monitor genome-wide changes in gene expression that became established during adaptation to fluconazole in four populations of *C. albicans* and persisted in the absence of the drug (23a). Although specific resistance determinants in *C. albicans* are well characterized, we did not anticipate the magnitude and extent of changes in genome-wide gene expression occurring with the evolution of drug resistance. Gene-expression profiling identified 301 open reading frames with expression levels that were significantly changed relative to the ancestor. Among the genes of altered expression, there were three distinct patterns underlying adaptation to the drug. One pattern was unique to one population and included overexpression of *CDR2*. A second pattern occurred at a late stage of adaptation in three populations. For two of these populations profiled earlier, as well as later, in their evolution, a third pattern was observed at an early stage of adaptation. Both the early- and late-stage patterns of gene expression included overexpression of *MDR1*. The succession of the early- and late-stage patterns of gene expression must represent a common program of adaptation to fluconazole. The same three patterns of gene expression were also identified in fluconazole-resistant clinical isolates of *C. albicans*.

Bearing in mind that mutations and persistent changes in genome-wide gene-expression patterns are recruited through natural selection, the question is whether a drug-resistant genotype will persist or go extinct. This depends on the relative fitness of resistant and sensitive genotypes. In bacteria and viruses investigations of the fitness cost of drug resistance have focused on the relative rates of growth, survival, and competitive performance of sensitive and resistant genotypes by means of pairwise competition experiments in vitro or in animal models. In most cases resistance confers a fitness cost in the absence of the drug (13). During subsequent evolution, however, the cost of resistance tends to decline as the organisms adapt to the environment (with or without the drug), compensating for the fitness cost of their drug resistance determinants (66).

Evidence from our experimental populations of *C. albicans* indicates that adaptations that reduce the fitness cost of drug resistance may also be common in fungi (23). The populations that were evolved with fluconazole diverged in fitness, measured by direct competition with a genetically marked version of the ancestor, and any cost of resistance was eliminated with further evolution. Two drug-resistant populations were fitter than the ancestor even in the absence of the drug. Compensatory evolution mitigating the cost of resistance is expected, rather than the reversion to sensitivity, when there are more possible mutations that improve fitness of the organism than back mutation alone (64). Compensatory evolution would be thwarted by frequent recombination that separates the compensatory mutation from the resistance mutation provided that the loci involved are not tightly linked. If fungal pathogens do not regularly undergo genetic exchange and recombination, for example when they are growing in the host, then compensatory evolution should proceed.

Pathogens showing a higher MIC of a drug are assumed fitter in the presence of the drug than those with a lower MIC. Despite the logic of this expectation, in the populations of *C. albicans* evolved with fluconazole we found cases of high fitness in the presence of the drug with low MIC, as well as of low fitness with high MIC [see Figure 1 in (22)]. Three of the six populations evolved with fluconazole decreased in MIC during continued evolution at the high drug concentration (24). In no case was this reduction in MIC accompanied by a reduction in fitness measured at the high drug concentration (23); in fact, in one case the drop in MIC was accompanied by a significant increase in fitness. The population with the largest fitness advantage in the presence of the highest concentration of drug used in this experiment had only an intermediate MIC. Finally, two of the three populations with the highest MIC did not show a fitness advantage in the presence of the drug. Because azoles are fungistatic, not fungicidal, the drug-sensitive ancestor grows well at concentrations above its MIC. We have also observed this phenomenon in *S. cerevisiae* (Figures 3 and 4). In *C. albicans* competitive fitness is not predicted by MIC or by growth parameters, such as exponential growth rate and stationary phase cell density, of either competitor in isolation. The discordance between MIC

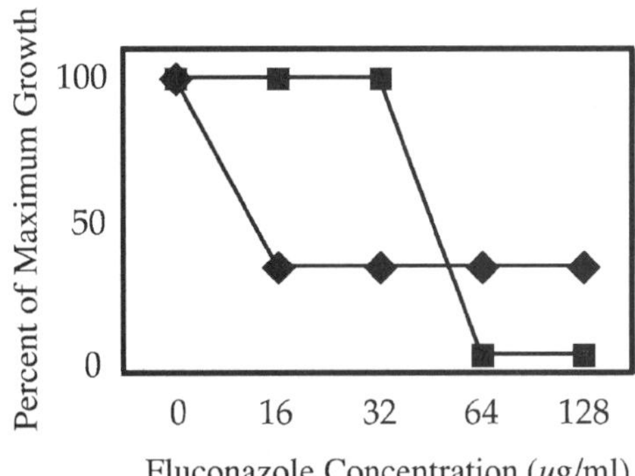

Figure 3 Minimum inhibitory concentration (MIC) is not always predictive of growth at high drug concentration. One hypothetical strain (*diamonds*) has a low MIC (16 μg/ml fluconazole) according the National Committee for Clinical Laboratory Standards protocol but continues to grow at the highest drug concentration. Another hypothetical strain (*squares*) has a high MIC (64 μg/ml) according the National Committee for Clinical Laboratory Standards protocol but grows poorly at the highest drug concentration.

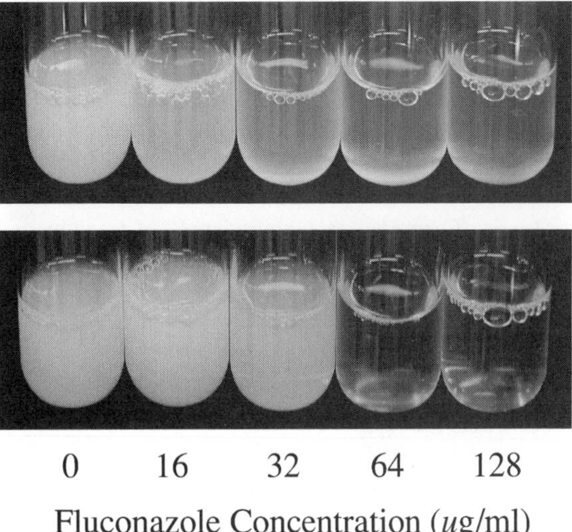

Figure 4 Growth of two strains of *S. cerevisiae* in separate cultures with various concentrations of fluconazole. One strain (*top panel*) with a low MIC (16 μg/ml fluconazole) continues to reproduce even at the highest concentration of the drug. Another strain (*bottom panel*) with a higher MIC (64 μg/ml) shows less reproduction at the highest concentration of the drug.

and fitness may be due to conditions in the tests. Our MIC tests use one starting genotype, static conditions, and small populations. Our tests of relative fitness use more than one starting genotype, constant agitation, and large populations. In the end, we suspect that interactions between mixed genotypes may also affect fitness in competition, but these interactions are not a factor in MIC tests.

The next step will be to study the evolution of drug resistance, with emphasis on fitness, in animal models. This has not yet been done with fungi, for which work with animal models has focused on measuring and comparing the virulence of mutants. Goldstein & McCusker used *S. cerevisiae* as a model pathogen to identify highly conserved signal transduction and metabolic pathways critical for survival in the host environment (44). Dominant drug-resistance markers were used to delete selected genes involved in metabolism and dimorphism in a clinically derived *S. cerevisiae* strain. Competitions were conducted in mice between a genetically marked control strain and two mutants, each homozygous for a deletion of the same gene but with different dominant drug-resistance markers. In addition to determining the importance of the selected genes for survival in the host, the fitness effects of a mutation conferring resistance to 5-FC were determined. Resistance to 5-FC was conferred by deletion of cytosine deaminase, encoded by *FCY1*. Deletion of *FCY1* was not deleterious but was actually advantageous in the mouse (44). This model could be applied to measure the fitness of *S. cerevisiae* mutants resistant to other antifungal drugs. Treating mice with the drug during the course of the competition experiment can test the relationship between MIC and fitness in vivo.

A system is also being developed to study genomic changes in *C. albicans* in a mouse model. Starting with *C. albicans* strains in which both copies of *GAL1* were deleted, this same gene was placed at various locations in the genome as a selectable marker. Loss of the introduced *GAL1* allele is expected with several different mechanisms of mitotic recombination, including gene conversion, crossing over, and other kinds of genome rearrangements. *GAL1* can be used for in vivo studies because it has no effect on virulence. The marked strains of *C. albicans* can now be passaged in mice, or in laboratory medium, in order to measure rates of mitotic recombination both in vivo and in vitro (P.T. Magee, personal communication).

PREDICTING THE EVOLUTION OF ANTIFUNGAL DRUG RESISTANCE

Mathematical models are usually the first step toward quantitative description of a complex system. If the assumptions of the model are realistic, the model will have predictive power. Most of the models of resistance to antifungal agents have been developed for fungicides in plant disease epidemics (epiphytotics). These models share common features with models of resistance to antibacterial (5, 14, 63) and antiviral (15) drugs in pathogens of humans and are useful in thinking about models of antifungal drug resistance. We consider two models of fungicide resistance based on different assumptions. The first model applies both deterministic and stochastic formulations to analyze the dynamics of competition between resistant and sensitive strains, the supply of susceptible host tissue, the rate of application of the fungicide, and the effects of the fungicide (46). Two strains of the pathogen are considered, one sensitive and one resistant. The model assumes that there is spatially homogeneous mixing of the pathogen in the crop and no mutation or recombination. The model derives thresholds for the invasion of fungicide resistance in a pathogen population within a single season. The threshold depends on both the relative fitness of the sensitive and resistant strains and on the fungicide efficacy; fungicide efficacy is affected by the fungicide decay rate, the amount of fungicide applied, and the period between applications. This model predicts a lower probability of resistance and a longer lag time before resistance appears with the application of lower amounts of fungicide.

The second model applies stochastic formulations to relate pathogen population size, the probability of fungicide-resistant individuals occurring, and the strength of selection for resistance, or fungicide efficacy (77). Again, there are only two genotypes, one resistant and one sensitive. Unlike the first model, pathogen growth is assumed to be exponential (availability of susceptible host tissue is not considered), and single mutations to resistance are allowed to occur (mutations to resistance in multiple individuals are ignored). The model also assumes that resistant and sensitive genotypes have equal fitness in the absence of the fungicide and that the fungicide has no inhibitory effect on resistant mutants. With small pathogen population size and slow pathogen growth rate, the model predicts that the probability of resistant individuals occurring is low. With a low mutation rate, resistant individuals are unlikely to appear until the population reaches a large size later in an epidemic. When the first resistant mutants occur late in the epidemic, there is little time for selection to act, and thus resistance remains at a low frequency. This model also predicts that when a population is kept small by an effective fungicide, the probability of resistance mutations occurring is low. In contrast to the first model, the second model predicts that given an initially small pathogen population size at the beginning of an outbreak of infection, the probability of developing resistance is lower when fungicide is applied intensively (implying both high fungicide concentration and frequency of application).

Models are necessarily simplifications of complex systems. While the effect of simplifying assumptions on predicted outcomes can be tested (67), the predictive power of any model also depends on accurate estimates of parameters. These parameters are often estimated from epidemiological studies, ideally on long temporal and broad geographic scales (6). Studies of experimental evolution are also especially useful here: population size, generation time, rates of different mutations conferring resistance, and the relative fitness of resistant and sensitive genotypes can be accurately measured under a variety of controlled laboratory conditions.

In addition to providing an empirical basis for setting parameters in modeling, experimental evolution can actually measure evolutionary potential. The potential for evolution of antifungal resistance can be measured at both the level of the gene and of the genome by evolving DNA sequences or microbial populations (22). Evolutionary potential at the level of the gene—of fine-scale interactions of primary resistance mutations within a gene—depends on the DNA sequence space (33) defined by the fitness effects of all possible mutations in candidate resistance genes (i.e., drug targets) in all possible combinations. This sequence space can be explored by evolving candidate resistance genes in vitro with sexual polymerase chain reaction (PCR), or DNA shuffling (47, 114), and then measuring the fitness effects of single and various combinations of multiple mutations. Sexual PCR explores sequence space far surpassing the range of mutational possibilities likely to be explored in natural populations. It is the mutational neighborhood (17) a small number of mutational steps away from the ancestral sequence that is key to predicting the resistance mutations that might arise in natural populations.

Evolutionary potential at the level of the genome—of large-scale interactions with many genes—depends on the composite of genome-wide changes in genotype and gene expression that accompany the evolution of resistance, as well as the impact of these changes on the fitness of the pathogen. Both rare and frequent adaptive changes can be identified through experimental evolution of fungal populations with inhibitory concentrations of any antifungal drug. Genes that are always associated with resistance to particular drugs, either through altered expression or through mutation, may represent keystones in the evolution of drug resistance. Whether or not these candidate keystone genes are critical to the development of drug resistance can be tested by knocking them out individually and measuring the ability of the resulting genotype to evolve resistance. Genes whose deletion results in impaired evolvability are potential targets for companion drugs designed to minimize the evolution of resistance to existing antifungals.

The most compelling rationale for investigating the evolution of drug resistance is to find ways to prolong the efficacy of existing drugs and to strategically develop and deploy novel drugs. Managing the evolution of resistance in the course of drug deployment will require a convergence of traditionally disparate disciplines, which must include evolutionary biology and experimental evolution. Mathematical modeling is the instrument for evaluating the effect of intervention strategies, including reducing pathogen transmission, increasing prophylactic use of drugs to keep pathogen populations small, cycling drugs with different targets, and minimizing the use of antimicrobial drugs by finding minimum effective dosages. In tandem with modeling, experimental evolution provides biologically realistic values for parameters used in models, bearing in mind that experimental populations can be reared in test tubes or in animals. These parameters will only be approximated from epidemiological sampling of patient populations, although such samples can survey long periods of time and large geographical areas. Adaptive changes in genes and genomes can be directly observed only in experimental populations. The challenge in interpreting genome-wide changes will be teasing out those changes that are directly related to drug resistance from those that are collateral. The ultimate challenge will be in developing models that are predictive in treating patients.

ACKNOWLEDGMENTS

L.M. Kohn and J.B. Anderson were supported by research grants from the Natural Sciences and Engineering Research Council (NSERC) of Canada. L.E. Cowen was supported by an NSERC postgraduate scholarship.

The *Annual Review of Microbiology* is online at http://micro.annualreviews.org

LITERATURE CITED

1. Alexander BD, Perfect JR. 1997. Antifungal resistance trends towards the year 2000. Implications for therapy and new approaches. *Drugs* 54:657–78
2. Anderson JB, Kohn LM. 1998. Genotyping, gene genealogies, and genomics bring fungal population genetics above ground. *Trends Ecol. Evol.* 13:444–49
3. Anderson JB, Wickens C, Khan M, Cowen LE, Federspiel N, et al. 2001. Infrequent genetic exchange and recombination in the mitochondrial genome of *Candida albicans*. *J. Bacteriol.* 183:865–72
4. Arjan JA, Visser M, Zeyl CW, Gerrish PJ, Blanchard JL, Lenski RE. 1999. Diminishing returns from mutation supply rate in asexual populations. *Science* 283:404–6
5. Austin DJ, Kakehashi M, Anderson RM. 1997. The transmission dynamics of antibiotic-resistant bacteria: the relationship between resistance in commensal organisms and antibiotic consumption. *Proc. R. Soc. London Ser. B* 264:1629–38
6. Austin DJ, Kristinsson KG, Anderson RM. 1999. The relationship between the volume of antimicrobial consumption in human communities and the frequency of resistance. *Proc. Natl. Acad. Sci. USA* 96: 1152–56
7. Avise JC, Wollenberg K. 1997. Phylogenetics and the origin of species. *Proc. Natl. Acad. Sci. USA* 94:7748–55
8. Baldauf SL, Roger AJ, Wenk-Siefert I, Doolittle WF. 2000. A kingdom-level phylogeny of eukaryotes based on combined protein data. *Science* 290:972–77
9. Barchiesi F, Calabrese D, Sanglard D, Falconi Di Francesco L, Caselli F, et al. 2000. Experimental induction of fluconazole resistance in *Candida tropicalis* ATCC 750. *Antimicrob. Agents Chemother.* 44:1578–84
10. Bektic J, Lell CP, Fuchs A, Stoiber H, Speth C, et al. 2001. HIV protease inhibitors attenuate adherence of *Candida albicans* to epithelial cells in vitro. *FEMS Immunol. Med. Microbiol.* 31:65–71
11. Bergstrom CT, McElhany P, Real LA. 1999. Transmission bottlenecks as determinants of virulence in rapidly evolving pathogens. *Proc. Natl. Acad. Sci. USA* 96: 5095–100
12. Binswanger HP. 2001. Public health. HIV/ AIDS treatment for millions. *Science* 292: 221–23
13. Björkman J, Andersson DI. 2000. The cost of antibiotic resistance from a bacterial perspective. *Drug Resist. Updates* 3:237–45
14. Bonhoeffer S, Lipsitch M, Levin BR. 1997. Evaluating treatment protocols to prevent antibiotic resistance. *Proc. Natl. Acad. Sci. USA* 94:12106–11
15. Bonhoeffer S, May RM, Shaw GM, Nowak MA. 1997. Virus dynamics and drug therapy. *Proc. Natl. Acad. Sci. USA* 94:6971–76
16. Bull JJ, Badgett MR, Wichman HA. 2000. Big-benefit mutations in a bacteriophage inhibited with heat. *Mol. Biol. Evol.* 17:942–50
17. Burch CL, Chao L. 2000. Evolvability of an RNA virus is determined by its mutational neighbourhood. *Nature* 406:625–28
18. Calabrese D, Bille J, Sanglard D. 2000.

A novel multidrug efflux transporter gene of the major facilitator superfamily from *Candida albicans* (*FLU1*) conferring resistance to fluconazole. *Microbiology* 146: 2743–54

19. Cassone A, De Bernardis F, Torosantucci A, Tacconelli E, Tumbarello M, Cauda R. 1999. In vitro and in vivo anticandidal activity of human immunodeficiency virus protease inhibitors. *J. Infect. Dis.* 180:448–53

19a. Chandra J, Kuhn DM, Mukherjee PK, Hoyer LL, McCormick T, Ghannoum MA. 2001. Biofilm formation by the fungal pathogen *Candida albicans*: development, architecture, and drug resistance. *J. Bacteriol.* 183:5385–94

20. Coleman DC, Rinaldi MG, Haynes KA, Rex JH, Summerbell RC, et al. 1998. Importance of *Candida* species other than *Candida albicans* as opportunistic pathogens. *Med. Mycol.* 36:156–65

21. Cooper VS, Lenski RE. 2000. The population genetics of ecological specialization in evolving *Escherichia coli* populations. *Nature* 407:736–39

22. Cowen LE. 2001. Predicting the emergence of resistance to antifungal drugs. *FEMS Microbiol. Lett.* 204:1–7

23. Cowen LE, Kohn LM, Anderson JB. 2001. Divergence in fitness and evolution of drug resistance in experimental populations of *Candida albicans*. *J. Bacteriol.* 183:2971–78

23a. Cowen LE, Nantel A, Whiteway MS, Thomas DY, Tessier DC, et al. 2002. Population genomics of drug resistance in *Candida albicans*. *Proc. Natl. Acad. Sci. USA* 99:9284–89

24. Cowen LE, Sanglard D, Calabrese D, Sirjusingh C, Anderson JB, Kohn LM. 2000. Evolution of drug resistance in experimental populations of *Candida albicans*. *J. Bacteriol.* 182:1515–22

25. Cowen LE, Sirjusingh C, Summerbell RC, Walmsley S, Richardson S, et al. 1999. Multilocus genotypes and DNA fingerprints do not predict variation in azole resistance among clinical isolates of *Candida albicans*. *Antimicrob. Agents Chemother.* 43:2930–38

26. Crill WD, Wichman HA, Bull JJ. 2000. Evolutionary reversals during viral adaptation to alternating hosts. *Genetics* 154:27–37

27. De Backer MD, Magee PT, Pla J. 2000. Recent developments in molecular genetics of *Candida albicans*. *Annu. Rev. Microbiol.* 54:463–98

28. DiDomenico B. 1999. Novel antifungal drugs. *Curr. Opin. Microbiol.* 2:509–15

29. Dismukes WE. 2000. Introduction to antifungal drugs. *Clin. Infect. Dis.* 30:653–57

30. Dupont B, Crewe Brown HH, Westermann K, Martins MD, Rex JH, et al. 2000. Mycoses in AIDS. *Med. Mycol.* 38:259–67

31. Dybul M, Chun TW, Yoder C, Hidalgo B, Belson M, et al. 2001. Short-cycle structured intermittent treatment of chronic HIV infection with highly active antiretroviral therapy: effects on virologic, immunologic, and toxicity parameters. *Proc. Natl. Acad. Sci. USA* 98:15161–66

32. Edwards JE Jr. 2001. Management of severe candidal infections: integration and review of current guidelines for treatment and prevention. *Curr. Clin. Top. Infect. Dis.* 21:135–47

33. Eigen M. 1987. New concepts for dealing with the evolution of nucleic acids. *Cold Spring Harbor Symp. Quant. Biol.* 52:307–20

34. Forche A, Schönian G, Gräser Y, Vilgalys R, Mitchell TG. 1999. Genetic structure of typical and atypical populations of *Candida albicans* from Africa. *Fungal Genet. Biol.* 28:107–25

35. Franz R, Kelly SL, Lamb DC, Kelly DE, Ruhnke M, Morschhauser J. 1998. Multiple molecular mechanisms contribute to a stepwise development of fluconazole resistance in clinical *Candida albicans* strains. *Antimicrob. Agents Chemother.* 42:3065–72

36. Garcia-Lerma JG, Heneine W. 2001. Resistance of human immunodeficiency virus type 1 to reverse transcriptase and protease inhibitors: genotypic and phenotypic testing. *J. Clin. Virol.* 21:197–212
37. Georgopapadakou NH. 1998. Antifungals: mechanism of action and resistance, established and novel drugs. *Curr. Opin. Microbiol.* 1:547–57
38. Georgopapadakou NH. 2001. Update on antifungals targeted to the cell wall: focus on beta-1,3-glucan synthase inhibitors. *Expert Opin. Invest. Drugs* 10:269–80
39. Georgopapadakou NH, Walsh TJ. 1994. Human mycoses: drugs and targets for emerging pathogens. *Science* 264:371–73
40. Gerrish PJ, Lenski RE. 1998. The fate of competing beneficial mutations in an asexual population. *Genetica* 103:127–44
41. Ghannoum MA. 1996. Is antifungal susceptibility testing useful in guiding fluconazole therapy? *Clin. Infect. Dis.* 22: S161–65
42. Ghannoum MA, Rex JH, Galgiani JN. 1996. Susceptibility testing of fungi: current status of correlation of in vitro data with clinical outcome. *J. Clin. Microbiol.* 34:489–95
43. Giaever G, Shoemaker DD, Jones TW, Liang H, Winzeler EA, et al. 1999. Genomic profiling of drug sensitivities via induced haploinsufficiency. *Nat. Genet.* 21:278–83
44. Goldstein AL, McCusker JH. 2001. Development of *Saccharomyces cerevisiae* as a model pathogen. A system for the genetic identification of gene products required for survival in the mammalian host environment. *Genetics* 159:499–513
45. Gräser Y, Volovsek M, Arrington J, Schönian G, Presber W, et al. 1996. Molecular markers reveal that population structure of the human pathogen *Candida albicans* exhibits both clonality and recombination. *Proc. Natl. Acad. Sci. USA* 93:12473–77
46. Gubbins S, Gilligan CA. 1999. Invasion thresholds for fungicide resistance: deterministic and stochastic analyses. *Proc. R. Soc. London Ser. B* 266:2539–49
47. Hall BG. 1999. Toward an understanding of evolutionary potential. *FEMS Microbiol. Lett.* 178:1–6
48. Hartman JL, Garvik B, Hartwell L. 2001. Principles for the buffering of genetic variation. *Science* 291:1001–4
49. Hermans P. 2001. Current review and clinical management of patients with primary HIV-1 infection: limits and perspectives. *Biomed. Pharmacother.* 55:301–7
50. Hull CM, Raisner RM, Johnson AD. 2000. Evidence for mating of the "asexual" yeast *Candida albicans* in a mammalian host. *Science* 289:307–10
51. Klepser ME, Lewis RE, Pfaller MA. 1998. Therapy of *Candida* infections: susceptibility testing, resistance, and therapeutic options. *Ann. Pharmacother.* 32:1353–61
52. Kolaczkowska A, Goffeau A. 1999. Regulation of pleiotropic drug resistance in yeast. *Drug Resist. Updates* 2:403–14
53. Kondrashov FA, Kondrashov AS. 2001. Multidimensional epistasis and the disadvantage of sex. *Proc. Natl. Acad. Sci. USA* 98:12089–92
54. Korona R. 1999. Unpredictable fitness transitions between haploid and diploid strains of the genetically loaded yeast *Saccharomyces cerevisiae*. *Genetics* 151:77–85
55. Korting HC, Schaller M, Eder G, Hamm G, Bohmer U, Hube B. 1999. Effects of the human immunodeficiency virus (HIV) proteinase inhibitors saquinavir and indinavir on in vitro activities of secreted aspartyl proteinases of *Candida albicans* isolates from HIV-infected patients. *Antimicrob. Agents Chemother.* 43:2038–42
56. Lamb D, Kelly D, Kelly S. 1999. Molecular aspects of azole antifungal action and resistance. *Drug Resist. Updates* 2:390–402
57. Lata Panwar S, Krishnamurthy S, Gupta V, Alarco AM, Raymond M, et al. 2001. Ca*ALK8*, an alkane assimilating cytochrome P450, confers multidrug

resistance when expressed in a hypersensitive strain of *Candida albicans*. *Yeast* 18: 1117–29

58. Lengeler KB, Davidson RC, D'Souza C, Harashima T, Shen WC, et al. 2000. Signal transduction cascades regulating fungal development and virulence. *Microbiol. Mol. Biol. Rev.* 64:746–85
59. Lengeler KB, Wang P, Cox GM, Perfect JR, Heitman J. 2000. Identification of the MATa mating-type locus of *Cryptococcus neoformans* reveals a serotype A MATa strain thought to have been extinct. *Proc. Natl. Acad. Sci. USA* 97:14455–60
60. Lenski RE. 1993. Assessing the genetic structure of microbial populations. *Proc. Natl. Acad. Sci. USA* 90:4334–36
61. Lenski RE, Mongold JA, Sniegowski PD, Travisano M, Vasi F, et al. 1998. Evolution of competitive fitness in experimental populations of *E. coli*: What makes one genotype a better competitor than another? *Antonie Van Leeuwenhoek* 73:35–47
62. Lenski RE, Travisano M. 1994. Dynamics of adaptation and diversification: a 10,000-generation experiment with bacterial populations. *Proc. Natl. Acad. Sci. USA* 91:6808–14
63. Levin BR. 2001. Minimizing potential resistance: a population dynamics view. *Clin. Infect. Dis.* 33:S161–69
64. Levin BR, Bergstrom CT. 2000. Bacteria are different: observations, interpretations, speculations, and opinions about the mechanisms of adaptive evolution in prokaryotes. *Proc. Natl. Acad. Sci. USA* 97:6981–85
65. Levin BR, Lipsitch M, Bonhoeffer S. 1999. Population biology, evolution, and infectious disease: convergence and synthesis. *Science* 283:806–9
66. Levin BR, Perrot V, Walker N. 2000. Compensatory mutations, antibiotic resistance and the population genetics of adaptive evolution in bacteria. *Genetics* 154:985–97
67. Lipsitch M, Bergstrom CT, Levin BR. 2000. The epidemiology of antibiotic resistance in hospitals: paradoxes and prescriptions. *Proc. Natl. Acad. Sci. USA* 97:1938–43
68. Lorenz MC, Fink GR. 2001. The glyoxylate cycle is required for fungal virulence. *Nature* 412:83–86
69. Lott TJ, Effat MM. 2001. Evidence for a more recently evolved clade within a *Candida albicans* North American population. *Microbiology* 147:1687–92
70. Lott TJ, Holloway BP, Logan DA, Fundyga R, Arnold J. 1999. Towards understanding the evolution of the human commensal yeast *Candida albicans*. *Microbiology* 145:1137–43
71. Luria SE, Delbruck M. 1943. Mutations of bacteria from virus sensitivity to virus resistance. *Genetics* 28:491–511
72. Luu LN, Cowen LE, Sirjusingh C, Kohn LM, Anderson JB. 2001. Multilocus genotyping indicates that the ability to invade the bloodstream is widespread among *Candida albicans* isolates. *J. Clin. Microbiol.* 39:1657–60
73. Mable BK, Otto SP. 2001. Masking and purging mutations following EMS treatment in haploid, diploid and tetraploid yeast (*Saccharomyces cerevisiae*). *Genet. Res.* 77:9–26
74. Magee BB, Magee PT. 2000. Induction of mating in *Candida albicans* by construction of MTLa and MTLalpha strains. *Science* 289:310–13
75. Marton MJ, DeRisi JL, Bennett HA, Iyer VR, Meyer MR, et al. 1998. Drug target validation and identification of secondary drug target effects. *Nat. Med.* 4:1293–301
76. McNeil MM, Nash SL, Hajjeh RA, Phelan MA, Conn LA, et al. 2001. Trends in mortality due to invasive mycotic diseases in the United States, 1980–1997. *Clin. Infect. Dis.* 33:641–47
77. Milgroom MG. 1990. A stochastic model for the initial occurrence and development of fungicide resistance in plant pathogen populations. *Phytopathology* 80:410–16

78. Mitchell AP. 1998. Dimorphism and virulence in *Candida albicans*. *Curr. Opin. Microbiol.* 1:687–92
79. Morison L. 2001. The global epidemiology of HIV/AIDS. *Br. Med. Bull.* 58:7–18
80. Nguyen MH, Peacock JJ, Morris AJ, Tanner DC, Nguyen ML, et al. 1996. The changing face of candidemia: emergence of non-*Candida albicans* species and antifungal resistance. *Am. J. Med.* 100:617–23
81. Odds FC. 2000. Pathogenic fungi in the 21st century. *Trends Microbiol.* 8:200–1
82. Odds FC, Gow NA, Brown AJ. 2001. Fungal virulence studies come of age. *Genome Biol.* 2: reviews 1009.1–1009.4
83. Orr HA. 1999. The evolutionary genetics of adaptation: a simulation study. *Genet. Res.* 74:207–14
84. Otto SP, Barton NH. 2001. Selection for recombination in small populations. *Evolution* 55:1921–31
85. Paquin CE, Adams J. 1983. Relative fitness can decrease in evolving asexual populations of *S. cerevisiae*. *Nature* 306:368–71
86. Pennisi E. 2001. The push to pit genomics against fungal pathogens. *Science* 292:2273–74
87. Perepnikhatka V, Fischer FJ, Niimi M, Baker RA, Cannon RD, et al. 1999. Specific chromosome alterations in fluconazole-resistant mutants of *Candida albicans*. *J. Bacteriol.* 181:4041–49
88. Perfect JR, Cox GM. 1999. Drug resistance in *Cryptococcus neoformans*. *Drug Resist. Updates* 2:259–69
89. Pfaller MA. 1998. The epidemiology of invasive mycoses—narrowing the gap. *Clin. Infect. Dis.* 27:1148–50
90. Pfaller MA, Diekema DJ, Jones RN, Sader HS, Fluit AC, et al. 2001. International surveillance of bloodstream infections due to *Candida* species: frequency of occurrence and in vitro susceptibilities to fluconazole, ravuconazole, and voriconazole of isolates collected from 1997 through 1999 in the SENTRY antimicrobial surveillance program. *J. Clin. Microbiol.* 39:3254–59
91. Pfaller MA, Jones RN, Messer SA, Edmond MB, Wenzel RP. 1998. National surveillance of nosocomial bloodstream infection due to *Candida albicans*: frequency of occurrence and antifungal susceptibility in the SCOPE Program. *Diagn. Microbiol. Infect. Dis.* 31:327–32
92. Pfaller MA, Lockhart SR, Pujol C, Swails WJ, Messer SA, et al. 1998. Hospital specificity, region specificity, and fluconazole resistance of *Candida albicans* bloodstream isolates. *J. Clin. Microbiol.* 36:1518–29
93. Pfaller MA, Messer SA, Hollis RJ, Jones RN, Doern GV, et al. 1999. Trends in species distribution and susceptibility to fluconazole among bloodstream isolates of *Candida* species in the United States. *Diagn. Microbiol. Infect. Dis.* 33:217–22
94. Pinner RW, Teutsch SM, Simonsen L, Klug LA, Graber JM, et al. 1996. Trends in infectious diseases mortality in the United States. *JAMA* 275:189–93
95. Pujol C, Reynes J, Renaud F, Raymond M, Tibayrenc M, et al. 1993. The yeast *Candida albicans* has a clonal mode of reproduction in a population of infected human immunodeficiency virus-positive patients. *Proc. Natl. Acad. Sci. USA* 90:9456–59
96. Rainey PB, Buckling A, Kassen R, Travisano M. 2000. The emergence and maintenance of diversity: insights from experimental bacterial populations. *Trends Ecol. Evol.* 15:243–47
97. Rainey PB, Travisano M. 1998. Adaptive radiation in a heterogeneous environment. *Nature* 394:69–72

97a. Ramage G, VandeWalle K, Wickes BL, Lopez-Ribot JL. 2001. Characteristics of biofilm formation by *Candida albicans*. *Rev. Iberoam. Micol.* 18:163–70

97b. Ramage G, Wickes BL, Lopez-Ribot JL. 2001. Biofilms of *Candida albicans* and their associated resistance to antifungal agents. *Am. Clin. Lab.* 20:42–44

98. Rangel-Frausto MS, Wiblin T, Blumberg

HM, Saiman L, Patterson J, et al. 1999. National epidemiology of mycoses survey (NEMIS): variations in rates of bloodstream infections due to *Candida* species in seven surgical intensive care units and six neonatal intensive care units. *Clin. Infect. Dis.* 29:253–58

99. Raymond M, Dignard D, Alarco AM, Clark KL, Weber S, et al. 2000. Molecular cloning of the *CRM1* gene from *Candida albicans*. *Yeast* 16:531–38
100. Rex JH, Pfaller MA, Galgiani JN, Bartlett MS, Espinel-Ingroff A, et al. 1997. Development of interpretive breakpoints for antifungal susceptibility testing: conceptual framework and analysis of in vitro/in vivo correlation data for fluconazole, itraconazole, and *Candida* infections. *Clin. Infect. Dis.* 24:235–47
101. Rex JH, Pfaller MA, Walsh TJ, Chaturvedi V, Espinel-Ingroff A, et al. 2001. Antifungal susceptibility testing: practical aspects and current challenges. *Clin. Microbiol. Rev.* 14:643–58
102. Rex JH, Rinaldi MG, Pfaller MA. 1995. Resistance of *Candida* species to fluconazole. *Antimicrob. Agents Chemother.* 39:1–8
103. Rex JH, Walsh TJ, Sobel JD, Filler SG, Pappas PG, et al. 2000. Practice guidelines for the treatment of candidiasis. Infectious Diseases Society of America. *Clin. Infect. Dis.* 30:662–78
104. Reyes G, Ghannoum MA. 2000. Antifungal susceptibility testing of yeasts: uses and limitations. *Drug Resist. Updates* 3:14–19
105. Rose MR, Lauder GV, eds. 1996. *Adaptation.* San Diego: Academic. 511 pp.
106. Rosewich U, Kistler H. 2000. Role of horizontal gene transfer in the evolution of fungi. *Annu. Rev. Phytopathol.* 38:325–63
107. Ruiz-Diez B, Martinez V, Alvarez M, Rodriguez-Tudela JL, Martinez-Suarez JV. 1997. Molecular tracking of *Candida albicans* in a neonatal intensive care unit: long-term colonizations versus catheter-related infections. *J. Clin. Microbiol.* 35:3032–36
108. Sanglard D. 2001. Integrated antifungal drug discovery in *Candida albicans*. *Nat. Biotechnol.* 19:212–13
109. Sanglard D, Ischer F, Calabrese D, de Micheli M, Bille J. 1998. Multiple resistance mechanisms to azole antifungals in yeast clinical isolates. *Drug Resist. Updates* 1:255–65
110. Sheehan DJ, Hitchcock CA, Sibley CM. 1999. Current and emerging azole antifungal agents. *Clin. Microbiol. Rev.* 12:40–79
111. Sniegowski P. 1999. The genomics of adaptation in yeast. *Curr. Biol.* 9:R897–98
112. Sniegowski PD, Gerrish PJ, Johnson T, Shaver A. 2000. The evolution of mutation rates: separating causes from consequences. *Bioessays* 22:1057–66
113. Sniegowski PD, Gerrish PJ, Lenski RE. 1997. Evolution of high mutation rates in experimental populations of *E. coli*. *Nature* 387:703–5
114. Stemmer WP. 1994. Rapid evolution of a protein in vitro by DNA shuffling. *Nature* 370:389–91
115. Talibi D, Raymond M. 1999. Isolation of a putative *Candida albicans* transcriptional regulator involved in pleiotropic drug resistance by functional complementation of a *pdr1 pdr3* mutation in *Saccharomyces cerevisiae*. *J. Bacteriol.* 181:231–40
116. Taylor JW, Geiser DM, Burt A, Koufopanou V. 1999. The evolutionary biology and population genetics underlying fungal strain typing. *Clin. Microbiol. Rev.* 12:126–46
117. Taylor JW, Jacobson DJ, Kroken S, Kasuga T, Geiser DM, et al. 2000. Phylogenetic species recognition and species concepts in fungi. *Fungal Genet. Biol.* 31: 21–32
118. Taylor M, Feyereisen R. 1996. Molecular biology and evolution of resistance to toxicants. *Mol. Biol. Evol.* 13:719–34
119. Tibayrenc M. 1997. Are *Candida albicans* natural populations subdivided? *Trends Microbiol.* 5:253–54

120. Tkacz JS, DiDomenico B. 2001. Antifungals: What's in the pipeline. *Curr. Opin. Microbiol.* 4:540–45
121. Travisano M. 2001. Evolution: towards a genetical theory of adaptation. *Curr. Biol.* 11:R440–42
122. Travisano M, Mongold JA, Bennett AF, Lenski RE. 1995. Experimental tests of the roles of adaptation, chance, and history in evolution. *Science* 267:87–90
123. Tzung KW, Williams RM, Scherer S, Federspiel N, Jones T, et al. 2001. Genomic evidence for a complete sexual cycle in *Candida albicans*. *Proc. Natl. Acad. Sci. USA* 98:3249–53
124. Vanden Bossche H, Dromer F, Improvisi I, Lozano CM, Rex JH, Sanglard D. 1998. Antifungal drug resistance in pathogenic fungi. *Med. Mycol. Suppl.* 1:119–28
125. Vanden Bossche H, Willemsens G, Marichal P. 1987. Anti-*Candida* drugs—the biochemical basis for their activity. *Crit. Rev. Microbiol.* 15:57–72
126. Verduyn Lunel FM, Meis JF, Voss A. 1999. Nosocomial fungal infections: candidemia. *Diagn. Microbiol. Infect. Dis.* 34:213–20
127. Vilgalys R, Gräser Y, Schönian G, Presber W. 1997. Response to Tibayrenc. *Trends Microbiol.* 5:254–57
128. Voss A, Pfaller MA, Hollis RJ, Rhine-Chalberg J, Doebbeling BN. 1995. Investigation of *Candida albicans* transmission in a surgical intensive care unit cluster by using genomic DNA typing methods. *J. Clin. Microbiol.* 33:576–80
129. Wahl LM, Krakauer DC. 2000. Models of experimental evolution: the role of genetic chance and selective necessity. *Genetics* 156:1437–48
130. Walsh TJ, Viviani MA, Arathoon E, Chiou C, Ghannoum M, et al. 2000. New targets and delivery systems for antifungal therapy. *Med. Mycol.* 38:335–47
131. White KP. 2001. Functional genomics and the study of development, variation and evolution. *Nat. Rev. Genet.* 2:528–37
132. White TC, Marr KA, Bowden RA. 1998. Clinical, cellular, and molecular factors that contribute to antifungal drug resistance. *Clin. Microbiol. Rev.* 11:382–402
133. Whiteway M. 2000. Transcriptional control of cell type and morphogenesis in *Candida albicans*. *Curr. Opin. Microbiol.* 3:582–88
134. Wichman HA, Badgett MR, Scott LA, Boulianne CM, Bull JJ. 1999. Different trajectories of parallel evolution during viral adaptation. *Science* 285:422–24
135. Wichman HA, Scott LA, Yarber CD, Bull JJ. 2000. Experimental evolution recapitulates natural evolution. *Philos. Trans. R. Soc. London Ser. B* 355:1677–84
136. Wloch DM, Szafraniec K, Borts RH, Korona R. 2001. Direct estimate of the mutation rate and the distribution of fitness effects in the yeast *Saccharomyces cerevisiae*. *Genetics* 159:441–52
137. Xu J, Boyd CM, Livingston E, Meyer W, Madden JF, Mitchell TG. 1999. Species and genotypic diversities and similarities of pathogenic yeasts colonizing women. *J. Clin. Microbiol.* 37:3835–43
138. Xu J, Mitchell TG, Vilgalys R. 1999. PCR-restriction fragment length polymorphism (RFLP) analyses reveal both extensive clonality and local genetic differences in *Candida albicans*. *Mol. Ecol.* 8:59–73
139. Xu J, Ramos AR, Vilgalys R, Mitchell TG. 2000. Clonal and spontaneous origins of fluconazole resistance in *Candida albicans*. *J. Clin. Microbiol.* 38:1214–20
140. Xu J, Vilgalys R, Mitchell TG. 1999. Lack of genetic differentiation between two geographically diverse samples of *Candida albicans* isolated from patients infected with human immunodeficiency virus. *J. Bacteriol.* 181:1369–73
141. Zeyl C. 2000. Budding yeast as a model organism for population genetics. *Yeast* 16:773–84
142. Zeyl C, DeVisser JA. 2001. Estimates of the rate and distribution of fitness effects of spontaneous mutation in *Saccharomyces cerevisiae*. *Genetics* 157:53–61
143. Zeyl C, Mizesko M, de Visser JA. 2001. Mutational meltdown in laboratory yeast populations. *Evolution* 55:909–17

Annu. Rev. Microbiol. 2002. 56:167–85
doi: 10.1146/annurev.micro.56.012302.160616

First published online as a Review in Advance on April 19, 2002

BIOTERRORISM: From Threat to Reality

Ronald M. Atlas
Department of Biology, University of Louisville, Louisville, Kentucky 40292; e-mail: r.atlas@louisville.edu

Key Words biological weapon, anthrax, smallpox, microorganism, toxin

■ **Abstract** The fears and predictions of attacks with biological weapons, which were increasing at the close of the twentieth century, were transformed into reality not long after September 11, 2001, when several anthrax-laden letters were sent through the U.S. postal system. The attack challenged our medical preparedness and scientific understanding of the epidemiology of biothreat agents. It is fortunate that this was not a massive aerosol release that could have exposed hundreds of thousands. Rapid diagnoses and medical treatments limited casualties and increased survival rates, but tragically some individuals died of inhalational anthrax. Even as physicians tested new treatment regimes and scientists employed new ways of detecting anthrax and decontaminating the mail, new predictions were made for potentially even more devastating attacks with anthrax, smallpox, plague, tularemia, botulism, or hemorrhagic fever viruses. Fear gripped the nation. Law enforcement sought to find the villain(s) who sent the anthrax letters and to deter future bioterrorist attacks. The biomedical community began to seek new ways of protecting against such future threats of bioterrorism.

CONTENTS

THE THREAT OF BIOTERRORISM AND BIOLOGICAL WARFARE

At the dawn of the twenty-first century former Secretary of Defense William Cohen proclaimed that the United States faced a "superpower paradox": The unrivaled supremacy of the United States in the conventional military arena had prompted adversaries to seek unconventional, asymmetric means to strike at the United States. This then raised the prospect that weapons of terror would be used by individuals or groups of fanatical terrorists or by self-proclaimed apocalyptic prophets (17). What would have happened if Timothy McVeigh had used anthrax instead of explosives in his attack on the Oklahoma City federal building (8)? Were we at the brink of a new age of catastrophic terrorism? Was Michael Osterholm, former state epidemiologist for Minnesota, right when he said, "I do not believe it is a question of whether a lone terrorist or terrorist group will use infectious disease agents to kill unsuspecting citizens; I'm convinced it's really just a question of when and where" (40)? And is he right when he now predicts even more devastating bioterrorism events—perhaps with smallpox? Are we facing the reality of his *Living Terrors* (41), that is, the specter that a smart terrorist could use smallpox, anthrax, bubonic plague, or other deadly diseases to wreak havoc on millions of Americans?

Several facts contributed to the growing danger of bioterrorism: Minute amounts of select biological agents can cause mass casualties; agents for biological weapons are easily acquired; technology for production and weaponization is readily available; only limited financing and training are needed to establish a biological weapons program; biological weapons have low visibility and are relatively easy to deliver; and the motivations of terrorists have changed (15). According to the Stockholm International Peace Research Institute, the cost to inflict civilian casualties is \$2000 per square kilometer with conventional weapons, \$800 with nuclear weapons, and \$1 with biological weapons (44). Recipes for homebrews of botulinum toxin could easily be found on the World Wide Web, and terrorist groups have attempted to acquire biological agents and expertise from governmental biological weapons programs. The estimates of casualties from a potential biological weapons attack range into the millions (57).

Whereas some argued that few terrorist groups possessed the scientific-technical resources required for the successful large-scale release of a biological agent, citing the example of Aum Shinrikyo, which had considerable wealth and scientific expertise, but failed in 10 separate attempts to carry out open-air attacks against urban targets with aerosolized anthrax spores or botulinum toxin (54), others said bioterrorism was inevitable. Kortepeter & Parker proposed that a wide range of groups or individuals might use biological agents as instruments of terror, claiming that well-funded and possibly state-supported organizations could be expected to cause the greatest harm because of their access to scientific expertise, biological agents, and most importantly, dissemination technology, which includes the capability to produce refined dry agent, deliverable in milled particles of the proper size for aerosol dissemination (34).

Siegrist contended that two of three prerequisites for bioterrorism, vulnerability and capability, have been in place and that only the third, the intent of a perpetrator to carry out an attack with biological weapons, has been lacking (50). He stated that previously terrorists made political statements through violence in order to influence an audience, but "postmodern" or "superterrorism" aims to maximize the number of casualties, reflecting a shift in the goal of the terrorists, from making political statements to maximizing damage to the target as an end in itself. These terrorists are often motivated by ethnic or religious considerations. Siegrist believes that a superpower's strength provides the incentive for a rogue state to launch a clandestine attack. He claims that if a rogue regime were to mount an unconventional asymmetric attack, they would choose biological weapons because their extreme destructive potential is concentrated in a relatively small and unremarkable package with virtually no detectable sensor signature; because of the agent's incubation period, the perpetrators might be gone before anyone knew that an attack had been made; and because biological agents, unlike ballistic missiles, lend themselves to clandestine dissemination. He joined others who view biological weapons, which kill people but leave infrastructure intact, as the "poor man's neutron bomb."

In *Toxic Terror* Tucker (55) presented a series of case studies of terrorist groups that have attempted to develop and in some cases to use biological weapons. By comparing and contrasting 12 cases in which terrorist groups and individuals who, from 1946 to 1998, allegedly acquired or employed chemical and biological weapons agents, the book identifies characteristic motivations and patterns of behavior associated with chemical and biological terrorism. The motivations range from apocalyptic prophecies to environmental ideology to revenge.

The activities of the Japanese cult Aum Shinrikyo greatly raised the level of reasonable concern about the threat of bioterrorism (31, 39). The Aum experimented with various biological agents, including botulinum toxin, anthrax, cholera, and Q fever. It sought to acquire Ebola virus, and in early 1994 cult doctors were quoted on Russian radio as discussing the possibility of using Ebola as a biological weapon. The Aum actually attempted several unsuccessful acts of biological terrorism with anthrax and botulinum toxin in Japan between 1990 and 1995.

Although not owing to bioterrorism, an accidental release of spores of *Bacillus anthracis* from a biological weapons plant in 1979 resulted in a major outbreak of human anthrax in Sverdlovsk (now Ekaterinburg, Russia). Soviet officials claimed that the outbreak was due to the consumption of contaminated meat, but Western governments believed it resulted from inhalation of spores accidentally released from a nearby military research facility (56), a conclusion sustained by an in-depth investigation by Meselson et al. (37). This conclusion was based upon epidemiological data showing that most victims worked or lived in a narrow zone extending from the military facility to the southern city limit. Farther south, livestock died of anthrax along the zone's extended axis. The zone paralleled the northerly wind that prevailed shortly before the outbreak. Guillemin (20) also conducted a thorough epidemiological investigation that included interviews with individuals who witnessed the outbreak and its progression. Abramova et al. (1) performed detailed studies that documented the pathology of inhalational anthrax in the victims. Analysis of tissues from 11 persons who died during the epidemic revealed the presence of *B. anthracis* toxin and capsular antigen genes required for pathogenicity; genes from at least four of the five known strain categories defined by this region were present (29). The Sverdlovsk incident taught us a great deal about the deadly nature of inhalational anthrax and the forensic evidence that should be examined in cases of bioterrorist attacks with anthrax.

The Sverdlosk incident revealed the Soviet Union's biological weapons program, which was conducted in violation of the Biological Weapons Convention. Tens of thousands of people worked as bioweaponeers, and tons of bioweapons were produced (3). The United States suspects several terrorist states have biological warfare programs (34). It has accused Iraq, North Korea, Iran, Libya, Syria, and Sudan, of having offensive biological weapons programs in violation of the Biological Weapons Convention, which prohibits the development and stockpiling of biological weapons (9).

Clearly there are those who want to use biological weapons despite formidable political risks: groups that can acquire the agent and a dissemination device, however crude, and groups whose organizational structure enables them to deliver or disseminate the agent covertly. This set of groups is small but growing, especially for low-technology attacks such as contaminating food or disseminating biological agents in an enclosed space. The most important lesson of the last decade of the twentieth century may have been that biological weapons pose a threat, particularly in conflicts involving Third World countries and/or terrorist groups (45). As we entered the twenty-first century, major attacks with biological weapons were becoming more likely (51, 52).

THE MAJOR BIOTHREAT AGENTS

The Centers for Disease Control and Prevention (CDC) has divided biological agents that are critical biothreat agents into categories based upon their risks for causing mass casualties in the event of a bioterrorist attack (32). Category A agents, the highest priority, represent organisms that (*a*) pose a risk to national security because they can be easily disseminated or transmitted person-to-person; (*b*) cause high mortality, with potential for major public health impact; (*c*) might cause public panic and social disruption; and (*d*) require special action for public health preparedness. Category A agents include variola major (smallpox); *B. anthracis*; *Yersinia pestis*; *Clostridium botulinum* toxin; *Francisella tularensis*; filoviruses, Ebola hemorrhagic fever, Marburg hemorrhagic fever; arenaviruses, Lassa (Lassa fever), Junin (Argentine hemorrhagic fever); and related viruses. Category B includes agents that are moderately easy to disseminate, cause moderate morbidity and low mortality, and require specific enhancements of CDC's diagnostic capacity and enhanced disease surveillance. They include *Coxiella burnetti* (Q fever), *Brucella* species (brucellosis), *Burkholderia mallei* (glanders), alphaviruses, Venezuelan encephalomyelitis, eastern and western equine encephalomyelitis, ricin toxin from *Ricinus communis* (castor beans), epsilon toxin of *Clostridium perfringens*, and *Staphylococcus* enterotoxin B. A subset of category B agents includes pathogens that are food- or water-borne. These pathogens include but are not limited to *Salmonella* species, *Shigella dysenteriae*, *Escherichia coli* O157:H7, *Vibrio cholerae*, and *Cryptosporidium parvum.* Category C agents include emerging pathogens that could be engineered for mass dissemination in the future because of availability, ease of production and dissemination, and potential for high morbidity and mortality and major health impact. They include Nipah virus, hantaviruses, tick-borne hemorrhagic fever viruses, tick-borne encephalitis viruses, yellow fever, and multidrug-resistant tuberculosis. Preparedness for category C agents requires ongoing research to improve disease detection, diagnosis, treatment, and prevention.

Most concern is placed on the category A biothreat agents. Eitzen et al. (18) prepared a handbook for the medical management of biological casualties, and Franz et al. (19) at USAMRIID produced a primer on 10 classical biological warfare agents. These references are invaluable resources for physicians in the recognition of diseases that might result from a bioterrorist attack.

Anthrax

Anthrax is a biological warfare agent that has been produced and weaponized. Studies of animals suggest that inhalational anthrax results from inspiration of 8000–50,000 spores of *B. anthracis* (28). The World Health Organization estimated that 50 kg of anthrax spores spread over a 2-km path of a major city could result in 100,000 casualties and 95,000 deaths, making anthrax a formidable bioterrorism threat (57). The U.S. Department of Defense considers anthrax the greatest potential biological warfare danger to U.S. Service personnel (17); anthrax also is a major threat to civilian populations.

Prior to September 11, 2001, the following information was available for diagnosing anthrax (19, 28). Inhalational anthrax begins with a prodrome featuring fever, malaise, and fatigue; nonproductive cough and vague chest discomfort also may be present. After an incubation period (typically 1–6 days but as long as 43 days has been reported) a nonspecific flu-like illness ensues, characterized by fever, myalgia, headache, a nonproductive cough, and mild chest discomfort (13). A brief intervening period of improvement sometimes follows 1–3 days of these prodromal symptoms, but rapid deterioration follows. This second phase is marked by high fever, dyspnea, stridor, cyanosis, and shock. Chest wall edema and hemorrhagic meningitis (present in up to 50% of cases) may be seen late in the disease course. Physical findings are usually nonspecific. The chest X-ray film is typically without infiltrates but may reveal a widened mediastinum with pleural effusions, which may be hemorrhagic. Chest radiographs may show pleural effusions and a widened mediastinum, although true pneumonitis is typically absent. Blood smears in the later stages of illness may contain the characteristic gram-positive spore-forming bacilli. Death is universal in untreated cases and may occur in as many as 95% of treated cases if therapy is begun more than 48 h after the onset of symptoms.

For treatment the U.S. Food and Drug Administration had approved ciprofloxacin, and for protection they licensed anthrax vaccine adsorbed (AVA), which is prepared from a cell-free filtrate of *B. anthracis* culture. The strain used (V770-NP1-R) is toxigenic and nonencapsulated. The filtrate, containing the toxins and other cellular products, is adsorbed to aluminum hydroxide. Achieving immunity requires six administrations of vaccine, and maintenance of immunity requires annual boosters. This vaccine, however, is only available to the military. More than 487,000 service members have received over 1.9 million vaccinations.

To understand the problem a deliberate release of anthrax spores would present to the medical community, Inglesby (25) presented a hypothetical scenario. During a football game with 74,000 in attendance a truck drives by, releasing an aerosol of anthrax spores. Approximately 16,000 of the 74,000 fans and another 4000 in the business and residential districts downwind of the region are infected. Two days later hundreds of people become ill with flu-like symptoms. Some self-administer with over-the-counter cold remedies; some seek phone advice from medical personnel; others are seen in clinics, doctors' offices, and emergency departments. A few get chest radiographs to exclude pneumonia. Only in retrospect, would the widened mediastinum seen on a number of chest radiographs be recognized as diagnostic for anthrax. Soon many are stricken and human casualties of bioterrorism become a reality. Intensive care units and isolation beds soon are completely full. Patients are febrile, hypotensive, and seem to be in septic shock; some have meningitis. Patients are dying. This scenario is ominous but realistically reflects the diagnostic challenge that would be faced by the healthcare community in the event of an aerosolized anthrax bioterrorist attack.

Plague

Yersinia pestis can also be aerosolized and is considered a serious threat for a bioterrorist attack. The most likely clinical presentation for the disease would be primary pneumonic plague (19, 26). After an incubation period of 2–3 days patients would appear with pneumonia featuring the acute and often fulminant onset of malaise, high fever, chills, headache, myalgia, cough with production of a bloody sputum, and clinical sepsis. Chest X-rays would show patchy or consolidated bronchopneumonia. Pneumonic plague progresses rapidly, resulting in dyspnea, stridor, and cyanosis. The terminal course may feature respiratory failure, shock, and ecchymoses. A presumptive diagnosis can be made by identifying a gram-negative coccobacillus and safety pin bipolar staining organisms in gram-stained or Wright-Giemsa-stained smears from peripheral blood, lymph node needle aspirate, sputum, or other clinical specimens. Immunofluorescent staining for the capsule is diagnostic. Plague pneumonia is almost always fatal if treatment is not initiated within 24 h of the onset of symptoms.

A large-scale simulated bioterrorist attack involving plague, exercise TOPOFF (named for the engagement of top officials of the U.S. government), was carried out in May 2000 (27). The exercise, directed by the U.S. Department of Justice, sought to highlight public health system weaknesses by staging a mock bioweapons event, the release of an aerosol of *Y. pestis* in Denver, Colorado (23). As the exercise proceeded, it became clear that unless controlling the spread of the disease and triage and treatment of patients in hospitals receive equal effort, the demand for healthcare services would not diminish. Importantly, the exercise suggested that both healthcare facilities and local and state public health agencies are poorly prepared for such attacks. For example, preassembled stockpiles of supplies, pharmaceuticals, and medical equipment intended for emergency delivery within 12 h arrived at the desired locations on time but were inefficiently broken down and distributed within the community (46).

Tularemia

Tons of *Francisella tularensis* were produced as a biological weapon by the former Soviet Union and hence is considered a likely biological weapon (3). Typhoidal or septicemic tularemia manifests as fever, prostration, and weight loss, but without adenopathy (16, 19). Respiratory symptoms, substernal discomfort and a nonproductive cough may also be present. Radiological evidence of pneumonia, with associated pleural effusion in some cases, may be present in all forms of tularemia, but it is most common with typhoidal disease. The case-fatality rate with all forms of untreated typhoidal disease is approximately 35%. Diagnosis of primary typhoidal tularemia is difficult because signs and symptoms are nonspecific and the patient is frequently lacking a suggestive exposure history. The diagnosis is best established retrospectively by serologic testing.

The Hanuman Redux 2001 exercise, staged by the CDC and run in Louisville, Kentucky, a few weeks prior to September 11, 2001, used tularemia as the biological agent (24). A suspicious device was found in an area where hotel guests had gathered as a result of a fire at the hotel. Some illnesses among first responders raised suspicion of a possible biological attack. Although these illnesses were found to be unrelated, an epidemiological investigation detected the bacterium that causes tularemia on the device. The device was designed to release an aerosol, and it was quickly determined that the agent would have been carried for several miles from the site of release by the prevailing wind, potentially exposing over 200,000 people who live in the area. Modeling indicated that 4800 individuals would have become ill and 1700 would have died without medical intervention. The national pharmaceutical stockpile was activated and shipped to Louisville, arriving within 9 h. It took an additional 41 h and round-the-clock work by 500 individuals to get doxycycline to the 209,000 individuals who might have been exposed. This would have been within the timeframe needed to prevent mass casualties. The exercise showed that regions with adequate bioterrorism response plans can respond to even a sizeable attack and that local, state, and federal health, law enforcement, and other agencies can work together effectively.

Botulism

Botulism, which is caused by a neurotoxin produced from the anaerobic, spore-forming bacterium *C. botulinum*, also is a likely candidate for use by bioterrorists. The Aum Shinrikyo, for example, on several occasions tried to spread aerosolized botulinum toxin. Botulinum toxins are the most toxic compounds known, with an estimated toxic dose (serotype A) of only 0.001 micrograms/kg of body weight; botulinum toxin is 15,000 times more toxic than the nerve agent VX and 100,000 times more toxic than sarin (5). Botulism is characterized by symmetric, descending, flaccid paralysis of motor and autonomic nerves, usually beginning with the cranial nerves. Blurred vision, dysphagia, and dysarthria are common initial complaints. Skeletal muscle paralysis follows, manifested as a symmetrical, descending, and progressive weakness that may culminate abruptly in respiratory failure. Deep tendon reflexes may be present or absent. With severe respiratory muscle paralysis, the patient may become cyanotic or exhibit narcosis from carbon dioxide retention (5, 48, 49). The occurrence of an epidemic of cases of a descending and progressive bulbar and skeletal paralysis in afebrile patients typical of classical botulism points to the diagnosis of botulinum intoxication. Onset usually occurs 18–36 h after exposure (range, 6 h to 8 days). In severe cases extensive respiratory muscle paralysis leads to ventilatory failure and death unless supportive care is provided. Ventilatory support and 24-h nursing care is most commonly needed for 2–8 weeks, but some patients require such support for up to 7 months before the return of muscular function. Respiratory failure secondary to paralysis of respiratory muscles is the most serious complication and, generally, the cause of death. With tracheostomy or endotracheal intubation and ventilatory assistance, fatalities should be less than 5%.

In the event of a large outbreak of botulism caused by an enteric or aerosolized route of exposure, the primary means of treating victims would be supportive care through the rapid mobilization of mechanical ventilators. However, no municipality has sufficient ventilators to provide intensive care for mass victims of an aerosolized botulinum toxin bioterrorist attack. Rapid administration of botulism antitoxin is the only pharmacologic treatment available and would probably reduce mortality rates. In U.S. Army experiments, equine F(ab′) botulism antitoxin given therapeutically to rhesus monkeys as late as 24 h after an aerosol challenge with a lethal dose of type A toxin resulted in high rates of survival. Without mechanical ventilation, however, the toxin was uniformly lethal if antitoxin administration was delayed until clinical signs had occurred (29–46 h after exposure).

Smallpox

Smallpox is regarded as the greatest threat if acquired by bioterrorists (21). The international black market trade in weapons of mass destruction is probably the only means of acquiring the smallpox virus. Thus, only a terrorist supported by the resources of a rogue state would be able to procure and deploy smallpox. Any attack with smallpox virus would certainly have to be very sophisticated (42). The potential of smallpox as a biological weapon was most dramatically illustrated by two European smallpox outbreaks in the 1970s (21). The first occurred in Meschede, Germany, in 1970 and demonstrated that smallpox virus in an aerosol suspension can spread widely and infect at low doses. The second outbreak occurred in Yugoslavia in February 1972 and showed that even with routine vaccination the disease spread; each infected individual infected 11–13 more people. In the event of an actual release of smallpox and subsequent epidemic, early detection, isolation of infected individuals, surveillance of contacts, and a focused selective vaccination program could be the essential components of an effective control program (22).

Smallpox is contagious. Patients typically become febrile and have severe aching pains and prostration 12 to 14 days after infection (21, 22). The abrupt onset of clinical manifestations is marked by systemic toxicity with prominent malaise, fever, rigors, vomiting, headache, and backache; 15% of patients develop delirium. Some 2–3 days later a papular rash develops over the face and spreads to the extremities. The rash soon becomes vesicular and later, pustular. Lesions are more abundant on the extremities and face; this centrifugal distribution is an important diagnostic feature. In distinct contrast to varicella, lesions on various segments of the body remain generally synchronous in their stage of development. The patient remains febrile throughout the evolution of the rash and customarily experiences considerable pain as the pustules grow and expand. Gradually scabs form, which eventually separate, leaving pitted scars. In the 30% of the cases that result in mortality, death usually occurs during the second week. In 5%–10% of smallpox patients the disease is more rapidly progressive and malignant and almost always fatal within 5–7 days. In such patients lesions are so densely confluent that the skin looks like crepe rubber; some patients exhibit bleeding into the skin and intestinal tract. Such cases are exceedingly infectious.

To appreciate the difficulty that physicians would face in diagnosing smallpox, O'Toole (42) presented a hypothetical scenario of how a smallpox bioterrorist attack might present itself. Initially patients are thought to have influenza (later they are diagnosed as cases of adult chickenpox) but smallpox is not recognized. The disease continues to spread throughout the United States and to numerous other countries. Tens of thousands are afflicted; 30% of them die. In analyzing this scenario, Bardi (7) pointed out that at the start of the epidemic, two weeks after the bioterrorist attack, confusion reigns. There is uncertainty as to what the infection is and reluctance to diagnose smallpox even when it is suspected, because it was eradicated in the late 1970s.

The impact of a covert smallpox attack was further highlighted in an exercise called Dark Winter (43). Key lessons from this exercise were (*a*) leaders are unfamiliar with the nature of bioterrorism and lack an understanding of policy options, (*b*) public health officials and physicians will play a key role in the response, (*c*) individual actions will determine the breadth of epidemic spread, (*d*) there is a lack of surge capacity to deal with mass casualties, and (*e*) the lack of sufficient vaccine limits management options for smallpox.

Recognizing that a new national stockpile of vaccinia vaccine is urgently needed to respond to the possible threat of a deliberate release of smallpox virus, the United States is trying to acquire sufficient doses of smallpox vaccine for every American by the end of 2002. Were an attack with smallpox to occur now, the current national stockpile (fewer than seven million doses of vaccinia virus vaccine) might prove insufficient. The CDC has released an interim smallpox response plan and guidelines for containment and vaccination in the event of a smallpox bioterrorist attack. The plan calls for isolation of confirmed or suspected smallpox patients to limit the potential exposure of nonimmune persons (2, 11). Vaccinia vaccine would be recommended for persons who were exposed to the initial release of the virus; persons who had face-to-face, household, or close-proximity contact (<6.5 feet or 2 meters) with a confirmed or suspected smallpox patient at any time from the onset of the patient's fever until all scabs have separated; personnel involved in the direct medical or public health evaluation, care, or transportation of confirmed or suspected smallpox patients; laboratory personnel involved in the collection or processing of clinical specimens from confirmed or suspected smallpox patients; and other persons who have an increased likelihood of contact with infectious materials from a smallpox patient (e.g., personnel responsible for medical waste disposal, linen disposal or disinfection, and room disinfection in a facility where smallpox patients are present). Vaccination might also be initiated for other groups whose unhindered function is deemed essential to the support of response activities (e.g., selected law enforcement, emergency response, or military personnel) and those who are not otherwise engaged in patient care activities but who have a reasonable probability of contact with smallpox patients or infectious materials.

Smallpox vaccination is associated with some risk for adverse reactions; the two most serious are postvaccinal encephalitis and progressive vaccinia (2). Postvaccinal encephalitis occurs at a rate of three per million primary vaccines; 40% of the cases are fatal, and some patients are left with permanent neurologic damage. Progressive vaccinia occurs among those who are immunosuppressed because of a congenital defect, malignancy, radiation therapy, or AIDS. The vaccinia virus simply continues to grow, and unless these patients are treated with vaccinia-immune globulin, they may not recover. Pustular material from the vaccination site may also be transferred to other parts of the body, sometimes with serious results.

At this time the risks of smallpox vaccination seem to outweigh the benefits. Thus, whereas sufficient stocks of smallpox vaccine with vaccinia virus will be produced, no mandatory smallpox vaccination program is yet planned. A new, safer vaccine as well as antiviral drugs will be sought to enhance preparedness against smallpox. Given the potential threat of smallpox being used as a biological weapon, the U.S. has decided to maintain the remaining stocks of smallpox (variola) virus at the CDC for the foreseeable future. Having live variola virus is viewed as critical for biodefense and efficacy testing of new vaccines and antiviral drugs. Should a credible threat of smallpox be verified or an actual attack occur, vaccine could be rapidly delivered to surround cases and limit the spread of the disease. Currently such a containment strategy is the planned response to combat the potential horrors of bioterrorist use of smallpox.

And if the threat from natural smallpox virus is not dangerous enough, genetic engineering potentially could be used to make an even more dangerous smallpox virus. The former Soviet Union reportedly explored forming a chimera between Marburg and smallpox viruses. The potential for creating horrific biological weapons through genetic engineering was highlighted when a virulent mousepox virus was accidentally created through insertion of an IL-4 gene (38); the resultant recombinant destroyed the cell-mediated response of infected mice and caused 100% mortality within nine days. Although the modified mousepox virus does not affect humans, it is closely related to smallpox, raising fears that the technology could be used to create a smallpox virus that would cause 100% mortality.

We clearly are unprepared now to deal with a major bioterrorist attack from a genetically engineered smallpox virus; in fact, we may never be. Consider the following scenario from the Center for Strategic and International Studies (14). No signs or symptoms of an attack manifested themselves during the incubation period following the covert release of a biological agent. The first cases of illness occurred among the portion of the population with the weakest immune systems: children, the elderly, AIDS patients, and patients undergoing chemotherapy. As the biological weapon produced person-to-person contagion, the victims infected their family and friends. After some days the CDC determined that the outbreak was due to a genetically altered strain of smallpox. Victims—and people believing themselves to be ill—crowded the hospitals, depleting the supply of beds and equipment. Hoarding of medication by medical staffs across the country increased sharply. Antivirals were flown in but, without a distribution mechanism in place, failed to reach the public. Widespread illness resulted in significant shortages of personnel, thereby disrupting critical services including telecommunications, electric power, and air traffic control. Quarantine was instituted, but it was too late. The announcement of the quarantine sparked panic and people tried to flee. Governors deployed the National Guard to prevent people from crossing state lines. Unable to enter surrounding states, while unwilling to return to their homes, thousands of citizens became refugees. Civil order collapsed.

THE REALITY OF BIOTERRORISM

Terrorist Uses of Biological Agents Prior to September 11, 2001

Prior to the post-September 11, 2001 anthrax attack, the largest bioterrorist attack in the United States occurred in 1984 when a cult intentionally contaminated salad bars with *Salmonella typhimurium* in The Dalles, Oregon (53). Epidemiological investigations identified the vehicles of transmission as foods on multiple self-service salad bars and probable times when contamination occurred. A criminal investigation found that members of a nearby religious commune, followers of Bhagwan Shree Rajneesh, had intentionally contaminated the salad bars on multiple occasions. Commune members believed that the outcome of the November 6, 1984, elections for Wasco County commissioners would have an important impact on land-use decisions and plotted to keep voters at home by spreading *Salmonella*. The outbreak of salmonellosis affected at least 751 persons. Despite the extensive epidemiological investigation, the source of *S. typhimurium* initially went unrecognized. It was not until more than a year after the outbreak that sufficient evidence had accumulated to link the religious commune with the outbreak and to identify the outbreak strain of *S. typhimurium* as originating from a clinical laboratory operated by the religious commune. Records showed that it was purchased before the outbreak, and laboratory testing during the following months demonstrated that the isolate matched the outbreak strain. As in many areas of our open society, current practices are inadequate to prevent deliberate contamination of food items by customers.

An outbreak of shigellosis at a medical center in Texas in 1996 also was most likely due to intentional contamination (33). From October 29 through November 1, 12 laboratory workers experienced severe gastrointestinal illness after eating muffins and doughnuts anonymously left in their break room between the night and morning shifts of October 29. The source of the organism was most likely the medical center's own stock culture of *S. dysenteriae* type 2. Although the motive and method of contamination are still unknown, the most likely hypothesis is that it was done by someone who had access to the freezer, had the laboratory skills to culture the organism from the beads and inoculate the pastries, and had access to the locked break room. Cases of deliberate contamination at a hospital, most likely carried out by a hospital worker, have occurred before. During the mid-1960s, several outbreaks of typhoid fever and dysentery in Japanese hospitals were traced to food contaminated by a research bacteriologist who later infected family members and neighbors (4).

The Post-September 11, 2001, Anthrax Attacks

As soon as terrorists attacked the United States on September 11, 2001, the CDC began warning about the possibility of a bioterrorist attack and asking physicians to be alert. In early October a case of inhalational anthrax was detected in a man in Florida. Bob Stevens, who worked for American Media, presented with signs of meningitis, and high numbers of *B. anthracis* were detected in his cerebrospinal fluid. The case proved fatal. Other exposures followed, and envelopes containing spores of *B. anthracis* sent through the U.S. Postal Service were identified in New York and Washington, DC. Anthrax-tainted letters had been sent to several people in the news media and to U.S. Senators. Hoaxes concerning envelopes filled with harmless powders and threatening notes exacerbated the problems facing investigators and the medical community, as did hysteria among Americans as they opened their mail.

The victims included four workers from the postal facility where an anthrax-tainted letter to Senator Daschle had been processed; later a similar letter sent to Senator Leahy was also found. Both letters had been sent from Trenton, New Jersey. Other workers in the New Jersey postal system contracted cutaneous (skin) anthrax and others developed inhalational anthrax. One woman in New York with no known contact with any of the letters developed a fatal case of inhalational anthrax, as did a 94-year-old woman in Connecticut. By the middle of November, there were 18 cases of anthrax—7 cutaneous and 11 inhalational, 5 of which were fatal. Of the 11 inhalational anthrax cases, 7 occurred in postal employees in New Jersey and the District of Columbia who were likely exposed to letters known to be contaminated with *B. anthracis* spores (47). Two cases were employees of a media company in Florida: One is believed to have received contaminated mail, the other to have sorted and distributed that mail. One case was the woman in New York; the nature of her exposure to *B. anthracis* is currently unknown but may have come from contaminated mail. The final case was the 94-year-old woman in Connecticut. The most likely source of her infection was a letter that may have been cross-contaminated. Because of her age she was probably particularly susceptible to inhalational anthrax. In fact there may be an age relationship to inhalational anthrax: All cases of inhalational anthrax involved individuals over 50 years old.

Presumptive isolation of *B. anthracis* in these patients was confirmed by gamma phage lysis, presence of a capsule, detection of capsule and cell wall antigens by direct fluorescent antibody, and *B. anthracis*–specific polymerase chain reaction. All isolates were confirmed by state and other laboratory facilities of the National Bioterrorism Laboratory Response Network and by the laboratories of the National Center for Infectious Diseases (NCID) at the CDC. Other tests performed at NCID for confirming the diagnosis of inhalational anthrax included immunohistochemical testing of clinical specimens by using *B. anthracis* capsule and cell wall antibody, *B. anthracis*–specific polymerase chain reaction, and serologic detection of immunoglobulin G to *B. anthracis*–protective antigen.

Following detection of the bioterrorism attack, the CDC issued interim guidelines for the management of exposed individuals and antimicrobial therapy regimes for those with cutaneous and inhalational anthrax (12). Antimicrobial susceptibility patterns showed that the *B. anthracis* isolates all were susceptible to ciprofloxacin, doxycycline, chloramphenicol, clindamycin, tetracycline, rifampin, vancomycin, penicillin, and amoxicillin. The CDC recommended that persons with an exposure or contact with an item or environment known, or suspected, to have been contaminated with *B. anthracis*—regardless of laboratory tests results—be offered antimicrobial prophylaxis. Antimicrobial prophylaxis was to be initiated pending additional information when (*a*) a person was exposed to an air space where a suspicious material may have been aerosolized or (*b*) a person shared the air space likely to be the source of an inhalational anthrax case. Antimicrobial prophylaxis was recommended to be continued for 60 days for cases (*a*) and (*b*) above and for persons along the transit path of an envelope or other vehicle containing *B. anthracis* that may have been aerosolized and unvaccinated laboratory workers exposed to confirmed *B. anthracis* cultures. The CDC also stated that antimicrobial prophylaxis was not indicated for prevention of cutaneous anthrax, for autopsy personnel when appropriate isolation precautions and procedures are followed, for hospital personnel, or for persons who routinely handle mail in the absence of a suspicious letter or credible threat.

At first ciprofloxacin was recommended as the treatment of choice for prophylaxis, but it was replaced with doxycycline because it is less expensive and has fewer side effects. Because of the mortality associated with inhalational anthrax, two or more antimicrobial agents predicted to be effective were recommended in suspected cases. Other agents with in vitro activity used in conjunction with ciprofloxacin or doxycycline include rifampin, vancomycin, imipenem, chloramphenicol, penicillin and ampicillin, clindamycin, and clarithromycin. Except for penicillin, limited or no data exist regarding the use of these agents in the treatment of this infection. Cephalosporins and trimethoprim-sulfamethoxazole should not be used.

Jernigan et al. (30) detailed the clinical presentation and course of the first 10 confirmed cases of inhalational anthrax caused by intentional release of *B. anthracis*. The median incubation period from time of exposure to onset of symptoms, when known, was four days. Symptoms at initial presentation included fever and chills, sweats, fatigue or malaise, minimal or nonproductive cough, dyspnea, and nausea or vomiting. There was greater fatigue, sweats, and abdominal symptoms than in previously described outbreaks of inhalational anthrax. Median white blood cell counts were 9.9×10^3 /mm^3, often with increased neutrophils and band forms. Seven patients had elevated serum transaminase levels and five were hypoxic. All 10 had abnormal chest X-rays; abnormalities included infiltrates, pleural effusion, and mediastinal widening. Computed tomography of the chest, when performed, showed mediastinal lymphadenopathy. Pleural effusions were a consistent clinical feature. The pleural effusions were often small on presentation, but in the surviving patients effusions were characterized by progressive enlargement and persistence. Drainage of the pleural cavity was required in 7 of the 10. The characteristics of the pleural fluid were all similar: hemorrhagic, with a high protein concentration and relatively few leukocytes. Immunohistochemistry demonstrated large quantities of *B. anthracis* capsule and cell wall antigens in pleural tissue or pleural fluid cell blocks. Blood cultures grew *B. anthracis* even in the initial phase of the illness in all who had not received prior antibiotic therapy. Blood cultures rapidly became sterile after initiation of antibiotic therapy, suggesting that prior antibiotic treatment may substantially decrease the sensitivity of blood cultures as a diagnostic test.

Eight of first 10 patients were in the initial phase of illness when they first sought care. Of these eight, six received antibiotics with activity against *B. anthracis* on the same day, and all six survived. Four patients, including one with meningitis, were exhibiting fulminant signs of illness when they first received antibiotics with activity against *B. anthracis* and died. Mayer et al. (36) described in greater detail the clinical presentation of inhalational anthrax for two surviving patients; both were postal employees working in the Washington, DC facility where the anthrax-laden letter sent to Senator Daschle was processed. Both were treated with ciprofloxacin, rifampin, and clindamycin and did not develop the classic findings of high-grade fever, dyspnea, profound respiratory distress, and shock. Timely intervention with antibiotic therapy and supportive therapy slowed the progression of inhalational anthrax and resulted in improved clinical outcome.

The survival rate with aggressive multidrug antibiotic regimens and supportive care of the 11 patients with inhalational anthrax was nearly 60%; the survival rate in previous occurrences was 5%–15%. The apparent improvement in survival suggests that the antibiotic combinations may have therapeutic advantage over previous regimens. Other explanations for the improved survival rate include earlier recognition and initiation of treatment, better supportive care, differences in the pathogenesis of bioterrorism-related anthrax, differences in susceptibility of the hosts, or a combination of the these factors.

Borio et al. (10) described in detail the cases of the two postal workers who died of inhalational anthrax. Both had nonspecific prodromal illnesses; both sought medical care for apparently mild, nonspecific illnesses and were sent home. One developed predominantly gastrointestinal symptoms, including nausea, vomiting, and abdominal pain. The other had an influenza-like illness associated with myalgias and malaise. Only after the news media reported cases of inhalational anthrax involving two postal workers from the local mail facility did their physicians consider the possibility of inhalational anthrax. The courses of their illness are similar to those reported in the outbreak of anthrax that occurred in Sverdlovsk in 1979. Despite aggressive medical therapy, their disease conditions progressed rapidly, and both died. The duration of the illnesses were 7 days and 5 days from onset of symptoms to death; both patients died within 24 h of hospitalization. The cause of death was certified as inhalational anthrax. The manner of death was certified as homicide.

These cases emphasize that in the event of serious outbreaks of infectious diseases, rapid communication of epidemiologic data to front-line medical care providers (especially emergency physicians and primary care clinicians) is essential so that they may initiate appropriate diagnostic procedures and therapies. To prevent similar deaths and cases of anthrax, more than 30,000 people received antibiotics as a consequence of possible exposure (35). Brookmeyer & Blades (10a) estimated that this prophylactic treatment prevented 10–28 cases of inhalational anthrax. The need for continual reevaluation of conventional wisdom regarding this disease as well as other potential bioterrorist threats has been made clear from these recent experiences.

Just as the medical profession has struggled to develop appropriate prophylactics and treatments, the postal service has also faced difficult decisions about how to safely operate the mail delivery system. Following the discovery of the letter to Senator Daschle, mail at the Senate and various other federal offices was sealed. Thus, it was weeks later that a contaminated letter sent to Senator Leahy was discovered. As postal workers became ill, the postal service tested postal facilities for possible contamination. Some facilities tested positive and had to be temporarily closed and decontaminated. It became clear that spores had passed through the pores of envelopes, contaminating postal equipment and cross-contaminating other letters. Tens of thousands of additional letters may have been contaminated with low levels of spores. These were thought to pose little risk for inhalational anthrax, but the case of the 94-year-old Connecticut woman indicates that the risk is not zero. To protect the mail from such spread of anthrax, the postal service has been in some cases implementing irradiation with electron beams. A radiation dose of 41.5 kGy will cause a 99.9% kill of anthrax spores. The White House Office of Science and Technology has been overseeing tests to achieve greater kill levels. Irradiation can ensure that the mail is free of viable anthrax, but various products, including biological diagnostic specimens, films, and so forth, will have to be shipped by alternate means if widespread irradiation is implemented to protect the mail against the threat of bioterrorism.

CONCLUDING REMARKS

There is no reason to believe that the fall 2001 anthrax attack represents an isolated act of bioterrorism (35). In fact, it is likely that additional attacks involving *B. anthracis*, and perhaps other pathogens, will occur. Each will present the healthcare community, law enforcement, mail delivery, and other segments of society with a new set of challenges and a need for rapid dissemination of reliable up-to-date information. Sharing of information among law enforcement authorities, public health officials, and front-line healthcare providers will continue to be essential. The alertness, open-mindedness, and sound clinical judgment of physicians and other healthcare professionals will be critical to the successful public health response to current and future threats.

A new awareness of the threat of biological weapons and the preventative role that bioscientists can play is imperative (6). Scientists and physicians have an important role to play in informing the public and policymakers concerning actions that can be taken to prevent biological warfare and bioterrorism and to minimize the impact of any use of biological weapons. Scientists and physicians have played important roles in developing biological weapons. They now must seek to deter and defend against bioterrorism, and along with clinicians and researchers they must help diagnose and mitigate a bioterrorist attack.

The *Annual Review of Microbiology* is online at http://micro.annualreviews.org

LITERATURE CITED

1. Abramova F, Grinberg L, Yampolskaya O, Walker D. 1993. Pathology of inhalational anthrax in forty-two cases from the Sverdlovsk outbreak of 1979. *Proc. Natl. Acad. Sci. USA* 90:2291–94
2. Advisory Committee on Immunization Practices. 2001. Vaccinia (smallpox) vaccine. Recommendations of the Advisory Committee on Immunization Practices (ACIP). *MMWR* 50:1–25

3. Alibek K. 1999. *Biohazard.* New York: Random House
4. Anonymous. 1966. Deliberate spreading of typhoid in Japan. *Sci. J.* 2(10):11, 13
5. Arnon SS, Schechter R, Inglesby TV, Henderson DA, Bartlett JG, et al. (Working Group on Civilian Biodefense). 2001. Botulinum toxin as a biological weapon: medical and public health management. *JAMA* 285:1059–70
6. Atlas R. 1998. Biological weapons pose challenge for microbiology community: Microbiologists should help shape policies protecting against biological weapons but safeguarding legitimate research. *ASM News* 64:383–89
7. Bardi J. 1999. Aftermath of a hypothetical smallpox disaster. *Emerg. Infect. Dis.* 5:547–51
8. Betts RK. 1998. The new threat of mass destruction. *Foreign Aff.* 77:26–42
9. Bolton JR. 2001. *5th Biological Weapons Convention Rev Con Meeting, Biological Weapons Convention Remarks*, Geneva November 19. http://www.state.gov/t/us/rm/2001/index.cfm?docid=6231
10. Borio L, Frank D, Mani V, Chiriboga C, Pollanen M, et al. 2001. Death due to bioterrorism-related inhalational anthrax report of 2 patients. *JAMA* 286:2554–59

10a. Brookmeyer R, Blades N. 2002. Prevention of inhalational anthrax in the U.S. outbreak. *Science* 295:1861

11. Centers for Disease Control and Prevention. 2001. Executive summary for interim smallpox response plan & guidelines plus ACIP vaccine recommendations. http://www.cdc.gov/nip/diseases/smallpox
12. Centers for Disease Control and Prevention. 2001. Update: investigation of bioterrorism-related anthrax and interim guidelines for exposure management and antimicrobial therapy. *MMWR* 50:909–19
13. Cieslak TJ, Eitzen EM. 1999. Clinical and epidemiologic principles of anthrax. *Emerg. Infect. Dis.* 5:552–55
14. Ciluffo FJ, Cardash SL, Lederman GN. 2000. *Combating Chemical, Biological, Radiological and Nuclear Terrorism: A Comprehensive Strategy.* Washington, DC: Cent. Strategic Int. Studies
15. Danzig R, Berkowsky PB. 1997. Why should we be concerned about biological warfare? *JAMA* 278:431–32
16. Dennis DT, Inglesby TV, Henderson DA, Bartlett JG, Ascher MS, et al. (Working Group on Civilian Biodefense). 2001. Tularemia as a biological weapon: medical and public health management. *JAMA* 285:2763–73
17. Department of Defense. 2001. *Proliferation: Threat and Response.* http://www.defenselink.mil/news/Jan2001/b01102001_bt010-01.html
18. Eitzen E, Pavlin J, Cieslak T, Christopher G, Culpepper R, eds. 1998. *Medical Management of Biological Casualties Handbook.* Fort Dietrick, MD: USAMRIID
19. Franz DR, Jahrling PB, Friedlander AM, McClain DJ, Hoover DL, et al. 1997. Clinical recognition and management of patients exposed to biological warfare agents. *JAMA* 278:399–411
20. Guillemin J. 1999. *Anthrax: The Investigation of a Deadly Outbreak.* Berkeley: Univ. Calif. Press
21. Henderson DA. 1999. Smallpox: clinical and epidemiologic features. *Emerg. Infect. Dis.* 5:537–39
22. Henderson DA, Inglesby TV, Bartlet JG, Ascher MS, Eitzen E, et al. 1999. Smallpox as a biological weapon: medical and public health management. *JAMA* 281: 2127–37
23. Hoffman RE, Norton JE. 2000. Lessons learned from a full-scale exercise. *Emerg. Infect. Dis.* 6:652–53
24. Humbaugh KE. 2001. Discussion of the lessons learned from the recent CDC bioterrorism exercise held in Louisville. http://www.louisville.edu/medschool/cme/bioterrorismwebcast.html
25. Inglesby TV. 1999. Anthrax: a possible case history. *Emerg. Infect. Dis.* 5:556–60
26. Inglesby TV, Dennis DT, Henderson DA, Bartlett JG, Ascher MS, et al. (Working

Group on Civilian Biodefense). 2000. Plague as a biological weapon: medical and public health management. *JAMA* 283:2281–90
27. Ingleby T, Grossman R, O'Toole T. 2001. A plague on your city: observations from TOPOFF. *Clin. Infect. Dis.* 32:436–45
28. Inglesby TV, Henderson DA, Bartlett JG, Ascher MS, Eitzen E, et al. (Working Group on Civilian Biodefense). 1999. Anthrax as a biological weapon: medical and public health management. *JAMA* 281: 1735–45
29. Jackson PJ, Hugh-Jones ME, Adair DM, Green G, Hill KK, et al. 1998. PCR analysis of tissue samples from the 1979 Sverdlovsk anthrax victims: the presence of multiple *Bacillus anthracis* strains in different victims. *Proc. Natl. Acad. Sci. USA* 95:1224–29
30. Jernigan JA, Stephens DS, Ashford DA, Omenaca C, Martin S, et al. 2001. Bioterrorism-related inhalational anthrax: the first 10 cases reported in the United States. *Emerg. Infect. Dis.* 7:933–44
31. Kaplan DE. 2000. Aum Shinrikyo. See Ref. 55, pp. 207–26
32. Khan AS, Sage MJ. 2000. Biological and chemical terrorism: strategic plan for preparedness and response. *MMWR* 49:1–14
33. Kolavic S, Kimura A, Simons S, Slutsker L, Barth S, Haley C. 1997. An outbreak of *Shigella dysenteriae* Type 2 among laboratory workers due to intentional contamination. *JAMA* 278:396–98
34. Kortepeter MG, Parker GW. 1999. Potential biological weapons threats. *Emerg. Infect. Dis.* 5:523–27
35. Lane CL, Fauci AS. 2001. Bioterrorism on the home front: a new challenge for American medicine. *JAMA* 286:2595–98
36. Mayer TA, Bersoff-Matcha S, Murphy C, Earls J, Harper S, et al. 2001. Clinical presentation of inhalational anthrax following bioterrorism exposure report of 2 surviving patients. *JAMA* 286:2549–53
37. Meselson M, Guillemin J, Hugh-Jones M, Langmuir A, Popova I, et al. 1994. The Sverdlovsk anthrax outbreak of 1979. *Science* 266:1202–8
38. Nowak R. 2001. Disaster in the making. *New Sci.* 169:4
39. Olson KB. 1999. Aum Shinrikyo: once and future threat? *Emerg. Infect. Dis.* 5: 513–16
40. Osterholm M. 1997. The silent killers. *Newsweek* 130(20):32–33
41. Osterholm MT, Schwartz J. 2000. *Living Terrors: What America Needs to Know to Survive the Coming Bioterrorist Catastrophe.* New York: Delacorte
42. O'Toole T. 1999. Smallpox: an attack scenario. *Emerg. Infect. Dis.* 5:540–46
43. O'Toole T, Inglesby T. 2001. *Shining Light on Dark Winter.* www.hopkins-biodefense.org/lessons.html
44. Pearson GS. 1998. *The Threat of Deliberate Disease in the 21st Century.* Henry L. Stimson Center Report, No.24. Washington, DC: Stimson Cent.
45. Poupard JA, Miller LA. 1992. History of biological warfare: catapults to capsomeres. *Ann. NY Acad. Sci.* 666:9–20
46. Russo E. 2001. Bioterrorism preparedness. *Scientist* 15:1
47. Schupp JM, Klevytska AM, Zinser G, Price LB, Keim P. 2000. vrrB, a hypervariable open reading frame in *Bacillus anthracis*. *J. Bacteriol.* 182:3989–97
48. Shapiro R, Hatheway C, Becher J, Swerdlow D. 1997. Botulism surveillance and emergency response. *JAMA* 278:433–35
49. Shapiro R, Hatheway C, Swerdlow D. 1998. Botulism in the United States: a clinical and epidemiologic review. *Ann. Intern. Med.* 129:221–28
50. Siegrist DW. 1999. The threat of biological attack: why concern now? *Emerg. Infect. Dis.* 5:505–8
51. Stern J. 1999. *The Ultimate Terrorists.* Boston: Harvard Univ. Press
52. Stern J. 1999. The prospect of domestic bioterrorism. *Emerg. Infect. Dis.* 5:517–22

53. Török T, Tauxe R, Wise R, Livengood J, Sokolow R, et al. 1997. A large community outbreak of salmonellosis caused by intentional contamination of restaurant salad bars. *JAMA* 278:389–95
54. Tucker JB. 1999. Historical trends related to bioterrorism: an empirical analysis. *Emerg. Infect. Dis.* 5:498–504
55. Tucker JB, ed. 2000. *Toxic Terror: Assessing Terrorist Use of Chemical and Biological Weapons*. Boston: MIT Press
56. Wade N. 1980. Death at Sverdlovsk: a critical diagnosis. *Science* 209:1501–2
57. WHO Group of Consultants. 1970. *Health Aspects of Chemical and Biological Weapons*. Geneva: WHO

Annu. Rev. Microbiol. 2002. 56:187–209
doi: 10.1146/annurev.micro.56.012302.160705

First published online as a Review in Advance on April 19, 2002

BIOFILMS AS COMPLEX DIFFERENTIATED COMMUNITIES

P. Stoodley,[1] K. Sauer,[1,2] D. G. Davies,[2] and J. W. Costerton[1]
[1]*Center for Biofilm Engineering, Montana State University, Bozeman, Montana 59717; e-mail: bill_c@erc.montana.edu*
[2]*Department of Biology, State University of New York, Binghamton, Binghampton, New York 13902*

Key Words cell-cell signaling, biofilm phenotype, prokaryotic development, microbial communities, integrated communities, hydrodynamics

■ **Abstract** Prokaryotic biofilms that predominate in a diverse range of ecosystems are often composed of highly structured multispecies communities. Within these communities metabolic activities are integrated, and developmental sequences, not unlike those of multicellular organisms, can be detected. These structural adaptations and interrelationships are made possible by the expression of sets of genes that result in phenotypes that differ profoundly from those of planktonically grown cells of the same species. Molecular and microscopic evidence suggest the existence of a succession of *de facto* biofilm phenotypes. We submit that complex cell-cell interactions within prokaryotic communities are an ancient characteristic, the development of which was facilitated by the localization of cells at surfaces. In addition to spatial localization, surfaces may have provided the protective niche in which attached cells could create a localized homeostatic environment. In a holistic sense both biofilm and planktonic phenotypes may be viewed as integrated components of prokaryote life.

CONTENTS

INTRODUCTION

In just two decades we have learned that biofilms comprise highly structured matrix-enclosed communities (10) whose cells express genes in a pattern (51) that differs profoundly from that of their planktonic counterparts. Because direct observations show that biofilms constitute the majority of bacteria in most natural (8) and pathogenic (11) ecosystems, it seems unwise to continue to extrapolate from planktonic cultures in studies of these systems. A new mindset is clearly required because direct observations of structural complexity and unequivocal demonstrations of one or more distinct biofilm phenotypes presage a new concept in which biofilms are seen as complex differentiated communities. Observations of biofilms formed in pure cultures of the gamma proteobacteria group of bacteria and of mixed species biofilms in natural ecosystems show a basic organization in which cells grow in matrix-enclosed microcolonies separated by a network of open water channels. The importance of these complex structures, which are seen in direct observations of living biofilms by scanning confocal laser microscopy (SCLM), is that they demonstrate a level of differentiation that requires a sophisticated system of cell-cell signals and a degree of cellular specialization. The simple maintenance of open water channels in multispecies biofilms requires interspecies signaling to direct growth and exo-polysaccharide production away from the channels. We have barely begun to decipher the system of environmental cues and phenotypic responses that shape the multispecies microbial communities that predominate in most ecosystems, but the communities themselves bear witness to a remarkably complex developmental process. This developmental process is unique in biology in that it involves the coordinated activity of several relatively small prokaryotic genomes, rather than one or more large and coordinated eukaryotic genomes, to produce a functional multicellular community. This notion alters the perceived position of bacteria in the hierarchy of living things because the single cells we have studied so acidulously in planktonic cultures are actually members of coordinated multicellular communities whose complexity and sophistication are only now being appreciated.

STRUCTURE AND DIFFERENTIATION IN BIOFILMS

Claude Zobell first noted the preference of marine bacteria for growth on surfaces (66), and Costerton's group has extended this observation to freshwater systems and to a variety of microbial ecosystems, including those on the surfaces of eukaryotic tissues (11). This concept of preferential growth on surfaces included no implications of complex biofilm structure, and as late as 1987 (7), biofilms were perceived (and pictured) as simple "slabs" of matrix material in which sessile bacterial cells were randomly embedded. In this early biofilm era, the question that was often asked and never answered was, "How do the deeply embedded cells in the biofilm have access to nutrients, including oxygen?" This pivotal question was answered, and the modern biofilm era began, when the first SCLM images of living biofilms (35) showed that sessile bacteria grow in matrix-enclosed microcolonies interspersed between open water channels as shown in Figure 1. The basic biofilm structure presented in Figure 1 is a computer-assisted compilation of confocal images of living fully hydrated biofilms formed by pure cultures of several proteobacteria and by natural mixed-species populations in natural ecosystems, such as mountain streams. This basic microcolony and water channel architecture is affected by shear forces and by many other factors discussed below, but its main consequence is that water from the bulk phase is entrained into channels and can therefore deliver nutrients deep within the complex community. Even in

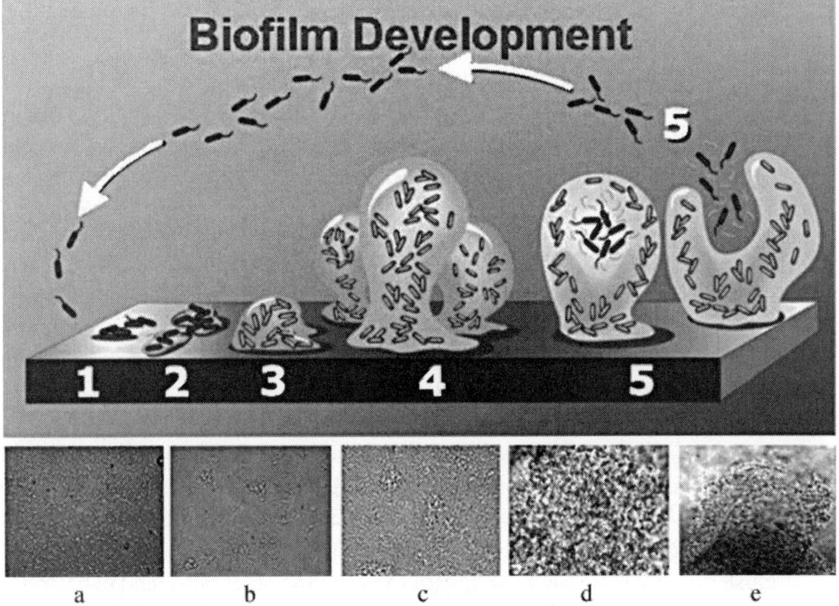

Figure 1 Diagram showing the development of a biofilm as a five-stage process. Stage 1: initial attachment of cells to the surface. Stage 2: production of EPS resulting in more firmly adhered "irreversible" attachment. Stage 3: early development of biofilm architecture. Stage 4: maturation of biofilm architecture. Stage 5: dispersion of single cells from the biofilm. The *bottom panels* (a-e) show each of the five stages of development represented by a photomicrograph of *P. aeruginosa* when grown under continuous-flow conditions on a glass substratum.

thick biofilms, such as the subgingival plaque seen in periodontitis, the functional microcolony architecture is maintained because each structurally distinct microcolony is anchored on the tooth surface and each moves independently in response to shear forces. The exchange of nutrients facilitiated by this biofilm architecture enables biofilm communities to develop considerable thickness and complexity while keeping individual cells (some of which are physiologically specialized) in optimal nutrient situations in many locations within the biofilm.

Initial Processes in Biofilm Formation and Subsequent Structural Differentiation

Biofilm formation can occur by at least three mechanisms. One is by the redistribution of attached cells by surface motility (12, 30). Results from O'Toole & Kolter (45) on studies of *Pseudomonas aeruginosa* mutants suggest that type IV pili-mediated twitching motility plays a role in surface aggregation for this organism. A second mechanism is from the binary division of attached cells (25). As cells divide, daughter cells spread outward and upward from the attachment surface to form cell clusters, in a similar manner to colony formation on agar plates. A time-lapse video illustrating this type of aggregation can be found at the ASM MicrobeLibrary ("Growth and Detachment of Biofilm Cell Cluster in Turbulent Flow," www.microbelibrary.org/Visual/page1.htm). A third mechanism of aggregation is the recruitment of cells from the bulk fluid to the developing biofilm (58a). The relative contribution of each of these mechanisms will depend on the organisms involved, the nature of the surface being colonized, and the physical and chemical conditions of the environment. In oligotrophic environments mature biofilms may consist of little more than a sparse covering of cells with relatively little structural complexity. Biofilms can take over 10 days to reach structural maturity, based on microscopically measured physical dimensions and visual comparison (25, 53). For this reason we must be careful not to misinterpret biofilm retardation in the initial events of attachment, which is caused by a particular gene knockout, with the total suppression of biofilm development.

In *P. putida* and *Escherichia coli* it has been shown that the activity of cells in the centers of the cell clusters diminished as the clusters grew larger, but their activity could be restored by the addition of a more readily utilizable carbon source, indicating that cell activity in the interior of the clusters may be controlled by nutrient availability (52). DeBeer et al. (17) demonstrated that the channels surrounding the cell clusters could increase the supply of oxygen (and other nutrients) to bacteria within the biofilm, thus relating structure to function. The biofilm structure appears to be largely determined by the production of slime-like matrix of extracellular polymeric substances (EPS), which provides the structural support for the biofilm (21). The EPS is composed of polysaccharides, proteins, and nucleic acids. Nivens et al. (42) found that mucoid strains of *P. aeruginosa* produced more structurally differentiated biofilms than nonmucoid strains and more specifically that O-acetylation of alginate (a principal component of the EPS of mucoid *P. aeruginosa* strains) was also required for structural development. Henzter et al. (24) found that they could induce structural complexity in a normally flat undifferentiated wild-type non-mucoid strain of *P. aeruginosa* biofilm by causing the overexpression of alginate. These alginate overproducing biofilms formed mound- and mushroom-shaped cell clusters separated by water channels similar to those formed by the wild-type mucoid strains (42). The nonmucoid PAO1 wild-type strain produced a flat biofilm in the Hentzer (24) study, but a heterogeneous biofilm structurally differentiated into mushroom-shaped cell clusters in the Davies (16) study. Hentzer et al. attribute this difference to differing nutrient conditions. The connection between EPS formation and differentiated biofilm structure is supported by Danese et al. (13) and Watnick et al. (60), who reported that colonic acid and EPS were required for the development of complex structures in *E. coli* K-12 and *Vibrio cholerae* O139 biofilms, respectively. These studies also showed that disruption of EPS not only resulted in less structural complexity but also conferred greater susceptibility to antimicrobial agents, while induced EPS overproduction had the opposite effect.

Influence of Hydrodynamics on Biofilm Structure and Material Properties

While the structure of biofilms is clearly influenced by a number of biological factors such as twitching motility, growth rate, cell signaling, and EPS production, the physical growth environment may also play a significant role in the determination of biofilm structure. Most laboratory biofilms are grown under low laminar flow conditions and the towers and mushroom-shaped microcolonies, and the patterns they form on the surface are generally isotropic i.e., there is no obvious evidence of directionality (Figure 1*a,c*). However, under higher unidirectional flows the influence of the increased shear becomes apparent and the biofilm cell clusters become elongated in the downstream direction to form filamentous streamers (55). Filamentous biofilms, or mats, commonly occur in both archeal and bacterial biofilms growing in fast-flowing environments such as hot springs (49b) or acid mine drainage runoff (20a). Filamentous biofilm streamers can also be formed by nonsheathed species such as *P. aeruginosa*, which grow as single cells in flask culture (49b, 55). The streamers are attached to the substratum by an upstream "head," while the downstream "tails" can freely oscillate in the flow (54a). In vitro studies of the development of biofilm streamers showed that the streamers became increasingly elongated over time, and scanning electron micrographs of in vitro *P. aeruginosa* biofilms show that the streamers thin along the tail until there is only a small chain of single cells at their tips ("*Pseudomonas aeruginosa*—Biofilm Streamers Growing in Turbulent Flow," ASM MicrobeLibrary, http://www.microbelibrary.org).

In addition to influencing biofilm structure, fluid shear also influences the physical properties of biofilms such as density and strength. A study by Liu & Tay (37) found that biofilms grown at higher shear were smoother and denser than those grown at low shear. Stoodley et al. (54) reported that *Desulfovibrio* spp.

(an anaerobic sulfate-reducing bacteria) and *P. aeruginosa* biofilms grown at higher shear stresses were more rigid and stronger than those biofilms grown at lower shear. It is yet unclear if the increased density and strength of biofilms exposed to higher shear stresses are regulated at the genetic level, occur through selection processes (i.e., only those cells that produce strong EPS remain attached and divide), or are determined by purely physical mechanisms (i.e., the alignment of EPS polymers that would be expected to occur when subjected to unidirectional shear). The deformation of laboratory-grown biofilms in response to normal or shear loading demonstrate that biofilms are viscoelastic and the viscoelastic properties were influenced by multivalent cations, presumably due to cross-linking of polymers in the EPS matrix (31, 54, 55). Preliminary evidence suggests that when biofilms are subjected to loads over short periods of time (seconds) they behave elastically, but over longer time periods they behave like viscous fluids. This may explain the fluid-like flow of biofilm ripples along the walls of a glass flow cell that was revealed by time-lapse microscopy (53, 56). These data and behaviors can be modeled using principles of associated polymer viscoelastic systems (29a), suggesting that the material properties of biofilm are largely determined by the EPS. Thus it appears that the EPS determines both the structure and cohesive strength of biofilms (21). Since it has been shown that alginate can protect mucoid *P. aeruginosa* FRD1 biofilm cells from exposure to ultra-violet radiation, it is possible that there was early selection pressures for biofilms colonizing wetted surfaces exposed to ultra-violet radiation to produce protective EPS. The biomechanics of biofilms are an important factor in linking structure to function and may ultimately help answer the question of whether biofilms are coordinated entities that actively shape their structure or whether they are merely aggregates of cells that are passively shaped by the chemistry and physics of the environment.

PHENOTYPIC DIFFERENTIATION DURING BIOFILM DEVELOPMENT

As in many processes of differentiation in many other biological systems, the differentiation process that transforms small groups of adherent bacteria into a thick matrix-enclosed biofilm community on a colonized surface was first perceived as a series of morphological changes. The individual adherent cells that initiate biofilm formation on a surface are surrounded by only small amounts of exopolymeric material, and many are capable of independent movement (45) by means of pilus-mediated twitching or gliding (Figure 1, *stage 1*). These adherent cells are not yet "committed" to the differentiation process leading to biofilm formation, and many may actually leave the surface to resume the planktonic lifestyle. During this stage of reversible adhesion (38) the bacteria exhibit several species-specific behaviors, which include rolling, creeping, aggregate formation, and "windrow" formation (30), before they begin to exude exopolysaccharide and adhere irreversibly (Figure 1, *stage 2*). Davies & Geesey (15) showed that in *P. aeruginosa* the cluster of genes responsible for alginate production is upregulated within 15 min of the cell's initial contact with the colonized surface and that this genetic event initiates the process of biofilm formation. As biofilms mature they develop the basic microcolony/water channel architecture that is now well recognized in natural and in in vitro biofilms (Figure 1, *stage 3*), and many cells alter their physiological processes (e.g., grow anaerobically) in response to conditions in their particular niches. Individual microcolonies may detach from the surface or may give rise to planktonic revertants that swim or float away from these matrix-enclosed structures, leaving hollow remnants of microcolonies or empty spaces that become parts of the water channels (Figure 1, *stages 4 and 5*). Additionally, whole microcolonies may naturally break away from the biofilm (detach without any obvious perturbation to the system), although the mechanisms behind this phenomenon are yet unclear. These processes are not necessarily synchronized throughout the whole biofilm but are often localized so that at any one time a small area on the surface may contain biofilm at each developmental stage.

Differentiation in biofilm development has been explored in increasing detail since the 1980s. Studies from this period have resulted in various models that characterize biofilm development as a process of adaptation and changing genetic regulation. In a recent review, O'Toole et al. (44) describe biofilm development as a process of microbial development, not unlike that observed in fruiting-body formation by *Myxococcus xanthus* and sporulation in *Bacillus subtilis*. A number of investigations have been directed at determining the degree to which gene regulation during biofilm development controls the switch from planktonic to biofilm growth. Brözel and coworkers (3) monitored changes in global gene expression in attached *P. aeruginosa* cells and found more than 11 genes whose levels were altered during various stages of attachment. Whiteley et al. (61) used DNA microarrays to analyze gene expression of *P. aeruginosa* grown in chemostats and as a biofilm on submerged gravel substrata, and 73 genes that showed differences in regulation when compared with planktonic bacteria were identified. Prigent-Combaret et al. (48) carried out a screen in *E. coli* K-12 and found attachment-related changes in the regulation of 38% of the generated *lacZ* gene fusions (out of 446 clones). In a separate study by Sauer et al. subtractive hybridization of *P. putida* biofilms grown in continuous culture in tubes and planktonic bacteria in chemostats revealed more than 30 operons that were altered within 6 h following attachment (50). It is difficult to extrapolate to a universality of these changes in phenotypic expression in response to quite different growth strategies, but several examples are now available to illustrate these changes in several species.

Irreversible Attachment

Once attachment to a surface has been affected, by reversible attachment, the bacteria must maintain contact with the substratum and grow in order to develop a mature biofilm. This change from reversible to irreversible attachment was noted as early as 1943 by Zobel, and it has been characterized by Characklis (5) as the transition from a weak interaction of the cell with the substratum to a permanent bonding, frequently mediated by the presence of extracellular polymers. Early investigators did not appreciate the possibility of surface transduction as a mechanism for inducing irreversible attachment, but recent investigations have gone a long way to suggest that profound physiological changes may accompany the transition to permanent attachment at a surface. One means of transition from reversible to irreversible attachment is mediated by type IV pili. Twitching motility is a mode of locomotion used by *P. aeruginosa* in which type IV polar pili are believed to extend and retract, propelling bacteria across a surface. In *P. putida*, irreversible attachment to a surface was shown by Sauer & Camper (50) to induce a surface-regulated switch from flagella-based motility to type IV pili–based twitching motility, as shown by differential gene expression and immunoblot analyses. Twitching motility is speculated by O'Toole & Kolter (45) to be involved in the formation of microcolonies. The authors suggest that interactions of bacteria with one another at a surface, forming groups of cells, help to strengthen the degree of attachment to the surface. Working with *Staphylococcus epidermis*, Gerke et al. (22) showed that adherent cells produce a polysaccharide intercellular adhesin that bonds the cells together and facilitates the formation of microcolonies and the maturation of the biofilm.

The hallmark of bacterial biofilms that segregates them from bacteria that are simply attached to a substratum is that biofilms contain EPS that surround the resident bacteria. Microbial EPS are biosynthetic polymers that can be highly diverse in chemical composition and may include substituted and unsubstituted polysaccharides, substituted and unsubstituted proteins, nucleic acids, and phospholipids (64). Among the best-characterized of all EPS is the bacterial product alginate, which has been shown to be involved in biofilm formation by *P. aeruginosa* in pulmonary infections and in industrial water systems. The production of alginate by *P. aeruginosa* has been regulated in response to a variety of environmental factors. The activation of a critical alginate promoter, *algD*, has been show to take place during nitrogen limitation, membrane perturbation induced by ethanol, and when cells were exposed to media of high osmolarity (20). The *algC* promoter is activated by environmental signals such as high osmolarity, and this activation is dependent on the presence of the response regulator protein AlgR1 (65). These experiments have been performed in liquid medium and on agar-based medium and have not been duplicated for biofilm bacteria, but they hint that environmental activation of alginate genes may take place in bacterial biofilms. Because the activities of alginate synthesis enzymes are difficult to detect, reporter constructs have been used to detect alginate production, and the *algC* gene has been shown to be regulated within 15 min of attachment to a substratum (15). This indicates that the production of alginate is an early event in the formation of a biofilm by *P. aeruginosa*, and we presume that it forms the structural and mechanical framework required for biofilm maturation. Although the term "irreversible attachment" was originally used to distinguish processes in the early events of biofilm formation, it implys that biofilms are rigidly "cemented" to surfaces. Time-lapse microscopy showing the dynamic motion of single cells over surfaces by twitching motility, within biofilm microcolonies by flagellar motility, the flow of entire microcolonies along surfaces, and the continual detachment of single cells and entire microcolonies from mature biofilms (each of which are discussed in this review) indicate that this term may need to be revised.

Biofilm Maturation

The next phase of biofilm development, maturation, results in the generation of complex architecture, channels, pores, and a redistribution of bacteria away from the substratum (16). In a recent study, mature biofilms of *P. aeruginosa* were shown to have a radically different protein profile from planktonic bacteria grown in chemostats (51). As much as 50% of the detectable proteome (over 800 proteins) was shown to have a sixfold or greater difference in expression. Of these, more than 300 proteins were detectable in mature biofilm samples that were undetectable in planktonic bacteria. The identified proteins fall into five major classes: metabolism, phospholipid and LPS-biosynthesis, membrane transport and secretion, as well as adaptation and protective mechanisms (51). In a separate study, Whiteley et al. (61) used DNA microarray technology to evaluate mature biofilms and compared these to chemostat cultures of *P. aeruginosa*. Just over 70 genes were shown to undergo alterations in expression in this study. Among those genes detected to be upexpressed in mature biofilms were genes encoding proteins involved in translation, metabolism, membrane transport and/or secretion, and gene regulation. These researchers also observed that the sigma factor *rpoH* was upexpressed and that the sigma factor *rpoS* was downexpressed in mature biofilms.

Cell-Cell Communication During Biofilm Formation

McLean et al. (39) have shown that acyl HSL autoinducers are detectable in naturally occurring biofilms, suggesting that biofilm communities in nature contain populations that are able to undergo cell density-dependent gene regulation. In 1998, Davies et al. (16) showed that *P. aeruginosa* PAO1 requires the *lasI* gene product $3OC_{12}$-HSL in order to develop a normal differentiated biofilm. In this study, it was observed that bacteria knocked out in the quorum-sensing inducer gene *lasI* produced biofilms that were only 20% as thick as those produced by the wild-type organism. In addition, these mutants grew as continuous sheets on the substratum, lacking differentiation and not demonstrating evidence of matrix polymer. By contrast, the wild-type organism formed characteristic microcolonies composed of groups of cells that were separated by intervening matrix polymer and separated from one another by water channels. When the autoinducer $3OC_{12}$-HSL was added to the medium of growing *lasI* mutant bacteria, these cells developed biofilms that were indistinguishable from the wild-type organism. These results indicated that $3OC_{12}$-HSL was responsible for the complex architecture observed in mature biofilms produced by *P. aeruginosa*. Spatial analysis revealed that the genes *lasI* and *rhlI* were maximally expressed in biofilm cells located at the substratum (18). Whereas *lasI* gene expression was found to decrease over time with increasing biofilm height, *rhlI* expression remained steady throughout biofilm development but occurred in a lower percentage of cells.

Since biofilm architecture is presumed to be influenced by matrix polymer, we can hypothesize that autoinduction is at least partly responsible for regulation of EPS synthesis in *P. aeruginosa*. In support of this hypothesis, it has been demonstrated that the alginate genes *algC* and *algD* are induced in *P. aeruginosa* strain 8830 by $3OC_{12}$-HSL and by $3OC_4$-HSL (D.G. Davies & J.W. Costerton, unpublished data). These observations indicate the possible role of quorum sensing as a signal transduction system by which *P. aeruginosa* may initiate the production of alginate and possibly other types of EPS. In a separate study Olvera et al. (43) have observed that *algC* is necessary for the production of rhamnolipid. The transcription of rhamnolipid has been shown to be dependent on activation by the quorum-sensing gene product *rhlI* (46), further indicating that *algC* transcription is regulated at some level by bacterial quorum sensing. In addition to activation of multiple factors by autoinduction, it has recently been shown that *P. aeruginosa* is able to repress the transcription of *lasI* in response to RsaL, an 11-kDa protein whose gene lies downstream from *lasR* (19). This protein has been shown to suppress LasB production and presumably all other factors regulated by the lasI quorum-sensing system. This finding demonstrates a further level of control of quorum sensing in *P. aeruginosa* and suggests that cell-to-cell communication is highly complex.

Altogether, *P. aeruginosa* is known to have 39 genes that are under the regulation of the *lasI*/*rhlI* quorum-sensing systems (62). In addition to cell-to-cell communication via acyl HSL-mediated quorum sensing, it has recently come to light that other communication systems are used by *P. aeruginosa*. Holden et al. (26) have discovered that cell-free extracts of cultures in which *P. aeruginosa* were grown contain diketopiperazines (DKPs), a cyclic dipeptide known to participate in cell-to-cell communication mostly in gram-positive bacteria. This compound has been shown to be capable of activating a LuxR biosensing system and swarming motility in *Serratia marcescens*. In addition, a third extracellular sensing system has been discovered in *P. aeruginosa* by Pesci et al. (47). In this system, *P. aeruginosa* produces a quinolone signaling molecule (2-heptyl-3-hydroxy-4-quinolone) that regulates LasB production via the *lasI*/*rhlI* quorum-sensing system. This observation hints at a super-regulatory function for quinolones in bacterial communication.

Sauer et al. (51) have recently shown that quorum sensing does not account for all biofilm-specific protein production in *P. aeruginosa*. Analysis of protein patterns of irreversible attached (1 day) and planktonic cells revealed the presence of quorum-sensing independent but probably surface-induced differential expression of proteins. Fifty-seven unique protein spots were identified in 1-day biofilm protein patterns for both *P. aeruginosa* strain PAO1 and *LasI* minus strain PAO-JP1 that were absent in planktonic protein patterns and 48 protein spots that were unique for planktonic growth. These results imply that quorum sensing accounts for only a portion of the total number of genes whose regulation is altered during the irreversible stage of biofilm development and that the physiological change in attached bacteria is not due solely to induction by PAI-1 quorum-sensing autoinducer (see 50). The implication of these results is that undiscovered biofilm regulons probably exist, that the role of quorum-sensing and other intercellular signaling mechanisms is unclear, and that these complex systems are under intercellular as well as extracellular control. The existence of these systems further suggests that interspecies communication is possible, as is evident from studies in which bioassays have been used to detect the presence of autoinducers from multiple species.

RpoS Activity During Biofilm Development

During stationary phase, gram-negative bacteria develop stress-response resistance that is coordinately regulated through the induction of a stationary-phase sigma factor known as RpoS (23). Biofilm bacteria are generally considered to show physiological similarity to stationary-phase bacteria in batch cultures. Thus, it is presumed that the synthesis and export of stationary-phase autoinducer-mediated exoproducts occur generally within biofilms. The stationary-phase behavior of biofilm bacteria may be explained by the activity of accumulated acyl HSL within cell clusters. The mechanism causing biofilm bacteria to demonstrate stationary-phase behavior is hinted at by the discovery that RpoS is produced in response to accumulation of the *rhlI* gene product in *P. aeruginosa* cultures (34). In a recent report, a similar relation was made, but rather than the RhlR-RhlI system influencing the expression *of* rpoS, it was demonstrated that RpoS regulated *rhlI* (63). In another study, Suh et al. (58) have shown that in *P. aeruginosa* PAO1 RpoS is responsible for a decrease in the production of exotoxin A, elastase A, LasA protease, and twitching motility. Additionally, it was found that *P. aeruginosa* FRD1 demonstrated a 70% loss in alginate synthesis when *rpoS* was inactivated.

Detachment as a Component of Biofilm Development

Detachment is a generalized term used to describe the release of cells (either individually or in groups) from a biofilm or substratum. Active detachment is a physiologically regulated event, but only a few studies (57) have been performed to demonstrate a biological basis for this process. Allison et al. (1) showed that following extended incubation, *P. fluorescens* biofilms experienced detachment, coincident with a reduction in EPS. In *Clostridium thermocellum* the onset of stationary phase has been correlated with increased detachment from the substratum (33). It has been postulated that starvation may lead to detachment by an unknown mechanism that allows bacteria to search for nutrient-rich habitats (44). This hypothesis is in agreement with recent observations by Sauer et al. (51) who compared two-dimensional-gel protein patterns to show that dispersing cells of *P. aeruginosa* are more similar to planktonic than to mature biofilm cells. This finding indicated that dispersing biofilm cells revert to the planktonic mode of growth; thus, the biofilm developmental life cycle comes full circle (51).

The transition from a flowing system to a batch culture system has been observed by many laboratories to result in biofilm detachment (14). We surmise that an increase in the concentration of an inducer molecule is responsible for the release of matrix polymer-degrading enzymes, which results in detachment of cells from the biofilm. One such example is found with the gram-positive organism *Streptococcus mutans*, which produces a surface protein releasing enzyme (SPRE) that mediates the release of cells from biofilms (36). Boyd & Chakrabarty (2) showed that overexpression of alginate lyase causes the degradation of alginate and produces biofilms that can be removed from surfaces by gentle rinsing. Cell density–dependent regulation may also trigger the release of matrix-degrading enzymes, allowing bacteria to disperse from a biofilm, when cell density reaches a high level in biofilm microcolonies (D.G. Davies, J.W. Costerton & H.M. Lappin-Scott, unpublished observations). Other investigators have demonstrated that homoserine lactones may play a role in detachment. Puckas et al. (49) reported that homoserine lactone production was negatively correlated with cell cluster formation in *Rhodobacter sphaeroides*, and Allison et al. (1) reported that the addition of *N*-acyl-C_6 homoserine lactone to *P. fluorescens* biofilms caused a reduction in biofilm and loss of exopolymer. The regulation of detachment events in microbial biofilms represents an important area for future research, and the new methods of analysis of gene expression promise to speed this process.

Biofilm Formation as a Developmental Process

As demonstrated above, biofilm development can be partitioned into at least four distinct stages. These are (*a*) reversible attachment, (*b*) irreversible attachment, (*c*) maturation, and (*d*) detachment. Detached cells are believed to return to the planktonic mode of growth, thus closing the biofilm developmental life cycle. A schematic overview is shown in Figure 1. These four biofilm developmental stages indicate significant episodes in the formation of a bacterial biofilm. Bacteria within each of the stages of biofilm development are generally believed to be physiologically distinct from cells in other stages. In a recent study, biofilm development in *P. aeruginosa* has been partitioned into four stages, and each stage was analyzed separately by two-dimensional polyacrylamide gel electrophoresis (51). The average difference in protein production between each developmental episode was 35% of detectable proteins. The transition from planktonic growth to the stage of irreversible attachment resulted in a 29% change in the production of detectable proteins. The transition from irreversibly attached cells to the stage of mature biofilms caused a change in the protein production of 40%, with the majority of proteins showing an increase in concentration. In contrast, the transition from mature-stage biofilm to the dispersion stage resulted in a reduction in 35% of detectable proteins. Cells during this stage of development had protein profiles that were more similar to planktonic cells than to mature-stage biofilm cells. The most profound differences were observed when planktonic cells were compared to mature biofilm cells, with more than 800 detectable proteins showing more than a sixfold change, or to dispersing biofilm cells. Differences in the protein patterns of planktonic cells and mature and dispersing biofilm cells of a clonal population of *P. aeruginosa* were as profound as the difference for different but related *Pseudomonas* species grown at the same stage of development. Therefore, the authors conclude that *P. aeruginosa* displays at least three phenotypes, (*a*) planktonic, (*b*) mature biofilm, and (*c*) dispersion, during biofilm development (51).

SALIENT CHARACTERISTICS OF BIOFILM COMMUNITIES

Biofilms as Physiologically Integrated Microbial Communities

When we examine the mature communities that dominate particular microbial ecosystems, with some emphasis on systems with high levels of physiological efficiency (like the bovine rumen), we find that these communities are composed of complex multispecies biofilms. The biofilm mode of growth is optimal when bacterial communities must colonize insoluble nutrient substrates (e.g., cellulose), and it is especially useful when multistage digestive processes involve many different species whose cells can be stably juxtaposed within a communal matrix. The bovine rumen provides a particularly cogent example of a biofilm-driven ecosystem because fresh nutrient substrate is ingested continually, the primary cellulose degraders colonize it avidly, and physiologically cooperative species accrete to form an efficient digestive consortium. The efficiency of the digestive consortium is as dependent on microbial teamwork as it is on the activity of the primary cellulose degraders because the removal of the products of primary digestion (e.g., butyrate) drives the whole cellulolytic process (6). In the rumen ecosystem, some of the organisms that drive cellulose digestion by product removal live within the biofilms on the cellulose fibers, whereas others are highly mobile cells (e.g., Treponema) that graze on "hot spots" of high butyrate concentration on the biofilm surfaces (32).

Because our speculations on the formation of digestive biofilms in the rumen predate modern perceptions of the role of cell-cell signaling in biofilm formation (16), all the mechanisms suggested to date have involved the attraction of cooperative species by nutrient advantage. The basic concept is that primary cellulose degraders attach to their preferred nutrient substrate and that secondary organisms are chemotactically attracted by butyrate and other volatile products of anaerobic cellulose digestion. However, as we emphasize in subsequent sections, nothing in the cellulolytic ecosystem precludes a possible role of cell-cell signaling in the development of these physiologically integrated biofilms. The notion that virtually all aspects of bacterial behavior are controlled by intercellular signals has a liberating effect on ecological speculations concerning the origin and organization of the complex physiologically integrated biofilms that actually dominate most ecosystems. We are no longer constrained by the necessity of explaining the recruitment of particular bacteria into particular ecological niches in terms of immediate physiological advantage. We now understand molecular mechanisms, in both bacteria and archea, by which one prokaryotic cell can attract another prokaryote (or even a eukaryote) by means of a simple signal molecule. Furthermore, the cells that then find themselves in stable juxtaposition can manipulate each other's physiology, to mutual advantage, and produce efficient integrated physiological consortia. We can now adopt an approach similar to that of the embryologists who explain the developmental changes in multicellular eukaryotic organisms in terms of hormonal signals, sent and received, rather than in terms of immediate nutrient advantage to participating cells. Logically we can view the complex microbial communities that develop in a calf's rumen, initiated by bacterial genomes from its mother's microbial population, with the same awe we usually reserve for the development of its tissues and organs from its mother's fertilized egg.

Biofilms as Behaviorally Integrated Microbial Communities

The fruiting bodies formed by Myxobacteria comprise one of many examples of macroscopic biofilms that are produced by individual cells of many bacterial species in response to specific environmental cues, such as starvation and other adverse factors. In nutrient-sufficient conditions these same myxobacterial cells move through their environments, by a form of gliding motility that depends on the coordinated activity of retractable pili (27), in large swarms that react to nutrient gradients. This complex pattern of integrated behavior is facilitated by the production of surfactants and is controlled by cell-cell signals [acyl homoserine lactones (AHLs)] in a manner that is described in detail in a review by Kim et al. (28). When environmental conditions deteriorate, in terms of drying or of nutrient deprivation, swarms of myxobacteria form microcolonies in which individual cells pile up on each other and produce large amounts of extracellular matrix material to produce macroscopic fruiting bodies. The individual bacterial cells within these *de facto* biofilms differentiate to form cysts, and the biofilms themselves differentiate to form elaborate shapes, in which simple mushroom shapes may morph into complex structures resembling jester's caps or ornate crowns. The net result of this signal-controlled process of differentiation is that the cells of a particular species have formed a drying-resistant biofilm that is resistant to the environmental conditions that threatened its planktonic way of life in more halcyon times. Thus, in these mobile biofilms, control mechanisms based on cell-cell signaling may be integrated with cellular strategies for nutrient acquisition, and particular cells may burgeon when both processes bring them into a biofilm microniche, in which conditions are optimal (9).

Biofilms as Highly Structured Microbial Communities

Microbial biofilms are often sufficiently thick and extensive to be visible to the unaided eye, and they may contain millions of prokaryotic (and sometimes some eukaryotic) cells in arrangements that facilitate the stable juxtaposition of physiologically cooperative organisms. Cells of particular species are found consistently in certain locations, near the colonized surface or at the apices of towers or mushrooms; the sessile cells that comprise single-species biofilms are located within the microcolonies in species-specific distribution patterns. In these single-species biofilms, the preponderance of the sessile cells may be found in the caps of mushrooms in a highly organized pattern with relatively regular cell-cell spacing, and the stalks of mushrooms formed by some species are virtually devoid of cells. We have up to this time explained these distribution patterns on the basis of nutrient advantage, as in cases where cellulolytic organisms occurred next to their substrate and fastidious aerobes occurred next to open water channels, but these explanations now seem to be unsatisfactory. They contain no provision for the movement of cells within the biofilm matrix or for the establishment of uniform cell-cell distance between sister cells, and it is difficult to see how some areas of the matrix are heavily colonized while others are devoid of cells. We hereby predict that a mechanism will be discovered whereby sessile cells within biofilms can move within the matrix, perhaps by the activity of pilus-like structures, attached to two or more cells, that can shorten or elongate to position the cells with some degree of precision. The high levels of lateral gene and plasmid transfer seen in biofilms may indicate that these or other structures may facilitate genetic exchange between neighboring cells, and the matrix may be structured in a manner that facilitates signaling between adjacent cells. Physiological cooperation between cells of the same or different species would be facilitated if the secondary structure of the EPS favored the preferential diffusion of particular molecules or even electrons. The random extrusion of EPS, with consequent cell displacement, is simply insufficient to account for the sophistication of the cell distribution patterns that we see in biofilms. Therefore we suggest that biofilms are surface-associated microbial communities, within which the position and spatial relationship(s) of each component cell are predetermined by means of a coordinated developmental cycle that is mediated (at least in part) by signal molecules and some type of positioning mechanism.

Biofilms as Self-Assembling Microbial Communities

The bacterial genome is expressed in many different phenotypes, as are the genomes of virtually all living things, and the proteomic data in the "Biofilm Formation as a Developmental Process" section of this review establishes the profound nature of the phenotypic changes that accompany the transition from planktonic to biofilm growth. Similar changes in gene expression occur when planktonic bacteria react to starvation (29). As a consequence of this phenotypic plasticity the genomes of thousands of bacterial species exist, in varying patterns of expression, in virtually all ecosystems in which environmental conditions allow bacterial growth or persistence. Recent examinations of oligotrophic environments, including the deep subsurface and the deep oceans (29), have revealed that approximately 1×10^5 bacterial cells are present in each milliliter of water and that most of these cells are in a dormant starved state. For this reason, the biosphere can be visualized as being a continuum of fluids in which the genomes of thousands of species of bacteria exist in forms that range from virtually dormant to active in either the planktonic or the biofilm patterns of gene expression.

This ubiquity of the genomes of so many different species of bacteria provides the mechanistic basis for the self-assembly of microbial communities. In modern times, the ocean reservoir of bacterial genomes routinely serves as the genetic source for the development of self-assembled microbial communities that develop when black smokers suddenly deliver volcanic gasses (notably hydrogen and H_2S) into the marine environment. Dormant bacteria are resuscitated into a physiologically active state, and these planktonic cells undertake biofilm formation on available surfaces in a nutrient-rich environment. As H_2S is oxidized to form thiosulfate and other sulfur compounds, more species of sulfur-cycle organisms will be resuscitated and these organisms will be recruited into the developing black smoker biofilm community. A complex microbial biofilm community will develop in response to this new nutrient opportunity in a pattern that is remarkably similar in virtually all ocean regions, and the new availability of organic carbon compounds will recruit heterotrophic prokaryotes and macrophytes. Eventually, the anabolic processes of the heterotrophic organisms will balance the catabolic processes of the chemolithotrophs, and a stable climax community will have developed. Black smoker communities are representative of bacterial communities that may have developed in the primitive earth by the recruitment of hundreds of bacterial genomes and by the subsequent differentiation of these genomes to produce physiologically integrated communities. Because black smoker communities have thrived in the deep oceans for eons, the reservoir of prokaryotic genomes that float in the depths may be especially rich in potential members for the sessile communities that develop quickly in response to each new opportunity.

It may be useful to contrast the phenotypic plasticity and ubiquity of the bacterial genome with the same properties of higher plants and animals in order to examine their relative success in responding to transient nutrient availability. In the interests of homeostasis and physiological integration, plants and animals have amassed much larger genomes so that they can build large and complex multicellular organisms from a condensed and coordinated genetic base. Their genomes are usually kept isolated and protected, except during gamete release by some species, and many of the higher forms are dependent on seasonal sexual cycles for their reproduction. It is axiomatic that the standing crop of insects in a stream is roughly proportional to the availability of organic nutrients in the system, but a huge input of particulate organics would have no effect at all on the mayfly population if it did not coincide with a feeding larval stage of that insect. Similarly, the sudden availability of large amounts of soluble nutrients on the bank of the same stream would have no effect on the colonization of that environment by bullrushes, if the bullrush genome was not present or if rainfall was insufficient. In contrast, the sudden availability of either soluble or particulate nutrients would have a profound effect on the bacteria in both the stream and the soil, and both would produce complex biofilm communities to make maximum use of this largesse to produce large microbial populations. Significantly, the phenotypic plasticity of the bacteria allows these organisms to process all the nutrients in stationary biofilm communities and then to disperse into both the stream and soil ecosystems as persistent genomes that are capable of responding to the next fortuitous nutrient event.

THE ROLE(S) OF SURFACES IN THE ORIGIN OF CELLS AND OF MULTICELLULAR COMMUNITIES

Our perception of prokaryotic cells has evolved from the notion of a membrane-enclosed bag of randomly seething protein and nucleic acid molecules to the concept of organized cytoplasmic elements enclosed within a highly organized membrane (and an even more structured cell wall). The concept of randomly floating or swimming bacterial cells has evolved into the thesis of this review, which is that most ecosystems operate on the basis of highly structured and coordinated communities of prokaryotic cells in which many members have specialized functions. The eukaryotic cell is also increasingly seen as a highly structured entity with a complex cytoskeleton that holds every organelle, and virtually every structural molecule, in a predetermined location that dictates its contribution to the coordinated function of the whole cell. Multicellular eukaryotic organisms are now known to coordinate the activities of their cells in patterns much more complex than those that were visualized when biologists first sat down to contemplate the origins of life. The role of random interactions of molecules or of cells has generally taken a beating in the past two decades, and it may be salutary to take a fresh look at the origins of life in light of these new concepts.

If an ordered structure lies at the root of our present concept of even the simplest cell, and of even the simplest multicellular community, it seems futile to look for this basic order in the planktonic state in which molecules and cells move about in a roiling mass in constant Brownian movement. In the nonliving world, high levels of order are found in crystals, and even the humble and ubiquitous clays display a highly structured series of alternating bands of different compositions and charges. Cairns-Smith (4) has speculated that the assembly of enzyme molecules might have been facilitated by the use of nonliving clay templates. This rudimentary replicating system may have later incorporated RNA, which eventually replaced the crystalline clay as the information template. Others have speculated that this association may have stabilized these pioneer nucleic acids against the entropy imposed by ionizing radiation. Our modern concept of prokaryotic cell membranes, with their complex arrays of protein molecules and their control over the replication of DNA, are barely recognizable from the early notion of a simple lipid bilayer, and logic seems to demand their original assembly on an ordered surface. These considerations would seem to suggest that the first organization of bacterial or archeal cells would have taken place at a surface: the surface acting as a locus for cells and absorbed nutrients to interact. The physical proximity of one cell to another may have provided a selective pressure to develop complex interactions within these early communities.

The theme of this review is the notion that complex interactions within prokaryotic communities evolved in surface-associated biofilms. The basis for our speculation is the fact that the earliest fossil records of life occur in surface-associated microbial mats in ancient hyperthermal vents and hot springs environments (49a, 49b) and that biofilm mat formation is found in the most ancient linages of archaeal and bacterial lines, the Korarchaeota and Aquificales (20a, 27a, 49b). A stable structural juxtaposition seems absolutely necessary as cells developed the physiological cooperation and reciprocal signaling so implicit in the origin of bacterial biofilms. To the extent that it is literally mind boggling to think that any such interactive community could develop as individual prokaryotic cells wheel past each other in the endless and frantic Brownian dance that all of us have seen under our microscope lenses. The fact that physiological coordination and signal-based cooperation are both based on the orderly diffusion of organic molecules demands the stable juxtaposition of cells and makes biofilm development a *sine qua non* for the major step from single cells to increasingly coordinated communities. Our current observation that structured and physiologically integrated communities comprise the predominant forms of growth in both bacteria and archea imply convergent evolution in two kingdoms that have evolved both independently and successfully.

As the first bacterial and archeal cells arrived at an evolutionary state that allowed replication, perhaps through the intermediate step of functional replicating nanobacteria (59), their immediate need would be for a protective homeostatic environment. The primitive earth was undoubtedly hostile, in chemical and physical terms, and these pioneer organisms would only survive if they could avoid floating from a permissive niche into a hotspring or an acid pool. Morita has suggested (40) that the energy source that supported the first successful prokaryotes was hydrogen, and we can easily imagine a matrix-enclosed community of primitive prokaryotes attached to a surface adjacent to a hydrogen source. In this way, the first multicellular communities of both bacteria and archea would develop in the same protected nutrient-rich niches in which, over millions of years, their cell machinery had been assembled to produce the first replicating bacterial and archeal cells. Cells of cooperative species would be found in stable juxtaposition within biofilm communities, and bacteria could cooperate with archea in multispecies communities that would be the primitive equivalent of modern black smoker communities. In this thesis, biofilms are the predominant form of prokaryotic life because basic biological processes, including the first assembly of cells and the first assembly of multicellular communities, can draw their basic order and their stable juxtaposition from the inherent order of inanimate surfaces. Once formed, these cells and these communities are protected from antibacterial influences (first chemical and physical and then viral and eukaryotic) by their specialized biofilm phenotype and by their production of matrix materials that further condition their selected niche. The concept of highly structured and metabolically integrated bacterial communities brings the prokaryotic domains much closer to the eukaryotic domain and provides a conceptual framework within which the transition of bacteria to mitochondria and of blue-green bacteria to chloroplasts may have occurred.

If we accept this notion that prokaryotic cells and prokaryotic communities developed preferentially on surfaces, we soon encounter the primary disadvantage of growth in biofilms, in that the conditioned niches become crowded with cells and access to nutrients is compromised. The first adaptation in response to this problem was probably the development of the structures inherent in the microcolony and water channel architecture of biofilms, which increase substrate access in sessile communities. The next adaptation in response to crowding in primitive biofilms may well have been the development of lateral surface–associated motility, which takes the form of twitching if it is mediated by type IV pili (45) and swarming if it is mediated by retractable pili (27). This twitching motility enables bacteria to colonize adjacent areas of the surface, and it is also involved in cellular rearrangements prior to biofilm formation during the colonization of newly available surfaces. Swarming motility involves a newly discovered mechanism that depends on retractable pili and can achieve amazing feats of coordinated cellular movement, so that its most efficient proponents (the Myxobacteria) can travel long distances and form elaborate structures. Some of the microbial biofilm communities that developed in these ways were macroscopic, and their early development is evident in the earliest fossil records as stromatolites.

We propose, for discussion, that the planktonic phenotype may have actually developed in the relative homeostatic environment afforded by the biofilm. The planktonic phenotype differs profoundly from the biofilm phentoype, and it involves the expression of a large number of genes that allow bacterial and archeal cells to detach themselves from biofilms and assume a floating or swimming mode of growth. Many planktonic cells are characterized by a heavy investment of genetic capability in detachment, in the formation of flagella, and in chemotactic mechanisms that allow them to find new and favorable niches for the development of stable protected biofilms. In the development of the Plant Kingdom primitive forms concentrated on growth and lateral spread in favorable niches, and higher forms only developed seeds when the available ecosystems became crowded and dispersal became a preferred strategy. In the same way, bacteria may have evolved their dispersal mechanisms late in their evolution, in response to the basic problem of crowding and to the opportunity of colonizing new favorable niches. We suggest that the evolution of the highly specialized planktonic phenotype has allowed certain species of bacteria and archea to colonize new environments, even at considerable distances from their original niches, and that this late-developing phenotype accounts for the phenomenal ecological success of these species.

ACKNOWLEDGMENTS

Financial support of this work has been provided by the National Institutes of Health RO1 GM60052 (P. Stoodley) and RO1 DC04173 (J.W. Costerton, K. Sauer, and D.G. Davies) and the cooperative agreement ##C-8907039 between the NSF and MSU.

The *Annual Review of Microbiology* is online at http://micro.annualreviews.org

LITERATURE CITED

1. Allison DG, Ruiz B, SanJose C, Jaspe A, Gilbert P. 1998. Extracellular products as mediators of the formation and detachment of *Pseudomonas fluorescens* biofilms. *FEMS Microbiol. Lett.* 167:179–84
2. Boyd A, Chakrabarty AM. 1995. Role of alginate lyase in cell detachment of *Pseudomonas aeruginosa*. *Appl. Environ. Microbiol.* 60:2355–59
3. Brözel VS, Strydom GM, Cloete TEE. 1995. A method for the study of *de novo* protein synthesis in *Pseudomonas aeruginosa* after attachment. *Biofouling* 8:195–210
4. Cairns-Smith AG. 1983. *Genetic Takeover and The Mineral Origins of Life*. New York: Cambridge Univ. Press
5. Characklis WG. 1990. Biofilm processes. In *Biofilms*, ed. WG Characklis, KC Marshall, pp. 195–231. New York: Wiley
6. Cheng K-J, Fay JP, Howarth RE, Costerton JW. 1980. Sequence of events in the digestion of fresh legume leaves by rumen bacteria. *Appl. Environ. Microbiol.* 40: 613–25
7. Costerton JW, Cheng K-J, Geesey GG, Ladd TI, Nickel JC, et al. 1987. Bacterial biofilms in nature and disease. *Annu. Rev. Microbiol.* 41:435–64
8. Costerton JW, Geesey GG, Cheng K-J. 1978. How bacteria stick. *Sci. Am.* 238: 86–95
9. Costerton JW, Lewandowski Z, DeBeer D, Caldwell D, Korber D, James G. 1994. Minireview: biofilms, the customized micronich. *J. Bacteriol.* 176(8):2137–42
10. Costerton JW, Stewart PS. 2001. Battling biofilms. *Sci. Am.* 285:75–81
11. Costerton JW, Stewart PS, Greenberg EP. 1999. Bacterial biofilms: a common cause of persistent infections. *Science* 284: 1318–22
12. Dalton HM, Goodman AE, Marshall KC. 1996. Diversity in surface colonization behavior in marine bacteria. *J. Ind. Microbiol.* 17:228–34
13. Danese PN, Pratt LA, Kolter R. 2000. Exopolysaccharide production is required for development of *Escherichia coli* K-12 biofilm architecture. *J. Bacteriol.* 182:3593–96
14. Davies DG. 1999. Regulation of matrix polymer in biofilm formation and dispersion. In *Microbial Extracellular Polymeric Substances*, ed. J Wingender, TR Neu, H-C Fleming, pp. 93–112. Berlin: Springer
15. Davies DG, Geesey GG. 1995. Regulation of the alginate biosynthesis gene *algC* in *Pseudomonas aeruginosa* during biofilm development in continuous culture. *Appl. Environ. Microbiol.* 61:860–67
16. Davies DG, Parsek MR, Pearson JP, Iglewski BH, Costerton JW, Greenberg EP. 1998. The involvement of cell-to-cell signals in the development of a bacterial biofilm. *Science* 280:295–98
17. de Beer D, Stoodley P, Roe F, Lewandowski Z. 1994. Effects of biofilm structures on oxygen distribution and mass transport. *Biotechnol. Bioeng.* 43:1131–38
18. De Kievit TR, Gillis R, Marx S, Brown C, Iglewski BH. 2001. Quorum-sensing genes in *Pseudomonas aeruginosa* biofilms: their role and expression patterns. *Appl. Environ. Microbiol.* 67:1865–73
19. De Kievit TR, Seed PC, Nezezon J, Passador L, Iglewski BH. 1999. RsaL, a novel repressor of virulence gene expression in

Pseudomonas aeruginosa. J. Bacteriol. 181:2175–84

20. DeVault JD, Berry A, Misra TK, Chakrabarty AM. 1989. Environmental sensory signals and microbial pathogenesis: *Pseudomonas aeruginosa* infection in cystic fibrosis. *BioTechnology* 7:352–57

20a. Edwards JK, Bond PL, Gihring TM, Banfield JF. 2000. An archaeal iron-oxidizing extreme acidophile important in acid mine drainage. *Science* 287(5459):1731–32

21. Flemming HC, Wingender J, Mayer C, Korstgens V, Borchard W. 2000. Cohesiveness in biofilm matrix polymers. In *Community Structure and Cooperation in Biofilms. Cambridge: SGM Symposium Series 59*, ed. D Allison, P Gilbert, HM Lappin-Scott, M Wilson, pp. 87–105. Cambridge, UK: Cambridge Univ. Press

22. Gerke C, Kraft A, Sussmuth R, Schweitzer O, Gotz F. 1998. Characterization of the N-acetylglucosaminyltransferase activity involved in the biosynthesis of the *Staphylococcus epidermidis* polysaccharide intercellular adhesin. *J. Biol. Chem.* 273:18586–93

23. Hengge-Aronis R. 1993. Survival of hunger and stress: the role of *RpoS* in early stationary phase regulation in *Escherichia coli. Cell* 72:165–68

24. Hentzer M, Teitzel GM, Balzer GJ, Heydorn A, Molin S, et al. 2001. Alginate overproduction affects *Pseudomonas aeruginosa* biofilm structure and function. *J. Bacteriol.* 183:5395–401

25. Heydorn A, Nielsen AT, Hentzer M, Sternberg C, Givskov M, et al. 2000. Quantification of biofilm structures by the novel computer program COMSTAT. *Microbiology* 146:2395–407

26. Holden MT, Ram Chhabra S, de Nys R, Stead P, Bainton NJ, et al. 1999. Quorum-sensing cross talk: isolation and chemical characterization of cyclic dipeptides from *Pseudomonas aeruginosa* and other gram-negative bacteria. *Mol. Microbiol.* 33:1254–66

27. Hong S, Zusman D, Shi W. 2000. Type IV pilus of *Myxococcus xanthus* is a motility apparatus controlled by the frz chemotaxis homologs. *Curr. Biol.* 10:1143–46

27a. Jahnke LL, Eder W, Huber R, Hope JM, Hinrichs KU, et al. 2001. Signature lipids and stable carbon isotope analyses of octopus spring hyperthermophilic communities compared with those of aquificales representatives. *Appl. Env. Microbiol.* 67:5179–89

28. Kim SK, Kaiser D, Kuspa A. 1992. Control of cell density and pattern by intercellular signaling in Myxococcus development. *Annu. Rev. Microbiol.* 46:117–39

29. Kjelleberg S. 1993. *Starvation in Bacteria*. New York: Plenum. 277 pp.

29a. Klapper I, Rupp CJ, Cargo R, Purevdorj B, Stoodley P. 2002. A viscoelastic fluid description of bacterial biofilm material properties. *Biotechnol. Bioeng.* In press

30. Korber DR, Lawrence JR, Lappin-Scott HM, Costerton JW. 1995. Growth of microorganisms on surfaces. In *Microbial Biofilms*, ed. HM Lappin-Scott, JW Costerton, pp. 15–45. Cambridge, UK: Cambridge Univ. Press

31. Körstgens V, Flemming H-C, Wingender J, Borchard W. 2001. Uniaxial compression measurement device for investigation of the mechanical stability of biofilms. *J. Microbiol. Methods* 46:9–17

32. Kudo H, Cheng K-J, Costerton JW. 1987. Interactions between *Treponema bryantii* and cellulolytic bacteria in the *in vitro* degradation of straw cellulose. *Can. J. Microbiol.* 33:244–48

33. Lamed R, Bayer EA. 1986. Contact and cellulolysis in *Clostridium thermocellum* via extensive surface organelles. *Experientia* 42:72–73

34. Latifi A, Foglino M, Tanaka K, Williams P, Lazdunski A. 1996. A hierarchical quorum-sensing cascade in *Pseudomonas aeruginosa* links the transcriptional activators LasR and RhlR (VsmR) to expression of the stationary-phase sigma factor RpoS. *Mol. Microbiol.* 21:1137–46

35. Lawrence JR, Korber DR, Hoyle BD,

Costerton JW, Caldwell DE. 1991. Optical sectioning of microbial biofilms. *J. Bacteriol.* 173:6558–67
36. Lee SF, Li YH, Bowden GH. 1996. Detachment of *Streptococcus mutans* biofilm cells by an endogenous enzymatic activity. *Infect. Immun.* 64:1035–38
37. Liu Y, Tay JH. 2001. Metabolic response of biofilm to shear stress in fixed-film culture. *J. Appl. Microbiol.* 90:337–42
38. Marshall KC, Stout R, Mitchell R. 1971. Mechanisms of the initial events in the sorption of marine bacteria to surfaces. *J. Gen. Microbiol.* 68:337–48
39. McLean RJ, Whiteley M, Stickler DJ, Fuqua WC. 1997. Evidence of autoinducer activity in naturally occurring biofilms. *FEMS Microbiol. Lett.* 154:259–63
40. Morita RY. 2000. Is hydrogen the universal energy source for long-term survival? *Microbiol. Ecol.* 38:307–20
41. Deleted in proof
42. Nivens DE, Ohman DE, Williams J, Franklin MJ. 2001. Role of alginate and its O acetylation in formation of *Pseudomonas aeruginosa* microcolonies and biofilms *J. Bacteriol.* 183:1047–57
43. Olvera C, Goldberg JB, Sanchez R, Soeron-Chavez G. 1999. The *Pseudomonas aeruginosa algC* gene product participates in rhamnolipid biosynthesis. *FEMS Microbiol. Lett.* 179:85–90
44. O'Toole GA, Kaplan HB, Kolter R. 2000. Biofilm formation as microbial development. *Annu. Rev. Microbiol.* 54:49–79
45. O'Toole GA, Kolter R. 1998. Flagellar and twitching motility are necessary for *Pseudomonas aeruginosa* biofilm development. *Mol. Microbiol.* 30:295–304
46. Passador L, Cook JM, Gambello MJ, Rust L, Iglewski BH. 1993. Expression of *Pseudomonas aeruginosa* virulence genes requires cell-to-cell communication. *Science* 260:1127–30
47. Pesci EC, Milbank JB, Pearson JP, McKnight S, Kende AS, et al. 1999. Quinolone signaling in the cell-to-cell communication system of *Pseudomonas aeruginosa*. *Proc. Natl. Acad. Sci. USA* 96:11229–34
48. Prigent-Combaret C, Vidal O, Dorel C, Lejeune P. 1999. Abiotic surface sensing and biofilm-dependent regulation of gene expression in *Escherichia coli*. *J. Bacteriol.* 181:5993–6002
49. Puckas MR, Greenberg EP, Kaplan S, Schaefer AL. 1997. A quorum-sensing system in the free-living photosynthetic bacterium *Rhodobacter sphaeroides*. *J. Bacteriol.* 179:7530–37
49a. Rasmussen B. 2000. Filamentous microfossils in a 3,235-million-year-old volcanogenic massive sulphide depost. *Nature* 405:676–79
49b. Reysenbach AL, Cady SL. 2001. Microbiology of ancient and modern hydrothermal systems. *Trends Microbiol.* 9:79–86
50. Sauer K, Camper AK. 2001. Characterization of phenotypic changes in *Pseudomonas putida* in response to surface-associated growth. *J. Bacteriol.* 183:6579–89
51. Sauer K, Camper AK, Ehrlich GD, Costerton JW, Davies DG. 2002. *Pseudomonas aeruginosa* displays multiple phenotypes during development as a biofilm. *J. Bacteriol.* 184:1140–54
52. Sternberg C, Christensen BB, Johansen T, Nielsen AT, Andersen JB, et al. 1999. Distribution of growth activity in flow-chamber biofilms. *Appl. Environ. Microbiol.* 65:4108–17
53. Stoodley P, Dodds I, Boyle JD, Lappin-Scott HM. 1999. Influence of hydrodynamics and nutrients on biofilm structure. *J. Appl. Microbiol.* 85:19S–28S
54. Stoodley P, Jacobsen A, Dunsmore BC, Purevdorj B, Wilson S, et al. 2001. The influence of fluid shear and $AlCl_3$ on the material properties of *Pseudomonas aeruginosa* PAO1 and *Desulfovibrio* sp. EX265 biofilms. *Water Sci. Technol.* 43:113–20
54a. Stoodley P, Lewandowski Z, Boyle JD, Lappin-Scott HM. 1998. Oscillation characteristics of biofilm streamers in

flowing water as related to drag and pressure drop. *Biotechnol. Bioeng.* 57:536–44
55. Stoodley P, Lewandowski Z, Boyle JD, Lappin-Scott HM. 1999. Structural deformation of bacterial biofilms caused by short term fluctuations in flow velocity: an in-situ demonstration of biofilm viscoelasticity. *Biotechnol. Bioeng.* 65:83–92
56. Stoodley P, Lewandowski Z, Boyle JD, Lappin-Scott HM. 1999. The formation of migratory ripples in a mixed species bacterial biofilm growing in turbulent flow. *Environ. Microbiol.* 1:447–57
57. Stoodley P, Wilson S, Hall-Stoodley L, Boyle JD, Lappin-Scott HM, Costerton JW. 2001. Growth and detachment of cell clusters from mature mixed species biofilms. *Appl. Environ. Microbiol.* 67: 5608–13
58. Suh SJ, Silo-Suh L, Woods DE, Hasset DJ, West SE, Ohman EE. 1999. Effect of *rpoS* mutation on the stress response and expression of virulence factors in *Pseudomonas aeruginosa. J. Bacteriol.* 181:3890–97
58a. Tolker-Nielson T, Brinch UC, Ragas PC, Andersen JB, Jacobsen CS, Molin S. 2000. Development and dynamics of *Pseudomonas* sp. biofilms. *J. Bacteriol.* 182:6482–89
59. Trevors JT, Psenner R. 2001. From self-assembly of life to present-day bacteria: a possible role for nanocells. *FEMS Microbiol. Rev.* 25:573–82
60. Watnick PI, Lauriano CM, Klose KE, Croal L, Kolter R. 2001. The absence of a flagellum leads to altered colony morphology, biofilm development and virulence in *Vibrio cholerae* O139. *Mol. Microbiol.* 39:223–35
61. Whiteley M, Bangera MG, Bumgarner RE, Parsek MR, Teitzel GM, et al. 2001. Gene expression in *Pseudomonas aeruginosa* biofilms. *Nature* 413:860–64
62. Whiteley M, Lee KM, Greenberg EP. 1999. Identification of genes controlled by quorum sensing in *Pseudomonas aeruginosa. Proc. Natl. Acad. Sci. USA* 96: 13904–9
63. Whiteley M, Parsek MR, Greenberg EP. 2000. Regulation of quorum sensing by RpoS in *Pseudomonas aeruginosa. J. Bacteriol.* 182:4356–60
64. Wingender J, Neu TR, Flemming H-C. 1999. What are bacterial extracellular polymeric substances? In *Microbial Extracellular Polymeric Substances*, ed. J Wingender, TR Neu, H-C Fleming, pp. 93–112. Berlin: Springer
65. Zielinski NA, Chakrabarty AM, Berry A. 1991. Characterization and regulation of the *Pseudomonas aeruginosa algC* gene encoding phosphomannomutase. *J. Biol. Chem.* 266:9754–63
66. Zobel CE. 1943. The effect of solid surfaces upon bacterial activity. *J. Bacteriol.* 46:39–56

Annu. Rev. Microbiol. 2002. 56:211–36
doi: 10.1146/annurev.micro.56.012302.161120

First published online as a Review in Advance on April 19, 2002

MICROBIAL COMMUNITIES AND THEIR INTERACTIONS IN SOIL AND RHIZOSPHERE ECOSYSTEMS

Angela D. Kent[1] and Eric W. Triplett[2]
[1]Center for Limnology and [2]Department of Agronomy, University of Wisconsin-Madison, Madison, Wisconsin 53706; e-mail: adkent@facstaff.wisc.edu; triplett@facstaff.wisc.edu

Key Words culture-independent community analysis, 16S rRNA, community fingerprint, microbial diversity, plant-microbe interactions, microbial ecology

■ **Abstract** Since the first estimate of prokaryotic abundance in soil was published, researchers have attempted to assess the abundance and distribution of species and relate this information on community structure to ecosystem function. Culture-based methods were found to be inadequate to the task, and as a consequence a number of culture-independent approaches have been applied to the study of microbial diversity in soil. Applications of various culture-independent methods to descriptions of soil and rhizosphere microbial communities are reviewed. Culture-independent analyses have been used to catalog the species present in various environmental samples and also to assess the impact of human activity and interactions with plants or other microbes on natural microbial communities. Recent work has investigated the linkage of specific organisms to ecosystem function. Prospects for increased understanding of the ecological significance of particular populations through the use of genomics and microarrays are discussed.

CONTENTS

INTRODUCTION TO THE PROBLEM

From the classic paper of Torsvik et al. (128) came the first culture-independent estimate of the number of prokaryotic genomes in soil. That estimate of 4600 distinct genomes per gram of soil was determined by the reassociation time of total community DNA compared with a standard curve of reassociation kinetics of a known number of cultured genomes. The classical ecological approach for the description of an ecosystem is to first characterize the community structure by identification and enumeration of the species present, and then to assign roles in ecosystem function to species or groups. This strategy, typically employed by ecosystem and population ecologists, has not always been practical for microbial ecologists.

Traditionally, taxonomic classification of bacteria has been determined based on metabolic, morphologic, and physiological traits (42, 55). This approach emulates the methodological approach of botanists and zoologists; however, it requires the isolation and cultivation of individual bacterial species. Assessments of bacterial communities from a number of environments have found that the fraction of cells that may be cultured is not representative of the abundance or diversity of the microbial community present in the environment; it is often observed that direct microscopic counts exceed viable cell counts by several orders of magnitude [reviewed in (4, 54, 94)]. Clearly, culture-based methodology is inadequate to serve the needs of microbial ecologists seeking to describe the diversity of bacterial communities in environmental samples.

CULTURE-INDEPENDENT ASSESSMENTS OF MICROBIAL COMMUNITIES

The rRNA molecules have long been recognized for their utility as molecular chronometers (136). These molecules occur in all organisms and possess a high degree of structural and functional conservation. The larger rRNA molecules contain many domains with independent rates of sequence change (related to their structural and functional conservation). Examination of these changes over time allows phylogenetic relationships to be measured. A number of methods have been developed that exploit this sequence divergence among taxa to examine microbial community structure. These culture-independent methods for microbial community analysis most often utilize polymerase chain reactions (PCR) to amplify phylogenetic markers from DNA extracted from the microbial community. Methodologies commonly used for microbial community analyses are summarized below. For more detailed information on these methods and their limitations, the reader is referred to recent reviews (46, 76, 103, 127, 130).

Early Analyses of Microbial Diversity in Soils by Molecular Means

In the first assessments of soils by culture-independent means, *Proteobacteria* were found to dominate 16S rDNA clone libraries using template DNA from a Queensland soil (68, 124). A more diverse population was found in a Japanese soybean field (129). However, in these papers, few clones were sequenced. In a larger study, three bacterial divisions, the *Proteobacteria*, the *Fibrobacter*, and the low G + C gram-positive bacteria, were represented in nearly 60% of the 16S rDNA clones with a Wisconsin pasture soil as the source of the template DNA (14). In sharp contrast to these agricultural soils, analysis of 16S rDNA clones from a Siberian tundra soil showed that over 60% of the clones belonged to the *Proteobacteria* and 16% to the *Fibrobacter* (144).

The publication of this early work encouraged analyses of soil microbial diversity under a wide variety of conditions, which are summarized below. These comparative studies must still be considered preliminary, as no known methods are capable of efficiently assessing the fate of over 10,000 distinct organisms per gram of soil over time and space. We have learned what divisions of bacteria commonly dominate soil, but we do not know their ecological significance. When changes are observed with soil treatments, we still do not know whether these changes affect ecosystem function. The current era of investigation can be viewed as the descriptive phase, which is necessary prior to a testing phase where we will learn the role and perhaps the functional redundancy of the perhaps hundreds of millions of operational taxonomic units in soils on earth.

Given the breadth of the current literature it is difficult for any review of this length to be comprehensive. For those important works that we have neglected to mention, we must apologize in advance. The reader is also referred to a number of important recent reviews by Amann (2), DeLong & Pace (29), Hughes et al. (56), Johnsen et al. (60), Øvreås (90), and Pace (92). Each of these covers topics that are now mentioned here and in some cases offers more detail on particular issues.

Prior to continuing this discussion of microbial diversity in soils, a description of the methods used for such assessments is necessary.

CULTURE-INDEPENDENT METHODS OF ASSESSING MICROBIAL DIVERSITY

PCR-Based Methods

Community analyses based on PCR have a number of steps that may introduce biases, starting with DNA extraction. Bacterial cell structure varies among taxonomic groups, with some bacteria being more easily disrupted than others. In addition, environmental factors require special consideration for both sample collection and DNA extraction. Inhibition of PCR by environmental compounds has been reviewed by Wilson et al. (135). Methods for sample collection and DNA extraction must take into account such factors as coextraction of humic substances from soil and low bacterial cell density in some environments, and at the same time optimize lysis of structurally different cells. Niemi et al. (86) demonstrated that soil bacterial community profiles differed depending on the DNA extraction and purification method utilized. Methods that include mechanical lysis using a bead beater were found to yield the most consistent results.

Despite these caveats, PCR-based community analysis methods are commonly used because of the ease with which many samples can be analyzed and the ability to tailor the analysis to examine particular organisms or taxa of interest through the use of universal or group-specific primers (19, 47, 66, 107). A number of community "fingerprint" methods are commonly used to assess differences in community composition between samples or treatments or to assess changes in microbial communities over time. Such techniques as ribosomal intergenic spacer analysis (RISA) (15, 100, 101, 107), denaturing gradient gel electrophoresis (DGGE) (83), temperature gradient gel electrophoresis (TGGE) (49), single-strand-conformation polymorphism (SSCP) (115, 116), ITS-restriction fragment length polymorphism (ITS-RFLP) (27), random amplified polymorphic DNA (RAPD) (44, 142), or amplified ribosomal DNA restriction analysis (ARDRA) (78) yield complex community profiles that do not directly offer phylogenetic information but do allow analysis and comparisons of community composition. Differences in electrophoretic profiles between samples reflect differences in community composition and abundance of individual microbial populations in a community. Although the fingerprint obtained from an environmental sample cannot reveal the taxonomic composition of a microbial community, phylogenetic information about particular community members may be obtained by isolation and sequence analysis of bands of interest.

A number of approaches have been developed to improve the detection and resolution of fragment analysis, including automated ribosomal intergenic spacer analysis (ARISA) (39, 102), length heterogeneity PCR (LH-PCR) (106, 126), and terminal restriction fragment length polymorphism (T-RFLP) (65, 70, 77, 89). For details of these methods, the reader is referred to the original literature. These methods utilize a fluorescently labeled oligonucleotide primer for PCR amplification and an automated system such as the Applied Biosystems capillary or gel electrophoresis instruments for separation and detection of PCR fragments. Automation of the procedure increases sample throughput and allows the rapid analysis of bacterial community structure. The high resolution offered by automated electrophoresis instruments and the high sensitivity of fluorescence detection increase the number of peaks detected compared to methods that use standard gel electrophoresis and detection. In addition, the band intensity can be measured more precisely by fluorescence detection methods, which allows a more accurate comparison of community profiles. Ranjard et al. (102) point out that this level of sensitivity may have undesirable aspects, as it may introduce variability in community profiles that has no biological origin. When the fluorescent fragment analysis techniques are used, information on the relative abundance of individual fragments (presumed to represent different bacterial taxa) is collected. These data are analogous to typical ecological data about species composition and abundance, and as such can be used to express the diversity of a community using indices of ecological diversity such as the Shannon-Weaver diversity index, Sorenson's similarity index, or the Bray-Curtis similarity index, as well as measures such as richness or evenness (73).

Of the fragment analysis methods listed above, only T-RFLP offers phylogenetic information directly without further sequencing of the fragments. Fragment length obtained from T-RFLP analysis of a microbial community may be compared to the expected terminal restriction fragment length obtained from analysis of known 16S rRNA gene sequences (77). In practice, however, the complexity of T-RFLP profiles obtained from environmental samples can hinder phylogenetic assignment of individual fragments (65, 70, 76). When more specific phylogenetic information is desired, researchers employ the more laborious strategy of constructing a clone library from the amplified phylogenetic markers (6, 16, 30, 51, 131). This approach exchanges rapid analysis of microbial community composition for fine-scale taxonomic assignment of dominant community members. The logistics of sequencing sufficient and statistically significant numbers of clones to describe the diversity of an environment make this a cumbersome technique for the comparison of microbial communities. Clone libraries are most useful for identification and characterization of previously undescribed species or for augmenting molecular fingerprinting techniques (65, 80). Approaches utilizing DNA microarrays are being developed to increase the application of clone libraries for comparisons of microbial communities (121).

While correlations between the distribution of PCR-amplified phylogenetic markers and species distribution have limitations owing to the presence of multiple rRNA operons in bacteria (35, 98) and PCR and cloning biases (23, 97, 105, 126), molecular methods for community analysis can reveal the presence of microorganisms that remain intractable to traditional cultivation techniques. Fernández et al. (37) found that the effects of these biases are minimized when relative changes are studied in the same ecosystem and when replicate community profiles are produced.

The majority of studies that utilize these PCR-based techniques are carrying out community analyses using the ribosomal RNA operon, typically the 16S rRNA gene, though the 5S rRNA gene is sometimes used (53). Marsh (76) summarized several bacterial housekeeping genes with potential as phylogenetic markers. These include genes for heat shock proteins, glutamine synthetase, ATPases, and topoisomerases (76). When specific traits or functional characteristics are under investigation, phylogenetic markers other than the rRNA genes can be used to characterize microbial communities. For soil microorganisms, functions associated with nitrogen metabolism have been widely used for community analysis. The phylogenetic markers used in these studies include a structural gene for nitrogenase (*nifH*) [studies reviewed by (81, 95)], nitrous oxide reductase (*nosZ*) (111, 112), and nitrite reductase genes (*nirK* and *nirS*) (18). Genes involved in methane oxidation (*pmoA*, *mmoB*, and *mxaF*) have also been used to characterize soil microbial communities (47). When a particular function is restricted to specific bacterial taxa, 16S rRNA sequence may be used to differentiate these community members. This approach is used to study autotrophic ammonia oxidizers as well as methane-oxidizing bacteria; PCR primers specific for the 16S rDNA of the closely related organisms capable of these functions can be used to carry out the community analyses described above (19, 47).

Alternatives to PCR Approaches

Methods that examine physiological or metabolic characteristics of microbial communities are alternatives to PCR-based approaches. Fatty acid methyl ester (FAME) profiles and phospholipid fatty acid analysis have been used extensively to characterize the composition of soil microbial communities [(57, 106) and references therein].

Direct microscopic examinations are also important for analysis of microbial communities. Fluorescence in situ hybridization (FISH) can be used to evaluate the distribution and function of microorganisms in situ (3, 5). This method uses oligonucleotide hybridization probes complementary to regions of the 16S rRNA gene for determination of in situ abundance. Like the PCR-based methods, this technique can be customized to target specific groups of organisms. To place the dynamics of important populations within the context of community-level phenomena, FISH can be used in combination with DAPI (4′,6′-diamidino-2-phenylindole), 2-(*p*-iodophenyl)-3-(*p*-nitrophenyl)-5-phenyltetrazolium chloride (INT)-formazan, or 5-cyano-2,3-ditolyl tetrazolium chloride (CTC) staining (64, 96) to determine the contribution made by the populations of interest to total abundance or active cell count. However, the low throughput of FISH limits its application for comparison of large numbers of samples.

Careful morphological analysis of bacterial cells can provide powerful information on the diversity, microbial abundance, and two-dimensional spatial distribution of microbial community members. A computer-aided system has been developed by the Center for Microbial Ecology at Michigan State University to assist in such assessments. CMEIAS (Center for Microbial Ecology Image Analysis System) is a semi-automated analysis tool that uses digital-image processing and pattern-recognition techniques in conjunction with microscopy to gather size and shape measurements of digital images of microorganisms to classify them into their appropriate morphotype, allowing culture-independent quantitative analysis of the diversity and distribution of complex microbial communities (69). This tool holds much promise for automating a tedious but important evaluation of microbial communities.

Methods to Assess Community Function in Soil

As microbial ecology involves the study of both the structure and function of an ecosystem, meaningful assessments of microbial communities must consider not only the abundance and distribution of species but also the functional diversity and redundancy present in a microbial community. Gaston (41) has described functional diversity as the number of distinct processes (functions) that can potentially be performed by a community, whereas functional redundancy is measured as the number of different species within the functional groups present in a community. The diversity of metabolic functions possessed by microbial communities is often examined using BIOLOG GN substrate utilization assays (40, 50, 59, 122), which assess the ability of the community as a whole to utilize select carbon substrates. This method has the inherent biases of other culture-based approaches; however, the resulting metabolic fingerprint may not be an accurate representation of the functional diversity of the natural microbial community (91, 122). An alternative technique that avoids misrepresentation of functional diversity due to culture bias assesses community response (measured as CO_2 respiration) after the addition of selected carbon substrates directly to the soil environment (28).

To gain better insight to the microbial processes within an ecosystem, it is essential to study functional diversity in combination with taxonomic diversity. Recent studies have attempted to characterize the portion of the microbial community that responds to nutrient availability by comparing community fingerprints after incubation in individual BIOLOG wells (122) or by isolating DNA from microbial community populations that responded to nutrient addition by uptake and incorporation of a thymidine nucleotide analog, bromodeoxyuridine (BrdU) (13). Molecular fingerprint analysis of the responsive portion of the microbial community (as defined by BrdU labeling) was also used to assess the functional redundancy of bacterial communities along a vegetation gradient (143).

A DNA microarray technique for the simultaneous identification of ecological function and phylogenetic affiliation of microbial populations has been recently developed (11). The approach combines a community-specific 16S rDNA-based oligonucleotide array (functional diversity array) with incubations of the microbial community with various radiotracer substrates. Total RNA extracted from the in situ incubations is hybridized to the microarray of species-specific probes, which allows the identification of populations that were active in the metabolism of the labeled substrates. This approach permits the assessment of growth rate and substrate utilization of individual microbial populations within a community. In addition, whole-genome DNA microarrays specific for a single organism can be used to analyze related organisms and may ultimately prove useful for community analysis (1, 31, 82). These arrays can be used for both gene discovery as well as for analysis of gene expression in the environment.

CHANGES IN UNCULTURED BACTERIAL COMMUNITIES WITH DISTURBANCE

The methods available to assess the effects of pesticides on bacterial diversity in soils were recently reviewed by Johnsen et al. (60). That discussion is not repeated here. Rather, we describe the results to date of the efforts made to assess how agricultural management, pesticides, and pollutants have altered the microbial landscape.

Heavy Metals

A soil treated with sludge containing either high or low amounts of heavy metals was analyzed for soil bacterial diversity of three subdivisions of the Proteobacteria: the Cytophaga-Flavobacterium division, the gram-positive high G + C division, and the gram-positive low G + C division. All were measured using a dot blot hybridization procedure (109). Heavy metal treatment had the greatest influence on two of these taxa. Sandaa et al. (109) found that the abundance of the α-Proteobacteria more than doubled while the Cytophaga-Flavobacterium division abundance declined by more than two thirds with the high heavy metal treatment. Other taxa seemed to decline with high heavy metal treatment, but their low abundance with the low heavy metal treatment made a quantitative assessment of the decline difficult. In another study, Sandaa et al. (110) showed that the number of prokaryotic genomes per gram of wet weight of soil declined eightfold following many years of heavy metal treatment. With the exception of the α-Proteobacteria, all phylogenetic taxa examined declined as a percentage of the total number of prokaryotes in the soil. The percentage of α-Proteobacteria more than doubled with the heavy metal treatment.

Addition of Hg(II) to a silt loam caused an increase in abundance of two RISA bands (100). These bands were excised, sequenced, and identified as having originated from a *Clostridium*-like gram-positive organisms and a *Ralstonia*-like

β-Proteobacterium. Verification of the identity of these bacteria was done by hybridization. This is an excellent example of the identification of uncultured bacteria following a given treatment.

Addition of Pollutants to Soil

During a pentachlorophenol enrichment in a reactor containing a soil slurry, the dominant organisms found after a period of enrichment were related to the genus *Sphingomonas* (8). Incubation of a soil sample with methane resulted in the enrichment of a group of putative methylotrophic α-Proteobacteria distantly related to known methylotrophs as well as an increase in type II methanotrophs (59a). Øvreås & Torsvik (91) found that methane treatment reduced overall bacterial diversity as measured by DGGE while enriching methanotrophs.

In assessments of bacterial diversity in soil microcosms using TGGE, bacterial diversity declined following treatment with chlorinated benzoates compared to the untreated control (99). *Burkholderia*-like organisms were found to increase with the addition of chlorinated benzoates in these experiments.

The DGGE profile of 16S rDNA from a polyaromatic hydrocarbon-contaminated sandy loam was considerably less diverse than those of noncontaminated soils (84). However, a direct assessment of the effects of polyaromatic hydrocarbons cannot be made since there was no uncontaminated sample available from the same site.

Few pollutants in soils have been examined for their effects on soil microflora from a culture-independent perspective. As the techniques to make such assessments continue to improve and simplify, such analyses may become a routine part of environmental assessments required by governmental agencies prior to the use of a new compound in the environment. However, with such assessments must come ideas on how such results can be properly interpreted for risk assessment analysis. Does an impact on microbial diversity have a significant effect on soil function? Does the functional redundancy of microbial processes render microbial diversity analyses based on phylogenetics meaningless with regard to ecosystem function? If that is the case in some circumstances, should microbial processes be measured along with the taxonomic diversity assays?

Pesticide Treatment

El Fantroussi et al. (34) examined the effect of three phenyl urea herbicides on microbial communities in soils over an 11-year period. All three herbicides significantly decreased the number of culturable heterotrophic bacteria. BIOLOG GN fingerprint analysis also showed that the treated communities differed significantly compared to the control. A striking result of this work is the apparent decline of uncultured *Acidobacterium* upon treatment with any of the three herbicides. Uncultured *Acidobacterium* are commonly found in culture-independent analyses of soils. It is not clear whether the decline is caused directly by the herbicides or as a consequence of the changes in the macroflora community resulting from herbicide use. Treatment of soil with the fungicide triadimefon caused a decline in organic carbon and soil microbial biomass but no decline in microbial DNA diversity as measured with RAPD random primer amplification (142). This can be explained by the common contradiction that although fungi can comprise a large proportion of soil biomass, fungal DNA concentrations in soil are low (45).

Xia et al. (138) evaluated microbial community response to the experimental application of 2,4-dichlorophenoxyacetic acid (2,4-D) using RAPD fingerprints. No changes in community structure were observed in response to 2,4-D application to three different soils. Hybridization studies indicated that application of 2,4-D at the recommended application rates did not select for bacterial populations capable of 2,4-D degradation.

Two culture-dependent studies on the effects of herbicides on soil bacterial diversity present conflicting results. Nicholson & Hirsch (85) showed an increase in culturable bacterial populations in soils treated with herbicides such as glyphosate. The authors thought that the increased crop yield resulting from the herbicide treatment might have contributed to higher bacterial numbers. In contrast, Busse et al. (22) found lower bacterial numbers in a pine plantation treated with glyphosate compared to the untreated control. Culture-independent analyses are needed to resolve this question.

Fumigants are used widely in high-value crops for the control of eukaryotic soil-borne pests such as fungal pathogens, nematodes, and weeds. Ibekwe et al. (58) studied their effect on soil prokaryotic communities from a culture-independent perspective. Of four fumigants used, methyl bromide caused the greatest and longest-lasting impact on soil bacterial diversity. Chloropicrin had virtually no impact.

The few studies published to date from a culture-independent perspective suggest that pesticides have little impact on soil bacterial diversity. However, so little has been done in so few locations with so few pesticides, that no conclusions can be drawn at this time. This is an area that is ripe for more investigation.

Agricultural Management

Through the use of ribosomal intergenic space analysis (RISA), deforestation in Amazonia was shown to have a profound, qualitative impact on soil bacterial diversity (15). This early work on a culture-independent assessment of the effects of land management on microbial diversity has since been followed by several more quantitative measurements of the effects of land use. Another analysis of community changes in tropical soils with deforestation was done by Nüsslein & Tiedje (88). The $G + C$ content of the pasture soil DNA was significantly higher than that of the forest soil DNA. Whereas the Fibrobacter were dominant in the forest soil, the β- and α-Proteobacteria dominated the pasture soil.

Improved and unimproved Scottish grasslands differing in fertilizer regimes and plant cover were assessed for microbial diversity using 16S rDNA clone libraries (79). Both pastures were dominated by α-Proteobacteria (about 40% of the total clones) followed by the actinomycetes (13.3% of the total). Indices of diversity including the Shannon-Weaver index as well as evenness and dominance measurements were similar between the two pastures.

Grasslands in the Netherlands taken out of agricultural production over a period of 30 years were examined for changes in microbial diversity (36). The multiple competitive RT-PCR procedure used did not have sufficient resolution to distinguish those pastures currently in agricultural production from those taken out of production 30 years earlier. Similarly, the application of sewage sludge to a grassland site for over 100 years failed to confer a measurable change in soil microbial diversity as determined by fatty acid methyl ester patterns and carbon substrate utilization by the community (67).

Through a culture-independent analysis of soils collected from the Kellogg Biological Station's Long Term Ecological Research project of Michigan State University, Buckley & Schmidt (21) found that the microbial diversity of cultivated fields differed little from each other regardless of the specific agricultural management regime. However, the bacterial diversity of the managed soils were significantly different from soils of nearby fields that had never been cultivated. This analysis was done using 16S rDNA taxa-specific probes and T-RFLP analysis of amplified 16S rDNA. This is an excellent site for such analyses because the Long Term Ecological Research sites have long-term data on the temporal and spatial variability of a wide range of physical, chemical, and biological properties of the experimental location. These long-term data allow the investigator to correlate changes in microbial communities with ecosystem processes. An interesting follow-up to the work of Buckley & Schmidt (21) would be to address the question of the temporal variability of bacterial diversity in these soils and determine whether any of the observed variability is correlated to any of the physical and chemical characteristics of these soils.

MORE DIRECT COMPARISONS BETWEEN CULTURE-DEPENDENT AND CULTURE-INDEPENDENT ASSAYS OF SOIL MICROBIAL DIVERSITY

The microbial communities of four arid soils from northern Arizona were compared by identifying cultured isolates and by restriction fragment-length polymorphism and sequence analysis of 16S rDNA clones derived from soil DNA (32). Seven bacterial divisions were represented among the clone libraries while only three were found among the isolates. *Acidobacterium*-related organisms comprised nearly half the organisms identified in the clone libraries, while nearly 80% of the isolates were gram-positive strains. As expected, the culture bias failed to identify most of the organisms observed in the clone libraries.

Similar results were obtained in a wheat field from Holland where *Acidobacterium* and the Proteobacteria dominated the uncultured organisms and gram-positive organisms dominated the culture collection (123). In addition, Smit et al. (123) looked at seasonal changes in these populations and found that samples taken in July were significantly different than those taken during other times of the year.

Analysis of cultured isolates from a sandy loam and an organic soil from Norway suggested that the bacterial diversity in the two soils was similar (91). However, as the DGGE profiles were not digitized, the authors were unable to assess differences in diversity by culture-independent means. Nevertheless, one of the hallmark assays from the Torsvik group was performed. Using thermal denaturation and reassociation of community DNA, the authors showed that the organic soil possessed 10–62-fold higher genome complexity than the sandy loam soil. This result confirms the need to assess diversity by culture-independent means.

SOIL PARTICLE SIZE

Ranjard et al. (101) used RISA to show that microbial diversity varies with soil particle size. Although it is not surprising that different organisms can occupy niches of different size, this is the first paper to demonstrate this from a culture-independent perspective. Sessitsch et al. (117) took these ideas one step further and showed that microbial diversity, as determined by T-RFLP profiles, increases with decreasing particle size. Larger particles were dominated by the α-Proteobacteria while the *Holophaga/Acidobacterium* were most common in clay particles.

MICROBIAL DIVERSITY OF THE SOIL-ROOT INTERFACE: THE RHIZOSPHERE

The rhizosphere is defined as the soil surrounding the roots that is influenced by living roots. This influence may occur by root exudation of carbon substrates that affect microbial communities. Shortly after the reports of culture-independent analyses of bulk soil were published, many investigators around the world turned their attention to the rhizosphere where so many interactions between microorganisms and plants take place. The number of issues that can be studied is limited only by the imagination. We review a set of these below.

Influence of the Host Plant on Rhizosphere Bacterial Communities

Smalla et al. (122a) used DGGE to distinguish microbial communities in bulk soils versus those in the rhizospheres of strawberry, canola, and potato. Rhizosphere communities differed significantly from bulk soil communities. Canola and potato rhizosphere communities were more similar to each other than they were to strawberry. Sequencing of some of the DGGE bands excised from rhizosphere sample gels revealed that most were derived from gram-positive strains. Plant species, root zone, and soil type all influence the rhizosphere bacterial community in the DGGE analysis of 16S rDNA by Marschner et al. (75). However, no data were provided concerning the identity of the specific organisms affected by these treatments. Similarly, the presence of rye or alfalfa roots in the soil influenced the rhizosphere community more strongly than did soil type. Kaiser et al. (63) found that the rhizosphere of canola was dominated by the α-Proteobacteria subdivision and the Cytophaga-Flavobacterium-Bacteroides division. This was in contrast to the cultured isolates from the rhizosphere that were dominated by organisms from the β- and γ-Proteobacteria subdivisions.

Normander & Prosser (87) assessed barley phytosphere bacterial diversity by using DGGE profiles. They found that plant age up to 36 days had little influence on the rhizosphere communities. The rhizosphere community was more similar to the bulk soil community than it was to the endophytic community. This suggests that rhizosphere bacteria are of soil origin, whereas many of the endophytic bacteria are seed borne.

Marilley & Aragno (74) sequenced 16S rDNA clones prepared from DNA templates collected from bulk soil, the rhizosphere, and the interior of roots. The γ-Proteobacteria increased along the gradient toward the interior of the plant while the *Holophaga/Acidobacterium* group decreased.

Chelius & Triplett (26) discovered that the interior of maize roots is inhabited by six bacterial and two archaeal divisions. In agreement with Marilley & Aragno (74), the Proteobacteria dominated the interior of the root. Several independent isolations of *Klebsiella pneumoniae* have been made from the interior of maize roots (24, 25, 31, 93), also in agreement with the observation that γ-Proteobacteria are enriched in the plant interior (74). The cultured collection from maize roots was also diverse with members from four bacterial divisions, which included a new bacterial genus and species in the Flexibacter group, *Dyadobacter fermentens* (24, 25).

Clearly the plant species strongly influences rhizosphere bacterial diversity. However, much more remains to be done to understand these relationships including whether the origin of some rhizosphere bacteria may be the plant seed.

Transgenic Plants and Microbial Diversity

The microbial diversity of the rhizospheres of field-grown T4-lysozyme-expressing potatoes, control transgenic potatoes that possess only the marker gene, and nontransgenic parental potatoes were not significantly different as measured with 16S rDNA DGGE profiles. In addition, a T4-lysozyme-tolerant Pseudomonad used as an inoculum strain did not become dominant in the community of any of the rhizospheres, including the rhizospheres of the T4-lysozyme-expressing plants. As many bacteria are sensitive to T4-lysozyme, this was a surprising result. The half-life of T4-lysozyme in nonsterile soil may be so brief that the concentration of this protein never builds to a high enough level to affect the bacterial community. Similarly, the rhizosphere communities of Barnase/Barstar transgenic potato differed little from those of the nontransgenic parent plant (71).

Herbicide-tolerant transgenic canola and wheat plants can harbor a different bacterial community either in the rhizosphere or in the root interior compared to nontransgenic varieties (33, 118, 119), but these differences cannot be attributed solely to the transgene because the plant varieties tested were not otherwise isogenic.

Bacterial Rhizosphere Communities and Plant Disease

Soils that suppress plant disease have been known for many years. However, the cause of the plant disease suppression in many cases remains unknown. One area in which much progress has been made is the study of natural supression of "take-all," a wheat root disease caused by the fungus *Gaeumannomyces graminis* var. *tritici*. This natural disease suppression, a phenomenon known as "take-all decline" (TAD) is manifested as a spontaneous reduction of disease after an extended period of barley or wheat monoculture. Many studies have reported the association of antibiotic-producing fluorescent *Pseudomonas* species with disease suppression [summarized by Raaijmakers et al. (96b, 96c)]. Further studies demonstrated that production of the antibiotic compound 2,4-diacetylphloroglucinol by fluorescent *Pseudomonas* spp. was a critical component of this natural disease suppression (96a, 96c). Inoculation with the causal agent of take-all disease of wheat, *Gueumannomyces graminis* var. tritici, results in a noticeable increase in the culturable Pseudomonads in the rhizosphere bacterial community (80). Several other groups of bacteria increased in these rhizospheres as well. Some of these were cultured but were not able to significantly influence the growth of the fungal pathogen in vitro, but they did vary in their ability to be antagonistic toward a *Pseudomonas* strain suppressive toward the disease.

Yang et al. (140) discovered that the bacterial diversity of *Phytophthora*-infected avocado roots is much greater than that of uninfected roots. Roots inoculated with the disease-suppressive bacterium *Pseudomonas fluorescens* 513 were disease-free and had a rhizosphere community similar to that of plants not inoculated with the pathogen. A likely explanation of this result is that diseased roots probably release more nutrients into the rhizosphere as a result of their own decay. These increased nutrients attract bacteria that might not normally be competitive in the rhizosphere.

Bacterial Antibiotic Production and Rhizosphere Microbial Diversity

In the first culture-independent analysis of the effects of bacterial antibiotic production on a natural community, Robleto et al. (107) showed that the production of the peptide antibiotic by *Rhizobium etli* CE3(pT2TFXK) resulted in a dramatic reduction in the diversity of trifolitoxin-sensitive α-Proteobacteria in the rhizosphere of *Phaseolus vulgaris*. This reduction did not occur in the rhizosphere of plants inoculated with the isogenic, nontrifolitoxin-producing strain, *R. etli* CE3(pT2TX3K). None of the treatments caused a detectable decline in total bacterial diversity of the rhizosphere, which was expected, as most bacteria are trifolitoxin-resistant.

Glandorf et al. (43) engineered *Pseudomonas putida* WCS358r to produce the antifungal compound phenazine-1-carboxylic acid (PCA). The engineered strains decreased the fungal diversity of the wheat rhizosphere more consistently and for a longer period.

Though the study by McSpadden Gardener et al. (80) indicates that the interaction of the fungal pathogen with the plant alters the diversity and composition of the wheat rhizosphere, the 2,4-diacetylphlorglucinol produced by *Pseudomonas* spp. recovered from such assays may play a role in restructuring the rhizosphere microbial community in the study discussed above (80). This compound is known to have antibacterial properties and has been recovered from soils naturally suppressive for take-all decline (96a).

Rhizosphere Microbial Diversity and Plant Nutrient Status

Few papers have examined the effects of mineral nutrient status on rhizosphere microbial diversity. Yang & Crowley (139) grew barley plants in low- and high-iron soils. Using amplified 16S rDNA from various locations on the roots and separated by DGGE, the authors found qualitative differences between the treatments but did not define which organisms comprised the differences. In a study of low-nitrogen conditions on bacterial diversity in the rhizosphere of bean plants, Schallmach et al. (113) found that low N increased the proportion of α-Proteobacteria relative to the entire bacterial population near the root tip. This may be caused by the accumulation of root-nodulating rhizobia near the root tip. Throughout the root, low-N status increased the proportion of high $G + C$ gram-positive bacteria. Both studies used methodology that has inherently low resolution. As a result, differences between treatments were difficult to observe.

Epiphytic Bacterial Diversity

Little has been done to describe epiphytes from a culture-independent perspective. As with other environments, leaf surfaces are inhabited by a wide variety of bacteria not known in culture collections. But to date this has only been studied on citrus (141) and the seagrass *Halophia stipulacea* (133, 134).

NONMETHANOGENIC ARCHAEA IN SOIL

Organisms within the domain Archaea have been classified among three divisions: Crenarchaeota, Euryarchaeota, and Korarchaeota (7). Methanogens are Euryarchaeota from soil and were among the earliest organisms recognized as members of the Archaea (137). Many methanogens have been cultured and characterized. A culture-independent study of Archaea in a Finnish forest soil reported the presence of *Halobacterium*-like Euryarchaeota (61). To date, none of the Korarchaeota has been cultured, and the only culture-independent evidence of their existence comes from extreme environments, particularly hot springs (52, 104). Prior to their first report in a temperate soil, the Crenarchaeota were thought to be present only in extreme environments. Ueda et al. (129) first reported culture-independent evidence for Crenarchaeota in a nonextreme soil environment. Two

16S rDNA clones were sequenced and found to be phylogenetically most similar to 16S rDNA sequences of Crenarchaeotal origin. Their template DNA was derived from a soybean field in Japan. Using an Archaeal-specific primer for the amplification of 16S rDNA clones, Bintrim et al. (12) found a wider diversity of Crenarchaeota in a Wisconsin soil. Culture-independent evidence of Crenarchaeota in lake sediments, tropical soils, and agricultural soils from Germany, Indiana, Michigan, and Norway suggests that the diversity of Crenarchaeota in terrestrial environments is broad (15, 20, 48, 72, 84, 110, 114).

Although not one of these soil Crenarchaeota has been cultured, their abundance can be determined using molecular tools with some interesting results. The first suggestion of the abundance of Crenarchaeota in soil relative to the Bacteria came from a culture-independent analysis of 16S rDNA clones from two adjacent Amazonian soils (15), where just 2 of 100 clones were of Archaeal origin. A more thorough analysis of Crenarchaeal abundance in soil by probing total community rRNA with a Crenarchaeota-specific probe showed that their abundance in a native soil ($0.37\% \pm 0.13\%$) was much lower than in a cultivated soil ($1.42\% \pm 0.59\%$). Jurgens & Saano (62) found that soil Crenarchaeota are sensitive to deforestation. Different Crenarchaeota were found in the control uncut-forest soil compared to soils from forests that were recently cleared with or without prescribed burning. Although the Crenarchaeota in the cut-forest soils were different from the native soil, their phylogenetic diversity was much higher than in the native soil. Sandaa et al. (110) found that the abundance of Crenarchaeota declined from $1.3\% + 0.3\%$ of all DAPI-stained cells to a level that was below detection following heavy metal contamination of the soil.

Some of the work summarized in the above paragraph strongly suggests that the plant community in the soil may influence the diversity and abundance of soil Crenarchaeota. Simon et al. (120) showed root colonization of soil Crenarchaeota. Chelius & Triplett (26) presented the first evidence for the interior colonization of plant roots by Crenarchaeota and Euryarchaeota.

Clearly more work is needed on the interactions of Archaea in soil and plants, especially in regard to their roles in these environments. Rapid progress will require culturing of these organisms so that their metabolism and physiology can be assessed rigorously through mutagenesis.

FUTURE NEEDS IN UNDERSTANDING MICROBIAL DIVERSITY

Current methodologies struggle to describe the vast microbial diversity in soil. For this reason we have recently turned to studying freshwater microbial diversity, as the microbial diversity of this habitat is considerably less complex than is soil (38, 39). Our hope is that once we have learned to comprehensively analyze microbial diversity in lakes, we can then begin to understand how to scale up our methods to understand soil communities.

In the meantime, all methods continue to improve. In particular, genomic approaches can be expected to improve our understanding of the role of uncultured microbes in the environment. The first genomic libraries of environmental DNA were created by DeLong and coworkers (10, 125, 132). This work resulted in the discovery of a photoactive proteorhodopsin in marine organisms that might play an important role in phototrophy in oceans (9). Similar analyses of uncultured organisms are in progress with soil as the source of DNA. Rondon et al. (108) reported the construction of a BAC library containing inserts of soil microbial DNA. Brady et al. (17) prepared a cosmid library of soil DNA and identified and sequenced a gene cluster for the biosynthesis of violacein, a potent broad-spectrum antibiotic. These works illustrate the ability to use these libraries in the identification of new secondary products, which would be difficult to discover by first culturing the producing organism.

Genomics of environmental DNA can lead to the discovery of important physiological processes in uncultured microorganisms. Efforts are already underway in a number of laboratories to obtain a significant amount of genome sequence from uncultured soil organisms, particularly from the *Acidobacterium* group, which are so common in soils. These data will provide clues regarding the role of these organisms in soil. With enough sequence information, the metabolic pathways of these organisms can be constructed, leading to effective strategies for the culturing of these organisms. The sequence data will also permit the construction of microarrays containing all known open reading frames that can be used to determine gene expression over time and space in the environment.

However, given the large number of organisms in soil, the target DNA to be sequenced must be chosen with care. For example, although the *Acidobacterium* group appears to be common in soil, no one has yet correlated its presence or absence to any microbial process in soil or any chemical or physical properties in soil. Some strong correlations between the presence of an organism with microbial processes or soil structure and function would be helpful in making wise choices prior to making large investments in such sequencing projects. As a result, linkage of specific organisms to ecosystem function over time and space is fundamental work that must go forward if we are to understand the role of these uncultured organisms.

The field also needs more definitive descriptions of those organisms that appear or disappear with a given treatment. Seeing patterns of diversity changing with treatments using DGGE, TGGE, ARISA, or T-RFLP is less satisfying if no attempt is made to identify the organisms affected.

We have also found that snapshots of diversity where the microbial diversity is assessed at one time in one location is not terribly informative. Where diversity over time and space has been measured, the dynamic nature of microbial communities is observed and the presence or absence of individual populations can be compared with changes in ecosystem processes.

The future in this area is bright, particularly as methods and data analysis improve. What we know today is still very much dwarfed by what we do not know.

Systems approaches, whether they be at the level of genomics, ecosystem function, or biocomplexity, will no doubt bring us new insights over the next decades that cannot now be imagined.

ACKNOWLEDGMENTS

The authors wish to acknowledge the support of those agencies that have supported our research in microbial diversity. These agencies include the National Science Foundation (grants DEB 9632853 and 9977903), the USDA (NRI grant 9802884), the Consortium for Plant Biotechnology Research, and the College of Agricultural and Life Sciences of the University of Wisconsin-Madison.

The *Annual Review of Microbiology* is online at http://micro.annualreviews.org

LITERATURE CITED

1. Akman L, Aksoy S. 2001. A novel application of gene arrays: *Escherichia coli* array provides insight into the biology of the obligate endosymbiont of tsetse flies. *Proc. Natl. Acad. Sci. USA* 98:7546–51
2. Amann R. 2000. Who is out there? Microbial aspects of biodiversity. *Syst. Appl. Microbiol.* 23:1–8
3. Amann R, Fuchs BM, Behrens S. 2001. The identification of microorganisms by fluorescence in situ hybridisation. *Curr. Opin. Biotechnol.* 12:231–36
4. Amann RI. 1995. Fluorescently labeled, ribosomal RNA-targeted oligonucleotide probes in the study of microbial ecology. *Mol. Ecol.* 4:543–53
5. Amann RI, Ludwig W, Schleifer KH. 1995. Phylogenetic identification and insitu detection of individual microbialcells without cultivation. *Microbiol. Rev.* 59:143–69
6. Bahr M, Hobbie JE, Sogin ML. 1996. Bacterial diversity in an arctic lake: a freshwater SAR11 cluster. *Aquat. Microb. Ecol.* 11:271–77
7. Barns SM, Delwiche CF, Palmer JD, Pace NR. 1996. Perspectives on archaeal diversity, thermophily and monophyly from environmental rRNA sequences. *Proc. Natl. Acad. Sci. USA* 93:9188–93
8. Beaulieu M, Bécaert V, Deschênes L, Villemur R. 2000. Evolution of bacterial diversity during enrichment of PCP-degrading activated soils. *Microb. Ecol.* 40:345–55
9. Béjà O, Aravind L, Koonin EV, Suzuki MT, Hadd A, et al. 2000. Bacterial rhodopsin: evidence for a new type of phototrophy in the sea. *Science* 289:1902–6
10. Beja O, Suzuki MT, Koonin EV, Aravind L, Hadd A, et al. 2000. Construction and analysis of bacterial artificial chromosome libraries from a marine microbial assemblage. *Environ. Microbiol.* 2:516–29
11. Bertilsson A, Polz M. 2001. Application of a diversity array to study specific substrate utilization in individual populations of aquatic heterotrophic bacteria. *9th Int. Symp. Microb. Ecol.* Amsterdam, The Netherlands
12. Bintrim SB, Donohue TJ, Handelsman J, Roberts GP, Goodman RM. 1997. Molecular phylogeny of archaea from soil. *Proc. Natl. Acad. Sci. USA* 94:277–82
13. Borneman J. 1999. Culture-independent identification of microorganisms that

respond to specified stimuli. *Appl. Environ. Microbiol.* 65:3398–400

14. Borneman J, Skroch PW, Osullivan KM, Palus JA, Rumjanek NG, et al. 1996. Molecular microbial diversity of an agricultural soil in Wisconsin. *Appl. Environ. Microbiol.* 62:1935–43
15. Borneman J, Triplett EW. 1997. Molecular microbial diversity in soils from eastern Amazonia: evidence for unusual microorganisms and microbial population shifts associated with deforestation. *Appl. Environ. Microbiol.* 63:2647–53
16. Bowman JP, McCammon SA, Rea SM, McMeekin TA. 2000. The microbial composition of three limnologically disparate hypersaline Antarctic lakes. *FEMS Microbiol. Lett.* 183:81–88
17. Brady SF, Chao CJ, Handelsman J, Clardy J. 2001. Cloning and heterologous expression of a natural product biosynthetic gene cluster from eDNA. *Org. Lett.* 3:1981–84
18. Braker G, Zhou J, Wu L, Devol AH, Tiedje JM. 2000. Nitrite reductase genes (*nirK* and *nirS*) as functional markers to investigate diversity of denitrifying bacteria in pacific northwest marine sediment communities. *Appl. Environ. Microbiol.* 66:2096–104
19. Bruns MA. 1999. Comparitive diversity of ammonia oxidizer 16S rRNA gene sequences in native, tilled, and successional soils. *Appl. Environ. Microbiol.* 65:2994–3000
20. Buckley DH, Graber JR, Schmidt TM. 1998. Phylogenetic analysis of nonthermophilic members of the kingdom Crenarchaeota and their diversity and abundance in soils. *Appl. Environ. Microbiol.* 64:4333–39
21. Buckley DH, Schmidt TM. 2001. The structure of microbial communities in soil and the lasting impact of cultivation. *Microb. Ecol.* 42:11–21
22. Busse MD, Ratcliff AW, Shestak CJ, Powers RF. 2001. Glyphosate toxicity and the effects of long-term vegetation control on soil microbial communities. *Soil Biol. Biochem.* 33:1777–89
23. Chandler DP, Fredrickson JK, Brockman J. 1997. Effect of PCR template concentration on the composition and distribution of total community 16S rDNA clone libraries. *Mol. Ecol.* 6:475–82
24. Chelius MK, Triplett EW. 2000. *Dyadobacter fermentans* gen. nov., sp nov., a novel gram-negative bacterium isolated from surface-sterilized *Zea mays* stems. *Int. J. Syst. Evol. Microbiol.* 50:751–58
25. Chelius MK, Triplett EW. 2000. Immunolocalization of dinitrogenase reductase produced by *Klebsiella pneumoniae* in association with *Zea mays* L. *Appl. Environ. Microbiol.* 66:783–87
26. Chelius MK, Triplett EW. 2001. The diversity of archaea and bacteria in association with the roots of *Zea mays* L. *Microb. Ecol.* 41:252–63
27. Cho J-C, Tiedje JM. 2000. Biogeography and degree of endemicity of fluorescent *Pseudomonas* strains in soil. *Appl. Environ. Microbiol.* 66:5448–56
28. Degens BP, Harris JA. 1997. Development of a physiological approach to measuring the catabolic diversity of soil microbial communities. *Soil Biol. Biochem.* 29:1309–20
29. DeLong EE, Pace NR. 2001. Environmental diversity of Bacteria and Archaea. *Syst. Biol.* 50:470–78
30. Dojka MA, Harris JK, Pace NR. 2000. Expanding the known diversity and environmental distribution of an uncultured phylogenetic division of bacteria. *Appl. Environ. Microbiol.* 66:1617–21
31. Dong YM, Glasner JD, Blattner FR, Triplett EW. 2001. Genomic interspecies microarray hybridization: rapid discovery of three thousand genes in the maize endophyte, *Klebsiella pneumoniae* 342, by microarray hybridization with *Escherichia coli* K-12 open reading frames. *Appl. Environ. Microbiol.* 67:1911–21
32. Dunbar J, Takala S, Barns SM, Davis

JA, Kuske CR. 1999. Levels of bacterial community diversity in four arid soils compared by cultivation and 16S rRNA gene cloning. *Appl. Environ. Microbiol.* 65:1662–69

33. Dunfield KE, Xavier LJC, Germida JJ. 1999. Identification of *Rhizobium leguminosarum* and *Rhizobium* sp. (Cicer) strains using a custom fatty acid methyl ester (FAME) profile library. *J. Appl. Microbiol.* 86:78–86
34. El Fantroussi S, Verschuere L, Verstraete W, Top EM. 1999. Effect of phenylurea herbicides on soil microbial communities estimated by analysis of 16S rRNA gene fingerprints and community-level physiological profiles. *Appl. Environ. Microbiol.* 65:982–88
35. Farrelly V, Rainey FA, Stackebrandt E. 1995. Effect of genome size and *rrn* gene copy number on PCR amplification of 16S rRNA genes from a mixture of bacterial species. *Appl. Environ. Microb.* 61:2798–801
36. Felske A, Wolterink A, Van Lis R, De Vos WM, Akkermans ADL. 2000. Response of a soil bacterial community to grassland succession as monitored by 16S rRNA levels of the predominant ribotypes. *Appl. Environ. Microbiol.* 66: 3998–4003
37. Fernández A, Huang SY, Seston S, Xing J, Hickey R, et al. 1999. How stable is stable? Function versus community composition. *Appl. Environ. Microbiol.* 65: 3697–704
38. Fisher MM, Klug JL, Lauster G, Newton M, Triplett EW. 2000. Effects of resources and trophic interactions on freshwater bacterioplankton diversity. *Microb. Ecol.* 40:125–38
39. Fisher MM, Triplett EW. 1999. Automated approach for ribosomal intergenic spacer analysis of microbial diversity and its application to freshwater bacterial communities. *Appl. Environ. Microbiol.* 65:4630–36
40. Garland JL, Mills AL. 1991. Classification and characterization of heterotrophic microbial communities on the basis of patterns of community-level sole-carbon-source utilization. *Appl. Environ. Microbiol.* 57:2351–59
41. Gaston KJ. 1996. *Biodiversity: A Biology of Numbers and Difference.* Oxford, UK: Blackwell Sci.
42. Gerhardt P, ed. 1981. *Manual of Methods for General Bacteriology.* Washington, DC: ASM
43. Glandorf DCM, Verheggen P, Jansen T, Jorritsma JW, Smit E, et al. 2001. Effect of genetically modified *Pseudomonas putida* WCS358r on the fungal rhizosphere microflora of field-grown wheat. *Appl. Environ. Microbiol.* 67:3371–78
44. Hadrys H, Balick M, Schierwater B. 1992. Application of random amplified polymorphic DNA (RAPD) in molecular ecology. *Mol. Ecol.* 1:55–63
45. Harris D. 1994. Analyses of DNA extracted from microbial communities. In *Beyond the Biomass*, ed. K Ritz, J Dighton, KE Giller, pp. 111–18. Chichester, UK: Wiley
46. Head IM, Saunders JR, Pickup RW. 1998. Microbial evolution, diversity, and ecology: a decade of ribosomal RNA analysis of uncultivated microorganisms. *Microb. Ecol.* 35:1–21
47. Henckel T, Friedrich M, Conrad R. 1999. Molecular analyses of the methane-oxidizing microbial community in rice field soil by targeting the genes of the 16S rRNA, particulate methane monooxygenase, and methanol dehydrogenase. *Appl. Environ. Microbiol.* 65:1980–90
48. Hershberger KL, Barns SM, Reysenbach AL, Dawson SC, Pace NR. 1996. Wide diversity of Crenarchaeota. *Nature* 384: 420
49. Heuer H, Smalla K. 1997. Application of denaturing gradient gel electrophoresis and temperature gradient gel electrophoresis for studying soil microbial communities. In *Modern Soil Microbiology*, ed. JD van Elsas, EMH Wellington,

JT Trevors, pp. 353–74. New York: Marcel Dekker
50. Heuer H, Smalla K. 1997. Evaluation of community-level catabolic profiling using BIOLOG GN microplates to study microbial community changes in potato phyllosphere. *J. Microbiol. Methods* 30:49–61
51. Hiorns WD, Methe BA, Nierzwicki-Bauer SA, Zehr JP. 1997. Bacterial diversity in Adirondack Mountain lakes as revealed by 16S rRNA gene sequences. *Appl. Environ. Microbiol.* 63:2957–60
52. Hjorleifsdottir S, Skirnisdottir S, Hreggvidsson GO, Holst O, Kristjansson JK. 2001. Species composition of cultivated and noncultivated bacteria from short filaments in an Icelandic hot spring at 88 degrees C. *Microb. Ecol.* 42:117–25
53. Höfle MG, Haas H, Dominik K. 1999. Seasonal dynamics of bacterioplankton community structure in a eutrophic lake as determined by 5S rRNA analysis. *Appl. Environ. Microbiol.* 65:3164–74
54. Holben WE, Harris D. 1995. DNA-based monitoring of total bacterial community structure in environmental samples. *Mol. Ecol.* 4:627–31
55. Holt JG, Krieg NR, Sneath PHA, Staley JT, Williams ST. 1994. *Bergey's Manual of Determinative Bacteriology*. Baltimore, MD: Wilkins & Wilkins
56. Hughes JB, Hellmann JJ, Ricketts TH, Bohannan BJM. 2001. Counting the uncountable: statistical approaches to estimating microbial diversity. *Appl. Environ. Microbiol.* 67:4399–406
57. Ibekwe AM, Kennedy AC. 1998. Fatty acid methyl ester (FAME) profiles as a tool to investigate community structure of two agricultural sails. *Plant Soil* 206:151–61
58. Ibekwe AM, Papiernik SK, Gan J, Yates SR, Yang CH, Crowley DE. 2001. Impact of fumigants on soil microbial communities. *Appl. Environ. Microbiol.* 67:3245–57
59. Insam H. 1997. Substrate utilization tests in microbial ecology—a preface to the special issue of the *Journal of Microbiological Methods*. *J. Microbiol. Methods* 30:1–2
59a. Jensen S, Øvreås L, Daae FL, Torsvik V. 1998. Diversity in methane enrichments from agricultural soil revealed by DGGE separation of PCR amplifed 16S rDNA fragments. *FEMS Microbiol. Ecol.* 26:17–26
60. Johnsen K, Jacobsen CS, Torsvik V, Sørenson J. 2001. Pesticide effects on bacterial diversity in agricultural soils—a review. *Biol. Fertil. Soils* 33:443–53
61. Jurgens G, Lindstrom K, Saano A. 1997. Novel group within the kingdom Crenarchaeota from boreal forest soil. *Appl. Environ. Microbiol.* 63:803–5
62. Jurgens G, Saano A. 1999. Diversity of soil Archaea in boreal forest before, and after clear-cutting and prescribed burning. *FEMS Microbiol. Ecol.* 29:205–13
63. Kaiser O, Pühler A, Selbitschka W. 2001. Phylogenetic analysis of microbial diversity in the rhizoplane of oilseed rape (*Brassica napus* cv. Westar) employing cultivation-dependent and cultivation-independent approaches. *Microb. Ecol.* 42:136–49
64. Karner M, Fuhrman JA. 1997. Determination of active marine bacterioplankton: a comparison of universal 16S rRNA probes, autoradiography, and nucleoid staining. *Appl. Environ. Microbiol.* 63: 1208–13
65. Kitts CL. 2001. Terminal restriction fragment patterns: a tool for comparing microbial communities and assessing community dynamics. *Curr. Issues Intest. Microbiol.* 2:17–25
66. Lane DJ. 1991. 16S/23S rRNA sequencing. In *Nucleic Acid Techniques in Bacterial Systematics*, ed. E Stackebrandt, M Goodfellow, pp. 131–73. Chichester, UK: Wiley
67. Lawlor K, Knight BP, Barbosa-Jefferson VL, Lane PW, Lilley AK, et al. 2000. Comparison of methods to investigate

microbial populations in soils under different agricultural management. *FEMS Microbiol. Ecol.* 33:129–37

68. Liesack W, Stackebrandt E. 1992. Occurrence of novel groups of the domain bacteria as revealed by analysis of genetic material isolated from an Australian terrestrial environment. *J. Bacteriol.* 174:5072–78
69. Liu J, Dazzo FB, Glagoleva O, Yu B, Jain AK. 2001. CMEIAS: a computer-aided system for the image analysis of bacterial morphotypes in microbial communities. *Microb. Ecol.* 41:173–94
70. Liu WT, Marsh TL, Cheng H, Forney LJ. 1997. Characterization of microbial diversity by determining terminal restriction fragment length polymorphisms of genes encoding 16S rRNA. *Appl. Environ. Microbiol.* 63:4516–22
71. Lukow T, Dunfield PF, Liesack W. 2000. Use of the T-RFLP technique to assess spatial and temporal changes in the bacterial community structure within an agricultural soil planted with transgenic and non-transgenic potato plants. *FEMS Microbiol. Ecol.* 32:241–47
72. MacGregor BJ, Moser DP, Alm EW, Nealson KH, Stahl DA. 1997. Crenarchaeota in Lake Michigan sediment. *Appl. Environ. Microbiol.* 63:1178–81
73. Magurran AE. 1988. *Ecological Diversity and Its Measurement.* Princeton, NJ: Princeton Univ. Press
74. Marilley L, Aragno M. 1999. Phylogenetic diversity of bacterial communities differing in degree of proximity of *Lolium perenne* and *Trifolium repens* roots. *Appl. Soil Ecol.* 13:127–36
75. Marschner P, Yang C-H, Lieberei R, Crowley DE. 2001. Soil and plant specific effects on bacterial community composition in the rhizosphere. *Soil Biol. Biochem.* 33:1437–45
76. Marsh T. 1999. Terminal restriction fragment length polymorphism (T-RFLP): an emerging method for characterizing diversity among homologous populations of amplification products. *Curr. Opin. Microbiol.* 2:323–27
77. Marsh TL, Saxman P, Cole J, Tiedje J. 2000. Terminal restriction fragment length polymorphism analysis program, a web-based research tool for microbial community analysis. *Appl. Environ. Microbiol.* 66:3616–20
78. Massol-Deya AA, Odelson DA, Hickey RF, Tiedje JM. 1995. Bacterial community fingerprinting of amplified 16S and 16-23S ribosomal DNA gene sequences and restriction endonuclease analysis (ARDRA). In *Molecular Microbial Ecology Manual*, ed. ADL Akkermans, JD van Elsas, FJ de Bruijn, 2:1–8. Dordrecht/Boston/London: Kluwer
79. McCaig AE, Glover LA, Prosser JI. 1999. Molecular analysis of bacterial community structure and diversity in unimproved and improved upland grass pastures. *Appl. Environ. Microbiol.* 65:1721–30
80. McSpadden Gardener B, Schroeder K, Kalloger S, Raaijmakers J, et al. 2000. Genotypic and phenotypic diversity of phlD-containing *Pseudomonas* isolated from the rhizosphere of wheat. *Appl. Environ. Microbiol.* 66:1939–46
81. Mergel A, Schmitz O, Mallman T, Bothe H. 2001. Relative abundance of denitrifying and dinitrogen-fixing bacteria in layers of a forest soil. *FEMS Microbiol. Ecol.* 36:33–42
82. Murray AE, Lies D, Li G, Nealson K, Zhou J, Tiedje JM. 2001. DNA/DNA hybridization to microarrays reveals gene-specific differences between closely related microbial genomes. *Proc. Natl. Acad. Sci. USA* 98:9853–58
83. Muyzer GA, de Waal EC, Uitterlinden AG. 1993. Profiling of complex microbial populations by denaturing gradient gel electrophoresis analysis of polymerase chain reaction-amplified genes coding for 16S rRNA. *Appl. Environ. Microbiol.* 59:695–700
84. Nakatsu CH, Torsvik V, Øvreås L. 2000.

Soil community analysis using DGGE of 16S rDNA polymerase chain reaction products. *Soil Sci. Soc. Am. J.* 64:1382–88

85. Nicholson PS, Hirsch PR. 1998. The effects of pesticides on the diversity of culturable soil bacteria. *J. Appl. Microbiol.* 84:551–58
86. Niemi RM, Heiskanen I, Wallenius K, Lindstrom K. 2001. Extraction and purification of DNA in rhizosphere soil samples for PCR-DGGE analysis of bacterial consortia. *J. Microbiol. Methods* 45:155–65
87. Normander B, Prosser JI. 2000. Bacterial origin and community composition in the barley phytosphere as a function of habitat and presowing conditions. *Appl. Environ. Microbiol.* 66:4372–77
88. Nüsslein K, Tiedje JM. 1999. Soil bacterial community shift correlated with change from forest to pasture vegetation in a tropical soil. *Appl. Environ. Microbiol.* 65:3622–26
89. Osborne AM, Moore ERB, Timmis KN. 2000. An evaluation of terminal-restriction fragment length polymorphism (T-RFLP) analysis for the study of microbial community structure and dynamics. *Environ. Microbiol.* 2:39–50
90. Øvreås L. 2000. Population and community level approaches for analysing microbial diversity in natural environments. *Ecol. Lett.* 3:236–51
91. Øvreås L, Torsvik V. 1998. Microbial diversity and community structure in two different agricultural soil communities. *Microb. Ecol.* 36:303–15
92. Pace NR. 2000. Community interactions: towards a natural history of the microbial world. *Environ. Microbiol.* 2: 7–8
93. Palus JA, Borneman J, Ludden PW, Triplett EW. 1996. A diazotrophic bacterial endophyte isolated from stems of *Zea mays* L and *Zea luxurians* Iltis and Doebley. *Plant Soil* 186:135–42
94. Pickup R. 1991. Detection and study of microorganisms in the environment—new approaches. 8:499–503
95. Poly F, Ranjard L, Nazaret S, Gourbiere F, Monrozier LJ. 2001. Comparison of *nifH* gene pools in soils and soil microenvironments with contrasting properties. *Appl. Environ. Microbiol.* 67:2255–62
96. Posch T, Pernthaler J, Alfreider A, Psenner R. 1997. Cell-specific respiratory activity of aquatic bacteria studied with the tetrazolium reduction method, cyto-clear slides, and image analysis. *Appl. Environ. Microbiol.* 63:867–73

96a. Raaijmakers JM, Bonsall RE, Weller DM. 1999. Effect of population density of *Pseudomonas fluorescens* on production of 2,4-diacetylphloroglucinol in the rhizosphere of wheat. *Phytopathology* 89:470–75

96b. Raaijmakers JM, Weller DM. 1998. Natural plant protection by 2,4-diacetylphloroglucinol-producing *Pseudomonas* spp. in take-all decline soils. *Mol. Plant-Microb. Interact.* 11:144–52

96c. Raaijmakers JM, Weller DM, Thomashow LS. 1997. Frequency of antibiotic-producing *Pseudomonas* spp. in natural environments. *Appl. Environ. Microbiol.* 63:881–87

97. Rainey FA, Ward N, Sly LI, Stackebrandt E. 1994. Dependence on the taxon composition of clone libraries for PCR amplified, naturally occurring 16S rDNA, on the primer pair and the cloning system used. *Experientia* 50:796–97
98. Rainey FA, Ward-Rainey NL, Stackebrandt E. 1996. *Clostridium paradoxum* DSM 7308T contains multiple 16S rRNA genes with heterogeneous intervening sequences. *Microbiology* 142: 2087–91
99. Ramirez-Saad HC, Sessitsch A, de Vos WM, Akkermans ADL. 2000. Bacterial community changes and enrichment of *Burkholderia*-like bacteria induced by chlorinated benzoates in a peat-forest soil-microcosm. *Syst. Appl. Microbiol.* 23:591–98

100. Ranjard L, Brothier E, Nazaret S. 2000. Sequencing bands of ribosomal intergenic spacer analysis fingerprints for characterization and microscale distribution of soil bacterium populations responding to mercury spiking. *Appl. Environ. Microbiol.* 66:5334–39
101. Ranjard L, Poly F, Combrisson J, Richaume A, Gourbiere F, et al. 2000. Heterogeneous cell density and genetic structure of bacterial pools associated with various soil microenvironments as determined by enumeration and DNA fingerprinting approach (RISA). *Microb. Ecol.* 39:263–72
102. Ranjard L, Poly F, Lata JC, Mougel C, Thioulouse J, Nazaret S. 2001. Characterization of bacterial and fungal soil communities by automated ribosomal intergenic spacer analysis fingerprints: biological and methodological variability. *Appl. Environ. Microbiol.* 67:4479–87
103. Ranjard L, Richaume AS. 2001. Quantitative and qualitative microscale distribution of bacteria in soil. *Res. Microbiol.* 152:707–16
104. Reysenbach AL, Ehringer H, Hershberger K. 2000. Microbial diversity at 83 degrees C in Calcite Springs, Yellowstone National Park: another environment where the aquificales and "Korarchaeota" coexist. *Extremophiles* 4: 61–67
105. Reysenbach AL, Giver LJ, Wickham GS, Pace NR. 1992. Differential amplification of rRNA genes by polymerase chain reaction. *Appl. Environ. Microbiol.* 58:3417–18
106. Ritchie NJ, Schutter ME, Dick RP, Myrold DD. 2000. Use of length heterogeneity PCR and fatty acid methyl ester profiles to characterize microbial communities in soil. *Appl. Environ. Microbiol.* 66:1668–75
107. Robleto EA, Borneman J, Triplett EW. 1998. Effects of bacterial antibiotic production on rhizosphere microbial communities from a culture-independent perspective. *Appl. Environ. Microbiol.* 64: 5020–22
108. Rondon MR, August PR, Bettermann AD, Brady SF, Grossman TH, et al. 2000. Cloning the soil metagenome: a strategy for accessing the genetic and functional diversity of uncultured microorganisms. *Appl. Environ. Microbiol.* 66:2541–47
109. Sandaa R-A, Torsvik V, Enger Ø. 2001. Influence of long-term heavy-metal contamination on microbial communities in soil. *Soil Biol. Biochem.* 33:287–95
110. Sandaa R-A, Torsvik V, Enger Ø, Daae FL, Castberg T, et al. 1999. Analysis of bacterial communities in heavy metal-contaminated soils at different levels of resolution. *FEMS Microbiol. Ecol.* 30: 237–51
111. Scala DJ, Kerkhof LJ. 1999. Diversity of nitrous oxide reductase (*nosZ*) genes in continental shelf sediments. *Appl. Environ. Microb.* 65:1681–87
112. Scala DJ, Kerkhof LJ. 2000. Horizontal heterogeneity of denitrifying bacterial communities in marine sediments by terminal restriction fragment length polymorphism analysis. *Appl. Environ. Microbiol.* 66:1980–86
113. Schallmach E, Minz D, Jurkevitch E. 2000. Culture-independent detection of changes in root-associated bacterial populations of common bean (*Phaseolus vulgaris* L.) following nitrogen depletion. *Microb. Ecol.* 40:309–16
114. Schleper C, Holben W, Klenk HP. 1997. Recovery of Crenarchaeotal ribosomal DNA sequences from freshwater-lake sediments. *Appl. Environ. Microbiol.* 63: 321–23
115. Schmalenberger A, Schwieger F, Tebbe CC. 2001. Effect of primers hybridizing to different evolutionarily conserved regions of the small-subunit rRNA gene in PCR-based microbial community analyses and genetic profiling. *Appl. Environ. Microbiol.* 67:3557–63
116. Schwieger F, Tebbe CC. 1998. A new approach to utilize PCR-single-strand

conformation polymorphism for 16s rRNA gene-based microbial community analysis. *Appl. Environ. Microbiol.* 64: 4870–76

117. Sessitsch A, Weilharter A, Gerzabek MH, Kirchmann H, Kandeler E. 2001. Microbial population structures in soil particle size fractions of a long-term fertilizer field experiment. *Appl. Environ. Microbiol.* 67:4215–24
118. Siciliano SD, Germida JJ. 1999. Taxonomic diversity of bacteria associated with the roots of field-grown transgenic *Brassica napus* cv. Quest, compared to the non-transgenic *B. napus* cv. Excel and *B. rapa* cv. Parkland. *FEMS Microbiol. Ecol.* 29:263–72
119. Siciliano SD, Theoret CM, de Freitas JR, Hucl PJ, Germida JJ. 1998. Differences in the microbial communities associated with the roots of different cultivars of canola and wheat. *Can. J. Microbiol.* 44: 844–51
120. Simon HM, Dodsworth JA, Goodman RM. 2000. Crenarchaeota colonize terrestrial plant roots. *Environ. Microbiol.* 2:495–505
121. Small J, Call DR, Brockman FJ, Straub TM, Chandler DP. 2001. Direct detection of 16S rRNA in soil extracts by using oligonucleotide microarrays. *Appl. Environ. Microbiol.* 67:4708–16
122. Smalla K, Wachtendorf U, Liu W-T, Forney L. 1998. Analysis of BIOLOG GN substrate utilization patterns by microbial communities. *Appl. Environ. Microbiol.* 64:1220–25

122a. Smalla K, Wieland G, Buchner A, Zock A, Parzy J, et al. 2001. Bulk and rhizosphere soil bacterial communities studied by denaturing gradient gel electrophoresis: plant-dependent enrichment and seasonal shifts revealed. *Appl. Environ. Microbiol.* 67:4742–51

123. Smit E, Leeflang P, Gommans S, van den Broek J, van Mil S, Wernars K. 2001. Diversity and seasonal fluctuations of the dominant members of the bacterial soil community in a wheat field as determined by cultivation and molecular methods. *Appl. Environ. Microbiol.* 67: 2284–91
124. Stackebrandt E, Liesack W, Goebel BM. 1993. Bacterial diversity in a soil sample from a subtropical Australian environment as determined by 16S rDNA analysis. *FASEB J.* 7:232–36
125. Stein JL, Marsh TL, Wu KY, Shizuya H, DeLong EF. 1996. Characterization of uncultivated prokaryotes: isolation and analysis of a 40-kilobase-pair genome fragment from a planktonic marine archaeon. *J. Bacteriol.* 178:591–99
126. Suzuki MT, Giovannoni SJ. 1996. Bias caused by template annealing in the amplification of mixtures of 16S rRNA genes by PCR. *Appl. Environ. Microbiol.* 62:625–30
127. Tiedje JM, Asuming-Brempong S, Nusslein K, Marsh TL, Flynn SJ. 1999. Opening the black box of soil microbial diversity. *Appl. Soil Ecol.* 13:109–22
128. Torsvik V, Salte K, Sorheim R, Goksoyr J. 1990. Comparison of phenotypic diversity and DNA heterogeneity in a population of soil bacteria. *Appl. Environ. Microbiol.* 56:776–81
129. Ueda T, Suga Y, Matsuguchi T. 1995. Molecular phylogenetic analysis of a soil microbial community in a soybean field. *Eur. J. Soil Sci.* 46:415–21
130. van Elsas JD, Duarte GF, Rosado AS, Smalla K. 1998. Microbiological and molecular biological methods for monitoring microbial inoculants and their effects in the soil environment. *J. Microb. Methods* 32:133–54
131. Vergin KL, Rappe MS, Giovannoni SJ. 2001. Streamlined method to analyze 16S rRNA gene clone libraries. *Biotechniques* 30:938–42
132. Vergin KL, Urbach E, Stein JL, DeLong EF, Lanoil BD, Giovannoni SJ. 1998. Screening of a fosmid library of marine environmental genomic DNA fragments

reveals four clones related to members of the order Planctomycetales. *Appl. Environ. Microbiol.* 64:3075–78

133. Weidner S, Arnold W, Puhler A. 1996. Diversity of uncultured microorganisms associated with the seagrass *Halophila stipulacea* estimated by restriction fragment length polymorphism analysis of PCR-amplified 16S rRNA genes. *Appl. Environ. Microbiol.* 62:766–71
134. Weidner S, Arnold W, Stackebrandt E, Puhler A. 2000. Phylogenetic analysis of bacterial communities associated with leaves of the seagrass *Halophila stipulacea* by a culture-independent small-subunit rRNA gene approach. *Microb. Ecol.* 39:22–31
135. Wilson MJ, Weightman AJ, Wade WG. 1997. Applications of molecular ecology in the characterization of uncultured microorganisms associated with human disease. *Rev. Med. Microbiol.* 8:91–101
136. Woese C. 1987. Bacterial evolution. *Microbiol. Rev.* 51:221–71
137. Woese CR, Kandler O, Wheelis ML. 1990. Towards a natural system of organisms—proposal for the domains Archaea, Bacteria, and Eucarya. *Proc. Natl. Acad. Sci. USA* 87:4576–79
138. Xia XQ, Bollinger J, Ogram A. 1995. Molecular genetic analysis of the response of three soil microbial communities to the application of 2,4-D. *Mol. Ecol.* 4:17–28
139. Yang C-H, Crowley DE. 2000. Rhizosphere microbial community structure in relation to root location and plant iron nutritional status. *Appl. Environ. Microbiol.* 66:345–51
140. Yang C-H, Crowley DE, Menge JA. 2001. 16S rDNA fingerprinting of rhizosphere bacterial communities associated with healthy and *Phytophthora* infected avocado roots. *FEMS Microbiol. Ecol.* 35:129–36
141. Yang C-H, Crowley DE, Borneman J, Keen NT. 2001. Microbial phyllosphere populations are more complex than previously realized. *Proc. Natl. Acad. Sci. USA* 98:3889–94
142. Yang Y-H, Yao J, Hu S, Qi Y. 2000. Effects of agricultural chemicals on DNA sequence diversity of soil microbial community: a study with RAPD marker. *Microb. Ecol.* 39:72–79
143. Yin B, Crowley D, Sparovek G, De Melo WJ, Borneman J. 2000. Bacterial functional redundancy along a soil reclamation gradient. *Appl. Environ. Microbiol.* 66:4361–65
144. Zhou J, Davey ME, Figueras JB, Rivkina E, Gilichinsky D, Tiedje JM. 1997. Phylogenetic diversity of a bacterial community determined from Siberian tundra soil DNA. *Microbiology* 143:3913–19

Annu. Rev. Microbiol. 2002. 56:237–61
doi: 10.1146/annurev.micro.56.012302.160847

First published online as a Review in Advance on April 29, 2002

TRANSITION METAL TRANSPORT IN YEAST

Anthony Van Ho,[1] Diane McVey Ward,[2] and Jerry Kaplan[2]
Departments of [1]Internal Medicine and [2]Pathology, University of Utah, School of Medicine, Salt Lake City, Utah 84132; e-mail: jerry.kaplan@path.utah.edu

Key Words regulation, iron, oxygen, high-affinity, copper, reductase

■ **Abstract** All eukaryotes and most prokaryotes require transition metals. In recent years there has been an enormous advance in our understanding of how these metals are transported across the plasma membrane. Much of this understanding has resulted from studies on the budding yeast *Saccharomyces cerevisiae*. A variety of genetic and biochemical approaches have led to a detailed understanding of how transition metals such as iron, copper, manganese, and zinc are acquired by cells. The regulation of metal transport has been defined at both the transcriptional and posttranslational levels. Results from studies on *S. cerevisiae* have been used to understand metal transport in other species of yeast as well as in higher eukaryotes.

CONTENTS

0066-4227/02/1013-0237$14.00

INTRODUCTION

Transition metals comprise the center portion of the Periodic Table. Transition metals are chemically defined as any element that forms at least one ion with a partially filled subshell of d electrons. The transition elements that have the most physiological relevance are iron, copper, manganese, cobalt, and zinc[1]. All eukaryotes and most prokaryotes require these elements. Iron, copper, cobalt, and to a lesser extent manganese participate in a variety of redox reactions. Zinc, although not a redox active metal, plays a role in defining protein structure. All organisms have developed mechanisms for utilizing and storing transition metals.

Within the past ten years enormous progress has been made in identifying the genes that mediate transition metal acquisition. Most of this progress has been made in identifying genes that mediate plasma membrane transport through the genetics and molecular biology of the budding yeast *Saccharomyces cerevisiae*. The tractable genetics and the sequenced genome have made this organism the model eukaryote. The identification of genes for transition metal transporters in *S. cerevisiae* has had enormous significance. First, we know more about how metals cross the plasma membrane of this organism than of any other organism. Second, results from *S. cerevisiae* have proven applicable to the study of metal transport in other species of yeast or fungi. Third, the transport mutants and the genes for yeast transporters have been used to identify transporters in higher eukaryotes including plants and humans. The ability to complement mutant phenotypes in yeast has led to the identification of families of transporters. In examining the chronology of the field of transition metal transport, it is extraordinary to realize that the first transporter genes were identified as recently as 1994. Since then, the budding yeast has been used as a vehicle to identify whole families of transporters. Although there are still gaps in our understanding of how many transporters work, we speculate that most of the transporters used to accumulate transition metals have been identified. The following review focuses on the characterization of transition metal transporters and their regulation in *S. cerevisiae* and where applicable in other yeast as well.

IRON

Iron is an element required by virtually all organisms. Only a few species of bacteria do not utilize iron. The facile ability of iron to gain and lose electrons renders this metal an essential cofactor in redox reactions. In addition, the high affinity of iron for oxygen has made iron the active site in heme, which is commonly involved in oxygen-binding and oxygen-based enzymatic reactions. Iron is abundant in the earth's crust, but it is available primarily in the ferric (Fe^{3+}) form, which

---

[1]The name of the metal will be written when it is referred to generically or when the valence state is unknown.

is insoluble. Furthermore, iron reacts with oxygen-containing compounds such as hydrogen peroxide to generate oxygen radicals, most particularly hydroxyl radicals and superoxide anions, which have a variety of toxic effects on cells. Hence, all organisms are faced with the problem of obtaining iron as well as regulating its concentration in biological fluids and cellular compartments.

Organisms as disparate as prokaryotes and mammals have developed a variety of mechanisms to overcome the problem of iron bioavailability and to regulate the concentration of iron in solution. Higher eukaryotes regulate cytosolic-free iron by controlling iron uptake through regulation of iron entry into the body and into individual cells. Within cells iron may be stored in the cytosolic protein ferritin. These processes are regulated by the iron-sensing regulatory proteins IRP1 and IRP2 [for review see (20, 68)].

Ferrireductase

In yeast, iron uptake is facilitated by multiple transport systems, all of which appear to require ferrous (Fe^{2+}) iron as a substrate (19, 49). Ferrous iron is much more soluble at physiological pH than ferric iron, but under atmospheric conditions ferrous iron is unstable and is oxidized rapidly to ferric iron. Thus, the initial step in iron transport is the reduction of ferric iron by two known transmembrane metallo-reductases encoded by the *FRE1* and *FRE2* genes (10, 26).[2] "Nutritional" studies demonstrated that cells would accumulate iron more readily when the iron was presented in a reduced form (49). Use of impermeable iron compounds also provided evidence for a cell surface ferric reductase. But compelling evidence for the presence of surface transmembrane oxido-reductases awaited the identification of genes that encode such activity. Identification of the ferric reductase gene, *FRE1*, came from a series of experiments in *S. cerevisiae* performed over a decade ago. A strain isolated from a mutagenesis screen lacking ferric reductase activity was identified using a colorimetric assay (10). The mutant, when compared to wild-type strains, was extremely sensitive to iron deprivation in the growth medium and was deficient in the uptake of ferric iron but not ferrous iron. Ferric reductase activity correlated with ferric iron uptake, and genetic analysis indicated that deficiencies in both resulted from a single mutation. The *FRE1* gene was identified by transformation of a genomic library that complemented the ferric reductase and ferric uptake activities in the mutant. Transcription of the *FRE1* gene and ferric reductase activity is induced by iron deprivation. Deletion of *FRE1* in wild-type cells resulted in decreased reductase activity, deficient ferric iron uptake, and impaired growth in iron-depleted medium. The decrease in ferric reductase activity in Δ*fre1* cells was only partial, and a substantial level (40%–60%) of activity remained.

[2]In this manuscript we use standard nomenclature for *S. cerevisiae* and *S. pombe* genes and proteins. *S. cerevisiae* wild-type genes are capitalized and italicized (*FET3*). *S. pombe* wild-type genes are in lower case, italicized, with a superscripted "+" at the end (*fio1*$^+$). *S. cerevisiae* and *S. pombe* protein nomenclature is the same and is written with a "p" at the end (Fet3p, fio1p).

A gene encoding a second transmembrane ferric reductase, *FRE2*, was identified on chromosome XI through the *S. cerevisiae* genome project (26). The *FRE2* gene product exhibited significant amino acid similarity to that of *FRE1*. Disruption of *FRE2* in Δ*fre1* yeast resulted in the almost-complete absence of ferric reductase activity and resulted in impaired growth in iron-depleted medium.

Although both *FRE1* and *FRE2* are highly homologous, have seemingly redundant functions, and are transcriptionally regulated, their transcription is controlled by different factors. The expression of *FRE1* is controlled both by iron and copper through the action of the iron transcription factor Aft1p (77) and by the copper transcription factor Mac1p (27). *FRE2* is only regulated by iron deprivation through the action of Aft1p (77). The products of the *FRE* genes are not specific to iron reduction and can also reduce copper from Cu^{2+} to Cu^{+}. While *FRE1* and *FRE2* appear to account for almost all of the cell surface ferric-reductase and cupric-reductase activity, there are five other *FRE* homologs in the *S. cerevisiae* genome. Fre3p, which is localized on the cell surface, has been suggested to reduce siderophore-iron complexes (81). Neither the subcellular location nor function(s) are known for the other *FRE* genes. Transcription of *FRE3*, *4*, *5*, and *6* is controlled by iron, while transcription of *FRE7* is controlled by copper (57). It is thought that these other *FRE* genes play roles in intracellular iron/copper metabolism.

High-Affinity Elemental Iron Transport

Analysis of the concentration dependency of iron transport reveals that iron uptake occurs through two distinct systems: a high-affinity uptake system (Km of 0.15 μM) and a low-affinity system (Km of 30–40 μM) (19). The iron content of the media inversely regulates the high-affinity uptake system, and cells grown in high iron media only exhibit the low-uptake system. Conversely, iron deprivation results in the expression of the high-affinity iron transport system. The existence of two physiologically separable iron transport systems led to the creation of a genetic screen that targeted the high-affinity iron transport system, which is required for growth on low iron media (2). Yeast, with mutations in the high-affinity iron transport system, can grow on high iron through the activity of the low-affinity iron transport system. Genetic inactivation of the high-affinity system was accomplished through use of the drug streptonigrin. This antibiotic diffuses into cells where it becomes reduced. In the presence of iron and oxygen, reduced streptonigrin generates toxic oxygen radicals. Yeast grown in low iron media to induce the high-affinity iron transport system were incubated with both iron and streptonigrin for 2 h. The yeast were then plated on both low and high iron–containing media. Yeast unable to grow on low iron but which could grow on high iron were selected for further study. One such mutant *fet3* was shown to have a normal ferrireductase but was unable to transport iron through the high-affinity system. The gene responsible for the defect was identified by transformation of a genomic library into mutant cells. A gene that permitted cells to grow on low iron media by restoring iron transport was shown by rigorous yeast genetics to be allelic to the mutant *fet3* gene.

The protein encoded by *FET3* is a plasma membrane protein with a single transmembrane domain (2, 13, 80). This latter feature suggested that Fet3p itself was not a transmembrane permease. Its role in iron transport was suggested by the presence of a signature motif denoting its membership in a small family of proteins termed multicopper oxidases (2). These proteins catalyze the oxidation of four moles of substrate in concert with the reduction of molecular oxygen. One member of the family, the vertebrate plasma protein ceruloplasmin, was shown to be a ferroxidase that utilized iron as a substrate (23, 47).

Several lines of evidence confirmed that Fet3p was also a ferroxidase. Genetic studies showed that high-affinity iron transport relied on copper homeostasis, as expected of an enzyme that requires copper (2, 11). High-affinity iron uptake is an oxygen-consuming reaction and Fet3p does not function in the absence of oxygen (13, 33). The isolation of both Fet3p and a secreted form, lacking the transmembrane domain, revealed that the protein contained four atoms of copper (14, 34). Spectroscopic studies revealed that the copper atoms were in the expected configuration: a type I Cu that imbues the protein with its deep blue color, a type II Cu that is EPR active, and two type III Cu that are electronically silent (34). The type I Cu is reduced upon oxidation of the substrate. Reduction involves the transfer of an electron from iron to the protein-bound copper. The protein oxidizes four substrates in sequential fashion, storing the extracted electrons. When the fourth substrate has been oxidized, the protein then reduces molecular oxygen to water. This reduction process does not generate partial oxygen products precluding the formation of reactive oxygen intermediates. The reaction catalyzed by Fet3p is $4Fe^{2+} + H^+ + O_2^- \rightarrow 4Fe^{3+} + 2H_2O$.

Fet3p was hypothesized to function in conjunction with an iron permease (Figure 1). The permease was identified by genetic studies that took advantage of a clever selection system. Klausner and colleagues (77) engineered a yeast strain in which the promoter for the iron-sensitive *FRE1* gene was placed in front of the *HIS3* gene. This construct was transformed into a *his3* strain. The strain was then mutagenized and grown in low iron media. Cells that could transport iron and were iron replete remained histidine auxotrophs and died on selective media that lacked histidine. A defect in iron transport, however, would result in transcription of the *HIS3* gene and permit cells to grow on media lacking histidine. The genes responsible for such mutants could be identified by complementation of a low iron growth defect. Additionally, in cells containing an *FRE1*-β galactosidase construct, a complementing gene would turn "blue" colonies back into "white" colonies. Using this approach Stearman et al. identified the transmembrane iron permease encoded by the *FTR1* gene (71). The Ftr1p is a multitopic protein localized to the cell surface. The protein has a putative ferric-binding site, which is similar to that found in ferritin light chain. The most persuasive evidence that Ftr1p is a partner for Fet3p comes from genetic studies that demonstrate that in the absence of Ftr1p, Fet3p is not localized to the cell surface. Conversely, in the absence of Fet3p, Ftr1p does not localize to the cell surface. Furthermore, both genes are regulated by the iron-sensing transcription factor *AFT1*.

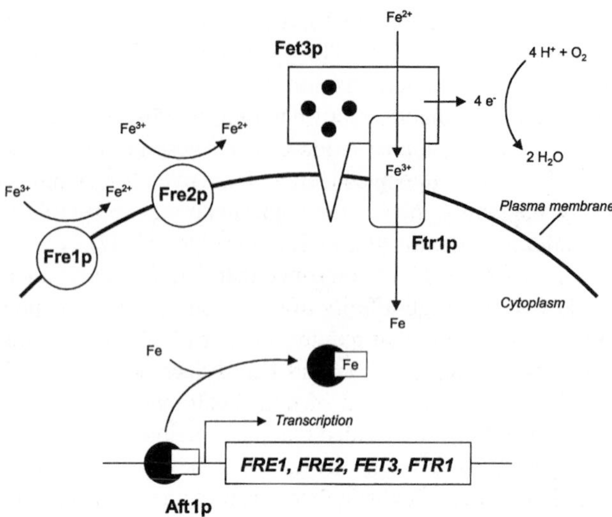

Figure 1 Model of high-affinity iron transport. Fe^{3+} is reduced to Fe^{2+} by the activity of the reductases Fre1p and Fre2p. This Fe^{2+} is oxidized to Fe^{3+} by the Fet3p ferroxidase subunit of the high-affinity complex Fet3p/Ftr1p. (Sequential oxidation of $4Fe^{2+}$ ions is followed by a single four-electron reduction of O_2 to $2H_2O$). Fe^{3+} is transported into the cell by the Ftr1p permease. *FRE1*, *FRE2*, *FET3*, and *FTR1* are transcribed according to iron need by the transcriptional activator Aft1p. (•) Copper atoms bound to Fet3p.

Other genetic data indicating that these two proteins work together came from an examination of iron transport in other species of yeast. *Schizosaccharomyces pombe*, the fission yeast, also has *FET3* and *FTR1* homologs (3). Interestingly, these genes are transcribed in the opposite direction using the same promoter. Expression of the *S. pombe* oxidase, *fio1*$^+$ in *S. cerevisiae* will not permit a *S. cerevisiae* Δ*fet3* strain to grow on low iron. This result suggests that Fio1p cannot partner with the endogenous *S. cerevisiae* Ftr1p. If the *S. pombe FTR1* homolog *fip1*$^+$ is transformed into the strain Δ*fet3* that also contains *Fio1*$^+$, then iron transport is restored. This data confirms that the oxidase and permease function together, and strongly suggests that no other plasma membrane protein is required to reconstitute high-affinity iron transport. It is curious that a physical interaction between Fet3p and Ftr1p has not yet been shown. There is a report of an interaction based on coimmunoprecipitation between an oxidase (Fet5p) and a permease homolog (Fth1p) present in yeast vacuole (74). In addition to *S. pombe* and *S. cerevisiae*, other yeast [*Candida albicans* (17), *Pischia pastoris* (5)] utilize an oxidase/permease transport system for high-affinity iron uptake. It is not known whether fungi other than yeast also utilize this system.

Both Fet3p and Ftr1p are plasma membrane glycoproteins and as such are synthesized as precursors in the endoplasmic reticulum, which are then transported to the Golgi apparatus. The Fet3p/Ftr1p complex is transported through a post-Golgi compartment prior to being transported to the cell surface. Proper vesicular transport is of particular importance for Fet3p, as it obtains its copper in a post-Golgi compartment, most likely the prevacuole (80). The function of Fet3p requires proper copper homeostasis, and defects in either plasma membrane or intracellular copper transport result in an inactive high-affinity iron transport system and the appearance of an apoFet3p on the cell surface. In some yeast strains the lack of Fet3p leads to no growth on glycerol-ethanol, which reflects the higher iron demand for respiratory activity. A lack of growth due to an apoFet3p can be differentiated from a mutation in the *FET3* gene. Mutations in genes required for vesicular traffic result in the appearance of apoFet3p on the cell surface. ApoFet3p can be copper loaded on the cell surface (80). Reconstitution of surface apoFet3p can occur at 0°C in the presence of Cu^+ and Cl^- (12). A number of mutations in the vesicular transport pathway from *trans*-Golgi to vacuole (termed vacuolar protein sorting) show a defect in the copper loading of Fet3p (67, 79).

Three genes are specific for the copper loading of apoFet3p, as mutations in these genes do not have generic effects on vesicular traffic. The first gene *CCC2* encodes an ATP-dependent copper transporter that is located in the intracellular compartment in which apoFet3p is copper loaded (79). This gene is a homolog for the Menkes/Wilsons gene in mammals. The absence of the Wilsons gene results in an inability to copper load ceruloplasmin as well as excessive copper accumulation in liver cells. The absence of the Menkes gene product results in a severe disease, which leads to death usually in the first year. All clinical symptoms result from an inability to transport copper from the intestinal absorptive cell into the plasma. Both the Menkes and the Wilsons gene can replace *CCC2* in *S. cerevisiae*, indicating that the genes are orthologous (39, 63).

While Ccc2p mediates copper transport into the vesicle, the substrate for the transporter is not copper but copper bound to a small-molecular-weight protein, Atx1p (53). This protein was the first of the copper chaperones to be discovered and is one of a family of three proteins. Each of the proteins "captures" cytosolic copper and delivers it to its target (8). In the case of Atx1p, the target is Ccc2p; copper bound to Atx1p is transferred to Ccc2p, which then transports it into the lumen of the vesicle where apoFet3p resides. As Ccc2p has mammalian homologs so does Atx1p, and the human homolog can replace Atx1p (44).

The third gene required for the copper loading of apoFet3p is *GEF1*, which encodes a vesicular Cl^- channel. *GEF1* was identified through a mutant screen as a mutant that could not grow on glycerol-ethanol without high iron media (30). Based on homology *GEF1* is a member of the family of voltage-regulated chloride channels. In the presence of a voltage gradient the channel is opened permitting the movement of Cl^- in response to an electrochemical gradient. The Gef1p is located in the same compartment as apoFet3p. In the absence of Gef1p, apoFet3p is transported to the cell surface. A deletion in *GEF1* has two effects on the copper loading of apoFet3p. The first effect results from an inability to transport cations in the absence of a counter ion, which dissipates an unfavorable voltage potential (25). The second effect reflects a direct requirement for Cl^- to copper load apoFet3p (12). Purified apoFet3p cannot be copper loaded in vitro in the absence of Cl^-.

Siderophore Iron Transport

While yeast have the ability to take up elemental iron they can also accumulate iron bound to organic molecules termed siderophores. Siderophores are low-molecular-weight ferric chelators produced by both bacteria and fungi [for review see (60, 61)]. Nearly all fungi, with the exception of the budding and fission yeast, make siderophores. Siderophores have an extraordinarily high affinity for iron. It has been reported that some siderophores can extract iron from stainless steel pots. Desferrioxamine, a siderophore produced by *Streptomyces pilosus*, has an affinity for iron of 10^{34}. In comparison, the mammalian iron-binding protein transferrin only has an affinity of 10^{18}. Siderophores serve two purposes. First, the high-affinity constant essentially solubilizes the normally insoluble Fe^{3+}. Second, special mechanisms are required to extract iron from the iron-siderophore complex. Microorganisms that possess these transport mechanisms have a decided growth advantage. Yeast that do not make siderophores can utilize siderophores produced by other organisms. Most of what is known about siderophore uptake comes from studies in *S. cerevisiae*. Although *S. cerevisiae* does not make siderophores, it has the ability to take up iron from a variety of siderophore-iron complexes (50).

There are two mechanisms that yeast use to extract iron from siderophores. The first mechanism can effect iron uptake from a variety of chemically diverse siderophores. Iron uptake is due to reduction of the siderophore-bound Fe^{3+} to Fe^{2+}, which then dissociates from the siderophore (48) and is transported by the high-affinity iron transport system composed of Fet3p/Ftr1p (50, 82). Reduction of ferrioxamine-iron complexes occurs through the cell surface ferric reductase Fre3p. High concentrations of rhodotorulic acid iron could be reduced by Fre4p (81). There is evidence that other fungi also utilize a reductase-based mechanism for siderophore-iron uptake (38). The second mechanism leading to uptake of iron-siderophore complexes requires the expression of specific transporters that are members of the facilitator-diffusion superfamily. Transporter-based siderophore uptake only becomes rate limiting under conditions in which the reductase-based uptake system is inoperative. The first siderophore transporter to be identified was discovered in a Δ*fet3*Δ*fet4* strain that is unable to grow on low iron unless the media is supplemented with siderophore-iron complexes (50). This strain was mutagenized and a mutant unable to utilize siderophores was identified. The lack of growth on siderophore-iron supplemented media was used to clone a siderophore transporter out of a genomic library. A siderophore transporter, *SIT1* specific for ferrioxamine-iron complexes was identified in this manner. At the same time these studies were performed, four highly homologous iron-regulated genes that are members of the facilitator-diffusion superfamily were identified by microarray analysis (82). One of these genes, *ARN3*, turned out to be identical to *SIT1*. The other genes were shown to encode transporters with specificity for different siderophores: Enb1p transports enterobactin (37); Taf1p/Arn2p triactylfusarinine C (36); and Arn1p ferrichrome. The specificity of the transporters is not absolute, as Arn2p can also take up ferrichrome to some degree. The motive force behind siderophore transport is unknown. What is known is that the transporters are highly regulated by the iron-sensing transcription factor *AFT1*.

Most of our knowledge on siderophore transport in fungi comes from studies on just one species, the budding yeast *S. cerevisiae*. There is reason, however, to believe that other species of yeast utilize similar mechanisms and homologous transporters. The pathogenic yeast *C. albicans* can also accumulate iron from siderophores (41). Older studies indicate that *C. albicans* can secrete both phenolate and hydroxymate siderophores (40, 59). *C. albicans* is expected to utilize a reductase-mediated uptake of siderophore iron, as it possesses both cell surface ferrireductases and a high-affinity elemental iron transport system composed of *FET3* and *FTR1* homologs. In addition to the elemental reductase-based transport system, *C. albicans* contains a ferrichrome transporter that is highly homologous to the *S. cerevisiae* ferrichrome transporter *ARN1* (1). The level of mRNA for this transporter is inversely correlated with media iron levels, suggesting that transcription is increased in response to iron need. Expression of the *C. albicans* ferrichrome transporter in *S. cerevisiae* cells that have mutations in both the high-affinity iron transport system (Δ*fet3*) and all four siderophore transporters (Δ*arn1–4*) permits this cell to accumulate and grow on ferrichrome iron. This transporter is highly specific for hydroxymate siderophores and will not transport ferrioxamine or other siderophores. It appears that *C. albicans* only has one siderophore transporter gene, which is specific to the siderophore that it secretes. *S. cerevisiae*, which does not secrete siderophores, has four highly homologous genes that appear to result from gene duplication. It has been speculated that *S. cerevisiae* lost the ability to secrete siderophores, which led to evolutionary pressure to accumulate siderophores made by other organisms. Because it could make siderophores, *C. albicans* was not subjected to the same pressure and retained the ancestral gene. This hypothesis is supported by examination of the fission yeast *S. pombe*, which also does not make siderophores. Examination of its genome reveals the presence of three genes highly homologous to the *S. cerevisiae* siderophore transporters. This observation suggests that the loss of siderophore biosynthesis led to evolutionary pressure to diversify siderophore transporter genes, resulting in an increased ability to accumulate chemically diverse siderophores.

Low-Affinity Iron Transport

The high-affinity iron transport system is required for growth on low iron media but not on high iron media. Furthermore, as might be expected from a transport system that requires oxygen, the high-affinity iron transport system is not functional under anaerobic conditions and is not transcribed (33). Iron uptake under anaerobic conditions requires the activity of other iron transport systems. That such systems exist is supported by two observations: Cells deleted for *FET3* can grow in iron-replete media (2), and the concentration dependency of iron uptake in *S. cerevisiae* is biphasic showing high- and low-affinity uptake processes (19). Genetic identification of a low-affinity iron transport system encoded by *FET4* came from studies in which the growth deficit of a Δ*fet3* strain was complemented by a cDNA library under the control of the galactose promoter (15). The complementing gene *FET4* encodes a membrane protein with six transmembrane domains. Deletion of this gene results in the loss of low-affinity iron transport and in an inability to grow on moderate iron under anaerobic conditions (15, 16).

Like the Fet3p/Ftr1p transport system, the substrate for the Fet4p transport system is Fe^{2+}. As opposed to the *FET3/FTR1* transport system, the species of iron transported by the *FET4* transport system is most probably Fe^{2+}. This conclusion is based on the observation that other transition metals, Cu^{+}, Mn^{2+}, Zn^{2+}, and Co^{2+} can compete with Fe^{2+} for transport (15, 16). Thus, Fet4p is a transition metal transport system that is not specific for iron. This conclusion reinforces a common property of Fe^{2+} transport systems: They have broad metal specificity and will transport other transition metals. Fe^{3+} transport systems, Fet3p/Ftr1p in yeast, and ceruloplasmin/transferrin in vertebrates are highly iron specific.

Transcription of *FET4* mRNA is affected by iron level but not through the activity of Aft1p (16). Under anaerobic conditions Fet3p is nonfunctional because it relies on oxygen for its catalytic activity and because it is not transcribed. In anaerobiosis, Fet4p is the major iron transporter and levels of *FET4* mRNA increase enormously (73). It is suspected that *FET4* transcription may be controlled by transcription factors activated by anaerobiosis.

Other low-affinity iron transport systems must exist, as a Δ*fet3*Δ*fet4* deletion strain can grow in high iron media. *SMF1/SMF2*, which are H^{+}/M^{2+} symporters, can also transport iron. Furthermore, a homolog of *SMF1/SMF2/SMF3* is a vacuolar iron transporter, which is regulated by iron. Similar to plasma membrane iron transport, the vacuole has both high- and low-affinity iron transporters. An oxidase/permease high-affinity system is composed of the *FET5/FTH1* gene products, which are homologs of *FET3/FTR1* (70, 74). *SMF3* may function as a low-affinity Fe^{2+}/Mn^{2+} transporter. These transport systems may serve to scavenge iron that has accumulated through endocytosis or through storage of cytosolic iron. A Fe^{2+}/Mn^{2+}-specific transporter Ccc1p has recently been shown to mediate cytosol to vacuolar Fe^{2+} and Mn^{2+} storage (51).

Regulation of Iron Transport

The same screen that resulted in the identification of *FTR1* also led to the identification of *AFT1*, the iron-sensing transcriptional regulator (77). The original mutant was constitutive (*aft1^{up}*) and led to high levels of transport activity even in the face of high-iron media. As the promoter for the *FRE1-HIS3* construct was subject to iron regulation, cells expressing *aft1^{up}* were histidine prototrophs. The gene was cloned through a complementation phenotype, using a genomic library obtained from the mutant cells. The cloned gene encodes a 78-kDa protein. The absence of the protein results in no transcription of target genes, suggesting that the apoAft1p is a transcriptional activator and that iron prevents transcription. Footprinting analysis demonstrated that the protein binds to sequences in the promoter region of target genes. All of the genes required for high-affinity iron transport are part of the iron regulon controlled by Aft1p. These genes include six of the eight *FRE* genes, the high-affinity iron transport system (*FET3*/*FTR1*), two genes required for the copper loading of Fet3p, *CCC2*, *ATX1*, the four siderophore iron transporters (*ARN1-4*), the high-affinity vacuolar iron transport uptake system composed of a multicopper oxidase *FET5* and the transmembrane permease *FTH1*, and *SMF3*, a vacuolar H^+/Fe^{2+}, Mn^{2+} transporter (22, 66). The genes that show the highest degree of regulation by *AFT1* encode four homologous GPI-linked cell surface proteins (22, 66). Deletion of these genes, however, has only minor (twofold) effects on siderophore transport, leaving their role in iron transport unclear.

COPPER

Copper is a redox-active metal and can easily transition between two oxidation states, Cu^+ and Cu^{2+}. Like iron, copper is both essential and potentially dangerous due to its ability to engage in Fenton chemistry and generate reactive oxygen intermediates. The number of copper-containing proteins is less than that of iron but includes essential proteins such as cytochrome oxidase, multicopper oxidases (Fet3p, laccase), and superoxide dismutase.

High-Affinity Copper Transport

Copper uptake, like the uptake of other transition metals, occurs through both high- and low-affinity transport systems (Figure 2). Like iron, copper can exist in two different valence states, and the lower valence form is the substrate for both high- and low-affinity transport systems (32). Reduction of the more commonly occurring Cu^{2+} is effected by the plasma membrane reductases encoded by *FRE1* and *FRE2* (27). Like the transport of iron, the requirement for the reductases can be bypassed by chemical reduction of Cu^{2+} by ascorbate, demonstrating that reductase and transport activities are separable biochemical events. Transcription of *FRE1* is regulated by intracellular copper through the action of the copper-dependent transcription factor Mac1p (27, 32).

High-affinity copper uptake is mediated by two plasma membrane transporters encoded by *CTR1* and *CTR3* (11, 45). *CTR1* was initially identified by a genetic selection for mutants defective in high-affinity iron uptake. This screen utilized the *FRE1-HIS3* construct, which as described above selects for cells with a defect in iron transport. This approach led to the discovery of a mutant (*ctr1*) that had decreased transport of both Fe^{2+} and Cu^+, in contrast to the *fet3* mutant, which is defective only in ferrous iron uptake. The deficiency of ferrous iron uptake in the

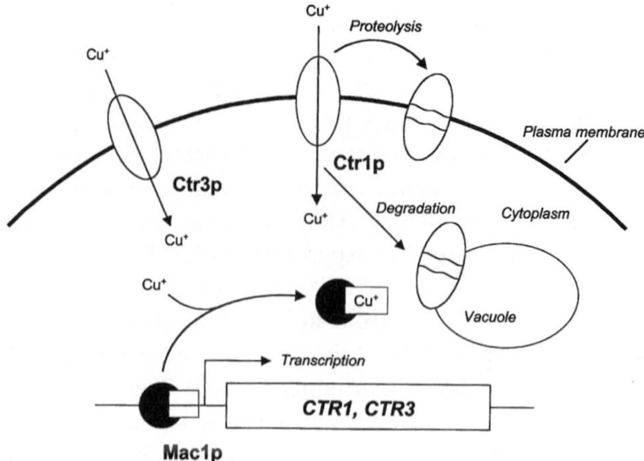

Figure 2 Model of copper transport. Cu^{2+} is reduced to Cu^{+} by the reductases Fre1p and Fre2p (as in Figure 1). Ctr1p or Ctr3p then transports Cu^{+} into the cell. These membrane proteins are transcriptionally regulated by the activity of Mac1p. Regulation of copper transport can also occur by a posttranslational process in which Ctr1p in the presence of moderate levels of Cu^{+} is endocytosed and degraded in the vacuole by a protease at the plasma membrane or Ctr1p.

CTR1 mutant, but not the *fet3* mutant, was corrected by growth in the presence of high concentrations of copper. The connection between the strict requirement for copper and high-affinity ferrous iron uptake in the *ctr1* mutant can be explained by the fact that Fet3p is a multicopper oxidase [(2); see above]. Copper transport, either through the high-affinity system requiring expression of *CTR1* or through a separate low-affinity system, supplies Fet3p with copper. The *CTR1* gene was cloned through complementation of the low iron phenotype by a plasmid containing a genomic insert. The *CTR1* gene encodes a plasma membrane protein of 406 amino acids and 3 potential transmembrane domains (11). Its amino terminus is unusually rich in methionine and serine residues and contains multiple repeats of a motif found in prokaryotic proteins involved in the processing of copper. In the plasma membrane of cells, Ctr1p is a homomultimer (9). This was proven by transforming cells with two different epitope-tagged constructs of *CTR1*. Western analysis of immunoprecipitates of Ctr1p using antibodies against one of the epitopes revealed the presence of the second epitope. This result suggests that Ctr1p must at least be a dimer.

In most yeast strains a deletion in *CTR1* does not lead to copper deprivation, whereas the same gene deletion in a small fraction of yeast strains does lead to copper deprivation (45). It was discovered that the majority of yeast strains contains another plasma membrane copper transporter, termed *CTR3*. In a minority of yeast strains the *CTR3* gene is inactivated by a transposon insertion. Ctr3p shows low sequence similarity to Ctr1p despite localizing to the plasma membrane and having a similar function. Deletion of both *CTR1* and *CTR3* results in severe copper deprivation. *CTR3* encodes a protein of 241 amino acids and 3 potential transmembrane domains. Using cross-linking agents Ctr3p was found to be a homotrimer (64).

Although most of what is known about transition metal transport comes from studies on *S. cerevisiae*, there are extensive studies regarding copper transport and storage in other species of yeast. In particular a copper transporter from *S. pombe* was identified through its ability to complement a $\Delta ctr1 \Delta ctr3$ *S. cerevisiae* strain (86). Gene *ctr4*$^{+}$ was identified as being necessary for Cu^{+} transport in *S. pombe*, as a deletion of this gene abrogated Cu^{+} transport. Although the ctr4p was expressed in *S. cerevisiae*, it could not complement the copper defect of a $\Delta ctr1 \Delta ctr3$ strain. Complementation required a second *S. pombe* gene, *ctr5*$^{+}$. The protein encoded by *ctr5*$^{+}$ has 41% amino acid identity with Ctr4p, and like Ctr4p has three potential transmembrane domains. In the absence of each other, neither Ctr4p nor Ctr5p localizes to the cell surface. Immunoprecipitation experiments demonstrate that the Ctr4p and Ctr5p are found in a high-molecular-weight complex.

What is most intriguing is a sequence comparison of the yeast *CTR* genes with *CTR* genes from organisms as disparate as humans and invertebrates (64). All of the copper transporters have three transmembrane domains and all of the fungal transporters are higher multimers. With the exception of *CTR1*, all the CTRs have high amino acid similarity in their transmembrane domains. The ectodomains of all known and projected copper transporters have the potential copper-binding domain of MX2MXM. Ctr3p lacks this domain, although it has a high concentration of cysteines, many of which were shown critical for its function. An intriguing hypothesis is that CTR from other organisms may have resulted from a fusion of the yeast *CTR1* and *CTR3* genes.

Low-Affinity Copper Transport

There appears to be a multiplicity of low-affinity Cu^{+} transporters, with at least three different gene products capable of transporting Cu^{+}. Functional complementation in yeast mutants defective for copper uptake led to the discovery of a copper transporter in *Arabidopsis thaliana*, which in turn revealed high sequence similarity to a potential low-affinity copper transporter in *S. cerevisiae* termed *CTR2* (43). Yeast strains with a $\Delta ctr2$ displayed an increased resistance to toxic copper concentrations. Overexpression of *CTR2* resulted in an increased sensitivity to copper toxicity and an increased resistance to copper depletion. The amino acid sequence of *CTR2* also contains three potential transmembrane domains.

Further studies demonstrated that Fet4p could also function as a low-affinity copper transporter (33). Overexpression of Fet4p could complement a $\Delta ctr1 \Delta ctr3$ cell and permit the copper loading of apoFet3p. Cu^{+} could compete with Fe^{2+} for Fet4p-mediated transport. Mutant forms of Fet4p defective in Fe^{2+} transport were similarly defective in Cu^{+} transport. Smf1p, which plays a critical role in Mn^{2+} transport, was also identified as a low-affinity Cu^{+} transporter.

Transcriptional Regulation of Copper Transport

Under copper-limiting conditions *CTR1*, *CTR3*, and *FRE1* are highly expressed, whereas under copper-replete conditions these genes are downregulated (32, 46, 57, 78). A copper-responsive transcription activator encoded by the *MAC1* gene controls expression of these genes. Mac1p resides within the nucleus in both copper-deficient and copper-replete cells. Regulation by Mac1p requires copper-responsive *cis*-acting elements (CuREs) 5′-TTTGCTC-3′, which are arranged either in tandem or inverted repeats in the promoters of *CTR1*, *CTR3*, and *FRE1* (46, 78). Footprinting analysis demonstrated that the CuREs are occupied under copper-limiting conditions in which *CTR1*, *CTR3*, and *FRE1* are expressed and unoccupied under copper-replete conditions in which transcription of these genes is inactivated. Copper inhibition of Mac1p arises from a copper-induced intramolecular interaction that represses both in vivo DNA binding and transactivation activities (42). DNA binding and transactivation activities map to distinct regions of Mac1p. The DNA-binding domain resides in the amino-terminal segment of the polypeptide and the activation domain maps to the carboxyl-terminal region. A recent study demonstrated that Mac1p must be phosphorylated to bind to CuREs (35). In the presence of moderate amounts of copper, transcription of *MAC1*-responsive genes does not occur, presumably because Mac1p is not phosphorylated. High concentrations of copper (>10 mM) lead to the degradation of Mac1p (87). Given that copper prevents Mac1p transcription, it is not clear why Mac1p need be degraded in the presence of copper. While the mechanism of Mac1p degradation is unknown, genetic studies showed that Mac1p degradation promotes resistance to copper toxicity. This observation suggests that there must be some residual transcription by Mac1p even in the presence copper, in which case the degradation of Mac1p would prevent copper intoxication.

Posttranslational Regulation of Copper Transport

While both *CTR1* and *CTR3* are regulated transcriptionally, Ctr1p also shows posttranslational regulation. In the presence of high concentrations of copper, plasma membrane Ctr1p is degraded (62). The degradation of Ctr1p occurs through two routes. At low concentrations Ctr1p is endocytosed and transported to the vacuole where it is degraded by vacuolar proteases. At high concentrations of copper (>10 mM) Ctr1p is degraded by a mechanism that does not involve either endocytosis or vacuolar proteases. A variant Ctr1p lacking the cytosolic domain cannot be internalized and degraded in lysosomes. This variant, however, can be degraded in the presence of high concentrations of copper. The details of either route of degradation are unknown. Questions to be resolved include what protein(s) senses high copper and induces the internalization and/or degradation of Ctr1p. Only Ctr1p is subject to posttranslational regulation, Ctr3p is unaffected (64). Since both proteins accumulate copper, it is not clear what advantage there is for a cell to only "inactivate" one of the two copper transporters.

ZINC

Although zinc is not a redox-active metal, it plays an essential role in protein structure. Hundreds of proteins require zinc for proper function. In most instances Zn^{2+} plays a structural role in defining protein shape. The concentration of Zn^{2+} in cells is extremely high and has been reported to be on the same order of magnitude as that of nucleic acid bases.

High-Affinity Zinc Transport

Analysis of the concentration dependence of Zn^{2+} uptake by yeast cells shows the presence of at least two uptake systems (Figure 3). One system has high affinity for substrate (Km of 1 μM) and is induced in Zn^{2+}-deficient cells (83). The second system has lower affinity (Km of 10 μM) and is not highly controlled by Zn^{2+} availability (84). The gene encoding the high-affinity zinc transporter, *ZRT1* (for zinc-regulated transporter), was identified because of its significant homology to *IRT1*, an iron-regulated transporter gene from the plant *A. thaliana* (18). *IRT1* was cloned by functional expression in a *S. cerevisiae* strain defective for iron uptake (Δ*fet3*/Δ*fet4*). The amino acid sequence of Irt1p is 30% identical and 50% similar to that of Zrt1p. Both proteins have eight potential transmembrane domains located in the same positions and the greatest degree of sequence similarity among these proteins is in these domains. Both transporters contain a histidine-rich domain suggested to be on the cytoplasmic surface of the membrane. This domain is thought to bind metal, as the imidazole ring nitrogens of histidine may serve as coordinating ligands for metal ions.

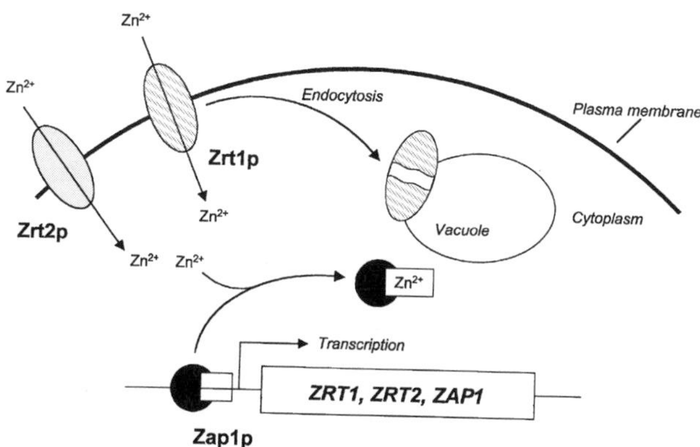

Figure 3 Model of zinc transport. Zrt1p (high-affinity) and Zrt2p (low-affinity) transport Zn^{2+} into the cell. The genes encoding Zrt1p and Zrt2p are transcriptionally regulated by the activity of Zap1p, which activates transcription under zinc-limiting conditions. At the posttranslational level, Zrt1p undergoes endocytosis and is degraded within the vacuoles.

Despite its significant similarity to the Irt1p, Zrt1p does not play a role in iron uptake in yeast (83). A deletion strain Δ*zrt1*, however, did not grow on media containing a high EDTA concentration. The addition of 500 μM Zn^{2+} rescued the growth of *zrt1* mutant in this EDTA media, whereas supplementation with 500 μM Co^{2+}, Cu^{+}, Fe^{2+}, Mg^{2+}, or Mn^{2+} did not rescue growth. The growth of both wild-type and Δ*zrt1* strains was severely inhibited in the absence of Zn^{2+}. Growth was recovered when Zn^{2+} was added to both strains, but the mutant strain required a 75-fold increase in the zinc concentration compared to the wild-type strain. The growth defect in Δ*zrt1* could be complemented fully by a genomic clone of the *ZRT1* gene. Zinc uptake assays in Δ*zrt1* cells grown in Zn^{2+}-depleted and Zn^{2+}-rich media demonstrated only low-affinity uptake activity. In Zn^{2+}-replete cells, expression of *ZRT1* from the *GAL1* promoter resulted in high rates of Zn^{2+} accumulation. Expression studies show that the Zrt1p protein is specific for Zn^{2+} and will not transport any other metal.

Low-Affinity Zinc Transport

The existence of a separate, low-affinity zinc transport system was suggested by the observation that Δ*zrt1* cells can grow in Zn^{2+}-replete media. Analysis of the concentration dependence of Zn^{2+} uptake in Δ*zrt1* cells showed a low-affinity uptake mechanism that was concentration- and temperature-dependent (84). Uptake assays demonstrated that Cu^{+} and Fe^{2+} could potentially be substrates, as they inhibited Zn^{2+} uptake, whereas Co^{2+}, Mn^{2+}, Mg^{2+}, and Ni^{2+} had no effect on Zn^{2+} uptake. The gene responsible for low-affinity Zn^{2+} transport is *ZRT2*, and it was first identified owing to its sequence homology to *ZRT1* (84). High copy expression of *ZRT2* could suppress the Zn^{2+}-depleted growth defect of Δ*zrt1*. Overexpression of *ZRT2* increased Zn^{2+} uptake rates in Δ*zrt1* cells. Conversely, disruption of the *ZRT2* gene eliminated low-affinity uptake without an effect on the high-affinity system. The amino acid sequence of Zrt2p has a 44% and 35% identity to Zrt1p and Irt1p, respectively. Interestingly cells that have a deletion in both *ZRT1* and *ZRT2* can grow in Zn^{2+}-replete media, indicating other low-affinity Zn^{2+} transporters exist. Fet4p may also be a low-affinity Zn^{2+} transporter (52).

Transcriptional Regulation of Zinc Transport

The high-affinity Zn^{2+} transport system is regulated in response to changes in intracellular Zn^{2+} pools (83). Evidence for transcriptional regulation is based on the fact that *ZRT1* mRNA levels were inversely regulated in response to cellular Zn^{2+} levels; Zn^{2+}-depleted cells had 10-fold more *ZRT1* mRNA than did Zn^{2+}-replete cells. A *ZRT1*-β galactosidase reporter construct, using the promoter region of the *ZRT1* gene, was used to show that regulation of mRNA levels occurred through control of transcription rather than mRNA stability. In wild-type cells, β-galactosidase activity was highest in Zn^{2+}-limited cells and decreased with increasing Zn^{2+} concentrations. In the Δ*zrt1* mutant strain, however, β-galactosidase activity remained at its maximum level in cells, even when cells were grown with high concentrations of Zn^{2+}.

To identify genes that effect Zn^{2+}-responsive transcriptional regulation, Zhao & Eide (85) implemented a genetic screen similar to that used to identify *FTR1* and *AFT1*. The promoter of the Zn^{2+}-sensitive *ZRT1* gene was inserted upstream of the coding region of the *HIS3* gene. The *ZRT1-HIS3* construct was integrated into a Δ*his3* strain. The screening method identified mutants with an elevated level of *ZRT1* transcription during growth in Zn^{2+}-replete media. Cells that could transport Zn^{2+} and were Zn^{2+} replete remained histidine auxotrophs and did not grow on selective media lacking histidine, whereas cells with defective Zn^{2+} transport and that were Zn^{2+} depleted would trigger transcription of the *ZRT1-HIS3* fusion gene and allow cells to grow on media lacking histidine (His^+ cells). One such His^+ strain was isolated and shown to have high Zn^{2+} uptake activity but was Zn^{2+} replete. Addition of Zn^{2+} did not lead to a decrease in *ZRT1* transcript nor Zn^{2+} acquisition. A genomic library was generated from this mutant strain, transformed into the Δ*his3* Δ*ZRT1-HIS3* strain, and then transformants were screened for histidine prototrophy. Using this approach, a constitutive mutant allele of the *ZAP1* gene (*ZAP1-1^{up}*) was isolated.

The cloned *ZAP1* gene encodes a protein that has 880 amino acids and a molecular mass of 93 kDa (85). Deletion of the *ZAP1* gene resulted in no expression of the target genes and, hence, Zn^{2+} deficiency. Sequence analysis of *ZAP1* revealed that the gene product shares similarity with many transcriptional activators: The carboxyl-terminal region contains five C2H2 zinc finger domains and the amino-terminal domain contains two acidic regions that could act as transcriptional activation domains. The five zinc fingers are necessary for high-affinity and sequence-specific DNA binding to sites, called zinc-responsive elements (ZREs), which are found in the promoters of genes targeted by Zap1p (56). The current consensus ZRE sequence, derived from mutational studies as well as comparison of many such elements from potential Zap1 target gene promoters, is 5′-ACCTTNAAGGT-3′ (85). Also, the constitutive *ZAP1-1^{up}* gene caused expression of the target genes in both zinc-limited and zinc-replete cells. Finally, the transcription of *ZAP1* is regulated by zinc, potentially through a positive autoregulatory mechanism. This type of regulatory circuitry would allow a rapid amplified response to changes in Zn^{2+} levels. Transcriptional autoregulation of Zap1p, however, is a minor component of Zn responsiveness; most regulation of Zap1p activity occurs at the posttranslational level (56). *ZRT2* is regulated transcriptionally by Zap1p in a similar fashion to *ZRT1*. The expression of *ZRT2* is not as tightly regulated as *ZRT1*, i.e., it takes a lower concentration of Zn^{2+} to activate *ZRT1*. The promoter region of *ZRT1* contains three ZREs, while *ZRT2* contains two ZREs, and the promoter region of *ZAP1*, which is the least sensitive of the three genes, contains one ZRE.

Posttranslational Regulation of Zinc Transport

Zrt1p activity is also regulated by Zn^{2+} at the posttranslational level (4, 28, 29). *S. cerevisiae* has a two-tiered regulatory system for Zn^{2+} transport in which transcriptional regulation can respond to moderate changes in Zn^{2+} availability and posttranslational control responds to more extreme variations. At extreme Zn^{2+}

concentrations, Zrt1p undergoes ubiquitin conjugation and is recruited into clathrin-coated pits as they form at the plasma membrane. Internalization then occurs as the endosome pinches off into the cytoplasm. The Zrt1p-containing endosomes fuse with the late endosome, where the transporter is packaged into vesicles for delivery to the vacuole where it is degraded by vacuolar proteases. Zinc-induced endocytosis is specific to Zrt1p and the endocytic response is relatively specific for Zn^{2+}; Cd^{2+} and Co^{2+} can trigger the response but less effectively than Zn^{2+}. The critical step in the endocytosis of Zrt1p is its ubiquitination, mediated by the ubiquitin-protein ligase and Ubc4p and Ubc5p, ubiquitin-conjugating enzymes. Ubiquitination and endocytosis require a specific lysine residue in the cytosolic loop of the protein. Posttranslational regulation of *ZRT2* has not been established, but it is unlikely that such a mechanism exists because the low-affinity system need not be as highly controlled as *ZRT1* by extreme variations in zinc availability.

MANGANESE

Manganese is considered only partially redox active. While the number of manganese-containing proteins is low, in some instances Mn^{2+} can substitute for Ca^{2+} in which case it can bind to a greater number of proteins. The ability to bind Mn^{2+} instead of Ca^{2+} may contribute to Mn^{2+} toxicity. Among the important proteins that bind Mn^{2+} physiologically are superoxide dismutase, glycosyl transferases, and endonucleases.

High-Affinity Manganese Transport

Manganese uptake occurs through both high-affinity (Km of 0.3 μM) and low-affinity (Km of 62 μM) systems. Opposed to transport of other transition metals, the high-affinity Mn^{2+} system has broad metal specificity, as transport may be competitively inhibited by the divalent metal ions Cd^{2+}, Co^{2+}, Mg^{2+}, and Zn^{2+} (24). High-affinity Mn^{2+} uptake is mediated through the *SMF1* and *SMF2* gene products. Smf1p was identified through a screen that selected mutants unable to grow in the presence of the transition metal chelator EGTA (72). A specific mutant was identified that was unable to grow in the presence of EGTA but could grow when Mn^{2+} was added back. Transformation of a genomic library into the mutant cells led to the identification of *SMF1* as a gene that could complement the mutant phenotype. This gene had previously been identified as a multicopy suppressor of a temperature-sensitive mutation in a mitochondrial processing peptidase (76). Overexpression of Smf1p or the presence of 1 mM manganese chloride permitted the Ts mutant to grow at the restrictive temperature. These results suggest that increased expression of Smf1p aided cellular manganese uptake. It was subsequently recognized that the mitochondrial processing protease was a manganese-dependent enzyme.

Examination of the yeast genome revealed the presence of two homologs of *SMF1*, *SMF2* and *SMF3*. These genes are part of an evolutionarily conserved gene

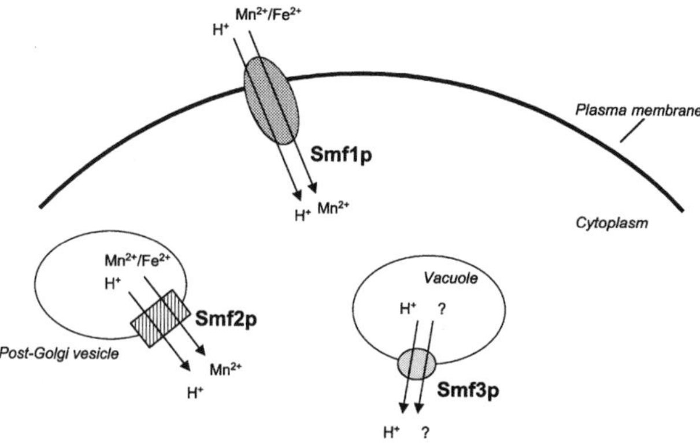

Figure 4 Model of manganese transport. High-affinity Mn^{2+} uptake is mediated through Smf1p and Smf2p. Smf1p and Smf2p can also transport Fe^{2+} and Co^{2+} as well as other transition metals. The role of Smf3p in Mn^{2+} transport is unclear. Smf1p and Smf2p are H^+/Mn^{2+} symporters. The metal specificity of Smf3p is unknown.

family, with members of the family found in all biological kingdoms. The family is often referred to as the NRAMP family, named after a gene required by mice for resistance to intracellular bacteria (75). The *SMF* proteins show 50% identity with each other and encode proteins of 59 to 51 KD. The proteins are membrane proteins, have either 10 or 12 transmembrane domains, and are localized in different membranes (7, 65) (Figure 4). Smf1p is found in the plasma membrane, Smf2p in a post-Golgi vesicle population, and Smf3p in the vacuole. The metal specificity and function of the *SMF* transporters were defined by examining the phenotype deletion strains (i.e., Δ*smf1*, Δ*smf1/2*, Δ*smf1-3*) in which a single *SMF* gene was overexpressed. The triple null mutant Δ*smf1-3* showed a dramatic decrease in Mn^{2+} accumulation (7). Overexpression of either Smf1p or Smf2p resulted in Mn^{2+} uptake that was higher than wild-type cells, but overexpression of Smf3p showed no significant uptake above that of the triple null mutant. Other metals taken up by Smf1p/Smf2p included Fe^{2+}, Cu^{+}, and Co^{2+} (72). Smf2p appears to have much greater preference for Co^{2+} than does Smf1p (55).

Although there is little data on the mechanism of transport for the high- or low-affinity Fe^{2+}/Fe^{3+}, Zn^{2+} or Cu^{2+}, transporters, there is a good understanding of how Smf1p1/Smf2p mediate Mn^{2+}or Fe^{2+} transport. These proteins are H^+/transition metal co-transporters. The involvement of pH in transport was identified first for the mammalian homolog DMT1 (also referred to as NRAMP2/DCT1). This protein is found in the apical plasma membrane of the intestinal epithelial cell as well as in the endosome of most vertebrate cell types (21, 31). Identified as a Fe^{2+} transport, DMT1 can also transport Mn^{2+} as well as other transition metals. Both the intestinal and endosomal lumens are acidic compartments, and disruption in pH inhibits iron transport. Information about DMT1 and SMF1/2 resulted from the ectopic expression of these proteins in *Xenopus* oocytes (6, 31, 69). The expression of these proteins in oocytes permits the use of both radioactive tracers and electrophysiological techniques to monitor metal transport. Transport of Fe^{2+} or Mn^{2+} by DMT1 and Smf1p/Smf2p is electrogenic and requires H^+ ions. It is of interest that Na^+, which is not transported by Smf1p, can inhibit metal transport. Incubation of cells in EDTA and high Na^+ inhibits cell growth by preventing the uptake of Mn^{2+}.

Regulation of Manganese Transport

In yeast, regulation of Mn^{2+} uptake at the transcriptional level has not been observed. Rather Mn^{2+} uptake is regulated by the posttranslational control of Smf1p, Smf2p trafficking, and degradation. Incubation of cells in low concentrations of transition metals results in increased levels of Smf1p and Smf2p (54, 55). Addition of Mn^{2+} or Fe^{2+}, but not Zn^{2+}, Cu^+, or Co^{2+}, led to the rapid disappearance of Smf1p and Smf2p. The disappearance of these proteins is due to a metal-induced change in intracellular trafficking. In the presence of Mn^{2+} or Fe^{2+}, Smf1p and Smf2p are trafficked to the vacuole where they are degraded by vacuolar proteases (54, 65). Liu et al. (55) identified a role for the *BSD2* gene product in regulating manganese accumulation by *SMF1* and *SMF2*. Mutations in *BSD2* caused yeast cells to accumulate elevated levels of copper that rendered cells hypersensitive to Cu^+. Further analysis revealed that *bsd2* mutant cells showed increased accumulation of Mn^{2+}, Co^{2+}, and Cd^{2+}. Whereas supplementation of wild-type yeast cells with Mn^{2+} led to the rapid degradation of Smf1/2p, *bsd2* cells showed high levels of Smf1/2p that were unaffected by Mn^{2+}. Bsd2p is a membrane protein localized to the endoplasmic reticulum that appears to direct Smf1p to the vacuole in response to Mn^{2+} supplementation. When cells are starved for Mn^{2+}, Smf1p does not enter the vacuole, and the transporter localizes to the plasma membrane. Similarly in the absence of Bsd2p, Smf2p is found in an intracellular vesicle. The mechanism by which Bsd2p controls the intracellular traffic of the Smf proteins is unknown.

While *SMF3* is highly homologous to *SMF1/2*, its role in Mn^{2+} transport is unclear, as deletion of *SMF3* does not affect Mn^{2+} accumulation (7, 65). As Smf3p is located in the vacuole, its affect on metal metabolism may not be as straightforward to analyze as surface transporters. Furthermore, Smf3p is regulated differently than Smf1/Smf2p. Bds2p does not regulate Smf3p; rather *SMF3* mRNA transcription is regulated as part of the iron regulon by *AFT1* (65). The current view of Smf3p is that it may be a low-affinity Fe^{2+}/Mn^{2+} transporter that transports these metals from vacuole to cytosol.

ACKNOWLEDGMENT

This work is supported by National Institute of Health Grant NIDDK-30534 to J. Kaplan.

The *Annual Review of Microbiology* is online at http://micro.annualreviews.org

LITERATURE CITED

1. Ardon O, Bussey H, Philpott C, Ward DM, Davis-Kaplan S, et al. 2001. Identification of a *Candida albicans* ferrichrome transporter and its characterization by expression in *Saccharomyces cerevisiae*. *J. Biol. Chem.* 276:43049–55
2. Askwith C, Eide D, Van Ho A, Bernard PS, Li L, et al. 1994. The FET3 gene of *S. cerevisiae* encodes a multicopper oxidase required for ferrous iron uptake. *Cell* 76:403–10
3. Askwith C, Kaplan J. 1997. An oxidase-permease-based iron transport system in *Schizosaccharomyces pombe* and its expression in *Saccharomyces cerevisiae*. *J. Biol. Chem.* 272:401–5
4. Bird AJ, Zhao H, Luo H, Jensen LT, Srinivasan C, et al. 2000. A dual role for zinc fingers in both DNA binding and zinc sensing by the Zap1 transcriptional activator. *EMBO J.* 19:3704–13
5. Bonaccorsi di Patti MC, Bellenchi GC, Bielli P, Calabrese L. 1999. Release of highly active Fet3 from membranes of the yeast *Pichia pastoris* by limited proteolysis. *Arch. Biochem. Biophys.* 372:295–99
6. Chen XZ, Peng JB, Cohen A, Nelson H, Nelson N, Hediger MA. 1999. Yeast SMF1 mediates H(+)-coupled iron uptake with concomitant uncoupled cation currents. *J. Biol. Chem.* 274:35089–94
7. Cohen A, Nelson H, Nelson N. 2000. The family of SMF metal ion transporters in yeast cells. *J. Biol. Chem.* 275:33388–94
8. Culotta VC, Lin SJ, Schmidt P, Klomp LW, Casareno RL, Gitlin J. 1999. Intracellular pathways of copper trafficking in yeast and humans. *Adv. Exp. Med. Biol.* 448:247–54
9. Dancis A, Haile D, Yuan DS, Klausner RD. 1994. The *Saccharomyces cerevisiae* copper transport protein (Ctr1p). Biochemical characterization, regulation by copper, and physiologic role in copper uptake. *J. Biol. Chem.* 269:25660–67
10. Dancis A, Klausner RD, Hinnebusch AG, Barriocanal JG. 1990. Genetic evidence that ferric reductase is required for iron uptake in *Saccharomyces cerevisiae*. *Mol. Cell. Biol.* 10:2294–301
11. Dancis A, Yuan DS, Haile D, Askwith C, Eide D, et al. 1994. Molecular characterization of a copper transport protein in *S. cerevisiae*: an unexpected role for copper in iron transport. *Cell* 76:393–402
12. Davis-Kaplan SR, Askwith CC, Bengtzen AC, Radisky D, Kaplan J. 1998. Chloride is an allosteric effector of copper assembly for the yeast multicopper oxidase fet3p: an unexpected role for intracellular chloride channels. *Proc. Natl. Acad. Sci. USA* 95:13641–45
13. De Silva DS, Askwith CC, Eide D, Kaplan J. 1995. The FET3 gene product required for high affinity iron transport in yeast is a cell surface ferroxidase. *J. Biol. Chem.* 270:1098–101
14. De Silva DS, Davis-Kaplan SR, Fergestad J, Kaplan J. 1997. Purification and characterization of Fet3 protein, a yeast homologue of ceruloplasmin. *J. Biol. Chem.* 272: 14208–13
15. Dix DR, Bridgham JT, Broderius MA, Byersdorfer CA, Eide DJ. 1994. The FET4 gene encodes the low affinity Fe(II) transport protein of *Saccharomyces cerevisiae*. *J. Biol. Chem.* 269:26092–99
16. Dix DR, Bridgham JT, Broderius M, Eide D. 1997. Characterization of the FET4 protein of yeast. Evidence for a direct role in the transport of iron. *J. Biol. Chem.* 272: 11770–77
17. Eck R, Hundt S, Hartl A, Roemer E, Kunkel W. 1999. A multicopper oxidase gene from *Candida albicans*: cloning, characterization and disruption. *Microbiology* 145: 2415–22
18. Eide D, Broderius M, Fett J, Guerinot ML. 1996. A novel iron-regulated metal

transporter from plants identified by functional expression in yeast. *Proc. Natl. Acad. Sci. USA* 93:5624–28

19. Eide D, Davis-Kaplan SR, Jordan I, Sipe D, Kaplan J. 1992. Regulation of iron uptake in *Saccharomyces cerevisiae*. The ferrireductase and Fe(II) transporter are regulated independently. *J. Biol. Chem.* 267:20774–81
20. Eisenstein RS, Blemings DB. 1998. Iron regulatory proteins, iron responsive elements and iron homeostasis. *J. Nutr.* 128:2295–98
21. Fleming MD, Trenor CC, Su MA, Foernzler D, Beier DR, et al. 1997. Microcytic anaemia mice have a mutation in Nramp2, a candidate iron transporter gene. *Nat. Genet.* 16:383–86
22. Foury F, Talibi D. 2001. Mitochondrial control of iron homeostasis. A genome wide analysis of gene expression in a yeast frataxin-deficient strain. *J. Biol. Chem.* 276:7762–68
23. Frieden E, Hsieh HS. 1976. The biological role of ceruloplasmin and its oxidase activity. *Adv. Exp. Med. Biol.* 74:505–29
24. Gadd GM, Laurence OS. 1996. Demonstration of high-affinity Mn2+ uptake in *Saccharomyces cerevisiae*: specificity and kinetics. *Microbiology* 142:1159–67
25. Gaxiola RA, Yuan DS, Klausner RD, Fink GR. 1998. The yeast CLC chloride channel functions in cation homeostasis. *Proc. Natl. Acad. Sci. USA* 95:4046–50
26. Georgatsou E, Alexandraki D. 1994. Two distinctly regulated genes are required for ferric reduction, the first step of iron uptake in *Saccharomyces cerevisiae. Mol. Cell. Biol.* 14:3065–73
27. Georgatsou E, Mavrogiannis LA, Fragiadakis GS, Alexandraki D. 1997. The yeast Fre1p/Fre2p cupric reductases facilitate copper uptake and are regulated by the copper-modulated Mac1p activator. *J. Biol. Chem.* 272:13786–92
28. Gitan RS, Eide DJ. 2000. Zinc-regulated ubiquitin conjugation signals endocytosis of the yeast ZRT1 zinc transporter. *Biochem. J.* 346(Pt 2):329–36
29. Gitan RS, Luo H, Rodgers J, Broderius M, Eide DJ. 1998. Zinc-induced inactivation of the yeast ZRT1 zinc transporter occurs through endocytosis and vacuolar degradation. *J. Biol. Chem.* 273:28617–24
30. Greene JR, Brown NH, DiDomenico BJ, Kaplan J, Eide DJ. 1993. The GEF1 gene of *Saccharomyces cerevisiae* encodes an integral membrane protein; mutations in which have effects on respiration and iron-limited growth. *Mol. Gen. Genet.* 241:542–53
31. Gunshin H, Mackenzie B, Berger UV, Gunshin Y, Romero MF, et al. 1997. Cloning and characterization of a mammalian proton-coupled metal-ion transporter. *Nature* 388:482–88
32. Hassett R, Kosman DJ. 1995. Evidence for Cu(II) reduction as a component of copper uptake by *Saccharomyces cerevisiae. J. Biol. Chem.* 270:128–34
33. Hassett RF, Romeo AM, Kosman DJ. 1998. Regulation of high affinity iron uptake in the yeast *Saccharomyces cerevisiae*. Role of dioxygen and Fe. *J. Biol. Chem.* 273:7628–36
34. Hassett RF, Yuan DS, Kosman DJ. 1998. Spectral and kinetic properties of the Fet3 protein from *Saccharomyces cerevisiae*, a multinuclear copper ferroxidase enzyme. *J. Biol. Chem.* 273:23274–82
35. Heredia J, Crooks M, Zhu Z. 2001. Phosphorylation and Cu^+ coordination-dependent DNA binding of the transcription factor Mac1p in the regulation of copper transport. *J. Biol. Chem.* 276:8793–97
36. Heymann P, Ernst JF, Winkelmann G. 1999. Identification of a fungal triacetylfusarinine C siderophore transport gene (TAF1) in *Saccharomyces cerevisiae* as a member of the major facilitator superfamily. *Biometals* 12:301–6
37. Heymann P, Ernst JF, Winkelmann G. 2000. A gene of the major facilitator superfamily encodes a transporter for enterobactin (Enb1p) in *Saccharomyces cerevisiae. Biometals* 13:65–72

38. Howard DH. 1999. Acquisition, transport, and storage of iron by pathogenic fungi. *Clin. Microbiol. Rev.* 12:394–404
39. Hung IH, Suzuki M, Yamaguchi Y, Yuan DS, Klausner RD, Gitlin JD. 1997. Biochemical characterization of the Wilson disease protein and functional expression in the yeast *Saccharomyces cerevisiae*. *J. Biol. Chem.* 272:21461–66
40. Ismail A, Bedell GW, Lupan DM. 1985. Siderophore production by the pathogenic yeast, *Candida albicans*. *Biochem. Biophys. Res. Commun.* 130:885–91
41. Ismail A, Lupan DM. 1986. Utilization of siderophores by *Candida albicans*. *Mycopathologia* 96:109–13
42. Jensen LT, Winge DR. 1998. Identification of a copper-induced intramolecular interaction in the transcription factor mac1 from *Saccharomyces cerevisiae*. *EMBO J.* 17:5400–8
43. Kampfenkel K, Kushnir S, Babiychuk E, Inze D, Van Montagu M. 1995. Molecular characterization of a putative *Arabidopsis thaliana* copper transporter and its yeast homologue. *J. Biol. Chem.* 270:28479–86
44. Klomp LW, Lin SJ, Yuan DS, Klausner RD, Culotta VC, Gitlin JD. 1997. Identification and functional expression of HAH1, a novel human gene involved in copper homeostasis. *J. Biol. Chem.* 272:9221–26
45. Knight SA, Labbe S, Kwon LF, Kosman DJ, Thiele DJ. 1996. A widespread transposable element masks expression of a yeast copper transport gene. *Genes Dev.* 10:1917–29
46. Labbe S, Zhu Z, Thiele DJ. 1997. Copper-specific transcriptional repression of yeast genes encoding critical components in the copper transport pathway. *J. Biol. Chem.* 272:15951–58
47. Lee GR, Nacht S, Lukens JN, Cartwright GE. 1968. Iron metabolism in copper-deficient swine. *J. Clin. Invest.* 47:2058–69
48. Lesuisse E, Crichton RR, Labbe P. 1990. Iron-reductases in the yeast *Saccharomyces cerevisiae*. *Biochim. Biophys. Acta* 1038:253–59
49. Lesuisse E, Raguzzi F, Crichton RR. 1987. Iron uptake by the yeast *Saccharomyces cerevisiae*: involvement of a reduction step. *J. Gen. Microbiol.* 133:3229–36
50. Lesuisse E, Simon-Casteras M, Labbe P. 1998. Siderophore-mediated iron uptake in *Saccharomyces cerevisiae*: the SIT1 gene encodes a ferrioxamine B permease that belongs to the major facilitator superfamily. *Microbiology* 144:3455–62
51. Li L, Chen OS, Ward DM, Kaplan J. 2001. CCC1 is a transporter that mediates vacuolar iron storage in yeast. *J. Biol. Chem.* 276:29515–19
52. Li L, Kaplan J. 1998. Defects in the yeast high affinity iron transport system result in increased metal sensitivity because of the increased expression of transporters with a broad transition metal specificity. *J. Biol. Chem.* 273:22181–87
53. Lin SJ, Culotta VC. 1995. The ATX1 gene of *Saccharomyces cerevisiae* encodes a small metal homeostasis factor that protects cells against reactive oxygen toxicity. *Proc. Natl. Acad. Sci. USA* 92:3784–88
54. Liu XF, Culotta VC. 1999. Post-translation control of Nramp metal transport in yeast. Role of metal ions and the BSD2 gene. *J. Biol. Chem.* 274:4863–68
55. Liu XF, Supek F, Nelson N, Culotta VC. 1997. Negative control of heavy metal uptake by the *Saccharomyces cerevisiae* BSD2 gene. *J. Biol. Chem.* 272:11763–69
56. Lyons TJ, Gasch AP, Gaither LA, Botstein D, Brown PO, Eide DJ. 2000. Genome-wide characterization of the Zap1p zinc-responsive regulon in yeast. *Proc. Natl. Acad. Sci. USA* 97:7957–62
57. Martins LJ, Jensen LT, Simon JR, Keller GL, Winge DR, Simons JR. 1998. Metalloregulation of FRE1 and FRE2 homologs in *Saccharomyces cerevisiae*. *J. Biol. Chem.* 273:23716–21
58. Deleted in proof
59. Minnick AA, Eizember LE, McKee JA, Dolence EK, Miller MJ. 1991. Bioassay for siderophore utilization by *Candida albicans*. *Anal. Biochem.* 194:223–29

60. Neilands JB. 1995. Siderophores: structure and functions of microbial iron transport compounds. *J. Biol. Chem.* 270:26723–26
61. Neilands JB, Leong SA. 1986. Siderophores in relation to plant growth and disease. *Annu. Rev. Plant Physiol.* 37:187–208
62. Ooi CE, Rabinovich E, Dancis A, Bonifacino JS, Klausner RD. 1996. Copper-dependent degradation of the *Saccharomyces cerevisiae* plasma membrane copper transporter Ctr1p in the apparent absence of endocytosis. *EMBO J.* 15:3515–23
63. Payne AS, Gitlin JD. 1998. Functional expression of the Menkes disease protein reveals common biochemical mechanisms among the copper-transporting P-type ATPases. *J. Biol. Chem.* 273:3765–70
64. Pena MM, Puig S, Thiele DJ. 2000. Characterization of the *Saccharomyces cerevisiae* high affinity copper transporter Ctr3. *J. Biol. Chem.* 275:33244–51
65. Portnoy ME, Liu XF, Culotta VC. 2000. *Saccharomyces cerevisiae* expresses three functionally distinct homologues of the nramp family of metal transporters. *Mol. Cell. Biol.* 20:7893–902
66. Protchenko O, Ferea T, Rashford J, Tiedeman J, Brown PO, et al. 2001. Three cell wall mannoproteins facilitate the uptake of iron in *Saccharomyces cerevisiae*. *J. Biol. Chem.* 276:49244–50
67. Radisky DC, Snyder WB, Emr SD, Kaplan J. 1997. Characterization of VPS41, a gene required for vacuolar trafficking and high-affinity iron transport in yeast. *Proc. Natl. Acad. Sci. USA* 94:5662–66
68. Rouault T, Klausner R. 1997. Regulation of iron metabolism in eukaryotes. *Curr. Top. Cell. Regul.* 35:1–19
69. Sacher A, Cohen A, Nelson N. 2001. Properties of the mammalian and yeast metal-ion transporters DCT1 and Smf1p expressed in *Xenopus laevis* oocytes. *J. Exp. Biol.* 204:1053–61
70. Spizzo TC, Byersdorfer S, Duesterhoeft S, Eide DJ. 1997. The yeast FET5 gene encodes a FET3-related multicopper oxidase implicated in iron transport. *Mol. Gen. Genet.* 256:547–56
71. Stearman RD, Yuan DS, Yamaguchi-Iwai Y, Klausner RD, Dancis A. 1996. A permease-oxidase complex involved in high-affinity iron uptake in yeast. *Science* 271:1552–57
72. Supek F, Supekova L, Nelson H, Nelson N. 1996. A yeast manganese transporter related to the macrophage protein involved in conferring resistance to mycobacteria. *Proc. Natl. Acad. Sci. USA* 93:5105–10
73. ter Linde JJ, Liang JH, Davis RW, Steensma HY, van Dijken JP, Pronk JT. 1999. Genome-wide transcriptional analysis of aerobic and anaerobic chemostat cultures of *Saccharomyces cerevisiae*. *J. Bacteriol.* 181:7409–13
74. Urbanowski JL, Piper RC. 1999. The iron transporter Fth1p forms a complex with the Fet5 iron oxidase and resides on the vacuolar membrane. *J. Biol. Chem.* 274:38061–70
75. Vidal SM, Pinner E, Lepage P, Gauthier S, Gros P. 1996. Natural resistance to intracellular infections: Nramp1 encodes a membrane phosphoglycoprotein absent in macrophages from susceptible (Nramp1 D169) mouse strains. *J. Immunol.* 157:3559–68
76. West AH, Clark DJ, Martin J, Neupert W, Hartl FU, Horwich AL. 1992. Two related genes encoding extremely hydrophobic proteins suppress a lethal mutation in the yeast mitochondrial processing enhancing protein. *J. Biol. Chem.* 267:24625–33
77. Yamaguchi-Iwai Y, Dancis A, Klausner RD. 1995. AFT1: a mediator of iron regulated transcriptional control in *Saccharomyces cerevisiae*. *EMBO J.* 14:1231–39
78. Yamaguchi-Iwai Y, Serpe YM, Haile D, Yang W, Kosman DJ, et al. 1997. Homeostatic regulation of copper uptake in yeast via direct binding of MAC1 protein to upstream regulatory sequences of FRE1 and CTR1. *J. Biol. Chem.* 272:17711–18
79. Yuan DS, Dancis A, Klausner RD. 1997.

Restriction of copper export in *Saccharomyces cerevisiae* to a late Golgi or post-Golgi compartment in the secretory pathway. *J. Biol. Chem.* 272:25787–93

80. Yuan DS, Stearman R, Dancis A, Dunn T, Beeler T, Klausner RD. 1995. The Menkes/Wilson disease gene homologue in yeast provides copper to a ceruloplasmin-like oxidase required for iron uptake. *Proc. Natl. Acad. Sci. USA* 92:2632–36
81. Yun CW, Bauler M, Moore RE, Klebba PE, Philpott CC. 2001. The role of the FRE family of plasma membrane reductases in the uptake of siderophore-iron in *Saccharomyces cerevisiae. J. Biol. Chem.* 276: 10218–23
82. Yun CW, Ferea T, Rashford J, Ardon O, Brown PO, et al. 2000. Desferrioxamine-mediated iron uptake in *Saccharomyces cerevisiae*. Evidence for two pathways of iron uptake. *J. Biol. Chem.* 275:10709–15
83. Zhao H, Eide DJ. 1996. The yeast ZRT1 gene encodes the zinc transporter protein of a high-affinity uptake system induced by zinc limitation. *Proc. Natl. Acad. Sci. USA* 93:2454–58
84. Zhao H, Eide DJ. 1996. The ZRT2 gene encodes the low affinity zinc transporter in *Saccharomyces cerevisiae. J. Biol. Chem.* 271:23203–10
85. Zhao H, Eide DJ. 1997. Zap1p, a metalloregulatory protein involved in zinc-responsive transcriptional regulation in *Saccharomyces cerevisiae. Mol. Cell. Biol.* 17:5044–52
86. Zhou H, Thiele DJ. 2001. Identification of a novel high affinity copper transport complex in the fission yeast *Schizosaccharomyces pombe. J. Biol. Chem.* 276:20529–35
87. Zhu Z, Labbe S, Pena MM, Thiele DJ. 1998. Copper differentially regulates the activity and degradation of yeast Mac1 transcription factor. *J. Biol. Chem.* 273: 1277–80

Annu. Rev. Microbiol. 2002. 56:263–87
doi: 10.1146/annurev.micro.56.012302.160741

First published online as a Review in Advance on April 29, 2002

INTEINS: Structure, Function, and Evolution

J. Peter Gogarten,[1] Alireza G. Senejani, Olga Zhaxybayeva, Lorraine Olendzenski, and Elena Hilario[2]

Department of Molecular and Cell Biology, University of Connecticut, 75 North Eagleville Road, Storrs, Connecticut 06269-3044; e-mail: gogarten@uconn.edu; ali@carrot.mcb.uconn.edu; olga@carrot.mcb.uconn.edu; lorraine@carrot.mcb.uconn.edu; elena@carrot.mcb.uconn.edu

Key Words homing endonuclease, splicing, protein introns, selfish genes, parasitic genes

■ **Abstract** Inteins are genetic elements that disrupt the coding sequence of genes. However, in contrast to introns, inteins are transcribed and translated together with their host protein. Inteins appear most frequently in Archaea, but they are found in organisms belonging to all three domains of life and in viral and phage proteins. Most inteins consist of two domains: One is involved in autocatalytic splicing, and the other is an endonuclease that is important in the spread of inteins. This review focuses on the evolution and technical application of inteins and only briefly summarizes recent advances in the study of the catalytic activities and structures of inteins. In particular, this review considers inteins as selfish or parasitic genetic elements, a point of view that explains many otherwise puzzling aspects of inteins.

CONTENTS

[1]corresponding author.
[2]current address: HortResearch, 120 Mt. Albert Road, Private Bag 92 169, Mt. Albert, Auckland, New Zealand.

INTRODUCTION

Definitions and History

Inteins (internal proteins) are genetic elements similar to self-splicing introns; however, inteins are transcribed and translated together with their host protein. Only at the protein level do the inteins excise themselves from the host protein. The two portions of the host protein separated by the intein are called exteins (external proteins) (16, 25, 69). During the splicing process the intein is excised, the two exteins are joined by a peptide bond, and the host protein assumes its normal folding and function. The first intein was discovered in 1987 when the carrot and *Neurospora crassa* vacuolar ATPases were compared with a putative Ca^{2+}-pumping ATPase. The latter had been isolated as a gene whose mutation made yeast resistant against the calmodulin antagonist trifluoperazine (91). The beginning and end of the encoded protein was very similar to the vacuolar ATPase subunits whose sequences had been submitted to the databanks at the same time. However, the central region of the putative calcium pump had no similarity to any known ATPase. Rather, this portion showed weak similarity to endonucleases. Anraku's lab (41) isolated the cDNA for the yeast vacuolar ATPase A-subunit and found the same sequence, including the central region, that had been earlier described as the trifluoperazine resistance gene (91). Surprisingly, denaturing polyacrylamide gel electrophoresis of the isolated protein demonstrated that the catalytic subunit of the functioning yeast V-ATPase had a molecular weight of only 70 kDa, as expected for a subunit without the insertion. Subsequently Kane et al. (48) showed that the insertion was still present in the mRNA, that the whole protein including the insertion was translated, and that the insertion spliced itself out of the protein during posttranslational processing.

Nomenclature

Inteins are named after the organism and host protein in which they reside (68, 69). If there is more than one intein in the host protein, the different inteins are designated by numbers. For example, *Tfu* Pol-2 denotes the second intein in the DNA polymerase of *Thermococcus fumicolans*. Inteins that occupy a homologous site in the host protein in a different organism are called intein alleles. For example, the third intein in the *Thermococcus aggregans* DNA polymerase, *Tag* Pol-3, is located in a position corresponding to the *Tfu* Pol-2 insertion site. These two inteins therefore are considered alleles (83).

Many inteins contain an endonuclease domain, and thus also receive a name following the conventions for naming homing endonucleases (2). The name of these endonucleases begins with PI (for protein insert) to denote them as an intein, followed by a three letter species indicator and a roman numeral specifying the different PI endonucleases present in an organism. Other prefixes used for endonucleases are I for intron and F for freestanding (2). For example, PI *Mga*I denotes the endonuclease activity of the intein in an ABC transporter from *Mycobacterium gastri* (*Mga* Pps1) (85); PI-*Sce*I denotes endonuclease activity of the intein in the yeast vacuolar ATPase catalytic subunit, *Sce* VMA1; and I-*Sce*I denotes the endonuclease in the 23S mitochondrial rRNA intron.

INTEIN DISTRIBUTION, TYPES, AND STRUCTURE

Distribution of Inteins

The intein database, InBase (68), provides information on all described inteins. Among other data it lists the inteins' sequences, conserved motifs, host organisms, and host proteins. More than 130 inteins are known in 34 different types of proteins (68, 76, 77). The inteins are between 134 and 608 amino acids long, and they are found in members of all three domains of life: Eukaryotes, Bacteria, and Archaea (Table 1) (Figure 1). Inteins are found in proteins with diverse functions, including metabolic enzymes, DNA and RNA polymerases, proteases, ribonucleotide reductases, and the vacuolar-type ATPase. However, enzymes involved in DNA replication and repair appear to dominate (55).

The ratio of an intein's size to that of its host protein varies widely. For instance, the size of the three inteins found in the *Methanococcus jannaschii* replication factor C (*Mja* RFC 1-3) are more than four times the size of their host protein, whereas the *Filobasidiella neoformans* pre-mRNA splicing factor (*Fne* pRP8) intein is less than one tenth of its host protein size (68).

TABLE 1 Number of inteins reported for the three domains of life (68)

	Number of species	Number of inteins
Eukaryotes	7	7
Eubacteria	25	44
Archaea	16	79
Total	48	130

Position of Inteins within Host Proteins

Within the host protein, inteins appear to prefer conserved regions, for example, nucleotide-binding domains (76). Figure 2 analyzes three families of conserved host proteins with respect to intein and intron insertion points. The ATPase catalytic subunits have two intein insertion sites: The vacuolar type ATPases of *Saccharomyces cerevisiae* and *Candida tropicalis* have an intein in location "a" (77); the archaeal ATPases of *Pyrococcus abyssi*, *P. furiosus*, *P. horikoshii*, *Thermoplasma acidophilum*, and *T. volcanium* have an intein in location "b" (Figure 2*a*). Replication factor C, which is less than 300 amino acids long, accommodates inteins in three different sites (Figure 2*b*). Each intein is twice the length of its host. CDC21 (cell division control protein 21) harbors six inteins: three in location "a" and three

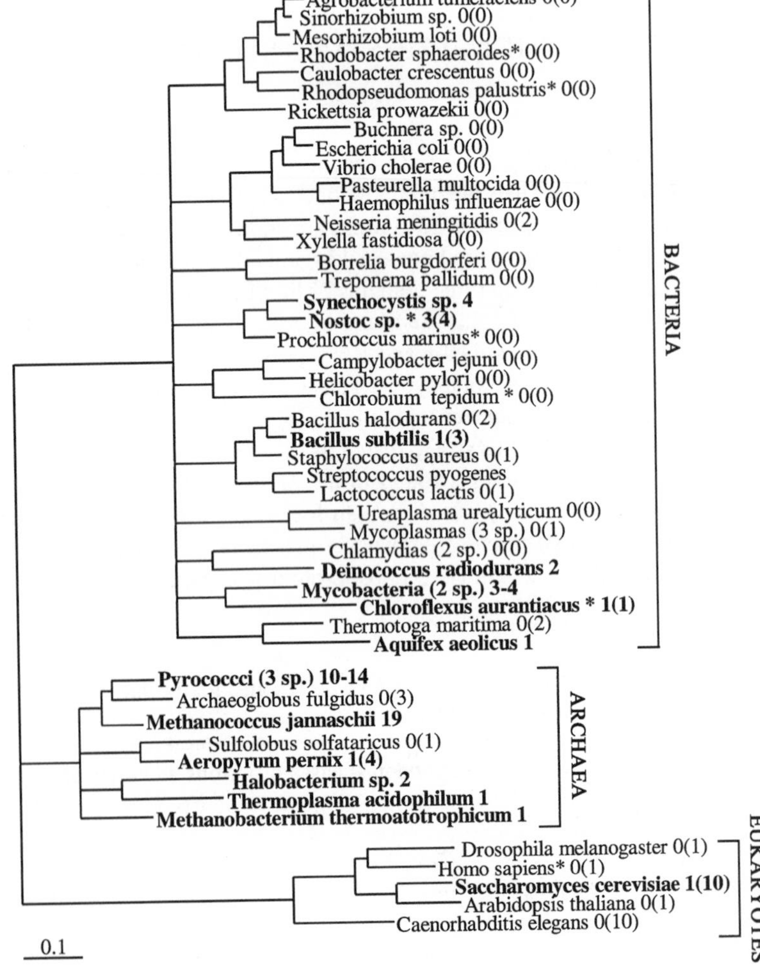

in location "b." All the inteins are inserted in the most conserved parts of the host protein. In contrast, the ATPase and replication factor C intron insertion sites do not appear to be restricted to conserved parts of the host protein (Figure 2*a*,*b*).

Large and Mini-Inteins

Most reported inteins consist of two domains (55): a self-splicing domain and an endonuclease domain. These are known as large inteins. The N- and C-terminal regions of the large inteins contain the elements necessary for splicing. Some inteins, known as mini-inteins, consist of only the self-splicing domain. Deletion of the endonuclease domain from a large intein does not affect protein splicing (13, 22, 92). The endonuclease found in large inteins is thought to play a crucial role in the spread of inteins (see below). Of the 4 subfamilies of homing endonucleases, the LAGLIDADG family is the largest, with more than 150 members reported to date (17, 99). With two exceptions (see below), all endonucleases found in inteins belong to the LAGLIDADG family.

Split Inteins

Two split inteins capable of protein trans-splicing were identified in DnaE, the catalytic subunit alpha of DNA polymerase III, in the cyanobacteria *Synechocystis* sp. strain PCC6803 and *Nostoc punctiforme* (28, 68, 104). The *Ssp* DnaE intein

←

Figure 1 Distribution of inteins across the three domains of life. Only organisms whose genome sequences are complete or nearly complete are included. Organisms that harbor inteins are highlighted in bold. The distribution of inteins is often depicted showing all organisms that contain inteins, giving the impression that inteins are widely distributed. However, among those Bacteria and Eukaryotes whose genomes have been sequenced, only a minority harbors inteins. Genomes were screened for the presence of inteins using PSI-BLAST (1, 86) profiles of 114 known inteins (68). Genomes marked with * are in progress and have not yet been annotated. Profiles were calculated using five iterations of PSI-BLAST, the known inteins as queries and the NCBI's nonredundant database as a target. Each of the 114 profiles was used to search the indicated genomes. The numbers of PSI-BLAST hits with E-value lower than 10^{-5} are given in parentheses after the organism's name. If no number is given in parentheses, then the number reflects that given in InBase for this organism (68). The hits in the individual genomes were carefully examined for the presence or absence of inteins. Many of the hits were homing endonucleases of introns and self-splicing regulatory proteins. The number of identified inteins is given without parentheses. Multiple species of the same genus are represented by a single branch with number of species examined indicated. For those genera the number of inteins and the number of PSI-BLAST hits is given per species. The tree was calculated in TREE-PUZZLE 5.0 (95) using the HKY85 substitution model and an alignment of small subunit rRNA genes from the European ribosomal RNA database (100).

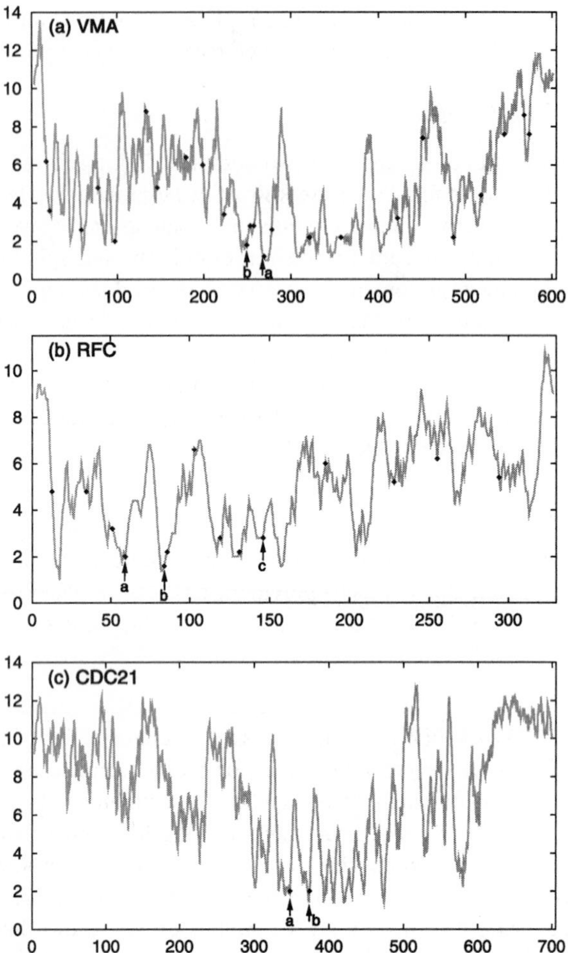

Figure 2 Positions of inteins and introns along the coding sequence of the host protein. The graph represents the conservation of the sites in protein alignments of homologs of subunit A of the vacuolar/archaeal ATPase (*panel a*), replication factor C (*panel b*), and cell division control protein 21 (*panel c*). The abscissa indicates the amino acid position along the alignment, and the ordinate gives the number of different amino acids present in that position averaged over a window of size 5. Sites in the alignment where more than 50% of the sequences had a gap were excluded from the analysis; all other gaps were treated as twenty-first amino acid. The positions of inteins are indicated as *dots with arrows*. Positions of introns from *Drosophila melanogaster*, *Arabidopsis thaliana*, and *Schizosaccharomyces pombe* are indicated as *dots without arrows*. Inteins are situated in the most conserved parts of the proteins, whereas this does not hold true for intron locations in general.

was identified as a naturally occurring split mini-intein capable of protein trans-splicing. The intein and DnaE are encoded by two separate genes, *dna*E-n and *dna*E-c, located more than 700 kbp apart in the genome and on opposite DNA strands (104). *dna*E-n encodes the amino-terminal part of the DNA polymerase subunit and the amino-terminal part of the intein; *dna*E-c encodes the carboxy-terminal parts of the intein and of the DNA polymerase subunit. Splicing and cleavage activity of the split intein result in the functional DNA polymerase and have been described in detail in (57).

Conserved Motifs

Comparative analysis of intein sequences reveals conserved motifs that can be used to identify and characterize inteins (68, 70, 72). Four conserved motifs, blocks A, B, F, and G, are found in all known inteins (Table 2). Blocks A and B are present at the intein N terminus, and blocks F and G are present at the C-terminal end of the intein. Pietrokovski (75) characterized two additional motifs (N2 and N4) that are located close to the N-terminal (Table 2). Inteins with an endonuclease domain have another four conserved motifs (blocks C, D, E, and H) (70). Studies using site-directed mutagenesis and comparative sequence and structure analyses (18, 22, 49, 70, 75) indicate that the N- and C-terminal motifs (blocks A, N2, B, N4, F, and G) are involved in protein splicing, whereas the endonuclease activity involves the central blocks C, D, E, and H (Table 2). The amino acid of the extein following the intein insertion site and block A at the N terminus of the intein contain residues chemically essential for splicing (15, 70, 105). Blocks C and E are the dodecapeptide motifs required for endonuclease activity (32, 42).

TABLE 2 Comparison of names used for conserved intein motifs. The first column gives the abbreviation used in the older literature (68, 70, 72); the second column gives the names suggested in (75)

Perler et al. (69)	Pietrokovski (75)	Other names
A	N1	N-terminal splicing motif
—	N2	—
B	N3	—
—	N4	—
C	EN1	DOD, LAGLIDADG motif
D	EN2	—
E	EN3	DOD, LAGLIDADG motif
H	EN4	—
—	HNH	HNH endonuclease motif
F	C2	—
G	C1	C-terminal splicing motif

Comparisons of Three-Dimensional Structures

The structure of the single, well-characterized intein in *S. cerevisiae* (*Sce* VMA1 or PI *Sce*I) (24, 34, 40, 43, 78) was determined by X-ray crystallography (24). The two-domain structure of PI *Sce*I is clearly visible (Figure 3, see color insert). The self-splicing domain is similar in structure to the mini-intein in the *Mycobacterium xenopi* gyrase (*Mxe* GyrA) (51). As expected, PI-*Sce*I makes target sequence-specific contacts using residues from the endonuclease domain (93). However, a part of the other domain that is distant from the PI-*Sce*I cleavage site also contributes to the recognition of the target sequence (24, 43). These additional interactions were determined using photo-crosslinking and affinity cleavage (43).

Figure 3 compares these two intein structures to the structure of the autoprocessing domain of the hedgehog protein from *Drosophila melanogaster* (39). Hedgehog proteins undergo autocatalytic cleavage and esterification, and they play an important role in the development of multicellular animals (26, 54, 80). The splicing and autoprocessing domains are all-beta structures and contain two homologous subdomains related by a pseudo two-fold axis of symmetry [see (39) and below for a discussion of evolutionary implications].

MECHANISM OF PROTEIN SPLICING

Substantial information about the chemical reactions involved in protein splicing is available (12, 15, 58, 66, 71, 79, 90, 94, 105) and was recently reviewed in (66). Briefly, protein splicing involves the following four steps (Figure 4):

Step 1 The amino-terminal splice junction of the intein is activated by an N-O or N-S shift that leads to an ester or thioester intermediate. As a result of this rearrangement, the N-extein binds to the oxygen of a serine or to the sulfur of a cysteine residue at the amino-terminal splice junction.

Step 2 Cleavage of the ester at the amino-terminal splice junction occurs through attack of a nucleophilic residue located at the carboxy-terminal splice junction. This transesterification results in a branched protein intermediate.

Step 3 The cleavage proceeds through asparagine cyclization, which causes intein excision and splicing of the two exteins by an ester bond. Several inteins have glutamine rather than asparagine residues at their C-terminal end, suggesting that cleavage in these inteins might occur via an aminoglutarimide rather than an aminosuccinimide intermediate (66, 74).

Step 4 A spontaneous rearrangement results in formation of a peptide bond between the two exteins.

INTEINS AS PARASITIC GENES

Endonucleases and Homing

Homing is the transfer of a parasitic genetic element to a cognate allele that lacks that element (9, 30, 33, 47). The result of homing is the duplication of the parasitic

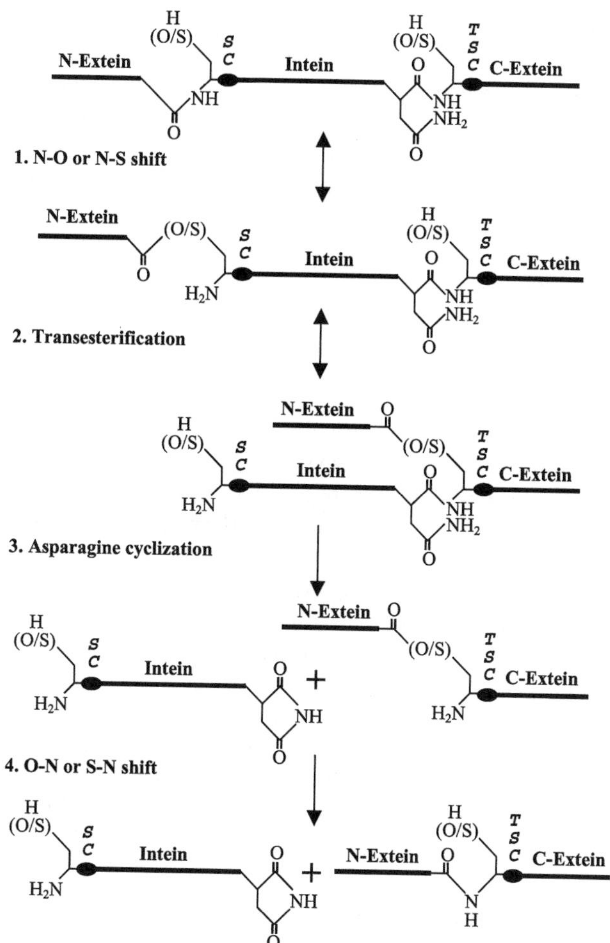

Figure 4 Splicing mechanism of inteins. Intein splicing takes place in four reaction steps (for further discussion see text).

genetic element (47). Homing allows for super-Mendelian inheritance and guarantees the rapid spread of the genetic element in a population. Homing was first described for introns that harbor a homing endonuclease, but inteins appear to spread by the same mechanism (21, 30). When an allele that contains an intron or intein with a functional homing endonuclease coexists in the same cell as a gene that contains an allele without the parasitic element, homing can be initiated by the endonuclease, and the intein/intron-free allele can be converted into an intein/intron-containing allele.

The structure and function of homing endonucleases were recently reviewed (2, 47). Briefly, the homing endonucleases recognize sites of 14–40 residues and usually do not require a complete match with the target sequence (9, 47). The

Sce VMA-1 intein recognizes a site of 31 bp (34, 40) residues, but only 9 of these residues are essential. The recognition site is so large that the homing endonuclease often cleaves a genome in a single site only. During the repair of the double-strand break, the gene encoding the homing endonuclease and surrounding sequences are copied into the cleavage site. This copying can be the result of legitimate or illegitimate recombination. The former case results in homing: The parasitic genetic element is copied into an intein/intron-free version of the allele. The latter might result in a new parasitic element if the endonuclease integrates into an existing intron or intein, or an existing parasitic element might be copied into a new target gene (2).

Selfish and Parasitic Genes

Many aspects of inteins, in particular their evolution, properties, and distribution, are better understood if one views them as parasitic genetic elements. There is an ongoing debate about which genes should be labeled as selfish [e.g., (7)]. Dawkins considers all genes as selfish (20). His gene-centered view considers the organisms' cells and bodies as vessels constructed by the genes to ensure the gene's future existence and their propagation into the next generation. Although all genes are selfish, in most instances the gene's interests coincide with the interest of the organism: A gene increases its survival chances by increasing the fitness of the organism. In contrast, the genetic elements that utilize homing are selfish in an egoistic sense. The proliferation of a parasitic genetic element via homing will continue even if the presence of this element does not contribute to the fitness of the organisms carrying the affected allele. To clearly separate these egoistic genes from Dawkins' more benign selfishness (20), we denote the former as parasitic genetic elements.

The Cyclic Reinvasion Model for Endonuclease Maintenance

Homing will lead to the rapid propagation of the genetic element that employs homing, and ultimately the allele that contains the element will be fixed in the population. At this point, however, there will be no more selection for functioning endonuclease activity. The endonuclease is only under selection for function during the super-Mendelian spreading phase; once fixed in a population, the endonuclease is expected to decay owing to random genetic drift or the deletion bias of the organism (59). Following this reasoning Goddard & Burt (35) studied the distribution and sequence of a self-splicing intron in the mitochondrial rRNA gene of different yeasts. They found that the intron frequently jumped between different yeast species and concluded that there is a cycle beginning with (*a*) the invasion of the empty site by an endonuclease-containing parasitic genetic element through horizontal transfer, homing, and fixation of the invaded allele in the population, followed first by (*b*) degeneration and loss of the endonuclease open reading frame and then by (*c*) loss of the parasitic genetic element through precise deletion, followed again by (*a*). The resulting cycle is depicted on the left side of Figure 5. This

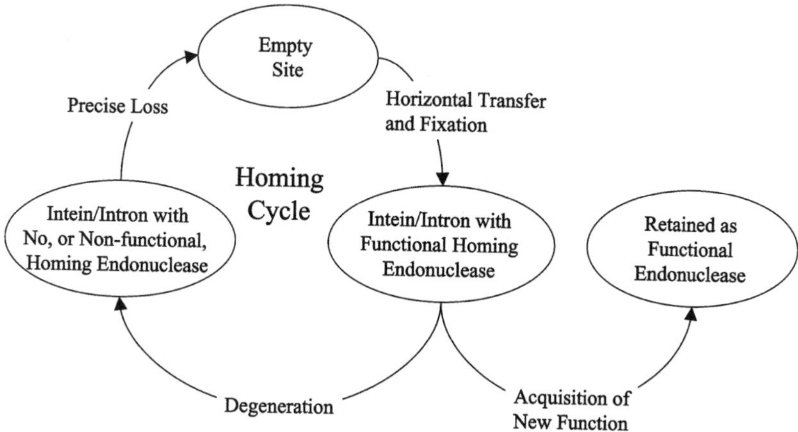

Figure 5 Homing cycle of parasitic genetic elements [modified from (35)].

homing cycle provides an explanation of how a functional homing endonuclease activity can be maintained over long periods of time through selection. Without re-invasion of a population, i.e., without the cycle operating, the parasitic genetic element would lose its endonuclease activity after the allele became fixed in the population. Not surprisingly, many inteins do not have a functioning endonuclease domain [see Figure 6 and (55, 68, 70, 75, 76)].

HOMING, SEX, AND HORIZONTAL TRANSFER In eukaryotes the intein-free and intein-containing alleles can be brought together through sex. Gimble & Thorner (33) reintroduced an intein-free version of the *vma*-1 gene into *S. cerevisiae*. After meiosis all *vma*-1 alleles contained the intein. Horizontal gene transfer is an important process that brings intein/intron-free copies together with those that contain the parasitic genetic element. Many inteins and introns with homing endonuclease were discovered in organellar genomes. For example, the Clp protease in *Chlamydomonas eugametos* choroplasts and the DNA helicase B in red algal and chryptophyte plastids (23a, 73, 81) contain inteins; one of the most thorough studies on the population dynamics of an intron with homing endonuclease was performed on a group I intron in the large subunit RNA in yeast mitochondria (35). At least in the latter case the data clearly indicate frequent re-invasion events. It is remarkable that these parasitic genetic elements are transferred with a frequency sufficient to survive inside organellar DNA, which is usually considered to avoid recombination and therefore thought to be subject to Muller's ratchet, i.e., the accumulation of slightly deleterious mutations in asexual populations (3, 82).

ENDONUCLEASE LOSS AND HORIZONTAL TRANSFER In the case of inteins there is insufficient data to estimate the relative times for each step of the homing cycle (Figure 5). For the intron in the large mitochondrial ribosomal subunit Goddard &

Burt (35) estimated the timescale of the homing cycle including the horizontal transmission to be 10^6–10^7 years, and perhaps much faster. Cho et al. (10) estimated that the intron in the plant mitochondrial *cox1* genes was transferred between species over 1000 times during angiosperm evolution. The precise loss of an intein occurs less frequently than precise loss of introns. Similar endonuclease-free inteins are found in related species, suggesting the possibility that these inteins might be exclusively maintained by vertical inheritance over long periods of time (e.g., the small inteins in *vma*-1 in the genus *Thermoplasma*) (Figure 6). However, transfer between divergent organisms is evident for several inteins with an endonuclease domain. The best-documented instance to date is the intein in the *dnaB* gene in *Rhodothermus marinus*. This gene is similar to the intein in *Synechocystis* sp. (56) and *N. punctiforme*. In *R. marinus* the intein has a different codon usage than the surrounding gene, and the intein sequences show higher similarity between *Synechocystis* sp. and *R. marinus* than the exteins. Taken together these findings indicate a recent invasion of the *dnaB* gene in *R. marinus* (56).

Many other instances of horizontal transfer are suggested by the sporadic distribution of intein alleles among divergent organisms. Similar inteins often occupy homologous sites in related proteins, but the exteins are not each other's closest relatives (56, 89). For example, the *Pyrococcus furiosus* DNA topoisomerase I and the *M. jannaschii* reverse gyrase exteins are structurally related and harbor an intein in corresponding sites, but the two enzymes have different functions (14). Intein homing has been observed under laboratory conditions (33); in contrast, the transfer of an intein to a species that did not previously harbor this intein or the invasion of a new, previously intein-free gene can at present only be inferred from comparative phylogenetic analyses (30). However, given that inteins are parasitic genetic elements, the alternative assumption of exclusive vertical inheritance with multiple parallel and convergent intein losses is untenable.

How Did Inteins Originate?

The splicing domain of inteins is similar to the autoprocessing domains of regulatory proteins that undergo autocatalytic excision [see (52) and Figure 3]. It is not yet clear which came first: inteins as parasitic genetic elements (70) or the autocatalytic splicing domain in regulatory proteins (76). Liu (55) described a scenario

Figure 6 Phylogenetic tree of 128 known inteins and *Sce* HO. Amino acid sequences were kindly provided by F.B. Perler, curator of InBase (68), and aligned using SAM (44). The tree was calculated using parsimony as implemented in PAUP* v. 4.0beta8 (97). Numbers denote branches with Bremer decay indices (6) smaller than 4. Unlabeled branches do not decay even if three additional steps are allowed. The notation of names is as in InBase (68). "*" indicates inteins without endonuclease domain; "#" denotes inteins in which a DOD family endonuclease is present in a different reading frame; and "?" specifies those inteins for which the presence of an endonuclease domain is questionable.

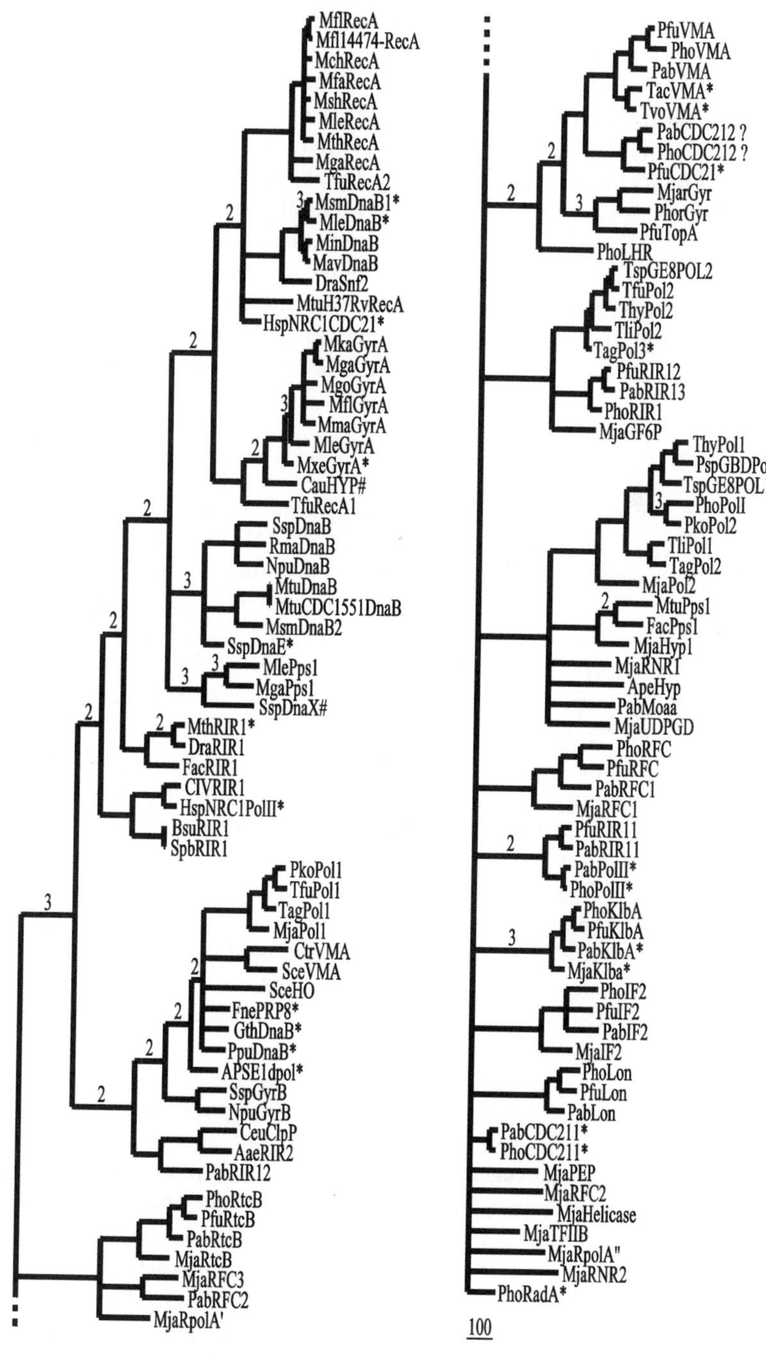
MflRecA
Mfl14474-RecA
MchRecA
MfaRecA
MshRecA
MleRecA
MthRecA
MgaRecA
TfuRecA2
MsmDnaB1*
MleDnaB*
MinDnaB
MavDnaB
DraSnf2
MtuH37RvRecA
HspNRC1CDC21*
MkaGyrA
MgaGyrA
MgoGyrA
MflGyrA
MmaGyrA
MleGyrA
MxeGyrA*
CauHYP#
TfuRecA1
SspDnaB
RmaDnaB
NpuDnaB
MtuDnaB
MtuCDC1551DnaB
MsmDnaB2
SspDnaE*
MlePps1
MgaPps1
SspDnaX#
MthRIR1*
DraRIR1
FacRIR1
CIVRIR1
HspNRC1PolII*
BsuRIR1
SpbRIR1
PkoPol1
TfuPol1
TagPol1
MjaPol1
CtrVMA
SceVMA
SceHO
FnePRP8*
GthDnaB*
PpuDnaB*
APSE1dpol*
SspGyrB
NpuGyrB
CeuClpP
AaeRIR2
PabRIR12
PhoRtcB
PfuRtcB
PabRtcB
MjaRtcB
MjaRFC3
PabRFC2
MjaRpolA'
PfuVMA
PhoVMA
PabVMA
TacVMA*
TvoVMA*
PabCDC212 ?
PhoCDC212 ?
PfuCDC21*
MjarGyr
PhorGyr
PfuTopA
PhoLHR
TspGE8POL2
TfuPol2
ThyPol2
TliPol2
TagPol3*
PfuRIR12
PabRIR13
PhoRIR1
MjaGF6P
ThyPol1
PspGBDPol
TspGE8POL1
PhoPolI
PkoPol2
TliPol1
TagPol2
MjaPol2
MtuPps1
FacPps1
MjaHyp1
MjaRNR1
ApeHyp
PabMoaa
MjaUDPGD
PhoRFC
PfuRFC
PabRFC1
MjaRFC1
PfuRIR11
PabRIR11
PabPolII*
PhoPolII*
PhoKlbA
PfuKlbA
PabKlbA*
MjaKlba*
PhoIF2
PfuIF2
PabIF2
MjaIF2
PhoLon
PfuLon
PabLon
PabCDC211*
PhoCDC211*
MjaPEP
MjaRFC2
MjaHelicase
MjaTFIIB
MjaRpolA"
MjaRNR2
PhoRadA*
100

in which a domain with self-cleaving activity undergoes a duplication resulting in a protein that autocatalytically cleaves itself into three separate peptides. A loop exchange is postulated to have integrated the two self-cleaving units, producing an intein that splices rather than cleaves the protein. The homing endonucleases were likely to invade such an element at a later time. The latter hypothesis is supported by finding different types of endonucleases in inteins (17, 70, 75) and by finding the intein-typical LAGLIDADG endonucleases also in introns and as open reading frames without clear association with parasitic elements (2). The idea of homing endonucleases invading self-splicing introns and inteins after these elements originated is also supported by the following reasoning (21): If endonucleases are mobile in genomes, genetic elements that remove themselves from the gene product would constitute preferred integration sites because in these locations the functions encoded by the surrounding DNA would not be disrupted.

Multiple Origins?

Inteins can be readily identified using position-specific iterated BLAST (PSI BLAST) or profile hidden Markov model (HMM) searches (1, 17, 72) (Figure 1). Protein space is so huge that it appears unlikely that protein folds and domains recognized by these approaches originated through convergent evolution (50); however, this possibility, albeit unlikely, cannot yet be ruled out. Both intein domains, the splicing and the endonuclease domain, probably had unique and independent origins. However, an intein that functions as an effective parasitic element only results from the combination of both domains. The homing cycle predicts that the endonuclease domain should be lost from the intein before the splicing domain; Gimble's (31) study of inteins in different yeasts illustrates that the endonuclease domain indeed has a tendency to decay, and several inteins without endonuclease domain or with an endonuclease that probably is nonfunctional have been described (68, 98). It is reasonable to assume that most of the mini-inteins that only contain the splicing domain evolved from inteins that had a functional endonuclease domain (30, 55). This view is supported by the finding that the mini-inteins do not form a coherent group in phylogenetic reconstructions as do the hedgehog proteins, but rather that inteins without an endonuclease are found for several different intein alleles (Figure 6) (18, 70).

Most inteins contain endonucleases of the LAGLIDADG type (also called DOD-type), and the relationship between splicing and endonuclease domains is similar in the different inteins (66–68, 70, 72, 75); however, an intein in the gyrase of two cyanobacteria (*Ssp* GyrB and *Npu* GyrB) contains an HNH-type endonuclease (17, 70, 75). (These endonucleases are named after the conserved amino acids sequence motifs "LAGLIDADG" and "HNH.") The observations that the endonuclease is always inserted between the same splicing motifs argues for only a single or a few recombination events giving rise to inteins; however, only one region in the splicing domain might be successfully invaded by an endonuclease without disruption of the excision and splicing activity. The fact that two inteins (*Ssp* GyrB

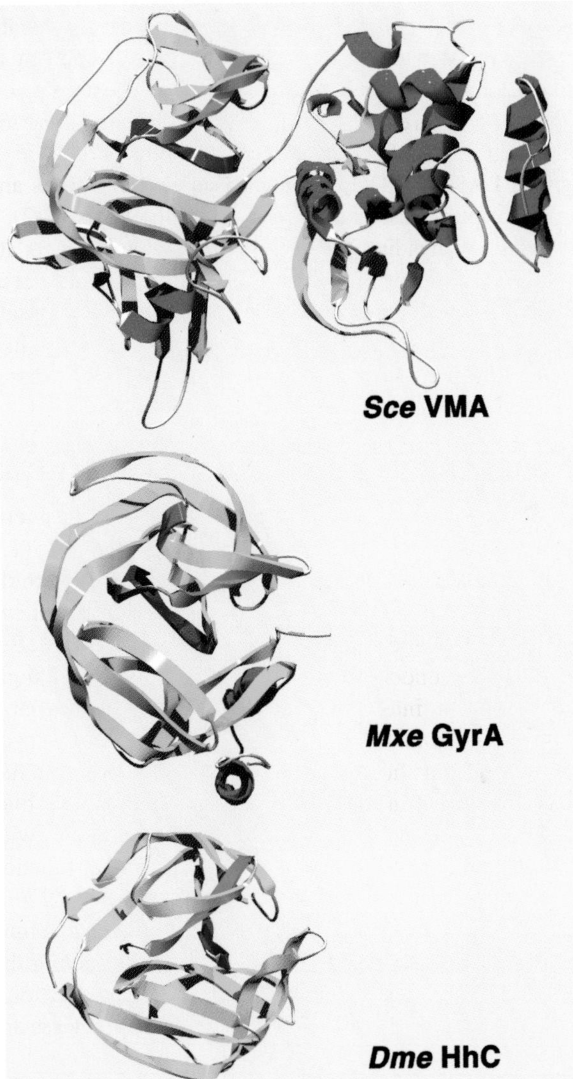

Figure 3 Comparison of intein and hedgehog protein structures. The structures of the *Sce* VMA, *Mxe* GyrA, and the autoprocessing domain of the hedgehog protein from *Drosophila melanogaster* have been determined by X-ray crystallography (24, 39, 51). The pdb-files were retrieved from the Protein Data Bank (4) and processed using the Swiss-PdbViewer (38). The *orange arrows* and *red helices* indicate β-sheets and α-helices. The amino-terminal β-sheets are colored in *green*, and the carboxy-terminal β-sheets are shown as *blue arrows*. The endonuclease domain present in *Sce* VMA (*top panel, right*) forms a domain clearly distinct from the self-splicing domain (*top panel, left*). The part of the *Sce* VMA structure that is not part of the endonuclease domain, but partakes in DNA binding, is depicted in *light blue*.

and *Npu* GyrB) contain a different endonuclease type shows that the combination between the endonuclease and splicing domains occurred at least twice.

Little phylogenetic information is retained in the primary intein sequences (18, 70) (Figure 6). Intein alleles group together in most instances. Inteins that occupy different integration sites do not group together reproducibly in different phylogenetic reconstructions. Whereas the comparison between intein, host protein, and assumed organismal phylogeny often provides evidence for the horizontal transfer of the intein as a parasitic genetic element (see above), this information is insufficient to assess the frequency of loss or gain of the endonuclease domain.

Pietrokovski (76) recently argued that the distribution of inteins among organisms from the three domains of life demonstrated their ancient character. Given the frequency of genetic exchange across even domain boundaries (23, 61), and the fact that inteins are parasitic genetic elements whose long-term survival as functional units depends on horizontal transfer, this conclusion seems unwarranted. To the contrary, the scarcity of inteins in organisms that link recombination to sex argues that the survival of inteins over evolutionary timescales indeed depends on frequent interspecies transfer. The diverse but extremely sporadic distribution of inteins (Figure 1) is probably due to their frequent transfer and does not necessarily imply an ancient origin.

Breaking the Homing Cycle: Acquisition of Nonparasitic Functions

Inteins are not strongly counterselected because they do not interrupt the function of the host protein they invade. Their rate of loss through excision is low because they are located in those parts of the host protein that are most important for function and therefore are under most selection pressure. An imprecise deletion would disrupt the host protein's function and is strongly counterselected. A precise deletion of the intein restores the restriction site, and the homing cycle can start again. One way for a parasitic genetic element to escape the cycle of deletion and re-invasion is to acquire a function that provides a positive selective advantage to the host organism. Among several different positive contributions that have been suggested, the one most frequently mentioned is regulation of expression [see (21, 55, 76) for recent summaries].

REGULATION OF EXPRESSION AND DEVELOPMENT The use of autocatalytic protein splicing or hydrolysis in regulation is a potential contribution of the splicing domain. Clearly, any obligatory posttranslational processing step provides opportunity for further regulation. The split inteins in the DNA polymerase of *N. punctiforme* and *Synechocystis* sp. (37, 68, 103) provide an example in which a parasitic genetic element became nearly essential for the survival of the organism. Without the intein the two exteins that are encoded in different parts of the genome would not be joined together and the organism could not survive. The more elaborate posttranslational processing makes it even more difficult for the organisms to delete the intein; however, even in this case it has not yet been demonstrated that the posttranslational splicing is actually utilized to fine-tune expression.

Paulus (66) reviewed several regulatory processes that utilize protein splicing or hydrolysis. In most instances it was not demonstrated that these processes are homologous to the splicing activity in inteins. The exceptions are the hedgehog proteins (26, 54). These proteins play an important role in animal development. During posttranslational processing the precursor is cleaved into the autoprocessing domain and the amino-terminal regulatory domain is esterified with cholesterol and secreted (80). The autoprocessing domain of the hedgehog proteins clearly is homologous to the splicing domain of inteins (52) (Figure 3). In phylogenetic analyses all hedgehog proteins group together, separate from the inteins, indicating that the transition between intein and hedgehog protein occurred only once (18). However, at present it is not clear if the ancestor was an intein that lost its endonuclease domain and whose protein-splicing domain was utilized for regulation (18), or if the common ancestor functioned in regulation and later was invaded by an endonuclease generating an intein that spread into different host proteins (76). The finding that the distribution of hedgehog proteins appears to be restricted to multicellular eukaryotes suggests the former, but given that inteins are parasitic genetic elements their wider distribution does not provide a strong argument for them being more ancient.

PARASITES HELPING ONE ANOTHER The homing endonuclease cleaves alleles that do not contain a parasitic genetic element that disrupts the recognition site. This feature might help phages and viruses that harbor an endonuclease in one of their genes to out-compete close relatives in mixed infections (21, 36). Support for this idea is provided by the *Bacillus subtilis* phages SPO1 and SP82. Both phages contain introns with homing endonucleases, but surprisingly both endonucleases prefer the heterologous DNA as a target. The SP82-encoded endonuclease is responsible for excluding the SPO1 intron and flanking genetic markers from the progeny of mixed infections (36).

OTHER FUNCTIONS OF ENDONUCLEASES WITH COMPLEX RECOGNITION SITES The most striking example of an intein acquiring a new function is that of the HO endonuclease in yeast. The role of this endonuclease is to initiate a gene conversion event that results in a switch of the mating type. The HO endonuclease catalyzes a double-strand break in the MAT locus, which initiates recombination with one of two unlinked loci that interconvert MATa and MAT-α (53, 101). The HO endonuclease is homologous to the homing endonucleases of the LAGLIDADG-type (17, 69); in particular, the HO endonuclease is similar to the intein in the yeast *vma*-1 gene. In phylogenetic analysis the HO endonuclease groups with the yeast's *vma*-1 inteins (Figure 6) (77). Although it does not undergo autocatalytic splicing, the HO endonuclease contains self-splicing motifs (70) (Figure 7). The presence of a nonfunctioning self-splicing domain clearly indicates that the mating-type switching endonuclease evolved from an intein ancestor.

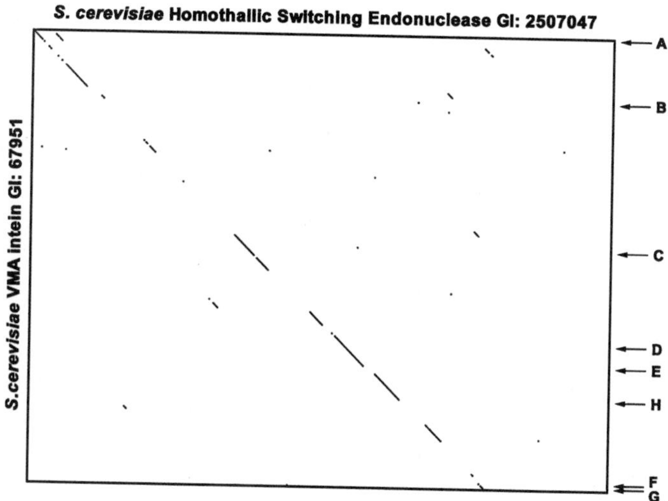

Figure 7 Dot plot of *Saccharomyces cerevisiae* V-ATPase intein (*Sce* VMA) against the *Saccharomyces cerevisiae* HO endonuclease (*Sce* HO). *Arrows* indicate the location of the intein-specific motifs abbreviated *A* through *G* according to Perler (68, 70). *Sce* HO shows a high degree of similarity to *Sce* VMA. The similarity includes both the endonuclease and the self-splicing domains. The dot plot was generated using the Dotlet Applet (46).

Nishioka et al. (60) characterized two endonucleases that are part of the inteins in the DNA polymerase of *Pyrococcus kodakarensis* (a.k.a. *Thermococcus kodakarensis*). Both of these endonucleases have characteristics typically found in homing endonucleases. One of two enzymes (PI Pko2 or Tko Pol2) has a minimal recognition sequence of only 16 bp and cleaves the intein-free allele of the DNA polymerase. However, the enzyme also cleaves a site at the junction between intein and DNA encoding the host protein and digests chromosomal DNA from *P. kodakarensis*. The authors suggested that this endonuclease might play a role in chromosomal rearrangement similar to the role of HO endonuclease in yeast.

Why Are Inteins Located Where They Are?

Inteins are found in conserved regions of conserved proteins (Figure 2). Many, but not all, of the host proteins are involved in nucleic acid metabolism and DNA replication. The following explanations for the selection of host proteins and integration sites have been proposed, but at present the discussion remains controversial (55, 76):

1. Many inteins are located in conserved regions because an intein in these regions cannot be easily deleted from the host protein. The regions are conserved because any mutation in this region leads to a nonfunctional protein,

and therefore changes are strongly counterselected. Any imprecise deletion of the intein will disrupt the function of the host protein and therefore will be counterselected.

2. Inteins survive in conserved regions of conserved proteins because only these regions are sufficiently conserved across species boundaries to allow the homing cycle to operate (Figure 5). In a less conserved region, populations could become immune to a parasitic genetic element through substitution of a site essential for endonuclease recognition. Although these substitutions have been observed in nature (84), presumably, they would be more frequent for an intein occupying a site that is not under strong selection.
3. Inteins would have a selective advantage if they occupied genes that are frequently transferred horizontally. Phage and viral genomes often encode their own enzymes used in DNA replication. An intein in these genes has a better chance of jumping across species boundaries and invading new populations (21, 74).
4. Homing endonucleases cleave genomic DNA. It is advantageous for the homing endonuclease to be expressed only when the machinery for DNA replication and repair is expressed as well. The homing process relies on the cell's machinery to repair the double-strand break using the intein-containing allele, and the host cell will be damaged if it is not prepared to repair double-strand brakes caused by the endonuclease (55).

APPLICATION OF INTEINS IN BIOTECHNOLOGY

Inteins are fascinating tools from a biotechnological point of view. Recombinant protein purification has benefited most from the applications of protein splicing, but many other tools based on inteins are being developed (29).

Traditional protein purification techniques incorporate unique tags (e.g., histidine tags) into a gene to be expressed. These tags are used to bind the desired protein onto an affinity chromatography column and then release the desired protein by the addition of external agents (ions, proteases). These methods are often inconvenient because it is necessary to further purify the recombinant protein from contaminants, lowering the yield and compromising its stability, solubility, and activity.

By mutating either the N or C terminus of an intein gene, a desired gene can be ligated and expressed in frame to the intein tag (11, 12). After expression the fusion protein is extracted from the cell and purified utilizing properties of the intein. The recombinant protein is released from the mutant intein through a change in reducing conditions. This technique has been commercialized by New England Biolabs (IMPACT^{TM}). However, the release of the recombinant protein from the affinity column–intein complex can be slow, and the purified recombinant protein can be chemically modified by the reducing agent. Improvements have been made by engineering a naturally occurring intein to produce a mini-intein with compromised activity. Through random mutation and selection a mutant mini-intein with pH-sensitive C-terminal cleavage was obtained and incorporated into a recombinant protein-affinity purification protocol (102). These inteins have been incorporated into a commercial purification kit (IMPACT-TWIN^{TM}, New England Biolabs) [for a review of current techniques see (5)]. The use of a green fluorescent protein mini-intein fusion system for protein expression (106) allows for the direct correlation between whole-cell fluorescence and protein yield. This greatly simplifies the tedious and difficult task of optimizing expression of fusion proteins.

An interesting application of inteins is the semisynthetic synthesis of cytotoxic proteins (27). Part of a cytotoxic protein is fused to an intein and expressed in vivo. Subsequent in vitro processing of the fusion protein results in cleavage of the intein, and addition of a synthetic peptide then yields the full-length and active cytotoxic proteins.

In vitro trans-splicing allows introducing nuclear magnetic resonance (NMR) labels into only a part of a large protein (62, 63). This approach promises to permit structural analysis of proteins over 50 kDa by NMR.

Daugelat & Jacobs (19) reported use of inteins in epitope mapping and antigen screening. In ORFTRAP, open reading frame fragments are selectively cloned into the *Mycobacterium tuberculosis recA* intein inserted in-frame into a kanamycin resistance gene. Only the clones containing a DNA fragment without stop codons and frameshifts (an open reading frame fragment) allow the correct intein splicing reaction and the kanamycin resistance phenotype to be expressed. Although this technique is in its preliminary stages, it could be further developed to overcome some of the initial drawbacks detected (e.g., only small inserts up to 425 bp can be cloned).

Peptide libraries are important tools in the search for ligand diversity in pharmacology. An intein-based method to biosynthesize backbone cyclic peptide libraries has been developed using the *Synechocystis* sp. PCC6803 *dnaE* split intein (28, 87, 88). Libraries of small, stable cyclic peptides were generated through site-directed mutagenesis. Libraries with 10^7–10^8 primary transformants were generated for cyclic peptides with five variable residues and either one or four fixed residues.

Split inteins also form the basis for several studies of protein-protein interaction and active-site dissection. An optical probe to study protein-protein interactions has been made from the amino and carboxy termini of the split intein in *Synechocystis* sp. PCC6803 *dnaE*, fused to the amino and carboxy fragments of firefly luciferase. The luciferase fragments in turn are linked to the proteins of interest (64). Protein-protein interaction triggers folding and trans-splicing of the *dnaE* intein, thereby recovering the luciferase enzymatic activity. Phosphorylation cascades, integral membrane protein interactions, and high-throughput drug screening are some of the potential applications of this technique. The N and C terminus of the green fluorescent protein have also been used as a reporter system for protein-protein interactions, using an artificial split intein engineered from PI *Sce*I (65), as well as the PI *Pfu*I intein from *P. furiosus* (45).

The use of split inteins in transgenic plants might allow engineering of plants whose transgenes cannot be easily transferred to other species (8, 96). Using the *dnaE* intein from *Synechocystis* sp. PCC6803 coupled to two unlinked fragments of a herbicide resistance protein might result in nontransferable genes because each of the split exteins fused to the N or C terminus of a mini-intein could be inserted into unlinked regions of a genome or into two different cellular compartments.

CONCLUDING REMARKS

Many features of inteins make sense when considered in an evolutionary context. However, a few puzzles remain controversial: How and how often did inteins originate? What is their relation to existing regulatory proteins? Why are some protein families preferred hosts? The endonuclease and the self-splicing domains of the large inteins are utilized in nature for nonparasitic purposes, and both domains are also being used in biotechnological applications. Many of these are still in their infancy, and future advancements will depend on a better understanding and utilization of the forces that shaped inteins in vivo.

ACKNOWLEDGMENTS

Support in the authors' lab was provided through the NASA Exobiology Program and through the NASA Astrobiology Institute at Arizona State University.

The *Annual Review of Microbiology* is online at http://micro.annualreviews.org

LITERATURE CITED

1. Altschul SF, Madden TL, Schaffer AA, Zhang J, Zhang Z, et al. 1997. Gapped BLAST and PSI-BLAST: a new generation of protein database search programs. *Nucleic Acids Res.* 25:3389–402
2. Belfort M, Roberts RJ. 1997. Homing endonucleases: keeping the house in order. *Nucleic Acids Res.* 25:3379–88
3. Bergstrom CT, Pritchard J. 1998. Germline bottlenecks and the evolutionary maintenance of mitochondrial genomes. *Genetics* 149:2135–46
4. Berman HM, Westbrook J, Feng Z, Gilliland G, Bhat TN, et al. 2000. The Protein Data Bank. *Nucleic Acids Res.* 28:235–42
5. Blaschke UK, Silberstein J, Muir TW. 2000. Protein engineering by expressed protein ligation. *Methods Enzymol.* 328: 478–96
6. Bremer K. 1988. The limits of amino acid sequence data in angiosperm phylogenetic reconstruction. *Evolution* 42:795–803
7. Cavalier-Smith T. 2001. Obcells as protoorganisms: membrane heredity, lithophosphorylation, and the origins of the genetic code, the first cells, and photosynthesis. *J. Mol. Evol.* 53:555–95
8. Chen L, Pradhan S, Evans TC Jr. 2001. Herbicide resistance from a divided EPSPS protein: the split *Synechocystis* DnaE intein as an in vivo affinity domain. *Gene* 263:39–48
9. Chevalier BS, Stoddard BL. 2001. Homing endonucleases: structural and functional insight into the catalysts of intron/intein mobility. *Nucleic Acids Res.* 29: 3757–74
10. Cho Y, Qiu YL, Kuhlman P, Palmer JD.

1998. Explosive invasion of plant mitochondria by a group I intron. *Proc. Natl. Acad. Sci. USA* 95:14244–49

11. Chong S, Mersha FB, Comb DG, Scott ME, Landry D, et al. 1997. Single-column purification of free recombinant proteins using a self-cleavable affinity tag derived from a protein splicing element. *Gene* 192:271–81
12. Chong S, Montello GE, Zhang A, Cantor EJ, Liao W, et al. 1998. Utilizing the C-terminal cleavage activity of a protein splicing element to purify recombinant proteins in a single chromatographic step. *Nucleic Acids Res.* 26:5109–15
13. Chong S, Xu MQ. 1997. Protein splicing of the *Saccharomyces cerevisiae* VMA intein without the endonuclease motifs. *J. Biol. Chem.* 272:15587–90
14. Chute IC, Hu Z, Liu XQ. 1998. A topA intein in *Pyrococcus furiosus* and its relatedness to the r-gyr intein of *Methanococcus jannaschii. Gene* 210:85–92
15. Cooper AA, Chen YJ, Lindorfer MA, Stevens TH. 1993. Protein splicing of the yeast TFP1 intervening protein sequence: a model for self-excision. *EMBO J.* 12:2575–83
16. Cooper AA, Stevens TH. 1995. Protein splicing: self-splicing of genetically mobile elements at the protein level. *Trends Biochem. Sci.* 20:351–56
17. Dalgaard JZ, Klar AJ, Moser MJ, Holley WR, Chatterjee A, Mian IS. 1997. Statistical modeling and analysis of the LAGLI-DADG family of site-specific endonucleases and identification of an intein that encodes a site-specific endonuclease of the HNH family. *Nucleic Acids Res.* 25:4626–38
18. Dalgaard JZ, Moser MJ, Hughey R, Mian IS. 1997. Statistical modeling, phylogenetic analysis and structure prediction of a protein splicing domain common to inteins and hedgehog proteins. *J. Comput. Biol.* 4:193–214
19. Daugelat S, Jacobs WR Jr. 1999. The *Mycobacterium tuberculosis* recA intein can be used in an ORFTRAP to select for open reading frames. *Protein Sci.* 8:644–53
20. Dawkins R. 1976. *The Selfish Gene*. Oxford, UK: Oxford Univ. Press
21. Derbyshire V, Belfort M. 1998. Lightning strikes twice: intron-intein coincidence. *Proc. Natl. Acad. Sci. USA* 95:1356–57
22. Derbyshire V, Wood DW, Wu W, Dansereau JT, Dalgaard JZ, Belfort M. 1997. Genetic definition of a protein-splicing domain: functional mini-inteins support structure predictions and a model for intein evolution. *Proc. Natl. Acad. Sci. USA* 94:11466–71
23. Doolittle WF. 1999. Phylogenetic classification and the universal tree. *Science* 284:2124–29

23a. Douglas SE, Penny SL. 1999. The plastid genome of the cryptophyte alga, *Guillardia theta*: complete sequence and conserved synteny groups confirm its common ancestry with red algae. *J. Mol. Evol.* 48:236–44

24. Duan X, Gimble FS, Quiocho FA. 1997. Crystal structure of PI-SceI, a homing endonuclease with protein splicing activity. *Cell* 89:555–64
25. Dujon B. 1989. Group I introns as mobile genetic elements: facts and mechanistic speculations—a review. *Gene* 82:91–114
26. Echelard Y, Epstein DJ, St-Jacques B, Shen L, Mohler J, et al. 1993. Sonic hedgehog, a member of a family of putative signaling molecules, is implicated in the regulation of CNS polarity. *Cell* 75:1417–30
27. Evans TC Jr, Benner J, Xu MQ. 1998. Semisynthesis of cytotoxic proteins using a modified protein splicing element. *Protein Sci.* 7:2256–64
28. Evans TC Jr, Martin D, Kolly R, Panne D, Sun L, et al. 2000. Protein trans-splicing and cyclization by a naturally split intein from the *dnaE* gene of *Synechocystis* species PCC6803. *J. Biol. Chem.* 275:9091–94

29. Evans TC Jr, Xu MQ. 1999. Intein-mediated protein ligation: harnessing nature's escape artists. *Biopolymers* 51:333–42
30. Gimble FS. 2000. Invasion of a multitude of genetic niches by mobile endonuclease genes. *FEMS Microbiol. Lett.* 185:99–107
31. Gimble FS. 2001. Degeneration of a homing endonuclease and its target sequence in a wild yeast strain. *Nucleic Acids Res.* 29:4215–23
32. Gimble FS, Stephens BW. 1995. Substitutions in conserved dodecapeptide motifs that uncouple the DNA binding and DNA cleavage activities of PI-SceI endonuclease. *J. Biol. Chem.* 270:5849–56
33. Gimble FS, Thorner J. 1992. Homing of a DNA endonuclease gene by meiotic gene conversion in *Saccharomyces cerevisiae*. *Nature* 357:301–6
34. Gimble FS, Wang J. 1996. Substrate recognition and induced DNA distortion by the PI-SceI endonuclease, an enzyme generated by protein splicing. *J. Mol. Biol.* 263:163–80
35. Goddard MR, Burt A. 1999. Recurrent invasion and extinction of a selfish gene. *Proc. Natl. Acad. Sci. USA* 96:13880–85
36. Goodrich-Blair H, Shub DA. 1996. Beyond homing: competition between intron endonucleases confers a selective advantage on flanking genetic markers. *Cell* 84:211–21
37. Gorbalenya AE. 1998. Non-canonical inteins. *Nucleic Acids Res.* 26:1741–48
38. Guex N, Peitsch MC. 1997. SWISS-MODEL and the Swiss-PdbViewer: an environment for comparative protein modeling. *Electrophoresis* 18:2714–23
39. Hall TM, Porter JA, Young KE, Koonin EV, Beachy PA, Leahy DJ. 1997. Crystal structure of a Hedgehog autoprocessing domain: homology between Hedgehog and self-splicing proteins. *Cell* 91:85–97
40. He Z, Crist M, Yen H, Duan X, Quiocho FA, Gimble FS. 1998. Amino acid residues in both the protein splicing and endonuclease domains of the PI-SceI intein mediate DNA binding. *J. Biol. Chem.* 273:4607–15
41. Hirata R, Ohsumk Y, Nakano A, Kawasaki H, Suzuki K, Anraku Y. 1990. Molecular structure of a gene, VMA1, encoding the catalytic subunit of H(+)-translocating adenosine triphosphatase from vacuolar membranes of *Saccharomyces cerevisiae*. *J. Biol. Chem.* 265: 6726–33
42. Hodges RA, Perler FB, Noren CJ, Jack WE. 1992. Protein splicing removes intervening sequences in an archaea DNA polymerase. *Nucleic Acids Res.* 20:6153–57
43. Hu D, Crist M, Duan X, Quiocho FA, Gimble FS. 2000. Probing the structure of the PI-SceI-DNA complex by affinity cleavage and affinity photocross-linking. *J. Biol. Chem.* 275:2705–12
44. Hughey R, Karplus K, Krogh A. 2000. http://www.cse.ucsc.edu/research/compbio/sam.html
45. Iwai H, Lingel A, Pluckthun A. 2001. Cyclic green fluorescent protein produced in vivo using an artificially split PI-PfuI intein from *Pyrococcus furiosus*. *J. Biol. Chem.* 276:16548–54
46. Junier T, Pagni M. 1999. http://www.isrec.isb-sib.ch/java/dotlet/Dotlet.html
47. Jurica MS, Stoddard BL. 1999. Homing endonucleases: structure, function and evolution. *Cell Mol. Life Sci.* 55:1304–26
48. Kane PM, Yamashiro CT, Wolczyk DF, Neff N, Goebl M, Stevens TH. 1990. Protein splicing converts the yeast TFP1 gene product to the 69-kD subunit of the vacuolar H(+)-adenosine triphosphatase. *Science* 250:651–57
49. Kawasaki M, Nogami S, Satow Y, Ohya Y, Anraku Y. 1997. Identification of three core regions essential for protein splicing of the yeast Vma1 protozyme. A random mutagenesis study of the entire Vma1-derived endonuclease sequence. *J. Biol. Chem.* 272:15668–74
50. Keefe AD, Szostak JW. 2001. Functional

proteins from a random-sequence library. *Nature* 410:715–18
51. Klabunde T, Sharma S, Telenti A, Jacobs WR Jr, Sacchettini JC. 1998. Crystal structure of GyrA intein from *Mycobacterium xenopi* reveals structural basis of protein splicing. *Nat. Struct. Biol.* 5:31–36
52. Koonin EV. 1995. A protein splice-junction motif in hedgehog family proteins. *Trends Biochem. Sci.* 20:141–42
53. Kostriken R, Strathern JN, Klar AJ, Hicks JB, Heffron F. 1983. A site-specific endonuclease essential for mating-type switching in *Saccharomyces cerevisiae*. *Cell* 35:167–74
54. Lee JJ, Ekker SC, von Kessler DP, Porter JA, Sun BI, Beachy PA. 1994. Autoproteolysis in hedgehog protein biogenesis. *Science* 266:1528–37
55. Liu XQ. 2000. Protein-splicing intein: genetic mobility, origin, and evolution. *Annu. Rev. Genet.* 34:61–76
56. Liu XQ, Hu Z. 1997. A DnaB intein in *Rhodothermus marinus*: indication of recent intein homing across remotely related organisms. *Proc. Natl. Acad. Sci. USA* 94: 7851–56
57. Martin DD, Xu MQ, Evans TC Jr. 2001. Characterization of a naturally occurring trans-splicing intein from *Synechocystis* sp. PCC6803. *Biochemistry* 40:1393–402
58. Mathys S, Evans TC, Chute IC, Wu H, Chong S, et al. 1999. Characterization of a self-splicing mini-intein and its conversion into autocatalytic N- and C-terminal cleavage elements: facile production of protein building blocks for protein ligation. *Gene* 231:1–13
59. Mira A, Ochman H, Moran NA. 2001. Deletional bias and the evolution of bacterial genomes. *Trends Genet.* 17:589–96
60. Nishioka M, Fujiwara S, Takagi M, Imanaka T. 1998. Characterization of two intein homing endonucleases encoded in the DNA polymerase gene of *Pyrococcus kodakaraensis* strain KOD1. *Nucleic Acids Res.* 26:4409–12
61. Olendzenski L, Liu L, Zhaxybayeva O, Murphey R, Shin DG, Gogarten JP. 2000. Horizontal transfer of archaeal genes into the Deinococcaceae: detection by molecular and computer-based approaches. *J. Mol. Evol.* 51:587–99
62. Otomo T, Ito N, Kyogoku Y, Yamazaki T. 1999. NMR observation of selected segments in a larger protein: central-segment isotope labeling through intein-mediated ligation. *Biochemistry* 38:16040–44
63. Otomo T, Teruya K, Uegaki K, Yamazaki T, Kyogoku Y. 1999. Improved segmental isotope labeling of proteins and application to a larger protein. *J. Biomol. NMR* 14:105–14
64. Ozawa T, Kaihara A, Sato M, Tachihara K, Umezawa Y. 2001. Split luciferase as an optical probe for detecting protein-protein interactions in mammalian cells based on protein splicing. *Anal. Chem.* 73:2516–21
65. Ozawa T, Nogami S, Sato M, Ohya Y, Umezawa Y. 2000. A fluorescent indicator for detecting protein-protein interactions in vivo based on protein splicing. *Anal. Chem.* 72:5151–57
66. Paulus H. 2000. Protein splicing and related forms of protein autoprocessing. *Annu. Rev. Biochem.* 69:447–96
67. Perler FB. 1999. InBase, the New England Biolabs Intein Database. *Nucleic Acids Res.* 27:346–47
68. Perler FB. 2000. InBase, the Intein Database. *Nucleic Acids Res.* 28:344–45
69. Perler FB, Davis EO, Dean GE, Gimble FS, Jack WE, et al. 1994. Protein splicing elements: inteins and exteins—a definition of terms and recommended nomenclature. *Nucleic Acids Res.* 22:1125–27
70. Perler FB, Olsen GJ, Adam E. 1997. Compilation and analysis of intein sequences. *Nucleic Acids Res.* 25:1087–93
71. Perler FB, Xu MQ, Paulus H. 1997. Protein splicing and autoproteolysis mechanisms. *Curr. Opin. Chem. Biol.* 1:292–99

72. Pietrokovski S. 1994. Conserved sequence features of inteins (protein introns) and their use in identifying new inteins and related proteins. *Protein Sci.* 3:2340–50
73. Pietrokovski S. 1996. A new intein in cyanobacteria and its significance for the spread of inteins. *Trends Genet.* 12:287–88
74. Pietrokovski S. 1998. Identification of a virus intein and a possible variation in the protein-splicing reaction. *Curr. Biol.* 8:R634–35
75. Pietrokovski S. 1998. Modular organization of inteins and C-terminal autocatalytic domains. *Protein Sci.* 7:64–71
76. Pietrokovski S. 2001. Intein spread and extinction in evolution. *Trends Genet.* 17: 465–72
77. Pietrokovski S. 2001. *Inteins—Protein Introns.* http://bioinfo.weizmann.ac.il/~pietro/inteins
78. Pingoud V, Thole H, Christ F, Grindl W, Wende W, Pingoud A. 1999. Photocross-linking of the homing endonuclease PI-SceI to its recognition sequence. *J. Biol. Chem.* 274:10235–43
79. Poland BW, Xu MQ, Quiocho FA. 2000. Structural insights into the protein splicing mechanism of PI-SceI. *J. Biol. Chem.* 275:16408–13
80. Porter JA, Young KE, Beachy PA. 1996. Cholesterol modification of hedgehog signaling proteins in animal development. *Science* 274:255–59
81. Reith ME, Munholland J. 1995. Complete nucleotide sequence of the *Porphyra purpurea* chloroplast genome. *Plant Mol. Biol. Rep.* 13:333–35
82. Saccone C, Gissi C, Lanave C, Larizza A, Pesole G, Reyes A. 2000. Evolution of the mitochondrial genetic system: an overview. *Gene* 261:153–59
83. Saves I, Eleaume H, Dietrich J, Masson JM. 2000. The thy pol-2 intein of *Thermococcus hydrothermalis* is an isoschizomer of PI-TliI and PI-TfuII endonucleases. *Nucleic Acids Res.* 28:4391–96
84. Saves I, Ozanne V, Dietrich J, Masson JM. 2000. Inteins of *Thermococcus fumicolans* DNA polymerase are endonucleases with distinct enzymatic behaviors. *J. Biol. Chem.* 275:2335–41
85. Saves I, Westrelin F, Daffe M, Masson JM. 2001. Identification of the first eubacterial endonuclease coded by an intein allele in the pps1 gene of mycobacteria. *Nucleic Acids Res.* 29:4310–18
86. Schaffer AA, Aravind L, Madden TL, Shavirin S, Spouge JL, et al. 2001. Improving the accuracy of PSI-BLAST protein database searches with composition-based statistics and other refinements. *Nucleic Acids Res.* 29:2994–3005
87. Scott CP, Abel-Santos E, Jones AD, Benkovic SJ. 2001. Structural requirements for the biosynthesis of backbone cyclic peptide libraries. *Chem. Biol.* 8:801–15
88. Scott CP, Abel-Santos E, Wall M, Wahnon DC, Benkovic SJ. 1999. Production of cyclic peptides and proteins in vivo. *Proc. Natl. Acad. Sci. USA* 96:13638–43
89. Senejani AG, Hilario E, Gogarten JP. 2001. The intein of the *Thermoplasma* A-ATPase A subunit: structure, evolution and expression in *E. coli*. *BMC Biochem.* 2:13
90. Shao Y, Paulus H. 1997. Protein splicing: estimation of the rate of O-N and S-N acyl rearrangements, the last step of the splicing process. *J. Pept. Res.* 50:193–98
91. Shih CK, Wagner R, Feinstein S, Kanik-Ennulat C, Neff N. 1988. A dominant trifluoperazine resistance gene from *Saccharomyces cerevisiae* has homology with F0F1 ATP synthase and confers calcium-sensitive growth. *Mol. Cell. Biol.* 8:3094–103
92. Shingledecker K, Jiang SQ, Paulus H. 1998. Molecular dissection of the *Mycobacterium tuberculosis* RecA intein: design of a minimal intein and of a trans-splicing system involving two intein fragments. *Gene* 207:187–95
93. Silva GH, Dalgaard JZ, Belfort M, Van Roey P. 1999. Crystal structure of the

thermostable archaeal intron-encoded endonuclease I-DmoI. *J. Mol. Biol.* 286: 1123–36
94. Southworth MW, Amaya K, Evans TC, Xu MQ, Perler FB. 1999. Purification of proteins fused to either the amino or carboxy terminus of the *Mycobacterium xenopi* gyrase A intein. *Biotechniques* 27: 110–14, 116, 118–20
95. Strimmer K, von Haeseler A. 1996. Quartet puzzling: a quartet maximum-likelihood method for reconstructing tree topologies. *Mol. Biol. Evol.* 13:964–69
96. Sun L, Ghosh I, Paulus H, Xu MQ. 2001. Protein trans-splicing to produce herbicide-resistant acetolactate synthase. *Appl. Environ. Microbiol.* 67:1025–29
97. Swofford D. 1998. *PAUP* 4.0 beta version, Phylogenetic Analysis Using Parsimony (and Other Methods)*. Sunderland, MA: Sinauer
98. Telenti A, Southworth M, Alcaide F, Daugelat S, Jacobs WR Jr, Perler FB. 1997. The *Mycobacterium xenopi* GyrA protein splicing element: characterization of a minimal intein. *J. Bacteriol.* 179:6378–82
99. Turmel M, Otis C, Cote V, Lemieux C. 1997. Evolutionarily conserved and functionally important residues in the I-CeuI homing endonuclease. *Nucleic Acids Res.* 25:2610–19
100. Van de Peer Y, De Rijk P, Wuyts J, Winkelmans T, De Wachter R. 2000. The European small subunit ribosomal RNA database. *Nucleic Acids Res.* 28:175–76
101. Wang R, Jin Y, Norris D. 1997. Identification of a protein that binds to the HO endonuclease recognition sequence at the yeast mating type locus. *Mol. Cell. Biol.* 17:770–77
102. Wood DW, Wu W, Belfort G, Derbyshire V, Belfort M. 1999. A genetic system yields self-cleaving inteins for bioseparations. *Nat. Biotechnol.* 17:889–92
103. Wu H, Hu Z, Liu XQ. 1998. Protein trans-splicing by a split intein encoded in a split DnaE gene of *Synechocystis* sp. PCC6803. *Proc. Natl. Acad. Sci. USA* 95: 9226–31
104. Wu H, Xu MQ, Liu XQ. 1998. Protein trans-splicing and functional mini-inteins of a cyanobacterial dnaB intein. *Biochim. Biophys. Acta* 1387:422–32
105. Xu MQ, Perler FB. 1996. The mechanism of protein splicing and its modulation by mutation. *EMBO J.* 15:5146–53
106. Zhang A, Gonzalez SM, Cantor EJ, Chong S. 2001. Construction of a mini-intein fusion system to allow both direct monitoring of soluble protein expression and rapid purification of target proteins. *Gene* 275:241–52

Annu. Rev. Microbiol. 2002. 56:289–314
doi: 10.1146/annurev.micro.56.012302.160938

First published online as a Review in Advance on April 29, 2002

TYPE IV PILI AND TWITCHING MOTILITY

John S. Mattick
ARC Special Research Centre for Functional and Applied Genomics, Institute for Molecular Bioscience, University of Queensland, Brisbane Qld. 4072, Australia; e-mail: j.mattick@imb.uq.edu.au

Key Words type 4 fimbriae, *Pseudomonas*, *Myxococcus*, *Neisseria*, colonization

■ **Abstract** Twitching motility is a flagella-independent form of bacterial translocation over moist surfaces. It occurs by the extension, tethering, and then retraction of polar type IV pili, which operate in a manner similar to a grappling hook. Twitching motility is equivalent to social gliding motility in *Myxococcus xanthus* and is important in host colonization by a wide range of plant and animal pathogens, as well as in the formation of biofilms and fruiting bodies. The biogenesis and function of type IV pili is controlled by a large number of genes, almost 40 of which have been identified in *Pseudomonas aeruginosa*. A number of genes required for pili assembly are homologous to genes involved in type II protein secretion and competence for DNA uptake, suggesting that these systems share a common architecture. Twitching motility is also controlled by a range of signal transduction systems, including two-component sensor-regulators and a complex chemosensory system.

CONTENTS

INTRODUCTION

The term twitching motility was first coined by Lautrop in 1961 (63) to describe flagella-independent surface motility in *Acinetobacter calcoaceticus*. The term derived from the observation that cells in this mode of motility appeared to move in a somewhat jerky fashion when viewed in suspension, which resembled twitching. Twitching motility has been shown to occur in a wide range of bacteria, of which the best studied are *Pseudomonas aeruginosa*, *Neisseria gonorrhoeae*, and *Myxococcus xanthus*, where it is referred to as "social gliding motility." Twitching motility appears to be principally a means of rapid colonization by bacterial communities of new surfaces under conditions of high nutrient availability. It occurs on wet surfaces and is important for host colonization and other forms of complex colonial behavior, including the formation of biofilms and fruiting bodies (85, 106, 127, 129). Twitching motility is mediated by type IV pili located at one or both poles of the cell (15, 45, 55) and is distinct from swimming motility (such as in *Escherichia coli* and *P. aeruginosa*), which is mediated by the rotation of unipolar flagella, and from swarming motility (such as in *Proteus mirabilis*), which is mediated by peritrichous flagella, although the term swarming is used somewhat indiscriminately to describe various forms of organized bacterial motility.

Twitching motility appears to be largely if not completely restricted to gram-negative bacteria, mainly the β, γ, and δ subdivisions of the Proteobacteria (44, 77), although there is one report of twitching motility and polar pili in the gram-positive species *Streptococcus sanguis* (46), not yet confirmed by molecular data. Based on a twitching motility phenotype, polar pili, and/or the presence of genes encoding type IV pili, twitching motility also occurs in *Aeromonas hydrophila*, *Azoarcus* spp., *Bacteroides ureolyticus*, *Branhamella catarrhalis*, *Comomonas testosteroni*, *Dichelobacter nodosus*, *Eikenella corrodens*, *Kingella denitrificans*, *K. kingae*, *Legionella pneumophila*, *Moraxella bovis*, *M. lacunata*, *M. nonliquefaciens*, *M. kingii*, *N. meningitidis*, *Pasteurella multocida*, *Pseudomonas stutzeri*, *P. putida*, *P. syringae*, *Ralstonia solanacearum*, *Shewanella putrefaciens*, *Suttonella indologenes*, *Synechocystis* sp. PCC 6803, *Vibrio cholerae*, and *Wolinella* spp., among others, many of which are important pathogens of animals, plants, and fungi (10, 25, 33, 38, 44, 65, 66, 77, 92, 104). Related pili termed type IVB occur in pathogenic *E. coli* and *V. cholerae* (35, 120) and may impart a restricted form of twitching motility (12). Type IV pili are also encoded by the IncI plasmid R64, where they are required for liquid mating (142).

The principal model for the genetic and functional analysis of twitching motility has been *P. aeruginosa*, which has the advantages of being an easily cultured aerobe, with a strong history of genetic and more recently genomic analysis, good host-vector systems for gene cloning, expression, and allelic exchange, and importantly an easily scored macroscopic phenotype for identifying twitching-impaired mutants (5, 105). *P. aeruginosa* is also an important opportunistic pathogen, being the major cause of lung damage in those suffering cystic fibrosis as well as of opportunistic infections in immunocompromised individuals, such as burn victims or those undergoing chemotherapy (69). *P. aeruginosa* is also capable of infecting a wide range of plant and animal hosts. These include model organisms such as mice, fruit fly (*Drosophila melanogaster*), nematode worm (*Caenorhabditis elegans*), and the mustard plant (*Arabidopsis thaliana*) (91), which affords the opportunity of exploring the genetics and genomics of host-pathogen interactions as the next stage in understanding the natural dynamics of bacterial infections. Genes affecting twitching motility have been shown to be important for infection by *P. aeruginosa* (19, 22), as well as for biofilm formation, for which *P. aeruginosa* is also an important model (85, 135) and which appears to be involved in chronic infections (108).

The other well-studied species in relation to the genetics and functional analysis of twitching motility are *M. xanthus* and *N. gonorrhoeae/N. meningitidis*, the former because of its prominence as a model for bacterial social behavior and development (fruiting body formation) (79, 110, 124, 127), and the latter because of their importance as obligate human pathogens and their high natural competence for transformation (34).

The genomic arrangement of the genes that are involved in twitching motility is quite idiosyncratic in different species, as are the regulatory pathways involved (5, 123, 124). At the core, however, are a set of proteins required for the assembly and mechanical function of type IV pili (also called tfp or type 4 fimbriae), a term coined by Ottow (86) in his 1975 review on different types of fimbriae and pili to designate that class of extracellular filaments that are located on the poles of bacteria which exhibit twitching motility.

PHENOTYPIC DESCRIPTION OF TWITCHING MOTILITY

Twitching motility takes place on both organic and inorganic surfaces, including agar gels, epithelial cells, plastics, glass, and metals. It appears to be a means for bacteria to travel in environments with a low water content and/or to colonize hydrated surfaces, as opposed to free living in fluids. At the macroscopic level, the manifestation of twitching motility can be variable depending on the species and culture conditions. In *P. aeruginosa*, mutants incapable of twitching motility produce smooth domed colonies on agar plates, whereas wild-type colonies form flat, spreading colonies with a characteristic rough appearance and a small peripheral twitching zone consisting of a thin layer of cells, and are quite easy to distinguish (78, 105). This fine serrated or "ground glass" edge around flat colonies on agar is characteristic of twitching motility in many species, including *M. xanthus* (124) (see below), although in *N. gonorrhoeae* and *N. meningitidis* twitching zones are only visible microscopically (137), which has hampered genetic screens for twitching motility mutants in these species.

Although small twitching motility zones are frequently discernable at the edge of colonies on the surface of agar plates, it has become evident in recent years that some species, specifically *M. bovis* (80) and *P. aeruginosa* (3, 23), exhibit active twitching motility at the interstitial surface between agar or other nutrient gel and plastic or glass with radial rates of expansion approaching 1 mm/hr, which results in large but fine twitching zones (halos) approaching 2–3 cm in diameter after overnight growth (105) (Figure 1*a*, see color insert). This is not a function of the agar-plastic/glass interface per se, or of a microaerophilic environment, but rather the smoothness of the surface, as similar fast rates of twitching zone expansion are observed on the surface if the agar is inverted, which implies that the normal air-dried surface of agar in plates is an inhibiting factor, presumably due to irregularities in the surface that do not occur when the gel is solidified against or in intimate contact with a smooth solid surface (105). The zone may be more easily visualized by staining, for example with Coomassie Blue, and is completely absent in twitching-deficient mutants (3). The size of the zone is also an excellent semi-quantitative measure of the activity of twitching motility in twitching-impaired mutants, which has led to the identification of a number of genes that have more subtle effects on twitching motility (see below). The twitching zone also shows characteristic concentric rings, reflecting periodic rounds of twitching-mediated colony expansion, wherein the expression of type IV pili and twitching activity is restricted to the outermost zone (80, 105) (Figure 1).

Twitching motility is colonial in nature and usually involves cell-cell contact. Isolated cells rarely move, except when within a certain distance (up to several μm) of others that corresponds to the length of type IV pili (83, 105, 124). Using time-lapse video microscopy at a gel-glass interface in vitro, twitching motility in *P. aeruginosa* occurs initially via the movement outward from the colony center of rafts or spearhead-like clusters of aggregated cells, typically 10–50 cells in width (105) (Figure 1*b*, see color insert). Such formations are quite characteristic of twitching motility in different species (15, 43) and are also observed in social gliding motility in *M. xanthus* (124). The rafts generally move radially outward following the long axis of the cells, which are highly aligned in tight cell-cell contact, although the rafts do meander and individual cells within them may reverse direction. Behind these advancing rafts, groups of cells appear to stretch out and break up into smaller aggregates that move off in different directions and cross-connect with other groups, accompanied by frequent reversals in cell movement (Figure 1). Cells from one group will join into another only when in close proximity. Such cells at first move end on toward the other cells until they touch another cell with their pole, then rapidly snap into an aligned position (105), which accounts for the characteristic jerky "twitching" motion observed with this form of motility (15, 43, 45). This process ultimately leads to the formation of a dynamic lattice-like network of trails (Figure 1*b*) within which cells move rapidly up and down as individual units or small groups, often reversing direction but always following the long axis of the cell (105). Similar trails and cell reversals are also observed in social gliding motility in *M. xanthus* and may involve the production of extracellular slime that contains lubricating properties or developmental cues (106). Once a mature lattice has been formed, the cells appear to settle and to form a three-dimensional microcolony structure that resembles a biofilm (105) (Figure 1*b*).

Although twitching motility is primarily a social activity, it has also been demonstrated that individual cells can show limited movement by twitching motility when in contact with inert substrata such as glass, quartz, or polystyrene (83, 109, 117), or on agar at low concentrations (117). Cells are capable of crawling over surfaces by this method either to aggregate into microcolonies or to move out of microcolonies (83, 85). It is important to appreciate that twitching motility can be employed both to bring cells together into complex structures such as biofilms or fruiting bodies under conditions of nutrient depletion, and to promote rapid expeditionary movement of colonies over new surfaces during vegetative growth under nutrient-rich conditions (85, 105, 107, 127), processes which are regulated by complex signal transduction systems (5, 124, 127, 128) (see below).

There has been debate for many years about whether twitching motility and gliding motility are equivalent. This debate has been confused by a number of issues, including the fact that gliding motility, at least in *M. xanthus*, consists of two components, adventurous and social, which are required for coordinated cell movement and aggregation into fruiting bodies (124). There are also clearly different forms of "gliding" motility that occur in different species (79). Adventurous gliding motility in *M. xanthus* involves more than 37 genes and is cell contact independent (110, 124). The mechanism is unknown (79, 110). Recently, however, it has been shown that *M. xanthus* and cyanobacteria possess small polar nozzles that secrete a polyelectrolyte slime from the rear of the cell and provide forward jet propulsion (139). Such systems may also underlie the gliding observed in some other bacteria (79, 110, 139). On the other hand, social gliding motility has many similarities to twitching motility in *P. aeruginosa*, not just phenotypically, but in its absolute dependence on type IV pili (105, 124), although the regulation of the system may be somewhat different. The issue is further complicated by the fact that social gliding motility in *M. xanthus* is also at least partly dependent on peritrichous carbohydrate-protein fibrils that link cells together and facilitate the cohesive movement ("swarming") of large groups of cells (58, 107, 110, 124, 139).

In his 1972 review on different types of bacterial motility, Henrichsen (43) had concluded that twitching motility and gliding motility were distinctive processes based on observations that the former involved small intermittent jerks that often change the direction of cell movement, whereas the latter involved smooth organized motions of cells directed along the long axis of the cell. This distinction has been repeated in some recent reviews on social gliding motility, albeit also acknowledging that social gliding motility does represent a variation of twitching motility (110, 124). Nonetheless it is now clear from studies such as that described above (105) that twitching motility and social gliding motility are mechanistically equivalent and that the apparent differences in the details of twitching motility and social gliding motility mainly reflect artifactual differences observed in vitro owing to suboptimal or unnatural observational conditions and/or species-specific idiosyncrasies in pili structure or the regulation and communal integration of the process. It is also clear that although twitching motility can be relatively aimless in isolated cells (83), the process is highly organized within a community of cells that are twitching together, which appears to be the normal state of affairs (50, 56, 105, 107).

Perhaps the best description of the process as it might occur in a natural growth setting was given by Henrichsen (43), who stated in reference to gliding motility in fruiting and nonfruiting myxobacteria, such as *Myxococcus* and *Cytophaga* spp., that "under conditions optimal for gliding, the colonies will be seen as 'completely flat, rapidly spreading almost invisible swarms' or as a spreading, rhizoid growth with a honeycomb appearance. Movement takes place mainly in 'spearheads' (i.e., spearhead-shaped cell aggregates at the edge of the colony), single isolated cells very rarely being motile, and the picture is one of a 'changing dispersed border' with interlacing bands being continuously rearranged.... The cells are arranged in a loose pattern of interlacing bands of rafts and cells.... Groups of cells resembling spearheads are seen projecting outward. The locomotion, which is principally seen in the groups of cells, i.e., rafts and spearheads, and takes the direction of the longitudinal axis of the bacteria, gives rise to a constantly changing picture, steadily gliding groups of cells uniting or dividing...." This description also fits well that of twitching motility under optimal culture conditions in *P. aeruginosa* (105) (Figure 1*b*), wherein there is no evidence that adventurous gliding motility can occur or that there are cohesive fibrils produced, although either remains possible.

It therefore seems that the historical distinction between social gliding motility and twitching motility is now largely semantic, as has been suggested by others (79). This holds true despite the fact that twitching motility may have somewhat different manifestations and functional roles under different environmental conditions, such as in nutrient-rich or -poor conditions, and in different species, for example in *N. gonorrhoeae* wherein twitching motility is apparently not controlled by a chemosensory system, unlike *M. xanthus* and *P. aeruginosa* (79). Gliding motility in the cyanobacterium *Synechocystis* has also been shown to be dependent on type IV pili (10, 143). Therefore, given that the term "gliding motility" is used broadly and somewhat ambiguously, whereas twitching motility is not, it may be time to either settle on the latter to refer to that motility mediated by type IV pili, or perhaps to invent a new term, such as "retractile motility" (see below), that is more neutral and descriptively accurate.

MECHANISM OF TWITCHING MOTILITY

The motive force for twitching motility is pili retraction. This was first deduced by Bradley on the basis of electron micrographic studies of twitching mutants and of the binding of pilus-specific bacteriophages [see (15) and references therein]. Bradley showed that phage-resistant mutants of *P. aeruginosa* fell into two categories, those lacking pili and those that were hyperpiliated, both of which were also defective in twitching motility. In wild-type cells these phages were located on the cell surface at the junction of the pili and the cell pole. Phages did not bind to the cell surface in nonpiliated mutants. Importantly, in the hyperpiliated mutants, phages were found not at the cell surface but bound at random points along the pili, whose average length was twice that observed in the wild type. Bacteriophage infection and twitching motility could also be inhibited by treatment with pilus-specific antibodies, which also resulted in a large increase in the number of visible pili per cell. On this basis, Bradley (15) concluded that pili retraction was required for both phage infectivity and twitching motility. The hyperpiliated mutants were later shown to have suffered small deletions in a gene termed *pilT* (133, 134), whose product is a NTP-binding protein required for pili retraction (83, 109, 117, 133) probably by filament disassembly (74, 138) (see below). Interestingly, while mutations in genes that are absolutely required for pili assembly or retraction result in phage resistance, a number of other twitching motility mutants retain sensitivity to (some) pilus-specific bacteriophages (23, 78, 134), suggesting that in these cases pili can be formed but their function is aberrant.

Confirmation that pili retraction is the basis of twitching motility has been recently made by three elegant studies in *N. gonorrhoeae* (83), *M. xanthus* (117), and *P. aeruginosa* (109) bound to various surfaces. In *N. gonorrhoeae*, optical tweezers were used to confirm and measure the retractile force generated by pili on isolated cells positioned near a microcolony or on cells tethered to latex beads coated with anti-gonococcal antibodies positioned near smaller latex beads coated with anti-pilus antibodies. In both cases cells were pulled toward a nearby microcolony or the bead pulled toward a tethered cell at speeds of around 1 μm/sec, similar to the rate measured for twitching motility–mediated movement, and with forces ranging up to 90 pN, which showed peaks in displacement histograms, possibly representing multiples of the force generated by the retraction of individual pili. Retraction events were sporadic, separated by 1–20 sec. No tethering or movement was observed in nonpiliated mutants, whereas *pilT* mutants could form static tethers but were unable to generate retractile forces (83).

Studies of fluorescently labeled pili in *P. aeruginosa* showed that pili extend and retract at approximately 0.5 μm/sec (109), similar to the rates of twitching motility observed in culture (105). Moreover, individual pili on the same cell appeared to extend and retract independently. Extension per se was not associated with cell movement, presumptively because the pili are too flexible to push cells forward, but cells were moved by retraction of pili after they had attached to the substratum at their distal tip. The force exerted by retraction of individual pili was estimated to be approximately 10pN (109).

Similar observations have been made in *M. xanthus* tethered to polystyrene or glass covered with methylcellulose, where rates of twitching motility of around 0.4 μm/sec were observed (117). The *M. xanthus* strain studied was defective in adventurous gliding motility to remove the complication of this form of motility in the analysis. Tethering to these surfaces was again dependent on pili and associated with a jiggling motion absent in *pilT* mutants. The number of tethered cells was higher in *pilT* mutants than in the wild type, indicating that retraction is accompanied by subsequent release of the tethering (117), as was also observed in *N. gonorrhoeae* (83). Tethered wild-type cells appeared to be retracted toward the surface via one pole, then to lie down parallel with the surface and be pulled forward (away from the initial attachment site) by extrusion of pili from the other pole and then retraction after tethering of these pili to the surface at their distal end. This was proposed to be the means by which cells in advancing rafts can move forward over new surfaces during twitching/social gliding motility (117). In addition, these studies indicated that movement reversals involve alternating the activity of pili from one cell pole to the other, the frequency of which was correlated with attachment time, and which is controlled by the *frz* chemosensory system (see below).

This model explains how cells might move forward to colonize new surfaces by twitching motility. However, as noted above, twitching motility is normally a social activity and involves cells moving together with others in rafts and trails during expeditionary expansion as well as coming together to form microcolonies in biofilms or fruiting bodies under nutrient-depleted conditions. This requires contact and the retraction of pili to bring cells into close alignment, and presumably involves specific recognition and possibly sensing by the pili of receptors on the surface of neighboring cells, although such receptors have not been identified. However, social gliding motility in *M. xanthus* is dependent on the product of a gene called *tgl* (transient gliding), which can stimulate pili extrusion on other cells in trans (125). A similar protein (PilF) is also required for twitching motility in *P. aeruginosa* (105, 129). Mutants that are *tgl− pilA+* do not express pili nor show motility, but they can be induced to do both by contact with other cells that are *tgl+ pilA−*, although the latter themselves cannot produce pili or twitch (55, 125). The *tgl* gene product has a type II (lipoprotein) signal sequence, followed by six tandem degenerate 34–amino acid repeats, which are involved in protein-protein interactions (125). Tgl stimulation of pili appears to involve end-to-end (polar) interactions between donor and recipient cells (123). These results also suggest that the regulation of the activity of twitching motility, as opposed to the direction, is (at least partly) at the level of pili assembly/extrusion, which fits well with results obtained from other regulatory mutants (see below), and raises the possibility that although pili retraction provides the force for twitching motility, it may occur somewhat reflexively. Retraction of individual pili appears to occur independently of whether pili are attached in *P. aeruginosa* (109), although this does not preclude sensing and triggering of the process by an initial attachment event, perhaps signaled through tensional or torsional stress on the filament.

The mechanism of pili retraction is thought to be filament disassembly mediated by PilT (49, 56, 74), a process that has been estimated to occur at around 1000 pilin subunits/sec (83). PilT is a nucleotide-binding protein that is homologous to another (PilB) required for the assembly of pili, but which has the opposite effect (133). PilT is unique to type IV pili, whereas PilB homologs exist in other systems. PilT mutants are unable to retract their pili, leading to hyperpiliation (or ingrown pili if the PilQ-export pore is blocked) and loss of twitching motility (15, 133, 138). PilT is also required for pilin degradation (138). Because retracted pili are normally broken down into subunits, which may be recycled into the assembly of new pili (109), these observations suggest that the normal role of PilT is to provide the means for the disassembly of pili filaments into subunits and that this process underlies the observed retraction of these structures. In addition, because the filament is a helical polymer (see below), the assembly/disassembly process must occur either by rotation of the point of subunit insertion or removal (with a rotary motor driving these processes), or the filament itself must rotate as a consequence of pilin insertion or removal from a fixed point. If the latter, the pilus may operate as a screw ratchet, which could create torsional stress on attachment, as well as being capable of transporting attached macromolecules (DNA and protein) into and out of the cell (74) (see below). On the other hand, as a member of the AAA family of motor proteins, it has also been suggested that PilT may form a hexameric ATPase surrounding the base of the pilus, which would provide an axial rotary power stroke analogous to that provided by F_1 ATPase (56).

P. aeruginosa, *P. stutzeri*, and *N. gonorrhoeae* also contain a gene directly downstream of *pilT* called *pilU* (38, 134, 138), which when mutated also leads to hyperpiliation and loss of visible twitching motility, but not phage sensitivity (38, 134) or ingrown pili in *pilQ* mutants (138). This suggests that loss of PilQ may impair but not prevent PilT-mediated pilus disassembly, and reciprocally that the PilU may cooperate to enhance, but is not absolutely required for, the function of PilT.

ADHESION AND TARGET SPECIFICITY OF TYPE IV PILI

Type IV pili can bind to a variety of surfaces, including inert surfaces, other bacterial cells, and eukaryotic cells, where they can mediate both colonization of the surface and intimate contact through pili retraction (82, 117). The attachment of pili to inert surfaces appears to occur via relatively nonspecific adhesion at their tip (109, 117). Pili on cells are always observed to attach to surfaces at their distal end, and broken (free) pili also only attach via an end (109).

Pili appear to bind via their tip to specific receptors on mammalian epithelial cells and other cell types (39, 82). In *P. aeruginosa* pilin a C-terminal disulfide-bonded region of 12–17 semiconserved amino acid residues, which is otherwise buried within the filament (see below), is exposed at the tip of the pilus and binds to the carbohydrate moiety of the glycosphingolipids asialo-GM1 and asialo-GM2 on epithelial cells (39, 41, 64). This C-terminal disulfide-bonded loop is also partly conserved in many other type IV piliated bacteria, including *N. gonorrhoeae* and *N. meningitidis* (21, 41, 120).

In *N. gonorrhoeae* and *N. meningitidis* binding of pili to epithelial and endothelial cells requires the product of a gene *pilC* (101, 137), which appears to function as a tip adhesin (95), although it is also found in the cell membrane. Most strains of *N. gonorrhoeae* have two variant copies of *pilC* (*pilC1* and *pilC2*), the expression of at least one of which is required for the appearance of extracellular pili (93, 97), but whose loss can be overcome with a compensating mutation in *pilT* (137). One interpretation of the latter finding is that PilC may be required to cap or stabilize the pili, in the absence of which they are retracted by PilT (137). A homolog of PilC (PilY1) is also found in *P. aeruginosa*, associated with both the cell membrane and extracellular pili, and is also required for the appearance of extracellular pili and twitching motility (2). Soluble pilin has also been found to bind specifically to human but not rodent cells, apparently through the CD46 receptor (99). Variations in the primary structure of the pilin can also alter the binding characteristics of *Neisseria* pili (53), and it appears that different pili may have different binding specificities for different types of cells mediated both by structural features of the pilin itself and/or by the presence of tip-exposed binding sites or tip-associated adhesins, which may be quite specific to the species and its ecology.

STRUCTURE OF TYPE IV PILI

Type IV pili are typically 5–7 nm in diameter and can extend to several μm in length. They are composed primarily of a single small protein subunit, usually termed PilA or pilin, which is arranged in helical conformation with 5 subunits per turn and which may be glycosylated and/or phosphorylated in different species (see below). Pilins from different species are usually 145–160 aa in length and have a highly distinct primary structure, notably a short positively charged leader sequence, and a highly conserved and highly hydrophobic amino-terminal domain (Figure 2, see color insert) with consensus sequence FTLIELMIVVAIIGILAA-IALPAYQDYTARSQ, which forms the core of the pilus fiber. The N-terminal residue of mature pilins is normally a methylated phenylalanine, although there are naturally occurring exceptions, and site-directed mutagenesis has established that a variety of hydrophobic amino acids can be tolerated at this position (114, 115). The leader sequence is cleaved and the resulting N-terminal residue is methylated by a specific leader peptidase (prepilin peptidase or PilD), which also cleaves a wide variety of substrates with prepilin-like leader sequences and type IV amino-terminal domains, including those involved in protein secretion (67, 116) (see below). A variety of studies suggest that the invariant glycine at -1 is required for prepilin cleavage and assembly and that the glutamate at $+5$ is essential for assembly and efficient N-terminal methylation but not cleavage (61, 70, 114, 115).

Beyond the conserved amino-terminal region the sequences of the major pilin diverge, although the region from residues 30–55/56 (which is usually a glycine residue) is semiconserved, as is a region near the carboxy terminus at the distal end containing a disulfide loop that may be involved in recognition of receptors (21, 39, 41, 120). The central and carboxy-terminal two thirds of the protein are relatively hydrophilic, often stabilized by one or two disulfide bridges, and contain the major sites of structural and antigenic variation (77, 120). This variation has been best characterized in *D. nodosus* pilin, wherein there are four major regions of hypervariability as well other variable sites that together define the serogroups and serotypes of this organism (75, 77) and are usually considered to be the result of host-immune pressure. It is interesting, however, that although vaccination with purified pili can protect against infection by related strains of the same *D. nodosus* serogroup, natural infections are immunologically remote and do not even prompt amnestic responses after prior infection or vaccination (27). This suggests that the reason for structural variation in type IV pili, at least for some epithelial pathogens, may not be host-immune pressure, but other factors, perhaps selection pressure by pili-specific bacteriophages, adaptation to different environments/receptors, or simply evolutionary drift. In *M. bovis* and *M. lacunata* the pilin sequence is switched between two forms by an invertible segment that alternates the orientation of an incomplete pilin sequence with respect to an external promoter and conserved N-terminal coding sequences (32). On the other hand, *N. gonorrhoeae* and *N. meningitidis*, which are not remote from the immune system, have active mechanisms for varying their pilin sequence by recombinational exchange of cassettes of pilin sequences from silent loci into the expression locus via pilin-dependent DNA uptake from lysed cells in the population (34, 81, 102, 103). In addition, these species produce large amounts of soluble pilin (S-pilin, cleaved after residue 39) apparently as antigenic decoys (99).

The three-dimensional crystal structure of *N. gonorrhoeae* MS11 pilin was solved in 1995 (87) and showed that pilin adopts a highly asymmetrical three-dimensional α-β roll fold (Figure 2) comprising an 85 Å α-helical spine formed by the conserved hydrophobic amino-terminal region of the protein, a 24–amino acid "sugar loop," a four-stranded anti-parallel β-sheet, and a C-terminal β-sheet and extended loop, which are stabilized by a disulfide bridge and interactions between other key residues, but which also allow domains of high sequence flexibility. Analysis of this structure in conjunction with earlier biophysical measurements carried out on *P. aeruginosa* pili (29, 131) indicates that pilins are packed into a conserved helical structure, composed of 5 subunits per turn, with a 41 Å pitch and an outer diameter of 60 Å (31). The core of the fiber is a parallel, overlapping, coiled-coil composed of the hydrophobic N-terminal α-helices (which penetrate over twice the length of the helical rise). Because this region is a highly conserved feature of type IV pilins and many related proteins, this helical hydrophobic core must be a common feature of the structures formed by these proteins. The outside of the fiber is formed by a highly organized scaffold of β-sheets packed flat against the core, with an outermost layer composed of an exposed hypervariable region, covalently attached saccharide, and an extended C-terminal tail (31). The latter are not integral parts of the fiber structure and can accommodate extreme amino acid changes and even C-terminal epitope additions by protein engineering (52) without disrupting the normal assembly of the fiber. The subsequent publication of the crystal structure of (truncated) *P. aeruginosa* PAK and K122-4 pilins showed that this general structure is conserved, albeit with some structural variations, notably the absence of two β-strands that occur in the sugar loop of MS11 pilin, and other differences in the flexible C-terminal tail (41, 59). However, these studies also suggest that the pilus fiber may have the opposite polarity to that suggested previously (31), i.e., that the N-terminal hydrophobic α-helix is directed upward toward the pilus tip, rather than downward toward the pilus base (41, 59).

Conservation of the general structure of type IV pili is also reflected in their interchangeability. Many type IV pilins may be artificially transferred from one species to another, wherein they are happily assembled into extracellular pili that function relatively normally in twitching motility and phage sensitivity in heterologous hosts (7, 76, 130). However, cells that are induced to coexpress two distinct pilins produce two types of pili, composed exclusively of one or the other pilin type, suggesting either that there is some specificity of interaction between homologous pilins in pilus assembly or that pilin translation and pili assembly are tightly coupled (28). The exchange of subunit genes also appears to have occurred naturally, as some strains of *P. aeruginosa*, *D. nodosus*, and *E. corrodens* have unusual pilins that have almost certainly been derived by lateral transfer from other species (41, 48, 75, 120, 121). Thus, whereas the central and C-terminal regions of pilins can vary substantially, it seems that all type IV pili have the same quaternary structure and a similar tertiary structure, albeit with some primary and secondary structural differences at the external surface of the fiber in different species and strains. In addition, the same core helical structure appears to have been adapted to other purposes in conjunction with quite different external domains (see below).

A number of posttranslational modifications of pilin have been observed. In *N. gonorrhoeae* pilin is phosphorylated at Ser68, which does not appear to affect the function (twitching motility or DNA uptake) or epithelial cell adhesion of the pili, but does modify their surface characteristics, including solubility, aggregation (bundling), and filament topography (curliness) (30). In *N. meningitidis*, and possibly also *N. gonorrhoeae*, pilin is also modified by a novel α-glycerophosphate linkage to Ser93/94 (111), which is exposed toward one end of the pilus fiber (30) and would be consistent with the suggestion that glycerophosphate may be a substrate for lipid attachment (111) either to help anchor the pili in the outer membrane or to modify its surface or binding properties.

N. gonorrhoeae and *N. meningitidis* pili may also be glycosylated at Ser63, via O-linkage to a variety of di- and trisaccharides, the identity of which differs between species and strains (30, 71, 88, 112). Again this modification does not appear to affect overtly the function or cell adhesion properties of the pili, but it does affect the solubility characteristics of the fiber (51, 71). In addition, the terminal Gal α1–3 Gal linkage found on many meningococcal pilins is a target for human anti-gal antibodies that block complement-mediated killing (40). Pilin glycosylation has also been observed in some strains of *P. aeruginosa* such as 1244, which has a trisaccharide substituent containing xylopyranosyl and furanosyl residues, but not in others including the widely studied PAK and PAO strains (17).

ASSEMBLY OF TYPE IV PILI

Genetic analysis in *P. aeruginosa* has also revealed the existence of a cluster of genes encoding a number of minor pilin-like proteins (PilE, PilV, PilW, PilX, FimT, and FimU) that contain the signature hydrophobic amino-terminal α-helical region, and which (except for FimT, which can substitute for FimU) are required for pili assembly, twitching motility, and infection by pilus-specific phage, apparently in correct stoichiometry (2–4, 97). These subunits also each have similarity with particular pilin-type proteins required for protein secretion (2, 4, 96), and at least some (PilV, W, and X) are located exclusively in the cell membrane fraction, suggesting that they may form the base structure for the pilus fiber, similar to the pseudopilus structures that appear to be involved in type II protein secretion (49, 74, 89) (see below). This includes one protein (PilX/GspK) that lacks the normally obligate Glu5 required for subunit assembly (2), which may function as an anchor or terminator of the pilus. However this does not rule out the possibility that one or more of these minor pilins may be incorporated elsewhere in the structure, for example at the tip, where they may be involved in initiation or stabilization of pilus assembly. Although not yet identified by genetic screens,

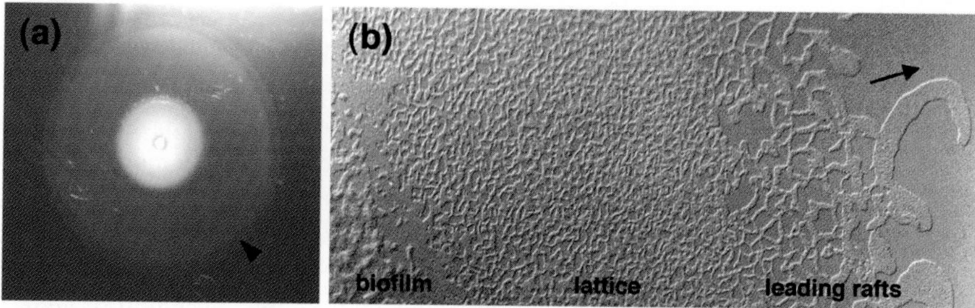

Figure 1 Twitching motility zones. (*a*) Macroscopic view of the twitching zone formed by *P. aeruginosa* at the interface between the agar gel and plastic at the bottom of a petri dish. Note the fine outer zone (indicated by *arrowhead*), which represents the active twitching zone. (*b*) Microscopic view of the twitching zone of *P. aeruginosa*, showing the outermost rafts, lattice-like network, and microcolonies in the biofilm behind the twitching zone. The *arrow* indicates the general direction of twitching movement away from the colony center. The panels in this figure are adapted from (104) and are reproduced with the permission of *Microbiology*.

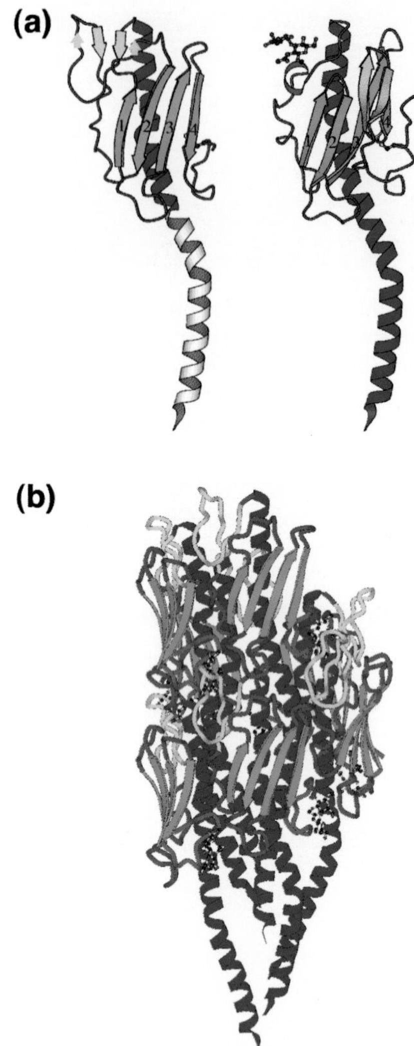

Figure 2 Model of the structure of the type IV pilin monomer and pilus fiber. (*a*) Crystal structures of *P. aeruginosa* PAK and *N. gonorrhoeae* MS11 pilins. (*b*) Fiber model of *P. aeruginosa* PAK pilin. Reproduced from (41) with the permission of the authors and the *Journal of Molecular Biology*.

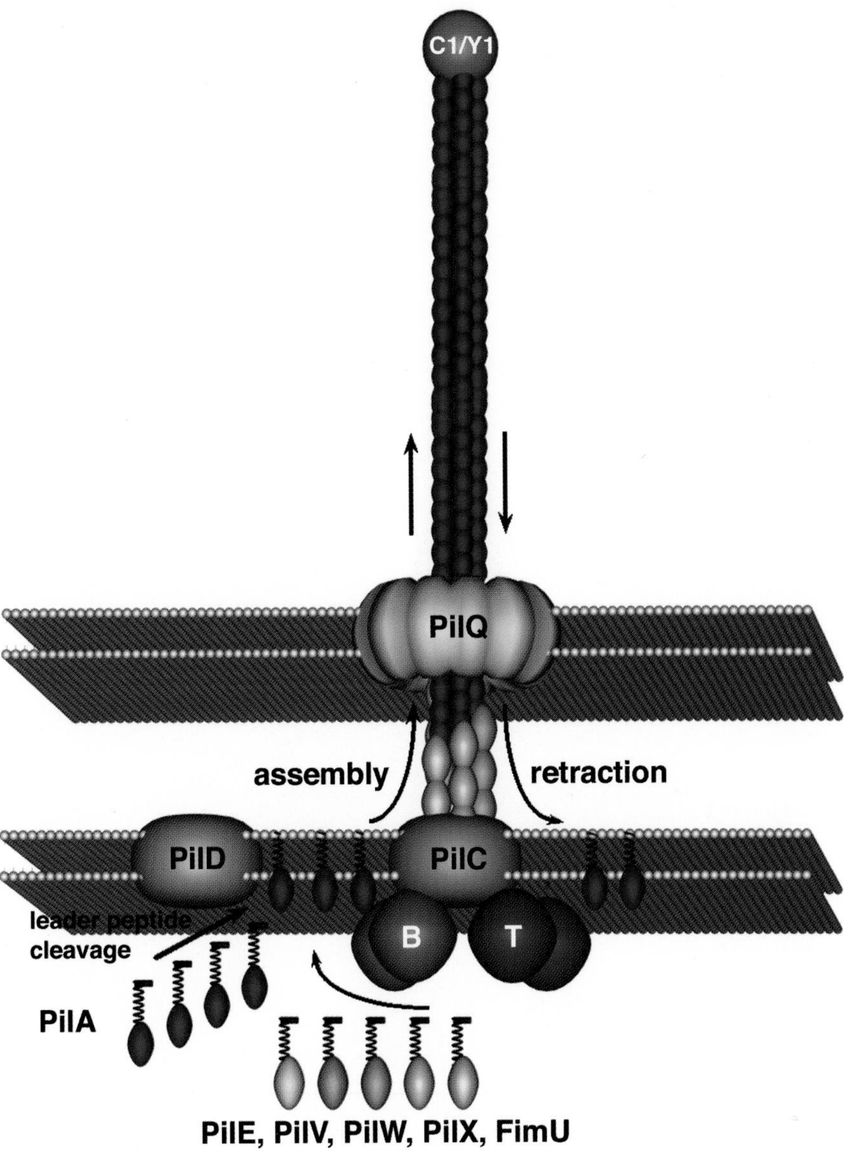

Figure 3 Model of type IV pili assembly and retraction, adapted from (49, 56, 73, 137). Prepilin leader sequences are cleaved (and the pilin N-methylated) by PilD. Processed PilA is assembled on a base of minor pilins (PilE, V, W, X, and FimU) by the action of the cytoplasmic membrane protein PilC and the NTP-binding protein PilB, and the pilus extruded through the outer membrane via a pore composed of multimeric PilQ, stabilized by the lipoprotein PilP. Pili are retracted by the action of PilT (aided by PilU).

analogous proteins probably occur in other type IV piliated species; for example, putative PilV homologs have been annotated in the genomes of *N. meningitidis* and *R. solanacearum*. It is also interesting and perhaps significant that the gene *pilY1*, which encodes a homolog of the *Neisseria* tip adhesin PilC (and another gene *pilY2*, which encodes a small protein that appears to be associated with extracellular pili), is located within the same cluster as those encoding the minor pilins in *P. aeruginosa*, between *pilX* and *pilE* (2).

In all species examined to date, the assembly of pili also requires a nucleotide-binding protein (PilB) similar to PilT, a polytopic inner membrane protein (PilC, called PilF in *N. gonorrhoeae*), the prepilin peptidase and methylase (PilD) discussed earlier, and a multimeric outer membrane protein (PilQ) that is a member of a large family of proteins that form gated pores in the outer membrane, through which the pilus is thought to extrude (5, 49, 62, 67, 72, 84, 96, 116) (Figure 3, see color insert). Analysis of the quaternary structure of PilQ by electron microscopy shows that this protein forms a dodecameric doughnut-shaped complex whose internal cavity diameter (53 Å and 65 Å) closely matches that of the diameter of the corresponding pilus (52 Å and 60 Å) in *P. aeruginosa* (13, 59) and *N. meningitidis* (18). PilQ homologs are also involved in the extrusion of filamentous phage (96), whose diameter resembles type IV pili, and in type II protein secretion where they also form multimeric pores, albeit of larger internal diameter (13) perhaps consistent with the wider diameter of the pseudopilus formed in the latter system (100) (see below).

In *P. aeruginosa* the *pilQ* gene is located in an operon that includes four other genes (*pilMNOPQ*) also required for pili assembly, twitching motility, and phage sensitivity (73). *M. xanthus*, *N. gonorrhoeae*, and other type IV piliated bacteria also have similar clusters with the same gene order (26, 73, 124, 143). Mutations in *pilM* are dominant negative, indicating that PilM, like PilQ, exists as a multimeric complex (73). PilM contains motifs conserved in the ATP-binding domain of actin and shows homology with the rod shape–determining protein MreB from *E. coli* as well as the *Bacillus subtilis* and *E. coli* cell division protein FtsA, which suggests that the assembly of type IV pili may be linked to cell shape and possibly directed through PilM to the cell pole (73).

In this context it is worth noting that a penicillin-binding protein gene is located immediately upstream of *pilM* in *P. aeruginosa* and another is located within the same operon as the pilin gene in *D. nodosus*, suggesting that pili assembly may also be linked with local cell wall remodeling (73). In addition, pili assembly and twitching motility also require the product of a gene *fimV*, which encodes a highly acidic protein that is probably located in the cytoplasmic membrane and which contains a putative peptidoglycan-binding domain (104). The function of FimV is dose-sensitive and overexpression causes dramatic elongation of the cells. Homologs of FimV have only thus far been found in other type IV piliated bacteria, with the strongest region of homology being around the peptidoglycan-binding domain (104). A link to the cell cycle is also indicated by the fact that pili assembly also requires a gene, *pilZ*, which is transcriptionally coupled to *holB*, which encodes the δ subunit of DNA polymerase III (1).

PilN and PilO encode proteins with long N-terminal stretches of hydrophobic residues that may act as cytoplasmic membrane–anchoring domains. PilN homologs are also required for type II protein secretion. PilP encodes a lipoprotein that appears to be required to stabilize PilQ multimer formation (26, 73). Interestingly in *N. gonorrhoeae*, both *pilP* and *pilQ* mutants shed the tip adhesin PilC to the medium in a soluble form (26).

It is now clear that type IV pili are a subset of a much larger and more widely distributed system of cell surface complexes that are involved in macromolecular transport in both gram-negative and gram-positive bacteria (49, 74, 133). There are multiple sets of homologies between the assembly proteins PilB, PilC, PilD, PilN, PilQ, the five minor pilins (PilE, PilV, PilW, PilX and FimU), and similar proteins involved in type II protein secretion in a wide variety of bacteria (49, 74, 89, 96). In *P. aeruginosa* PilD (also known as XcpA) is required to process the leader sequence from prepilin-like proteins for both pili assembly and protein secretion (67, 116). A similar set of proteins is also required for competence for DNA uptake in *B. subtilis* (49), which lacks PilQ, as gram-positive bacteria do not have an outer membrane. Interestingly, type IV pili are themselves required for DNA uptake in many bacteria, including *N. gonorrhoeae* (see below). PilA is also required for optimal protein secretion in *P. aeruginosa* and may interact with the pseudopilin proteins involved (68). Taken together, this suggests that these components form similar complexes at the cell surface, which have a common evolutionary origin and core architecture, but which have been adapted to different functions.

Since all of these complexes involve components that contain the conserved N-terminal hydrophobic domain, it seems likely that they are all formed of helical pseudopilus-like structures (49, 89) whose assembly is driven or at least controlled by PilB-like nucleotide-binding proteins (74). This was recently confirmed by overexpression in *E. coli* of the protein secretion pseudopilin PulG (together with other assembly proteins), which resulted in the production of extracellular pili-like structures (100). However, there is some debate about exactly how such structures might mediate transport of (a wide variety) of globular proteins. However, in the case of twitching motility it seems that the base complex has been modified by the addition of PilA, which is overexpressed and extruded as a filament and can be retracted by the antagonistic action of PilT/U (74) (Figure 3). Thus PilA is not a substrate of the secretion system in the normal sense but rather an infrastructural component. PilB and PilT appear to be associated with both the cytoplasmic and inner membrane fractions, perhaps extending into the periplasmic space (16, 89). The assembly/disassembly process must occur from the base, as the hydrophobic α-helical core of the type IV pilus is too tightly packed to allow macromolecular transport through the pilus (31, 59).

REGULATION OF TWITCHING MOTILITY

In *P. aeruginosa* almost 40 genes have now been identified whose products are required for twitching motility. Apart from those involved in the biogenesis and mechanical function of pili (*pilA, B, C, D, E, F, M, N, O, P, Q, T, U, V, W, X, Y1, Y2, Z,* and *fimT, U, V*), which have been discussed above, the others (*pilG, H, I, K, L-chpA, B, C, D, E, pilS, R, rpoN, fimS, algR, algU,* and *vfr*) are involved in transcriptional regulation and chemosensory pathways that control the expression or activity of the system[1] (5, 8).

Transcription of PilA in *P. aeruginosa* is tightly controlled by a classical RpoN-dependent, two-component sensor-regulator pair, PilS/PilR (47, 116). PilS is a transmembrane protein located at the pole of the cell, but the signal to which it responds is as yet unknown (14). Twitching motility in *P. aeruginosa* is also controlled by an atypical sensor-regulator pair FimS/AlgR (132), mutants of which show limited raft production at the colony edge but no development of the lattice network typical of normal twitching motility. AlgR, but not FimS, also regulates the production of the exopolysaccharide alginate in mucoid strains (132). AlgR appears to be a DNA-binding protein that interacts with the sigma factor AlgU, overexpression of which can also restore twitching motility in *fimS* mutants. FimS and AlgR do not affect PilA expression and so must be regulating some other aspect of the system, as both *fimS* and *algR* mutants lack extracellular pili. FimS appears to be incapable of autophosphorylation, and so is presumed either to be phosphorylated by an upstream signaling pathway, or to be capable of dephosphorylating AlgR, which is phosphorylated by another pathway (132). AlgR can be phosphorylated in vitro by histidine kinases such as CheA (see below) and by small-molecular-weight phosphate donors, such as carbamoyl phosphate and acetyl phosphate. In addition it has been recently discovered that twitching motility is also controlled by Vfr, the *P. aeruginosa* homolog of the catabolite repressor protein Crp (8). Vfr may bind both cAMP and cGMP (8), and there is evidence that *P. aeruginosa* widely utilizes guanine nucleotides for global physiological regulation (20).

In *M. xanthus*, genes encoding two sets of sensor-regulator pairs *pilS/pilR* and *pilS2/pilR2* are located in the *pil* gene cluster. PilA expression requires PilR, but not PilS, which appears to function as a negative rather than a positive regulator (140). PilA expression is also autoregulated (140), and there is some evidence that this may also occur in *P. aeruginosa* (28), which might be expected if cells are to control the size of the pool of pilins in the cell membrane. PilS2 is not required for twitching motility and may also be a negative regulator. The role of PilR2 is currently unknown (124). In *M. xanthus* motility also requires the products of a cluster of three genes encoding ABC transporters (*pilG, H,* and *I*), which are located within the main *pil* gene cluster (124), as well as that of a gene (*frzS*) that is located near the *frz* gene cluster (see below) and encodes a protein with a N-terminal response regulator domain connected to a C-terminal repeat domain that appears to form a long coiled-coil structure, similar to those found in fibrous proteins such as myosin (79, 126). Homologs of these proteins have not yet been identified, functionally or genomically, in other type IV piliated bacteria.

[1]C.B. Whitchurch, M.D. Young, H.D. Kennedy, A.J. Leech, J.L. Sargent, A.B.T. Semmler, S.A. Beatson, J.C. Commolli, L. Nguyen, J.N. Engel, P.R. Martin, A.A. Mellick, R.A. Alm, M. Hobbs, A.A. Darzins & J.S. Mattick, manuscript in preparation.

In both *M. xanthus* and *P. aeruginosa* twitching motility is also controlled by chemosensory phosphotransfer signal transduction systems (*frz* and *chp*) that are similar to, but substantially more complex than, the chemotactic (*che*) system, which controls flagella rotation in swimming motility in a variety of bacteria (5, 124, 127, 128). This system operates via methyl-accepting chemotaxis proteins (MCPs), which can induce autophosphorylation of a central histidine kinase (CheA) which in turn phosphorylates a response regulator (CheY) that reverses the direction of rotation of the flagella motor. The other proteins involved (CheW, CheR, CheB, and CheZ) act as adaptors between the MCPs and CheA, or are involved in modulating the methylation state of the MCPs and in the re-setting (dephosphorylation) of CheY. Interestingly, *P. aeruginosa* actually has four complete chemosensory systems encoded in its genome (20), one of which controls swimming motility, one of which controls twitching motility, and two of which have unknown function.

The chemosensory system that controls pili-dependent motility was first discovered in *M. xanthus* and is termed the *frizzy* (*frz*) system because of the phenotype of mutant colonies. A second chemosensory system encoded by the *dif* genes appears to control the expression of fibrils that play a role in other aspects of social motility (110, 124, 141). Homologs of all the Che proteins (except CheZ) are encoded by the *frz* cluster, with the additional complexity that there are three CheY domains, two of which are located in one protein (FrzZ) and one of which is located at the C terminus of the CheA homolog (FrzE). The *frz* system controls the frequency of cell reversal and therefore the direction and pattern of twitching motility, both in normal vegetative growth and during fruiting body formation (127). Most mutants in *frz* genes are impaired in cell reversal, but gain-of-function mutations in *frzCD* (which encodes an MCP homolog) increase the frequency of cell reversal (127). FrzCD is a soluble cytoplasmic protein, unlike other members of the methylated chemotaxis protein family, which are integral membrane proteins. FrzCD is also methylated during fruiting body formation and vegetative growth when cells undergo directed movement, and demethylated when cells are exposed to repellants (107).

The *chp* gene cluster in *P. aeruginosa* is similar to the *frz* cluster in that it also contains three CheY-like response receiver domains, two encoded by adjacent genes (*pilG* and *pilH*, which are probably equivalent to *frzZ*) and another at the C terminus of the CheA homolog (*pilL*/*chpA*, which were previously reported to encode two separate proteins but in fact encode one large protein of almost 2500 amino acids)[1] (5, 23, 24). This protein (henceforth referred to as ChpA) is far more complicated than its *M. xanthus* counterpart, as it contains not one but seven histidine phosphotransfer (Hpt) domains, one of which contains a serine rather than a histidine at the active site of the Hpt domain[1] (5). Mutations affecting PilG (CheY), PilI (CheW), and PilJ (MCP) are all defective in twitching motility and extracellular pili (although the cells remain phage sensitive), whereas those affecting PilH (the second CheY) show an aberrant pattern of twitching motility reminiscent perhaps of the *frzD* alleles of the dual CheY protein FrzZ) (23, 24). Different mutations in *chpA* show different phenotypes from "frizzy" colony morphology to a complete lack of overt twitching motility,[1] and site-directed mutagenesis has shown that some but not others of the Hpt domains in ChpA, as well as the C-terminal CheY-like domain, are required for twitching motility (A.J. Leech & J.S. Mattick, unpublished results). Like *frz* mutants, *chp* mutants also appear to control the speed and frequency of cell reversal in twitching motility.[1]

ChpA appears to be the most sophisticated signal transduction protein yet defined in nature[1] (5). The fact that it has multiple Hpt domains suggests that this protein is receiving and integrating multiple inputs and/or transmitting multiple outputs, which may involve not just the three CheY domains, but also other systems including conventional two-component sensor-regulator systems that also contain Hpt transmitter and/or CheY-like receiver domains. Indeed some two-component signal transduction systems have been shown to be part of larger sensory relays that involve multiple His-Asp-His-Asp phosphotransfer cascades [see (128)]. One of the targets of the Chp system may well be FimS or AlgR, as the phenotypes of some Chp mutants are similar.[1] In addition, *chpA* mutants show defects in expression of a range of virulence factors, suggesting that this protein is involved in the complex regulation and integration of a range of cellular responses during surface and host colonization[1] (22).

The environmental signals that control twitching motility are not well understood. There is evidence that twitching motility is influenced by nutritional status, including certain amino acids, as well as by cell density, and may also involve self-generated soluble and cell contact-dependent intercellular signals during vegetative swarming and microcolony/fruiting body development (107, 124, 127, 128). A recent report suggests that twitching motility in *P. aeruginosa* is responsive to phosphatidylethanolamine (PE), possibly chemokinetically rather than chemotactically (57). PE has also been shown to enhance social gliding motility in *M. xanthus*, but in this case at least the effect appears to be transduced through the fibrils (58).

It has also been reported that twitching motility in *P. aeruginosa* is dependent on quorum sensing by the *las* and *rhl* systems (36). However, this is not the case, as *lasI*, *lasR*, *rhlI*, and *rhlR* mutants all exhibit apparently normal twitching motility (9). However, it also seems that quorum-sensing mutants may be susceptible to compensatory second site mutations in genes encoding other regulators that (among other things) are required for twitching motility, specifically *algR* in the case of *rhlI* and *vfr* in the case of *lasI*, the latter of which result in a specific microdeletion in the cyclic nucleotide-binding domain (8, 9).

P. aeruginosa and *M. xanthus* are free-living bacteria in which twitching motility is responsive to a variety of environmental and nutritional signals, whereas other species live in different environments and, not surprisingly, regulate twitching motility differently. In *Synechocystis* sp. PCC 6803, light has been shown to control motility and production of extracellular type IV pili via a phototactic gene cluster where signaling is linked to chromophore-binding photoreceptor domains (11). In contrast, *N. gonorrhoeae* and *N. meningitidis* are obligate pathogens and do not have any chemosensory systems encoded in their genomes. Not a great deal is known about how twitching motility is regulated in these organisms, other than the fact that pilin (called PilE in these species) and PilC expression is influenced by two genes, *pilB* and *pilA*, that appear to encode an unusual two-component regulatory system responsive to GTP (6, 118, 119). Pilin gene expression is also subject to phase variation and exchange of cassettes of variable pilin sequences, presumably to avoid host immune responses (34, 102). *pilC* is also subject to phase variation (54).

FUNCTIONS OF TWITCHING MOTILITY

Twitching motility is involved in rapid expeditionary colonization of surfaces during vegetative growth as well as in the formation of complex colonial structures in biofilms and fruiting bodies (79, 85, 105, 124, 127). Apart from cell movement, functional type IV pili are also required for a wide variety of other processes, including transformation (see below), conjugation (142), and bacteriophage infection (15, 130, 134).

Twitching motility is also required for host colonization and pathogenesis, including the activation of host cell responses (82). *pilA* (pilin-deficient) mutants are not pathogenic (or have reduced virulence, depending on the model examined) (19, 42), and vaccination with purified pili can protect against infection with serologically related strains (27, 120). However, this is not simply a function of pilus-mediated adhesion to host cells, although this is important in its own right (101), but also of twitching motility (pilus retraction). For example, *pilT* mutants of *N. gonorrhoeae* are unable to make intimate contact with, or form attaching effacing lesions on, epithelial cells (90). *P. aeruginosa pilT* mutants are not infective in corneal tissue (42) and exhibit reduced cytotoxicity to epithelial cells in culture (19). A recent screen for *P. aeruginosa* mutants that are impaired in virulence in *Drosophila* showed that all strains that were strongly impaired in fly killing also lacked twitching motility, with the majority of mutations occurring in the *pilGHIJKL-chpABCDE* chemosensory gene cluster, although this appeared to be a consequence of *chp* control of other virulence factors as other (apparently) nontwitching variants had normal virulence (22). In addition, *pilQ* and *pilT* mutants of the phytopathogen *R. solanacearum* cause slower and less severe wilting on tomato plants (66).

Finally, in many bacteria, notably the *Neisseriaceae* but also including *D. nodosus*, *E. corrodens*, *L. pneumophila*, *P. stutzeri*, and *Synechocystis* sp. PCC 6803, functional type IV pili are required for competence for DNA transformation (34, 37, 60, 94, 113, 122, 143). PilA, PilC, PilQ, and importantly PilT are all required for this process (38, 94, 98, 136). *pilU* mutants are severely impaired but not lacking in transformation (38). Interestingly, competence for DNA transformation in *pilA* mutants of *P. stutzeri* can be restored by complementation with a heterologous PilA subunit from *P. aeruginosa* (37), which is itself not naturally competent. There is insufficient space within the core of the pilus filament for nucleic acids to pass through (31, 59), so the mechanism by which type IV pili mediate natural transformation must involve another process, which is still unknown. However, it has been suggested that minimal pilin assembly structures and/or pili retraction may allow the PilQ pore to open for DNA entry (38).

Thus twitching motility is involved in various aspects of bacterial biology, based around a core process of assembly and retraction of type IV pili. This system has been modified and adapted in surprisingly sophisticated ways and underlies a number of complex behaviors in bacteria. However, much remains to be learned about how the process is regulated in different organisms and in different environmental conditions.

ACKNOWLEDGMENTS

I thank the members of my laboratory and my colleagues in the type IV pili/twitching community for many valuable discussions and insights over many years. My sincere apologies are extended to those authors whose primary work has only been quoted indirectly through reference to recent reviews because of space constraints on this article.

The *Annual Review of Microbiology* is online at http://micro.annualreviews.org

LITERATURE CITED

1. Alm RA, Bodero AJ, Free PD, Mattick JS. 1996. Identification of a novel gene, *pilZ*, essential for type 4 fimbrial biogenesis in *Pseudomonas aeruginosa*. *J. Bacteriol.* 178:46–53
2. Alm RA, Hallinan JP, Watson AA, Mattick JS. 1996. Fimbrial biogenesis genes of *Pseudomonas aeruginosa*: *pilW* and *pilX* increase the similarity of type 4 fimbriae to the GSP protein-secretion systems and *pilY1* encodes a gonococcal PilC homologue. *Mol. Microbiol.* 22:161–73
3. Alm RA, Mattick JS. 1995. Identification of a gene, *pilV*, required for type 4 fimbrial biogenesis in *Pseudomonas aeruginosa*, whose product possesses a pre-pilin-like leader sequence. *Mol. Microbiol.* 16:485–96
4. Alm RA, Mattick JS. 1996. Identification of two genes with prepilin-like leader sequences involved in type 4 fimbrial biogenesis in *Pseudomonas aeruginosa*. *J. Bacteriol.* 178:3809–17
5. Alm RA, Mattick JS. 1997. Genes involved in the biogenesis and function of type-4 fimbriae in *Pseudomonas aeruginosa*. *Gene* 192:89–98
6. Arvidson CG, So M. 1995. The Neisseria transcriptional regulator PilA has a GTPase activity. *J. Biol. Chem.* 270:26045–48
7. Beard MK, Mattick JS, Moore LJ, Mott MR, Marrs CF, Egerton JR. 1990. Morphogenetic expression of *Moraxella bovis* fimbriae (pili) in *Pseudomonas aeruginosa*. *J. Bacteriol.* 172:2601–7
8. Beatson SA, Whitchurch CB, Sargent JL, Levesque RC, Mattick JS. 2002. Differential regulation of twitching motility and elastase production by Vfr in *Pseudomonas aeruginosa*. *J. Bacteriol.* 184:3605–13
9. Beatson SA, Whitchurch CB, Semmler ABT, Mattick JS. 2002. Quorum sensing is not required for twitching motility in *Pseudomonas aeruginosa*. *J. Bacteriol.* 184:3598–604
10. Bhaya D, Bianco NR, Bryant D,

Grossman A. 2000. Type IV pilus biogenesis and motility in the cyanobacterium *Synechocystis* sp. PCC6803. *Mol. Microbiol.* 37:941–51

11. Bhaya D, Takahashi A, Grossman AR. 2001. Light regulation of type IV pilus-dependent motility by chemosensor-like elements in *Synechocystis* PCC6803. *Proc. Natl. Acad. Sci. USA* 98:7540–45
12. Bieber D, Ramer SW, Wu CY, Murray WJ, Tobe T, et al. 1998. Type IV pili, transient bacterial aggregates, and virulence of enteropathogenic *Escherichia coli. Science* 280:2114–18
13. Bitter W, Koster M, Latijnhouwers M, de Cock H, Tommassen J. 1998. Formation of oligomeric rings by XcpQ and PilQ, which are involved in protein transport across the outer membrane of *Pseudomonas aeruginosa. Mol. Microbiol.* 27:209–19
14. Boyd JM. 2000. Localization of the histidine kinase PilS to the poles of *Pseudomonas aeruginosa* and identification of a localization domain. *Mol. Microbiol.* 36:153–62
15. Bradley DE. 1980. A function of *Pseudomonas aeruginosa* PAO pili: twitching motility. *Can. J. Microbiol.* 26:146–54
16. Brossay L, Paradis G, Fox R, Koomey M, Hebert J. 1994. Identification, localization, and distribution of the PilT protein in *Neisseria gonorrhoeae. Infect. Immun.* 62:2302–8
17. Castric P, Cassels FJ, Carlson RW. 2001. Structural characterization of the *Pseudomonas aeruginosa* 1244 pilin glycan. *J. Biol. Chem.* 276:26479–85
18. Collins RF, Davidsen L, Derrick JP, Ford RC, Tonjum T. 2001. Analysis of the PilQ secretin from *Neisseria meningitidis* by transmission electron microscopy reveals a dodecameric quaternary structure. *J. Bacteriol.* 183:3825–32
19. Comolli JC, Hauser AR, Waite L, Whitchurch CB, Mattick JS, Engel JN. 1999. *Pseudomonas aeruginosa* gene products PilT and PilU are required for cytotoxicity in vitro and virulence in a mouse model of acute pneumonia. *Infect. Immun.* 67:3625–30
20. Croft L, Beatson SA, Whitchurch CB, Huang B, Blakeley RL, Mattick JS. 2000. An interactive web-based *Pseudomonas aeruginosa* genome database: discovery of new genes, pathways and structures. *Microbiology* 146:2351–64
21. Dalrymple B, Mattick JS. 1987. An analysis of the organization and evolution of type 4 fimbrial (MePhe) subunit proteins. *J. Mol. Evol.* 25:261–69
22. D'Argenio DA, Gallagher LA, Berg CA, Manoil C. 2001. Drosophila as a model host for *Pseudomonas aeruginosa* infection. *J. Bacteriol.* 183:1466–71
23. Darzins A. 1993. The *pilG* gene product, required for *Pseudomonas aeruginosa* pilus production and twitching motility, is homologous to the enteric, single-domain response regulator CheY. *J. Bacteriol.* 175:5934–44
24. Darzins A. 1994. Characterization of a *Pseudomonas aeruginosa* gene cluster involved in pilus biosynthesis and twitching motility: sequence similarity to the chemotaxis proteins of enterics and the gliding bacterium *Myxococcus xanthus. Mol. Microbiol.* 11:137–53
25. Dorr J, Hurek T, Reinhold-Hurek B. 1998. Type IV pili are involved in plant-microbe and fungus-microbe interactions. *Mol. Microbiol.* 30:7–17
26. Drake SL, Sandstedt SA, Koomey M. 1997. PilP, a pilus biogenesis lipoprotein in *Neisseria gonorrhoeae*, affects expression of PilQ as a high-molecular-mass multimer. *Mol. Microbiol.* 23:657–68
27. Egerton JR, Cox PT, Anderson BJ, Kristo C, Norman M, Mattick JS. 1987. Protection of sheep against footrot with a recombinant DNA-based fimbrial vaccine. *Vet. Microbiol.* 14:393–409
28. Elleman TC, Peterson JE. 1987. Expression of multiple types of N-methyl Phe pili in *Pseudomonas aeruginosa. Mol. Microbiol.* 1:377–80
29. Folkhard W, Marvin DA, Watts TH,

Paranchych W. 1981. Structure of polar pili from *Pseudomonas aeruginosa* strains K and O. *J. Mol. Biol.* 149:79–93
30. Forest KT, Dunham SA, Koomey M, Tainer JA. 1999. Crystallographic structure reveals phosphorylated pilin from *Neisseria*: phosphoserine sites modify type IV pilus surface chemistry and fibre morphology. *Mol. Microbiol.* 31:743–52
31. Forest KT, Tainer JA. 1997. Type-4 pilus structure: outside to inside and top to bottom—a minireview. *Gene* 192:165–69
32. Fulks KA, Marrs CF, Stevens SP, Green MR. 1990. Sequence analysis of the inversion region containing the pilin genes of *Moraxella bovis*. *J. Bacteriol.* 172:310–16
33. Fullner KJ, Mekalanos JJ. 1999. Genetic characterization of a new type IV-A pilus gene cluster found in both classical and El Tor biotypes of *Vibrio cholerae*. *Infect. Immun.* 67:1393–404
34. Fussenegger M, Rudel T, Barten R, Ryll R, Meyer TF. 1997. Transformation competence and type-4 pilus biogenesis in *Neisseria gonorrhoeae*—a review. *Gene* 192:125–34
35. Giron JA, Gomez-Duarte OG, Jarvis KG, Kaper JB. 1997. Longus pilus of enterotoxigenic *Escherichia coli* and its relatedness to other type-4 pili—a minireview. *Gene* 192:39–43
36. Glessner A, Smith RS, Iglewski BH, Robinson JB. 1999. Roles of *Pseudomonas aeruginosa* las and rhl quorum-sensing systems in control of twitching motility. *J. Bacteriol.* 181:1623–29
37. Graupner S, Wackernagel W. 2001. *Pseudomonas stutzeri* has two closely related *pilA* genes (type IV pilus structural protein) with opposite influences on natural genetic transformation. *J. Bacteriol.* 183:2359–66
38. Graupner S, Weger N, Sohni M, Wackernagel W. 2001. Requirement of novel competence genes *pilT* and *pilU* of *Pseudomonas stutzeri* for natural transformation and suppression of *pilT* deficiency by a hexahistidine tag on the type IV pilus protein PilAI. *J. Bacteriol.* 183:4694–701
39. Hahn HP. 1997. The type-4 pilus is the major virulence-associated adhesin of *Pseudomonas aeruginosa*—a review. *Gene* 192:99–108
40. Hamadeh RM, Estabrook MM, Zhou P, Jarvis GA, Griffiss JM. 1995. Anti-Gal binds to pili of *Neisseria meningitidis*: the immunoglobulin A isotype blocks complement-mediated killing. *Infect. Immun.* 63:4900–6
41. Hazes B, Sastry PA, Hayakawa K, Read RJ, Irvin RT. 2000. Crystal structure of *Pseudomonas aeruginosa* PAK pilin suggests a main-chain-dominated mode of receptor binding. *J. Mol. Biol.* 299:1005–17
42. Hazlett LD, Moon MM, Singh A, Berk RS, Rudner XL. 1991. Analysis of adhesion, piliation, protease production and ocular infectivity of several *P. aeruginosa* strains. *Curr. Eye Res.* 10:351–62
43. Henrichsen J. 1972. Bacterial surface translocation: a survey and a classification. *Bacteriol. Rev.* 36:478–503
44. Henrichsen J. 1975. The occurrence of twitching motility among gram-negative bacteria. *Acta Pathol. Microbiol. Scand. B* 83:171–78
45. Henrichsen J. 1983. Twitching motility. *Annu. Rev. Microbiol.* 37:81–93
46. Henriksen SD, Henrichsen J. 1975. Twitching motility and possession of polar fimbriae in spreading *Streptococcus sanguis* isolates from the human throat. *Acta Pathol. Microbiol. Scand. B* 83:133–40
47. Hobbs M, Collie ES, Free PD, Livingston SP, Mattick JS. 1993. PilS and PilR, a two-component transcriptional regulatory system controlling expression of type 4 fimbriae in *Pseudomonas aeruginosa*. *Mol. Microbiol.* 7:669–82
48. Hobbs M, Dalrymple BP, Cox PT, Livingston SP, Delaney SF, Mattick JS. 1991. Organization of the fimbrial gene region of *Bacteroides nodosus*: class I and class II strains. *Mol. Microbiol.* 5:543–60

49. Hobbs M, Mattick JS. 1993. Common components in the assembly of type 4 fimbriae, DNA transfer systems, filamentous phage and protein-secretion apparatus: a general system for the formation of surface-associated protein complexes. *Mol. Microbiol.* 10:233–43
50. Jelsbak L, Sogaard-Andersen L. 2000. Pattern formation: fruiting body morphogenesis in *Myxococcus xanthus*. *Curr. Opin. Microbiol.* 3:637–42
51. Jennings MP, Virji M, Evans D, Foster V, Srikhanta YN, et al. 1998. Identification of a novel gene involved in pilin glycosylation in *Neisseria meningitidis*. *Mol. Microbiol.* 29:975–84
52. Jennings PA, Bills MM, Irving DO, Mattick JS. 1989. Fimbriae of *Bacteroides nodosus*: protein engineering of the structural subunit for the production of an exogenous peptide. *Protein Eng.* 2:365–69
53. Jonsson AB, Ilver D, Falk P, Pepose J, Normark S. 1994. Sequence changes in the pilus subunit lead to tropism variation of *Neisseria gonorrhoeae* to human tissue. *Mol. Microbiol.* 13:403–16
54. Jonsson AB, Nyberg G, Normark S. 1991. Phase variation of gonococcal pili by frameshift mutation in *pilC*, a novel gene for pilus assembly. *EMBO J.* 10:477–88
55. Kaiser D. 1979. Social gliding is correlated with the presence of pili in *Myxococcus xanthus*. *Proc. Natl. Acad. Sci. USA* 76:5952–56
56. Kaiser D. 2000. Bacterial motility: How do pili pull? *Curr. Biol.* 10:R777–80
57. Kearns DB, Robinson J, Shimkets LJ. 2001. *Pseudomonas aeruginosa* exhibits directed twitching motility up phosphatidylethanolamine gradients. *J. Bacteriol.* 183:763–67
58. Kearns DB, Shimkets LJ. 2001. Lipid chemotaxis and signal transduction in *Myxococcus xanthus*. *Trends Microbiol.* 9:126–29
59. Keizer DW, Slupsky CM, Kalisiak M, Campbell AP, Crump MP, et al. 2001. Structure of a pilin monomer from *Pseudomonas aeruginosa*: implications for the assembly of pili. *J. Biol. Chem.* 276: 24186–93
60. Kennan RM, Dhungyel OP, Whittington RJ, Egerton JR, Rood JI. 2001. The type IV fimbrial subunit gene (*fimA*) of *Dichelobacter nodosus* is essential for virulence, protease secretion, and natural competence. *J. Bacteriol.* 183:4451–58
61. Koomey M, Bergstrom S, Blake M, Swanson J. 1991. Pilin expression and processing in pilus mutants of *Neisseria gonorrhoeae*: critical role of Gly-1 in assembly. *Mol. Microbiol.* 5:279–87
62. Lauer P, Albertson NH, Koomey M. 1993. Conservation of genes encoding components of a type IV pilus assembly/two-step protein export pathway in *Neisseria gonorrhoeae*. *Mol. Microbiol.* 8:357–68
63. Lautrop H. 1961. *Bacterium anitratum* transferred to the genus *Cytophaga*. *Int. Bull. Bacteriol. Nomencl.* 11:107–8
64. Lee KK, Sheth HB, Wong WY, Sherburne R, Paranchych W, et al. 1994. The binding of *Pseudomonas aeruginosa* pili to glycosphingolipids is a tip-associated event involving the C-terminal region of the structural pilin subunit. *Mol. Microbiol.* 11:705–13
65. Liles MR, Viswanathan VK, Cianciotto NP. 1998. Identification and temperature regulation of *Legionella pneumophila* genes involved in type IV pilus biogenesis and type II protein secretion. *Infect. Immun.* 66:1776–82
66. Liu H, Kang Y, Genin S, Schell MA, Denny TP. 2001. Twitching motility of *Ralstonia solanacearum* requires a type IV pilus system. *Microbiology* 147:3215–29
67. Lory S, Strom MS. 1997. Structure-function relationship of type-IV prepilin peptidase of *Pseudomonas aeruginosa*—a review. *Gene* 192:117–21
68. Lu HM, Motley ST, Lory S. 1997. Interactions of the components of the general secretion pathway: role of *Pseudomonas*

aeruginosa type IV pilin subunits in complex formation and extracellular protein secretion. *Mol. Microbiol.* 25:247–59
69. Lyczak JB, Cannon CL, Pier GB. 2000. Establishment of *Pseudomonas aeruginosa* infection: lessons from a versatile opportunist. *Microbes Infect.* 2:1051–60
70. Macdonald DL, Pasloske BL, Paranchych W. 1993. Mutations in the fifth-position glutamate in *Pseudomonas aeruginosa* pilin affect the transmethylation of the N-terminal phenylalanine. *Can. J. Microbiol.* 39:500–5
71. Marceau M, Forest K, Beretti JL, Tainer J, Nassif X. 1998. Consequences of the loss of O-linked glycosylation of meningococcal type IV pilin on piliation and pilus-mediated adhesion. *Mol. Microbiol.* 27: 705–15
72. Martin PR, Hobbs M, Free PD, Jeske Y, Mattick JS. 1993. Characterization of *pilQ*, a new gene required for the biogenesis of type 4 fimbriae in *Pseudomonas aeruginosa*. *Mol. Microbiol.* 9:857–68
73. Martin PR, Watson AA, McCaul TF, Mattick JS. 1995. Characterization of a five-gene cluster required for the biogenesis of type 4 fimbriae in *Pseudomonas aeruginosa*. *Mol. Microbiol.* 16:497–508
74. Mattick JS, Alm RA. 1995. Common architecture of type 4 fimbriae and complexes involved in macromolecular traffic. *Trends Microbiol.* 3:411–13
75. Mattick JS, Anderson BJ, Cox PT, Dalrymple BP, Bills MM, et al. 1991. Gene sequences and comparison of the fimbrial subunits representative of *Bacteroides nodosus* serotypes A to I: class I and class II strains. *Mol. Microbiol.* 5:561–73
76. Mattick JS, Bills MM, Anderson BJ, Dalrymple B, Mott MR, Egerton JR. 1987. Morphogenetic expression of *Bacteroides nodosus* fimbriae in *Pseudomonas aeruginosa*. *J. Bacteriol.* 169:33–41
77. Mattick JS, Hobbs M, Cox PT, Dalrymple BP. 1993. Molecular biology of the fimbriae of *Dichelobacter* (prev. *Bacteroides*) *nodosus*. In *Genetics and Molecular Biology of Anaerobic Bacteria*, ed. M Sebald, pp. 517–45. New York: Springer
78. Mattick JS, Whitchurch CB, Alm RA. 1996. The molecular genetics of type 4 fimbriae in *Pseudomonas aeruginosa*—a review. *Gene* 179:147–55
79. McBride MJ. 2001. Bacterial gliding motility: multiple mechanisms for cell movement over surfaces. *Annu. Rev. Microbiol.* 55:49–75
80. McMichael JC. 1992. Bacterial differentiation within *Moraxella bovis* colonies growing at the interface of the agar medium with the petri dish. *J. Gen. Microbiol.* 138:2687–95
81. Mehr IJ, Long CD, Serkin CD, Seifert HS. 2000. A homologue of the recombination-dependent growth gene, *rdgC*, is involved in gonococcal pilin antigenic variation. *Genetics* 154:523–32
82. Merz AJ, So M. 2000. Interactions of pathogenic neisseriae with epithelial cell membranes. *Annu. Rev. Cell. Dev. Biol.* 16:423–57
83. Merz AJ, So M, Sheetz MP. 2000. Pilus retraction powers bacterial twitching motility. *Nature* 407:98–102
84. Nunn D, Bergman S, Lory S. 1990. Products of three accessory genes, *pilB*, *pilC*, and *pilD*, are required for biogenesis of *Pseudomonas aeruginosa* pili. *J. Bacteriol.* 172:2911–19
85. O'Toole GA, Kolter R. 1998. Flagellar and twitching motility are necessary for *Pseudomonas aeruginosa* biofilm development. *Mol. Microbiol.* 30:295–304
86. Ottow JC. 1975. Ecology, physiology, and genetics of fimbriae and pili. *Annu. Rev. Microbiol.* 29:79–108
87. Parge HE, Forest KT, Hickey MJ, Christensen DA, Getzoff ED, Tainer JA. 1995. Structure of the fibre-forming protein pilin at 2.6 Å resolution. *Nature* 378:32–38
88. Power PM, Roddam LF, Dieckelmann M, Srikhanta YN, Tan YC, et al. 2000. Genetic characterization of pilin glycosylation in *Neisseria meningitidis*. *Microbiology* 146:967–79

89. Pugsley AP. 1993. The complete general secretory pathway in gram-negative bacteria. *Microbiol. Rev.* 57:50–108
90. Pujol C, Eugene E, Marceau M, Nassif X. 1999. The meningococcal PilT protein is required for induction of intimate attachment to epithelial cells following pilus-mediated adhesion. *Proc. Natl. Acad. Sci. USA* 96:4017–22
91. Rahme LG, Ausubel FM, Cao H, Drenkard E, Goumnerov BC, et al. 2000. Plants and animals share functionally common bacterial virulence factors. *Proc. Natl. Acad. Sci. USA* 97:8815–21
92. Roine E, Nunn DN, Paulin L, Romantschuk M. 1996. Characterization of genes required for pilus expression in *Pseudomonas syringae* pathovar *phaseolicola*. *J. Bacteriol.* 178:410–17
93. Rudel T, Boxberger HJ, Meyer TF. 1995. Pilus biogenesis and epithelial cell adherence of *Neisseria gonorrhoeae pilC* double knock-out mutants. *Mol. Microbiol.* 17:1057–71
94. Rudel T, Facius D, Barten R, Scheuerpflug I, Nonnenmacher E, Meyer TF. 1995. Role of pili and the phase-variable PilC protein in natural competence for transformation of *Neisseria gonorrhoeae*. *Proc. Natl. Acad. Sci. USA* 92:7986–90
95. Rudel T, Scheurerpflug I, Meyer TF. 1995. Neisseria PilC protein identified as type-4 pilus tip-located adhesin. *Nature* 373:357–59
96. Russel M. 1998. Macromolecular assembly and secretion across the bacterial cell envelope: type II protein secretion systems. *J. Mol. Biol.* 279:485–99
97. Russell MA, Darzins A. 1994. The *pilE* gene product of *Pseudomonas aeruginosa*, required for pilus biogenesis, shares amino acid sequence identity with the N-termini of type 4 prepilin proteins. *Mol. Microbiol.* 13:973–85
98. Ryll RR, Rudel T, Scheuerpflug I, Barten R, Meyer TF. 1997. PilC of *Neisseria meningitidis* is involved in class II pilus formation and restores pilus assembly, natural transformation competence and adherence to epithelial cells in PilC-deficient gonococci. *Mol. Microbiol.* 23:879–92
99. Rytkonen A, Johansson L, Asp V, Albiger B, Jonsson AB. 2001. Soluble pilin of *Neisseria gonorrhoeae* interacts with human target cells and tissue. *Infect. Immun.* 69:6419–26
100. Sauvonnet N, Vignon G, Pugsley AP, Gounon P. 2000. Pilus formation and protein secretion by the same machinery in *Escherichia coli*. *EMBO J.* 19:2221–28
101. Scheuerpflug I, Rudel T, Ryll R, Pandit J, Meyer TF. 1999. Roles of PilC and PilE proteins in pilus-mediated adherence of *Neisseria gonorrhoeae* and *Neisseria meningitidis* to human erythrocytes and endothelial and epithelial cells. *Infect. Immun.* 67:834–43
102. Seifert HS. 1996. Questions about gonococcal pilus phase- and antigenic variation. *Mol. Microbiol.* 21:433–40
103. Seifert HS, Ajioka RS, Marchal C, Sparling PF, So M. 1988. DNA transformation leads to pilin antigenic variation in *Neisseria gonorrhoeae*. *Nature* 336:392–95
104. Semmler AB, Whitchurch CB, Leech AJ, Mattick JS. 2000. Identification of a novel gene, *fimV*, involved in twitching motility in *Pseudomonas aeruginosa*. *Microbiology* 146:1321–32
105. Semmler AB, Whitchurch CB, Mattick JS. 1999. A re-examination of twitching motility in *Pseudomonas aeruginosa*. *Microbiology* 145:2863–73
106. Shi W, Ngok FK, Zusman DR. 1996. Cell density regulates cellular reversal frequency in *Myxococcus xanthus*. *Proc. Natl. Acad. Sci. USA* 93:4142–46
107. Shimkets LJ. 1999. Intercellular signaling during fruiting-body development of *Myxococcus xanthus*. *Annu. Rev. Microbiol.* 53:525–49
108. Singh PK, Schaefer AL, Parsek MR, Moninger TO, Welsh MJ, Greenberg EP. 2000. Quorum-sensing signals indicate

that cystic fibrosis lungs are infected with bacterial biofilms. *Nature* 407:762–64
109. Skerker JM, Berg HC. 2001. Direct observation of extension and retraction of type IV pili. *Proc. Natl. Acad. Sci. USA* 98:6901–4
110. Spormann AM. 1999. Gliding motility in bacteria: insights from studies of *Myxococcus xanthus*. *Microbiol. Mol. Biol. Rev.* 63:621–41
111. Stimson E, Virji M, Barker S, Panico M, Blench I, et al. 1996. Discovery of a novel protein modification: alpha-glycerophosphate is a substituent of meningococcal pilin. *Biochem. J.* 316:29–33
112. Stimson E, Virji M, Makepeace K, Dell A, Morris HR, et al. 1995. Meningococcal pilin: a glycoprotein substituted with digalactosyl 2,4-diacetamido-2,4,6-trideoxyhexose. *Mol. Microbiol.* 17:1201–14
113. Stone BJ, Kwaik YA. 1999. Natural competence for DNA transformation by *Legionella pneumophila* and its association with expression of type IV pili. *J. Bacteriol.* 181:1395–402
114. Strom MS, Lory S. 1991. Amino acid substitutions in pilin of *Pseudomonas aeruginosa*. Effect on leader peptide cleavage, amino-terminal methylation, and pilus assembly. *J. Biol. Chem.* 266:1656–64
115. Strom MS, Lory S. 1992. Kinetics and sequence specificity of processing of prepilin by PilD, the type IV leader peptidase of *Pseudomonas aeruginosa*. *J. Bacteriol.* 174:7345–51
116. Strom MS, Lory S. 1993. Structure-function and biogenesis of the type IV pili. *Annu. Rev. Microbiol.* 47:565–96
117. Sun H, Zusman DR, Shi W. 2000. Type IV pilus of *Myxococcus xanthus* is a motility apparatus controlled by the *frz* chemosensory system. *Curr. Biol.* 10:1143–46
118. Taha MK, Giorgini D. 1995. Phosphorylation and functional analysis of PilA, a protein involved in the transcriptional regulation of the pilin gene in *Neisseria gonorrhoeae*. *Mol. Microbiol.* 15:667–77
119. Taha MK, Giorgini D, Nassif X. 1996. The *pilA* regulatory gene modulates the pilus-mediated adhesion of *Neisseria meningitidis* by controlling the transcription of *pilC1*. *Mol. Microbiol.* 19:1073–84
120. Tennent JM, Mattick JS. 1994. Type 4 fimbriae. In *Fimbriae: Aspects of Adhesion, Genetics, Biogenesis and Vaccines*, ed. P Klemm, pp. 127–46. Boca Raton: CRC
121. Tonjum T, Weir S, Bovre K, Progulske-Fox A, Marrs CF. 1993. Sequence divergence in two tandemly located pilin genes of *Eikenella corrodens*. *Infect. Immun.* 61:1909–16
122. Villar MT, Hirschberg RL, Schaefer MR. 2001. Role of the *Eikenella corrodens pilA* locus in pilus function and phase variation. *J. Bacteriol.* 183:55–62
123. Wall D, Kaiser D. 1998. Alignment enhances the cell-to-cell transfer of pilus phenotype. *Proc. Natl. Acad. Sci. USA* 95:3054–58
124. Wall D, Kaiser D. 1999. Type IV pili and cell motility. *Mol. Microbiol.* 32:1–10
125. Wall D, Wu SS, Kaiser D. 1998. Contact stimulation of Tgl and type IV pili in *Myxococcus xanthus*. *J. Bacteriol.* 180:759–61
126. Ward MJ, Lew H, Zusman DR. 2000. Social motility in *Myxococcus xanthus* requires FrzS, a protein with an extensive coiled-coil domain. *Mol. Microbiol.* 37:1357–71
127. Ward MJ, Zusman DR. 1997. Regulation of directed motility in *Myxococcus xanthus*. *Mol. Microbiol.* 24:885–93
128. Ward MJ, Zusman DR. 1999. Motility in *Myxococcus xanthus* and its role in developmental aggregation. *Curr. Opin. Microbiol.* 2:624–29
129. Watson AA, Alm RA, Mattick JS. 1996. Identification of a gene, *pilF*, required for type 4 fimbrial biogenesis and twitching motility in *Pseudomonas aeruginosa*. *Gene* 180:49–56
130. Watson AA, Mattick JS, Alm RA. 1996. Functional expression of heterologous type 4 fimbriae in *Pseudomonas aeruginosa*. *Gene* 175:143–50

131. Watts TH, Kay CM, Paranchych W. 1983. Spectral properties of three quaternary arrangements of *Pseudomonas* pilin. *Biochemistry* 22:3640–46
132. Whitchurch CB, Alm RA, Mattick JS. 1996. The alginate regulator AlgR and an associated sensor FimS are required for twitching motility in *Pseudomonas aeruginosa*. *Proc. Natl. Acad. Sci. USA* 93: 9839–43
133. Whitchurch CB, Hobbs M, Livingston SP, Krishnapillai V, Mattick JS. 1991. Characterisation of a *Pseudomonas aeruginosa* twitching motility gene and evidence for a specialised protein export system widespread in eubacteria. *Gene* 101:33–44
134. Whitchurch CB, Mattick JS. 1994. Characterization of a gene, *pilU*, required for twitching motility but not phage sensitivity in *Pseudomonas aeruginosa*. *Mol. Microbiol.* 13:1079–91
135. Whitchurch CB, Tolker-Nielsen T, Ragas P, Mattick JS. 2002. Extracellular DNA required for bacterial biofilm formation. *Science* 295:1487
136. Wolfgang M, Lauer P, Park HS, Brossay L, Hebert J, Koomey M. 1998. PilT mutations lead to simultaneous defects in competence for natural transformation and twitching motility in piliated *Neisseria gonorrhoeae*. *Mol. Microbiol.* 29:321–30
137. Wolfgang M, Park HS, Hayes SF, van Putten JP, Koomey M. 1998. Suppression of an absolute defect in type IV pilus biogenesis by loss-of-function mutations in *pilT*, a twitching motility gene in *Neisseria gonorrhoeae*. *Proc. Natl. Acad. Sci. USA* 95:14973–78
138. Wolfgang M, van Putten JPM, Hayes SF, Dorward D, Koomey M. 2000. Components and dynamics of fiber formation define a ubiquitous biogenesis pathway for bacterial pili. *EMBO J.* 19:6408–18
139. Wolgemuth C, Hoiczyk E, Kaiser D, Oster G. 2002. How myxobacteria bacteria glide. *Curr. Biol.* 12:369–77
140. Wu SS, Kaiser D. 1997. Regulation of expression of the *pilA* gene in *Myxococcus xanthus*. *J. Bacteriol.* 179:7748–58
141. Yang Z, Geng Y, Xu D, Kaplan HB, Shi W. 1998. A new set of chemotaxis homologues is essential for *Myxococcus xanthus* social motility. *Mol. Microbiol.* 30: 1123–30
142. Yoshida T, Kim SR, Komano T. 1999. Twelve *pil* genes are required for biogenesis of the R64 thin pilus. *J. Bacteriol.* 181:2038–43
143. Yoshihara S, Geng X, Okamoto S, Yura K, Murata T, et al. 2001. Mutational analysis of genes involved in pilus structure, motility and transformation competency in the unicellular motile cyanobacterium *Synechocystis* sp. PCC 6803. *Plant Cell Physiol.* 42:63–73

Annu. Rev. Microbiol. 2002. 56:315–44
doi: 10.1146/annurev.micro.56.012302.160950
First published online as a Review in Advance on May 7, 2002

THE CLASS MESOMYCETOZOEA: A Heterogeneous Group of Microorganisms at the Animal-Fungal Boundary*

Leonel Mendoza,[1] John W. Taylor,[2] and Libero Ajello[3]
[1]Medical Technology Program, Department of Microbiology and Molecular Genetics, Michigan State University, East Lansing Michigan, 48824-1030; e-mail: mendoza9@msu.edu
[2]Department of Plant and Microbial Biology, University of California, Berkeley, California 94720-3102; e-mail: jtaylor@socrates.berkeley.edu
[3]Centers for Disease Control and Prevention, Mycotic Diseases Branch, Atlanta Georgia 30333; e-mail: lial@cdc.gov

Key Words Protista, Protozoa, Neomonada, DRIP, Ichthyosporea

■ **Abstract** When the enigmatic fish pathogen, the rosette agent, was first found to be closely related to the choanoflagellates, no one anticipated finding a new group of organisms. Subsequently, a new group of microorganisms at the boundary between animals and fungi was reported. Several microbes with similar phylogenetic backgrounds were soon added to the group. Interestingly, these microbes had been considered to be fungi or protists. This novel phylogenetic group has been referred to as the DRIP clade (an acronym of the original members: *Dermocystidium*, rosette agent, *Ichthyophonus*, and *Psorospermium*), as the class Ichthyosporea, and more recently as the class Mesomycetozoea. Two orders have been described in the mesomycetozoeans: the Dermocystida and the Ichthyophonida. So far, all members in the order Dermocystida have been pathogens either of fish (*Dermocystidium* spp. and the rosette agent) or of mammals and birds (*Rhinosporidium seeberi*), and most produce uniflagellated zoospores. Fish pathogens also are found in the order Ichthyophonida, but so are saprotrophic microbes. The Ichthyophonida species do not produce flagellated cells, but many produce amoeba-like cells. This review provides descriptions of the genera that comprise the class Mesomycetozoea and highlights their morphological features, pathogenic roles, and phylogenetic relationships.

CONTENTS

INTRODUCTION

Led by Haeckel's proposal that the metazoans may have had an ancestor within the unicellular protists, numerous studies utilizing morphology as well as molecular and phylogenetic analyses have supported his concept (18, 48, 51, 56). Those studies were validated in 1993 when Wainright et al. (95) used phylogenetic analyses of small subunit ribosomal DNA (18S SSU rDNA) sequences to conclude that the metazoans were a monophyletic group that shared ancestry with the choanoflagellates. They also reported that animals and fungi might have had a more recent common ancestor than either group had with plants, alveolates, or stramenopiles (14, 95). First Spanggaard et al. (84) and then Ragan et al. (77) reported that a new group of parasitic and saprotrophic protists had been found near the animal and fungal divergence. Later, others verified their findings (8, 18, 19, 30, 39, 44). These investigators confirmed that previously unclassified animal parasites and saprotrophic microbes grouped together as a new protistan monophyletic clade located near the point where the animals had diverged from the fungal boundary (Figure 1). Early on Ragan et al. referred to those microorganisms as the DRIP clade (an acronym for *Dermocystidium*, rosette agent, *Ichthyophonus* and *Psorospermium*). Later Cavalier-Smith placed them in the class Ichthyosporea (18), and more recently Mendoza et al. (57) established the class Mesomycetozoea to accommodate them. The location of this group at the divergence between animals and fungi was significant because it indicated that this unique group of microorganisms arose near the time that animals had diverged from fungi (Figure 1), providing additional organisms for comparative studies that could reveal the nature of the progenitor of the animal and the fungi, two of the kingdoms of multicellular and macroscopic organisms (17, 39, 77).

Examination of Figure 1 shows that the class Mesomycetozoea is a monophyletic group composed of two strongly supported clades, the orders Ichthyophonida and Dermocystida. However, as often is the case, the relationships of the Mesomycetozoea to other broad taxa are poorly supported in molecular phylogenetic analyses. Incorporating phenotypical data, Cavalier-Smith (17, 18) held that the ancestors of the animals and fungi were not mesomycetozoeans but unicellular flagellate organisms in the choanoflagellates. He based his conclusions on the facts that the mesomycetozoeans did not possess chitin or flagellate stages, a concept that recently was proven to be incorrect.

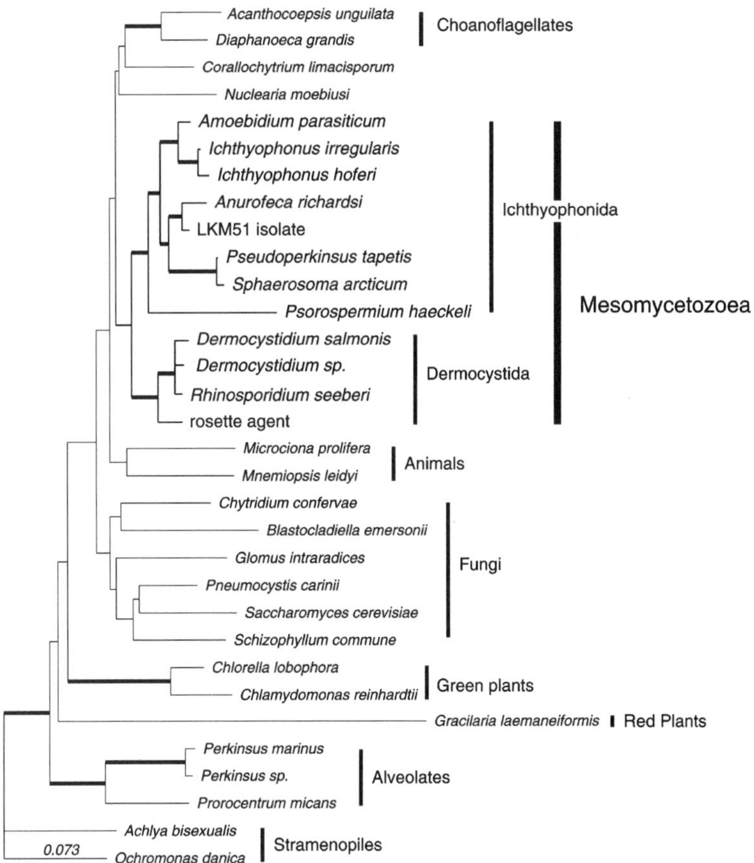

Figure 1 A phylogenetic analysis made by neighbor joining in Phylogenetic Analysis Using Parsimony (PAUP) using distances estimated by maximum likelihood (transitions/transversions estimated by ML; emperical nulceotide frequencies among site variation allowed comparable to the HKY85 model). The thickened branches are supported by 90% or greater of 1000 bootstrapped datasets. The topologies and support are similar to those found with parsimony analysis, with the exception of the placement of *Nuclearia moebiusi*, which can move as far as the branch subtending animals + choanoflagellaes + Mesomycetozoea. The scale for percent nucleotide substitution per nucleotide is given on the branch to *Ochromonas danica*.

Originally, Cavalier-Smith included the mesomycetozoeans in the class Ichthyosporea (18) (Table 1). This class was based on the fact that the original members of the DRIP clade described by Ragan et al. (77) were all fish parasites. However, new members of the group including *Amoebidium parasiticum*, *Anurofeca richardsi*, *Pseudoperkinsus tapetis*, *Rhinosporidium seeberi*, *Sphaerosoma*

TABLE 1 Current classification of the class Mesomycetozoea including new members

Kingdom Protozoa (Cavalier-Smith 1997)	
Subkingdoms	
1-Archezoa (*Giardia, Chilomastix, Retortamonas*)	
2-Eozoa (*Trichomonas*)	
3-Neozoa (four Infrakingdoms)	
a) Sarcodina (*Amoeba, Acanthamoeba, Entamoeba, Dictyostelium*)	
b) Alveolata (*Perkinsus, Eimeria, Plasmodium, Babesia, Paramecium, Tetrahymena*)	
c) Actinopoda (*Acanthocystis*, Radiozoa)	
d) Neomonada, Cavalier-Smith 1997	
Phylum Neomonada	
Subphyla	
a) Apusozoa (*Apusomonas*)	
b) Isomita (*Nephromyces*)	
c) Mesomycetozoa (four Classes) stat. nov.	
Class 1 Choanoflagellatea (*Sphaeroeca*)	
Class 2 Corallochytrea (*Corallochytrium*)	
Class 3 Mesomycetozoea (em. Mendoza et al. 2001)–(Ichthyosporea, Cavalier-Smith 1998)	
Order Dermocystida	Order Ichthyophonida
a) *Dermocystidium* spp.	a) *Amoebidium parasiticum*
b) *Rhinosporidium seeberi*	b) *Anurofeca richardsi*
c) Rosette agent	c) *Ichthyophonus* spp.
	d) *Pseudoperkinsus tapetis*
	e) *Psorospermium haeckeli*
	f) *Sphaerosoma articum*
	g) LKM51
Class 4 Cristidiscoidea, Order Nucleariida (*Nuclearia*)	

arcticum, and the isolate LKM51 are not fish parasites, which renders the term Icthyosporea inappropriate. In addition, the finding that *Ichthyophonus hoferi* has chitin in its cell walls (83) and that *R. seeberi* possesses at least one of the chitin synthase genes (40) suggested that this group of microbes may also have chitin in its cell walls, as do the fungi and some stramenopilans (60). This result did not come as a surprise because several investigators had previously suggested the presence of this polymer in the cell walls of some mesomycetozoeans (40, 83). The main problem has been that this polymer is not abundant in Mesomycetozoeans' cell walls. Thus it was difficult to demonstrate its presence in this group of protista (61). Moreover, chitin has always been associated with fungi but not with the protista, which in part discouraged some investigators from seeking it in organisms other than fungi (9). Mulisch (61) defied this trend by reporting that several protists possess chitin in their cell walls including the filamentous stramenopilans (60, 84). This finding strongly supported the reports of the presence of this polymer in *I. hoferi* and *R. seeberi*, and by extrapolation in all of the other mesomycetozoeans (9, 40, 83). Based on these facts, we recently emended Cavalier-Smith's original proposal and introduced the new class Mesomycetozoea (57), a name more suitable for this group of microbes (Figure 1). Accordingly, we use this term throughout our review. The subphylum Choanozoa was also emended to subphylum Mesomycetozoa. Because the choanoflagellates were the only group between animals and fungi at that time, the epithet Chonozoa was previously introduced (17). With the inclusion of related nearby microbes, the term Mesomycetozoa (between animals and fungi), originally proposed by Herr et al. (39), was considered to be more appropriate.

Based on phylogenetic analyses of 18S small subunit rDNA genes, the class Mesomycetozoea comprises 10 different parasitic and saprophytic microbes. They are members of the genera *Amoebidium*, *Anurofeca*, *Dermocystidium*, *Ichthyophonus*, *Pseudoperkinsus*, *Psorospermium*, *Rhinosporidium*, *Sphaerosoma*, and two as yet unnamed microbes, the "rosette agent" and "clone LKM51" (Figure 1). Each of the species in these genera has several morphological characteristics in common with species in the other genera of the clade, but each has unique characteristics as well (Table 2). Although members of this group are typically thought to be aquatic pathogens of fish, there are exceptions: *R. seeberi*, the only member of the class

TABLE 2 Members of orders Dermocystida and Ichthyophonida and the only species of the class Corallochytrea (Cora) with highlights of their key attributes

Taxa	Mitocondrial cristae	Life cycle traits	Hosts
Dermocystida			
Dermocystidium spp	Flat	Cysts with endospores uniflagellate zoospores	Fish
Rosette agent	Flat?	Spherules with endospores uniflagellate zoospores	Fish
Rhinosporidium seeberi	Flat	Sporangium with endospores uniflagellate zoospores?	Mammals, Birds
Ichthyophonida			
Amoebidium parasiticum	Flat	Sporangium, sporangiospores amoebic stage	Insects Crustaceans
Ichthyophonus hoferi	Tubular	Hyphae, plasmodium, spores amoebic stage	Fish
Psorospermium haeckeli	Flat	Ovoid shell-bearing spores amoebic stage	Crayfish
Anurofeca richardsi	Flat?	Spherules with endospores	Anural larvae
Pseudoperkinsus tapetis	Unknown	Spherules with spores uniflagellate zoospores?	Clams
Sphaerosoma articum	Flat	Spherules with endospores	Saprotrophic?
Isolate LKM51	Unknown	Unknown	Saprotrophic?
Corallochytrea			
Corallochytrium limacisporum	Flat?	Spherules with spores amoebic stage	Saprotrophic

? = not clear.

known to cause infections in mammals and birds; *Amoebidium* species, a genus of saprotrophic organisms; *A. richardsi*, whose role in diseases of anuran larvae is not clear; *Ps. tapetis*, a possibly nonpathogenic species associated with clams; *S. arcticum*, usually a saprotrophic organism; and the isolate termed LKM51, a putatively saprotrophic eukaryote found in phytoplankton.

Nothing is known regarding Mesomycetozoea's geographical distribution or their relationships with their natural environments. Thus, their epidemiological features and their interactions with other microbes in their ecological niches are also largely unknown. Because most mesomycetozoeans are animal parasites, what we do know about their cell cycles was learned from studies conducted on their parasitic stages. These studies have been of pivotal importance for the partial construction of their life cycles (5, 64, 83, 92, 97) (Figure 2). For instance, it was demonstrated that the species of the genera *Amoebidium*, *Ichthyophonus*, and *Psorospermium* developed amoeba-like cells in vitro. Based on these studies it was speculated that the amoeba-like cells could be the infecting propagules in nature (83, 92, 97) (Figure 2). Likewise, in vitro the *Dermocystidium* spp. and the rosette agent developed uniflagellated zoospores, indicating that they could serve as the infectious propagules [(37, 65) K.D. Arkush, personal communication] (Figure 2). The major contribution of these studies is the finding that during their parasitic stages, at least one phenotypic form of the mesomycetozoean species could initiate a new cycle outside their hosts.

In spite of these findings, their true life cycles in nature remain a mystery. Sexual development in the Mesomycetozoea has yet to be reported. This is due in part to the fact that most mesomycetozoeans have only been studied in their parasitic stages rather than in culture. Thus, sexual fusion, gamete formation, meiosis, and other major important genetic traits have yet to be found or induced. It is important to note that some investigators reported the presence of multiple nuclei during the parasitic stages of some mesomycetozoeans during the formation of new spores. In addition, little is known about their feeding habits. It has been found, however, that during their parasitic stages *D. salmonis* (65), *I. hoeferi* (45), *P. haeckeli* (89, 91), and *R. seeberi* (59) absorb nutrients from the hosts through the mesomycetozoean's cell walls, a finding that might suggest a similar behavior during their environmental stages. However, more studies are necessary to validate this assumption.

The epithets used to identify the phenotypic stages of the members of the class Mesomycetozoea varied according to the type of microorganism the investigators thought them to be. For instance, mycological terminology was used to identify the structures of *A. parasiticum*, *A. richardsi*, *I. hoferi*, and *R. seeberi*, all of them studied by mycologists (5, 21, 38, 62, 98, 99). The terms they used included endospores, hyphae, sporangia, spores, sporangiospores, thalli, and others. In contrast, protozoological names such as amoeba, cyst, plasmodium, sporocyst, and zoospore were used by protistologists to identify the structures formed by *A. parasiticum*, *D. salmonis*, *I. hoferi*, *P. haeckeli*, and the rosette agent (14, 24, 42, 62, 64, 75). With their inclusion in the class Mesomycetozoea, standardization of the names

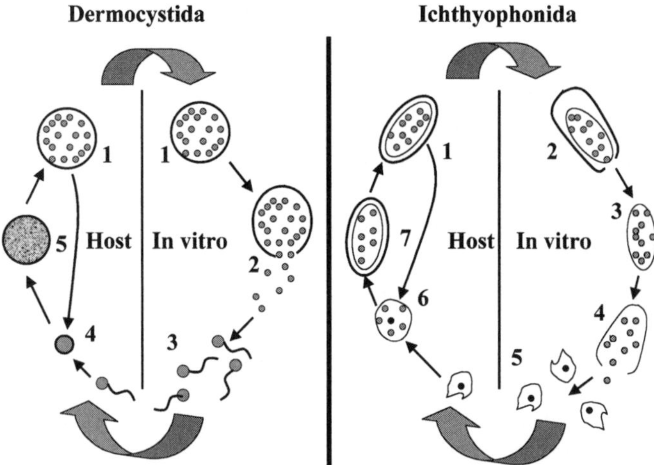

Figure 2 Depiction of the putative life cycle of members of the orders Dermocystida (*left*) and Ichthyophonida (*right*). Members of the order Dermocystida develop spherical cells with endospores (stage 1). In vitro the released endospores (stage 2) give rise to uniflagellated zoospores (stage 3). When the zoopores (infecting units) infect the host, they encyst (stage 4) and increase in size (stages 4, 5) and undergo cleavage into endospores (stage 1). The endospores can also be directly released within the host's infected tissues, and the cycle is repeated inside the hosts (stages 1, 4, 5, 1). Members of the order Ichthyophonida develop spherical (*Ichthyophonus* spp.) or ovoid cells in infected tissues (*Psorospermium haekeli*) or on its hosts (*Amoebidium parasiticum*) (stage 1). In vitro (stage 1) the hatching of spore receptacles occurs from the ovoid cells (stage 2); the receptacles containing spores (stage 3) then rupture and release their endospores (stage 4), which develop into motile amoeboid cells (infecting units) (stage 5). The amoeboid cell reach the hosts and develops into a small receptacle (stage 6) that later generates a hard cell wall with endospores (stages 7, 1). The endospores can also be released within their hosts, repeating the cycle (stages 1, 6, 7, 1). Note that in the genus *Ichthyophonus*, the development of hyphae that produce spherical cells with endopores (in vitro) is a feature, so far, not encountered in the other members of the order (62, 64, 83). For the order Ichthyophonida the cell cycle was adapted from Vogt & Rugg (92).

used to describe their phenotypic stages would be of importance in studying the members of this class. In this review, however, we continue to use the traditional nomenclature until a consensus is reached regarding their terminology.

We provide a brief description of the genera that are currently in the class Mesomycetozoea, highlighting their similarities and differences, with the intention of covering the latest known developments of the mesomycetozoeans and introducing the microbiological community to this novel class of microorganisms and their morphological, pathogenic, and phylogenetic relationships.

MEMBERS OF THE CLASS MESOMYCETOZOEA

Order Ichthyophonida

AMOEBIDIUM PARASITICUM The genus *Amoebidium* (21) is composed of four species: *A. australiense* (52), *A. colluviei* (53), *A. parasiticum* (98), and *A. recticola* (54). One of the well-studied species of this genus is *A. parasiticum*. This arthropodophilous symbiont is frequently found on crustaceans and insects that inhabit fresh water ponds (21, 52, 98). *A. parasiticum* shares with the other members of the class Mesomycetozoea its historical inclusion within the kingdom Fungi. The inclusion of this apparently nonpathogenic microorganism within the fungi was based on its production of a unicellular fungal-like thallus (sporangium), which is externally attached to the host. Like some mesomycetozoeans, *A. parasiticum* has been difficult to isolate in culture. It also possesses morphological characteristics different from most members of the group (Figure 3) (96–98). For instance, most mesomycetozoeans have spherical spores, but *A. parasiticum* has cigar-shaped spores (sporangiospores), from which amoeba arise, and also elongated thalli (sporangium) when attached to crustacea and insects (Figure 3).

Similarly, early morphological studies of *A. parasiticum* isolates showed that this microbe possesses unique features not found within the fungal Trichomycetes (Zygomycota), where it had long been classified (86, 97) (Figure 2). These facts

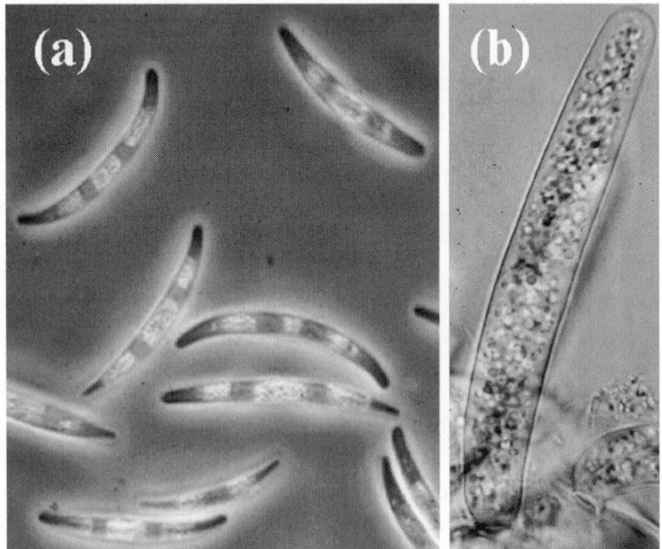

Figure 3 (*a*) Colorless cigar-shaped sporangiospores of *Amoebidium parasiticum* from which the amoeba phase developed (X 1000). (*b*) Sporangium-like thallus detached from its host's body (X 1250) (courtesy of R.W. Lichtwardt).

prompted some investigators to hypothesize that *A. parasiticum* represents a nonfungal evolutionary lineage possibly derived from a protozoan ancestor (54, 96). The finding of Whisler (97) that *A. parasiticum* forms amoebal cysts that give rise to several motile amoebas, which later encyst to form sporangium-like structures, provided the strongest evidence for this hypothesis. Although other investigators supported this new position (74, 85), those studies did not resolve the phylogenetic connection of this genus with other living organisms.

The phylogenetic affinities of *A. parasiticum* were almost simultaneously resolved when first Ustinova et al. (86) and later Benny & O'Donnell (12) sequenced the 18S SSU rDNA of *A. parasiticum*. Using phylogenetic analyses, these two independent groups of researchers found that *A. parasiticum* is a member of the class Mesomycetozoea and far from the fungal class Trichomycetes of the kingdom Fungi. In their phylogenetic trees, *A. parasiticum* was always closely related to *I. hoferi*. Interestingly, ultrastructural studies have indicated that *A. parasiticum* possesses mitochondria with flat cristae (9, 98), a finding that contrasts with the tubular cristae found on *I. hoferi*'s mitochondria (77). Our phylogenetic analyses confirm that *A. parasiticum* is the sister clade to *I. hoferi* and *I. irregularis* (Figure 1). These independent analyses showed that *A. parasiticum* belongs to the order Ichthyophonida, family Ichthyophonae in the class Mesomycetozoea (Table 1). Unfortunately, the other three species of the genus *Amoebidium* have not been phylogenetically investigated and are yet to be classified.

ANUROFECA RICHARDSI *Anurofeca richardsi* was originally held to be an alga and described under the name *Prototheca richardsi* (11, 99). It was first reported by Richards in 1958 (79) in feces of the larval state of *Rana pipiens* and implicated as a gut parasite of anural larvae (10, 79). Indeed, *A. richardsi* is isolated only from the guts of anuran larvae (99). Although the occurrence of *A. richardsi* is correlated with the presence of amphibian larvae in ponds, its role as a pathogen of anuran larvae is still unclear (8).

The genus *Anurofeca* was proposed by Baker et al. in 1999 (8). *Anurofeca richardsi* produces nonpigmented spherical cells 2–20 μm in diameter. Several small, round endospore-like cells develop within the spherical cells and are released after the cell wall breaks open. The inclusion of *A. richardsi* within the algal genus *Prototheca* was originally motivated by its morphological resemblance to members of that genus (Figure 4). Some investigators, however, noticed several inconsistencies. For instance, *A. richardsi* grows weakly on *Prototheca* isolation media, whereas the *Protheca* species can be easily cultured on that medium (73). It was also noticed that while other *Prototheca* species antigenically cross-reacted with each other, *A. richardsi* showed weak or no cross-reactions (99). Nonetheless, the final position of this algal-like organism was resolved by neither its morphological nor by its antigenic features.

A. richardsi's phylogenetic connection with the class Mesomycetozoea recently was revealed by Baker et al. (8). Using molecular approaches, they found that *A. richardsi* formed the sister clade to *I. hoferi* and was closely related to *P. haeckeli*,

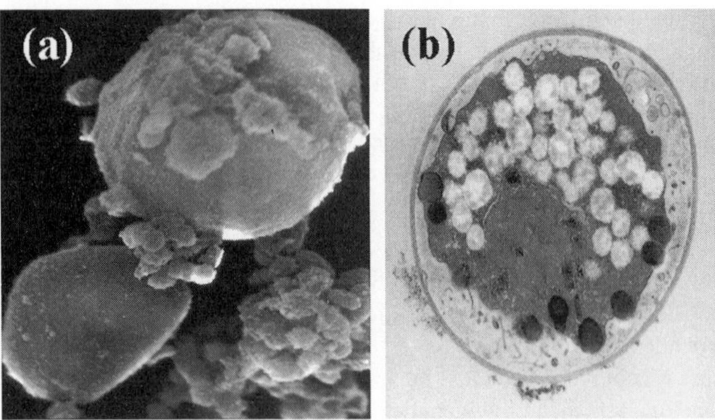

Figure 4 (*a*) Scanning electron microscopy of *Anurofeca richardsi* showing its spherical phenotype. Bacteria and debris are observed around the cells (X 5000). (*b*) Transmission electron microscopy of *A. richardsi* showing numerous vesicles within its cytoplasm (X 8720) (courtesy of T.J.C. Beebee).

far away from the algae. In our phylogenetic tree *A. richardsi* is closely related to the clone LKM51 isolate, and together they are the sister clade of *Pseudoperkinsus tapetis* plus *Sphaerosoma arcticum*. Mendoza et al. (57) placed *A. richardsi* within the phylum Neomonada, subphylum Mesomycetozoa (previously known as subphylum Choanozoa), class Mesomycetozoea, order Ichthyophonida, family Ichthyophonae (Table 1) (Figure 1).

ICHTHYOPHONUS HOFERI AND *I. IRREGULARIS* *Ichthyophonus hoferi* (72) was until recently considered to be a species of unknown taxonomic placement. Owing to its spherical in vivo morphologic features, *I. hoferi* was considered to be either a fungus or a protozoan, a taxonomic history that it shares with the other members of the class Mesomycetozoea. At least three other species, *I. gastrophilum* (20), *I. intestinalis* (50), and *I. lotae* (49), have also been described. More recently, however, *I. gastrophilum* was found to be synonymous with *I. hoferi*, whereas *I. intestinalis* and *I. lotae* were found to be related to the order Enthomophthorales in the kingdom Fungi (49, 62). This placement partly explains why early investigators considered *I. hoferi* to be a zygomycetous fungus.

Since the early twentieth century many reports of infections caused by *I. hoferi* in freshwater and marine fishes have been published. At least 14 marine and 6 freshwater fishes were found infected with *I. hoferi* worldwide. This microorganism frequently affects the internal organs of the infected animals such as the heart, liver, muscles, and spleen. *I. hoferi* occurs in infected tissues as spherical, thick-walled, multinucleated cells referred to as cysts, spores, and/or resting spores (45) (Figure 5). This organism can easily be recovered in pure culture from tissue samples of infected hosts (Figure 5).

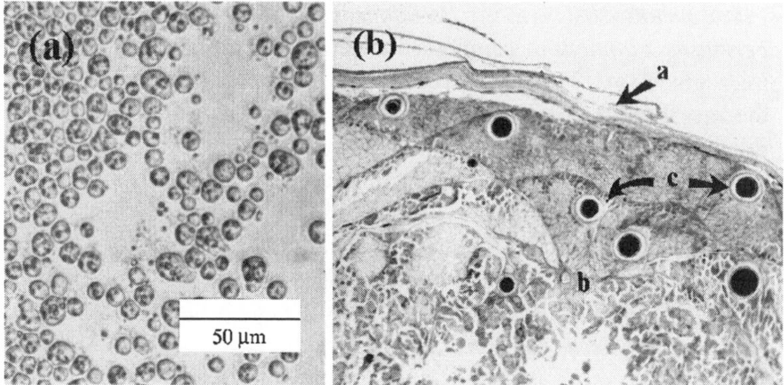

Figure 5 (A) *Ichthyophonus hoferi*'s "microspores" released from the tips of its hyphae. (*B*) Histological section of herring infected with *I. hoferi* depicting (*a*) epidermis, (*b*) body musculature, and (*c*) *I. hoferi* spores (X 100) (courtesy of R.M. Kocan).

Ichthyophonus hoferi has been the only species of the genus *Ichthyophonus* recognized to cause disease in fish. However, it had been reported for some time that a variety of fishes were infected with an *Ichthyophonus*-like organisms (38, 75). Based on morphological criteria and phylogenetic DNA sequences, Rand et al. (76) found that an unusual species of *Ichthyophonus*, morphologically different from *I. hoferi*, had been recovered from yellowtail flounders (*Limanda feruginea*) in Nova Scotia. When the 18S SSU rDNA from this particular isolate was sequenced, it was found to be distinct from two previously sequenced and geographically distant isolates of *I. hoferi* recovered from infected fish. These investigators proposed that their isolates were a new species, which they named *Ichthyophonus irregularis*. This revealed that the genus *Ichthyophonus* comprises more than one species and that all may be fish pathogens.

The controversy over the phylogenetic placement of *I. hoferi* came to an end when Spanggaard et al. (84) isolated the 18S SSU rDNA from this pathogen and found that it was related to the choanoflagellates. Later, when Ragan et al. (77) included more taxa in their analysis, it was found that *I. hoferi* was not a choanoflagellate but a member of the class Mesomycetozoea. In contrast to some members of this class, *I. hoferi* can be easily cultured (64, 75, 76, 83), and it possesses mitochondria with tubular cristae (9, 77). Cultivation studies have shown that at a low pH the organism developed hyphal forms, but when shifted to higher pHs (7–9) it developed motile amoeba-like forms (64). The studies were later confirmed by others (83). This finding indicated that the infecting units of this pathogen could be its amoeba-like form. The ability to develop motile amoeba-like forms was also found in some other members of the order Ichthyophonida (89, 91, 92, 97), but so far not in members of the order Dermocystida.

PSEUDOPERKINSUS TAPETIS *Pseudoperkinsus tapetis* was originally described as *Perkinsus atlanticus* in samples isolated from clams infected with a *Perkinsus*-like microorganism [(29) B. Novoa & A. Figueras, personal communication]. Although the morphological features that separate *Ps. tapetis* from *Perkinsus* spp. were relatively minor, with the use of molecular techniques it was evident that *Ps. tapetis* belongs to the new class Mesomycetozoea, phylogenetically far from the alveolate genus *Perkinsus* (9, 17, 18, 71) (Table 1). With this information, Figueras et al. (29) recommended a new genus to accommodate the organism originally named *P. atlanticus* by his group. He proposed the genus and the species *Pseudoperkinsus tapetis* (29). Confirming evidence that *Ps. tapetis* is distinct from the members of the genus *Perkinsus* came also from molecular studies using 5.8S ribosomal RNA (7, 31, 41, 46) on *P. marinus*, *P. atlanticus*, and other *Perkinsus* species. Those studies revealed that the *Perkinsus* spp. clustered together with the alveolates and that they were all different from the DNA sequences of *Ps. tapetis* deposited in the GenBank by Figueras et al. (29).

Although morphologically the *Ps. tapetis* and *Perkinsus* spp. are indistinguishable, the fact that *Ps. tapetis* did not cause high mortality in species of clams susceptible to *Perkinsus* spp. provided a phenotypic difference that correlated with the molecular divergence (B. Novoa & A. Figueras, personal communication). However, the virulence of *Ps. tapetis* to several other species of clams is still under investigation (68, 69). *P. tapetis* is spherical with several round vesicles in its cytoplasm (Figure 6). It can be recovered easily from clams and cultivated in synthetic media. It is interesting to note that in thioglycolate, this organism increases in size and produces spherical inclusions within its cytoplasm, similar to those observed in the hypnospores of the species classified in the genus *Perkinsus*. The production

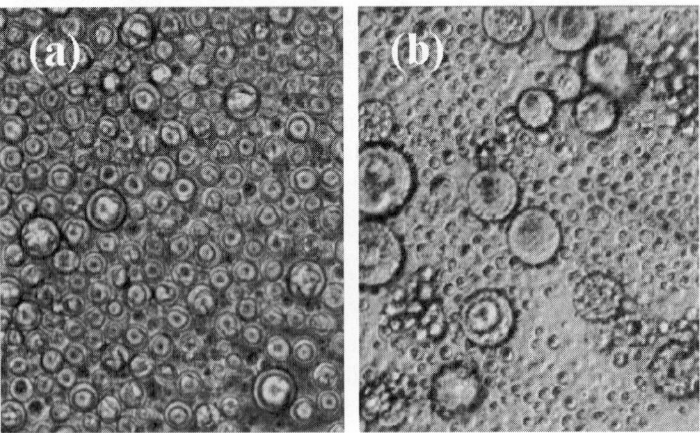

Figure 6 (*a*) Fresh mount preparation of *Pseudoperkinsus tapetis* from Spanish clam cultures (X 250). (*b*) Detail of the spherules developed by *Ps. tapetis* from culture samples (X 400) (courtesy of A. Figueras & B. Novoa).

of amoeba-like cells, typical of the order Ichthyophonida, has not yet been found in *Ps. tapetis* (B. Novoa & A. Figueras, personal communication). However, Ordás & Figueras (67) reported that one of their isolates, identified initially as *P. atlanticus* and later considered to be *Ps. tapetis*, produced uniflagellated zoospores. If confirmed, this report would be the first evidence for flagellated cells in the order Icthyophonida. The observation of flagella in *Ps. tapetis*, however, is controversial because *Perkinsus* spp. are well known to develop uniflagellated zoospores, raising the possibility that the *Ps. tapetis* cultures had been contaminated with a *Perkinsus* sp. Contamination is plausible because, as mentioned above, both microbes are morphologically indistinguishable, and they often have been recovered simultaneously in culture (B. Novoa, personal communication). Nonetheless, the finding of uniflagellated cells in this microbe (67) warrants further study.

Pseudoperkinsus tapetis was found to belong in the class Mesomycetozoea when its 18S SSU rDNA sequence was used for phylogenetic analysis (29). Figueras and colleagues found that *Ps. tapetis* is part of the class Ichthyophonida in the phylum Neomonada and the sister taxon of *S. arcticum*, as is also shown in our analysis (Table 1) (Figure 1). More importantly, its DNA sequences were found to possess a 99.5% identity with *S. arcticum*, a finding that may indicate that *Ps. tapetis* is a member of the genus *Sphaerosoma* and not in a separate genus. Additional data, however, from other molecules are needed to validate this claim.

PSOROSPERMIUM HAECKELI Haeckel (33) first reported crayfish infections caused by *P. haeckeli* in 1857 (33), but he did not propose a name for the etiologic agent. In 1883 Hilgendorf (42) named the organism *P. haeckeli* as a tribute to Haeckel. This organism mainly infects crayfish and has been reported in the Americas, the majority of European countries, as well as in Siberia (23, 47, 62, 90, 93). Originally *P. haeckeli* was considered to be either a fungus, a protist, or an alga based on its morphological similarities with fungi and algae (63). Owing to the taxonomic uncertainties of *P. haeckeli* it has always been referred to as "the enigmatic parasite of freshwater crayfish."

Psorospermium haeckeli may not be the only member of the genus. In addition, *Psorospermium orconectis* (24) and a *Psorospermium* sp. from Siberia (93) have been described, although the morphological features of *P. orconectis* were similar to those of *P. haeckeli* and the two may be conspecific. Based on size, morphology, and histology, Vogt (89) recently stated that the genus *Psorospermium* is probably composed of at least six morphotypes, each one perhaps representing different pathogenic species.

Psorospermium haeckeli affects the connective tissues around the gut of infected crayfish. It also has been reported to affect muscles and other tissues. In the infected areas *P. haeckeli* appears as a 45×90 μm–diameter ovoid and elongated spore with a refractive cell wall enclosing refringent globules of different sizes in their cytoplasm (Figure 7). Its oval morphology contrasts with the other members of the class, which have spherical phenotypes. The development of a binucleate amoeba-like stage has been used in the past to classify this parasite within the Protozoa

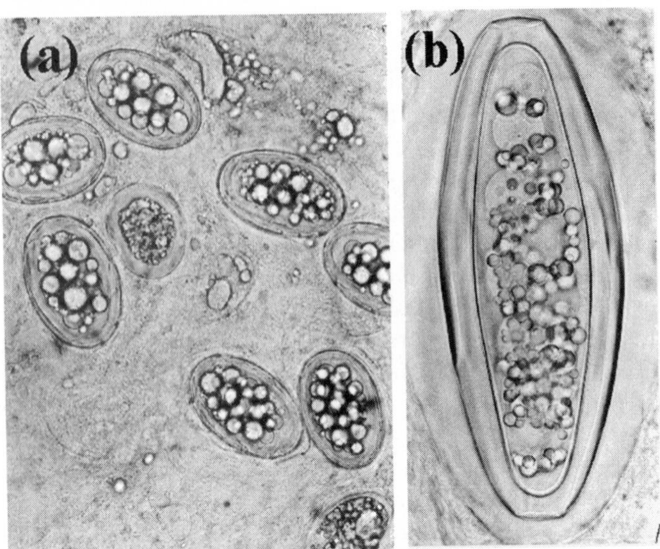

Figure 7 (*a*) Fresh mount preparation of connective tissue from crayfish infected with the ovoid cells of *Psorospermiun haekeli* (X 400). Note several globular vesicles within the cell's cytoplasm. (*b*) Elongate spore of *P. haekeli* from the connective tissue of a crayfish (X 1000) (courtesy of G. Vogt).

(81, 32). This feature was corroborated by Vogt (89), who also induced the hatching of slowly moving amoeboid cells from oval spores obtained from crayfish tissue infected with *P. haeckeli*. *P. haeckeli* in all constructed phylogenetic trees has been found to be closely related to *I. hoferi*. The two species have amoeba-like bodies and pathogenicity for marine animals. They differ, however, in that *P. haeckeli* possesses flat mitochondrial cristae rather than the tubular form found in *I. hoferi* (92). Mitochondrial ultrastructure was considered to be a key characteristic prior to the development of molecular phylogenetics, but it seems that the morphological character is not as stable as was thought, judging from the apparent shift from flat to tubular cristae in the case of *I. hoferi* (87, 101).

The mystery of its taxonomic affinities was resolved when Ragan et al. (77) amplified the 18S SSU rDNA molecule from this pathogen. They found that *P. haeckeli* was not a fungus but that it was part of the DRIP clade. In that study *P. haeckeli* clustered close to *I. hoferi*, a finding confirmed later by Cavalier-Smith (18), Baker et al. (8), Herr et al. (39), Fredericks et al. (30), and others. In our phylogenetic tree *P. haeckeli* lies at the base of the order Ichthyophonida, with the whole order resting on a well-supported branch (Figure 1).

SPHAEROSOMA ARCTICUM *Sphaerosoma arcticum* is a new member of the Mesomycetozoea. The story of how this organism was first found is an example of the rapid expansion of our understanding of the Mesomycetozoea. During an expedition to the high Arctic Dr. Bjarne Landfald and collaborators, while investigating unusual bacteria in cold marine habitats, isolated a microorganism from the amphipod *Gammarus setosus*. They recovered in culture a spherical eukaryotic microorganism (B. Landfald, personal communication). Because the host from which the organism had been isolated did not show pathological changes, it was not clear if *S. arcticum* was a parasite, a harmless protist, or that perhaps *G. setosus* had only an accidental association with this spherical organism, e.g., a food organism.

In culture *S. arcticum* develops spherical structures containing several individual cells, which at maturity release ~100 smaller cells (B. Landfald, personal communication). Electron microscopic studies of *S. arcticum* showed that the spherical bodies, with well-defined internal spherical cells, were similar to the morphological features of the algal species of *Chlorella* and *Prototheca* (Figure 8). The internal cells contained mitochondria with flat cristae and a well-defined nucleus. Attempts to infect related amphipod species with *S. arcticum* were unsuccessful (B. Landfald, personal communication). However, the fact that *S. arcticum*'s 18S SSU rDNA sequence is 99.5% similar to the *Ps. tapetis* sequence suggests

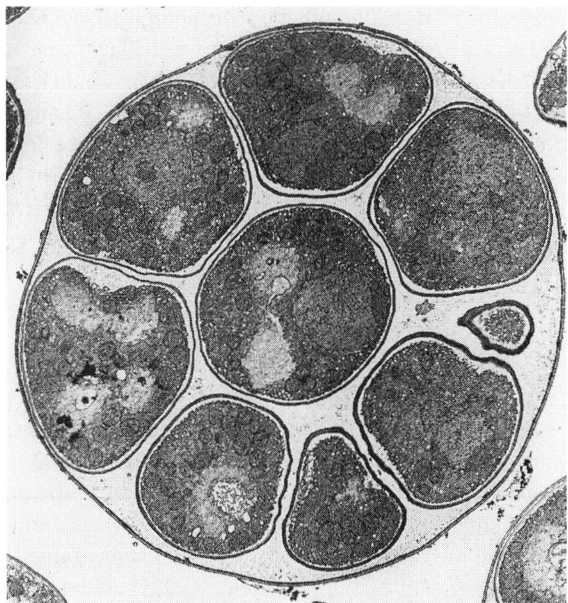

Figure 8 Transmission electron microscopy section of *Sphaerosoma arcticum* from cultured samples. The presence of numerous spores within the cytoplasm is its main phenotypic characteristic. The morphological features that *S. arcticum* shares with the species of the algal genera *Prototheca* and *Chlorella* are strikingly similar (X 26,400) (courtesy of B. Landfald).

not only that *Ps. tapetis* may be perhaps a species of the genus *Spaerosoma* but also that *S. arcticum* may be a microbe associated with clams. Therefore, one can speculate that *S. arcticum* may be a nonpathogenic species associated with clams, as in the case of *Ps. tapetis*. This might explain in part why Landfald could not infect amphipods with *S. arcticum*. Phylogenetic analysis of the *S. arcticum*'s 18S SSU rDNA showed that this organism is a member of the class Mesomycetozoea (B. Landfald, personal communication). In our analysis *S. arcticum* and *Ps. tapetis* form the sister clade to *A. richardsi* plus the LKM51 isolate, and both groups are well supported within the order Ichthyophonida (Figure 1).

ISOLATE LKM51 Isolate LKM51 is known only from an 18S SSU rDNA sequence obtained from DNA isolated from phytoplankton (86a). Thus, mesomycetozoeans have joined the many archaea, eubacteria, and uncultivated eukaryotic microbes whose presence is known only from environmental DNA, but which could well comprise 70% of the earth's microbiota (3, 43, 80, 94). Recently van Hannen et al. (86a), while investigating the correlation between the biomass of bacteria and the biomass of protozoans in phytoplankton samples, found that the affiliation of ~20% of the 18S SSU rDNA sequences investigated during this study could not be quickly determined. One of these sequences, clone LKM51, was later grouped with the mesomycetozoeans. Although the morphological features and other characteristics of the organism from which the DNA was isolated were not investigated, this was the first evidence that some mesomycetozoeans could be planktonic. Whether isolate LKM51 represents a free-living planktonic organism or simply part of its life cycle is not known, but were it shown to be free living, it would be the only mesomycetozoean that is not an obligate parasite. Phylogenetic analysis of the LKM51 clone showed that it is the sister clade of *A. richardsi* and well supported within the order Ichthyophonida (86a). Our analysis gave the same results (Figure 1).

Order Dermocystida

DERMOCYSTIDIUM SPP. The genus *Dermocystidium* is the sister taxon of *Rhinosporidium*. The latter genus comprises approximately 12 species, all of which cause deadly infections in fish and other marine animals (22, 27, 37, 66). The *Dermocystidium* spp. are characterized by their spherical sporangia (cysts) and the production of endospore-like structures (Figure 9). Such features are also encountered in the other members of the order Dermocystida (Table 1). These similarities could well explain why early investigators called attention to the morphological relationship between the *Dermocystidium* spp. and their sister taxon *R. seeberi* (39). Owing to their morphological features, the *Dermocystidium* spp. were previously considered to be either members of the haplosporeans, the fungi, or the apicomplexa groups (9, 26, 62, 70, 77, 88, 100).

Because of their intractability to cultivation, the ecological distribution of *Dermocystidium* spp. is poorly known. It is believed that fish are infected through

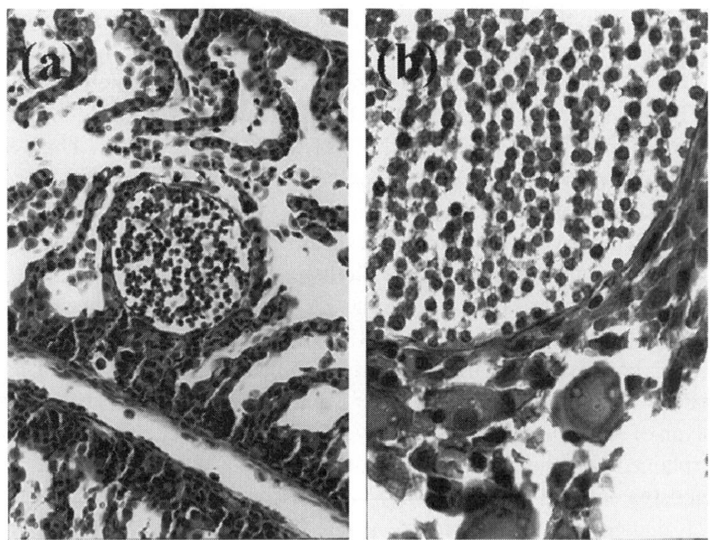

Figure 9 (*a*) Histological section of a fish's gills infected with a sporangium of *Dermocystidium salmonis* containing numerous endospores (X 400). (*b*) Histological section of an enlargement of *D. salmonis*'s sporangium. Note the presence of well-developed endospores, a feature also observed in *Rhinosporidium seeberi* (X 800) (courtesy of K.D. Arkush).

contact with the endospores released from infected fish. The spherical sporangia of *Dermocystidium* measure 200–400 μm in diameter and contain hundreds of endospores. They are readily observed in infected gills (Figure 9). These spherical sporangia are produced in great numbers to the point of making it impossible for the infected fish to take up oxygen and they finally succumb to their infections. However, the presence of uniflagellate zoospores in *D. salmonis* (65) suggests that motile zoospores could be the infective propagules of the *Dermocystidium* species. Recently it has also been found that the rosette agent develops uniflagellate zoospores (K.D. Arkush, personal communication). The rosette agent is also a member of the order Dermocystida and also causes a disease that primarily affects fish gills.

Early reports of *Dermocystidium* spp. suggested that the species of that genus could cause infections not only in fish but also in amphibians (14, 15, 25). Based on its spherical parasitic stage and the ecological distribution of its hosts, however, other investigators speculated that the true etiology of the spherical structures in amphibians were not the *Dermocystidium* species but members of a new genus (14). Accordingly, the genus *Dermosporidium* was created to accommodate the *Dermocystidium* spp. in infected amphibians (14). However, based on morphological characteristics, Herr et al. (39) speculated that *Dermosporidium granulosum* (14) and *Dermosporidium ranarum* (15) should be identified as *R. seeberi* in amphibians rather than as species of the newly proposed genus *Dermosporidium* or species of the genus *Dermocystidium*. Nevertheless, the phylogenetic connection of the amphibian parasite with *R. seeberi* has yet to be established by comparison of DNA sequences. It is important to note that the species of *Dermocystidium* are also phenotypically similar to those of the genus *Perkinsus*. Thus, the *Perkinsus* species have been studied in the past as members of the genus *Dermocystidium*. A good example of misidentification is *P. marinus*, which was once erroneously included in the genus *Dermocystidium* as *D. marinus* (71). We now know, however, that members of both genera are phylogenetically distant (71).

The taxonomic affinities of the *Dermocystidium* spp. were resolved when Ragan et al. (77) sequenced their 18S SSU rDNAs. Those studies revealed that these fish pathogens possess mitochondria with flat cristae and were related to other fish parasites. Later, Herr et al. (39) found that the genus *Dermocystidium* was the sister taxon of *R. seeberi* and was closely related to the rosette agent. That study, in part, explained why the phenotypic features of the *Dermocystidium* spp. had been previously confused with *R. seeberi* in the tissues of their infected hosts (14, 15, 25). Cavalier-Smith (18) placed the genus *Dermocystidium* in the order Dermocystida along with the rosette agent; *R. seeberi* was added later (57) (Figure 1) (Table 1).

RHINOSPORIDIUM SEEBERI *Rhinosporidium seeberi* Seeber (82) is the etiologic agent of rhinosporidiosis, a cutaneous and subcutaneous disease of humans, other mammals, and birds, which is characterized by the formation of polypoidal masses in mucous membranes. *R. seeberi* appears in infected tissue as spherical structures referred to as sporangia. As is the case in the species of the genus *Dermocystidium*, these sporangia can grow up to 450 μm in diameter and can hold as many as several thousand endospores (Figure 10). Like some mesomycetozoeans, *R. seeberi* cannot be isolated in culture, and until recently it was classified both as a fungus and as

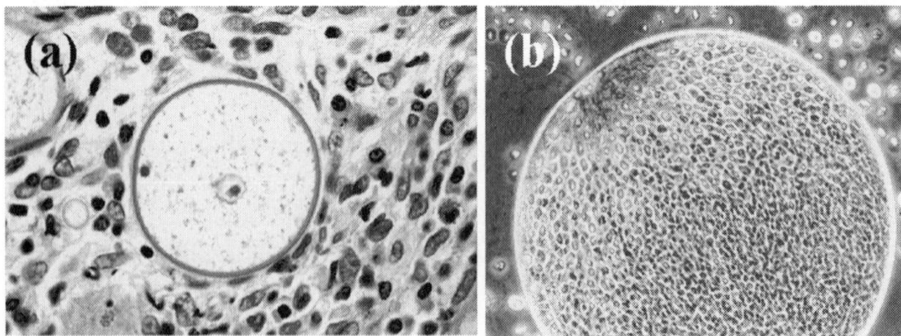

Figure 10 (*a*) Histological section of polypoidal tissue from a Sri Lankan man infected with *Rhinosporidium seeberi*, depicting an immature sporangium and inflammatory cells (X 400). (*b*) In vitro release of endospores from a mature sporangium. Note the large number of endospores within the sporangium (X 800).

a protozoan (5, 6, 23). The possible development of uniflagellated zoospores by *R. seeberi*, similar to those of *Dermocystidium* spp. and the rosette agent, is currently under investigation (D. McMeekin & L. Mendoza, unpublished data).

Light and electron microscopic analyses in the past 100 years have indicated that *R. seeberi* has a complex in vivo life cycle (5, 22) that is initiated with the release of endospores into its host's tissues from spherical bodies (100–450 μm) referred to as sporangia (Figure 10). Once implanted, the endospores increase in size and progressively develop into juvenile, intermediate, and finally mature sporangia with endospores (59). The endospores are then released and the in vivo cycle is reinitiated. These analyses, however, did not provide clues regarding *R. seeberi*'s taxonomic affinities. In 1999 a team of investigators from Michigan State University, Emory University, University of California, Berkeley, and the University of Paradeniya, Sri Lanka (39), using phylogenetic analysis, found that the 18S SSU rDNA of *R. seeberi* from humans with rhinosporidiosis clustered with the DRIPs, then a recently discovered group of fish parasites. Their data showed that *R. seeberi* was the sister taxon of two *Dermocystidium* spp. and that this trio was close to the rosette agent. This study was later corroborated by Fredericks et al. (30), whose 18S SSU rDNA sequence from a dog with rhinosporidiosis proved to be identical to that of the human isolate sequenced by Herr et al. (39). That finding strongly supported the view that *R. seeberi* may be a monotypic genus. Although Frederick et al. (30) reported that *R. seeberi* possessed mitochondria with tubular cristae, Herr et al. and Mendoza and colleagues (39, 58) demonstrated that *R. seeberi* did indeed have mitochondria with flat cristae.

The phylogenetic home of *R. seeberi* has been controversial for over a century. When Seeber described the first known infection caused by *R. seeberi* in 1900 (82), he believed that this spherical microorganism was a coccidium. Later, other investigators, using morphological and staining procedures, suggested that the pathogen was more closely related to members of the kingdom Fungi than to members of the kingdom Protoctista. Investigations mainly focused on the in vivo histopathological features of *R. seeberi* because it could not be cultivated. None of those investigations, however, provided clues as to the true nature of this pathogen. This frustration led other investigators to propose extreme views such as that *R. seeberi* was a carbohydrate waste product resulting from the ingestion of tapioca (1) or a cyanobacterium in the genus *Microcystis* (2). The recent finding that *R. seeberi*'s 18S SSU rDNA clustered with a novel clade of fish parasites in the divergence between animals and fungi ended 100 years of taxonomic uncertainties. *R. seeberi* differs from the other mesomycetozoeans in that it is the only member pathogenic to mammals and birds. Correspondingly, *R. seeberi* has several features in common with the other mesomycetozoeans: (*a*) It was previously classified as a fungus or a different type of protozoan; (*b*) it was associated with aquatic environments; (*c*) it produced spherical structures containing several daughter cells (endospores); and (*d*) *R. seeberi* and some other mesomycetozoeans are intractable to culture (Table 1).

ROSETTE AGENT This organism is an obligate intracellular fish parasite. The infections it causes were first described by Harrell et al. (35) in net-pen-reared chinook salmon (*Onchorhynchus tshawytscha*). No generic name was proposed at that time. It was only referred to as the rosette agent because of the false impression that it clustered as a six-celled organism, resembling a rosette, in infected tissues (28). Based on morphological findings, Harrell et al. (35) pointed out that this unique salmon pathogen could either be a fungus, a protozoan, or an alga.

Infected salmon develop severe anemia and lymphocytosis. Swollen kidneys and spleens occur. Gram-stained smears of the infected tissue showed gram-positive spherical structures ~5–7 μm in diameter. The spherical organisms were found within the macrophages of infected kidneys and spleens. Transmission electron microcopy of the infected areas revealed spherical cells with multilayered cell walls, vacuoles, and a prominent nucleus (Figure 11). More recently, morphological and molecular studies of several isolates of the rosette agent showed that their 18S SSU rDNAs were identical. That study suggested that the rosette agent represented a new protozoan genus and species. Details of this proposal will be published elsewhere (K. D. Arkush, personal communication).

Experimental infections using rosette agent endospores taken from infected salmon tissues failed. Recently, however, Arkush at the University of California, Davis, induced the production of uniflagellated zoospores by the cells

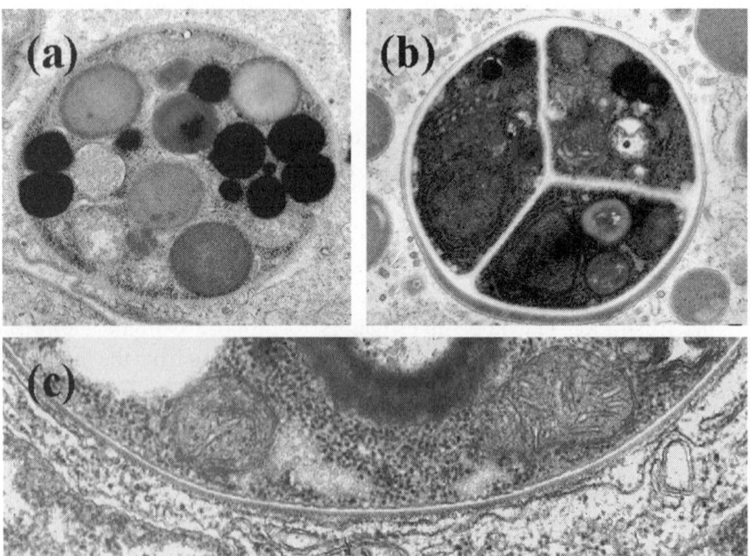

Figure 11 (*a*) Transmission electron microscopy section of a salmon kidney's intersticium infected with the rosette agent (X 27250). (*b*) Transmission electron microscopy showing the rosette agent undergoing division by progressive internal cleavage of its cytoplasm (X 24500). (*c*) Enlargement of a rosette agent's spherule showing two mitochondria with flat cristae (X 34024) (courtesy of R.A. Elston & K.D. Arkush).

of the rosette agent. The zoospores were approximately 1–2 μm in diameter and nearly spherical. They had only one flagellum ~10 μm long with a typical 9 + 2 configuration of microtubules. When the zoospores were used to infect salmon, the animals developed lesions comparable to those in natural infections (K. D. Arkush, personal communication). In addition, the rosette agent has been successfully cultured using the chinook salmon embryo cell line CHSE-214 at 15°C (4).

The rough phylogenetic relationships of the rosette agent were first revealed when Kerk et al. (44) characterized its18S SSU rDNA. Spanggaard et al. (84) corroborated this finding. Both studies revealed that the rosette agent was closely related to the choanoflagellates. Ragan et al. (77) established that this pathogen was part of a novel clade of parasites located between the divergence of fungi and animals, a finding corroborated by others (8, 12, 19, 29, 30, 39, 57, 101). In our phylogenetic analyses the rosette agent is the sister clade to *R. seeberi* and *Dermocystidium* spp., all possessing flat mitochondrial cristae and all closely related to other members of the order Ichthyophonida (Figure 1). Interestingly, Arkush et al. (4) found that the rosette agent possessed tubular mitochondrial cristae. However, a close evaluation of their electron microscopic figures and new photographic data indicated that the rosette agent might have mitochondria with flat cristae (Figure 11*c*).

RELATIONSHIPS OF THE MESOMYCETOZOEANS WITH THE OTHER CLADES ARISING NEAR THE ANIMAL-FUNGAL DIVERGENCE

The report of Ragan et al. (77) that the mesomycetozoeans form a monophyletic clade near the animal-fungal divergence was an unexpected finding. The exact position of this group, however, remains controversial. Ragan et al. (77) found that, depending on the sequences added to their phylogenetic trees, the mesomycetozoeans are either the sister group to the Animal kingdom plus the choanoflagellates or the sister group to the animals plus the choanoflagellates plus the fungi. Cavalier-Smith (18) argued that the mesomycetozoeans are the sister group to choanoflagellates plus *Corallochytrium*, and not to choanoflagellates plus both *Corallochytrium* and animals as inferred by Ragan et al. (77). Broader phylogenetic studies tend to confirm Cavalier-Smith's position (30, 39). Phylogenetic studies including additional members of the order Ichthyophonida, suggested that the mesomycetozoeans are the sister clade to the choanoflagellates and that together they are the sister group to animals and, with the inclusion of the animals, to the fungi (12, 29, 86). Our own phylogenetic analyses, using all the microbes in previous studies, indicated that mesomycetozoeans are indeed the sister group to the choanoflagellates plus *Corallochytrium* and the *Nuclearia* (Figure 1), thus supporting previous interpretations (18). However, it must be kept in mind that none of the deeper branches relating the Mesomycetozoea to other clades are well supported.

Based on the data published by several investigators (44, 77), Cavalier-Smith (18) proposed two different orders within the mesomycetozoeans, the

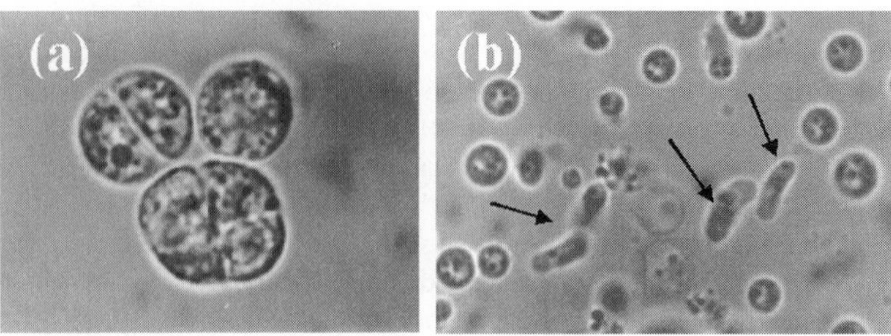

Figure 12 (*a*) Single, diad, and tetrad spherical vegetative cells of *Corallochytrium limacisporum* (X 1500). (*b*) Vegetative and elongated limax amoeba cells of *C. limacisporum* from cultures (*arrows*) (2000) (courtesy of S. Raghu-Kumar).

Dermocystida and Ichthyphonida. These two phylogenetic groups were found to differ from each other, not only on the basis of their 18S SSU rDNAs and internal transcriber spacers sequences but also on the morphological stages of their life cycles (Figure 12). For instance, the production of uniflagellate zoospores was the main feature of the *Dermocystidium* species and the rosette agent, both in the order Dermocystida. Flagellated cells are not known in *R. seeberi*, but as mentioned above, the matter is currently under investigation (Table 2). In contrast, most members of the order Ichthyophonida (*Amoebidium*, *Ichthyophonus*, and *Psorospermium* spp.) develop motile, amoeba-like cells. No such simple dichotomy exists when it comes to the morphology of mitochondrial cristae. Both *R. seeberi* and *Dermocystidium* spp. in the Dermocystida have mitochondria with flat cristae. The rosette agent was reported to have tubular mitochondrial cristae (4). However, the electron micrographs of the rosette agent's mitochondria described by Arkush et al. (4) had features compatible with flat mitochondrial cristae. Recent morphological studies of the rosette agent's mitochondria confirmed that it possesses flat cristae (Figure 11*c*). Alone among the mesomycetozoeans, *I. hoferi* seems to have tubular mitochondrial cristae. It may be that the morphology of cristae easily changed over evolutionary time or that interpretation of mitochondrial morphology is difficult in this group. As Zettler et al. (101) argued, the finding of different mitochondrial cristae types among microbes considered to be monophyletic should be carefully evaluated together with other morphological and physiological characteristics to avoid misinterpretations.

CLOSE RELATIVES OF THE MESOMYCETOZOEA AND THE EVOLUTION OF TRAITS COMMON TO ALL

The closest members of the mesomycetozoeans are *Corallochytrium limacisporum* and *Nuclearia* spp. in the classes Corallochytrea and Cristidiscoidea, and the choanoflagellates (Table 1). Study of these taxa may help us understand the origins of the Mesomycetozoea. Although *C. limacisporum* is not a mesomycetozoean, it is included in this review for its phylogenetic proximity to that group.

Corallochytrium limacisporum is a novel type of saprotrophic marine protist (78). This organism was originally isolated from a coral reef in the Indian Ocean (78). Like some members of the class Mesomycetozoea, it is a spherical, single-celled organism, 4.5–20.0 μm in diameter that undergoes several binary fissions to later release numerous elongated daughter cells (up to 32 daughters per single cell) (Figure 12). *C. limacisporum* releases its endospores through one or more pores in its cell wall, recalling the behavior of *R. seeberi*. However, in *R. seeberi* there is only one exit pore (59). The elongated released spores are amoeba-like and have a slow sinusoidal movement (Figure 12). The production of an amoebic stage is a characteristic shared with mesomycetozoeans in the order Ichthyophonida. In that order *Amoebidium parasiticum* also produces elongated spores and has an amoeba-like stage. *C. limacisporum* apparently possess mitochondria with flat cristae (19), but photomicrographs depicting these organelles have not been published, and their morphology needs verification (Table 2).

Using phylogenetic analyses of 18S SSU rDNA, Cavalier-Smith & Allsopp (19) reported that *C. limacisporum* was closely related to the choanoflagellates and to the only member of the mesomycetozoeans known at that time, the rosette agent; it was not a thraustochytrid, the group in which it had previously been placed. Based on this result, Cavalier-Smith later created a new class, the Corallochytrea (18). Our phylogenetic analyses confirmed Cavalier-Smith's placement of this microbe outside of the mesomycetozoeans and close to the choanoflagellates (Table 1) (Figure 1).

Cavalier-Smith (16–18) placed the nucleariid amoeba in the phylum Neomonada, subphylum Mesomycetozoa (new subphylum that replaces the subphylum Choanozoa), class Cristidiscoidea, order Nucleariia, closely associated with *C. limacisporum* and the mesomycetozoeans (Table 1). According to the taxonomic classification of Cavalier-Smith (18), they all share the same phylum and subphylum (Table 1). This was confirmed by Zettler et al. (101) and by our own phylogenetic analyses (Figure 1). Members of the nucleariid amoeba are classified according to their morphological characteristics, especially those related to locomotion styles and feeding habits. Mitochondrial cristae morphology is important in the classification of nucleariid amoebae, which contain flat and discoidal mitochondrial cristae, the least common of the mitochondrial types. Based on phylogenetic analyses, Zettler et al. (101) speculated that the nucleariid amoeba might have developed discoidal mitochondrial cristae independently from other microbes that possess this type of cristae, again pointing out the evolutionary plasticity of this characteristic. The recent placement of the nucleariid amoeba near the mesomycetozoeans, choanoflagellates, and corallochytrians, and the rate at which new groups are being found at the animal-fungal boundary, suggest that more microbes at the same location will be found. Both *Corallochytrium* and *Nuclearia* spp. have amoeba or amoeba-like cells and neither have flagella, whereas animals, choanoflagellates, some fungi, and some Mesomycetozoea retain flagella.

Assuming that it is much easier to lose a flagellum than to gain one, it seems that the *Corallochytrium* and *Nuclearia* spp. must have lost flagella independently of similar losses in the Ichthyophonida and several phyla of the kingdom Fungi (87).

Amoebal motility, like the flagellar type, appears to be an ancestral trait found in amoebal-flagellates like the plasmoidal slime molds and their relatives (9, 18, 87). Even in the kingdom Fungi, when flagellated zoospores, especially those of the Blastocladiales, are trapped under cover glass and microscope slides, amoebal motility is observed (55). Cell walls also appear to be ancestral, being found in the choanoflagellates, fungi, the Mesomycetozoea, the *Corallochytrium*, and *Nuclearia* spp., but not in the animals. Surely, cell walls were lost in animals. The ancestors of all groups at the animal-fungal boundary must have been unicellular, and only in the animals and fungi did multicellularity develop and become widespread. In the Mesomycetozoea *I. hoferi* is filamentous, but the filaments are cenocytic and lack regular septa. They resemble the fungal hyphae in the Blastocladiales, Entomophthorales, Mucorales, Glomales, and other zygomycete orders. The members of those fungal orders similarly lack the multicellularity and macroscopic morphology typical of fungi in the Ascomycota and Basidiomycota. Multicellularity and macroscopic structures clearly developed independently in several lineages, including the green, brown, and red plants, in addition to fungi and animals. Further study may find macroscopic Mesomycetozoea. Most Mesomycetozoea are associated with animals as parasites or commensals. We speculate that *Sphaerosoma* and the LKM51 isolate are saprotrophic and free-living. It seems likely that other saprotrophic Mesomycetozoea will be found when methods of cultivating these organisms are applied to samples taken from nature.

FUTURE DIRECTIONS OF RESEARCH IN THE MESOMYCETOZOEA

What can we expect from the Mesomycetozoea in the near future, in addition to the discovery of more members? The cultivation of more members seems likely, and with it the possibility of more detailed comparisons of mesomycetozoean development with that of the other groups at the animal-fungal boundary. For example, the problem of mitochondrial form might be solved, as well as discovery of more flagellated species in the order Dermocystida. Cultivation also increases the possibility of gathering sequences from additional DNA regions in hopes of resolving the relationships among Mesomycetozoea, fungi, animals, and choanoflagellates. For example, Loytynoja & Milikovitch (55) have compared sequences of mitochondrial ADP-ATP carriers to question the relationship of fungi and animals as sister taxa, albeit without including the Mesomycetozoea or choanoflagellates. It would be worth extending the analysis to include these groups. The geological timing of divergence of the Mesomycetozoea, fungi, and animals is also an important topic. Several estimates of the date when fungi diverged from animals have been made, ranging from ~1 billion (13) to 1.6 billion yeas ago (36). Again, the class

Mesomycetozoea is not represented in these studies, and additional sequences might lead to new estimates of the timing of the divergence of these taxa. Cultivation also introduces the possibility of genomic analyses, and given the obvious interest in the origin of the Animal Kingdom, a strong case can be made for obtaining the genome of a mesomycetozoean for comparative genomics. Important contributions can be made, however, with traditional approaches. For example, the life cycle of no member of the Mesomycetozoea is known completely. Investigations to determine if gametes are produced, and the nuclear division needed to produce them, as well as the fusion of gametes to form zygotes, would be most welcome, as they could bring genetics to the class.

ACKNOWLEDGMENTS

We thank all the investigators who shared their unpublished and in press information with us. We also thank Matías J. Cafaro and Roger A. Herr for their contributions to some sections of this review.

The *Annual Review of Microbiology* is online at http://micro.annualreviews.org

LITERATURE CITED

1. Ahluwalia KB. 1992. New interpretations in rhinosporidiosis, enigmatic disease of the last nine decades *J. Submicrosc. Cytol. Pathol.* 24:109–14
2. Ahluwalia KB. 1999. Culture of the organism that causes rhinosporidiosis. *J. Laryngol. Otol.* 113:523–28
3. Amann RI, Ludwing W, Schleifer KH. 1995. Phylogenetic identification and *in situ* detection of individual microbial cells without cultivation. *Microbiol. Rev.* 59:143–69
4. Arkush KD, Frasca S Jr, Hedrick RP. 1998. Pathology associated with the rosette agent, a systemic protist infecting salmonid fishes. *J. Aquat. Anim. Health* 10:1–11
5. Arseculeratne SN, Ajello L. 1998. *Rhinosporidium seeberi*. In *Topley and Wilson's Microbiology and Microbial Infections*, ed. L Ajello, RJ Hay, 4:595–643. London/Sydney/Auckland: Arnold
6. Ashworth JH. 1923. On *Rhinosporidium seeberi* (Wernecki, 1903) with special reference to its sporulation and affinities. *Trans. R. Soc. London Edinb.* 53:301–42
7. Azevedo C. 1989. Fine structure of *Perkinsus atlanticus* n. sp. (Apicomplexa, Perkinsea) parasite of the clam *Ruditapes decussates* from Portugal. *J. Parasitol.* 75:627–35
8. Baker GC, Beebee TJC, Ragan MA. 1999. *Prototheca richardsi*, a pathogen of anuran larvae, is related to a clade of protistan parasites near the animal-fungal divergence. *Microbiology* 145:1777–84
9. Beakes GW. 1998. Relationships between lower fungi and protozoa. See Ref. 22a, pp. 351–73
10. Beebee TJC. 1991. Purification of an agent causing growth inhibition in anuran larvae and its identification as a unicellular unpigmented alga. *Can. J. Zool.* 69:2146–53
11. Beebee TJC, Wong ALC. 1993. Stimulation of cell division and DNA replication in *Prototheca richardsi* by passage through larval amphibian guts. *Parasitology* 107:119–24

12. Benny GL, O'Donnell K. 2000. *Amoebidium parasiticum* is a protozoan, not a trichomycete (Zygomycota). *Mycologia* 92:1133–37
13. Berbee ML, Taylor JW. 2001. Fungal molecular evolution: gene trees and geologic time. In *The Mycota: Systematics and Evolution*, ed. D McLaughlin, E McLaughlin P Lemke, VII, part B:229–45. Berlin: Springer-Verlag
14. Broz O, Privora M. 1951. Two skin parasites of *Rana temporaria: Dermocystidium ranae* Guyenot & Naville and *Dermosporidium granulosum* n. sp. *Parasitology* 24:65–69
15. Carini A. 1940. Sobre um pasasito semelhante ao "*Rhinosporidium*" encontrado em quistos da pele de uma "hyla." *Arq. Inst. Biol. Sao Paulo* 11:93–96
16. Cavalier-Smith T. 1997. Amoeboflagellates and mitochondrial cristae in eukaryote evolution: megasystematics of the new protozoan subkingdoms Eozoa and Neozoa. *Arch. Protistenkünde.* 147:237–58
17. Cavalier-Smith T. 1998. A revised six-kingdom system of life. *Biol. Rev.* 73:203-66
18. Cavalier-Smith T. 1998. Neomonada and the origin of animal and fungi. See Ref. 22a, pp. 375–407
19. Cavalier-Smith T, Allsopp MTEP. 1996. *Corallochytrium*, an enigmatic non-flagellated protozoan related to choanoflagellates. *Eur. J. Protistol.* 32:306–10
20. Caullery M, Menil F. 1905. Sur les haplosporidies parasites de poisons marins. *Cr. Séanc. Soc. Biol.* 58:640–43
21. Cienkowski L. 1861. Ueber parasitische Schläuche auf Crustaceen und einigen Insektenlarven (*Amoebidium parasiticum* m.). *Bot. Ztg.* 19:169–74
22. Chuan E, Kannan-Kutty M. 1975. *Rhinosporidium seeberi*: spherules and their significance. *Pathology* 7:133–37
22a. Coombs GH, Vickerman K, Sleigh MA, Warren A, eds. 1998. *Evolutionary Relationships Among Protozoa*, Vol. 56. Dordrecht, The Netherlands: Kluwer
23. Daniels RA. 1980. Distribution and status of crayfish in the Pit River drainage, California. *Crustaceana* 38:131–38
24. de la Herran R, Garrido MA, Navas RJ, Rejon CR, Rejon MR. 2000. Molecular characterization of the ribosomal RNA gene region of *Perkinsus atlanticus*: its use in phylogenetic analysis and as a target for a molecular diagnosis. *Parasitology* 120:345–53
25. Dukerly JS. 1914. *Dermocystidium pusula* Perez, parasitic on trutta fario. *Zool. Anz.* 44:179–82
26. Dykova L, Lom J. 1992. New evidence of fungal nature of *Dermocystidium koi* Hoshina and Sahara, 1950. *J. Appl. Ichthyol.* 8:180–85
27. Eiras JC, Silva-Souza AT. 2000. A *Dermocystidium* infection in *Trichomycterus* sp. (Osteichthyes, Trichomycteridae). *Parasite* 7:323–26
28. Elston RA, Harrell L, Wilkinson MT. 1986. Isolation and in vitro characterization of Chinook salmon (*Oncorhyncus tshawytscha*) rosette agent. *Aquaculture* 56:1–21
29. Figueras A, Lorenzo G, Ordas MC, Gouy M, Novoa B. 2000. Sequence of the small subunit ribosomal RNA gene of *Perkinsus atlanticus*-like isolated from carpet shell clam in Galicia, Spain. *Mar. Biotechnol.* 2:419–28
30. Fredericks DN, Jolley JA, Lepp PW, Kosek JC, Relman DA. 2000. *Rhinosporidium seeberi*: a human pathogen from a novel group of aquatic protistan parasites. *Emerg. Infect. Dis.* 6:273–82
31. Goggin CL. 1994. Variation in the internal transcribed spacers and 5.8S ribosomal RNA from 5 isolates of marine parasites *Perkinsus* (Protist, Apicomplexa). *Mol. Biochem. Parasitol.* 65:179–82
32. Günter V, Rug M. 1999. Life stages and tentative life cycle of *Psorospermium haeckeli*, a species of the novel DRIPs clade from the animal-fungal dichotomy. *J. Exp. Zool.* 283:31–42
33. Haeckel E. 1857. Ueber die gewebe des

flusskrebses. *Arch. Anat. Physiol. Wiss. Med.* 24:469–568
34. Deleted in proof
35. Harrell LW, Elston RA, Scott TM, Wilkinson MT. 1986. A significant new systemic disease of net-pen reared chinook salmon (*Oncorhyncus tshawytscha*) brook stock. *Aquaculture* 55:249–62
36. Heckman DS, Geiser DM, Eidell BR, Stauffer RL, Kardos NL, Hedges SB. 2001. Molecular evidence for the early colonization of land by fungi and plants. *Science* 293:1129–33
37. Hedrick RP, Friedman CS, Modin J. 1989. Systemic infection in Atlantic salmon *Salmo salar* with a *Dermocystidium*-like species. *Dis. Aquat. Org.* 7:171–77
38. Herman RL. 1984. *Ichthyophonus*-like infection in newts (*Notophthalmus viridescens* rafinisque). *J. Wildl. Dis.* 20:55–56
39. Herr AR, Ajello L, Taylor JW, Arseculeratne SN, Mendoza L. 1999. Phylogenetic analysis of *Rhinosporidium seeberi*'s 18S small-subunit ribosomal DNA groups this pathogen among members of the protoctistant mysomycetozoan clade. *J. Clin. Microbiol.* 37:2750–54
40. Herr AR, Ajello L, Mendoza L. 1999. Chitin synthase class 2 (*CHS2*) gene from the human and animal pathogen *Rhinosporidium seeberi*. *Proc. 99th Gen. Meet. Am. Soc. Microbiol.* p. 296 (Abstr.)
41. Deleted in proof
42. Hilgendorf F. 1883. Bemerkungen über die sogenanntekrebspest, insbesondere über *Psorospermium haeckeli* spec. nova. *Sitz. Gess. Nat. Freun. Berlin* 9:179–83
43. Hugenholtz P, Goebel BM, Pace NR. 1998. Impact of culture-independent studies on the emerging phylogenetic view of bacterial diversity. *J. Bacteriol.* 180: 4765–74
44. Kerk D, Gee A, Standish M, Wainwright PO, Drum AS, et al. 1995. The rosette agent of Chinook salmon (*Oncorhyncus tshawytscha*) is closely related to choanoflagellates, as determined by phylogenetic analyses of its small ribosomal subunit RNA. *Mar. Biol.* 122:187–92
45. Kocan RM, Hershberger P, Mehl T, Elder N, Bradley M, et al. 1999. Pathogenicity of *Ichthyosphonus hoferi* for laboratory-reared pacific herring *Clupea pallasi* and its early appearance in wild Puget Sound herring. *Dis. Aquat. Org.* 35:23–29
46. Kotob SI, McLaughlin SM, van Berkum P, Faisal M. 1999. Discrimination between two *Perkinsus* spp. isolated from the softshell clam, *Mya arenaria*, by sequence analysis of two internal transcribed spacer region and the 5.8S ribosomal RNA gene. *Parasitology* 119:363–68
47. Krucinska J, Simon E. 1968. On the parasites and epibionts of the branchial cavity in crayfish at Wrocaw and vicinity. *Przegald. Zool.* 12:288–90
48. Leeddale GF. 1974. How many are the kingdoms of organisms? *Taxon* 32:261–70
49. Léger L. 1924. Sur un organisme du type *Ichthyophonus* parasite du tube digestif de la Lote d'eau douce. *Cr. Hebd. Seanc. Acad. Sci. Paris* 197:785–87
50. Léger L, Hesse E. 1923. Sur un champignon du type *Ichthyophonus* parasite de l'intestin de la truite. *Cr. Hebd. Séanc. Acad. Sci. Paris* 176:420–22
51. Levine ND. 1970. Taxonomy of the sporozoa. *J. Parasitol.* 56:208–9
52. Lichtwardt RW. 1997. Costa Rican gut fungi (Trichomycetes) infecting lotic insect larvae. *Rev. Biol. Trop.* 45:1339–83
53. Lichtwardt RW, Arenas J. 1996. Trichomycetes in aquatic insects from southern Chile. *Micologia* 88:844–57
54. Lichtwardt RW, Williams MC. 1990. Trichomycete gut fungi in Australian aquatic insect larvae. *Can. J. Bot.* 1057–74
55. Loytynoja A, Milikovitch MC. 2001. Molecular phylogenetic analyses of the mitochondrial ADP-ATP carriers: the Plantae/Fungi/Metazoa trichotomy revisited *Proc. Natl. Acad. Sci. USA* 98:10202–7
56. Margulis L. 1996. Archeal-eubacterial

mergers in the origin of eukarya: phylogenetic classification of life. *Proc. Natl. Acad. Sci. USA* 93:1071–76

57. Mendoza L, Ajello L, Taylor JW. 2001. The taxonomic status of *Lacazia loboi* and *Rhinosporidium seeberi* has been finally resolved with the use of molecular tools. *Rev. Iberoam. Micol.* 18:95–98
58. Mendoza L, Herr AR, Ajello L. 2001. Causative agent of rhinosporidiosis. *J. Clin. Microbiol.* 39:413–15
59. Mendoza L, Herr RA, Arseculeratne SN, Ajello L. 1999. *In vitro* studies on the mechanism of endospore release by *Rhinosporidium seeberi*. *Mycopathologia* 148: 9–15
60. Mort-Bontemps M, Gay L, Fevre M. 1997. *CHS2*, a chitin synthase gene from the Oomycete *Saprolegnia monoica*. *Microbiol.* 143:2009–20
61. Mulisch M. 1993. Chitin in protistan organisms. *Eur. J. Protistol.* 29:1–18
62. Neish GA, Hughes GC. 1980. Fungal diseases of fishes, Book 6. In *Diseases of Fishes*, ed. SF Snieszco, HR Axelrod, pp. 61–100. USA/New Jersey/Gt. Britain/ Surrey: TFH
63. Nylund V, Westman K, Laintmaa K. 1983. Ultrastructure and taxonomic position of the crayfish parasite *Psorospermium haeckeli* Heilgendorf. *Freshw. Crayfish* 5:307–14
64. Okamoto N, Nakase K, Suzuki H, Nakai Y, Fujii K, Sano T. 1985. Life history and morphology of *Ichthyophonus hoferi* in vitro. *Fish Pathol.* 20:273–85
65. Olson RE, Dungan CF, Holt RA. 1991. Water-borne transmission of *Dermocystidium salmonis* in the laboratory. *Dis. Aquat. Org.* 12:41–48
66. Olson RE, Holt RA. 1995. The gill pathogen *Dermocystidium salmonis* in Oregon salmonids. *J. Aquat. Anim. Health* 7:111–17
67. Ordás MC, Figueras A. 1998. *In vitro* culture of *Perkinsus atlanticus*, a parasite of the carpet shell clam *Ruditapes decussates*. *Dis. Aquat. Org.* 33:129–36
68. Ordás MC, Novoa B, Faisal M, Mclaughlin S, Figueras A. 2001. Proteolytic activity of cultured *Pseudoperkinsus tapetis* extracellular products. *Comp. Biol. Physiol. B* 130:199–206
69. Ordás MC, Novoa B, Figueras A. 2000. Modulation of the chemiluminescence response of Mediterranean mussel (*Mytilus galloprovencialis*) haematocytes. *Fish Shellfish Immunol.* 10:611–22
70. Paperna I, Kim SH. 1996. Ultrastructure of the etiological agent of systemic granuloma in goldfish. *Dis. Aquat. Org.* 26:43–47
71. Perkins FO. 1996. The structure of *Perkinsus marinus* (Mackin, Owen and Collier, 1950) Levine, 1978 with comments on the taxonomy and phylogeny of *Perkinsus* spp. *J. Shellfish Res.* 15:67–87
72. Plehn M, Mulsow K. 1911. Der erreger der "taumelkrankheir" der salmoniden. *Zentbl. Bakt. ParasitKde, Abt.* 58:63–68
73. Pore RS. 1985. *Prototheca* taxonomy. *Mycopathologia* 90:129–39
74. Porter D, Smiley R. 1979. Ribosomal RNA molecular weights of trichomycetes and zygomycetes. *Exp. Mycol.* 3:188–93
75. Rand TG. 1994. An unusual form of *Ichthyophonus hoferi* (Icthyophonales: Ichthyophonaceae) from yellowtail flounder *Limanda ferruginea* from the Nova Scotia shelf. *Dis. Aquat. Org.* 18:21–28
76. Rand TG, White K, Cannone JJ, Gutell RR, Murphy CA, Ragan MA. 2000. *Ichthyophonus irregularis* sp. nov. from the yellowtail flounder *Limanda ferruginea* from the Nova Scotia shelf. *Dis. Aquat. Organ.* 41:31–36
77. Ragan MA, Goggins CL, Cawthorn RJ, Cerenius L, Jamienson AVC, et al. 1998. A novel clade of protistan parasites near the animal-fungal divergence. *Proc. Natl. Acad. Sci. USA* 93:11907–12
78. Raghu-Kumar S. 1987. Occurrence of the traustochytrid, *Corallochytrium limacisperum* gen. et sp. nov. in the coral reef lagoons of the Lakshadweep Islands in the Arabian Sea. *Bot. Mar.* 30:83–89

79. Richards CM. 1958. The inhibition of growth in crowded *Rana pipiens* tadpoles. *Physiol. Zool.* 31:138–51
80. Rondon MR, Augus PR, Bettermann AD, Brady SF, Grossman TH, et al. 2000. Cloning the soil metagenome: a strategy for accessing the genetic and functional diversity of uncultured microorganisms. *Appl. Environ. Microbiol.* 66:2541–47
81. Rug M, Vogt G. 1994. Developmental stages of the crayfish parasite *Psorospermium haeckeli* in thoraxic arteries of *Astacus astacus*. *J. Invert. Pathol.* 64:153–55
82. Seeber GR. 1900. *Un nuevo sporozuario parasito del hombre. Dos casos encontrados en polypos nasals.* MD thesis, Univ. Nacional Buenos Aires, Imprenta Libreria "Boullosa"
83. Spanggaard B, Huss HH, Brescianni J. 1995. Morphology of *Ichthyophonus hoferi* assessed by light and scanning electron microscopy. *J. Fish Dis.* 18:567–77
84. Spanggaard B, Skouboe P, Rossen L, Taylor JW. 1996. Phylogenetic relationships of the intercellular fish pathogen *Ichthyophonus hoferi*, and fungi, choanoflagellates and the rosette agent. *Mar. Biol.* 126:109–15
85. Trotter MJ, Whistler HC. 1965. Chemical composition of the cell wall of *Amoebidium parasiticum*. *Can. J. Bot.* 43:869–76
86. Ustinova I, Kienitz L, Huss VAR. 2000. *Hyaloraphidium curvatum* is not a green alga, but a lower fungus; *Amoebidium parasiticum* is not a fungus, but a member of the DRIPs. *Protist* 151:253–62
86a. van Hannen EJ, Mooij W, van Agterveld MP, Gons HJ, van Laanbroek P. 1999. Detritus-dependent development of the microbial community in an experimental system: quality analysis by denaturing gradient gel electrophoresis. *Appl. Environ. Microbiol.* 65:2478–84
87. Vickerman K. 1998. Revolution among the protozoa. See Ref. 22a, pp. 1–24
88. Vítovec SCJ, Lom J, Kubu JHF. 1974. Dermocystidiosis—a gill disease of the carp due to *Dermocystidium cyprini* n.sp. *J. Fish Biol.* 6:689–99
89. Vogt G. 1999. In vitro induction of further development of sporocysts of the crayfish parasite *Psorospermium haeckeli*. *Freshw. Crayfish* 12:319–34
90. Vogt G, Keller M, Brandis D. 1996. Occurrence of *Psorospermium haeckeli* in the stone crayfish *Austropotamobius torrentium* from a population naturally mixed with the noble crayfish *Astacus astacus*. *Dis. Aquat. Org.* 25:233–38
91. Vogt G, Rug M. 1995. Microscopic anatomy and histochemistry of the crayfish parasite *Psorospermium haeckely*. *Dis. Aquat. Org.* 21:79–90
92. Vogt G, Rug M. 1999. Life stages and tentative life cycle of *Psorospermium haeckeli*, a species of a novel DRIPs clade from the animal-fungal dichotomy. *J. Exp. Zool.* 283:31–42
92a. von Dietheilm-Scheer D. 1979. *Psorospermium orconectis* n. sp. a new parasite in *Orconectes limosus*. *Arch. Protistenkd.* 121:381–91
93. Voronin VN. 1975. On findings of *Psorospermium lacustris*. *Izvest. Gos. Nauch. Instit. Oz. Irech. Ryb. Khoz.* (Leningrad) 93:125–26
94. Ward DM, Weller R, Bateson MM. 1990. 16S rDNA sequences reveal numerous uncultured microorganisms in natural community. *Nature* 345:63–65
95. Wainright PO, Hinkle G, Sogin ML, Stickel SK. 1993. Monophyletic origins of the metazoans: an evolutionary link with fungi. *Science* 260:340–42
96. Whisler HC. 1962. Culture and nutrition of *Amoebidium parasiticum*. *Am. J. Bot.* 29:193–99
97. Whisler HC. 1968. Developmental control of *Amoebidium parasiticum*. *Dev. Biol.* 17:562–70
98. Whisler HC, Fuller M. 1968. Preliminary observations on the holdfast of *Amoebidium parasiticum*. *Micologia* 60:1068–79
99. Wong A, Beebee T. 1994. Identification of a unicellular, non-pigmented alga

that mediates growth inhibition in anuran tadpoles: a new species of the genus *Prototheca* (Chlorophyceae: Chlorococcales). *Hydrobiologia* 277:85–96

100. Wootten R, Mcvicar AH. 1982. *Dermocystidium* from cultured eels, *Anguilla anguilla* L., in Scotland. *J. Fish Dis.* 5:215–22

101. Zettler LAA, Nerad TA, O'Kelly CJ, Sogin ML. 2001. The nucleariid amoeba: more protists at the animal-fungal boundary. *J. Eukaryot. Microbiol.* 48:293–97

Annu. Rev. Microbiol. 2002. 56:345–69
doi: 10.1146/annurev.micro.56.012302.160749

First published online as a Review in Advance on May 7, 2002

METABOLIC DIVERSITY IN AROMATIC COMPOUND UTILIZATION BY ANAEROBIC MICROBES

Jane Gibson and Caroline S. Harwood
Department of Microbiology, 3-432 Bowen Science Building, The University of Iowa, Iowa City, Iowa 52242; e-mail: caroline-harwood@uiowa.edu

Key Words biodegradation, dehalorespiration, humic substances, bacteria, aromatic hydrocarbons

■ **Abstract** A vast array of structurally diverse aromatic compounds is continually released into the environment due to the decomposition of green plants and as a consequence of human industrial activities. Increasing numbers of bacteria that utilize aromatic compounds in the absence of oxygen have been brought into pure culture in recent years. These include most major metabolic types of anaerobic heterotrophs and acetogenic bacteria. Diverse microbes utilize aromatic compounds for diverse purposes. Chlorinated aromatic compounds can serve as electron acceptors in dehalorespiration. Humic substances serve as electron shuttles to enable the use of inorganic electron acceptors, such as insoluble iron oxides, that are not always easily reduced by microbes. Substituents that are attached to aromatic rings may serve as carbon or energy sources for microbes. Examples include acyl side chains and methyl groups. Finally, aromatic compounds can be completely degraded to serve as carbon and energy sources. Routes by which various types of aromatic compounds, including toluene, ethylbenzene, phenol, benzoate, and dihydroxylated compounds, are degraded have been elucidated in recent years. Biochemical strategies employed by microbes to destabilize the aromatic ring in preparation for degradation have become apparent from this work.

CONTENTS

INTRODUCTION

About 25% of the Earth's biomass is composed of compounds that have a benzene ring as the main structural constituent. Green plants synthesize the overwhelming proportion of these aromatic compounds and assemble them to form lignin, a polymer that is highly stabilized by a complexity of ether and other linkages. The processes that lead to lignin depolymerization and the ultimate breakdown of the aromatic compounds generated during depolymerization are poorly understood. In addition to natural sources, human ingenuity and activities have added a plethora of new aromatic chemicals to the environment. Although the input of these xenobiotic compounds is less in total amount than that of plant materials, their novel structures pose major challenges to the microbial communities that are the major recyclers of natural products. Concerns about environmental pollution have resulted in many attempts to define conditions under which aromatic compound degradation can occur, as this would help in the development of strategies for accelerating rates of biodegradation.

Many of the aromatic nucleus-containing materials that are released into the environment by humans or by the degradation of plant material find their way into anaerobic sediments. This has prompted studies of aromatic compound degradation in the absence of oxygen, where the already well-understood aerobic reactions catalyzed by mono- or dioxygenases, or the peroxidases that initiate attack on lignins cannot occur. The past decade has seen major advances in our understanding of the range and rates at which aromatic compounds of varying degrees of complexity can be degraded in natural environments that are devoid of oxygen. More importantly, a number of bacterial strains that degrade aromatic compounds in pure culture have recently been isolated, opening the door to more detailed physiological, biochemical, and molecular genetic studies. Many of these microbes and the compounds they degrade are listed in Table 1, which is provided as supplementary material on the web (follow the Supplemental Material link in the online version of this chapter or at http://www.annualreviews.org). Several factors have facilitated these isolations. Among the most significant is the recognition that a broader range of electron acceptors can be used for anaerobic respiration than had been previously thought. The use of nitrate and sulfate for this purpose has been known for a long time, but the significance of Fe(III) reduction and chlorate respiration has only recently been recognized. Ferric iron is widespread and abundant in most environments, and bacteria that can benefit from using this compound as a terminal electron acceptor are widely distributed (77). Chlorate used as a bleaching agent, disinfectant, and component of rocket propellants has been a significant anthropogenic addition to natural environments only for a small number of decades. Even so, it is possible to isolate strains from many habitats that use chlorate, as well as nitrate, as electron acceptors in the absence of oxygen (22). Use of inert carriers such as heptamethylnonane (93) or Amberlite XAD7 (3, 85) to mediate the slow release of toxic organic compounds into growth media has enabled the cultivation of microbes on substrates that may be tolerated over only a narrow concentration range. Bacteria thought to grow only as obligate syntrophs in coculture with a hydrogen-utilizing microbe such as a methanogen were found to grow in pure cultures on crotonate (6, 57). Finally, many small improvements in handling and culturing strict anaerobes have made the long incubations required for slow-growing cultures technically feasible.

Important new concepts have emerged recently. These include recognition that interspecies transfers of the kind first defined for hydrogen by Wolin and colleagues (56) occur with other metabolic intermediates, including acetate, formate, and probably many others in natural ecosystems. Also that some reactions thought to be the province of organic chemists can be carried out by enzymes. Examples include the biological Birch reduction involved in anaerobic benzoate ring reduction and a Kolbe-Schmidt-type reaction for aromatic dehydroxylation reactions. In addition there is an appreciation that enzyme radical chemistry is critical in overcoming the stability of aromatic as well as of alkyl hydrocarbons, as demonstrated by the proposed reaction mechanisms for initial attack on ethylbenzene and toluene.

This review considers the physiological roles that aromatic compounds play in the lives of facultatively or obligately anaerobic microorganisms, and how these lead to the partial or complete degradation annually of large quantities of biomass and industrial by-products. This viewpoint naturally restricts our focus to studies with isolated pure strains or with defined cocultures. We have concentrated on work published during the past five years and restricted the range of aromatic compounds considered almost entirely to monocyclic compounds. Many excellent reviews on various aspects of the anaerobic utilization of aromatics give more detailed accounts, especially of the biochemistry (47, 52, 53, 77, 78, 97, 115).

AROMATIC COMPOUNDS AS ELECTRON ACCEPTORS AND SHUTTLES

General Considerations

Many microorganisms can utilize substituted and complex aromatic compounds in the absence of oxygen in ways that do not perturb the benzene nucleus itself (Figure 1). Aromatic compounds can participate in energy metabolism by serving

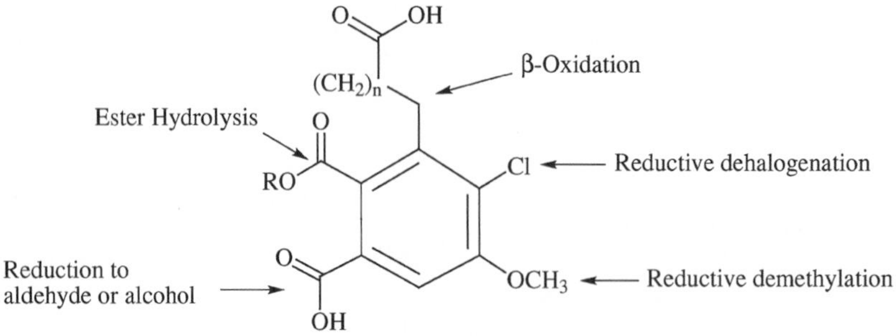

Figure 1 Commonly encountered attacks on aromatic ring substituents that are mediated by microbes.

as electron acceptors, generally with accompanying modifications of ring substituents. An example would be the use of chlorinated aromatic compounds as electron acceptors in dehalorespiration. Aromatic compounds can also serve as electron shuttles. For instance, extracellular quinones can bridge the activities of different kinds of microorganisms by transferring electrons from the respiratory chain of one microorganism to an insoluble electron acceptor or even to another microorganism (81). This provides a mechanism for extending the possibilities for interspecies transfer to the level of a single electron.

Humic Substances

These are complex mixtures of partial lignin degradation products that can be extracted from soils of many kinds and that contain a number of quinonoid compounds. Lovley et al. (80) showed that humic substances could serve as electron acceptors for natural sediment populations and for pure cultures of *Geobacter* or *Shewanella* species. Amounts of acetate that were well in excess of the Fe(III) content of the medium were oxidized when humic substances were present. EPR measurements showed that quinonoid molecules in the mixture were reduced. Subsequently, it has been found that humic substances could be replaced by the convenient analog anthraquinone disulfonate (AQDS).

Humics can serve as electron acceptors. In natural situations, however, it is likely that they more commonly serve as electron shuttles. Microbes reduce quinones to hydroquinones, which in turn react spontaneously to reduce insoluble iron oxides. The oxidized quinone can then shuttle back to the microbe to be reduced again (Figure 2). Substantially higher rates of respiration, and therefore growth, are thus achieved in the presence of humic substances. The Lovley group (81) has also emphasized that humic substances can be reduced by a number of facultatively anaerobic heterotrophic bacteria such as *Paracoccus denitrificans* under anaerobic conditions. This opens the way for ferric iron in the environment to act as terminal electron acceptor for microorganisms that lack the ability to reduce Fe(III) directly (104). Reduced humic substances or anthrahydroquinone disulfonate (the product

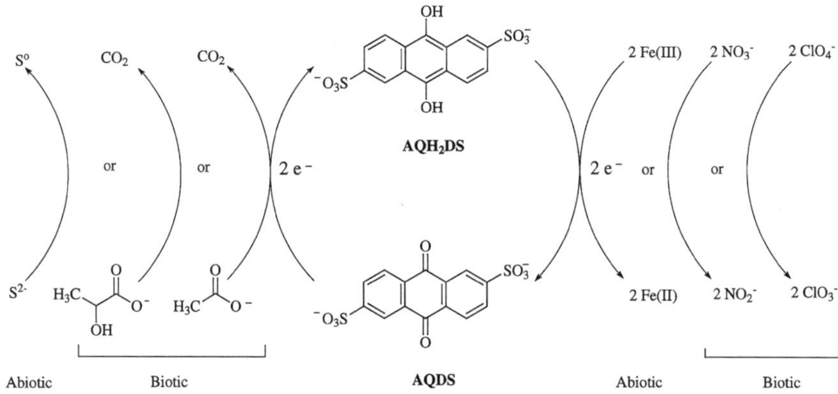

Figure 2 Use of the humic substance analog anthraquinone disulfonate (AQDS) as an electron shuttle to mediate the reduction of inorganic compounds. This enables the oxidation of organic or inorganic compounds by microbes.

of AQDS reduction) can also be used as electron donor for reduction of other terminal electron acceptors such as nitrate, fumarate, arsenate, or selenate (81) (Figure 2). These and related studies suggest that naturally occurring aromatic derivatives can play an important role in interspecies distribution of reducing equivalents.

Compounds produced by the microorganisms themselves may also facilitate electron transfer to abundant but poorly soluble potential acceptors. An intriguing report by Newman & Kolter (88) demonstrated that mutants of *Shewanella putrefaciens* that were unable to reduce humic substance or AQDS differed from the wild type in failing to secrete a small redox-active menaquinone derivative, which was required to mediate the reduction.

Dehalogenation

Public concern about environmental contamination, particularly by halogenated industrial products, has led to extensive investigations of the fates of such compounds in anaerobic environments. Many studies have shown that reductive dehalogenation of aromatic as well as aliphatic compounds occurs fairly rapidly in anaerobic sediments (116), and a large number of pure cultures of anaerobes that carry out reductive dehalogenations have been isolated (54).

Understandably, major emphasis has been placed on the dehalogenation reactions themselves, and less attention has been given to analyzing in detail the advantages a microorganism might derive from catalyzing these reactions. Dehalogenations without further conversions of products are carried out by many different groups of strictly anaerobic and facultatively anaerobic microbes. Under anaerobic conditions, dehalogenations are almost invariably reductive, and consequently haloaryl as well as haloalkyl compounds can serve as electron acceptors for the process of dehalorespiration. Usually those bacteria that have this potential grow more rapidly with alternative electron acceptors such as nitrate or sulfate, or even by fermentation of a readily usable substrate such as pyruvate. It is only when they are required to use a more reduced carbon substrate, for example lactate or acetate, and are also constrained by the availability of potential electron acceptors that their ability to use halogenated compounds as an electron acceptor becomes evident. The isolates that can dehalogenate aromatic compounds appear to have moderately relaxed substrate specificity, but a few highly specialized isolates, such as strain 2CP-1 (23) and several *Desulfitobacterium* strains (54), that dehalogenate only haloaromatic substrates have been described.

Some strictly anaerobic bacteria, such as *Desulfitobacterium frappieri* (16), that dechlorinate a range of aromatic chlorobenzoates or phenols also use other electron acceptors such as sulfite, thiosulfate, or nitrate when the carbon and energy source is pyruvate. A strain of *Desulfovibrio* TBP-1 grows anaerobically on lactate with a number of mono- or dibromophenols in place of oxidized sulfur compounds, but it is unable to use chlorinated or fluorinated or iodinated phenols (17). A strong preference for the positions that are dehalogenated has been evident in many, but not all, isolates. *Desulfomonile tiedgei* (26) preferentially catalyzes *m*-dechlorinations of substituted benzoates, while *Desulfitobacterium dehalogenans* (108, 109) reacts almost exclusively with chlorophenols that have *o*-substituents. *D. frappieri*, however, dechlorinates aromatic compounds with substituents in the para, meta, and ortho positions (16). The number of different compounds that are dehalogenated anaerobically in enrichment sediment samples is much larger than the number of compounds found to be dehalogenated by pure cultures. One must expect therefore that many additional dehalogenating strains or defined consortia of strains exist and have yet to be brought into captivity.

The discovery that halogenated compounds could replace sulfate as electron acceptors opened the possibility that such reactions might contribute to the microorganisms' energy metabolism as well as redox balance. The enzymes involved in dehalogenation of tetrachlorethylene have been extensively investigated, leading to the suggestion that a short electron transport chain is involved, which can develop a proton gradient (54). Energy conservation accompanying dehalogenation of aromatic compounds does not appear to be universally possible. It has, however, been convincingly demonstrated in a number of strains, in the first place during the use of 3-chlorobenzoate by *Desulfomonile tiedgei* [(29, 84); see (54) for a full review of the evidence]. Careful comparisons of growth yields in cultures of *D. dehalogenans* with limited carbon source, or with electron donors such as H_2 or formate, indicated that the increment generated from dehalorespiration is fairly modest (110). Moreover, energy was conserved from dehalogenation only when substrate-level phosphorylation was precluded. The major physiological advantage of reductive dehalogenation is likely to arise less from the generation of an electrochemical gradient than from the use of chlorinated compounds as electron acceptors. This frees cells from having to use any of the available carbon as electron acceptors in fermentation reactions. However, the enzymes that carry out dehalorespiration of aromatic derivatives resemble aliphatic dehalogenases in being membrane-bound, and a comparable energy conservation system based on vectorial proton translocation has not been excluded.

The enzymes that catalyze dehalogenation reactions have repeatedly been found to be inducible by growth with halogenated aromatics, as in the very thorough study of Utkin et al. (108). Dehalogenases active with haloaromatics have been purified from *D. tiedgei* (89) and *Desulfitobacterium hafniense* (20), *D. dehalogens* (111), and *Desulfitobacterium chlororespirans* (64). Whereas the *D. tiedgei* 3-chlorobenzoate dehalogenase is a heme enzyme, all other reductive dehalogenases that have been purified are corrinoid-containing iron-sulfur proteins. The purified enzymes from *Desulfitobacterium* species all require a hydroxyl group ortho to the chlorine substituent in order for dehalogenation to occur. As indicated above, activities of whole cells suggest, however, that aromatic dehalogenases with different substrate specificities exist but have not yet been purified.

AROMATICS AS SOURCES OF ENERGY AND/OR CARBON

General Considerations

Microorganisms able to use the entire molecule of aromatic compounds as sources of carbon for growth, or as both carbon and energy sources, in the absence of oxygen are distributed among almost all groups of the Bacteria. Until this year, no Archaea with this ability had been described, although a strain of *Haloferax* could use phenylpropionic acid aerobically (41). Lovley's group (106) has demonstrated that the strictly anaerobic and hyperthermophilic archeon *Ferroglobus placidus*, originally isolated by others by incubating under nitrate-reducing conditions (45), can utilize a number of aromatic compounds, provided Fe(III) is added as an electron acceptor in place of nitrate. Molar growth yield measurements clearly indicate that the aromatic ring itself is attacked. Complete mineralization of aromatic molecules under anoxic conditions appears to be a trait restricted to prokaryotes (48), although fungi may carry out superficial attacks on complex molecules.

The rate of growth and the growth yield that is achieved by a particular microbe when it utilizes an aromatic compound as a source of both carbon and energy depends not only on the biochemistry of the degradation pathway but also on the fate of the reducing equivalents released. Clearly the use of any outside acceptor for electrons increases the proportion of substrate carbon that can be used for biosynthesis. The greatest energy conservation with a given substrate is obviously to be expected when nitrate or chlorate is the final electron acceptor (NO_3^-/NO_2^-, +430 mV; ClO_4^{2-}/ClO_3^{2-}, +1190 mV). Use of ferric iron (Fe^{III}/Fe^{II}, (~0–400 mV)) [see (104)] as an electron acceptor also allows considerable energy conservation. The midpoint potential of the SO_4^{2-}/SO_3^{2-} couple is only −250 mV, so the potential for energy conservation when sulfate is the electron acceptor is much more limited. Fermentative bacteria are even more tightly constrained energetically. So far, the only strains of fermentative bacteria that have been obtained in pure culture are those that use simple aromatic substrates, such as benzoate, as carbon and energy sources. These fermentative strains tend to be restricted to a syntrophic existence when they are growing on aromatic compounds. Typically, complete biodegradation becomes energetically feasible only when accompanying methanogens or sulfate-reducing bacteria promptly use metabolic end products such as hydrogen that are generated during aromatic compound breakdown. Photosynthetic bacteria avoid many of the constraints imposed on their nonphotosynthetic counterparts, since all their energy is derived from light and aromatic compounds are degraded to form intermediary metabolites such as acetyl-CoA, which are subsequently used in biosynthetic reactions. Carbon and energy sources are thus uncoupled from each other.

Biochemical Strategies for Aromatic Substrate Degradation

The outline and much of the biochemical details of anaerobic attack on an array of simple and substituted aromatic compounds are now known. The substituents attached to more complex aromatics may serve as carbon and energy, or even as nitrogen sources, for microorganisms that are unable to attack the benzene ring (Figure 1). These reactions may serve the additional purpose of making an attack on the ring feasible for those bacteria that can perform complete mineralization but cannot remove some blocking substituents. Benzoyl-CoA emerges as the most common intermediate in the degradation of a diversity of aromatic molecules that are halogenated, methoxylated, or have carbon side chains. Benzoyl-CoA is also a frequent intermediate in the degradation of monohydroxylated aromatic substrates and of some dihydroxylated compounds such as catechol (97).

A few common themes are recognizable in many of the pathways that have been studied in detail. One is that aromatic compounds that are destined to enter the benzoyl-CoA degradation pathway must either carry a carboxyl group (i.e., be an aromatic acid) to begin with or they must be carboxylated to form an aromatic acid in one of the initial metabolic steps. Reactions of this kind initiate attack on phenol, *o*-cresol (2-methylphenol), catechol, and hydroquinone (47, 97). All of the intermediates of the reductive benzoate degradation pathway are CoA thioesters, and intermediates in the funneling pathways that lead to benzoyl-CoA formation are themselves frequently thioesterified at a carboxyl group with coenzyme A (47). Initial attack on aryl hydrocarbons requires condensation with fumarate, followed by β-oxidation reactions that lead to formation of a thioesterified carboxyl group (115) (Figure 3).

Aromatic compounds with two or more hydroxyl groups are less stabilized by the resonance of the benzene ring and are more easily degraded (Figure 4). The degradation of these compounds does not necessarily include carboxylation as an initial step, and further hydroxylations or rearrangements provide alternative ways to decrease the stability of the ring (97). In fact, trihydroxylated aromatic acids, such as gallate, undergo a decarboxylation reaction prior to further degradation (70).

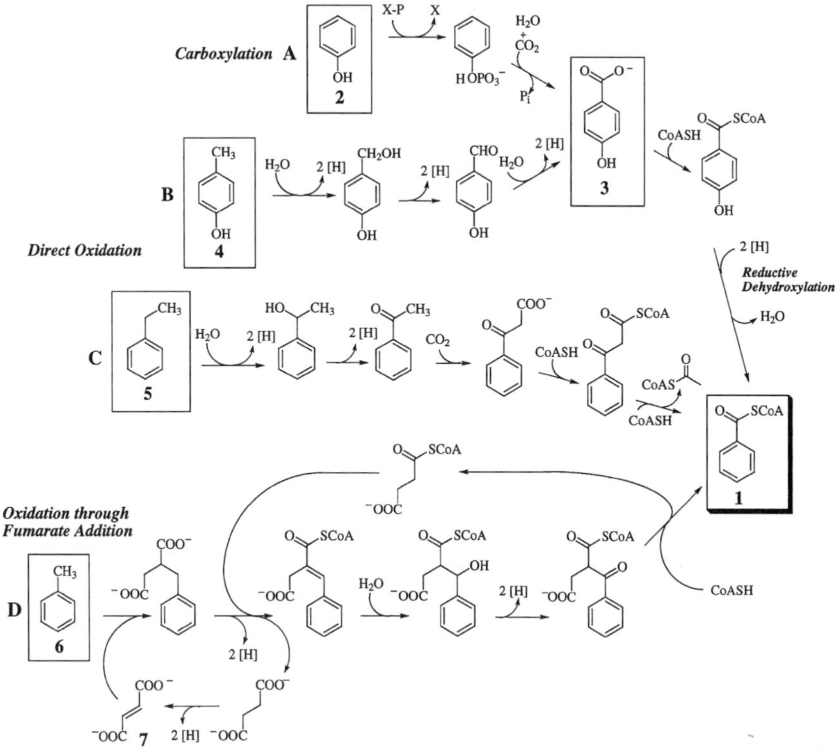

Figure 3 Representative routes of carboxylation and direct oxidation that are followed by bacteria in the conversion of phenol [2] and 4-cresol [4] to 4-hydroxybenzoate [3] and then to benzoyl-CoA [1] as a central intermediate. Ethylbenzene [5] is directly oxidized to benzoyl-CoA, and toluene [6] is oxidized to benzoyl-CoA following fumarate [7] addition.

REACTIONS THAT REMOVE RING SUBSTITUENTS AND BENEFIT THE MICROBE

Many microbes utilize aromatic ring substituents to their advantage while leaving the benzene ring untouched. Such modifications include acyl side chain removal, demethoxylation, and ester hydrolysis (Figure 1).

Acyl Side Chains

The side chains of phenylalkanoates, such as cinnamate, are readily degraded by β-oxidation, yielding acetyl groups for biosynthesis or for energy. Benzoate may appear transiently in enrichments (51) or in pure cultures of the phototroph *Rhodopseudomonas palustris* growing on phenylalkanoates (35), or it may accumulate as an end product (25, 44).

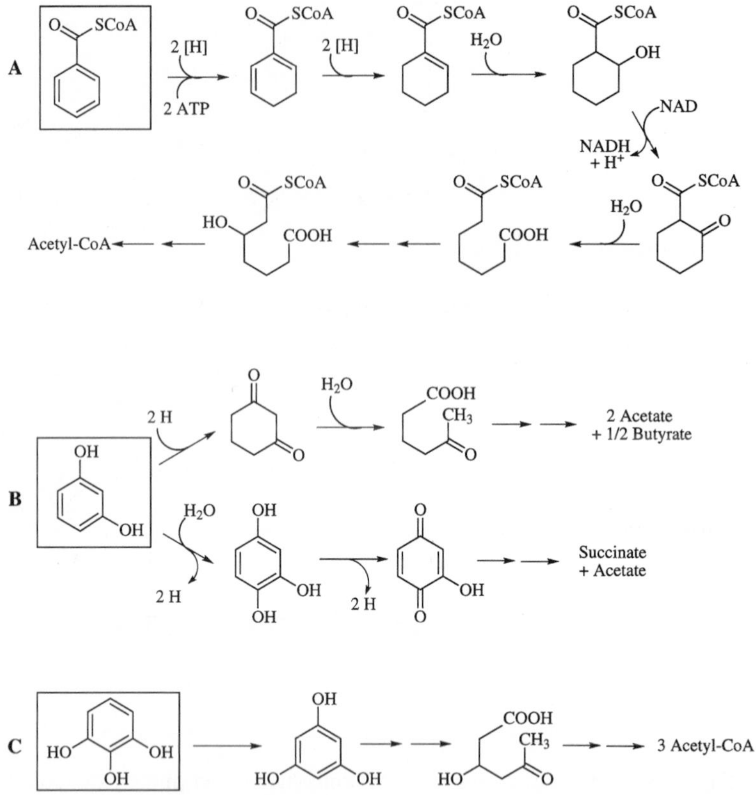

Figure 4 Anaerobic ring cleavage pathways. (*A*) Benzoyl-CoA pathway in *R. palustris*. A slightly different pathway is used by *T. aromatica*. Monohydroxylated and some dihydroxylated aromatic compounds are processed though benzoyl-CoA. (*B*) Dihydroxylated compounds such as resorcinol can be degraded either oxidatively or reductively by routes that do not involve benzoyl-CoA. (*C*) Initial steps in the degradation of trihydroxybenzene.

Aromatic Esters

Aromatic esters are hydrolyzed by enzymes from various sources. Among anaerobic fungi found in the rumen, such as *Neocallimastix* strain MC-2, *p*-coumaroyl, and feruloyl, esterases play an important part in the degradation of plant cell walls (14, 15). The physiological value of these enzymes to the producer lies in improving access to the xylanases that serve as the fermentable substrates for their growth. The aromatic ring does not appear to be attacked by these microorganisms.

The mammalian urine constituent hippurate (benzoylglycine) can be used anaerobically as well as aerobically by a range of microorganisms, among them nonsulfur purple photosynthetic bacteria. Whereas *Rhodobacter capsulatus* uses only the glycine resulting from hydrolysis by its hippuricase, as indicated by both recovery of benzoate from the culture supernatant and the molar growth yield of the culture, *R. palustris* uses both hydrolysis products (82).

Methoxylated Compounds

Methoxylated aromatic molecules are major components of lignins. A number of acetogenic and other bacteria can use the methyl group of phenylmethyl ethers for synthesis of acetic acid. The CH_3 portion is removed by reactions involving corrinoid and tetrahydrofolate cofactors, but the phenolic derivative of the reactions is not used (27, 65, 66, 114).

Aromatic Nitrogen Sources

Nitrotoluenes are readily reduced either partially or completely in anaerobic environments to corresponding amino compounds, which are considerably more stable. There is evidence that anaerobic enrichments of microbes can use these products as nitrogen source, but the identity of the microorganisms that benefit in this way has not yet been established (39, 71).

MOBILIZING THE BENZENE RING FOR COMPLETE DEGRADATION

As noted above, benzoyl-CoA enters a central pathway of ring reduction and cleavage that allows for the complete degradation of diverse aromatic compounds to acetyl-CoA and carbon dioxide under anaerobic conditions. Many natural and synthetic aromatics either lack the carboxyl group that is necessary for benzoyl-CoA to be ultimately formed or they carry ring substituents that must first be removed. Some examples of modifications of aromatic compounds that occur to prepare them for entry into central ring degradation pathways follow, and are shown in Figure 3.

Amino Groups

Anthranilate (2-aminobenzoate) supports anaerobic growth of denitrifying bacteria as sole carbon and nitrogen source (18). In *Thauera aromatica*, the molecule is first thioesterified and subsequently reductively deaminated to benzoyl-CoA (76). Aniline serves as carbon and nitrogen source for both sulfate-reducing (98) and denitrifying strains (61). The molecule is first carboxylated to 4-aminobenzoate, converted to the CoA thioester, and then reductively deaminated as in *T. aromatica* (61, 99).

Halogen Groups

Monohalogenated benzoates can be completely degraded and serve as carbon sources for a variety of anaerobes, for example, some denitrifiers of the *Thauera/Azoarcus* group (46, 102). Some but not all strains of the phototrophic bacterium

R. palustris also degrade chlorinated benzoates (112). A strain (RCB100) of *R. palustris* that can grow with 3-chlorobenzoate as sole carbon source forms 3-chlorobenzoyl-CoA as an initial intermediate, and this is then dehalogenated to yield benzoyl CoA (31). Most laboratory strains of *R. palustris* are unable to degrade 3-chlorobenzoate, but enrichments usually yield 3-chlorobenzoate-degrading strains (31). Moreover, strains that do not degrade 3-chlorobenzoate initially will often acquire this ability upon extended incubation with 3-chlorobenzoate as the sole utilizable carbon source. This suggests that just one or a few mutations may suffice for phototrophs to acquire the ability to convert chlorinated benzoates to benzoyl-CoA. A similar phenomenon may contribute to the long lags commonly observed in establishing nonphotosynthetic dehalogenating enrichments.

Phthalates

Phthalates are aromatic dicarboxylates that are widely used industrially. Various isomers have repeatedly been shown to be degraded under nitrate-reducing, sulfate-reducing, and methanogenic conditions, but few pure cultures have been studied. A denitrifying *Pseudomonas*-like strain accumulated benzoate transiently (90, 91) during growth on phthalate, indicating that one of the carboxyl groups was removed before the ring was degraded. The enzymology of phthalate degradation has not been defined.

Hydroxylated Compounds

At least four distinct pathways for the degradation of hydroxylated aromatic compounds have been described (97). The specific route taken appears to be dictated in large part by the number of hydroxyl groups and their relative positions on the ring. Compounds with a single hydroxyl group are processed through the benzoyl-CoA pathway, following introduction of a carboxyl group when this is not already present. Phenol carboxylation by *T. aromatica* has been extensively studied. Phenylphosphate, formed using an unidentified phosphoryl donor, is carboxylated to give 4-hydroxybenzoate, and this is thioesterified and then reductively dehydroxylated (72, 73). Rapid progress is being made in defining the genes and enzyme mechanisms involved (19). It is not yet clear whether the energy-requiring carboxylation reaction is also used by bacteria that have more restricted energy budgets, but phenol utilization under methanogenic or sulfate-reducing conditions has commonly been observed (5, 42).

The reversible 4- and 3,4-hydroxybenzoate carboxylases/decarboxylases from *Clostridium hydroxybenzoicum* (49, 50, 55, 118) have strikingly different properties from the phenol carboxylase of *T. aromatica*. *Cl. hydroxybenzoicum* uses amino acids for growth and does not completely degrade either phenol or 4-hydroxybenzoate. Thus it is not yet clear what physiological advantage this bacterium may derive from carrying out the carboxylation/decarboxylation reaction.

The pathways and the bacteria that can use aromatic substrates bearing two or more hydroxyl groups or a hydroxyl group in addition to other substituents have been comprehensively reviewed (97) and are only superficially described here (Figure 4). Hydroquinones and catechols are often fed into the benzoate pathway in sulfate-reducing bacteria and fermenters, using carboxylation reactions that have not been studied in detail. Dihydroxylated compounds can also be degraded, particularly by denitrifying bacteria, by routes that do not include benzoyl-CoA as an intermediate. When more than one hydroxyl is present, destabilization of the aromatic ring may be achieved by tautomerization, sometimes after adding another hydroxyl group. Tautomerized intermediates may then react either oxidatively or reductively (Figure 4), and subsequent ring opening is easily achieved. Although detailed studies have only been carried out with nitrate-reducing, sulfate-reducing, and some fermentative bacteria, it is probable that similar metabolic steps will be found in both iron and chlorate reducers. The availability and nature of the electron acceptor(s) used by the cultures govern the end products of the degradations.

Aromatic Hydrocarbons

As with halogenated compounds, potential toxicity and widely distributed contamination have prompted a plethora of studies on aromatic hydrocarbon degradation. Strategies used by anaerobic bacteria of various physiological types for the degradation of the mono aromatic compounds benzene, toluene, ethylbenzene, and xylene (BTEX) have been examined (53). So far, all studies indicate that the initial attack on these compounds is geared to converting them to benzoyl-CoA. Specific, inducible enzymes catalyze several unusual reactions in the sequences of reactions leading to benzoyl-CoA formation.

TOLUENE Of the BTEX components, toluene appears to be most readily degraded in anaerobic microcosms established from contaminated, or even pristine, sites (53, 79, 92, 95, 103). Benzylsuccinate, which appears extracellularly in generally small quantities (~1%) during toluene degradation, was at first believed to be a metabolic dead-end by-product (40). Benzylsuccinate is formed by the condensation of toluene with fumarate and this reaction is now recognized as the initial step in toluene degradation in strains of *Thauera* and *Azoarcus*, as well as a sulfate-reducing strain (7, 8, 10, 67). Benzylsuccinyl-CoA is then converted by β-oxidation reactions to form benzoyl-CoA (Figure 3). Many of the enzymes catalyzing toluene degradation have been purified and the corresponding genes cloned (1, 24, 74, 75). It is becoming clear that addition of fumarate is a general reaction used to activate hydrocarbons, and it occurs also in the initial reactions with other methylated aromatics such as xylenes (7, 9, 37, 67), *m*- and *p*-cresol (86, 87), and methyl naphthalene (3). It also has a counterpart in the first reaction undergone by alkanes (69, 94). So far, it appears that fumarate is the only cellular metabolite that can initiate the reaction sequence.

ETHYLBENZENE Although less rapidly removed from contaminated sites, ethylbenzene can yet serve as carbon and energy sources for nitrate- and sulfate-reducing bacteria (93, 96), provided steps are taken to keep its concentration in the growth medium low. Maximum growth rates are low; doubling times of 30–48 h seem to be common. Ethylbenzene is oxidized to 1-phenylethanol and then converted in a series of reactions to benzoyl-CoA (Figure 3). The steps leading to benzoyl-CoA formation have been defined, and the initiating enzyme, which converts ethylbenzene to 1-phenylethanol, has been purified, characterized, and the corresponding gene cloned (58, 59, 62, 63).

BENZENE Benzene has been shown to disappear from contaminated soils and aquifers with accompanying reduction of ferric iron, nitrate, or sulfate, and stable anoxic microcosms in which benzene was oxidized have also been studied (2, 53, 78). Despite intense efforts, pure cultures of benzene-degrading bacteria have not yet been obtained from such enrichments, which has precluded biochemical studies. This frustrating state of affairs is likely to change shortly. Coates et al. (21) have found that two strains of *Dechloromonas*, isolated under different conditions, could grow with benzene as the sole carbon source when nitrate was supplied as electron acceptor. Cells transferred to fresh benzene-containing medium grew as well as the original culture, and the doubling times appeared to be only a small number of hours. These striking observations provide obvious opportunities for elucidating the physiology and biochemistry of benzene degradation under anaerobic conditions.

NAPHTHALENE Naphthalene-degrading bacteria have been obtained in pure culture (43), but most work on mobilization of naphthalene and related compounds has been done with stable enrichments of microbes (3, 83, 105). There is evidence that a carboxyl group is added to the 2-carbon of one ring of naphthalene by a reaction(s) that has not been studied in detail (83, 117). The uncarboxylated ring is then reduced by a series of reactions that may be analogous to the benzoyl-CoA reduction outlined below. 2-Methylnaphthalene degradation has been shown to be initiated by fumarate addition (3).

BENZOATE REDUCTION

Benzoate is the aromatic compound that is most commonly degraded by anaerobic bacteria, and the processes involved have been more thoroughly studied than any other aspect of anaerobic aromatic compound degradation. Benzoate degradation pathways have been worked out in the phototroph *R. palustris* and the denitrifier *T. aromatica* K172 (Figure 4). Both are initiated by conversion of benzoate to benzoyl-CoA by specific CoA ligases (47), and this product is attacked reductively to give a cyclohexadienecarboxyl-CoA intermediate (13). Detailed studies with the four-subunit-enzyme benzoyl-CoA reductase from *T. aromatica* indicate that two

ATP are consumed in each two-electron reduction reaction (107). The subsequent steps of ring modification and ring cleavage differ in *T. aromatica* and in *R. palustris*, but each pathway eventually leads to the formation of 3-hydroxypimelyl-CoA, which is degraded to give three molecules of acetyl-CoA and one carbon dioxide. Genes encoding enzymes of anaerobic benzoate degradation have been cloned from both *R. palustris* and *T. aromatica*, and the gene sequences reflect the observed similarities and differences in the pathways used by each organism (47).

Whether the limited energy resources of benzoate-utilizing syntrophs and sulfate-reducing bacteria could support the same reductive enzymology as is utilized by the phototrophs and denitrifiers has been questioned (47, 101). In the syntrophic-fermenting bacteria, the generation of three ATP by substrate-level phosphorylation from conversion of benzoate to three molecules of acetate would be negated by the initial investment of three ATP (one for benzoate activation by a CoA ligase and two for benzoyl-CoA reduction). It has been suggested that some of the initial energy investment could be spared if benzoyl-CoA were reduced to the level of cyclohex-1-ene-carboxylate or cyclohexanecarboxylate by the concerted addition of four or six reducing equivalents in an ATP-independent reaction that has yet to be demonstrated. Evidence that is consistent with this idea is the observation that *Syntrophus aciditrophicus* transiently releases significant quantities (at least 20% of benzoate used) of cyclohexanecarboxylate into its growth medium during benzoate degradation when grown in a defined coculture with *Methanospirillum hungatei* (36). Cyclohexadienecarboxylate, the expected two-electron reduction product, was not released during benzoate degradation. Measurements of some appropriate enzyme activities (36), suggests that the steps following benzoyl-CoA ring reduction are the same in *S. aciditrophicus* and in *R. palustris* (34, 47).

In all syntrophs examined so far, benzoate is activated to its CoA derivative by an ATP-consuming ligase rather than by a CoA transferase. A proton-translocating pyrophosphatase that can theoretically recover one third of the ATP that is expended when benzoate is converted to benzoyl-CoA, pyrophosphate, and AMP, has been sought and identified in *Syntrophus gentianae* (100) and should be expected also in *S. aciditrophicus*.

Benzoate degradation intermediates appear to exist as CoA thioesters in *S. aciditrophicus* (36), and release of the free-acid forms of these compounds into the external medium thus requires that the CoA moiety first be cleaved off. This seems tantamount to the loss of a high-energy bond and accumulation of such a large amount of cyclohexanecarboxylate in the culture medium during benzoate degradation by *S. aciditrophicus* is therefore surprising. These considerations suggest that cyclohexanecarboxylate formation by syntrophs must confer some advantage to cells. The possibility that cyclohex-1-ene-carboxylate or its CoA derivative serves as an electron acceptor for *S. aciditrophicus* cells growing in pure culture on benzoate suggests one advantage and has recently received experimental support (38). Product analyses indicate that reducing equivalents generated during the fermentation of one molecule of benzoate to acetate and carbon dioxide are used to reduce a second benzoate molecule to cyclohexanecarboxylate. Pure cultures of *S.*

aciditrophicus do not utilize cyclohexanecarboxylate that is produced during benzoate fermentation. The observation that cyclohexanecarboxylate is both produced and then reutilized by cocultures of the syntroph and a hydrogen-utilizing microbe may reflect that *S. aciditrophicus* is physiologically biased toward a fermentative mode of growth on benzoate, even when a hydrogen-utilizing partner is present.

INSIGHTS FROM GENOMICS AND MOLECULAR APPROACHES

The use of traditional genetic approaches has lagged behind traditional biochemical investigations as a means of elucidating pathways of anaerobic aromatic compound degradation. This is partly because many of the microbes that have been used in studies are difficult to grow routinely as colonies on solid media. Also, relatively little effort has been put toward the development of genetic systems for use in those microbes that do grow well on plates. Molecular approaches have been successfully applied to *R. palustris* to identify regulatory genes involved in aromatic compound degradation (28) and to construct mutants for purposes of verifying proposed routes of degradation (34). Recently, an *Azoarcus* sp. strain T mutant with a disruption in one of the genes encoding benzylsuccinate synthase was constructed and found to be defective in growth on both toluene and *m*-xylene (1). This simple experiment provided the valuable information that the same enzyme that initiates toluene degradation initiates *m*-xylene degradation.

Genes encoding dozens of enzymes involved in various anaerobic aromatic degradation pathways have been sequenced. Clones carrying genes of interest have been identified by complementation of mutants (24), by immunoscreening with antisera raised against purified proteins (30), or most commonly, by using nucleotide probes that were designed from N-terminal amino acid sequences of purified enzymes. Inspection of gene sequences sometimes reveals features that are important for catalytic function but that were not immediately obvious from studies of purified enzymes. Based on sequence analysis it seemed probable that pyrogallol-phloroglucinol transhydroxylase, 4-hydroxybenzoyl-CoA reductase, and ethylbenzene dehydrogenase were molybdenum-containing enzymes (4, 28, 58, 62). Inferences about the mechanisms that govern efficient catalysis by benzoyl-CoA reductase, 4-hydroxybenzoate-CoA reductase, and benzylsuccinate synthase that were gleaned from gene sequence analysis have been confirmed by experimental studies (11, 12, 68).

As a final example, the N-terminal amino acid sequences of proteins extracted from two-dimensional polyacrylamide gels led to the identification of a cluster of genes from *T. aromatica* that encodes phenol-inducible proteins (19). The gene sequences suggest that the phenylphosphate carboxylation reaction may be mechanistically similar to 3-octaprenyl-4-hydroxybenzoate carboxy-lyase, an enzyme that catalyzes a decarboxylation reaction in ubiquinone biosynthesis. The reversible 4-hydroxybenzoate carboxylase from *Cl. hydroxybenzoicum* also bears some similarity to the same biosynthetic enzyme (55). Although phenylphosphate carboxylase and 4-hydroxybenzoate decarboxylase appear to be dissimilar enzymes, gene sequences suggest that the enzymes may have more in common than is apparent from a superficial comparison of their properties.

Molecular biology has revolutionized studies of microbial ecology by providing tools to examine the presence of specific genes, and therefore metabolic capabilities, in all sorts of soil and water environments. Typically, nucleotide primers specific to genes encoding particular enzymes are used to amplify DNA that has been extracted from environmental samples using polymerase chain reactions. This approach has not yet been widely applied to study aromatic compound degradation in anoxic environments, as there are as yet only a few gene sequences available from different organisms for any given enzyme. Without a set of related genes in hand, it is difficult to know which regions of the gene are the most highly conserved, and therefore the best suited for primer design. This situation is likely to change soon as more examples of gene sequences for enzymes like benzylsuccinate synthase, benzoyl-CoA reductase, and ethylbenzene dehydrogenase become available. Recently, primers specific to reductive dehalogenases have been developed and used to amplify reductive dehalogenase-like genes from dechlorinating microbial consortia (113).

At this writing the whole genome sequence of just one organism that can completely degrade aromatic compounds in the absence of oxygen is available; that of the phototroph *R. palustris*. The genome of the reductive dechlorinating bacterium *Desulfitobacterium hafniense* has also been sequenced (60). However, since it is now possible to sequence microbial genomes in a matter of weeks, one can anticipate that in the next few years the sequences of anaerobic aromatic degrading microbes representative of each metabolic type (denitrifiers, sulfate reducers, and Fe(III) reducers) will become available. Comparative genomics can be expected to yield valuable insights into aspects of aromatic compound degradation that are universal and aspects that are specific to a particular metabolic group. Approaches of functional genomics, such as DNA microarray analysis and whole-cell mass spectrometry, which take advantage of whole genome sequence information, provide the means for developing a comprehensive picture of the complete biodegradation potential of an individual microbial species. It should also be possible to determine how multiple degradation pathways are functionally coordinated at the level of a single microbe to operate simultaneously.

Another advantage of having the nucleotide sequences of large fragments of DNA is that one can more easily identify and study regulatory genes, transporters, and other ancillary genes that are important or essential for degradation but that do not directly contribute to catalysis. Several regulatory genes that control anaerobic benzoate and 4-hydroxybenzoate degradation were identified by examining a benzoate degradation gene cluster from *R. palustris* and by creating site-directed mutants to verify functions that were suspected based on nucleotide sequences (32, 33). A fragment of DNA encoding the ortho-chlorophenol reductive dehalogenase from *Desulfitobacterium dehalogenans* includes a gene for an integral membrane protein that may act as a membrane anchor for the dehalogenase (111).

FUTURE CHALLENGES AND PROSPECTS

We now have a better understanding of the microbial world's anaerobic degradation potential as a result of studies carried out in the past decade. Starting with initial observations of aromatic compound disappearance in anoxic sediments, metabolically diverse microbes have steadily been brought into captivity as pure cultures. These have served as the experimental subjects for a range of physiological and biochemical studies aimed mainly at elucidating the enzymatic basis for various aromatic compound degradation pathways. The number of microbes that are present in pure cultures is nevertheless still relatively small, and there is no reason to suppose that the full range of microorganisms that participate in anaerobic decomposition of aromatic compounds is known. In particular, defined groups of microorganisms that normally participate in a complex web of metabolite transfers between different species have only occasionally been cultivated.

To date, humans have tended to study aspects of biodegradation that are of interest to humans. Hence the strong emphasis on microbes and enzymes involved in BTEX degradation. Relatively little attention has been given to the microbial decomposition of green plant material in anaerobic environments. It may be useful practically to try to understand biodegradation from the point of view of the participating microbes. How does the possession and utilization of selected aromatic degradation pathways contribute to the success of that organism in the physical and biological environment in which it lives? With the possible exception of some halorespirers, aromatic substrates are among the less favored substrates for anaerobic bacteria. Degradation enzymes are induced only under appropriate conditions of absence of oxygen and presence of substrates, and the growth rate supported is almost always substantially lower than in the presence of more favored carbon and energy sources. Measurements of standing concentrations and fluxes of as large a number as possible of potential nutrients should help us understand factors that control the rates of aromatic compound degradation. We also need sensitive approaches for coaxing microbes to tell us more about how they perceive their environment. With the availability of complete genome sequences, differential displays of gene expression using microarrays can be used to provide insights into the effects of environmental changes. Such technical developments can be expected to have major impacts on our basic understanding of the degradation of aromatic substances in the absence of oxygen, as well as on the ability to manipulate the outcome of natural processes.

ACKNOWLEDGMENTS

We are extremely grateful to Dr. Janelle Torres Y Torres for extensive help with the Figures and with the Table. Work from the laboratory of C.S. Harwood was supported by the U.S. Department of Energy and the U.S. Army Research Office.

The *Annual Review of Microbiology* is online at http://micro.annualreviews.org

LITERATURE CITED

1. Achong GR, Rodriguez AM, Spormann AM. 2001. Benzylsuccinate synthase of *Azoarcus* sp. strain T: cloning, sequencing, transcriptional organization, and its role in anaerobic toluene and m-xylene mineralization. *J. Bacteriol.* 183:6763–70
2. Anderson RT, Lovley DR. 2000. Anaerobic bioremediation of benzene under sulfate-reducing conditions in a petroleum-contaminated aquifer. *Environ. Sci. Technol.* 34:2261–66
3. Annweiler E, Materna A, Safinowski M, Kappler A, Richnow HH, et al. 2000. Anaerobic degradation of 2-methylnaphthalene by a sulfate-reducing enrichment culture. *Appl. Environ. Microbiol.* 66:5329–33
4. Baas D, Retey J. 1999. Cloning, sequencing and heterologous expression of pyrogallol-phloroglucinol transhydroxylase from *Pelobacter acidigallici*. *Eur. J. Biochem.* 265:896–901
5. Bak F, Widdel W. 1986. Anaerobic degradation of phenol and phenol derivatives by *Desulfobacterium phenolicum* sp. nov. *Arch. Microbiol.* 146:177–80
6. Beaty PS, McInerney MJ. 1987. Growth of *Syntrophomonas wolfei* in pure culture on crotonate. *Arch. Microbiol.* 147:389–93
7. Beller HR, Spormann AM. 1997. Anaerobic activation of toluene and o-xylene by addition to fumarate in denitrifying strain T. *J. Bacteriol.* 179:670–76
8. Beller HR, Spormann AM. 1997. Benzylsuccinate formation as a means of anaerobic toluene activation by sulfate-reducing strain PRTOL1. *Appl. Environ. Microbiol.* 63:3729–31
9. Beller HR, Spormann AM. 1999. Substrate range of benzylsuccinate synthase from *Azoarcus* sp. strain T. *FEMS. Microbiol. Lett.* 178:147–53
10. Biegert T, Fuchs G, Heider J. 1996. Evidence that anaerobic oxidation of toluene in the denitrifying bacterium *Thauera aromatica* is initiated by formation of benzylsuccinate from toluene and fumarate. *Eur. J. Biochem.* 238:661–68
11. Boll M, Fuchs G, Meier C, Trautwein A, El Kasmi A, et al. 2001. Redox centers of 4-hydroxybenzoyl-CoA reductase, a member of the xanthine oxidase family of molybdenum-containing enzymes. *J. Biol. Chem.* 276:47853–62
12. Boll M, Fuchs G, Meier C, Trautwein A, Lowe DJ. 2000. EPR and Mossbauer studies of benzoyl-CoA reductase. *J. Biol. Chem.* 275:31857–68
13. Boll M, Laempe D, Eisenreich W, Bacher A, Mittelberger T, et al. 2000. Nonaromatic products from anoxic conversion of benzoyl-CoA with benzoyl-CoA reductase and cyclohexa-1,5-diene-1-carbonyl-CoA hydratase. *J. Biol. Chem.* 275: 21889–95
14. Borneman WS, Hartley RD, Akin DE, Ljungdahl LG. 1990. Feruloyl and *p*-coumaroyl esterase from anaerobic fungi in relation to plant cell wall degradation. *Appl. Microbiol. Technol.* 33:345–51
15. Borneman WS, Ljungdahl LG, Hartley RD, Akin DE. 1991. Isolation and characterization of *p*-coumaroyl esterase from the anaerobic fungus *Neocallimastix* strain MC-2. *Appl. Environ. Microbiol.* 57:2337–44
16. Bouchard B, Beaudet R, Villemur R, McSween G, Lepine F, Bisaillon JG. 1996. Isolation and characterization of *Desulfitobacterium frappieri* sp. nov., an anaerobic bacterium which reductively dechlorinates pentachlorophenol to 3-chlorophenol. *Int. J. Syst. Bacteriol.* 46: 1010–15
17. Boyle AW, Phelps CD, Young LY. 1999. Isolation from estuarine sediments of a *Desulfovibrio* strain which can grow on

lactate coupled to the reductive dehalogenation of 2,4,6-tribromophenol. *Appl. Environ. Microbiol.* 65:1133–40

18. Braun K, Gibson DT. 1984. Anaerobic degradation of 2-aminobenzoate (anthranilic acid) by denitrifying bacteria. *Appl. Environ. Microbiol.* 48:102–7
19. Breinig S, Schiltz E, Fuchs G. 2000. Genes involved in anaerobic metabolism of phenol in the bacterium *Thauera aromatica. J. Bacteriol.* 182:5849–63
20. Christiansen N, Ahring BK, Wohlfarth G, Diekert G. 1998. Purification and characterization of the 3-chloro-4-hydroxyphenylacetate reductive dehalogenase of *Desulfitobacterium hafniense. FEBS Lett.* 436:159–62
21. Coates JD, Chakraborty R, Lack JG, O'Connor SM, Cole KA, et al. 2001. Anaerobic benzene oxidation coupled to nitrate reduction in pure culture by two strains of *Dechloromonas. Nature* 411: 1039–43
22. Coates JD, Michaelidou U, Bruce RA, O'Connor SM, Crespi JN, Achenbach LA. 1999. Ubiquity and diversity of dissimilatory (per)chlorate-reducing bacteria. *Appl. Environ. Microbiol.* 65:5234–41
23. Cole JR, Cascarelli AL, Mohn WW, Tiedje JM. 1994. Isolation and characterization of a novel bacterium growing via reductive dehalogenation of 2-chlorophenol. *Appl. Environ. Microbiol.* 60: 3536–42
24. Coschigano PW, Wehrman TS, Young LY. 1998. Identification and analysis of genes involved in anaerobic toluene metabolism by strain T1: putative role of a glycine free radical. *Appl. Environ. Microbiol.* 64: 1650–56
25. Defnoun S, Labat M, Ambrosio M, Garcia JL, Patel BK. 2000. *Papillibacter cinnamivorans* gen. nov., sp. nov., a cinnamate-transforming bacterium from a shea cake digester. *Int. J. Syst. Evol. Microbiol.* 50:1221–28
26. DeWeerd KA, Mandelco L, Tanner RS, Suflita JM. 1991. *Desulfomonile tiedgei* gen. nov. and sp. nov., a novel anaerobic, dehalogenating, sulfate reducing bacterium. *Arch. Microbiol.* 154:23–30
27. DeWeerd KA, Saxena A, Nagle DP Jr, Suflita JM. 1988. Metabolism of the 18O-methoxy substituent of 3-methoxybenzoic acid and other unlabeled methoxybenzoic acids by anaerobic bacteria. *Appl. Environ. Microbiol.* 54:1237–42
28. Dispensa M, Thomas CT, Kim MK, Perrotta JA, Gibson J, Harwood CS. 1992. Anaerobic growth of *Rhodopseudomonas palustris* on 4-hydroxybenzoate is dependent on AadR, a member of the cyclic AMP receptor protein family of transcriptional regulators. *J. Bacteriol.* 174:5803–13
29. Dolfing J. 1990. Reductive dechlorination of 3-chlorobenzoate is coupled to ATP production and growth in an anaerobic bacterium DCB-1. *Arch. Microbiol.* 153: 264–66
30. Egland PG, Gibson J, Harwood CS. 1995. Benzoate-coenzyme A ligase, encoded by BadA, is one of three ligases able to catalyze benzoyl-coenzyme A formation during anaerobic growth of *Rhodopseudomonas palustris* on benzoate. *J. Bacteriol.* 177:6545–51
31. Egland PG, Gibson J, Harwood CS. 2001. Reductive, coenzyme A-mediated pathway for 3-chlorobenzoate degradation in the phototrophic bacterium *Rhodopseudomonas palustris. Appl. Environ. Microbiol.* 67:1396–99
32. Egland PG, Harwood CS. 1999. BadR, a new MarR family member, regulates anaerobic benzoate degradation by *Rhodopseudomonas palustris* in concert with AadR, an Fnr family member. *J. Bacteriol.* 181:2102–9
33. Egland PG, Harwood CS. 2000. HbaR, a 4-hydroxybenzoate sensor and FNR-CRP superfamily member, regulates anaerobic 4-hydroxybenzoate degradation by *Rhodopseudomonas palustris. J. Bacteriol.* 182:100–6
34. Egland PG, Pelletier DA, Dispensa M,

Gibson J, Harwood CS. 1997. A cluster of bacterial genes for anaerobic benzene ring biodegradation. *Proc. Natl. Acad. Sci. USA* 94:6484–89

35. Elder DJ, Morgan P, Kelly DJ. 1992. Anaerobic degradation of trans-cinnamate and omega-phenylalkane carboxylic acids by the photosynthetic bacterium *Rhodopseudomonas palustris*: evidence for a beta-oxidation mechanism. *Arch. Microbiol.* 157:148–54
36. Elshahed MS, Bhupathiraju VK, Wofford NQ, Nanny MA, McInerney MJ. 2001. Metabolism of benzoate, cyclohex-1-ene carboxylate, and cyclohexane carboxylate by *Syntrophus aciditrophicus* strain SB in syntrophic association with H(2)-using microorganisms. *Appl. Environ. Microbiol.* 67:1728–38
37. Elshahed MS, Gieg LM, McInerney MJ, Suflita JM. 2001. Signature metabolites attesting to the in situ attenuation of alkylbenzenes in anaerobic environments. *Environ. Sci. Technol.* 35:682–89
38. Elshahed MS, McInerney MJ. 2001. Benzoate fermentation by the anaerobic bacterium *Syntrophus aciditrophicus* in the absence of hydrogen-using microorganisms. *Appl. Environ. Microbiol.* 67:5520–25
39. Esteve-Núñez A, Caballero A, Ramos JL. 2001. Biological degradation of 2,4,6-trinitrotoluene. *Microbiol. Mol. Biol. Rev.* 65:335–52
40. Evans PJ, Ling W, Goldschmidt B, Ritter ER, Young LY. 1992. Metabolites formed during anaerobic transformation of toluene and o-xylene and their proposed relationship to the initial steps of toluene mineralization. *Appl. Environ. Microbiol.* 58:496–501
41. Fu W, Oriel P. 1999. Degradation of 3-phenylpropionic acid by *Haloferax* sp. D1227. *Extremophiles* 3:45–53
42. Gallert V, Winter J. 1992. Comparison of 4-hydroxybenzoate decarboxylase and phenol carboxylase in crude extracts in a defined 4-hydroxybenzoate- and phenol-degrading anaerobic consortium. *Appl. Microbiol. Biotechnol.* 37:119–24
43. Galushko A, Minz D, Schink B, Widdel F. 1999. Anaerobic degradation of naphthalene by a pure culture of a novel type of marine sulphate-reducing bacterium. *Environ. Microbiol.* 1:415–20
44. Grbic-Galic D. 1986. O-demethylation, dehydroxylation, ring-reduction and cleavage of aromatic substrates by *Enterobacteriaceae* under anaerobic conditions. *J. Appl. Bacteriol.* 61:491–97
45. Hafenbradl D, Keller M, Dirmeier R, Rachel R, Rossnagel P, et al. 1996. *Ferroglobus placidus* gen. nov., sp. nov., a novel hyperthermophilic archaeum that oxidizes Fe^{2+} at neutral pH under anoxic conditions. *Arch. Microbiol.* 166:308–14
46. Häggblom MM, Young LY. 1999. Anaerobic degradation of 3-halobenzoates by a denitrifying bacterium. *Arch. Microbiol.* 171:230–36
47. Harwood CS, Burchhardt G, Herrmann H, Fuchs G. 1999. Anaerobic metabolism of aromatic compounds via the benzoyl-CoA pathway. *FEMS Microbiol. Rev.* 22: 439–58
48. Harwood CS, Gibson J. 1997. Shedding light on anaerobic benzene ring degradation: a process unique to prokaryotes? *J. Bacteriol.* 179:301–9
49. He Z, Wiegel J. 1995. Purification and characterization of an oxygen-sensitive reversible 4-hydroxybenzoate decarboxylase from *Clostridium hydroxybenzoicum*. *Eur. J. Biochem.* 229:77–82
50. He Z, Wiegel J. 1996. Purification and characterization of an oxygen-sensitive, reversible 3,4-dihydroxybenzoate decarboxylase from *Clostridium hydroxybenzoicum*. *J. Bacteriol.* 178:3539–42
51. Healy JB Jr, Young LY, Reinhard M. 1980. Methanogenic decomposition of ferulic acid, a model lignin derivative. *Appl. Environ. Microbiol.* 39:436–44
52. Heider J, Fuchs G. 1997. Anaerobic metabolism of aromatic compounds. *Eur. J. Biochem.* 243:577–99

53. Heider JA, Spormann AM, Beller HR, Widdel F. 1999. Anaerobic bacterial metabolism of hydrocarbons. *FEMS Microbiol. Rev.* 22:459–73
54. Holliger C, Wohlfarth G, Diekert G. 1999. Reductive dechlorination in the energy metabolism of anaerobic bacteria. *FEMS Microbiol. Rev.* 22:383–98
55. Huang J, He Z, Wiegel J. 1999. Cloning, characterization, and expression of a novel gene encoding a reversible 4-hydroxybenzoate decarboxylase from *Clostridium hydroxybenzoicum*. *J. Bacteriol.* 181: 5119–22
56. Iannotti EL, Kafkewitz D, Wolin MJ, Bryant MP. 1973. Glucose fermentation products in *Ruminococcus albus* grown in continuous culture with *Vibrio succinogenes*: changes caused by interspecies transfer of H2. *J. Bacteriol.* 114:1231–40
57. Jackson BE, Bhupathiraju VK, Tanner RS, Woese CR, McInerney MJ. 1999. *Syntrophus aciditrophicus* sp. nov., a new anaerobic bacterium that degrades fatty acids and benzoate in syntrophic association with hydrogen-using microorganisms. *Arch. Microbiol.* 171:107–14
58. Johnson HA, Pelletier DA, Spormann AM. 2001. Isolation and characterization of anaerobic ethylbenzene dehydrogenase, a novel Mo-Fe-S enzyme. *J. Bacteriol.* 183:4536–42
59. Johnson HA, Spormann AM. 1999. In vitro studies on the initial reactions of anaerobic ethylbenzene mineralization. *J. Bacteriol.* 181:5662–68
60. Joint Genome Institute. 2001. http://www.jgi.doe.gov/JGI_microbial/html/index.html
61. Kahng HY, Kukor JJ, Oh KH. 2000. Characterization of strain HY99, a novel microorganism capable of aerobic and anaerobic degradation of aniline. *FEMS Microbiol. Lett.* 190:215–21
62. Kniemeyer O, Heider J. 2001. Ethylbenzene dehydrogenase, a novel hydrocarbon-oxidizing molybdenum/iron-sulfur/heme enzyme. *J. Biol. Chem.* 276:21381–86
63. Kniemeyer O, Heider J. 2001. (S)-1-Phenylethanol dehydrogenase of *Azoarcus* sp. strain EbN1, an enzyme of anaerobic ethylbenzene catabolism. *Arch. Microbiol.* 176:129–35
64. Krasotkina J, Walters T, Maruya KA, Ragsdale SW. 2001. Characterization of the B12- and iron-sulfur-containing reductive dehalogenase from *Desulfitobacterium chlororespirans*. *J. Biol. Chem.* 276:40991–97
65. Kreft JU, Schink B. 1993. Demethylation and degradation of phenylmethylethers by the sulfide-methylating homoacetogenic bacterium strain TMBS 4. *Arch. Microbiol.* 159:308–15
66. Kreft JU, Schink B. 1994. O-demethylation by the homoacetogenic anaerobe *Holophaga foetida* studied by a new photometric methylation assay using electrochemically produced cob(I) alamin. *Eur. J. Biochem.* 226:945–51
67. Krieger CJ, Beller HR, Reinhard M, Spormann AM. 1999. Initial reactions in anaerobic oxidation of m-xylene by the denitrifying bacterium *Azoarcus* sp. strain T. *J. Bacteriol.* 181:6403–10
68. Krieger CJ, Roseboom W, Albracht SP, Spormann AM. 2001. A stable organic free radical in anaerobic benzylsuccinate synthase of *Azoarcus* sp. strain T. *J. Biol. Chem.* 276:12924–27
69. Kropp KG, Davidova IA, Suflita JM. 2000. Anaerobic oxidation of n-dodecane by an addition reaction in a sulfate-reducing bacterial enrichment culture. *Appl. Environ. Microbiol.* 66:5393–98
70. Krumholz LR, Crawford RL, Hemling ME, Bryant MP. 1987. Metabolism of gallate and phloroglucinol in *Eubacterium oxidoreducens* via 3-hydroxy-5-oxohexanoate. *J. Bacteriol.* 169:1886–90
71. Krumholz LR, Li J, Clarkson WW, Wilber GG, Suflita JM. 1997. Transformations of TNT and related aminotoluenes in

groundwater aquifer slurries under different electron-accepting conditions. *J. Ind. Microbiol. Biotechnol.* 18:161–69
72. Lack A, Fuchs G. 1992. Carboxylation of phenylphosphate by phenol carboxylase, an enzyme system of anaerobic phenol metabolism. *J. Bacteriol.* 174:3629–36
73. Lack A, Fuchs G. 1994. Evidence that phenol phosphorylation to phenylphosphate is the first step in anaerobic phenol metabolism in a denitrifying *Pseudomonas* sp. *Arch. Microbiol.* 161:132–39
74. Leuthner B, Heider J. 2000. Anaerobic toluene catabolism of *Thauera aromatica*: the *bbs* operon codes for enzymes of beta oxidation of the intermediate benzylsuccinate. *J. Bacteriol.* 182:272–77
75. Leuthner B, Leutwein C, Schulz H, Horth P, Haehnel W, et al. 1998. Biochemical and genetic characterization of benzylsuccinate synthase from *Thauera aromatica*: a new glycyl radical enzyme catalysing the first step in anaerobic toluene metabolism. *Mol. Microbiol.* 28:615–28
76. Lochmeyer C, Koch J, Fuchs G. 1992. Anaerobic degradation of 2-aminobenzoic acid (anthranilic acid) via benzoyl-coenzyme A (CoA) and cyclohex-1-enecarboxyl-CoA in a denitrifying bacterium. *J. Bacteriol.* 174:3621–28
77. Lovley DR. 1997. Microbial Fe(III) reduction in subsurface environments. *FEMS Microbiol. Rev.* 20:305–13
78. Lovley DR. 2000. Anaerobic benzene degradation. *Biodegradation* 11:107–16
79. Lovley DR. 2001. Bioremediation. Anaerobes to the rescue. *Science* 293:1444–46
80. Lovley DR, Coates JD, Blunt-Harris EL, Phillips JP, Woodward JC. 1996. Humic substances as electron acceptors for microbial respiration. *Nature* 382:445–48
81. Lovley DR, Fraga JL, Coates JD, Blunt-Harris EL. 1999. Humics as electron donor for anaerobic respiration. *Environ. Microbiol.* 1:89–98
82. Madigan MT, Jung DO, Resnick SM. 2001. Growth of the purple bacterium *Rhodobacter capsulatus* on the aromatic compound hippurate. *Arch. Microbiol.* 175:462–65
83. Meckenstock RU, Annweiler E, Michaelis W, Richnow HH, Schink B. 2000. Anaerobic naphthalene degradation by a sulfate-reducing enrichment culture. *Appl. Environ. Microbiol.* 66:2743–47
84. Mohn WW, Tiedge JM. 1990. Strain DCB-1 conserves energy for growth from reductive dechlorination coupled to formate oxidation. *Arch. Microbiol.* 153:267–71
85. Morasch B, Annweiler E, Warthmann RJ, Meckenstock RU. 2001. The use of a solid adsorber resin for enrichment of bacteria with toxic substrates and to identify metabolites: degradation of naphthalene, o-, and m-xylene by sulfate-reducing bacteria. *J. Microbiol. Methods* 44:183–91
86. Müller JA, Galushko AS, Kappler A, Schink B. 1999. Anaerobic degradation of m-cresol by *Desulfobacterium cetonicum* is initiated by formation of 3-hydroxybenzylsuccinate. *Arch. Microbiol.* 172:287–94
87. Müller JA, Galushko AS, Kappler A, Schink B. 2001. Initiation of anaerobic degradation of p-cresol by formation of 4-hydroxybenzylsuccinate in *Desulfobacterium cetonicum. J. Bacteriol.* 183:752–57
88. Newman DK, Kolter R. 2000. A role for excreted quinones in extracellular electron transfer. *Nature* 405:94–97
89. Ni S, Fredrickson JK, Xun L. 1995. Purification and characterization of a novel 3-chlorobenzoate-reductive dehalogenase from the cytoplasmic membrane of *Desulfomonile tiedgei* DCB-1. *J. Bacteriol.* 177:5135–39
90. Nozawa T, Maruyama Y. 1988. Anaerobic metabolism of phthalate and other aromatic compounds by a denitrifying bacterium. *J. Bacteriol.* 170:5778–84
91. Nozawa T, Maruyama Y. 1988. Denitrification by a soil bacterium with phthalate

and other aromatic compounds as substrates. *J. Bacteriol.* 170:2501–5

92. Rabus R, Fukui M, Wilkes H, Widdle F. 1996. Degradative capacities and 16S rRNA-targeted whole-cell hybridization of sulfate-reducing bacteria in an anaerobic enrichment culture utilizing alkylbenzenes from crude oil. *Appl. Environ. Microbiol.* 62:3605–13
93. Rabus R, Widdel F. 1995. Anaerobic degradation of ethylbenzene and other aromatic hydrocarbons by new denitrifying bacteria. *Arch. Microbiol.* 163:96–103
94. Rabus R, Wilkes H, Behrends A, Armstroff A, Fischer T, et al. 2001. Anaerobic initial reaction of n-alkanes in a denitrifying bacterium: evidence for (1-methylpentyl)succinate as initial product and for involvement of an organic radical in n-hexane metabolism. *J. Bacteriol.* 183: 1707–15
95. Rabus R, Wilkes H, Schramm A, Harms G, Behrends A, et al. 1999. Anaerobic utilization of alkylbenzenes and n-alkanes from crude oil in an enrichment culture of denitrifying bacteria affiliating with the beta-subclass of *Proteobacteria*. *Environ. Microbiol.* 1:145–57
96. Rueter P, Rabus R, Wilkes H, Aeckersberg F, Rainey FA, et al. 1994. Anaerobic oxidation of hydrocarbons in crude oil by new types of sulphate-reducing bacteria. *Nature* 372:455–58
97. Schink B, Philipp B, Müller J. 2000. Anaerobic degradation of phenolic compounds. *Naturwissenschaften* 87:12–23
98. Schnell S, Bak F, Pfennig N. 1989. Anaerobic degradation of aniline and dihydroxybenzenes by newly isolated sulfate-reducing bacteria and description of *Desulfobacterium anilini*. *Arch. Microbiol.* 152:556–63
99. Schnell S, Schink B. 1991. Anaerobic aniline degradation via reductive deamination of 4-aminobenzoyl-CoA in *Desulfobacterium anilini*. *Arch. Microbiol.* 155: 183–90
100. Schöcke L, Schink B. 1998. Membrane-bound proton-translocating pyrophosphatase of *Syntrophus gentianae*, a syntrophically benzoate-degrading fermenting bacterium. *Eur. J. Biochem.* 256:589–94
101. Schöcke L, Schink B. 1999. Biochemistry and energetics of fermentative benzoate degradation by *Syntrophus gentianae*. *Arch. Microbiol.* 171:331–37
102. Song B, Palleroni NJ, Kerkhof LJ, Häggblom MM. 2001. Characterization of halobenzoate-degrading, denitrifying *Azoarcus* and *Thauera* isolates and description of *Thauera chlorobenzoica* sp. nov. *Int. J. Syst. Evol. Microbiol.* 51:589–602
103. Spormann AM, Widdel F. 2000. Metabolism of alkylbenzenes, alkanes, and other hydrocarbons in anaerobic bacteria. *Biodegradation* 11:85–105
104. Straub KL, Benz M, Schink B. 2001. Iron metabolism in anoxic environments at near neutral pH. *FEMS Microbiol. Ecol.* 34:181–86
105. Sullivan ER, Zhang X, Phelps C, Young LY. 2001. Anaerobic mineralization of stable-isotope-labeled 2-methylnaphthalene. *Appl. Environ. Microbiol.* 67:4353–57
106. Tor JM, Lovley DR. 2001. Anaerobic degradation of aromatic compounds coupled to Fe(III) reduction by *Ferroglobus placidus*. *Environ. Microbiol.* 3:281–87
107. Unciuleac M, Boll M. 2001. Mechanism of ATP-driven electron transfer catalyzed by the benzene ring-reducing enzyme benzoyl-CoA reductase. *Proc. Natl. Acad. Sci. USA* 98:13619–24
108. Utkin I, Dalton DD, Wiegel J. 1995. Specificity of reductive dehalogenation of substituted ortho-phenols by *Desulfitobacterium dehalogenans* JW/IU-DC1. *Appl. Environ. Microbiol.* 61:346–51
109. Utkin I, Woese C, Wiegel J. 1994. Isolation and characterization of *Desulfitobacterium dehalogenans* gen. nov., spec. nov., an anaerobic bacterium which reductively

dechlorinates chlorophenolic compounds. *Int. J. Syst. Bacteriol.* 44:612–19

110. van de Pas BA, Jansen S, Dijkema C, Schraa G, de Vos WM, Stams AJ. 2001. Energy yield of respiration on chloroaromatic compounds in *Desulfitobacterium dehalogenans*. *Appl. Environ. Microbiol.* 67:3958–63

111. van de Pas BA, Smidt H, Hagen WR, van der Oost J, Schraa G, et al. 1999. Purification and molecular characterization of ortho-chlorophenol reductive dehalogenase, a key enzyme of halorespiration in *Desulfitobacterium dehalogenans*. *J. Biol. Chem.* 274:20287–92

112. van der Woude BJ, de Boer M, van der Put NM, van der Geld FM, Prins RA, Gottschal JC. 1994. Anaerobic degradation of halogenated benzoic acids by photoheterotrophic bacteria. *FEMS Microbiol. Lett.* 119:199–207

113. von Wintzingerode F, Schlotelburg C, Hauck R, Hegemann W, Gobel UB. 2001. Development of primers for amplifying genes encoding CprA- and PceA-like reductive dehalogenases in anaerobic microbial consortia, dechlorinating trichlorobenzene and 1,2-dichloropropane. *FEMS Microbiol. Ecol.* 35:189–96

114. White GF, Russell NJ, Tidswell EC. 1996. Bacterial scission of ether bonds. *Microbiol. Rev.* 60:216–32

115. Widdel F, Rabus R. 2001. Anaerobic biodegradation of saturated and aromatic hydrocarbons. *Curr. Opin. Biotechnol.* 12: 259–76

116. Wiegel J, Wu Q. 2000. Microbial reductive dehalogenation of polychlorinated biphenyls. *FEMS Microbiol. Ecol.* 32:1–15

117. Zhang X, Sullivan ER, Young LY. 2000. Evidence for aromatic ring reduction in the biodegradation pathway of carboxylated naphthalene by a sulfate reducing consortium. *Biodegradation* 11:117–24

118. Zhang X, Wiegel J. 1994. Reversible conversion of 4-hydroxybenzoate and phenol by *Clostridium hydroxybenzoicum*. *Appl. Environ. Microbiol.* 60:4182–85

Annu. Rev. Microbiol. 2002. 56:371–402
doi: 10.1146/annurev.micro.56.012302.160654

First published online as a Review in Advance on July 15, 2002

THE MOLECULAR BIOLOGY OF WEST NILE VIRUS: A New Invader of the Western Hemisphere

Margo A. Brinton
Department of Biology, Georgia State University, Atlanta, Georgia 30303; e-mail: mbrinton@gsu.edu

Key Words WNV replication, WNV proteins, *cis*-acting sequences, conserved RNA structures, RNA-protein interactions, virus-host interactions

■ **Abstract** West Nile virus (WNV) is a mosquito-borne flavivirus that primarily infects birds but occasionally also infects humans and horses. In recent years, the frequency of WNV outbreaks in humans has increased, and these outbreaks have been associated with a higher incidence of severe disease. In 1999, the geographical distribution of WNV expanded to the Western hemisphere. WNV has a positive strand RNA genome of about 11 kb that encodes a single polyprotein. WNV replicates in the cytoplasm of infected cells. Although there are still many questions to be answered, a large body of data on the molecular biology of WNV and other flaviviruses has already been obtained. Aspects of virion structure, the viral replication cycle, viral protein function, genome structure, conserved viral elements, host factors, virus-host interactions, and vaccines are discussed in this review.

CONTENTS

0066-4227/02/1013-0371$14.00

INTRODUCTION

West Nile virus (WNV) is a mosquito-borne virus that was first isolated in 1937 from the blood of a woman participating in a malaria study in the West Nile region of Uganda (136). Endemic in parts of Africa, the Middle East, and western Asia, particularly India, WNV has periodically been the causative agent of brief epizootics in France, Romania, Russia, Algeria, Madagascar, Senegal, and South Africa and of infrequent disease outbreaks in humans. Because a number of wild bird species develop high levels of viremia after WNV infection and sustain viremic levels of WNV of at least 10^5 PFU/ml of serum (the minimum level estimated to be required to infect a feeding mosquito) for days to weeks, they are the main reservoir hosts in endemic regions and are also the usual source of the virus initiating epizootics outside endemic areas (11, 86, 113). For example, the sequence of the envelope gene of a WNV isolate (RO97-50) obtained from *Culex* mosquitoes in Bucharest in 1999 was identical to that of WNV isolates obtained in Kenya and Senegal from *Culex* mosquitoes (130). During the recent outbreaks in Israel and the United States, there has been a higher-than-normal death rate in birds (90).

Humans and horses are incidental hosts with low viremic levels and do not play a role in the transmission cycle. Although the incidence of clinical disease among WNV-infected humans is low, in recent outbreaks there has been an increase in the severity of disease among those that develop clinical symptoms (113). Fever is the most common symptom observed in humans. The course of the fever is sometimes biphasic, and a rash on the chest, back, and upper extremities often develops during or just after the fever (29). Symptoms also include headaches, muscle weakness, and disorientation. A few infected individuals develop encephalitis, meningoencephalitis, or hepatitis. The brainstem, particularly the medulla, is the primary CNS target (129). Humans age 60 and older have an increased risk of developing fatal disease (36a).

The largest known human epidemic of WNV occurred in 1974 in Cape Province, South Africa, with about 3000 human clinical cases (101). Recent outbreaks of WNV in humans and horses have been more frequent (Romania and Morocco in 1996; Tunisia in 1997; Italy in 1998; Russia and the United States in 1999; and Israel, France, and the United States in 2000) (113). Estimates from serological data obtained in Queens, New York, the epicenter of the U.S. outbreak in 1999, suggest that at least 1900 humans in Queens may have been infected with WNV. While the majority of the infected humans experienced either no symptoms or a

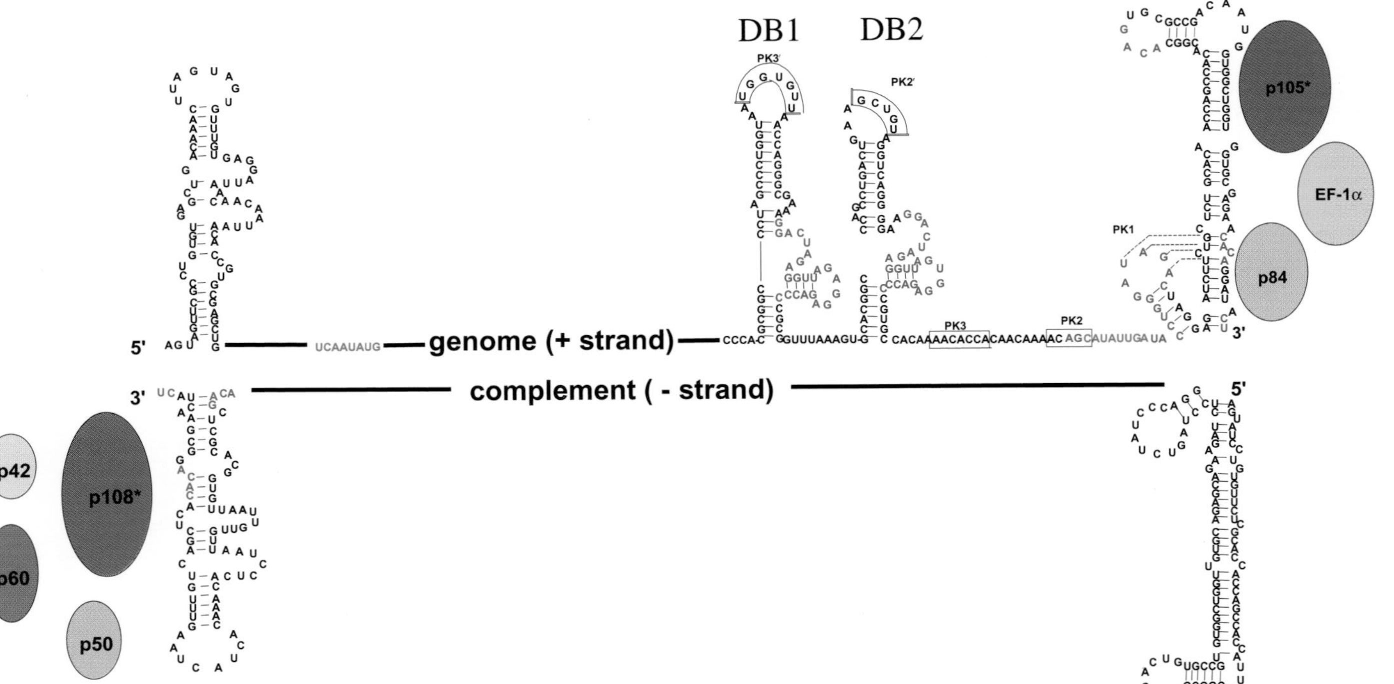

Figure 3 Conserved RNA structures and sequences in the WNV genome and the complementary minus-strand RNA and cell proteins that interact specifically with the viral 3′ RNAs. Conserved sequences are shown in *green*. The conserved 3′ and 5′ cyclization sequences are in *red*. PK, pseudoknot. The sequences involved in the tertiary interactions of PK2 and PK3 are *boxed*.

low-grade fever, 62 developed clinical disease and of these 7 died as a result of the infection (77).

WNV has been isolated from *Culex*, *Aedes*, *Anopheles*, *Minomyia*, and *Mansonia* mosquitoes in Africa, Asia, and the United States, but *Culex* species are the most susceptible to infection with WNV (29, 68). Also *Culex* mosquitoes feed on wild bird species that have high levels of viremia (145). Natural vertical transmission of WNV in *Culex* mosquitoes in Africa has been reported and is expected to enhance virus maintenance in nature (103). Mechanical transmission by ticks may also play a role in virus maintenance (29). The 1999 outbreak in New York represented the first introduction of WNV into the Western hemisphere (17). The mode of introduction of WNV into the United States is not known, but phylogenic analysis of the envelope gene of a WNV isolate from the New York outbreak indicated that it was most closely related to a goose isolate from Israel (WN-Israel 1998) (70, 90). WNV transmission reoccurred in New York during the summers of 2000 and 2001, and the epizootic spread to other states, indicating that this virus had become endemic in the United States. In future years, it is expected that viremic migratory birds will carry WNV to all parts of the United States as well as to Canada, the Carribean, and Central and South America (119). Mosquitoes capable of transmitting WNV to susceptible birds exist in all of these regions.

Environmental conditions such as heavy rains followed by flooding, irrigation, and high temperatures can cause an increase in mosquito populations and also in the incidence of mosquito-borne viruses such as WNV. However, in cities the lack of heavy rains, which periodically flush out the sewers, results in increased mosquito populations breeding in pools of stagnant water, as was the case for the 1999 outbreak in New York City. The convergence of birds at scarce pools of water also facilitates virus transmission.

VIRUS CLASSIFICATION

The family *Flaviviridae* consists of three genera: *Flavivirus*, *Pestivirus*, and *Hepacivirus*. Members of the different genera are distantly related but share a similar gene order and conserved nonstructural protein motifs. The ~70 viruses currently classified in the genus *Flavivirus* are further subdivided into twelve antigenic serogroups (65). WNV is a member of the Japanese encephalitis virus (JEV) serogroup, which also includes Cacipacore, Koutango, JE, Murray Valley encephalitis, St. Louis encephalitis, Usutu, and Yaounde viruses (65, 115). Based on sequence homology, Kunjin virus, which is endemic to Australia and Asia, is now considered a WNV subtype (65, 131). WNV isolates have been grouped into two genetic lineages (1 and 2) on the basis of signature amino acid substitutions or deletions in their envelope proteins (12). All the WNV isolates associated thus far with outbreaks of human disease have been in lineage 1 (70, 90). Lineage 2 viruses are restricted to endemic enzootic infections in Africa. Because of antigenic cross-reactivity between different flaviviruses, techniques such as in situ hybridization or sequence analysis of RT-PCR products are required to unequivocally identify WNV as the causative agent of an outbreak (17, 90, 139).

Because data are not available on all aspects of WNV and because it is likely that the properties of other flaviviruses will be similar to those of WNV, information obtained for other flaviviruses has been used when appropriate in this review. For additional information on flaviviruses, see (29, 64, 65, 96).

GENOME STRUCTURE

The WNV genome is a single-stranded RNA of positive polarity (mRNA sense). A type 1 cap structure (m^7GpppAmp) is present at the 5′ end (40), but the 3′ end terminates with CU_{OH} (26, 151). Flavivirus genomes are the only mammalian plus-strand RNA virus genomes that do not have a 3′ poly(A) tract. The 5′ noncoding region (NCR) of the WNV genome RNA is 96 nts in length, while the 3′ NCR is 631 nts. The WNV genome is 11,029 nt in length and contains a single open reading frame (ORF) of 10,301 nt (90). Ten mature viral proteins are produced via proteolytic processing of the single polyprotein by the viral serine protease (NS2B-NS3) and various cellular proteases (96, 108) (Figure 1). The three viral structural proteins, capsid (C), membrane (prM/M), and envelope (E), are encoded within the 5′ portion of the genomic ORF, while the seven nonstructural proteins (NS1, NS2A, NS2B, NS3, NS4A, NS4B, and NS5) are encoded within the 3′ portion (124).

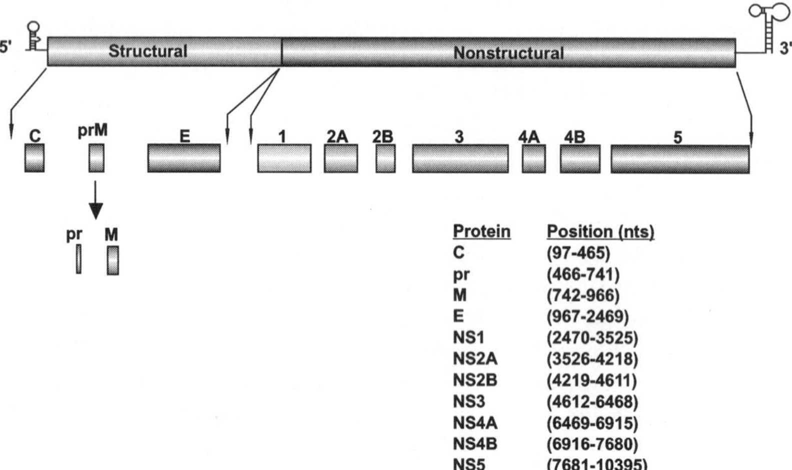

Protein	Position (nts)
C	(97-465)
pr	(466-741)
M	(742-966)
E	(967-2469)
NS1	(2470-3525)
NS2A	(3526-4218)
NS2B	(4219-4611)
NS3	(4612-6468)
NS4A	(6469-6915)
NS4B	(6916-7680)
NS5	(7681-10395)

Figure 1 The WNV genome. Noncoding regions with their 3′ terminal structures are indicated by black lines. The single open reading frame encodes a polyprotein that is processed by the viral NS2B-NS3 protease and cell proteases to the mature viral proteins. Structural proteins are shown in *blue* and nonstructural proteins are shown in *red*. The positions of each of the viral proteins in the nucleotide sequence of the genome of WNV, strain EG101, are shown. The genome is not drawn to scale.

VIRION MORPHOLOGY AND COMPOSITION

WN virions are small (~50 nm in diameter), spherical, enveloped, and have a buoyant density of ~1.2 g/cm^3. The spherical nucleocapsid is ~25 nm in diameter and is composed of multiple copies of the C protein. Cryo-electron microscopy data suggest that the virion envelope and capsid have icosahedral symmetry (65). Recent data indicate that the symmetry is conferred on the virus particle by interactions between E proteins rather than by interactions between capsid proteins (88a, 114). A precursor of C, designated anchored C, contains a hydrophobic region at its C terminus that is cleaved to generate the mature virion C protein. The encapsidation signal on the flavivirus genomic RNA recognized by the C protein has not yet been definitively mapped. Glutathione-S-transferase fusion proteins made from different fragments of the Kunjin virus C protein bound strongly to RNA probes from either the 3′ or 5′ noncoding regions of the genomic RNA (84). In infected cells, the C protein is located in the nucleus as well as in the cytoplasm late in the infection cycle (153a).

The two viral envelope proteins, E and M, are both type I integral membrane proteins with C-terminal membrane anchors. The E proteins of some strains of WNV have no N-linked glycosylation sites, whereas others have a single glycosylation site (2, 12, 156). The acquisition of a carbohydrate residue on the WNV E protein after passage in mosquito cells was not linked to attenuation of the virus (33). Cysteine residues in the E protein ectodomain are strictly conserved and all of them form intramolecular disulfide bonds (109). Crystallographic analysis of the three-dimensional structure of a soluble fragment of a flavivirus E protein revealed dimers composed mostly of β-sheets arranged in a head-to-tail orientation and suggested that the distal ends of each monomer were anchored in the membrane (64, 122). The E protein dimers in mature virions lie fairly flat against the lipid bilayer. The E protein has both receptor-binding and pH-dependent fusion activities and is composed of three domains (64). Domain III contains a fold typical of an immunoglobulin constant domain and has been postulated to contain the receptor-binding region; the lateral surfaces of the Domain III structures of mosquito- and tick-borne flavivirus E proteins differ significantly from each other and mutations in Domain III alter virulence (64, 91). A study on yellow fever virus suggested that a region of Domain II may also be involved in the binding of virus to cells in monkey brains (106).

WNV REPLICATION CYCLE

WNV replicates in a wide variety of cell cultures, including primary chicken, duck, and mouse embryo cells and continuous cell lines from monkeys, humans, pigs, rodents, amphibians, and insects, but does not cause obvious cytopathology in many cell lines (22). Glycosaminoglycans play a role in flavivirus entry (36, 91). However, Murray Valley virus mutants with substitutions in the hydrophilic region (FG loop) of the E protein that resulted in an increased dependence on glycosaminoglycans during entry of cultured cells showed decreased neurovirulence in mice

(91). The identification of two glycoproteins as putative receptors of dengue 4 virus (128) suggest that additional host cell surface molecules are necessary for flavivirus entry. Because flaviviruses are transmitted between insect and vertebrate hosts during their natural transmission cycle, it is likely that the cell receptor(s) they utilize is a highly conserved protein.

After binding to an unknown cell receptor(s), virions enter cells via receptor-mediated endocytosis followed by low-pH fusion of the viral membrane with the endosomal vesicle membrane releasing the nucleocapsid into the cytoplasm (64) (Figure 2*A*). The genome RNA is released and translated into a single polyprotein (Figure 2*B*). The viral serine protease, NS2B-NS3, and several cell proteases then cleave the polyprotein at multiple sites to generate the mature viral proteins (Figure 2*C*). The viral RNA-dependent RNA polymerase (RdRp), NS5, in conjunction with other viral nonstructural proteins and possibly cell proteins, copies complementary minus strands from the genomic RNA template (Figure 2*D*), and these minus-strand RNAs in turn serve as templates for the synthesis of new genomic RNAs (Figure 2*E*). Flaviviral RNA synthesis is semiconservative and asymmetric. Genome RNA synthesis is about 10 times more efficient than minus-strand RNA synthesis (38, 41). Data obtained by Chu & Westaway (38) with a Kunjin

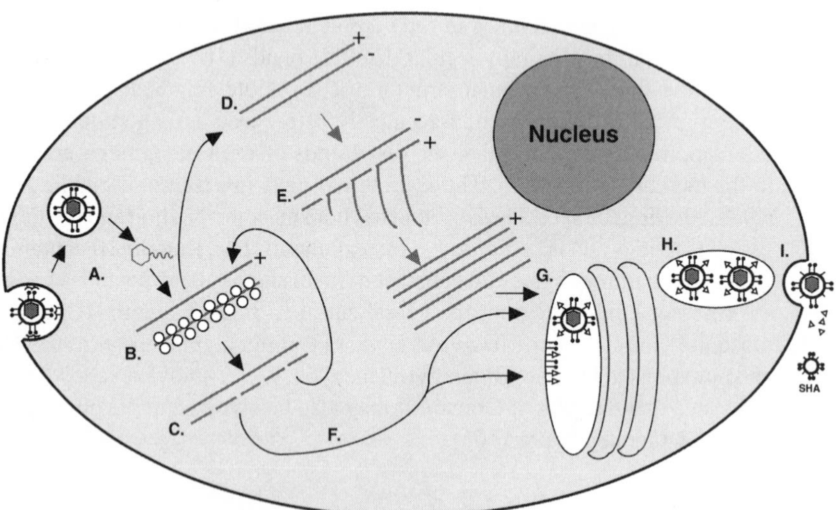

Figure 2 The WNV replication cycle. *A*. Attachment and entry of the virion. *B*. Uncoating and translation of the virion RNA. *C*. Proteolytic processing of the polyprotein. *D*. Synthesis of the minus-strand RNA from the virion RNA. *E*. Synthesis of nascent genome RNA from the minus-strand RNA. *F*. Transport of structural proteins to cytoplasmic vesicle membranes. *G*. Encapsidation of nascent genome RNA and budding of nascent virions. *H*. Movement of nascent virions to the cell surface. *I*. Release of nascent virions. SHA, slowly sedimenting hemagglutinin, a subviral particle that is also sometimes released.

virus suggested that only a single, nascent (−) RNA is copied from a plus-strand template at a time (replicative form, RF), while the minus-strand template is efficiently reinitiated so that multiple, nascent plus strands are simultaneously copied from a single minus-strand template (replicative intermediate, RI). Once established, both plus- and minus-strand viral RNA synthesis can continue even in the absence of protein synthesis, indicating that transient viral polyprotein precursors are not required (38, 41, 154). Extensive reorganization and proliferation of cytoplasmic, perinuclear endoplasmic reticular (ER) membranes are observed in infected cells (96). Nascent genome RNAs could function as templates for translation and transcription and as substrates for encapsidation. Data obtained with a Kunjin replicon suggest that translation is a prerequisite for replication of nascent RNAs (81) and that replication is a prerequisite of encapsidation (83a). At the beginning of the replication cycle, nascent genome RNAs may alternate between replication and translation because a sufficient pool of structural proteins has not yet accumulated.

Virion assembly occurs in association with rough ER membranes (Figure 2*F*,*G*). Little is known about this process because budding intermediates and free nucleocapsids in the cytoplasm have rarely been observed by electron microscopy (EM), but they may have been observed by cryo-EM (107). Intracellular immature virions, which contain heterodimers of E and prM (the precursor of M), accumulate in vesicles and are then transported through the host secretory pathway where glycans on E and prM are modified (66, 149) (Figure 2*H*). The glycosylated and hydrophilic N-terminal portion of prM is cleaved in the *trans*-Golgi network by cellular furin or a related protease (138). The C-terminal portion (M) remains inserted in the envelope of the mature virion and contains a short ectodomain and two membrane-spanning domains (104). The prM-E interaction may maintain the E protein in a stable, fusion-inactive conformation during the assembly and release of new virions (64) because virions with prM-E complexes are less infectious than mature virions. Virions are transported to the plasma membrane in vesicles and are released by exocytosis (99, 108) (Figure 2*I*). Progeny virions are released from infected mammalian cells starting at 10 to 12 h after infection, and maximal extracellular virus titers are not usually observed until approximately 24 hr. Infected cells also produce noninfectious, but antigenic, subviral particles called slowly sedimenting hemagglutinin (SHA), which are composed of cell membranes inserted with E and M proteins as well as some prM (126) (Figure 2).

VIRAL NONSTRUCTURAL PROTEINS

The majority of the WNV nonstructural proteins are multifunctional. All seven of the flavivirus nonstructural proteins appear to be directly or indirectly involved in viral RNA synthesis. However, little is known about the interactions between the various viral nonstructural proteins and between these proteins and cell proteins that may also be required for the formation of active viral RNA replication complexes.

NS2A, NS2B, NS4A, and NS4B

NS2A, NS2B, NS4A and NS4B are small, hydrophobic proteins that contain no conserved motifs characteristic of known enzymes. Some or all of these four proteins may facilitate the assembly of viral replication complexes and/or their localization on cytoplasmic membranes. Two forms of NS2A are produced in infected cells, the full-length protein and a C-terminal truncated fragment (96). Mutation of either of the C-terminal sites in the yellow fever virus NS2A was lethal for virus replication (123). NS2B has been shown to complex with and function as the cofactor for the viral serine protease (NS3) (34, 50). The conserved hydrophilic domain of NS2B is essential for its cofactor activity (34, 50), whereas the hydrophobic regions of NS2B are required for cotranslational insertion of the NS2B-NS3 precursor into ER membranes and for the interaction between NS3 and NS2B (18, 42). Although the NS4A protein has been identified in Kunjin virus–infected cells (137), only uncleaved NS3-NS4A was found in cells infected with other flaviviruses (96). Since the size of NS4B estimated on gels is higher than predicted, it is likely that this membrane-spanning protein is posttranslationally modified. In infected cells, NS4B is enriched in the perinuclear endoplasmic reticular membranes but later in the infection cycle is also dispersed through the nucleoplasm (153a). Attempts to complement NS2A, NS2B, NS4A, and NS4B in *trans* were unsuccessful (83). Overexpression of recombinant NS2B, NS4A, or NS4B proteins from Japanese encephalitis virus in an inducible bacteria system permeabilized the cells to hygromycin B (35). NS4B was the least efficient of these three proteins at permeabilizing bacteria. Successful expression of NS2A in bacteria was not achieved. In mammalian cells, expression of NS2A, NS2B, or NS4B, but not NS4A, decreased reporter protein expression from two different cotransfected plasmids. Although the authors suggest that this observed decrease in reporter protein expression was due to the modification of cellular membranes by the expressed viral proteins, no data on the localization of the expressed viral proteins in cells nor the mechanism by which these proteins alter cell permeability were presented.

NS1

NS1 is a glycoprotein with 2 conserved N-linked glycosylation sites and 12 conserved cysteines that are essential for virus viability (96, 123). NS1 is translocated into the lumen of the ER prior to cleavage at the E/NS1 junction by a cellular signalase (32). NS2A is cleaved from the C-terminal end of NS1 by a membrane-bound host protease associated with the ER (49). NS1 can be detected within as well as on the surface of infected cells and is secreted by mammalian cells but not by mosquito cells (135). NS1 monomers are soluble but homodimers are hydrophobic and membrane-associated (135, 157, 158). The mechanism by which NS1 associates with cell membranes is not yet known. A proline-to-leucine substitution at amino acid 250 in the Kunjin NS1 was shown to completely inhibit NS1 dimer formation (63). The mutant virus grew to 100-fold lower titers in Vero cells than did wild-type virus and was 10-fold less infectious for weanling mice, but NS1 was secreted efficiently from cells infected with this mutant (63). These data suggest that NS1 dimerization enhances, but is not essential for, its function or secretion. The normal form of secreted NS1 has recently been identified as a hexamer consisting of three homodimers (51). In the dengue 1 NS1, one of the two oligosaccharide chains is protected in the dimer from cellular maturation enzymes and remains as a high-mannose chain, while the other exposed chain is processed into a multibranched endo-beta-N-acetylglucosaminidase F-resistant complex sugar. Complete processing of exposed complex sugar chains appears to be essential for NS1 secretion; NS1 is not secreted by mammalian cells treated with glycosylation inhibitors nor by mosquito cells that produce glycoproteins with high-mannose chains rather than complex sugars (51).

Both indirect immunofluorescence and cryo-immuno-electron microscopy studies with dengue 1 virus–infected cells have shown that NS1 associates with intracellular membranes but not with mature virions (96). Dual labeling with anti-NS1 and anti-dsRNA antisera suggested colocalization of NS1 and viral double-stranded RNA to vesicle packets in infected Vero cells, as well as to the luminal face of and to membrane fragments within large cytoplasmic vacuoles in infected C6/36 mosquito cells. In cells infected with a mutant of yellow fever virus that had a temperature-sensitive mutation in NS1, RNA accumulation was not observed at the nonpermissive temperature (105). No temperature-dependent differences were observed in NS1 stability or secretion, in viral polyprotein processing nor in the efficiency of virus release, suggesting that NS1 plays a critical role in flavivirus RNA synthesis. Successful *trans*-complementation of a defective yellow fever virus RNA, which contained a large in-frame deletion in NS1, with wild-type NS1 expressed from a Sindbis replicon allowed the uncoupling of the NS1 function from expression and processing of the viral polyprotein (94). An advantage of this system is that the Sindbis replicon RNA was not packaged so that pure stocks of the NS1 deletion mutant could be obtained. Plus- and minus-strand RNA synthesis was analyzed in cells infected with the mutant virus using an RNase-protection assay. Neither positive- nor negative-strand viral RNA accumulated unless NS1 was supplied in *trans*, suggesting that NS1 functions prior to or early in minus-strand synthesis, the first type of RNA synthesis to occur in infected cells (Figure 2). The NS1 of a different flavivirus, dengue, was unable to *trans*-complement the yellow fever NS1 deletion mutant RNA, which suggests that NS1 may interact with another viral protein (95). Three independent genetic screens performed to select yellow fever virus variants that could utilize the dengue NS1 yielded variants with a single base mutation in the NS4A gene. These data indicate that an interaction between NS1 and NS4A is required for viral RNA synthesis.

A recent study with dengue virus suggests a role for NS1 in a cell activation mechanism (69). A glycosyl-phosphatidylinositol-linked form of NS1 was detected on the surface of infected cells as well as on cells expressing only recombinant NS1. Addition of anti-NS1 antibody induced signal transduction as detected by tyrosine phosphorylation of cellular proteins. The role of a virus-induced cell activation mechanism during the flavivirus replication cycle is not yet known.

NS3

NS3 is a highly conserved protein. The N-terminal third of NS3 encodes a serine protease that is a member of the trypsin superfamily (8, 9, 54, 150). However, NS3 is not an active protease until it forms a stable complex with NS2B. The N-terminal 184 amino acid residues of NS3 are sufficient for the interaction with NS2B (3). The heterodimerization of NS3 and NS2B is thought to generate additional specific interactions with the P2 and P3 residues of the substrate cleavage sites (165). Membrane association of the NS3-NS2B complex is required for efficient polyprotein processing. The NS3 protease cleaves the anchor region from the C protein and also cleaves at multiple sites within the nonstructural polyprotein. Cleavage sites consist of two basic amino acids followed by an amino acid with a short side chain (96, 116, 165). Cleavage within the NS3 protein has been observed in mammalian cells but not in mosquito cells (3, 152). A putative structure for the dengue 2 virus NS3 protease that was modeled by homology on the structural coordinates of the crystallized hepatitis C NS3 protease predicted some differences between the two viral proteases, particularly in the substrate-binding site (18). The NS2B-NS3 protease complex upregulates the processing of the C-prM precursor by cell signalase and promotes the efficient secretion of prM-E expressed from a vector (159).

The C-terminal portion of NS3 contains regions with homology to the DEAD family of RNA helicases (55), to an RNA-stimulated nucleoside triphosphatase (NTPase), and to an RNA triphosphatase (148, 152, 153). The RNA triphosphatase is thought to function in the synthesis of the 5′ cap structure added to the viral RNA in the cytoplasm of infected cells (153). Overlapping functional domains for both the protease and NTPase were mapped in the dengue 2 NS3 to a region located between amino acid residues 160 and 180 (92). A C-terminal fragment of NS3 purified from WNV-infected cells or culture media had both helicase and NTPase activity (16). Although the helicase utilizes energy from NTP hydrolysis for RNA unwinding, the NTPase activity was not directly coupled to the helicase-unwinding reaction (16). When deletions or point mutations that inactivate either the helicase activity alone or both the helicase and NTPase activities of NS3 were engineered into an infectious clone of the related pestivirus, bovine viral diarrhea virus, no detectable minus-strand RNA was synthesized and no infectious virus was produced (59). A study with a dengue virus infectious clone also showed that an active helicase was essential for virus viability (100). Although the exact functions of these two enzymes during the virus replication cycle have not yet been demonstrated, the helicase may be required to unwind the 3′ terminal secondary structure of the viral genome prior to the initiation of minus-strand synthesis, to facilitate polymerase (NS5) processivity during elongation of nascent strands, and/or to free the 3′ end of the completed minus-strand from the plus-strand template for the initiation of positive-strand synthesis.

NS3 not only interacts with NS2B, but also with NS5. A protein-protein interaction between NS3 and the polymerase, NS5, has been demonstrated both in vivo and in vitro by coimmunoprecipitation and immunoblotting (76). This interaction occurred in the absence of other viral proteins but was dependent on the phosphorylation state of NS5. For dengue virus, the regions participating in this protein-protein interaction were mapped to the C terminus of NS3 (residues 303 to 618) and the N terminus of NS5 (residues 320 to 368). In vitro NTPase activity of a recombinant dengue NS3 was stimulated twofold by the addition of recombinant NS5 (43), which suggests NS3 activity might be regulated by the interaction between these two proteins. An interaction between a recombinant dengue NS3 protein and regions of the viral 3′ noncoding region RNA has also been reported (35a, 43). However, although apparently specific, this RNA-protein interaction was relatively weak, which may explain why an interaction between the WNV 3′ RNA and NS3 in an infected cytoplasmic extract was not detected under more-stringent assay conditions (13). The interaction between NS3 and NS5 would facilitate the coordination of the helicase, polymerase, and capping activities.

NS5

NS5 is located at the C terminus of the viral polyprotein (Figure 1). It is the largest as well as the most highly conserved of the flavivirus proteins (96). In vivo *trans*-complementation of a defective Kunjin virus replicon with NS5 has been demonstrated (79, 81, 82). Although NS5 is a basic protein, it does not contain long hydrophobic regions characteristic of *trans*-membrane domains. The C-terminal portion of NS5 contains motifs characteristic of all RNA-dependent RNA polymerases (RdRps), including the highly conserved GDD (73, 87), and homology modeling on the crystal structure of the poliovirus polymerase showed that these two polymerases contained the same structural elements in a similar order (111). The N-terminal region of NS5 contains a methyltransferase domain that is thought to function in the methylation of the type I cap (88). Site-directed mutagenesis studies confirmed that both active polymerase and methyltransferase are required for virus replication (79, 82). In the cell nucleus, three activities, an RNA triphosphatase, a guanylyltransferase, and a methyltransferase, are required to synthesize a cap structure. Since flaviviruses replicate in the cytoplasm, the synthesis of the viral type I cap must be accomplished by viral enzymes. The flavivirus NS3 contains the RNA triphosphatase, while NS5 contains the methyltransferase, but the guanylyltransferase has not yet been identified.

NS5 is phosphorylated only on serine residues by cellular serine/threonine kinases (76, 121). After SDS-PAGE, two forms of NS5 were detected in the cytoplasm of infected cells grown in media containing 2% FCS. Both forms of NS5 were phosphorylated, but the protein in the upper band was hyperphosphorylated (76). In cells grown in media containing 10% FCS, only the hyperphosphorylated form of NS5 was detectable in the cytoplasm. The phosphorylation state of NS5 may regulate the association between NS5 and NS3 (76). NS5 has been detected in both the cytoplasm and the nucleus of infected cells (28, 76). The hyperphosphorylated form of NS5 is located primarily in the nucleus of infected cells, whereas the hypophosphorylated form is found only in the cytoplasm (76). The dengue NS5 nuclear localization signal (NLS) has been mapped to a region located between the methyltransferase and the polymerase domains (residues 369 to 405) (52). This region was capable of targeting a β-galactosidase fusion protein to the nucleus and was recognized by the β-subunit of the NLS-binding importin receptor complex with an affinity similar to that of the bipartite NLS of the retinoblastoma protein. Further experiments indicated that both the NLS as well as differential phosphorylation may be involved in targeting NS5 to the nucleus in a regulated manner during the infection cycle. Whether nuclear localization of NS5 plays a role in the viral replication cycle and/or is part of a mechanism used by the virus to alter a cell function is currently unknown. The same N-terminal region of NS5 required for interaction with NS3 was also shown to be recognized by importin-beta, suggesting that competition between NS3 and importin-beta may also play a role in regulating the localization of NS5 to the nucleus (71). A number of cellular phosphoproteins as well as a cellular kinase(s) were observed to be coimmunoprecipitated with NS5, suggesting that cellular proteins may also complex with NS5 (121). Association of cell proteins with purified, transcriptionally active viral replication complexes has previously been reported; three cell proteins, translation elongation factor (EF)-Tu, EF-Ts, and S-1, that copurified with the polymerase of Qβ phage were shown to be required for polymerase activity (15). The polymerase (L) of the minus-strand RNA virus, vesicular stomatitis virus, expressed from a recombinant baculovirus in insect Sf21 cells was complexed with the α-, β-, and γ-subunits of EF-1, and association with these cell proteins was shown to be required for polymerase activity (45).

POLYMERASE ACTIVITY OF INFECTED CELL CYTOPLASMIC EXTRACTS

In vitro polymerase activity in cytoplasmic extracts prepared from infected cells has been demonstrated for WNV (56) and Kunjin virus (38). Enzyme activity was inhibited by high concentrations of NaCl and was dependent on the presence of Mg^{2+}, but activity was also observed if Mn^{2+} instead of Mg^{2+} was present (56). The RdRp activity was associated with "heavy" cytoplasmic membrane fractions (39, 57), but polymerase activity was retained after solubilization with detergent (58). The heavy membrane fractions were enriched in NS3, NS2A, and NS2B/NS4A (39), but they contained only about 10% of the NS5 (57). Subsequent metabolic labeling and colocalization experiments indicated that nascent viral RNA replication intermediates and the viral nonstructural proteins, NS1, NS2A, NS3, and NS4A, are associated with the perinuclear vesicle membranes, whereas NS3 and NS2B colocalize with convoluted membranes (96, 154, 155). These data suggest there are separate compartments for viral polyprotein translation and processing and for viral RNA replication after virus-induced membrane reorganization has occurred in infected cells. Also, although many of the viral nonstructural proteins form a complex on the proliferated ER membranes, NS5 appears to be brought into this complex only when a RNA template is initiated or reinitiated. Both anti-NS5 and anti-NS3 antibody blocked Kunjin–in vitro polymerase activity (7), which suggests that both of these proteins are required for flavivirus RNA synthesis.

Elongation and release of nascent plus strands from already initiated, endogenous minus-strand templates constituted the majority of the in vitro polymerase activity in experiments done with cytoplasmic extracts containing endogenous templates, and only a limited amount of reinitiation was observed (38, 56). Results from an in vitro polymerase assay that consisted of nuclease-treated dengue virus–infected cell extracts and exogenous RNA templates indicated that the primary product was twice the size of the template (163). Such a product would be produced by self-priming due to the 3′ end of the template RNA folding back on itself. Only templates that contained both the viral 3′ and 5′ terminal sequences could prime RNA synthesis. In addition, RNA synthesis occurred when the 5′ sequence was present either in *cis* or in *trans*.

Both uridylyl and adenylyl terminal transferase activities were shown to be associated with the cellular cytoplasmic fractions that contained WNV polymerase activity (56). Recent studies with purified recombinant RdRps from viruses in the other two genera of the *Flaviviridae* family, hepatitis C and bovine viral diarrhea virus (BVDV), showed that both of these RdRps contain an intrinsic terminal transferase activity (118). The nucleotides added by the viral terminal transferase differed depending on the 3′ terminal sequence of the template RNA, and it was postulated that the function of this activity was to repair the 3′ termini of viral RNA templates. It is likely that the WNV polymerase also contains a similar activity.

IN VITRO ACTIVITY OF RECOMBINANT NS5

Tan et al. (142) reported that a recombinant dengue NS5 protein, expressed in *E. coli* as a GST fusion protein and then cleaved by thrombin, exhibited in vitro RdRP activity with synthetic RNA templates consisting of the terminal 3′ 629 or 3200 nts of the dengue type 1 genome RNA. However, these investigators may have been measuring only terminal transferase addition to the template rather than de novo initiation followed by polymerization of a nascent strand because after ^{32}P-UTP incorporation only template-sized products were observed on nondenaturing gels. Steffens et al. (140) reported that baculovirus-expressed WNV NS5 produced RNA hairpins via self-priming from both WNV as well as from other template RNAs. RNA polymerization from a homopolymeric template [poly (C)] that could not base pair with itself was observed only when a primer was present. A baculovirus-expressed Kunjin NS5 copied only full-length, copy-back products from both a 9035 nt Kunjin replicon and a 8300 nt Semliki Forest replicon template (60). In contrast, two products, one of template size that was double stranded until denatured and another twice the length of the template, were observed in reactions with a recombinant dengue virus NS5 produced in *E. coli* (1), suggesting that both template self-priming as well as de novo initiation occurred in these reactions. The replicon template (770 nts) used in these reactions contained 3′ and 5′ terminal sequences that included both the 3′ and 5′ cyclization sequences. The authors suggested that two alternative conformations of the 3′ terminal sequence form and that these conformations determined which of the two types of initiation occurred. The 5′ sequences present and the temperature of the transcription reaction were also postulated to affect the RNA conformation. Models of the RNA structures were proposed, but proof of their existence was not provided.

The intrinsic activities of all the recombinant flavivirus polymerases were low and none of them were template specific. However, template specificity is a property of the viral polymerase complexes found in infected cells (75). Although an NS5-deficient Kunjin replicon could be complemented in *trans* by NS5 expressed alone, complementation was much more efficient when all seven of the viral nonstructural proteins were expressed (81, 83). Cell proteins that interact specifically with the 3′ terminal regions of the WNV RNA have also been detected (Figure 3, see color insert). In addition, RNA replication complexes in infected cells are associated with membranes. Which of these interactions confers template specificity to the WNV polymerase is currently not known.

Initiation of RNA synthesis was shown to occur on the 3′ nt of a short template consisting of the first 21 3′ nts of the BVDV minus-strand RNA (74). This initiation was primer independent (de novo), template specific, and required a 3′ pyrimidine and a +1 cytidylate (75). Consistent with these requirements, both the WNV genomic and minus-strand RNAs have a U as the 3′ terminal nt and the +1 nt is a C. To initiate at the 3′ terminal nt, the catalytic site of the viral polymerase must interact with the first two nts of the template RNA and also with the two nts that will become the first and second nts of the nascent strand.

The development of in vitro polymerase assays are of interest for the screening of antiviral drug candidates that target viral RNA synthesis. Polyoxometalate HPA-23 and several structurally related compounds inhibited the elongation of Kunjin RNA in vitro in a dose-dependent fashion (6). Ribavirin inhibited WNV replication in human neural cell cultures, but high doses were required (72).

CONSERVED SECONDARY AND TERTIARY TERMINAL RNA STRUCTURES

Stable secondary structures are located at both the 3′ and 5′ termini of the WNV genome RNA (25, 26, 133, 134) (Figure 3). The sizes and shapes of the terminal RNA structures are conserved among divergent flaviviruses even though the majority of the sequences composing them are not conserved (25, 26, 120). The presence of a 3′ terminal structure on the WNV genomic RNA was demonstrated in solution by chemical digestion under partially denaturing conditions (26). The 5′ stem loop (SL) is smaller and less stable than the 3′ SL and has a different shape (25).

Terminal SL structures were subsequently predicted in the complementary minus-strand RNA (Figure 3). Circular dichroism spectroscopic, thermal melting curve and RNase probing data supported the existence of an RNA structure at the 3′ end of the WNV minus-strand RNA (134; P.-Y. Shi & M.A. Brinton, unpublished data). RNase probing also indicated that the structure of this SL was as predicted by version 3.1 of M-fold (M.M. Emara & M.A. Brinton, unpublished data). The data also suggested that nts beyond the terminal of the structure might be involved in tertiary interactions (P.-Y. Shi & M.A. Brinton, unpublished data). The formation of the 3′ (−) SL structure may facilitate the release of nascent minus-strands from plus-strand templates and may also facilitate the formation of active replication complexes at the 3′ end of the minus-strand template. Since minus strands are present in infected cells only in RF and RI structures, it is unknown whether the predicted 5′ SL of the minus strand ever forms (Figure 3). Such a structure might form transiently upon the release of a nascent plus strand and serve to protect the 5′ end of the minus-strand template from digestion by cellular nucleases.

Deletion of the sequences composing either the genomic 3′ or 5′ terminal structures in a dengue infectious clone was lethal (31, 89, 102). Deletion of sequences in the 3′ NCR also decreased the replication efficiency of a Kunjin replicon (85). Studies with a BVDV infectious clone (164), dengue-WNV chimeric infectious clones (166), and dengue-WNV chimeric replicons (162, 163) indicated that short regions within the 3′ terminal structure of the genome RNA had *cis*-acting function.

The sequence located just 5′ of the genomic 3′ terminal SL was shown to fold into a small conserved SL structure. The loop of this structure could form a tertiary interaction (pseudoknot) with nts in an unstable region on the 5′ side of the 3′ terminal SL (111, 133; P.-Y. Shi & M.A. Brinton, unpublished data) (Figure 3). The feasibility of this pseudoknot interaction was demonstrated in solution by RNase and chemical probing and by melting experiments (133). Data from an dengue in vitro RNA replication system suggested that this pseudoknot may be functionally important for in vitro initiation of minus-strand synthesis from the genomic RNA (162, 163). Two additional conserved SL (dumbbell) structures within the 3′ NCR, each with loop sequences predicted to participate in tertiary interactions with downstream sequences (indicated by *boxes* in Figure 3), have recently been predicted in mosquito-borne flavivirus genome RNAs (110). Various deletions made within the 3′ NCR of dengue and tick-borne encephalitis virus infectious clones in the region containing these two predicted pseudoknots reduced the replication efficiency of the mutant virus produced, and in some cases highly attenuated virus and/or virus with an altered phenotype was produced (89, 98, 102). These deletions were designed in the absence of information on the structures formed by the 3′ NCR RNA but are predicted to result in the rearrangement of one or both pseudoknots (110, 117). These data suggest that similar to some types of plant RNA virus genomes, multiple pseudoknots are also located near the 3′ end of the flavivirus genomic RNA (110). The 3′ NCR of flavivirus genomic RNAs is relatively long (about 500–600 nts). Each of the three conserved 3′ tertiary structures may represent separate local folding domains that serve to stabilize and compact this region of the RNA. These structures could also be important for the formation of binding sites for viral and/or cellular proteins.

CONSERVED SEQUENCES

Several short, conserved sequences have been identified within the 3′ NCR of flavivirus genome RNAs (indicated in *green* in Figure 3), but the functional significance of these conserved sequences is not yet known (26, 61, 110, 123). In addition to the first two 3′ nts 3′-UC-5′, which are conserved in both the genome and minus-strand RNAs, there are two additional conserved sequences within the 3′ terminal SL. One sequence, 3′-ACAC-5′, is located on the 3′ side of the stem across from the 3′ pseudoknot and is also found near the 3′ terminus of the minus-strand RNA (22) (Figure 3). The other sequence, 3′-GACA-5′, is located in the second top loop (26) (Figure 3). Within the 3′ NCR of the WNV genome RNA, there is one copy of a sequence designated conserved sequence 1 (CS1) and two copies of a second conserved sequence designated CS2 and RCS2 (61, 110) (indicated in *green* in Figure 3). CS2 and RCS2 are located within the dumbbell structures that form the second and third pseudoknots (Figure 3). An 8-nt sequence, designated the cyclization sequence (indicated in *red* in Figure 3), is located in the 5′ portion of CS1 and is immediately adjacent to the first pseudoknot (61). CS1 is complementary to 5′ CS (indicated in *red* in Figure 3), which is located in the capsid-coding region near the 5′ end of the genome (61). Base pairing between the 3′ and 5′ cyclization sequences, even though this interaction would span more than 10,000 nts, was calculated to be thermodynamically feasible (61). Data from experiments using a Kunjin replicon in an in vivo system (80) and a dengue replicon in an in vitro system (162) suggested that base pairing between the 5′ and 3′ cyclization sequences does occur and is necessary for in vitro RNA replication. Base pairing between the 3′ and 5′ sequences, not the sequences themselves, was required for this function. Data obtained with the dengue replicon suggested that the 5′ NCR and additional regions of the 3′ NCR may be functionally important for initiation of minus-strand RNA synthesis from the genome RNA (162). However, the means by which these various elements facilitate virus RNA replication is not currently understood. Within CS1 the 3′ cyclization sequence and the sequence predicted to be involved in the tertiary interaction of the second 3′ pseudoknot overlap by 1 nt, and the 3′ side of the cyclization sequence is adjacent to the first pseudoknot (Figure 3). The data obtained thus far suggest that the 3′ NCR of the WNV RNA participates in complex and possibly alternative RNA-RNA interactions.

HOST CELL PROTEINS THAT INTERACT WITH THE WNV 3′ TERMINAL SLS

Genomic 3′ Terminal SL

Cell proteins with molecular masses of 52, 84, and 105 kDa that bind specifically to the 3′ terminal SL (83 nts) of the WNV genome RNA were detected in RNA-protein interaction assays with mammalian S100 cell cytoplasmic extracts (13) (Figure 3).

Competition gel mobility shift assays indicated that p105, but not p84 or p52, also interacts specifically with the WNV 3′ (−) RNA (13) (Figure 3, p. 108). Calf intestinal alkaline phosphatase treatment had no effect on the abilities of the 84- or 105-kDa proteins to bind to the WNV 3′ (+) SL RNA, but it completely abolished the ability of p52 to bind to this RNA (data not shown). Three protein bands, p52, p84, and p105, with identical molecular masses were detected by Northwestern blotting assays in BHK S100 extracts with probes consisting of the 3′ SLs of dengue virus, yellow fever virus, tick-borne encephalitis virus, and WNV. Also, each of these four 3′ SLs competed to the same extent with a WNV probe, indicating that the same set of cell proteins binds to the 3′ SLs of each of these divergent flaviviruses (W.G. Davis, A.A. Perelygin, J.L. Blackwell, B. Smith, V.F. Yamshchikov & M.A. Brinton, manuscript submitted). An additional broad protein band, ~p34, was sometimes detected in the RNA-protein interaction assays, but because this protein band could be competed away by nonspecific competitor RNAs, this RNA-protein interaction was considered to be of low affinity or nonspecific (13). The 36-kDa Mov34 protein was subsequently reported to bind to the 3′ SL RNA of Japanese encephalitis virus (141).

The 52-kDa protein was identified as eukaryotic elongation factor 1 alpha (EF-1α) (14). The primary cellular function of EF-1α is to ferry aminoacylated tRNAs to the ribosomes. EF-1α binds specifically to all of the charged tRNAs and to ribosomal RNA (125). Anti-EF-1α antibody supershifted RNA-protein complexes formed between proteins in S100 extracts and the WNV 3′ (+) SL RNA probe and also coimmunoprecipitated WNV genome RNA from infected cells (14). The dissociation constant (K_d) for the EF-1α-WNV 3′ (+) SL RNA interaction was calculated to be 2.1×10^{-9} M, which is similar to that for the interaction between EF-1α and aminoacylated tRNA (10^{-9} to 10^{-10} M). This suggests that an in vivo interaction between the viral 3′ (+) RNA and EF-1α is feasible (14).

The major binding site on the WNV 3′ (+) RNA recognized by EF-1α was first mapped by footprinting to a region in the middle of the 5′ side of the terminal SL of the WNV 3′ (+) RNA and confirmed in filter-binding assays using various mutated or truncated WNV 3′ (+) SL RNAs and purified EF-1α (14); mutation of the site detected by footprinting reduced binding by about 60%. Two additional weaker binding sites were also identified, indicating that EF-1α interacts with the viral 3′ RNA at multiple sites. One minor binding site was located in a single-stranded loop at the top of the 3′ terminal SL and the other was inactivated by deletion of the second small SL. Mutation of the major binding site for EF-1α in different structural contexts in a WNV infectious clone (160) produced virus that replicated inefficiently and produced small plaques. Upon passage, virus that produced larger plaques arose spontaneously and the genomes of these viruses contained a partial reversion of the mutated sequence (W.G. Davis, A.A. Perelygin, J.L. Blackwell, B. Smith, V.F. Yamshchikov & M.A. Brinton, manuscript submitted). These data suggest that the binding of EF-1α to the WNV 3′ SL plays a functionally important role in virus replication.

Minus-Strand 3′ Terminal SL

Four cell proteins, p108, p60, p50, and p42, that bind specifically to the 3′ terminal SL (75 nts) of the WNV complementary minus-strand RNA were detected in RNA-protein interaction assays with BHK S100 cell cytoplasmic extracts (134) (Figure 3). Calf intestinal alkaline phosphatase treatment of the cell extracts had no effect on the binding of these proteins to the WNV 3′ (−) SL RNA, indicating that protein phosphorylation is not important for the binding of these proteins. The 42-kDa protein was identified as T cell–restricted intracellular antigen-1 (TIA)-related protein (TIAR) and as the closely related protein TIA-1 (W. Li, Y. Li, N. Kedersha, P. Anderson, M. Emara, K. Swiderek, T. Moreno & M.A. Brinton, manuscript submitted). These two multifunctional proteins belong to the RRM family of RNA-binding proteins and are expressed in most tissues. They are normally present both in the cytoplasm and in the nucleus of cells, although their concentrations are usually higher in the nucleus (10, 47, 143, 144). Recombinant TIAR bound specifically to the WNV 3′ (−) RNA, and the K_d for this interaction was estimated to be greater than 10^{-8} M (W. Li, Y. Li, N. Kedersha, P. Anderson, M. Emara, K. Swiderek, T. Moreno & M.A. Brinton, manuscript submitted). The observation that the replication of WNV, but not that of Sindbis virus, vesicular stomatitis virus, herpes simplex virus, type 1, or vaccinia virus, was less efficient in TIAR-knockout murine cell lines than in the wild-type cells suggests a functional role for these proteins during WNV replication.

EFFECT OF THE GENOMIC 3′ SL RNA ON TRANSLATION

According to a current model of translation initiation of cell mRNAs, the 3′ and 5′ ends of the mRNA are brought into close proximity in a closed-loop complex via interactions between proteins that bind to the 3′ poly A and the 5′ cap structure (127) (Figure 4). mRNAs that are in a closed-loop configuration show a significant increase in the efficiency with which they recruit and recycle ribosomes. The WNV genomic RNA does not have a 3′ poly A tail and therefore cannot utilize poly A–binding protein to facilitate a 3′-5′ interaction. Other types of viral mRNAs that do not have 3′ poly A tails utilize alternative proteins to bind to their 3′ sequences and facilitate interactions with the 5′ cap–binding complex. For instance, the 3′ end of the plant positive-strand RNA virus, brome mosaic virus, binds to the cell protein Lsm1p (48), and the 3′ ends of rotavirus mRNAs bind to the viral protein NSP3 (147). Both these proteins facilitate the formation of closed-loop complexes (Figure 4). The translational efficiencies of chimeric mRNAs, which consisted of a CAT reporter gene sequence flanked by either viral or nonviral 3′ and 5′ sequences, were reduced when the WNV 3′ (+) SL was present at the 3′ end of the RNA (93). Reduced translation was observed with both capped and uncapped chimeric mRNAs and with chimeric mRNAs that had either a WNV or nonviral 5′ NCR, suggesting that the cell proteins interacting with the WNV 3′ (+) SL did not facilitate the formation of a 3′–5′ closed-loop complex. In vitro, the WNV 3′ (+)

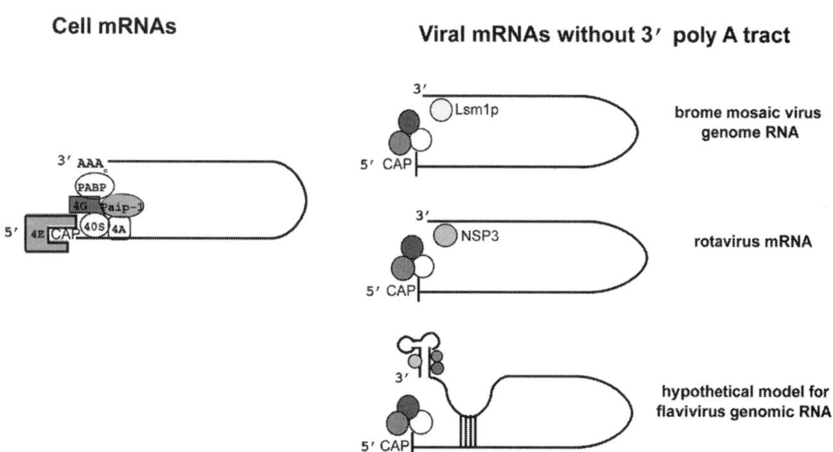

Figure 4 Closed-loop complexes formed by viral and cellular mRNAs. The 3′ regions of viral mRNAs that do not have a 3′ poly A tract interact with alternative cell (brome mosaic virus) or viral proteins (rotavirus), which facilitate interaction with 5′ cap–binding proteins. A 3′-5′ interaction for the WNV genome RNA may be facilitated by RNA-RNA interactions.

SL reduced translation both in *cis* and in *trans* in a dose-dependent manner, which is consistent with competition for translation factors. At least one cell translation–associated protein, EF-1α, has been shown to bind to the viral 3′ (+) SL and to charged tRNA with similar affinities (14). Base pairing between the 3′ and 5′ cyclization sequences (indicated in *red* in Figure 3) instead of protein-protein interactions may facilitate a 3′-5′ interaction in the WNV genomic RNA (Figure 4). Neither cyclization sequence was present in the chimeric mRNAs tested. Alternatively, a viral protein could facilitate such an interaction; weak binding of NS3 and NS5 to the viral 3′ NCR has been reported (35a, 43). These results suggest that the viral 3′ structures, the cell proteins that interact with them and long-distance RNA-RNA interactions, may be involved in regulating translation as well as minus-strand RNA synthesis.

VIRUS-HOST INTERACTIONS

Little is known about the mechanisms involved in WNV-induced cytopathology. Apoptosis was observed in WNV-infected K562 and Neuro-2a cells and was shown to be bax dependent (112). Virus replication appears to be required since UV-inactivated virus failed to induce apoptosis. WNV infection has also been reported to induce expression of ICAM-1 in infected cells (132). WNV-specific, interferon-independent induction of ICAM-1 was observed within 2 h after infection in quiescent but not replicating fibroblasts, suggesting involvement of the cell cycle.

Infection of diploid vertebrate cells with WNV has been reported to increase cell surface expression of MHC-1, which results from increased MHC-1 mRNA transcription activated by NF-κB (78). Activation of NF-κB appeared to be mediated via virus-induced phosphorylation of inhibitor κB.

Variation in the response of individuals to flavivirus infection has been observed in humans as well as in other host species. In mice, the alleles of a single Mendelian dominant gene, *Flv*, can determine whether an infection is lethal [reviewed in (22)]. Most of the currently used inbred mouse strains are susceptible; only BRVR, BSVR, C3H.RV, Det, PRI, and most wild *Mus musculus domesticus* are resistant. Resistant animals and cell cultures can be infected with WNV, but they produce lower yields of virus than do susceptible ones. The *Flv* gene was mapped to a region of murine chromosome 5 (146) and has been tentatively identified as Oas1b using a positional cloning approach (A.A. Perelygin, S.V. Scherbik, I.B. Zhulin, B.M. Stockman, Y. Li & M.A. Brinton, manuscript submitted).

WNV defective interfering (DI) particles were detected during serial undiluted passage of virus in mouse embryofibroblasts (21, 22, 44). The production of and interference by WNV DI particles is difficult to detect during acute infections of permissive cells, which produce high levels of virus and in which cytopathology develops rapidly. Because the production of standard virus is lower in genetically resistant mouse cells and little cytopathology is observed after WNV infection, DI particles represent a larger portion of the virus population and DI interference can be readily demonstrated. Analysis of the genomes of the DI particles suggested that DI particles with different-sized genomes were simultaneously replicated (21).

Persistence of flaviviruses in mice, monkeys, and humans has been reported [reviewed in (22)]. Persistent infections of WNV were successfully established in mouse embryo fibroblast cultures, and multiple virus mutant populations were shown to evolve and replicate simultaneously in these cultures. Temperature-sensitive (TS) mutants, nonplaquing mutants, DI particles, and one replication-efficient mutant were isolated (19, 20, 22, 24). Most isolated TS mutants had a RNA-minus phenotype. Complementation between individual TS mutants was not observed.

Prolonged passage of WNV in mosquito cells resulted in virus that was no longer neuroinvasive (33). Although these viruses had a mutation in the E protein that produced a N-linked glycosylation site, this mutation was shown not to correlate directly with attenuation (33, 130a). Additional mutations in E as well as mutations in other regions of the genome appeared to also be required for attenuation. Studies with yellow fever virus chimeras containing Japanese encephalitis virus E and prM showed that, although residue 138 of the E protein had a dominant effect on virulence, multiple mutations in the E protein were required to convert an attenuated virus to a virulent one (4). Analysis of nts that differed between a dengue vaccine candidate and its virulent parent showed that the mutations that primarily affected virulence in mice were located in the 5′ NCR and NS1, but again multiple mutations were required to restore the virulent phenotype (30).

DNA vectors expressing noncytopathic Kunjin replicons that encode a heterologous gene have been developed (146a). These vectors directed high-level, long-term expression of the heterologous gene in a number of mammalian cell lines and for eight weeks in the lung epithelium after delivery to mice by the intranasal route, indicating that such vectors would be useful in gene therapy applications.

IMMUNE RESPONSE AND VACCINE STRATEGIES

The majority of the antibodies elicited in flavivirus-infected hosts are directed to epitopes in the E, NS1, and NS3 proteins, although antibodies to most other viral proteins can also be detected (27). The dominant cytotoxic T cell epitopes are located in NS3, but epitopes have also been demonstrated in NS1-NS2A, E, NS4A, and NS4B (27, 67). Partial protective immunity has been achieved by vaccination with purified E or NS1 (62), but a stronger response was elicited following expression of C-prM-E-NS1-NS2A from a vaccinia recombinant (62). West Nile vaccines currently under development include DNA vaccines expressing either WNV prM-E or C (46, 161), a live attenuated WNV (97), and an attenuated live virus chimera consisting of yellow fever 17D vaccine with the prM and E genes replaced with those of WNV (5).

CONCLUSIONS

Although much has been learned about the molecular biology of flaviviruses, there are still many unanswered questions. A growing body of literature suggests that viruses utilize cell proteins during many steps of their replication cycles, such as attachment, entry, translation, transcription/replication, and assembly. Because WNV alternates between insect vectors and vertebrates in nature, any cellular proteins that this virus uses during replication would be expected to be evolutionarily conserved. Of particular interest will be the identification of the cell protein(s) used for virus attachment and entry and the elucidation of the molecular mechanisms involved in these processes. Further study of the viral envelope proteins and their novel role in imposing symmetry on virions will broaden our understanding of virus virulence.

Little is currently known about the molecular mechanisms utilized to initiate the synthesis of flavivirus RNA in a template-specific manner. Cell proteins may facilitate specific template recognition by and/or enhance the processivity of the viral polymerase. The mechanisms utilized for the initiation and regulation of flavivirus RNA synthesis and for the regulation of translation are likely to depend on interactions between viral nonstructural proteins and between cellular and viral proteins, on interactions between viral RNA and cell and/or viral proteins, and on secondary and tertiary RNA interactions. For WNV, as well as for most other RNA viruses, our understanding of these types of complex interactions and their functions in viral RNA replication is still rudimentary. Both local and long-distance RNA-RNA interactions may help to provide the appropriate three-dimensional shape required for the polymerase complex to specifically recognize and initiate synthesis at the 3′ end of a viral RNA template. Cell and/or viral proteins may gain an ability to interact with each other only after binding to different regions of the viral RNA. Interactions between the 3′ and 5′ ends of the genomic RNA may play a role in regulating the switch between translation and RNA replication (53). Although the general aspects of RNA synthesis may be similar for different groups of viruses, the molecular interactions utilized by the different groups of RNA viruses to accomplish the same tasks are likely to be unique.

There is a significant need for the development of safe and effective WNV vaccines for humans, especially the elderly, as well as for horses and birds. Data from previous studies on vaccines for other flaviviruses should help to speed the development of WNV vaccines.

The development of low-toxicity, efficacious antiviral therapies is needed to treat patients who develop central nervous system diseases after WNV infection.

ACKNOWLEDGMENTS

The author thanks Taronna R. Maines and William G. Davis for producing the graphics.

The *Annual Review of Microbiology* is online at http://micro.annualreviews.org

LITERATURE CITED

1. Ackermannn M, Padmanabhan R. 2001. De novo synthesis of RNA by the dengue virus RNA-dependent RNA polymerase exhibits temperature dependence at the initiation but not elongation phase. *J. Biol. Chem.* 276:39926–37
2. Adams SC, Broom AK, Sammels LM, Harnett AC, Howard M, et al. 1995. Glycosylation and antigenic variation among Kunjin virus isolates. *Virology* 206:40–56
3. Arias CF, Preugschat F, Strauss JH. 1993. Dengue 2 virus NS2B and NS3 form a stable complex that can cleave NS3 within the helicase domain. *Virology* 193:888–99
4. Arroyo J, Guirakhoo F, Fenner S, Zhang ZX, Monath TP, et al. 2001. Molecular basis for attenuation of a yellow fever virus/Japanese encephalitis virus chimera vaccine (ChimeriVax-JE). *J. Virol.* 75:934–42
5. Arroyo J, Miller CA, Catalan J, Monath TP. 2001. Yellow fever vector live-virus vaccines: West Nile virus vaccine development. *Trends Mol. Med.* 7:350–54
6. Bartholomeusz A, Tomlinson E, Wright PJ, Birch C, Locarnini S, et al. 1994. Use of a flavivirus RNA-dependent RNA polymerase assay to investigate the antiviral activity of selected compounds. *Antiviral Res.* 24:341–50
7. Bartholomeusz AI, Wright PJ. 1993. Synthesis of dengue virus RNA in vitro: initiation and the involvement of proteins NS3 and NS5. *Arch. Virol.* 128:111–21
8. Bazan JF, Fletterick RJ. 1989. Detection of trypsin-like serine protease domain in flaviviruses and pestiviruses. *Virology* 171:637–39

9. Bazan JF, Fletterick RJ. 1990. Structural and catalytic models of trypsin-like proteases. *Semin. Virol.* 1:311–22
10. Beck AR, Medley QG, O'Brien S, Anderson P, Streuli M. 1996. Structure, tissue distribution and genomic organization of the murine RRM-type RNA binding proteins TIA-1 and TIAR. *Nucleic Acids Res.* 24:3829–35
11. Bernard KA, Maffei JG, Jones SA, Kauffman EB, Ebel GD, et al. 2001. West Nile virus infection in birds and mosquitoes, New York State, 2000. *Emerg. Infect. Dis.* 7:679–85
12. Berthet FX, Zeller HG, Drouet M-T, Rauzier J, Digoutte J-P, Deubel V. 1997. Extensive nucleotide changes and deletions within the envelope glycoprotein gene of Euro-African West Nile viruses. *J. Gen. Virol.* 78:2293–97
13. Blackwell JL, Brinton MA. 1995. BHK cell proteins that bind to the 3′ stem-loop structure of the West Nile virus genome RNA. *J. Virol.* 69:5650–58
14. Blackwell JL, Brinton MA. 1997. Translation elongation factor-1α interacts with the 3′ stem-loop of West Nile virus genomic RNA. *J. Virol.* 71:6433–44
15. Blumenthal T, Carmichael GG. 1979. RNA replication: function and structure of the Qβ replicase. *Annu. Rev. Biochem.* 48:525–48
16. Borowski P, Niebuhr A, Mueller O, Bretner M, Felczak K, et al. 2001. Purification and characterization of West Nile virus nucleoside triphosphatase (NTPase)/helicase: evidence for dissociation of the NTPase and helicase activities of the enzyme. *J. Virol.* 75:3220–29
17. Briese T, Jia X-Y, Huang C, Grady LJ, Lipkin WI. 1999. Identification of a Kunjin/West Nile-like flavivirus in brains of patients with New York encephalitis. *Lancet* 354:1261–62
18. Brinkworth RI, Fairlie DP, Leung D, Young PR. 1999. Homology model of the dengue 2 virus NS3 protease: putative interactions with both substrate and NS2B cofactor. *J. Gen. Virol.* 80:1167–77
19. Brinton MA. 1981. Isolation of a replication-efficient mutant of West Nile virus from a persistently infected genetically resistant mouse cell culture. *J. Virol.* 39:413–21
20. Brinton MA. 1982. Characterization of West Nile virus persistent infections in genetically resistant and susceptible mouse cells. I. Generation of defective nonplaquing virus particles. *Virology* 116:84–98
21. Brinton MA. 1983. Analysis of extracellular West Nile virus particles produced by cell cultures from genetically and resistant and susceptible mice indicated enhanced amplification of defective interfering particles by resistant cultures. *J. Virol.* 46:860–70
22. Brinton MA. 1986. Replication of flaviviruses. In *Togaviridae and Flaviviridae, The Viruses*, ed. S Schlesinger, M Schlesinger, pp. 329–76. New York: Plenum
23. Brinton MA. 2001. Host factors involved in virus replication. *Ann. NY Acad. Sci.* 951:207–19
24. Brinton MA, Davis J, Schaefer D. 1985. Characterization of West Nile virus persistent infections in genetically susceptible and resistant mouse cells. II. Generation of temperature sensitive mutants. *Virology* 140:152–58
25. Brinton MA, Dispoto JH. 1988. Sequence and secondary structure analysis of the 5′ terminal region of flavivirus genome RNA. *Virology* 61:3641–44
26. Brinton MA, Fernandez AV, Dispoto JH. 1986. The 3′-nucleotides of flavivirus genomic RNA form a conserved secondary structure. *Virology* 153:113–21
27. Brinton MA, Kurane I, Mathew A, Zeng L, Shi PY, et al. 1998. Immune mediated and inherited defences against flaviviruses. *Clin. Diagn. Virol.* 10:129–39
28. Buckley A, Gaidamovich S, Turchinskaya A, Gould EA. 1992. Monoclonal

antibodies identify the NS5 yellow fever virus non-structural protein in the nuclei of infected cells. *J. Gen. Virol.* 73:1125–30

29. Burke DS, Monath TP. 2001. Flaviviruses. See Ref. 85a, pp. 1043–125
30. Butrapet S, Huang CY, Pierro DJ, Bhamarapravati N, Gubler DJ, et al. 2000. Attenuation markers of a candidate dengue type 2 vaccine virus, strain 16681 (PDK-53), are defined by mutations in the 5′ noncoding region and nonstructural proteins 1 and 3. *J. Virol.* 74:3011–19
31. Cahour A, Pletnev A, Vazeille-Falcoz M, Rosen L, Lai CJ. 1995. Growth-restricted dengue virus mutants containing deletions in the 5′ noncoding region of the RNA genome. *Virology* 207:68–76
32. Chambers TJ, Hahn CS, Galler R, Rice CM. 1990. Flavivirus genome organization, expression, and replication. *Annu. Rev. Microbiol.* 44:649–88
33. Chambers TJ, Halevy M, Nestorowicz A, Rice CM, Lustig S. 1998. West Nile virus envelope proteins: nucleotide sequence analysis of strains differing in mouse neuroinvasiveness. *J. Gen. Virol.* 79:2375–80
34. Chambers TJ, Nestorowicz A, Amberg SM, Rice CM. 1993. Mutagenesis of the yellow fever virus NS2B protein: effects on proteolytic processing, NS2B-NS3 complex formation, and viral replication. *J. Virol.* 67:6797–807
35. Chang YS, Liao CL, Tsao CH, Chen MC, Liu CI, et al. 1999. Membrane permeabilization by small hydrophobic nonstructural proteins of Japanese encephalitis virus. *J. Virol.* 73:6257–64

35a. Chen CJ, Kuo MD, Chien LJ, Hsu SL, Wang YM, et al. 1997. RNA-protein interactions: involvement of NS3, NS5, and 3′ noncoding regions of Japanese encephalitis virus genomic RNA. *J. Virol.* 71:3466–73

36. Chen Y, Maguire T, Hileman RE, Fromm JR, Esko JD, et al. 1997. Dengue virus infectivity depends on envelope protein binding to target cell heparan sulfate. *Nat. Med.* 3:866–71

36a. Chowers MY, Lang R, Nasser F, Ben-David D, Giladi M. 2001. Clinical characteristics of the West Nile fever outbreak, Israel 2000. *Emerg. Infect. Dis.* 7: 611–14

37. Chu PW, Westaway EG. 1987. Characterization of Kunjin virus RNA-dependent RNA polymerase: reinitiation of synthesis in vitro. *Virology* 157:330–37
38. Chu PWG, Westaway EG. 1985. Replication strategy of Kunjin virus: evidence for recycling role of replicative form RNA as template in semiconservative and asymmetric replication. *Virology* 140:68–79
39. Chu PWG, Westaway EG. 1992. Molecular and ultrastructural analysis of heavy membrane fractions associated with the replication of Kunjin virus RNA. *Arch. Virol.* 125:177–91
40. Cleaves GR, Dubin DT. 1979. Methylation status of intracellular dengue type 2 40S RNA. *Virology* 96:159–65
41. Cleaves GR, Ryan TE, Schlesinger RW. 1981. Identification and characterization of type 2 dengue virus replicative intermediate and replicative form RNAs. *Virology* 111:73–83
42. Clum S, Ebner KE, Padmanabhan R. 1997. Cotranslational membrane insertion of the serine proteinase precursor NS2B-NS3 (Pro) of dengue virus type 2 is required for efficient in vitro processing and is mediated through the hydrophobic regions of NS2B. *J. Biol. Chem.* 272:30715–23
43. Cui T, Sugrue RJ, Xu Q, Lee AK, Chan YC, et al. 1998. Recombinant dengue virus type 1 NS3 protein exhibits specific viral RNA binding and NTPase activity regulated by the NS5 protein. *Virology* 246:409–17
44. Darnell MB, Koprowski H. 1974. Genetically determined resistance to infection with group B arboviruses. II. Increased production of interfering particles in cell

cultures from resistant mice. *J. Infect. Dis.* 129:248–56

45. Das T, Mathur M, Gupta AK, Janssen GM, Banerjee AK. 1998. RNA polymerase of vesicular stomatitis virus specifically associates with translation elongation factor-1 alphabetagamma for its activity. *Proc. Natl. Acad. Sci. USA* 95:1449–54
46. Davis BS, Chang G-JJ, Cropp B, Roehrig JT, Martin DA, et al. 2001. West Nile virus recombinant DNA vaccine protects mouse and horse from virus challenge and expresses in vitro a noninfectious recombinant antigen that can be used in enzyme-linked immunosorbent assays. *J. Virol.* 75:4040–47
47. Dember LM, Kim ND, Liu KQ, Anderson P. 1996. Individual RNA recognition motifs of TIA-1 and TIAR have different RNA binding specificities. *J. Biol. Chem.* 271:2783–88
48. Diez J, Ishikawa M, Kaido M, Ahlquist P. 2000. Identification and characterization of a host protein required for efficient template selection in viral RNA replication. *Proc. Natl. Acad. Sci. USA* 97:3913–18
49. Falgout B, Markoff L. 1995. Evidence that flavivirus NS1-NS2A cleavage is mediated by membrane-bound host protease in the endoplasmic reticulum. *J. Virol.* 69:7232–43
50. Falgout B, Miller RH, Lai C-J. 1993. Deletion analysis of dengue virus type 4 nonstructural protein NS2B: identification of a domain required for NS2B-NS3 proteinase activity. *J. Virol.* 67:2034–42
51. Flamand M, Megret F, Mathieu M, Lepault J, Rey FA, et al. 1999. Dengue virus type 1 nonstructural glycoprotein NS1 is secreted from mammalian cells as a soluble hexamer in a glycosylation-dependent fashion. *J. Virol.* 73:6104–10
52. Forwood JK, Brooks A, Briggs LJ, Xiao CY, Jans DA, et al. 1999. The 37–amino acid interdomain of dengue virus NS5 protein contains a functional NLS and inhibitory CK2 site. *Biochem. Biophys. Res. Commun.* 257:731–37
53. Gamarnik AV, Andino R. 1998. Switch from translation to RNA replication in a positive-stranded RNA virus. *Genes Dev.* 12:2293–304
54. Gorbalenya AE, Donchenko AP, Koonin EV, Blinov VM. 1989. N-terminal domains of putative helicases of flavi- and pestiviruses may be serine proteases. *Nucleic Acids Res.* 17:3889–97
55. Gorbalenya AE, Koonin EV, Donchenko AP, Blinov VM. 1989. Two related superfamilies of putative helicases involved in replication, recombination, repair and expression of DNA and RNA genomes. *Nucleic Acids Res.* 17:4713–30
56. Grun JB, Brinton MA. 1986. Characterization of West Nile viral RNA-dependent RNA polymerase and cellular terminal adenylyl and uridylyl transferases in cell-free extracts. *J. Virol.* 60:1113–24
57. Grun JB, Brinton MA. 1987. Dissociation of NS5 from cell fractions containing West Nile virus-specific polymerase activity. *J. Virol.* 61:3641–44
58. Grun JB, Brinton MA. 1988. Separation of functional West Nile virus replication complexes from intracellular membrane fragments. *J. Gen. Virol.* 12:3121–27
59. Gu B, Liu C, Lin-Goerke J, Maley DR, Gutshall LL, et al. 2000. The RNA helicase and nucleotide triphosphatase activities of bovine viral diarrhea virus NS3 protein are essential for viral replication. *J. Virol.* 74:1794–800
60. Guyatt KJ, Westaway EG, Khromykh AA. 2001. Expression and purification of enzymatically active recombinant RNA-dependent RNA polymerase (NS5) of the flavivirus Kunjin. *J. Virol. Methods* 92:37–44
61. Hahn CS, Hahn YS, Rice CM, Lee E, Dalgarno L, et al. 1987. Conserved elements in the 3′ untranslated region of flavivirus RNAs and potential cyclization sequences. *J. Mol. Biol.* 198:33–41
62. Hall RA, Brand TN, Lobigs M, Sangster

MY, Howard MJ, et al. 1996. Protective immune responses to the E and NS1 proteins of Murray Valley encephalitis virus in hybrids of flavivirus-resistant mice. *J. Gen. Virol.* 77:1287–94

63. Hall RA, Khromykh AA, Mackenzie JM, Scherret JH, Khromykh TI, et al. 1999. Loss of dimerisation of the nonstructural protein NS1 of Kunjin virus delays viral replication and reduces virulence in mice, but still allows secretion of NS1. *Virology* 264:66–75
64. Heinz FX, Allison SL. 2000. Structures and mechanisms in flavivirus fusion. *Adv. Virus Res.* 55:231–69
65. Heinz FX, Purcell MS, Gould EA, Howard CR, Houghton M, et al. 2000. Family Flaviviridae. In *Virus Taxonomy*, ed. MHV Regenmortel, CM Fauquet, DHL Bishop, EB Carstens, MK Estes, et al., pp. 860–78. San Diego: Academic
66. Heinz FX, Stiasny K, Puschner-Auer G, Holzmann H, Allison SL, et al. 1994. Structural changes and functional control of the tick-borne encephalitis virus glycoprotein E by the heterodimeric association with protein prM. *Virology* 198:109–17
67. Hill AB, Mullbacher A, Parrish C, Coia G, Westaway EG, et al. 1992. Broad cross-reactivity with marked fine specificity in the cytotoxic T cell response to flaviviruses. *J. Gen. Virol.* 73:115–23
68. Ilkal MA, Mavale MS, Prasanna Y, Jacob PG, Geevarghese G, et al. 1997. Experimental studies on the vector potential of certain *Culex* species to West Nile virus. *Indian J. Med. Res.* 106:225–28
69. Jacobs MG, Robinson PJ, Bletchly C, Mackenzie JM, Young PR. 2000. Dengue virus nonstructural protein 1 is expressed in a glycosyl-phosphatidylinositol-linked form that is capable of signal transduction. *FASEB J.* 14:1603–10
70. Jia X-Y, Briese T, Jordan I, Rambaut A, Chi HC, et al. 1999. Genetic analysis of West Nile New York 1999 encephalitis virus. *Lancet* 354:1971–72
71. Johansson M, Brooks AJ, Jans DA, Vasudevan SG. 2001. A small region of the dengue virus-encoded RNA-dependent polymerase, NS5, confers interaction with both the nuclear transport receptor importin-beta and the viral helicase, NS3. *J. Gen. Virol.* 82:735–45
72. Jordan I, Briese T, Fischer N, Lau JY, Lipkin WI. 2000. Ribavirin inhibits West Nile virus replication and cytopathic effect in neural cells. *J. Infect. Dis.* 182: 1214–17
73. Kamer G, Argos P. 1984. Primary structural comparison of RNA-dependent polymerases from plant, animal, and bacterial viruses. *Nucleic Acids Res.* 12: 7269–82
74. Kao CC, Del Vecchio AM, Zhong W. 1999. De novo initiation of RNA synthesis by a recombinant Flaviviridae RNA-dependent RNA polymerase. *Virology* 253:1–7
75. Kao CC, Singh P, Ecker DL. 2001. De novo initiation of viral RNA-dependent RNA synthesis. *Virology* 287:251–60
76. Kapoor M, Zhang L, Ramachandra M, Kusukawa J, Ebner KE, et al. 1995. Association between NS3 and NS5 proteins of dengue virus type 2 in the putative RNA replicase is linked to differential phosphorylation of NS5. *J. Biol. Chem.* 270:19100–6
77. Kaye K, Mojica AM. 2000. West Nile virus—a briefing (http://www.ci.nyc.ny.us/html/doh/pdf/chi/chi19–1.pdf). *City Health Inf. NY City Dep. Health* 19:1–6
78. Kesson AM, King NJ. 2001. Transcriptional regulation of major histocompatibility complex class I by flavivirus West Nile is dependent on NF-kappa B activation. *J. Infect. Dis.* 15:947–54
79. Khromykh AA, Kenney MT, Westaway EG. 1998. *trans*-complementation of flavivirus RNA polymerase gene NS5 by using Kunjin virus replicon-expressing BHK cells. *J. Virol.* 72:7270–79
80. Khromykh AA, Meka H, Guyatt KJ,

Westaway EG. 2001. Essential role of cyclization sequences in flavivirus RNA replication. *J. Virol.* 75:6719–28

81. Khromykh AA, Sedlak PL, Guyatt KJ, Hall RA, Westaway EG. 1999. Efficient trans-complementation of the flavivirus Kunjin NS5 protein but not of the NS1 protein requires its coexpression with other components of the viral replicase. *J. Virol.* 73:10272–80
82. Khromykh AA, Sedlak PL, Westaway EG. 1999. *trans*-complementation analysis of the flavivirus Kunjin NS5 gene reveals an essential role for translation of its N-terminal half in RNA replication. *J. Virol.* 73:9247–55
83. Khromykh AA, Sedlak PL, Westaway EG. 2000. *cis*- and *trans*-acting elements in flavivirus RNA replication. *J. Virol.* 74:3253–63

83a. Khromykh AA, Varnavski AN, Sedlak PL, Westaway EG. 2001. Coupling between replication and packaging of flavivirus RNA: evidence derived from the use of DNA-based full-length cDNA clones of Kunjin virus. *J. Virol.* 75:4633–40

84. Khromykh AA, Westaway EG. 1996. RNA binding properties of core protein of the flavivirus Kunjin. *Arch. Virol.* 141: 685–99
85. Khromykh AA, Westaway EG. 1997. Subgenomic replicons of the flavivirus Kunjin: construction and applications. *J. Virol.* 71:1497–505

85a. Knipe DM, Howley PM, eds. 2001. *Fields Virology*. Philadelphia: Lippincott, Williams & Wilkins. 4th ed.

86. Komar N. 2000. *Sentinel live bird surveillance*. Second Natl. Plan. Meet. Surveill. Prev. Control West Nile virus U.S. (http://www.CDC.gov/ncidod/dvbib/westnile/misc/slides/index.htm)
87. Koonin EV. 1991. The phylogeny of RNA-dependent RNA polymerases of positive-strand RNA viruses. *J. Gen. Virol.* 72:2197–206
88. Koonin EV. 1993. Computer-assisted identification of a putative methyltransferase domain in NS5 protein of flaviviruses and lambda 2 protein of reovirus. *J. Gen. Virol.* 74:733–40

88a. Kuhn RJ, Zhang W, Rossman MG, Pletnev SV, Corver J, et al. 2002, Structure of dengue virus. Implications for flavivirus organization, maturation, and fusion. *Cell* 108:717–25

89. Lai C-J, Men R, Pethel M, Bray M. 1992. Infectious RNA transcribed from stably cloned full-length cDNA: construction of growth-restricted dengue virus mutants. In *Vaccines 92*, ed. F Brown, RM Chanock, HS Ginsberg, RA Lerner, pp. 265–70. Cold Spring Harbor: Cold Spring Harbor Lab. Press
90. Lanciotti RS, Roehrig JT, Deubel V, Smith J, Parker M, et al. 1999. Origin of the West Nile virus responsible for the outbreak of encephalitis in the Northeastern United States. *Science* 286:2333–37
91. Lee E, Lobigs M. 2000. Substitutions at the putative receptor-binding site of an encephalitic flavivirus alter virulence and host cell tropism and reveal a role for glycosaminoglycans in entry. *J. Virol.* 74: 8867–75
92. Li H, Clum S, You S, Ebner KE, Padmanabhan R. 1999. The serine protease and RNA-stimulated nucleoside triphosphatase and RNA helicase functional domains of dengue virus type 2 NS3 converge within a region of 20 amino acids. *J. Virol.* 73:3108–16
93. Li W, Brinton MA. 2001. The 3′ stem loop of the West Nile virus genome RNA can suppress translation of chimeric mRNAs. *Virology* 287:49–61
94. Lindenbach BD, Rice CM. 1997. *trans*-complementation of yellow fever virus NS1 reveals a role in early RNA replication. *J. Virol.* 71:9608–17
95. Lindenbach BD, Rice CM. 1999. Genetic interaction of flavivirus nonstructural proteins NS1 and NS4A as a determinant of replicase function. *J. Virol.* 73:4611–21

96. Lindenbach BD, Rice CM. 2001. Flaviridae: the viruses and their replication. See Ref. 85a, pp. 991–1041
97. Lustig S, Olshevsky U, Ben-Nathan D, Lachmi BE, Malkinson M, et al. 2000. A live attenuated West Nile virus strain as a potential veterinary vaccine. *Viral Immunol.* 13:401–10
98. Mandl CW, Holzmann H, Meixner T, Rauscher S, Stadler P, et al. 1998. Spontaneous and engineered deletions in the 3′ noncoding region of tick-borne encephalitis virus: construction of highly attenuated mutants of a flavivirus. *J. Virol.* 72:2132–40
99. Mason PW. 1989. Maturation of Japanese encephalitis virus glycoproteins produced by infected mammalian and mosquito cells. *Virology* 169:354–64
100. Matusan AE, Pryor MJ, Davidson AD, Wright PJ. 2001. Mutagenesis of the dengue virus type 2 NS3 protein within and outside helicase motifs: effects on enzyme activity and virus replication. *J. Virol.* 75:9633–43
101. McIntosh BM, Jupp PG, Dos Santos I, Meenehan GM. 1976. Epidemics of West Nile and Sindbis viruses in South Africa with *Culex* (Culex) *univittatus* Theobald as vector. *S. Afr. J. Sci.* 72:295–300
102. Men R, Bray M, Clark D, Chanock RM, Lai CJ. 1996. Dengue type 4 virus mutants containing deletions in the 3′ noncoding region of the RNA genome: analysis of growth restriction in cell culture and altered viremia pattern and immunogenicity in Rhesus monkeys. *J. Virol.* 70:3930–37
103. Miller BR, Nasci RS, Godsey MS, Savage HM, Lutwama JJ, et al. 2000. First field evidence for natural vertical transmission of West Nile virus in *Culex univittatus* complex mosquitoes from Rift Valley province, Kenya. *Am. J. Trop. Med. Hyg.* 62:240–46
104. Murray JM, Aaskov JG, Wright PJ. 1993. Processing of the dengue virus type 2 proteins prM and C-prM. *J. Gen. Virol.* 74:175–82
105. Muylaert IR, Galler RG, Rice CM. 1997. Genetic analysis of yellow fever virus NS1 protein: identification of a temperature-sensitive mutation which blocks RNA accumulation. *J. Virol.* 71:291–98
106. Ni H, Ryman KD, Wang H, Saeed MF, Hull R, et al. 2000. Interaction of yellow fever virus French neurotropic vaccine strain with monkey brain: characterization of monkey brain membrane receptor escape variants. *J. Virol.* 74:2903–6
107. Ng ML, Tan SH, Chu JJH. 2001. Transport and budding at two distinct sites of visible nucleocapsids of West Nile (Sarafend) virus. *J. Med. Virol.* 65:758–64
108. Nowak T, Farber PM, Wengler G, Wengler G. 1989. Analyses of the terminal sequences of West Nile virus structural proteins and of the in vitro translation of these proteins allow the proposal of a complete scheme of the proteolytic cleavages involved in their synthesis. *Virology* 169:365–76
109. Nowak T, Wengler G. 1987. Analysis of disulfides present in the membrane proteins of the West Nile flavivirus. *Virology* 156:127–37
110. Olsthoorn RCL, Bol JF. 2001. Sequence comparison and secondary structure analysis of the 3′ noncoding region of flavivirus genomes reveals multiple pseudoknots. *RNA* 7:1370–77
111. O'Reilly EK, Kao CC. 1998. Analysis of RNA-dependent RNA polymerase structure and function as guided by known polymerase structures and computer predictions of secondary structure. *Virology* 252:287–303
112. Parquet MC, Kumatori A, Hasebe F, Morita K, Igarashi A. 2001. West Nile virus-induced bax-dependent apoptosis. *FEBS Lett.* 29:17–24
113. Petersen LR, Roehrig JT. 2001. West Nile Virus: a reemerging global pathogen. *Emerg. Infect. Dis.* 7:611–14

114. Pletnev SV, Zhang W, Mukhopadhyay S, Fisher BR, Hernandez R, et al. 2001. Locations of carbohydrate sites on alphavirus glycoproteins show that E1 forms an icosahedral scaffold. *Cell* 105:127–36
115. Poidinger M, Hall RA, Mackenzie JS. 1996. Molecular characterization of the Japanese encephalitis serocomplex of the flavivirus genus. *Virology* 218:417–21
116. Preugschat F, Yao C-W, Strauss JH. 1990. In vitro processing of dengue virus type 2 nonstructural proteins NS2A, NS2B, and NS3. *J. Virol.* 64:4364–74
117. Proutski V, Gritsun TS, Gould EA, Holmes EC. 1999. Biological consequences of deletions within the 3′-untranslated region of flaviviruses may be due to rearrangements of RNA secondary structure. *Virus Res.* 64:107–23
118. Ranjith-Kumar CT, Gajewski J, Gutshall L, Maley D, Sarisky RT, et al. 2001. Terminal nucleotidyl transferase activity of recombinant Flaviviridae RNA-dependent RNA polymerases: implication for viral RNA synthesis. *J. Virol.* 75:8615–23
119. Rappole JH, Derrickson SR, Hubalek Z. 2000. Migratory birds and spread of West Nile in the Western Hemisphere. *Emerg. Infect. Dis.* 6:319–28
120. Rauscher S, Flamm C, Mandl CW, Heinz FX, Stadler PF. 1997. Secondary structure of the 3′-noncoding region of flavivirus genomes: comparative analysis of base pairing probabilities. *RNA* 3:779–91
121. Reed KE, Gorbalenya AE, Rice CM. 1998. The NS5A/NS5 proteins of viruses from three genera of the family Flaviviridae are phosphorylated by associated serine/threonine kinases. *J. Virol.* 72:6199–206
122. Rey FA, Heinx FX, Mandl C, Kunz C, Harrison SC. 1995. The envelope glycoprotein from tick-borne encephalitis virus at 2 Å resolution. *Nature* 375:291–98
123. Rice CM. 1996. Flaviviridae: the viruses and their replication. In *Fields Virology*, ed. BN Fields, DM Knipe, PM Howley, pp. 931–59. Philadelphia: Lippincott-Raven. 3rd ed.
124. Rice CM, Lenches EM, Eddy SR, Shin SJ, Sheets RL, et al. 1985. Nucleotide sequence of yellow fever virus: implications for flavivirus gene expression and evolution. *Science* 229:726–35
125. Riis B, Rattanm SI, Clark BF, Merrick WC. 1990. Eukaryotic protein elongation factors. *Trends Biochem. Sci.* 15:429–24
126. Russell PK, Brandt WE, Dalrymple JM. 1980. Chemical and antigenic structure of flaviviruses. In *The Togaviruses: Biology, Structure, Replication*, ed. RW Schlesinger, pp. 503–29. New York: Academic
127. Sachs AB, Sarnow P, Hentze MW. 1997. Starting at the beginning, middle, and end: translation initiation in eukaryotes. *Cell* 89:831–38
128. Salas-Benito JS, del Angel RM. 1997. Identification of two surface proteins from C6/36 cells that bind dengue type 4 virus. *J. Virol.* 71:7246–52
129. Sampson BA, Ambrosi C, Charlot A, Reiber K, Veress JF, et al. 2000. The pathology of human West Nile virus infection. *Hum. Pathol.* 31:527–31
130. Savage HM, Ceianu C, Nicolescu G, Karabatsos N, Lanciotti R, et al. 1999. Entomologic and avian investigations of an epidemic of West Nile fever in Romania in 1996, with serologic and molecular characterization of a virus isolate from mosquitoes. *Am. J. Trop. Med. Hyg.* 61:600–11
130a. Scherret JH, Mackenzie JS, Khromykh AA, Hall RA. 2001. Biological significance of glycosylation of the envelope protein of Kunjin virus. *Ann. NY Acad. Sci.* 953:361–63
131. Scherret JH, Poidinger M, Mackenzie JS,

Broom AK, Deubel V, et al. 2001. The relationships between West Nile and Kunjin viruses. *Emerg. Infect. Dis.* 7:697–705

132. Shen J, Devery JM, King NJ. 1995. Early induction of interferon-independent virus-specific ICAM-1 (CD54) expression by flavivirus in quiescent but not proliferating fibroblasts—implications for virus-host interactions. *Virology* 208: 437–49
133. Shi P-Y, Brinton MA, Veal JM, Zhong YY, Wilson WD. 1996. Evidence for the existence of a pseudoknot structure at the 3′ terminus of the flavivirus genomic RNA. *Biochemistry* 34:4222–30
134. Shi P-Y, Li W, Brinton MA. 1996. Cell proteins bind specifically to West Nile virus minus-strand 3′ stem-loop RNA. *J. Virol.* 70:6278–87
135. Smith GW, Wright PJ. 1985. Synthesis of proteins and glycoproteins in dengue type 2 virus-infected *Vero* and *Aedes albopictus* cells. *J. Gen. Virol.* 66:559–71
136. Smithburn KC, Hughes TP, Burke AW, Paul JH. 1940. A neurotropic virus isolated from the blood of a native of Uganda. *Am. J. Trop. Med. Hyg.* 20:471–92
137. Speight G, Westaway EG. 1989. Positive identification of NS4A, the last of the hypothetical nonstructural proteins of flaviviruses. *Virology* 170:299–301
138. Stadler K, Allison SL, Schalich J, Heinz FX. 1997. Proteolytic activation of tick-borne encephalitis virus by furin. *J. Virol.* 71:8475–81
139. Steele KE, Linn MJ, Schoepp RJ, Komar N, Geisbert TW, et al. 2000. Pathology of fatal West Nile virus infections in native and exotic birds during the 1999 outbreak in New York City, New York. *Vet. Pathol.* 37:208–24
140. Steffens S, Thiel H-J, Behrens S-E. 1999. The RNA-dependent RNA polymerases of different members of the family Flaviviridae exhibit similar properties in vitro. *J. Gen. Virol.* 80:2583–90
141. Ta M, Vrati S. 2000. Mov34 protein from mouse brain interacts with the 3′ noncoding region of Japanese encephalitis virus. *J. Virol.* 74:5108–15
142. Tan B-H, Fu J, Sugrue RJ, Yap E-H, Chan Y-C, Tan YH. 1996. Recombinant dengue type 1 virus NS5 protein expressed in *Escherichia coli* exhibits RNA-dependent RNA polymerase activity. *Virology* 216:317–25
143. Taupin JL, Tian Q, Kedersha N, Robertson M, Anderson P. 1995. The RNA-binding protein TIAR is translocated from the nucleus to the cytoplasm during Fas-mediated apoptotic cell death. *Proc. Natl. Acad. Sci. USA* 92:1629–33
144. Tian Q, Taupin J, Elledge S, Robertson M, Anderson P. 1995. Fas-activated serine/threonine kinase (FAST) phosphorylates TIA-1 during Fas-mediated apoptosis. *J. Exp. Med.* 182:865–74
145. Turell MJ, O'Guinn M, Oliver J. 2000. Potential for New York mosquitoes to transmit West Nile virus. *Am. J. Trop. Med. Hyg.* 62:413–14
146. Urosevic N, Mansfield JP, Mackenzie JS, Shellem GR. 1995. Low resolution mapping around the flavivirus locus (Flv) on mouse chromosome 5. *Mamm. Genome* 6:454–58

146a. Varnavski AN, Young PR, Khromykh AA. 2000. Stable high-level expression of heterologous genes in vitro and in vivo by noncytoplasmic DNA-based Kunjin virus replicon vectors. *J. Virol.* 74:4394–403

147. Vende P, Piron M, Castagne N, Poncet D. 2000. Efficient translation of rotavirus mRNA requires simultaneous interaction of NSP3 with eukaryotic translation initiation factor eIF4G and mRNA 3′ end. *J. Virol.* 74:7064–71
148. Warrener P, Tamura JK, Collett MS. 1993. An RNA-stimulated NTPase activity associated with yellow fever virus NS3 protein expressed in bacteria. *J. Virol.* 67:989–96
149. Wengler G. 1989. Cell-associated West

Nile flavivirus is covered with E+ pre-M protein heterodimers which are destroyed and reorganized by proteolytic cleavage during virus release. *J. Virol.* 63:2521–26

150. Wengler G, Czaya G, Farber PM, Hegemann JH. 1991. In vitro synthesis of West Nile virus proteins indicates that the amino-terminal segment of the NS3 protein contains the active centre of the protease which cleaves the viral polyprotein after multiple basic amino acids. *J. Gen. Virol.* 72:851–58
151. Wengler G, Wengler G. 1981. Terminal sequences of the genome and replicative-form RNA of the flavivirus West Nile virus: absence of poly (A) and possible role in RNA replication. *Virology* 113:544–55
152. Wengler G, Wengler G. 1991. The carboxy-terminal part of the NS3 protein of the West Nile flavivirus can be isolated as a soluble protein after proteolytic cleavage and represents an RNA-stimulated NTPase. *Virology* 184:707–15
153. Wengler G, Wengler G. 1993. The NS3 nonstructural protein of flaviviruses contains an RNA triphosphatase activity. *Virology* 197:265–73

153a. Westaway EG, Khromykh AA, Kenny MT, Mackenzie JM, Jones MK. 1997. Protein C and NS4B of the flavivirus Kunjin translocate independently into the nucleus. *Virology* 234:31–41

154. Westaway EG, Khromykh AA, Mackenzie JM. 1999. Nacent flavivirus RNA colocalized in situ with double-stranded RNA in stable replication complexes. *Virology* 258:108–17
155. Westaway EG, Mackenzie JM, Kenney MT, Mackenzie JM, Jones MK. 1997. Ultrastructure of Kunjin virus-infected cells: colocalization of NS1 and NS3 with double-stranded RNA, and of NS2B with NS3 in virus-induced membrane structures. *J. Virol.* 71:6650–61
156. Winkler G, Heinz FX, Kunz C. 1987. Studies on the glycosylation of flavivirus E proteins and the role of carbohydrate in antigenic structure. *Virology* 171:237–43
157. Winkler G, Maxwell SE, Ruemmler C, Stollar V. 1989. Newly synthesized dengue-2 virus nonstructural protein NS1 is a soluble protein but becomes partially hydrophobic and membrane-associated after dimerization. *Virology* 171:302–5
158. Winkler G, Randolph VB, Cleaves GR, Ryan TE, Stollar V. 1988. Evidence that the mature form of the flavivirus nonstructural protein NS1 is a dimer. *Virology* 162:187–96
159. Yamshchikov VF, Trent DW, Compans RW. 1997. Upregulation of signalase processing and induction of prM-E secretion by the flavivirus NS2B-NS3 protease: roles of protease components. *J. Virol.* 71:4364–71
160. Yamshchikov VF, Wengler G, Perelygin AA, Brinton MA, Compans R. 2001. An infectious clone of the West Nile flavivirus. *Virology* 281:294–304
161. Yang JS, Kim JJ, Hwang D, Choo AY, Dang K, et al. 2001. Induction of potent Th1-type immune responses from a novel DNA vaccine for West Nile virus New York isolate (WNV-NY1999). *J. Infect. Dis.* 184:809–16
162. You S, Falgout B, Markoff L, Padmanabhan R. 2001. In vitro RNA synthesis from exogenous dengue viral RNA templates requires long-range interactions between 5′- and 3′-terminal regions that influence RNA structure. *J. Biol. Chem.* 276:15581–91
163. You S, Padmanabhan R. 1999. A novel in vitro replication system for dengue virus. Initiation of RNA synthesis at the 3′-end of exogenous viral RNA templates requires 5′- and 3′-terminal complementary sequence motifs of the viral RNA. *J. Biol. Chem.* 274:33714–22
164. Yu H, Grassmann CW, Behrens S-E. 1999. Sequence and structural elements at the 3′ terminus of bovine viral diarrhea virus genomic RNA: functional role

during RNA replication. *J. Virol.* 73: 3638–48

165. Yusof R, Clum S, Wetzel M, Murthy HM, Padmanabhan R. 2000. Purified NS2B/NS3 serine protease of dengue virus type 2 exhibits cofactor NS2B dependence for cleavage of substrates with dibasic amino acids in vitro. *J. Biol. Chem.* 275:9963–69
166. Zeng L-L, Falgout B, Markoff L. 1998. Identification of specific nucleotide sequences within the conserved 3′-SL in the dengue type 2 virus genome required for replication. *J. Virol.* 72:7510–22

Annu. Rev. Microbiol. 2002. 56:403–32
doi: 10.1146/annurev.micro.56.012302.160838

Microbial Degradation of Polyhydroxyalkanoates*

Dieter Jendrossek and René Handrick
Institut für Mikrobiologie, Allmandring 31, D-70550 Stuttgart, Germany; e-mail: imbdj@po.uni-stuttgart.de; imbrh@po.uni-stuttgart.de

Key Words polyhydroxybutyrate, intracellular PHB depolymerase, extracellular PHB depolymerase, *Paucimonas lemoignei*

■ **Abstract** Polyesters such as poly(3-hydroxybutyrate) (PHB) or other polyhydroxyalkanoates (PHA) have attracted commercial and academic interest as new biodegradable materials. The ability to degrade PHA is widely distributed among bacteria and fungi and depends on the secretion of specific extracellular PHA depolymerases (e-PHA depolymerases), which are carboxyesterases (EC 3.1.1.75 and EC 3.1.1.76), and on the physical state of the polymer (amorphous or crystalline). This contribution provides a summary of the biochemical and molecular biological characteristics of e-PHA depolymerases and focuses on the intracellular mobilization of storage PHA by intracellular PHA depolymerases (i-PHA depolymerases) of PHA-accumulating bacteria. The importance of different assay systems for PHA depolymerase activity is also discussed.

CONTENTS

*Dedicated to Frank Mayer, Göttingen in recognition of his outstanding ability to teach and to visualize subcellular biological structures.

0066-4227/02/1013-0403$14.00

INTRODUCTION

Poly(*R*)-hydroxyalkanoic acids (PHA) are a group of storage compounds of carbon and energy that are accumulated during unbalanced growth by many bacteria, i.e., in the presence of an excess of a carbon source and if growth is limited by another nutrient (e.g., nitrogen). PHA are deposited intracellularly in the form of inclusion bodies ("granules," visible as brilliant globules, 100–500 nm) in the phase contrast microscope and may account for up to 90% of the cellular dry weight [for reviews on PHA see (3, 13, 63, 72, 89)]. Poly(*R*)-3-hydroxybutyric acid (PHB) is the most abundant polyester in bacteria. It consists of only one type of monomer, 3-hydroxybutyrate (3HB). However, about 150 hydroxyalkanoic acids (HA) other than 3HB have been identified as constituents of microbial polyesters (100, 102). Depending on the number of carbon atoms of the monomers, PHA are classified as short-chain-length PHA (PHA$_{SCL}$; 3 to 5 C-atoms) and medium-chain-length PHA (PHA$_{MCL}$; 6 or more C-atoms). In many bacteria PHA is composed of more than one type of monomer (copolyesters). The high number of monomers and the variable monomeric composition of PHA results in an enormous variation of the physical and chemical characteristics of different PHA. Despite the high numbers of different PHA, only PHB and copolymers of 3HB and 3-hydroxyvaleric acid (3HV) have been commercialized (e.g., Biopol®). PHA differ from chemosynthetically produced polymers such as polyethylene or polypropylene by their synthesis from renewable resources and by their easy biodegradability to water and carbon dioxide. In this contribution the key enzymes of PHA degradation, the PHA depolymerases, are reviewed.

DEFINITIONS

Any research on the biodegradation of PHA should clearly distinguish between intracellular PHA degradation and extracellular PHA degradation. Intracellular degradation is the active degradation (mobilization) of an endogeneous storage reservoir by the accumulating bacterium itself. Enzymes catalyzing the intracellular degradation of PHA are intracellular PHA depolymerases (i-PHA depolymerases). Conversely, extracellular degradation is the utilization of an exogeneous polymer by a not-necessarily accumulating microorganism that secretes extracellular PHA depolymerases (e-PHA depolymerases). The source of extracellular polymer is PHA released by accumulating cells after death and cell lysis. It is necessary to differentiate between extracellular and intracellular degradation because PHA in vivo and outside of the bacteria is present in two different biophysical states. In intracellullar PHA (so-called native PHA granules) the polyesters exist in the amorphous state and the molecules are mobile (2). Native PHA granules have a particular surface layer consisting of proteins and phospholipids that is sensitive to physical or chemical stress. The polymers of damaged granules crystallize and adopt an ordered helical structure (11a) (Figure 1). Extracellular PHB is a partially crystalline polymer (typical degree of crystallinity is 50%–60%) with an amorphous fraction characterized by the glass-to-rubber transition temperature ($T_g \sim 0$–$2°C$) and by a crystalline fraction that melts in the range 170–180°C. In this contribution PHA in the native (amorphous) state (i.e., intracellular granules with intact surface layer) are indicated as nPHA, whereas the same polymers in

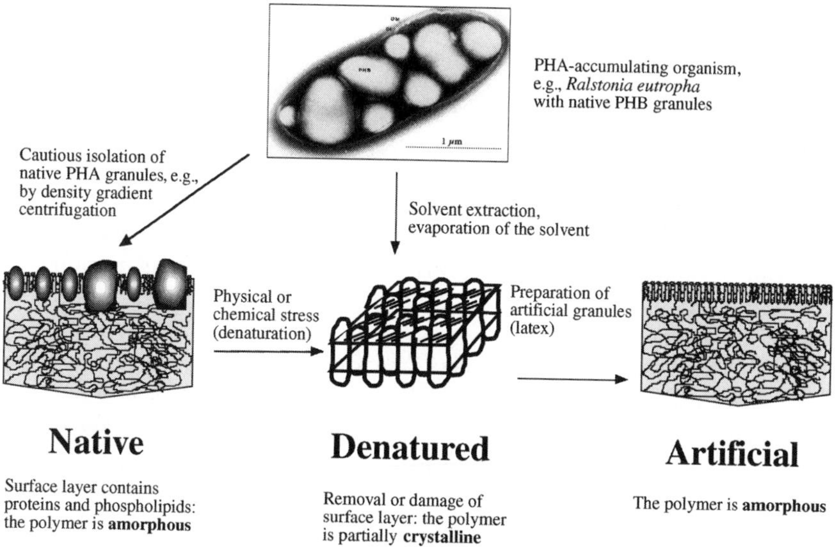

Figure 1 Physical state of PHA. *Top*: Electron micrograph of a PHB-accumulating bacterium. *Bottom left*: Schematic view of a cross-section through a native PHB granule; structural and functional proteins and phospholipids constituing the surface layer are indicated. *Middle*: Schematic view of isolated, denatured, partially crystalline PHB. *Bottom right*: Schematic view of a cross-section through an artificial PHB granule with surfactant molecules at the surface [modified from (42)].

the partially crystalline form are denoted as denatured, dPHA. The same notation is used to differentiate PHA depolymerases according to their ability to hydrolyze nPHA (nPHA depolymerases) or dPHA (dPHA depolymerases).

EXTRACELLULAR DEGRADATION OF PHA

General Aspects on the Isolation of dPHA-Degrading Microorganisms

Microorganisms that utilize extracellular dPHA as a sole source of carbon and energy can be enriched by inoculation of mineral salts solutions containing the desired polyester as the sole source of carbon and energy, with material from the environment (e.g., soil, sludge). This procedure generally results in the enrichment of those microbial species that have the shortest doubling times under the chosen laboratory conditions. However, these organisms might not necessarily represent the most efficient PHA-degrading strains in nature. A more suitable method for the assessment of the distribution of microbial PHA degraders is to plate dilutions of material from the ecosystem of interest on solid agar media that contain the polymer as a sole source of carbon in an opaque overlay prepared from dPHA. dPHA-degrading microorganisms secrete dPHA depolymerases, which hydrolyze the polymer extracellularly to water-soluble products and can be recognized by the appearance of transparent clearing zones around the colonies. This clear zone technique can be performed aerobically, anaerobically, or in agar shake tubes for the selection of aerobic, anaerobic, or microaerophilic microorganisms. The first study on PHA degradation was performed in 1963 in H.G. Schlegel's department by A.A. Chowdhury (11).

Unfortunately, only a few short-chain-length dPHA ($dPHA_{SCL}$), e.g., PHB, can be prepared as a milky suspension of denatured granules. Many other PHA, e.g., all medium-chain-length PHA (PHA_{MCL}), form rubber-like aggregations and cannot be used for the clear zone technique directly. Mineral agar plates containing solution-cast films of PHA_{MCL} have been successfully applied for the isolation of PHA_{MCL}-degrading bacteria (92). A significant improvement was achieved by the development of PHA emulsions (lattices) (64, 82), which have a milky appearance and can be used for the more sensitive clear zone technique. Growth of bacteria expressing poly(3-hydroxyoctanoate) (PHO) depolymerase activity on a PHO latex emulsified in the agar is shown in Figure 2. The different diameters of the clearing zones and the different degree of opaqueness are due to mutations of the PHO-depolymerase gene (see below) and indicate the sensitivity of the method. Other methods for latex preparation were developed by Horowitz and coworkers (35, 36): They prepared emulsions of PHA by dissolving the polymer in chloroform, adding a surfactant, and emulsifying with water by sonication. The solvent was removed by dialysis or evaporation, and an opaque, stable suspension of granules coated by the surfactant was obtained (artificial PHA granules) (Figure 1, bottom right).

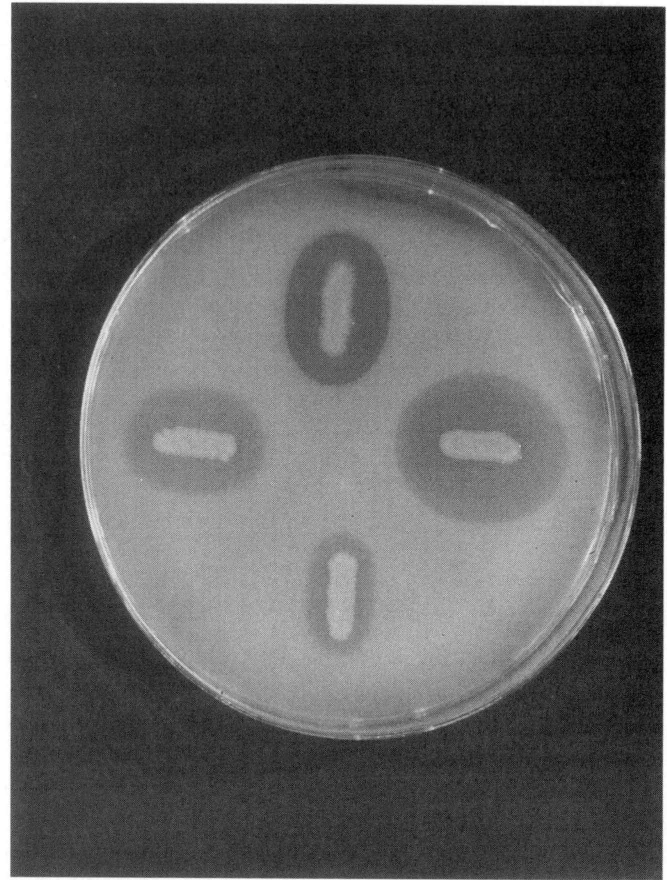

Figure 2 Growth of recombinant *E. coli* harboring the PHO depolymerase gene of *P. fluorescens* on solid medium with opaque latex of PHO. Note large diameter of clearing zones and complete clearing of wild-type colonies (*top*) and different diameters and variable degree of clearing zone formation in strains harboring mutations in the PHO depolymerase gene (*right*, *bottom*, *left*); for details see text (from 42).

Characterization of Extracellular dPHA-Degrading Microorganisms

The ability to degrade extracellular PHA is widely distributed among bacteria and fungi: Aerobic and anaerobic PHA-degrading microorganisms have been isolated from various ecosystems such as soil, compost, aerobic and anaerobic sewage sludge, fresh and marine water (including deep sea), estuarine sediment, and air. An overview on PHA-degrading microorganisms is given in Table 1 of (42, 43). Apparently, PHA-degrading microorganisms, in particular dPHB-degrading microorganisms, are widespread in the environment. For a survey of ecological and/or taxonomical studies on the abundance and diversity of PHA-degrading bacteria see (8, 47, 52, 78, 93) and references within (67).

PHA-degrading bacteria differ in respect to the type of polyester they can degrade. Most characterized bacteria are specific to either $dPHA_{SCL}$ or $dPHA_{MCL}$. However, some bacteria reveal a rather broad polyester specificity and are able to utilize a large variety of polymers including PHA_{SCL} and PHA_{MCL} (93). Such bacteria synthesize more than one PHA depolymerase, which differ in respect to the substrate specificity for PHA_{SCL} or PHA_{MCL}.

The ability to degrade PHA is not restricted to bacteria, and many PHA-degrading fungi have been identified [(24, 27) and references within (65)]. In a comparative study 95 genera of fungi have been identified that include at least one species of a $dPHA_{SCL}$- or $dPHA_{MCL}$-degrading fungus (74). Whether bacteria or fungi contribute more to the PHA-degrading activities in the environment is not known. Preliminary experiments using ciliates such as *Paramecium caudatum* indicated that accumulated PHA in food bacteria of ciliates can be utilized by ciliates (J. Gierlich & D. Jendrossek, unpublished results).

Biochemical Properties of Extracellular dPHA Depolymerases

The microbial degradation of PHA proceeds by enzymatic surface erosion of the polymer to water-soluble monomers and/or oligomers. For a survey of electron-microscopical analysis of surface-eroded PHA_{SCL} and PHA_{MCL} see (70, 75). The e-dPHA depolymerases of many bacteria and several fungi have been purified and/or characterized [for details see Table 1 in (42, 43)]. Most e-dPHA depolymerases share several characteristics: (*a*) high stability at a wide range of pH, temperature, and ionic strength (as far as it has been tested e-dPHA depolymerases are highly resistant to and even active in solvents); (*b*) a relatively small M_r (<70 kDa) with most depolymerases consisting of only one polypeptide; (*c*) alkaline pH optimum (7.5 to 9.8); (*d*) many $dPHA_{SCL}$ depolymerases are inhibited by reducing agents and by serine hydrolase inhibitors such as diisopropyl-fluorophosphate (DFP) or acylsulfonylderivates; (*e*) most PHA depolymerases, as far as it has been tested, do not bind to anion exchangers such as DEAE (at neutral pH) but have a strong affinity to hydrophobic materials. Therefore, many purification protocols include a hydrophobic interaction chromatography step.

The most intensively studied PHA-degrading bacterium is *Paucimonas* [*Pseudomonas*] *lemoignei* (44). It belongs to the β-subclass of the Proteobacteria. *P. lemoignei* is unique among PHA-degrading bacteria because it synthesizes at least seven e-PHA depolymerases [PhaZ1 to PhaZ7 (see below)]. Six of them degrade dPHA and are therefore e-dPHA depolymerases. A seventh e-PHA depolymerase of *P. lemoignei* with specificity to amorphous PHB is described below. At present, the selection advantage for having as many as seven e-PHA depolymerases is not known. A cooperative effect on PHB hydrolysis in the presence of two or more PHB depolymerases with different substrate specificities and

k_M-values for different chain-length molecules might exist. The presence of several isoenzymes of a respective polymer hydrolase within a single degrading strain seems to be a common theme for the degradation of natural polymers such as cellulose or chitin.

All purified dPHA depolymerases are specific for either $dPHA_{SCL}$ (EC 3.1.1.75) or $dPHA_{MCL}$ (EC 3.1.1.76). Even a dPHB depolymerase of *Streptomyces exfoliatus* K10, a strain that degrades both dPHB and dPHO, is specific for $dPHA_{SCL}$, indicating at least one additional depolymerase with specificity for $dPHA_{MCL}$ (55). Experiments with copolymers consisting of 3HB and 3-hydroxyhexanoate and *Alcaligenes faecalis* T1 dPHB depolymerase are in agreement with results obtained with $dPHA_{SCL}$ and $dPHA_{MCL}$: The depolymerase was not able to hydrolyze ester bonds between 3HB and 3-hydroxyhexanoate (15). Several lipases are able to hydrolyze PHA if the polymer consists of ω-hydroxyalkanoic acids, i.e., if there are no side chains in the carbon backbone (40).

In contrast to most other bacterial dPHA depolymerases, the dPHB depolymerases of *P. lemoignei* and those of fungi are glycosylated. Glycosylation in *P. lemoignei* dPHB depolymerases is not well understood. Although several data indicate the presence of carbohydrate (discrepancy between experimentally determined and DNA-deduced M_r, positive in periodic acid stain, proof of glucose and N-acetylglucosamine), all attempts to deglycosylate the purified enzyme by deglycosylases or by chemical means (trifluoromethansulfonic acid or alkali treatment) failed (5, 9). It might be that the carbohydrate moiety becomes noncovalently attached to the depolymerase during secretion of the enzyme across the cell wall. The presence of carbohydrate is not essential for activity of *P. lemoignei* depolymerases because recombinant *Escherichia coli* or *Bacillus subtilis* strains harboring the corresponding depolymerase gene synthesized the nonglycosylated form of the enzyme but had almost the same specific activity. It is possible that glycosylation enhances the resistance of the exoenzyme to elevated temperature and/or to hydrolytic cleavage by proteases of competing microorganisms (9, 46). PHB depolymerases apparently catalyze the reverse reaction, i.e., the synthesis of esters or transesterification if the reaction is performed in the absence of water (e.g., in solvents) (59, 105).

Molecular Biology and Functional Analysis of $dPHA_{SCL}$ Depolymerases

Twenty genes for extracellular bacterial dPHA depolymerases (*phaZ*) have been cloned and analyzed since 1989 (Table 1): *Acidovorax* sp. (58), *A. faecalis* (strains AE122 and T1) (54, 76, 87), *Delftia* (*Comamonas*) *acidovorans* strain YM1069 (50), *Comamonas testosteroni* (97), *Comamonas* sp. (45), *Leptothrix* sp. strain HS (107), *Pseudomonas* sp. strain GM101 (acession no. AF293347), *P. fluorescens* strain GK13 (49, 91, 93), *P. lemoignei* (six genes) (9, 46, 48, 94), *P. stutzeri* (79), *R. pickettii* (strains A1 and K1, acession no. JO4223, D25315), *S. exfoliatus* K10 (55), and *Streptomyces hygroscopicus* (112). Nineteen genes coded for depolymerases

TABLE 1 Properties of cloned bacterial PHA depolymerases

Strain	Gene	Length (bp)	*Mr* (mature protein)	T-opt [°C]	pH-opt	Carbohydrate content	Lipase box	Arrangement and types of domains[a]	Reference
Extracellular PHA$_{SCL}$ depolymerases									
P. lemoignei	*phaZ1$_{Ple}$*	1242	39.5	60[b]	8.0[b]	Yes	GLSAG	C1-Thr-SBD2	(48)
	phaZ2$_{Ple}$	1299	41.8	55	8.0	Yes	GLSAG	C1-Thr-SBD1	(9)
	phaZ3$_{Ple}$	1257	41.2				GLSAG	C1-Thr-SBD1	(9)
	phaZ4$_{Ple}$	1701	57.5	50[b]	8.0[b]		GLSAG	C1-Fn3-SBD2	(46)
	phaZ5$_{Ple}$	1299	42.2	55	8.0		GLSSG	C1-Thr-SBD1	(46)
	phaZ6$_{Ple}$	1254	42.8	55	8.0	Yes	GLSSG	C1-Thr-SBD1	(94)
Alcaligenes faecalis T1	*phaZ$_{Afa}$*	1464	46.8		7.5	Yes	GLSSG	C1-Fn3-SBD1	(87)
Alcaligenes faecalis AE122	*phaZ$_{AfaAE122}$*	1905	62.5				GLSSG	C1-Fn3-SBD1-SBD2	(54)
Ralstonia picketti	*phaZ$_{Ppi}$*	1467	46.8				GLSSG	C1-Fn3-SBD1	Accession no. JO4223
R. picketti strain K1	*phaZ$_{PpiK1}$*	1479	47.5				GLSSG	C1-Fn3-SBD1	Accession no. D25315
Pseudomonas stutzeri	*phaZ$_{Pst}$*	1731	57.5				GLSSG	C1-Cad-SBD1-SBD1	(79)
Pseudomonas sp. strain GM101	*phaZ$_{Psp}$*	1731	57.5				GLSSG	C1-Cad-SBD1-SBD1	Accession no. AF293347
Acidovorax sp.	*phaZ$_{Asp}$*	1476	48.5	<40	7.5		GLSSG	C2-Fn3-SBD1	(58)
Comamonas sp.	*phaZ$_{Csp}$*	1542	50.1	29–35	9.4		GLSSG	C2-Fn3-SBD1	(45)
Delftia acidovorans	*phaZ$_{Dac}$*	1485	48.6	~37	9.0	No	GLSSG	C2-Fn3-SBD1	(50)
C. testosteroni YM1004	*phaZ$_{Cte}$*	1539	50	9.5–10			GLSSG	C2-Fn3-SBD1	(97)
Leptothrix sp.	*phaZ$_{Lsp}$*	1488	49.6				GLSSG	C2-Fn3-SBD1	(107)
Streptomyces exfoliatus K10	*phaZ$_{Sex}$*	1467	47.9	40	8.5–9.0	No	GLSSG	C2-Fn3-SBD1	(55)
Streptomyces hygroscopicus	*phaZ$_{Shy}$*	1560	47.9				GLSAG	C1-Fn3-SBD1	(112)
Extracellular PHA$_{SCL}$ depolymerases									
P. lemoignei	*phaZ7$_{Ple}$*	1143	36.2	65	9.0	No	AHSMG		(29)
Extracellular PHA$_{MCL}$ depolymerases									
P. fluorescens GK13	*phaZ$_{Pfl}$*	837	25	45	8.5	No	GISSG		(91)

[a]Catalytic domain type 1 (C1), catalytic domain type 2 (C2), threonine-rich region (Thr), fibronectin 3 domain (Fn3), Cadherin-like domain (Cad), substrate-binding domain type 1 (SBD1), substrate-binding domain type 2 (SBD2).

[b]Value from protein purified from recombinant *E. coli*. Empty spaces indicate that the parameter is not known.

with specificity for only PHA_{SCL}, especially dPHB. Some of these depolymerase proteins have significant activity with denatured poly(3-hydroxyvalerate) (dPHV) homopolyester (PhaZ6 > PhaZ1 > PhaZ2, all from *P. lemoignei*), but none of them have significant activity with $dPHA_{MCL}$. In contrast, the *P. fluorescens* dPHO depolymerase is specific for dPHO and related $dPHA_{MCL}$ (see below). The dPHO depolymerase is completely inactive with $dPHA_{SCL}$. In conclusion, dPHA depolymerases are highly specific with respect to the length of the monomer carbon side chain of the PHA substrate.

All $dPHA_{SCL}$ depolymerase proteins have a composite domain structure (Figure 3) (Table 1) and consist of

1) A 22-to-58 amino acid–long signal peptide. The signal peptide is necessary for secretion of the polypetide across the cytoplasmic membrane and is cleaved off during the secretion process by signal peptidases.
2) A large catalytic domain at the N terminus of the mature protein. Three strictly conserved amino acids, serine, aspartate, and histidine, constitute the active center of the catalytic domain. The serine is part of a lipase-box pentapeptide Gly-Xaa_1-Ser-Xaa_2-Gly, which is present in almost all known serine hydrolases such as lipases, esterases, and serine proteases [for reviews on lipases see (38, 39)]. Site-directed mutagenesis of the lipase-box amino acids from two PHA_{SCL} and one PHA_{MCl} depolymerase resulted in inactive proteins and confirmed the requirement of the catalytic triad amino acids for activity (76, 93, 96, 105). A second histidine residue is also conserved in dPHB depolymerases (Figure 3). The region around this histidine resembles the oxyanion hole known from lipases [see references within (39)]. The oxyanion amino acid stabilizes the transient state of the hydrolysis reaction by allowing the formation of a hydrogen bridge from the histidine peptide bond to the negatively charged enzyme-bound tetrahedral intermediate (oxyanion). Two types of catalytic domains can be differentiated by the arrangement of the catalytic active amino acids within the primary amino acid sequence: In many PHA depolymerases and lipases the sequential order is histidine (oxyanion)-serine-aspartate-histidine (type 1, Figure 3) (supplemental Figure 1*A*; follow the Supplemental Material link on the Annual Reviews homepage at http://www.annualreviews.org/). In *S. exfoliatus* and in several *Comamonas* and *Leptothrix* species, the depolymerases have the putative histidine (oxyanion) C-terminal of the catalytic triad (type 2, Figure 3) (supplemental Figure 1*B*; follow the Supplemental Material link on the Annual Reviews homepage at http://www.annualreviews.org/), and the sequential order is serine-aspartate-histidine-histidine (oxyanion). This indicates a different primary structure of the protein.
3) A linking domain that links the catalytic domain to the C-terminal domain. The function of the linking region between the catalytic and the C-terminal domain in PHA_{SCL} depolymerases is unknown. In five dPHB depolymerases, namely PhaZ1, PhaZ2, PhaZ3, PhaZ5, and PhaZ6 of *P. lemoignei*, this

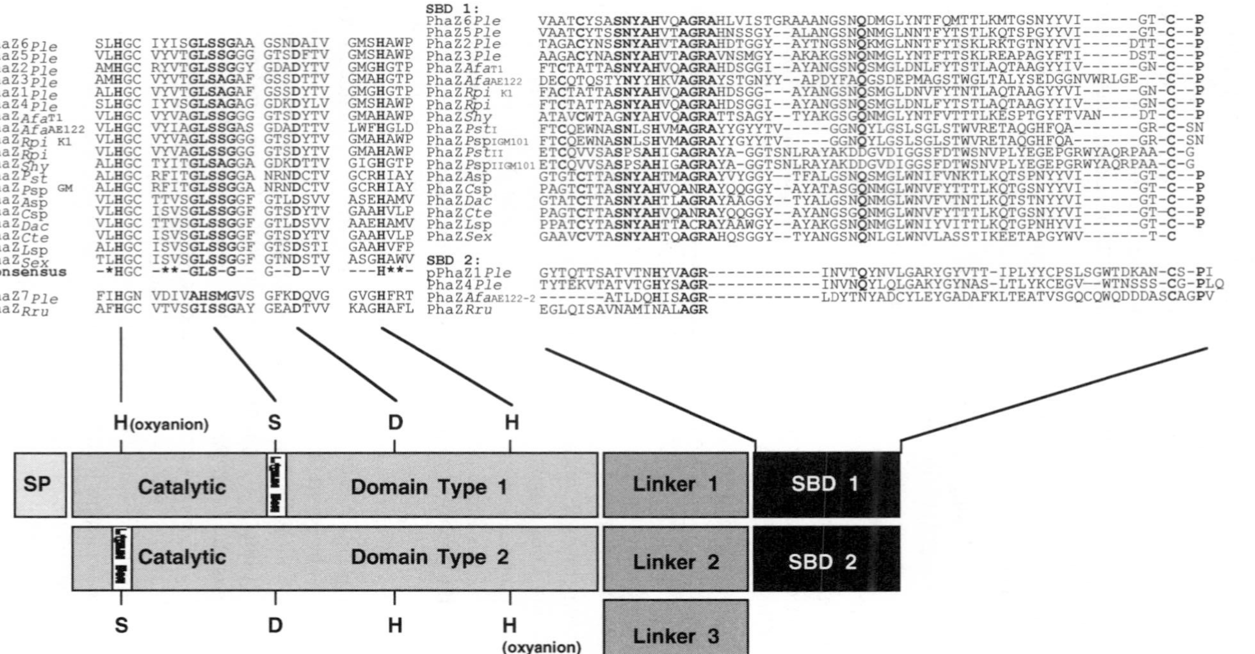

Figure 3 Alignment of PHA depolymerase catalytic active amino acids (*top*, *left*) and of substrate binding domains (*top*, *right*). Amino acids of the catalytic triad and the conserved histidine residue of the putative oxyanion are shown in bold and big letters. Amino acids conserved in all or more than 13 depolymerases are indicated by capital or lowercase letters in the consensus sequence, respectively. *Asterisks* (*) indicate amino acids with hydrophobic side chain. The lipase boxes are indicated by bold letters. A domain model of PHA_{SCL} depolymerases is shown below the alignment. Depending on the microbial origin the depolymerase consists of (*a*) a signal peptide (SP), (*b*) one of two possible types (type 1 or 2) of catalytic domains, (*c*) one of three possible types of linking domains (Fn3, Thr, or Cad), and (*d*) one or two of two types of substrate-binding domains (SBD 1 or 2).

region consists of about 40 amino acids with repeating clusters of two to six threonine residues. This threonine (Thr)-rich region is replaced by a fibronectin type III (Fn3) fingerprint in PhaZ4 of *P. lemoignei* and in most other sequenced $dPHA_{SCL}$ depolymerases. Fn3-sequences have been described for many eukaryotic extracellular matrix proteins and in several prokaryotic polymer-hydrolyzing proteins such as chitinases, cellulases, and several glucoamylases (61). The Fn3-sequence of the *Bacillus circulans* chitinase is not essential for activity (111). However, the Fn3-domain of the *A. faecalis* dPHB depolymerase was essential because its deletion resulted in a protein that had lost the depolymerase activity but not the 3HB dimer hydrolase activity (76). Interestingly, the Fn3-domain of the *A. faecalis* depolymerase could be functionally replaced by a Thr-rich region of the dPHB depolymerase A (PhaZ5) of *P. lemoignei*. These findings support the hypothesis that Fn3-sequences and/or the Thr-rich regions are necessary to provide a proper distance (linker) between the substrate-binding domain (SBD) and the catalytic center of the protein, but the nature of this linker appears to be less important. Recently, a third type of linker consisting of an ∼100–amino acid–long cadherin-like sequence (Cad) has been described for the dPHB depolymerase of *P. stutzeri* (79). Cad sequences have also been found in other polymer hydrolases, e.g., in chitinases.

4) C-terminal SBD $dPHA_{SCL}$ depolymerases bind to dPHA-granules. The C-terminal domains of the depolymerases (40 to 60 amino acids) (Figure 3) are responsible for this ability and constitute a SBD. The SBD apparently has some specificity for PHA because dPHB depolymerases do not bind to chitin or to (crystalline) cellulose (6, 51, 79). The dPHB-binding ability can be used for affinity purification of dPHA depolymerases. The dPHB-binding ability is lost in proteins that lack the C-terminal domain (60 amino acids), and these truncated enzymes do not hydrolyze dPHB anymore. However, the ability to hydrolyze soluble esters (e.g., *p*-nitrophenylesters) is unaffected (6, 7, 22, 34, 76). Obviously SBD is responsible and sufficient for dPHB binding. The function of SBD for binding the water-insoluble substrate was confirmed by analysis of fusion proteins consisting of the maltose-binding protein MalE or the glutathione-S-transferase (GST) with SBD (7, 50, 51, 79, 97). The fusion proteins were able to bind specifically to dPHA in all cases. Recently, the dPHB depolymerase sequences of *A. faecalis* AE122, *P. stutzeri*, and *Pseudomonas* sp. GM101 (the latter being identical to the *P. stutzeri* depolymerase), which contain two instead of only one dPHB-binding domains were published (54, 79).

Two types of SBD can be differentiated (types 1 and 2, Figure 3). However, several amino acids including positively charged amino acids such as histidine (His_{-49}), arginine (Arg_{-44}), and a cysteine residue (Cys_{-2}) are conserved in both types of binding domains. In type 1 a second cysteine at position −57 or −59 is present. It is not known whether the conserved amino acids are necessary to constitute a particular three-dimensional structure or whether these amino acids are directly involved in binding to the polymer chain, e.g., by hydrophobic, hydrophilic, and/or electrostatic interaction [see also (113)]. Computer analysis of the putative secondary structures of SBD predicted the absence of any α-helices and the presence of mainly coiled structures with one or two β-sheets. The sensitivity of dPHB depolymerases to DTT suggests that at least some of the cysteine residues participate in a disulfide bridge(s).

PhaZ7: A New Type of Thermoalkalophilic Hydrolase of *P. lemoignei* with Specificity for Amorphous PHA_{SCL}

All dPHA depolymerases hydrolyze crystalline dPHA, but many of them are inactive with intact amorphous nPHA granules. However, dPHB depolymerases are able to hydrolyze nPHB granules if the surface layer of the nPHB granules has been damaged or partially removed (e.g., by protease treatment). Once the surface layer is damaged, the hydrolysis of nPHA granules by dPHA depolymerases is three- to fourfold higher compared to that of crystalline dPHA. Some $dPHA_{SCL}$ depolymerases are even able to hydrolyze intact nPHB granules [e.g., recombinant PHB depolymerase B (PhaZ2) of *P. lemoignei*; R. Handrick & D. Jendrossek, unpublished observation]. Recently, the hydrolysis of atactic PHB, which is a synthetic copolyester of (*S*)-3HB and (*R*)-3HB, was analyzed (29). Due to the presence of both enantiomers being randomly distributed along the polymer, crystallization of atactic PHB is impossible and the polymer is completely amorphous. Conventional dPHB depolymerases such as PHB depolymerase A or B (PhaZ5, PhaZ2) of *P. lemoignei* have no or only little activity on atactic PHB. However, unexpected high levels of a novel extracellular enzymatic activity of *P. lemoignei* that hydrolyzed atactic PHB were detected (29): The respective depolymerase was purified and characterized. The new depolymerase (PhaZ7) was specific for amorphous PHB, such as nPHB granules with an intact surface layer or atactic PHB, but was inactive with dPHA. Therefore, the depolymerase is the first described extracellular $nPHA_{SCL}$ depolymerase. The purified enzyme is active and stable at high temperatures (59°C) and at alkaline pH (pH 9–12). The depolymerase PhaZ7 is highly active on nPHB granules (9000 μmol cleaved ester bonds $\times$ min^{-1} $\times$ mg^{-1}) but has no detectable lipase, cutinase, amidase/nitrilase, protease, DNAse, nor any significant RNAse activity. Only low activity was found with soluble *p*-nitrophenylacyl-esters ($\leq$0.38 U/mg). Therefore, PhaZ7 is not a conventional esterase or lipase (29). A summary of the biochemical properties of selected PHA_{SCL} depolymerases is given in Table 2.

Analysis of the DNA-deduced amino acid sequence of PhaZ7 showed no significant sequence similarity to any PHB depolymerase. Only low similarities to several hypothetical proteins of *Caenorhabditis elegans*, to lipase LipB of *B. subtilis*, and to some other lipase-like proteins were found (supplemental Figure 2; follow the Supplemental Material link on the Annual Reviews homepage at http://www.annualreviews.org/). The similarities were restricted to a modified lipase-box pentapeptide sequence Ala-His-Ser_{136}-Met-Gly and to short regions

TABLE 2 Properties of selected (mature) $dPHA_{SCL}$ and $nPHA_{SCL}$ depolymerases

	e-dPHB depolymerase PhaZ*Afa* ***A. faecalis* T1**	**e-dPHB depolymerase PhaZ2** ***P. lemoignei***	**e-dPHB depolymerase PhaZ5** ***P. lemoignei***	**e-nPHB depolymerase PhaZ7** ***P. lemoignei***	**i-nPHB depolymerase PhaZ*Rru*** ***R. rubrum***
Mr (SDS-PAGE) [kDa]	50	67	54–67[a]	36	34
Mr (MALDI-MS) [kDa]			46.8	36.2	35.3
Mr (DNA-deduced) [kDa]	46.8	41.8	42.2	36.2	35.3
Glycosylation		+	+	–	–
IEP (experimental) [pH]	8.6	9–10	7.6	9.2	9–10
pH optimum	7.5	8	8	9–10	9–10
T opt [°C]		51	55	65	50
Hydrolysis of nPHA granules isolated from:					
R. eutropha H16	–	+	–	++	++[b]
B. megaterium			–	++	++[b]
E. coli (phaCAB + phaP)		+	–	++	++[b]
n-PHV granules:					
C. violaceum			–	++	+[b]
n-PHO granules:					
P. oleovorans		–	–	–	–[b]
atactic PHB [Poly(*R*-, *S*-3HB)]		–	–	25[c]	–
Denatured (paracrystalline) PHA:					
d-PHB	+	+	+	–	–
d-PHV		(+)	–	–	–
d-PHO, d-PHO/HD[d]	–	–	–	–	–

(Continued)

TABLE 2 (*Continued*)

	e-dPHB depolymerase PhaZ*Afa* *A. faecalis* T1	**e-dPHB depolymerase PhaZ2 *P. lemoignei***	**e-dPHB depolymerase PhaZ5 *P. lemoignei***	**e-nPHB depolymerase PhaZ7 *P. lemoignei***	**i-nPHB depolymerase PhaZ*Rru* *R. rubrum***
Artificial PHA granules[e]:					
PHB		+	+	++	++
PHV			–	++	+
Main products[f]:	R2 (R3, R1)	R2 (R3, R1)	R2 (R3, R1)	R5, (R1–R8)	R1, R2, R3
Activity to 3HB oligomers (μmol acid/min × mg):					
R2	0	<0.01	<0.01	<5	<0.1
R3	33	130	110	<5	350
R4	635	3400	3800	7	1400
R5	707	1800	3900	7	2000
R6	654	2000	5000	25	1200
R7	572				
R8	712	1100	4000	500	900
S4	0	<0.01	<0.01	<0.01	<0.01
Hydrolysis of other substrates:					
Poly(6-hydroxyhexanoate)	–	–	–	–	–
Triolein, Tributyrin	–	–	–	–	–
Casein/N-α-benzoyl-L-arginine-nitranilide		–	–	–	–

***p*-nitrophenylalkanoates:**					
C_2			0.58	0.019	<0.001
C_4			0.50	0.048	<0.001
C_6			0.05	0.18	<0.001
C_8			–	0.38	<0.001
C_{10}			–	0.12	
C_{16}			–	0.055	
Dependence on ions:					
Ca^{2+}	–	+/–[g]	++	+[g]	+/–[g]
Inhibition (%):					
DFP (1 mM)	>99	>99	>99	>60[h]	>98
PMSF (10 mM)	>96		>99	10	<5
DDSC (10 mM)			>99	10	<5
EDTA (10 mM)		>64	>99	70	80
DTT (4 mM)		>99	>99	<1	>90
SDS (0.05%)			40	activates	55
TritonX100 (0.02%)	++ (>94% with 0.1% detergent)	92	94	<1	>80

[a]Value variable, depends on degree of glycosylation.

[b]Measured in the presence of "activator" at pH 9, activity >100 U/mg (++); <10 U/mg (–); 1 U = 1 μmol released acid min^{-1} (titration).

[c]Weight loss from defined films [mg cm^{-2} h^{-1} mg $protein^{-1}$].

[d]Copolymer consisting of 3-hydroxyoctanoate and 3-hydroxydecanoate.

[e]Prepared by coating with SDS or cholate.

[f]Monomer (R1), dimer (R2), trimer (R3).

[g]Ca^{2+} addition not necessary for activity however EDTA inhibits reaction.

[h]>90% inhibition at 10 mM.

Abbreviations: diisopropylfluorophosphate (DFP); phenylmethylsulfonylfluoride (PMSF); dodecylsulfonylchloride (DDSC); dithioerythritol (DDT).

Empty space indicates that the value is not known.

around a putative oxyanion pocket at His_{47}. This result, in combination with the observed sensitivity of the PhaZ7 to the serine esterase inhibitor DFP and secondary structure prediction *in silicio*, suggests that PhaZ7 is different from known PHA depolymerases and lipases, but apparently is a member of the serine hydrolase family having a catalytic triad of Ser_{136}, Asp_{242}, His_{306} (39, 80).

Enantioselectivity and Hydrolysis Products of PHA Depolymerases

The specificity of PHA depolymerases with respect to the enantiomeric composition of the polymer has been studied mainly with dPHB depolymerase of *A. faecalis* T1 (1, 4, 16) and with nPHB depolymerase PhaZ7 of *P. lemoignei* (29). All PHA depolymerases analyzed so far are specific for 3HA-esters in the natural (*R*)-configuration. Polymers or oligomers of (*S*)-3HA or ester bonds between a (*R*)- and (*S*)-3HA unit are not cleaved. However, copolymers of (*R*)- and (*S*)-3HA were hydrolyzed between two adjacent (*R*)-3HA (4). Such (*R*)-(*R*)-ester bonds statistically appear in atactic PHB [Poly(*R*, *S*-3HB)]. Therefore, the hydrolysis products of microbial-degraded atactic PHB are higher oligomers compared to the hydrolysis products of natural PHB with 100% (*R*)-configuration (18).

Hydrolysis of end-labeled 3HB oligomers by purified *A. faecalis* dPHB depolymerase showed that the enzyme mainly cleaved the second and third ester linkage from the hydroxy terminus (73). Since the enzyme also hydrolyzes cyclic 3HB oligomers, the *A. faecalis* depolymerase has endohydrolase activity in addition to exohydrolase activity (4, 98). Similar conclusions were drawn by de Koning et al. (11b), who demonstrated that covalently cross-linked $dPHA_{MCL}$ was hydrolyzed by *P. fluorescens* GK13. It is assumed that most if not all extracellular PHA depolymerases including fungal enzymes have endo- and exohydrolase activity (90). Depending on the depolymerase the end products are only monomers, monomers and dimers, or a mixture of oligomers. If oligomers are the end products of PHA depolymerases, bacteria have oligomer hydrolases or dimer hydrolases that cleave the oligomers/dimers to the monomer [see references within (85)].

Molecular Biology and Functional Analysis of $dPHA_{MCL}$ Depolymerases

In contrast to a large variety of isolated and more or less well-characterized $dPHA_{SCL}$ depolymerases, only few $dPHA_{MCL}$ depolymerases have been described so far. The dPHO depolymerase of *P. fluorescens* GK13 is the only one that has been purified and studied at the molecular level (49, 91, 92). For a survey of gram-positive and gram-negative $dPHA_{MCL}$-degrading bacteria see (93). The dPHO depolymerase of *P. fluorescens* significantly differs from $dPHA_{SCL}$ depolymerases: It is specific for PHA_{MCL} and for soluble esters such as *p*-nitrophenylacylesters with six or more carbon atoms in the fatty acid moiety. dPHB and other $dPHA_{SCL}$ are not hydrolyzed. The enzyme is not inhibited by DTT or EDTA and is not dependent on Ca^{2+} or other divalent cations. Recently, a PHA_{MCL} depolymerase was isolated from a *Xanthomonas*-like strain that hydrolyzed a copolyester of 3-hydroxy-5-phenoxyvalerate and 28% 3-hydroxy-7-phenoxyheptanoate (53). The purified enzyme could not hydrolyze PHA_{SCL}; the ability to hydrolyze PHO was not tested. Another bacterium with extracellular PHA_{MCL} activity is *Comamonas* sp. P37C (81). Cell-free culture fluid of PHO-grown *Comamonas* sp. P37C cells showed high PHO depolymerase and weak PHB depolymerase activity. It is not known whether the bacterium synthesizes two enzymes with different substrate specificities for PHA_{SCL} and PHA_{MCL} or only one enzyme with broad specificity for both types of polyesters.

The deduced amino acid sequence of the cloned dPHO depolymerase gene of *P. fluorescens* GK13 has no significant homology to PHA_{SCL} depolymerases except for small regions in the neighborhood of a lipase-box (Gly-Xaa_1-Ser_{172}-Xaa_2-Gly), an aspartate (Asp_{228}), and a histidine (His_{260}) residue. Site-directed mutagenesis of these amino acids confirmed that these amino acids are essential for activity (91, 93) and that they presumably constitute a catalytic triad (serine hydrolase). Interestingly, the three catalytic triad amino acids (Ser_{172}, Asp_{228}, His_{260}) are located in the C-terminal region of the protein, which is different to all known PHA_{SCL} depolymerases. It is assumed that the N-terminal part of the protein constitutes a polymer-binding site. This assumption was supported by PCR-induced mutagenesis of the dPHO depolymerase gene and selection of mutants that were impaired in dPHO depolymerase activity but still had almost-normal levels of esterase activity with *p*-nitrophenyloctanoate. DNA sequencing of 30 mutants revealed that most of the mutations were located in the N-terminal region of the mature depolymerase [amino acid 1-65 (49)]. In particular, amino acids near Leu_{15}, Phe_{50}, and Phe_{63} represented hot spots because these or neighboring amino acids were mutated in 16 independently isolated mutants (supplemental Figure 3; follow the Supplemental Material link on the Annual Reviews homepage at http://www.annualreviews.org/). Because the esterase activity of these mutant dPHO depolymerase proteins was not significantly changed, it is assumed that these amino acids are involved in the interaction of the enzyme with its polymeric substrate. The reason for the observed phenotypic differences between mutants with residual opaqueness of the clearing zones and mutants with clear halos is unknown (Figure 1).

Regulation of PHA Depolymerase Synthesis

The synthesis of PHA depolymerases in bacteria is generelly repressed if suitable soluble carbon sources such as glucose or organic acids are present. However, after exhaustion of the soluble nutrients synthesis of PHA depolymerases is derepressed in many strains (47). At least in some bacteria dPHB depolymerase is expressed even in the absence of the polymer after cessation of growth. Therefore, an induction mechanism by the polymer itself is not necessary.

In most known dPHA-degrading bacteria high levels of dPHA depolymerase are produced only during growth on dPHA but are repressed on succinate. In contrast, production of PHA_{SCL} depolymerases by *P. lemoignei* is maximal during growth in batch culture on succinate. Therefore, isolation of PHA_{SCL} depolymerases from *P. lemoignei* is usually performed from succinate-grown cells (12, 71, 73). Synthesis of dPHB depolymerase on succinate is pH-dependent and occurs only above pH 7 (103). Recently, the relationship between depolymerase synthesis and growth on/uptake of succinate has been elucidated (110): It was shown that transport of succinate into the bacteria is pH-dependent and does not work well above pH 7 in *P. lemoignei*. As a consequence the bacteria starve even in the presence of residual succinate at pH above 7, and depolymerase synthesis is derepressed. Analysis of the succinate transport system of *P. lemoignei* revealed that it utilizes only the monocarboxylate form of succinate (H-succinate^{1-}) but is not able to take up the dicarboxylate (succinate^{2-}). The pH of the culture fluid increases during growth of *P. lemoignei* in batch culture on succinate owing to the uptake of succinic acid. As a consequence, the concentration of the H-succinate^{1-} ion ($pK_A2 = 5.6$) decreases. The inition of dPHB depolymerase synthesis above pH 7 can be considered as induced by carbon starvation because of insufficient uptake of H-succinate^{1-} at high pH (110).

Analysis of the DNA region adjacent to depolymerase genes revealed a putative transcriptional regulator gene, *phaR*, in front of *phaZ2* in *P. lemoignei*. Based on analysis of *phaZ::lacZ* transcriptional fusions in *P. lemoignei* wild-type and in *phaR* null mutants, it was demonstrated that PhaR is a negative regulator of *phaZ2* (dPHB depolymerase B) but does not affect the expression of other depolymerases (109).

Regulation of $dPHA_{MCL}$ depolymerase synthesis apparently is similar to that of dPHB depolymerases: High levels of dPHO depolymerase activity were found during growth of *P. fluorescens* GK13 on $dPHA_{MCL}$ and on low concentrations of HA_{MCL}-monomers. The presence of sugars or fatty acids repressed synthesis of dPHO depolymerase (92). Similar results have been reported for a dPHO depolymerase of *P. maculicola* (21).

Influence of Physicochemical Properties of the Polymer on Its Biodegradability

The physicochemical properties of a polyester have a strong impact on its biodegradability. The most important factors are (*a*) stereoregularity, (*b*) crystallinity, (*c*) monomeric composition, and (*d*) accessibility of the polymer surface. For a recent summary of the influence of physicochemical properties of the polymer on its biodegradability see (42, 43).

INTRACELLULAR DEGRADATION OF PHA

Mobilization of PHA in Bacteria

The mechanism and regulation of the intracellular degradation of PHA (i.e., the mobilization of previously accumulated polyester) is poorly understood. Lemoigne (60) observed already 80 years ago that 3HB appeared upon anaerobic incubation of *Bacillus*. Macrae & Wilkinson (62) described a reduction of the PHB content of *Bacillus megaterium* after aerobic incubation of PHB-rich cells in phosphate buffer. The authors found that autolysis of PHB-rich cells occurred later and to a minor extent compared to PHB-poor cells, and proposed that PHB might function as a storage compound. Hayward et al. (31) observed that the intracellular content of PHA in *Rhizobium*, *Spirillum*, and *Pseudomonas* species had a maximum followed by a decrease in the stationary growth phase. Similar reports have been published for *Micrococcus halodenitrificans* and *Ralstonia eutropha* H16 [(32) and references within (33)]. Survival of bacteria in the absence of exogeneous carbon sources was dependent on the intracellular PHB content. A cyclic metabolic route (the PHB circle) from acetyl-CoA via acetoacetyl-CoA, 3HB-CoA to PHB (PHB biosynthesis-sequence); and from PHB via 3HB and acetoacetate to acetoacetyl-CoA and acetyl-CoA by i-PHB depolymerase, 3HB dehydrogenase, acetoacetate:succinyl-CoA transferase, and ketothiolase (PHB degradation-sequence) was formulated (77, 95). Doi and coworkers (17) reported on the cyclic nature of PHA metabolism. The assumption of a simultaneous synthesis and degradation of PHB was confirmed by ^{14}C-glucose pulse experiments (106): The bacteria incorporated the label into PHB at a high rate even after the end of net PHB accumulation, i.e., during PHB degradation. However, the physiological reason for the energy-consuming cycle of degradation and synthesis of PHA remains unknown. One could imagine that PHB functions as a carbon buffer similar to a water reservoir that is continuously filled up by river water (PHA accumulation) but is also continuously utilized by taking drinking water from the reservoir (PHA mobilization). Depending on the volumes of influx and efflux the sum of both processes is either positive (≈PHB net accumulation) or negative (≈PHB net mobilization).

Rapid intracellular mobilization of PHB was shown also for *Legionella pneumophila* and *Hydrogenophaga pseudoflava* in the absence of an exogeneous carbon source (10, 41). For *L. pneumophila* a correlation between the amount of accumulated PHB and the long-term survival (up to 600 days) in the absence of an exogeneous carbon source was found. Recently, it was shown that *R. eutropha* H16 could even grow in the absence of any exogeneous carbon source by utilizing previously accumulated PHB (30, 49). The key enzymes of intracellular mobilization of PHA are intracellularly located PHA depolymerases, which are able to hydrolyze nPHA. These enzymes differ from e-dPHA depolymerases in their inability to hydrolyze crystalline dPHA (see below). It is interesting that an e-PHB depolymerase (PhaZ7) with specificity for nPHB has been described for *P. lemoignei* [see above, (29)]. However, this enzyme was located extracellularly and cannot be responsible for intracellular PHA mobilization.

Structure and Properties of nPHA Granules

It is generally accepted that intracellular PHB granules are amorphous and are covered by a surface layer consisting of proteins and phospholipids (nPHB) (26, 66). The surface layer of nPHB granules can be destroyed or removed by exposure to chemicals such as solvents, detergents, and alkali [(25) and references within]. Physical stress such as freezing-thawing cycles or repeated centrifugation also leads to denaturation of nPHB granules (68) (Figure 1). Denaturation of PHB can be avoided by careful disruption of the cells (French press or enzymatic lysis) and subsequent cautious purification of the granules by density gradient centrifugation.

The major fraction of the proteins of the granule surface layer consists of relatively small amphiphilic polypeptides that presumably have a structural function by mediating between the water-insoluble polymer core and the hydrophilic cytoplasm. These structural proteins are called phasins and have a high affinity to PHA granules [(99) and references within (101)]. Phasins may also participate in regulation of PHA synthesis and deposition (114). The surface layer contains additional proteins responsible for biosynthesis (PHB synthase) and polymer degradation (i-PHB depolymerase).

Assay Systems for i-PHA Depolymerases

i-PHA depolymerases are specific for the native, amorphous form of the polymer, and dPHA are not a substrate for these enzymes. Therefore, any form of amorphous PHA may be suited for assay of i-PHA depolymerases. The best way to obtain amorphous PHB is to purifiy PHA granules in the native form. nPHA granules, at least those of *Zoogloea ramigera*, *P. oleovorans*, and *R. eutropha*, contain a significant portion of endogeneous PHB depolymerase attached to the surface of the granules. Therefore, nPHA granules have a slow but significant autohydrolysis rate if incubated under appropriate conditions (19, 20, 30, 49, 86). Alternatively, nPHA granules can be isolated from recombinant *E. coli* strains harboring the PHA-biosynthetic genes. Because *E. coli* does not have any PHA depolymerase genes, nPHA granules from recombinant *E. coli* are free of any endogeneous self-hydrolyzing activity (30).

Another way to prepare amorphous PHA is to prepare a PHA latex by emulsifying the chloroform-solubilized polymer with water in the presence of a detergent (SDS, cholate, oleate) and evaporating the solvent afterward (36). The obtained PHA latex (artificial PHA) remains amorphous for a long time and is a suitable substrate for many i- and some e-PHA depolymerases. However, it should be noted that such PHA lattices contain an artificial surface layer without phasin proteins. These proteins in vivo may have an impact on the accessibility and degradability of the polymer, and therefore depolymerase activities with artificial granules cannot be compared to those obtained with nPHA granules. Another method to obtain a substrate for i-PHA depolymerases is the treatment of isolated PHA granules with lipases and/or proteases in the presence or absence of surfactants (e.g., oleate) (88). It should be emphasized that the physiological relevance of an i-PHA depolymerase activity that has been determined with artificial granules remains uncertain unless the same activity has been confirmed with the in vivo substrate (nPHA granules) or by physiological analysis of the respective knock-out mutants.

i-PHA depolymerase activity can be determined by (*a*) following the decrease of the optical density in buffered solution (turbidometric method) or (*b*) by determination of the amount of NaOH required to keep the decreasing pH constant in an aqueous, nonbuffered medium (titristat method). The turbidometric method is quick, easy to perform, and can be automated in a microtiter plate reader. However, this method is sensitive to experimental errors such as artificial coalescence of the granules, change of optical density, and change in surface area due to changes of the granule's diameter during the reaction. It is not possible to correlate the degree of reduction in optical density with the number of ester bonds cleaved directly. The titristat method is more sophisticated, time-consuming, and needs to be performed in distilled water (problem of enzyme stability), but it provides quantitative data on the number of ester bonds cleaved. The assay of released product, e.g., 3HB or 3HB-oligomers by 3HB dehydrogenase or by HPLC, is possible but higher oligomers are hardly soluble in water and may not be detected quantitatively. Because 3HA oligomers generally are sensitive to treatment with heat, acid, or alkali, the reaction mixtures should not be stopped by these treatments unless appropriate controls on the stability of the oligomers have been performed.

The i-PHB Depolymerase System of *Rhodospirillum rubrum*

The first biochemical study on intracellular PHB degradation was done by Merrick & Doudoroff (68). nPHB granules isolated from *B. megaterium* were used as a substrate for the depolymerase assay. The granules themselves had only low self-hydrolysis activity but were rapidly hydrolyzed by crude extracts of *Rhodospirillum rubrum* to 3HB and oligomers. Remaining dimers and oligomers could be hydrolyzed to 3HB by a soluble 3HB dimer hydrolase [(85) and references within]. Interestingly, crude extracts of *R. rubrum* contained two soluble components that were necessary for hydrolysis of the polymer. One compound was heat-sensitive and was the i-PHB depolymerase itself. The second component was heat-stable and was named activator (68). Because the effect of the activator on nPHB granules could be replaced by incubation with trace amounts of trypsin (25) or other unspecific proteases (108), one would expect that the activator is a protease that removes some protein of the surface layer of nPHA thus making the polyester molecules accessible for the depolymerase. This assumption was supported by the finding that hydrolysis of artificial surfactant-coated PHB granules (see above) did not require treatment with activator (28, 69). However, the activator could not be inhibited by any protease inhibitor. The activator appeared in the high molecular fraction during gel filtration experiments and was sensitive to proteases; the activator therefore should have at least one peptide or ester bond. Trypsin-activated nPHB were rapidly hydrolyzed by dPHB depolymerase A (PhaZ5) from *P. lemoignei* but the same enzyme was inactive with nPHB that had been incubated with the activator compound of *R. rubrum* (U. Technow, R. Handrick & D. Jendrossek, unpublished result). This indicated that the mechanism of activation by trypsin is different from activation by the activator. The activator was resistant to many solvents including phenol and chloroform. It is assumed that the activator somehow interacts with the surface of nPHB, making the polymer accessible for the depolymerase. All attempts to purify the activator and to elucidate its structure were not successful (68, 108). It must be mentioned that we cannot exclude that the activator compound is an artifact of the in vitro test system.

Properties of the i-PHB Depolymerase of *R. rubrum*

The i-PHB depolymerase of *R. rubrum* has been purified (28). It consists of one polypeptide of 35 kDa and has a pH and temperature optimum of pH 9 and 50°C. In contrast to e-PHA depolymerases the purified i-PHB depolymerase is highly unstable. A high specific activity of 3100 μmol cleaved ester bonds min^{-1} mg^{-1} protein was determined (at 45°C) independent whether the nPHB granules were activated by trypsin or by activator. The purified enzyme was inactive with dPHB and had no lipase, protease, or esterase activity with *p*-nitrophenylacyl esters (2 to 8 carbon atoms). nPHV granules (trypsin-treated) were hydrolyzed but nPHO granules were not indicating a high substrate specificity similar to e-$dPHB_{SCL}$ depolymerases (Table 2).

Sequence analysis of the cloned i-PHB depolymerase gene of *R. rubrum* revealed that the depolymerase had high homologies to e-dPHB depolymerases with a catalytic triad and a type II catalytic domain (Figure 3; supplemental Figure 1*B*; follow the Supplemental Material link on the Annual Reviews homepage at http://www.annualreviews.org/, unpublished result), and it had a putative two-dimensional structure related to α/β-proteins. However, the polypeptide was significantly smaller compared to e-dPHB depolymerases (35 kDa compared to 40–60 kDa) and did not contain a linking domain or SBD. Interestingly, a typical signal peptide sequence was found at the N terminus of the predicted polypeptide, and the experiment-determined N terminus of the purified depolymerase coincided with the predicted signal peptidase cleavage site. Therefore, the depolymerase should be located extracellularly, but there is evidence that it is located intracellularly and no activity could be detected in cell-free culture supernatant (R. Handrick & D. Jendrossek, unpublished result). One might speculate that the signal peptide directs the depolymerase into the surface layer of the PHB granule; however, so far there is no experimental support for this assumption. It will be necessary to identify the true location of the depolymerase.

The i-PHA_{SCL} Depolymerases of *Ralstonia eutropha* and of *Paracoccus denitrificans*

Isolated n-PHB granules of *R. eutropha* have a low rate of self-hydrolysis, which is about two orders of magnitude lower compared to the hydrolysis rates obtained by *R. rubrum* extracts. The endogeneous activity of *R. eutropha* granules can be enhanced about threefold if the PHB-rich bacteria have been exposed to carbon starvation before cell harvest (i.e., mobilization conditions) (30). The pH optimum of this endogeneous i-PHB depolymerase activity was at pH 7. It was concluded that this activity represents the i-PHB depolymerase and was responsible for mobilization of the storage material during starvation. No significant i-PHB depolymerase activity could be detected in soluble cell extracts of *R. eutropha* if nPHB granules were used as a substrate (30). However, a soluble i-PHB depolymerase activity and a second pH optimum at pH 9 have been reported for *R. eutropha* (88), as high level trypsin-treated and oleate-coated PHB granules were used. This finding demonstrates the dependence of the results on the assay system. Recently, the DNA sequence of a putative i-PHB depolymerase of *R. eutropha* was published (84). The heterologically (*E. coli*) expressed depolymerase hydrolyzed artificial amorphous PHB granules but not freeze-dried dPHB. The depolymerase was located in the PHB granules fraction. Surprisingly, the DNA-deduced amino acid sequence (47.3 kDa) did not contain a lipase-box fingerprint and had no similarities to the *R. rubrum* i-PHB depolymerase. Chromosomal knock-out mutants in the i-PHB depolymerase gene showed a reduced but still significant self-hydrolyis activity. It was concluded that the gene could be partially responsible for PHB mobilization and that other (so far unkown) depolymerase isoenzymes must be present in *R. eutropha* (30).

Recently, a putative i-PHB depolymerase gene of *P. denitrificans* was found adjacent but in opposite orientation to the PHB synthase gene *phaC* (23). The heterologically expressed (*E. coli*) gene product hydrolyzed artificial protease- and oleate-treated PHB granules. The DNA-deduced amino acid sequence had no similarities to the *R. rubrum* i-PHB depolymerase but had high similarities to the *R. eutropha* i-PHB depolymerase and to several hypothetical proteins, which have been identified by sequencing of several bacterial genomes (supplemental Figure 4; follow the Supplemental Material link on the Annual Reviews homepage at http://www.annualreviews.org/). None of the DNA-deduced genes contained a lipase box or any other significant homologies to proteins of databases. More biochemical and physiological data on i-PHB depolymerase are necessary for a better understanding of the in vivo function of these genes.

i-PHA$_{MCL}$ Depolymerases

Self-hydrolysis of n-PHA granules has been described for n-PHO granules of *P. oleovorans* (19, 20, 104). The pH optimum was at pH 8. Inhibition by serine esterase inhibitors indicated that the active center of i-PHA$_{MCL}$ depolymerases might be related to that of e-PHA depolymerases and other serine esterases. A putative i-PHA$_{MCL}$ depolymerase gene (*phaD*, suggested to be renamed as *phaZ*) was identified first for *P. oleovorans* (37). Interestingly, *phaZ* was located colinear between two copies of PHA synthase genes (*phaC1* and *phaC2*). Mutants in *phaZ* were defective in PHA$_{MCL}$ degradation. Therefore, it is likely that the open reading frame between the two synthase genes represents the i-PHO depolymerase gene. The deduced amino acid sequence contained a potential lipase box (GVSWG) but no significant homologies to any i-PHA$_{SCL}$ depolymerases. Recently, physiological data confirmed the involvement of this gene for survival during starvation (83). The same arrangement of two PHO synthase genes separated by a putative depolymerase gene with high sequence similarity to the putative *P. oleovorans* i-PHA$_{MCL}$ depolymerase was also found for several other pseudomonads and related bacteria (see supplemental Figure 5; follow the Supplemental Material link on the Annual Reviews homepage at http://www.annualreviews.org/).

ACKNOWLEDGMENTS

We greatly acknowledge the support by the Deutsche Forschungsgemeinschaft, the Max-Buchner Forschungsstiftung, the Fonds der Chemischen Industrie, and the Studienstiftung des Deutschen Volkes.

The *Annual Review of Microbiology* is online at http://micro.annualreviews.org

LITERATURE CITED

1. Abe H, Doi Y. 1996. Enzymatic and environmental degradation of racemic poly(3-hydroxybutyric acid)s with different stereoregulatoris. *Macromolecules* 29:8683–88
2. Amor SR, Rayment T, Sanders JKM. 1991. Poly(3-hydroxybutyrate) in vivo: NMR and X-ray characterization of the elastomeric state. *Macromolecules* 24:4583–88
3. Anderson AJ, Dawes EA. 1990. Occurrence, metabolism, metabolic role, and industrial use of bacterial polyhydroxyalkanoates. *Microbiol. Rev.* 54:450–72
4. Bachmann BM, Seebach D. 1999. Investigation of the enzymatic cleavage of diastereomeric oligo(3-hydroxybutanoates) containing two to eight HB units. A model for the stereoselectivity of PHB depolymerase from *Alcaligenes faecalis* T1. *Macromolecules* 32:1777–84
5. Behrends A. 1994. *Untersuchungen zur Glykoproteinnatur von Polyhydroxyalkanoat-Depolymerasen aus* Pseudomonas lemoignei. Diploma thesis. Univ. Göttingen. 132 pp.
6. Behrends A, Klingbeil B, Jendrossek D. 1996. Poly(3-hydroxybutyrate) depolymerases bind to their substrate by a C-terminal located substrate binding site. *FEMS Microbiol. Lett.* 143:191–94
7. Briese B-H, Jendrossek D. 1998. Biological basis of enzyme-catalyzed polyester degradation: 59 C-terminal amino acids of poly(3-hydroxybutyrate) (PHB) depolymerase A from *Pseudomonas lemoignei* are sufficient for PHB-binding. *Macromol. Symp.* 130:205–16
8. Briese B-H, Jendrossek D, Schlegel HG. 1994. Degradation of poly(3-hydroxybutyrate-*co*-3-hydroxyvalerate) by aerobic sewage sludge. *FEMS Microbiol. Lett.* 117:107–12
9. Briese B-H, Schmidt B, Jendrossek D. 1994. *Pseudomonas lemoignei* has five poly(hydroxyalkanoic acid) (PHA) depolymerase genes: a comparative study of bacterial and eukaryotic depolymerases. *J. Environ. Polym. Degrad.* 2:75–87
10. Choi MH, Yoon SC, Lenz RW. 1999. Production of poly(3-hydroxybutyric acid-co-4-hydroxybutyric acid) and poly(4-hydroxybutyric acid) without subsequent degradation by *Hydrogenophaga pseudoflava*. *Appl. Environ. Microbiol.* 65:1570–77
11. Chowdhury AA. 1963. Poly-ß-hydroxybuttersäure abbauende Bakterien und Exoenzym. *Arch. Mikrobiol.* 47:167–200

11a. de Koning GJM, Lemstra PJ. 1992. The amorphous state of bacterial poly[(*R*)-3-hydroxyalkanoate)] in vivo. *Polymer* 33:3304–6

11b. de Koning GJM, van Bilsen HMM, Lemstra PJ, Hazenberg W, Witholt B, et al. 1994. A biodegradable rubber by cross linking polyhydroxyalkanoates from

Pseudomonas oleovorans. *Polymer* 35: 2090–97
12. Delafield FP, Doudoroff M, Palleroni NJ, Lusty CJ, Contopoulos R. 1965. Decomposition of poly-ß-hydroxybutyrate by pseudomonads. *J. Bacteriol.* 90:1455–66
13. Doi Y. 1990. *Microbial Polyesters*. New York: VHC
14. Deleted in proof
15. Doi Y, Kitamura S, Abe H. 1995. Microbial synthesis and characterization of poly(3-hydroxybutyrate-*co*-3-hydroxyhexanoate). *Macromolecules* 28:4822–28
16. Doi Y, Kumagai Y, Tanahashi N, Mukai K. 1992. Structural effects on biodegradation of microbial and synthetic poly(hydroxyalkanoates). In *Biodegradable Polymers and Plastics*, ed. M Vert, pp. 139–48. London: R. Soc. Chem.
17. Doi Y, Segawa A, Kawaguchi Y, Kunioka M. 1990. Cyclic nature of poly(3-hydroxyalkanoate) metabolism in *Alcaligenes eutrophus*. *FEMS Microbiol. Lett.* 67:165–70
18. Focarete ML, Scandola M, Jendrossek D, Adamus G, Sikorska W, Kowalczuk M. 1999. Bioassimilation of oligomers of atactic poly[(*R*,*S*)-3-hydroxybutyrate] by selected bacterial strains. *Macromolecules* 32:4814–18
19. Foster LJR, Lenz RW, Fuller RC. 1994. Quantitative determination of intracellular depolymerase activity in *Pseudomonas oleovorans* inclusions containing poly-3-hydroxyalkanoates with long alkyl substituents. *FEMS Microbiol. Lett.* 118: 279–82
20. Foster LJR, Stuart ES, Tehrani A, Lenz RW, Fuller RC. 1996. Intracellular depolymerase and polyhydroxyoctanoate granule integrity in *Pseudomonas oleovorans*. *Int. J. Biol. Macromol.* 19:177–83
21. Foster LJR, Zervas SJ, Lenz RW, Fuller RC. 1995. The biodegradation of poly-3-hydroxyalkanoates, PHAs, with long alkyl substituents by *Pseudomonas maculicola*. *Biodegradation* 6:67–73
22. Fukui T, Narikawa T, Miwa K, Shirakura Y, Saito T, Tomita K. 1988. Effect of limited tryptic modification of a bacterial poly(3-hydroxybutyrate) depolymerase on its catalytic activity. *Biochim. Biophys. Acta* 952:164–71
23. Gao S, Maehara A, Yamane T, Ueda S. 2001. Identification of the intracellular polyhydroxyalkanoate depolymerase gene of *Paracoccus denitrificans* and some properties of the gene product. *FEMS Microbiol. Lett.* 196:159–64
24. Gonda KE, Jendrossek D, Molitoris HP. 2000. Fungal degradation of thermoplastic polymers under simulated deep sea conditions. *Hydrobiologia* 426:73–183
25. Griebel RJ, Merrick JM. 1971. Metabolism of poly-ß-hydroxybutyrate: effect of mild alkaline extraction on native poly-ß-hydroxybutyrate granules. *J. Bacteriol.* 108:782–89
26. Griebel RJ, Smith Z, Merrick JM. 1968. Metabolism of poly-ß-hydroxybutyrate. I. Purification, composition, and properties of native poly-ß-hydroxybutyrate granules from *Bacillus megaterium*. *Biochemistry* 7:3676–81
27. Han J-S, Son Y-J, Chang C-S, Kim M-N. 1998. Purification and properties of extracellular poly(3-hydroxybutyrate) depolymerase produced by *Penicillium pinophilum*. *J. Microbiol.* 36:67–73
28. Handrick R, Jendrossek D. 1998. Extracellular and intracellular polyhydroxyalkanoate depolymerases: homologies and differences. In *Biochemical Principles and Mechanisms of Biodegradation and Biodegradation of Polymers*, ed. A. Steinbüchel, pp. 57–67. Weinheim: Wiley-VCH
29. Handrick R, Reinhardt S, Focarete ML, Scandola M, Adamus G, et al. 2001. A new type of thermoalkalophilic hydrolase of *Paucimonas lemoignei* with high specificity for amorphous polyesters of short-chain-length hydroxyalkanoic acids. *J. Biol. Chem.* 276:36215–24
30. Handrick R, Reinhard S, Jendrossek D.

2000. Mobilization of poly(3-hydroxybutyrate) in *Ralstonia eutropha*. *J. Bacteriol.* 182:5916–18

31. Hayward AC, Forsyth WGC, Roberts JB. 1959. Synthesis and breakdown of poly-ß-hydroxybutyric acid by bacteria. *J. Gen. Microbiol.* 20:510–18
32. Hippe H. 1967. Aufbau und Wiederverwertung von Poly-ß-hydroxybuttersäure durch *Hydrogenomonas* H16. *Arch. Mikrobiol.* 56:248–77
33. Hippe H, Schlegel HG. 1967. Hydrolyse von PHBS durch intrazelluläre Depolymerase von *Hydrogenomonas* H16. *Arch. Mikrobiol.* 56:278–99
34. Hiraishi T, Ohura T, Ito S, Kassuya K-I, Doi Y. 2000. Function of the catalytic domain of poly(3-hydroxybutyrate) depolymerase from *Pseudomonas stutzeri*. *Biomacromolecules* 1:320–24
35. Horowitz DM, Brennan EM, Koon JJ, Gerngross TU. 1999. Novel thermal route to an amorphous, film-forming polymer latex. *Macromolecules* 32:3347–52
36. Horowitz DM, Sanders JKM. 1995. Biomimetic, amorphous granules of polyhydroxyalkanoates: composition, mobility, and stabilization in vivo by proteins. *Can. J. Microbiol.* 41(Suppl. 1):115–23
37. Huisman GW, Wonink E, Meima R, Terpstra B, Witholt B. 1991. Metabolism of poly(3-hydroxyalkanoates) (PHA) by *Pseudomonas oleovorans*. *J. Biol. Chem.* 266:2191–98
38. Jaeger K-E, Dijkstra BW, Reetz MT. 1999. Bacterial biocatalysts: molecular biology, three-dimensional structures, and biotechnological applications. *Annu. Rev. Microbiol.* 53:315–35
39. Jaeger K-E, Ransac S, Dijkstra BW, Colson D, van Heuvel M, Misset O. 1994. Bacterial lipases. *FEMS Microbiol. Rev.* 15:29–63
40. Jaeger K-E, Steinbüchel A, Jendrossek D. 1995. Substrate specificities of bacterial polyhydroxyalkanoate depolymerases and lipases: bacterial lipases hydrolyze poly(ω-hydroxyalkanoates). *Appl. Environ. Microbiol.* 61:3113–18
41. James BW, Mauchline WS, Dennis PJ, Keevil CW, Wait R. 1999. Poly-3-hydroxybutyrate in *Legionella pneumophila*, an energy source for survival in low-nutrient environments. *Appl. Environ. Microbiol.* 65:822–27
42. Jendrossek D. 2001. Extracellular PHA depolymerases: the key enzymes of PHA degradation. See Ref. 100, pp. 41–83
43. Jendrossek D. 2001. Microbial degradation of polyesters. See Ref. 89, pp. 293–325
44. Jendrossek D. 2001. Transfer of [*Pseudomonas*] *lemoignei*, a gram-negative rod with restricted catabolic capacity, to *Paucimonas* gen. nov. with one species, *Paucimonas lemoignei* comb. nov. *Int. J. Syst. Evol. Microbiol.* 51:905–8
45. Jendrossek D, Backhaus M, Andermann M. 1995. Characterization of the extracellular poly(3-hydroxybutyrate) depolymerase of *Comamonas* sp. and of its structural gene. *Can. J. Microbiol.* 41(Suppl. 1):160–69
46. Jendrossek D, Frisse A, Behrends A, Andermann M, Kratzin HD, et al. 1995. Biochemical and molecular characterization of the *Pseudomonas lemoignei* depolymerase system. *J. Bacteriol.* 177:596–607
47. Jendrossek D, Knoke I, Habibian RB, Steinbüchel A, Schlegel HG. 1993. Degradation of poly(3-hydroxybutyrate), PHB, by bacteria and purification of a novel PHB depolymerase from *Comamonas* sp. *J. Environ. Polym. Degrad.* 1:53–63
48. Jendrossek D, Müller B, Schlegel HG. 1993. Cloning and characterization of the poly(hydroxyalkanoic acid)-depolymerase gene locus, *phaZ1*, of *Pseudomonas lemoignei* and its gene product. *Eur. J. Biochem.* 218:701–10
49. Jendrossek D, Schirmer A, Handrick R. 1997. Recent advances in characterization of bacterial PHA depolymerases.

In *International Symposium on Bacterial Polyhydroxyalkanoates*, ed. G Eggink, A Steinbüchel, Y Poirier, B Witholt, pp. 89–101. Ottawa, Can.: NRC Res.

50. Kasuya K, Inoune Y, Tanaka T, Akehata T, Iwata T, et al. 1997. Biochemical and molecular characterization of the polyhydroxybutyrate depolymerase of *Comamonas acidovorans*YM1609, isolated from freshwater. *Appl. Environ. Microbiol.* 63:4844–52
51. Kasuya K, Ohura T, Masuda K, Doi Y. 1999. Substrate and binding specificities of bacterial polyhydroxybutyrate depolymerases. *Int. Biol. Macromol.* 24:329–36
52. Kasuya K, Takagi K, Ishiwatari S, Yoshida Y, Doi Y. 1998. Biodegradabilities of various aliphatic polyesters in natural waters. *Polym. Degrad. Stab.* 59:327–32
53. Kim H, Ju H-S, Kim J. 2000. Characterization of an extracellular poly(3-hydroxy-5-phenylvalerate) depolymerase from *Xanthomonas* sp. JS02. *Appl. Microbiol. Biotechnol.* 53:323–27
54. Kita K, Mashiba S, Nagita M, Ishimaru K, Okamoto K, et al. 1997. Cloningme of poly(3-hydroxybutyrate) depolymerase from a marine bacterium, *Alcaligenes faecalis* AE122, and characterization of its gene product. *Biochim. Biophys. Acta* 1352:113–22
55. Klingbeil B, Kroppenstedt R, Jendrossek D. 1996. Taxonomical identification of *Streptomyces exfoliatus* K10 and characterization of its poly(3-hydroxybutyrate) depolymerase gene. *FEMS Microbiol. Lett.* 142:215–21
56. Deleted in proof
57. Deleted in proof
58. Kobayashi T, Sugiyama A, Kawase Y, Saito T, Mergaert J, Swings J. 1999. Biochemical and genetic characterization of an extracellular poly(3-hydroxybutyrate) depolymerase from *Acidovorax* sp. strain TP4. *J. Environ. Polym. Degrad.* 7:9–18
59. Kumar A, Gross RA, Jendrossek D. 2000. Poly(3-hydroxybutyrate)-depolymerase from *Pseudomonas lemoignei*: catalysis of esterifications in organic media. *J. Org. Chem.* 65:7800–6
60. Lemoigne M. 1925. Etudes sur l'autolyse microbienne—acidification par formation d'acide β oxybutyrique. *Ann. Inst. Pastuer* 39:144–55
61. Little E, Bork P, Doolittle RF. 1994. Tracing the spread of fibronectin type III domains in bacterial glycohydrolases. *J. Mol. Evol.* 39:631–43
62. Macrae RM, Wilkinson JF. 1958. Poly-ß-hydroxybutyrate metabolism in washed suspensions of *Bacillus cereus* and *Bacillus megaterium*. *J. Gen. Microbiol.* 19: 210–22
63. Madison LL, Huisman GW. 1999. Metabolic engineering of poly(3-hydroxyalkanoates): from DNA to plastic. *Microbiol. Mol. Biol. Rev.* 63:21–53
64. Marchessault RH, Morin FG, Wong S, Saracovan I. 1995. Artificial granule suspensions of long side chain poly(3-hydroxyalkanoate). *Can. J. Microbiol.* 41 (Suppl. 1):138–42
65. Matavulj M, Molitoris HP. 1992. Fungal degradation of polyhydroxyalkanoates and a semiquantitative assay for screening their degradation by terrestrial fungi. *FEMS Microbiol. Rev.* 103:323–32
66. Mayer F, Madkour MH, Pieper-Fürst U, Wieczorek R, Liebergesell M, Steinbüchel A. 1996. Electron microscopic observations on the macromolecular organization of the boundary layer of bacterial PHA inclusion bodies. *J. Gen. Appl. Microbiol.* 42:445–55
67. Mergaert J, Schirmer A, Hauben L, Mau M, Jendrossek D, Swings J. 1996. Isolation and identification of poly(3-hydroxyvalerate) degrading strains of *Pseudomonas lemoignei*. *Int. J. Syst. Bacteriol.* 46:769–73
68. Merrick JM, Doudoroff M. 1964. Depolymerisation of poly-ß-hydroxybutyrate by an intracellular enzyme system. *J. Bacteriol.* 88:60–71
69. Merrick JM, Steger R, Dombroski D.

1999. Hydrolysis of native poly(hydroxybutyrate) granules (PHB), crystalline PHB, and artificial amorphous PHB granules by intracellular and extracellular depolymerases. *Int. J. Biol. Macromol.* 25: 129–34
70. Molitoris KP, Moss ST, de Koning G, Jendrossek D. 1996. SEM analysis of polyhydroxyalkanoate degradation by bacteria. *Appl. Microbiol. Biotechnol.* 46:570–79
71. Müller B, Jendrossek D. 1993. Purification and properties of a poly(3-hydroxyvalerate) depolymerase from *Pseudomonas lemoignei. Appl. Microbiol. Biotechnol.* 38:487–92
72. Müller HM, Seebach D. 1993. Poly(hydroxyalkanoates): a fifth class of physiologically important organic biopolymers? *Angew. Chem. Int. Ed. Engl.* 32:477–502
73. Nakayama K, Saito T, Fukui T, Shirakura Y, Tomita K. 1985. Purification and properties of extracellular poly(3-hydroxybutyrate) depolymerases from *Pseudomonas lemoignei. Biochim. Biophys. Acta* 827:63–72
74. Neumeier S. 1994. *Abbau thermoplastischer Biopolymere auf Poly-β-Hydroxyalkanoat-Basis durch terrestrische und marine Pilze.* Diplomarbeit. Univ. Regensburg. 99 pp.
75. Nobes GAR, Marchessault RH, Chanzy H, Briese BH, Jendrossek D. 1996. Splintering of poly(3-hydroxybutyrate) single crystals by PHB-depolymerase A from *Pseudomonas lemoignei. Macromolecules* 29:8330–33
76. Nojiri M, Saito T. 1997. Structure and function of poly(3-hydroxybutyrate) depolymerases from *Alcaligenes faecalis* T1. *J. Bacteriol.* 179:6965–70
77. Oeding V, Schlegel HG. 1973. ß-Ketothiolase from *Hydrogenomonas eutropha* H16 and its significance in the regulation of poly-ß-hydroxybutyrate metabolism. *Biochem. J.* 134:239–48
78. Ohura T, Aoyagi Y, Takagi K, Yoshida Y, Kasuya K, Doi Y. 1999. Biodegradation of poly(3-hydroxyalkanoic acids) fibers and isolation of poly(3-hydroxybutyric acid)-degrading microorganisms under aquatic environments. *Polym. Degrad. Stab.* 63: 23–29
79. Ohura T, Kasuya K, Doi Y. 1999. Cloning and characterization of the polyhydroxybutyrate depolymerase gene of *Pseudomonas stutzeri* and analysis of the function of substrate-binding domains. *Appl. Environ. Microbiol.* 65:189–97
80. Ollis DL, Cheah E, Cygler M, Dijkstra B, Frolow F, et al. 1992. The α/β hydrolase-fold. *Protein Eng.* 5:197–211
81. Quinteros R, Goodwin S, Lenz RW, Park WH. 1999. Extracellular degradation of medium chain length poly(beta-hydroxyalkanoates) by *Comamonas* sp. *Int. J. Biol. Macromol.* 25:135–43
82. Ramsay BA, Saracovan I, Ramsay JA, Marchessault RH. 1994. A method for the isolation of microorganisms producing extracellular long-side-chain poly(ß-hydroxyalkanoate) depolymerase. *J. Environ. Polym. Degrad.* 2:1–7
83. Ruiz JA, Lopez NI, Fernandez RO, Mendez BS. 2001. Polyhydroxyalkanoate degradation is associated with nucleotide accumulation and enhances stress resistance and survival of *Pseudomonas oleovorans* in natural water microcosms. *Appl. Environ. Microbiol.* 67:225–30
84. Saegusa H, Shiraki M, Kanai C, Saito T. 2001. Cloning of an intracellular poly [D(−)-3-hydroxybutyrate] depolymerase gene from *Ralstonia eutropha* H16 and characterization of the gene product. *J. Bacteriol.* 183:94–100
85. Saito T, Kobayashi T. 2001. Intracellular degradation of PHAs. See Ref. 100, pp. 23–40
86. Saito T, Saegusa H, Miyata Y, Fukui T. 1992. Intracellular degradation of poly(3-hydroxybutyrate) granules of *Zoogloea ramigera* I-16-M. *FEMS Microbiol. Rev.* 103:333–38
87. Saito T, Suzuki K, Yamamoto J, Fukui T, Miwa K, et al. 1989. Cloning, nucleotide

sequence, and expression in *Escherichia coli* of the gene for poly(3-hydroxybutyrate) depolymerase from *Alcaligenes faecalis*. *J. Bacteriol.* 171:184–89

88. Saito T, Takizawa K, Saegusa H. 1995. Intracellular poly(3-hydroxybutyrate) depolymerase in *Alcaligenes eutrophus*. *Can. J. Microbiol.* 41(Suppl. 1):187–91

89. Scheper T. 2001. *Biopolyesters: Advances in Biochemical Engineering/Biotechnology*. Berlin/Heidelberg: Springer

90. Scherer TM, Fuller C, Goodwin S, Lenz RW. 2000. Enzymatic hydrolysis of oligomeric models of poly-3-hydroxybutyrate. *Biomacromolecules* 1:577–83

91. Schirmer A, Jendrossek D. 1994. Molecular characterization of the extracellular poly(3-hydroxyoktanoic acid) [P(3HO)] depolymerase gene of *Pseudomonas fluorescens* GK13 and of its gene product. *J. Bacteriol.* 176:7065–73

92. Schirmer A, Jendrossek D, Schlegel HG. 1993. Degradation of poly(3-hydroxyoctanoic acid), [P(3HO)], by bacteria. Purification and properties of a P(3HO) depolymerase of *Pseudomonas fluorescens* GK13 biovarV. *Appl. Environ. Microbiol.* 59:1220–27

93. Schirmer A, Matz C, Jendrossek D. 1995. Substrate specificities of PHA-degrading bacteria and active site studies on the extracellular poly(3-hydroxyoctanoic acid) [P(3HO)] depolymerase of *Pseudomonas fluorescens* GK13. *Can. J. Microbiol.* 41(Suppl. 1):170–79

94. Schöber U, Thiel C, Jendrossek D. 2000. Poly(3-hydroxyvalerate) depolymerase of *Pseudomonas lemoignei*. *Appl. Environ. Microbiol.* 66:1385–92

95. Senior PJ, Dawes EA. 1973. The regulation of poly-ß-hydroxybutyrate metabolism in *Azotobacter beijerinkii*. *Biochem. J.* 134:225–38

96. Shinohe T, Nojiri M, Saito T, Stanislawski T, Jendrossek D. 1996. Determination of the active sites serine of the poly(3-hydroxybutyrate) depolymerases of *Pseudomonas lemoignei* (PhaZ5) and of *Alcaligenes faecalis*. *FEMS Microbiol. Lett.* 141:103–9

97. Shinomiya M, Iwata T, Kasuya K, Doi Y. 1997. Cloning of the gene for poly(3-hydroxybutyric acid) depolymerase of *Comamonas testosteroni* and functional analysis of its substrate-binding domain. *FEMS Microbiol. Lett.* 154:89–94

98. Shirakura Y, Fukui T, Saito T, Okamoto Y, Narikawa T, et al. 1986. Degradation of poly(3-hydroxybutyrate) by poly(3-hydroxy-butyrate) depolymerase from *Alcaligenes faecalis* T_1. *Biochim. Biophys. Acta* 880:46–53

99. Steinbüchel A, Aerts K, Babel W, Föllner C, Liebergesell M, et al. 1995. Considerations on the structure and biochemistry of bacterial polyhydroxyalkanoic acids inclusions. *Can. J. Microbiol.* 41(Suppl. 1):94–105

100. Steinbüchel A, Doi Y, eds. 2001. *Biopolymers, Polyesters II, Properties and Chemical Synthesis*. Weinheim: Wiley-VCH. 468 pp.

101. Steinbüchel A, Hein S. 2001. Biochemical and molecular basis of microbial synthesis of polyhydroxyalkanoates in microorganisms. See Ref. 89, pp. 81–124

102. Steinbüchel A, Valentin HE. 1995. Diversity of bacterial polyhydroxyalkanoic acids. *FEMS Microbiol. Lett.* 128:219–28

103. Stinson MW, Merrick JM. 1974. Extracellular enzyme secretion by *Pseudomonas lemoignei*. *J. Bacteriol.* 119:152–61

104. Stuart ES, Foster LJR, Lenz RW, Fuller RC. 1996. Intracellular depolymerase functionality and location in *Pseudomonas oleovorans* inclusions containing polyhydroxyalkanoate. *Int. J. Biol. Macromol.* 19:171–76

105. Suzuki Y, Taguchi S, Saito S, Toshima K, Matsumura S, Doi Y. 2001. Involvement of catalytic amino acid residues in enzyme-catalyzed polymerization for the synthesis of polyesters. *Biomacromolecules* 2:541–44

•106. Taidi B, Mansfield DA, Anderson AJ.

1995. Turnover of poly(3-hydroxybutyrate) (PHB) and its influence on the molecular mass of the polymer accumulated by *Alcaligenes eutrophus* during batch culture. *FEMS Microbiol. Lett.* 129: 201–6
107. Takeda M, Kitashima K, Adachi K, Hanaoka Y, Suzuki I, Koizumi J. 2000. Cloning and expression of the gene encoding thermostable poly(3-hydroxybutyrate) depolymerase. *J. Biosci. Bioeng.* 90:416–21
108. Technow U. 1999. *Isolierung und Substratbindedomäne der extrazellulären PHB Depolymerase A von* Pseudomonas lemoignei *aus rekombinanten* Escherichia coli *und Anreicherung und Charakterisierung des Aktivators im intrazellulären PHB-Abbau von* Rhodospirillum rubrum. Diplomarbeit. Univ. Göttingen. 108 pp.
109. Terpe K. 1999. *Physiologische und molekularbiologische Untersuchungen zur Regulation extrazellulärer PHB-Depolymerasen in* P. lemoignei. PhD thesis. Univ. Göttingen. 158 pp.
110. Terpe K, Kerkhoff K, Pluta E, Jendrossek D. 1999. Relationship between succinate transport and production of extracellular poly(3-hydroxybutyrate) depolymerase in *Pseudomonas lemoignei. Appl. Environ. Microbiol.* 65:1703–9
111. Watanabe T, Ito Y, Yamada T, Hashimoto M, Sekine S, Tanaka H. 1994. The roles of the C-terminal domain and type III domains of chitinases A1 from *Bacillus circulans* WL-12 in chitin degradation. *J. Bacteriol.* 176:4465–72
112. Wu K, Chung L, Revill WP, Katz L, Reeves CD. 2000. The FK520 gene cluster of *Streptomyces hygroscopicus* var. *ascomyceticus* (ATCC 14891) contains genes for biosynthesis of unusual polyketide extender units. *Gene* 251:81–90
113. Yamashita K, Aoyagi Y, Abe H, Doi Y. 2001. Analysis of adsorption function of polyhydroxybutyrate depolymerase from *Alcaligenes facalis* T1 by using a quartz crystal microbalance. *Biomacromolecules* 2:25–28
114. York GM, Junker BH, Stubbe JA, Sinskey AJ. 2001. Accumulation of the PhaP phasin of *Ralstonia eutropha* is dependent on production of polyhydroxybutyrate in cells. *J. Bacteriol.* 183:4217–26

Annu. Rev. Microbiol. 2002. 56:433–55
doi: 10.1146/annurev.micro.56.012302.160625

First published online as a Review in Advance on July 15, 2002

MENACING MOLD: The Molecular Biology of *Aspergillus fumigatus*

Axel A. Brakhage and Kim Langfelder
Institut für Mikrobiologie, Universität Hannover, Schneiderberg 50, 30167 Hannover, Germany; e-mail: brakhage@ifmb.uni-hannover.de

Key Words *Aspergillus fumigatus*, molecular techniques, genomics, virulence

■ **Abstract** Infections with mold pathogens have emerged as an increasing risk faced by patients under sustained immunosuppression. Species of the *Aspergillus* family account for most of these infections, and in particular *Aspergillus fumigatus* may be regarded as the most important airborne pathogenic fungus. The improvement in transplant medicine and the therapy of hematological malignancies is often complicated by the threat of invasive aspergillosis. Specific diagnostic methods are still limited as are the possibilities of therapeutic intervention, leading to the disappointing fact that invasive aspergillosis is still associated with a high mortality rate that ranges from 30% to 90%. In recent years considerable progress has been made in understanding the genetics of *A. fumigatus*, and molecular techniques for the manipulation of the fungus have been developed. Molecular genetics offers not only approaches for the detailed characterization of gene products that appear to be key components of the infection process but also selection strategies that combine classical genetics and molecular biology to identify virulence determinants of *A. fumigatus*. Moreover, these methods have a major impact on the development of novel strategies leading to the identification of antimycotic drugs. This review summarizes the current knowledge on the biology, molecular genetics, and genomics of *A. fumigatus*.

CONTENTS

0066-4227/02/1013-0433$14.00

INTRODUCTION

In the past 20 years the deuteromycete *Aspergillus fumigatus* has gone from being a saprophytic fungus of minor interest to scientists to becoming one of the most important fungal pathogens. The main reason for the rise in systemic infections lies in the steady increase in the number of immunocompromised individuals, the main risk group for such infections (51). Diseases caused by *A. fumigatus* can be divided into three categories: (*a*) allergic reactions and (*b*) colonization with restricted invasiveness are observed in immunocompetent individuals, whereas (*c*) systemic infections with high mortality rates occur in immunocompromised patients. Added to this is the lack of effective therapy, resulting in a high mortality rate between 30 and 90% (32).

Although *A. fumigatus* only makes up a small proportion of all aerial spores, around 0.3% in the air of a particular hospital, it causes roughly 90% of systemic *Aspergillus* infections (73). This suggests that *A. fumigatus* possesses certain factors that allow it to become an opportunistic human pathogen in immunocompromised patients. In recent years great progress has been made in understanding the genetics of *A. fumigatus*; molecular genetic techniques have been developed that allow a detailed characterization of the fungus. Molecular biology provides effective techniques for understanding the mechanisms underlying *A. fumigatus* virulence and for investigating the key components of the infectious process.

In this review we present an overview of the molecular biological technology available for studying *A. fumigatus* and for identifying possible virulence determinants. Most importantly, gene transfer methods for *A. fumigatus* are available, some of which were first developed for other *Aspergillus* species such as *A. nidulans* and *A. niger* (55). Unlike its relative *A. nidulans*, *A. fumigatus* has not revealed a sexual reproduction cycle; however, some classical genetic studies are possible. In addition, a large range of molecular biological tools is now available for investigating *A. fumigatus*. The most important of these tools are described here. In conjunction with the genome sequence of *A. fumigatus*, these should eventually lead to the discovery of the key elements of *A. fumigatus* virulence and eventually to drugs for treating systemic *A. fumigatus* infections. Further general information can be obtained from the *Aspergillus* web site (http://www.aspergillus.man.ac.uk/). A comprehensive overview about putative virulence determinants of *A. fumigatus* has been given in several reviews (10, 11, 45, 51, 52).

BIOLOGY AND GENETICS OF *ASPERGILLUS FUMIGATUS*

The natural ecological niche of the saprobic fungus *A. fumigatus* is the soil. By contributing to the degradation of decaying organic matter such as compost and hay, *A. fumigatus* plays an important role in recycling carbon and nitrogen sources (24, 58, 82).

A. fumigatus conidia (spores) are gray-green in color and only 2.5–3.0 μm in diameter, allowing them to reach the lung alveoli (65, 69, 70) (Figure 1, see color

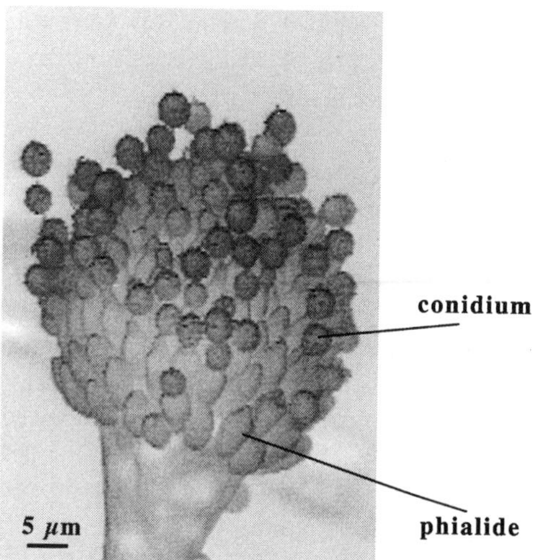

Figure 2 Scanning electron microscope image of an *Aspergillus fumigatus* conidiophore showing the stem cell, phialides, and conidia.

insert). Conidia contain a single haploid nucleus. The conidiophore extends from a foot cell at right angles to the mycelium and culminates in a broadly clavate vesicle (20–30 μm in diameter). A single layer of cells (phialides) is attached to the apical side of the vesicle. The conidia result from a constriction of an elongated portion of the phialide (13) (Figure 2).

A. fumigatus is able to grow rapidly on minimal agar plates containing a carbon source (e.g., glucose), a nitrogen source (e.g., nitrate), and trace elements. In addition to being relatively flexible with regard to growth medium, *A. fumigatus* is also able to withstand high temperatures. Growth occurs up to 55°C, and conidia are able to survive at temperatures up to 70°C [reviewed in (51, 70)].

The genome of *A. fumigatus* has not yet been fully characterized, but its size is estimated at around 27.8 Mb (http://www.tigr.org/), about the same as calculated for *A. nidulans* (14). Pulse-field gel electrophoresis has shown the presence of 5 chromosomal DNA bands ranging in size from 1.7 to 4.8 Mb (80). A first approximation suggests that the genomes of *A. nidulans* and *A. fumigatus* are in fact similar (2). A full genome sequencing project is under way for *A. fumigatus* (see section Genomics).

CLASSICAL GENETIC TECHNIQUES

The fact that *A. fumigatus* has haploid uninucleate conidia facilitates the isolation of clones and the analysis of mutants by classical or molecular techniques. The existence of a teleomorph form of *A. fumigatus* is still under debate. However,

DNA-DNA reassociation values less than 70% were found for *A. fumigatus* and *Neosartorya*, proving conclusively that *Neosartorya* is not the teleomorph form of *A. fumigatus* (64). Other analyses using different methods produced identical results (38). Unlike *A. nidulans*, which has a well-characterized sexual cycle, no sexual cycle has been described for *A. fumigatus*. Because a sexual cross is necessary to generate linkage data for different genetic loci, the amount of linkage information available for *A. fumigatus* has been restricted. However, in the early 1960s parasexuality was described for *A. fumigatus* (5, 77). The parasexual capability of *A. fumigatus*, in which two fungal strains growing in close proximity can be induced to form a heterokaryon (the frontier method), results in diploid nuclei that can be maintained as a stable diploid form. Treatment with a destabilizing agent (e.g., benomyl) results in haploid strains [reviewed in (63)]. Reports by de Lucas et al. (27) and Brookman & Denning (15) suggest that parasexuality in *A. fumigatus* can be a powerful technique. It could be used to exchange marker genes between strains. New auxotrophic mutants of *A. fumigatus* could help to establish this technique.

MOLECULAR GENETICS

Transformation Systems

TRANSFORMATION PROCEDURES Most of the transformation protocols currently used for *A. fumigatus* are based on methods developed for transformation of *A. nidulans* and *A. niger* in the early 1980s (4, 18, 34). The most frequently used method is the transformation of protoplasts following cell wall degradation by a lytic enzyme preparation. Both Glucanex (Novo Nordisk) and Panzym (Begerow) have been used successfully. Protoplasts are maintained in an osmotically stabilized solution containing PEG/$CaCl_2$ in order to induce the uptake of DNA into the cells. Subsequently protoplasts are regenerated on osmotically stabilized medium containing the relevant antibiotic or lacking the necessary nutritional supplement, depending on the selection marker being used. Transformation efficiencies are in the range of 1–100 transformants/μg DNA. Although this method is reliable, it does have several drawbacks. The low transformation efficiency can be a problem for complementing mutant alleles, while the method is time consuming. Unfortunately *A. fumigatus* protoplasts are not amenable to freezing (K. Langfelder, unpublished observation).

Electroporation of germinating conidia is another method that can be used to transform *A. fumigatus*. This method had previously been described for other filamentous fungi, including *A. oryzae* (20, 21), *A. niger* (61), and *A. nidulans* (71). Recently Weidner et al. (85) demonstrated that this technique can also be used for transformation of *A. fumigatus*. In *A. oryzae* the method is significantly quicker than the transformation of protoplasts because conidia do not require treatment with lytic enzymes and transformant colonies appear sooner (61). Pretreatment with cell wall–degrading enzymes increased transformation efficiencies to 100 transformants/μg DNA. In *A. fumigatus* transformation efficiencies are

2- to 10-fold times higher using electroporation than for the protoplast-forming method (31). However, it appears that the percentage of ectopic integrations into the genome is significantly higher when the electroporation method is used. This means that electroporation is not the method of choice when site-directed integration or single-copy integration is required, unless it is possible to select for specific integration sites.

de Groot et al. (25) demonstrated that it is also possible to use *Agrobacterium tumefaciens* to transform filamentous fungi. In *Aspergillus awamori* they showed an increase of 600-fold relative to conventional transformation techniques (i.e., protoplast formation and electroporation), with integration in a manner similar to that seen in plants (i.e., a single copy of the Ti-plasmid integrates at random at a chromosomal locus). They also demonstrated that the technique is applicable to filamentous fungi in general by successfully transforming *A. niger*, *Trichoderma reesei*, and *Neurospora crassa*, among others. Despite the utility of such a method few publications report the actual use of the method to transform filamentous fungi (40); not one describes its use in *A. fumigatus*.

DOMINANT SELECTABLE MARKERS *A. fumigatus* and a number of other filamentous fungi are sensitive to several antibiotics such as phleomycin/bleomycin and hygromycin B. Bacterial genes that produce resistance to such compounds can be used as dominant selectable markers in filamentous fungi. For *A. fumigatus* the most frequently used resistance gene is the hygromycin B phosphotransferase (*hph*) gene of *Escherichia coli*. Hygromycin B is an aminoglycoside antibiotic produced by *Streptomyces hygroscopicus*, which inhibits protein synthesis in prokaryotic and eukaryotic cells (39). The protein encoded by the *hph* gene phosphorylates hygromycin B molecules, which results in complete loss of biological activity (62). In order to express this gene in *A. fumigatus* and other filamentous fungi, expression cassettes were designed in which expression of the *hph* gene was placed under the control of fungal promoters. One particular cassette in which the *hph* gene is under the control of the strong, constitutive glyceraldehyde-3-phosphate dehydrogenase gene (*gpdA*) promoter of *A. nidulans* and the *trpC* terminator region of *A. nidulans* (67) has been used frequently for transforming a variety of different filamentous fungi (68). The *ble* gene, which confers resistance to phleomycin/bleomycin, has been used far less frequently for transformation of *A. fumigatus*. Bleomycin acts by intercalating in DNA, which leads to degradation of DNA (35). The Ble protein binds to bleomycin with high affinity and thus prevents it from interacting with DNA (35). On such occasions where bleomycin has been used, it was often required to generate a second gene deletion in a strain already carrying the *hph* gene (57, 74).

Avalos et al. (3) described using the *bar* gene of *S. hygroscopicus*, which confers resistance to the herbicide bialaphos, as a selectable marker in *N. crassa*. However, hygromycin B resistance is still the dominant marker of choice for *A. fumigatus* (31). Although the availability of other dominant selectable markers would provide useful tools for generating multiple deletions in a single strain, in the absence of such markers other techniques can be used, such as "*ura*-blaster" and auxotrophic markers.

AUXOTROPHIC MARKERS Auxotrophic markers provide an alternative to using dominant selectable markers for transformation of *A. fumigatus*. Researchers working on *A. fumigatus* do not have access to the large collection of auxotrophic mutants that exist for *A. nidulans*. However, new sequence data from the *A. fumigatus* sequencing project should make it possible to delete specific genes (e.g., *argB*) known to be essential for *A. nidulans*. In this way, a collection of auxotrophic strains can be generated that can be complemented with the corresponding *A. fumigatus*, *A. nidulans*, or *N. crassa* genes. In addition to providing additional selectable markers, transformation of auxotrophic mutants also makes other techniques possible (e.g., site-directed integration).

One example of an auxotrophic marker that is available in *A. fumigatus* is the *pyrG* gene, which encodes the orotidine-5′-monophosphate decarboxylase enzyme essential for the synthesis of uracil. *pyrG* mutants, auxotrophic for uracil, have been identified among strains that are not sensitive to 5-fluoro-orotic acid (FOA) (8, 29). *pyrG* homologs have been identified in a number of other species as well, including *A. nidulans* (*pyrG*) (60), *A. niger* (*pyrG*) (87), *Candida albicans* (*ura3*) (53), and *Schizosaccharomyces pombe* (*ura4*) (41). van Hartingsveldt et al. (84) developed *pyrG* as a selectable marker for transformation of *A. niger*. Later, Weidner et al. (85) cloned the *A. fumigatus pyrG* gene and established a homologous transformation system for *A. fumigatus* using *pyrG*. This means that *A. fumigatus pyrG* deletion strains can be utilized for both homologous and heterologous transformations with *A. fumigatus* or *A. niger pyrG*. In general, the background in transformations using uracil prototrophy as selection marker is relatively low in *A. fumigatus* (S. Gattung, K. Langfelder & A.A. Brakhage, unpublished results).

Recently a further homologous recombination system was developed for *A. fumigatus* based on the *sC* gene, which encodes ATP sulfurylase (26). The system had been established previously for *A. niger* (19). Fungal strains lacking the gene are incapable of utilizing sulfate as the sulfur source. Instead, they require reduced sulfur (e.g., methionine) as the sulfur source. In addition, they are no longer sensitive to selenate (which is taken up by the same system as sulfur but then becomes toxic after reduction inside the cell). *A. fumigatus sC* mutant strains were isolated by selection for selenate resistance. Such strains could be transformed successfully with a plasmid encoding the intact *sC* gene as the selectable marker gene. Transformants regained the ability to grow on minimal media using sulfate as the sole sulfur source. The authors found that transformation efficiencies were similar to those observed using *pyrG* as the selectable marker gene (26).

A further metabolic gene that has been used successfully as a selectable marker in filamentous fungi, although not yet in *A. fumigatus*, is the *niaA* gene (23, 72, 81). Its product, nitrate reductase, is required for growth when nitrate is the sole nitrogen source. This marker has the added advantage that strains lacking the *niaA* gene are not sensitive to chlorate, whereas strains carrying a functional *niaA* gene are sensitive. Thus it is possible to select for spontaneous *niaA* deletions, which is an advantage. *niaA* has in fact been developed as a selectable marker for *A. niger*

(81), and this system could probably be adapted in order to provide an additional selectable marker in *A. fumigatus*. The *A. fumigatus pabaA* gene, encoding the *para*-aminobenzoic acid synthase enzyme, has been cloned (16) and could provide a useful selectable marker. This and a number of other metabolic genes could also be suitable as selectable markers for *A. fumigatus*. Among them is the *A. nidulans argB* gene, which has been used as a heterologous marker for transformation of *A. niger* (18) and other genes involved in essential processes.

One possible disadvantage in using auxotrophic markers is complications in testing virulence of deletion mutants. Often auxotrophic strains show a significantly reduced virulence. For example, a strain auxotrophic for uracil (i.e., *pyrG* deletion strain) shows a marked growth defect and was shown to be apathogenic in a murine infection model (29). Likewise, Tang et al. (79) showed that *A. nidulans pabaA* deletion strains are completely apathogenic in a murine infection model. However, they also showed that by supplying the mice with *para*-aminobenzoic acid in their drinking water it was possible to fully restore virulence of the *pabaA* deletion strain. This knowledge could also be applied in order to make "safe" hypervirulent strains of *A. fumigatus*. Using a *pabaA* deletion strain and supplementing growth media and/or drinking water of test animals with *para*-aminobenzoic acid, one can carry out risk-free experiments with potentially hypervirulent *A. fumigatus* strains.

In summary, a number of selectable markers, both dominant and auxotrophic, are available for the transformation of *A. fumigatus*. As a result *A. fumigatus* is accessible to a variety of molecular biology techniques in order to define and investigate potential virulence factors. Even so, it would be advantageous to have a collection of auxotrophic mutants of a single wild-type strain in order to make better use of auxotrophic markers.

COUNTER-SELECTION SYSTEMS Although in theory there are a number of selection markers available for transformation of *A. fumigatus*, in practice it is difficult to generate strains with multiple gene deletions. To date only two such strains have been published (57, 74). In both cases hygromycin B and phleomycin were used as selectable markers. But what if three transformations are required?

One solution is to make use of a *ura*-blaster to transform a uracil-auxotrophic strain that has a nonfunctional *pyrG* gene (Figure 3). A *ura*-blaster cassette is similar to a conventional deletion vector in that it contains a selectable marker, in this case the *pyrG* gene, flanked by upstream and downstream sequences of the gene of interest. In addition, the blaster cassette must contain direct repeats in front of and behind the *pyrG* gene to allow recombination to occur. A functional *pyrG*-blaster for *A. fumigatus* was established using the *A. niger pyrG* gene as selectable marker and two copies of the gene-encoding neomycin phosphotransferase (*neo*) as direct repeats (28). Following transformation with the *pyrG*-blaster, transformants are selected that are uracil prototroph. These are analyzed to find strains carrying a deletion of the gene of interest. Such strains are plated on medium containing 5-FOA and supplemented with uracil. Only strains that have lost the *pyrG* gene through recombination between the direct repeats will grow under these conditions

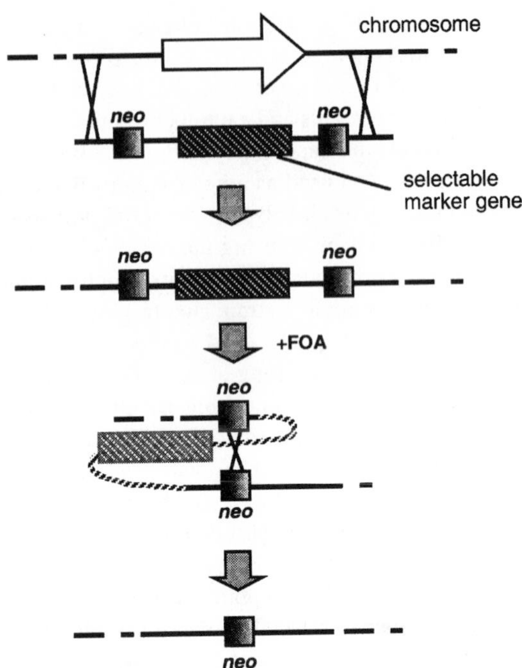

Figure 3 Schematic representation of the gene deletion method using a *ura*-blaster. The *arrow* marks a region of interest on a fungal chromosome. The *hatched box* represents the selectable marker gene *pyrG*, and the *small shaded boxes* represent the *Escherichia coli* neomycin phosphotransferase gene (*neo*). Following a double homologous recombination, the region of interest is replaced by the selectable marker gene and the flanking *neo* genes. After a transformant has been selected, which carries the required deletion, a negative selection is carried out. This strain is grown in a medium that contains 5′-fluoro-orotic acid (FOA) in addition to uridine. This allows selection of strains that have lost the *pyrG* gene through homologous recombination between the two *neo* gene copies flanking the *pyrG* gene. In such strains the *pyrG* gene can be used as selectable marker gene for further rounds of transformation.

because a functional orotidine-5′-monophosphate decarboxylase enzyme converts 5-FOA to a toxic product. Such strains can be used for a second round of transformation to delete further genes or redundant genes that might otherwise mask the effect of a single gene deletion (Figure 3). Under these conditions recombination events occur at a frequency as high as 10^{-4} in *A. fumigatus* and *A. nidulans* (28, 30).

An additional counter-selection system that allows multiple rounds of transformation using the same selectable marker gene is the *sC* system developed in *A. fumigatus* by de Lucas et al. (26). As described previously, the lack of a functional *sC* gene (encoding ATP sulfurylase) results in strains that are unable to utilize sulfate as a source of sulfur and therefore require a reduced sulfur source such as methionine. They are also not sensitive to selenate. These properties allow the *sC* selection system to be used in the same way as the *ura*-blaster in order to carry out repeated selection rounds with the *sC* gene as selectable marker. All that would be required to establish the system is a blaster construct analogous to the *ura*-blaster (i.e., with the *sC* gene flanked by two direct repeats, such as the *neo* gene). With this system, *sC* mutants could be transformed using the *sC* gene as the selectable marker gene. Then, transformants could be plated on selenate to select for strains that had lost the *sC* gene again through homologous recombination. These strains can be used in subsequent transformation rounds using the *sC* gene as the selectable marker gene. This should be another useful tool to assist in the functional analysis of the *A. fumigatus* genome.

MODE OF INTEGRATION OF DNA In order to produce genetically stable *A. fumigatus* transformants, DNA must be integrated at a chromosomal locus. If the recombinant DNA does not contain homologous sequences, then it can integrate at random in the genome (i.e., ectopically). Although no extensive studies have been carried out, initial experiments in *A. nidulans* suggest that some loci are more favored for homologous recombination than others (7). If the recombinant DNA carries homologous sequences, then integration can occur by homologous recombination. Even if the transforming DNA contains homologous sequences, it still integrates ectopically at high frequency. Bird & Bradshaw (7) showed that in *A. nidulans* the efficiency of homologous recombination correlates with the size of the homologous fragment. In addition, strategies exist that can increase the targeting efficiency at defined genetic loci to almost 50% (85).

AUTONOMOUSLY REPLICATING PLASMIDS Under normal circumstances stable transformation of *A. fumigatus* results from a recombination event taking place between the transforming DNA and one of the fungal chromosomes. The transforming DNA becomes integrated in the genome. This is useful because it allows the construction of genetically stable transformants. However, this is one of the reasons why the transformation efficiency is relatively low for *A. fumigatus*.

No replicating plasmids have been isolated from any *Aspergillus* species to date. However, Gems et al. (37) detected a sequence of chromosomal origin, AMA1, that allows plasmids carrying this sequence to replicate in *A. nidulans*. The AMA1 sequence is 6.1 kb in length and contains an inverted repeat separated by a unique sequence. The presence of AMA1 on a vector leads to a significant increase in the transformation efficiency (around 250-fold increase over normal transformation efficiencies). AMA1 was also functional in other filamentous fungi including *A. oryzae*, *A. niger*, and *Penicillium chrysogenum* (33). In *A. fumigatus* cotransformation of cosmids and a plasmid containing the AMA1 sequence increased transformation efficiency 10-fold (49). de Lucas et al. (26) also reported a 10-fold increase in transformation efficiency when a cotransformation with the plasmid pHELP, which carries the AMA1 sequence, was carried out. This effect can probably be explained by recombination between plasmid and cosmid. Unfortunately cotransformation with plasmids containing the AMA1 sequence frequently leads to rearrangements on the cotransformed cosmid (J. Van den Brulle & A.A. Brakhage, unpublished results). This prevents an otherwise useful tool being employed more often. Nevertheless plasmids containing the AMA1 sequence can be used to clone genes from gene libraries (36) and for other applications (1). Solving the side effects of the AMA1 sequence (genetic instability, rearrangements) would provide an additional tool for investigating *A. fumigatus*.

Complementation of Mutants

Several techniques exist to create a large number of *A. fumigatus* mutants, e.g., restriction enzyme-mediated integration (REMI) (17) or UV-mutagenesis (46). Complementation of mutant phenotypes with libraries of *A. fumigatus* or other filamentous fungi can be used to characterize mutations. For example, an *A. fumigatus* mutant with white spores could be complemented with the *A. nidulans wA* gene (9). More recently a mutant with an identical phenotype was generated by UV-mutagenesis (46) and complemented with an *A. fumigatus* wild-type cosmid library (49). The complementing gene, *pksP*, was indeed found to be the homolog of the *A. nidulans wA* gene, both of which encode a polyketide synthase involved in conidial pigment biosynthesis (56). This example shows that complementation of mutants is a powerful tool for analyzing the functions of specific *A. fumigatus* genes given that mutants display a recognizable phenotype.

Other methods for generating a large number of mutants include REMI and transposon mutagenesis. REMI was established in *A. fumigatus* by Brown et al. (17) as a method to generate a large population of roughly random mutants. Integrations were distributed throughout the genome, but clustering was observed to some extent. By including the restriction enzymes *Kpn*I or *Xho*I in the transformation, they obtained a large frequency of single-copy integrations in the genome. Transposons have been described for a number of fungi (47). Transposon tagging has been demonstrated successfully in filamentous fungi (48); recently a transposon-mediated mutagenesis was reported in *A. fumigatus* (6). This can be a powerful technique for generating a large library of mutant strains, with the added advantage that the site of integration is labeled and can easily be identified.

Knock-Out Versus Gene Disruption

Among the most important methods for determining the function of *A. fumigatus* genes and their possible involvement in virulence is the creation of strains in which the relevant gene is no longer functional. In this manner one can study the role of the gene based on the phenotype that results when it is not functional. A second method would be the overexpression of the relevant gene, but this is not always conclusive. Three different approaches can be used to generate strains in which the relevant gene is not functional.

The simplest method is called gene disruption (Figure 4*A*). Here an internal fragment of the gene of interest, often lying toward the 5′-region of the gene, is cloned into a vector carrying a selectable marker. The resulting plasmid is used

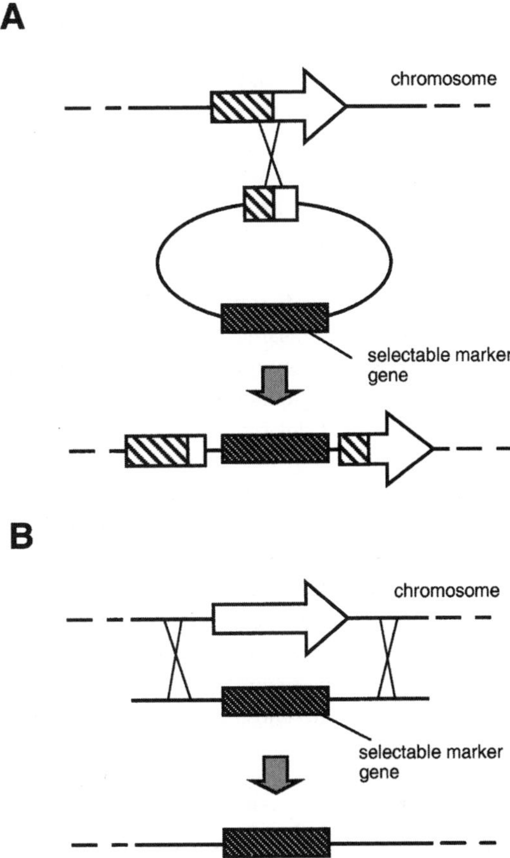

Figure 4 (*A*) Schematic representation of a gene disruption event. The *open arrow* marks a gene of interest on a fungal chromosome. The *dark hatched box* represents a selectable marker gene on a plasmid. Following a single homologous recombination, the plasmid integrates in the gene of interest causing a disruption of the gene. (*B*) Schematic representation of a gene deletion event. The *open arrow* marks a region of interest on a fungal chromosome. The *hatched box* represents a selectable marker gene. Following a double homologous recombination, the region of interest is replaced by the selectable marker gene.

to transform *A. fumigatus*. Homologous recombination with the gene of interest via a single crossover results in two mutant copies of the gene separated by the selectable marker. Neither of the genes should encode functional proteins, since both copies are truncated (one lacks the 3′ region while the other lacks the promoter and 5′ regions). Although gene disruption is relatively quick because few DNA cloning steps are involved, it does have several drawbacks. One problem is that gene disruption results in a truncated peptide, which may still be functional, particularly if the internal fragment chosen is located toward the 3′ region of the gene (Figure 4*A*). A second problem can arise from the fact that gene disruption creates a tandem repeat, which can be unstable under nonselective conditions. In our laboratory a strain in which the adenylate cyclase gene (*acyA*) was disrupted was only stable as long as hygromycin B was present in the medium. As soon as this selection pressure was removed, the disruption strains, which showed a significant growth defect, were outgrown by revertants, which had jettisoned the selection marker (B. Liebmann & A.A. Brakhage, unpublished observation). This result suggests that recombination is a relatively frequent event in *A. fumigatus* and argues against using gene disruption for studying gene function. In addition, gene disruption might not be feasible for small genes because relatively large internal fragments (0.5–1.0 kb) are required in order to achieve a reasonable gene disruption efficiency. This might simply not be possible for small genes.

As a result, most publications report using gene deletion rather than gene disruption to study gene function in *A. fumigatus* (31). The method of choice for studying gene function is called gene deletion or gene knock-out (Figure 4*B*). Here, a plasmid is constructed in which the two regions flanking the gene of interest are separated by a selection marker. The flanking regions should be at least 1.0 kb in length in order to achieve a reasonably high recombination efficiency. The vector is linearized (and dephosphorylated) prior to transformation in order to prevent ectopic integrations and to promote double homologous recombination. Recombination events between the homologous flanking regions lead to the replacement of the chromosomal gene by the selection marker. This is a "clean" method and there is no risk of reversion because the gene is completely removed. A number of cloning steps are required to construct the deletion vector. Also, the bigger the flanking regions are the higher the recombination efficiency is, meaning that fewer transformants need to be screened in order to find one with the required gene deletion. Chaveroche et al. (22) reported a new method for generating gene deletions in *A. nidulans*. We have shown that this method also works for *A. fumigatus* (48a) (Figure 5). The method is based on the ability of *E. coli* strains expressing the λ *red* genes to carry out recombination with >50 bp of flanking sequence. A similar method making use of the *E. coli rec* genes was described by Zhang et al. (88) [reviewed in (59)]. These methods can be used to replace the gene of interest on a cosmid with a selectable marker. The resulting cosmids have large flanking regions and lead to high-efficiency homologous recombination. This means that fewer transformants need to be screened in order to detect a strain carrying a deletion in the relevant gene. At the moment, the only selectable marker available for this system is the *pyrG* gene of *A. fumigatus*, which encodes orotidine-5′-monophosphate decarboxylase. The addition of other selectable markers will make this method even more powerful and will mean that *A. fumigatus* deletion strains can be generated more quickly and more easily than before.

Sometimes gene deletions can result in a lethal phenotype. In such cases no transformants will be obtained and it is difficult to study the effect of the gene. A solution to this problem used for e.g., *A. nidulans* can be to place the gene

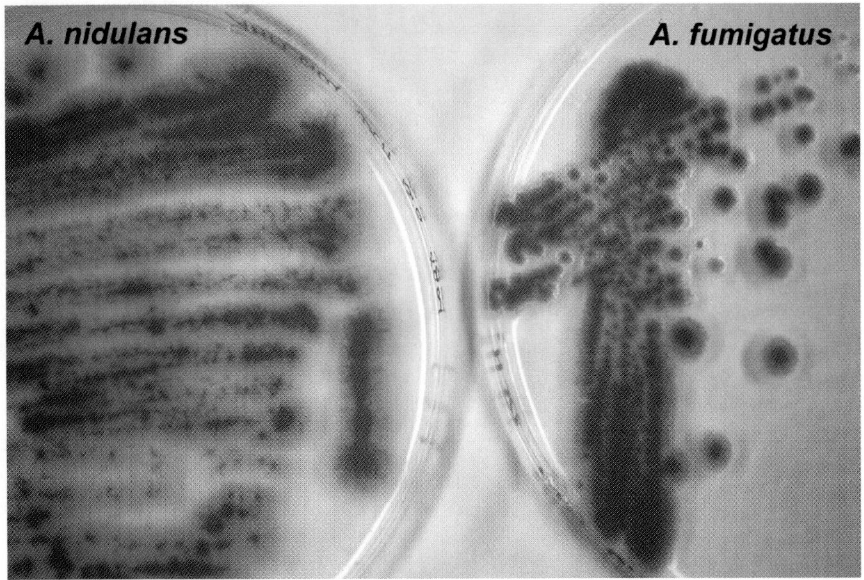

Figure 1 Sporulating colonies of *Aspergillus fumigatus* (*right*) and *Aspergillus nidulans* (*left*) growing on *Aspergillus* minimal medium agar plates. The characteristic *gray-green* color of *A. fumigatus* and the *green* color of *A. nidulans* result from conidial pigment.

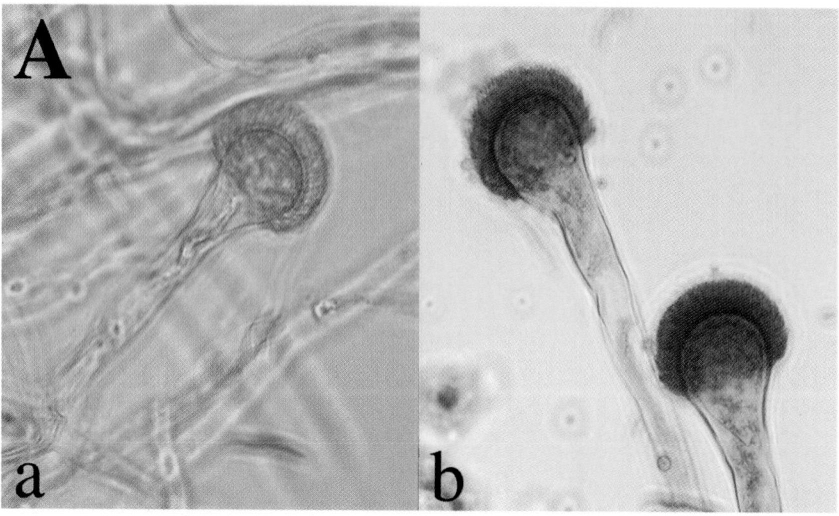

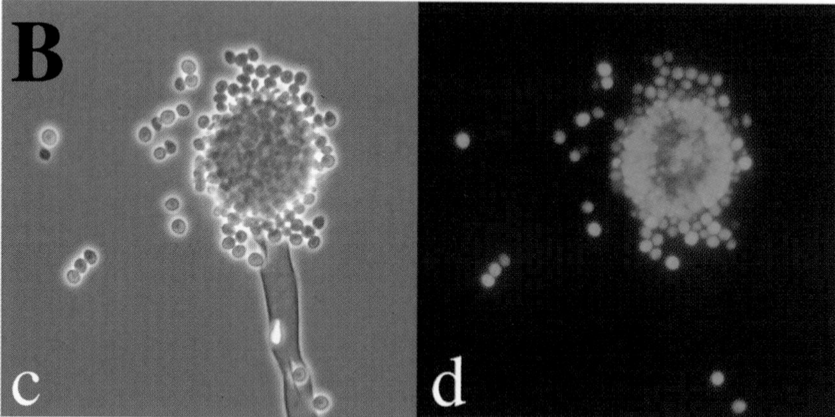

Figure 7 Reporter gene systems in *Aspergillus fumigatus*. (*A*) The *Escherichia coli lacZ* gene can be used as a reporter gene in *A. fumigatus*. This can be visualized by addition of x-gal to fungal conidiophores. *Panel a*: The wild type of *A. fumigatus* shows no β-galactosidase activity. *Panel b*: Fusion of the *lacZ* gene with a developmentally regulated gene results in a specific expression pattern (S. Gattung & A. A. Brakhage, unpublished observation). (*B*) The *egfp* gene encoding the enhanced green fluorescent protein has been used successfully to study gene expression patterns in *A. fumigatus* (50). *Panel c* shows the light microscopy image of a fungal conidiophore. *Panel d* shows the same conidiophore using a filter for green fluorescence. In this *A. fumigatus* strain the *egfp* gene is fused with a developmentally regulated gene, resulting in the observed expression pattern (K. Langfelder, unpublished result).

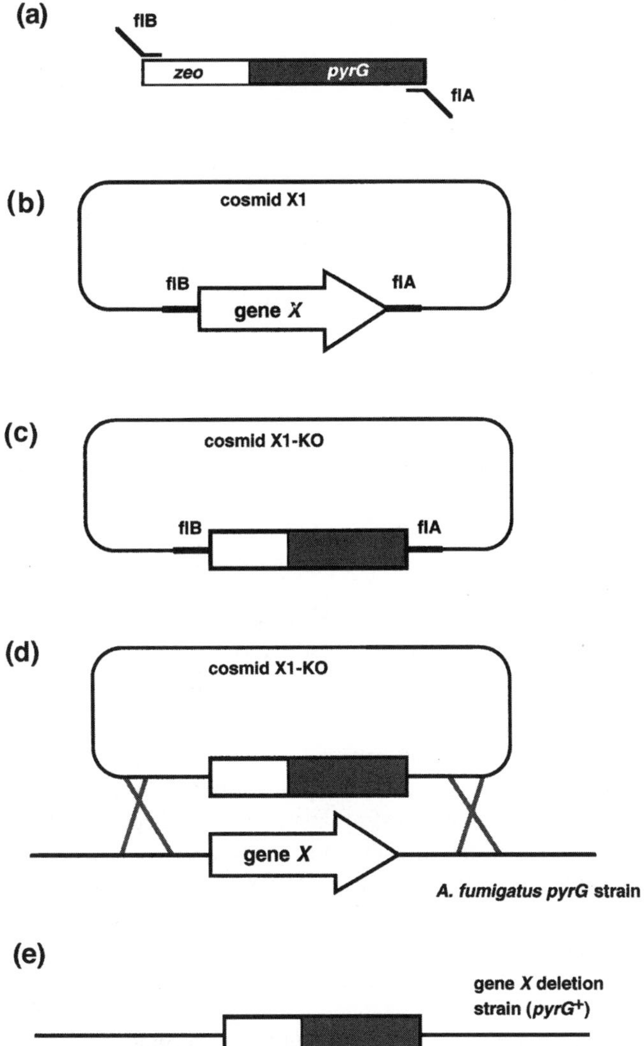

Figure 5 Schematic representation of the ET-cloning technique (22, 48a). (*a*) A region encoding a bifunctional marker (*zeo* and *pyrG*) is amplified by PCR using primers that contain approximately 60 bp of sequence flanking the gene of interest (gene X). (*b*) An *Escherichia coli* strain carrying a cosmid (cosmid X1) that contains the gene X is transformed with the PCR product by electroporation. The expression of the λ *red* operon in the *E. coli* cells results in homologous recombination between the cosmid and the flanking sequences on the PCR product. (*c*) Cosmids in which the gene X has been replaced by the bifunctional marker can be selected for zeocin resistance in addition to the marker previously present on the cosmid. (*d*) The cosmid X1-KO is used to transform an *A. fumigatus pyrG* strain. The large flanking regions lead to a high rate of homologous recombination so that fewer transformants need to be screened. (*e*) The method produces deletion strains that are uracil prototrophic.

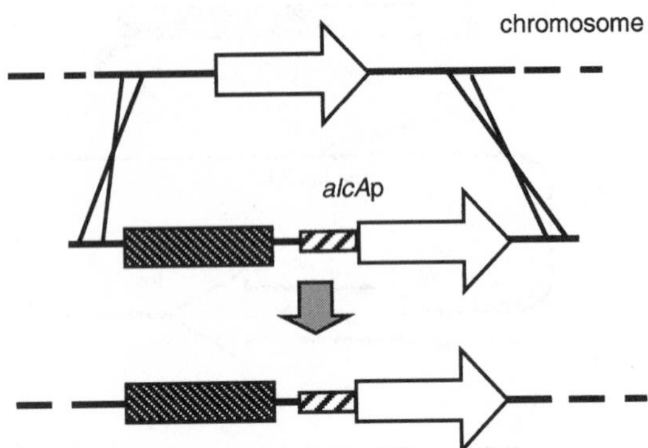

Figure 6 Schematic representation of a method to artificially induce a gene of interest via the *alcA* promoter region. The *open arrow* marks a gene of interest on a fungal chromosome. The *dark hatched box* represents a selectable marker gene. Following a double homologous recombination, the gene of interest is replaced by a copy of the same gene under the control of the *alcA* promoter and the selectable marker gene. In this strain the gene of interest can be induced by carbon sources such as threonine and repressed by glucose. It is unlikely, however, that the *alcA* promoter without the presence of the *alcR* gene is active in *A. fumigatus*.

under the control of an inducible promoter such as the *A. nidulans alcA* gene promoter (Figure 6). For transformation a construct similar to the deletion construct is required, except that in addition to the selection marker the gene of interest is placed under the control of an inducible promoter that is located between the two flanking regions. In *A. nidulans* the *alcA* promoter can be induced by e.g., L-threonine and is repressed by glucose. In this manner transformants can be selected under conditions where the gene is active and the effect of a nonfunctional gene can be observed under other conditions. One drawback of this system is that the *alcA* promoter is slightly "leaky," allowing a small amount of expression even under repressing conditions. To date no strong inducible promoters are known for *A. fumigatus*. This problem needs to be addressed.

Reporter Gene Technology

What mechanisms regulate gene expression in *A. fumigatus*? More specifically, how are genes involved in virulence regulated? Answers to these questions would add significantly to our understanding of *A. fumigatus* and the factors responsible for making it an opportunistic pathogen. In other organisms reporter genes have been established as important genetic tools for analyzing the expression of particular genes under various conditions (12, 50, 66, 78, 83). Reporter genes have also been used to identify regions in the upstream sequence of a gene, which are important for regulating gene expression (e.g., 76, 86). Application of the

β-galactosidase-encoding *E. coli lacZ* gene in *A. fumigatus* was demonstrated for the first time by Smith et al. (74). They fused the *A. fumigatus* alkaline protease (*alp*) gene with the *E. coli lacZ* gene to demonstrate that the *alp* gene is expressed during lung colonization in a murine infection model. An example showing the use of the *lacZ* gene of *E. coli* in *A. fumigatus* is shown in Figure 7*A* (see color insert). A reporter gene used extensively in *A. nidulans* (66) is the *uidA* gene encoding ß-glucuronidase. It could also be functional in *A. fumigatus* (31).

Recently Langfelder et al. (50) showed that another popular reporter gene, *egfp*, derived from the jellyfish *Aequorea victoria*, can also be used to study gene expression in *A. fumigatus* (Figure 7*B*). *egfp* encodes the enhanced green fluorescent protein (EGFP). *egfp* is especially well suited to studying protein localization and expression during infection because EGFP requires no additional cofactors. In the study the authors used *egfp* gene fusions to demonstrate differential expression and the expression of a virulence determinant during the infection process in a murine infection model. Taken together, these two reporter genes represent powerful tools for investigating gene expression in *A. fumigatus* and should provide new insights into the regulation of potential virulence determinants.

Ultimately the aim is to understand all the mechanisms controlling the expression of a gene. This includes a detailed analysis of the gene's upstream region, potential regulatory elements, and proteins binding to these elements. It is possible to use reporter genes, such as *lacZ*, to measure promoter activity (more difficult with *egfp* than with *lacZ*) by fusing the upstream region of the gene of interest with *lacZ*. In order to compare different upstream regions, possibly with deletions and/or point mutations, it is essential to integrate the constructs at a defined chromosomal locus. A system allowing the integration at the *pyrG* locus of *A. fumigatus* was developed by Weidner et al. (85) (Figure 8). The method relies upon using two mutant alleles of the *pyrG* gene, *pyrG1* and *pyrG2*, neither of which encodes a functional protein. The mutations in *pyrG1* and *pyrG2* are located at opposite ends of the gene. A uracil-auxotrophic *A. fumigatus* strain bearing the *pyrG1* allele is transformed with a plasmid bearing the *pyrG2* allele, and the selected transformants are uracil prototroph (i.e., do not require additional uracil). Such strains will have a functional *pyrG* gene, which can either result from a gene conversion of the *pyrG1* mutation or by integration of the plasmid via single crossover at the *pyrG* locus (Figure 8). The latter case occurs in about 45% of transformants and results in a targeted integration of the plasmid at the *pyrG* locus (85). Another locus that could be used for targeting gene fusions is the *sC* locus (26).

A useful tool that can be used in conjunction with the techniques described above is a method for generating point mutations in promoter regions by PCR. The method pioneered by Higuchi et al. (44) uses two sets of oligonucleotide primers that bind at each end of the promoter region to be studied as well as in the region where one or several point mutations are to be made. The oligonucleotide primers binding at the mutation site overlap and are oriented in opposite directions. They must be designed to contain the necessary mutation. Using these primers two independent PCR reactions are carried out. Both products will carry the point mutation. The

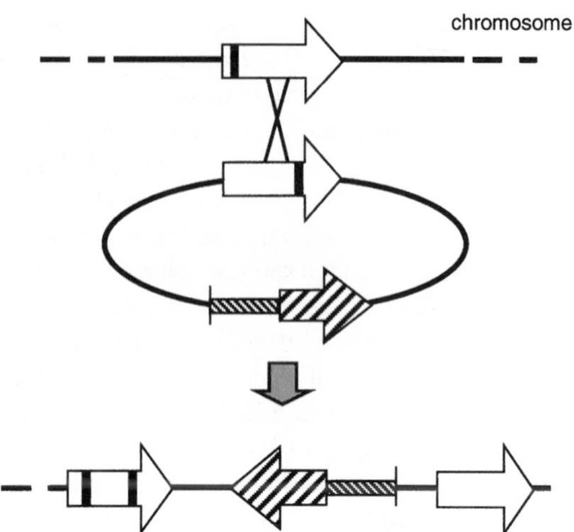

Figure 8 Schematic representation of site-directed integration. The *open arrow* represents a selectable gene on one of the fungal chromosomes (e.g., *pyrG*). The same gene is present on the plasmid used for transformation. Both the chromosomal copy and the plasmid copy of the selectable gene carry different mutations (indicated by *black bars*). Neither copy encodes a functional protein. Integration of the plasmid at the site of the selectable gene (e.g., *pyrG*) by single homologous recombination results in two copies of the gene, one of which is intact while the other carries both mutations. The *hatched arrow* marks a gene of interest, e.g., a reporter gene fusion with a promoter region to be analyzed.

PCR products are purified and mixed together. The mixture of PCR products is then denatured by heating to 94°C and allowed to cool slowly to room temperature. As the two PCR products overlap slightly a small proportion of fragments will form heterologous pairs (by far the greater number will bind to their complementary sequence). A second PCR reaction using the primers lying on the outside of the template will amplify only those PCR fragments that have formed heterologous pairs. The result of this PCR will be a full-length product that contains the point mutation required. This can then be cloned upstream of the *lacZ* reporter gene and transformed into *A. fumigatus* to measure the effect on gene expression. Although this method has not yet been described for *A. fumigatus*, it was used in *A. nidulans*, e.g., to analyze the promoter of the *hapB* gene (76).

Applied Transformation Technologies

SIGNATURE TAGGED MUTAGENESIS Signature tagged mutagenesis (STM) is a method used to screen for genes essential for infectious growth in pathogenic bacteria (42, 43). In STM large numbers of mutants are created by insertional mutagenesis using, e.g., REMI. Each DNA molecule inserted carries an individual signature tag, which can be identified by hybridization. Brown et al. (16) adapted the method to study virulence determinants in *A. fumigatus*. They created several pools of mutants and used these to infect immunosuppressed mice. The mutant strains were then reisolated from the lungs and compared with the strains used to infect the mice. Strains that could not be reisolated potentially carried a deletion in a gene essential for infectious growth of *A. fumigatus*. Using this technique Brown et al. (16) were able to identify the *A. fumigatus pabaA* gene that encodes *para*-aminobenzoic acid synthetase. Furthermore, they showed that the gene is essential for invasive growth. However, *pabaA* can hardly be termed a virulence factor, and no virulence factors have been identified to date using STM. One explanation could be that *A. fumigatus* does not have any major virulence factors but does have a number of physiological characteristics (or "virulence determinants") such as the *pksP* gene enabling it to be an opportunistic pathogen (52). Alternatively, the STM method described here is simply not sensitive enough to identify genes that might have only a small effect on virulence. It is possible that a coinfection model of aspergillosis would lead to better results (51), allowing this method to be used more effectively on *A. fumigatus*. For coinfection mice are infected simultaneously with equal doses of wild-type and mutant strains, resulting in the death of the mice. Fungal cells are then isolated from the mice and the ratio of wild-type to mutant colony-forming units is observed. A significant difference would indicate that the mutant strain is defective in some function involved in pathogenicity.

IN VIVO EXPRESSION TECHNOLOGY In vivo expression technology (IVET) was originally developed in bacteria to identify promoter regions that would lead to efficient expression during the infection process (54). This technique could be adapted for use in *A. fumigatus*, as was accomplished by Staib et al. (75) for *Candida albicans*. Two different approaches can be used for IVET. The first approach investigates a specific promoter region, as was done with *C. albicans* (75). In this case the promoter of interest is fused in frame with a marker gene that allows the identification of strains expressing this reporter gene during infection. Possible markers for *A. fumigatus* are the genes *pyrG*, *pabaA*, and *sC* in the relevant deletion strains. Following transformation of the relevant strain (using a different selectable marker), transformants are used in an *A. fumigatus* infection model. Strains grow only if the promoter fused with the marker gene is active during the infection process. Promoters that are not active under normal conditions but are specifically activated during infection could be involved in virulence.

The second approach is to generate a library of *A. fumigatus* mutants in which random fragments of fungal DNA are fused with the relevant marker gene (31). The strains are used in an infection model of aspergillosis. Again, only those strains can survive where the DNA region fused with the marker gene leads to expression during the infection process. If such DNA fragments do not lead to expression under normal conditions, they could indicate that the attached gene is involved in the infection process. It is advisable to use targeted integration (e.g., to the chromosomal *pyrG* locus) for both approaches to ensure that no external factors influence marker expression. A possible problem could result from the sensitivity of this method because it distinguishes only between growth and no growth.

GENOMICS

The complete genome of *A. fumigatus* is currently being sequenced (see http://www.tigr.org/; http://www.sanger.ac.uk/). Based on preliminary sequencing data, the genome of *A. fumigatus* was estimated to consist of 28.7 Mb (http://www.tigr.org/). In parallel to the progress of the sequencing, data have been already released. The available sequence data can be expected to have a major impact on *A. fumigatus* research. These data will allow comparative genomics to answer questions about the differences between pathogenic and nonpathogenic *Aspergillus* species. Furthermore, they will allow large-scale biological studies (transcriptome and proteome analyses) to identify genes/proteins that are specifically transcribed/produced during the infectious process and may represent factors unique to *A. fumigatus*. The genome information can also be expected to help to identify targets for the generation of novel drugs.

CONCLUSION

As we have shown here, the past years have seen a dramatic increase in the range of tools available for studying the opportunistic fungal pathogen *A. fumigatus*. Methods have been established for transforming the fungus. These can be used in conjunction with the existing selection systems for *A. fumigatus* or with systems that are currently used in other fungi such as *A. nidulans* and *A. niger* and can be adapted for *A. fumigatus*. It might be possible to establish a library of auxotrophic *A. fumigatus* mutants in order to facilitate working with the parasexuality of *A. fumigatus*. Recently developed tools for *A. fumigatus* include techniques for the rapid generation of deletion strains and new reporter genes. Working with these tools, improving the techniques available, and developing new approaches will enable researchers to carry out a detailed study of *A. fumigatus*. The genomic sequence of *A. fumigatus* that will soon become available will provide the *A. fumigatus* community with an additional powerful tool and allow for a better understanding of the fungus in general and factors involved in virulence in particular. Eventually this will allow the identification of new targets for antifungal substances and the development of new treatments to prevent the high mortality rates associated with a systemic aspergillosis.

ACKNOWLEDGMENTS

We gratefully acknowledge Bernhard Jahn for many helpful discussions. We thank Burghard Liebmann, and Stephanie Gattung for providing data prior to publication. The work in the authors' laboratory is supported by the Deutsche Forschungsgemeinschaft (DFG).

The *Annual Review of Microbiology* is online at http://micro.annualreviews.org

LITERATURE CITED

1. Aleksenko A, Clutterbuck AJ. 1997. Autonomous plasmid replication in *Aspergillus nidulans*: AMA1 and MATE elements. *Fungal. Genet. Biol.* 21:373–87
2. Amaar YG, Moore MM. 1998. Mapping of the nitrate-assimilation gene cluster (*crnA-niiA-niaD*) and characterization of the nitrite reductase gene (*niiA*) in the opportunistic fungal pathogen *Aspergillus fumigatus*. *Curr. Genet.* 33:206–15
3. Avalos J, Geever RF, Case ME. 1989. Bialaphos resistance as a dominant selectable marker in *Neurospora crassa*. *Curr. Genet.* 16:369–72
4. Ballance DJ, Turner G. 1985. Development of a high-frequency transforming vector for *Aspergillus nidulans*. *Gene* 36: 321–31
5. Berg CM, Garber ED. 1962. A genetic analysis of color mutants of *Aspergillus fumigatus*. *Genetics* 47:1139–46
6. Birch M, Rickers A, Brookman JL. 2001. Transposon mediated mutagenesis in *Aspergillus fumigatus*. *Symp. Physiol. Yeasts Filamentous Fungi*. Denmark. p. 96. (Abstr.)
7. Bird D, Bradshaw R. 1997. Gene targeting is locus dependent in the filamentous fungus *Aspergillus nidulans*. *Mol. Gen. Genet.* 255:219–25
8. Boeke JD, LaCroute F, Fink GR. 1984. A positive selection for mutants lacking orotidine-5′-phosphate decarboxylase activity in yeast: 5-fluoro-orotic acid resistance. *Mol. Gen. Genet.* 197:345–46
9. Borgia PT, Dodge CL, Eagleton LE, Adams TH. 1994. Bidirectional gene transfer between *Aspergillus fumigatus* and *Aspergillus nidulans*. *FEMS Microbiol. Lett.* 122:227–31
10. Brakhage AA, Jahn B. 2002. Molecular mechanisms of pathogenicity of *Aspergillus fumigatus*. In *Molecular Biology of Fungal Development*, ed. HD Osiewacz. New York: Dekker. In press
11. Brakhage AA, Jahn B, Schmidt A, eds. 1999. *Aspergillus fumigatus*. Biology, clinical aspects and molecular approaches to pathogenecity. In *Contributions to Microbiology, Vol. 2*. Basel: Karger. 221 pp.
12. Brakhage AA, Van den Brulle J. 1995. Use of reporter genes to identify recessive *trans*-acting mutations specifically involved in the regulation of *Aspergillus nidulans* penicillin biosynthesis genes. *J. Bacteriol.* 177:2781–88
13. Braude AI. 1991. The aspergilli. In *Infectious Disease and Medical Microbiology*, ed. AI Braude, CE Davis, J Fierer, 2:592–97. Philadelphia: Saunders
14. Brody H, Carbon J. 1989. Electrophoretic karyotype of *Aspergillus nidulans*. *Proc. Natl. Acad. Sci. USA* 86:6260–63
15. Brookman JL, Denning DW. 2000. Molecular genetics in *Aspergillus fumigatus*. *Curr. Opin. Microbiol.* 3:468–74
16. Brown JS, Aufauvre-Brown A, Brown J, Jennings JM, Arst HN Jr, Holden DW. 2000. Signature-tagged and directed mutagenesis identify PABA synthetase as essential for *Aspergillus fumigatus* pathogenicity. *Mol. Microbiol.* 36:1371–80
17. Brown JS, Aufauvre-Brown A, Holden DW. 1998. Insertional mutagenesis of *Aspergillus fumigatus*. *Mol. Gen. Genet.* 259:327–35
18. Buxton FP, Gwynne DI, Davies RW. 1985. Transformation of *Aspergillus niger* using the *argB* gene of *Aspergillus nidulans*. *Gene* 37:207–14
19. Buxton FP, Gwynne DI, Davies RW. 1989. Cloning of a new bidirectionally selectable marker for *Aspergillus* strains. *Gene* 84:329–34
20. Chakraborty BN, Kapoor M. 1990. Transformation of filamentous fungi by electroporation. *Nucleic Acids Res.* 18:6737

21. Chakraborty BN, Patterson NA, Kapoor M. 1991. An electroporation-based system for high-efficiency transformation of germinated conidia of filamentous fungi. *Can. J. Microbiol.* 37:858–63
22. Chaveroche MK, Ghigo JM, d'Enfert C. 2000. A rapid method for efficient gene replacement in the filamentous fungus *Aspergillus nidulans. Nucleic Acids Res.* 28:E97, 1–6
23. Daboussi MJ, Djeballi A, Gerlinger C, Blaiseau PL, Bouvier I, et al. 1989. Transformation of seven species of filamentous fungi using the nitrate reductase gene of *Aspergillus nidulans. Curr. Genet.* 15:453–56
24. Debeaupuis JP, Sarfati J, Chazalet V, Latgé JP. 1997. Genetic diversity among clinical and environmental isolates of *Aspergillus fumigatus. Infect. Immun.* 65: 3080–85
25. de Groot MJ, Bundock P, Hooykaas PJ, Beijersbergen AG. 1998. *Agrobacterium tumefaciens*-mediated transformation of filamentous fungi. *Nat. Biotechnol.* 16: 839–42
26. de Lucas JR, Domínguez AI, Higuero Y, Martinez O, Romero B, et al. 2001. Development of a homologous transformation system for the opportunistic human pathogen *Aspergillus fumigatus* based on the *sC* gene encoding ATP sulfurylase. *Arch. Microbiol.* 176:106–13
27. de Lucas JR, Domínguez AI, Mendoza A, Laborda F. 1998. Use of flow-cytometry to distinguish between haploid and diploid strains of *Aspergillus fumigatus. Fung. Genet. Newsl.* 45:7–9
28. d'Enfert C. 1996. Selection of multiple disruption events in *Aspergillus fumigatus* using the orotidine-5′-decarboxylase gene, *pyrG*, as a unique transformation marker. *Curr. Genet.* 30:76–82
29. d'Enfert C, Diaquin M, Delit A, Wuscher N, Debeaupuis JP, et al. 1996. Attenuated virulence of uridine-uracil auxotrophs of *Aspergillus fumigatus. Infect. Immun.* 64:4401–5
30. d'Enfert C, Fontaine T. 1997. Molecular characterization of the *Aspergillus nidulans treA* gene encoding an acid trehalase required for growth on trehalose. *Mol. Microbiol.* 24:203–16
31. d'Enfert C, Weidner G, Mol PC, Brakhage AA. 1999. Transformation systems of *Aspergillus fumigatus*: new tools to investigate fungal virulence. See Ref. 11, 2:149–66
32. Ellis M. 1999. Therapy of *Aspergillus fumigatus*-related diseases. See Ref. 11, 2:105–29
33. Fierro F, Kosalkova K, Gutierrez S, Martin JF. 1996. Autonomously replicating plasmids carrying the AMA1 region in *Penicillium chrysogenum. Curr. Genet.* 29: 482–89
34. Fincham JR. 1989. Transformation in fungi. *Microbiol. Rev.* 53:148–70
35. Gatignol A, Durand H, Tiraby G. 1988. Bleomycin resistance conferred by a drug-binding protein. *FEBS Lett.* 230:171–75
36. Gems DH, Clutterbuck AJ. 1993. Co-transformation with autonomously-replicating helper plasmids facilitates gene cloning from an *Aspergillus nidulans* gene library. *Curr. Genet.* 24:520–24
37. Gems DH, Johnstone IL, Clutterbuck AJ. 1991. An autonomously replicating plasmid transforms *Aspergillus nidulans* at high frequency. *Gene* 98:61–67
38. Girardin H, Monod M, Latgé JP. 1995. Molecular characterization of the food-borne fungus *Neosartorya fischeri* (Malloch and Cain). *Appl. Environ. Microbiol.* 61:1378–83
39. Gonzalez A, Jimenez A, Vazquez D, Davies JE, Schindler D. 1978. Studies on the mode of action of hygromycin B, an inhibitor of translocation in eukaryotes. *Biochim. Biophys. Acta* 521:459–69
40. Gouka RJ, Gerk C, Hooykaas PJ, Bundock P, Musters W, et al. 1999. Transformation of *Aspergillus awamori* by *Agrobacterium tumefaciens*-mediated homologous recombination. *Nat. Biotechnol.* 17:598–601

41. Grimm C, Kohli J, Murray J, Maundrell K. 1988. Genetic engineering of *Schizosaccharomyces pombe*: a system for gene disruption and replacement using the *ura4* gene as a selectable marker. *Mol. Gen. Genet.* 215:81–86
42. Hensel M, Holden DW. 1996. Molecular genetic approaches for the study of virulence in both pathogenic bacteria and fungi. *Microbiology* 142:1049–58
43. Hensel M, Shea JE, Gleeson C, Jones MD, Dalton E, Holden DW. 1995. Simultaneous identification of bacterial virulence genes by negative selection. *Science* 269:400–3
44. Higuchi R, Krummel B, Saiki RK. 1988. A general method of *in vitro* preparation and specific mutagenesis of DNA fragments: study of protein and DNA interactions. *Nucleic Acids Res.* 16:7351–67
45. Holden DW, Tang CM, Smith JM. 1994. Molecular genetics of *Aspergillus* pathogenicity. *Antonie van Leeuwenhoek* 65: 251–55
46. Jahn B, Koch A, Schmidt A, Wanner G, Gehringer H, et al. 1997. Isolation and characterization of a pigmentless-conidium mutant of *Aspergillus fumigatus* with altered conidial surface and reduced virulence. *Infect. Immun.* 65:5110–17
47. Kempken F, Jacobsen S, Kück U. 1998. Distribution of the fungal transposon Restless: full-length and truncated copies in closely related strains. *Fungal Genet. Biol.* 25:110–18
48. Kempken F, Kück U. 2000. Tagging of a nitrogen pathway-specific regulator gene in *Tolypocladium inflatum* by the transposon Restless. *Mol. Gen. Genet.* 263:302–8
48a. Langfelder K, Gattung S, Brakhage AA. 2002. A novel method used to delete a new *Aspergillus fumigatus* ABC transporter-encoding gene. *Curr. Genet.* In press
49. Langfelder K, Jahn B, Gehringer H, Schmidt A, Wanner G, Brakhage AA. 1998. Identification of a polyketide synthase gene (*pksP*) of *Aspergillus fumigatus* involved in conidial pigment biosynthesis and virulence. *Med. Microbiol. Immunol.* 187:79–89
50. Langfelder K, Philippe B, Jahn B, Latgé JP, Brakhage AA. 2001. Differential expression of the *Aspergillus fumigatus pksP* gene detected *in vitro* and *in vivo* with green fluorescent protein. *Infect. Immun.* 69:6411–18
51. Latgé JP. 1999. *Aspergillus fumigatus* and aspergillosis. *Clin. Microbiol. Rev.* 12: 310–50
52. Latgé JP. 2001. The pathobiology of *Aspergillus fumigatus*. *Trends Microbiol.* 9: 382–89
53. Losberger C, Ernst JF. 1989. Sequence and transcript analysis of the *C. albicans URA3* gene encoding orotidine-5′-phosphate decarboxylase. *Curr. Genet.* 16:153–58
54. Mahan MJ, Tobias JW, Slauch JM, Hanna PC, Collier RJ, Mekalanos JJ. 1995. Antibiotic-based selection for bacterial genes that are specifically induced during infection of a host. *Proc. Natl. Acad. Sci. USA* 92:669–73
55. Martinelli SD. 1994. Progress in industrial microbiology. In Aspergillus: *50 Years On*, Vol. 29, ed. RD Martinelli, JR Kinghorn. London: Elsevier. 851 pp.
56. Mayorga ME, Timberlake WE. 1992. The developmentally regulated *Aspergillus nidulans wA* gene encodes a polypeptide homologous to polyketide and fatty acid synthases. *Mol. Gen. Genet.* 235:205–12
57. Mellado E, Aufauvre-Brown A, Gow NA, Holden DW. 1996. The *Aspergillus fumigatus chsC* and *chsG* genes encode class III chitin synthases with different functions. *Mol. Microbiol.* 20:667–79
58. Mullins J, Harvey R, Seaton A. 1976. Sources and incidence of airborne *Aspergillus fumigatus* (Fres). *Clin. Allergy* 6:209–17
59. Muyrers JP, Zhang Y, Stewart AF. 2001. Techniques: recombinogenic engineering—new options for cloning and manipulating DNA. *Trends Biochem. Sci.* 26: 325–31

60. Oakley BR, Rinehart JE, Mitchell BL, Oakley CE, Carmona C, et al. 1987. Cloning, mapping and molecular analysis of the *pyrG* (orotidine-5′-phosphate decarboxylase) gene of *Aspergillus nidulans*. *Gene* 61:385–99
61. Ozeki K, Kyoya F, Hizume K, Kanda A, Hamachi M, Nunokawa Y. 1994. Transformation of intact *Aspergillus niger* by electroporation. *Biosci. Biotechnol. Biochem.* 58:2224–27
62. Pardo JM, Malpartida F, Rico M, Jimenez A. 1985. Biochemical basis of resistance to hygromycin B in *Streptomyces hygroscopicus*—the producing organism. *J. Gen. Microbiol.* 131:1289–98
63. Peberdy JF. 1979. Fungal protoplasts: isolation, reversion, and fusion. *Annu. Rev. Microbiol.* 33:21–39
64. Peterson SW. 1992. *Neosartorya pseudofischeri* sp. Nov. and its relationship to other species in *Aspergillus* section Fumigati. *Mycol. Res.* 96:547–54
65. Pitt JI. 1994. The current role of *Aspergillus* and *Penicillium* in human and animal health. *J. Med. Vet. Mycol.* 32 (Suppl.)1:17–32
66. Punt PJ, Greaves PA, Kuyvenhoven A, van Deutekom JC, Kinghorn JR, et al. 1991. A twin-reporter vector for simultaneous analysis of expression signals of divergently transcribed, contiguous genes in filamentous fungi. *Gene* 104:119–22
67. Punt PJ, Oliver RP, Dingemanse MA, Pouwels PH, van den Hondel CAMJJ. 1987. Transformation of *Aspergillus* based on the hygromycin B resistance marker from *Escherichia coli*. *Gene* 56:117–24
68. Punt PJ, van den Hondel CAMJJ. 1992. Transformation of filamentous fungi based on hygromycin B and phleomycin resistance markers. *Methods Enzymol.* 216:447–57
69. Rüchel R, Reichard U. 1999. Pathogenesis and clinical presentation of aspergillosis. See Ref. 11, 2:21–43
70. Samson RA. 1999. The genus *Aspergillus* with special regard to the *Aspergillus fumigatus* group. See Ref. 11, 2:25–20
71. Sanchez O, Navarro RE, Aguirre J. 1998. Increased transformation frequency and tagging of developmental genes in *Aspergillus nidulans* by restriction enzyme-mediated integration (REMI). *Mol. Gen. Genet.* 258:89–94
72. Sanchez-Fernandez R, Unkles SE, Campbell EI, Macro JA, Cerda-Olmedo E, Kinghorn JR. 1991. Transformation of the filamentous fungus *Gibberella fujikuroi* using the *Aspergillus niger niaD* gene encoding nitrate reductase. *Mol. Gen. Genet.* 225:231–33
73. Schmitt HJ, Blevins A, Sobeck K, Armstrong D. 1990. *Aspergillus* species from hospital air and from patients. *Mycoses* 33:539–41
74. Smith JM, Tang CM, Van Noorden S, Holden DW. 1994. Virulence of *Aspergillus fumigatus* double mutants lacking restrictocin and an alkaline protease in a low-dose model of invasive pulmonary aspergillosis. *Infect. Immun.* 62:5247–54
75. Staib P, Kretschmar M, Nichterlein T, Kohler G, Michel S, et al. 1999. Host-induced, stage-specific virulence gene activation in *Candida albicans* during infection. *Mol. Microbiol.* 32:533–46
76. Steidl S, Hynes MJ, Brakhage AA. 2001. The *Aspergillus nidulans* multimeric CCAAT binding complex AnCF is negatively autoregulated via its *hapB* subunit gene. *J. Mol. Biol.* 306:643–53
77. Stroemnaes I, Garber ED. 1962. Heterocaryosis and the parasexual cycle in *Aspergillus fumigatus*. *Genetics* 48:653–62
78. Suelmann R, Sievers N, Fischer R. 1997. Nuclear traffic in fungal hyphae: *in vivo* study of nuclear migration and positioning in *Aspergillus nidulans*. *Mol. Microbiol.* 25:757–69
79. Tang CM, Smith JM, Arst HN Jr, Holden DW. 1994. Virulence studies of *Aspergillus nidulans* mutants requiring lysine or *p*-aminobenzoic acid in invasive

pulmonary aspergillosis. *Infect. Immun.* 62:5255–60

80. Tobin MB, Peery RB, Skatrud PL. 1997. An electrophoretic molecular karyotype of a clinical isolate of *Aspergillus fumigatus* and localization of the MDR-like genes *AfuMDR1* and *AfuMDR2*. *Diagn. Microbiol. Infect. Dis.* 29:67–71
81. Unkles SE, Campbell EI, Carrez D, Grieve C, Contreras R, et al. 1989. Transformation of *Aspergillus niger* with the homologous nitrate reductase gene. *Gene* 78:157–66
82. Vanden Bossche H, Mackenzie DWR, Cauwenbergh G, eds. 1988. Aspergillus *and Aspergillosis*. New York: Plenum
83. van Gorcom RF, Pouwels PH, Goosen T, Visser J, van den Broek HW, et al. 1985. Expression of an *Escherichia coli* beta-galactosidase fusion gene in *Aspergillus nidulans*. *Gene* 40:99–106
84. van Hartingsveldt W, Mattern IE, van Zeijl CM, Pouwels PH, van den Hondel CAMJJ. 1987. Development of a homologous transformation system for *Aspergillus niger* based on the *pyrG* gene. *Mol. Gen. Genet.* 206:71–75
85. Weidner G, d'Enfert C, Koch A, Mol PC, Brakhage AA. 1998. Development of a homologous transformation system for the human pathogenic fungus *Aspergillus fumigatus* based on the *pyrG* gene encoding orotidine 5′-monophosphate decarboxylase. *Curr. Genet.* 33:378–85
86. Weidner G, Steidl S, Brakhage AA. 2001. The *Aspergillus nidulans* homoaconitase gene *lysF* is negatively regulated by the multimeric CCAAT-binding complex AnCF and positively regulated by GATA sites. *Arch. Microbiol.* 175:122–32
87. Wilson LJ, Carmona CL, Ward M. 1988. Sequence of the *Aspergillus niger pyrG* gene. *Nucleic Acids Res.* 16:2339
88. Zhang Y, Muyrers JP, Testa G, Stewart AF. 2000. DNA cloning by homologous recombination in *Escherichia coli*. *Nat. Biotechnol.* 18:1314–17

Annu. Rev. Microbiol. 2002. 56:457–87
doi: 10.1146/annurev.micro.56.012302.160634

First published online as a Review in Advance on May 10, 2002

WHAT ARE BACTERIAL SPECIES?

Frederick M. Cohan
Department of Biology, Wesleyan University, Middletown, Connecticut 06459-0170; e-mail: fcohan@wesleyan.edu

Key Words speciation, ecotype, species concept, genetic exchange, systematics, MLST

■ **Abstract** Bacterial systematics has not yet reached a consensus for defining the fundamental unit of biological diversity, the species. The past half-century of bacterial systematics has been characterized by improvements in methods for demarcating species as phenotypic and genetic clusters, but species demarcation has not been guided by a theory-based concept of species. Eukaryote systematists have developed a universal concept of species: A species is a group of organisms whose divergence is capped by a force of cohesion; divergence between different species is irreversible; and different species are ecologically distinct. In the case of bacteria, these universal properties are held not by the named species of systematics but by ecotypes. These are populations of organisms occupying the same ecological niche, whose divergence is purged recurrently by natural selection. These ecotypes can be discovered by several universal sequence-based approaches. These molecular methods suggest that a typical named species contains many ecotypes, each with the universal attributes of species. A named bacterial species is thus more like a genus than a species.

CONTENTS

INTRODUCTION

These are formative times for systematics. Owing to recent technological developments, the study of biological diversity has matured into a powerful science. Systematists have invented robust methods for deriving evolutionary trees (35) as well as powerful computer algorithms for implementing these methods (91). Also, the technology for DNA sequencing is readily accessible, so that systematic research today is typically based on sequence variation and sequence-derived phylogenies. Together, the new methods of systematics and a wealth of sequence data are allowing systematists to reconstruct the history of life and the origins of biodiversity with confidence.

For microbiologists, the new systematics has been particularly fruitful. Microbial systematists have built a universal tree of life, such that any newly discovered organism can be placed near its closest relatives (18, 104). Sequence surveys have fostered discovery of new bacterial taxa at all levels. Sequence data have frequently turned up organisms with no known close relatives; these organisms represent new divisions within the bacterial world (18). At the other extreme, sequence surveys have fostered the discovery of new species. For example, sequence data revealed the Lyme disease spirochete *Borrelia burgdorferi* (sensu lato) to consist of several species, each with its own etiology (7, 8). Sequence-based approaches have allowed systematists to characterize the species diversity even among uncultured bacteria, and species names have now been given to many uncultured organisms, pending further characterization (62).

Despite these remarkable successes, however, bacterial systematics has not yet reached a consensus for defining the fundamental unit of biological diversity, the species. Bacterial species exist—on this much bacteriologists can agree. Bacteriologists widely recognize that bacterial diversity is organized into discrete phenotypic and genetic clusters, which are separated by large phenotypic and genetic gaps, and these clusters are recognized as species (85). Beyond agreeing on the existence of species, however, bacteriologists differ on operational procedures for identifying species most appropriately (32). Moreover, we fail to agree on whether a bacterial species should be conceived simply as a cluster of phenotypically and genetically similar organisms, or whether we should also expect a species to have special genetic, ecological, evolutionary, or phylogenetic properties.

I argue that there are bacterial taxa called "ecotypes" (15), which share the quintessential set of dynamic properties held by all eukaryotic species. I demonstrate that, alas, the species generally recognized in bacterial systematics do not have these universal properties: Each named "species" appears to contain many ecotypes, each with the dynamic properties of a species. I present several universal sequence-based approaches for discovering ecotypes and recommend a means of incorporating ecotypes into bacterial taxonomy.

BACTERIAL SPECIES AS PHENOTYPIC AND GENETIC CLUSTERS

Bacterial systematics began in much the same way as the systematics of animals and plants. Before there was a widely accepted theory of species, systematists of all organisms identified species simply as phenotypic clusters (87). While macrobiologists generally surveyed morphological characters and microbiologists generally investigated metabolic characters, systematists of all major groups were successful in carving biological diversity into the phenotypic clusters they identified as species (4).

Macrobial and microbial systematics split profoundly with Mayr's publication of the Biological Species Concept in 1944 (57), in which evolutionary theory was incorporated into systematics. The Biological Species Concept and several later concepts of species changed zoologists' and botanists' views of what a species should represent. A species was no longer merely a cluster of similar organisms; a species was now viewed as a fundamental unit of ecology and evolution, with certain dynamic properties. In the case of Mayr's Biological Species Concept, a species was viewed as a group of organisms whose divergence is opposed by recombination between them. However, as we shall see, the mainstream of bacterial systematics has not incorporated theory-based concepts of species.

Instead, the past half-century of bacterial systematics has been characterized by improvements in methods for demarcating species as clusters. Technological advances in assaying phenotypes have increased our ability to discern closely related species. For example, gas chromatographic techniques for characterizing a strain's fatty acid content (81) frequently provide important diagnostic phenotypic characters (67). Automated equipment for assaying metabolism, such as the microtiter plate reader, allows many more strains to be assessed for many more metabolic traits (10, 53).

Bacterial systematists have also developed improved statistical methods for demarcating phenotypic clusters. Numerical taxonomy, developed by Sneath & Sokal (86), was designed as an objective, mathematical approach for demarcating clusters. In this method, large numbers of strains are assayed for many phenotypic traits, including degradation or metabolism of certain chemicals, the ability to produce and survive various antibiotics, the ability to grow on various carbon or nitrogen sources, staining reactions, and morphology (32). The multidimensional space of phenotypic diversity can then be collapsed onto two or three axes, for example, by principal component analysis. Practitioners of numerical taxonomy have clearly illustrated the discrete nature of phenotypic clusters within many bacterial genera (9). Note that phenotype-based numerical taxonomy is ultimately much more than a method for delimiting species; it is also a venture into the natural history of a bacterial group. By studying the phenotypic variation within a taxon, we learn about the ecological diversity that gives meaning to the taxonomic enterprise.

Over the past three decades, bacterial systematists have added molecular techniques to their arsenal for demarcating clusters. In the early 1970s bacteriologists adopted a genomic approach for discovering clusters. With whole-genome DNA-DNA hybridization, systematists could assay the genomic similarity of two strains, as measured by the fraction of their genomes that are homologous. Johnson (44) determined that strains from the same species, as defined by phenotypic clustering, nearly always shared 70% or more of their genomes and that strains from different species nearly always shared less than 70%. A 70% level of homology over the genome was thus adopted as a gold standard for determining whether two strains should be considered different species (100).

Systematists have more recently utilized DNA sequence divergence data, particularly divergence in the 16S rRNA genes, for demarcating species. Stackebrandt & Goebel (88) found that strains that are more than 3% divergent in 16S rRNA are nearly always members of different species, as determined by DNA-DNA hybridization, whereas strains that are less than 3% divergent may or may not be generally members of different species. A cutoff of 3% divergence was therefore recommended as a conservative criterion for demarcating species.

Molecular approaches have made bacterial systematics much more accessible. The existence of multiple species within a group can now be inferred by general molecular techniques, even when we are unaware of the phenotypic and ecological differences between the clusters. In the case of uncultivated bacteria, molecular methods are all that is available for identifying the diversity of species.

Nevertheless, I argue that molecular approaches have been under-utilized in bacterial systematics. This is because the molecular cutoffs for demarcating species have been calibrated to yield the species groupings already determined by phenotypic clustering. The 70% cutoff for DNA-DNA hybridization was calibrated to yield the phenotypic clusters previously recognized as separate species, and the 3% cutoff for 16S rRNA divergence was calibrated to yield the species previously determined by DNA-DNA hybridization and phenotypic clustering. Because bacterial systematics is lacking a theory-based concept of species, all we can do is calibrate each new molecular technique to yield the clusters previously determined by phenotypic criteria. I demonstrate that, if we adopt an ecological and evolutionary theory of species, molecular approaches will give us much more than the phenotypic clusters of yore; we will be able to identify nearly every ecologically distinct population of bacteria.

THE UNIVERSAL DYNAMIC PROPERTIES OF SPECIES

Evolutionary biologists and systematists of animals and plants have widely believed that species are more than just clusters of closely related and similar organisms. Species are believed to have some quintessential, dynamic properties as well. Most fundamentally, genetic diversity within a species is thought to be constrained by one or more forces of cohesion.

The principal insight of Mayr's (57) Biological Species Concept is that there is a reason why organisms form the tight clusters discovered by systematists. Every species has a force of cohesion that hinders genetic divergence among its members, and (at least for highly sexual animal and plant species) this cohesive force is genetic exchange. So long as organisms can successfully interbreed, argued Mayr, they will remain phenotypically and genetically similar, but when they lose the ability to interbreed, they become free to diverge without bound. Mayr thus defined species as a reproductive community, a group of organisms with the potential to interbreed and produce viable and fertile offspring.

The Biological Species Concept was widely accepted by zoologists and botanists for two compelling reasons, one based in theory and the other empirical. First, population genetic theory predicts that recurrent interbreeding between populations tends to homogenize the populations at all gene loci (105). Thus, Mayr (58) argued that even low levels of genetic exchange between populations should limit their genetic divergence [although probably not in the way Mayr envisioned; see (30)]. Second, Mayr pointed out a striking empirical correspondence between the phenotypic clusters previously recognized as species and groups of organisms with the potential to interbreed.

Systematists have pointed out that the Biological Species Concept does not felicitously accommodate hybridization between species of plants (89) and species of bacteria (75, 76). That is, there are many pairs of species (as defined by phenotypic clustering) that occasionally exchange genes yet retain their integrity as distinct phenotypic clusters. In these cases, genetic exchange is clearly not acting effectively as a force of cohesion.

The problem of hybridization has been accommodated by the Cohesion Species Concept by Meglitsch (61) and Templeton (92). According to the Cohesion Species Concept, a species is a group of organisms whose divergence is capped by one or more forces of cohesion. In the case of sexual species, the predominant cohesive force is understood to be genetic exchange. But the cohesion species concept takes into account that genetic exchange between two groups is not always sufficient to restrict their divergence. In cases where two clusters of organisms retain their separate adaptations, in spite of occasional genetic exchange between them, the clusters are considered separate species, according to the Cohesion Species Concept. With regard to highly sexual species, the Cohesion Species Concept may be understood as a paraphrasing of the Biological Species Concept: A species is a group of organisms whose divergence is constrained by genetic exchange. We see that the Cohesion Species Concept is especially useful in accommodating bacterial groups that form separate phenotypic clusters despite recurrent recombination between them.

Another limitation of the Biological Species Concept is that it does not accommodate asexual species. Again, the Cohesion Species Concept proves to be general enough to take into account species that fall outside the box built by the Biological Species Concept. Templeton (92) argues that asexual species are subject to their own powerful force of cohesion. This force is natural selection, which can purge all genetic diversity from an asexual population. Consider the fate of an adaptive mutation, which grants its bearer superior competitive ability compared to all other members of the population. For example, Ferea et al. (29) found that in asexual experimental populations of *Saccharomyces cerevisiae*, evolutionary adaptation to an oxygenated environment involved many mutations, some suppressing anaerobic metabolism and others augmenting oxidative phosphorylation. Each of these mutations was brought to fixation (i.e., to a frequency of 100%) by natural selection. In the absence of recombination, the entire genome of the successful mutant is brought to fixation, and so all the genetic diversity within the population (at all loci) is purged to zero. An asexual species may then be understood as a group of organisms whose divergence is constrained and recurrently reset to zero by intermittent bouts of natural selection. [This diversity-purging process is called "periodic selection" (5).]

What, then, constitutes speciation in the asexual world? A new species is formed when an asexual lineage evolves into a new ecological niche (i.e., uses a different set of resources or microhabitats), such that the new species cannot be extinguished by adaptive mutants from its former population (Figure 1). For example, within

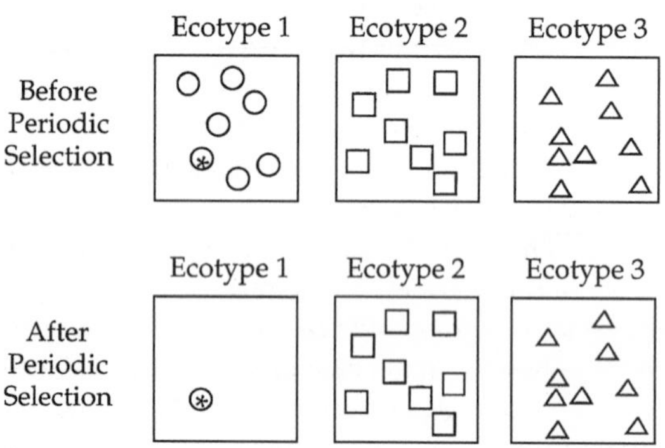

Figure 1 The transience of diversity within an asexual (or rarely sexual) species and the permanence of divergence among species. Each individual symbol represents an individual organism, and the distance between symbols represents the genetic divergence between organisms. The *asterisk* represents an adaptive mutation in one individual of Species 1. Because of the absence (or rarity) of recombination, the adaptive mutant and its clonal descendants replace the rest of the genetic diversity within the species. Genetic diversity within an asexual (or rarely sexual) species is thus transient, awaiting its demise with the next periodic selection event. Because asexual (or rarely sexual) species differ in the resources they use, the adaptive mutant from one species (e.g., Species 1) does not out-compete the organisms of other species, and the genetic diversity within these other species remains untouched. Once populations of organisms are divergent enough to escape one another's periodic selection events, these populations are free to diverge permanently and have reached the status of distinct species. Used with permission from the American Society of Microbiology (14a).

an experimental population of *Escherichia coli* inoculated with a single clone, a mutant population arose that fed on a waste metabolite from the original (95). While the original population and the cross-feeding population continued to coexist, each population endured its own periodic selection events. That is, adaptive mutants appearing in each population purged the diversity within their own populations. Owing to the differences in ecology between populations, however, periodic selection from one population failed to affect the diversity within the other. When two asexual populations reach the point that they can survive each other's periodic selection events, a force of cohesion no longer caps their divergence, and so they may be considered separate species.

In summary, any species, whether highly sexual (such as animals and plants) or completely asexual (such as experimental microbial populations engineered to lack sex), is subject to forces of cohesion. Within highly sexual populations, genetic exchange is a powerful force of cohesion, although genetic exchange between groups that recombine only rarely (as in hybridizing species) does not constrain divergence. Within asexual species, periodic selection is a powerful force of cohesion that recurrently resets the genetic diversity within the species to zero.

While cohesion is the fundamental attribute claimed for all species (13, 15, 61, 92), several corollary attributes follow from cohesion. Because different species are not bound together by any cohesive force, they are free to diverge without constraint from one another. Different species are thus understood to be irreversibly separate, with distinct evolutionary fates [the Evolutionary Species Concept by Simpson (83) and Wiley (103)].

Another corollary is that species occupy different ecological niches either by utilizing different kinds of resources or by utilizing the same kinds of resources at different times or within different microhabitats [the Ecological Species Concept by van Valen (98)]. This is most clearly true for asexual species, as asexual organisms must diverge ecologically before they can escape one another's periodic selection events. However, it is also true for any pair of species, if they are to coexist without constraint from one another. Early twentieth-century ecologists demonstrated that two species cannot coexist unless they differ at least somewhat in the resources they consume [the competitive exclusion principle (31)]. Thus, in the highly sexual world of animals and plants, speciation requires both reproductive divergence and ecological divergence (25, 31, 54, 59).

Although the various species concepts I have discussed emphasize different attributes of species, all modern species concepts endow species with one quintessential property: that species are evolutionary lineages that are irreversibly separate, each with its own evolutionary tendencies and historical fate (19).

Systematists have thus delineated general attributes of species that apply well for organisms at the extremes of sexuality, including species that exchange genes every generation as well as species that never exchange genes. As we shall see, bacteria fall into neither extreme, but nevertheless they fall into species that fit all the universal attributes of species I have delineated.

THE PECULIAR SEXUAL HABITS OF BACTERIA

Genetic exchange in bacteria differs profoundly from that in the highly sexual eukaryotes, the animals and plants, for which species concepts were invented. First, recombination in bacteria is extremely rare in nature. Several laboratories have taken a retrospective approach to determining the historical rate of recombination in nature (27, 28, 79). Based on surveys of diversity in allozymes, restriction recognition sites, and DNA sequences, recombination rates have been estimated from the degree of association between genes or parts of genes [(37, 39, 40); for limitations of this approach, see (16, 56)]. Survey-based approaches have shown that in most cases a given gene segment is involved in recombination at about the same rate or less, as mutation (27, 28, 56, 79, 82, 102). Less commonly, as in the cases of *Helicobacter pylori* and *Neisseria gonorrhoeae*, recombination occurs at least one order of magnitude more frequently than mutation, although survey methods do not allow us to determine by how much (66, 90).

Also in contrast to the case for animals and plants, recombination in bacteria is promiscuous. Whereas animal groups typically lose the ability to exchange genes entirely by the time their mitochondrial DNA sequences are 3% divergent (6), bacteria can undergo homologous recombination with organisms at least as divergent as 25% in DNA sequence (21, 52, 99).

There are, nevertheless, some important constraints on bacterial genetic exchange. Ecological differentiation between populations may prevent them from inhabiting the same microhabitat at the same time, as may be the case, for example, for *Streptococcus pyogenes* populations adapted for throat versus skin infection at different times of year (26). Recombination that depends on vectors, such as transduction or conjugation, is limited by the host ranges of the respective phage and plasmid vectors. Also, restriction-endonuclease activity can greatly reduce the rate of recombination by transduction (24, 60), although not by transformation (17, 94). Finally, homologous recombination is limited by the resistance to integration of divergent DNA sequences because mismatch repair tends to reverse integration of a mismatched donor-recipient heteroduplex (77, 99) and because integration requires a 20–30-bp stretch of nearly perfectly matched DNA (51, 52, 74).

In addition, recombination events in bacteria are localized to a small fraction of the genome. Segments transduced or transformed in the laboratory are frequently less than several kilobases in length (60, 107). Surveys of sequences in nature show that recombination in nature is likewise highly localized within the chromosome (27, 28, 50, 55).

Finally, recombination in bacteria is not limited to the transfer of homologous segments. Bacteria can acquire new gene loci from other organisms, which in some cases are extremely distantly related. This may occur as a side effect of homologous recombination, whereby a heterologous gene from a donor is integrated along with flanking homologous DNA (34, 52). Alternatively, heterologous genes may be integrated along with a transposable element brought into the recipient on a plasmid or phage (97). Genomic analyses have recently shown that a sizeable fraction of bacterial species' genomes (frequently 5%–15%) has typically been acquired from other species (65).

WHY THE BIOLOGICAL SPECIES CONCEPT IS INAPPROPRIATE FOR BACTERIOLOGY

Systematists of animals and plants are indebted to Mayr's (57) Biological Species Concept for the infusion of evolutionary theory into systematics. Even though animal and plant species are usually discovered as phenotypic clusters (87), systematists have understood that there is a force of cohesion that holds each cluster together and that organisms in different clusters are no longer bound by a force of cohesion (92).

Why has bacterial systematics not been similarly transformed by evolutionary theory? It is not for lack of wrestling with the concept. For example, in the early 1960s, Ravin (75, 76) attempted to apply the Biological Species Concept to bacteria. Recognizing that bacteria are sexual and that bacteria can exchange genes even with distant relatives, Ravin (76) defined "genospecies" as groups of bacteria that could exchange genes and "taxospecies" as the phenotypic clusters of mainstream bacterial systematics. In contrast to the case for animals and plants, the genospecies and taxospecies so defined did not correspond well: Many clusters retained their phenotypic distinctness despite their inclusion within a genospecies (76). The lack of correspondence between genospecies and taxospecies suggested that the ability to exchange genes had little effect on the evolution of phenotypic divergence in bacteria. Perhaps discouraged, bacteriologists did not attempt to apply the Biological Species Concept (or any other theory-based concept of species) for another three decades.

Dykhuizen & Green (23) proposed in 1991 to classify bacteria into species according to the Biological Species Concept, delimiting bacterial species as groups of strains that recombine with one another but not with strains from other such groups. They suggested a phylogenetic approach using sequence data, which would identify groups that have and have not been exchanging genes.

One minor problem is that this proposal does not recognize the intrinsic promiscuity of genetic exchange: Bacteria do exchange genes both within and between the clusters we recognize as named species (48, 55). Nevertheless, in practice it is often the case that phylogenetic evidence for recombination is much more common within named bacterial species than between them.

My principal objection to applying the Biological Species Concept to bacteria is that there is no biological motivation for doing so. The Biological Species Concept is appropriate for the highly sexual animals and plants because divergence between two closely related animal or plant populations cannot be permanent until the rate of recombination between the populations is severely reduced compared to the rate of recombination within populations. Owing to the high rate of recombination within animal and plant populations, interpopulation recombination would rapidly

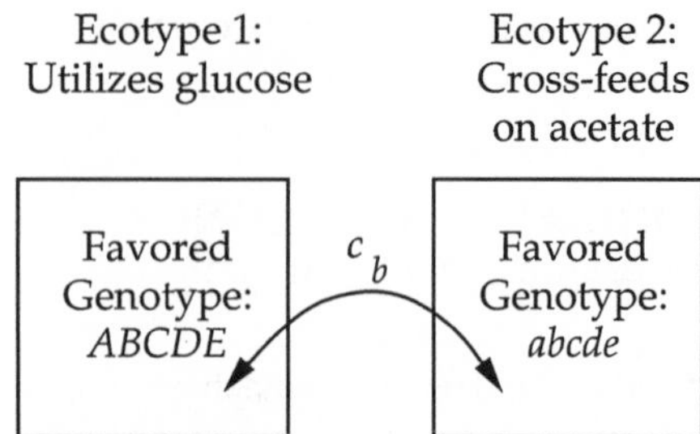

Figure 2 Ecological divergence between ecotypes is stable with respect to recurrent recombination. The genetic basis of ecological divergence among ecotypes is assumed to be due to differences in alleles at several gene loci (as in the figure), or divergence may be due to acquisition of different gene loci. Recombination occurs between ecotypes at rate c_b. The fitness penalty for recombination at any of the genes responsible for ecological divergence is s, such that the fitness of each nonrecombinant genotype (e.g., *ABCDE*) is 1, and the fitness of a single-locus recombinant (e.g., *ABCDe*) is $1 - s$. A mathematical model shows the equilibrium frequency of maladaptive foreign alleles in each ecotype to be c_b/s, which, given the low rate of recombination in bacteria, is a negligible frequency (14). Used with permission from American Society for Microbiology (14a).

eliminate interpopulation divergence if it were to proceed at the same rate as recombination within populations.

In contrast, because recombination in bacteria is so rare, recurrent recombination between bacterial species cannot hinder their divergence (14). Even if recombination between species were to occur at the same rate as recombination within them, natural selection against interspecies recombinants could easily limit the frequency of recombinant genotypes to negligible levels (Figure 2). While the evolution of sexual isolation is an important milestone in the origin of animal and plant species, it is irrelevant to the evolution of permanent divergence in the bacterial world. The Biological Species Concept is thus a red herring for bacterial systematics.

Nevertheless, bacteriologists need not envy the macrobial world for its tidy application of the Biological Species Concept. It turns out that there is an appropriate species concept for bacteria. Moreover, bacteria and eukaryotes both fit comfortably within a universal concept of species.

BACTERIA FORM SPECIES LIKE EVERYONE ELSE

Let us begin by defining a bacterial "ecotype" with respect to the fate of an adaptive mutant (14, 15): An ecotype is a set of strains using the same or similar ecological resources, such that an adaptive mutant from within the ecotype out-competes to extinction all other strains of the same ecotype; an adaptive mutant does not, however, drive to extinction strains from other ecotypes (Figure 1). For example, an adaptive mutant from an ecotype of *Streptococcus pyogenes* that is genetically adapted to infecting our throats would out-compete to extinction other members of its own ecotype but would not out-compete closely related ecotypes genetically adapted to infecting our skin.

If they were entirely asexual, bacterial ecotypes defined in this way would have the universal properties of species. I earlier discussed how asexual populations defined by the domains of periodic selection have all the attributes of species. Each such asexual ecotype would be subject to a force of cohesion (the diversity-purging effect of its own periodic selection events), different ecotypes would be irreversibly separate (free to diverge from one another indefinitely), and the ecotypes would be ecologically distinct.

Let us now add the reality of rare but promiscuous genetic exchange to these ecotypes. Would they still retain the universal qualities of species? We might imagine that periodic selection would not be an effective force of cohesion within a rarely recombining population. Perhaps even rare genetic exchange would rapidly place the adaptive mutation into many genetic backgrounds within the ecotype, such that selection would fail to purge sequence diversity at all loci. However, this is not the case. Under rates of recombination typical of bacteria, selection will purge each locus, on average, of 99.9% of its sequence diversity (13). Thus, natural selection does act as a potential force of cohesion within rarely recombining ecotypes.

In contrast, there is no effective force of cohesion binding different ecotypes. Ecotypes are defined to be free to diverge without the constraint of one another's periodic selection events; moreover, as we have seen, the rare recombination occurring in bacteria is unable to prevent adaptive divergence between ecotypes.

In summary, the bacterial ecotypes defined here share the fundamental properties of species. They are each subject to an intense force of cohesion. Once different ecotypes have diverged to the point of escaping one another's periodic selection events, there is no force that can prevent their divergence; and bacterial ecotypes are ecologically distinct. Bacterial ecotypes are therefore evolutionary lineages that are irreversibly separate, each with its own evolutionary tendencies and historical fate (19, 83, 103). A species in the bacterial world may be understood as an evolutionary lineage bound together by ecotype-specific periodic selection.

BACTERIAL SPECIATION AS AN EVERYDAY PROCESS

How frequently do bacterial populations split irreversibly into lineages with separate evolutionary fates? By applying the principles of population genetics and ecology, we can predict an enormous potential for speciation in the bacterial world, much greater than that in the highly sexual world of animals and plants.

First, speciation in highly sexual eukaryotes requires both reproductive (59) and ecological (25, 31) divergence, but speciation in bacteria requires only ecological divergence (13).

Second, speciation in highly sexual eukaryotes requires allopatry (i.e., that the incipient species inhabit different geographical regions) (58), or at least microallopatry (i.e., that they inhabit different microhabitats) (11, 101). This is because highly sexual populations cannot diverge as long as they are exchanging genes at a high rate; allopatry (or microallopatry) provides the only mechanism for reducing genetic exchange between populations in early stages of speciation. In contrast, as I have shown, genetic exchange is too rare to hinder divergence between bacterial populations, and so the need for allopatry in bacterial speciation is greatly reduced (but not necessarily eliminated, as we shall see).

Third, the extremely large population sizes of bacteria make rare mutation and recombination events much more accessible to a bacterial population than is the case for macroorganisms.

Fourth, whereas each animal and plant species is genetically closed to all other species (except for hybridization with closely related species), a bacterial species is open to gene transfer from many other species, even those that are distantly related (65, 106). So, while animal and plant species must evolve all their adaptations on their own, bacteria can take up existing adaptations from a great diversity of other species. Homologous recombination can substitute an adaptive allele from another species into an existing gene in the recipient (55); recombination can also introduce entirely novel genes and operons from other species (3, 33, 46, 47, 65). By granting an entirely new metabolic function, heterologous gene transfer has the potential to endow a strain with a new resource base, such that the strain and its descendants are instantaneously a new species—beyond the reach of periodic selection within the strain's former population. Since 5%–15% of the genes in a typical bacterial genome have been acquired from other species (65), it is possible that many speciation events in the past have been driven by the acquisition of new genes.

The transfer of adaptations across species is facilitated by the peculiar characteristics of bacterial genetic exchange. Incorporation of highly divergent DNA is fostered not only by the promiscuous nature of bacterial genetic exchange, but also by the localized nature of bacterial recombination, whereby only a small fraction of the donor's genome is integrated. This allows for the transfer of a generally useful adaptation (i.e., useful in the genetic backgrounds and the ecological niches of both the donor and recipient), without the co-transfer of narrowly adapted donor segments that would be deleterious for the recipient (107). This is in contrast to the case for most eukaryotes, where the processes of meiosis and fertilization yield hybrids that are a 1:1 mix of both parents' genomes.

Finally, genetic exchange between ecotypes (48) may enhance speciation by preventing a nascent ecotype from being extinguished by an adaptive mutant from the parental ecotype (15) (Figure 3). This can occur if ecological divergence between incipient ecotypes involves several mutational steps. In the early stages of such divergence, nearly every periodic selection event may be limited to purging the diversity within its own ecotype. Occasionally, however, an extraordinarily fit adaptive mutant from the parental ecotype might out-compete all strains from

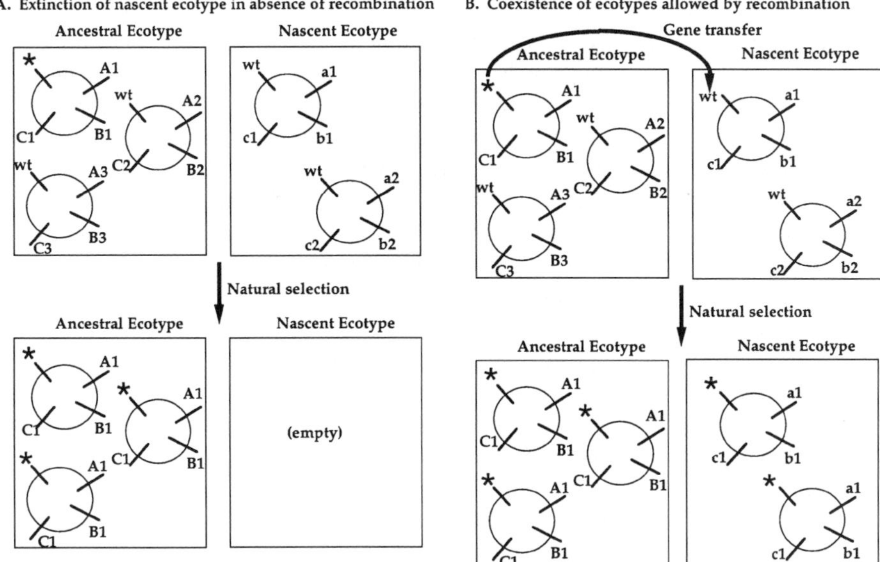

Figure 3 Facilitation of speciation by recombination among ecotypes. It is assumed that each newly divergent ecotype in the figure has already undergone several private periodic selection events. However, in the figure we suppose that an extraordinarily competitive adaptive mutant (with *asterisk*) has appeared in the ancestral ecotype, such that this mutant would out-compete the membership of the nascent ecotype as well as its own ecotype membership. (*A*) When there is no recombination between the newly divergent ecotypes, the adaptive mutant could extinguish the membership of the other ecotype, and the speciation process would be terminated. (*B*) When the adaptive mutation can be transferred from one ecotype to the other, periodic selection is less likely to cause extinction of one ecotype by another. The transfer of the adaptive mutation would cause a private periodic selection event within the nascent ecotype. Because the two ecotypes would then share the adaptive mutation, one ecotype would not be able to extinguish the other. Used with permission from the Society for Systematic Biology (15).

the nascent ecotype (as well as all the other strains from its own ecotype). In this case, the speciation process would be quashed by a periodic selection event before the two incipient ecotypes had diverged sufficiently to be completely free of one another (Figure 3*A*). However, recombination between two incipient species could potentially prevent this (Figure 3*B*). Through genetic exchange, the adaptive mutation could be transferred from the parental ecotype to a recipient in the other ecotype, and the new ecotype would lose its disadvantage.

Although recurrent recombination is not sufficient to prevent ecological divergence between ecotypes (13), recombination should be sufficient to allow an adaptive mutation to pass between ecotypes and enable the recipient ecotype to become fixed for the adaptation by natural selection. Preventing divergence between ecotypes requires recurrent recombination at a high rate, but initiating a natural selection event in a recipient population requires only a single recombinational transfer into the recipient ecotype. Given the enormous population sizes of bacterial populations, such a transfer event is not unlikely.

In summary, population genetic principles suggest that the rare but promiscuous nature of bacterial genetic exchange, as well as the large population sizes of bacteria, should foster a much higher rate of speciation in bacteria than is possible in plants and animals. Nevertheless, important questions remain unresolved.

It is not clear, for example, whether bacterial speciation can proceed without allopatry. As I have discussed, allopatry is unnecessary for evolution of sexual isolation between incipient ecotypes because sexual isolation is not a necessary step in the origin of bacterial ecotypes. However, allopatry may be necessary to give a nascent ecotype a chance to gradually build up its ecological distinctness from the parental ecotype before being exposed to periodic selection from the parental ecotype.

In addition, we do not know the typical source of adaptations that enable invasion of new niches: Is it mutational change in existing genes, or acquisition of new gene loci from other species? Finally, it is not clear how frequently speciation in bacteria actually occurs in nature.

Fortunately, these issues can now be addressed by model experimental systems developed for studying the origins of ecological diversity in the bacterial world. In these model systems, a clone and its descendants are cultured in liquid in the laboratory and are allowed to evolve on their own. In one system, using *E. coli*, bacteria are cultured in a chemostat (95); in another (also using *E. coli*), the bacteria are maintained in serial batch culture (43, 80). In yet another system (using *Pseudomonas fluorescens*), the culture medium is neither replenished nor stirred (72). In all these systems, no extrinsic source of DNA is provided, so novel genes cannot be introduced by horizontal transfer. Moreover, all vectors of recombination have been eliminated from the *E. coli* systems.

From research in all these systems, it appears inevitable that a bacterial clone can evolve into multiple ecotypes by mutation alone. Treves et al. (95) found replicable evolution of a new ecotype, which utilized acetate secreted by the original clone. In experiments in a nonstirred environment, as performed by Rainey & Travisano (72), ecotypes have replicably arisen that are specifically adapted to different parts of the structured environment (i.e., the surface, the bottom, and the water column). In other experiments, molecular markers have demonstrated the existence of a diversity of ecotypes (43). Here, each periodic selection event has purged the diversity in only a subset of the population, indicating that multiple ecotypes are present. In some cases, the putative ecotypes have coexisted over years of evolution (80, 95).

The rate at which new ecotypes can be formed is striking. In the case of the physically structured environment, new ecotypes originated with high replicability in the course of several days. In the unstructured environments of the chemostat and serial batch culture, ecotypes originated within several weeks.

These experiments also demonstrate that allopatry is not required for ecotypes to gradually build up their ecological distinctness. Because some incipient ecotypes have coexisted for years (80, 95), it appears that nascent ecotypes have evolved to escape periodic selection from the parental ecotype without the benefit of allopatry.

The work by Treves et al. is especially notable in demonstrating the vast potential for speciation even in the simplest of environments: The chemostat environment does not have daily or seasonal fluctuations; the stirring eliminated the possibility of adaptation to different microhabitats; no extrinsic DNA was present; and only one carbon source was introduced into the system. The bacterial metabolism itself created a diversity of resources (by secreting acetate), and this was all that was needed to foster speciation. Thus, the evolution of new ecotypes would appear to be an ineluctable process in the bacterial world.

SEQUENCE-BASED APPROACHES TO IDENTIFYING BACTERIAL SPECIES

Discovery of Ecotypes as Sequence Clusters

The theory of evolutionary genetics provides a compelling rationale for using sequence data to characterize bacterial diversity (69). Given enough time, each bacterial ecotype is expected to be identifiable as a sequence cluster, where the average sequence divergence between ecotypes is much greater than the average sequence divergence within them, for any gene shared by the ecotypes. In addition, each ecotype is expected to be identifiable as a monophyletic group in a phylogeny based on DNA sequence data (Figure 4).

The rationale can be outlined from a phylogenetic perspective. Suppose a new ecotype is derived clonally from one mutant cell that is adapted to a new ecological niche. The nascent ecotype constitutes a monophyletic group descending from this original recombinant [i.e., the ecotype consists of all and only the descendants of the original mutant (Figure 4*A*)]. However, this ecotype is not yet a sequence cluster; one would not conclude from the sequence-based phylogeny that two populations exist within this group. After periodic selection, however, the diversity within the new ecotype is purged (Figure 4*B*). Likewise, periodic selection events within the ancestral ecotype will purge diversity within that ecotype as well (Figures 4*C*,*D*). Note that owing to the diversity-purging effect of periodic selection within each ecotype, the ecotypes eventually appear as separate sequence clusters and each is a monophyletic group (Figure 4*E*). Although this result is seen most clearly in the case of no recombination, Palys et al. (69) showed that under the extremely low rates of recombination occurring in bacteria different ecotypes are nevertheless expected to fall eventually into different sequence clusters for any gene shared across ecotypes.

Conversely, ecotypes are not expected to split into two or more sequence clusters (22, 69) because multiple clusters within an ecotype would be unstable with respect to periodic selection (Figure 5). Each adaptive mutant within the ecotype would

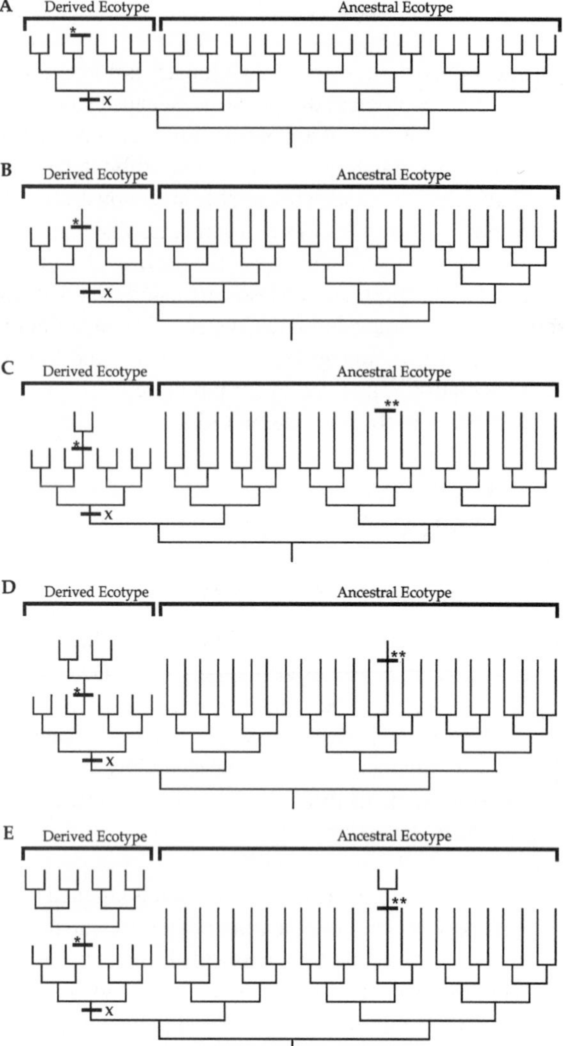

Figure 4 A phylogenetic perspective on periodic selection. As demonstrated here, two ecotypes will become distinct sequence clusters. (*A*) The derived ecotype consists of the descendants of a mutant (X) capable of utilizing a new ecological niche. The adaptive mutant in the derived ecotype (*) is capable of out-competing all other members of the derived ecotype. (*B*) The adaptive mutant (*) has driven all the other lineages within the derived ecotype to extinction. (*C*) With time, the derived ecotype becomes more genetically diverse. One cell in the ancestral ecotype (**) has developed a mutation that allows it to out-compete other members of its ecotype. (*D*) The adaptive mutant (**) has out-competed other members of the ancestral ecotype. (*E*) The ancestral ecotype is becoming more genetically diverse. At this point, each ecotype is a distinct sequence cluster as well as a monophyletic group. Used with permission from the Society for Systematic Biology (15).

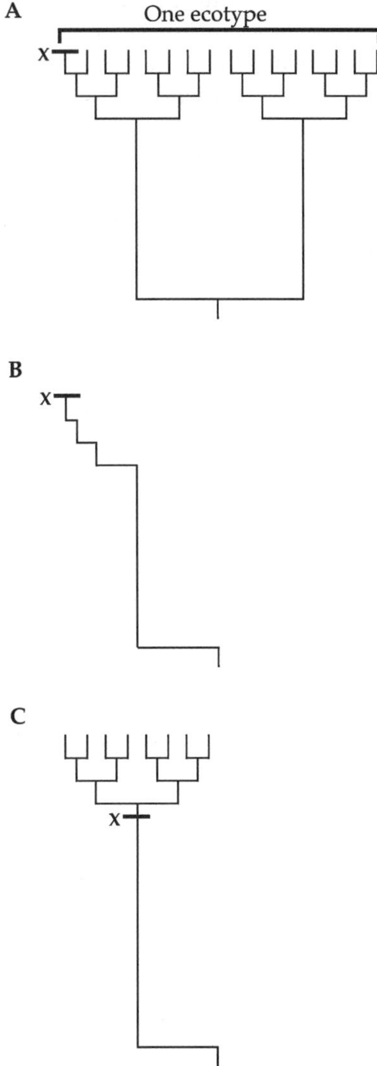

Figure 5 Sympatric members of a single ecotype cannot be split among multiple sequence clusters. (*A*) The ecotype initially contains two distinct sequence clusters. Then an adaptive mutation occurs in lineage X. (*B*) Because the adaptive mutant can purge diversity from the entire ecotype, only one cluster survives periodic selection. (*C*) After periodic selection, variation within the ecotype is re-established, but the population now forms a single sequence cluster. Used with permission from the American Society for Microbiology (69).

drive to extinction cells from all the clusters within the ecotype, and the cluster bearing the adaptive mutant would be all that survives this purge of diversity. If two highly divergent clusters have coexisted long enough to survive periodic selection, then the clusters must belong to different ecotypes.

There is one exception to this conclusion: Geographically isolated populations could diverge into separate sequence clusters, even if they are members of the same ecotype. In this case, an adaptive mutant from one geographical region would not be able to compete with subpopulations from other regions, so sequence divergence between the geographically isolated subpopulations could proceed indefinitely. Divergence among geographically isolated members of the same ecotype would be especially likely for bacteria with low mobility (perhaps pathogens of nonmobile hosts) but would not be possible for highly mobile organisms like *Bacillus*, where intercontinental migration of spores occurs extremely frequently (79). In any case, we can be sure that two highly divergent sequence clusters from the same geographic region (i.e., within migration range) must represent different ecotypes.

In summary, sequence clusters are expected to correspond, more or less, to ecotypes. Indeed, surveys of sequence diversity show a good correspondence between sequence clusters and groups known to be ecologically distinct (69, 96).

The correspondence between ecologically distinct populations and sequence clusters has proven useful for bacterial systematics in several ways (68). First, previously characterized sequence differences between taxa can be used diagnostically to identify unknown isolates.

Second, the correspondence between ecotypes and sequence clusters has enabled us to discover ecological diversity among uncultured bacteria. Increasingly often, uncultured taxa are being described on the basis of forming sequence clusters for 16S rRNA (20, 41, 62). For example, David Ward and coworkers have found that sequence clusters of uncultured *Synechococcus* strains from Yellowstone hot springs correspond to populations inhabiting distinct microenvironments defined by temperature, photic zone, and stage of ecological succession [(73); D. Ward, personal communication].

Finally, the correspondence between ecotypes and sequence clusters is useful for discovering cryptic ecological diversity within a named species. In several cases, a survey of sequence diversity within a named species has revealed multiple sequence clusters that were later found to be ecologically distinct (69). For example, a survey of sequence diversity of 16S rRNA in the genus *Frankia* uncovered previously unknown taxa with unique host specificities (64).

Discovery of Ecotypes as Star Clades

While sequence clusters provide a useful criterion for discovering ecotypes, a serious problem remains. A sequence-based phylogeny from almost any named bacterial species reveals a hierarchy of clusters, subclusters, and sub-subclusters. This raises the possibility that a typical named bacterial species may contain many cryptic and uncharacterized ecotypes, each corresponding to some small subcluster. The challenge is to determine which level of subcluster, if any, corresponds to ecotypes. Fortunately, the peculiar population dynamics of bacteria allows us to identify the clusters that correspond to ecotypes.

Jason Libsch and I have developed a model for identifying the clusters corresponding to ecotypes [(15); J. Libsch & F.M. Cohan, unpublished results]. Our "star clade" approach assumes that the sequence diversity within an ecotype is constrained largely by periodic selection and much less by genetic drift (random fluctuation in gene frequencies within a population, most notably within populations of small size). This assumption is correct if the population size of a bacterial ecotype is 10^{10} or greater. [If sequence diversity in populations of this size were limited only by genetic drift, sequence diversity would be far greater than the 0.5%–1.0% generally seen within sequence clusters (14)].

Consider next the consequences of periodic selection on the phylogeny of an ecotype. Nearly all strains randomly sampled from an ecotype should trace their ancestries directly back to the adaptive mutant that caused and survived the last periodic selection event. Thus, the phylogeny of an ecotype should be consistent with a star clade, with only one ancestral node, such that all members of the ecotype are equally closely related to one another (Figure 6). In contrast, a population whose sequence diversity is limited by genetic drift will have a phylogeny with many nodes.

In an asexual ecotype, a sequence-based phylogeny would yield a perfect star clade, with only minor exceptions due to homoplasy (i.e., convergent nucleotide

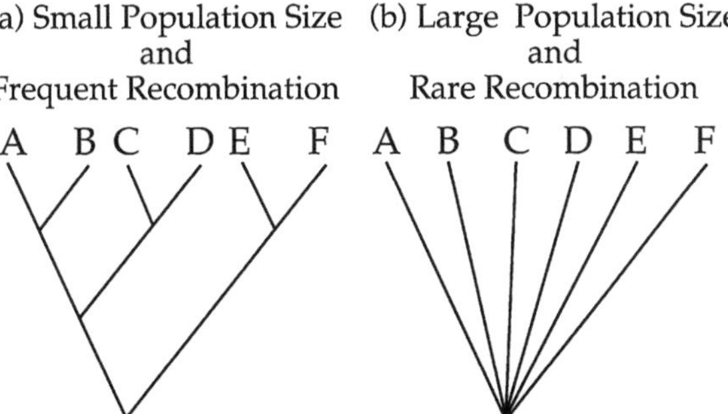

Figure 6 The phylogenetic signatures of populations whose diversity is controlled by periodic selection versus genetic drift. (*a*) In a population of small size, genetic drift causes coalescence of many pairs of lineages. Moreover, if recombination is frequent, there is no opportunity for genome-wide purging of diversity. Consequently, the phylogeny has many nodes. (*b*) In a bacterial population, characterized by large population size and rare recombination, the population's phylogeny is expected to resemble a star. Following periodic selection, each strain traces its ancestry directly back to the adaptive mutant that precipitated the periodic selection event. In addition, population sizes are too large for genetic drift to create coalescences between pairs of strains with appreciable frequency.

substitutions in different lineages and nucleotide substitutions reversing to a former state). However, in an ecotype subject to high rates of recombination, particularly with other ecotypes, the sequence-based phylogeny can deviate significantly from a perfect star clade. For example, suppose that a large segment from a divergent ecotype is recombined into one recipient within the ecotype and that this recipient later donates this foreign segment to another member of the ecotype (Figure 7). In a sequence-based phylogeny, these two strains would appear as closest relatives, a deviation from a perfect star clade.

We have used a computer simulation to determine how closely an ecotype's sequence-based phylogeny should resemble a perfect star clade (15). In general, within groups recombining only rarely (e.g., *Staphylococcus aureus*) (E. Feil, personal communication), the phylogeny of an ecotype is expected to closely

A. The actual history of an ecotype

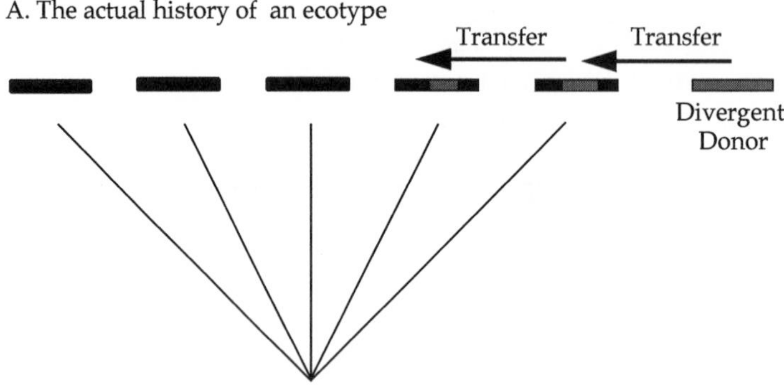

B. The history of ecotype, as determined by phylogeny

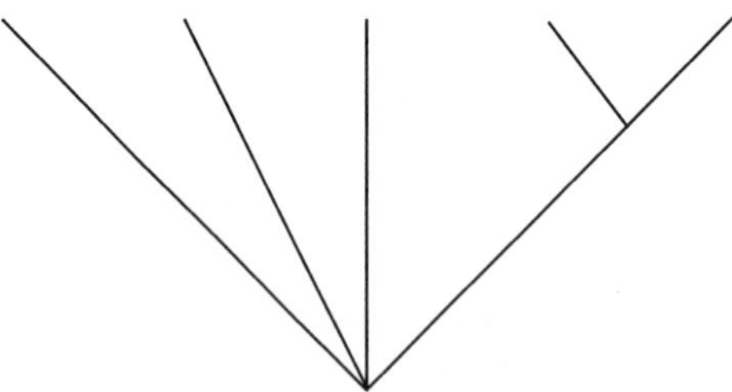

Figure 7 Recombination causes a bacterial ecotype's phylogeny to deviate from a star. Here a member of the ecotype has received a divergent donor's sequence in a gene on which the phylogeny is based, and the divergent sequence is in turn transferred to another member of the ecotype. Each of these strains then appears to be the other's closest relative, and a node is added to the phylogeny. The star clade computer simulation determines how many nodes an ecotype's phylogeny is expected to have, taking into account the recombination and mutation parameters of the taxon.

resemble a perfect star clade; within a taxon with more frequent recombination [e.g., *Neisseria meningitidis* (27)], the phylogeny of an ecotype is expected to deviate to a greater extent from the star form. Our approach is to determine, for a given taxon, how closely an ecotype's sequence-based phylogeny should resemble a star clade and then to identify the largest groups of strains that are each consistent with what is expected for an ecotype. Here the phylogenies are based on a concatenation of several gene loci, usually seven.

The number of nodes within a tree quantifies the degree of resemblance of an ecotype's phylogeny to a star clade: A perfect star has one node, and each additional coalescence of two or more lineages yields an additional node (Figure 8). In the case of *S. aureus*, where individual alleles are subject to mutation three times more frequently than recombination per gene (E. Feil, personal communication),

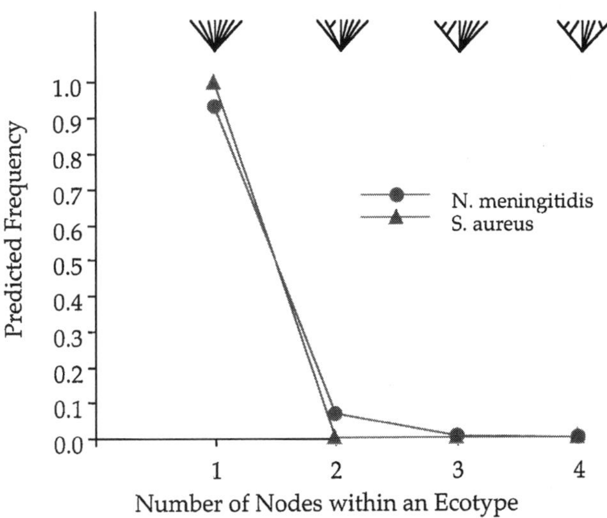

Figure 8 The number of nodes predicted to occur within an ecotype's phylogeny. The Star computer program simulated sequence evolution within an ecotype, at the seven loci used by MLST. The program took into account the recombination and mutation parameters for *N. meningitidis* and *S. aureus*, as estimated from sequence data. Sequence evolution was simulated over many replicate runs. From each replicate run, a 95% bootstrap-supported phylogeny of the ecotype was determined using PAUP*'s heuristic parsimony algorithm (91), based on a concatenation of the seven loci. The figure indicates for each taxon the fraction of times a particular number of nodes was found in the ecotype's phylogeny, over all replicate runs. Owing to the low frequency of recombination in *S. aureus*, an ecotype in this taxon usually has just one significant node; therefore, any collection of strains from *S. aureus* that has more than one node likely contains more than one ecotype. In the more frequently recombining *N. meningitidis*, an ecotype can have up to two significant nodes.

our simulations have shown that an ecotype is only rarely, by chance, expected to have more than one significant node in its phylogeny (F.M. Cohan, unpublished results) (Figure 8). Ecotypes may be identified, then, as the largest groups of strains whose phylogeny contains one significant node. On the other hand, the greater recombination rates within *N. meningitidis* [where recombination occurs four times more frequently than mutation per gene (28)] allow for greater deviation from a perfect star clade (F.M. Cohan, unpublished results) (Figure 8). An ecotype within *N. meningitidis* is expected to contain at most one or two significant nodes. Accordingly, we may tentatively identify ecotypes within *N. meningitidis* as the largest clusters whose phylogenies contain at most two significant nodes. As we shall see, there appear to be many ecotypes within each of these species and perhaps within most named species of bacteria.

Although the star clade approach produces a theory-based criterion for testing whether a set of strains are members of the same ecotype (i.e., the maximum number of nodes expected within an ecotype's phylogeny), this approach does not help us choose the groups of strains to be tested for membership within an ecotype. As we shall see, the MLST approach developed by B. Spratt and coworkers (50) produces accurate hypotheses for demarcating strains into ecotypes.

Discovery of Ecotypes Through Multilocus Sequence Typing

In MLST, strains of a named species are surveyed for partial sequences (usually ~450 bp) of seven gene loci that produce "housekeeping proteins" (proteins that are not involved in niche-specific adaptations and are presumably interchangeable between ecologically distinct groups). The evolutionary distance between strains is quantified in MLST as the number of loci that are different. Two strains are scored as different for a locus whether they differ by one nucleotide substitution or by scores of nucleotides (possibly due to a recombination event). Strains are then classified into "clonal complexes": All strains that are identical with a particular strain at five or more loci (in some cases, six or more loci) are deemed members of a clonal complex (Figure 9). The MLST website (http://www.mlst.net/new/index.htm) provides the "Burst" computer algorithm developed by E. Feil for assigning strains into clonal complexes according to criteria set by the user.

The clonal complexes defined by MLST correspond remarkably well to ecologically distinct clusters. The various hypervirulent lineages within *N. meningitidis* (1, 12, 50, 109) and within *Streptococcus pneumoniae* (36, 108) have been distinguished by MLST as separate clonal complexes. For example, one clonal complex of serogroup A in *N. meningitidis* causes pandemic and epidemic meningitis, particularly in sub-Saharan Africa; the other clonal complex within serogroup A is not associated with disease. One clonal complex of serogroup C causes localized outbreaks in primary schools, university dormitories, and prisons, where conditions are crowded; and one clonal complex of serogroup B causes disease more sporadically (1, 12, 50, 109). It is especially impressive that recombination within *N. meningitidis*, which occurs at a higher rate than in most bacteria, has not prevented

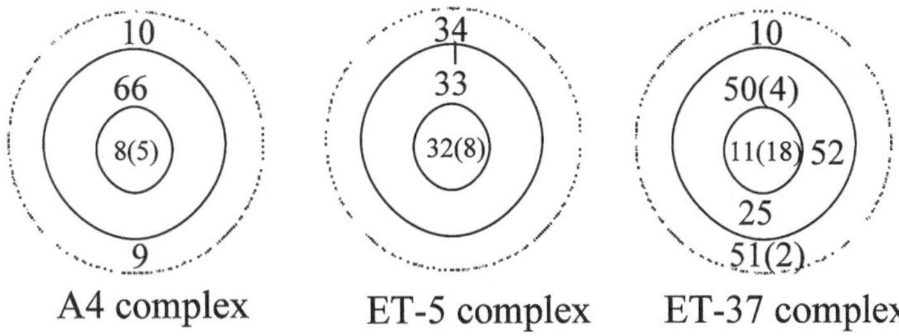

Figure 9 Demarcation of clonal complexes by MLST. The figure indicates three of the ten clonal complexes identified within *N. meningitidis* (27). In the case of *N. meningitidis*, the Burst computer algorithm by E. Feil has identified groups of strains that are identical to a "central" strain at five or more loci; in other species, a criterion of identity at six or more loci has been used (28). The numbers indicate multilocus sequence types, as listed at the MLST website. The numbers in parentheses indicate the number of strains with a given sequence type. Within the inside circle is the "central" sequence type that is identical to the rest of the clonal complex at five or more loci. In the next circle are sequence types identical to the central strain at exactly six loci. In the outside circle are sequence types identical to the central strain at five loci. The straight line indicates identity between peripheral strains at six loci. The clonal complexes identified in this way have been shown to be ecologically distinct, and the star clade approach shows almost every clonal complex to have the phylogenetic properties of an ecotype.

MLST from correctly identifying ecologically distinct groups (50). It will be interesting to see how well MLST holds up in analysis of the most frequently recombining bacteria, including *H. pylori* (90) and *N. gonorrhoeae* (66).

Why does MLST work so well? I have previously hypothesized that the clonal complexes identified by MLST are actually ecotypes, as defined here (16). Because each periodic selection event is recurrently purging the diversity within an ecotype, ecotypes are expected to accumulate little sequence diversity. It was the intuitive insight of E. Feil and B. Spratt that ecotypes would have only enough time between periodic selection events for a given strain to accumulate divergence at one or two loci out of seven, whether by mutation or recombination—this yields the 6/7 and 5/7 criteria used in MLST. Because MLST's 6/7 and 5/7 criteria are intuitively based, we should test analytically whether MLST's clonal complexes are indeed ecotypes.

Fortunately, the star clade approach can test whether the clonal complexes identified by MLST have phylogenies consistent with ecotypes, taking into account the recombination and mutation parameters estimated for the particular taxon. To this end, I have tested the strains of each of the ten clonal complexes within *N. meningitidis* for inclusion within a single ecotype. As shown in Figure 8, using recombination rates and other parameters estimated for *N. meningitidis*, an ecotype is expected to have at most two significant nodes in its phylogeny. As it turns out, the phylogenies of all the MLST clonal complexes within *N. meningitidis* contain one or two nodes, with the exception of the ET-37 complex, which has three. Because an *N. meningitidis* ecotype is so unlikely to contain three nodes, these results suggest that the ET-37 complex contains two ecotypes.

We may conversely address whether more than one MLST clonal complex may be subsumed within a single ecotype. In all but one case, the pool of strains from different *N. meningitidis* clonal complexes contained too many nodes to fit within a single ecotype.

The same pattern has emerged from analysis of ecotypical diversity within *S. aureus*: With few exceptions the clonal complexes demarcated by MLST are each consistent with the phylogeny of one ecotype, and each pair of complexes represents more than one ecotype.

It will be interesting to determine how MLST's criterion for demarcating clonal complexes should change with the rate of recombination. Intuitively, one might expect that a taxon with extremely frequent recombination (such as *H. pylori*) might undergo changes in more than two out of seven loci between periodic selection events. It will be interesting to simulate evolution within an ecotype (using the simulation developed for the star clade algorithm) to determine how the optimal criterion for demarcating MLST complexes changes with recombination rate.

In summary, the star clade approach demonstrates that, in most cases, the clonal complexes yielded by MLST have phylogenies consistent with ecotypes, at least within *S. aureus* and *N. meningitidis*. The clonal complexes produced by MLST thus produce reliable hypotheses about the membership of ecotypes, and these hypotheses can be tested by the star clade approach. These hypotheses can of course also be tested directly by an ecological investigation of the putative ecotypes.

What is striking is that each named species studied by MLST has so many clonal complexes, 10 in the case of *N. meningitidis* (27), 26 in *S. aureus* (E. Feil, personal communication), and 28 in *S. pneumoniae*. Because each of these clonal complexes appears to be a distinct ecotype, with the universal properties of a species, a named bacterial species may actually be more like a genus than a species.

RECOMMENDATIONS FOR BACTERIAL SYSTEMATICS

A principal aim of systematics is to discover, describe, and classify the diversity of living organisms. Systematists have concluded that the basic unit of biological diversity is the species, with these quintessential properties: Species are groups of organisms whose divergence is hindered by one or more forces of cohesion; they are ecologically distinct from one another; and they are irreversibly separate. In the case of bacteria, these universal properties of species are held not by the named species of bacterial systematics but by ecotypes. These are populations of organisms occupying the same ecological niche, whose divergence is purged recurrently by natural selection. Named species appear to contain many such ecotypes.

It should not come as a surprise that named species contain this magnitude of ecological diversity. For decades, systematists have known that there is considerable variation in metabolic traits within named species (49); also, DNA-DNA hybridization experiments have demonstrated a great diversity in genomic content within named species (63). More recently the sequencing of multiple genomes within species has shown considerable variation in the genes contained (3, 70, 71, 78, 93).

We have before us an urgent but accessible goal: to characterize the ecotypical diversity within our most familiar and important named species. A comprehensive study of ecological diversity would demand no less. From a clinical point of view, identifying a pathogen with its ecotype, and thus its distinct virulence properties, will be invaluable; indeed, this was a primary motivation in developing MLST (50). Finally, from an evolutionary genetic point of view, statistical techniques for estimating dynamic properties [such as recombination (37) and migration (84) rates] from sequence survey data require that the strains sampled come from a single ecotype.

A practical first step toward identifying ecotypes is to use a universal, molecular approach to form hypotheses about ecotypical diversity within named species. The clonal complexes yielded by MLST give a good first approximation for ecotype demarcation. The star clade method can then test whether the clonal complexes obtained are consistent with ecotypes. Apparently, in most cases they will be.

Next, one can test whether the ecotypes suggested by sequence-based clustering are indeed ecologically distinct, and the ecological differences can be characterized. Subtractive hybridization is a promising method for discovering the sets of genes not shared by two ecotypes and may suggest the nature of their ecological differences (2, 45). Also, ecological differences between ecotypes can be characterized by differences in the expression levels of all the genes they share by using microarray technology (29).

Sequence-based approaches are particularly important for discovering the ecological diversity among uncultured bacteria, which we now know to constitute the great majority of bacterial diversity (18). The most discerning sequence-based methods must be utilized, or we will likely underestimate the ecological diversity among uncultured bacteria. For example, limiting sequence surveys of uncultured bacteria to 16S rRNA data would likely miss many closely related ecotypes because multilocus sequence typing has revealed multiple ecotypes within named species of cultivated bacteria, and these ecotypes are typically identical or nearly identical in their 16S rRNA sequences.

We are left, then, with the practical, taxonomic matter of classifying the ecotypical diversity within named species. I recommend that when putative ecotypes demarcated by MLST and/or the star clade approach are confirmed to be ecologically distinct, we should recognize their existence with a Latin trinomial, giving the genus, species, and ecotype name. For example, the virulent serogroup A clonal complex and ecotype of *N. meningitidis* might be named *N. meningitidis* ecotypus *africana* for its role in epidemics and pandemics in sub-Saharan Africa. The clusters we identify in this fashion are the fundamental units of ecology and evolution. They deserve our attention and they deserve a name.

ACKNOWLEDGMENTS

I am grateful to E. Feil for sharing unpublished data about *S. aureus* and for many enlightening discussions about MLST. I am grateful to M. Dehn for help in clarifying the manuscript. This review was supported by NSF grant DEB-9815576 and research grants from Wesleyan University.

The *Annual Review of Microbiology* is online at http://micro.annualreviews.org

LITERATURE CITED

1. Achtman M, van der Ende A, Zhu P, Koroleva IS, Kusucek B, et al. 2001. Molecular epidemiology of serogroup A meningitis in Moscow, 1969 to 1997. *Emerg. Infect. Dis.* 7:420–27
2. Akopyants NS, Fradkov A, Diatchenko L, Hill JE, Siebert PD, et al. 1998. PCR-based subtractive hybridization and differences in gene content among strains of *Helicobacter pylori*. *Proc. Natl. Acad. Sci. USA* 95:13108–13
3. Alm RA, Ling LL, Moir DT, King BL, Brown ED, et al. 1999. Genomic-sequence comparison of two unrelated isolates of the human gastric pathogen *Helicobacter pylori*. *Nature* 397:176–80
4. American Society for Microbiology. 1957. *Bergey's Manual of Determinative Bacteriology*. Baltimore: Williams & Wilkins. 7th ed.
5. Atwood KC, Schneider LK, Ryan FJ. 1951. Periodic selection in *Escherichia coli*. *Proc. Natl. Acad. Sci. USA* 37:146–55
6. Avise JC. 2000. *Phylogeography: The History and Formation of Species*. Cambridge, MA: Harvard Univ. Press
7. Balmelli T, Piffaretti J. 1996. Analysis of the genetic polymorphism of *Borrelia burgdorferi* sensu lato by multilocus enzyme electrophoresis. *Int. J. Syst. Bacteriol.* 46:167–72
8. Baranton G, Postic D, Saint Girons I, Boerlin P, Piffaretti JC, et al. 1992. Delineation of *Borrelia burgdorferi* sensu stricto, *Borrelia garinii* sp. nov., and group VS461 associated with Lyme borreliosis. *Int. J. Syst. Bacteriol.* 42:378–83
9. Barrett SJ, Sneath PH. 1994. A numerical phenotypic taxonomic study of the genus *Neisseria*. *Microbiology* 140:2867–91
10. Bochner BR. 1989. Sleuthing out bacterial identities. *Nature* 339:157–58
11. Bush GL. 1994. Sympatric speciation in animals: new wine in old bottles. *Trends Ecol. Evol.* 9:285–88
12. Chan MS, Maiden MC, Spratt BG. 2001. Database-driven multi locus sequence typing (MLST) of bacterial pathogens. *Bioinformatics* 17:1077–83
13. Cohan FM. 1994. Genetic exchange and evolutionary divergence in prokaryotes. *Trends in Ecol. Evol.* 9:175–80
14. Cohan FM. 1994. The effects of rare but promiscuous genetic exchange on evolutionary divergence in prokaryotes. *Am. Nat.* 143:965–86

14a. Cohan FM. 1996. The role of genetic exchange in bacterial evolution. *ASM News* 62:631–36

15. Cohan FM. 2001. Bacterial species and speciation. *Syst. Biol.* 50:513–24
16. Cohan FM. 2002. Clonal structure: an overview. In *Encyclopedia of Evolution*, ed. M Pagel, pp. 159–61. Oxford: Oxford Univ. Press

17. Cohan FM, Roberts MS, King EC. 1991. The potential for genetic exchange by transformation within a natural population of *Bacillus subtilis*. *Evolution* 45:1393–421
18. DeLong E, Pace N. 2001. Environmental diversity of Bacteria and Archaea. *Syst. Biol.* 50:470–78
19. de Queiroz K. 1998. The general lineage concept of species, species criteria, and the process of speciation. In *Endless Forms: Species and Speciation*, ed. DJ Howard, SH Berlocher, pp 57–75. Oxford: Oxford Univ. Press
20. Dojka MA, Harris JK, Pace NR. 2000. Expanding the known diversity and environmental distribution of an uncultured phylogenetic division of Bacteria. *Appl. Environ. Microbiol.* 66:1617–21
21. Duncan KE, Istock CA, Graham JB, Ferguson N. 1989. Genetic exchange between *Bacillus subtilis* and *Bacillus licheniformis*: variable hybrid stability and the nature of species. *Evolution* 43:1585–609
22. Dykhuizen DE. 1998. Santa Rosalia revisited: Why are there so many species of bacteria? *Antonie Van Leeuwenhoek* 73:25–33
23. Dykhuizen DE, Green L. 1991. Recombination in *Escherichia coli* and the definition of biological species. *J. Bacteriol.* 173:7257–68
24. Edwards RA, Helm RA, Maloy SR. 1999. Increasing DNA transfer efficiency by temporary inactivation of host restriction. *Biotechniques* 26:892–98
25. Eldredge N. 1985. *Unfinished Synthesis: Biological Hierarchies and Modern Evolutionary Thought*. New York: Oxford Univ. Press
26. Enright MC, Spratt BG, Kalia A, Cross JH, Bessen DE. 2001. Multilocus sequence typing of *Streptococcus pyogenes* and the relationships between emm type and clone. *Infect. Immun.* 69:2416–27
27. Feil EJ, Maiden MC, Achtman M, Spratt BG. 1999. The relative contributions of recombination and mutation to the divergence of clones of *Neisseria meningitidis*. *Mol. Biol. Evol.* 16:1496–502
28. Feil EJ, Smith JM, Enright MC, Spratt BG. 2000. Estimating recombinational parameters in *Streptococcus pneumoniae* from multilocus sequence typing data. *Genetics* 154:1439–50
29. Ferea TL, Botstein D, Brown PO, Rosenzweig RF. 1999. Systematic changes in gene expression patterns following adaptive evolution in yeast. *Proc. Natl. Acad. Sci. USA* 96:9721–26
30. Futuyma DJ. 1987. On the role of species in anagenesis. *Am. Nat.* 130:465–73
31. Gause GF. 1934. *The Struggle for Existence*. Baltimore: Williams & Wilkins
32. Goodfellow M, Manfio GP, Chun J. 1997. Towards a practical species concept for cultivable bacteria. In *Species: The Units of Biodiversity*, ed. MF Claridge, HA Dawah, MR Wilson, pp 25–29. London: Chapman & Hall
33. Groisman EA, Ochman H. 1997. How *Salmonella* became a pathogen. *Trends Microbiol.* 5:343–49
34. Hamilton CM, Aldea M, Washburn BK, Babitzke P, Kushner SR. 1989. New method for generating deletions and gene replacements in *Escherichia coli*. *J. Bacteriol.* 171:4617–22
35. Hennig W. 1966. *Phylogenetic Systematics*. Urbana: Univ. Illinois Press
36. Henriques Normark B, Kalin M, Ortqvist A, Akerlund T, Liljequist BO, et al. 2001. Dynamics of penicillin-susceptible clones in invasive pneumococcal disease. *J. Infect. Dis.* 184:861–69
37. Hey J, Wakeley J. 1997. A coalescent estimator of the population recombination rate. *Genetics* 145:833–46
38. Deleted in proof
39. Hudson RR. 1987. Estimating the recombination parameter of a finite population model without selection. *Genet. Res.* 50:245–50
40. Hudson RR, Kaplan NL. 1985. Statistical properties of the number of recombination

events in the history of a sample of DNA sequences. *Genetics* 111:147–64
41. Hugenholtz PB, Goebel M, Pace NR. 1998. Impact of culture-independent studies on the emerging phylogenetic view of bacterial diversity. *J. Bacteriol.* 180: 4765–74
42. Deleted in text
43. Imhof M, Schlotterer C. 2001. Fitness effects of advantageous mutations in evolving *Escherichia coli* populations. *Proc. Natl. Acad. Sci. USA* 98:1113–17
44. Johnson JL. 1973. Use of nucleic-acid homologies in the taxonomy of anaerobic bacteria. *Int. J. Syst. Bacteriol.* 23:308–15
45. Lai YC, Yang SL, Peng HL, Chang HY. 2000. Identification of genes present specifically in a virulent strain of *Klebsiella pneumoniae. Infect. Immun.* 68:7149–51
46. Lan R, Reeves PR. 1996. Gene transfer is a major factor in bacterial evolution. *Mol. Biol. Evol.* 13:47–55
47. Lawrence JG, Ochman H. 1998. Molecular archaeology of the *Escherichia coli* genome. *Proc. Natl. Acad. Sci. USA* 95: 9413–17
48. Linz B, Schenker M, Zhu P, Achtman M. 2000. Frequent interspecific genetic exchange between commensal Neisseriae and *Neisseria meningitidis. Mol. Microbiol.* 36:1049–58
49. Logan NA, Berkeley RC. 1984. Identification of *Bacillus* strains using the API system. *J. Gen. Microbiol.* 130:1871–82
50. Maiden MC, Bygraves JA, Feil E, Morelli G, Russell JE, et al. 1998. Multilocus sequence typing: a portable approach to the identification of clones within populations of pathogenic microorganisms. *Proc. Natl. Acad. Sci. USA* 95:3140–45
51. Majewski J, Cohan FM. 1998. The effect of mismatch repair and heteroduplex formation on sexual isolation in *Bacillus. Genetics* 148:13–18
52. Majewski J, Cohan FM. 1999. Adapt globally, act locally: the effect of selective sweeps on bacterial sequence diversity. *Genetics* 152:1459–74
53. Mauchline WS, Keevil CW. 1991. Development of the BIOLOG substrate utilization system for identification of *Legionella* spp. *Appl. Environ. Microbiol.* 57:3345–49
54. May RM. 1973. On relationships among various types of population models. *Am. Nat.* 107:46–57
55. Maynard Smith J, Dowson CG, Spratt BG. 1991. Localized sex in bacteria. *Nature* 349:29–31
56. Maynard Smith J, Smith NH, O'Rourke M, Spratt BG. 1993. How clonal are bacteria? *Proc. Natl. Acad. Sci. USA* 90:4384–88
57. Mayr E. 1944. *Systematics and the Origin of Species from the Viewpoint of a Zoologist.* New York: Columbia Univ. Press
58. Mayr E. 1963. *Animal Species and Evolution.* Cambridge: Belknap Press/Harvard Univ. Press
59. Mayr E. 1982. *The Growth of Biological Thought: Diversity, Evolution, and Inheritance*, Chapter 6. Cambridge: Harvard Univ. Press
60. McKane M, Milkman R. 1995. Transduction, restriction and recombination patterns in *Escherichia coli. Genetics.* 139: 35–43
61. Meglitsch PA. 1954. On the nature of species. *Syst. Zool.* 3:491–503
62. Murray RGE, Stackebrandt E. 1995. Taxonomic note: implementation of the provisional status *Candidatus* for incompletely described procaryotes. *Int. J. Syst. Bacteriol.* 45:186–87
63. Nakamura LK, Roberts MS, Cohan FM. 1999. Relationship of *Bacillus subtilis* clades associated with strains 168 and W23: a proposal for *Bacillus subtilis* subsp. *subtilis* subsp. nov. and *Bacillus subtilis* subsp. *spizizenii* subsp. nov. *Int. J. Syst. Bacteriol.* 49:1211–15
64. Normand P, Orso S, Cournover B, Jeannin P, Chapelon C, et al. 1996. Molecular phylogeny of the genus *Frankia* and

related genera and emendation of the family Frankiaceae. *Int. J. Syst. Bacteriol.* 46:1–9

65. Ochman H, Lawrence JG, Groisman EA. 2000. Lateral gene transfer and the nature of bacterial innovation. *Nature* 405:299–304
66. O'Rourke M, Spratt BG. 1994. Further evidence for the non-clonal population structure of *Neisseria gonorrhoeae*: extensive genetic diversity within isolates of the same electrophoretic type. *Microbiology* 140:1285–90
67. Palmisano MM, Nakamura LK, Duncan KE, Istock CA, Cohan FM. 2001. *Bacillus sonorensis* sp. nov., a close relative of *Bacillus licheniformis*, isolated from soil in the Sonoran Desert, Arizona. *Int. J. Syst. Evol. Microbiol.* 51:16717–19
68. Palys T, Berger E, Mitrica I, Nakamura LK, Cohan FM. 2000. Protein-coding genes as molecular markers for ecologically distinct populations: the case of two *Bacillus* species. *Int. J. Syst. Evol. Microbiol.* 50:1021–28
69. Palys T, Cohan FM, Nakamura LK. 1997. Discovery and classification of ecological diversity in the bacterial world: the role of DNA sequence data. *Int. J. Syst. Bacteriol.* 47:1145–56
70. Parkhill J, Achtman M, James KD, Bentley SD, Churcher C, et al. 2000. Complete DNA sequence of a serogroup A strain of *Neisseria meningitidis* Z2491. *Nature* 404:502–6
71. Perna NT, Plunkett G 3rd, Burland V, Mau B, Glasner JD, et al. 2001. Genome sequence of enterohaemorrhagic *Escherichia coli* O157:H7. *Nature* 409:529–33
72. Rainey PB, Travisano M. 1998. Adaptive radiation in a heterogeneous environment. *Nature* 394:69–72
73. Ramsing NB, Ferris MJ, Ward DM. 2000. Highly ordered vertical structure of *Synechococcus* populations within the one-millimeter-thick photic zone of a hot spring cyanobacterial mat. *Appl. Environ. Microbiol.* 66:1038–49
74. Rao BJ, Chiu SK, Bazemore LR, Reddy G, Radding CM. 1995. How specific is the first recognition step of homologous recombination? *Trends Biochem. Sci.* 20: 102–13
75. Ravin AW. 1960. The origin of bacterial species: genetic recombination and factors limiting it between bacterial populations. *Bacteriol. Rev.* 24:201–20
76. Ravin AW. 1963. Experimental approaches to the study of bacterial phylogeny. *Am. Nat.* 97:307–18
77. Rayssiguier C, Thaler DS, Radman M. 1989. The barrier to recombination between *Escherichia coli* and *Salmonella typhimurium* is disrupted in mismatch repair mutants. *Nature* 342:396–401
78. Read TD, Brunham RC, Shen C, Gill SR, Heidelberg JF, et al. 2000. Genome sequences of *Chlamydia trachomatis* MoPn and *Chlamydia pneumoniae* AR39. *Nucleic Acids Res.* 28:1397–406
79. Roberts MS, Cohan FM. 1995. Recombination and migration rates in natural populations of *Bacillus subtilis* and *Bacillus mojavensis*. *Evolution* 49:1081–94
80. Rozen DE, Lenski RE. 2000. Long-term experimental evolution in *Escherichia coli*. VIII. Dynamics of a balanced polymorphism. *Am. Nat.* 155:24–35
81. Sasser M. 1990. Identification of bacteria by gas chromatography of cellular fatty acids. *Tech. Note 101. Microbial ID, Inc.*, Newark, Del.
82. Selander RK, Musser JM. 1990. Population genetics of bacterial pathogenesis. In *Molecular Basis of Bacterial Pathogenesis*, ed. BH Iglewski, VL Clark, pp. 11–36. San Diego: Academic
83. Simpson GG. 1961. *Principles of Animal Taxonomy*. New York: Columbia Univ. Press
84. Slatkin M, Maddison WP. 1989. A cladistic measure of gene flow inferred from the phylogenies of alleles. *Genetics* 123:603–13
85. Sneath PHA. 1985. Future of numerical taxonomy. In *Computer-Assisted*

Bacterial Systematics, ed. M Goodfellow, D Jones, FG Priest, pp. 415–31. Orlando: Academic

86. Sneath PHA, Sokal RR. 1973. *Numerical Taxonomy: The Principles and Practice of Numerical Classification*. San Francisco: Freeman
87. Sokal RR, Crovello TJ. 1970. The biological species concept: a critical evaluation. *Am. Nat.* 104:127–53
88. Stackebrandt E, Goebel BM. 1994. Taxonomic note: a place for DNA:DNA reassociation and 16S rRNA sequence analysis in the present species definition in bacteriology. *Int. J. Syst. Bacteriol.* 44:846–49
89. Stebbins GL. 1957. *Variation and Evolution in Plants*. Irvington: Columbia Univ. Press
90. Suerbaum S, Maynard Smith J, Bapumia K, Morelli G, Smith NH, et al. 1998. Free recombination within *Helicobacter pylori*. *Proc. Natl. Acad. Sci. USA* 95: 12619–24
91. Swofford DL. 1998. *Phylogenetic Analysis Using Parsimony (*And Other Methods)*. Version 4. Sunderland, MA: Sinauer
92. Templeton AR. 1989. The meaning of species and speciation: a genetic perspective. In *Speciation and Its Consequences*, ed. D Otte, JA Endler, pp. 3–27. Sunderland, MA: Sinauer
93. Tettelin H, Saunders NJ, Heidelberg J, Jeffries AC, Nelson KE, et al. 2000. Complete genome sequence of *Neisseria meningitidis* serogroup B strain MC58. *Science* 287:1809–15
94. Trautner TA, Pawlek B, Bron S, Anagnostopoulos C. 1974. Restriction and modification in *B. subtilis*. *Mol. Gen. Genet.* 131:181–91
95. Treves DS, Manning S, Adams J. 1998. Repeated evolution of an acetate cross-feeding polymorphism in long-term populations of *Escherichia coli*. *Mol. Biol. Evol.* 15:789–97
96. Vandamme P, Pot B, Gillis M, de Vos P, Kersters K, Swings J. 1996. Polyphasic taxonomy, a consensus approach to bacterial systematics. *Microbiol. Rev.* 60:407–38
97. Van Spanning RJ, Reijnders WN, Stouthamer AH. 1995. Integration of heterologous DNA into the genome of *Paracoccus denitrificans* is mediated by a family of IS1248-related elements and a second type of integrative recombination event. *J. Bacteriol.* 177:4772–78
98. Van Valen L. 1976. Ecological species, multispecies, and oaks. *Taxon* 25:233–39
99. Vulic M, Dionisio F, Taddei F, Radman M. 1997. Molecular keys to speciation: DNA polymorphism and the control of genetic exchange in enterobacteria. *Proc. Natl. Acad. Sci. USA* 94:9763–67
100. Wayne LG, Brenner DJ, Colwell RR, Grimont PAD, Kandler O, et al. 1987. Report of the ad hoc committee on reconciliation of approaches to bacterial systematics. *Int. J. Syst. Bacteriol.* 37:463–64
101. White MJD. 1978. *Modes of Speciation*. San Francisco: Freeman
102. Whittam TS, Ake SE. 1993. Genetic polymorphisms and recombination in natural populations of *Escherichia coli*. In *Molecular Paleopopulation Biology*, ed. N Takahata, AG Clark, pp. 223–45. Tokyo: Jpn. Sci. Soc. Press
103. Wiley EO. 1978. The evolutionary species concept reconsidered. *Syst. Zool.* 27:17–26
104. Woese CR. 2000. Interpreting the universal phylogenetic tree. *Proc. Natl. Acad. Sci. USA* 97:8392–96
105. Wright S. 1931. Evolution in Mendelian populations. *Genetics* 16:97–159
106. Young JPW. 1989. The population genetics of bacteria. In *Genetics of Bacterial Diversity*, ed. DA Hopwood, KF Chater, pp. 417–38. London: Academic
107. Zawadzki P, Cohan FM. 1995. The size and continuity of DNA segments integrated in *Bacillus* transformation. *Genetics* 141:1231–43
108. Zhou J, Enright MC, Spratt BG. 2000.

Identification of the major Spanish clones of penicillin-resistant pneumococci via the Internet using multilocus sequence typing. *J. Clin. Microbiol.* 38:977–86

109. Zhu P, van der Ende A, Falush D, Brieske N, Morelli G, et al. 2001. Fit genotypes and escape variants of subgroup III *Neisseria meningitidis* during three pandemics of epidemic meningitis. *Proc. Natl. Acad. Sci. USA* 98:5234–39

Annu. Rev. Microbiol. 2002. 56:489–520
doi: 10.1146/annurev.micro.56.012302.160916

First published online as a Review in Advance on May 10, 2002

GENOME REMODELING IN CILIATED PROTOZOA

Carolyn L. Jahn[1] and Lawrence A. Klobutcher[2]
[1]*Department of Cell and Molecular Biology, Northwestern University Medical School, Chicago, Illinois 60611; e-mail: jahn@casbah.acns.nwu.edu*
[2]*Department of Biochemistry, University of Connecticut Health Center, Farmington, Connecticut 06032; e-mail: klobutcher@nso2.uchc.edu*

Key Words DNA rearrangement, chromosome fragmentation, DNA splicing, gene scrambling

■ **Abstract** The germline genomes of ciliated protozoa are dynamic structures, undergoing massive DNA rearrangement during the formation of a functional macronucleus. Macronuclear development involves chromosome fragmentation coupled with de novo telomere synthesis, numerous DNA splicing events that remove internal segments of DNA, and, in some ciliates, the reordering of scrambled gene segments. Despite the fact that all ciliates share similar forms of DNA rearrangement, there appears to be great diversity in both the nature of the rearranged DNA and the molecular mechanisms involved. Epigenetic effects on rearrangement have also been observed, and recent work suggests that chromatin differentiation plays a role in specifying DNA segments either for rearrangement or for elimination.

CONTENTS

INTRODUCTION

Cells have evolved complex systems to ensure the faithful replication, correction, and transmission of genetic information from one generation to the next. Nonetheless, there are numerous cases in both prokaryotic and eukaryotic organisms where

0066-4227/02/1013-0489$14.00

programmed genetic changes occur. In some instances, these are reversible, so that there is no permanent change to the genome. In other cases, the changes are not reversible. Such rearrangements are typically limited to terminally differentiated cells and do not affect the germline genome. The existence of such processes has been evident since the pioneering studies of Boveri (13) more than a century ago. Using cytological methods alone, he observed that somatic cells of the parasitic nematode *Parascaris* fragmented their chromosomes and eliminated large heterochromatic terminal regions. More recent studies indicate that 80%–90% of the genome is eliminated by this process, which has come to be known as "chromatin diminution" (107). Chromatin diminution has been observed in a number of other organisms, including copepods, insects, fish, and marsupials, and takes a variety of forms (78). These include the excision of interstitial segments of DNA with rejoining of flanking sequences, as well as the loss of entire chromosomes. A common feature of many of these processes is that the eliminated DNA appears condensed and intensely stained by DNA dyes, features typically associated with heterochromatin.

Other DNA rearrangements affect only single or a limited number of genes in the genome. Examples include the rearrangement of immunoglobulin and T-cell receptor genes in the vertebrate immune system, and DNA inversion systems regulating flagellar proteins in *Salmonella* or tail fiber proteins in bacteriophage Mu [reviewed in (8)]. For many of these more limited systems, a good deal of information is available on molecular mechanisms. For at least some of these systems, the proteins that mediate rearrangement are related to proteins encoded by transposons and/or viruses, leading to suggestions that the rearrangement systems evolved from such mobile genetic elements (110).

The relationship of the large-scale chromatin diminution systems to the more limited forms of programmed DNA rearrangement is unclear. In this article we discuss the dramatic genome reorganization process that occurs in ciliated protozoa. These unicellular organisms undergo multiple forms of DNA rearrangement and in some cases eliminate >95% of their germline genome during the process. Work on a number of ciliates has provided information on the molecular mechanisms of rearrangement, including the involvement of transposons in the process. In addition, recent studies have both documented epigenetic effects that influence rearrangement and suggested that chromatin structure is involved in specifying regions of the genome for rearrangement. While the characterization of these systems is far from complete, the work to date indicates these organisms will prove quite informative in investigating the relationship of chromatin diminution processes to the more limited types of DNA rearrangement.

NUCLEAR DUALISM IN CILIATED PROTOZOA

Unicellular ciliates display "nuclear dimorphism." This feature allows them to undergo major genome reorganization, yet maintain the constancy of the genome from generation to generation. Each cell contains at least one micronucleus (MIC) and one macronucleus (MAC) (numbers vary in different species). MIC possess features of typical eukaryotic nuclei, such as a diploid genome composed of conventional chromosomes and division by mitosis. However, they are unusual in that they are transcriptionally inactive during asexual, or vegetative, growth of the organism. Nuclear transcription is limited to the MAC during this phase of the life cycle. The MAC genome is composed of a subset of the sequences present in the MIC. Moreover, the DNA molecules in the MAC are shorter in length than in the MIC and are present in multiple copies [e.g., *Tetrahymena thermophila* has 45 copies of each MAC DNA molecule, while the spirotrichs often have 1000 or more copies (21, 113)]. Both nuclei replicate their genomes and divide during vegetative reproduction; the MIC divides by mitosis, whereas the MAC is partitioned by essentially pinching in half, a process termed amitosis.

During sexual reproduction (conjugation) the MAC is destroyed, and a new MAC is generated from a copy of the MIC (Figure 1). Conjugation is typically induced by starvation conditions and begins with the pairing of cells of compatible mating types. The nuclear events of conjugation vary in different ciliate species. Figure 1 presents a simplified scheme of the key events shared by most species. Initially, the MIC undergoes meiotic divisions to generate haploid products. Haploid nuclei are then exchanged between members of the mating pair and fuse with a resident meiotic product to generate a new diploid nucleus (the zygotic nucleus). The zygotic nucleus in each cell replicates its genome and divides by mitosis to generate two identical diploid zygotic nuclei. One of these becomes the new

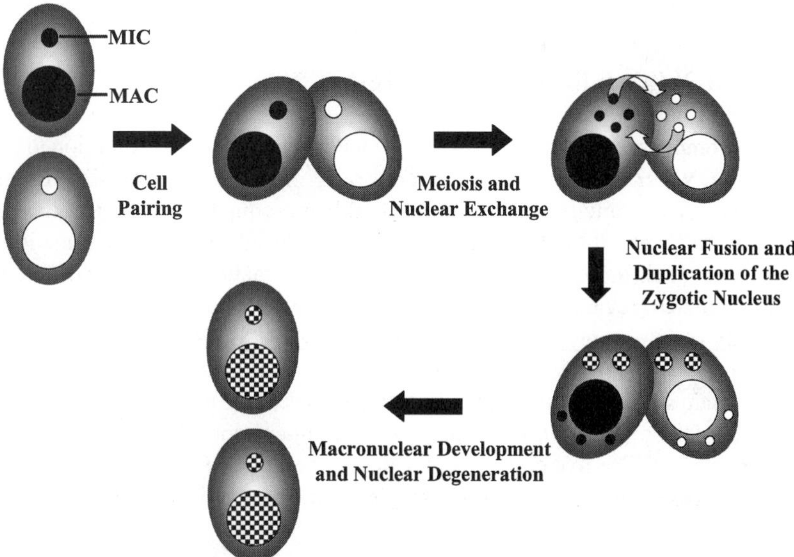

Figure 1 A simplified and generic scheme for conjugation and MAC development in ciliates. See text for details.

MIC while the other undergoes a series of genomic changes to become the new MAC. During this process, the old MAC becomes pycnotic and fragmented, and depending on the species, the fragments are either resorbed or they are lost during subsequent divisions. The unused meiotic products are also destroyed.

The process of MAC development involves a series of changes to the MIC genome. We discuss the substantial variability in the details of these processes among the various groups of ciliates. This is not entirely unexpected, as the ciliates are an ancient group of organisms that diverged from the eukaryotic lineage more than one billion years ago, and the major ciliate subgroups established soon thereafter (143). Nonetheless, there are a number of basic features of MAC development that are shared by all ciliates examined to date. One of these is endoreplication of the genome, which ultimately results in the multiple copies of the MAC chromosomes. During the duplication of the MIC chromosomes, two forms of DNA rearrangement also occur. The first is the site-specific fragmentation of the MIC chromosomes and the addition of simple telomeric repeat sequences to the DNA ends of the resulting MAC chromosomes. Second, interstitial segments of DNA, often called internal eliminated sequences (IESs), are excised from DNA that will be retained in the MAC (MAC-destined sequences) with rejoining of flanking DNA. During these processes, significant amounts of the MIC genome are discarded (10% to >95% depending on the species). This review discusses these different forms of rearrangement in the ciliates that have been best characterized at the molecular level, which are the oligohymenophorens (Class Oligohymenophorea) *Tetrahymena* and *Paramecium* and a number of members of the Class Spirotrichea (*Euplotes*, *Oxytricha*, and *Stylonychia*).[1]

CHROMOSOME FRAGMENTATION

Chromosome fragmentation, coupled with de novo telomere addition to the resulting DNA ends, is one of the DNA rearrangement processes of MAC development shared by all ciliates. The process is highly reproducible, in that identical or nearly identical sets of MAC chromosomes are produced following conjugation, but the extent of chromosome fragmentation differs dramatically between different groups of ciliates [reviewed in (21, 113)]. At one extreme are the spirotrichs, where the MIC chromosomes are fragmented into 10,000–20,000 MAC chromosomes with an average size of ~2 kbp. The majority of the spirotrich MAC chromosomes contain single open reading frames (ORFs) (i.e., one gene), but a few possess multiple ORFs (127). Other ciliates undergo much more limited chromosome fragmentation. The five chromosomes that make up the *T. thermophila* MIC genome are broken at ~200 positions to yield multigene MAC chromosomes with an

[1]There have been a number of recently proposed changes in species names among the spirotrichs. We have chosen to refer to these organisms by their long-standing designations but provide alternative designations in parentheses when the species is first mentioned.

average size of about 600 kbp, whereas *Paramecium* produce MAC chromosomes of an average size of 300 kbp.

Sequence elements that specify chromosome fragmentation have been identified in two groups of ciliates. In *T. thermophila* and related species, a 15-bp sequence termed the "chromosome breakage sequence" (Cbs; 5′-AAAGAGGTTGGTTTA-3′) resides at positions subject to fragmentation (23, 146) (Figure 2*a*). Transformation-based studies have provided strong evidence that the Cbs is the key *cis*-acting sequence for chromosome fragmentation; deletion of the Cbs eliminates fragmentation, whereas moving it to a novel position creates a new chromosome fragmentation site (145). The Cbs sequence is highly conserved, but an in vitro

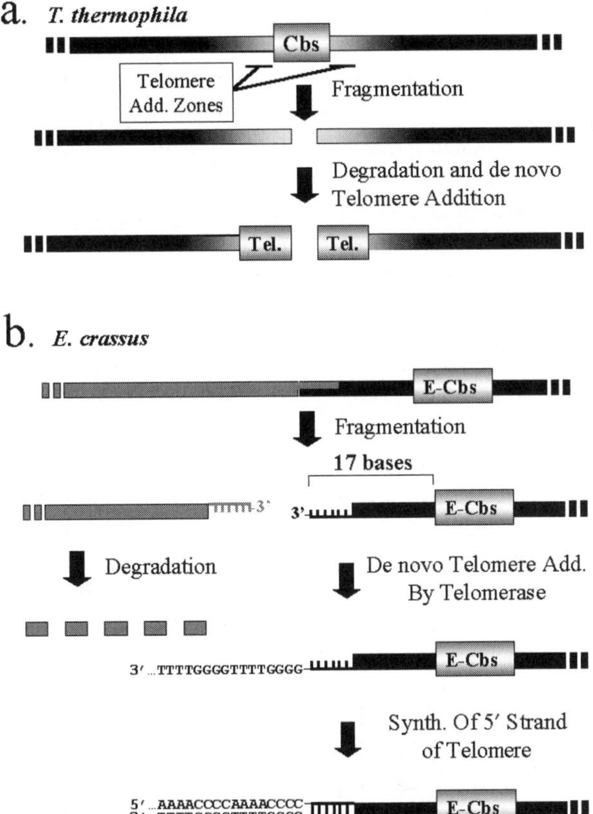

Figure 2 Models of chromosome fragmentation and de novo telomere formation in (*a*) *T. thermophila* and (*b*) *E. crassus*. MAC-destined sequences are shown in *black* and eliminated spacer DNA in *gray*. Tel., telomere; Cbs, *Tetrahymena* chromosome breakage sequence; E-Cbs, *Euplotes* chromosome breakage sequence. The models are based on those in references (21, 72).

mutagenesis study (37), as well as a broad survey of natural fragmentation sites in *T. thermophila* (E. Hamilton, D. Cassidy-Hanley, P. Bruns & E. Orias, unpublished results), indicates that one or two single-bp differences at a limited number of positions within the Cbs are tolerated (23, 146, 149). The Cbs are not retained within the MAC genome. Instead, two MAC chromosome ends are formed with their telomeric repeats added at variable positions within 5–30 bp of each side of the Cbs (36) (Figure 2*a*). This has led to a model (21) that suggests that the Cbs directs a single double-stranded break in the DNA. The two resulting DNA ends would then be degraded by nucleases to variable extents prior to the initiation of de novo telomere synthesis.

The spirotrich *Euplotes crassus* (*Moneuplotes crassus*) employs a somewhat different system of chromosome fragmentation. In this organism, "spacer" DNA segments eliminated during MAC development typically separate the precursors of the small MAC DNA molecules, and chromosome fragmentation/telomere addition is precise to the nucleotide (6, 106). A conserved 10-bp sequence termed the *Euplotes*-Cbs (E-Cbs; 5′-HATTGAAaHH-3′, H = A, C, or T) resides near each chromosome fragmentation site (6, 73) (Figure 2*b*). The E-Cbs is typically found within DNA retained in the MAC, with the highly conserved core 5′-TTGAA-3′ positioned 17 bp from the site of telomere addition. However, for about one third of the chromosome fragmentation sites, the E-Cbs is instead found in inverted orientation in the eliminated spacer DNA, with the core 5′TTGAA3′ positioned 11 bp from the telomere addition site. Based on this difference in spacing, it was proposed that the E-Cbs serves as a recognition and positioning sequence for the protein(s) that carries out chromosome fragmentation (6, 72) (Figure 2*b*). Fragmentation would generate a staggered, double-stranded break in the DNA to generate a 6-base, single-stranded overhang 17 bases in the 5′ direction from the E-Cbs core. The overhangs would then be filled in during the process of de novo telomere addition, accounting for the apparent difference in spacing of the E-Cbs when positioned in MAC-destined versus -eliminated spacer DNA. Support for this model has been provided by PCR-based experiments, which detected the predicted fragmentation intermediates with 6-base 3′ overhangs in DNA from cells during the period of chromosome fragmentation, but not in DNA from vegetative cells or developing cells prior to chromosome fragmentation (72). The ends of the fragmentation intermediates possessed 5′-phosphate and 3′-hydroxyl groups, consistent with a hydrolytic cleavage mechanism. This system of chromosome fragmentation is likely shared by a wide range of Euplotids, as the E-Cbs sequence is frequently seen at the ends of MAC DNA molecules from many different *Euplotes* species (76). However, the E-Cbs sequence does not entirely explain the specificity of chromosome fragmentation. Sequences quite similar to the E-Cbs can be found within the MAC DNA molecules of *E. crassus*. Thus, there may be additional *cis*-acting sequence elements near true chromosome fragmentation sites, or the internal E-Cbs might somehow be masked from the fragmentation machinery.

Less information is available for the fragmentation systems in other ciliates, but they appear to differ from the *Tetrahymena* Cbs or *Euplotes* E-Cbs recognition mechanisms, as credible matches to these sequence elements are not consistently found at similar positions relative to fragmentation/telomere addition sites. All other ciliates examined display heterogeneity in the position of telomere addition to MAC DNA ends [reviewed in (11, 21)], and two forms of variability have been described. First, there is microheterogeneity, such that the telomeres are added at positions spanning tens of bp in the non-Euplotid spirotrichs, and over a region of a several hundred bp in *Paramecium*. Second, chromosome fragmentation sites are sometimes used in an alternative manner in both *Oxytricha* and *Paramecium*, resulting in MAC chromosomes differing by multiple kbp in size (15, 38). While no conserved sequence elements have been consistently noted, a number of studies have provided information relevant to the process. Transformation-based studies on *Stylonychia lemnae* indicate that sequences flanking MAC-destined DNA are not required for proper fragmentation but that sequences within ~300 bp of the ends of MAC-destined segments are essential (69, 70). Intriguingly, good matches to the E-Cbs core sequence are found within the essential region, and a single-base mutation in this element abolished correct fragmentation (69). This suggests that an E-Cbs-like element may be important in *S. lemnae* but that its position is variable compared to *Euplotes*. It is also noteworthy that Prescott & Dizick (116) have found that the ends of non-Euplotid spirotrich MAC DNA molecules have an unusual base composition. The first 50 subtelomeric bases display a significant strand asymmetry in the ratio of purines to pyrimidines (~60:40). A broader survey of >2400 random MAC-chromosome ends of *Oxytricha trifallax* has confirmed this unusual base composition (T. Doak, R. Weiss, D. Dunn & G. Herrick, personal communication). It has been suggested that these terminal regions might be important in specifying chromosome fragmentation not by virtue of a specific DNA sequence, but because of the unusual base composition, which could influence the structure of either the DNA or chromatin (116).

Overall, the data strongly suggest that a good deal of diversity exists in the chromosome fragmentation systems of various ciliate groups. Whether these differences are merely superficial and represent evolutionary diversification of a common process, or are substantially different and a reflection of multiple origins for chromosome fragmentation, is yet unclear. Information on the cellular components that mediate chromosome fragmentation in the various ciliate groups could help resolve these issues as well as provide new information on the mechanism(s). Unfortunately, attempts to reconstruct chromosome fragmentation in vitro have been unsuccessful, and there are yet no known mutations specifically affecting chromosome fragmentation.

DE NOVO TELOMERE ADDITION

In most organisms breaks in DNA are either repaired or generate DNA rearrangements. Ciliate MAC development represents one of the few cases where breaks in the DNA are efficiently healed through the de novo synthesis of telomeres onto the

DNA ends. De novo telomere addition is tightly coupled to chromosome fragmentation, as it has been possible to detect fragmented forms that lack telomeres only through the use of sensitive, PCR-based methods (72). The telomeres of ciliates consist of simple, tandem repeats, such as 5′-GGGGTTTT-3′ (G_4T_4 repeats) in the spirotrichs and 5′-GGGGTT-3′ (G_4T_2 repeats) in *Tetrahymena* on the DNA strand representing the 3′-end of the telomere [reviewed in (11)]. In vegetative cells, the ribonucleoprotein telomerase maintains telomere length by synthesizing new repeats onto the existing repeats at the 3′-ends of telomeres. Studies in *Tetrahymena* provide a clear indication that telomerase is also responsible for de novo telomere synthesis during MAC development. Yu & Blackburn (149) incorporated a mutant form of the gene encoding the RNA component of telomerase, which provides the template for telomeric repeat synthesis, into *Tetrahymena* cells and allowed them to proceed through MAC development. Newly formed MAC telomeres contained repeats templated by the mutant form of telomerase RNA, indicating that it is involved in de novo telomere formation.

It is not entirely clear whether a specific sequence signal is required for de novo telomere addition, but most studies indicate that it may be a nonspecific process. In *Paramecium*, any DNA molecule introduced into the vegetative MAC can have telomeric repeats synthesized onto its ends (43). This does not appear to be the case in *Tetrahymena*, as introduction of an rDNA transformation construct containing ends generated by restriction endonuclease digestion into developing cells did not generate viable transformants (36). However, in vitro studies on telomerase isolated from vegetative or developing *Tetrahymena* indicate that it is capable of adding telomeric repeats onto the 3′-ends of nontelomeric oligonucleotides that are single stranded or have a 3′-overhang of at least 20 bases (137). A similar activity has been reported for the *E. crassus* telomerase (7, 94). In this case, the substrate oligonucleotide must have a telomeric repeat-like sequence near its 5′-end, which is suggested to be required for binding to telomerase, but there are apparently no specific sequence requirements for the 3′-end (94). Moreover, only telomerase from *E. crassus* cells undergoing MAC development has the ability to carry out this type of de novo telomere addition, suggesting that telomerase is modified during development. Indeed, telomerase in developing cells has been found to exist in higher-molecular-weight complexes compared to vegetative cells (47), and there is evidence for a de novo telomere addition factor that associates with telomerase during development (7). Although the in vitro studies do not entirely recapitulate the in vivo process (e.g., it is not clear that the in vivo substrate is single stranded), they are most consistent with the view that a specialized form of telomerase recognizes any free DNA end and initiates de novo telomere addition.

A final question is whether de novo telomere addition plays a role in differentiating between sequences to be retained in the MAC and those destined for degradation. One of the functions of telomeric sequences, in conjunction with telomere-binding proteins, is to protect natural chromosome ends from recombination and degradation [reviewed in (12)]. Thus, addition of telomeres to MAC-destined sequences could conceivably be the event that prevents their destruction during development. Although an attractive model, a recent study on *E. crassus* argues against this mechanism (106). Using PCR-based methods, eliminated spacer DNAs with telomeric repeats were transiently observed during development at the time of chromosome fragmentation. In contrast to the MAC DNA ends, the spacer DNAs displayed variability in telomere addition sites; telomeres were often added at positions expected to be at the end or within the 6-base 3′-overhangs of the fragmentation intermediates, but telomere addition onto sites initially within the spacer were also observed. The existence of spacer DNAs with telomeres makes it unlikely that differential telomere addition is responsible for discriminating between eliminated and retained sequences. Intriguingly, the variable telomere addition sites observed for *Euplotes* spacer DNAs resembles the situation for MAC-destined DNA in other ciliates. As a result, the precise telomere addition sites observed for the *Euplotes* MAC-destined sequences may prove to be only a minor variation of a more general process.

REMOVAL OF INTERNAL ELIMINATED SEQUENCES BY DNA BREAKAGE AND REJOINING

A second form of DNA rearrangement shared by all ciliates is the removal of interstitial DNA segments [internal eliminated sequences (IESs)], with the rejoining of flanking DNA. Again, there is diversity in the IES excision processes of different ciliates, but two general classes of IESs can be defined. The first consists of relatively short DNA sequences. This type of IES was first identified in *T. thermophila* (14, 144), where about 6000 IESs, ranging in size from 600 bp to a few kbp, are removed from the MIC genome. These IESs are typically AT-rich elements with direct repeats of 1–8 bp near their ends, and many are members of, or contain, moderately repetitive sequence families. The excision boundaries of *T. thermophila* IESs are variable, and none have yet been found to reside within coding regions. Analogous IESs exist in members of the spirotrich group and *Paramecium*, but with a number of significant differences. A substantial fraction of the spirotrich and *Paramecium* IESs are shorter than 100 bp, and there are many more present in the genome [e.g., *Paramecium aurelia* is estimated to have ~65,000 IESs (32) while *Oxytricha nova* (*Sterkiella nova*) has ~60,000 (120)]. Moreover, the IESs in these organisms are precisely excised with one copy of their short terminal direct repeats retained within the MAC DNA (133, 136). In all likelihood, precise excision accounts for the location of many IESs within the coding regions of genes in these organisms.

In contrast to these relatively small and noncoding IESs, the second class of IESs contain open reading frames (ORFs) and display features of transposable elements. These include the 4.1-kbp TBE1 elements of *O. fallax* and *O. trifallax* [*Sterkiella histriomuscorum* (53, 140)] and the 5.3-kbp Tec1 and Tec2 elements of *E. crassus* (5, 59, 62, 81), which are present in ~2000 and ~5000 copies per haploid

MIC genome. All these elements are bounded by short direct repeats, have long inverted terminal repeats, and contain multiple ORFs (25, 26, 27, 59, 142). For each element, one of the ORFs encodes a protein with similarity to the transposases of the Tc1/*mariner*/IS*630* transposons (26, 59), indicating that they are members of this phylogentically widespread transposon family. All these elements are eliminated during MAC development, and at least some interrupt coding regions and are removed as a unit.

Two transposon-like elements have also been identified in the MIC of *Tetrahymena*: the Tel-1 elements (19) and the more recently identified Tlr elements [(41, 138); J. Wuitschick, J. Gershan, A. Lochowicz, S. Li & K. Karrer, manuscript submitted]. The Tlr elements are clearly distinct from the spirotrich Tc1/*mariner*/IS*630*-like transposons. The main body of the element is 22 kbp in length, and it possesses long terminal inverted repeats that vary in composition and size among element family members. Tlrs have approximately 15 ORFs, and a number of these encode conceptual proteins related to those of double-stranded DNA viruses and retrotransposons, including a retroviral integrase (41). All Tlr elements are eliminated during MAC development, but it is not clear if the element can be excised as a unit. The deletion boundaries of the best-studied element, Tlr1, are variable and do not coincide with the ends of the element (138), and other Tlr family members appear to reside within other developmentally eliminated DNA.

The presence of both short IESs and longer transposon-like elements has led to suggestions that the excision process may have developed from transposable elements [see (75) and references therein]. One scenario is that the MIC genome was invaded by mobile elements that encoded functions that allowed for their excision during MAC development. Like mobile introns and inteins, such elements would be well tolerated in the genome because any detrimental effects are reversed at one of the steps of gene expression, in this case by excision at the DNA level. Over time, random mutations within elements would create a selective pressure for the host to assume the excision functions, perhaps by rare events that place element-encoded genes under the control of host promoters. Once this occurred, the elements would be free to mutate, except for the *cis*-acting sequences required for excision, forming the smaller, nontransposon-like IESs.

Although this scenario is attractive, various types of analyses on ciliate IESs provide mixed evidence relating to this model. First, orthologous regions of the MIC genome in different ciliate species (29, 55, 56), or alternate alleles of a MIC gene within the same species (53, 140), have been found to differ in the presence or absence of IESs. This indicates that the pattern of IESs in the MIC genome has changed during ciliate evolution. These results could be explained by the loss of IESs in some species/alleles, but it seems more likely that IESs have been mobile in the MIC genome.

Second, a number of studies have found that some of the genes of transposon-like IES have recently been under selection for function of the encoded proteins (25, 41, 142). Selection of element-encoded functions is not expected following the typical invasion of an organism's genome by a transposon, but it would be expected in ciliates if transposon IES genes encoded proteins that were initially involved in development excision (75). Nonetheless, while element-encoded genes appear to have been under selection in the past, it seems unlikely that they now encode proteins that participate in excision. Transcripts from the *E. crassus* Tec1 and Tec2 elements are barely detectable during MAC development (64), and no transcripts were detected for the retroviral integrase-like genes of the *T. thermophila* Tlr elements (41), making it unlikely that they encode the proteins necessary for the massive amount of IES excision. The TBE elements of *O. trifallax* produce abundant heterogeneous RNAs during development, but few polyadenylated transcripts are present, and immunological analyses have failed to detect the element-encoded transposase protein [(25); K. Williams & G. Herrick, unpublished results]. Thus, there is little evidence that transposon IESs are producing the proteins required for development excision; however, these results do not rule out the possibility that the genes of these transposon IESs, or others, have already been assumed by their ciliate hosts to mediate excision.

A third area bearing on the possible transposon origin of IES excision concerns the *cis*-acting sequences that specify excision. The key *cis*-acting sequences for transposition are typically located near the ends of elements, so that if IES excision is a related process, one might expect similarly positioned regulatory elements. This is likely to be the case for the IESs of *Euplotes* and *Paramecium*. In these organisms, essentially all IESs (both small IESs and transposon-like IESs) are bounded by a 5′-TA-3′ direct repeat. Moreover, this dinucleotide forms part of a longer, more loosely conserved inverted repeat at the ends of the smaller IESs. The consensus sequences for the *Euplotes* and *Paramecium* small IESs ends are 5′-TATrGCRN-3′ and 5′TAYAGYNR-3′ (Y = pyrimidine, R = purine) (33, 57, 74). These sequences are quite similar to the ends of the *E. crassus* Tec1 and Tec2 transposon IESs (5′-TATAGAGG-3′) (65), as well as the broader Tc1/*mariner* family of elements (5′-TACAGTKS-3′; K = G or T, S = C or G), supporting the notion that the smaller IESs are derived from the transposon IESs.

A number of lines of evidence indicate these terminal sequences are important for excision. In *Paramecium*, single-base mutations in the conserved terminal inverted repeat of an IES can result in the failure to excise that particular IES during development (51, 92, 93, 121). Moreover, orthologous IESs have been identified in different strains of *Paramecium* where either the deletion boundaries or efficiency of excision are altered, and these correlate with base changes in the terminal inverted repeats (30, 31). For *E. crassus*, there is no direct evidence for the involvement of the conserved terminal inverted repeat, but the recent discovery of a third type of transposon IES, Tec3, supports a role in excision (M. Jacobs, A. Sanchez-Blanco, L. Katz & L. Klobutcher, unpublished results). The 4.48-kbp Tec3 elements are excised at the same stage of development as the Tec1 and Tec2 transposon IESs. They share no detectable sequence similarity with Tec1 and Tec2 elements, with the exception of their ends. The terminal 9 bp of the Tec3 element precisely matches the ends of the Tec1 and Tec2 elements, and 16 of the terminal 23 bp are identical.

While the terminal sequences of IESs in *Paramecium* and *Euplotes* play a role in defining the boundaries of excision, sequences flanking IESs have also been implicated in some instances. In transformation-based deletion analysis of a small 28-bp IES in *Paramecium*, removal of a portion of the 72-bp flanking one end of the IES reduced the efficiency of excision, and complete removal of the 72-bp flanking region abolished excision (82). This result parallels studies in *Tetrahymena*, where flanking regions have been shown to be important for the excision of a number of IESs (16, 44, 45, 84, 111). The best-studied IES in this regard is the M element. In this case, a 10-bp polypurine tract with the sequence 5′-AAAAAGGGGG-3′ (A_5G_5) is located about 45 bp away from each deletion boundary. The A_5G_5 sequences play a major role in potentiating deletion as well as defining the deletion endpoints, as mutation of the element abolishes deletion, while repositioning of the element moves the deletion boundaries to new positions that remain about 45 bp from the A_5G_5 element (44, 45). Curiously, other *Tetrahymena* IESs lack the flanking A_5G_5 elements, including those where flanking sequences have been shown to be important for excision. This suggests that there are multiple sequence elements that serve the same role in specifying IES excision. In addition to flanking sequences, there is evidence that multiple, and possibly redundant, sequences within the main body of *Tetrahymena* IESs appear to enhance the efficiency of excision (21, 44, 45). Thus, the *cis*-acting sequence requirements for *Tetrahymena* IES removal do not readily fit the expectations for a transposon-like excision system.

Finally, studies on intermediates and/or products of IES excision have produced some insights into the molecular mechanism(s) mediating the process. A variety of transposition mechanisms exist, but most share the common steps of DNA strand cleavage followed by one, or more, direct transesterification reactions to link DNA ends (24). Some of the ciliate IES excision processes have been proposed to occur by a similar mechanism. In *Tetrahymena*, the primary product of excision appears to be a linear IES (124). PCR-based studies on DNA from developing cells have detected putative cleavage intermediates with four nucleotide 5′-overhangs that coincide with the observed excision boundaries (122, 123). Moreover, there is evidence that IES excision may be initiated by cleavage of DNA at only one end (123), leading to a transposition-like model of IES excision (122, 123) (Figure 3*a*). The 3′-hydroxyl group generated by cleavage at one end of the IES is proposed to serve as a nucleophile that attacks a phosphodiester bond at the opposite boundary of the IES, resulting in a strand-transfer reaction that links one DNA strand of the flanking sequences. The other strand of the DNA at the newly created junction would then be cleaved to release the IES and repaired to form the MAC-destined junction. A recent study characterizing the ends of free, linear IESs has provided support for this model (124).

In other ciliates, IES excision intermediates have not been characterized, but information on excision products exists for three species. In each case excision results in a MAC-destined sequence that retains one copy of the direct repeat that bounds the IES and a free circular form of the excised IESs (Figure 3*b*,*c*). For

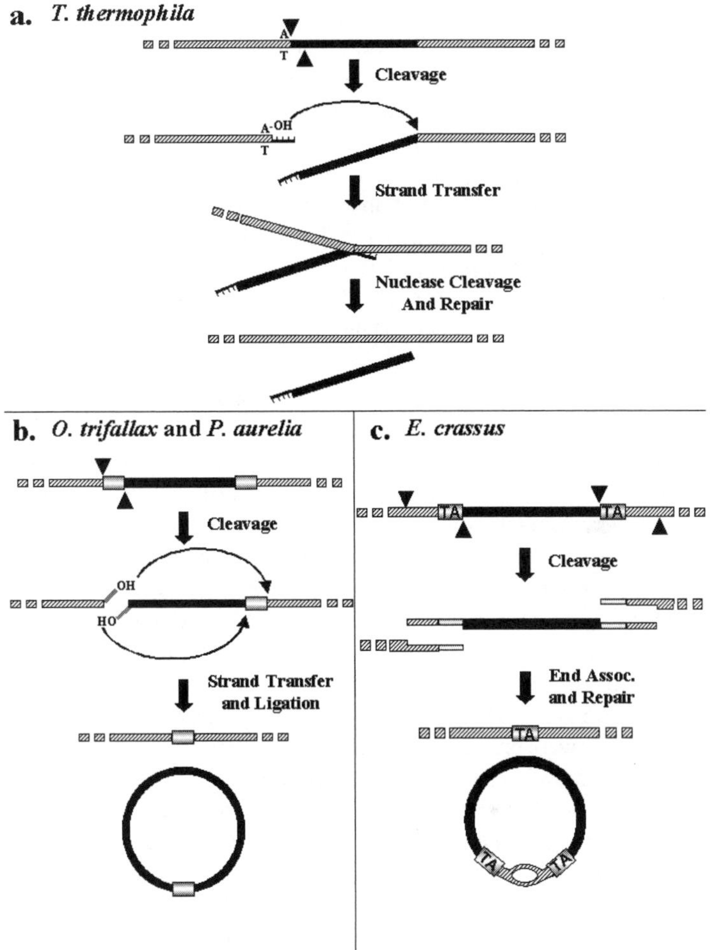

Figure 3 Products, intermediates, and models of IES excision in (*a*) *T. thermophila*, (*b*) *O. trifallax* and *Paramecium*, and (*c*) *E. crassus*. MAC-destined sequences are *cross-hatched*, and the IESs are indicated in *black* with the *shaded rectangles* representing their short terminal direct repeats. *Arrowheads* denote positions of DNA cleavage, and 3′-hydroxyl groups (−OH) involved in transesterification reactions and their targets are linked by *arrows*. For details, see text and references (46, 65, 77, 124, 140).

the *O. trifallax* TBE1 elements and the *Paramecium* IESs, the free circular IESs have a single direct repeat (5′-ANT-3′ and 5′-TA-3′) at their junctions (10, 140). These excision products suggest a transposition-like model of excision (46, 140) (Figure 3*b*) in which excision is initiated by a staggered, double-stranded break at one of the direct repeats of the IES. The hydroxyl groups of the two resulting 3′-ends would then serve as nucleophiles to attack phosphodiester bonds immediately adjacent to the other direct repeat, generating the free circular IES and precisely joined MAC junction. Alternatively, it has been suggested that the circular forms of the IESs may not be the primary products of excision and that IES removal in these organisms could initially generate linear molecules, perhaps by a process similar to that proposed for *Tetrahymena* (Figure 3*a*), with subsequent joining of the IES ends to form circles (46).

Circular forms of both Tec transposons and small IESs also have been found in *E. crassus* (60, 81, 135). However, despite the similarity in IES terminal sequence structure between *Euplotes* and *Paramecium*, the *E. crassus* circular IESs have a unique junction structure (65, 77). The junctions have two copies of the terminal 5′-TA-3′ direct repeat separated by 10 bases, part of which is a heteroduplex (i.e., the bases on the two DNA strands are not complementary). For the circular forms of the small IESs, the 10 bases are known to be derived from the sequences immediately flanking the unexcised IES, with the central 6 bases forming the heteroduplex region (77). Although the transposon Tn*916* is thought to form a free circular form with a heteroduplex following excision (126), this type of structure is extremely rare, and it is difficult to incorporate the *E. crassus* excision products into a model that involves a transesterification mechanism. Existing models propose that excision is initiated by staggered cuts at both ends of the IES (Figure 3*c*), with the resulting pairs of ends filled in and ligated in somewhat different manners to generate the two excision products [for details see (65, 77)].

Overall there is little doubt that transposons have come to be involved in the IES excision processes of at least some ciliates; however, it remains unclear as to whether they were primarily responsible for generating this form of rearrangement. If IES excision did originate with transposons, the bulk of the current data indicates that the excision functions are no longer encoded by the transposons and that they must have been assumed by the host ciliates. The analyses of excision products and intermediates suggest that IES removal may be occurring by a transposition-like mechanism in some, but not all, ciliates, but direct confirmation of the reaction mechanism(s) is still lacking. The ultimate answer to the relationship between IES removal and transposition will likely require the isolation and characterization of the proteins that directly mediate this process, and this remains an elusive goal.

DNA SCRAMBLING IN OXYTRICHIDS

A third form of DNA rearrangement, DNA unscrambling, has been found in the spirotrichs *Oxytricha* and *Stylonychia* [reviewed in (114, 115)]. The MIC copies of scrambled genes are not only interrupted by developmentally eliminated DNA, but the sequences that form the MAC chromosome are disordered (Figure 4) and in some cases inverted relative to others. The Prescott laboratory identified the first example of a scrambled gene, the actin gene in *O. nova* (48, 117), and this gene is now known to be scrambled in a number of *Oxytricha* and *Stylonychia* species (55). In addition, the MAC DNA molecules containing the α-telomere-binding

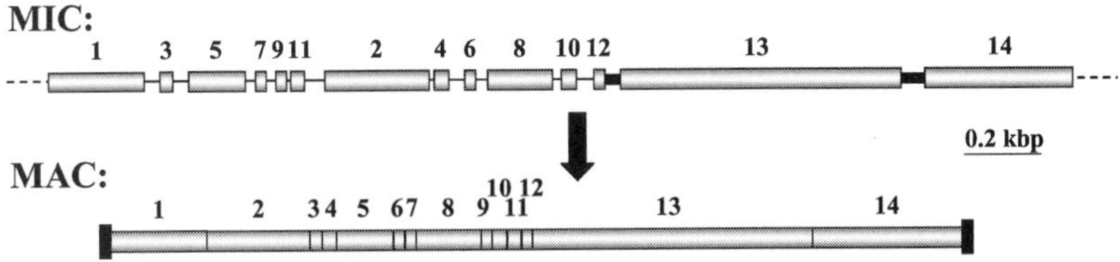

Figure 4 MIC and MAC organization of the scrambled gene encoding the α-subunit of the telomere-binding protein in *O. nova*. *Rectangles* represent MAC-destined DNA segments and are numbered to emphasize their differing linear arrangement in the MIC and MAC. Conventional IESs are shown as *thick lines*, eliminated DNA separating scrambled MAC-destined segments as *thin lines*, and telomeres as *black rectangles*. The figure is based on (105).

protein (119) and DNA polymerase α genes (54, 83) are also scrambled. As the MIC organizations of only 12 genes in total have been analyzed in these species, a scrambled arrangement may well be common (115). The DNA polymerase α gene represents the most highly segmented gene observed to date; it is split into >45 separate MAC-destined segments in the MIC. The MAC-destined segments of the actin gene give the appearance of being randomly scrambled, but the arrangement of MAC-destined sequences in the α-telomere-binding protein (Figure 4) and DNA polymerase α genes is decidedly nonrandom. For these genes, alternating blocks of MAC-destined sequences tend to be clustered together in the MIC (Figure 4). While the pattern of MAC-destined sequences is similar for the three genes in different species, there are differences, indicating that scrambling has occurred during recent evolution and may be an ongoing process.

Little is known about the timing or mechanism of unscrambling, although it almost certainly occurs prior to chromosome fragmentation, as the fragmentation/telomere addition sites often reside within the locus in the unrearranged MIC configuration. The unscrambling process also appears to involve recombination between short repeated sequence elements. The ends of pairs of MAC-destined sequence blocks that are joined together during unscrambling are bracketed by short blocks of identical sequence, averaging about 11 bp in length (115). The sequence of the repeats varies between pairs of MAC-destined ends, with no obvious similarities. The repeat length is longer than that for conventional IESs (i.e., where excision of the IES is linked to the joining of the immediately flanking sequences) and is nearly sufficient to provide an explanation of the specificity of the unscrambling process. That is, one expects to find only a limited number of identical sequences of this length in the entire genome. The nonrandom order of MAC-destined segments may facilitate the formation of "folded" structures at scrambled loci, which would enhance the specificity of the unscrambling process (115).

The evolutionary origin of scrambled genes is unclear, but a number of models have been proposed (83, 114, 115). A recent model (114) suggests that the interruption of MAC-destined sequences by conventional IESs served as a prelude to the scrambling process. Once a MAC-destined sequence contained conventional IESs, germline recombination between different IESs would result in both inversion and disordering of the MAC-destined DNA segments. To assess the validity of this model, the MIC arrangements of genes that are scrambled in *Oxytricha* and *Stylonychia* have been examined in more distantly related spirotrichs. The MIC actin genes of *Engelmanniella mobilis* and *Urostyla grandis*, species that diverge early in the spirotrich lineage, are not scrambled, but they are interrupted by conventional IESs (55). Similarly, the DNA polymerase α gene is not scrambled in *U. grandis* but contains 41 conventional IESs (W.-J. Chang, E. Curtis & L. Landweber, manuscript in preparation). These results are consistent with the proposal that conventional IESs could have potentiated the development of the scrambling system. Although they also suggest that scrambling developed late in evolution and may be limited to only a handful of ciliate species, it should be kept in mind that the MIC organizations for only a small number of genes are known for even the best-studied ciliates. As a result, gene scrambling could yet prove to be more widespread among ciliates.

EPIGENETIC EFFECTS ON DNA REARRANGEMENT

Epigenetic effects on DNA rearrangement have been documented in *Paramecium* and *Tetrahymena*. Because epigenetic effects in many cases involve heterochromatin formation as their underlying mechanism, we first describe the epigenetic effects in ciliates and then consider the evidence for chromatin structure as a determinant of DNA processing. The term "epigenetic" refers to changes in gene expression that arise outside of the gene itself or are not the result of nucleotide changes in the gene (52). Epigenetic effects are frequently seen as heritable changes in gene expression that can be reversible. For example, "position effect variegation" in *Drosophila* is an inactivation or silencing of genes due to chromosomal translocations that juxtapose the gene with heterochromatin. Because the determinants of heterochromatin structure can vary, the expression of the juxtaposed gene can differ from cell to cell (and its clonal progeny), leading to a variegated phenotype. In *Paramecium* and *Tetrahymena* the epigenetic effects are such that they result in heritable changes in the MAC that are not determined by the DNA sequence content of the germline MIC. Thus, in spite of the fact that the MAC is destroyed and regenerated from the MIC at each conjugation, the old MAC somehow directs the outcome of the formation of a new MAC. A recent review describes the epigenetic phenomena in detail (100), and we briefly describe these results and their implications.

The first case of an epigenetic effect on DNA processing was uncovered through the analysis of a *P. tetraurelia* mutant (*d48*) that failed to express the A immobilization antigen [a cell surface antigen that when recognized by antibodies causes immobilization of the cells (35)]. This mutation showed non-Mendelian behavior: The MIC was wild type, possessing normal copies of the gene (*A+*), but no matter what type of cell was mated to it, the progeny from the *d48* cell were *A−*, which

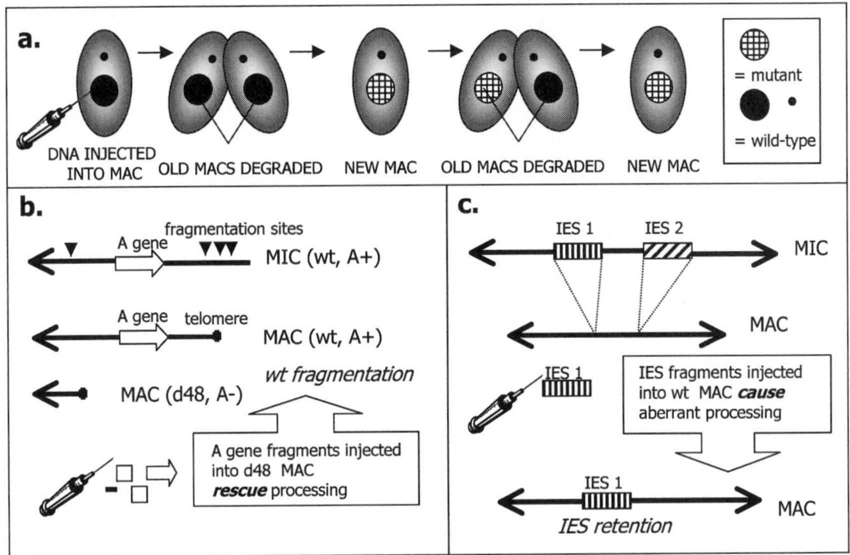

Figure 5 Microinjection of DNA and epigenetic effects. (*a*) The "maternal" inheritance pattern of a phenotype produced by injection of DNA into the vegetative MAC is diagrammed. (*b*) The structure of the *P. tetraurelia A* gene in the MIC, wild-type MAC, and *d48* mutant MAC is illustrated. "Curing" the *d48* defect through microinjection of *A* gene sequences is also shown. (*c*) An example of a sequence-specific epigenetic effect produced by introducing DNA containing an IES into the wild-type MAC. See text for details. wt, wild type.

was the same phenotype of MAC gene expression as the parental cell (Figure 5). This inheritance pattern resembles maternal inheritance in higher organisms, where the maternal contribution to the cytoplasm of the egg determines the phenotypic outcome regardless of progeny genotype. Intriguingly, injection of plasmids containing the *A* gene and various subregions of the gene and its immediately adjacent 5′ sequence into the vegetative MAC rescue the defect during the next round of MAC development (68, 71, 125, 147, 148). As shown in Figure 5, the *A* gene resides close to a set of fragmentation sites that generate a MAC chromosome. In the *d48* mutant, the MAC chromosomes were fragmented at a site that was 5′ to the *A* gene, thus deleting the gene. Injection of *A* gene sequences results in normal fragmentation and retention of the *A* gene in the new MAC. Thus, the presence or absence of a sequence in the old MAC determines the outcome of DNA processing in the developing MAC (anlagen).

Epigenetic changes in phenotype are also seen when DNA is introduced into the MAC of wild-type cells, but the effects can differ substantially from what was observed for the *d48* mutant. Instead of correcting a DNA-processing defect, the injected DNA can elicit abnormal DNA processing in a heritable manner (17, 32, 33, 97, 99, 101). In the first case studied, plasmids containing the G immobilization antigen gene of *P. primaurelia* (i.e., a MAC telomere-adjacent immobilization antigen gene similar to the *P. tetraurelia* A gene) were injected into the MAC (96). Following autonomy (conjugation without sexual exchange), progeny of these cells failed to express the *G* gene because they fragmented their chromosomes within a region that was 5′ to the gene instead of the normal heterogeneous fragmentation sites that are 3′ to the gene. Like the *A* gene rescue in *d48*, the injected fragments that caused the altered processing were sequence specific: Regions that elicited the new fragmentation site were restricted to *G* gene sequences and its immediate 5′ region. IES excision also can be altered by the sequence composition of the old MAC. In both *Tetrahymena* and *Paramecium*, introduction of high copy numbers of an IES into the MAC results in a failure to remove that IES during development with little or no effect on other IESs (Figure 5*c*). The conclusions from a large number of studies are that these are highly sequence-specific effects that are being transmitted from the old MAC to the developing MAC during conjugation. The sequence-specific nature of these processes suggests that DNA or RNA is somehow mediating the effect on processing.

As noted by others, the epigenetic effects on DNA processing resemble epigenetic effects seen in other organisms (17, 18, 32, 98). The alterations in processing induced by injected DNA are sensitive to dosage, which in this case is the amount of injected DNA (copy number) that is maintained in the vegetative MAC. This is similar to the "position effect variegation" results seen with integrations of transgenes in *Drosophila*, where higher copy numbers of tandem repeats of a gene result in increased silencing (28), or "repeat induced silencing" in *Neurospora* (128). These effects are elicited by heterochromatin formation, and DNA methylation with heterochromatin formation. Thus, the effect of increased copy number of injected sequences on ciliate DNA processing may be similar. The injected sequences may alter heterochromatin formation in the vicinity of sequences that are homologous to the injected DNA and thereby alter the DNA processing. Heterochromatin formation in these other organisms functions to silence genes. In contrast, in the ciliates, silencing, or heterochromatin formation, may be a signal for elimination. Based on what is known from other organisms on the mechanics of heterochromatin formation, two models come to mind with respect to these epigenetic effects in ciliates: One involves DNA pairing as a trigger for heterochromatin formation while the other involves RNA as the trigger (17, 18, 32, 98). In the next section we review what is known about chromatin and heterochromatin in the ciliates and how this relates to DNA processing, and then we discuss models in light of what is currently known.

CHROMATIN DIFFERENTIATION AND MACRONUCLEAR DEVELOPMENT

DNA elimination in ciliates has long been recognized as a chromatin elimination phenomenon. Microscopic studies of MAC development in the spirotrichs explored the relationship between chromosome structure and DNA elimination prior to any knowledge of the molecular events (79, 102, 103, 118, 130). In these organisms, MAC development begins with multiple rounds of DNA replication that result in the formation of polytene chromosomes. During the period of DNA elimination, the polytene chromosomes become compartmentalized into structures referred to as "vesicles" (a misnomer because they are proteinaceous structures that are not membrane bound). Thymidine labeling of DNA and analysis by EM autoradiography clearly show that chromatin is eliminated throughout the chromosomes because the amount of chromatin and labeled DNA subsequently decreases in the majority of compartments (79, 118). This indicates that the elimination process differs from other cases of chromatin diminution where elimination occurs in large chunks of chromosomes [for instance loss of large heterochromatic regions in nematodes (107)].

A distinction between eliminated and retained chromatin was observed in chromatin spreads of polytene chromosomes from *Stylonychia* (102, 103). Nucleosomal fibers (12 nm) were distinguishable from larger 30-nm fibers. The 30-nm fibers were seen as circles that were removed from the chromosomes and eventually disappeared, whereas the nucleosomal fibers remained and eventually were seen in lengths typical of the MAC linear DNA molecules. This suggests that a fairly large amount of DNA in *Stylonychia* is organized in a fiber structure that does not readily unravel to the nucleosomal structure during spreading and that these sequences are eliminated as chromatin circles. The size range of the chromatin circles suggests that this eliminated DNA is not IES DNA since it ranges in size from 18–150 kb (102). If *Stylonychia* is similar to *Oxytricha* and *Euplotes*, then these chromatin rings are most likely derived from the large amount of eliminated DNA that is outside the clustered MAC-destined sequences (62, 63). At present little is known about how these sequences are eliminated. However, in *E. crassus*, approximately two thirds of the Tec transposons are known to reside within these sequences. Analysis of the timing of DNA replication and elimination for this subclass of Tec elements demonstrated that these regions of the genome are "late replicating" and only replicate in the second period of DNA replication that forms the polytene chromosomes (39). Because "late replication" is a property of heterochromatic sequences in a diversity of organisms (141), this may be an indication of the heterochromatic nature of these sequences.

More recent studies have provided additional evidence that eliminated DNA in *Tetrahymena* and *Euplotes* is associated with a heterochromatic chromatin structure. In *Tetrahymena*, proteins have been identified that are synthesized specifically during conjugation and deposited into the anlagen (90). Three of these proteins have now been characterized by immunofluorescence and EM studies of their localization and by the isolation of the genes encoding them (89, 108, 129). The proteins are called Pdd1, 2, and 3 (Pdd = programmed DNA degradation). These proteins are found in both the old MAC and the anlagen and are thought to be associated with the elimination of the old MAC as well as the MIC-limited sequences in the anlagen, hence their designation as DNA degradation proteins. The Pdd proteins are initially diffusely localized in the developing MAC and then coalesce into large heterochromatic structures that resemble nucleoli (90, 129). These structures appear at the time of DNA rearrangement and subsequently become associated with the nuclear periphery. In situ hybridization demonstrated that eliminated DNA is associated with these structures. Likewise, chromatin immunoprecipitation showed an association of the Pdd proteins with IESs (108, 129). Differences in the timing of deposition and the localization of Pdd3 versus Pdd1 and 2 were apparent: Pdd3 is synthesized later and locates to the center and periphery of the structures, whereas Pdd1 or 2 are more peripheral. Thus, the Pdd3 protein may play a different role in the formation of the elimination structures.

Pdd1 (65 kD) and Pdd3 (32 kD) contain chromodomains (three in Pdd1 and one in Pdd3). The chromodomain is a conserved aa sequence present in heterochromatin-associated proteins from a wide range of organisms (80). In several cases the domain is associated with other domains, such as histone acetyltransferase, methyltransferase, and ATPase domains that modulate repressive chromatin effects. Mutational studies of the chromodomain of the *Drosophila* HP1 and polycomb proteins suggest that the chromodomain is not involved in DNA binding but is important for protein-protein interactions (95, 112). By analogy, the Pdd proteins are not thought to be DNA-binding proteins, and it is unknown how the Pdd proteins recognize the eliminated DNA to initiate heterochromatin formation. The recent finding that a chromodomain can bind RNA (1) suggests that the transcription of an IES prior to excision in *Tetrahymena* (described below) could play a role in heterochromatin assembly (18).

The functions of Pdd1 and Pdd2 have been examined in MAC knockouts [this disrupts all the ~45 copies of the MAC genes but leaves the MIC genes intact (22, 109)]. Thus, the initial expression of these *Pdd* genes, which occurs in the old MAC during the meiotic stage at about 3 h of development, is eliminated and transcription of the genes does not begin until ~7–10 h, when the normal copies of the genes present in the anlagen become active. This results in a decrease in the amount of Pdd proteins during the early stages of MAC development. These "knockdowns" result in a failure to complete MAC development. Assays of DNA processing indicate that IES excision is prevented, but DNA fragmentation proceeds normally in *Pdd1* MAC knockouts, whereas the *Pdd2* knockouts do not carry out IES excision or fragmentation and telomere addition. These disruptions in IES excision parallel the lack of formation of the large heterochromatic structures that contain the eliminated DNA and the Pdd proteins. The localization of these proteins instead remains diffuse. The old MACs are condensed and destroyed normally in the somatic mutants, but the cells do not eliminate one of the two MICs, a process that occurs late in development in wild-type cells. Likewise, the final endoreplication of the new MAC from 4C to 8C is blocked.

The formation of Pdd structures in *Tetrahymena* occurs in conjunction with many other changes in chromatin proteins that warrant description, as these well-documented changes set the stage for models of how chromatin structure relates to DNA processing and epigenetic effects. The Allis and Gorovsky labs have

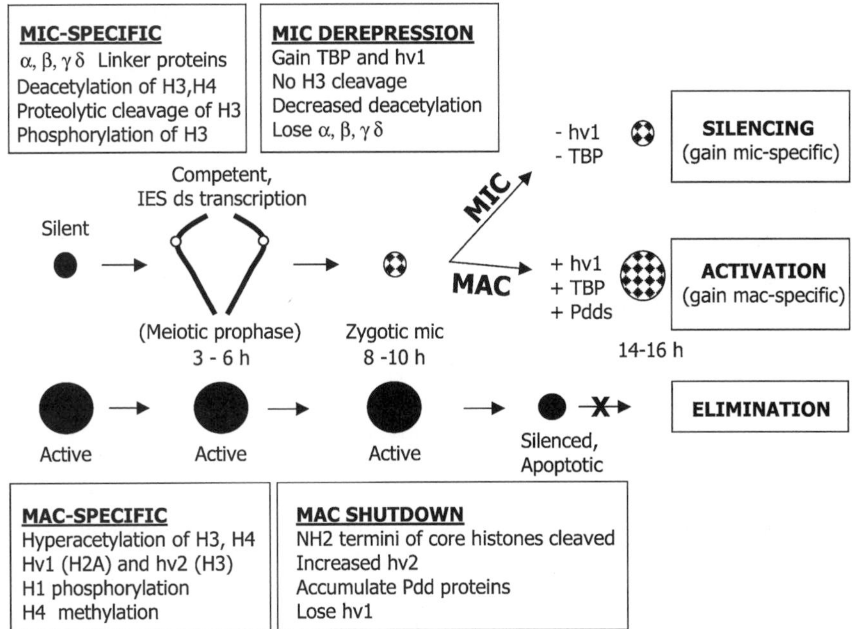

Figure 6 Summary of changes in histone proteins and their modifications in cellular nuclei during the timecourse of conjugation (times in hours are noted in the middle).

extensively characterized histones and their modifications in the MIC and MAC (Figure 6). Many of these change from a MIC-specific state to a MAC-specific state during conjugation (2–4, 20, 86, 87, 131, 132, 134, 139). The timing of the changes suggests a paradigm of chromatin differentiation that involves a multi-step process of reversal of MIC silencing followed by a progressive build up of the necessary structures for transcriptional activity. A "transcriptionally competent" state is apparent in the MIC at an early period in conjugation, corresponding to prophase of meiosis, as evidenced by the appearance of TATA-binding protein (TBP) and the hv1 histone H2A variant normally associated with the active MAC (131, 132). Interestingly, this corresponds to when transcripts of IESs are first detected (18). Like the aforementioned transcripts from *O. trifallax* TBE1 elements [(25); K. Williams & G. Herrick, unpublished results), these IES transcripts are heterogeneous in size and are produced from both strands of the IES. They are not polyadenylated, and transcription initiates outside of the IESs. Although it has not been formally tested, this early transcriptional activity of the MIC does not provide functional transcripts of genes.

During this early period of conjugation, some "deactivation" of the old MAC is also apparent, with loss of H1 histone phosphorylation and cleavage of the amino-terminal tails of the core histones [this results in loss of the major acetylation and methylation sites (86)]. In addition, the Pdd1 and Pdd2 proteins appear in the old MAC by 3 and 6 h (22, 109). At the 8–10 h time period, when copies of the zygotic nucleus first differentiate into MACs and MICs, further changes become evident: The old MAC condenses and becomes pycnotic and the hv1 and TBP proteins become restricted to the developing MAC (131, 132). Late in development, when the new MACs change in DNA content from 4C to 8C, the new MICs gain the MIC-specific linker proteins, and the new MACs gain histone H1 and the H3 variant hv2 (3, 131, 132). Changes in acetylation and deacetylation of histones H3 and H4 accompany these other chromatin changes (2, 20). Histone deacetylation predominates in the vegetative MIC, whereas acetylation predominates in the MAC. Both activities occur in the developing MAC, with the new MAC eventually becoming hyperacetylated late in development. Taken together these changes point to a complex series of chromatin remodeling steps in each nucleus. Thus, specific regulation of each developmental program of nuclear differentiation and coordination of the programs between the different nuclei is likely. Histone modifications are clearly a mechanism of regulating the differentiation of euchromatin and heterochromatin [or essentially transcriptional activation and silencing (67)]. The impact of the changes in histone modifications on DNA processing is only beginning to be addressed.

In *Euplotes*, studies of histone proteins and genes are less extensive, but the major differences uncovered to date involve anlagen-specific histone variants. Two abundant histone variants appear during the polytenization of chromosomes in the early anlagen of *E. crassus*: an H3 histone variant and an H2A variant [(42, 61) C.L. Jahn, unpublished observations]. The H3 histone variant has an extended amino terminus (+12 aa) with the potential to be modified at more lysine residues than a typical histone H3. The H2A variant is longer (+13 aa) at its carboxyl end, a conserved region of ubiquitination in typical H2A histones. The crystal structure of the nucleosome predicts that the extended ends of these two histones project from the same region of the nucleosome, specifically, the space occupied by DNA entering and exiting the core structure (88). Thus, the extended tails could play a role in modulating the opening and closing of the nucleosomal structure. Immunolocalization studies of the H3 variant indicate that it is associated with the anlagen throughout development and does not completely disappear with the eliminated DNA (42). This makes it unlikely that these histone variants are differentially localized with respect to eliminated and retained sequences. Nevertheless, they may provide additional control via modifications that differ between eliminated and retained DNA.

Chromatin structural changes associated with DNA rearrangement in *Euplotes* also have been documented. First, the nucleosomal spacing differs between the anlagen and the MIC (61). Throughout meiosis and until the time that the anlagen first begins to differentiate, the spacing of nucleosomes on the MIC-limited Tec elements is ~150 bp per repeat. As the polytene replications begin, the spacing becomes larger (~190 bp) and resembles the spacing in the MAC (~180 bp). In addition to this change in spacing, the Tec elements, which will all be eliminated, become incorporated into a chromatin structure that has highly unusual micrococcal nuclease digestion properties (58). Although the nucleosomal repeat length is the same as the MAC-destined sequences in the anlagen, the sizes of the multimeric micrococcal nuclease digestion products for the Tec element chromatin are shifted +70 bp from the ~200-bp multimeric repeats, and the minimal digestion product is ~300 bp in size instead of 150–200 bp. This suggests that digestion within linker regions between nucleosomes is suppressed and that the only digestion occurring is within the DNA associated with the nucleosomal core (58). Upon excision, the chromatin structure of the circular Tec element DNA becomes highly resistant to nuclease digestion, indicating that they have highly compacted structures.

The nuclease resistance of the Tec element chromatin, combined with its resistance to disruption by NaCl extraction, allowed the identification of an 85-kd protein as one of the few proteins that remains with the Tec element DNA (S. Sharp & C.L. Jahn, manuscript in preparation). Immunofluorescence with anti-p85 antibody demonstrated that the protein colocalizes with Tec element DNA and is specific to the developing MAC. It first appears in the anlagen during the replicative phase that forms the polytene chromosomes and disappears during the period when DNA is eliminated. The extraction properties of p85 are typical of a nuclear matrix or chromosomal scaffold (104). Thus, it was not unexpected to find that p85 coextracts and colocalizes with topoisomerase II, which is known to be associated with the nuclear matrix/chromosome scaffold in a wide range of organisms (9, 34, 40). This suggests that topoisomerase II may play a role in generating the compacted structure of the Tec element chromatin. Sequence analysis of the p85 gene indicates that it is a novel protein with alternating patches of acidic and basic amino acids: a property that may allow it to interact with both DNA and histones.

Overall, it is clear that macronuclear development entails a major reorganization of chromatin structure, and there is a growing amount of evidence that heterochromatin is a characteristic of eliminated DNA in ciliates. However, it is still unclear how heterochromatin relates to DNA processing. For instance, many of the sequences eliminated (particularly in spirotrichs) are extremely short stretches of DNA that would not encompass a single nucleosome and thus are unlikely to be organized into heterochromatin. Studies of other processes of DNA rearrangement and recombination suggest that the enzymology of these events requires a highly accessible chromatin structure similar to that required for transcription. In several cases, the actual sites of recombination and rearrangement occur in DNA that has a chromatin structure that is "nuclease hypersensitive" (85). Sensitivity to nucleases is a measurement of increased accessibility and, in some cases, distortion of DNA sequences (49). A logical question is whether the actual sites of DNA processing in ciliates (associated with the known consensus sequences) are recognizable as "hypersensitive" DNA structures, which would serve as the preferred targets of the DNA processing machinery. Based on preliminary data, the *E. crassus* Tec element inverted repeats may fit this scenario because the terminal ~100 bp of the repeats remain in a nucleosome-free, micrococcal nuclease "hypersensitive" structure throughout development (C.L. Jahn, unpublished data). In a similar vein, one hypothesis concerning the role of IES transcription in *Tetrahymena* is that this may keep these sequences in an open chromatin structure that is accessible to DNA processing (18).

One way to rationalize the need for accessible sites of processing with the finding of heterochromatic structures is that processing sites could occur at boundaries of chromatin domains in the cases where the eliminated DNA is a large region capable of forming heterochromatin. For instance, the Tec element IESs clearly differ in chromatin structure from the surrounding MAC-destined sequences, with the terminal inverted repeats forming some type of boundary (58). In contrast, the smaller IESs might simply represent nucleosome-free regions accessible to the rearrangement machinery. The observed epigenetic effects on DNA processing can also be accommodated under this view, if one assumes that they operate by altering either the state of hypersensitive sites or the formation of boundaries between euchromatin and heterochromatin. Injected DNA sequences could result in the aberrant formation of heterochromatin in regions that normally form a boundary or a hypersensitive site, thus masking a DNA processing site. Likewise, inducing different sites of heterochromatin formation could result in new boundaries, leading to new DNA processing sites. A variety of tools are available to address these issues, and it seems likely that we will soon know more about the chromatin structure of specific eliminated sequences as well as whether conditions that result in epigenetic effects also result in sequence-specific alterations in chromatin structure.

CONCLUDING REMARKS

The study of genome remodeling in ciliates has uncovered enormous variety in these processes. In some organisms, transposons form part of the DNA that is excised and eliminated, and many of the rearrangement processes are highly specific, as one might expect for a site-specific recombination system. Nonetheless, the observed epigenetic effects and changes in chromatin structure suggest a system more complex than the more limited DNA rearrangement systems observed in other organisms (e.g., V(D)J recombination or DNA inversion; see Introduction).

It seems likely that these rearrangement events represent an ongoing process of genome evolution within the special circumstances that nuclear duality allows. Thus, it is important to keep in mind the numerous genomic stability functions that a eukaryotic host of mobile elements would possess, and how such functions might have evolved under the special circumstances of ciliates. The involvement of heterochromatin in DNA processing suggests that ciliates are using an evolutionarily conserved genetic mechanism in a new way. DNA processing in ciliates likely arose within a system that already controlled transcription via chromatin structure, and it is carried out within the context of the transcriptional activation of an entire genome. The differences in histone modifications first noted in the ciliates by virtue of the differences seen between the transcriptionally silent MIC and the transcriptionally active MAC have proven to be conserved throughout all eukaryotes as a control mechanism for transcriptional activation and silencing (67). Similarly, chromodomain proteins are found throughout eukaryotes (80). Thus other aspects of MAC development and DNA processing may be conserved as well. It is frequently speculated that heterochromatin functions to silence transposons (66, 91). Perhaps it is not surprising that the ciliates would take these mechanisms one step further and simply use chromatin differentiation to purge transposons and their diverged remnants from the genome in order to generate a highly efficient "euchromatic" transcriptional state in the MAC. While many features of ciliates have initially been viewed as unusual and bizarre, investigations of these same features have often uncovered universal molecular mechanisms (e.g., self-splicing RNA, telomerase, and control by histone modification). Understanding how sequences are recognized for elimination and what type of enzymology carries out DNA rearrangement will also illuminate other universal processes.

ACKNOWLEDGMENTS

We sincerely thank the many colleagues who provided us with unpublished manuscripts and allowed us to cite their work, and we apologize to those whose work we were unable to include because of space limitations. C.L. Jahn is supported by NSF grant MCB-078182, and L.A. Klobutcher is supported by the National Science Foundation (MCB-9816765) and The Robert Leet & Clara Guthrie Patterson Trust.

The *Annual Review of Microbiology* is online at http://micro.annualreviews.org

LITERATURE CITED

1. Akhtar A, Zink D, Becker PB. 2000. Chromodomains are protein-RNA interaction modules. *Nature* 407:405–9
2. Allis CD, Chicoine LG, Richman R, Schulman IG. 1985. Deposition-related histone acetylation in micronuclei of conjugating *Tetrahymena. Proc. Natl. Acad. Sci. USA* 82:8048–52
3. Allis CD, Wiggins JC. 1984. Histone rearrangements accompany nuclear differentiation and dedifferentiation in *Tetrahymena. Dev. Biol.* 101:282–94
4. Allis CD, Wiggins JC. 1984. Proteolytic processing of micronuclear H3 and histone phosphorylation during conjugation in *Tetrahymena thermophila. Exp. Cell Res.* 153:287–98
5. Baird SE, Fino GM, Tausta SL, Klobutcher LA. 1989. Micronuclear genome organization in *Euplotes crassus*: a transposon-like element is removed during macronuclear development. *Mol. Cell. Biol.* 9:3793–807
6. Baird SE, Klobutcher LA. 1989. Characterization of chromosome fragmentation in two protozoans and identification of a candidate fragmentation sequence in *Euplotes crassus. Genes Dev.* 3:585–97
7. Bednenko J, Melek M, Greene EC, Shippen DE. 1997. Developmentally regulated initiation of DNA synthesis by telomerase: evidence for factor-assisted *de novo* telomere formation. *EMBO J.* 16:2507–18
8. Berg DE, Howe MM, eds. 1989. *Mobile DNA*. Washington, DC: ASM. 972 pp.
9. Berrios M, Osheroff N, Fisher PA. 1985. In situ localization of DNA topoisomerase II, a major polypeptide component of the

Drosophila nuclear matrix fraction. *Proc. Natl. Acad. Sci. USA* 82:4142–46
10. Betermier M, Duharcourt S, Seitz H, Meyer E. 2000. Timing of developmentally programmed excision and circularization of *Paramecium* internal eliminated sequences. *Mol. Cell. Biol.* 20:1553–61
11. Blackburn EH. 1995. Developmentally programmed healing of chromosomes. See Ref. 12, pp. 193–218
12. Blackburn EH, Greider CW, eds. 1995. *Telomeres*. Cold Spring Harbor, NY: Cold Spring Harbor Laboratory Press. 396 pp.
13. Boveri T. 1887. Ueber Differenzierung der Zellkerne wahrend der Furchung des Eies von *Ascaris megalocephala*. *Anat. Anz.* 2:688–93
14. Callahan RC, Shalke G, Gorovsky MA. 1984. Developmental rearrangements associated with a single type of expressed alpha-tubulin gene in Tetrahymena. *Cell* 36:441–45
15. Cartinhour SW, Herrick GA. 1984. Three different macronuclear DNAs in *Oxytricha fallax* share a common sequence block. *Mol. Cell. Biol.* 4:931–38
16. Chalker D, La Terza A, Wilson A, Kroenke C, Yao M-C. 1999. Flanking regulatory sequences of the *Tetrahymena* R deletion element determine the boundaries of DNA rearrangement. *Mol. Cell. Biol.* 19:5631–41
17. Chalker DL, Yao M-C. 1996. Non-Mendelian, heritable blocks to DNA rearrangement are induced by loading the somatic nucleus of *Tetrahymena* with germline-limited DNA. *Mol. Cell. Biol.* 16:3658–67
18. Chalker DL, Yao M-C. 2001. Nongenic, bidirectional transcription precedes and may promote developmental DNA deletion in *Tetrahymena thermophila*. *Genes Dev.* 15:1287–98
19. Cherry JM, Blackburn EH. 1985. The internally located telomeric sequences in the germ-line chromosomes of Tetrahymena are at the ends of transposon-like elements. *Cell* 43:747–58
20. Chicoine LG, Allis CD. 1986. Regulation of histone acetylation during macronuclear differentiation in *Tetrahymena*: evidence for control at the level of acetylation and deacetylation. *Dev. Biol.* 116:477–85
21. Coyne RS, Chalker DL, Yao M-C. 1996. Genome downsizing during ciliate development: nuclear division of labor through chromosome restructuring. *Annu. Rev. Genet.* 30:557–78
22. Coyne RS, Nikiforov MA, Smothers JF, Allis CD, Yao M-C. 1999. Parental expression of the chromodomain protein Pdd1p is required for completion of programmed DNA elimination and nuclear differentiation. *Mol. Cell* 4:865–72
23. Coyne RS, Yao M-C. 1996. Evolutionary conservation of sequences directing chromosome breakage and rDNA palindrome formation in Tetrahymenine ciliates. *Genetics* 144:1479–87
24. Craig NL. 1995. Unity in transposition reactions. *Science* 270:253–54
25. Doak TG. 2001. *Transposons in ciliated protozoa*. PhD thesis. Univ. Utah, Salt Lake City. 141 pp.
26. Doak TG, Doerder FP, Jahn CL, Herrick G. 1994. A proposed superfamily of transposase genes: transposon-like elements in ciliated protozoa and a common "D35E" motif. *Proc. Natl. Acad. Sci. USA* 91:942–46
27. Doak TG, Witherspoon DJ, Doerder FP, Williams K, Herrick G. 1997. Conserved features of TBE1 transposons in ciliated protozoa. *Genetica* 101:75–86
28. Dorer DR, Henikoff S. 1994. Expansions of transgene repeats cause heterochromatin formation and gene silencing in Drosophila. *Cell* 77:993–1002
29. DuBois ML, Prescott DM. 1997. Volatility of internal eliminated segments in germ line genes of hypotrichous ciliates. *Mol. Cell. Biol.* 17:326–37
30. Dubrana K, Amar L. 2001. Control of DNA excision efficiency in *Paramecium*. *Nucleic Acids Res.* 29:4654–62

31. Dubrana K, Le Mouel A, Amar L. 1997. Deletion endpoint allele-specificity in the developmentally regulated elimination of an internal sequence (IES) in *Paramecium. Nucleic Acids Res.* 25:2448–54
32. Duharcourt S, Butler A, Meyer E. 1995. Epigenetic self-regulation of developmental excision of an internal eliminated sequence in *Paramecium tetraurelia. Genes Dev.* 9:2065–77
33. Duharcourt S, Keller AM, Meyer E. 1998. Homology-dependent maternal inhibition of developmental excision of internal eliminated sequences in *Paramecium tetraurelia. Mol. Cell. Biol.* 18:7075–85
34. Earnshaw WC, Heck MM. 1985. Localization of topoisomerase II in mitotic chromosomes. *J. Cell Biol.* 100:1716–25
35. Epstein LM, Forney JD. 1984. Mendelian and non-Mendelian mutations affecting surface antigen expression in *Paramecium tetraurelia. Mol. Cell. Biol.* 4:1583–90
36. Fan Q, Yao M-C. 1996. New-telomere formation coupled with site-specific chromosome breakage in *Tetrahymena thermophila. Mol. Cell. Biol.* 16:1267–74
37. Fan Q, Yao M-C. 2000. A long stringent sequence for programmed chromosome breakage in *Tetrahymena thermophila. Nucleic Acids Res.* 28:895–900
38. Forney JD, Blackburn EH. 1988. Developmentally controlled telomere addition in wild-type and mutant *Paramecia. Mol. Cell. Biol.* 8:251–58
39. Frels JS, Tebeau CM, Doktor SZ, Jahn CL. 1996. Differential replication and DNA elimination in the polytene chromosomes of *Euplotes crassus. Mol. Biol. Cell* 7:755–68
40. Gasser SM, Laroche T, Falquet J, Boy de la Tour E, Laemmli U. 1986. Metaphase chromosome structure. Involvement of topoisomerase II. *J. Mol. Biol.* 188:613–29
41. Gershan JA, Karrer KM. 2000. A family of developmentally excised DNA elements in *Tetrahymena* is under selective pressure to maintain an open reading frame encoding an integrase-like protein. *Nucleic Acids Res.* 28:4105–12
42. Ghosh S, Klobutcher LA. 2000. A developmental-specific histone H3 localizes to the developing macronucleus of *Euplotes. Genesis* 26:179–88
43. Gilley D, Preer JJ, Aufderheide KJ, Polisky B. 1988. Autonomous replication and addition of telomere-like sequences to DNA microinjected into *Paramecium tetraurelia* macronuclei. *Mol. Cell. Biol.* 8:4765–72
44. Godiska R, James C, Yao M-C. 1993. A distant 10-bp sequence specifies the boundaries of a programmed DNA deletion in *Tetrahymena. Genes Dev.* 7:2357–65
45. Godiska R, Yao M-C. 1990. A programmed site-specific DNA rearrangement in Tetrahymena thermophila requires flanking polypurine tracts. *Cell* 61:1237–46
46. Gratias A, Betermier M. 2001. Developmentally programmed excision of internal DNA sequences in *Paramecium aurelia. Biochimie* 83:1009–22
47. Greene EC, Bednenko J, Shippen DE. 1998. Flexible positioning of the telomerase-associated nuclease leads to preferential elimination of nontelomeric DNA. *Mol. Cell. Biol.* 18:1544–52
48. Greslin AF, Prescott DM, Oka Y, Loukin SH, Chappel JC. 1989. Reordering of nine exons is necessary to form a functional actin gene in *Oxytricha nova. Proc. Natl. Acad. Sci. USA* 86:6264–48
49. Gross DS, Garrard WT. 1988. Nuclease hypersensitive sites in chromatin. *Annu. Rev. Biochem.* 57:159–97
50. Deleted in proof
51. Haynes WJ, Ling K-Y, Preston RR, Saimi Y, Kung C. 2000. The cloning and molecular analysis of *pawn-B* in *Paramecium tetraurelia. Genetics* 155:1105–17
52. Henikoff S, Matzke MA. 1997. Exploring and explaining epigenetic effects. *Trends Genet.* 13:293–95
53. Herrick G, Cartinhour S, Dawson D, Ang D, Sheets R, et al. 1985. Mobile elements

bounded by C4A4 telomeric repeats in Oxytricha fallax. *Cell* 43:759–68
54. Hoffman DC, Prescott DM. 1996. The germline gene encoding DNA polymerase α in the hypotrichous ciliate *Oxytricha nova* is extremely scrambled. *Nucleic Acids Res.* 24:3337–40
55. Hogan DJ, Hewitt EA, Orr KE, Prescott DM, Muller KM. 2001. Evolution of IESs and scrambling in the actin I gene in hypotrichous ciliates. *Proc. Natl. Acad. Sci. USA* 98:15101–6
56. Huvos P. 1995. Developmental DNA rearrangements and micronucleus-specific sequences in five species within the *Tetrahymena pyriformis* complex. *Genetics* 141:925–36
57. Jacobs ME, Klobutcher LA. 1996. The long and the short of developmental DNA deletion in *Euplotes crassus*. *J. Eukaryot. Microbiol.* 43:442–52
58. Jahn CL. 1999. Differentiation of chromatin during DNA elimination in *Euplotes crassus*. *Mol. Biol. Cell* 10:4217–30
59. Jahn CL, Doktor SZ, Frels JS, Jaraczewski JW, Krikau MF. 1993. Structures of the *Euplotes crassus* Tec1 and Tec2 elements: identification of putative transposase coding regions. *Gene* 133:71–78
60. Jahn CL, Krikau MF, Shyman S. 1989. Developmentally coordinated en masse excision of a highly repetitive element in E. crassus. *Cell* 59:1009–18
61. Jahn CL, Ling Z, Tebeau CM, Klobutcher LA. 1997. An unusual histone H3 specific for early macronuclear development in *Euplotes crassus*. *Proc. Natl. Acad. Sci. USA* 94:1332–37
62. Jahn CL, Nilles LA, Krikau MF. 1988. Organization of the *Euplotes crassus* micronuclear genome. *J. Protozool.* 35:590–601
63. Jahn CL, Prescott KE, Waggener MW. 1988. Organization of the micronuclear genome of *Oxytricha nova*. *Genetics* 120:123–34
64. Jaraczewski JW, Frels JS, Jahn CL. 1994. Developmentally regulated, low abundance Tec element transcripts in *Euplotes crassus*—implications for DNA elimination and transposition. *Nucleic Acids Res.* 22:4535–42
65. Jaraczewski JW, Jahn CL. 1993. Elimination of Tec elements involves a novel excision process. *Genes Dev.* 7:95–105
66. Jensen S, Gassama MP, Heidman T. 1999. Taming of transposable elements by homology-dependent gene silencing. *Nat. Genet.* 21:209–12
67. Jenuwein T, Allis CD. 2001. Translating the histone code. *Science* 293:1074–79
68. Jessop-Murray H, Martin LD, Gilley D, Preer JJ, Polisky B. 1991. Permanent rescue of a non-Mendelian mutation of *Paramecium* by microinjection of specific DNA sequences. *Genetics* 129:727–34
69. Jonsson F, Steinbruck G, Lipps HJ. 2001. Both subtelomeric regions are required and sufficient for specific DNA-fragmentation during macronuclear development in *Stylonychia lemnae*. *Genome Biol.* 2:research0005.1–.11
70. Jonsson F, Wem JP, Fetzer CP, Lipps HJ. 1999. A subtelomeric DNA sequence is required for correct processing of the macronuclear DNA sequences during macronuclear development in the hypotrichous ciliate *Stylonychia lemnae*. *Nucleic Acids Res.* 27:2832–41
71. Kim CS, Preer JJ, Polisky B. 1994. Identification of DNA segments capable of rescuing a non-Mendelian mutant in *Paramecium*. *Genetics* 136:1325–28
72. Klobutcher LA. 1999. Characterization of in vivo developmental chromosome fragmentation intermediates in *Euplotes crassus*. *Mol. Cell* 4:695–704
73. Klobutcher LA, Gygax SE, Podoloff JD, Vermeesch JR, Price CM, et al. 1998. Conserved DNA sequences adjacent to chromosome fragmentation sites in *Euplotes crassus*. *Nucleic Acids Res.* 26:4230–40
74. Klobutcher LA, Herrick G. 1995. Consensus inverted terminal repeat sequence of *Paramecium* IESs: resemblance to termini of Tc1-related and *Euplotes* Tec

transposons. *Nucleic Acids Res.* 23:2006–13
75. Klobutcher LA, Herrick G. 1997. Developmental genome reorganization in ciliated protozoa: the transposon link. *Prog. Nucleic Acid Res. Mol. Biol.* 56:1–62
76. Klobutcher LA, Swanton MT, Donini P, Prescott DM. 1981. All gene-sized DNA molecules in four species of hypotrichs have the same terminal sequence and an unusual 3′ terminus. *Proc. Natl. Acad. Sci. USA* 78:3015–19
77. Klobutcher LA, Turner LR, LaPlante J. 1993. Circular forms of developmentally excised DNA in *Euplotes crassus* have a heteroduplex junction. *Genes Dev.* 7:84–94
78. Kloc M, Zagrodzinska B. 2001. Chromatin elimination—an oddity or a common mechanism in differentiation and development? *Differentiation* 68:84–91
79. Kloetzel JA. 1970. Compartmentalization of the developing macronucleus following conjugation in *Stylonychia* and *Euplotes. J. Cell Biol.* 47:395–407
80. Koonin EV, Zhou S, Lucchesi JC. 1995. The chromo superfamily: new members, duplication of the chromodomain and possible role in delivering transcriptional regulators to chromatin. *Nucleic Acids Res.* 23:4229–33
81. Krikau MF, Jahn CL. 1991. Tec2, a second transposon-like element demonstrating developmentally programmed excision in *Euplotes crassus. Mol. Cell. Biol.* 11:4751–59
82. Ku M, Mayer K, Forney JD. 2000. Developmentally regulated excision of a 28-base-pair sequence from the *Paramecium* genome requires flanking DNA. *Mol. Cell. Biol.* 20:8390–96
83. Landweber LF, Kuo T-C, Curtis EA. 2000. Evolution and assembly of an extremely scrambled gene. *Proc. Natl. Acad. Sci. USA* 97:3298–303
84. Li J, Pearlman RE. 1996. Programmed DNA rearrangement from an intron during nuclear development in *Tetrahymena thermophila*: molecular analysis and identification of potential *cis*-acting sequences. *Nucleic Acids Res.* 24:1943–49
85. Lichten M, Goldman AS. 1995. Meiotic recombination hotspots. *Annu. Rev. Genet.* 29:423–44
86. Lin R, Cook RG, Allis CD. 1991. Proteolytic removal of core histone amino termini and dephosphorylation of histone H1 correlate with the formation of condensed chromatin and transcriptional silencing during *Tetrahymena* macronuclear development. *Genes Dev.* 5:1601–10
87. Lin R, Leone JW, Cook RG, Allis CD. 1989. Antibodies specific to acetylated histones document the existence of deposition- and transcription-related histone acetylation in *Tetrahymena. J. Cell Biol.* 108:1577–88
88. Luger K, Mader AW, Richmond RK, Sargent DF, Richmond TJ. 1997. Crystal structure of the nucleosome core particle at 2.8 Å resolution. *Nature* 389:251–60
89. Madireddi MT, Coyne RS, Smothers JF, Mickey KM, Yao M-C, Allis CD. 1996. Pdd1p, a novel chromodomain-containing protein, links heterochromatin assembly and DNA elimination in Tetrahymena. *Cell* 87:75–84
90. Madireddi MT, Davis MC, Allis CD. 1994. Identification of a novel polypeptide involved in the formation of DNA-containing vesicles during macronuclear development in *Tetrahymena. Dev. Biol.* 165:418–31
91. Matzke MA, Mette MF, Aufsatz W, Jakowitsch J, Matzke AJ. 1999. Host defenses to parasitic sequences and the evolution of genetic control mechanisms. *Genetica* 107:271–87
92. Mayer KM, Forney JD. 1999. A mutation in the flanking 5′-TA-3′ dinucleotide prevents excision of an internal eliminated sequence from the *Paramecium tetraurelia* genome. *Genetics* 151:597–604
93. Mayer KM, Mikami K, Forney JD. 1998. A mutation in *Paramecium tetraurelia* reveals functional and structural features of

developmentally excised DNA elements. *Genetics* 148:139–49
94. Melek M, Greene EC, Shippen DE. 1996. Processing of nontelomeric 3′ ends by telomerase: default template alignment and endonucleolytic cleavage. *Mol. Cell. Biol.* 16:3437–45
95. Messmer S, Fanke A, Paro R. 1992. Analysis of the functional role of the Polycomb chromodomain in *Drosophila melanogaster. Genes Dev.* 6:1241–54
96. Meyer E. 1992. Induction of specific macronuclear developmental mutations by microinjection of a cloned telomeric gene in *Paramecium primaurelia. Genes Dev.* 6:211–22
97. Meyer E, Butler A, Dubrana K, Duharcourt S, Caron F. 1997. Sequence-specific epigenetic effects of the maternal somatic genome on developmental rearrangements of the zygotic genome in *Paramecium primaurelia. Mol. Cell. Biol.* 17:3589–99
98. Meyer E, Duharcourt S. 1996. Epigenetic programming of developmental genome rearrangement in ciliates. *Cell* 87:9–12
99. Meyer E, Duharcourt S. 1996. Epigenetic regulation of programmed genomic rearrangements in *Paramecium aurelia. J. Eukaryot. Microbiol.* 43:453–61
100. Meyer E, Garnier O. 2002. Non-Mendelian inheritance and homology-dependent effects in ciliates. *Adv. Genet.* 46:305–38
101. Meyer E, Keller A-M. 1996. A Mendelian mutation affecting mating-type determination also affects developmental genomic rearrangements in *Paramecium tetraurelia. Genetics* 143:191–202
102. Meyer GF, Lipps HJ. 1980. Chromatin elimination in the hypotrichous ciliate *Stylonychia mytilus. Chromosoma* 77:285–97
103. Meyer GF, Lipps HJ. 1981. The formation of polytene chromosomes during macronuclear development of the hypotrichous ciliate *Stylonychia mytilus. Chromosoma* 82:309–14
104. Mirkovitch J, Mirault M-E, Laemmli UK. 1984. Organization of the higher-order chromatin loops: specific DNA attachment sites on nuclear scaffold. *Cell* 39:223–32
105. Mitcham JL, Lynn AJ, Prescott DM. 1992. Analysis of a scrambled gene: the gene encoding alpha-telomere-binding protein in *Oxytricha nova. Genes Dev.* 6:788–800
106. Mollenbeck M, Klobutcher LA. 2002. *De novo* telomere addition to spacer sequences prior to their developmental degradation in *Euplotes crassus. Nucleic Acids Res.* 30:523–31
107. Muller F, Tobler H. 2000. Chromatin diminution in the parasitic nematodes *Ascaris suum* and *Parascaris univalens. Int. J. Parisitol.* 30:391–99
108. Nikiforov MA, Gorovsky MA, Allis CD. 2000. A novel chromodomain protein, pdd3p, associates with internal eliminated sequences during macronuclear development in *Tetrahymena thermophila. Mol. Cell. Biol.* 20:4128–34
109. Nikiforov MA, Smothers JF, Gorovsky MA, Allis CD. 1999. Excision of micronuclear-specific DNA requires parental expression of Pdd2p and occurs independently from DNA replication in *Tetrahymena thermophila. Genes Dev.* 13:2852–62
110. Oettinger MA. 1999. V(D)J recombination: on the cutting edge. *Curr. Opin. Cell Biol.* 11:325–29
111. Patil N, Karrer K. 2000. A developmentally regulated deletion element with long terminal repeats has *cis*-acting sequences in the flanking DNA. *Nucleic Acids Res.* 28:1465–72
112. Platero SJ, Hartnett T, Eissenberg JC. 1995. Functional analysis of the chromodomain of HP1. *EMBO J.* 14:3977–86
113. Prescott DM. 1994. The DNA of ciliated protozoa. *Microbiol. Rev.* 58:233–67
114. Prescott DM. 1999. The evolutionary scrambling and developmental unscrambling of germline genes in hypotrichous ciliates. *Nucleic Acids Res.* 27:1243–50

115. Prescott DM. 2000. Genome gymnastics: unique modes of DNA evolution and processing in ciliates. *Nat. Rev. Genet.* 1:191–98
116. Prescott DM, Dizick SJ. 2000. A unique pattern of intrastrand anomolies in base composition of the DNA in hypotrichs. *Nucleic Acids Res.* 28:4679–88
117. Prescott DM, Greslin AF. 1992. Scrambled actin I gene in the micronucleus of *Oxytricha nova*. *Dev. Genet.* 13:66–74
118. Prescott DM, Murti KG. 1973. Chromosome structure in ciliated protozoans. *Cold Spring Harbor Symp. Quant. Biol.* 38:609–18
119. Prescott JD, DuBois ML, Prescott DM. 1998. Evolution of the scrambled germline gene encoding alpha-telomere binding protein in three hypotrichous ciliates. *Chromosoma* 107:293–303
120. Ribas-Aparicio RM, Sparkowski JJ, Proulx AE, Mitchell JD, Klobutcher LA. 1987. Nucleic acid splicing events occur frequently during macronuclear development in the protozoan *Oxytricha nova* and involve the elimination of unique DNA. *Genes Dev.* 1:323–36
121. Ruiz F, Krzywicka A, Klotz C, Keller A-M, Cohen J, et al. 2000. The *SM19* gene, required for duplication of basal bodies in *Paramecium*, encodes a novel tubulin, η-tubulin. *Curr. Biol.* 10:1451–54
122. Saveliev SV, Cox MM. 1995. Transient DNA breaks associated with programmed genomic deletion events in conjugating cells of *Tetrahymena thermophila*. *Genes Dev.* 9:248–55
123. Saveliev SV, Cox MM. 1996. Developmentally programmed DNA deletion in *Tetrahymena thermophila* by a transposon-like reaction pathway. *EMBO J.* 15:2858–69
124. Saveliev SV, Cox MM. 2001. Product analysis illuminates the final steps of IES deletion in *Tetrahymena thermophila*. *EMBO J.* 20:3251–61
125. Scott JM, Mikami K, Leeck CL, Forney JD. 1994. Non-Mendelian inheritance of macronuclear mutations is gene specific in *Paramecium tetraurelia*. *Mol. Cell. Biol.* 14:2479–84
126. Scott JR, Churchward GG. 1995. Conjugative transposition. *Annu. Rev. Microbiol.* 49:367–97
127. Seegmiller A, Williams KR, Herrick G. 1997. Two two-gene macronuclear chromosomes of the hypotrichous ciliates *Oxytricha fallax* and *O. trifallax*. *Dev. Genet.* 20:348–57
128. Selker EU. 1999. Gene silencing: repeats that count. *Cell* 97:157–60
129. Smothers JF, Mizzen CA, Tubbert MM, Cook RG, Allis CD. 1997. Pdd1p associates with germline-restricted chromatin and a second novel anlagen-enriched protein in developmentally programmed DNA elimination structures. *Development* 124:4537–45
130. Spear BB, Lauth MR. 1976. Polytene chromosomes of *Oxytricha*: biochemical and morphological changes during macronuclear development in a ciliated protozoan. *Chromosoma* 54:1–13
131. Stargell LA, Bowen J, Dadd CA, Dedon PC, Davis M, et al. 1993. Temporal and spatial association of histone H2A variant hv1 with transcriptionally competent chromatin during nuclear development in *Tetrahymena thermophila*. *Genes Dev.* 7:2641–51
132. Stargell LA, Gorovsky MA. 1994. TATA-binding protein and nuclear differentiation in *Tetrahymena thermophila*. *Mol. Cell. Biol.* 14:723–34
133. Steele CJ, Barkocy GG, Preer LB, Preer JJ. 1994. Developmentally excised sequences in micronuclear DNA of *Paramecium*. *Proc. Natl. Acad. Sci. USA* 91: 2255–59
134. Sweet MT, Jones K, Allis CD. 1996. Phosphorylation of linker histone is associated with transcriptional activation in a normally silent nucleus. *J. Cell Biol.* 135:1219–28
135. Tausta SL, Klobutcher LA. 1989. Detection of circular forms of eliminated DNA

during macronuclear development in E. crassus. *Cell* 59:1019–26
136. Tausta SL, Turner LR, Buckley LK, Klobutcher LA. 1991. High fidelity developmental excision of Tec1 transposons and internal eliminated sequences in *Euplotes crassus. Nucleic Acids Res.* 19: 3229–36
137. Wang H, Blackburn EH. 1997. *De novo* telomere addition by *Tetrahymena* telomerase *in vitro. EMBO J.* 16:866–79
138. Wells JM, Ellingson JL, Catt DM, Berger PJ, Karrer KM. 1994. A small family of elements with long inverted repeats is located near sites of developmentally regulated DNA rearrangement in *Tetrahymena thermophila. Mol. Cell. Biol.* 14:5939–49
139. Wenkert D, Allis CD. 1984. Timing of the appearance of macronuclear-specific histone variant hv1 and gene expression in developing new macronuclei of *Tetrahymena thermophila. J. Cell Biol.* 98:2107–17
140. Williams K, Doak TG, Herrick G. 1993. Developmental precise excision of *Oxytricha trifallax* telomere-bearing elements and formation of circles closed by a copy of the flanking target duplication. *EMBO J.* 12:4593–601
141. Wintersberger E. 2000. Why is there late replication? *Chromosoma* 109:300–7
142. Witherspoon DJ, Doak TG, Williams KR, Seegmiller A, Seger J, Herrick G. 1997. Selection on the protein-coding genes of the *TBE1* family of transposable elements in the ciliates *Oxytricha fallax* and *O. trifallax. Mol. Biol. Evol.* 14:696–706
143. Wright A-DG, Lynn DH. 1997. Maximum ages of ciliate lineages estimated using a small subunit rRNA molecular clock: crown eukaryotes date back to the paleoproterozoic. *Arch. Protistenkd.* 148:329–41
144. Yao M-C, Choi J, Yokoyama S, Austerberry CF, Yao C-H. 1984. DNA elimination in Tetrahymena: a developmental process involving extensive breakage and rejoining of DNA at defined sites. *Cell* 36:433–40
145. Yao M-C, Yao C-H, Monks B. 1990. The controlling sequence for site-specific chromosome breakage in Tetrahymena. *Cell* 63:763–72
146. Yao M-C, Zheng K, Yao C-H. 1987. A conserved nucleotide sequence at the sites of developmentally regulated chromosomal breakage in Tetrahymena. *Cell* 48:779–88
147. You Y, Aufderheide K, Morand J, Rodkey K, Forney J. 1991. Macronuclear transformation with specific DNA fragments controls the content of the new macronuclear genome in *Paramecium tetraurelia. Mol. Cell. Biol.* 11:1133–37
148. You Y, Scott J, Forney J. 1994. The role of macronuclear DNA sequences in the permanent rescue of a non-Mendelian mutation in *Paramecium tetraurelia. Genetics* 136:1319–24
149. Yu GL, Blackburn EH. 1991. Developmentally programmed healing of chromosomes by telomerase in Tetrahymena. *Cell* 67:823–32

Annu. Rev. Microbiol. 2002. 56:521–38
doi: 10.1146/annurev.micro.56.012302.160643

First published online as a Review in Advance on May 21, 2002

Common Principles in Viral Entry

Minna M. Poranen,[1] Rimantas Daugelavičius,[1,2] and Dennis H. Bamford[1]

[1]*Institute of Biotechnology and Department of Biosciences, University of Helsinki, Finland; e-mail:minna.poranen@helsinki.fi; dennis.bamford@helsinki.fi*
[2]*Department of Biochemistry and Biophysics, Vilnius University, LT-2009 Vilnius, Lithuania; e-mail: rimantas.daugelavicius@gf.vu.lt*

Key Words virus, bacteriophage, uncoating, energetics, virus cell interactions

■ **Abstract** Viruses occur throughout the biosphere. Cells of Eukarya, Bacteria, and Archaea are infected by a variety of viruses that considerably outnumber the host cells. Although viruses have adapted to different host systems during evolution and many different viral strategies have developed, certain similarities can be found. Viruses encounter common problems during their entry process into the host cells, and similar strategies seem to ensure, for example, that the movement toward the site of replication and the translocation through the host membrane occur. The penetration of the host cell's external envelope involves, across the viral world, either fusion between two membranes, channel formation through the host envelope, disruption of the membrane vesicle, or a combination of these events. Endocytic-type events may occur during the entry of a bacterial virus as well as during the entry of an animal virus; the same applies for membrane fusion.

CONTENTS

0066-4227/02/1013-0521$14.00

INTRODUCTION

Viruses are intracellular parasites that commonly have an extracellular phase between their reproduction cycles. This implies that viruses must have ways of transporting their genetic material and other needed components through the host cell barriers into the site of replication in the beginning of each new infection cycle. The structure and the properties of the cellular barriers are among the critical factors restricting the viral host range.

Viruses are diverse in their structure. The structure in turn has clear implications for the function of the virion i.e., how to deliver the viral genome into the host cell. However, the molecular mechanisms employed during virus entry obviously must obey the common physicochemical rules of cells. Consequently, the viral entry strategies follow certain principles, and similarities are found among viruses infecting highly divergent host cells.

We discuss the virus entry by describing similarities that exist between the entry strategies of viruses infecting organisms of different origins—even from viruses infecting cells from different domains of life: Eukarya, Bacteria, and Archaea. We intend to raise interesting viral examples equally from different lineages of life. Our aim is not to cover the details and variations, as excellent recent reviews are available (45, 52, 64, 67, 70, 80, 86, 100). Instead, we attempt to unify the discussion by covering the field using examples across the viral world.

Viruses Are Spread to All Domains of Life

Viruses are spread throughout the whole biosphere; cells of Eukarya, Bacteria, and Archaea are predisposed to viral infections. The number of viruses in the biosphere outnumber the host cells by at least an order of magnitude (9, 102). Evidence is accumulating that viruses have evolved with their hosts and have adapted to the different cellular systems during their common evolution (11). However, structurally similar viruses have also been discovered infecting highly divergent cell types, even infecting cells from different domains of life (3, 8, 42), indicating that viral functions may remain conserved for a long evolutionary period. Often such structural similarities also have functional consequences on the entry process.

Two groups of viruses have been under considerable investigation. Animal viruses have been extensively studied owing to their medical importance. Bacterial viruses, especially those infecting *Escherichia coli*, have commonly been in the forefront when basic concepts of biology and virology have been developed. Although many plant viruses have immense agricultural importance, they are less studied than bacterial and animal viruses. Also new information on algal viruses is accumulating. In contrast to the large number of known viruses for bacterial and eukaryotic hosts, only some 45 viruses have been described infecting archaeal hosts (1), and few of these have been studied in detail. This biased sampling obviously has affected our understanding and thinking of the origin and evolution of viruses, as well as of the ways they interact with their host cells.

Host Cell Barriers

Different cell types expose different external envelope structures, which the entering viruses have to penetrate in the beginning of each infection cycle. The plasma membrane, the limiting membrane of the cell, is the basic barrier all viruses have to encounter during their way from one cell to another.

Most plant, fungal, bacterial, and archaeal cells gain structural support from the cell wall located outside the plasma membrane, and these rigid structures make an additional barrier for the entering virus. Practically all bacteria have a peptidoglycan-containing cell wall. The peptidoglycan of gram-negative cells is uni- to triple-layered with a thickness of 2.5–7.5 nm (50), whereas in gram-positive bacteria the peptidoglycan can be about 25 nm thick (10). In Archaea a variety of cell wall types have been described from cell walls totally lacking a polysaccharide component to those that closely resemble peptidoglycan. In plants the major component of the cell wall is cellulose, while in algae the structure and the chemistry of the cell wall is more diverse (55). The fungal cell wall is commonly composed of chitin (55). Generally the cell wall severely restricts the passage of macromolecular complexes. For example, the cutoff value for the passage of proteins through the bacterial peptidoglycan is about 50 kDa (26). In addition to the peptidoglycan, gram-negative bacteria contain a second lipid bilayer that covers the peptidoglycan and comprises an additional layer around the cell. This bacterial outer membrane restricts the passage of large complexes, but it is relatively permeable to small metabolites owing to the porin channels (55). Furthermore, many prokaryotic cells have an external capsule or a slime layer that may play a role in attachment to surfaces or in resistance to desiccation.

Eukaryotic cells have the ability to take large macromolecular complexes into the cell using endocytosis. Several different endocytic processes exist. The receptor-mediated endocytosis via clathrin-coated pits is the best understood (57). In addition, animal cells can uptake extracellular material via caveolae, phagocytosis, and pinocytosis (76). Bacterial cells do not commonly employ such endocytic processes. Instead, pathways ancestrally related to bacterial conjugation systems allow the transport of nucleic acids and proteins through specific protein complexes spanning the envelope (21).

Many animal and some plant viruses use the host cell nucleus as their site of replication and thus have evolved mechanisms for nuclear entry. The normal transport of macromolecules between the nucleoplasm and the cytosol is controlled by nuclear pore complexes, which are tunnel-like structures at the nuclear envelope (71). The functional diameter of the nuclear pore complex is around 25 nm (31), which is clearly less than the diameter of most viral capsids.

Viral Strategies

The virion has two main tasks. First, it provides a stable protective coat for the genome against the physical insults during the passage from one cell to another. Second, it recognizes the correct host and fulfills the genome delivery role. The virion structure is a compromise between efficient assembly, stability, and entry requirements and can thus give indications of the entry mechanisms. For example, enzymes (e.g., polymerases and lytic-enzymes) packaged into the virion are needed during the early stages of infection, prior to the initiation of viral protein production.

The nature of the genome has many implications on the viral life cycle. As cells do not have machineries to cope with dsRNA or negative-strand (−) ssRNA, the dsRNA and (−) ssRNA viruses have to deliver not only their genome but also their own polymerase into the cell in order to replicate. Many dsRNA viruses deliver their genome within a protein capsid, and the dsRNA is never exposed to cell cytosol. On the other hand most DNA and (+) ssRNA viruses can rely on host functions in genome replication and transcription, and the delivery of viral nucleic acid into the cell is often sufficient to initiate a productive infection.

The initial recognition of the host cell is accomplished by specific interactions between the virus and its host. Receptor–binding proteins on the virion surface are adapted to specifically bind to structures normally exposed on the surface of a susceptible host cell (37, 41). The subsequent uncoating and host envelope penetration are dependent on fusogenic or pore-forming proteins, and the virions of both eukaryotic and prokaryotic viruses may contain enzymes that facilitate cell wall penetration.

Sometimes viruses may avoid normal cellular barriers by infecting cells only during certain phases in the cell cycle. Many retroviruses do not have a mechanism to access the nucleus, but they reside in the cytosol until the nuclear envelope temporarily disassembles during mitosis (100). Likewise, EsV-1 (Ectocarpus siliculosus Virus-1) of a brown algae *Ectocarpus siliculosus* cannot enter cell wall–containing vegetative cells and infects only the host's motile gametes or spores that lack the cell wall (65).

Many viruses have adapted to utilize normal cellular functions to access the cell's interior. This is best exemplified by animal viruses that employ cellular endocytic pathways in order to reach the cytoplasm. In addition viruses have adapted to move along cellular filaments and to utilize cellular transcriptional and translational machineries for uncoating.

MAKING THE APPROACH

Viruses are inert structures without the ability to move. However, directional moving could be advantageous for viruses to reach their host or an appropriate location for replication within the host. Indeed several viruses, both prokaryotic and eukaryotic, have evolved mechanisms to ride along cellular filaments during their entry processes.

Numerous viruses of gram-negative bacteria use a bacterial pilus (a long helical protein filament with an ability to polymerize and depolymerize) to dock themselves to the outer membrane of the bacterium. These include filamentous ssDNA (e.g., M13, fl and fd) (99) and icosahedral ssRNA phages (e.g., MS2 and Qβ) of *E. coli* (96), and enveloped bacteriophage φ6 of *Pseudomonas syringae* (78). The filamentous phages interact specifically with the tip of the host cell F-pilus via the minor coat protein (pIII) located at one end of the phage particle (44). Instead bacteriophage $\varphi 6$ and icosahedral ssRNA phages attach to the side of the pili (4, 96). The pilus allows efficient capture of the phage at a distance, and the retraction of the pilus translocates the bound virions to the bacterial surface (78, 96, 99). Such strategy also facilitates the penetration of the possible slime layer on the cell surface (78).

Practically all animal DNA viruses, except poxviruses, replicate inside the nucleus. Also several RNA viruses, such as influenza virus and retroviruses, need to access the nucleus (100). The viscose cytosol efficiently restricts the diffusion of virus-size particles (53), and consequently several viruses replicating within the nucleus are dependent on an active cytosolic transport. These viruses make use of cytoskeletal components, microtubules, and actin filaments. Adenovirus and herpes simplex virus capsids associate with dynein, a microtubule-dependent motor (88, 92). On the other hand, the baculovirus nucleocapsid binds directly to actin inducing its polymerization, and the assembled actin cables are involved in the transport of the viral capsid toward the nucleus (51). The movement proteins responsible for the cell-to-cell movement of plant viruses also interact with microtubules, suggesting a microtubule-dependent transport (12).

CELL WALL PASSAGE

Most of the cellular life forms are covered with a cell wall; animal and protozoal cells are an exception. The fungal viruses have solved the problem caused by the rigid cell wall by persistently infecting the host and never escaping from the cell. Thus fungal viruses do not need a mechanism to penetrate the chitin wall, rather they are spread by cell-cell fusion occurring during the mating process (101). The plant, algae, bacterial, and archaeal viruses have an extracellular phase between the reproduction cycles and thus need a strategy to penetrate the rigid barrier made of the host cell wall. The viruses of higher plants are either transmitted through seeds or vegetative propagation or directly introduced into the host cytosol via feeding of invertebrate vectors, or via mechanical action that involves partial destruction of the cell wall and perforation of the plasma membrane (58). Thus plant viruses, in a way similar to fungal ones, do not have specialized mechanisms to enter the host and cross the cell wall. However, viral-encoded movement proteins allow infection to spread to neighboring plant cells via plasmodesmata connections (22).

Active cell wall penetration mechanisms are employed by bacterial, some algal, and likely by archaeal viruses. Bacterial viruses have genes that encode peptidoglycan-modifying enzymes, transglycosylases, endopeptidases, and lysozyme-like lytic enzymes. The protein products of such genes are often structural components of the virion, suggesting a role in cell wall penetration during the entry. The baseplate of tailed bacteriophage T4 carries a protein (gp5) with lysozyme activity, which is putatively needed to facilitate the tail penetration through the peptidoglycan layer (46, 66). Enveloped bacteriophage $\varphi 6$ has nucleocapsid-associated endopeptidase (P5), which locally digests the peptidoglycan layer and thus allows the nucleocapsid to reach the plasma membrane (5, 16). Recently, the enzymatic activity of bacteriophage PRD1 virion-associated transglycosylase was demonstrated. In the absence of this activity the PRD1 DNA entry is delayed (81). Similarly the putative transglycosylase activity of bacteriophage T7 is needed to facilitate successful infection of host cells grown to stationary phase (62).

The algal virus PBCV-1, which infects *Chlorella* microalgae, encodes five proteins, including two chitinases, a chitosanase, a β-1,3-glucanase, and an amidase, which are all potentially involved in degrading the *Chlorella* cell wall (97). At least two of these enzymes, one chitinase and one chitosanase, are assembled into the virions (90, 91). Such enzymes are likely to be involved in the digestion of the host cell wall at the point of virion attachment (97).

Two major approaches seem to prevail in cell wall penetration. Either the viruses are unable to cope with the cell wall, as in the case of plants and fungi, or specific degradative enzymes are assembled into virions in a way that their action can be directed to the outer layers of the host cells. The activity of such enzymes needs to be controlled and directed strictly to the site of genome entry, as uncontrolled activity could cause the lysis of the cell. The fact that fungal and plant viruses do not have cell wall–degradating enzymes could reflect the more stable and complex structure of these cell walls. In addition, the genomes of most known plant viruses are far too small to code specialized cell wall–degrading enzymes.

GENOME UNCOATING

The initial steps in the infection are connected to conformational changes in the viral capsid proteins and sometimes to structural rearrangements in the entire virion. These changes are triggered by the signals from the cell and are induced by receptor interaction, changes in pH, ionic environment, or by the action of proteases. Agents that stabilize the virion structure can efficiently block virus uncoating and are potential antiviral drugs. The anti-influenza agents in clinical use, amantadine and rimantadine, interfere with the activity of viral M2 channels, which prime the influenza virus nucleocapsid for dissociation upon penetration into the cytosol (38). Furthermore, potential antiviral drugs for several picornaviruses (WIN compounds) bind in the viral capsid in a hydrophobic pocket and either inhibit attachment or block the virus from uncoating (80).

In many viral systems the entry process leads to the total dissociation of the virion, while in others the capsid remains intact and the genome is released via a specialized channel located at one of the vertices of the icosahedral virion. In the case of dsRNA viruses, such as bacteriophage φ6 and rotavirus, the genome is maintained and replicated within the innermost viral capsid during the entire viral life cycle to avoid the dsRNA-induced host responses.

The uncoating of the virion can occur at different locations: at the plasma membrane, within an intracellular vesicle, in the cytosol, at the nuclear membrane, or even in the nucleus. The uncoating may occur in a sequential manner during the virus passage into its site of replication as exemplified by adenovirus (36) and bacteriophage $\varphi 6$ (5, 78, 79).

Genome Delivery Through an Icosahedral Vertex

The rigid multilayered envelope of prokaryotic cells is a tight barrier for macromolecules and large structures like viral capsids. Most bacterial viruses have therefore adopted an entry mechanism where the capsid is never internalized into the host cell but remains at the outer surface of the envelope. The genome-containing capsid (or head) is stable, and the genome is delivered from the capsid through a genome delivery apparatus located at one of the vertices. The empty capsid stays unaltered outside the cell. This strategy is employed by all known tailed dsDNA bacteriophages (e.g., λ, T4, T7), lipid-containing bacteriophage PRD1, icosahedral ssRNA bacteriophages (e.g., MS2, $Q\beta$), and also likely applies for those archaeal viruses that have the appearance of the tailed bacteriophages (e.g., φH, ψM1).

The tailed dsDNA bacteriophages pack and deliver their genome through a circular portal complex located at one of the vertices of the icosahedral capsid. The portal also connects the tail to the capsid (95). The tail is a specialized structure with which its appendages perform the entry functions of the phage including host recognition, attachment, and initial stages of the genome delivery. Permeability changes of the host plasma membrane have been described during the infection of various tailed bacteriophages, suggesting that DNA delivery involves the formation of a pore or a channel into the host envelope (30). The channel for phage DNA delivery across the host envelope can be of different origin. Bacteriophage λ induces the formation or opening of a pore produced by cellular proteins (77). Instead, phage T5 carries a pore-forming protein (pb2), as a part of its tail, which can form water-filled transmembrane channels into lipid membranes (32). In the case of bacteriophage T7, internal core proteins (gp14, gp15, and gp16) are ejected from the virion and probably constitute a channel across the cell envelope (64). The channel for the delivery of bacteriophage T4 genome appears to be formed by phage-induced fusion of the plasma and the outer membranes (93).

The dsDNA bacteriophage PRD1 has a lipid bilayer between the dsDNA genome and the capsid but has no tail. Receptor interaction induces the release of the viral receptor-binding spike complex from one of the capsid vertices, resulting in a formation of an opening with a diameter of 14 nm (15, 83). Subsequently part of the protein-rich phage membrane, which encloses the genome, is transformed into a tubular structure that crosses the rigid capsid through the opening at the vertex. This "tail-tube" appendage penetrates the cell envelope with the assistance of several viral membrane proteins. The interiors of the cell and the virus are connected and the genome is delivered into the cytoplasm (A.M. Grahn, R. Daugelavičius and D.H. Bamford, submitted).

Not only the bacterial viruses but also plant and algal viruses have to penetrate a multilayered cell envelope. Like many bacteriophages, the algal virus PBCV-1 infects its host cell by attaching to the cell wall, degrading the wall at the attachment point, and then releasing the viral DNA into the cell leaving the empty protein capsid on the cell surface (97). The structure of the PBCV-1 virion resembles that of PRD1 (including a lipid bilayer beneath the capsid). Similar to PRD1, PBCV-1 might utilize its internal membrane as a genome delivery device.

Although many animal viruses penetrate the host plasma membrane, the small icosahedral ssRNA viruses, such as poliovirus and nodaviruses, seem to follow an entry pathway where the protein capsid is never internalized and only the ssRNA genome is delivered into the cytosol (20, 45). It is suggested that the genome release occurs either at the plasma membrane immediately after the receptor binding or after the receptor-virion complex has been internalized into an endocytic vesicle. Receptor binding induces a conformational transition in poliovirus virion resulting in so-called 135S particle, which lack capsid protein VP4 and have exposed the amino termini of protein VP1 near the virion vertices (7, 20). The externalized components confer membrane-binding properties (33), and a putative transmembrane pore is formed through which the viral RNA genome enters into the host cytosol (7, 28). Like tailed dsDNA bacteriophages, the poliovirus capsids can induce the formation of ion-permeable channels to target membranes (94).

The final step in the entry pathway of herpes virus is the penetration of the nuclear envelope. The herpes virus capsid, delivered into the host cytosol by fusion between the viral and host membranes, binds to the nuclear pore complex so that one of the vertices is facing the pore. The genome is released through the pore and the apparently intact empty capsid, devoid of DNA, remains associated with the nucleus (69, 88). Recently, Newcomb et al. (68) showed that herpes simplex virus has a portal structure located at a unique vertex of the capsid, which forms a channel through which the viral DNA could pass. Such a finding suggests that the genome delivery from the herpes virus capsid could occur, as in tailed bacteriophages, through the portal. The apparent similarities in virion structure and assembly with tailed bacteriophages have led to a suggestion of evolutionary connection between these two viral groups. Thus the mechanism of genome delivery could also utilize similar principles. The nuclear envelope and the envelope of gram-negative bacteria resemble each other topologically; both are composed of two membrane layers. While herpes appears to utilize a preexisting channel (the nuclear pore complex) through which the genome is delivered, the tailed phages appear to induce the formation or opening of such a channel.

The entry strategy "genome delivery through an icosahedral vertex" seems to be widespread in the viral world. A common theme is the metastable structure built into the virion vertices. Such a structure is designed to be triggered and is actively capable of functioning in genome translocation. This metastability is frequently achieved by symmetry mismatch and is carried out by building a multimeric spike or tail structure, which does not obey fivefold symmetry, at the fivefold capsid vertex. Due to the symmetry mismatch such structures are not firmly bound to the virion and are thus able to undergo structural rearrangements. Examples of symmetry mismatch include the portals of dsDNA bacteriophages and herpes simplex virus (which obey 12- or 13-fold symmetry) (68, 95), and the trimeric spikes of adenovirus and bacteriophage PRD1 (14, 83).

Coat Dissociation at the Cell Envelope

The outermost layer of a virion can be utilized as a device for the penetration of the cell envelope. Similar events are observed in the viruses infecting hosts as diverse as gram-negative bacteria and humans. The examples given are bacteriophage $\varphi 6$ and human immunodeficiency virus type-1 (HIV-1). Both viruses are enveloped and contain oligomeric extensions on the virion surface, which are built of two protein species: the receptor-binding spike (P3 of $\varphi 6$ and gp120 of HIV-1) and the fusogenic transmembrane protein (P6 of $\varphi 6$ and gp41 of HIV-1) anchoring the spike (5, 60, 70).

The primary receptor for bacteriophage $\varphi 6$ is the bacterial pilus to which the virion is absorbed via the spike protein P3 (5, 78). Removal of the spike after pilus interaction activates the fusogenic transmembrane protein (P6) and pH-independent fusion between the bacterial outer membrane, and the phage envelope commences leading to the mixing of the viral envelope and bacterial outer membrane (5).

The HIV-1 entry is also multistage process where cellular proteins are involved in virus binding and fusion activation. The spike protein (gp120) of HIV-1 envelope first binds to its primary receptor, CD4. This induces a conformational change in gp120, which leads to the exposure of a previously hidden coreceptor-binding site in the gp120. Binding to the coreceptor (chemokine receptor) causes an additional conformational change. These events transform the transmembrane protein (gp41) into a fusion-active state and lead to gp41-mediated pH-independent fusion between the HIV-1 envelope and the host plasma membrane (49, 70). Viral membrane fusion with the host plasma membrane is also characteristic of other members of the Retroviridae family as well as for viruses belonging to the Paramyxoviridae family. Unlike the case of HIV-1, however, the binding and fusion activation are often carried out by a single cell surface protein (70).

The uncoating at the cell envelope can also occur during the entry of nonenveloped viruses, as exemplified by filamentous bacteriophages such as fd. The major capsid protein (pVIII) has a helical hydrophobic central region and is associated with the host plasma membrane during both virus assembly and entry (99). The insertion of the coat protein of fd into the host membrane during the entry process logistically resembles the insertion of the viral envelope components (proteins and lipids) into the host membrane. As in HIV-1, the infection of fd is dependent on a sequential two-way docking mechanism involving a primary receptor that captures the virus to the cell surface and a coreceptor that induces the uncoating event (23, 75). As discussed earlier, filamentous phages use the host cell pili as their primary receptor, and this interaction is mediated by capsid protein pIII. Interaction with the pilus is not, however, sufficient to trigger infection, but a secondary binding site at the cell envelope is needed. The obligatory coreceptor for fd is the C-terminal domain of the periplasmic protein TolA (75), which binds to the N terminus of pIII. In the absence of the pilus the binding of the virus to the secondary receptor is hindered, suggesting that interaction with the primary receptor induces a conformational change in pIII, which exposes the binding site for TolA (75). The interaction with the secondary receptor triggers the virion uncoating, and the viral major capsid protein is inserted into the bacterial plasma membrane (24).

As described above, the fusion-type uncoating and entry mechanism is utilized by both bacterial and eukaryotic viruses. The fusion is performed by a viral protein that is transformed to a fusion-active state owing to the interaction with the cellular receptor(s). When such a fusion occurs at the outer surface of the cell, it is essential to conserve the cellular membrane integrity by formation of a nonleaky junction between the viral and the host components. The uncoating at the cell surface may accomplish the entire entry process (fd), or additional membrane penetration and uncoating events may be required to reach the site of replication (HIV-1, $\varphi 6$).

Internalization of Viral Particles Into The Host Cell

Often the viral particle is internalized from the cell surface without major changes in the virion structure. This is typical for numerous animal viruses. Many times the virion does not, however, reach the host cytosol, but the uncoating occurs during the escape from the intracellular vesicle, as with semliki forest virus and influenza virus. The viral fusion protein undergoes low pH-dependent conformational change, which triggers the fusion of viral and endosomal membranes. The genome, together with some viral proteins, is delivered into the cytosol (48, 86). In addition, nonenveloped viruses, such as picornaviruses and adenovirus, are apparently taken up via endocytosis. However, they cannot escape from the endocytic vesicle by fusion; they have to either disrupt the membrane, so that the capsid can enter the cytosol (as proposed for adenovirus) (67), or induce the formation of a pore (as proposed for poliovirus) (7, 28) through which the genome and possible accessory factors can be delivered to the cytosol.

Semliki forest virus, adenovirus, and a number of other animal viruses are dependent on the clathrin-coated vesicles for their entry (27, 29, 98). The entry of these viruses into the host cell is characteristically dependent on the low-pH step in the endosomal compartment and can be inhibited with agents raising the endosomal pH (40), blocking the action of the vacuolar H^+-ATPase (73), or with antibodies against clathrin (29).

Clathrin-mediated endocytosis is not, however, the only possible endocytic pathway for virion internalization. Studies on Simian virus 40 (SV40), a member of Papovaviridae, have shown that alternative endocytic pathways exist. On the cell surface SV40 particles are trapped into tight caveolin-containing vesicles (47, 89). These vesicles are directed to caveosomes, (membranous organelles distributed throughout the cytosol). From the caveosomes viral particles are transported within tubular membrane vesicles toward smooth endoplasmic reticulum (72). This entry route can be blocked by treatment of cells with compounds that disrupt or prevent the formation of caveolae (2) or by expression of a dominant-negative mutant of caveolin (34).

The internalization of an entire viral capsid is common in eukaryotic viral systems but occurs rarely in bacteria. However, as a dsRNA virus, the bacteriophage $\varphi 6$ must have a mechanism to deliver its virion core (protein capsid surrounding the genome and having polymerase activity) into the host cell cytosol. The entry of $\varphi 6$ involves viral envelope fusion with the host cell's outer membrane (5), cell wall degradation inside the periplasmic space by a viral enzyme (16, 61), and endocytic-like host plasma membrane penetration (74, 79). As bacterial cells do not generally possess endocytic uptake of periplasmic material, it is likely that the formation of the membrane curvature and the subsequent pinching off of the vesicle from the plasma membrane are driven by viral proteins. The transcriptionally active core (47 nm in diameter) is delivered from the vesicle into the host cytosol likely by a virion-induced membrane disruption.

The endocytic route of internalization is widely exploited by eukaryotic viruses, and different endocytic routes appear to be employed by the entering viruses. In addition to clathrin- and caveolin-dependent routes, endocytosis through uncoated vesicles has been described [e.g., canine parvovirus and polyomavirus (6, 54)]. The uptake of polyomavirus appears to occur in a clathrin-, caveolin-, and dynamin I–independent manner, suggesting utilization of some type of not-well-characterized vesicles to enter the cell (34). The endocytic-type entry mechanism of bacterial virus $\varphi 6$ highlights the universal usage of this type of entry mechanisms.

ENERGY CONSIDERATIONS

Virus entry and genome uncoating are energy-dependent processes. Structural proteins of the virion are commonly found in metastable conformations. Interaction with the cell leads to a more stable, lower-energy conformation. Such a change constitutes an irreversible step in virus entry and primes the stable extracellular virion for uncoating or genome delivery. For example, poliovirus virions undergo externalization of the capsid protein VP4 and the amino termini of protein VP1 after interaction with the receptor molecule. The prerequisite for this transition is the tight binding between the virion and the receptor, which introduces a transition state and allows the conversion in the virion structure. These changes, in turn, prime the virion for genome delivery (20). A well-documented structural rearrangement occurs in the influenza virus envelope protein hemagglutinin (86). Exposure to the low pH in the endosomal compartment induces major irreversible changes that include the exposure of the fusion peptide from the stem of the hemagglutinin molecule to the surface (13, 17). Such changes in the structure of the molecule are critical for its fusogenic activity. Similar, albeit not as well understood, changes have been documented in the early events of bacteriophage entry [(25); A.M. Grahn, R. Daugelavičius and D.H. Bamford, submitted].

Another type of energy captured in the virion is in the form of internal pressure caused by the ATP-driven dense genome packaging. During the phage φ29 dsDNA genome packaging, a high internal pressure is built up, and this force is suggested to be used for the initiation of genome ejection from the capsid (87). This force is likely to play an important role during the initial stages of DNA delivery of viruses applying an entry mechanism where the genome is delivered from the capsid through a portal-like opening.

Viruses can often enter and deliver their genome only into cells that are metabolically active, indicating that the cellular energy status is important for successful virus entry. The energy available can be in the form of the high-energy bond of ATP (or any NTP) or proton motive force (Δp) composed of membrane voltage ($\Delta\psi$, transmembrane difference of electrical potential) and pH gradient (ΔpH, transmembrane pH difference).

The transfer of viral particles along cellular filaments is dependent on cellular energy. The assembly of bacterial pili is an ATP-dependent process (21, 84). In addition, agents reducing transmembrane Δp cause pilus retraction or otherwise inhibit normal interaction between the phage and the host cell and thus prevent the adsorption of pilus-dependent bacteriophages (79, 105, 106). Likewise, the polymerization and depolymerization of eukaryotic microtubules and actin filaments are controlled by GTP and ATP, respectively, and microtubule-dependent motors (e.g., dynein), employed by several viruses, utilize the energy of ATP hydrolysis to move unidirectionally along a tubule. Thus the riding of bacteriophages φ6 and M13 toward the cell envelope on a bacterial pilus as well as adenovirus and herpes virus toward the nuclear envelope on microtubules utilizes a similar energy source.

The production of membrane curvature, which is necessary for the formation of endocytic vesicles, is not spontaneous. In eukaryotic cells, vesicle coat proteins, such as clathrin and caveolin, force the membrane to bend. The polymerization-depolymerization cycle of clathrin is dependent on ATP, and dynamin catalyzes the GTP-dependent fission of both caveolin- (59) and clathrin-coated vesicles from the plasma membrane (57). In bacterial cells there are no parallels for eukaryotic vesicle coat molecules. In accordance the endocytic-like entry process of bacteriophage φ6 is not dependent on ATP. Instead, in this case, the membrane vesicle formation is membrane voltage dependent (74). The endocytic-type virus entry appears to be dependent on host cell energetics in eukaryotic as well as in prokaryotic hosts.

A number of bacterial viruses that employ the entry strategy "genome delivery through an icosahedral vertex" are dependent on proton motive force (30, 52). Several studies on semliki forest virus and poliovirus have also indicated that membrane voltage or transmembrane proton gradient might be required for successful entry of these animal viruses (39, 43, 56, 73). Such energy is generated in the eukaryotic endosomes by the activity of the vacuolar H^+-ATPase, and in bacteria by the plasma membrane-associated respiratory chain and/or H^+-ATPase. It has been proposed that proton motive force during the animal virus entry could be used to promote virus uncoating and to drive the viral genome through the lipid barrier of the endosomal membrane (18, 19). The utilization of energy in a similar form during the phage genome delivery has been well documented for several tailed bacteriophages (30, 52). The actual role of Δp has been a matter of debate, but in most cases it appears to be needed for the formation of an opening or a pore into the host envelope through which the genome can be delivered into the cytosol (30, 52). It is interesting that the appearance of poliovirus-induced channels is also controlled by membrane voltage (94). In addition, the herpes simplex virus type 1 genome delivery at the nuclear pore is dependent on a cellular energy source (69).

The viral genome uncoating is in some cases dependent on the host cell activities, such as transcription and translation. The initial 850 bp of bacteriophage T7 dsDNA genome is first delivered into the cytosol in a Δp-dependent manner, the remainder of the genome is then mechanically pulled into the cell by active transcription. *E. coli* RNA polymerase pulls the first 19% of the T7 genome into the cell, and the rest is drawn by the activity of phage-encoded RNA polymerase (63). Ribosomes play an important role during semliki forest virus nucleocapsid uncoating by releasing the coat proteins from the viral ssRNA genome (85). Likewise the initial uncoating of the helical tobacco mosaic virus virion occurs by a cotranslational mechanism. As the ribosome reads along the viral RNA in the 5′ to 3′ direction, it releases the bound viral capsid proteins from approximately 75% of the viral ssRNA (104). The uncoating of the rest of the genome occurs in the 3′ to 5′ direction and is probably accomplished by a coreplicational mechanism (103). Thus synthesis of either RNA or polypeptides can drive the uncoating process in several apparently unrelated viral systems.

CONCLUSIONS

There are common themes and mechanisms employed by viruses infecting a vast variety of host cells. An intriguing question is whether the observed similarities are convergent evolution or whether the early viruses developed mechanisms that are utilized by the contemporary viruses. If so, this would mean that viruses were developed early and form lineages that reach to all domains of life.

Several examples of structural resemblance have been described between viruses from different domains of life. The recognition of structural similarities between tailed bacteriophages and herpes simplex virus has led to the description of a portal complex and a scaffolding-dependent assembly pathway for an eukaryotic virus. On the other hand the apparent evolutional connection between adenovirus and bacteriophage PRD1 (8, 42) has directed the studies on PRD1 to recognize, for example, capsid-cementing proteins and a vertex structure similar to those found in adenovirus (14, 82, 83). Comparative analyses on virus entry mechanisms will likely bring new ideas and hypotheses, which will lead to new experiments and approaches. The increasing understanding of viral entry mechanisms is important not only to combat viral diseases, but also because such studies have often led to the discovery of cellular functions and mechanisms not previously described.

ACKNOWLEDGMENTS

We acknowledge Professor J. Valkonen and Docent K. Mäkinen for helpful discussions. This work has been supported by research grants from the Academy of Finland [to D. H. Bamford; Finnish Center of Excellence Program (2000-2005)] and from the European Union (to R. Daugelavičius; ICA1-CT-2000-70027).

The *Annual Review of Microbiology* is online at http://micro.annualreviews.org

LITERATURE CITED

1. Ackermann HW. 2001. Frequency of morphological phage descriptions in the year 2000. Brief review. *Arch. Virol.* 146:843–57
2. Anderson HA, Chen Y, Norkin LC. 1996. Bound simian virus 40 translocates to caveolin-enriched membrane domains, and its entry is inhibited by drugs that selectively disrupt caveolae. *Mol. Biol. Cell* 7:1825–34
3. Bamford DH, Burnett RG, Stuart DI. 2002. Evolution of viral structure. *Theor. Popul. Biol.* In press
4. Bamford DH, Palva ET, Lounatmaa K. 1976. Ultrastructure and life cycle of the lipid-containing bacteriophage φ6. *J. Gen. Virol.* 32:249–59
5. Bamford DH, Romantschuk M, Somerharju PJ. 1987. Membrane fusion in prokaryotes: bacteriophage φ6 membrane fuses with the *Pseudomonas syringae* outer membrane. *EMBO J.* 6:1467–73
6. Basak S, Turner H. 1992. Infectious entry pathway for canine parvovirus. *Virology* 186:368–76
7. Belnap DM, Filman DJ, Trus BL, Cheng N, Booy FP, et al. 2000. Molecular tectonic model of virus structural transitions: the putative cell entry states of poliovirus. *J. Virol.* 74:1342–54
8. Benson SD, Bamford JK, Bamford DH, Burnett RM. 1999. Viral evolution revealed by bacteriophage PRD1 and human adenovirus coat protein structures. *Cell* 98:825–33
9. Bergh O, Borsheim KY, Bratbak G, Heldal M. 1989. High abundance of viruses found in aquatic environments. *Nature* 340:467–68
10. Beveridge TJ, Graham LL. 1991. Surface layers of bacteria. *Microbiol. Rev.* 55:684–705
11. Blaisdell BE, Campbell AM, Karlin S. 1996. Similarities and dissimilarities of phage genomes. *Proc. Natl. Acad. Sci. USA* 93:5854–59
12. Boyko V, Ferralli J, Ashby J, Schellenbaum P, Heinlein M. 2000. Function of microtubules in intercellular transport of plant virus RNA. *Nat. Cell Biol.* 2:826–32
13. Bullough PA, Hughson FM, Skehel JJ, Wiley DC. 1994. Structure of influenza hemagglutinin at the pH of membrane fusion. *Nature* 371:37–43
14. Burnett RM. 1997. The structure of adenovirus. See Ref. 19a, pp. 209–38
15. Butcher SJ, Bamford DH, Fuller SD. 1995. DNA packaging orders the membrane of bacteriophage PRD1. *EMBO J.* 14:6078–86
16. Caldentey J, Bamford DH. 1992. The lytic enzyme of the *Pseudomonas* phage φ6. Purification and biochemical characterization. *Biochim. Biophys. Acta.* 1159:44–50
17. Carr CM, Kim PS. 1993. A spring-loaded

mechanism for the conformational change of influenza hemagglutinin. *Cell* 73:823–32

18. Carrasco L. 1994. Entry of animal viruses and macromolecules into cells. *FEBS Lett.* 350:151–54
19. Carrasco L. 1995. Modification of membrane permeability by animal viruses. *Adv. Virus. Res.* 45:61–112

19a. Chiu W, Burnett RM, Garcea RL, eds. 1997. *Structural Biology of Viruses*. NY: Oxford Univ. Press

20. Chow M, Basavappa R, Hogle JM. 1997. The role of conformational transitions in poliovirus pathogenesis. See Ref. 19a, pp. 157–83
21. Christie PJ. 2001. Type IV secretion: intercellular transfer of macromolecules by systems ancestrally related to conjugation machines. *Mol. Microbiol.* 40:294–305
22. Citovsky V. 1999. Tobacco mosaic virus: a pioneer of cell-to-cell movement. *Philos. Trans. R. Soc. London B. Biol. Sci.* 354: 637–43
23. Click EM, Webster RE. 1997. Filamentous phage infection: required interactions with the TolA protein. *J. Bacteriol.* 179:6464–71
24. Click EM, Webster RE. 1998. The TolQRA proteins are required for membrane insertion of the major capsid protein of the filamentous phage f1 during infection. *J. Bacteriol.* 180:1723–28
25. Crowther RA, Lenk EV, Kikuchi Y, King J. 1977. Molecular reorganization in the hexagon to star transition of the baseplate of bacteriophage T4. *J. Mol. Biol.* 116: 489–523
26. Demchick P, Koch AL. 1996. The permeability of the wall fabric of *Escherichia coli* and *Bacillus subtilis*. *J. Bacteriol.* 178:768–73
27. DeTulleo L, Kirchhausen T. 1998. The clathrin endocytic pathway in viral infection. *EMBO J.* 17:4585–93
28. Dimitrov DS. 2000. Cell biology of virus entry. *Cell* 101:697–702
29. Doxsey SJ, Brodsky FM, Blank GS, Helenius A. 1987. Inhibition of endocytosis by anti-clathrin antibodies. *Cell* 50:453–63
30. Dreiseikelmann B. 1994. Translocation of DNA across bacterial membranes. *Microbiol. Rev.* 58:293–316
31. Feldherr CM, Akin D. 1990. The permeability of the nuclear envelope in dividing and nondividing cell cultures. *J. Cell Biol.* 111:1–8
32. Feucht A, Schmid A, Benz R, Schwarz H, Heller KJ. 1990. Pore formation associated with the tail-tip protein pb2 of bacteriophage T5. *J. Biol. Chem.* 265:18561–67
33. Fricks CE, Hogle JM. 1990. Cell-induced conformational change in poliovirus: externalization of the amino terminus of VP1 is responsible for liposome binding. *J. Virol.* 64:1934–45
34. Gilbert JM, Benjamin TL. 2000. Early steps of polyomavirus entry into cells. *J. Virol.* 74:8582–88
35. Deleted in proof
36. Greber UF, Willetts M, Webster P, Helenius A. 1993. Stepwise dismantling of adenovirus 2 during entry into cells. *Cell* 75:477–86
37. Haywood AM. 1994. Virus receptors: binding, adhesion strengthening, and changes in viral structure. *J. Virol.* 68:1–5
38. Helenius A. 1992. Unpacking the incoming influenza virus. *Cell* 69:577–78
39. Helenius A, Kielian M, Wellsteed J, Mellman I, Rudnick G. 1985. Effects of monovalent cations on semliki forest virus entry into BHK-21 cells. *J. Biol. Chem.* 260: 5691–97
40. Helenius A, Marsh M, White J. 1982. Inhibition of semliki forest virus penetration by lysosomotropic weak bases. *J. Gen. Virol.* 58:47–61
41. Heller KJ. 1992. Molecular interaction between bacteriophage and the gram-negative cell envelope. *Arch. Microbiol.* 158:235–48
42. Hendrix RW. 1999. Evolution: the long evolutionary reach of viruses. *Curr. Biol.* 9:R914–17
43. Irurzun A, Carrasco L. 2001. Entry of

poliovirus into cells is blocked by valinomycin and concanamycin A. *Biochemistry* 40:3589–600

44. Jacobson A. 1972. Role of F pili in the penetration of bacteriophage fl. *J. Virol.* 10:835–43
45. Johnson JE, Rueckert RR. 1997. Packaging and release of viral genome. See Ref. 19a, pp. 269–87
46. Kao SH, McClain WH. 1980. Baseplate protein of bacteriophage T4 with both structural and lytic functions. *J. Virol.* 34: 95–103
47. Kartenbeck J, Stukenbrok H, Helenius A. 1989. Endocytosis of simian virus 40 into the endoplasmic reticulum. *J. Cell. Biol.* 109:2721–29
48. Kielian M, Jungerwirth S. 1990. Mechanisms of enveloped virus entry into cells. *Mol. Biol. Med.* 7:17–31
49. Kristiansen TB, Knudsen TB, Eugen-Olsen J. 1998. Chemokine receptors and their crucial role in human immunodeficiency virus infection: major breakthroughs in HIV research. *Scand. J. Immunol.* 48:339–46
50. Labischinski H, Goodell EW, Goodell A, Hochberg ML. 1991. Direct proof of a "more-than-single-layered" peptidoglycan architecture of *Escherichia coli* W7: a neutron small-angle scattering study. *J. Bacteriol.* 173:751–56
51. Lanier LM, Volkman LE. 1998. Actin binding and nucleation by *Autographa california* M nucleopolyhedrovirus. *Virology* 243:167–77
52. Letellier L, Plancon L, Bonhivers M, Boulanger P. 1999. Phage DNA transport across membranes. *Res. Microbiol.* 150: 499–505
53. Luby-Phelps K. 1994. Physical properties of cytoplasm. *Curr. Opin. Cell Biol.* 6:3–9
54. Mackay RL, Consigli RA. 1976. Early events in polyomavirus infection: attachment, penetration, and nuclear entry. *J. Virol.* 19:620–36
55. Madigan MT, Martinko JM, Paker J, eds. 2000. *Brock Biology of Microorganisms*. Upper Saddle River: Prentice-Hall. 991 pp.
56. Madshus IH, Olsnes S, Sandvig K. 1984. Requirements for entry of poliovirus RNA into cells at low pH. *EMBO J.* 3:1945–50
57. Marsh M, McMahon HT. 1999. The structural era of endocytosis. *Science* 285:215–20
58. Matthews REF. 1992. *Fundamentals of Plant Virology*. San Diego: Academic. 835 pp.
59. Matveev S, Li X, Everson W, Smart EJ. 2001. The role of caveolae and caveolin in vesicle-dependent and vesicle-independent trafficking. *Adv. Drug. Deliv. Rev.* 49:237–50
60. Mindich L, Bamford DH. 1988. Lipid-containing bacteriophages. In *The Bacteriophages*, ed. R Calendar, pp. 475–519. New York: Plenum
61. Mindich L, Lehman J. 1979. Cell wall lysin as a component of the bacteriophage φ6 virion. *J. Virol.* 30:489–96
62. Moak M, Molineux IJ. 2000. Role of the Gp16 lytic transglycosylase motif in bacteriophage T7 virions at the initiation of infection. *Mol. Microbiol.* 37:345–55
63. Molineux I. 1999. T7 bacteriophages. In *Encyclopedia of Molecular Biology*, ed. TE Creighton, pp. 2495–507. New York: Wiley
64. Molineux IJ. 2001. No syringes please, ejection of phage T7 DNA from the virion is enzyme driven. *Mol. Microbiol.* 40:1–8
65. Muller DG, Kawai H, Stache B, Lanka S. 1990. A virus infection in marine brown alga *Ectocarpus siliculosus* (Phaeophyceae). *Bot. Acta.* 103:72–82
66. Nakagawa H, Arisaka F, Ishii S. 1985. Isolation and characterization of the bacteriophage T4 tail-associated lysozyme. *J. Virol.* 54:460–66
67. Nemerow GR. 2000. Cell receptors involved in adenovirus entry. *Virology* 274: 1–4
68. Newcomb WW, Juhas RM, Thomsen DR, Homa FL, Burch AD, et al. 2001. The UL6 gene product forms the portal for entry of

DNA into the herpes simplex virus capsid. *J. Virol* 75:10923–32
69. Ojala PM, Sodeik B, Ebersold MW, Kutay U, Helenius A. 2000. Herpes simplex virus type 1 entry into host cells: reconstitution of capsid binding and uncoating at the nuclear pore complex in vitro. *Mol. Cell. Biol.* 20:4922–31
70. Overbaugh J, Miller AD, Eiden MV. 2001. Receptors and entry cofactors for retroviruses include single and multiple transmembrane-spanning proteins as well as newly described glycophosphatidylinositol-anchored and secreted proteins. *Microbiol. Mol. Biol. Rev.* 65:371–89
71. Pante N, Aebi U. 1996. Molecular dissection of the nuclear pore complex. *Crit. Rev. Biochem. Mol. Biol.* 31:153–99
72. Pelkmans L, Kartenbeck J, Helenius A. 2001. Caveolar endocytosis of simian virus 40 reveals a new two-step vesicular-transport pathway to the ER. *Nat. Cell Biol.* 3:473–83
73. Perez L, Carrasco L. 1994. Involvement of the vacuolar H(+)-ATPase in animal virus entry. *J. Gen. Virol.* 75:2595–606
74. Poranen MM, Daugelavičius R, Ojala PM, Hess MW, Bamford DH. 1999. A novel virus-host cell membrane interaction. Membrane voltage-dependent endocytic-like entry of bacteriophage φ6 nucleocapsid. *J. Cell Biol.* 147:671–82
75. Riechmann L, Holliger P. 1997. The C-terminal domain of TolA is the coreceptor for filamentous phage infection of *E. coli*. *Cell* 90:351–60
76. Riezman H, Woodman PG, van Meer G, Marsh M. 1997. Molecular mechanisms of endocytosis. *Cell* 91:731–38
77. Roessner CA, Ihler GM. 1986. Formation of transmembrane channels in liposomes during injection of lambda DNA. *J. Biol. Chem.* 261:386–90
78. Romantschuk M, Bamford DH. 1985. Function of pili in bacteriophage φ6 penetration. *J. Gen. Virol.* 66:2461–69
79. Romantschuk M, Olkkonen VM, Bamford DH. 1988. The nucleocapsid of bacteriophage φ6 penetrates the host cytoplasmic membrane. *EMBO J.* 7:1821–29
80. Rossmann MG, Greve JM, Kolatkar PR, Olson NH, Smith TJ, et al. 1997. Rhinovirus attachment and cell entry. See Ref. 19a, pp. 105–33
81. Rydman PS, Bamford DH. 2000. Bacteriophage PRD1 DNA entry uses a viral membrane-associated transglycosylase activity. *Mol. Microbiol.* 37:356–63
82. Rydman PS, Bamford JK, Bamford DH. 2001. A minor capsid protein P30 is essential for bacteriophage PRD1 capsid assembly. *J. Mol. Biol.* 313:785–95
83. Rydman PS, Caldentey J, Butcher SJ, Fuller SD, Rutten T, Bamford DH. 1999. Bacteriophage PRD1 contains a labile receptor-binding structure at each vertex. *J. Mol. Biol.* 291:575–87
84. Sandkvist M. 2001. Biology of type II secretion. *Mol. Microbiol.* 40:271–83
85. Singh I, Helenius A. 1992. Role of ribosomes in Semliki Forest virus nucleocapsid uncoating. *J. Virol.* 66:7049–58
86. Skehel JJ, Wiley DC. 2000. Receptor binding and membrane fusion in virus entry: the influenza hemagglutinin. *Annu. Rev. Biochem.* 69:531–69
87. Smith DE, Tans SJ, Smith SB, Grimes S, Anderson DL, Bustamante C. 2001. The bacteriophage φ29 portal motor can package DNA against a large internal force. *Nature* 413:748–52
88. Sodeik B, Ebersold MW, Helenius A. 1997. Microtubule-mediated transport of incoming herpes simplex virus 1 capsids to the nucleus. *J. Cell. Biol.* 136:1007–21
89. Stang E, Kartenbeck J, Parton RG. 1997. Major histocompatibility complex class I molecules mediate association of SV40 with caveolae. *Mol. Biol. Cell* 8:47–57
90. Sun L, Adams B, Gurnon JR, Ye Y, Van Etten JL. 1999. Characterization of two chitinase genes and one chitosanase gene encoded by *Chlorella* virus PBCV-1. *Virology* 263:376–87
91. Sun L, Gurnon JR, Adams BJ, Graves MV, Van Etten JL. 2000. Characterization of a

beta-1,3-glucanase encoded by *Chlorella* virus PBCV-1. *Virology* 276:27–36

92. Suomalainen M, Nakano MY, Keller S, Boucke K, Stidwill RP, Greber UF. 1999. Microtubule-dependent plus- and minus end-directed motilities are competing processes for nuclear targeting of adenovirus. *J. Cell Biol.* 144:657–72
93. Tarahovsky YS, Khusainov AA, Deev AA, Kim YV. 1991. Membrane fusion during infection of *Escherichia coli* cells by phage T4. *FEBS Lett.* 289:18–22
94. Tosteson MT, Chow M. 1997. Characterization of the ion channels formed by poliovirus in planar lipid membranes. *J. Virol.* 71:507–11
95. Valpuesta JM, Carrascosa JL. 1994. Structure of viral connectors and their function in bacteriophage assembly and DNA packaging. *Q. Rev. Biophys.* 27:107–55
96. van Duin J. 1988. The single-stranded RNA bacteriophages. In *The Bacteriophages*, ed. R Calendar, pp. 117–67. New York: Plenum
97. Van Etten JL, Meints RH. 1999. Giant viruses infecting algae. *Annu. Rev. Microbiol.* 53:447–94
98. Varga MJ, Weibull C, Everitt E. 1991. Infectious entry pathway of adenovirus type 2. *J. Virol.* 65:6061–70
99. Webster RE. 1996. Biology of filamentous bacteriophage. In *Phage Display of Peptides and Proteins*, ed. BK Kay, pp. 1–20. San Diego: Academic
100. Whittaker GR, Kann M, Helenius A. 2000. Viral entry into the nucleus. *Annu. Rev. Cell Dev. Biol.* 16:627–51
101. Wickner RB. 1995. Viruses of yeast, fungi and parasitic microorganisms. In *Fields Virology*, ed. BN Fields, DM Knipe, PM Howley, pp. 557–85. New York: Raven
102. Wommack KE, Colwell RR. 2000. Virioplankton: viruses in aquatic ecosystems. *Microbiol. Mol. Biol. Rev.* 64:69–114
103. Wu X, Shaw JG. 1997. Evidence that a viral replicase protein is involved in the disassembly of tobacco mosaic virus particles in vivo. *Virology* 239:426–34
104. Wu X, Xu Z, Shaw JG. 1994. Uncoating of tobacco mosaic virus RNA in protoplasts. *Virology* 200:256–62
105. Yamamoto M, Kanegasaki S, Yoshikawa M. 1980. Effects of temperature and energy inhibitors on complex formation between *Escherichia coli* male cells and filamentous phage fd. *J. Gen. Microbiol.* 119:87–93
106. Yamamoto M, Kanegasaki S, Yoshikawa M. 1981. Role of membrane potential and ATP in complex formation between *Escherichia coli* male cells and filamentous phage fd. *J. Gen. Microbiol.* 123:343–49

Annu. Rev. Microbiol. 2002. 56:539–65
doi: 10.1146/annurev.micro.56.012302.161110

First published online as a Review in Advance on July 15, 2002

CROSS-SPECIES INFECTIONS AND THEIR ANALYSIS

Man-Wah Tan

Department of Genetics, and Department of Microbiology and Immunology, Stanford University School of Medicine, Stanford, California 94305; e-mail: mwtan@stanford.edu

Key Words multihost pathogens, nonvertebrate hosts, host-pathogen interactions, *Caenorhabditis elegans*, *Drosophila*

■ **Abstract** The ability of certain pathogens to infect multiple hosts has led to the development of genetically tractable nonvertebrate hosts to elucidate the molecular mechanisms of interactions between these pathogens and their hosts. The use of plant, insect, nematode, and protozoan hosts to study human pathogens has facilitated the elucidation of molecular nature of pathogenesis and host responses. Analyses of virulence of multihost pathogens on their respective hosts revealed that pathogens utilize many universal offensive strategies to overcome host defenses, irrespective of the evolutionary lineage of the host. Likewise, genetic dissections of the defense response of the nonvertebrate hosts have also shown that key features underlying host defense responses are highly conserved. This review summarizes how the information gained from the analysis of cross-species infections contributes to our understanding of host-pathogen interactions.

CONTENTS

0066-4227/02/1013-0539$14.00

INTRODUCTION

Many pathogens are capable of infecting more than one host species. In a comprehensive survey of literature that identified 1415 species of pathogenic organisms that are infectious to humans, Taylor et al. (130) showed that 61% of these are zoonotic; that is, they can be transmitted between humans and animals. The ability of a human pathogen to infect hosts from different orders and classes within the subphylum Vertebrata, especially among mammalian species, has led to the development of many mammalian models to study pathogenesis. This has resulted in the characterization of many microbial virulence factors and the elucidation of the mechanisms of pathogenesis. However, it has not been feasible to perform systematic and comprehensive genetic dissection of the host pathways targeted by bacterial virulence factors because the mammalian hosts are not genetically tractable. Recent studies on the defense mechanisms of nonmammalian hosts have provided compelling evidence that some of the key features underlying host defense responses may have ancient origins and are conserved at the molecular level across phylogeny (74, 122). Similarly, the offensive strategies employed by pathogenic microbes, including mechanisms for exploitation of host surface molecules, systems to deliver bacterial virulence proteins, and mechanisms to subvert host defenses, are highly conserved among plant and animal pathogens (23, 45, 50, 108). These findings suggest that pathogenic interactions deciphered using a nonmammalian host could greatly facilitate our understanding of these universal mechanisms underlying host-pathogen interactions (88).

Some human pathogens cross a larger evolutionary gulf to naturally or experimentally infect organisms from other phyla or kingdoms (Table 1). For example, a clinical isolate, *Pseudomonas aeruginosa* strain PA14, can elicit severe soft-rot-like symptoms and proliferate when infiltrated into *Arabidopsis* leaves (109), cause lethal sepsis in a mouse full-skin-thickness burn model (109), kill larvae of the wax moth *Galleria mellonella*, kill the nematode *Caenorhabditis elegans*, and infect the soil amoeba *Dictyostelium discoideum* (68, 89, 104a, 128, 129). Taking advantage of the extensive host range of PA14, Ausubel and colleagues pioneered the use of nonmammalian hosts to directly screen for virulence-attenuated mutants of *P. aeruginosa*. An important discovery from the "multihost pathogenesis system" is that there is significant overlap among the PA14 virulence factors required for pathogenesis in plants, nematodes, insects, and mice. These studies demonstrate that there is an extensive conservation in the virulence mechanisms used by *P. aeruginosa* to infect evolutionarily divergent hosts and that many of these fundamental mechanisms can be efficiently identified using nonmammalian hosts. The analyses of cross-species infections can be extended to other human pathogens that infect nonmammalian hosts.

The use of genetically tractable nonmammalian hosts, for which complete genome sequences are available, has great potential in advancing the field of microbial pathogenesis in another respect. It allows us to readily apply and integrate traditional genetics approaches with recent functional genomics innovations to

TABLE 1 Human pathogens and some of their nonvertebrate hosts. Pathogenesis in some of these hosts is discussed in the text.

Pathogen	Plants	Insecta	Nematoda	Protozoa
P. aeruginosa	*Arabidopsis*, *thaliana*, lettuce, rutabaga, brussel sprouts, celery, carrot, tomato, potato, and cucumber	Grasshoppers (*Melanoplus bivattatus*, *Camnula pellucida*), locusts, greater wax moth (*Galleria mellonella*), silkworm (*Bombyx mori*), tent caterpillars (*Malacosoma americanum*), cutworms, hornworms (*Manduca* spp.), diamondback moth (*Plutella xylostella*), *D. melanogaster*	*C. elegans*, *C. briggsae*, *C. vulgaris*	*D. discoideum*, *Acanthamoeba*
S. marcescens	Curcubits	Anopheline mosquito, *D. melanogaster*	*C. elegans*	
B. pseudomallei			*C. elegans*	*A. astronyxis*, *A. castellani*, *A. polyphaga*
B. cepacia	Onion		*C. elegans*	*Acanthamoeba*, *Tetrahymena pyriformis*
E. carotovora	Carrot, potato, turnip	*D. melanogaster*	*C. elegans*	
S. enterica		*G. mellonella*, *D. melanogaster*	*C. elegans*, *C. briggsae*	
E. faecalis		*G. mellonella*	*C. elegans*	*A. polyphaga*, *A. castellanii*, *D. discoideum*

identify and characterize components of the host immune response. In this review, I focus on various human pathogens that can infect nonmammalian hosts. I discuss current knowledge of the molecular basis of host-pathogen interactions based on the analyses of cross-species infections by these pathogens. I also discuss the utility of *Drosophila*, *C. elegans*, and *Dictyostelium* as model hosts for the study of the innate immune system.

DISSECTION OF VIRULENCE DETERMINANTS OF MULTIHOST PATHOGENS

The gram-negative and gram-positive bacterial pathogens that infect humans and their nonvertebrate hosts are given in Table 1, and the analyses of their cross-species infections are discussed below.

Extracellular Gram-Negative Bacterial Pathogens

PSEUDOMONAS AERUGINOSA *P. aeruginosa* is one of the most versatile multihost pathogens. In addition to being an important human pathogen capable of infecting virtually all tissues, *P. aeruginosa* also infects plants, insects, nematodes, and protozoa (45, 88, 104a, 108, 127) (Table 1). The basis for this versatility is an impressive array of enzymes and excreted compounds that it produces; the former allow *P. aeruginosa* to use a large variety of compounds as nutrients and the latter, also called virulence determinants, allow the bacterium to survive and thrive in hostile host environments.

P. aeruginosa ***as a human pathogen*** *P. aeruginosa* is an important pathogen in cystic fibrosis patients, and in patients whose immune system is compromised by medical intervention, infection, or burn. *P. aeruginosa* can infect burns or surgical wounds, the urinary tract, the gastrointestinal tract, the respiratory tract, eyes, ears, and meninges. This ability to infect a huge array of tissues lies in part to the large variety of virulence factors it produces (86).

P. aeruginosa ***as a plant pathogen*** Many species of plants are susceptible to *P. aeruginosa* (27, 114) (Table 1). In *Arabidopsis thaliana*, *P. aeruginosa* strain PA14 causes severe soft-rot symptoms that correspond to bacterial proliferation in *Arabidopsis* leaves (101, 108). PA14 can invade *A. thaliana* leaves directly through the stomata without the requirement of wounding. PA14 primarily colonizes the intercellular spaces, causing disruption of plant cell walls and membrane structures, which ultimately leads to a systemic infection that results in rotting of the petiole and central bud and death of the plant. PA14 also causes soft-rot symptoms to develop when inoculated into the midrib of lettuce leaf stem (110). The severity of symptom development in lettuce directly correlates with bacterial growth in *Arabidopsis*. Because of the ease of testing several strains on a single lettuce leaf, Rahme et al. (110) used lettuce to screen for *P. aeruginosa* PA14 transposon-insertion mutants that failed to elicit disease symptoms. The lettuce screen led to the identification of 9 bacterial mutants out of 2500 prototrophic mutants tested. Importantly, all nine mutants identified from the plant screen exhibited reduced pathogenicity when tested in a burned-mouse pathogenicity model at a dose of 5×10^3 bacteria (110). Even at the higher inoculum of 5×10^5 bacteria, 7 of the 9 mutants caused significantly lower mouse mortality than the wild-type. The identity and function of each of the known genes has been the subject of recent reviews (23, 108) and is summarized in Table 2.

P. aeruginosa ***as a pathogen of nematodes*** The bacteriophagic nematode *C. elegans* and *P. aeruginosa* inhabit the same habitat, the soil, and associate with each other naturally (56). Some, but not all, *P. aeruginosa* strains are pathogenic to *C. elegans*. For example, *C. elegans* feeding on monogenic lawns of *P. aeruginosa* strains PAK or PO37 grown in low-salt slow-killing medium (SKM) develop and

TABLE 2 Identity of *P. aeruginosa* mutants identified from plant and nematode screens and their pathogenicity in mice [modified from (89, 108, 110, 129)]

Strain	Gene identity	Gene product function or other phenotypes	% Mouse mortality[a]
Mutants identified from lettuce assay			
1D7	*gacA*[b]	A response regulator in the GacS/GacA two-component system	50
pho34B12	*mvfR*	LysR-like transcriptional regulator (24)	56
pho15	*dsbA*[b]	Periplasmic disulfide bond–forming enzyme	62
33C7	no matches[c]	Unknown	0
25A12	no matches[c]	Unknown	87
33A9	no matches[d]	Unknown, increased attachment but reduced motility	0
25F1	*PA0596*	Unknown, 41% similar to *orfT* of *C. tepidum*	20
34H4	no matches[c]	Contains a bipartite nuclear localization signal; translocates into nucleus of host cells (23)	33
16G12	PA0268	Transcriptional regulator with a *gntR* family signature	100
Mutants identified in the *C. elegans* fast-killing assay			
36A4	*mdoH*[b]	Involves in periplasmic glucan synthesis in *E. coli* and virulence in *P. syringae*	0
3E8; 6A6	*PA1900*	*phzB2*, phenazine biosynthesis	18
8C12	PA0997	Unknown	63
23A2	*mexA*	Multidrug transporter of the *mexAB-oprM* operon	85
1G2	*crbA*	Two-component system of the NtrB-NtrC family; modulates catabolism of natural substrate in response to C:N ratio (98)	100
Mutants identified in the *C. elegans* slow-killing assay			
50E12	*ptsP*[b]	phosphoenolpyruvate phosphotransferase, involved in a complex two-component sensing and regulatory phosphate transfer system	0
35H7	*gacA*[b]	A response regulator in the GacS/GacA two-component system	NT
48D9	*gacS*[b]	A sensor kinase in the GacS/GacA two-component system; known previously as *lemA*	50
12A1	*lasR*[b]	Quorum-sensing regulator	50
35A9	*PA2020*	Also known as *mexZ* and *amrR*; a transcriptional regulator of multidrug transporter with a tetR family signature	53
44B1	*no matches*[d]	Unknown	56
41C1	*PA5022*	Similar to *aefA*, also known as *kefA*, a cation-specific channel (92a)	81
41A5	*PA0328*	Putative autotransporter protein (61)	100

[a]Six-week-old male ARK/J inbred strain mice (from Jackson Laboratories), weighing between 20 and 30 g were inoculated with 5×10^5 cells; 100% of mice die when inoculated with wild-type PA14 at this dose. When 25A12, 16G12 were tested for virulence in mice at an inoculum of 5×10^3 cells, they caused significantly lower mortality than wild-type at the same inoculum (110). The number of animals that died of sepsis was monitored each day for seven days. NT, not tested. Another *gacA* mutant 1D7 was independently isolated from a plant screen. Mutant 1D7 has been tested on mice and showed 50% mortality.

[b]These genes are also required for full virulence in other plant and/or animal bacterial pathogens.

[c]It is not clear if the homolog is present in the completed genome of strain PA01.

[d]No matches to the completed genome of strain PA01 and appears to be unique to strain PA14.

reproduce normally (128). Similarly, several environmental isolates of *P. aeruginosa* could support growth and reproduction of *C. elegans* for several generations (8, 56). In contrast, several strains, including several clinical isolates of *P. aeruginosa* kill *C. elegans* strain Bristol (N2).

Three largely distinct mechanisms of killing, which are dependent on the media used to grow the bacteria and the bacterial genotype, have been defined. When grown in SKM, strain PA14 and PA29 kill worms over a period of 2–3 days (slow killing) by an infection-like process that correlates with the accumulation of bacteria in the worm gut (126, 128). The mechanism by which live *P. aeruginosa* PA14 kills *C. elegans* is not clear. When strain PA14 is grown in a peptone-glucose-sorbitol (PGS) medium, it kills worms within 4–24 h (fast killing) by the production of low-molecular-weight toxins, including a family of related redox-active compounds, the phenazines (89, 126). The sequenced strain of *P. aeruginosa*, PA01, kills by yet another mechanism. When grown on brain-heart infusion (BHI) agar, worms become paralyzed within 4 h upon contact with the bacterial lawn. This "paralytic killing" is mediated primarily by hydrogen cyanide, which is under the control of LasR and RhlR quorum-sensing regulators (35, 51). Unlike fast killing, phenazines do not appear to play a role in paralytic killing (51). Strain PA14, however, does not cause the level of paralysis seen with strain PA01 when grown on BHI.

Due to the antagonistic nature of the association between *P. aeruginosa* and *C. elegans*, it is reasonable to hypothesize that the two organisms have developed an arsenal of weapons to combat each other and that some of these strategies may have been conserved during evolution. Indeed, when individual clones of mutagenized PA14 were screened in the *C. elegans* fast- and slow-killing assays, mutations in 5 and 8 genes were identified that were attenuated in *C. elegans* fast- or slow-killing (Table 2). Importantly, at least 10 of the 13 *P. aeruginosa* genes required for killing *C. elegans* are also required for virulence in a burned-mouse model when tested at the inoculum of 5×10^5 bacteria (28, 89, 129) (Table 2), indicating that *P. aeruginosa* uses common virulence determinants to overwhelm its mammalian and nematode hosts. In a separate screen for mutants defective in paralytic killing, 21 genes were identified that included the *hcnC* gene that encodes a subunit of hydrogen cyanide synthase, regulatory genes, genes encoding metabolic enzymes, a probable metal transporter, and five genes with previously known or postulated virulence functions (51). With the exception of *gacS* and *mvrR*, which have been shown to be multihost virulence factors (Table 2), the function of the other genes in pathogenesis has yet to be determined in a mammalian model. The role of cyanide in *P. aeruginosa* pathogenesis in human is also unexplored.

P. aeruginosa ***as a pathogen of insects*** *P. aeruginosa* is also a potent pathogen of insects (20, 68) (Table 1). For example, the 50% lethal dose (LD_{50}) of *P. aeruginosa* when injected into the hemolymph of the greater wax moth *Galleria mellonella* larvae can be as low as one bacterium (68). Various strains of *P. aeruginosa*, including strains PA14 and PA01, can cause lethal infection when injected into

Drosophila melanogaster [(15, 36); G. Lau, S. Mahajan-Miklos, F.M. Ausubel & L.G. Rahme, personal communication].

Jander et al. (68) tested PA14 transposon-insertion mutants isolated from the plant and nematode screens and several isogenic mutants generated by marker exchange, and showed a positive correlation between virulence of PA14 mutants in *Galleria* and mouse. By using *Drosophila* as a model host, D'Argenio et al. (36) screened approximately 1500 transposon-insertion mutants of PA01 and identified 33 mutants that are impaired in fly killing. All mutants strongly impaired in fly killing also lacked twitching motility, a form of solid surface translocation that occurs in a wide range of bacteria and is dependent on the presence of functional type IV fimbriae or pili (92). However, twitching motility per se is not required for full virulence in the fly because an additional set of twitching-motility mutants generated in that study was not impaired in fly killing. It is likely that genes required for twitching motility and fly pathogenesis, such as the *pilGHIJKL chpABCDE* gene cluster, control the expression of yet unknown virulence factors. One of these genes, *chpA*, was required for *P. aeruginosa*–mediated cytotoxicity of mammalian epithelial cells (70). It would be informative to test all the mutants identified from the fly-killing assay for virulence in a mammalian infection model.

SERRATIA MARCESCENS Another versatile bacterium with a broad host range is *Serratia marcescens*. It is a hospital-acquired pathogen associated with a number of specific outbreaks, particularly in critically ill neonates (60). Like *P. aeruginosa*, it is most commonly isolated in contact lens–associated keratitis (6).

S. marcescens can infect insects, such as anopheline mosquitoes, *Drosophila*, and *G. mellonella* (48, 116) (Table 1). Strain DB11 is a spontaneous streptomycin-resistant derivative of strain DB10 originally isolated from moribund *Drosophila* flies, and it establishes an infection that subsequently kills *C. elegans*. In contrast, a pleiotropic protease-deficient derivative of DB11 is less pathogenic in *Drosophila* and *C. elegans* (49, 104). Interestingly, protease is also an important virulence factor in mammalian keratitis. It is responsible for cornea damage in a rabbit eye model (75, 87). With the exception of protease, the identity of virulence factors required for *C. elegans* or *Drosophila* killing is unknown. However, screens of a *S. marcescens* transposon insertion library in *C. elegans* suggest that a significant number of genes are involved in virulence (78), and it is likely that many of these are important virulence determinants in mammalian infections.

S. marcescens can cause yellow vine disease in cucurbits. This bacterium is transmitted to plants by the squash bug *Anasa tristis*. *S. marcescens* colonizes the plant phloem, an infection that leads to chlorosis, rapid wilting, and death (14, 19). It is not known if plant, nematode, and insect pathogenic strains are capable of infecting humans or other mammals. Nonetheless, these findings open the possibility of using several nonvertebrate hosts to study *S. marcescens* pathogenesis.

BURKHOLDERIA CEPACIA *B. cepacia* is another opportunistic human pathogen that also infects plants and animals. The taxonomy of *B. cepacia* species is complex.

The *B. cepacia* complex (Bcc) consists of several species of closely related and extremely versatile gram-negative bacteria found naturally in soil, water, and the rhizosphere of plants. Strains of Bcc have been used in biological control of plant diseases and bioremediation. Some strains are plant pathogens. Indeed, *B. cepacia* was first described in 1950 as a pathogen causing soft-rot on onion bulbs (21). Over the past two decades it has emerged as an important pulmonary pathogen in individuals with chronic granulomatous disease, immunocompromised patients, and patients with cystic fibrosis (96). The original species *B. cepacia* has been split into eight genetic species (genomovars), but taxonomic distinctions have not enabled biological control strains to be clearly distinguished from human pathogenic strains (100). For genomovars II, IV, V, VII, and a new unnumbered genomovar, there are sufficient phenotypic characteristics for them to be assigned specific names: *B. multivorans*, *B. stabilis*, *B. vietnamiensis*, *B. ambifaria*, and *B. pyrrocinia*. Genomovars I, III, and VI have not yet received species names. Of these, only *B. pyrrocinia* has not been reported to be pathogenic in humans.

B. cepacia genomovars I and IV can kill *C. elegans* (99, 127). It is not yet known if *B. cepacia* infects insects. *B. cepacia* was present together with other bacterial species in the midgut of anopheline mosquitoes (105). Although *B. cepacia* is a phytopathogen, plants have not been exploited as a host for studying virulence factors important for *B. cepacia* pathogenesis in mammals. It is not known if the ability of *B. cepacia* to infect plant lies in the genetic material that is unique to plant pathogenic strains. It has been reported that a plant pathogenic strain of *B. cepacia* contains a plasmid-borne *pehA* gene, which encodes secreted endopolygalacturonase (enzymes capable of macerating plant tissues). In contrast, a soil isolate and a clinical isolate did not produce endopolygalacturonase and were incapable of macerating onion tissue (55). However, this study is based on a small number of strains and does not completely rule out the utility of plants as a surrogate to dissect *B. cepacia* virulence determinants. It will be of interest to determine if clinical isolates can infect a genetically tractable plant species, such as *Arabidopsis*.

Several reports have demonstrated that *B. cepacia* can invade and survive within epithelial cells and macrophages in vitro (22, 91, 113). The invasive ability of this bacillus may explain the persistence of infection in the face of antibiotic therapy and its propensity to cause bacteremic infections in cystic fibrosis patients. Speert and colleagues (26) showed that *B. cepacia* is capable of attaching to and invading murine respiratory tract epithelial cells and pulmonary macrophages. The ability of *B. cepacia* to survive and grow within the free-living amoeba *Acanthamoeba* species (82, 90) and *Tetrahymena pyriformis* (106) suggests that these protozoan, and perhaps *Dictyostelium*, may be used as surrogates to study the intracellular pathogenesis of this bacterium.

ERWINIA Members of the genus *Erwinia* are universally recognized as causing plant diseases and have been reported to be pathogenic in humans and other animals, including *C. elegans* (16, 42, 121). Human and animal isolates of *Erwinia* spp. are capable of infecting plants (81). A variety of factors influences the ability of this organism to attack plant tissue. Of major importance are a number of secreted pectinases and degradative enzymes, such as cellulase and protease. Whether these factors are also required for mammalian pathogenesis is not known. *E. carotovora* and *E. paradisiaca*, the causal agents of soft-rot in a number of plant species, can cause infections and trigger an immune response when ingested by *Drosophila* larvae (11). Together, these indicate that species of *Erwinia* are bona fide multihost pathogens.

Intracellular Gram-Negative Bacterial Pathogens

SALMONELLA ENTERICA *S. enterica* is one of the most extensively characterized bacterial pathogens and the leading cause of gastroenteritis and enteric fever in humans. Unlike the pathogens discussed above, *Salmonella* serotypes are known to be specialized or host-adapted pathogen. For example, epidemiological studies have limited the host range of *S. enterica* subspecies I serotypes to warm-blooded vertebrates (12). Surprisingly several *S. enterica* subspecies I serovars, including serovars Typhimurium, Dublin, and Enteritidis, kill *C. elegans* (3, 80). *S. enterica* serovar Typhimurium proliferates and establishes a persistent infection in the intestine of *C. elegans*. Colonization of the intestine also leads to increased cell death in the *C. elegans* germline, a process dependent on the host apoptotic machinery and requires the bacterial PhoP-PhoQ two-component system, which regulates the expression of at least 40 virulence genes (1). However, unlike the infection of mammalian hosts, there is no evidence that the bacteria invade intestinal cells of *C. elegans* (3, 80). Nevertheless, *S. enterica* serovar Typhimurium pathogenesis of nematode and mammals requires several factors in common, such as (*a*) PhoP-PhoQ; (*b*) OmpR, part of the OmpR-EnvZ two-component system that is responsible for both activation and repression of gene expression in response to changes in osmolarity and pH; (*c*) Fur-1, an iron-regulated protein required for acid tolerance, and (*d*) RpoS, the stationary-phase sigma factor (3, 80). In contrast, a *ssaV* mutant, which is less virulent in mammalian hosts, is not defective in *C. elegans* infection (13, 80). *ssaV* encodes an apparatus of the Type III secretion system that makes up the *Salmonella* pathogenicity island 2 (SPI2), indicating that SPI2 is critical for *Salmonella* pathogenesis in mice but not in nematodes. These results indicate that *C. elegans* is a good in vivo model system to dissect certain aspects of *Salmonella* pathogenesis. The ability of *S. enterica* serovar Typhimurium to infect *G. mellonella* (77) and *Drosophila* suggests that insects can also be used to study *Salmonella* pathogenesis.

LEGIONELLA PNEUMOPHILA Another important intracellular human pathogen is *Legionella pneumophila*, a gram-negative bacterium that is ubiquitous in aquatic environments and is the causal agent of Legionnaires' disease (124). Pathogenesis of legionellosis is largely due to the ability of *L. pneumophila* to invade and grow within alveolar macrophages, and it is widely believed that this ability results from a prior adaptation to intracellular niches in nature. Since the pioneering work of Rowbotham, it is now well established that *L. pneumophila* flourishes as an intracellular parasite of freshwater amoebae. At least 13 species of amoebae and 2 species of ciliated protozoa that allow intracellular bacterial replication are potential environmental hosts for legionellae (44).

The multihost capacity of *L. pneumophila* has led to the study of this bacterium in both protozoan and mammalian hosts. An important result from these cross-species investigations is that the life cycle of *L. pneumophila* in amoebae strongly resembles that observed in macrophages. There is also strong genetic evidence that the molecular basis for *L. pneumophila* invasion of protozoa and entry into human macrophages are similar. Several studies have shown that many of the *L. pneumophila* factors that promote its survival and replication within U937 macrophage-like cells are also required for growth in amoebae. For example, among 121 distinct miniTn*10::kan* insertion mutants with various degrees of defects in survival and replication within U937 cells, 89 are classified as protozoan and macrophage infectivity (*pmi*) mutants because they also exhibit similar phenotypic defects in *Acanthamoeba polyphaga* (52). In a separate study, 13 intracellular multiplication (*icm*) genes, identified by insertion mutations that conferred defective bacterial growth in U937 cells, are also required for replication within *Acanthamoeba castellanii* (115). The *icm* genes are also known as *dot* genes, because mutants in these genes are defective in organelle trafficking (*dot*). Likewise, seven transposon mutants isolated as defective for flagellar production are avirulent in both *Hartmannella vermiformis* and in U937 cells (103). However, there are genes that are specific for invasion of mammalian cells, such as the macrophage infectivity loci [*mil* (53)], and these would not be identified using amoebae as hosts.

Although much has been learned about the pathogenesis of *L. pneumophila* from the use of murine models, macrophage-like cell lines, and freshwater amoebae, the ability to analyze host factors essential for *L. pneumophila* growth is limited because these hosts are not amenable to genetic analysis. Recently, two groups have extended the concept of using genetically amenable hosts to study human pathogens by investigating the interaction between the free-living protozoa, *Dictyostelium discoideum*, and *L. pneumophila* (57, 119, 120). In addition, they showed that *L. pneumophila* grew by the same mechanism within *D. discoideum* as it does in mammalian macrophages and in two amoebae commonly used to study *L. pneumophila* phagocytosis, *H. vermiformis* and *A. castellanii*. For example, mutants in each of the *dotH/icmK*, *dotI/icmL*, and *dotO/icmB* genes that had severe intracellular growth defects in mouse bone marrow–derived macrophages (9) and in *A. castellanii* (115) also failed to grow intracellularly in *D. discoideum* (120).

The development of *D. discoideum* as a model host is an important contribution to the study of *L. pneumophila*–host interactions because it brings to bear the well-developed cell biological and biochemical techniques, and the powerful genetics tools for use in this organism. The *Dictyostelium* genome, consisting of 34 Mb carried on six chromosomes, can easily be manipulated by means of recombinant DNA techniques. Because it has a haploid genome, genes could be targeted for disruption by homologous recombination and as such, large pools of defined mutants are already available. Mutant *Dictyostelium* cells for systematic screens can be generated quickly and efficiently using techniques such as restriction enzyme–mediated integration [REMI (4)]. It should be possible to identify host factors that interact with *L. pneumophila* by directly screening the REMI-generated mutants for increased susceptibility or resistance to *L. pneumophila* infection. Finally, the anticipated completion of the full genome sequence of *D. discoideum* will further bring to bear many functional genomics tools, such as genome-wide expression analysis, for studying *L. pneumophila-D. discoideum* interactions.

BURKHOLDERIA PSEUDOMALLEI *B. pseudomallei* is a soil saprophyte endemic to northern Australia and Southeast Asia that causes melioidosis, a potentially fatal disease. The clinical outcomes include rapid-onset septicemia and relapsing and delayed-onset infections. It is primarily acquired by inoculation of compromised surface tissues by *B. pseudomallei*–contaminated water or soil (17, 133). Molecular typing has found identical *B. pseudomallei* bacteria from animals, humans, and soil (29). Recently, *C. elegans* has been used to model *B. pseudomallei* pathogenesis. *B. pseudomallei* causes paralysis in *C. elegans*, with a LT_{50} (a calculated time at which 50% of the nematode population exposed to the pathogen was observed to die) ranging between 16–23 h, depending on the strain used (99). The virulence factor(s) that causes *C. elegans* mortality is not yet known. Flagella, proteases, and lipases that are exported by the general secretion machinery, or exoproducts transported by the AmrAB-OprA multidrug efflux system do not appear to be important in causing *C. elegans* mortality (99). Analyses of known *C. elegans* mutants, however, indicate that the virulence factor(s) acts upon the Ca^{2+} signaling pathway mediated by a L-type Ca^{2+} channel (99).

Although *B. pseudomallei* is a facultative intracellular pathogen in humans, it is not known whether it is able to invade *C. elegans* cells. *B. pseudomallei* survives and proliferates within free-living acanthamoebae (*Acanthamoeba astronyxis*, *A. castellanii*, and *A. polyphaga*) in vitro. The interaction between *B. pseudomallei* and the amoebae—entry by coiling phagocytosis, survival within amoebic vacuoles, and eventual escape from vacuoles into amebic cytoplasm and to the surrounding medium—is similar to its interactions with mammalian cells (66). This raises the possibility that similar intracellular mechanisms are involved in the interactions between *B. pseudomallei* and its mammalian and protozoan hosts. The ability of *B. pseudomallei* to infect *C. elegans* and acantamoebae suggests that these hosts can be used to directly screen for bacterial virulence factors. These mutants can then be tested in mammalian hosts, such as *B. pseudomallei*-susceptible SWISS or BALB/c mice, to determine the role of these factors in mammalian pathogenesis.

Gram-Positive Bacterial Pathogens

ENTEROCOCCUS Gram-positive enterococcal infections are caused by at least 12 species of *Enterococcus*, two of which account for over 95% of the clinically important strains, *E. faecalis* (85%–90%) and *E. faecium* (5%–10%). *E. faecalis* (previously known as *Streptococcus faecalis*) is a commensal organism well suited for survival in intestinal and vaginal tracts and the oral cavity. It is increasingly gaining importance as community- and hospital-acquired infectious agents in humans, capable of infecting the blood, endocardium, genitourinary tract, abdomen, wounds, and skin and soft tissues. The natural ability of enterococci to readily acquire, accumulate, and share extrachromosomal elements encoding virulence traits or antibiotic resistance genes lends advantages to their survival under unusual environmental stresses and in part explains their ability to infect multiple hosts. Several clinical isolates of *E. faecalis*, including strains OG1RF (38) and V583 (112), can effectively colonize and kill *C. elegans* with a LT_{50} of approximately four days (54). *E. faecalis* also infects insects; when injected in moderate doses *E. faecalis* kills *G. mellonella* larvae (20). Importantly, several of the virulence factors required for *C. elegans* pathogenesis, such as *fsrB*, *scrB*, and cytolysin, are also required for mammalian pathogenesis (54). FsrB is similar to sensor kinases of the two-component regulatory system known to regulate many virulence factors, including gelatinase (107). Cytolysin is related to a family of antibacterial peptides termed lantibiotics, and it disrupts the membrane of eukaryotic cells, thus contributing to tissue damage in a rabbit model of endocarditis (28). Garsin et al. (54) also screened for *E. faecalis* mutants that are nonpathogenic on *C. elegans*. They showed that a mutant with a *Tn*917 insertion in a gene homologous to *scrB*, which encodes sucrose-6-phosphate hydrolase, is also attenuated in mouse pathogenesis.

In contrast to *E. faecalis*, *E. faecium* can neither form a persistent infection nor kill *C. elegans* even though it can accumulate to a high titer in the worm lumen. Analysis of virulence factors of *E. faecalis* and *E. faecium* cultured from blood showed that hemagglutinin and lipase may represent additional virulence factors of *E. faecalis* that are absent in *E. faecium* (39). Whether this contributes to the lack of virulence of *E. faecium* on *C. elegans* is unclear.

OTHER GRAM-POSITIVE BACTERIAL PATHOGENS Two other important gram-positive human pathogens *Streptococcus pneumoniae* and *Staphylococcus aureus* also kill *C. elegans*, with a LT_{50} of approximately 1 and 2 days (54). However, it is not known if there is any overlap in the virulence determinants required for nematode and mammalian pathogenesis, and the validity of *C. elegans* as a model for these pathogens remains to be demonstrated.

UTILITY AND LIMITATIONS OF CROSS-SPECIES ANALYSIS

P. aeruginosa mutants isolated from the plant and nematode screens have been grouped into classes based on their defects in pathogenicity in plant, nematode, and mice (129). This revealed that some virulence factors are species- and model-dependent (Table 3). This result suggests that in order to increase the efficiency for

TABLE 3 Classification of *P. aeruginosa* transposon-insertion mutants based on their defects in pathogenicity in nematode, plant, and mammalian hosts

Class	Exclusively less pathogenic in:	PA14 mutant strain or *gene*	# out of 21
I	Nematodes, plants, and mice	*gacA*, 25F1 (*orfT*), 34H4, pho15 (*dsbA*), pho34B12 (*mvfR*), *lasR*, 48D9 (*gacS*), 50E12 (*pstP*), 3E8 (*PA1900*), 36A4 (*mdoH*), 8C12 (*PA0997*), 41C1 (*aefA* or *kefA*)	12
II	Nematodes and mice	35A9 (*PA2020*), 44B1	2
III	Plants and mice	33A9, 33C7, 25A12, 16G12 (*PA0268*)	4
IV	Plants and nematodes	41A5 (*PA0328*), 23A2 (*mexA*)	2
V	Nematodes only	IG2 (*crbA*)	1

1. For most of the mutants, virulence in mice was determined at an inoculum of 5×10^5 cells, a relatively high dose that results in 100% mortality with PA14 wild-type. Under this condition, mutants that caused higher than 85% mouse mortality are considered to show no significant decrease in mouse pathogenicity in this classification. It is conceivable that some of the class IV and V are considered to mutants would be classified as being attenuated in mouse pathogenesis if a lower inoculation dose of 5×10^3 cells had been used, as demonstrated for mutants 25A12, 16G12 (see Table 2).

2. In creating these categories, the mutants affecting either fast or slow *C. elegans* killing were combined for the sake of simplicity.

identifying novel virulence determinants required for mammalian pathogenesis, it would be advantageous to use more than one nonvertebrate host. As exemplified by class II mutants of *P. aeruginosa* that are less pathogenic in nematodes and mice but not in plants, and by class III mutants of *P. aeruginosa* that are less pathogenic in plants and mice but not in nematodes, a significant proportion of the mutants that exhibited attenuated virulence in the burned-mouse model would have been missed if only one screen in a single nonvertebrate host had been used. From an evolutionary perspective, the use of multiple hosts provides extremely insightful information about conserved and host-specific factors. The conserved factors, i.e., genes required for pathogenesis in all hosts, are likely to exert their effect by targeting conserved components within diverse hosts. Conversely, species-specific virulence factors likely target host factors unique to the species, analyses of which would shed light on the question of species specificity in host-pathogen interactions.

Studies from different mammalian models have shown that many different virulence determinants are required for the manifestation of disease in any particular tissue, but the set of factors may differ from one tissue type to the other. For example, for *P. aeruginosa* pathogenesis, the hemolytic phospholipase C is important in causing mortality in burned mice (109) but is not essential for corneal infection in mice (102). Thus, in determining which virulence factors identified from nonvertebrate hosts are essential for mammalian pathogenesis, it is important to test these genes in a variety of mammalian models.

The use of model hosts to decipher molecular mechanisms of mammalian pathogenesis is likely to be limited to the identification and characterization of virulence determinants that target evolutionarily conserved cellular mechanisms. For each pathogen, the identification of mammal-specific virulence determinants still relies on mammalian models. Various techniques such as in vivo expression technology (IVET) and signature-tagged mutagenesis (STM) have been developed for this purpose (25, 93, 117). Another potential limitation in using *D. discoideum*, *Caenorhabditis*, and *Drosophila* as model hosts is that experiments using these animals have to be performed at temperatures less than 30°C, whereas for macrophages and amoeba they can be performed at 35–37°C. Because some virulence factors are known to be activated at 37°C, it is unclear if the effect of these virulence factors on host-pathogen interactions can be effectively elucidated using these model organisms.

ANALYSIS OF HOST DEFENSE RESPONSE IN NONVERTEBRATE ANIMAL HOSTS

Conservation of the Innate Immune System

The ability to detect an invading pathogen and activate an instantaneous defense is crucial for survival. The innate immune system is the first line of defense, acting effectively without previous exposure to a pathogen and conferring broad protection. In higher vertebrates, the innate immune system controls and assists the acquired immune system, which takes days or weeks to develop maximum efficacy. A wealth of compelling evidence indicates that major molecular components of the innate immune system are conserved in invertebrates and humans (74). The conservation between mammalian and nonvertebrate immune responses should not be surprising because pathogens invading these organisms have common features and utilize similar offensive strategies to overcome host defenses (23, 45, 46, 88, 108). For example, the virulence factors *gacA*, *gacS*, *lasR*, and *mucD* are required by *P. aeruginosa* in order to successfully infect nematode, insect, plant, and mice. Coupled with the demand for survival that is inescapably shared by these hosts, evolution would have conserved essential elements in these organisms to effectively combat pathogens.

A striking example of phylogenetic conservation of the innate immune system is the signaling pathway mediated by Toll-like receptor (TLR) proteins (65). The TLR family encodes transmembrane proteins containing extracellular leucine-rich repeats (LRR) and an intracellular domain that is now known as the Toll/interleukin receptor (TIR) domain because of its significant homology to the intracellular domain of the interleukin-1 receptor. First elucidated in *Drosophila*, the Toll pathway functions in dorso-ventral patterning during embryonic development and in activating defense response in adult flies upon challenge by infectious agents (83). In mammals, the LRR domains of TLRs are involved in the recognition of conserved molecular patterns on pathogens, leading to the activation of innate immune response (7, 76). Molecules that bear similarity to the TLRs and some of the downstream components are also present in nematodes and plants (43, 122, 127).

C. elegans has a Toll homolog, *tol-1*, that is essential both for development and for recognition of a bacterial pathogen (104).

Recent work suggests that the mitogen-activated protein kinase (MAPK) cassette is another innate immune pathway that is conserved in phylogenetically diverse organisms. The MAPK cassette has a core unit consisting of a three-member protein cascade. MAPKs are activated by MAPK kinases (MKKs), dual specificity kinases that catalyze the phosphorylation of MAPKs on both tyrosine and threonine residues. The MKKs are themselves phosphorylated and activated by serine/threonine kinases that function as MKK kinases (MAP3Ks). The evolutionarily conserved MAPKs can be divided into three subgroups: the p38 kinases, the c-Jun N-terminal (JNK, also known as stress-activated protein kinase), and the extracellular signal–regulated kinase (ERK). Typically the p38 and JNK pathways are activated in response to stress stimuli or pathogens (79), whereas the ERK pathway transduces the signals from growth factors or mitogens. A wealth of cell biological and biochemical data has implicated a critical role for mammalian p38 MAPK kinase signaling in cellular immune response (79). Cell biological studies have also implicated p38 and JNK MAP kinases in the modulation of *Drosophila* immune response (59, 118). Using *C. elegans* as a host, we have recently obtained direct genetic evidence that the p38 MAPK pathway is required in the immune response of a whole animal (D. Kim, R. Feinbaum, M.-W. Tan & F. M. Ausubel, unpublished data).

Genetic and Genomic Dissections of the Innate Immune System

Because the innate immune system is phylogenetically conserved, genetically tractable nonvertebrate animal hosts can be used to identify components of the innate immune system, many of which are likely to be conserved in mammals. In addition to being genetically facile, complete genome sequences are available for hosts such as *C. elegans* and *Drosophila* (5, 33). Thus, traditional genetics approaches and recent functional genomics innovations can readily be applied and integrated in these organisms to identify and characterize components of the innate immune system. A direct approach to identify host factors that are involved in defense response to pathogens is to screen for host mutants that are either more resistant or more susceptible to pathogen attack. This screen does not rest on any assumptions about which signaling pathway or downstream components are important for defense against pathogens, thus allowing the identification of novel innate immune mechanisms. This approach has been used successfully to isolate *Drosophila* and *C. elegans* mutants that are more susceptible to pathogens in order to elucidate defense pathways in insects (84, 131) and nematodes (73).

Transcriptional control of genes has been shown to play a key role in host-pathogen interactions (34, 123). Microarray technology can be used to monitor gene expression in the host at the whole-genome level when it is interacting with pathogens. Typically hundreds or even thousands of genes show altered expression in a microarray experiment. To determine the function of this large number of differentially expressed genes for defense against pathogens, an efficient, economical, and logistically feasible technology to inactivate each gene is essential. Recently, a powerful targeted-gene inactivation method called double-stranded (ds) RNA-mediated interference (RNAi) was established. The RNAi machinery uses the sequence information in the dsRNA to generate a protein-RNA complex that destroys the corresponding mRNA (58, 135). Thus with RNAi sequence of a particular gene can now be used to disrupt the function of that gene, generating in most cases animals or cells that phenocopy the loss-of-function phenotype due to chromosomal mutation for that gene (47). This method of gene inactivation in the whole organism works most efficiently in *C. elegans*, and it is becoming an important tool in the study of other invertebrates, including *Drosophila* (72, 95). Mutants generated by RNAi can then be assayed for their susceptibility to pathogens; this targeted gene knock-out approach therefore directly connects susceptible or resistant mutant phenotypes with known genes. Although genome-wide surveys of host gene expression in response to pathogen can also be performed for the human and mouse genomes, it is difficult to determine the function of the differentially expressed genes for host defense because the technology for efficient and economical gene knock-out for these organisms is not yet available.

Another important future contribution of the use of simple model organisms to the study of host-pathogen interactions will be the functional analysis of virulence factors at specific stages of interaction with the host. For example, because slow killing involves an infection process, *C. elegans* can be used to analyze the functional contribution of each *P. aeruginosa* virulence factor for colonization, survival, and replication within the host, an analysis that is difficult to perform using mammalian models.

The conservation of core parts of the innate immune pathways, such as the Toll and MAPK pathways between vertebrate and nonvertebrates, suggests that these pathways were under strong evolutionary constraints. However, the effector molecules used to directly combat pathogens were likely to be less constrained and hence could have freely evolved with each evolutionary lineage. Thus, the evolutionary distance between the nonvertebrate model organisms and humans can be an added advantage. Because they could be infected by a common set of pathogens, over the millions of years of independent evolution between these pathogens and their respective hosts, each host could have evolved unique effector molecules to directly kill or inhibit the invading pathogens. This would give rise to effector molecules that are unique to each lineage, despite being regulated by the same conserved pathways. For example, many antimicrobial peptides in *Drosophila* have no counterpart in humans. The identification of these effector molecules in the nonvertebrate hosts can potentially be used as a source for the development of novel anti-infective agents for the augmentation antimicrobial defense. In the following, I discuss the salient features of *Drosophila, C. elegans*, and *Dictyostelium* as model organisms for the dissection of the innate immune system.

Nonvertebrate Animal Hosts

DROSOPHILA *Drosophila* activates multiple cellular and humoral immune responses upon microbial infections (63, 64). Studies on bacterial infection of *Drosophila* have relied on direct introduction of bacteria into the body cavity by injection or pricking, or by ingestion of bacteria by the fly larvae. Interestingly, it appears that different pathways may be involved in activating antibacterial responses, depending on the mode of infection. For example, in the absence of physical injury, hemocytes play a significant role in activating a systemic antibacterial response, whereas introduction of bacteria by pricking triggers other pathways that bypass the requirement of hemocytes (11).

A combination of classical genetic screens and genomic approaches has revealed a wealth of information regarding the *Drosophila* defense responses. Characterization of mutations on chromosome X that resulted in adult fly that failed to survive infection by a bacterial pathogen has identified a *Dredd*, a caspase-encoding gene, and *dTAK1*, a homolog of the mammalian TAK1 [TGFβ–activated kinase (84, 131)], as essential components of defense. In a separate screen, 14 genes on the third chromosome have been identified based on their requirement to induce transcription of an antibacterial peptide gene in response to infection or to form melanized clots at the site of wounding. Among these are *scribble* and *kurtz/modulo*, which affect the cellular organization of the fat body, the tissue responsible for antimicrobial peptide production, a homolog of I kappa B kinase (*DmIkk beta*) and *Relish*, a Rel-family transcription factor (134). Genome-wide analyses of the *Drosophila* immune response to gram-negative or gram-positive bacteria or fungi carried out using oligonucleotide arrays have also revealed numerous novel components of the immune response (37, 67). These studies should provide a basis for a comprehensive understanding of the genetic control of immune responses in *Drosophila*. The striking similarity between *Drosophila* and mammalian innate immune responses make the fly a powerful model for elucidating the innate immune system (74, 97).

CAENORHABDITIS ELEGANS The use of *C. elegans* to elucidate the molecular nature of interactions between an animal host and bacterial pathogen is now an active field of research (2, 42, 78, 125, 127). Table 4 shows a list of bacteria species demonstrated to kill *C. elegans* strain Bristol (N2) and the media conditions used to grow the bacterial lawn. A key aspect of the *C. elegans* pathogenesis model is that the mode and extent of killing is dependent on a variety of genetic and environmental factors. One of the critical environmental factors is the medium on which the microorganism is grown. For example, *E. coli* OP50, which does not kill *C. elegans* when grown on NGM medium, is pathogenic when grown on BHI (54). Therefore, it is likely that the number of pathogens able to use *C. elegans* as a model host would increase when more strains and conditions are tested.

With the exception of *S. enterica*, the majority of the human pathogens that kill *C. elegans* survive well as soil saprophytes. Because *C. elegans* uses bacteria

TABLE 4 Strains of bacteria tested for the ability to kill *C. elegans* and the media conditions used in those tests

Bacterial species	Media condition	Reference
Strains that kill *C. elegans*		
P. aeruginosa	SKA, PG, BHI	35, 89, 126, 128
*P. fluorescens**	PG, BHI	35, 128
Burkholderia cepacia	SKA, PG	99, 127
*B. pseudomallei, B. thailandensis**	SKA, PG	99
Salmonella enterica serovar Typhimurium, Dublin, and Enteritidis	SKA, NGM	3, 80
Aeromonas hydrophila, Agrobacterium tumefaciens, Erwinia chrysanthemi, E. carotovora carotovora, Shewanella massalia, S. putrifaciens	NGM	42
Enterococcus faecalis, Streptococcus pneumoniae, Staphylococcus aureus	BHI	54
Strains that infect but do not kill *C. elegans*		
*Microbacterium nematophilum**	NGM	62
Strains that do not infect or kill *C. elegans*		
E. faecium, Strept. pyogenes, Bacillus subtilis	BHI	54
P. syringae, P. fluorescens,* P. circhorii,* Burkholderia mallei, B. vietnamensis*	SKA, PG	99, 128
Salmonella enterica serovar Typhi, Paratyphi	SKA	3

*Not known to be pathogenic in humans; recipes for media used are given in (125).

and fungi as food, some of these organisms could have evolved mechanisms to protect themselves against this predator. For those organisms that are also capable of infecting humans, as demonstrated by the analysis of cross-species infections of *P. aeruginosa*, it would not be surprising then to find common virulence determinants (weapons) that are used to inflict harm in the nematode and human hosts. Limited experimental evidence thus far suggests that *C. elegans* may not be a suitable host for obligate pathogens or host-specific pathogens. O'Quinn et al. (99) compared the ability of two phylogenetically closely related human pathogens and showed that *B. pseudomallei*, a soil saprophyte, kills *C. elegans* and that *B. mallei*, an obligate pathogen with no known environmental reservoir, does not kill *C. elegans*.

Unlike *Drosophila*, less is known about the *C. elegans* immune system. Recent studies on genome-wide gene expression of worms in response to bacterial infections have revealed that *C. elegans* has an inducible defense response (J.J. Ewbank, personal communications; M.-W. Tan, unpublished work). In addition, genetic screens to identify mutant worms that are either more susceptible or resistant to pathogens have been conducted. Analysis of ethyl methanesulfonate (EMS)-induced mutant worms that showed enhanced susceptibility to *P. aeruginosa* PA14 revealed that *C. elegans* uses the Sma pathway, which is homologous to the mammalian TGFβ pathway, and the p38 MAPK pathway to transduce signals required for antibacterial defense [(73); M.-W. Tan, unpublished work]. Darby et al. (35) screened F_2 progeny of EMS mutagenized *C. elegans* for resistance to *P. fluorescens* SE59-mediated paralysis. They identified two alleles of *egl-9* and showed that *egl-9* mutants are also resistant to cyanide-induced paralysis by *P. aeruginosa* PA01. EGL-9 is a 2-oxoglutarate-dependent dioxygenase that regulates the hypoxia inducible factor (HIF) by prolyl hydroxylation (40). However, its interaction with hydrogen cyanide remains to be elucidated. *C. elegans* also produces antimicrobial peptides, and the recombinant of one of these, ABF-2, has potent microbicidal activity against gram-positive and gram-negative bacteria and yeast (71). Together with the many tools available (125), these results indicate that *C. elegans* can also be a powerful model to comprehensively dissect the innate immune system.

DICTYOSTELIUM Several facultative bacterial intracellular pathogens, notably *L. pneumophila*, *Mycobacterium avium*, *Chlamydia pneumoniae*, and *Listeria monocytogenes*, survive as endosymbionts or infect free-living amoebae such as *Acanthamoeba*, *Hartmanella*, and *Naegleria* (32, 41, 85, 111). Amoebae in the environment have also been found to harbor several other pathogenic bacteria, including *E. coli* 0157, *P. aeruginosa*, *Vibrio*, *B. cepacia*, and *Franciscella* (10, 18, 90, 94, 132). Although direct evidence for the involvement of amoebic endosymbionts in the pathogenesis of human infection has yet to be found, amoebic endosymbiosis is known to augment the virulence of *L. pneumophila* and *M. avium* (30, 31, 32).

As discussed in several examples above, the molecular mechanisms by which amoebae and macrophages ingest microbes bear striking similarities. It is perhaps not surprising then that the strategies used by pathogens to parasitize macrophages overlap with those enabling survival within amoebae. *D. discoideum* is a unicellular amoeba that lives in soil and feeds on bacteria. Because of its genetic tractability as a haploid organism and a relatively small genome, cell biologists have used *Dictyostelium* as a model to study phagocytosis (69). The amoeba also shares many physiological functions seen in mammalian cells and is amenable to genetic manipulation. The successful use of *Dictyostelium* to study *L. pneumophila* pathogenesis suggests that there is great potential in developing this genetically facile organism to model interactions with other intracellular pathogens.

CONCLUDING REMARKS

A multihost pathogen is a generalist not only in host range but also in tissue range within a particular host. The majority of these multihost pathogens are opportunistic human pathogens, that is, they cause disease in hosts whose local or systemic immune attributes have been impaired, damaged, or are innately dysfunctional. The interaction of pathogens with their more common nonvertebrate hosts could be the main force that drives the evolution and maintenance of many of the virulence factors. These virulence factors could be identified by directly screening for mutants that are attenuated in virulence in their respective nonvertebrate hosts.

The use of genetically facile nonvertebrate hosts to study the pathogenesis of multihost human pathogen models also opens up the possibility of exploiting the power of genetics and functional genomics to examine the interplay that occurs between a pathogen and its host by manipulating host factors involved in disease resistance, as well as to identify and study bacterial virulence factors. Although these models do not perfectly reflect infections in humans, the conservation of many signaling pathways, particularly in innate immunity, suggests that the use of an invertebrate model will provide insights into general mechanisms of host-pathogen interactions.

ACKNOWLEDGMENTS

Work in the author's laboratory is supported in part by the Donald E. and Delia B. Baxter Foundation. I thank members of my laboratory for helpful comments on the manuscript.

The *Annual Review of Microbiology* is online at http://micro.annualreviews.org

LITERATURE CITED

1. Aballay A, Ausubel FM. 2001. Programmed cell death mediated by *ced-3* and *ced-4* protects *Caenorhabditis elegans* from *Salmonella typhimurium*-mediated killing. *Proc. Natl. Acad. Sci. USA* 98: 2735–39
2. Aballay A, Ausubel FM. 2002. *Caenorhabditis elegans* as a host for the study of host-pathogen interactions. *Curr. Opin. Microbiol.* 5:97–101
3. Aballay A, Yorgey P, Ausubel FM. 2000. *Salmonella typhimurium* proliferates and establishes a persistent infection in the intestine of *Caenorhabditis elegans*. *Curr. Biol.* 10:1539–42
4. Adachi H, Hasebe T, Yoshinaga K, Ohta T, Sutoh K. 1994. Isolation of *Dictyostelium discoideum* cytokinesis mutants by restriction enzyme-mediated integration of the blasticidin S resistance marker. *Biochem. Biophys. Res. Commun.* 205:1808–14
5. Adams MD, Celniker SE, Holt RA, Evans CA, Gocayne JD, et al. 2000. The genome sequence of *Drosophila melanogaster*. *Science* 287:2185–95
6. Alexandrakis G, Alfonso EC, Miller D. 2000. Shifting trends in bacterial keratitis in south Florida and emerging resistance to fluoroquinolones. *Ophthalmology* 107:1497–502
7. Anderson KV. 2000. Toll signaling pathways in the innate immune response. *Curr. Opin. Immunol.* 12:13–19
8. Andrew PA, Nicholas WL. 1976. Effect of bacteria on dispersal of *Caenorhabditis elegans* (Rhabditidae). *Nematologica* 22:451–61
9. Andrews HL, Vogel JP, Isberg RR. 1998. Identification of linked *Legionella pneumophila* genes essential for intracellular growth and evasion of the endocytic pathway. *Infect. Immun.* 66:950–58
10. Barker J, Humphrey TJ, Brown MW. 1999. Survival of *Escherichia coli* 0157 in a soil protozoan: implications for disease. *FEMS Microbiol. Lett.* 173:291–95

11. Basset A, Khush RS, Braun A, Gardan L, Boccard F, et al. 2000. The phytopathogenic bacteria *Erwinia carotovora* infects *Drosophila* and activates an immune response. *Proc. Natl. Acad. Sci. USA* 97:3376–81
12. Baumler AJ, Tsolis RM, Ficht TA, Adams LG. 1998. Evolution of host adaptation in *Salmonella enterica. Infect. Immun.* 66:4579–87
13. Beuzon CR, Meresse S, Unsworth KE, Ruiz-Albert J, Garvis S, et al. 2000. *Salmonella* maintains the integrity of its intracellular vacuole through the action of SifA. *EMBO J.* 19:3235–49
14. Bextine B, Wayadande A, Pair S, Bruton B, Mitchell F, Fletcher J. 2001. Parameters of *Serratia marcescens* transmission by the squash bug, *Anasa tristis. Phytopathology* 91:S8 (Abstr.)
15. Boman HG, Nilsson I, Rasmuson B. 1972. Inducible antibacterial defence system in *Drosophila. Nature* 237:232–35
16. Bottone E, Schneierson SS. 1972. *Erwinia* species: an emerging human pathogen. *Am. J. Clin. Pathol.* 57:400–5
17. Brett PJ, Woods DE. 2000. Pathogenesis of and immunity to melioidosis. *Acta Trop.* 74:201–10
18. Brown MR, Barker J. 1999. Unexplored reservoirs of pathogenic bacteria: protozoa and biofilms. *Trends Microbiol.* 7:46–50
19. Bruton B, Mitchell F, Bextine B, Wayadande A, Pair S, et al. 2001. Yellow vine of cucurbits: pathogenicity of *Serratia marcescens* and transmission by *Anasa tristis. Phytopathology* 91:S11 (Abstr.)
20. Bulla LA, Rhodes RA, St. Julian G. 1975. Bacteria as insect pathogens. *Annu. Rev. Microbiol.* 29:163–90
21. Burkholder W. 1950. Sour skin, a bacterial rot of onion bulbs. *Phytopathology* 40:115–18
22. Burns JL, Jonas M, Chi EY, Clark DK, Berger A, Griffith A. 1996. Invasion of respiratory epithelial cells by *Burkholderia (Pseudomonas) cepacia. Infect. Immun.* 64:4054–59
23. Cao H, Baldini RL, Rahme LG. 2001. Common mechanisms for pathogens of plants and animals. *Annu. Rev. Phytopathol.* 39:259–84
24. Cao H, Krishnan G, Goumnerov B, Tsongalis J, Tompkins R, Rahme LG. 2001. A quorum sensing-associated virulence gene of *Pseudomonas aeruginosa* encodes a LysR-like transcription regulator with a unique self-regulatory mechanism. *Proc. Natl. Acad. Sci. USA* 98:14613–18
25. Chiang SL, Mekalanos JJ, Holden DW. 1999. In vivo genetic analysis of bacterial virulence. *Annu. Rev. Microbiol.* 53:129–54
26. Chiu CH, Ostry A, Speert DP. 2001. Invasion of murine respiratory epithelial cells *in vivo* by *Burkholderia cepacia. J. Med. Microbiol.* 50:594–601
27. Cho JJ, Schroth MN, Kominos SD, Green SK. 1975. Ornamental plants as carriers of *Pseudomonas aeruginosa. Phytopathology* 65:425–31
28. Chow JW, Thal LA, Perri MB, Vazquez JA, Donabedian SM, et al. 1993. Plasmid-associated hemolysin and aggregation substance production contribute to virulence in experimental enterococcal endocarditis. *Antimicrob. Agents Chemother.* 37:2474–77
29. Choy JL, Mayo M, Janmaat A, Currie BJ. 2000. Animal melioidosis in Australia. *Acta Trop.* 74:153–58
30. Cirillo JD, Cirillo SL, Yan L, Bermudez LE, Falkow S, Tompkins LS. 1999. Intracellular growth in *Acanthamoeba castellanii* affects monocyte entry mechanisms and enhances virulence of *Legionella pneumophila. Infect. Immun.* 67:4427–34
31. Cirillo JD, Falkow S, Tompkins LS. 1994. Growth of *Legionella pneumophila* in *Acanthamoeba castellanii* enhances invasion. *Infect. Immun.* 62:3254–61

32. Cirillo JD, Falkow S, Tompkins LS, Bermudez LE. 1997. Interaction of *Mycobacterium avium* with environmental amoebae enhances virulence. *Infect. Immun.* 65:3759–67
33. *C. elegans* Sequencing Consortium. 1998. Genome sequence of the nematode *C. elegans*: a platform for investigating biology. *Science* 282:2012–18
34. Cotter PA, Miller JF. 1998. *In vivo* and *ex vivo* regulation of bacterial virulence gene expression. *Curr. Opin. Microbiol.* 1:17–26
35. Darby C, Cosma CL, Thomas JH, Manoil C. 1999. Lethal paralysis of *Caenorhabditis elegans* by *Pseudomonas aeruginosa*. *Proc. Natl. Acad. Sci. USA* 96: 15202–7
36. D'Argenio DA, Gallagher LA, Berg CA, Manoil C. 2001. *Drosophila* as a model host for *Pseudomonas aeruginosa* infection. *J. Bacteriol.* 183:1466–71
37. De Gregorio E, Spellman PT, Rubin GM, Lemaitre B. 2001. Genome-wide analysis of the *Drosophila* immune response by using oligonucleotide microarrays. *Proc. Natl. Acad. Sci. USA* 98:12590–95
38. Dunny GM, Brown BL, Clewell DB. 1978. Induced cell aggregation and mating in *Streptococcus faecalis*: evidence for a bacterial sex pheromone. *Proc. Natl. Acad. Sci. USA* 75:3479–83
39. Elsner HA, Sobottka I, Mack D, Claussen M, Laufs R, Wirth R. 2000. Virulence factors of *Enterococcus faecalis* and *Enterococcus faecium* blood culture isolates. *Eur. J. Clin. Microbiol. Infect. Dis.* 19:39–42
40. Epstein AC, Gleadle JM, McNeill LA, Hewitson KS, O'Rourke J, et al. 2001. *C. elegans* EGL-9 and mammalian homologs define a family of dioxygenases that regulate HIF by prolyl hydroxylation. *Cell* 107:43–54
41. Essig A, Heinemann M, Simnacher U, Marre R. 1997. Infection of *Acanthamoeba castellanii* by *Chlamydia pneumoniae*. *Appl. Environ. Microbiol.* 63: 1396–99
42. Ewbank JJ. 2002. Tackling both sides of the host-pathogen equation with *Caenorhabditis elegans*. *Microbes Infect.* 4: 247–56
43. Fallon PG, Allen RL, Rich T. 2001. Primitive Toll signalling: bugs, flies, worms and man. *Trends Immunol.* 22:63–66
44. Fields BS. 1996. The molecular ecology of legionellae. *Trends Microbiol.* 4:286–90
45. Finlay BB. 1999. Bacterial disease in diverse hosts. *Cell* 96:315–58
46. Finlay BB, Falkow S. 1997. Common themes in microbial pathogenicity revisited. *Microbiol. Mol. Biol. Rev.* 61:136–69
47. Fire A, Xu S, Montgomery MK, Kostas SA, Driver SE, Mello CC. 1998. Potent and specific genetic interference by double-stranded RNA in *Caenorhabditis elegans*. *Nature* 391:806–11
48. Flyg C, Kenne K, Boman HG. 1980. Insect pathogenic properties of *Serratia marcescens*: phage-resistant mutants with a decreased resistance to *Cecropia* immunity and a decreased virulence to *Drosophila*. *J. Gen. Microbiol.* 120:173–81
49. Flyg C, Xanthopoulos KG. 1983. Insect pathogenic properties of *Serratia marcescens*: passive and active resistance to insect immunity studied with protease-deficient and phage-resistant mutants. *J. Gen. Microbiol.* 129:453–64
50. Galan JE, Collmer A. 1999. Type III secretion machines: bacterial devices for protein delivery into host cells. *Science* 284:1322–28
51. Gallagher LA, Manoil C. 2001. *Pseudomonas aeruginosa* PAO1 kills *Caenorhabditis elegans* by cyanide poisoning. *J. Bacteriol.* 183:6207–14
52. Gao LY, Harb OS, Kwaik YA. 1997. Utilization of similar mechanisms by *Legionella pneumophila* to parasitize two evolutionarily distant host cells,

mammalian macrophages and protozoa. *Infect. Immun.* 65:4738–46
53. Gao LY, Harb OS, Kwaik YA. 1998. Identification of macrophage-specific infectivity loci (mil) of *Legionella pneumophila* that are not required for infectivity of protozoa. *Infect. Immun.* 66:883–92
54. Garsin DA, Sifri CD, Mylonakis E, Qin X, Singh KV, et al. 2001. A simple model host for identifying gram-positive virulence factors. *Proc. Natl. Acad. Sci. USA* 98:10892–97
55. Gonzalez CF, Pettit EA, Valadez VA, Provin EM. 1997. Mobilization, cloning, and sequence determination of a plasmid-encoded polygalacturonase from a phytopathogenic *Burkholderia (Pseudomonas) cepacia. Mol. Plant Microbe Interact.* 10:840–51
56. Grewal PS. 1991. Influence of bacteria and temperature on the reproduction of *Caenorhabditis elegans* (Nematoda: Rhabditidae) infesting mushrooms (*Agaricus bisporus*). *Nematologica* 37: 72–82
57. Hagele S, Kohler R, Merkert H, Schleicher M, Hacker J, Steinert M. 2000. *Dictyostelium discoideum*: a new host model system for intracellular pathogens of the genus *Legionella. Cell. Microbiol.* 2:165–71
58. Hammond SM, Bernstein E, Beach D, Hannon GJ. 2000. An RNA-directed nuclease mediates post-transcriptional gene silencing in *Drosophila* cells. *Nature* 404:293–96
59. Han ZS, Enslen H, Hu X, Meng X, Wu IH, et al. 1998. A conserved p38 mitogen-activated protein kinase pathway regulates *Drosophila* immunity gene expression. *Mol. Cell Biol.* 18:3527–39
60. Hejazi A, Falkiner FR. 1997. *Serratia marcescens. J. Med. Microbiol.* 46:903–12
61. Henderson IR, Navarro-Garcia F, Nataro JP. 1998. The great escape: structure and function of the autotransporter proteins. *Trends Microbiol.* 6:370–78
62. Hodgkin J, Kuwabara PE, Corneliussen B. 2000. A novel bacterial pathogen, *Microbacterium nematophilum*, induces morphological change in the nematode *C. elegans. Curr. Biol.* 10:1615–18
63. Hoffmann JA, Reichhart J. 1997. *Drosophila* immunity. *Trends Cell Biol.* 7:309–16
64. Imler JL, Hoffmann JA. 2000. Signaling mechanisms in the antimicrobial host defense of *Drosophila. Curr. Opin. Microbiol.* 3:16–22
65. Imler JL, Hoffmann JA. 2001. Toll receptors in innate immunity. *Trends Cell Biol.* 11:304–11
66. Inglis TJ, Rigby P, Robertson TA, Dutton NS, Henderson M, Chang BJ. 2000. Interaction between *Burkholderia pseudomallei* and *Acanthamoeba* species results in coiling phagocytosis, endamebic bacterial survival, and escape. *Infect. Immun.* 68:1681–86
67. Irving P, Troxler L, Heuer TS, Belvin M, Kopczynski C, et al. 2001. A genome-wide analysis of immune responses in *Drosophila. Proc. Natl. Acad. Sci. USA* 98:15119–24
68. Jander G, Rahme LG, Ausubel FM. 2000. Positive correlation between virulence of *Pseudomonas aeruginosa* mutants in mice and insects. *J. Bacteriol.* 182:3843–45
69. Janssen KP, Schleicher M. 2001. *Dictyostelium discoideum*: a genetic model system for the study of professional phagocytes. Profilin, phosphoinositides and the *lmp* gene family in *Dictyostelium. Biochim. Biophys. Acta* 1525:228–33
70. Kang PJ, Hauser AR, Apodaca G, Fleiszig SM, Wiener-Kronish J, et al. 1997. Identification of *Pseudomonas aeruginosa* genes required for epithelial cell injury. *Mol. Microbiol.* 24:1249–62
71. Kato Y, Aizawa T, Hoshino H, Kawano

K, Nitta K, Zhang H. 2002. *abf-1* and *abf-2*, ASABF-type antimicrobial peptide genes in *Caenorhabditis elegans*. *Biochem. J.* 361:221–30
72. Kennerdell JR, Carthew RW. 1998. Use of dsRNA-mediated genetic interference to demonstrate that frizzled and frizzled 2 act in the wingless pathway. *Cell* 95:1017–26
73. Deleted in proof
74. Kimbrell DA, Beutler B. 2001. The evolution and genetics of innate immunity. *Nat. Rev. Genet.* 2:256–67
75. Kreger AS, Griffin OK. 1975. Cornea-damaging proteases of *Serratia marcescens*. *Invest. Ophthalmol.* 14:190–98
76. Krutzik SR, Sieling PA, Modlin RL. 2001. The role of Toll-like receptors in host defense against microbial infection. *Curr. Opin. Immunol.* 13:104–8
77. Kurstak E, Vega CE. 1968. Bacterial infection due to *Salmonella typhimurium* in an invertebrate, *Galleria mellonella* L. *Can. J. Microbiol.* 14:233–37
78. Kurz CL, Ewbank JJ. 2000. *Caenorhabditis elegans* for the study of host-pathogen interactions. *Trends Microbiol.* 8: 142–44
79. Kyriakis JM, Avruch J. 2001. Mammalian mitogen-activated protein kinase signal transduction pathways activated by stress and inflammation. *Physiol. Rev.* 81:807–69
80. Labrousse A, Chauvet S, Couillault C, Kurz CL, Ewbank JJ. 2000. *Caenorhabditis elegans* is a model host for *Salmonella typhimurium*. *Curr. Biol.* 10:1543–45
81. Lakso JU, Starr MP. 1970. Comparative injuriousness to plants of *Erwinia* spp. and other enterobacteria from plants and animals. *J. Appl. Bacteriol.* 33:692–707
82. Landers P, Kerr KG, Rowbotham TJ, Tipper JL, Keig PM, et al. 2000. Survival and growth of *Burkholderia cepacia* within the free-living amoeba *Acanthamoeba polyphaga*. *Eur. J. Clin. Microbiol. Infect. Dis.* 19:121–23
83. Lemaitre B, Nicolas E, Michaut L, Reichhart JM, Hoffmann JA. 1996. The dorsoventral regulatory gene cassette spatzle/Toll/cactus controls the potent antifungal response in *Drosophila* adults. *Cell* 86:973–83
84. Leulier F, Rodriguez A, Khush RS, Abrams JM, Lemaitre B. 2000. The *Drosophila* caspase Dredd is required to resist gram-negative bacterial infection. *EMBO Rep.* 1:353–58
85. Ly TM, Muller HE. 1990. Ingested *Listeria monocytogenes* survive and multiply in protozoa. *J. Med. Microbiol.* 33:51–54
86. Lyczak JB, Cannon CL, Pier GB. 2000. Establishment of *Pseudomonas aeruginosa* infection: lessons from a versatile opportunist. *Microbes Infect.* 2:1051–60
87. Lyerly D, Gray L, Kreger A. 1981. Characterization of rabbit corneal damage produced by *Serratia* keratitis and by a serratia protease. *Infect. Immun.* 33:927–32
88. Mahajan-Miklos S, Rahme LG, Ausubel FM. 2000. Elucidating the molecular mechanisms of bacterial virulence using non-mammalian hosts. *Mol. Microbiol.* 37:981–88
89. Mahajan-Miklos S, Tan MW, Rahme LG, Ausubel FM. 1999. Molecular mechanisms of bacterial virulence elucidated using a *Pseudomonas aeruginosa-Caenorhabditis elegans* pathogenesis model. *Cell* 96:47–56
90. Marolda CL, Hauroder B, John MA, Michel R, Valvano MA. 1999. Intracellular survival and saprophytic growth of isolates from the *Burkholderia cepacia* complex in free-living amoebae. *Microbiology* 145:1509–17
91. Martin DW, Mohr CD. 2000. Invasion and intracellular survival of *Burkholderia cepacia*. *Infect. Immun.* 68:24–29
92. McBride MJ. 2001. Bacterial gliding motility: multiple mechanisms for cell movement over surfaces. *Annu. Rev. Microbiol.* 55:49–75

93. Merrell DS, Camilli A. 2000. Detection and analysis of gene expression during infection by in vivo expression technology. *Philos. Trans. R. Soc. London B Biol. Sci.* 355:587–99
93a. McLaggan D, Jones MA, Gouesbet G, Levina N, Lindey S, et al. 2002. Analysis of the *kefA2* mutation suggests that KefA is a cation-specific channel involved in osmotic adaptation in *Escherichia coli. EMBO J.* 43:521–36
94. Michel R, Burghardt H, Bergmann H. 1995. *Acanthamoeba*, naturally intracellularly infected with *Pseudomonas aeruginosa*, after their isolation from a microbiologically contaminated drinking water system in a hospital. *Zentralbl. Hyg. Umweltmed.* 196:532–44
95. Misquitta L, Paterson BM. 1999. Targeted disruption of gene function in *Drosophila* by RNA interference (RNA-i): a role for nautilus in embryonic somatic muscle formation. *Proc. Natl. Acad. Sci. USA* 96:1451–56
96. Mohr CD, Tomich M, Herfst CA. 2001. Cellular aspects of *Burkholderia cepacia* infection. *Microbes Infect.* 3:425–35
97. Mushegian A, Medzhitov R. 2001. Evolutionary perspective on innate immune recognition. *J. Cell. Biol.* 155:705–10
98. Nishijyo T, Haas D, Itoh Y. 2001. The CbrA-CbrB two-component regulatory system controls the utilization of multiple carbon and nitrogen sources in *Pseudomonas aeruginosa. Mol. Microbiol.* 40:917–31
99. O'Quinn AL, Wiegand EM, Jeddeloh JA. 2001. *Burkholderia pseudomallei* kills the nematode *Caenorhabditis elegans* using an endotoxin-mediated paralysis. *Cell Microbiol.* 3:381–94
100. Parke JL, Gurian-Sherman D. 2001. Diversity of the *Burkholderia cepacia* complex and implications for risk assessment of biological control strains. *Annu. Rev. Phytopathol.* 39:225–58
101. Plotnikova JM, Rahme LG, Ausubel FM. 2000. Pathogenesis of the human opportunistic pathogen *Pseudomonas aeruginosa* PA14 in *Arabidopsis. Plant Physiol.* 124:1766–74
102. Preston MJ, Fleiszig SM, Zaidi TS, Goldberg JB, Shortridge VD, et al. 1995. Rapid and sensitive method for evaluating *Pseudomonas aeruginosa* virulence factors during corneal infections in mice. *Infect. Immun.* 63:3497–501
103. Pruckler JM, Benson RF, Moyenuddin M, Martin WT, Fields BS. 1995. Association of flagellum expression and intracellular growth of *Legionella pneumophila. Infect. Immun.* 63:4928–32
104. Pujol N, Link EM, Liu LX, Kurz CL, Alloing G, et al. 2001. A reverse genetic analysis of components of the Toll signaling pathway in *Caenorhabditis elegans. Curr. Biol.* 11:809–21
104a. Pukatzki S, Kessin RH, Mekalanos JJ. 2002. The human pathogen *Pseudomonas aeruginosa* utilizes conserved virulence pathways to infect the social amoeba *Dictyostelium discoideum. Proc. Natl. Acad. Sci. USA* 99:3159–64
105. Pumpuni CB, Demaio J, Kent M, Davis JR, Beier JC. 1996. Bacterial population dynamics in three anopheline species: the impact on *Plasmodium* sporogonic development. *Am. J. Trop. Med. Hyg.* 54: 214–18
106. Pushkareva VI, Konstantinova ND, Litvin VI, Popov VL, Shustrova NM, Safutina GV. 1992. Pseudomonads as parasites of protozoa. *Zh. Mikrobiol. Epidemiol. Immunobiol.* 2:4–10
107. Qin X, Singh KV, Weinstock GM, Murray BE. 2001. Characterization of fsr, a regulator controlling expression of gelatinase and serine protease in *Enterococcus faecalis* OG1RF. *J. Bacteriol.* 183:3372–82
108. Rahme LG, Ausubel FM, Cao H, Drenkard E, Goumnerov BC, et al. 2000. Plants and animals share functionally common bacterial virulence factors. *Proc. Natl. Acad. Sci. USA* 97:8815–21

109. Rahme LG, Stevens EJ, Wolfort SF, Shao J, Tompkins RG, Ausubel FM. 1995. Common virulence factors for bacterial pathogenicity in plants and animals. *Science* 268:1899–902

110. Rahme LG, Tan MW, Le L, Wong SM, Tompkins RG, et al. 1997. Use of model plant hosts to identify *Pseudomonas aeruginosa* virulence factors. *Proc. Natl. Acad. Sci. USA* 94:13245–50

111. Rowbotham TJ. 1980. Preliminary report on the pathogenicity of *Legionella pneumophila* for freshwater and soil amoebae. *J. Clin. Pathol.* 33:1179–83

112. Sahm DF, Kissinger J, Gilmore MS, Murray PR, Mulder R, et al. 1989. In vitro susceptibility studies of vancomycin-resistant *Enterococcus faecalis. Antimicrob. Agents Chemother.* 33: 1588–91

113. Saini LS, Galsworthy SB, John MA, Valvano MA. 1999. Intracellular survival of *Burkholderia cepacia* complex isolates in the presence of macrophage cell activation. *Microbiology* 145:3465–75

114. Schroth MN, Cho JJ, Green SK, Kominos SD. 1977. Epidemiology of *Pseudomonas aeruginosa* in agricultural areas. In Pseudomonas aeruginosa: *Ecological Aspects and Patient Colonization*, ed. VM Young, pp. 1–29. New York: Raven

115. Segal G, Shuman HA. 1999. *Legionella pneumophila* utilizes the same genes to multiply within *Acanthamoeba castellanii* and human macrophages. *Infect. Immun.* 67:2117–24

116. Seitz HM, Maier WA, Rottok M, Becker-Feldmann H. 1987. Concomitant infections of *Anopheles stephensi* with *Plasmodium berghei* and *Serratia marcescens*: additive detrimental effects. *Zentralbl. Bakteriol. Mikrobiol. Hyg. A* 266:155–66

117. Shea JE, Santangelo JD, Feldman RG. 2000. Signature-tagged mutagenesis in the identification of virulence genes in pathogens. *Curr. Opin. Microbiol.* 3: 451–58

118. Sluss HK, Han Z, Barrett T, Davis RJ, Ip YT. 1996. A JNK signal transduction pathway that mediates morphogenesis and an immune response in *Drosophila. Genes Dev.* 10:2745–58

119. Solomon JM, Isberg RR. 2000. Growth of *Legionella pneumophila* in *Dictyostelium discoideum*: a novel system for genetic analysis of host-pathogen interactions. *Trends Microbiol.* 8:478–80

120. Solomon JM, Rupper A, Cardelli JA, Isberg RR. 2000. Intracellular growth of *Legionella pneumophila* in *Dictyostelium discoideum*, a system for genetic analysis of host-pathogen interactions. *Infect. Immun.* 68:2939–47

121. Starr MP, Chatterjee AK. 1972. The genus *Erwinia*: enterobacteria pathogenic to plants and animals. *Annu. Rev. Microbiol.* 26:389–426

122. Staskawicz BJ, Mudgett MB, Dangl JL, Galan JE. 2001. Common and contrasting themes of plant and animal diseases. *Science* 292:2285–89

123. Svanborg C, Godaly G, Hedlund M. 1999. Cytokine responses during mucosal infections: role in disease pathogenesis and host defence. *Curr. Opin. Microbiol.* 2:99–105

124. Swanson MS, Hammer BK. 2000. *Legionella pneumophila* pathogenesis: a fateful journey from amoebae to macrophages. *Annu. Rev. Microbiol.* 54:567–613

125. Tan M-W. 2002. Identification of host and pathogen factors involved in virulence using *C. elegans. Methods Enzymol.* In press

126. Tan M-W, Ausubel FM. 2002. Alternative models in microbial pathogens. In *Molecular Cellular Microbiology*, Vol. 31, ed. P Sansonetti, A Zychlinsky, pp. 461–75. London: Academic

127. Tan MW, Ausubel FM. 2000. *Caenorhabditis elegans*: a model genetic host to

study *Pseudomonas aeruginosa* pathogenesis. *Curr. Opin. Microbiol.* 3:29–34

128. Tan MW, Mahajan-Miklos S, Ausubel FM. 1999. Killing of *Caenorhabditis elegans* by *Pseudomonas aeruginosa* used to model mammalian bacterial pathogenesis. *Proc. Natl. Acad. Sci. USA* 96:715–20
129. Tan MW, Rahme LG, Sternberg JA, Tompkins RG, Ausubel FM. 1999. *Pseudomonas aeruginosa* killing of *Caenorhabditis elegans* used to identify *P. aeruginosa* virulence factors. *Proc. Natl. Acad. Sci. USA* 96:2408–13
130. Taylor LH, Latham SM, Woolhouse ME. 2001. Risk factors for human disease emergence. *Philos. Trans. R. Soc. London B Biol. Sci.* 356:983–89
131. Vidal S, Khush RS, Leulier F, Tzou P, Nakamura M, Lemaitre B. 2001. Mutations in the *Drosophila dTAK1* gene reveal a conserved function for MAPKKKs in the control of rel/NF-kappaB-dependent innate immune responses. *Genes Dev.* 15:1900–12
132. Winiecka-Krusnell J, Linder E. 1999. Free-living amoebae protecting *Legionella* in water: the tip of an iceberg? *Scand. J. Infect. Dis.* 31:383–85
133. Woods DE, DeShazer D, Moore RA, Brett PJ, Burtnick MN, et al. 1999. Current studies on the pathogenesis of melioidosis. *Microbes Infect.* 1:157–62
134. Wu LP, Choe KM, Lu Y, Anderson KV. 2001. *Drosophila* immunity: genes on the third chromosome required for the response to bacterial infection. *Genetics* 159:189–99
135. Zamore PD, Tuschl T, Sharp PA, Bartel DP. 2000. RNAi: double-stranded RNA directs the ATP-dependent cleavage of mRNA at 21 to 23 nucleotide intervals. *Cell* 101:25–33

Annu. Rev. Microbiol. 2002. 56:567–97
doi: 10.1146/annurev.micro.56.012302.160729

First published online as a Review in Advance on June 4, 2002

BACTERIAL CHROMOSOME SEGREGATION

Geoffrey C. Draper and James W. Gober
Department of Chemistry and Biochemistry, University of California, Los Angeles, Los Angeles, California 90095-1569; e-mail: gober@chem.ucla.edu

Key Words chromosome segregation, SMC proteins, ParA, ParB, Soj, Spo0J, SpoIIIE

■ **Abstract** Recent studies have made great strides toward our understanding of the mechanisms of microbial chromosome segregation and partitioning. This review first describes the mechanisms that function to segregate newly replicated chromosomes, generating daughter molecules that are viable substrates for partitioning. Then experiments that address the mechanisms of bulk chromosome movement are summarized. Recent evidence indicates that a stationary DNA replication factory may be responsible for supplying the force necessary to move newly duplicated DNA toward the cell poles. Some factors contributing to the directionality of chromosome movement probably include centromere-like-binding proteins, DNA condensation proteins, and DNA translocation proteins.

CONTENTS

INTRODUCTION

Bacterial cells possess a host of functions that replicate, maintain, and segregate their chromosomal DNA. Although the underlying mechanisms of bacterial DNA replication and repair are well known, our understanding of the mechanisms of chromosomal DNA partitioning is largely incomplete. While the relatively small size of the bacterial cell has hampered efforts to visualize specific structures within the cell that might constitute a mitotic apparatus, the advent of powerful immunocytological and genetic techniques has allowed the visualization of the subcellular localization of a variety of other structures within the bacterial cell. In particular, it is now known that the bacterial nucleoid has a specific orientation within the cell, with origin of replication (*oriC*) regions being located toward the cell poles for much of the cell cycle. The foundations of the bacterial DNA partitioning field has its roots in genetic screens that identified mutants defective in chromosome partitioning. Most of these mutants were later found to possess mutations in genes encoding DNA gyrase and topoisomerases, which do not function to move replicated chromosomes per se but facilitate the production of substrates that are viable for segregation.

For many years it was accepted that newly replicated chromosomes were segregated into their daughter cell compartments by a passive, membrane-bound process. This "replicon model" of chromosome segregation, which was originally applied to explain the fidelity of F-factor partitioning in *Escherichia coli*, suggested that the origin regions of the daughter nucleoids were anchored to the cell membrane at a central position of the cell. DNA replication would result in daughter chromosomes also positioned at the midcell. Cell elongation and addition of new cell membrane specifically between the attachment sites of the two chromosomes to the membrane were proposed to provide the motive force to segregate the nucleoids into the daughter cell compartments (63). It would later be shown that the F and P1 plasmids encode specific determinants, named *par* (or *sop*) genes, that facilitate their efficient partitioning [reviewed in (41)]. Chromosomally encoded homologs of the *par* system have since been discovered in a wide range of bacteria including *Bacillus subtilis* (Soj and Spo0J) and *Caulobacter crescentus* (ParA and ParB) and may influence the inheritance of newly replicated chromosomes (see section below). ParB and Spo0J bind specific regions of DNA located close to the origin of replication. Thus, this system appears to be functionally related to the eukaryotic kinetochore and centromere. ParB (Spo0J) has a dynamic localization pattern in vivo. Early in the cell cycle, ParB is localized to the poles of the cell. During chromosome replication another region of ParB localization can be seen (presumably from ParB binding to the duplicated origin region), which then migrates to the opposite pole of the cell. Upon completion of chromosome replication and just prior to cell division, ParB is localized to both poles of the cell. Division results in daughter cells with ParB localized to one of the cell poles.

Other factors that influence the partitioning of bacterial chromosomes include topoisomerases and recombinases that separate physically linked and dimerized chromosomes. Proteins homologous to the eukaryotic SMC (structural maintenance of chromosomes) proteins act to condense the nucleoid and show a dynamic pattern of localization through the cell cycle (see section below).

In this review we summarize recent findings that may lead to an understanding of bacterial chromosome segregation (i.e., the physical separation of newly replicated chromosomes). We then detail efforts to illuminate the spatial organization of the nucleoid within the bacterial cell, particularly recent exciting experiments in which the dynamic localization of proteins thought to contribute to partitioning are assayed in living cells. These same groundbreaking techniques have also been used to visualize the subcellular location of specific regions of the genome. We discuss factors likely to contribute to chromosome movement, including the DNA-condensing proteins, MukB and SMC, the process of DNA replication itself, and DNA translocation by cell division–related proteins. Last we report on the role of the cellular homologs of plasmid-partitioning proteins in coupling chromosome dynamics to developmentally or cell cycle–regulated pathways.

GENERATION OF MONOMERIC CHROMOSOME SUBSTRATES FOR PARTITIONING

The partitioning of newly replicated chromosomes can only take place after DNA segregation, which includes (*a*) resolution of dimeric chromosomes formed by odd numbers of recombination events and (*b*) successful decatenation of the two covalently closed circular DNA molecules. In *E. coli*, resolution of chromosome dimers is accomplished by the *dif*/XerCD system. The unlinking of the catenated chromosomes formed upon completion of DNA replication requires the concerted action of the DNA gyrase and topoisomerase IV.

dif and XerCD: Resolution of Chromosome Dimers Formed by Sister Chromosome Exchange

Daughter chromosomes are prone to recombination that produces a circular dimer (76). Such dimers must be resolved into monomeric chromosomes prior to segregation. In *E. coli*, resolution occurs at a specific locus, known as *dif* (deletion induced filamentation) (16, 76). *dif* is a 28-bp sequence located in the central region of the replication terminus (10, 76). Strains carrying a *dif* deletion fail to resolve chromosome dimers and elicit a characteristic Dif phenotype, typified by an induction of the SOS response, aberrant nucleoid morphology, filamentation, and reduced viability and growth rate (19, 76). Although the SOS response is induced, filamentation observed in *dif* mutants is independent of the SOS-induced division inhibitor, SulA, suggesting that it is the lack of resolution of chromosome dimers itself that blocks cell division, possibly by nucleoid occlusion of the division site (46, 76, 157). The core *dif* resolvase site functions in the absence of any accessory sequences (141). A 173-kb deletion of the terminus region can be replaced with a 33-bp *dif* sequence that is competent for resolution. However, the position of the *dif* site is important; the Dif phenotype is not suppressed when the only copy of *dif* is moved 30 kb from the normal *dif* locus (19, 75, 141).

Recombination at *dif* is independent of RecA and requires the XerC and XerD proteins (10, 11). The XerC and XerD resolvases belong to the lambda integrase family (37% identity) (10, 11) and bind to 11-bp sequences within the *dif* site opposite a 6- to 8-bp central region (130). Strains carrying mutations in one or both *xer* genes result in the Dif phenotype (10, 76, 141). Individually, the Xer proteins act as type I topoisomerases and relax supercoils by nicking one strand of the *dif* site. The rate of nicking is, however, low and is not thought to be of any biological importance (18). Homologs of XerC and XerD have been described from a number of organisms, including *Pseudomonas aeruginosa* (55), *Salmonella typhimurium* (44), *B. subtilis* (125, 132), and several species of *Enterobacteriaceae* (131), which suggests that this mechanism of dimer resolution is highly conserved.

Topoisomerases: Unlinking the Circles

Early investigations in *E. coli* gave rise to a number of mutants defective in chromosome partitioning (52, 54, 70, 71, 89, 124, 133). These *par* mutant strains exhibited a temperature-sensitive, filamentous cell morphology, which contains nonsegregating nucleoids at the nonpermissive temperature. These strains were later shown to carry mutations in genes encoding either the subunits of DNA gyrase or topoisomerase IV. The *parA* allele mapped to the *gyrB* gene, which encodes the B subunit of DNA gyrase (70). *parD* was originally thought to be a novel gene (60) but this was later shown to be unlikely, as the strain also carried an amber mutation in the gene that encodes the gyrase A subunit, *gyrA* (61). The *parC* allele mapped to a gene encoding a homolog of the A subunit of DNA gyrase (69). The *parE* gene was mapped and cloned, and the deduced protein was shown to have a significant degree of similarity to the B subunit of DNA gyrase (133). Crude cell extracts from strains overexpressing *parC* and *parE* were shown to relax supercoiled DNA. Increasing the copy number of both the *parC* and *parE* genes suppressed a mutation in *topA*, the gene-encoding topoisomerase I. Together, these experiments showed the presence of a second type II topoisomerase system in *E. coli*, encoded by *parC* and *parE*, which was named topoisomerase IV (69). Purified topoisomerase

IV was also demonstrated to be capable of catalyzing the efficient decatenation of circular DNA molecules in vitro (2, 72, 114, 115). Pulse-labeling experiments showed that nearly all newly replicated plasmid DNA is catenated under conditions in which topo IV is not active (168). These experiments also revealed that DNA gyrase activity could slowly process catenanes formed as a consequence of DNA replication in the absence of topo IV activity. Therefore, the combined activities of topo IV and gyrase are responsible for the decatenation of chromosomal DNA formed during DNA replication, although the rate of decatenation by gyrase alone is 100-fold less than in the presence of both proteins (168, 169). A possible function of DNA gyrase in the unlinking of chromosomes is to introduce negative supercoils into the DNA, creating a better substrate for decatenation by topo IV than for relaxed DNA (169).

In contrast, in *Caulobacter crescentus*, cells lacking topo IV (*parE*) did not form long filaments, but rather possessed multiple sites of pinching along the length of the cell, indicating a defect in a later stage of cell division (150). Furthermore, unlike enteric bacteria, *C. crescentus* topo IV mutants exhibited no defect in chromosome segregation. However, DNA segregation defects could be readily demonstrated if the *parE* mutation was combined with a conditional mutation in the early cell division gene, *ftsA*, thus providing a genetic link between the chromosome segregation and cell division (150). This difference in phenotypes between *C. crescentus* and enteric bacteria probably reflects the tight coupling of cell cycle–regulated events in *Caulobacter*.

Interestingly, ParC of *B. subtilis* has been localized to the cell poles (58). In the absence of ParE, ParC is localized to the nucleoid instead of the poles. ParE had no specific localization pattern and could be detected throughout the cytoplasm (58). The same authors also provided evidence that both the DNA gyrase A and B subunits are generally associated with the nucleoid, consistent with the established idea that DNA gyrase is required for higher-order DNA structure. The results of these localization experiments are consistent with the findings that topo IV functions to decatenate linked chromosomes prior to partitioning and, in agreement with Zechiedrich et al. (169), that DNA gyrase creates a DNA topology favorable for topo IV function. A previous study used immunogold-labeled antibodies to show that *E. coli* GyrA and GyrB were randomly localized in the cytoplasm (143). The difference in localization patterns may indicate that *B. subtilis* and *E. coli* use different mechanisms to segregate their nucleoids.

Other Mutations Affecting Chromosome Segregation in *Escherichia coli*

Many other mutants of *E. coli* that possess observed chromosomal DNA segregation or partitioning defects have been identified. Mutations in these genes result in either recombination defects or global cell cycle defects, which in turn generate a DNA segregation defect. None of these is likely to be directly involved with segregation.

1. One of the *parB* mutants described in the preceding section was found to be an allele of the gene-encoding DNA primase, *dnaG* (53, 111). Later experiments showed that the *parB* allele of *dnaG* retained the ability to synthesize functional RNA primers for DNA replication (147). There is a distinct possibility that the *par* phenotype observed in strains carrying the *parB* allele of *dnaG* could be due to a loss of interaction of DnaG with a putative mitotic apparatus (147).
2. *E. coli* strains bearing mutations in the *ruvC*, *recBC*, and *sbcCD* genes also form filamentous cells with chromosome segregation defects (167). The phenotype is due to incomplete recombination events initiated by RecA that cannot progress into mature recombinants owing to the loss of *ruvC* function. Because functional RuvABC is required for resolution of Holliday intermediates formed by RecA-mediated strand exchange, it follows that in these mutants the chromosomes are physically linked at multiple sites via Holliday intermediates (167).
3. *E. coli* mutants lacking the histone-like proteins H-NS and HU produce anucleate cells possibly because of the loss of higher-order structure of the nucleoid (31, 59, 64, 68, 148). These proteins bind and bend DNA and have been implicated in DNA replication, transposition, and DNA and chromosome segregation [reviewed in (32)]. An insertionally inactivated *hupA* gene, which encodes HU2 (one of the subunits of the heterodimeric HU protein), could not be combined with a null-allele of *mukB*, indicating that a global loss of chromosome organization and condensation is lethal (64).
4. In a genetic screen designed to detect anucleate cell production, a number of mutations were generated in the *min* locus (see section below) (3, 49, 22). The *min* locus consists of three genes, *minC*, *minD*, and *minE*, whose products exert control over cell division site selection in *E. coli* (25–27). Mutations in the *min* locus result in the loss of division site specificity and divisions can take place at both the central (normal) division site and at the cell poles, which are remnants of previous cell divisions. Polar divisions can lead to the production of anucleate minicells, and thus it is not unexpected that mutations in the *min* genes could be generated in the screen designed to isolate partitioning mutants (see *mukB* section below) (49). Interestingly, Mulder et al. (101) found that plasmids segregated into minicells show reduced levels of supercoiling, a phenomenon that could reflect decreased topoisomerase activity in minicells, exacerbating the partitioning defect.

SPATIAL ORGANIZATION OF THE BACTERIAL CHROMOSOME

In order for partitioning to operate with efficiency, there must exist mechanisms that function to orient each daughter chromosome to opposite poles of the cell. The circular chromosome of bacteria, although compacted by supercoiling and the binding of histone-like proteins, still occupies much of the internal volume of the cell. Surprisingly, however, a significant fraction of the bulk chromosomal DNA resides at specific locations within the cell. Below, we summarize experiments that first described the spatial organization of the bacterial chromosome.

Sporulating Cells Have a Stringent Requirement for Efficient Chromosome Partitioning

Initial indications that the bacterial chromosome might have a distinct orientation within the cell came from studies on sporulation in *Bacillus subtilis*. In response to starvation, *B. subtilis* undergoes a program of development that results in the formation of a dormant spore [reviewed in (35, 81)]. The initial morphogenetic event in this developmental pathway is the formation of an asymmetric septum near the pole of the cell. The smaller compartment thus formed, the prespore, eventually becomes the spore. Both the prespore compartment and the larger adjacent mother cell compartment each contain a complete chromosome, and differential programs of gene expression within each compartment directs the successful development of the spore. The SpoIIIE protein was shown to be required for complete segregation of chromosomal DNA into the prespore (159). During the asymmetrical cell division, the closure of the septum results in the bisection of the newly replicated chromosomes. At this point approximately 30% of the chromosome is located in the prespore with the remainder in the mother cell compartment. Mutations in *spoIIIE* prevent the completion of DNA segregation into the prespore (159). SpoIIIE is localized to the prespore septum where it is proposed to move the remaining 70% of the chromosome from the mother cell compartment into the prespore via a conjugation-like mechanism (160). Using elegant genetic and physical methods, it was determined that the DNA located in the prespore in a *spoIIIE* mutant always consisted of the same region of chromosome (159, 161). This region corresponded to a 500-kb region of DNA surrounding the origin of replication, *oriC* (127, 159, 160). The observation that a specific chromosomal region arrives first in the prespore compartment pointed toward the presence of a mitotic-like apparatus responsible for the orientation of the origin to the cell pole during asymmetric cell division.

The gene encoding SpoIIIE is expressed constitutively and has the effect of enhancing the fidelity of vegetative chromosome segregation in *B. subtilis* by translocating trapped DNA from the closing septum during cell division (126). Indeed, biochemical experiments have demonstrated that the conserved carboxyl-terminal domain of SpoIIIE possesses an ATPase activity that can be stimulated 10-fold by the presence of double-stranded DNA (7). Furthermore, by probing the topology of the DNA substrate following ATP hydrolysis, these experiments show that DNA tracks along SpoIIIE or vice versa (7). Taken together, these results indicate that SpoIIIE functions as a DNA pump that translocates unpartitioned chromosomal DNA across the prespore division septum. The homologs of SpoIIIE, such as FtsK, found in nonspore-forming bacterium may have a similar role in chromosome movement.

Localization of the Origin and Terminus Regions of the Bacterial Nucleoid

Specific orientation of the chromosome was subsequently discovered in *E. coli* and in vegetative cells of *B. subtilis* (40, 105, 128, 153, 154). An ingenious reporter system, originally used to visualize chromosomal regions in *Saccharomyces cerevisiae* (137), showed that the origin regions of both the *B. subtilis* (137) and *E. coli* (154) chromosomes are dynamically positioned. The technique relies on a tandem repeat of 256 *E. coli lac* operator (*lacO*) sites that are inserted into the chromosome at specific loci, in these cases near either the origin or terminus. A *lac* repressor-green fluorescent protein chimera (LacI-Gfp) expressed in these cells bound the operator sites, and the localization of the complex was detected by fluorescence microscopy (40, 153, 154). Fluorescent in situ hybridization (FISH) has also been employed in *E. coli* (105) and *C. crescentus* (65) to assay location of *oriC* in the cell. The reports had similar findings; in newborn *E. coli* cells with ongoing DNA replication (i.e., fast-growing cells), the two copies of the origin are located at opposite cell poles (Figure 1). Before DNA replication is reinitiated, the origin regions move from the cell pole toward the midcell (the presumptive division site). After replication, one of the origins from each pair of chromosomes moves toward the cell pole while the other remains localized toward the cell center. Formation of a septum and subsequent cell division results in a daughter cell with a chromosome origin at each pole. *B. subtilis* origin placement is similar to that observed in *E. coli* (153, 128). Origin regions of the *E. coli* chromosome detected by FISH are localized to the pole proximal border of the nucleoid (110, 128). The bulk-segregated nucleoids in the postseptational but predivisional cell occupy the quarter positions of the cell (29, 50). In *B. subtilis*, nucleoids are separated by a fixed distance that is not dependent on the cell division machinery (128). The presence of a hypothetical chromosomal "ruler" that dictates the positioning of segregated nucleoids is an attractive theory that could explain the regularity observed in nucleoid spacing (129).

The location of the terminus regions of *B. subtilis* and *E. coli* were also visualized by *lacO*/LacI-GFP (40, 154) or FISH in *E. coli* (105) and *C. crescentus* (65). The terminus initially appears at the center of the cell, and completion of DNA replication results in the formation of two termini that separate. The termini move toward the middle of the two nascent daughter cells as the second copy of origin migrates to the potential division site. Based on these observations, Niki & Hiraga (105) proposed the existence of an active mechanism that caused the movement of the terminus to the midcell. However, the replisome model of chromosome replication and dynamics could suggest a more passive mechanism of terminus mobility. The dynamics of the origin and terminus regions in slow-growing *E. coli* is somewhat different from that observed in fast-growing cells. Mononuclear newborn cells have their single origin and terminus regions at opposite poles of the cell (Figure 2). Later in the cell cycle the origin and terminus move to the center of the cell and DNA replication ensues. The daughter origin regions migrate to the cell poles and the terminus regions remain in the center of the cell. Cell division results in daughter cells with polar origin and terminus placement.

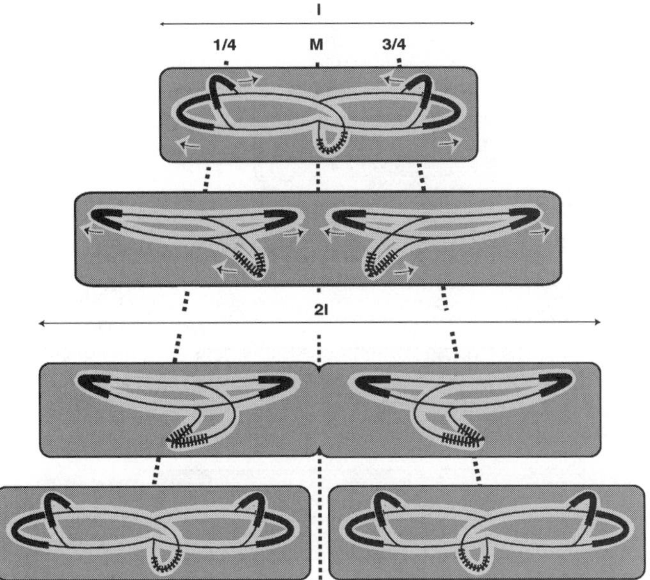

Figure 1 Origin of replication (*oriC*) and terminus localization in fast-growing *E. coli* cells. Depicted is a schematic showing a newly born *E. coli* cell in rich medium. The *bar* above the cell indicates the cell length (l) immediately following division. The midcell (M) and one-quarter and three-quarter positions are indicated, as are the *oriC* regions (highlighted as *dark*) and the terminus (*hatched line*). Immediately following division in fast-growing cells there are four *oriC* regions. Two are located closer to the cell poles than the other two, which reside near the one-quarter and three-quarter positions. At this stage, the terminus region is located at the midcell. As the cell grows and eventually reaches a length that is twice that of the newly born cell (2l), the *oriC* regions that occupied both the one-quarter and three-quarter positions, migrate to the midcell. The termini migrate to both quarter positions at this stage. Subsequent cell division results in the formation of daughter cells each containing *oriC* regions at their poles and a terminus at the midcell position.

CHROMOSOME MOBILITY AND PLACEMENT

There are two proposed opposing views on the mechanism and rate of bulk chromosome movement. One body of work reported that nucleoids partition slowly as the cell elongates (145, 158). This is thought to result from the coupled transcription/translation of membrane proteins. In growing cells these processes are proposed to link active regions of the chromosome to the cell membrane, which in turn slowly drag the chromosome apart as the cell elongates [reviewed in (156)]. This model of nucleoid partitioning implicitly requires a zonal, outward expansion of the cell membrane during cell elongation. In *B. subtilis*, there is evidence that DNA is associated with the peptidoglycan layer, and outward growth of the cell

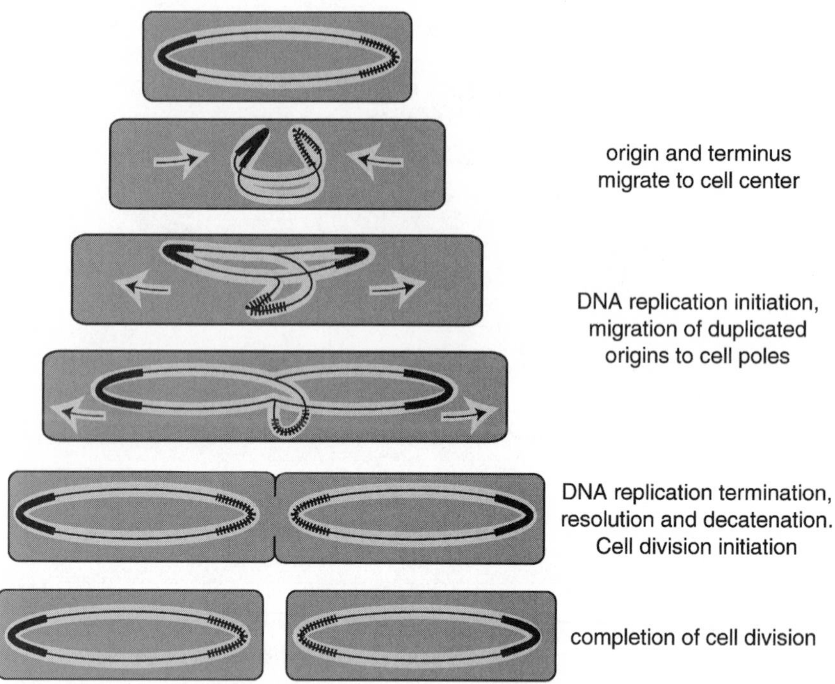

Figure 2 Origin of replication (*oriC*) and terminus localization in slow-growing *E. coli* cells. Depicted is a schematic view showing a newly born *E. coli* cell. The *oriC* regions (highlighted as *dark*) and the terminus (*hatched line*) are indicated. Immediately following division in slow-growing cells there is one *oriC* region and a single terminus. These migrate to the center of the cell, where replication initiates and duplicates the *oriC* region. The duplicated *oriC* regions rapidly migrate toward the pole of the cell. Replication then terminates, with subsequent resolution of chromosome dimers and the removal of catenanes. This is followed by the completion of cell division. Note that in at least one organism (*C. crescentus*) the localization of the origin region and ParB is required for the initiation of cell division (see text for details).

could lead to partitioning of the nucleoid via such attachments (122, 123). This, however, is not the case in *E. coli*, where extension of the cell occurs in a nonzonal manner (103, 128).

In contrast other experimental results suggest that chromosomes partition to predetermined sites in the daughter cell compartments quite rapidly, mediated by an active mechanism (9, 29, 50). Indeed, upon resumption of protein synthesis after a period of inhibition, fused chromosomes reoccupy DNA-free regions of filaments without any appreciable increase in cell length, supporting the idea of a mitotic apparatus that functions to actively segregate chromosomal DNA (145). Webb et al. (153) estimated the velocity of segregating origin regions of

B. subtilis to be 0.17 mm min^{-1}, whereas the rate of cell elongation was only 0.011–0.025 mm min^{-1}. As the rate of origin movement exceeds that of cell elongation, this finding is in conflict with those who propose that chromosome partitioning is mediated by nucleoid association with the cell membrane. Van Helvoort et al. (144) have attempted to reconcile their proposal that DNA partitioning is mediated by interactions of the transcription/translation machinery with the cell membrane and as such is a gradual process with the discovery of active DNA-partitioning proteins and centromere-like regions of DNA in *B. subtilis* (154) and *C. crescentus* (95). The authors suggest that the function of the Soj/Spo0J (*B. subtilis*) and ParA/ParB (*C. crescentus*) systems is to orient the newly replicated origin regions to different ends of the cell, after which the transcription-mediated partitioning system actually moves the bulk of the nucleoid. This is an insightful interpretation of their data, which relies on a gradual movement of nucleoids, but is not in harmony with the data of other groups that demonstrate active and rapid chromosome movements during DNA partitioning.

DNA Replication: Giving Chromosomes a Helpful Push

Replication of the circular bacterial chromosome initiates at the origin of replication and proceeds in a bidirectional manner, until halting in a region opposite the origin called the terminus. The loading of DnaA onto the origin results in localized unwinding, a process continued by DNA helicase. The unwinding of the origin permits formation of the DNA replication complex, or replisome. It was generally accepted that the replisome was mobile and replicated chromosomal DNA by moving along the template DNA in a processive manner. It has recently been shown, however, that the replisome is a structure that has a limited amount of mobility, as it is located toward the cell center (77). To demonstrate the localization of the replisome, the δ', τ, and PolC components of the DNA polymerase holoenzyme from *B. subtilis* were independently fused to GFP. The resultant fusion proteins were functional. Fluorescence microscopy revealed that the replisome is localized to the midcell for much of the cell cycle (77). Fast-growing cells with multiple chromosomes had additional foci, located at the cell one-quarter and three-quarter positions. Under conditions where DNA replication was blocked there were no observable replisome foci, supporting the proposal that the foci demarcated the position of the replisome and not just aggregations of excess replisome components. Furthermore, it has been demonstrated that the localized DNA polymerase is associated with replication forks and that replication effectively extrudes DNA following synthesis (78).

The authors propose that the spooling effect of the nascent daughter chromosomes from the replisome could impart a motive force on the DNA to aid segregation (77–79). This pushing action could be complemented by the pulling force of DNA condensation activity of the SMC proteins on the newly replicated regions of the chromosome (Figure 3). Other factors, perhaps the Par system, could impart the directionality of chromosome movement (i.e., each newly replicated origin toward opposite cell poles). Terminus mobility could also be directly

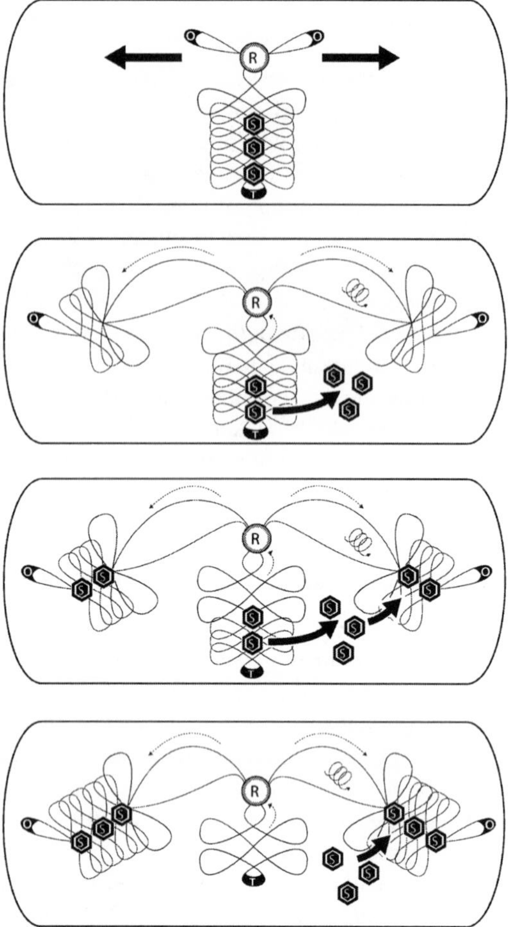

Figure 3 Model summarizing the factors contributing to bacterial chromosome partitioning. At the top of the diagram is a schematic view of a cell shortly following the initiation of DNA replication. A stationary DNA replication factory (replisome) located at the midcell is indicated (R). The newly duplicated *oriC* regions, indicated as (O), then migrate relatively rapidly toward the opposite poles of the cell. Data indicate that the driving force for movement at this stage may be the polymerization of DNA by the stationary replisome. Also shown is the dynamic role of DNA condensation proteins (S), such as MukB or SMC, in chromosome partitioning. These proteins probably disassemble from the DNA to permit the passage of replication forks and then possibly reassemble on the newly duplicated regions of the chromosome. This would lead to condensation and perhaps, in conjunction with a tethered *oriC*, result in a bulk movement of the newly replicated DNA toward opposite cell poles. The completion of DNA replication results in the localization of the terminus region at the midcell. At this location, chromosome dimers are resolved by XerCD homologs via a FtsK-stimulated recombination reaction (not shown) (see text for details), thus uniting cell division and chromosome segregation.

influenced by the replisome. As chromosomal DNA is spooled through the replication complex, the terminus would inevitably be dragged toward the midcell as the amount of initial template DNA decreases.

Experiments with *E. coli* indirectly indicate that the replisome may also be localized at the midcell. The localization of the SeqA protein was followed in synchronized populations of cells (48). SeqA is responsible for sequestering newly replicated hemimethylated DNA in *E. coli* (87). These experiments have demonstrated that foci of SeqA exhibit a bidirectional migration from the midcell distinct from the observed migration of the duplicated *oriC* regions (47, 48, 113). Because SeqA preferentially binds to newly replicated, hemimethylated DNA, this result suggests that newly replicated regions of the *E. coli* chromosome migrate from the midcell in a concerted fashion and suggests the possibility of a midcell localized replisome in *E. coli*. In contrast, in *C. crescentus*, the replisome migrates from the pole of the cell to the midcell shortly after the initiation of DNA replication (66). This variation could be a consequence of the fact that the *C. crescentus oriC* (*Cori*) remains localized at the cell pole following partitioning (65). The authors provide evidence that the replisome movement requires DNA replication, suggesting that migration to the midcell is passively mediated through displacement by newly replicated DNA.

ROLE OF CHROMOSOME CONDENSATION IN PARTITIONING

SMC and MukB are DNA Condensation Proteins

The SMC proteins of *B. subtilis* and *C. crescentus* and the MukB protein of *E. coli* are functionally and structurally similar proteins that participate in chromosome partitioning (14, 65, 94, 99, 109, 107). The two proteins belong to the larger SMC family of proteins whose functions include sister chromatid cohesion, DNA repair, and chromosome condensation and segregation in eukaryotes [reviewed in (138)]. Eukaryotic SMC proteins function as heterodimers, whereas their eubacterial counterparts are homodimers, as there is only one *smc* (*mukB*) gene per genome. Due to the relatively large size of the *B. subtilis* SMC (BsSMC) and *E. coli* MukB proteins (135 kD and 176 kD), it has been possible to view their structure using the electron microscope (94, 107). The dimeric complexes consist of two globular domains separated by an atypical antiparallel coiled-coil domain (94, 107). In the central portion of the coiled-coil domain a flexible hinge region allows the complex to adopt many conformations ranging from a straight, rod-like form to a V-like structure to a point where the two arms are completely closed (94, 107). Purified MukB binds DNA, and its amino-terminal region contains a nucleotide-binding domain that binds both ATP and GTP (94). The binding and hydrolysis of ATP is not required for MukB to bind DNA, but incubation of MukB with single-stranded DNA results in an increase in ATPase activity (107). The dynamic structure of MukB, coupled with its DNA and nucleotide-binding properties, has led to the suggestion that this protein acts as a motor that, in part, drives chromosome segregation (107, 109).

MukB, SMC, and Chromosome Segregation

The *mukB* gene was originally isolated in a genetic screen to detect mutants defective in chromosome segregation (49, 109). This screen employed a plasmid-borne copy of the *lacZ* gene, the expression of which was repressed by a chromosomally expressed *lac* repressor. Mutated cells were plated onto indicator plates containing X-Gal. Those mutants that elicited chromosome partitioning defects (i.e., anucleate cell production) produced a blue pigment due to the β-galactosidase expression from the randomly segregated *lacZ*-containing plasmid into the anucleate cells in which no *lac* repressor was being expressed. Another mutant generated in the same screen, *mukA*, was mapped to the *tolC* gene, mutations in which are pleiotropic and exhibit reduced resistance to colicins and antibiotics, abnormal OmpF and OmpC expression, and increased sensitivity to SDS (49, 108, 155). At present, the involvement of TolC in chromosome segregation remains unclear. The two genes immediately upstream of the *mukB* in *E. coli* are also involved in chromosome partitioning (163). These genes, *mukE* and *mukF*, are essential for accurate chromosome segregation, and MukB, MukE, and MukF interact, possibly via the carboxyl terminus of MukB (163, 164).

Mutations in *mukB* or *B. subtilis smc* result in a chromosome partitioning defect. MukB point mutations result in slow growth, 5% anucleate cell production at 22°C and highly restrict growth at 42°C in rich media. Both *mukB* and *smc* null-mutants do not form colonies in rich media at elevated temperatures (14, 99, 109). At 22°C the *mukB* null-mutant produces a mixture of anucleate cells, cells with bisected nucleoids and septated pairs of cells, one of which is anucleate (109). Incubation of this *mukB* mutant at 42°C causes the cells to elongate and chromosome positioning becomes erratic; extended incubation at 42°C results in cell division, with anucleate cells comprising 40%–60% of the population (109). Similarly, null mutations in both the *B. subtilis* (14, 99) and *C. crescentus* (66) *smc* genes result in the inability to form colonies in rich media at elevated temperatures. A significant fraction (10%–15%) of the cells in *B. subtilis* strains bearing *smc* null-alleles was anucleate (14, 99). Moriya et al. (99) described how a strain with the *smc* gene under control of the IPTG inducible *spac* promoter showed an increase in anucleate cell production that correlated with decreasing concentrations of inducer. A *C. crescentus* strain in which the *smc* gene was insertionally inactivated did not produce a significant amount of anucleate cells at either the permissive or the elevated temperatures (65). Instead, a majority of the cells were unable to divide, arresting at a predivisional stage, and did not elongate. Examination of stained chromosomes in these "stalled" cells showed that the distribution of the nucleoids was impaired.

In *B. subtilis*, SMC is localized to the nucleoid (14, 42). Britton et al. (14) also described 1 to 2 foci of BsSMC at polar, although chromosome associated, sites and suggested that these sites might represent regions where nucleoid condensation was active. In contrast, Graumann et al. (42) described these polar foci as being separate from the nucleoid. The same group also indicated that the polar foci were present in only newborn cells. In older cells with segregated chromosomes there were no detectable polar BsSMC foci, suggesting that BsSMC moves from the polar sites to the chromosome prior to or during nucleoid partitioning. The methods used to generate the above data could explain why similar experiments yielded (albeit slightly) different results. Britton et al. (14) used a BsSMC-GFP chimera to detect BsSMC localization, whereas Graumann et al. (42) employed immunofluorescence microscopy with BsSMC-specific antibodies. Either the BsSMC-GFP fusion protein has different localization properties than the wild-type protein, or the fixation protocol required for IFM resulted in a dissociation of the polar foci from the chromosome borders. The carboxyl-terminal region of BsSMC appears to be required for nucleoid association but not formation of polar foci (42). Both the polar and nucleoid-associated forms of BsSMC foci were visible in aseptate filaments generated by the depletion of the essential cell division protein FtsZ from the cell. BsSMC, therefore, does not require the cell division apparatus to dictate its pattern of localization.

Another phenotype associated with a disruption of MukB/BsSMC function is abberrant chromosome condensation [(42, 56, 99); reviewed in (138)]. *smc* knockouts in *B. subtilis* and *C. crescentus* result in a decondensation or relaxation of the nucleoid (14, 65, 99). Analysis of the localization of the *oriC*-associated Spo0J protein in *smc* mutants of *B. subtilis* revealed that the origin region was mislocalized in the absence of BsSMC (14, 99). It is possible that bacterial MukB/SMC proteins condense the nucleoid during or prior to segregation to provide tension that allows the chromosomes to partition into daughter cell compartments. The presence of a mitotic apparatus linking the origin region of the chromosome to a polar structure could act as an anchor to maintain the tension imparted by chromosome condensation, thus ensuring the direction the chromosomes will recoil upon decatenation and resolution. Alternatively, nascent replicated chromosomes emerging from the centrally positioned replisome (77) could be directed toward the cell poles by the origin region localization systems (the *par* system) and continuously condensed at the polar regions (see Figure 3). As discussed above, a *mukB* null-allele cannot be combined with an inactivated *hupA* gene, which encodes the HU2 subunit of the *E. coli* DNA-binding protein HU, again stressing the importance of nucleoid structure and condensation in chromosome segregation (64).

A *mukB* null mutation also cannot be combined with an *ftsK* allele that expresses only the amino terminus (cell division domain) of the protein in *E. coli* (165, 166). In *B. subtilis*, an *smc spoIIIE* double-deletion mutant grows poorly and accumulates suppressor mutations (13). In a *B. subtilis* strain with an inducible copy of *smc* and either a *spoIIIE* null-allele or the *spoIIIE36* allele, which harbors three missense mutations, growth ceased after several generations in media without inducer. BsSMC depletion also caused the appearance of cells with nucleoids bisected by invaginating septa (13). The trapping of nucleoids at the closing septum is associated with a loss of function of SMC and SpoIIIE. Britton & Grossman

(13) show that when the *spoIIIE* and *smc* mutations are combined, the proportion of cells with the bisected nucleoid phenotype is 38% compared to ~10% of cells with the *smc* knockout alone. The earlier work of Sharpe & Errington (126) also showed that in a *spoIIIE* mutant, chromosomes could be trapped by septa. Thus, in situations where the nucleoid is decondensed (by virtue of a mutation in *smc* or *mukB*) and cannot move straggling regions of the chromosome (because of the lack of functional SpoIIIE or the SpoIIIE-like domain of *E. coli* FtsK), the trapped nucleoid phenomenon is exacerbated, strengthening the case for a role for each protein in chromosome segregation.

Interestingly, a spontaneous mutation that suppressed the growth defect phenotype of *E. coli* cells bearing a *mukB* deletion was shown to map to the *topA* gene (encoding topoisomerase I) (121). This and other mutations in *topA* were found to be general suppressors in a variety of *mukB*, *mukE*, and *mukF* mutants. The authors propose a mechanism by which the increased negative supercoiling imparted on the chromosomes by the unrestricted activity of DNA gyrase in the *topA*/*muk* mutant strains is responsible for increasing the compactness of the nucleoid and thus aiding segregation into daughter cells. Inhibition of DNA gyrase activity by coumermycin reverses this suppression and a typical *mukB*$^-$ phenotype was observed. Some gyrase mutants are defective in segregating their chromosomal DNA (134). Possibly both MukBEF and negative supercoiling act in concert to promote a compact nucleoid structure (121) that favors efficient partitioning.

INTEGRATION OF CHROMOSOME PARTITIONING WITH THE CELL CYCLE

A Link Between Cell Division and Replication Termination Events

It has been demonstrated that resolution of chromosome dimers via the *dif*-XerCD system requires cell division (136). It was subsequently discovered that it was not the lack of cell division that resulted in the loss of dimer resolution but that the carboxyl terminus of the cell division protein FtsK was required for resolution of dimers (120, 135). FtsK is a member of the SpoIIIE family of DNA translocases that localize to the septum during cell division (8, 149, 159, 161, 166). The amino-terminal domain of FtsK contains a membrane-spanning region that is essential for cell division (8, 30, 149, 166), and the carboxyl terminus contains an ATP-binding site and is required for efficient chromosome partitioning (8, 86, 165). Strains carrying deletions in the carboxyl-terminal region of FtsK do not resolve chromosomal or plasmid-borne *dif* sites, but there is no effect on the resolution of the plasmid recombination sites *cer* and *psi*, which also require XerCD for resolution (19, 120, 135, 139). It has been demonstrated that FtsK functions to activate resolution of chromosome dimers via XerCD late in the cell cycle (120). Dimer resolution via XerCD is a sequential process where a Holliday junction intermediate between *dif* sites is formed by the XerC-mediated exchange of a first pair of DNA strands. The carboxyl-terminal domain of FtsK functions to activate the XerD-catalyzed exchange of the second pair of strands (6). The localization of FtsK to the midcell, through association with other cell division proteins, would effectively deliver it to dimer chromosomes that have been trapped or guillotined by the closing septum. This would insure that resolution takes place immediately prior to completion of cell division and explain the requirement for cell division for *dif*-mediated resolution of dimeric chromosomes (120, 135). Another perhaps complementary role for FtsK in nucleoid partitioning is in DNA translocation. The similarity of the carboxyl terminus of FtsK with SpoIIIE suggests that FtsK could remove trapped chromosomes into the daughter cell compartments at the last stages of septal closure (86, 120, 135).

A similar relationship between replication termination events and partitioning has been discovered in *B. subtilis* (80). As in *E. coli*, chromosome dimers are resolved by a XerD homolog, RipX. Mutations in the gene encoding RipX result in the generation of a significant fraction of anucleate cells. These authors discovered a synthetic effect when a *ripX* mutation was combined with a null mutation in *rtp*, a gene encoding a replication termination DNA-binding protein that blocks replication forks in the terminus region. A *ripX rtp* double mutant generated twice as many anucleate cells as the strain bearing the single *ripX* mutation. Notably a strain containing only a *rtp* null mutation did not exhibit a partitioning defect. Interestingly, the *rtp* null mutation, when combined with a *spoIIIE* mutation, also exhibited a synthetic partitioning defect (80). The authors suggest that these effects result from overreplication of the terminus region in the absence of *rtp*, which would generate a double-stranded end that is highly recombinogenic. In the *rtp spoIIIE* double mutant, an increase in anucleate cells is observed because SpoIIIE, in a role analogous to FtsK in *E. coli*, is probably required to enhance or facilitate the dimer resolution catalyzed by RipX.

ParA AND ParB: CELL CYCLE REGULATORS OR COMPONENTS OF A BACTERIAL MITOTIC APPARATUS?

The positioning of the origin regions and the regular mobility of newly replicated origins suggested that an active mechanism is responsible for the movement and positioning of the bacterial nucleoid. However, no structure similar to the mitotic spindle has been observed or detected in the eubacteria. The origin of replication is the first region of chromosomal DNA to be replicated (by definition) and it displays a dramatic dynamic localization pattern (see section above). Therefore, it is possible that the determinant for chromosomal DNA polarity is the origin. However, experiments have shown that *oriC* is not the actual determinant of bulk chromosome positioning. For example, in *E. coli oriC*-containing plasmids are not stably inherited (106, 112). Introduction of the partitioning functions of F-factor into such a plasmid, however, can confer stable inheritance (106). One fifth of the chromosome, centered around the origin region, colocalizes with the origin. Likewise one fifth of the chromosome, which contains the terminus, appears to take the same position as the terminus itself (110). Thus these extended regions of chromosomal origin and terminus DNA contain the regions partly responsible for chromosome placement. Given these findings, it is apparent that there must exist a mechanism that recognizes and partitions copies of the origin to the poles of the predivisional cell. The discovery that many species of bacterial cells contain homologs of the well-characterized partitioning proteins of low-copy-number plasmids (F-factor) and bacteriophage (P1) suggests the existence of a similar, simple mitotic apparatus. Below, we briefly review the role of these proteins in the partitioning of episomes. [For a more detailed review of partitioning mechanisms in these and other unit-copy plasmids, the reader is referred to (41)]. We then summarize the experimental evidence for the role of the cellular homologs of plasmid partitioning genes in chromosome partitioning.

Plasmid and Bacteriophage DNA Segregation

Low-copy-number plasmids, like chromosomes, must possess mechanisms to ensure that efficient replicon partitioning occurs following DNA replication. In particular, experiments with bacteriophage P1, which can exist as a single-copy episome when acting as a lysogen, and F-factor have provided insights into the function of the widely occurring partitioning proteins ParA and ParB. The plasmid-encoded *parA* (*sopA*) and *parB* (*sopB*) gene products from bacteriophage P1 and P7 (1, 4, 5, 88) or F-factor (97, 98, 112) have been demonstrated to be essential for efficient partitioning. Conserved homologs of *parA* have been sequenced from at least 10 different plasmids or prophages. The deduced amino acid sequence of *parA* shows that it is homologous to a large distinct family of ATPases (74, 100), including the *minD* gene product of *E. coli* and *B. subtilis*. MinD, in concert with other *min* gene products such as MinC and MinE (in *E. coli*)/DivIVA (in *B. subtilis*), specifies the topological position of the cell division septum (27, 82, 91, 146). The plasmid *parB*/*sopB* gene encodes a sequence-specific DNA-binding protein (23, 24, 98, 151). For example, the *parB* gene product from the prophage form of *E. coli* bacteriophage P1 binds to repeated sequences of dyad symmetry that lie immediately following the end of the *parB* coding region (23, 38, 93). This *cis*-acting sequence, called *parS* (sopC in F-factor), is required for efficient plasmid partitioning and is regarded to function as the plasmid equivalent of a centromere. Sequences of this type have been identified adjacent to the *parAB* operons of several different plasmids, and mutant plasmids with deletions in this *cis*-acting region cannot be stably maintained. In addition, P1 plasmids containing only the *parS* sequence do not partition efficiently unless the *parA* and *parB* gene products are supplied in *trans* (93). Therefore, these experiments demonstrate that the *parS* sequence functions as the plasmid equivalent of a centromere during the partitioning process; the binding of ParB to the *parS* centromere sequence is hypothesized to function in daughter replicon pairing as a prelude to partitioning (33, 92).

Experiments with purified ParA and SopA have demonstrated that they possess a weak ATPase activity in vitro (20, 24, 152). ATPase activity can be stimulated in the presence of purified *parB* gene product and *parS* DNA, indicating that these two proteins interact with each other in vivo (21, 24). Experiments with P1 and P7 proteins have identified the sequences in ParA and ParB that are essential for this interaction (118). The *parA* gene products of plasmid and phage bind to DNA sequences within their own promoter and function to regulate expression of the *parAB* operon (36, 37, 45, 51). This autoregulation is critical for proper partitioning, as overexpression of *parA* or *parB* leads to partitioning defects (1, 38). The precise biochemical function fulfilled by ParA ATPase activity is not clear. It has been proposed that the ATP hydrolysis functions to break apart paired sister replicons as a necessary prelude for partitioning. Alternatively, it has been suggested that the hydrolysis of ATP is required for movement of the sister plasmids toward the cell poles. Experiments have shown that both ATP and ADP can be bound to ParA (21). Biochemical experiments have demonstrated that the double-stranded DNA-binding activity is influenced by the nucleotide-bound state of ParA (12). When bound to ADP ParA represses transcription by binding to the *parAB* operon promoter region, and when bound to ATP it interacts with the ParB-*parS* complex (12). ParB presumably regulates this transition between these two activities by stimulating ATPase activity.

Subcellular localization experiments have demonstrated that both P1 (40) and F-factor (40, 104) are located at the midcell of newly formed cells. When they are duplicated in growing cells, each daughter episome migrates to the one-quarter and three-quarter positions of the predivisional cell. Plasmids without the partitioning proteins exhibited a somewhat random localization, occupying regions of the cell that lacked nucleoid material (104). Taken together these results show that ParA (SopA) and ParB (SopB) are responsible for directing the orientation and partitioning of bacteriophage P1 and F-factor. An important question is whether these proteins, in particular the centromere-binding protein ParB/SopB, bind to the *parS*/*sopC* sequence and travel with the segregating plasmid, or rather exist in preformed localized foci that function as receptors for the migrating daughter plasmids. In the case of P1, experiments have demonstrated that the formation of localized foci depends on the presence of *parS* sequences (34) thus supporting the idea that ParB migrates with the plasmid DNA during partitioning. The data are not as clear for F-factor. Experiments using immunofluorescence microscopy have shown that the formation of SopB foci depends on the plasmid centromere sequence (*sopC*) (51). However, using a SopB-Gfp fusion, Kim & Wang (73) have shown that foci of this fusion protein form near the poles of the cell even in the absence of *sopC* sequences. These authors even demonstrated that a fusion lacking the SopB DNA-binding sequences exhibited a similar pattern of localization (73), indicating the existence of a cellular receptor for SopB.

Role of the Cellular Homologs of ParA and ParB in the Partitioning of Chromosomal DNA

Chromosomally encoded homologs of the plasmid partitioning genes *parA* and *parB* have been identified in the genome sequences of at least 29 different eubacterial species (162). Most of our knowledge regarding the role of these proteins in partitioning is mainly derived from experiments with *B. subtilis* and *C. crescentus*. These proteins, Soj/Spo0J in *B. subtilis* and ParA/ParB in *C. crescentus*, possess biochemical properties that are similar to those described for the plasmid homologs.

Genetic experiments in *B. subtilis* were the first to implicate the chromosomally encoded *par* homologs in the partitioning of chromosomal DNA. The gene encoding the ParB homolog, *spo0J*, was originally identified as being required for an early stage of the sporulation pathway (102). Specifically, *spo0J* is required for the expression of relatively early-acting sporulation genes, which require the Spo0A transcription factor for expression. The gene encoding the ParA homolog, *soj*, can suppress the sporulation defect in *spo0J* mutant strains by restoring Spo0A-activated transcription (62). Mutations in *spo0J* result in a 100-fold increase in the generation of anucleate cells, suggesting an active role in chromosome partitioning (62). Like its plasmid homolog, Spo0J possesses double-stranded DNA-binding activity. Spo0J binds to relatively conserved 16-bp *parS* sequences, clustered around the *soj spo0J* operon (84). These sequences occur approximately ten times within the origin-proximal 20% of the *B. subtilis* genome (84).

Foci of Spo0J colocalize with the origin of replication region (39, 83, 85). Furthermore, experiments that assayed prespore-compartment-specific gene expression indicated that *spo0J* mutants had defects in the orientation of the prespore chromosome (127). One important experiment revealed that an unstable plasmid can be stabilized by the presence of a Spo0J-binding site. This stabilization was also dependent on the presence of Spo0J and Soj (84). These findings prompted the hypothesis that Spo0J was involved in spatial organization of the origin proximal regions of the chromosome and thus is one component of the *B. subtilis* mitotic apparatus. In support of this idea, Soj and Spo0J can function to partition a plasmid bearing *parS* sequences in *E. coli* (162). In contrast to these findings, which indicate a role for Spo0J in partitioning, it appears that Spo0J is not required for proper positioning or movement of the chromosomal origin of replication in *B. subtilis* (153). It is possible that this result indicates that *B. subtilis* may have an additional system for directing the *oriC* to the cell poles. Indeed, recent evidence has shown that the division site-selection protein (91) DivIVA has an additional role in orienting the *oriC* region of the *B. subtilis* chromosome (142). These authors found that *divIVA* mutants displayed a defect in both the expression of early sporulation genes and in DNA partitioning during sporulation. An alteration in prespore-specific transcription indicated that the *divIVA* mutant strain had a defect in the positioning of *oriC* at the prespore pole (142). The authors speculate that the widespread distribution of *divIVA* homologs in other gram-positive organisms, particularly those lacking a *min* system, possibly indicates a primary role for this protein in chromosome partitioning (142).

Deletion of *soj* (*parA* homolog) does not result in a DNA segregation defect, but it is required for the stability of *parS*-containing plasmids (162). Biochemical experiments have shown that Soj inhibits Spo0A-activated transcription by binding to single-stranded DNA in the transcription open complex (15). This unusual mechanism of repression is most evident in the absence of Spo0J; using in vivo crosslinking followed by immunoprecipitation, Soj preferentially associated with early (*spoII* class) sporulation promoters in a *spo0J* mutant (117). Soj also reinforces this negative regulation of early sporulation promoters by repressing the expression of *spo0A*, the gene encoding their transcriptional activator (116).

Thus the Soj/Spo0J system of *B. subtilis* operates a checkpoint that couples chromosome partitioning to developmental gene expression. When partitioning is incomplete, it is likely that Soj represses the expression of Spo0A-dependent promoters. The completion of partitioning results in Spo0J inactivating the repressive activities of Soj. Although the details are not yet known, this regulation involves the bipolar localization of Spo0J. In other words, partitioning is probably sensed by this signal transduction system when the chromosomal origin of replication and thus Spo0J arrives at the poles of the cell.

How does the spatial location of Spo0J regulate Soj activity? Soj displays a remarkable pattern of subcellular localization. Soj oscillates from one cell pole to the other within a period of approximately 20 sec. Such oscillation is dependent on Spo0J, and in its absence Soj localizes to the nucleoid (90, 117). It can be inferred from these observations that the oscillating form of Soj does not repress early-sporulation promoters and that the nonoscillating, nucleoid-associated form does direct repression. The conversion between these two forms may be linked to Soj ATPase activity. In the related ATPase, MinD, which regulates division site selection in *E. coli*, polar oscillation is regulated by the MinE protein (43, 57, 119), which stimulates MinD ATPase activity (57). In support of this idea, a Soj mutant predicted to be defective in ATPase activity failed to exhibit bipolar oscillation (117).

The ParA/ParB system has also been studied in some detail in *C. crescentus*. Like *B. subtilis*, the operon containing *parA* and *parB* is near the origin of replication region (*Cori*) (95). ParB binds to several *parS* sequences adjacent to the *par* operon and the *Cori* [(95); J. Easter, R.M. Figge & J.W. Gober, unpublished data). Cell cycle subcellular localization experiments have shown that a single focus of ParB is found at the poles of cells containing a single chromosome (95, 96). Initiation of DNA replication results in the relatively rapid formation of an additional ParB focus at the opposite pole of the cell. This dynamic pattern of localization reflects the movement of the newly duplicated *Cori* region during the cell cycle (65). ParA localizes at the cell poles; however, it has not been assayed for polar oscillation in living cells (95). Overexpression of either ParA or ParB causes cell filamentation and a marked chromosome partitioning defect. Simultaneous overproduction of both proteins has little effect on cell division but still results in a relatively severe partitioning defect (95). These results suggest that a proper balance of ParA and ParB concentration in the cell is required for proper progression of the cell cycle. In contrast to *B. subtilis*, inactivation of either *parA* or *parB* in *C. crescentus* is lethal to the cell. In ParB depletion experiments, loss of the protein leads to cell filamentation and eventual death (96). Under these conditions, filamentation is the result of the inability to form FtsZ rings, the earliest cytological event in bacterial cell division. An identical phenotype with regard to FtsZ ring formation is observed when ParA is overexpressed. Thus increasing the ratio of ParA to ParB inhibits cell division (96). Based on these results it has been hypothesized that ParA and ParB in *C. crescentus* operate a cell cycle checkpoint that couples partitioning to cytokinesis (96). With analogy to *B. subtilis*, it is proposed that a ParA cell division block is probably relieved upon the arrival of ParB to opposite poles of the cell. Indeed in this regard the formation of bipolar ParB foci precedes the formation of an FtsZ ring by 20 min (96). The role of ParA and ParB in directing chromosome partitioning still remains unclear.

PERSPECTIVES

The question of how the bacteria segregate their chromosomes with such a high degree of fidelity is now being addressed from a number of approaches. Origin regions are segregated and positioned by an active mechanism. The ParA/ParB system or an analogous but elusive centromere-binding protein may be partly responsible for the localization of the origin to the cell pole. Bulk chromosome polarity could be imparted simply by the order in which the chromosomal DNA is replicated; with the first regions to be duplicated, the origins, being pushed and pulled to the cell poles by the action of the replisome and SMC proteins (Figure 3). Postreplication, poleward mobility of the origin, has been described in many species and can be explained by the above mechanisms, but what of the movement of the origin to the center of the cell prior to replication in slow-growing cells? Does SMC play a role in relocalizing chromosomal DNA prior to DNA replication? Is there a scaffold protein present in the cell in which the origin, and perhaps bulk chromosomal DNA, can track along? What are the cues that trigger SMC to associate with the nucleoid, condense the nucleoid, and presumably permit decondensation of the nucleoid prior to DNA replication?

One additional persistent question is: What determines the location of the cell poles and the midcell? Insights from cell division studies in *E. coli* have implicated the *min* system in determining that the central potential division site should be used for division (rod-shaped organisms can be thought of as having three potential division sites, one at the cell center and one at either pole, which are relics of previous division events). This question remains: What is the descriptor that the cell employs to indicate a pole? The periseptal annulus, a ring-like structure positioned on either side of the midcell potential division site that, upon cell division, results in a single annulus at the cell pole, is a candidate for a pole "tagging" structure (17). Other landmarks such as polar peptidoglycan structure(s) or membrane domains could also help the cell determine its poles.

ACKNOWLEDGMENTS

We are grateful to J. Easter, J. England, and R. Muir for helpful comments regarding the preparation of this manuscript. Work in our laboratory is supported by grants from the National Institutes of Health (GM48417) and The National Science Foundation (MCB-9986127).

The *Annual Review of Microbiology* is online at http://micro.annualreviews.org

LITERATURE CITED

1. Abeles AL, Friedman SA, Austin SJ. 1985. Partition of unit-copy miniplasmids to daughter cells. III. The DNA sequence and functional organization of the P1 partition region. *J. Mol. Biol.* 185:261–72
2. Adams DE, Shekhtman EM, Zechiedrich EL, Schmid MB, Cozzarelli NR. 1992. The role of topoisomerase IV in partitioning bacterial replicons and the structure of catenated intermediates in DNA replication. *Cell* 71:277–88
3. Adler HI, Fisher WD, Cohen A, Hardigree AA. 1967. Miniature *Escherichia coli* cells deficient in DNA. *Proc. Natl. Acad. Sci. USA* 57:321–26
4. Austin S, Abeles A. 1983. Partition of unit-copy miniplasmids to daughter cells. I. P1 and F miniplasmids contain discrete, interchangeable sequences sufficient to promoter equipartition. *J. Mol. Biol.* 169:353–72
5. Austin S, Abeles A. 1983. Partition of unit-copy miniplasmids to daughter cells. II. The partition region of miniplasmid P1 encodes an essential protein and a centromere-like site at which it acts. *J. Mol. Biol.* 169:373–87
6. Barre FX, Aroyo M, Colloms SD, Helfrich A, Cornet F, Sherratt DJ. 2000. FtsK functions in the processing of a Holliday junction intermediate during bacterial chromosome segregation. *Genes Dev.* 14:2976–88
7. Bath J, Wu LJ, Errington J, Wang JC. 2000. Role of *Bacillus subtilis* SpoIIIE in DNA transport across the mother cell-prespore division septum. *Science* 290:995–97
8. Begg KJ, Dewar SJ, Donachie WD. 1995. A new *Escherichia coli* cell division gene, *ftsK*. *J. Bacteriol.* 177:6211–22
9. Begg KJ, Donachie WD. 1991. Experiments on chromosome separation and positioning in *Escherichia coli*. *New Biol.* 3:475–86
10. Blakely G, Colloms S, May G, Burke M, Sherratt D. 1991. *Escherichia coli* XerC recombinase is required for chromosomal segregation at cell division. *New Biol.* 3:789–98
11. Blakely G, May G, McCulloch R, Arciszewska LK, Burke M, et al. 1993. Two related recombinases are required for site-specific recombination at *dif* and *cer* in *E. coli* K12. *Cell* 75:351–61
12. Bouet JY, Funnell BE. 1999. P1 ParA interacts with the P1 partition complex at *parS* and an ATP-ADP switch controls ParA activities. *EMBO J.* 18:1415–24
13. Britton RA, Grossman AD. 1999. Synthetic lethal phenotypes caused by mutations affecting chromosome partitioning in *Bacillus subtilis*. *J. Bacteriol.* 181:5860–64
14. Britton RA, Lin DC, Grossman AD. 1998. Characterization of a prokaryotic SMC protein involved in chromosome partitioning. *Genes Dev.* 12:1254–59
15. Cervin MA, Spiegelman GB, Raether B, Ohlsen K, Perego M, Hoch JA. 1998. A negative regulator linking chromosome segregation to developmental transcription in *Bacillus subtilis*. *Mol. Microbiol.* 29:85–95
16. Clerget M. 1991. Site-specific recombination promoted by a short DNA segment of plasmid R1 and by a homologous segment in the terminus region of the *Escherichia coli* chromosome. *New Biol.* 3:780–88
17. Cook WR, Kepes F, Jseleau-Petit D, MacAlister TJ, Rothfield LI. 1987. Proposed mechanism for generation and localization of new cell division sites during the division cycle of *Escherichia coli*. *Proc. Natl. Acad. Sci. USA* 84:7144–48
18. Cornet F, Hallet B, Sherratt DJ. 1997. Xer

recombination in *Escherichia coli*. Site-specific DNA topoisomerase activity of the XerC and XerD recombinases. *J. Biol. Chem.* 272:21927–31

19. Cornet F, Louarn J, Patte J, Louarn JM. 1996. Restriction of the activity of the recombination site *dif* to a small zone of the *Escherichia coli* chromosome. *Genes Dev.* 10:1152–61
20. Davey MJ, Funnell BE. 1994. The P1 plasmid partition protein ParA. A role for ATP in site-specific DNA binding. *J. Biol. Chem.* 269:22908–13
21. Davey MJ, Funnell BE. 1997. Modulation of the P1 plasmid partition protein ParA by ATP, ADP, and P1 ParB. *J. Biol. Chem.* 272:15286–92
22. Davie E, Sydnor K, Rothfield LI. 1984. Genetic basis of minicell formation in *Escherichia coli* K-12. *J. Bacteriol.* 158: 1202–23
23. Davis M, Martin K, Austin S. 1992. Biochemical activities of the ParA partition protein of the P1 plasmid. *Mol. Microbiol.* 6:1141–47
24. Davis MA, Austin SJ. 1988. Recognition of the P1 plasmid centromere analog involves the binding of the ParB protein and is modified by a specific host factor. *EMBO J.* 7:1881–88
25. de Boer PA, Crossley RE, Rothfield LI. 1988. Isolation and properties of *minB*, a complex genetic locus involved in correct placement of the division site in *Escherichia coli*. *J. Bacteriol.* 170:2106–12
26. de Boer PA, Crossley RE, Rothfield LI. 1989. A division inhibitor and a topological specificity factor coded for by the minicell locus determine proper placement of the division septum in *Escherichia coli*. *Cell* 56:641–49
27. de Boer PAJ, Crossley RE, Hand AR, Rothfield LI. 1991. The MinD protein is a membrane ATPase required for the correct placement of the *Escherichia coli* cell division site. *EMBO J.* 10:4371–80
28. de Pedro MA, Quintela JC, Holtjë JV, Schwarz H. 1997. Murein segregation in *Escherichia coli*. *J. Bacteriol.* 179:2823–34
29. Donachie WD, Begg KJ. 1989. Chromosome partition in *Escherichia coli* requires postreplication protein synthesis. *J. Bacteriol.* 171:5405–9
30. Draper GC, McLennan N, Begg K, Masters M, Donachie WD. 1998. Only the N-terminal domain of FtsK functions in cell division. *J. Bacteriol.* 180:4621–27
31. Dri AM, Rouvière-Yaniv J, Moreau PL. 1991. Inhibition of cell division in *hupA hupB* mutant bacteria lacking HU protein. *J. Bacteriol.* 173:2852–63
32. Drlica K, Rouvière-Yaniv J. 1987. Histonelike proteins of bacteria. *Microbiol. Rev.* 51:301–19
33. Edgar R, Chattoraj DK, Yarmolinsky M. 2001. Pairing of plasmid partition sites by ParB. *Mol. Microbiol.* 42:1363–70
34. Erdmann N, Petroff T, Funnell BE. 1999. Intracellular localization of P1 ParB protein depends on ParA and *parS*. *Proc. Natl. Acad. Sci. USA* 96:14905–10
35. Errington J. 1993. *Bacillus subtilis* sporulation: regulation of gene expression and control of morphogenesis. *Microbiol. Rev.* 57:1–33
36. Friedman SA, Austin SJ. 1988. The P1 plasmid-partition system synthesizes two proteins from an autoregulated operon. *Plasmid* 19:103–12
37. Fung E, Bouet JY, Funnell BE. 2001. Probing the ATP-binding site of P1 ParA: Partition and repression have different requirements for ATP binding and hydrolysis. *EMBO J.* 20:4901–11
38. Funnell BE. 1988. Mini-P1 plasmid partitioning: Excess ParB protein destabilizes plasmids containing the centromere *parS*. *J. Bacteriol.* 170:954–60
39. Glaser P, Sharpe ME, Raether B, Perego M, Ohlsen K, Errington J. 1997. Dynamic, mitotic-like behavior of a bacterial protein required for accurate chromosome partitioning. *Genes Dev.* 11:1160–68
40. Gordon GS, Sitnikov D, Webb CD, Teleman A, Straight A, et al. 1997.

Chromosome and low copy plasmid segregation in *E. coli*: visual evidence for distinct mechanisms. *Cell* 90:1113–21

41. Gordon GS, Wright A. 2000. DNA segregation in bacteria. *Annu. Rev. Microbiol.* 54:681–708
42. Graumann PL, Losick R, Strunnikov AV. 1998. Subcellular localization of *Bacillus subtilis* SMC, a protein involved in chromosome condensation and segregation. *J. Bacteriol.* 180:5749–55
43. Hale CA, Meinhardt H, de Boer PA. 2001. Dynamic localization cycle of the cell division regulator MinE in *Escherichia coli*. *EMBO J.* 20:1563–72
44. Hayes F, Lubetzki SA, Sherratt DJ. 1997. *Salmonella typhimurium* specifies a circular chromosome dimer resolution system which is homologous to the Xer site-specific recombination system of *Escherichia coli*. *Gene* 198:105–10
45. Hayes F, Radnedge L, Davis MA, Austin SJ. 1994. The homologous operons for P1 and P7 partition are autoregulated from dissimilar operator sites. *Mol. Microbiol.* 11:249–60
46. Helmstetter CE, Pierucci O, Weinburger M, Holmes M, Tang M-S. 1979. Control of cell division in *Escherichia coli*. In *The Bacteria VII*, ed. PJR Sokatch, LN Ornston. 517–79. New York: Academic
47. Hiraga S, Ichinose C, Niki H, Yamazoe M. 1998. Cell cycle-dependent duplication and bidirectional migration of SeqA-associated DNA-protein complexes in *E. coli*. *Mol. Cell* 1:381–87
48. Hiraga S, Ichinose C, Onogi T, Niki H, Yamazoe M. 2000. Bidirectional migration of SeqA-bound hemimethylated DNA clusters and pairing of *oriC* copies in *Escherichia coli*. *Genes Cells* 5:327–41
49. Hiraga S, Niki H, Ogura T, Ichinose C, Mori H, et al. 1989. Chromosome partitioning in *Escherichia coli*: novel mutants producing anucleate cells. *J. Bacteriol.* 171:1496–505
50. Hiraga S, Ogura T, Niki H, Ichinose C, Mori H. 1990. Positioning of replicated chromosomes in *Escherichia coli*. *J. Bacteriol.* 172:31–39
51. Hirano M, Mori H, Onogi T, Yamazoe M, Niki H, et al. 1998. Autoregulation of the partition genes of the mini-F plasmid and the intracellular localization of their products in *Escherichia coli*. *Mol. Gen. Genet.* 257:392–403
52. Hirota Y, Jacob F, Ryter A, Buttin G, Nakai T. 1968. On the process of cellular division in *Escherichia coli*. I. Asymmetrical cell division and production of deoxyribonucleic acid-less bacteria. *J. Mol. Biol.* 35:175–92
53. Hirota Y, Mordoh J, Jacob F. 1970. On the process of cellular division in *Escherichia coli*. III. Thermosensitive mutants of *Escherichia coli* altered in the process of DNA initiation. *J. Mol. Biol.* 53:369–87
54. Hirota Y, Ryter A, Jacob F. 1968. Thermosensitive mutants of *E. coli* affected in the processes of DNA synthesis and cellular division. *Cold Spring Harbor Symp. Quant. Biol.* 33:677–93
55. Hofte M, Dong Q, Kourambas S, Krishnapillai V, Sherratt D, Mergeay M. 1994. The *sss* gene product, which affects pyoverdin production in *Pseudomonas aeruginosa* 7NSK2, is a site-specific recombinase. *Mol. Microbiol.* 14:1011–20
56. Hu KH, Liu E, Dean K, Gingras M, DeGraff W, Trun NJ. 1996. Overproduction of three genes leads to camphor resistance and chromosome condensation in *Escherichia coli*. *Genetics* 143:1521–32
57. Hu Z, Lutkenhaus J. 2001. Topological regulation of cell division in *E. coli*. Spatiotemporal oscillation of MinD requires stimulation of its ATPase by MinE and phospholipid. *Mol. Cell* 7:1337–43
58. Huang WM, Libbey JL, van der Hoeven P, Yu SX. 1998. Bipolar localization of *Bacillus subtilis* topoisomerase IV, an enzyme required for chromosome segregation. *Proc. Natl. Acad. Sci. USA* 95:4652–27
59. Huisman O, Faelen M, Girard D, Jaffe A, Toussaint A, Rouvière-Yaniv J. 1989.

Multiple defects in *Escherichia coli* mutants lacking HU protein. *J. Bacteriol.* 171:3704–12
60. Hussain K, Begg KJ, Salmond GP, Donachie WD. 1987. *parD*: a new gene coding for a protein required for chromosome partitioning and septum localization in *Escherichia coli. Mol. Microbiol.* 1:73–81
61. Hussain K, Elliott EJ, Salmond GP. 1987. The *parD*$^-$ mutant of *Escherichia coli* also carries a *gyrAam* mutation. The complete sequence of *gyrA*. *Mol. Microbiol.* 1:259–73
62. Ireton K, Gunther NW 4th, Grossman AD. 1994. *spo0J* is required for normal chromosome segregation as well as the initiation of sporulation in *Bacillus subtilis. J. Bacteriol.* 176:5320–29
63. Jacob F, Brenner S, Cuzin F. 1963. On the regulation of DNA replication in bacteria. *Cold Spring Harbor Symp. Quant. Biol.* 28:329–48
64. Jaffe A, Vinella D, D'Ari R. 1997. The *Escherichia coli* histone-like protein HU affects DNA initiation, chromosome partitioning via MukB, and cell division via MinCDE. *J. Bacteriol.* 179:3494–99
65. Jensen RB, Shapiro L. 1999. The *Caulobacter crescentus smc* gene is required for cell cycle progression and chromosome segregation. *Proc. Natl. Acad. Sci. USA* 96:10661–67
66. Jensen RB, Wang SC, Shapiro L. 2001. A moving DNA replication factory in *Caulobacter crescentus. EMBO J.* 20: 4952–63
67. Deleted in proof
68. Kaidow A, Wachi M, Nakamura J, Magae J, Nagai K. 1995. Anucleate cell production by *Escherichia coli* Dhns mutant lacking a histone-like protein, H-NS. *J. Bacteriol.* 177:3589–92
69. Kato J, Nishimura Y, Imamura R, Niki H, Hiraga S, Suzuki H. 1990. New topoisomerase essential for chromosome segregation in *E. coli. Cell* 63:393–404
70. Kato J, Nishimura Y, Suzuki H. 1989. *Escherichia coli parA* is an allele of the *gyrB* gene. *Mol. Gen. Genet.* 217:178–81
71. Kato J, Nishimura Y, Yamada M, Suzuki H, Hirota Y. 1988. Gene organization in the region containing a new gene involved in chromosome partition in *Escherichia coli. J. Bacteriol.* 170:3967–77
72. Kato J, Suzuki H, Ikeda H. 1992. Purification and characterization of DNA topoisomerase IV in *Escherichia coli. J. Biol. Chem.* 267:25676–84
73. Kim SK, Wang JC. 1998. Localization of F plasmid SopB protein to positions near the poles of *Escherichia coli* cells. *Proc. Natl. Acad. Sci. USA* 95:1523–27
74. Koonin EV. 1993. A superfamily of ATPases with diverse functions containing either classical or deviant ATP-binding motif. *J. Mol. Biol.* 229:1165–74
75. Kuempel P, Hogaard A, Nielsen M, Nagappan O, Tecklenburg M. 1996. Use of a transposon (Tn*dif*) to obtain suppressing and nonsuppressing insertions of the *dif* resolvase site of *Escherichia coli. Genes Dev.* 10:1162–71
76. Kuempel PL, Henson JM, Dircks L, Tecklenburg M, Lim DF. 1991. *dif*, a *recA*-independent recombination site in the terminus region of the chromosome of *Escherichia coli. New Biol.* 3:799–811
77. Lemon KP, Grossman AD. 1998. Localization of bacterial DNA polymerase: evidence for a factory model of replication. *Science* 282:1516–19
78. Lemon KP, Grossman AD. 2000. Movement of replicating DNA through a stationary replisome. *Mol. Cell* 6:1321–30
79. Lemon KP, Grossman AD. 2001. The extrusion-capture model for chromosome partitioning in bacteria. *Genes Dev.* 15: 2031–41
80. Lemon KP, Kurtser I, Grossman AD. 2001. Effects of replication termination mutants on chromosome partitioning in *Bacillus subtilis. Proc. Natl. Acad. Sci. USA* 98:212–17
81. Levin PA, Losick R. 2000. Asymmetric division and cell fate during sporulation

in *Bacillus subtilis*. In *Prokaryotic Development*, ed. YV Brun, LJ Shimkets, pp. 167–89. Washington, DC: ASM

82. Levin PA, Margolis PS, Setlow P, Losick R, Sun D. 1992. Identification of *Bacillus subtilis* genes for septum placement and shape determination. *J. Bacteriol.* 174:6717–28
83. Lewis PJ, Errington J. 1997. Direct evidence for active segregation of *oriC* regions of the *Bacillus subtilis* chromosome and co-localization with the SpoOJ partitioning protein. *Mol. Microbiol.* 25:945–54
84. Lin DCH, Grossman AD. 1998. Identification and characterization of a bacterial chromosome partitioning site. *Cell* 92:675–85
85. Lin DCH, Levin PA, Grossman AD. 1997. Bipolar localization of a chromosome partition protein in *Bacillus subtilis*. *Proc. Natl. Acad. Sci. USA* 94:4721–66
86. Liu G, Draper GC, Donachie WD. 1998. FtsK is a bifunctional protein involved in cell division and chromosome localization in *Escherichia coli*. *Mol. Microbiol.* 29:893–903
87. Lu M, Campbell JL, Boye E, Kleckner N. 1994. SeqA: a negative regulator of replication initiation in *E. coli*. *Cell* 77:413–26
88. Ludtke DN, Eichorn BG, Austin SJ. 1989. Plasmid-partition functions of the P7 prophage. *J. Mol. Biol.* 209:393–96
89. Luttinger AL, Springer AL, Schmid MB. 1991. A cluster of genes that affects nucleoid segregation in *Salmonella typhimurium*. *New Biol.* 3:687–97
90. Marston AL, Errington J. 1999. Dynamic movement of the ParA-like Soj protein of *B. subtilis* and its dual role in nucleoid organization and developmental regulation. *Mol. Cell* 4:673–82
91. Marston AL, Thomaides HB, Edwards DH, Sharpe ME, Errington J. 1998. Polar localization of the MinD protein of *Bacillus subtilis* and its role in selection of the mid-cell division site. *Genes Dev.* 12:3419–30
92. Martin KA, Davis MA, Austin S. 1991. Fine-structure analysis of the P1 plasmid partition site. *J. Bacteriol.* 173:3630–34
93. Martin KA, Friedman SA, Austin SJ. 1987. Partition site of the P1 plasmid. *Proc. Natl. Acad. Sci. USA* 84:8544–47
94. Melby TE, Ciampaglio CN, Briscoe G, Erickson HP. 1998. The symmetrical structure of structural maintenance of chromosomes (SMC) and MukB proteins: long, antiparallel coiled coils, folded at a flexible hinge. *J. Cell. Biol.* 142:1595–604
95. Mohl DA, Gober JW. 1997. Cell cycle-dependent polar localization of chromosome partitioning proteins in *Caulobacter crescentus*. *Cell* 88:675–84
96. Mohl DA, Easter J, Gober JW. 2001. The chromosome partitioning protein, ParB, is required for cytokinesis in *Caulobacter crescentus*. *Mol. Microbiol.* 42:741–55
97. Mori H, Kondo A, Ohshima A, Ogura T, Hiraga S. 1986. Structure and function of the F plasmid genes essential for partitioning. *J. Mol. Biol.* 192:1–15
98. Mori H, Mori Y, Ichinose C, Niki H, Ogura T, et al. 1989. Purification and characterization of SopA and SopB proteins essential for F plasmid partitioning. *J. Biol. Chem.* 264:15535–41
99. Moriya S, Tsujikawa E, Hassan AK, Asai K, Kodama T, Ogasawara N. 1998. A *Bacillus subtilis* gene-encoding protein homologous to eukaryotic SMC motor protein is necessary for chromosome partition. *Mol. Microbiol.* 29:179–87
100. Motellebi-Veshareh M, Rouch DA, Thomas CM. 1990. A family of ATPases involved in active partitioning of diverse bacterial plasmids. *Mol. Microbiol.* 4:1455–63
101. Mulder E, El'Bouhali M, Pas E, Woldringh CL. 1990. The *Escherichia coli minB* mutation resembles *gyrB* in defective nucleoid segregation and decreased negative supercoiling of plasmids. *Mol. Gen. Genet.* 221:87–93
102. Mysliwiec TH, Errington J, Vaidya AB, Bramucci MG. 1991. The *Bacillus*

subtilis spo0J gene: evidence for involvement in catabolite repression of sporulation. *J. Bacteriol.* 173:1911–19
103. Nanninga N, Wientjes FB, de Jonge BL, Woldringh CL. 1990. Polar cap formation during cell division in *Escherichia coli. Res. Microbiol.* 141:103–18
104. Niki H, Hiraga S. 1997. Subcellular distribution of actively partitioning F plasmid during the cell division cycle in *E. coli. Cell* 90:951–57
105. Niki H, Hiraga S. 1998. Polar localization of the replication origin and terminus in *Escherichia coli* nucleoids during chromosome partitioning. *Genes Dev.* 12: 1036–45
106. Niki H, Hiraga S. 1999. Subcellular localization of plasmids containing the *oriC* region of the *Escherichia coli* chromosome, with or without the *sopABC* partitioning system. *Mol. Microbiol.* 34:498–503
107. Niki H, Imamura R, Kitaoka M, Yamanaka K, Ogura T, Hiraga S. 1992. *E. coli* MukB protein involved in chromosome partition forms a homodimer with a rod-and-hinge structure having DNA binding and ATP/GTP binding activities. *EMBO J.* 11:5101–19
108. Niki H, Imamura R, Ogura T, Hiraga S. 1990. Nucleotide sequence of the *tolC* gene of *Escherichia coli. Nucleic Acids Res.* 18:5547
109. Niki H, Jaffe A, Imamura R, Ogura T, Hiraga S. 1991. The new gene *mukB* codes for a 177 kd protein with coiled-coil domains involved in chromosome partitioning of *E. coli. EMBO J.* 10:183–93
110. Niki H, Yamaichi Y, Hiraga S. 2000. Dynamic organization of chromosomal DNA in *Escherichia coli. Genes Dev.* 14:212–23
111. Norris V, Alliotte T, Jaffe A, D'Ari R. 1986. DNA replication termination in *Escherichia coli parB* (a *dnaG* allele), *parA*, and *gyrB* mutants affected in DNA distribution. *J. Bacteriol.* 168:494–504
112. Ogura T, Hiraga S. 1983. Partition mechanism of F plasmid: Two plasmid gene-encoded products and a *cis*-acting region are involved in partition. *Cell* 32:351–60
113. Onogi T, Niki H, Yamazoe M, Hiraga S. 1999. The assembly and migration of SeqA-Gfp fusion in living cells of *Escherichia coli. Mol. Microbiol.* 31:1775–82
114. Peng H, Marians KJ. 1993. Decatenation activity of topoisomerase IV during *oriC* and pBR322 DNA replication *in vitro. Proc. Natl. Acad. Sci. USA* 90:8571–75
115. Peng H, Marians KJ. 1993. *Escherichia coli* topoisomerase IV. Purification, characterization, subunit structure, and subunit interactions. *J. Biol. Chem.* 268: 24481–90
116. Quisel JD, Grossman AD. 2000. Control of sporulation gene expression in *Bacillus subtilis* by the chromosome partitioning proteins Soj (ParA) and Spo0J (ParB). *J. Bacteriol.* 182:3446–51
117. Quisel JD, Lin DC, Grossman AD. 1999. Control of development by altered localization of a transcription factor in *B. subtilis. Mol. Cell* 4:665–72
118. Radnedge L, Youngren B, Davis M, Austin S. 1998. Probing the structure of complex macromolecular interactions by homolog specificity scanning: the P1 and P7 plasmid partition systems. *EMBO J.* 17:6076–85
119. Raskin DM, de Boer PA. 1999. Rapid pole-to-pole oscillation of a protein required for directing division to the middle of *Escherichia coli. Proc. Natl. Acad. Sci. USA* 96:4971–76
120. Recchia GD, Aroyo M, Wolf D, Blakely G, Sherratt DJ. 1999. FtsK-dependent and -independent pathways of *xer* site-specific recombination. *EMBO J.* 18:5724–34
121. Sawitzke JA, Austin S. 2000. Suppression of chromosome segregation defects of *Escherichia coli muk* mutants by mutations in topoisomerase I. *Proc. Natl. Acad. Sci. USA* 97:1671–76
122. Schlaeppi JM, Karamata D. 1982. Cosegregation of cell wall and DNA in *Bacillus subtilis. J. Bacteriol.* 152:1231–40

123. Schlaeppi JM, Schaefer O, Karamata D. 1985. Cell wall and DNA cosegregation in *Bacillus subtilis* studied by electron microscope autoradiography. *J. Bacteriol.* 164:130–35
124. Schmid MB. 1990. A locus affecting nucleoid segregation in *Salmonella typhimurium*. *J. Bacteriol.* 172:5416–24
125. Sciochetti SA, Piggot PJ, Sherratt DJ, Blakely G. 1999. The *ripX* locus *of Bacillus subtilis* encodes a site-specific recombinase involved in proper chromosome partitioning. *J. Bacteriol.* 181:6053–62
126. Sharpe ME, Errington J. 1995. Postseptational chromosome partitioning in bacteria. *Proc. Natl. Acad. Sci. USA* 92:8630–34
127. Sharpe ME, Errington J. 1996. The *Bacillus subtilis soj-spo0J* locus is required for a centromere-like function involved in prespore chromosome partitioning. *Mol. Microbiol.* 21:501–9
128. Sharpe ME, Errington J. 1998. A fixed distance for separation of newly replicated copies of *oriC* in *Bacillus subtilis*: implications for co-ordination of chromosome segregation and cell division. *Mol. Microbiol.* 28:981–90
129. Sharpe ME, Errington J. 1999. Upheaval in the bacterial nucleoid. An active chromosome segregation mechanism. *Trends Genet.* 15:70–74
130. Sherratt DJ. 1993. Site-specific recombination and the segregation of sister chromosomes. In *Nucleic Acids and Molecular Biology*, ed. F Eckstein, DMJ Lilley, pp. 202–16. Heidelberg: Springer
131. Sirois S, Szatmari G. 1995. Detection of XerC and XerD recombinases in gram-negative bacteria of the family *Enterobacteriaceae*. *J. Bacteriol.* 177:4183–86
132. Slack FJ, Serror P, Joyce E, Sonenshein AL. 1995. A gene required for nutritional repression of the *Bacillus subtilis* dipeptide permease operon. *Mol. Microbiol.* 15:689–702
133. Springer AL, Schmid MB. 1993. Molecular characterization of the *Salmonella typhimurium parE* gene. *Nucleic Acids Res.* 21:1805–9
134. Steck TR, Drlica K. 1984. Bacterial chromosome segregation: evidence for DNA gyrase involvement in decatenation. *Cell* 36:1081–88
135. Steiner W, Liu G, Donachie WD, Kuempel P. 1999. The cytoplasmic domain of FtsK protein is required for resolution of chromosome dimers. *Mol. Microbiol.* 31: 579–83
136. Steiner WW, Kuempel PL. 1998. Cell division is required for resolution of dimer chromosomes at the *dif* locus of *Escherichia coli*. *Mol. Microbiol.* 27:257–68
137. Straight AF, Belmont AS, Robinett CC, Murray AW. 1996. GFP tagging of budding yeast chromosomes reveals that protein-protein interactions can mediate sister chromatid cohesion. *Curr. Biol.* 6: 1599–608
138. Strunnikov AV, Jessberger R. 1999. Structural maintenance of chromosomes (SMC) proteins: conserved molecular properties for multiple biological functions. *Eur. J. Biochem.* 263:6–13
139. Summers DK, Sherratt DJ. 1984. Multimerization of high copy number plasmids causes instability: ColE1 encodes a determinant essential for plasmid monomerization and stability. *Cell* 36:1097–103
140. Deleted in proof
141. Tecklenburg M, Naumer A, Nagappan O, Kuempel P. 1995. The *dif* resolvase locus of the *Escherichia coli* chromosome can be replaced by a 33-bp sequence, but function depends on location. *Proc. Natl. Acad. Sci. USA* 92:1352–56
142. Thomaides HB, Freeman M, El Karoui M, Errington J. 2001. Division site selection protein DivIVA of *Bacillus subtilis* has a second distinct function in chromosome segregation during sporulation. *Genes Dev.* 15:1662–73
143. Thornton M, Armitage M, Maxwell A, Dosanjh B, Howells AJ, et al. 1994. Immunogold localization of GyrA and GyrB

proteins in *Escherichia coli*. *Microbiology* 140:2371–82

144. Van Helvoort JM, Huls PG, Vischer NO, Woldringh CL. 1998. Fused nucleoids resegregate faster than cell elongation in *Escherichia coli* pbpB(Ts) filaments after release from chloramphenicol inhibition. *Microbiology* 144:1309–17
145. van Helvoort JM, Woldringh CL. 1994. Nucleoid partitioning in *Escherichia coli* during steady-state growth and upon recovery from chloramphenicol treatment. *Mol. Microbiol.* 13:577–83
146. Varley AW, Stewart GC. 1992. The *divIVB* region of the *Bacillus subtilis* chromosome encodes homologs of *Escherichia coli* septum placement (*minCD*) and cell shape (*mreBCD*) determinants. *J. Bacteriol.* 174:6729–42
147. Versalovic J, Lupski JR. 1997. Missense mutations in the 3′ end of the *Escherichia coli dnaG* gene do not abolish primase activity but do confer the chromosome-segregation defective (*par*) phenotype. *Microbiology* 143:585–94
148. Wada M, Kano Y, Ogawa T, Okazaki T, Imamoto F. 1988. Construction and characterization of the deletion mutant of *hupA* and *hupB* genes in *Escherichia coli*. *J. Mol. Biol.* 204:581–91
149. Wang L, Lutkenhaus J. 1998. FtsK is an essential cell division protein that is localized to the septum and induced as part of the SOS response. *Mol. Microbiol.* 29:731–40
150. Ward D, Newton A. 1997. Requirement of topoisomerase IV *parC* and *parE* genes for cell cycle progression and developmental regulation in *Caulobacter crescentus*. *Mol. Microbiol.* 26:897–910
151. Watanabe E, Inamoto S, Lee MH, Kim SU, Ogua T, et al. 1989. Purification and characterization of the *sopB* gene product which is responsible for stable maintenance of mini-F plasmid. *Mol. Gen. Genet.* 218:431–36
152. Watanabe E, Wachi M, Yamasaki M, Nagai K. 1992. ATPase activity of SopA, a protein essential for active partitioning of F plasmid. *Mol. Gen. Genet.* 234:346–52
153. Webb CD, Graumann PL, Kahana JA, Teleman AA, Silver PA, Losick R. 1998. Use of time-lapse microscopy to visualize rapid movement of the replication origin region of the chromosome during the cell cycle in *Bacillus subtilis*. *Mol. Microbiol.* 28:883–92
154. Webb CD, Teleman A, Gordon S, Straight A, Belmont A, et al. 1997. Bipolar localization of the replication origin regions of chromosomes in vegetative and sporulating cells of *B. subtilis*. *Cell* 88:667–74
155. Webster RE. 1991. The *tol* gene products and the import of macromolecules into *Escherichia coli*. *Mol. Microbiol.* 5:1005–11
156. Woldringh CL, Jensen PR, Westerhoff HV. 1995. Structure and partitioning of bacterial DNA: determined by a balance of compaction and expansion forces? *FEMS Microbiol. Lett.* 131:235–42
157. Woldringh CL, Mulder E, Huls PG, Vischer N. 1991. Toporegulation of bacterial division according to the nucleoid occlusion model. *Res. Microbiol.* 142:309–20
158. Woldringh CL, Zaritsky A, Grover NB. 1994. Nucleoid partitioning and the division plane in *Escherichia coli*. *J. Bacteriol.* 176:6030–38
159. Wu LJ, Errington J. 1994. *Bacillus subtilis* SpoIIIE protein required for DNA segregation during asymmetric cell division. *Science* 264:572–75
160. Wu LJ, Errington J. 1997. Septal localization of the SpoIIIE chromosome partitioning protein in *Bacillus subtilis*. *EMBO J.* 16:2161–19
161. Wu LJ, Lewis PJ, Allmansberger R, Hauser PM, Errington J. 1995. A conjugation-like mechanism for prespore chromosome partitioning during sporulation in *Bacillus subtilis*. *Genes Dev.* 9:1316–26
162. Yamaichi Y, Niki H. 2000. Active segregation by the *Bacillus subtilis* partitioning system in *Escherichia coli*. *Proc. Natl. Acad. Sci. USA* 97:14656–61

163. Yamanaka K, Ogura T, Niki H, Hiraga S. 1996. Identification of two new genes, *mukE* and *mukF*, involved in chromosome partitioning in *Escherichia coli*. *Mol. Gen. Genet.* 250:241–51
164. Yamazoe M, Onogi T, Sunako Y, Niki H, Yamanaka K, et al. 1999. Complex formation of MukB, MukE and MukF proteins involved in chromosome partitioning in *Escherichia coli*. *EMBO J.* 18:5873–84
165. Yu XC, Tran AH, Sun Q, Margolin W. 1998. Localization of cell division protein FtsK to the *Escherichia coli* septum and identification of a potential N-terminal targeting domain. *J. Bacteriol.* 180:1296–304
166. Yu XC, Weihe EK, Margolin W. 1998. Role of the C terminus of FtsK in *Escherichia coli* chromosome segregation. *J. Bacteriol.* 180:6424–28
167. Zahradka D, Vlahovic K, Petranovic M, Petranovic D. 1999. Chromosome segregation and cell division defects in *recBC sbcBC ruvC* mutants of *Escherichia coli*. *J. Bacteriol.* 181:6179–83
168. Zechiedrich EL, Cozzarelli NR. 1995. Roles of topoisomerase IV and DNA gyrase in DNA unlinking during replication in *Escherichia coli*. *Genes Dev.* 9:2859–69
169. Zechiedrich EL, Khodursky AB, Cozzarelli NR. 1997. Topoisomerase IV, not gyrase, decatenates products of site-specific recombination in *Escherichia coli*. *Genes Dev.* 11:2580–92

Annu. Rev. Microbiol. 2002. 56:599–624
doi: 10.1146/annurev.micro.56.012302.160925

First published online as a Review in Advance on July 15, 2002

IMPACT OF GENOMIC TECHNOLOGIES ON STUDIES OF BACTERIAL GENE EXPRESSION

Virgil Rhodius,[1] Tina K. Van Dyk,[2] Carol Gross,[1] and Robert A. LaRossa[2]
[1]*Department of Stomatology, University of California, San Francisco, San Francisco, California 94143; e-mail: rhodius@itsa.ucsf.edu; cgross@cgl.ucsf.edu*
[2]*DuPont Company, Central Research and Development, Wilmington, Delaware 19880-0173; e-mail: tina.k.van-dyk@usa.dupont.com; robert.a.larossa@usa.dupont.com*

Key Words mRNA profiles, gene arrays, gene fusions, bioinformatic inferences, data analyses

■ **Abstract** The ability to simultaneously monitor expression of all genes in any bacterium whose genome has been sequenced has only recently become available. This requires not only careful experimentation but also that voluminous data be organized and interpreted. Here we review the emerging technologies that are impacting the study of bacterial global regulatory mechanisms with a view toward discussing both perceived best practices and the current state of the art. To do this, we concentrate upon examples using *Escherichia coli* and *Bacillus subtilis* because prior work in these organisms provides a sound basis for comparison.

CONTENTS

INTRODUCTION

The means by which bacteria respond to their environment is most diverse. Included are changes in their patterns of motility, metabolism, and gene expression. Each of these, extensively studied for decades, is a cornerstone in our understanding of bacterial behavior. Nonetheless, the bedrock upon which we have studied bacteria is changing due to the availability of completed genomic sequences. Here we illustrate ways that genomic sequence information affects global analyses of gene expression.

Past work could be characterized as predominantly reductionist. Usually, a gene or enzyme was studied in great detail by defining its metabolic and regulatory roles. Nonetheless, genetic research continually reminded us that the cell was crowded, complicated, and surprising. At an early time, several avenues indicated that regulation of gene expression could be highly complex. For example, *his* operon control in *Salmonella typhimurium* (95) differed significantly from those of the paradigm *lac* and *trp* operons in that mutations in several loci, rather than a single regulatory gene, elevated *his* operon expression. Moreover, mutations in one of these loci, *hisT* encoding a tRNA modification enzyme, also elevated expression of two other amino acid biosynthetic operons (18) and conferred resistance to several inhibitory analogs of amino acids (52), indicating that the *hisT* product negatively affected many metabolic pathways. Conversely, mutations of *relA*, whose gene product is responsible for the synthesis of ppGpp and pppGpp, led to several amino acid analog-sensitive phenotypes, which suggests that these highly phosphorylated nucleotides were pleiotropic positive effectors of gene expression (12). Thus control of amino acid biosynthesis was thought to have both specific and global components. This thought was not limited to this single aspect of metabolism. Studies of *cya*, *crp*, and *crr* mutants indicated that catabolism was also subject to a general, global control mediated by the interplay between adenylate cyclase, the PTS system, and the catabolite activator protein (62) as well as specific sugar-repressor interactions.

These realizations, coupled with subsequent technological developments, ushered in many more detailed studies of gene expression on a global scale. Biochemically, methods used to measure individual components of small molecule and stable RNA pools enhanced our understanding of the roles of these molecules and two-dimensional gels, the forerunner of current proteomics efforts, (55) allowed identification and quantification of gene products coordinately regulated by either an environmental perturbation or a mutation. By using such methods about 1800 *Escherichia coli* K12 gene products were resolved (88). Sensitivity and reliable identification of polypeptide spots arising from the two-dimensional constellation limited this application; recent solutions to these obstacles portend rapid progress.

A second thrust identifying regulatory circuitry involved genetic screens; this approach was aided by the application of gene fusions. The original, cumbersome use of reporter genes to monitor transcription dates to the early 1960s (2) and accelerated with the development of streamlined methods to fuse the *lacZ* reporter to regulatory regions (11). Soon thereafter, the *lacZ* reporter was used for genome-wide transcriptional analysis, including the pioneering studies with gene fusions induced by DNA damage (35) or phosphate starvation in *E. coli* (89). At the time these and other such studies were conducted, only the subset of highly upregulated gene fusions were identified.

By the mid-1990s it became likely that the genomic sequences of many bacterial species would become available; that reality is evident today. These data have effected our approaches to genetics; for example, insertional mutations (54) and gene amplification events resulting in multicopy suppression or relief from inhibitor action (65) are identified by sequencing reactions rather than by traditional laborious and time-consuming mapping techniques. Genomics is also changing the study of gene expression. Today the separation and quantification of polypeptide constituents are being revisited within the context of proteomics. The resolution and identification problems of two-dimensional protein gel electrophoretic separations are being addressed with mass spectroscopic techniques that, in conjunction with complete genomic sequences, allow polypeptide fragments to be assigned to genes and quantified (31). In a similar vein, the availability of completely sequenced genomes allows sequence-based, precise mapping of entire gene fusion libraries (87). Furthermore, hybridization assays, a standard measure of gene expression, have been dramatically impacted by genomic sequence data. Complete sequences of microbial genomes allow multiplexing of this assay. Transcripts corresponding to many genes can be measured in parallel as first demonstrated with an ordered collection of specialized transducing λ phages representative of the entire *E. coli* chromosome (15) and is now done routinely for essentially all transcripts of several microbes. Such methodology can be used to define stimulons, which are ensembles of genes effected by an environmental perturbation, and regulons, which are collections of genes under the control of a single regulatory protein (51).

In this review we first describe a series of tools that provides not only several complementary views of comprehensive expression profiling but also an organizational framework to what might otherwise be seemingly large and amorphous datasets. We then detail both biochemical measures, based upon nucleic acid hybridization, and genetic measures, capitalizing on gene fusion technology, to comprehensively monitor global responses. We then illustrate how these approaches have been utilized to further our understanding of biological regulatory mechanisms.

DATABASES

These global measures of gene expression produce large amounts of information that must be organized and analyzed in a coherent fashion. Aiding in this matter are several publicly available databases that specialize in *E. coli* and *B. subtilis*, enabling one to access comprehensive information about their genes and their relationships with other organisms. The web addresses for the databases listed below are given in Table 1.

NCBI

NCBI has information on many of the published complete bacterial genomic sequences including *E. coli* K-12 MG1655, O157:H7, and O157:H7 EDL933, and *B. subtilis* 168. One can search for genes by name or position in the genome and perform BLAST searches against the genome or inferred protein sequences. There are several BLAST protein homolog databases: Cluster of Orthologous Groups (COGs), sequences with known structure (3D Structure), sequences grouped by superkingdom (TaxMap), three-way genome comparison (TaxPlot), and Conserved Domain Database (CDD). The site also provides the NCBI Gene Expression Omnibus (GEO), which is a repository for published microarray data.

KEGG

Kyoto Encyclopedia of Genes and Genomes (KEGG) contains many bacterial genome sequences including *E. coli* K-12 W3110 and MG1655, *E. coli* O157 EDL933 and O157 Sakai, and *B. subtilis*. The site supports a variety of standard query tools for gene analysis and also contains a repository for published microarray data. Importantly, explicit links between gene and function within the context of pathways are found within KEGG. Moreover, KEGG allows the comparison of genomes within the contexts provided by metabolic pathways.

EcoGene DB

EcoGene DB serves as a curated genome sequence annotation source for both for Colibri and SwissProt *E. coli* K-12 MG1655 records (68). One can download large datasets from EcoGene for database management and computational analysis. Both the SwissProt and Colibri databases enhance the EcoGene data with additional information, graphics, more powerful data query and retrieval systems, and WWW links. A useful feature of EcoGene DB is the additional information, including literature, available for every gene.

Colibri/SubtiList

The Colibri/SubtiList, created by the Institut Pasteur, are dedicated to *E. coli* K-12 MG1655 and *B. subtilis* 168 genome sequences. The sites provide a graphical genome browser enabling one to visualize the arrangement of specific genes on the chromosome. In addition to the standard query tools, it is possible to search based on codon usage, molecular weight, and isoelectric point. Also, one can download nucleotide and protein sequence data for specified genes.

GenProtEC

GenProtEC (64) assigns the genes of *E. coli* K-12 MG1655 into a multifunctional classification scheme devised by Serres & Riley (71). The classification scheme is hierarchical, providing a useful description of the known or predicted physiological roles for gene products. It is also informative to determine whether a list of genes identified from a microarray experiment are significantly enriched in gene members from one or more functional categories.

Micado

Micado has complete genome information for *B. subtilis*, including a hierarchical classification of genes into different metabolic pathways. The site enables one to extract the complete genomic sequence, specific nucleotide or protein sequences, and perform Blast searches.

RegulonDB

RegulonDB contains the known and predicted transcriptional and regulatory information of *E. coli* K-12 MG1655 (69). It includes annotations for known promoters, transcription factor-binding sites, operons and terminators, and predictions for promoter and operons. RegulonDB is also part of EcoCyc.

DBTBS

DBTBS is a database of *B. subtilis* promoters and transcription factors (32) with known binding site information for 90 transcription factors, predicted regulons, and 403 promoters including the organization of known *cis*-acting elements within each promoter region.

EcoCyc/MetaCyc

EcoCyc/MetaCyc contain literature-based genomic and biochemical information for microorganisms including *E. coli* and *B. subtilis* (34). EcoCyc is specific for *E. coli* and, in addition to functioning as a gene database, provides excellent

TABLE 1 Bioinformatic resources for microarray analysis

Resources and software	URL	Reference
Statistical analysis		
SAM	http://www-stat.stanford.edu/~tibs/SAM/index.html	(82)
Cluster analysis		
Cluster and TreeView	http://rana.lbl.gov/EisenSoftware.htm	(22)
GeneCluster	http://www-genome.wi.mit.edu/cancer/software/software.html	(75)
Genomic analysis		
NCBI	http://www.ncbi.nlm.nih.gov/PMGifs/Genomes/eub.html	
KEGG	http://www.genome.ad.jp/kegg/	
EcoGene	http://bmb.med.miami.edu/EcoGene/EcoWeb/index.html	(68)
Colibri	http://genolist.pasteur.fr/Colibri/	
SubtiList	http://genolist.pasteur.fr/SubtiList/	
GenProtEC	http://genprotec.mbl.edu/	(64)
Micado	http://locus.jouy.inra.fr/cgi-bin/genmic/madbase/progs/madbase.operl	
RegulonDB v3.2	http://www.cifn.unam.mx/Computational_Genomics/regulondb/	(69)
DBTBS	http://elmo.ims.u-tokyo.ac.jp/dbtbs/	(32)
EcoCyc	http://biocyc.org/ecoli/	(34)
Expression databases		
NCBI GEO	http://www.ncbi.nlm.nih.gov/geo/	
KEGG	http://www.genome.ad.jp/kegg/	
Regulatory site analysis		
AlignAce	http://atlas.med.harvard.edu/cgi-bin/alignace.pl	(67)
MEME	http://meme.sdsc.edu/meme/website/	(5)
CONSENSUS	http://bioweb.pasteur.fr/seqanal/interfaces/consensus.html	(29)
Gibbs Motif Sampler	http://bayesweb.wadsworth.org/gibbs/gibbs.html	(42)

information and graphical representations of enzymatic reactions, metabolic pathways, metabolites, and transporters. Submitted microarray expression data can be displayed on a metabolic pathway overview diagram, enabling interpretation of gene expression data in a pathway context.

COMPREHENSIVE EXPRESSION PROFILING WITH GENE ARRAYS

Methods and Some Theoretical Considerations

A variety of related techniques have been introduced to measure the distribution of transcripts found in a prokaryotic cell. Common steps within these methods are RNA preparation, reverse transcription of the RNA into cDNA, labeling of the cDNA, hybridization of the cDNA to an array, collection of a signal corresponding to the hybridization, and data manipulation. Different approaches, however, have been used to achieve each common step and have been recently reviewed (38, 60). When isolating RNA, it is critical to arrest transcription and prevent degradation so that the isolated RNA is representative of the population present at the time of culture sampling. To accomplish this some practitioners snap freeze cultures, while others collect azide-treated cells by rapid centrifugation through shaved ice.

Also an important factor is the RNA population to be studied. In contrast to studies of eukaryotic cells, a means to readily separate mRNA from stable RNA species is not available in bacteria because 3′ ends of the nucleic acid are not polyadenylated. In addition, only a single, core RNA polymerase is found in bacteria. This polymerase transcribes genes that specify final products that are either protein or RNA. Coupled with the realization that ribosomal and transfer RNAs make up as much as 90% of the total RNA pool, there is great risk in ignoring this stable fraction, as has been enunciated in studies of the stringent response. Nonetheless that is often precisely what is done because expression of genes whose final products are RNAs is typically disregarded, as genes encoding stable RNAs are not found on most arrays. Additionally, protocols that selectively remove these noncoding transcripts through tricks [such as forming double-stranded ribosomal RNA-DNA hybrids for enzymatic degradation or preventing copying of ribosomal RNA into cDNA through the design of a population of primers specific for each open reading frame (ORF) of a genome] focus the analysis on the distribution of RNA polymerase among promoters that drive expression of ORFs.

Another consideration is obtaining as faithful a cDNA population as possible following reverse transcription of the isolated RNA pool. To this end, both gene-specific and random hexamers have been used to prime the copying reaction. Use of gene-specific primers results in under-representation of the stable RNA pool. Moreover, a recent study suggests that the random hexamers may result in a superior cDNA product (3). Nucleotides, radioactively labeled with ^{33}P, fluorescently labeled with Cy3 or Cy5, or bearing a modification allowing subsequent label attachment, are incorporated into the polynucleotide product of the polymerase. This attachment point could be a polynucleotide terminus or distributed along the length of the chain. Postreplicative incorporation into cDNA of fluorescent labels eliminates problems associated with differential acceptance of fluorescent nucleotide triphosphates by polymerases. After the labeled polynucleotide is generated, it is hybridized to a surface that contains nucleic acids representative of each gene of an organism.

Currently, these highly parallel hybridization assays are performed in one of three basic formats. The first involves spotting of PCR-amplified ORFs onto nylon membranes yielding macroarrays (63). Radioactive cDNA, prepared from bulk RNA, is hybridized to the membrane. Alternatively, those same products, spotted onto glass slides, are hybridized with two cDNA samples labeled with different fluorescent dyes. This latter microarray method (63, 91, 99) uses greatly reduced amounts of PCR products. The third substratum used is one in which oligonucleotides are synthesized on a surface following a photolithographic protocol (46).

In this latter oligonucleotide array method, either bulk RNA or cDNA is terminally modified with a biotinylated nucleotide in vitro. Subsequently, a fluorescent label is incorporated into the polynucleotide.

Data Normalization

The difference in gene expression between two cell types when determined by cDNA microarrays is typically expressed as the ratio of measured expression levels. The calculated Cy3 (green, G) and Cy5 (red, R) fluorescence level hybridized to each gene on an array is corrected for background noise and then converted to a ratio: log (R/G). Both the Cy3 and Cy5 dyes are relatively unstable and differ in their quantum efficiencies, their efficiency of detection, and their ability, as components of modified nucleotides, to be incorporated into cDNA during labeling reactions. Consequently, in order to analyze gene expression ratios both within and across experiments, it is necessary to normalize the fluorescent intensities of the Cy3- and Cy5-labeled cDNAs relative to each other. Several normalization methods are used: (*a*) total mRNA normalization, (*b*) constant-mRNA normalization, and (*c*) dye-dependent normalization, discussed in more detail by Yang et al. (96).

TOTAL mRNA NORMALIZATION In many biological comparisons analyzed on microarrays, only a small fraction of genes are differentially expressed. As a result the total level of gene expression should be the same in the two conditions. Therefore, the remaining genes can be used as an indicator of the relative intensity of the two dyes. For example, Richmond et al. (63) corrected their data by summing the background-corrected Cy3 and Cy5 fluorescent intensities for all the genes on the array and then normalizing the totals to each other, thereby adjusting all the gene expression ratios. This assumption also holds true if there is a symmetric increase or decrease in gene expression.

CONSTANT-mRNA NORMALIZATION When a significant fraction of genes are differentially expressed, a subset of genes with constant expression can be used for normalization. This precludes applying a predetermined constraint, such as that imposed by global normalization, if it is inconsistent with the biological knowledge of the system under study. A predefined set of uniformly expressed housekeeping genes or genes identified from the dataset that exhibit constant expression can instead be used for normalization. For example, Tseng et al. (81) empirically identified constantly expressed genes within their datasets by first ranking the background-corrected fluorescent intensities of each gene for both Cy3 and Cy5. They then excluded the highest- and lowest-ranking genes from the subset to be normalized. Next they selected genes with similar ranking in both Cy3 and Cy5 fluorescence intensity to use for normalization.

DYE-DEPENDENT NORMALIZATION Data that have been normalized using one of the above methods can still exhibit dye-dependent effects. Low-intensity spots tend to exhibit R/G expression ratios less than 1 due to an excess of Cy3 fluorescence, whereas high-intensity spots tend to exhibit R/G expression ratios greater than 1 due to an excess of Cy5 fluorescence. This systematic variation of expression ratios as a function of gene expression level can be easily visualized by plotting the log intensity ratio, M ($\log_2$ R/G), versus the mean log intensity, A ($\log_2 \sqrt{RG}$). This *M* vs. *A* plot produces a "scatter" plot that enables artifacts in the data to be easily identified (21). In an ideal case, the data should scatter around the line, $M = 0$, indicating no bias in the log intensity ratios for all gene expression levels. Tseng et al. (81) show that for an individual hybridization, the scatter plot can be skewed with low-intensity spots scattered below the line and high-intensity spots scattered above the line. The simplest correction for this artifact is to perform a reverse labeling "dye-swap" experiment, which effectively cancels out the intensity-dependent dye effects (81, 91). The data from each experiment are then normalized and the expression ratios averaged.

Identifying Differentially Expressed Genes

When does a measured change reflect a biological change in gene expression? It is not sufficient to simply apply a twofold cutoff to identify differentially expressed genes, as a gene with low expression in one or both conditions has more variable expression ratios than a gene with a more substantial level of basal expression. In addition, the strain, a particular gene, the quality of the mRNA, cross-hybridization issues or systematic issues such as incorporation and scanning properties of the fluorescent labels, or spot quality due to print-tip or spatial effects can be sources of variation. Newton et al. (53) and Yang et al. (96) describe normalization methods to account for some of the systematic variations in microarray experiments.

Without replicate microarray experiments, it is not possible to know whether genes that are differentially expressed are real or spurious (43). Performing replicates enables a confidence level to be assigned to the gene expression ratios and gives a measure for nonsystematic variation between the experiments. In addition, by calculating probability scores, or *p* values for the gene expression ratios, the ratios can be progressively ranked, enabling one to prioritize further analysis. One approach is to normalize the data from replicated experiments and then perform the Student's *t*-test for every gene using the replicated log intensity ratios to derive a probability score. The *t*-test detects differences in the means between two experimental groups and uses the variance of the data as an indicator of precision in this difference. The null hypothesis H_0 is that there is no difference between the two datasets; hence, the resulting *p* values range from 1 for data that is identical to very small *p* values for data that is different at a highly significant level. In replicated microarray experiments where the comparison is condition *A* versus condition *B*, H_0 states that there is no differential gene expression, i.e., the log gene expression ratios are zero. In such an example, the *t*-test compares the replicated experimental log expression ratios with theoretical values of zero for every gene on the array. Richmond et al. (63) used the *t*-test to verify the reproducibility of genes that were more than fivefold differentially expressed in their datasets. A more stringent approach employed by Callow et al. (10) is to build a background model that more accurately describes the natural variability of the microarray system, such as gene expression, culture, and experimental variations. To do this, they performed replicates of control versus control experiments, as well as an equal number of control versus condition *A* experiments. Consequently, they employed the *t*-test for every gene to compare the log expression ratios of the control versus control against the control versus condition *A* datasets. Arfin et al. (3) utilized the *t*-test to address reproducibility of their nylon macroarrays. Note that because they are not dealing with gene expression ratios, they use the *t*-test for every gene to compare the replicates of background-corrected spot intensity in condition *A* against replicates of background-corrected spot intensity in condition *B*.

Analysis of microarray expression data using *t*-tests identifies genes that are consistently differentially expressed in replicated experiments by their low *p* values. A useful consequence is that low-intensity spots that tend to give highly variable expression ratios are more likely to result in high *p* values, making it unnecessary to filter data for low-intensity spots (10, 21). The use of confidence levels enables gene expression ratios with low *p* values to be termed significant and be selected for further analysis; a confidence level of 5% ($p < 0.05$) means that there is a false-positive rate of 5% of the total number of trials. The confidence level is also related to the total number of genes simultaneously tested, otherwise known as multiple testing. Consequently, using a confidence level of 5% for a 4000-gene array means that there would be a false prediction level of 200 genes just by chance alone! There are several correction methods to deal with multiple testing; the simplest is the Bonferroni method, which adjusts the *p* values in a single step. Simply, the calculated *p* values are multiplied by the number of genes in the array to give a new adjusted *p* value. For the calculated *p* value of gene$_j$ in an array of *N* genes, the adjusted *p* value, p_a, is given by:

$$\mathrm{p_{aj}} = \mathrm{Np_j}$$

(Note: calculated p_{aj} values > 1 are adjusted to 1).

A confidence level is then applied to filter the adjusted *p* values (e.g., $p_{aj} < 0.05$). Note that the unadjusted *p* values are required to be very small to score as significant, necessitating many replicate array experiments in order to obtain highly reproducible data with small *p* values. Also, the Bonferroni method tends to be conservative and assumes that the variables are independent. This is not true in microarray experiments because groups of genes have highly correlated expression levels due to coregulation. Callow et al. (10) use an estimation of adjusted *p* values by permutation proposed by Westfall & Young (93) that is less conservative and takes into account the dependent nature of gene expression, explained in detail by Dudoit et al. (21). Probably the most accessible measure is Significance Analysis of Microarrays (SAM) described by Tusher et al. (82), which also employs the *t*-test. SAM uses permutations of the repeated measurements to estimate the percentage of genes identified by chance (false discovery rate, FDR) for a particular cutoff threshold. The user can freely adjust the cutoff threshold to identify larger or smaller sets of genes, and the FDRs are calculated for each set. Their program is freely available to the academic community (Table 1) and runs as an addin within Excel on Windows® machines.

An alternative method to the *t*-test is the analysis of variance (ANOVA) used by Kerr et al. (37) in order to estimate the error variation associated with an estimated change in gene expression. They employ ANOVA methods to normalize replicated microarray data as an automatic correction for the extraneous effects in a microarray experiment, such as dye, slide, and gene effects. This provides normalized estimates for genes that are differentially expressed, which can then be used more reliably in downstream data analysis such as clustering. The different methods of statistical analysis of microarray data are under rapid development and are not yet widely accepted. However, it is clear that there are fundamental issues of reproducibility and the significance of differentially expressed genes that have to be addressed in order to analyze microarray data in a useful and meaningful way.

Clustering Methods

After identifying genes that are significantly differentially expressed, the next step is to use biological insight, clustering algorithms, and visualization programs to identify patterns and groups in the data that can be used to assign biological meaning to the expression profiles. Several different clustering algorithms have been developed [reviewed in (16, 72)]. The CLUSTER and TREEVIEW programs (22) (Table 1) employ a hierarchical tree-building method used in sequence and phylogenetic analyses to cluster and visualize the data. The expression profile for a particular gene in various conditions is compared to that of every other gene using the Pearson correlation coefficient, and the most highly correlated genes are identified. CLUSTER reiterates this procedure for every gene to build a similarity matrix. Using a hierarchical process, the two most-related genes are joined by a node/branch of a tree, removed from the dataset, and replaced by an item that represents the new branch. The most related gene to this new branch in the dataset is identified, joined by a node, removed from the dataset, and a new branch added. This process is reiterated until all the genes in the dataset have been computed. The program TREEVIEW is then used to visualize the relationship of the gene expression profiles as a dendritic tree that connects a color map of the gene expression data.

Hierarchical cluster analysis is best suited for exploring ranked similarities between gene expression data; however, expression patterns can be similar in many different ways. In order to capture the complexity of gene expression data, other clustering techniques have been used, such as self-organizing maps (SOMs) (75, 80), *k*-means clustering (78), and *QT_Clust* (30). These algorithms involve partitioning the data into homogeneous groups with similar expression profiles; both SOMs and *k*-means require the user to input the number of groups, something that has to be determined empirically or by trial and error. Tamayo et al. (75) give a good explanation of SOMs and have developed a SOM algorithm called GENECLUSTER for microarray analysis that is freely available (Table 1). The *QT_Clust* algorithm determines cluster groups by having a threshold, i.e., the gene members must have a minimum pairwise correlation in order to belong in a particular cluster. The advantages of *QT_Clust* are that the numbers of clusters do not have to be previously determined and that a cluster can contain single genes, preventing the disruption of otherwise homogeneous clusters with outliers. Many of the clustering algorithms assign genes into groups based on the correlation of their expression patterns. Kerr & Churchill (36) raise the important issue that these expression patterns are estimates with margins of error. They have used a bootstrapping cluster analysis algorithm to assess the reliability of cluster results in light of variation in the data.

The use of clustering algorithms with microarray data enables the identification of genes that are coexpressed under different environmental conditions. An increasingly popular approach is to analyze the upstream regulatory regions of coexpressed genes for common sequence motifs. This approach is based on the assumption that genes with similar patterns of expression are probably regulated by common factors, and therefore they are likely to share common binding sites for these factors. The regulatory regions of coexpressed genes can be searched for statistically over-represented motifs using a variety of algorithms such as Gibbs Motif Sampler (42), MEME (5), AlignAce (67), and Consensus (29), all of which run on servers accessible on the Web (see Table 1). These algorithms function by performing local multiple alignments of all the sequences, which can be computationally intensive. To overcome this, the algorithms employ slightly different strategies to identify over-represented motifs, such that it is often advantageous to use several of the algorithms when querying sequences. Results are significantly improved by careful selection of the sequences to be aligned, using statistical criteria for identifying differentially expressed genes and cluster analysis to identify coexpressed genes. The use of microarrays thus will greatly facilitate the identification of global regulatory networks by integrating the identification of blocks of coregulated genes under many different environmental conditions together with predictions of DNA-binding sites. In addition, it should be possible to correlate gene expression patterns with transcription factor-binding sites. The recent availability of many bacterial genome sequences has facilitated a new approach to identify conserved binding sites by aligning the regulatory regions of highly conserved genes from different genomes (48, 49). This method can be further developed with microarrays to identify highly conserved regulatory pathways by identifying coregulated orthologous genes that share similar expression patterns under the same stresses in different organisms.

Exploiting Operon and Regulatory Network Organization

These regulatory structures provide a conceptual biological framework for the analysis of comprehensive expression profiling experiments. An advantage inherent in bacterial studies of gene expression is the cotranscription of several genes that are organized into operons. This means that several measures of the transcription of a multigene operon are made in a whole-genome microarray, macroarray, or oligonucleotide array. Such measurements provide independent determinations of transcript quantity from a single RNA sample; a concordance of results provides a degree of confidence lacking if only a single measurement is made. Such analyses are aided by databases such as RegulonDB (69), which lists experimentally verified and predicted operons. Because the complexity of an individual operon's control and structure can be quite baroque (95), noncoordinate responses may also be observed. A similar quality check involves consideration of regulatory networks because the genes of a regulon (51) should respond in concert to a given perturbation. Differing combinations of multiple promoter elements controlling operon expression, however, can give rise to noncoordinate responses of regulon genes. Understanding of such transcriptional regulatory networks is furthest advanced in *E. coli*. Approximately 800 known transcription factor-binding sites are documented in RegulonDB and EcoCyc, and hundreds of predictions based on derived consensus binding sites of 55 known factors have been made by Robison et al. (66) and Thieffry et al. (79). However, Pérez-Rueda & Collado-Vides (58) estimate that there are approximately 350 regulatory macromolecules in *E. coli*, illustrating that our knowledge of regulatory networks is still limited.

GENOME-WIDE GENE FUSION ASSAYS

Ordered Collections

Gene fusion arrays represent an important alternative technology for genome-wide gene expression analysis in bacteria. The largest reported set of mapped gene fusions in bacteria was obtained beginning with a collection made by joining random segments of the *E. coli* chromosome to a bioluminescent reporter (83). Each of 8000 random gene fusions was sequenced to define the ends and orientation of the chromosomal segment upstream of the reporter genes yielding 5000 mapped gene fusions (87). Due to the formation of nonfunctional fusions and redundancies inherent in a random approach, this set contained gene fusions to 689 of the 2584 predicted transcriptional units in *E. coli*, or 27% coverage of the *E. coli* chromosome (85). An alternative to the random approach of finding gene fusions is directed construction. The known and predicted promoter regions deduced from genomic sequence data can be amplified by the polymerase chain reaction and cloned upstream of a reporter gene. Such an approach has been successfully applied in yeast to generate a large set of fusions to *gfp* (20), but to date the approach has been used only to generate relatively small panels of selected gene fusions in bacteria (7, 33, 57). The choice for development of reporter gene arrays depends on the relative ease and cost of DNA sequencing compared to cloning. Of course, a combination of random and directed methods is also possible, where the random approach can be used first and directed gene fusions are made for missing promoter regions.

Methods and Instrumentation

Once a set of reporter gene fusion strains has been developed, there are several methods to perturb and collect data in a parallel fashion to study genome-wide gene expression. One approach involves "printing" the set of strains containing the gene fusions in a high-density ordered array to solidified agar (20, 85). Subsequently, images of the signal generated from reporter constructs are collected in such a manner that the signal intensity can be quantified. This approach relies on the use of reporter genes that generate a visible signal. Thus, bioluminescence or fluoresence have been used. Conveniently, several software programs designed for quantitation of DNA array data can be applied to find the pixel density of the image at each spot of the reporter gene array. Other approaches for parallel analysis of reporter gene activity, such as in growth and assay in microplates (50, 84), are amenable to automation and thus provide an avenue applicable to reporters that require manipulation such as cell lysis and substrate addition to generate a detectable signal. Furthermore, several recent intriguing developments have potential for implementation in a highly parallel fashion with reporter systems that generate light as a signal. Patch biosensors have been made from bioluminescent bacterial reporter strains immobilized in latex copolymers (47). This technology is also applicable to the formation of three-dimensional microstructures by inkjet deposition, which could be used for construction of microarrays of reporter gene strains. Another development is direct connection of a reporter strain with the measurement instrumentation. A small (2.2 mm $\times$ 2.2 mm) integrated circuit combined with a bioluminescent reporter bacterial strain detects, processes, and communicates the signal from the reporter strain (73). The variety of options for implementation is an advantageous feature of reporter gene arrays.

Reporter Choice

The choice of reporter gene for development of gene fusion arrays is critical. The traditional reporter gene, *lacZ*, is not optimal for high-throughput applications because cells need to be lysed or permeabilized and substrate added to accurately measure β-galactosidase levels. The number of manipulations is a limitation for development of arrays where many measurements must be done in parallel. Thus, reporters that generate a visible signal without need for cell disruption or substrate addition are advantageous for the development of gene fusion arrays. Two widely used reporters in this category are the green fluorescent protein (GFP) and the bacterial bioluminescent operon *luxCDABE*. The advantage of GFP, that it is a single gene reporter for which expression can be mathematically modeled (44), must be balanced with the disadvantage that the autocatalytic protein modification required for fluorescence proceeds slowly in bacteria (1). GFP variants with improved folding efficiency and fluorescence intensity have been found following mutagenesis (17, 70); however, the rate of fluorescent signal development is still slow, taking more than 1 h to develop a full fluorescent signal within the bacterial cell. Other single-gene product fluorescent proteins, such as dsRed, also suffer from slow maturation kinetics (6, 28). These fluorescent reporters are best suited for steady-state analysis of gene expression.

For transient expression analysis following perturbation in highly parallel assays, the bacterial bioluminescence reporter is currently the best option. The bioluminescent signal can be directly detected in as short a time as 15 to 20 min following a perturbation that activates promoters controlling *luxCDABE* expression (8, 86). Light production can be accurately quantitated over a range of greater than seven orders of magnitude using several commercial instruments. The bioluminescent signal can also be quantified by image analysis. However, the light production pathway requires active cellular metabolism and expression of five genes (13). Thus, this reporter is only useful for analysis of gene expression alterations when stresses do not compromise energy and reductant pools. This is not a severe limitation because sublethal stress conditions are often of interest, and numerical correction of reporter signal diminution due to impairment of cellular metabolism can be readily accomplished when a large set of gene fusions in a reporter gene array is used (85). The need to have five proteins expressed to generate light implies that the bioluminescent signal is a report of the expression of the rate-limiting enzyme of the light production pathway. That the five genes are cotranscribed in light-producing bacteria allows a degree of coordinated expression in heterologous host strains. The *luxCDABE* operon from bacteria such as *Vibrio fischeri* and *Photorhabdus luminescens* are efficiently transcribed and translated in numerous gram-negative bacteria. Moreover, the temperature optima of the *P. luminescens* Lux proteins are well matched with the optimal temperature range for *E. coli* growth. Modifications to the translation initiation region of the *lux* operon have recently been made, which allow improved expression in gram-positive bacteria (24, 25).

Utility

When using a reporter gene fusion, promoter activity is inferred from the activity of the reporter gene product. Thus, reporter gene arrays provide an independent analysis method to that of hybridization experiments. To the extent that gene expression is controlled at the level of transcriptional initiation, as is typical in many prokaryotes, these methods should yield equivalent answers as has been demonstrated by analysis of gene expression responses to DNA damage (85, 87). Technology based upon large, genome-registered collections of Lux fusions provides several other advantages. The noninvasive monitoring of light production allows efficient determination of antimicrobial minimal inhibitory concentrations (9), induction of stress responses (86), and detailed, if necessary continuous, kinetic characterizations. Selected fusion-bearing strains can serve as whole-cell biosensors or as the basis of high-throughput assays (40). Consequently, these fusion libraries provide immediate advantages for kinetic monitoring and range finding, as well as components for downstream applications predicated upon the information found in gene expression profiles.

CASE STUDIES

In this section, we first review the studies that validate the use of microarrays to examine gene expression in bacteria and then show the variety of ways this technology has been used to elucidate aspects of microbial biology. We showcase a few important studies rather than present a comprehensive review of all results in the literature. In each of these studies, hybridization was to gene-sized ORFs on glass slides, and "control" and "experimental" RNAs were differentially labeled with either Cy3 or Cy5 unless otherwise noted.

Validation of the Technique

The *lac* operon and the *trp* regulon of *E. coli* are probably the two best-studied bacterial regulatory systems. Both have been examined with whole-genome expression analysis, and the results indicate the general utility of this approach to obtain a qualitatively accurate picture of gene expression. Three different studies (63, 91, 92) report that *lac* genes are induced approximately 30- to 50-fold upon addition of IPTG, a significant increase although somewhat less than the 500-fold induction observed in protein studies. Inability to accurately measure the basal value, or saturation of signal, may account for lower induction ratios. As expected from this specific inducer, the three to four other genes reported to be induced showed much lower induction ratios (two- to fivefold) and may be artifacts of the particular experiments; only one of these genes was induced in two of these studies and none were induced in all three. Examining expression of the *trp* regulon was a more stringent test of array analysis because increases in gene expression are smaller and because 15 genes, organized into 9 operons, were known to be in the regulon. Whole-genome expression analysis was carried out under a wide variety of conditions that should alter expression of this regulon including excess and limiting tryptophan and mutants inactivating the *trp* repressor (39). The microarray analyses, performed without dye-swapping, accurately reflect the regulatory patterns expected: Changes in expression measured by this method were within a factor of two of those determined from alterations in proteins encoded in the *trp* regulon. Given the ability of this analysis to reproduce known regulatory events, we are confident that it can be used to examine situations in which expectations are unknown. We note, however, that these experiments filtered out noise through the use of multiple replicates and that spurious signals can occur. This issue suggests that such analyses provide a starting point for further study rather than a definitive solution.

Use of Arrays to Define Regulons

The most straightforward use of whole-genome expression analysis is to define regulons of particular transcription factors. Such experiments can lead to a greater understanding of regulon function, allow development of novel physiological hypotheses for testing, and eventually yield a blueprint of the cellular transcriptional regulatory networks. In addition, such experiments will allow us to explore how the function of a particular group of regulators has diverged across the bacterial world.

In these experiments, RNA from cells expected to have low or no expression of regulon genes is compared to RNA from cells highly expressing regulon members. Of primary importance here is that both sets of cells grow under the same conditions and at approximately the same rate, so that as many changes as possible between the two cultures reflect the particular regulatory event being studied. This has been successfully accomplished in two ways: (*a*) comparing mutants that respectively either increase or decrease regulon expression, or (*b*) following gene expression as a function of time after either an environmental challenge that stimulates a regulatory circuit or overexpression of a regulator. In both cases control and experimental cultures are grown in the same conditions. Statistical criteria are developed to identify those genes whose behavior is consistent with membership in the regulon. Often additional data are used to bolster assignments including (*a*) direct experimental determination that the particular gene is controlled by that regulator, (*b*) identification of putative regulator-binding sites in the vicinity of the gene postulated to be under the control of the regulator, or (*c*) physiological classification of the differentially regulated genes.

Using the paired mutant approach, Kustu and collaborators (99) have defined *E. coli* genes controlled by NtrC, a transcription factor whose activity is increased upon nitrogen limitation and functions to activate σ^{54}. The experimental design incorporated multiple replicates but did not include dye-dependent normalization. About 2% of the genome is either directly under the control of NtrC or indirectly controlled by NtrC as a result of its activation of Nac, a σ^{70}-dependent transcriptional activator. It is interesting that almost two thirds of these genes encode members of transport operons, indicating that the first line of defense against nitrogen starvation is to scavenge nitrogen from the medium and the periplasmic space consistent with the broader presumption that import is preferable to biosynthesis.

Using other paired mutants with and without ultra-violet irradiation, Courcelle et al. (19) analyzed the LexA-dependent regulon of *E. coli*. This intensively studied system, known as the SOS response, is a coordinated reply to DNA insults that is induced following cleavage of the LexA repressor by the RecA coprotease activated by the genomic damage. In this study, a noncleavable version of the LexA repressor encoded by the *lexA1* allele was used. No replicates or dye swaps were conducted, instead RNA was isolated from control and *lexA1* strains at five time points following ultra-violet irradiation, as well as from unexposed cultures. Although this experimental protocol did not allow statistical analysis, kinetics of gene expression were obtained. Most of the 31 previously known SOS genes were upregulated at multiple time points in a *lexA*-dependent manner. The false negatives in this study can be defined as those known SOS genes for which the RNA levels did not increase following ultra-violet irradiation. The 23% rate of false negatives compares favorably with a 33% rate from another DNA array study of *E. coli* treated with mitomycin C, a DNA-damaging agent, where replicates and dye swaps were done (87). Of note is that these two studies, which used subsets of the same PCR products for array fabrication, had two false negatives in common, suggesting individual spotted PCR products as a possible source of error. This points out that the quality of individual components of the array is an inherent uncertainty in these large-scale analyses. One aim of this study was elucidating the extent of the LexA regulon. Accordingly, 17 genes with a similar pattern of gene expression as the known members were suggested to be additional constituents of the LexA regulon. Sequences related to the LexA-binding site were found upstream of some but not all of these genes. The DNA array analysis thus provided a starting point for further experimentation to confirm these findings and to determine if the observed upregulation is directly or indirectly dependent on LexA. Confirmation of the LexA-dependent DNA damage response was found for one such gene from a gene fusion array analysis, but for another gene, the LexA-dependence of the response was not observed (85). Thus, in spite of this genome-wide survey, the detailed enumeration of the LexA regulon remains to be finalized.

Using overexpression of *rpoE*, Gross and collaborators have defined *E. coli* genes in the σ^E regulon, which is believed to control the extra-cytoplasmic stress response (W. Suh, V. Rhodius & C. Gross, manuscript in preparation). It is satisfying that most of the genes identified were involved in processes that occurred in the extra-cytoplasmic space, including maintaining the barrier function of the outer membrane. In contrast, amplification of *sdiA*, a transcriptional activator of cell division genes, led to a complex expression pattern in which not only were the expected *sdiA* and *ftsZ* operon transcripts elevated but mRNAs corresponding to genes near the origin and a terminus of replication were over-represented (90).

Use of Arrays to Define an Organismal Response to Altered Conditions

Prokaryotic gene regulation is a means to optimize growth or survival as the environment changes. Using genetic, biochemical, and molecular biological techniques, a large data bank of many of the regulatory responses utilized by model organisms has been generated. However, even for the best-studied organisms, our information is incomplete because all regulon members have not been identified and there may be additional unidentified regulatory responses to a perturbation. Whole-genome expression analysis should fill in both of these blanks. However, going from the data to an understanding of the transcriptional regulation underlying the response can be difficult especially when expression of a large fraction of the genome changes in response to the perturbation. Several tools are helpful. First, the data-defining regulons can help identify the regulators involved in the response. Second, upstream regions of coregulated genes can be searched for common motifs to identify predicted binding sites for regulators as well as putative regulons. Finally, clustering methods assign groups of genes to units that are coregulated under a variety of different conditions, thereby identifying stimulons and potential new regulons. These studies should provide a greatly improved understanding of the transcriptional regulatory circuits of model organisms at the global level. In addition, given the extensive conservation of transcriptional regulators in prokaryotes, this information coupled with whole-genome expression analysis of less-studied organisms will allow rapid progress in understanding the transcriptional programs of these bacteria as well. Indeed such progress is apparent in work aimed at understanding processes such as growth rate control (76), infectious disease (4, 94), metabolic engineering (56, 77), and the bacterial cell cycle (41). The two examples below illustrate studies of cellular responses to perturbations. It is notable that even in the best-studied cases, microarray analysis has uncovered novel facets of the response.

Using two-dimensional gels to monitor protein expression after exposure to H_2O_2, Ames and colleagues (14) initially identified the cellular response to this chemical; mutational analysis then identified OxyR as the transcription factor regulating expression of a majority of the proteins. By examining both the OxyR-dependent and -independent responses to H_2O_2 addition using microarrays, Storz and colleagues (98) have added considerably to our understanding of the response. This study added several members with intriguing functions to the OxyR regulon and suggested, for the first time, that part of the response involves OxyR functioning as a repressor. Their findings also indicate that the OxyR-independent response to peroxide is complex, including induction of the superoxide regulator, SoxS (at high concentrations of H_2O_2), as previously suggested by a gene fusion study (7) and additional redox-sensing mechanism(s). Finally, this work, along with other studies investigating the response to ultra-violet light (19) and superoxides (61) establishes that *E. coli* has a predominant, distinct response to each stress, as had been inferred from earlier experiments. In contrast, in yeast a large group of the same proteins are induced in response to many different stresses (26).

Although the sporulation process in *B. subtilis* has been extensively studied, the global changes in gene expression during this process were poorly understood. Youngman and colleagues (23) have now remedied this situation with expression profiling using ORFs affixed to nylon membranes and probed with radioactively labeled cDNA. The SpoOA transcription factor and σ^F were both known to play a role in initiating sporulation. By examining gene expression in cells expressing a constitutively activated SpoOA, and comparing sporulation in wild-type cells with those lacking either SpoOA or σ^F, the authors identified 520 genes directly or indirectly dependent on SpoOA, of which about half were activated and the remainder repressed. An additional 66 genes were dependent on both SpoOA and σ^F. This study indicates that there are profound changes in the pattern of gene expression during sporulation. Two of the newly identified genes were disrupted and affect sporulation, and it is likely that many others will play essential roles in sporulation.

Global Studies of Repression

Much of the work reported to date has focused upon those genes whose expression is elevated by mutations or stress. Certain genes may also exhibit significant downregulation. A study in which an amino acid analog causes induction of the histidine operon provides some insight. Several genes are downregulated, including those encoding components of the translational apparatus. Surprisingly, several highly expressed biosynthetic genes, expected to be upregulated by amino acid starvation, are not significantly transcribed when cells are challenged with the amino acid analog. This suggests that the inhibitor has differential effects on biosynthetic gene expression; the mechanisms of this disparate regulation remain to be determined (74).

Transcriptome and Proteome Comparisons

Gene expression in prokaryotes is controlled principally at the level of transcription; thus, the expectation is that quantities of mRNA measured by arrays should correlate with protein titers. These correlations, however, may not hold for individual polypeptide-transcript pairs if other processes such as protein turnover and translational regulation have a significant impact. Indeed, several comparisons of transcriptome measurements from DNA arrays with proteomic measurements based upon two-dimensional gel electrophoretic separations have found only modest correlations between these two measures.

Most microarray experiments compare the transcript contents of two cultures differing in a single environmental or genetic variable. The output of such experiments is a ratio that communicates the fold induction or repression of mRNA content that is observed. Nonetheless, the distribution of mRNA species within *E. coli* under steady-state growth conditions can be described by using the simplifying assumption that copies of genomic DNA are equivalent to an equimolar mixture of all mRNAs encoded within a genome (91). Comparison of these measures for *E. coli* actively growing in defined medium with the distribution of polypeptides from proteomic data of Church and colleagues (45) for similar conditions yields a coarse correlation ($r^2 = 0.31$) (R.A. LaRossa, unpublished). Similarly, Gmuender et al. (27) found modest correlation from parallel transcriptome and proteome analyses of *Haemophilus influenzae* cultures. Hybridization signals on a duplicate high-density oligonucleotide array were compared to the average intensity of five measurements of 223 two-dimensional protein gel spots that had been assigned identity and had been normalized for methionine content. A correlation coefficient of 0.5 was observed, indicating weak agreement between these measures of mRNA content and newly synthesized proteins. However, when perturbation by two antibiotics was studied, the direction of change of RNA measures and protein measures was correlated, although the magnitude of the change differed. A similar correlation (0.61) between changes in mRNA and protein content for 289 yeast genes has been reported using a different measure of polypeptide content. Together, these results point to the danger of quantitative interpretations of the data from either method, but nevertheless they suggest that the validity of descriptive interpretations is bolstered when two independent methods yield comparable results.

The benefits of combined analyses for description of genes belonging to regulons or stimulons are evident from two recent studies in *B. subtilis*. Previously, proteomic, genetic, and bioinformatic screening has identified 75 genes belonging to the σ^B-controlled general stress response. Using a DNA array analysis, Hecker and coworkers (59) confirmed 62 of these and added 63 additional putative members of this regulon. Notably, all members of the regulon that had been identified by the proteomic approach were confirmed with the DNA array analyses; these were almost exclusively cytoplasmic proteins. In contrast, many of the newly identified putative members encoded proteins with predicted transmembrane regions, suggesting that DNA array analyses detect alterations in expression of membrane protein gene expression that are missed in proteomic analyses. However, DNA array analyses alone did not uncover all regulon members. *opuE*, an acknowledged member of the σ^B regulon, was not identified in this study, indicating at least one false-negative result. Likewise, using a combined approach to study glucose repression, Fujita and collaborators (97) found many more candidate genes from the transcriptome analysis than from the proteome analysis (97). The overall patterns emanating from the two methods were similar. Nonetheless, 2 of the 11 genes identified by proteomics were missed in the DNA array analyses. Thus, a combination of approaches is likely to allow the most comprehensive description of expression changes.

CONCLUDING REMARKS

How have these genomic technologies for studying gene expression changed the ways that microbiology is approached? They are evoking a renaissance; terms like pleiotropy, global regulatory mechanisms, and metabolic integration are beginning to reoccupy "center-stage." There is an emerging realization that these methods will provide essential insights for the analysis of any microbial species or physiological response. We share this optimistic view; however, we caution that even the best measures are estimates of transcript levels and that biological regulatory mechanisms work at the level of protein activity as well as at the level of gene expression. We thus look at the enhanced capability to monitor gene expression on a global basis as a means to scout out new and ill-defined territories. Only after settlers follow scouts, however, will the abundance of the newly discovered arenas be exploited. There, many things will be combined—classical and new technologies, holistic and reductionist approaches, genetics with cell biology and biochemistry—to gain a greater understanding of the microbial cell.

The *Annual Review of Microbiology* is online at http://micro.annualreviews.org

LITERATURE CITED

1. Albano CR, Randers-Eichhorn L, Chang Q, Bentley WE, Rao G. 1996. Quantitative measurement of green fluorescent protein expression. *Biotechnol. Tech.* 10:953–58
2. Ames BN, Hartman PE, Jacob F. 1963. Chromosomal alterations affecting the regulation of histidine enzymes in *Salmonella. Mol. Biol.* 7:23–42
3. Arfin SM, Long AD, Ito ET, Tolleri L, Riehle MM, et al. 2000. Global gene expression profiling in *Escherichia coli* K12. *J. Biol. Chem.* 275:29672–84
4. Arnold CN, McElhanon J, Lee A, Leonhart R, Siegele DA. 2001. Global analysis of *Escherichia coli* gene expression during the acetate-induced acid tolerance

stress response. *J. Bacteriol.* 183:2178–86

5. Bailey TL, Elkan C. 1994. Fitting a mixture model by expectation maximization to discover motifs in biopolymers. *Proc. Int. Conf. Intell. Syst. Mol. Biol.* 2:28–36
6. Baird G, Zacharias D, Tsien R. 2000. Biochemistry, mutagenesis, and oligomerization of DsRed, a red fluorescent protein from coral. *Proc. Natl. Acad. Sci. USA* 97:11984–89
7. Belkin S, Smulski DR, Dadon S, Vollmer AC, Van Dyk TK, LaRossa RA. 1997. A panel of stress-responsive luminous bacteria for the detection of selected classes of toxicants. *Wat. Res.* 31:3009–16
8. Belkin S, Smulski DR, Vollmer AC, Van Dyk TK, LaRossa RA. 1996. Oxidative stress detection with *Escherichia coli* harboring a *katG′::lux* fusion. *Appl. Environ. Microbiol.* 62:2252–56
9. Bulich AA. 1982. A practical and reliable method for monitoring the toxicity of aquatic samples. *Process Biochem.* 17:45–47
10. Callow MJ, Dudoit S, Gong EL, Speed TP, Rubin EM. 2000. Microarray expression profiling identifies genes with altered expression in HDL-deficient mice. *Genome Res.* 10:2022–29
11. Casadaban MJ, Cohen SN. 1980. Lactose genes fused to exogenous promoters in one step using a new-*lac* bacteriophage: *in vivo* probe for transcriptional control sequences. *Proc. Natl. Acad. Sci. USA* 76:4530–33
12. Cashel M, Gentry DR, Hernandez VJ, Vinella D. 1996. The stringent response. See Ref. 50a, pp. 1458–96
13. Chatterjee J, Meighen EA. 1995. Biotechnological applications of bacterial bioluminescence (*lux*) genes. *Photochem. Photobiol.* 62:641–50
14. Christman MF, Morgan RW, Jacobson FS, Ames BN. 1985. Positive control of a regulon for defenses against oxidative stress and some heat-shock proteins in *Salmonella typhimurium*. *Cell* 41:735–62
15. Chuang S-E, Daniels DL, Blattner FR. 1993. Global regulation of gene expression in *Escherichia coli*. *J. Bacteriol.* 175:2026–36
16. Claverie J-M. 1999. Computational methods for the identification of differential and coordinated gene expression. *Hum. Mol. Genet.* 8:1821–32
17. Cormack BP, Valdivia RH, Falkow S. 1996. FACS-optimized mutants of the green fluorescent protein (GFP). *Gene* 173:33–38
18. Cortese R, Landsberg RA, von der Haar RA, Umbarger HE, Ames BN. 1974. Pleiotropy of *hisT* mutants blocked in pseudouridine synthesis in tRNA: leucine and isoleucine-valine operons. *Proc. Natl. Acad. Sci. USA* 71:1857–61
19. Courcelle J, Khodursky A, Peter B, Brown PO, Hanawalt PC. 2001. Comparative gene expression profiles following UV exposure in wild-type and SOS-deficient *Escherichia coli*. *Genetics* 158:41–64
20. Dimster-Denk D, Rine J, Phillips J, Scherer S, Cundiff P, et al. 1999. Comprehensive evaluation of isoprenoid biosynthesis regulation in *Saccharomyces cerevisiae* utilizing the Genome Reporter Matrix. *J. Lipid Res.* 40:850–60
21. Dudoit S, Yang YH, Callow MJ, Speed TP. 2000. Statistical methods for identifying differentially expressed genes in replicated cDNA microarray experiments. Technical Report #578: http://www.stat.berkeley.edu/users/terry/zarray/Html/papersindex.html
22. Eisen MB, Spellman PT, Brown PO, Botstein D. 1998. Cluster analysis and display of genome-wide expression patterns. *Proc. Natl. Acad. Sci. USA* 95:14863–68
23. Fawcett P, Eichenberger P, Losick R, Youngman P. 2000. The transcriptional profile of early to middle sporulation in *Bacillus subtilis*. *Proc. Natl. Acad. Sci. USA* 97:8063–68
24. Francis K, Yu J, Bellinger-Kawahara C, Joh D, Hawkinson M, et al. 2001. Visualizing pneumococcal infections in the

lungs of live mice using bioluminescent *Streptococcus pneumoniae* transformed with a novel gram-positive *lux* transposon. *Infect Immun.* 69:3350–58

25. Francis KP, Joh D, Bellinger-Kawahara C, Hawkinson MJ, Purchio TF, Contag PR. 2000. Monitoring bioluminescent *Staphylococcus aureus* infections in living mice using a novel *luxABCDE* construct. *Infect. Immun.* 68:3594–600
26. Gasch AP, Spellman PT, Kao CM, Carmel-Harel O, Eisen MB, et al. 2000. Genomic expression programs in the response of yeast cells to environmental changes. *Mol. Biol. Cell* 11:4241–57
27. Gmuender H, Kuratli K, Di Padova K, Gray C, Keck W, Evers S. 2001. Gene expression changes triggered by exposure of *Haemophilus influenzae* to novobiocin or ciprofloxacin: combined transcription and translation analysis. *Genome Res.* 11:28–42
28. Heikal A, Hess S, Baird G, Tsien R, Webb W. 2000. Molecular spectroscopy and dynamics of intrinsically fluorescent proteins: coral red (dsRed) and yellow (citrine). *Proc. Natl. Acad. Sci. USA* 97:11996–2001
29. Hertz GZ, Stormo GD. 1999. Identifying DNA and protein patterns with statistically significant alignments of multiple sequences. *Bioinformatics* 15:563–77
30. Heyer LJ, Kruglyak S, Yooseph S. 1999. Exploring expression data: identification and analysis of coexpressed genes. *Genome Res.* 9:1106–15
31. Ideker T, Thorsson V, Ranish JA, Christmas R, Buhler J, et al. 2001. Integrated genomic and proteomic analyses of a systematically perturbed metabolic network. *Science* 292:929–34
32. Ishii T, Yoshida K, Terai G, Fujita Y, Nakai K. 2001. DBTBS: a database of *Bacillus subtilis* promoters and transcription factors. *Nucleic Acids Res.* 29:278–80
33. Kalir S, McClure J, Pabbaraju K, Southward C, Ronen M, et al. 2001. Ordering genes in a flagella pathway by analysis of expression kinetics from living bacteria. *Science* 292:2080–83
34. Karp PD, Riley M, Saier M, Paulsen IT, Paley SM, Pellegrini-Toole A. 2000. The EcoCyc and MetaCyc databases. *Nucleic Acids Res.* 28:56–59
35. Kenyon CJ, Walker GC. 1980. DNA damaging agents stimulate gene expression at specific loci in *Escherichia coli. Proc. Natl. Acad. Sci. USA* 77:2819–23
36. Kerr MK, Churchill GA. 2001. Bootstrapping cluster analysis: assessing the reliability of conclusions from microarray experiments. *Proc. Natl. Acad. Sci. USA* 98:8961–65
37. Kerr MK, Martin M, Churchill GA. 2000. Analysis of variance for gene expression microarray data. *J. Comp. Biol.* 7:819–37
38. Khodursky A, Bernstein JA, Peter BJ, Rhodius V, Wendisch VF, Zimmer DP. 2002. *Escherichia coli* spotted double strand DNA microarrays: RNA extraction, labeling hybridization, quality control and data management. In *Methods in Microbiology*, ed. M Brownstein, pp. In press. Totowa, NJ: Humana
39. Khodursky AB, Peter BJ, Cozzarelli NR, Botstein D, Brown PO, Yanofsky C. 2000. DNA microarray analysis of gene expression in response to physiological and genetic changes that affect tryptophan metabolism in *Escherichia coli. Proc. Natl. Acad. Sci. USA* 97:12170–75
40. LaRossa RA, Van Dyk TK. 2000. Applications of stress responses for environmental monitoring and molecular toxicology. In *Bacterial Stress Responses*, ed. G Storz, R Hengge-Aronis, pp. 453–68. Washington, DC: ASM
41. Laub MT, McAdams HH, Feldblyum T, Fraser CM, Shapiro L. 2000. Global analysis of the genetic network controlling a bacterial cell cycle. *Science* 290:2144–48
42. Lawrence CE, Altschul SF, Boguski MS, Liu JS, Neuwald AF, Wootton JC. 1993. Detecting subtle sequence signals: a Gibbs sampling strategy for multiple alignment. *Science* 262:208–14

43. Lee M-LT, Kuo FC, Whitmore GA, Sklar J. 2000. Importance of replication in microarray gene expression studies: statistical methods and evidence from repetitive cDNA hybridizations. *Proc. Natl. Acad. Sci. USA* 97:9834–39
44. Leveau J, Lindow S. 2001. Predictive and interpretative simulation of green fluorescent protein expression in reporter bacteria. *J. Bacteriol.* 183:6752–62
45. Link AJ, Robison K, Church GM. 1997. Comparing the predicted and observed properties of proteins encoded in the genome of *Escherichia coli* K-12. *Elect rophoresis* 18:1259–313; http://arep.med.harvard.edu/labgc/proteom.html
46. Lockhart DJ, Dong H, Byrne MC, Follettie MT, Gallo MV, et al. 1996. Expression monitoring by hybridization to high-density oligonucleotide arrays. *Nat. Biotechnol.* 14:1675–80
47. Lyngberg O, Stemke D, Schottel J, Flickinger M. 1999. A single-use luciferase-based mercury biosensor using *Escherichia coli* HB101 immobilized in a latex copolymer film. *J. Ind. Microbiol. Biotechnol.* 23:668–76
48. McCue AM, Thompson W, Carmack CS, Ryan MP, Liu JS, et al. 2001. Phylogenetic footprinting of transcription factor binding sites in proteobacterial genomes. *Nucleic Acids Res.* 29:774–82
49. McGuire AM, Hughes JD, Church GM. 2000. Conservation of DNA regulatory motifs and discovery of new motifs in microbial genomes. *Genome Res.* 10:744–57
50. Menzel R. 1989. A microtiter plate-based system for the semiautomated growth and assay of bacterial cells for β-galactosidases activity. *Anal. Biochem.* 181:40–50
50a. Neidhardt FC, ed. 1996. Escherichia coli *and* Salmonella*: Cellular and Molecular Biology*. Washington, DC: ASM 2822 pp.
51. Neidhardt FC, Savageau MF. 1996. Regulation beyond the operon. See Ref. 50a, pp. 1310–24
52. Neil RJ. 1979. *Mutations affecting regulation of amino acid, nucleotide, and carbon metabolism in* Salmonella typhimurium. PhD thesis. Univ. California, Berkeley. 174 pp.
53. Newton MA, Kendziorski CM, Richmond CS, Blattner FR, Tsui KW. 2001. On differential variability of expression ratios: improving statistical inference about gene expression changes from microarray data. *J. Comp. Biol.* 8:37–52
54. Nichols BP, Shafiq O, Meiners V. 1998. Sequence analysis of Tn*10* insertion sites in a collection of *Escherichia coli* strains used for genetic mapping and strain construction. *J. Bacteriol.* 180:6408–11
55. O'Farrell PH. 1975. High-resolution two-dimensional electrophoresis of proteins. *J. Biol. Chem.* 250:4007–21
56. Oh M-K, Liao JC. 2000. DNA microarray detection of metabolic responses to protein overproduction in *Escherichia coli. Metab. Eng.* 2:201–9
57. Orser CS, Foong FCF, Capaldi SR, Nalezny J, MacKay W, et al. 1995. Use of prokaryotic stress promoters as indicators of the mechanisms of chemical toxicity. *In Vitro Toxicol.* 8:71–85
58. Pérez-Rueda E, Collado-Vides J. 2000. The repertoire of DNA-binding transcriptional regulators in *Escherichia coli* K-12. *Nucleic Acids Res.* 28:1838–47
59. Petersohn A, Brigulla M, Haas S, Hoheisel J, Volker U, Hecker M. 2001. Global analysis of the general stress response of *Bacillus subtilis. J. Bacteriol.* 183:5617–31
60. Picataggio SK, Templeton LJ, Smulski DR, LaRossa RA. 2002. Comprehensive transcript profiling of *Escherichia coli* using high-density DNA microarrays. In *Bacterial Pathogenesis, Part C, Volume 358 of Methods in Enzymology*, ed. VL Clark, PM Bavoil, pp. In press. San Diego, CA: Academic
61. Pomposiello PJ, Bennik MHJ, Demple B. 2001. Genome-wide transcriptional profiling of the *Escherichia coli* responses to superoxide stress and sodium salicylate. *J. Bacteriol.* 183:3890–902

62. Potsma PW, Lengeler JW, Jacobson GR. 1996. Phosphoenolpyruvate: carbohydrate phosphotransferase systems. See Ref. 50a, pp. 1149–74
63. Richmond CS, Glaser JD, Mau R, Jin H, Blattner FR. 1999. Genome-wide expression profiling in *Escherichia coli* K-12. *Nucleic Acids Res.* 27:3821–35
64. Riley M. 1998. Genes and proteins of *Escherichia coli* K-12 (GenProtEC). *Nucleic Acids Res.* 26:54
65. Rine J, Hansen W, Hardeman E, Davis RW. 1983. Targeted selection of recombinant clones through gene dosage effects. *Proc. Natl. Acad. Sci. USA* 80:6750–54
66. Robison K, McGuire AM, Church GM. 1998. A comprehensive library of DNA-binding site matrices for 55 proteins applied to the complete *Escherichia coli* K-12 genome. *J. Mol. Biol.* 284:241–54
67. Roth FP, Hughes JD, Estep PW, Church GM. 1998. Finding DNA regulatory motifs within unaligned noncoding sequences clustered by whole-genome mRNA quantitation. *Nat. Biotechnol.* 16:939–45
68. Rudd KE. 2000. EcoGene: a genome sequence database for *Escherichia coli* K-12. *Nucleic Acids Res.* 28:60–64
69. Salgado H, Santos-Zavaleta A, Gama-Castro S, Millán-Zárate D, Díaz-Peredo E, et al. 2001. RegulonDB (version 3.2): transcriptional regulation and operon organization in *Escherichia coli* K-12. *Nucleic Acids Res.* 29:72–74
70. Scholz O, Thiel A, Hillen W, Niederweis M. 2000. Quantitative analysis of gene expression with an improved green fluorescent protein. *Eur. J. Biochem.* 267:1565–70
71. Serres MH, Riley M. 2000. MultiFun, a multifunctional classification scheme for *Escherichia coli* K-12 gene products. *Microb. Comp. Genomics* 5:205–22
72. Sherlock G. 2000. Analysis of large-scale gene expression data. *Curr. Opin. Immunol.* 12:201–5
73. Simpson ML, Sayler GS, Applegate BM, Ripp S, Nivens DE, et al. 1998. Bioluminescent-bioreporter integrated circuits form novel whole-cell biosensors. *TIBTECH* 16:332–38
74. Smulski DR, Huang LL, McCluskey MP, Reeve MJG, Vollmer AC, et al. 2001. Combined, functional genomic-biochemical approach to intermediary metabolism: interaction of acivicin, a glutamine amidotransferase inhibitor, with *Escherichia coli* K-12. *J. Bacteriol.* 183: 3353–64
75. Tamayo P, Slonim D, Mesirov J, Zhu Q, Kitareewan S, et al. 1999. Interpreting patterns of gene expression with self-organizing maps: methods and application to hematopoietic differentiation. *Proc. Natl. Acad. Sci. USA* 96:2907–12
76. Tao H, Bausch C, Richmond C, Blattner FR, Conway T. 1999. Functional genomics: expression analysis of *Escherichia coli* growing on minimal and rich media. *J. Bacteriol.* 181:6425–40
77. Tao H, Gonzalez R, Martinez A, Rodriguez M, Ingram LO, et al. 2001. Engineering a homo-ethanol pathway in *Escherichia coli*: increased glycolytic flux and levels of expression of glycolytic genes during xylose fermentation. *J. Bacteriol.* 183:2979–88
78. Tavazoie S, Hughes JD, Campbell MJ, Cho RJ, Church GM. 1999. Systematic determination of genetic network architecture. *Nat. Biotechnol.* 22:281–85
79. Thieffry D, Salgado H, Huerta AM, Collado-Vides J. 1998. Prediction of transcriptional regulatory sites in the complete genome sequence of *Escherichia coli* K-12. *Bioinformatics* 14:391–400
80. Törönen P, Kolehmainen M, Wong G, Castrén E. 1999. Analysis of gene expression data using self-organizing maps. *FEBS Lett.* 451:142–46
81. Tseng GC, Oh M-K, Rohlin L, Liao JC, Wong WH. 2001. Issues in cDNA microarray analysis: quality filtering, channel normalization, modes of variations and

assessment of gene effects. *Nucleic Acids Res.* 29:2549–57

82. Tusher VG, Tibshirani R, Chu G. 2001. Significance analysis of microarrays applied to the ionizing radiation response. *Proc. Natl. Acad. Sci. USA* 98:5116–21
83. Van Dyk TK. 1998. Stress detection using bioluminescent reporters of the heat shock response. In *Methods in Molecular Biology: Bioluminescence Methods and Protocols*, ed. RA LaRossa, pp. 153–60. Totowa, NJ: Humana
84. Van Dyk TK, Ayers BL, Morgan RW, LaRossa RA. 1998. Constricted flux through the branched-chain amino acid biosynthetic enzyme acetolactate synthase triggers elevated expression of genes regulated by *rpoS* and internal acidification. *J. Bacteriol.* 180:785–92
85. Van Dyk TK, DeRose EJ, Gonye GE. 2001. LuxArray, a high-density, genome-wide transcription analysis of *Escherichia coli* using bioluminescent reporter strains. *J. Bacteriol.* 183:5496–505
86. Van Dyk TK, Majarian WR, Konstantinov KB, Young RM, Dhurjati PS, LaRossa RA. 1994. Rapid and sensitive pollutant detection by induction of heat shock gene-bioluminescence gene fusions. *Appl. Environ. Microbiol.* 60:1414–20
87. Van Dyk TK, Wei Y, Hanafey MK, Dolan M, Reeve MJG, et al. 2001. A genomic approach to gene fusion technology. *Proc. Natl. Acad. Sci. USA* 98:2555–60
88. VanBogelen RA, Abshire KZ, Pertsemlidis A, Clark RL, Neidhardt FC. 1996. Gene-protein database of *Escherichia coli* K-12, edition 6. See Ref. 50a, pp. 2067–117
89. Wanner BL, McSharry R. 1982. Phosphate-controlled gene expression in *Escherichia coli* using Mud1-directed *lacZ* fusions. *J. Mol. Biol.* 158:347–63
90. Wei Y, Lee J-M, LaRossa RA. 2001. The Global impact of *sdiA* amplification revealed by comprehensive gene expression profiling of *Escherichia coli*. *J. Bacteriol.* 183:2265–72
91. Wei Y, Lee J-M, Richmond C, Blattner FR, Rafalski JA, LaRossa RA. 2001. High-density miroarrray mediated gene expression profiling of *Escherichia coli*. *J. Bacteriol.* 183:545–56
92. Wendisch VF, Zimmer DP, Khodursky AB, Peter B, Cozzarelli N, Kustu S. 2001. Isolation of *Escherichia coli* mRNA and comparison of expression using mRNA and total RNA on DNA microarrays. *Anal. Biochem.* 290:205–13
93. Westfall PH, Young SS. 1993. *Resampling-based Multiple Testing: Examples and Methods for p-Value Adjustment*. New York: Wiley. 340 pp.
94. Wilson M, DeRisi J, Kristensen H-K, Imboden P, Rane S, et al. 1999. Exploring drug-induced alterations in gene expression in *Mycobacteriun tuberculosis* by microarray hybridization. *Proc. Natl. Acad. Sci. USA* 96:12833–38
95. Winkler ME. 1996. Biosynthesis of histidine. See Ref. 50a, pp. 485–505
96. Yang YH, Dudoit S, Luu P, Speed TP. 2001. Normalization for cDNA microarray data. http://www.stat.berkeley.edu/users/terry/zarray/Html/papersindex.html
97. Yoshida K, Kobayashi K, Miwa Y, Kang C, Matsunaga M, et al. 2001. Combined transcriptome and proteome analysis as a powerful approach to study genes under glucose repression in *Bacillus subtilis*. *Nucleic Acids Res.* 29:683–92
98. Zheng M, Wang X, Templeton LJ, Smulski DR, LaRossa RA, Storz G. 2001. DNA microarray-mediated transcriptional profiling of the *Escehrichia coli* response to hydrogen peroxide. *J. Bacteriol.* 183:4562–70
99. Zimmer DP, Soupene E, Lee HL, Wendisch VF, Khodursky AB, et al. 2000. Nitrogen regulatory protein C-controlled genes of *Escherichia coli*: scavenging as a defense against nitrogen limitation. *Proc. Natl. Acad. Sci. USA* 97:14674–79

Annu. Rev. Microbiol. 2002. 56:625–56
doi: 10.1146/annurev.micro.56.012302.161103

First published online as a Review in Advance on June 4, 2002

CONTROL OF CHROMOSOME REPLICATION IN *CAULOBACTER CRESCENTUS*

Gregory T. Marczynski[1] and Lucy Shapiro[2]
[1]*Department of Microbiology and Immunology, McGill University, Montreal, Quebec, Canada H3A 2B4; e-mail: gmarczynski@microimm.mcgill.ca*
[2]*Department of Developmental Biology, Beckman Center, Stanford University School of Medicine, Stanford, California 94305-5329; e-mail: shapiro@cmgm.stanford.edu*

Key Words replication origin, cell cycle, response regulator, CtrA, DnaA

■ **Abstract** *Caulobacter crescentus* permits detailed analysis of chromosome replication control during a developmental cell cycle. Its chromosome replication origin (*Cori*) may be prototypical of the large and diverse class of alpha-proteobacteria. *Cori* has features that both affiliate and distinguish it from the *Escherichia coli* chromosome replication origin. For example, requirements for DnaA protein and RNA transcription affiliate both origins. However, *Cori* is distinguished by several features, and especially by five binding sites for the CtrA response regulator protein. To selectively repress and limit chromosome replication, CtrA receives both protein degradation and protein phosphorylation signals. The signal mediators, proteases, response regulators, and kinases, as well as *Cori* DNA and the replisome, all show distinct patterns of temporal and spatial organization during cell cycle progression. Future studies should integrate our knowledge of biochemical activities at *Cori* with our emerging understanding of cytological dynamics in *C. crescentus* and other bacteria.

CONTENTS

INTRODUCTION

Dimorphic Chromosome Replication

Caulobacter crescentus presents a bacterial model for cell cycle processes typically attributed to eukaryotes and their apparently more sophisticated multicellular development. *C. crescentus* asymmetrically divides into two different cell types (Figure 1), the swarmer cell and the stalked cell, which differ in morphology and behavior (21, 37, 42, 106, 140). This dimorphic cell division is intrinsic to the cell cycle, and *C. crescentus* cell division always produces a nonreplicating swarmer cell and a replicating stalked cell. The swarmer cell differentiates into a stalked cell in order to replicate and divide (28). This nonreplicating swarmer cell versus replicating stalked cell dichotomy presents the basic question that we address throughout this review: What are the mechanisms that allow chromosome replication in the stalked cell but not in the swarmer cell?

Chromosome Replication During the *Caulobacter crescentus* Cell Cycle

The swarmer versus stalked cell decision for chromosome replication occurs in the context of other staged cell cycle events. The *C. crescentus* cell cycle is

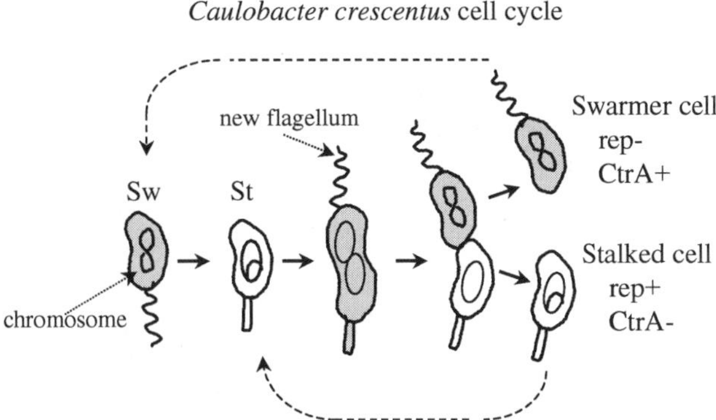

Figure 1 The dimorphic *C. crescentus* cell cycle. The condensed chromosome (*twisted circle*) in the swarmer (Sw) cell decondenses and initiates replication (*theta shape*) in the stalked cell (St). Subsequent asymmetric cell division produces a new polar flagellum and two chromosomes that are placed in either the nonreplicating (rep−) swarmer or the replicating (rep+) stalked cell. The shading indicates the presence of CtrA protein inside the flagellar cells.

operationally defined as beginning with the swimming swarmer cell because these cells are easily isolated and synchronized by a density gradient method (38). When released into fresh media, swarmer cells progress through the cell cycle and its intrinsic program of cell differentiation and asymmetric cell division. The swarmer cell is distinguished by its one polar flagellum and polar chemotaxis apparatus (117). In dilute nutrient environments, the swarmer cells swim to find preferred growth conditions. Swarmer-to-stalked cell differentiation can be delayed by nitrogen starvation (25); however, under standard laboratory conditions, when nutrients are in excess, the swarmer cells synchronously differentiate into stalked cells following a predictable delay that depends on prescribed growth conditions (34, 106).

The swarmer-to-stalked cell differentiation involves ejection of the polar flagellum and the targeted degradation of at least one flagellar protein in the basal body (59). In addition, the chemoreceptors at the swarmer cell pole are selectively degraded apparently because these motility systems are no longer needed in the sessile stalked cells (1).

During swarmer-to-stalked cell differentiation, the CtrA (cell cycle transcription regulator) protein is also selectively degraded (32). CtrA is a global transcription factor (108) and a selective chromosome replication inhibitor (109). Polar pili, which act as unique DNA phage receptors, also disappear at this time (120). A tubular extension of the cell wall and cytoplasm begins to grow at the base of the former flagellum (44, 126). This appendage, termed the "stalk," marks the new replicating cell type.

Chromosome replication occurs only in the stalked cell type that develops from these and many additional molecular events (28). Replication begins from one region of a circular chromosome and proceeds bidirectionally at a uniform rate

(31, 75). These considerations imply that replication control mechanisms act at or near the replication origin. Interestingly, the nucleoid appears to expand or decondense in the stalked cell type, as evidenced by a decreased sedimentation rate (42). We do not know what alters the nucleoid, but this phenomenon could reflect global controls on chromosome replication as well as transcription.

The stalked cell is committed to both chromosome replication and asymmetric cell division. The stalked cell essentially acts as a stem cell that releases swarmer cells. Asymmetric cell division requires both selective gene expression and protein localization (117). Most conspicuously, along with a new nonreplicating chromosome, a new flagellum and a new chemotaxis system must be synthesized and placed at the new swarmer pole every time the stalked cell divides. Over 40 genes are required for flagellar biogenesis, and these are transcribed in an ordered cascade (35, 140). Flagellar gene transcription and flagellar assembly are both staged processes that require checkpoint controls (140, 144). The earliest flagellar gene transcription initiates during the middle of chromosome replication. These genes require CtrA for transcription, and as illustrated in Figure 1, new CtrA protein is made at this time (32, 108). Subsequent stages of flagellar biogenesis, including flagellar protein synthesis, assembly, and activation (rotation), occur as chromosome replication ends and the chromosomes are positioned at opposite poles of the predivisional cell. CtrA protein accumulates as the cell divides. However, when the division-plane constricts and the two chromosomes acquire separate compartments, CtrA protein is degraded in the stalked cell compartment but not in the swarmer cell compartment (32). We present evidence that CtrA is a replication repressor that tracks the swarmer cell and therefore its nonreplicating chromosome (109, 118) (Figure 1).

Cell Cycle Phospho-Relay Signals

Staging and progression through the *C. crescentus* cell cycle are in part controlled by signals passed through phospho-relays composed of histidine kinase (HK) and aspartate response regulator (RR) modules (53, 97, 129). These two-component regulatory proteins dominate bacterial adaptation responses. For example, phospho-relays signal between chemoreceptors and the flagellar motor during chemotaxis. However, phospho-relays that are more typical alter transcription in response to changes in environmental conditions. Complex pathways are suggested by a recent microarray analysis of global transcription profiles that identified new HK and RR proteins whose transcript levels are temporally controlled during the *C. crescentus* cell cycle (71).

Genetic disruptions of the dimorphic *C. crescentus* cell cycle also identified HK and RR proteins that play critical roles during cell cycle progression (97, 108, 129). Cell division mutations identified the expected homologs to *E. coli fts* (filamentous temperature sensitive), cell wall synthesis, and topoisomerase genes (100). However, *C. crescentus* cell division mutations also identified the HK DivJ, the HK DivL (98), and the RR DivK (47, 48). Additional *C. crescentus* mutations that disrupt swarmer pole development, the so-called pleiotropic mutations affecting the flagellum, pili, and phage receptors, identified the HK PleC and the RR PleD (22, 122, 129).

The CtrA RR transcription factor is probably the end receiver of a phospho-relay composed of several HK and RR pairs (117, 141). This hypothesis, based on the *Bacillus subtilis* sporulation model, has been well reviewed (117, 141). The CckA HK is a likely cognate kinase for CtrA. CckA and CtrA were identified in a genetic screen for temperature-sensitive alleles, and one temperature-sensitive allele of *cckA* selectively blocks CtrA ~ P phosphorylation in vivo (55).

Experiments employing green fluorescent protein fusions to HK and RR proteins, and microscopic examination of synchronized cells, revealed that these two-component proteins have unique cell cycle programs of polar localization and delocalization (56, 134). For example, CckA is evenly dispersed throughout the cytoplasmic membrane in swarmer cells, but CckA aggregates preferentially toward the swarmer pole and then to the stalked pole in predivisional cells (55). If CckA kinase activity and aggregation are related, then this observation may help explain the repression of chromosome replication. Bipolar CckA aggregation may reflect CtrA ~ P phosphorylation and chromosome repression at both the swarmer and the stalked cell poles. PleC, DivJ, and DivK have similarly interesting but different localization programs (56, 134) that we discuss further in relation to polar localization of the replication origin. The chemoreceptors and CheA HK complexes are also polarly localized in *E. coli* (77) and *C. crescentus* (1), and this implies a conserved mechanism that has been adapted by *C. crescentus* for cell cycle control.

Additional novel uses and mechanisms for *C. crescentus* phospho-relays are suggested by several reports. For example, DivL represents a new tyrosine kinase class that resembles the typical HK proteins except that tryrosine has replaced the histidine as the phosphate acceptor within the H-box domain (142). In addition, typical bacterial HK and RR systems are nonessential, and their deletions selectively impair adaptation without impairing cell viability. In contrast, the best-studied *C. crescentus* HK and RR proteins, HK-like DivL (142), HK CckA (55), RR divK (48), and RR CtrA (108), are all essential for cell viability. Perhaps this reflects their many global regulatory roles, or an intimate and yet to be understood integration with essential cell cycle functions such as membrane synthesis, cell wall synthesis, and chromosome replication. The cell cycle cues that govern the kinase, phosphatase, as well as protein aggregation and polar localization activities of *C. crescentus* HK and RR proteins are as yet not understood. However, there is clearly a firm connection between the phospho-relays and the cell cycle control of the replication origin through the CtrA protein (55, 109).

The Scope and Perspective of this Review

We view replication origins as receivers and integrators of cell cycle signals, as well as DNA platforms for assembling the replication system, the "replisome."

We know a great deal about the *E. coli* chromosome replication origin (*oriC*) and how it serves as a platform for replisome assembly (8, 88), but we know comparatively little about the cell cycle signals and how they are processed (49). The exceptionally amenable *C. crescentus* cell cycle may provide this missing as well as novel information.

To address replication control, we first discuss replisome assembly at the best-studied replication origin, that of *E. coli*. We then compare the *C. crescentus* replication origin and draw attention to both common and unique features. We argue that signaling through the DnaA protein and RNA polymerase probably reflects common bacterial control mechanisms, whereas signaling through phospho-relay proteins (CtrA) and perhaps through CcrM DNA methylation may be unique to *C. crescentus* and related alpha-proteobacteria. We finally consider the whole chromosome and ask if additional global controls are needed to account for the precision of cell cycle–regulated chromosome replication observed in *C. crescentus*.

BACTERIAL REPLICATION ORIGINS

The *Escherichia coli oriC* Model for Chromosome Replication

Any model for *C. crescentus* chromosome replication, or that of any other organism, relies on precedents set by *E. coli*. Early *E. coli* studies provided our basic concepts of replication proteins acting through special DNA elements, the so-called replicon hypothesis. A search for temperature-sensitive mutants that are selective for chromosome replication identified mutations in the gene-encoding DnaA, which is a major replication control protein in *E. coli* (119), *B. subtilis* (91), *C. crescentus* (45, 149), and presumably most eubacteria (88).

The *E. coli* chromosome replication origin (*oriC*) was isolated as a small and unique piece of autonomously replicating DNA (145). This result was consistent with prior experiments demonstrating bidirectional replication starting from a single zone on the circular chromosome (86). The functional significance of *oriC* was further demonstrated through biochemical and physiological experiments. In vitro, *oriC* DNA uniquely supports replication in a crude *E. coli* protein preparation (8). Subsequently, fractionation and reconstitution of this crude replication system permitted the isolation of the individual proteins required to selectively initiate replication from *oriC* (63). Many of these proteins, such as DnaA, DnaB, DnaC, and DnaG, correspond to temperature-sensitive mutations that block in vivo chromosome replication (8). Additional physiological significance derives from in vivo *oriC*-plasmid experiments that demonstrated timed cell cycle DNA synthesis (73).

Escherichia coli oriC Transcription and Chromosome Replication

E. coli oriC replication and transcription are somehow interrelated (88). The linear DNA structure of *E. coli oriC* is diagrammed in Figure 2. The length of DNA

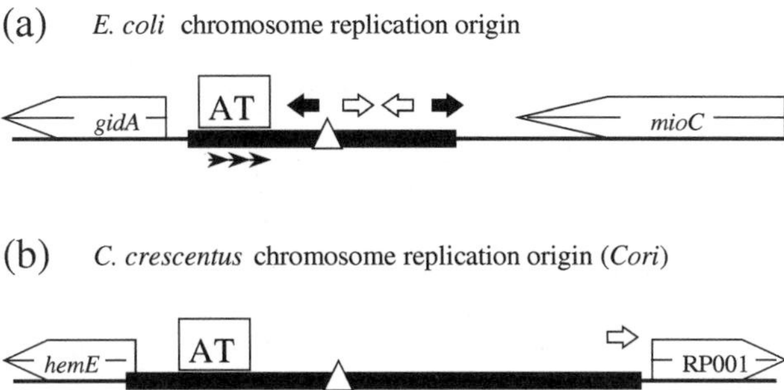

Figure 2 Similarities between (*a*) the *E. coli* chromosome replication origin and (*b*) the *C. crescentus* chromosome replication origin. The *thick lines* indicate the DNA required for autonomous replication in the respective bacteria. The *open triangle* marks the IHF-binding sites. The AT box marks approximately 40 bp of DNA that is exceptionally AT-rich. The *arrowheads* below the lines show 13-mer sequences, which in *E. coli* are unwinding elements but which in *C. crescentus* are only conspicuous sequence similarities. The *arrows* above the lines mark DnaA boxes. The two *solid arrows* in (*a*) mark the perfect *E. coli* consensus TTATCCACA sequence. The *open arrow* in (*b*) marks the best matching TGATCCACA DnaA box in the *C. crescentus* chromosome replication origin. The flanking coding sequences for *E. coli gidA* and *mioC*, as well as *C. crescentus hemE* and RP001, are marked by *pointed bars* to indicate polarity.

sequence required for autonomous replication is only about 250 bp. However, autonomous replication requires additional AT-rich sequences if sufficient leftward transcription from the *gidA* promoter does not occur (4). Transcription from *mioC* also promotes autonomous *oriC*-plasmid replication, but transcription from *gidA* and *mioC* promoters is not required. Promoter deletions failed to produce a detectable influence on chromosome replication when assayed in the context of the whole chromosome (13). A similar paradox is observed with the *C. crescentus* replication origin. *E. coli oriC* replication requires RNA polymerase transcription in vitro, but only when a crude *E. coli* replication protein preparation is employed (7). RNA polymerase does not prime DNA synthesis; this function belongs to DnaG primase (8). Instead, RNA polymerase may promote replication by twisting and unwinding DNA or by removing bound proteins that block replication. Transcription is also conspicuously involved in *C. crescentus* chromosome replication (83), and both similarities and differences are discussed below.

The *Escherichia coli* DnaA Protein: A Chromosome Replication Switch

DnaA is the major chromosome replication protein that binds four sites termed DnaA boxes inside *E. coli oriC* (39). A detailed understanding of DnaA and its multiple functions is perhaps fundamental to understanding how bacteria commit to chromosome replication. Biochemical analysis of the *oriC* in vitro replication system indicates that DnaA is an ATP/ADP (on/off) switch protein (69, 123). Only DnaA bound with ATP initiates *oriC* in vitro replication. DnaA bound with ADP is produced by an intrinsic ATPase activity, and this form of DnaA does not support in vitro replication from *oriC*. Interestingly, DnaA/ADP supports replication from DnaA-dependent plasmid replication origins (93). This suggests that plasmids use DnaA without taking advantage of its cell cycle switch. Although this point is controversial, apparently these plasmids replicate randomly throughout the cell cycle (73).

Potential Cell Cycle Signals for *Escherichia coli* DnaA

Fluid membranes and assembled replication forks may provide cell cycle signals for DnaA. A search for biochemical factors that restore DnaA/ATP from DnaA/ADP suggests that fluid membranes, particularly those rich in cardiolipin with unsaturated fatty acids, may flip the "on-switch" (26, 147). DnaA has a special amphipathic helix that mediates membrane interactions (46) and it is primarily a peripheral membrane protein (94). Likewise, a complementary search for biochemical factors that stimulate DnaA/ADP formation from DnaA/ATP suggests that the DnaN protein, the ring, or "sliding clamp" component of DNA polymerase III flips the "off-switch" (65). This biochemical property suggests that DnaN mediates a negative feedback that quenches *E. coli*'s capacity to initiate *oriC* replication. Both of these biochemically inferred on/off mechanisms are also supported by physiological observations.

The *Escherichia coli* DnaA Protein Interactions at *oriC*

DnaA protein binds *oriC* apparently throughout most of the *E. coli* cell cycle, except at the weakest binding DnaA box termed R3, which is transiently bound at the start of chromosome replication (23). Electron-microscopy suggests that a specific shape of clustered DnaA proteins is required to initiate *oriC* in vitro replication (27). Therefore, sufficient DnaA/*oriC* binding appears to complete a structure required for replication.

DnaA also cooperates with histone-like proteins at the start of chromosome replication. During in vitro *oriC* replication, DnaA protein bound to DnaA boxes makes secondary DNA contacts in the AT-rich side of *oriC* (17). This DNA bending requires cooperation with basic histone-like proteins, HU or IHF, and causes *oriC* to wrap around a cluster of 20 to 40 DnaA molecules (27). Subsequently the AT-rich side of *oriC* becomes unwound and accessible to replication proteins. In vitro footprints demonstrate that IHF protein has one, apparently strategic binding site inside *oriC* (107). In vivo footprints imply that IHF binds *oriC* uniquely at the start of chromosome replication (23, 24). Therefore, another cell cycle control point could involve IHF access to *oriC*, and a similar situation is suggested by the *C. crescentus* replication origin discussed below. The Fis histone-like protein also selectively binds *E. coli oriC* (40) and Fis represses in vitro *oriC* replication (137). Alternative IHF and Fis binding at *oriC* is another proposed on/off switch (88).

DnaA acts during several stages at the start of chromosome replication. During in vitro *oriC* replication, DnaA bound to ATP contacts DnaA boxes, and ATP creates an affinity for secondary DNA-binding sites (TGATCA) clustered in the AT-rich side of *oriC* (123). This property presumably contributes to the selective unwinding of the AT-rich "unwinding element," consisting of three tandem 13-mer motifs (17). This tandem arrangement apparently contributes to a zipper-like mechanism of unwinding the AT-element by DnaA/ATP. Further unwinding of *oriC* DNA requires the loading of two hexameric DnaB helicases (8). DnaA recruits DnaC to bring DnaB to *oriC*, and DnaA binds DnaB directly, perhaps during a loading and/or positioning process. A strategically loaded pair of hexameric DnaB helicases is presumed to commit *oriC* to bidirectional replication. Subsequent loading of primase and DNA polymerase III apparently does not require DnaA (88).

Escherichia coli DNA Methylation and Chromosome Replication

Although DNA methylation was implicated from the beginning of the *E. coli oriC* replication studies, solid support connecting DNA methylation with replication accumulated slowly and required a series of genetic as well as biochemical studies (16). *C. crescentus* research may be following a similar path (110). A comparison of six closely related enteric bacteria replication origins establish a "consensus" *oriC* with clearly conserved Dam methylation (GATC) sites as well as DnaA boxes (150). The precisely conserved spacing between these DNA sites reinforced their significance, which was only later supported by physiological studies. For example, *E. coli dam* mutants randomly initiate replication (too early as well as too late) and apparently lose a "pace-making" mechanism during exponential growth that keeps a constant period between successive initiations of chromosome replication (9, 15).

We now know that this *E. coli* pacemaking requires Dam methylation and the SeqA membrane protein. SeqA literally sequesters *oriC* from immediately starting a second round of replication (76). Fully methylated *E. coli* DNA (Dam methylated on both top and bottom strands) becomes hemi-methylated upon replication. SeqA shows low-affinity cooperative binding for fully methylated DNA and high-affinity (but less cooperative binding) for hemi-methylated DNA (121). These properties of SeqA provide plausible explanations for both positive and negative effects of Dam methylation on *oriC* replication (15). Conceivably, low-affinity cooperative binding delivers the fully methylated *oriC* to the membrane where DnaA is also present. Cooperative low-affinity SeqA binding is easily reversed and *oriC* DNA should remain readily available to DnaA whose replication activity is also stimulated by SeqA. Following replication, the high-affinity noncooperative SeqA binding for hemi-methylated *oriC* blocks DnaA and other replication proteins, apparently at several stages of the initiation process. Accurate resetting of a fully methylated *oriC* requires the precise expression of the Dam methyltransferase because too much or too little Dam causes premature replication initiation (15).

Extending these DNA methylation results to other bacteria is difficult because Dam DNA methylation and the *seqA* gene are restricted to bacteria that are closely related to enteric bacteria (52). The counterpart adenine DNA methyltransferase in *C. crescentus*, as well as most alpha-proteobacteria, is the CcrM (GANTC) methyltransferase (110). Although its direct role in chromosome replication has not been established, CcrM is an essential cell cycle protein, and its potential regulatory roles are considered below.

THE *CAULOBACTER CRESCENTUS* REPLICATION ORIGIN

The Criteria for Origin Identification and Isolation

To study the control of chromosome replication, the *C. crescentus* replication origin (*Cori*) was cloned (84). This project was aided by a whole-genome pulse field gel electrophoresis (PFGE) map (35, 36). Combined with DNA labeling experiments, PFGE analysis of synchronized cells permitted the tracking of replication forks across the entire ~4-Mb *C. crescentus* genome (31). This technique clearly demonstrated bidirectional replication, initiating from only one region of the chromosome. The hybridization of synchronized in vivo labeled DNA to specific chromosome fragments also established bidirectional chromosome replication and initiation from a narrow DNA zone (75).

Direct isolation of *Cori* from the earliest replicating *C. crescentus* DNA was accomplished by preparative-scale PFGE (84). This DNA was used to probe and isolate overlapping clones that spanned the earliest replicating region. A temperature-sensitive DNA synthesis mutant (99, 103) was also exploited to identify early replicating DNA (84). In synchronized swarmer cells, this mutation blocked chromosome replication at the start of S-phase (during the swarmer-to-stalked cell transition). Only a few small DNA bands were labeled under these conditions. These DNA bands were contained in the clones derived from the PFGE experiment and they comprised less than 30 kb of chromosome DNA. When these DNA bands were individually tested for autonomous replication, only one DNA fragment supported autonomous plasmid replication in *C. crescentus*, and it was designated *Cori*. Subsequent experiments that selected replicating DNA from pooled DNA clones always isolated DNA fragments that overlapped *Cori* (G.T. Marczynski, unpublished data). These results support the original hypothesis of one bidirectional replication origin.

Additional criteria argued that *Cori* is a cell cycle–regulated replication origin. Most importantly, *Cori*-directed plasmid replication occurred in stalked cells and not in swarmer cells (84). Therefore, the cloned *Cori* retains elements that direct cell type–specific DNA synthesis. Accordingly, this result provides access to this developmental problem by molecular techniques that are discussed further below.

This result also strongly argues against artifacts such as the cloning of an integrated phage or an integrated plasmid replication origin. Broad host-range plasmids, such as incP and incQ plasmids, replicate in both swarmer and stalked cells (82). Therefore, broad host-range plasmids have a cell cycle replication pattern that is distinct from *Cori*-plasmids.

Confirmation that *Cori* is the Actual Replication Origin

Replication origins can also be mapped by analyzing replication intermediates with the so-called Brewer and Fangman two-dimensional gel electrophoresis technique (127). This technique resolves nonlinear DNA fragments created by replication forks. Specific DNA regions are revealed by Southern-blot hybridization. Both the origin as well as the direction of the replication forks can be inferred by the DNA arc patterns visualized by this technique (127). Originally developed to confirm that yeast autonomously replicating sequences are in fact replication origins, this technique is rarely used and seldom successful in bacterial studies. Presumably, the relatively rapid replication fork movements in bacteria cause correspondingly few replication intermediates. Consequently, successful detection of replication forks requires precise replication synchrony. Germinating *B. subtilis* spores are sufficiently synchronized. Consistent with earlier radiolabel studies, the *B. subtilis* chromosome replication origin was mapped to the 3′ end of its *dnaA* gene by this two-dimensional gel technique (92). It is tacitly assumed that the same replication origin is used during germination and during vegetative growth.

C. crescentus swarmer cells are sufficiently synchronized to reveal replication intermediates during vegetative growth by the same two-dimensional gel technique (18). The *C. crescentus* chromosome replication origin was mapped to a 1.6-kb *Bam*HI fragment spanning the region between *hemE* and RP001. In wild-type swarmer cells, two-dimensional gels indicate that replication forks move outward from this 1.6-kb *Bam*HI fragment, whereas in a temperature-sensitive DNA synthesis mutant the replication forks stall and accumulate on both sides of this same fragment (18).

In summary, a consistent location for *Cori* was provided by independent methods. Most reported bacterial replication origins are not defined by rigorous criteria, and often only one piece of evidence such as autonomous replication defines a replication origin. The whole-genome pulsed field gel and in vivo radiolabel techniques localized *Cori* to ~30 kb. Inside this DNA region, the Brewer and Fangman two-dimensional gel technique localized *Cori* to ~1.6 kb. This DNA contains the unique autonomously replicating sequences. Deletions and the requirements for regulated autonomous plasmid replication further localize *Cori* to ~0.5 kb (83). These DNA sequences are the focus of replication studies discussed below.

A New Class of Replication Origins

The *C. crescentus* origin (*Cori*) is the prototype for a new class of chromosome replication origins within the alpha-proteobacteria (19). When *Cori* was isolated, its overlap with the *hemE* gene was without precedent (84). Most bacterial replication origins are linked to the *dnaA* gene, as for example *Pseudomonas putida* (146) and *B. subtilis* (90). The *E. coli* origin is linked to the *gidA* gene, but this position is viewed as a translocation away from an ancestral position at *dnaA*. Interestingly, *P. putida* and *P. aeruginosa* have two autonomously replicating sequences (146), one at *gidA* and one at *dnaA*. It is not clear whether either or both of these autonomously replicating sequences serve as the actual chromosome replication origins, although the one at *dnaA* was designated *oriC*.

In addition to their genetic contexts, chromosome replication origins have also been classified based on interspecies autonomous replication and their DNA sequence organization. For example, origins from *P. putida* and *P. aeruginosa* do not replicate in enteric bacteria species and vice versa. However, a distant enteric such as *Vibrio harveyi* has an origin that replicates in *E. coli* presumably because they share a similar consensus sequence (150). These enteric replication origins are relatively short (~250 bp), and they have precisely conserved DnaA boxes, Dam methylation sites, and an AT-rich sequence adjacent to the *gidA* promoter. The *P. putida* and *P. aeruginosa* origins are relatively poorly characterized. These origins are apparently larger and contain a cluster of DNA sequences matching the *E. coli* consensus DnaA box. The *B. subtilis* replication origin is even larger and "bipartite" because it requires two clusters of DnaA boxes separated by the *dnaA* gene (90). A critical spacing (~1 kb) and not the exact DNA sequences between the clusters of DnaA boxes are required for autonomous replication. Many of the *B. subtilis* DnaA boxes match the *E. coli* consesus DnaA box, and their respective DnaA proteins have similar affinities for their DnaA boxes (68). Related gram-positive bacteria, such as *Streptomyces* species, share this large bipartite DNA sequence organization (57, 90).

In respect to these criteria, the *C. crescentus* origin (*Cori*) is clearly different from these previously characterized origins. For example, the enteric replication origin of *Klebsiella pneumoniae* does not replicate in *C. crescentus* or in a related alpha-proteobacteria (101). Likewise, *E. coli oriC* does not replicate in *C. crescentus*, and *Cori* does not replicate in *E. coli* (G. T. Marczynski, unpublished data). In addition, except for one DnaA box (TGATCCACA), a clustering of *E. coli*-like DnaA boxes is conspicuously absent from *Cori*. Previous *Cori* diagrams were adjusted to fit the apparently more typical bacterial model of multiple DnaA boxes, and consequently weak sequence matches were counted as potential DnaA boxes (84, 85). Additional unique *Cori* sequence motifs are described below.

Potential Alpha-Proteobacteria Replication Origins

Conserved genetic organization may identify replication origins in related species. The *hemE*/RP001 genetic organization around the *C. crescentus* origin (*Cori*) is present in the sequenced genomes of several alpha-proteobacteria, further implying a new class of replication origins (19). This was first suggested by comparing the genomes of *C. crescentus* and *Rickettsia prowazekii*. The GC-strand bias technique, applied to the whole genome of *R. prowazekii*, suggested one replication origin between *hemE* and the first designated open reading frame (RP001) (2). Apparently, the leading DNA synthesis strand has an excess GC content, and therefore this bias switches strands at replication origins. However, this origin assignment should be considered tentative. The biochemical or biophysical basis of the GC-strand bias is debated (105), and there are few well-documented chromosome replication origins that serve as controls. Of these, the *E. coli* origin shows GC-strand bias but *B. subtilis* does not. Nonetheless, the retained (or conserved) *R. prowazekii hemE*/RP001 genetic organization is significant considering that this parasite has reduced its genome to only 1.1 Mb (2) compared to 4.0 Mb for free-living *C. crescentus* (95). In addition, *R. prowazekii* has retained the *dnaA* gene (128), and *E. coli*–like DnaA boxes are present inside this apparent replication origin (2).

Cori DNA Replication Assays

The *Cori* replication sequences were defined by deletions, insertions, and point mutations that were first assayed for autonomous *Cori*-plasmid replication (83, 84), and later in the whole chromosome context by homologous recombination and *Cori* DNA replacement (G.T. Marczynski, unpublished data). Mutations that permitted autonomous replication and did not perturb the cell cycle (swarmer to stalked cell) pattern of *Cori*-plasmid replication were considered outside the replication origin. Mutations that blocked autonomous replication or perturbed the cell cycle pattern of *Cori*-plasmid replication were considered inside the replication origin.

Some mutations that block or perturb *Cori*-plasmid replication but are tolerated on the chromosome were defined as in auxiliary elements. Mutations that were not tolerated on *Cori*-plasmids or in the whole chromosome context defined essential *Cori* DNA elements. The *Cori*-plasmid replication assay appears to be more sensitive because many mutations that abolish *Cori*-plasmid replication are tolerated when recombined into the chromosome. However, these chromosome mutations produce cells that have subtle phenotypes; a similar situation is also observed at the *E. coli* origin (3, 12). These mutant *C. crescentus* strains are described below for specific *Cori* elements.

Apparent Similarities Between the Replication Origins of *Caulobacter crescentus* and *Escherichia coli*

Figure 2 perhaps overemphasizes the apparent similarities between the *C. crescentus Cori* and *E. coli oriC*. For example, *Cori* also contains a 40-bp AT-rich sequence that is conspicuous and unusual in a CG-rich genome. *Cori* also contains a sequence, inside the AT-rich sequence, that resembles an *E. coli* 13-mer, suggesting that both are used by DnaA to unwind the DNA. Other potential bacterial origins of replication have been identified as clusters of *E. coli*-like DnaA boxes. Although *Cori* contains only one conspicuous DnaA box, with a 1-bp mismatch with the *E. coli* consensus, point mutations in this *Cori* DnaA box completely

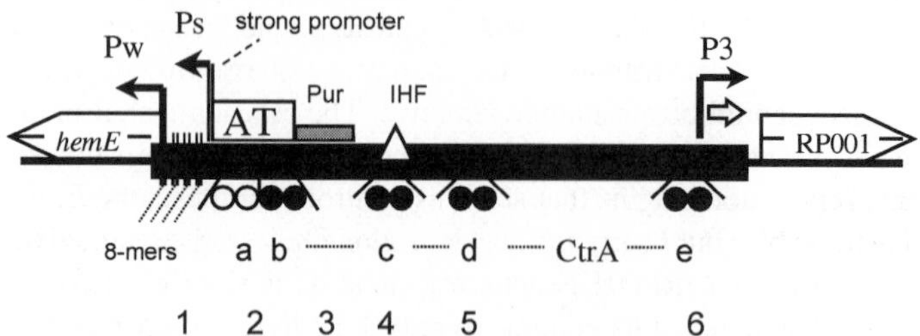

Figure 3 Extra sequence motifs superimposed over the *C. crescentus* replication origin presented in Figure 2*b*. The three *bent arrows*, labeled Pw, Ps, and P3, mark the RNA start sites for transcription promoters. The four *perpendicular bars* next to *hemE* mark the 8-mer motifs. The *shaded box* (Pur) next to the AT box marks the DNA sequences that are purine-rich in the top strand. The *filled double circles* mark the consensus CtrA-binding sites (b–e; TTAA-N7-TTAA), and the *open double circle* marks the weakest CtrA-binding site (a; TTAA-N7-CTAA). The numbers below these motifs divide *Cori* into six hypothetical elements, as detailed in the text.

abolish *Cori*-plasmid replication (84). The position of an IHF (integration host factor)–binding site also argues for an *E. coli*–like replication origin.

Cori DNA Replication Elements: Established and Hypothetical

A tentative organization of *Cori* DNA elements (#1–#6) is presented in Figure 3. These are composed of binding sites for proteins (DnaA, CtrA, IHF), transcription promoters, and enigmatic DNA sequences whose biochemical properties have not yet been identified. *Cori* may have more DNA elements and these represent a summary of current data. A replication model based on the *E. coli* model is presented below, but it is too simple and does not account for all six *Cori* elements. Note that the best-characterized eukaryotic chromosome replication origin, the yeast ARS1, is both smaller and conceptually simpler because ARS1 only has four distinct DNA elements (80).

Cori DNA element #1 is the weak transcription promoter (Pw) overlapped by four enigmatic 8-mer motifs (83). Element #1 is essential, and accordingly small deletions here are not tolerated on *Cori*-plasmids or on the chromosome. Pw matches the typical sigma-70 type promoter (79), and Pw is transcribed equally in both swarmer and stalked cells. This transcription pattern is consistent with *hemE* encoding a housekeeping heme biosynthetic gene, uroporphyrinogen decarboxylase, that is required by both cell types. Pw directs most of the HemE protein synthesis (83). This is surprising because Pw is weaker than the upstream "strong" promoter is. In addition, Pw mRNA starts on the A of the *hemE* AUG start codon and therefore lacks the typical ribosome-binding site. The *C. crescentus dnaX* gene also shares this unusual organization (136). By inference a ribosome-binding site downstream from the AUG must be used by *hemE* and *dnaX*.

Cori DNA element #2 is 40 bp of exceptionally AT-rich DNA containing the strong transcription promoter and two CtrA-binding sites ("a" and "b"). Unlike element #1, element #2 is auxiliary, and accordingly its deletion is tolerated on the chromosome but not on a *Cori*-plasmid (G.T. Marczynski, unpublished data). However, as argued below, element #2 is clearly a regulatory replication element.

The unusual *Cori* DNA element #3 is 40 bp that is exceptionally purine (A and G)-rich in the top strand. Like element #1, element #3 is essential for replication. All deletions or insertions in element #3 abolished *Cori*-plasmid replication and none were tolerated in the chromosome (G.T. Marczynski, unpublished data). There is no precedent for a purine-rich element in bacterial replication origins. However, purine-rich strands are essential elements of eukaryotic virus replication origins where leading strand DNA synthesis initiates (30).

The three additional *Cori* elements are composed of protein-binding sites. *Cori* DNA element #4 is a CtrA-binding site (site "c") overlapping an IHF-binding site. Like element #2, element #4 is auxiliary (R. Siam & G.T. Marczynski, unpublished data). However, although its deletion is tolerated on the chromosome, these *C. crescentus* strains are significantly impaired. *Cori* DNA element #5 is a CtrA-binding site ("d"), as is *Cori* DNA element #6 [CtrA-binding site ("e")], adjacent to a DnaA box (84).

Some but not all of these *Cori* DNA elements can be fitted into an *E. coli*–like (*oriC*) replication model (17, 88), superimposed by the swarmer cell–specific repressor, CtrA. In this view, DnaA protein binding (element #6) would contact and unwind the AT-rich region (element #2), and as in *E. coli*, IHF-induced DNA bending (element #4) might promote this reaction. Note that all these elements overlap CtrA-binding sites, and each step of the origin recognition and unwinding reaction can hypothetically be repressed by CtrA. The roles of the other elements are harder to reconcile with this model, whose merits and demerits are discussed further below.

Transcription and *Cori* Replication Control

Transcription and chromosome replication are inter-related (7, 88). In this case, RNA polymerase does not prime DNA synthesis, as in the ColE1-plasmid or mitochondrial mechanisms, but instead transcription promotes replication by altering the replication origin (5). For example, *E. coli gidA* transcription promotes *oriC*-plasmid replication apparently because RNA polymerase movement creates negative supercoiled DNA in its wake that promotes the unwinding of the AT-rich region (4, 6). In this view, *gidA* transcription aids DnaA protein during AT DNA unwinding, and transcription is not a primary replication signal. Also *gidA* transcription aids replication in the *oriC*-plasmid context, but not in the whole-chromosome context (13). When the *gidA* promoter was deleted from the *E. coli* chromosome, these strains were indistinguishable from control strains with respect to growth and replication, as assayed by fluorescent DNA cytometry.

The *Cori* Pw promoter, inside element #1, only suggests a function similar to the *gidA* promoter. Both promoters are similarly situated with respect to the AT region. However, unlike the *E. coli gidA* promoter, element #1 deletions are not tolerated in the chromosome. Interestingly, a DNA duplication inside element #1 that blocks *Cori*-plasmid replication, and also abolishes Pw transcription, is tolerated in the whole-chromosome context (G.T. Marczynski, unpublished data). This suggests that the four 8-mer motifs are the essential components of element #1. The 8-mers overlap the −10 and −35 sequences of Pw, but they do not appear to be Pw regulators because Pw is transcribed equally in all cell types (83). Instead it was suggested that the four 8-mers base pair to fold the RNA starting from Ps (83). The significance of this observation remains to be determined, but RNA structure is fundamental to plasmid replication and copy number control (29).

Current evidence favors an auxiliary role for Ps in *Cori* replication control. The *Cori* Ps promoter, inside element #2, also suggests a function similar to the *E. coli gidA* promoter because RNA polymerase could also conceivably unwind the AT-rich DNA in its wake. Unlike Pw, Ps is regulated by CtrA, and Ps is actively transcribed in stalked cells and not in swarmer cells (83). Although Ps is a stronger promoter and situated ~50 bp upstream from *hemE*, Ps transcription is uncoupled from HemE protein synthesis. Paradoxically, the "weak" Pw directs most HemE protein synthesis (83). The Ps RNA lacks a good ribosome-binding site and could fold through bases pairing between the enigmatic 8-mer motifs. On *Cori*-plasmids, Ps is a timing element that stimulates replication when the swarmer cells differentiate into stalked cells. A Ps mutation that affects CtrA binding, and that permits equal transcription in stalked and swarmer cells, also causes promiscuous *Cori*-plasmid replication (83). However, like the *gidA* promoter, a Ps deletion (34 out of the 40 AT-rich bp) is tolerated in the whole chromosome context. Nonetheless, these strains grow slower and are more sensitive to an oversupply of the CtrA repressor (G.T. Marczynski, unpublished data).

CtrA Cell Cycle Regulation and *Cori* Replication

CtrA connects chromosome replication with the *C. crescentus* cell cycle. CtrA is a global cell cycle regulator that, according to a recent gene microarray experiment, controls 26% of the transcripts that vary during the cell cycle (71). *C. crescentus* CtrA is a transcription activator for the early flagellar operons (108), a chemotaxis operon (62), the *pilA* gene in the pili biosynthetic biosynthetic cluster (120), and the *ccrM* DNA methyltransferase gene (108). CtrA also controls cell wall biosynthesis, since it is required for stalk growth (108) and cell division (20, 67, 116). CtrA activates the transcription of some cell division genes, but CtrA represses the transcription of the *ftsZ* gene (20), as it does the *Cori* Ps promoter in swarmer cells.

CtrA activity is itself regulated during the cell cycle by at least three basic mechanisms:

1. Transcription control: The *ctrA* gene is autoregulated such that CtrA activates one promoter while repressing another *ctrA* promoter (33).
2. Directed proteolysis: The ClpX,P ATP-dependent chaperone and protease pair degrade CtrA in stalked cells (32, 58).
3. Protein phosphorylation: Like all response regulators, CtrA is phosphorylated at a conserved aspartate (D51) within its receiver domain, and CtrA ~ P has a higher affinity for target DNA (32, 108, 111).

As would be expected for a response regulator protein, phosphorylated CtrA (CtrA ~ P) is the dominant cell cycle input (32, 111). Under laboratory conditions, extra CtrA regulatory mechanisms appear to be superfluous. Both regulated transcription and proteolysis can be circumvented by linking *ctrA* to a constitutive promoter and by adding a CtrA carboxy-terminal mutation that blocks its proteolysis. Under these conditions, cell cycle variations of CtrA ~ P phosphorylation, high CtrA ~ P in swarmer and low CtrA ~ P in stalked cells, alone appear to drive the dimorphic cell cycle, as assayed by cell morphology and chromosome DNA content (32). Accordingly, in the simplest view cell cycle variations in CtrA ~ P, and therefore variations in CtrA affinity for *Cori* CtrA-binding sites, could account for dimorphic replication control.

CtrA-Binding Sites, Binding Modes, and CtrA Binding in *Cori*

In vitro footprint experiments with pure CtrA protein demonstrated five *Cori* positions of selective CtrA binding centered over a consensus GTTAA-N7-TTAA motif (109). A similar but more divergent DNA motif was also observed over early flagellar operon promoters (108), and subsequently over the *ccrM* methyltransferase (125) and other promoters (53). Systematic point mutations, in the context of a flagellar promoter (*fliL*), demonstrated that the TTAA half-motif is the dominant recognition signal for CtrA to bind and activate transcription (104). Although additional recognition sequences were not apparent, the flanking DNA, as well as the N7 spacer DNA sequences, is also important for CtrA binding.

Cori CtrA-binding sites appear to be strategically organized (109). These five CtrA sites span the entire stretch of autonomously replicating DNA sequences (83), suggesting that CtrA regulates multiple replication steps. Strong binding TTAA-N7-TTAA consensus sites are rare, and the concentration of CtrA sites at *Cori* is unique. Considering the cell cycle variations of CtrA, *Cori* is probably the first place on the entire genome to bind CtrA, and *Cori* is likewise the last place to release CtrA.

The CtrA protein has at least three different modes of DNA binding that are physiologically significant, and at least two different modes are observed at *Cori* in vitro. In the first and most common binding mode, CtrA binds two perfect or imperfect TTAA half-sites (with N7 spacing) apparently as a CtrA dimer (111, 118), and binding affinity is increased (10 to 50-fold) upon CtrA ~ P phosphorylation.

The second DNA binding mode for CtrA is also cooperative and exhibited at adjacent *Cori* sites (a, b) (118). Here, relatively strong CtrA binding to site "b" (TTAA-N7-TTAA) aids otherwise weak CtrA binding to site "a" (TTAA-N7-CTAA). CtrA ~ P protein phosphorylation stimulates this second mode of binding as well as the first mode, but the protein-to-protein contacts may not be the same. A whole-genome scan for adjacent CtrA-binding sites failed to reveal a situation comparable to *Cori* Ps sites (a, b). This is surprising considering that *E. coli* response regulators NtrC (78, 143) and OmpR (14, 54) show cooperative DNA binding between two or more adjacent binding sites. Apparently, CtrA generally acts through only one binding site resembling TTAA-N7-TTAA, and although it is common in *E. coli*, the type of cooperative binding seen at Ps sites (a, b) is either rare or unique to *Cori*.

The third DNA binding mode for CtrA is independent binding to TTAA half-sites. Unlike the first and second binding modes, this binding is weak and the DNA binding affinity is not increased by CtrA ~ P phosphorylation. Presumably, one CtrA molecule binds one TTAA half-site without the benefit of additional protein to protein contacts. Although this binding mode is probably unphysiological in *Cori*, it is apparently used for *ctrA* autoregulation (R. Siam & G.T. Marczynski, unpublished data). The two promoters that drive *ctrA* transcription contain three TTAA motifs with atypical spacing (33). CtrA clearly binds and regulates transcription through these sites in vivo (33), but CtrA affinity is weak (Kd ~500 nM) and it is not increased by CtrA ~ P phosphorylation in vitro (R. Siam & G.T. Marczynski, unpublished data). This result implies that the effective in vivo CtrA concentration must exceed 500 nM in the dividing stalked cells and that all five *Cori* sites are bound to CtrA except at critical transitions of CtrA activity.

CtrA Homologs

CtrA homologs form a unique subfamily of response regulator proteins in alpha-proteobacteria (11). Like the *C. crescentus ctrA* gene, the *Sinorhizobium meliloti ctrA* homolog is an essential gene and therefore it too controls essential functions (11), and it appears to be autoregulated, at least as evidenced by *C. crescentus* CtrA protein binding to the promoter DNA (11). The *R. prowazekii* CtrA homolog, CzcR, is especially significant because it is only one of four response regulators retained in this degenerate genome (2). By comparison *C. crescentus* has at least 80 response regulators. (CzcR was originally named for a weaker homology to a different response regulator.) In vitro footprint experiments demonstrate that purified affinity-tagged CzcR protein selectively binds all five CtrA-binding sites in *Cori*. In vivo, *czcR* driven by the *C. crescentus* PxylX promoter (87) partially complements the growth defects of the *ctrA*401 temperature allele (A.K.C. Brassinga & G.T. Marczynski, unpublished data). This argues that CzcR and CtrA have similar DNA-binding specificity as well as interactions with RNA polymerase and perhaps additional replication proteins. Apparently, these CtrA homologs have similar global regulatory functions, including perhaps the control of chromosome replication.

However, comparisons between *C. crescentus* and related alpha-purple *Rhodobacter capsulatus* suggest an altered functional evolution for the CtrA protein. In *R. capsulatus*, the synthesis of the "gene transfer agent," a phage-like particle, is controlled by a CtrA homolog as well as the cognate CckA kinase homolog (70). The *R. capsulatus ctrA* as well as the *R. capsulatus cckA* genes are not essential and their disruptions only block production of the gene transfer agent. It is possible that *R. capsulatus* has a second essential *ctrA* homolog. However, two-component systems are extremely versatile, and it is possible that the highly conserved *ctrA* gene serves different functions in different bacteria.

A Potential Role for IHF in *Cori* Replication

IHF may have an auxiliary role for *C. crescentus Cori*, as discussed above for *E. coli oriC*. *C. crescentus* has an IHF protein that complements in vitro the lambda-phage integration reaction (41, 43). Therefore, both the *E. coli* and *C. crescentus* IHF proteins recognize the same target and bend the DNA through a comparable U-turn (112). In vitro footprint experiments demonstrate that *E. coli* IHF protein has one selective binding site in the middle of *Cori* overlapping CtrA-binding site "c" (element #3) (R. Siam & G.T. Marczynski, unpublished data). IHF binding antagonizes CtrA binding at site "c" and also weakens CtrA binding at nearby site "d." This antagonism is reasonable considering that the IHF-binding site consensus and the CtrA-binding site consensus sequences overlap. Presumably, IHF and CtrA compete for the same contacts on the DNA molecule.

A replication model is suggested by these in vitro binding and cell cycle studies. This model implies that CtrA regulates IHF *Cori* bending and that IHF selectively promotes *Cori* bending at the start of S-phase. *C. crescentus* IHF is preferentially transcribed just prior to S-phase (71), and IHF protein concentrations increase during the cell cycle as cells progress from swarmer-to-stalked and predivisional stages (41, 43). Therefore, removal of CtrA in stalked cells may allow IHF access to *Cori* at the start of S-phase, and subsequently, the return of CtrA may displace IHF and in a sense reset *Cori* in an unbent conformation. Nonetheless, element #3 is auxiliary, and despite growth defects, its deletion is tolerated in the *C. crescentus* chromosome (B. Yang & G.T. Marczynski, unpublished data). Apparently, this proposed bending and unbending mechanism adjusts or optimizes replication control.

Roles for DnaA in *Cori* Replication

C. crescentus DnaA protein is required to initiate chromosome replication (45). Although this was the anticipated result, it could not be assumed a priori that

C. crescentus DnaA has the same role at *Cori* as *E. coli* DnaA at *oriC*. As noted above, *Cori* is not a typical replication origin, since it has only one *E. coli*–like DnaA box (a 1-bp mismatch to the *E. coli* consensus) (Figure 2). Also, the DnaA protein in cyanobacteria is not required for chromosome replication or any obvious cell cycle function (113). Until now, only *E. coli* and *B. subtilis* (91) have provided physiological studies linking DnaA with chromosome replication.

To confirm that the *C. crescentus* DnaA protein is required to initiate chromosome replication, the *C. crescentus dnaA* gene was placed under the conditional control of the *C. crescentus* xylose promoter and ribosome-binding sites (45). Blocked *dnaA* expression prevented the initiation of chromosome replication, but not ongoing replication, as evidenced by DNA synthesis that lasted approximately one cell cycle period. These kinetics were expected if established forks completed replication without initiation. Consistent with this interpretation, single-cell DNA cytometry confirmed that these cells accumulated one complete chromosome.

Cell Cycle DNA Methylation and the *Caulobacter crescentus* DNA Methyltransferase, CcrM

C. crescentus DNA methylation is essential and required for cell viability as well as normal cell cycle progression (110, 124, 148). Although the mechanisms that use DNA methylation for cell cycle regulation have not been identified, circumstantial data suggest that DNA methylation is a part of a cell cycle control network. The original motivation for isolating the *C. crescentus* DNA methyltransferase, CcrM, was provided by the *E. coli* sequestration studies demonstrating SeqA protein binding to Dam (GATC) hemi-methylated DNA and the consequent repression of *E. coli oriC* replication. CcrM methylates the A within a similar recognition sequence (GANTC), but CcrM is not particularly related to Dam (148). The CcrM regulatory systems probably evolved separately from Dam. CcrM methylation and genetic homologs are restricted to the alpha-proteobacteria (64, 115, 138), whereas Dam methylation and the *seqA* gene are restricted to the gamma-proteobacteria (52).

CcrM is under strict cell cycle control so that its presence and activity are restricted to the end of DNA synthesis (the S-phase) but prior to cell division (124, 148). Interestingly, CtrA also activates *ccrM* transcription. CtrA protein binds the *ccrM* promoter, but relatively weakly, when compared to that of the early flagellar *fliQ* promoter (111, 125). This relatively weak CtrA binding presumably accounts for the relatively late transcription of *ccrM*. CcrM protein synthesis coincides with transcription of the *ccrM* gene, and in the absence of new transcription, CcrM protein is efficiently degraded by a homolog of the *E. coli* Lon protease (139). Bypassing either selective transcription or protein degradation impairs cell division and chromosome replication. Strains with a continuously transcribed *ccrM* gene (148), or strains with a *lon* genetic disruption (139), have a similarly altered morphology: elongated cells with extra chromosomes.

The *C. crescentus* chromosome alternates between fully methylated and hemi-methylated DNA during the cell cycle because of the restricted expression of *ccrM* (81, 148). In swarmer cells, the chromosome DNA is fully methylated (on both top and bottom strands), and later in stalked cells, semiconservative DNA replication produces hemi-methylated DNA that persists until the end of S-phase when CcrM returns and methylates the complementary strand.

This chromosome methylation cycle has both theoretical implications and practical applications. *E. coli oriC* also cycles between fully and hemi-methylated states, and although the mechanisms for maintaining hemi-methylated DNA differ, this parallel suggests a similar switch between active and inactive replication templates. However, unlike *E. coli* where SeqA binding maintains only short regions of hemi-methylated DNA, in *C. crescentus* most of the chromosome becomes hemi-methylated toward the end of S-phase (81). Also, the earliest replicated DNA remains hemi-methylated the longest. It is likely that DNA methylation has as yet unidentified consequences for DNA transactions, including replication, transcription, DNA repair, and chromosome organization.

Does asymmetric DNA methylation influence the genetic programs in the swarmer and stalked cell progeny? *C. crescentus* could produce two large and chemically distinct DNA molecules because during replication the top DNA strand remains methylated in one chromosome while the bottom DNA strand remains methylated in the other chromosome. In theory, this differential methylation could tag the chromosomes, but as yet there is no evidence for this idea. Two independent studies suggest that chromosome DNA strands are randomly partitioned to the progeny cells (82, 102). Either the top versus bottom methylated strands are picked at random, or homologous recombination scrambles the DNA prior to chromosome separation.

The pattern of CcrM DNA methylation also demonstrated that *C. crescentus* chromosome replication initiates exactly once per cell cycle (81). This concept is usually attributed to eukaryotes whose chromosomes are restricted, or "licensed" to replicate only once during S-phase. In *C. crescentus*, broad host-range plasmids produce unmethylated DNA (148) because they replicate promiscuously during the cell cycle (82). Unmethylated DNA is distinguishable from hemi-methylated and fully methylated DNA by selective restriction endonuclease digestion. However, when the same assays were directed to the chromosome DNA near *Cori*, no unmethylated DNA was detected above the background noise (81). Factoring the signal-to-noise ratio indicated that less than 1 per 1000 cells reinitiate chromosome replication before the end of S-phase.

Does *C. crescentus* have a mechanism similar to Dam/SeqA that represses replication? This is not known, but a CcrM-based repression mechanism is suggested by the autorepression of the *ccrM* promoter. Mutations in the CcrM methylation sites of the *ccrM* promoter increase transcription and allow transcription in the swarmer cells where this promoter is normally silent (125). However, if *C. crescentus* has DNA methylation-binding proteins, they are unlike SeqA because they must bind tightly the fully methylated DNA in the swarmer cells.

GLOBAL CHROMOSOME REPLICATION CONTROLS

Global Controls of Biosynthetic Reactions

A mechanism for adjusting general biosynthetic rates is reasonable considering the swarmer cell and stalked cell dichotomy. The nonreplicating swarmer cell only needs to compensate for turnover, while the replicating stalked cell must approximately duplicate its total macromolecules. Cell metabolism, as measured by bulk RNA and protein synthesis per cell, is three- to fourfold higher in the stalked cells than in the swarmer cells (82). As well as swarmer cell– and stalked cell–specific transcription, the expression of many biosynthetic genes that are required in both cell types must be increased in stalked cells. A general mechanism for adjusting rather than for turning on or off, the expression of many genes is analogous to chromosome dosage compensation in eukaryotes, but similar relatively subtle regulators have not yet been identified in *C. crescentus.*

A major portion of the lower biosynthetic rates in swarmer cells may be due to specific repressors. As well as the versatile CtrA protein, global transcription regulators that decrease swarmer cell transcription include the CIRCE element repressor of heat shock genes (10, 114) and the RRF (repression of replication factors) protein that binds upstream of DNA replication genes (66). The *dnaA*, *dnaX*, *dnaK*, and *gyrB* genes (135), as well as *parCE* (topoisomerase) (130, 131) and *holB*(*dnaC*) genes (99, 100), all have a unique consensus sequence between the −10 and −35 promoter elements that bind the proposed "RRF" repressor in swarmer cells (66). The RRF protein has yet to be purified, and its gene has yet to be identified and analyzed. However, promoter mutations that increase *dnaX* transcription in swarmer cells strongly suggest that RRF is responsible for increasing the transcription of many DNA replication genes at the swarmer-to-stalked cell transition and maintaining higher transcription in stalked cells.

Although it has not been directly confirmed that stalked cells have more replication proteins, this idea is certainly suggested by plasmid replication studies. Broad host-range incP and incQ plasmids require host-encoded replication proteins, and these plasmids have 5- to 10-fold higher replication rates in stalked cells versus swarmer cells (82). However, transcription rates do not necessarily correlate with protein abundance. For example, *holB*(*dnaC*) gene transcription increases in stalked cells (99), apparently regulated by RRF, but HolB protein, required for loading the clamp for DNA polymerase, is present in approximately equal abundance in swarmer and stalked cells (61). In this case, HolB activity and HolB protein localization vary during the cell cycle.

Polar Localization of the Replication Origin, Replication Proteins, and Kinases

Chromosome positioning is fundamental to the cell cycle control of chromosome replication. Essentially, an active chromosome must be placed into the proliferating stalked cell, and likewise an inactive chromosome must be placed in the nonproliferating swarmer cell. *E. coli* (96), *B. subtilis* (132, 133), and *C. crescentus* (60) replication origin regions are positioned at or near the cell poles, apparently as part of a mitosis-like process for efficient chromosome segregation. In *B. subtilis*, this positioning is not due to the replication origin, but instead a dispersed set of origin-proximal DNA motifs appears to position the chromosome (74). Likewise in *C. crescentus*, because *Cori*-plasmids are unstable, chromosome positioning sequences are probably not confined to *Cori*, and similar *Cori* proximal DNA sites were proposed to position the chromosome (89). Also, both *B. subtilis* (74) and *C. crescentus* (89) contain chromosome-encoded homologs of plasmid partition genes, *parA* and *parB*, that apparently act through these chromosome positioning sites. Despite much study of the plasmid-encoded ParA and ParB proteins, it is still not known how they act to evenly distribute DNA molecules between progeny cells (50, 51). Another observation that eludes a simple explanation is the rapid movement of the *C. crescentus Cori* region from the stalked pole to the future swarmer pole (60).

Cori proximal DNA is always associated with the cell poles (60), and accordingly the mechanisms that both activate and repress *Cori* replication must also be associated with the cell poles. Chromosome replication initiates at the stalked pole. Next, one replicated *Cori* region is placed at the future swarmer pole, while the other replicated *Cori* region remains at the stalked pole. As noted above, the top versus bottom methylated DNA strands are probably picked at random for this migration. Random choice supports the idea that the molecules at the destination, and not just the cargo molecules, dictate chromosome programs. Stalked pole–to–swarmer pole movement must be rapid because DNA fluorescent in situ hybridization experiments fail to detect intermediate *Cori* positions inside the cells (60). More importantly, because replication only initiates once per cell cycle, both poles must repress chromosome replication throughout most of the cell cycle. The stalked-pole *Cori* must rapidly switch from an active to an inactive (presumably repressed) state, while the swarmer-pole *Cori* must acquire and keep a repressed state. Subsequent *Cori* activation requires developmental and cell cycle cues that remain to be identified.

The polarly localized phospho-relay proteins, and especially the DivJ, PleC, and CckA protein kinases, may mediate the cues that activate and repress *Cori*. Fluorescent protein fusion experiments argue that each of these membrane proteins precisely aggregate and disaggregate during the cell cycle (117). DivJ shows stalked pole–specific aggregation (134), whereas PleC (134) and CckA (55) show swarmer pole–specific aggregation. Interestingly, some CckA also aggregates at the stalked pole, suggesting that CckA phosphorylation may both activate and repress *Cori*. Hypothetically, at the swarmer pole, aggregated kinase activity may focus phosphorylated CtrA ~ P directly onto *Cori* and maintain a hyper-repressed state. A similar process at the stalked pole may also prevent premature replication and help limit chromosome replication to only once per cell cycle. However these and similar ideas require additional experiments, and perhaps new methodologies, before they are substantiated.

The replication "factory model," as recently proposed for bacteria (72), also implies that chromosome positioning is fundamental to the initiation of chromosome replication. In *B. subtilis* (72) and later in *C. crescentus* (61), green fluorescence protein fusions with replication proteins demonstrated specific foci where replication proteins aggregate during chromosome replication. The replication factory model implies that a replication origin must be brought to a fixed replication factory where the replication proteins are gathered. This contrasts with the older view where replication proteins are brought to the replication origin. The recent time-lapse observations of *C. crescentus* replication factories accommodate both views.

C. crescentus has a single untethered replication factory that assembles at the stalked pole and migrates to the midline as replication progresses (61). As visualized by HolB, HolC, or DnaB fluorescent protein fusions, this replication factory assembles at the start of S-phase and disassembles when replication stops. In this experimental system, it is replication that dictates assembly, and not the other way around, because this replication factory disperses when replication is interrupted. HolB, HolC (both subunits of the clamp loading complex), and DnaB (helicase) are required for the elongation of DNA synthesis. It is possible that there exists a separate replication initiation factory that was not visualized and might be fixed or have a different movement. The stalked pole–to-midline movement is easily rationalized as the extrusion of chromosome DNA on either side of the replication factory (61). Therefore, there is no need to invoke a special tether. One interesting implication of the mobile replication factory is that it is first drawn toward and then away from the phospho-relay proteins, such as DivJ that focus at the poles. This provides yet another link, to be explored in the future, between polar development during the cell cycle and replication activity.

ACKNOWLEDGMENTS

Caulobacter research is supported by the National Institutes of Health Grant GM51426 to L. Shapiro, and by the Canadian Institutes of Health Research Grant MT-13453 to G.T. Marczynski.

The *Annual Review of Microbiology* is online at http://micro.annualreviews.org

LITERATURE CITED

1. Alley MRK, Maddock JR, Shapiro L. 1993. Requirement for the carboxy terminus of a bacterial chemoreceptor for its targeted proteolysis. *Science* 259: 1754–57
2. Andersson SGE, Zomorodipour A, Andersson JO, Sicheritz-Ponten T, Alsmark UCM, et al. 1998. The genome sequence of *Rickettsia prowazekii* and the origin of mitochondria. *Nature* 396:133–40
3. Asai T, Bates DB, Boye E, Kogoma T. 1998. Are minichromosomes valid model systems for DNA replication control? Lessons learned from *Escherichia coli*. *Mol. Microbiol.* 29:671–75
4. Asai T, Chen CP, Nagata T, Takanami

M, Imai M. 1992. Transcription in vivo within the replication origin of the *Escherichia coli* chromosome: a mechanism for activating initiation of replication. *Mol. Gen. Genet.* 231:169–78

5. Asai T, Kogoma T. 1994. D-loops and R-loops: alternative mechanisms for the initiation of chromosome replication in *Escherichia coli*. *J. Bacteriol.* 176:1807–12
6. Asai T, Takanami M, Imai M. 1990. The AT richness and *gidA* transcription determine the left border of the replication origin of the *E. coli* chromosome. *EMBO J.* 9:4065–72
7. Baker TA, Kornberg A. 1988. Transcription activation of initiation of replication from the *E. coli* chromosomal origin: an RNA-DNA hybrid near *oriC*. *Cell* 55: 113–23
8. Baker TA, Wickner SH. 1992. Genetics and enzymology of DNA replication in *Escherichia coli*. *Annu. Rev. Genet.* 26: 447–77
9. Bakker A, Smith DW. 1989. Methylation of GATC sites is required for precise timing between rounds of DNA replication in *Escherichia coli*. *J. Bacteriol.* 171: 5738–42
10. Baldini RL, Avedissian M, Gomes SL. 1998. The CIRCE element and its putative repressor control cell cycle expression of the *Caulobacter crescentus groESL* operon. *J. Bacteriol.* 180:1632–41
11. Barnett MJ, Hung DY, Reisenauer A, Shapiro L, Long SR. 2001. A homolog of the CtrA cell cycle regulator is present and essential in *Sinorhizobium meliloti*. *J. Bacteriol.* 183:3204–10
12. Bates DB, Asai T, Cao Y, Chambers MW, Cadwell GW, Boye E, Kogoma T. 1995. The DnaA box R4 in the minimal *oriC* is dispensable for initiation of *Escherichia coli* chromosome replication. *Nucleic Acids Res.* 23:3119–25
13. Bates DB, Boye E, Asai T, Kogoma T. 1997. The absence of an effect of *gidA* or *mioC* transcription on the initiation of chromosome replication in *Escherichia coli*. *Proc. Natl. Acad. Sci. USA* 94:12497–502
14. Bergstrom LC, Qin L, Harlocker SL, Egger LA, Inouye M. 1998. Hierarchical and co-operative binding of OmpR to a fusion construct containing the *ompC* and *ompF* upstream regulatory sequences of *Escherichia coli*. *Genes Cells* 3: 777–88
15. Boye E, Lobner-Olesen A. 1990. The role of *dam* methyltransferase in the control of DNA replication in *E. coli*. *Cell* 62:981–89
16. Boye E, Lobner-Olesen A, Skarstad K. 2000. Limiting DNA replication to once and only once. *EMBO Rep.* 1:479–83
17. Bramhill D, Kornberg A. 1988. A model for the initiation at origins of replication. *Cell* 54:915–18
18. Brassinga AKC, Marczynski GT. 2001. Replication intermediate analysis confirms that chromosome replication initiates from an unusual intergenic region in *Caulobacter crescentus*. *Nucleic Acids Res.* 29:4441–51
19. Brassinga AKC, Siam R, Marczynski GT. 2001. Conserved gene cluster at replication origins of the alpha-proteobacteria *Caulobacter crescentus* and *Rickettsia prowazekii*. *J. Bacteriol.* 183: 1824–29
20. Brun YV, Janakiraman R. 2000. The dimorphic life cycle of *Caulobacter* and stalked bacteria. See Ref. 117a, pp. 297–316
21. Brun YV, Marczynski GT, Shapiro L. 1994. The expression of asymmetry during *Caulobacter* cell differentiation. *Annu. Rev. Biochem.* 63:419–50
22. Burton GJ, Hecht GB, Newton A. 1997. Roles of the histidine kinase PleC in *Caulobacter crescentus* motility and chemotaxis. *J. Bacteriol.* 179:5849–53
23. Cassler M, Grimwade J, Leonard A. 1995. Cell cycle-specific changes in nucleoprotein complexes at a chromosome replication origin. *EMBO J.* 14:5833–41
24. Cassler M, Grimwade J, McGarry KC,

Mott RT, Leonard A. 1999. Drunken-cell footprints: Nuclease treatment of ethanol-permeablilized bacteria reveals an initiation-like nucleoprotein complex in stationary phase replication origins. *Nucleic Acids Res.* 27:4570–76

25. Chiaverotti TA, Parker G, Gallant J, Agabian N. 1981. Conditions that trigger guanosine teraphosphate accumulation in *Caulobacter crescentus. J. Bacteriol.* 145:1463–65
26. Crooke E, Castuma CE, Kornberg A. 1992. The chromosome origin of *Escherichia coli* stabilizes DnaA protein during rejuvenation by phospholipids. *J. Biol. Chem.* 267:16779–82
27. Crooke E, Thresher R, Hwang DS, Griffith J, Kornberg A. 1993. Replicatively active complexes of DnaA protein and the *Escherichia coli* chromosomal origin observed in the electron microscope. *J. Mol. Biol.* 233:16–24
28. Degnen ST, Newton A. 1972. Chromosome replication during development in *Caulobacter crescentus. J. Mol. Biol.* 64: 671–80
29. Del Solar G, Giraldo R, Ruiz-Echevarria MJ, Espinosa M, Diaz-Orejas R. 1998. Replication and control of circular bacterial plasmids. *Microbiol. Mol. Biol. Rev.* 62:434–64
30. DePamphilis ML. 1993. Eukaryotic DNA replication: anatomy of an origin. *Annu. Rev. Biochem.* 62:29–63
31. Dingwall A, Shapiro L. 1989. Rate, origin, and bidirectionality of *Caulobacter* chromosome replication as determined by pulsed-field gel electrophoresis. *Proc. Natl. Acad. Sci. USA* 86:119–23
32. Domian IJ, Quon KC, Shapiro L. 1997. Cell type-specific phosphorylation and proteolysis of a transcriptional regulator controls the G1 to S transition in a bacterial cell cycle. *Cell* 90:415–24
33. Domian IJ, Reisenauer A, Shapiro L. 1999. Feedback control of a master bacterial cell cycle regulator. *Proc. Natl. Acad. Sci. USA* 96:6648–53
34. Ely B. 1991. Genetics of *Caulobacter crescentus. Methods Enzymol.* 204:372–84
35. Ely B, Ely TW. 1989. Use of pulsed field gel electrophoresis to estimate the total number of genes required for motility in *Caulobacter crescentus. Genetics* 123: 649–54
36. Ely B, Ely TW, Gerardot CJ, Dingwall A. 1990. Circularity of the *Caulobacter crescentus* chromosome determined by pulsed-field gel electrophoresis. *J. Bacteriol.* 172:1262–66
37. Ely B, Shapiro L. 1984. Regulation of cell differentiation in *Caulobacter crescentus.* In *Microbial Development*, ed. R Losick, L Shapiro. pp. 1–26. Cold Spring Harbor Laboratory: Cold Spring Harbor
38. Evinger M, Agabian N. 1977. Envelope-associated nucleoid from *Caulobacter crescentus* stalked and swarmer cells. *J. Bacteriol.* 132:294–301
39. Fuller RS, Funnell BE, Kornberg A. 1984. The DnaA protein complex with the *E. coli* chromosomal replication origin (*oriC*) and other DNA sites. *Cell* 38: 889–900
40. Gille H, Egan JB, Roth A, Messer W. 1991. The FIS protein binds and bends the origin of chromosomal DNA replication, *oriC*, of *Escherichia coli. Nucleic Acids Res.* 19:4167–72
41. Gober JW, Shapiro L. 1990. Integration host factor is required for the activation of developmentally regulated genes in *Caulobacter. Genes Dev.* 4:1494–505
42. Gober JW, Shapiro L. 1991. Temporal and spatial regulation of developmentally expressed genes in *Caulobacter. BioEssays* 13:277–83
43. Gober JW, Shapiro L. 1992. A developmentally regulated *Caulobacter* flagellar promoter is activated by a 3′ enhancer and IHF binding elements. *Mol. Biol. Cell* 3:913–26
44. Gonin M, Quardokus EM, O'Donnol D, Maddock J, Brun YV. 2000. Regulation of stalk elongation by phosphate

in *Caulobacter crescentus*. *J. Bacteriol.* 182:337–47
45. Gorbatyuk B, Marczynski GT. 2001. Physiological consequences of blocked *Caulobacter crescentus* DnaA expression, an essential DNA replication gene. *Mol. Microbiol.* 40:485–97
46. Hase M, Yoshimi T, Ishikawa Y, Ohba A, Guo L, et al. 1998. Site-directed mutational analysis for the membrane binding of DnaA protein. *J. Biol. Chem.* 273: 28651–56
47. Hecht G, Newton A. 1995. Identification of a novel response regulator required for the swarmer-to-stalked-cell transition in *Caulobacter crescentus*. *J. Bacteriol.* 177:6223–29
48. Hecht GB, Lane T, Ohta N, Sommer JM, Newton A. 1995. An essential single domain response regulator required for normal cell division and differentiation in *Caulobacter crescentus*. *EMBO J.* 14:3915–24
49. Herrick J, Kohiyama M, Atlung T, Hansen FG. 1996. The initiation mess? *Mol. Microbiol.* 19:659–66
50. Hiraga S. 1992. Chromosome and plasmid partition in *Escherichia coli*. *Annu. Rev. Biochem.* 61:283–306
51. Hiraga S. 2000. Dynamic localization of bacterial and plasmid chromosomes. *Annu. Rev. Genet.* 34:21–59
52. Hiraga S, Ichinose C, Onogi T, Niki H, Yamazoe M. 2000. Bidirectional migration of SeqA-bound hemimethylated DNA clusters and pairing of *oriC* copies in *Escherichia coli*. *Genes Cells* 5: 327–41
53. Huang D, McAdams H, Shapiro L. 2000. Regulation of the *Caulobacter* cell cycle. See Ref. 117a, pp. 361–78
54. Huang K-J, Lan C-Y, Igo MM. 1997. Phosphorylation stimulates the cooperative DNA-binding properties of the transcription factor OmpR. *Proc. Natl. Acad. Sci. USA* 94:2828–32
55. Jacobs C, Domian IJ, Maddock JR, Shapiro L. 1999. Cell cycle-dependent polar localization of an essential bacterial histidine kinase that controls DNA replication and cell division. *Cell* 97: 111–20
56. Jacobs C, Hung D, Shapiro L. 2001. Dynamic localization of a cytoplasmic signal transduction response regulator controls morphogenesis during the *Caulobacter* cell cycle. *Proc. Natl. Acad. Sci. USA* 98:4095–100
57. Jakimowicz D, Majka J, Messer W, Speck C, Fernandez M, et al. 1998. Structural elements of the *Streptomyces oriC* region and their interactions with the DnaA protein. *Microbiology* 144:1281–90
58. Jenal U, Fuchs T. 1998. An essential protease involved in bacterial cell cycle control. *EMBO J.* 17:5658–69
59. Jenal U, Shapiro L. 1996. Cell cycle-controlled proteolysis of a flagellar moter protein that is asmmetrically distributed in the *Caulobacter* predivisional cell. *EMBO J.* 15:2393–406
60. Jensen RB, Shapiro L. 1999. The *Caulobacter crescentus smc* gene is required for cell cycle progression and chromosome segregation. *Proc. Natl. Acad. Sci. USA* 96:10661–66
61. Jensen RB, Wang SC, Shapiro L. 2001. A moving DNA replication factory in *Caulobacter crescentus*. *EMBO J.* 20: 4952–63
62. Jones SE, Ferguson NL, Alley MRK. 2001. New member of the *ctrA* regulon: The major chemotaxis operon in *Caulobacter* is CtrA dependent. *Microbiology* 147:949–58
63. Kaguni JM, Kornberg A. 1984. Replication initiation at the origin (*oriC*) of the *E. coli* chromosome reconstituted with purified enzymes. *Cell* 38:183–90
64. Kahng LS, Shapiro L. 2001. The CcrM DNA methyltransferase of *Agrobacterium tumefaciens* is essential, and its activity is cell cycle regulated. *J. Bacteriol.* 183:3065–75
65. Katayama T, Kubota T, Kurokawa K,

Crooke E, Sekimizu K. 1998. The initiator function of DnaA protein is negatively regulated by the sliding clamp of the *E. coli* chromosome replicase. *Cell* 94:61–71
66. Keiler K, Shapiro L. 2001. Conserved promoter motif is required for cell cycle timing of *dnaX* transcription in *Caulobacter*. *J. Bacteriol.* 183:4860–65
67. Kelly AJ, Sackett MJ, Din N, Quardokus E, Brun YV. 1998. Cell cycle-dependent transcriptional and proteolytic regulation of FtsZ in *Caulobacter*. *Genes Dev.* 15:880–93
68. Krause M, Ruckert B, Lurz R, Messer W. 1997. Complexes at the replication origin of *Bacillus subtilis* with homologous and heterologous DnaA protein. *J. Mol. Biol.* 274:365–80
69. Kubota T, Katayama T, Ito Y, Mizishima T, Sekimizu K. 1997. Conformational transition of DnaA protein by ATP: structural analysis of DnaA protein, the initiator of *Escherichia coli* chromosome replication. *Biochem. Biophys. Res. Commun.* 232:130–35
70. Lang AS, Beatty JT. 2000. Genetic analysis of a bacterial genetic exchange element: the gene transfer agent of *Rhodobacter capsulatus*. *Proc. Natl. Acad. Sci. USA* 97:859–64
71. Laub MT, McAdams HH, Fraser CM, Shapiro L. 2000. Global analysis of the genetic network controlling a bacterial cell cycle. *Science* 290:2144–48
72. Lemon KP, Grossman AD. 1998. Localization of bacterial DNA polymerase: evidence for a factory model of replication. *Science* 282:1516–19
73. Leonard AC, Helmstetter CE. 1988. Replication patterns of multiple plasmids coexisting in *Escherichia coli*. *J. Bacteriol.* 170:1380–83
74. Lin DC, Grossman AD. 1998. Identification and characterization of a bacterial chromosome partitioning site. *Cell* 92:675–85
75. Lott T, Ohta N, Newton A. 1987. Order of gene replication in *Caulobacter crescentus*: use of in vivo labeled genomic DNA as a probe. *Mol. Gen. Genet.* 210:543–50
76. Lu M, Campbell JL, Boye E, Kleckner N. 1994. SeqA: a negative modulator of replication initiation in *E. coli*. *Cell* 77:413–26
77. Maddock JR, Shapiro L. 1993. Polar localization of the chemoreceptor complex in the *Escherichia coli* cell. *Science* 259:1717–23
78. Magasanik B. 1996. Regulation of nitrogen utilization. See Ref. 93a, 1:1344–56
79. Malakooti J, Ely B. 1995. Principal sigma subunit of the *Caulobacter crescentus* RNA polymerase. *J. Bacteriol.* 177:6854–60
80. Marahrens Y, Stillman B. 1992. A yeast chromosome origin of DNA replication defined by multiple functional elements. *Science* 255:817–23
81. Marczynski GT. 1999. Chromosome methylation and the measurement of faithful, once and only once per cell cycle chromosome replication in *Caulobacter crescentus*. *J. Bacteriol.* 181:1984–93
82. Marczynski GT, Dingwall A, Shapiro L. 1990. Plasmid and chromosomal DNA replication and partitioning during the *Caulobacter crescentus* cell cycle. *J. Mol. Biol.* 212:709–22
83. Marczynski GT, Lentine K, Shapiro L. 1995. A developmentally regulated chromosomal origin of replication uses essential transcription elements. *Genes Dev.* 9:1543–57
84. Marczynski GT, Shapiro L. 1992. Cell-cycle control of a cloned chromosomal origin of replication from *Caulobacter crescentus*. *J. Mol. Biol.* 226:959–77
85. Marczynski GT, Shapiro L. 1993. Bacterial chromosome origins of replication. *Curr. Opin. Genet. Dev.* 3:775–82
86. Marsh RC, Worcel A. 1977. A DNA fragment containing the origin of replication of the *Escherichia coli* chromosome. *Proc. Natl. Acad. Sci. USA* 74:2720–24
87. Meisenzahl AC, Shapiro L, Jenal U.

1997. Isolation and characterization of a xylose-dependent promoter from *Caulobacter crescentus*. *J. Bacteriol.* 179:592–600
88. Messer W, Weigel C. 1996. Initiation of chromosome replication. See Ref. 93a, 1:1579–601
89. Mohl DA, Gober JW. 1997. Cell cycle-dependent polar localization of chromosome partitioning proteins in *Caulobacter crescentus*. *Cell* 88:675–84
90. Moriya S, Imai Y, Hassan AKM, Ogasawara N. 1999. Regulation of initiation of *Bacillus subtilis* chromosome replication. *Plasmid* 41:17–29
91. Moriya S, Kato K, Yoshikawa H, Ogasawara N. 1990. Isolation of a *dnaA* mutant of *Bacillus subtilis* defective in initiation of replication: amount of DnaA protein determines cells' initiation potential. *EMBO J.* 9:2905–10
92. Moriya S, Ogasawara N. 1996. Mapping of the replication origin of the *Bacillus subtilis* chromosome by the two-dimensional gel method. *Gene* 176:81–84
93. Mukhopadhyay G, Carr KM, Kaguni JM, Chattoraj DK. 1993. Open-complex formation by the host initiator, DnaA, at the origin of P1 plasmid replication. *EMBO J.* 12:4547–54
93a. Neidhardt FC, Curtiss R III, Ingraham JL, Lin ECC, Low KB, et al., eds. 1996. Escherichia coli *and* Salmonella, *Cellular and Molecular Biology*. Washington, DC: ASM
94. Newman G, Crooke E. 2000. DnaA, the initiator of *Escherichia coli* chromosome replication, is located at the cell membrane. *J. Bacteriol.* 182:2604–10
95. Nierman WC, Feldblyum TV, Laub MT, Paulsen IT, Nelson KE, et al. 2001. Complete genome sequence of *Caulobacter crescentus*. *Proc. Natl. Acad. Sci. USA* 98:4136–41
96. Niki H, Yamaichi Y, Hiraga S. 2000. Dynamic organization of chromosomal DNA in *Escherichia coli*. *Genes Dev.* 14:212–23
97. Ohta N, Grebe TW, Newton A. 2000. Signal transduction and cell cycle checkpoints in developmental regulation of *Caulobacter*. See Ref. 117a, pp. 341–59
98. Ohta N, Lane T, Ninfa EG, Sommer JM, Newton A. 1992. A histidine protein kinase homologue required for regulation of bacterial cell division and differentiation. *Proc. Natl. Acad. Sci. USA* 89:10297–301
99. Ohta N, Masurekar M, Newton A. 1990. Cloning and cell cycle-dependent expression of DNA replication gene *dnaC* from *Caulobacter crescentus*. *J. Bacteriol.* 172:7027–34
100. Ohta N, Ninfa AJ, Allaire A, Kulick L, Newton A. 1997. Identification, characterization, and chromosomal organization of cell division cycle genes in *Caulobacter crescentus*. *J. Bacteriol.* 179:2169–80
101. O'Neill EA, Bender RA. 1988. *Klebsiella pneumonia* origin of replication (*oriC*) is not active in *Caulobacter crescentus*, *Pseudomonas putida*, and *Rhodobacter spheroides*. *J. Bacteriol.* 170:3774–77
102. Osley MA, Newton A. 1974. Chromosome segregation and development in *Caulobacter crescentus*. *J. Mol. Biol.* 90:359–70
103. Osley MA, Newton A. 1977. Mutational analysis of developmental control in *Caulobacter crescentus*. *Proc. Natl. Acad. Sci. USA* 74:124–28
104. Ouimet MC, Marczynski GT. 2000. Analysis of a cell-cycle promoter bound by a response regulator. *J. Mol. Biol.* 302:761–75
105. Picardeau M, Lobry JR, Hinnebusch BJ. 1999. Physical mapping of an origin of bidirectional replication at the center of the *Borrelia burgdorferi* linear chromosome. *Mol. Microbiol.* 32:437–45
106. Poindexter JS. 1981. The *Caulobacters*: ubiquitous unusual bacteria. *Microbiol. Rev.* 45:123–79
107. Polaczek P. 1990. Bending of the origin

of replication of *E. coli* by binding of IHF at a specific site. *New Biol.* 2:265–71

108. Quon KC, Marczynski GT, Shapiro L. 1996. Cell cycle control by an essential bacterial two-component signal transduction protein. *Cell* 84:83–93
109. Quon KC, Yang B, Domian IJ, Shapiro L, Marczynski GT. 1998. Negative control of bacterial DNA replication by a cell cycle regulatory protein that binds at the chromosome origin. *Proc. Natl. Acad. Sci. USA* 95:120–25
110. Reisenauer A, Kahng LS, McCollum S, Shapiro L. 1999. Bacterial DNA methylation: a cell cycle regulator? *J. Bacteriol.* 181:5135–39
111. Reisenauer A, Quon K, Shapiro L. 1999. The CtrA response regulator mediates temporal control of gene expression during the *Caulobacter* cell cycle. *J. Bacteriol.* 181:2430–39
112. Rice PA, Yang SW, Mizuuchi K, Nash HA. 1996. Crystal structure of an IHF-DNA complex: a protein-induced DNA U-turn. *Cell* 87:1295–96
113. Richter S, Hagemann M, Messer W. 1998. Transcription analysis and mutation of a *dnaA*-like gene in *Synechocystis* sp. strain PCC6803. *J. Bacteriol.* 177:4245–51
114. Roberts RC, Toochinda C, Avedissian M, Baldini RL, Gomes SL, Shapiro L. 1996. Identification of a *Caulobacter crescentus* operon encoding *hrcA*, involved in negatively regulating heat-inducible transcription, the chaperone gene *grpE*. *J. Bacteriol.* 178:1829–41
115. Robertson GT, Reisenauer A, Wright R, Jensen RB, Jensen A, Shapiro L, Roop RM. 2000. The *Brucella abortus* CcrM DNA methyltransferase is essential for viability, and its overexpression attenuates intracellular replication in murine macrophage. *J. Bacteriol.* 182:3482–89
116. Sackett MJ, Kelly AJ, Brun YV. 1998. Ordered expression of *ftsQA* and *ftsZ* cell division genes during the *Caulobacter crescentus* cell cycle. *Mol. Microbiol.* 28:421–34
117. Shapiro L, Losick R. 2000. Dynamic spacial regulation in the bacterial cell. *Cell* 100:89–98

117a. Shimkets LJ, Brun YV, eds. 2000. *Prokaryotic Development*. Washington, DC: ASM

118. Siam R, Marczynski GT. 2000. Cell cycle regulator phosphorylation stimulates two distinct modes of binding at a chromosome replication origin. *EMBO J.* 19:1138–47
119. Skarstad K, Boye E. 1994. The initiator protein DnaA: evolution, properties and function. *Biochim. Biophys. Acta* 1217:111–30
120. Skerker JM, Shapiro L. 2000. Identification and cell cycle control of a novel pilus system in *Caulobacter crescentus*. *EMBO J.* 19:3223–34
121. Slater S, Wold S, Lu M, Boye E, Skarstad K, Kleckner N. 1995. *E. coli* SeqA protein binds *oriC* in two different methyl-modulated reactions appropriate to its roles in DNA replication initiation and origin sequestration. *Cell* 82:927–36
122. Sommer JM, Newton A. 1989. Turning off flagellum rotation requires the pleiotropic gene *pleD*: *pleA*, *pleC*, and *pleD* define two morphogenic pathways in *Caulobacter crescentus*. *J. Bacteriol.* 171:392–401
123. Speck C, Weigel C, Messer W. 1999. ATP- and ADP-DnaA protein, a molecular switch in gene regulation. *EMBO J.* 18:6169–76
124. Stephens CM, Reisenauer A, Wright R, Shapiro L. 1996. A cell cycle-regulated bacterial DNA methyltransferase is essential for viability. *Proc. Natl. Acad. Sci. USA* 93:1210–14
125. Stephens CM, Zweiger G, Shapiro L. 1995. Coordinate cell cycle control of a *Caulobacter* DNA methyltransferase and the flagellar genetic hierarchy. *J. Bacteriol.* 177:1662–69

126. Terrana B, Newton A. 1976. Requirement of a cell division step for stalk formation. *J. Bacteriol.* 128:456–62
127. van Brabant AJ, Hunt SY, Fangman WL, Brewer BJ. 1998. Identifying sites of replication initiation in yeast chromosomes: looking for origins in all the right places. *Electrophoresis* 19:1239–46
128. Waite RT, Shaw EI, Winkler HH, Wood DO. 1998. Isolation and characterization of the *dnaA* gene of *Rickettsia prowazekii*. *Acta Virol.* 42:95–101
129. Wang SP, Sharma PL, Shoenlein PV, Ely B. 1993. A histidine protein kinase is involved in polar organelle development in *Caulobacter crescentus*. *Proc. Natl. Acad. Sci. USA* 90:630–34
130. Ward D, Newton A. 1997. Requirement of topoisomerase IV *parC* and *parE* genes for cell cycle progression and developmental regulation in *Caulobacter crescentus*. *Mol. Microbiol.* 26:897–910
131. Ward DV, Newton A. 1999. Cell cycle expression and transcriptional regulation of DNA topoisomerase IV genes in *Caulobacter*. *J. Bacteriol.* 181:3321–29
132. Webb CD, Graumann PL, Kahana JA, Teleman AA, Silver PA, Losick R. 1998. Use of time-lapse microscopy to visualize rapid movement of the replication origin region of the chromosome during the cell cycle in *Bacillus subtilis*. *Mol. Microbiol.* 28:883–92
133. Webb CD, Teleman A, Gordon S, Straight A, Belmont A, et al. 1997. Bipolar localization of the replication origin regions of chromosomes in vegetative and sporulating cells of *B. subtilis*. *Cell* 88:667–74
134. Wheeler RT, Shapiro L. 1999. Differential localization of two histidine kinases controlling bacterial cell differentiation. *Mol. Cell* 4:683–94
135. Winzeler E, Shapiro L. 1996. A novel promoter motif for *Caulobacter* cell cycle controlled DNA replication genes. *J. Mol. Biol.* 264:412–25
136. Winzeler E, Shapiro L. 1997. Translation of the leaderless *Caulobacter dnaX* mRNA. *J. Bacteriol.* 179:3981–88
137. Wold S, Crooke E, Skarstad K. 1996. The *Escherichia coli* Fis protein prevents initiation of DNA replication from *oriC* in vitro. *Nucleic Acids Res.* 24:3527–32
138. Wright R, Stephens C, Shapiro L. 1997. The CcrM DNA methyltransferase is widespread in the alpha subdivision of proteobacteria, and its essential functions are conserved in *Rhizobium meliloti* and *Caulobacter crescentus*. *J. Bacteriol.* 179:5869–77
139. Wright R, Stephens C, Zweiger G, Shapiro L, Alley MRK. 1996. *Caulobacter* Lon protease has a critical role in cell-cycle control of DNA methylation. *Genes Dev.* 10:1532–42
140. Wu J, Newton A. 1997. Regulation of *Caulobacter* flagellar gene hierarchy: not just for motility. *Mol. Microbiol.* 24:233–39
141. Wu J, Ohta N, Newton A. 1998. An essential, multicomponent signal transduction pathway required for cell cycle regulation in *Caulobacter*. *Proc. Natl. Acad. Sci. USA* 95:1443–48
142. Wu J, Ohta N, Zhao JL, Newton A. 1999. A novel bacterial tyrosine kinase essential for cell division and differentiation. *Proc. Natl. Acad. Sci. USA* 96:13068–73
143. Wyman C, Rombel I, North AK, Bustamante C, Kustu S. 1997. Unusual oligomerization required for activity of NtrC, a bacterial enhancer-binding protein. *Science.* 275:1658–61
144. Xu H, Dingwall A, Shapiro L. 1989. Negative transcriptional regulation in the *Caulobacter* flagellar hierarchy. *Proc. Natl. Acad. Sci. USA* 86:6656–60
145. Yasuda S, Hirota Y. 1977. Cloning and mapping of the replication origin of *Escherichia coli*. *Proc. Natl. Acad. Sci. USA* 74:5458–62
146. Yee TW, Smith DW. 1990. *Pseudomonas* chromosome replication origins: a

bacterial class distict from *Escherichia coli*-type origins. *Proc. Natl. Acad. Sci. USA* 87:1278–82

147. Zheng W, Li Z, Skarstad K, Crooke E. 2001. Mutation in the DnaA protein suppress the growth arrest of acidic phospholipid-deficient *Escherichia coli* cells. *EMBO J.* 20:1164–72
148. Zweiger G, Marczynski GT, Shapiro L. 1994. A *Caulobacter* DNA methyltransferase that functions only in the predivisional cell. *J. Mol. Biol.* 235:472–85
149. Zweiger G, Shapiro L. 1994. Expression of *Caulobacter dnaA* as a function of the cell cycle. *J. Bacteriol.* 176:401–8
150. Zyskind JW, Cleary JM, Brusilow WSA, Harding NE, Smith DW. 1983. Chromosome replication origin from the marine bacterium *Vibrio harveyi* functions in *Escherichia coli*: *oriC* consensus sequence. *Proc. Natl. Acad. Sci. USA* 80:1164–68

Annu. Rev. Microbiol. 2002. 56:657–75
doi: 10.1146/annurev.micro.56.012302.160806

First published online as a Review in Advance on July 18, 2002

THE PREVALENCE AND MECHANISMS OF VANCOMYCIN RESISTANCE IN *STAPHYLOCOCCUS AUREUS*

Timothy R. Walsh[1] and Robin A. Howe[2]
[1]Department of Pathology and Microbiology, School of Medical Sciences, University of Bristol, Bristol BS8 1TD, United Kingdom; e-mail: t.r.walsh@bristol.ac.uk
[2]Bristol Centre for Antimicrobial Research and Evaluation, Southmead Hospital, Bristol BS10 5NB, United Kingdom; e-mail: robin.howe@north-bristol.swest.nhs.uk

Key Words VISA, hVISA, glycopeptide resistance

■ **Abstract** The emergence of *Staphylococcus aureus* resistant to vancomycin has caused considerable concern. Such strains are currently rare, although they have been isolated from many areas of the world. Considerable controversy surrounds strains of *S. aureus* displaying heterogeneous resistance to vancomycin regarding their definition and methods for detection. This has led to considerable variance in estimates of prevalence (0–1.3%–20% in Japan) and has hindered efforts to define the clinical relevance of these strains. The mechanism of resistance involves a complex reorganization of cell wall metabolism, leading to a grossly thickened cell wall with reduced peptidoglycan cross-linking. There may be many different ways in which strains achieve this endpoint. Current knowledge and theories are summarized.

CONTENTS

0066-4227/02/1013-0657$14.00

INTRODUCTION

Staphylococcus aureus is an important cause of serious infections in both hospitals and the community (43). Unfortunately this pathogen has been particularly efficient at developing resistance to antimicrobials. Methicillin-resistant strains (MRSA) in particular have spread across the world and are frequently resistant to many other agents (7). In the United States, the proportion of nosocomial intensive therapy unit *S. aureus* infections that are due to MRSA is now greater than 50% (National Nosocomial Infections Surveillance Data at http://www.cdc.gov/ncidod/hip/Aresist/aresist.htm). Until recently vancomycin was believed to have retained activity against all strains of *S. aureus*; therefore the spread of MRSA has led to increased vancomycin usage and hence increased selective pressure for the development of resistance.

The first clinical strain of *S. aureus* resistant to vancomycin (Mu50) was reported in 1997 from Japan (32), and although such strains appear rare, there have subsequently been reports of similar strains from most parts of the world. A resistance phenotype of hetero-resistance to vancomycin (strains sensitive by conventional testing but possessing a subpopulation of cells able to grow in the presence of higher vancomycin concentrations) has also been described in Japan (31), and similar strains have been reported from many countries, although there is controversy surrounding the definition of this type of resistance and the methods employed to detect it.

The mechanism of resistance to vancomycin in *S. aureus* appears novel. All strains tested so far are negative for the *van* genes that confer glycopeptide resistance in enterococci. Although there are some phenotypic differences between vancomycin-resistant strains from different geographical areas, a common characteristic of these strains appears to be a grossly thickened cell wall. Alterations in cell wall metabolism have been described in both clinically and laboratory-derived strains. We summarize current knowledge regarding the mechanism of resistance and present a theory for the mechanism underlying the expression of vancomycin resistance in *S. aureus*.

HISTORY

Vancomycin is a glycopeptide antibiotic that was introduced in the mid-1950s. At that time it had good activity against *S. aureus*, although, using the methods of the time, 16 of 31 strains tested had a minimal inhibitory concentration (MIC) of 8 or 16 mg/L (21). Since its introduction, clinical experience has suggested that the development of resistance to vancomycin by *S. aureus* was difficult, although a number of workers have selected strains in vitro with low-level resistance (9, 19, 24, 69). Teicoplanin is another glycopeptide antibiotic that is currently not licensed in the United States. A strain of *S. aureus* that developed resistance to teicoplanin during therapy for infective endocarditis has been reported, but the strain remained susceptible to vancomycin (39).

Nevertheless, the development of vancomycin resistance in clinical strains seemed likely to occur given the increased use of vancomycin and particularly after the development of vancomycin resistance in enterococci. The demonstration that the *van* resistance genes could be transferred from enterococci to *S. aureus* in vitro and expressed, conferring vancomycin resistance, caused concern (48). However, when vancomycin resistance was reported in a clinical strain of *S. aureus* in 1997, it was low-level resistance mediated by a novel mechanism (32).

TERMINOLOGY, DEFINITIONS, AND METHODS OF DETECTION

The area of vancomycin resistance in *S. aureus* is beset by problems regarding terminology and definitions that are largely related to methodological problems in their detection. Antimicrobial resistance or susceptibility is defined by interpretive criteria that are determined for MIC tests performed according to standardized methods. Many national institutions publish MIC interpretative criteria that give breakpoints above which an organism would be defined as resistant and below which an organism may be intermediately or fully susceptible. Unfortunately these criteria differ between countries (Table 1). Thus for example, in the United States and France there is an intermediate category for vancomycin susceptibility but not in the United Kingdom or Sweden. This has led to the imprecise use of the acronyms VISA (vancomycin intermediate *S. aureus*) and VRSA (vancomycin resistant *S. aureus*). We propose that the term VISA defined as *S. aureus* with a MIC of 8–16 mg/L when performed by a recognized method should be used in order to reduce confusion. The term VRSA should be reserved for strains of *S. aureus* with a MIC of ≥32 mg/L. The term GISA (also GRSA), meaning glycopeptide intermediate *S. aureus*, has been proposed to encompass strains that are resistant to

TABLE 1 Interpretative criteria (mg/L) for MIC testing of *S. aureus*

	Vancomycin			Teicoplanin		
Organization	**Susceptible**	**Intermediate**	**Resistant**	**Susceptible**	**Intermediate**	**Resistant**
BSAC	≤4	***	≥8	≤4	***	≥8
NCCLS	≤4	8–16	≥32	≤8	16	≥32
SFM	≤4	8–16	≥32	≤4	8–16	≥32
SRGA	≤4	***	≥8	≤4	***	≥8

BSAC, British Society for Antimicrobial Chemotherapy

NCCLS, National Committee for Clinical Laboratory Standards

SFM, Société Française de Microbiologie

SRGA, Swedish Reference Group for Antibiotics

***, no intermediate category defined

teicoplanin. Although this is an accurate description, not all teicoplanin-resistant strains are resistant to vancomycin, and the international differences in teicoplanin breakpoints are likely to lead to further confusion.

Detection of VISA requires MIC measurement; disc diffusion tests have been shown to be unreliable (64). NCCLS recommends incubation for a full 24 h to give a clearer endpoint. Some confirmed strains of VISA such as the New York isolate may sometimes give susceptible MICs on broth microdilution testing; therefore CDC has proposed three criteria that should be satisfied to identify VISA stains: broth microdilution MIC of 8–16 mg/L, Etest MIC of ≥6 mg/L, and growth on commercial BHI (brain heart infusion) agar screen plates containing 6 mg/L vancomycin at 24 h (63). Enriched media such as BHI and high-bacterial inocula should not be used for MIC testing, as these allow growth at higher vancomycin concentrations and therefore may give inaccurate results (35).

Hetero-resistance to vancomycin was first described by Hiramatsu et al. (31), who found that some strains sensitive to vancomycin by MIC estimation possessed subpopulations of cells that could grow in the presence of ≥4 mg/L vancomycin on an enriched medium, brain heart infusion agar (BHIA). The population heterogeneity in susceptibility can be seen if a full population analysis profile (PAP) is performed on such hetero-vancomycin intermediate *S. aureus* (hVISA). In this procedure an overnight broth is plated onto agar plates containing increasing concentrations of vancomycin. The colony counts on each concentration of vancomycin give a measure of the resistance of subpopulations of the test strain as shown in Figure 1. Mu3 (the archetypal hVISA) shows heterogeneity in vancomycin susceptibility not seen in MSSA and MRSA strains. VISA strains Mu50, MC963 (Michigan strain), and NJ992 (New Jersey strain) also display population heterogeneity, although at higher concentrations than Mu3 (Figure 1*b*).

The definition of hVISA strains is central to the study of their prevalence and clinical importance. The original definition as proposed by Hiramatsu et al. states that "Hetero-VRSA status was considered definite if the strain produced a subclone(s) with a vancomycin MIC of 8 mg/L or above upon selection with vancomycin, with the stability of the strain persisting beyond 9 days in a drug-free medium." The requirement for a strain to give rise to stably resistant subclones is a problem, as Mu3 does not fulfill this criterion (35). However, we believe that hVISA constitute a distinct group of strains that can be defined by their population heterogeneity, may be associated with treatment failure, and may be precursors of VISA. Our definition of such strains is *S. aureus* with a MIC of ≤4 mg/L, which shows population heterogeneity similar to Mu3 when subjected to a full PAP.

Many methods have been proposed for screening and confirming hVISA. The original method proposed by Hiramatsu et al. (31) was based on an abbreviated population analysis (APA). Essentially in this method an overnight culture is inoculated onto BHIA containing 4 mg/L vancomycin. Confluent growth after 24-h incubation at 37°C suggests a possible VISA (to be confirmed with by formal MIC measurement). A countable number of colonies growing after 48-h incubation indicates a resistant subpopulation of cells and therefore suggests a possible

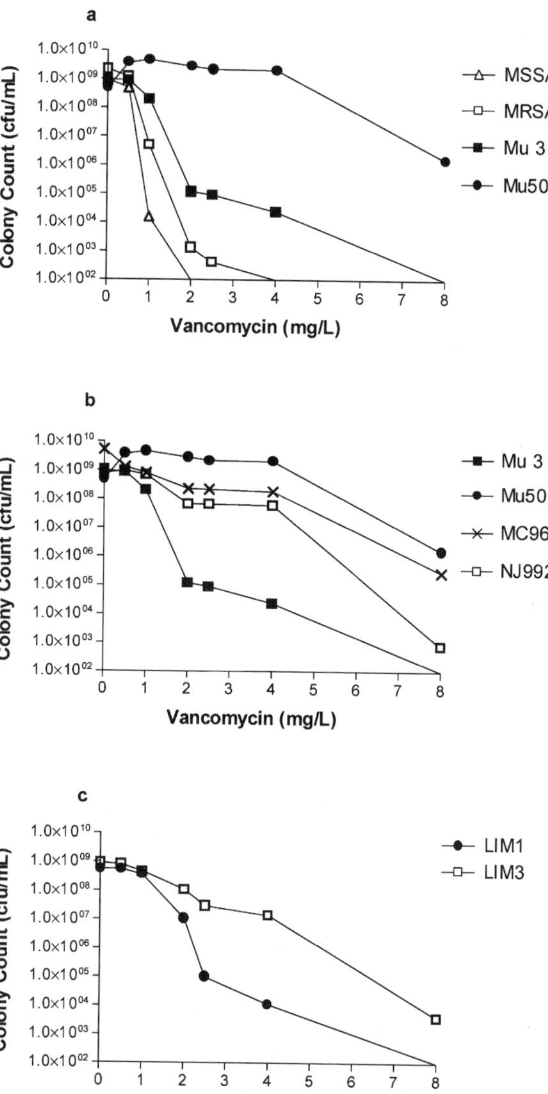

Figure 1 Population analysis profiles performed on Brain Heart Infusion Agar. (*a*) Strains Mu50 (VISA), Mu3 (hVISA), MRSA, MSSA. (*b*) Mu50, Mu3, MC963 (Michigan VISA), NJ992 (New Jersey VISA). (*c*) LIM1 (hVISA), LIM3 (VISA).

hVISA. A number of groups have assessed this screening method and found it to be unreliable. In one of our studies, Mu3 was detected on only 80% of occasions (74), and in another study the method gave a sensitivity of 71% and specificity of 88% when compared with PAP-AUC ratio (66). Attempts to improve sensitivity and specificity have explored the use of different concentrations of vancomycin, use of teicoplanin instead of vancomycin, and use of different media. However, it should be stressed that these APA methods are for screening strains and a subsequent confirmatory test is required.

An Etest method using a high inoculum (macro Etest) has been proposed in order to identify the low-frequency-resistant subpopulation seen in hVISA strains. Figure 2 shows a macro Etest with microcolonies representing the more resistant subpopulation growing within the main zone of inhibition. Criteria for a strain to be a possible hVISA or VISA are a vancomycin and teicoplanin MIC of ≥8 mg/L or a teicoplanin MIC of ≥12 mg/L. A study of 284 MRSA and 45 other staphylococci with reduced vancomycin susceptibility showed a sensitivity of 96% and specificity of 97% (66). Some workers have applied standard breakpoint criteria to macro Etest results, which is incorrect (16).

Antagonism between vancomycin and β-lactams was observed in many of the original strains of VISA and hVISA, and this has been proposed as a method for detecting such strains (26, 33). However, the potential problem with this test is that

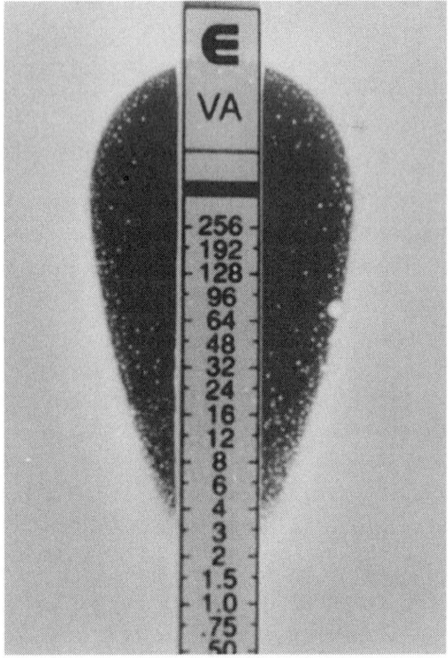

Figure 2 A macro E-test of Mu50 showing resistant microcolonies growing into the main zone of inhibition (by permission American Society for Microbiology).

the underlying mechanism is unknown; therefore it is not certain that the effect will be seen in all strains with reduced vancomycin susceptibility. Indeed some hVISA strains have now been reported that do not display antagonism (16).

The methods described above are for screening for hVISA or VISA strains and a confirmatory test is required. VISA status can be confirmed by performing an MIC using a recognized method. More stringent confirmation can be obtained by testing for all three criteria suggested by the CDC. Confirmation of hVISA status is more complicated, as the initially published criteria are not reliably fulfilled by the archetypal strain (Mu3). We suggest that the definitive method for confirmation of hVISA status is to demonstrate population heterogeneity in a full population analysis profile. Heterogeneity is difficult to define precisely, and to overcome this we calculate the area under the population analysis curve and compare it to Mu3 as a control. A ratio of $\geq$0.90 defines an hVISA (74). We have found this PAP-AUC ratio to be a reliable and reproducible method of defining population heterogenicity.

THE PREVALENCE AND CLINICAL IMPORTANCE OF VISA AND hVISA

Although VISA strains have now been isolated from many countries around the world, they remain rare and only 21 strains have so far been reported in the literature. (Details of these reports are given in an extensive table available online. See the Supplemental Material link in the online version of this chapter or at http://www.annualreviews.org/.) The level of resistance is low and no clinical VRSA (MIC of $\geq$32 mg/L) has been reported. Definitive statements regarding the clinical impact of vancomycin resistance are difficult given the paucity of information available. However, there are some characteristics of the patients infected and the type of infections that emerge. Patients tend to have an underlying illness (often end-stage renal disease, diabetes mellitus, or a malignancy) and have usually received a prolonged course of vancomycin prior to VISA isolation. Infections often involve biomedical devices, such as central venous catheters or peritoneal dialysis catheters, and do not respond well to vancomycin therapy.

The problem with determining the prevalence of hVISA is that the result depends largely on the definition and methodologies employed for screening and confirmation. A number of studies have aimed to determine the prevalence of strains with some degree of reduced-susceptibility vancomycin. When hVISA status is defined by the selection of subclones with a stable vancomycin MIC of 8 mg/L, reported frequencies among MRSA range from 1.3%–20% for different hospitals in Japan (31) to 2.7% in Hong Kong (72), 1.9% in Thailand (65), and 1.1% in Italy (44). An alternative study in Japan that tested 6625 strains from 278 hospitals found no hVISA using this definition and therefore casts some doubt on the original Japanese report (35). A survey in 1997 of 630 MRSA from 33 U.S. hospitals also failed to identify any hVISA (34). Studies examining the prevalence of hVISA defined by the existence of a resistant subpopulation of cells have not generally been well standardized and have given rates of hVISA of 0%–23% of MRSA with lower rates in the United Kingdom (5, 74, 70, 73) and higher rates in continental Europe (14, 16, 25). (Further details of screening studies are available online. See the Supplemental Material link in the online version of this chapter or at http://www.annualreviews.org/.) Taking all available evidence, it seems that the prevalence of hVISA in most parts of the world is approximately 0%–5% of MRSA, although it does appear to be more common in some areas of France, Germany, Spain, and possibly Japan.

The potential importance of hVISA is that it may be associated with treatment failure and it may be a precursor of VISA. Case reports of infections caused by hVISA show a similar picture to VISA infections. (Details of reports are available online. See the Supplemental Material link in the online version of this chapter or at http://www.annualreviews.org/.) Patients often have an underlying illness and have received prior vancomycin. Infections tend to respond poorly to vancomycin. However, it should be remembered that these cases are selected to some extent by their authors and in some cases have been selected for testing for hVISA status due to failure of vancomycin therapy (3). There has not been a prospective study of the clinical features and outcomes of hVISA infections, but Ariza et al. (3) in a retrospective study reported a failure rate of 85.7% for hVISA infections compared with a rate of 70% for comparable non-hVISA infections.

The suggestion that hVISA is a precursor of VISA was proposed in the first description by Hiramatsu et al. (31). Subsequently, evidence of the evolution of hVISA to VISA during an infection has been reported in the New York VISA strain (57). PAPs for the LIM strains from France [kindly donated by M.C. Ploy (51)] are shown in Figure 1*c*. The progression from LIM1 isolated early in infection, which shows a heterogeneous profile and has a vancomycin MIC of 2 mg/L (hence hVISA) to LIM3 isolated from the same patient that has an MIC of 8 mg/L (VISA), is clearly seen. Further evidence to support the proposition that hVISA are VISA precursors comes from the fact that they show similar thickening of the cell wall, particularly when exposed to vancomycin.

The term hVISA is not widely used in the United States, presumably because of the difficulties in producing a meaningful definition and a reliable methodology. However, the CDC has been gathering data on strains that have an MIC of 4 mg/L. These isolates are susceptible by NCCLS criteria but have a higher MIC than most U.S. strains. Fridkin et al. (23) have recently reported that patients with infections due to these strains have a significantly higher mortality compared with controls infected with vancomycin-susceptible MRSA (attributable death 63% versus 12%). They suggest that differentiating infection control recommendations for VISA from those for *S. aureus* with a MIC of 4 mg/L may need to be reassessed (23). Although the prevalence of such strains has not been widely studied, there appears to be considerable geographical variation. Of strains submitted to the CDC as possible VISA between 1999 and 2001, only 15 strains were found to have an MIC of 4 mg/L (23). However, it is reported that 15.2% of MRSA in the United Kingdom in 1999 had a vancomycin MIC of 4 mg/L (5). Recently a study of MRSA isolated from an intensive therapy unit in South Korea showed that 37 of 136 (27.2%) had a MIC of 4 mg/L (38).

CHARACTERISTICS OF hVISA AND VISA

VISA and hVISA strains have now been isolated from many areas of the world, and there are a number of characteristics shared by both clinically and laboratory-derived strains that can give some clues as to the mechanism of resistance. With the exception of the laboratory-derived strains of Sieradski & Tomasz (58), all strains have had only low-level resistance with a MIC of 8–16 mg/L. A striking characteristic is the abnormally thickened cell wall seen in most strains studied (19, 31, 32, 44, 49, 57, 58, 62). In addition, many strains display reduced susceptibility to lysostaphin and decreased autolysis (11, 50), and a slower rate of growth (11, 44, 50, 52). On a macro-phenotypic scale, VISA and hVISA strains often look mixed with large and small colony forms, and this has been suggested as a potentially useful diagnostic observation (10, 44, 45, 52, 68). A characteristic that may cause problems in the diagnostic laboratory is the fact that some strains show reduced/delayed coagulase activity (19, 52).

PEPTIDOGLYCAN BIOSYNTHESIS IN *S. aureus* AND THE ACTION OF VANCOMYCIN

Vancomycin is a glycopeptide antibiotic that has as its primary target the D-ala-D-ala subunits of the gram-positive cell wall, which causes cell death by inhibiting cell wall cross-linking. Vancomycin resistance in *S. aureus* is not mediated by a simple mechanism (e.g., target site modification, production of an antibiotic-modifying enzyme, drug efflux or impermeability), and the *vanA*, *vanB*, *vanC*, *vanD*, *vanE*, and *vanG* genes associated with resistance in enterococci have not been found in VISA or hVISA. The resistance mechanism appears to involve a complex reorganization of cell wall anabolism; therefore it would be prudent to review the mechanism of cell wall synthesis in order to appreciate the action of vancomycin and how the bacterium adjusts cell wall metabolism to confer vancomycin resistance.

The fundamental component (50%) of the cell wall of *S. aureus* is peptidoglycan, which is a polysaccharidic matrix composed of unbranched, β-linked chains containing alternating murein subunits of *N*-acetylglucosamine (GlcNAc) and *N*-acetylmuramic acid (MurNAc). The initial step in production of a murein monomer is the formation of glucosamine-1-phosphate from either glucose or glucosamine. Pyrimidine (UDP) biosynthesis occurs from glutamine via a complex set of reactions and is then linked to glucosamine-1-phosphate to form the first subunit UDP-*N*-acetyl-glucosamine (UDP-GlcNAc). The next stage is the formation of the second subunit from UDP-GlcNAc, which involves the initial production of phosphoenol pyruvate from glucose. The UDP-GlcNAc and phosphoenol pyruvate combine to form the second subunit UDP-N-acetyl-muramic acid (UDP-MurNAc). An L-alanine is then linked to the UDP-MurNAc residue followed by D-glutamate, L-lysine, and D-alanyl-D-alanine to form the UDP-MurNAc pentapeptide or Parks nucleotide. The Parks nucleotide then transfers to C55-undecaprenylphosphate to form LIPID-I of the cytoplasmic membrane peptidoglycan transporter (47). The next stage involves the sequential transfer of five glycine residues to the L-lysine component of the pentapeptide to form a pentaglycine side chain. The D-glutamate component of the pentapeptide is then amidated to D-glutamine using L-glutamine as a NH_4^+ donor (42). This complex is then transferred (LIPID II) across the membrane where it is joined to the nascent peptidoglycan chain by the action of transglycosylases, which link the sugar moieties GlcNAc-MurNAc-pentapeptide-pentaglycine. Transpeptidases then link the pentaglycine chain of one monomer to the penultimate D-alanine residue of the pentapeptide of the neighboring peptidoglycan chain, with the release of the D-alanine residue from the recipient pentapeptide (47). Theoretically, all D-ala-D-ala motifs should be consumed in the cross-bridging reaction; however, approximately 20% remains intact. It has been estimated that each *S. aureus* cell contains approximately 10^6 D-ala-D-ala exposed peptides.

Glycopeptides form a stoichiometric 1:1 complex with the peptidoglycan precursor UDP-MurNAc-pentapeptide (15) and inhibit transpeptidation and nascent peptidoglycan synthesis by binding the D-alanyl-D-alanine residues of the murein monomer. The transglycosylase enzyme that transfers the disaccharide of the peptidoglycan precursor to the growing glycan polymer is inhibited, presumably owing to the steric hindrance from the large glycopeptide-peptidoglycan precursor complex (1). Even if some glycan chain synthesis occurs, the transpeptidase enzyme reaction is also excluded as the acyl-D-ala-D-ala portion of the peptidoglycan precursor is enveloped by the bulky glycopeptide.

THE MECHANISM OF RESISTANCE OF VISA/hVISA: COMPLEX CHANGES TO PEPTIDOGLYCAN BIOSYNTHESIS AND ITS CONTROL

Evidence regarding the possible mechanism for vancomycin resistance in *S. aureus* derives from studies on clinical strains with low-level resistance and also from two highly resistant laboratory-generated strains, VM and TNM. Most clinically and laboratory-derived strains share characteristics of a thickened cell wall with reduced levels of peptidoglycan cross-linking, which may represent the final common pathway for the expression of vancomycin resistance. It is postulated that reduced peptidoglycan cross-linking leads to an increase in free D-ala-D-ala side chains to which vancomycin can bind. The D-alanyl-D-alanine/glycopeptide complexes at the outer edges of the peptidoglycan matrix are likely to hinder other glycopeptides from reaching the sites of cell wall biosynthesis at the plasma membrane (56). Thus whereas vancomycin binds to its target (D-ala-D-ala), it does not bind to the D-ala-D-ala side chains at the cytoplasmic membrane that are actively involved in the formation of cross-links. The decreased strength of the poorly cross-linked cell wall is presumably compensated for by the greatly increased thickness of peptidoglycan produced. Although this theory may provide an explanation for the observations from most clinically and laboratory-derived strains, the mechanism by which the cell re-organizes its cell wall metabolism to give this phenotype has yet to be established; it is possible that strains from the laboratory and different geographical sources have achieved this phenotype by different means. Two theories that have been proposed to explain the reduced cross-linking involve reduced amidation of muropeptides and/or reduced expression of PBP4.

During the pre-membrane transfer stages of peptidoglycan biosynthesis, the D-glutamate of the pentapeptide complex is converted to D-glutamine. However, Mu50 (original clinical vancomycin-resistant strain) has been reported as having an increased proportion of non-amidated muropeptides in its cell wall (18). The muropeptides containing glutamate instead of glutamine are thought to be poorer substrates for transpeptidases and consequently, fewer cross links are formed (18, 27, 46). The anabolism of peptidoglycan is highly dependent on the availability of L-glutamine. It acts as an amine donor converting fructose-6-phosphate to glucosamine-6-phosphate in the production of UDP-GlcNAc, and it is an initial component of phosphoenol pyruvate in the formation of UDP-MurNAc. In addition, it is required as an amine donor for the conversion of the pentapeptide D-glutamate to D-glutamine. The regeneration of L-glutamine from L-glutamate is catalyzed by glutamine synthetase, which is one of the most tightly regulated enzymes in the bacterial cell (51). The way in which glutamine synthetase is controlled in *S. aureus* has not been determined; however, as the basic intermediary metabolic pathways of *Escherichia coli* and *S. aureus* are similar it can be assumed that the glutamine synthetase regulation in *S. aureus* is also similar (13), i.e., inhibited by glucosamine-6-phosphate and activated by the presence of either 2-ketoglutarate or L-glutamate (37). In Mu50, there exists evidence that the cell preferentially uses glucose over glucosamine from the environment to produce peptidoglycan. Thus the L-glutamine pool is depleted owing to the increased glucosamine-6-phosphate production during the utilization of glucose to form glucosamine-1-phosphate and subsequently GlcNAc (18, 36). Therefore, the amount of L-glutamine available for the amidation of D-glutamate in the pentapeptide used to assemble peptidoglycan is also reduced (18, 42). Thus, it would appear that the extent of peptidoglycan cross-linking is intrinsically linked to L-glutamine levels and ultimately dependant on the tricarboxylic acid cycle.

Laboratory studies by Tomasz and colleagues resulted in the isolation of two highly glycopeptide-resistant mutants, VM and TNM, with a vancomycin MIC of 72 mg/L and a teicoplanin MIC of 200 mg/L, respectively. The mutants were obtained from serial passaging in liquid culture; therefore frequencies cannot be given for the rate at which mutants were generated. Resistance was shown in both cases to correlate with a thickened cell wall and the release of large amounts of peptidoglycan into the medium. This released peptidoglycan had a reduction in cross-linking of up to 85% and hence increased free D-ala-D-ala side chains that could bind and sequester large quantities of vancomycin in a biologically active but unavailable state (59, 60). In both VM and TNM the expression of PBP4 is either greatly reduced or inactivated. PBP4 has DD-carboxypeptidase activities as well as a secondary transpeptidase required for the extensive cross-linking of peptidoglycan (28, 75). Impaired PBP4 activity is therefore likely to reduce cross-linking and increase the number of vancomycin targets. The studies of Finan et al. (22) on clinically derived VISA showed a marked decrease or no PBP4 activity. However, passage-derived VISA from the same study with similar MIC levels to that of the clinical VISA still demonstrated detectable levels of PBP4. In accordance with these findings, the passage-derived mutant VM did not lose PBP4 activity until a vancomycin MIC of 50 mg/L had been obtained via a mutation in the structural gene (56). This suggests that there is at least a quantitative difference in the resistance mechanism between clinically and laboratory-derived strains. How PBP4 is downregulated in the clinically derived VISA is unclear. It is known that *pbp4* is transcribed divergently from a gene, *abcA*, possessing an ATP-binding cassette motif in its sequence. Studies examining the regulatory relationship of *pbp4* and *abcA* have, in some instances, given conflicting results (20, 29). However, Finan et al. (22) have demonstrated that loss of PBP4 could be complemented by the transformation of competent *pbp4* genes, suggesting that if *pbp4* is regulated, the regulation is not stringently controlled. It is also worth noting that there appears to be a link between *abcA* mutants and cell autolysis, a phenomenon that is often altered in hVISA/VISA (55).

The above theories provide some explanation for the final mechanism of vancomycin resistance, but it should be noted that a minority of VISA strains have been described that do not show these alterations in peptidoglycan composition (12). These strains have not been well studied but may have an alternative mechanism of resistance.

GENETIC MUTATIONS AND ALTERED GENE EXPRESSION IN VISA

Although reduced PBP4 expression or L-glutamine depletion due to upregulated cell wall anabolism may provide possible explanations for vancomycin resistance mediated by a thickened peptidoglycan cell wall with reduced cross-linking, the genetic changes required to effect these alterations remains unclear. Mu50 is the VISA strain that has been most thoroughly studied. Kuroda et al. (40) created a genomic library of Mu50 DNA and probed it with radioactive cDNA from Mu50, and the VSSA revertant Mu50w. Seven clones were identified that hybridized more Mu50 cDNA than Mu50w cDNA, suggesting that at least one gene in each clone is upregulated in the VISA strain. Further studies including sequence analysis and homology searches of databases identified these genes as *vraA-G* together with a putative two-component regulator pair, *vraS* and *vraR*. This study was preliminary, and unfortunately the *vra* gene sequences were not reported; rather *Bacillus cereus* gene homologs were given. Little work was done to determine the levels of expression of these genes in Mu50 or Mu3 with the exception of *vra*R, which was shown to be overexpressed in Mu50 (40). Interestingly, Aritaka et al. (2) have suggested that β-lactams induce increased transcription of the response regulator, *vraR*, perhaps giving a hint as to the mechanism behind the antagonism seen between β-lactams and glycopeptides with some VISA and hVISA strains.

In addition to the regulation analysis of Mu50, the recent publication by Kuroda et al. (41) on the genomic sequence of Mu50 provides a unique opportunity to study specific altered/truncated genes that are involved in VISA. The study compared the sequence of the archetype VISA, Mu50, and a related strain, N315. The comparative genetic analysis reported was superficial and did not highlight differences between the two sequences pertaining to vancomycin resistance. It seems likely that there could be major differences in genes that encode enzymes involved in cell wall synthesis between Mu50 and N315. We have performed an analysis of the genomic sequences and found a number of possible frameshift mutations in key genes involved in cell wall metabolism (6). However, the initially published sequence contained a number of sequencing errors and our analysis awaits confirmation.

GLOBAL REGULATORY CHANGES SEEN IN VISA

The obvious morphological difference between normal MRSA strains and hVISA/VISAs is the thickened cell wall, which in turn is a reflection of changes in the expression of peptidoglycan biosynthesis genes. Several extracellular proteins, as well as cell-bound adhesins, considered essential for the pathogenicity of *S. aureus*, may also be altered in their expression. The fact that hVISA can adhere more readily to artificial surfaces than its MRSA progenitor may reflect this (67, 71). Two regulatory loci, *agr* and *sar*, contribute to the differential regulation of virulence determinants in *S. aureus* (4). *sar* codes three overlapping transcripts termed *sarA*, *sarC*, and *sarB* (8). Recently, SarA has been shown to bind upstream of the *fbnA* gene and therefore mediate its expression. The regulator *agr* is thought to influence the expression of many proteins including δ-hemolysin, fibronectin-binding proteins, and protein A, as well as influencing biofilm production and invasive infection (17, 30). *agr* also controls the expression of *abcA*, suggesting a genetic linkage between *agr*, *abcA*, and the expression of *pbp4* (55). In vitro studies on an array of lytic enzymes show that in VISA and hVISA these enzymes are secreted less than their MRSA counterparts (53). Furthermore, Sakoulas et al. (54) indicate that δ-hemolysin expression by some clinical isolates is regulated by a group II *agr*. When further analyzed, *agr* was found to be mutated in their clinical VISA/hVISA isolates as well as in a laboratory-passaged hVISA. Our group has demonstrated an increase in adherence to artificial surfaces by hVISA compared with VSSA, particularly when challenged with vancomycin (67). These data are also supported by the findings of Williams et al. (71), which showed an increase in vancomycin MICs when *S. aureus* strains were examined from biofilms.

When TNM was mutated back to the teicoplanin susceptibility displayed by the parent strain (COL26) by the insertion of the transposon Tn*551*, the genetic insertions were mapped and found to locate to genes encoding cell wall synthesis and also two global regulatory genes *sar* and *sigma B* (61, 60).

Thus it would appear that mutations in *agr* and/or *sar*, or altered expression of these regulators, may lead to a VISA or hVISA phenotype and affect the organism's ability to express virulence factors and form biofilms. Thus, in VISA/hVISA there seems to be a clear link between pathogenicity and resistance.

CONCLUSION

Vancomycin resistance has at last emerged in *S. aureus*, albeit at a low level. VISA strains are rare but have been reported from many areas of the world. It seems possible that VISA and hVISA strains, which seem to share some phenotypic similarities, may constitute the parts of a spectrum of resistance caused by a common phenotype expressed to different levels in different strains. The fact that the range of resistance straddles current breakpoints leads to problems in definition and terminology. Further studies are required to define the clinical impact of these strains, and in order for these to be performed, reliable tests for their identification are required. Although the final mechanism of resistance appears the same for most clinical strains studied, i.e., a grossly thickened cell wall, there would seem to be many ways in which *S. aureus* may achieve this. Given that vancomycin resistance in *S. aureus* is likely to involve changes in global regulators, other factors such as virulence will be affected as well as the bacterium's fitness. The fact that laboratory-derived strains with similar mechanisms of resistance can achieve high levels of resistance (MIC of 72 mg/L) (58) may be a warning of what is to come in clinical strains.

ACKNOWLEDGMENTS

We wish to acknowledge the members of the VISA group at the Bristol Centre for Antimicrobial Research and Evaluation (M. Wooton, M.B. Avison, P.M. Bennett, A.P. MacGowan) for their contributions to our work on VISA and for stimulating discussions about the clinical role and mechanism of resistance in *S. aureus*.

The *Annual Review of Microbiology* is online at http://micro.annualreviews.org

LITERATURE CITED

1. Anderson JS, Matsuhashi M, Haskin MA, Strominger JL. 1967. Biosythesis of the peptidoglycan of bacterial cell walls. II. Phospholipid carriers in the reaction sequence. *J. Biol. Chem.* 242:3180–90
2. Aritaka N, Hanaki H, Cui L, Hiramatsu K. 2001. Combination effect of vancomycin and beta-lactams against a *Staphylococcus aureus* strain, Mu3, with heterogeneous resistance to vancomycin. *Antimicrob. Agents Chemother.* 45:1292–94
3. Ariza J, Pujol M, Cabo J, Pena C,

Fernandez N, et al. 1999. Vancomycin in surgical infections due to methicillin-resistant *Staphylococcus aureus* with heterogeneous resistance to vancomycin. *Lancet* 353:1587–88

4. Arvidson SK, Tegmark K. 2001. Regulation of virulence determinants in *Staphylococcus aureus. Int. J. Med. Microbiol.* 291: 159–70
5. Aucken HM, Warner M, Ganner M, Johnson AP, Richardson JF, et al. 2000. Twenty months of screening for glycopeptide-intermediate *Staphylococcus aureus. J. Antimicrob. Chemother.* 46:639–40
6. Avison MB, Bennett PM, Howe RA, Walsh TR. 2002. Preliminary analysis of the genetic basis for vancomcyin resistance in *Staphylococcus aureus* strain Mu50. *J. Antimicrob. Chemother.* 49:255–60
7. Ayliffe GA. 1997. The progressive intercontinental spread of methicillin-resistant *Staphylococcus aureus. Clin. Infect. Dis.* 24 (Suppl. 1):S74–79
8. Bayer MG, Heinrichs JH, Cheung AL. 1996. The molecular architecture of the sar locus in *Staphylococcus aureus. J. Bacteriol.* 178:4563–70
9. Biavasco F, Giovanetti E, Montanari, MP, Lupidi R, Varaldo PE. 1991. Development of in-vitro resistance to glycopeptide antibiotics: assessment in staphylococci of different species. *J. Antimicrob. Chemother.* 27:71–79
10. Bobin-Dubreux S, Reverdy ME, Nervi C, Rougier M, Bolmstrom A, et al. 2001. Clinical isolate of vancomycin-heterointermediate *Staphylococcus aureus* susceptible to methicillin and in vitro selection of a vancomycin-resistant derivative. *Antimicrob. Agents Chemother.* 45:349–52
11. Boyle-Vavra S, Berke SK, Lee JC, Daum RS. 2000. Reversion of the glycopeptide resistance phenotype in *Staphylococcus aureus* clinical isolates. *Antimicrob. Agents Chemother.* 44:272–77
12. Boyle-Vavra S, Labischinski H, Ebert CC, Ehlert K, Daum RS. 2001. A spectrum of changes occurs in peptidoglycan composition of glycopeptide-intermediate clinical *Staphylococcus aureus* isolates. *Antimicrob. Agents Chemother.* 45:280–87
13. Bruckner R, Bassias J. 2000. Carbohydrate catabolism pathways and regulation in Staphylococci. In *Gram Positive Pathogens*, ed. VA Fiscetti, RP Novick, JJ Ferretti, DA Portnoy, JI Rood. pp. 339-44. Washington, DC: ASM Press
14. Canton R, Mir N, Martinez-Ferrer M, Sanchez del Saz B, Soler I, Baquero F. 1999. Prospective study of *Staphylococcus aureus* with reduced susceptibility to glycopeptides. *Rev. Esp. Quim.* 12:48–53
15. Chatterjee AN, Perkins HR. 1966. Compounds formed between nucleotides related to the biosynthesis of bacterial cell wall and vancomycin. *Biochem. Biophys. Res. Commun.* 24:489–94
16. Chesneau O, Morvan A, Solh NE. 2000. Retrospective screening for heterogeneous vancomycin resistance in diverse *Staphylococcus aureus* clones disseminated in French hospitals. *J. Antimicrob. Chemother.* 45:887–90
17. Cheung AL, Schmidt K, Bateman B, Manna AC. 2001. SarS, a SarA homolog repressible by agr, is an activator of protein A synthesis in *Staphylococcus aureus. Infect. Immun.* 69:2448–55
18. Cui L, Murakami H, Kuwahara-Arai K, Hanaki H, Hiramatsu K. 2000. Contribution of a thickened cell wall and its glutamine nonamidated component to the vancomycin resistance expressed by *Staphylococcus aureus* Mu50. *Antimicrob. Agents Chemother.* 44:2276–85
19. Daum RS, Gupta S, Sabbagh R, Milewski WM. 1992. Characterization of *Staphylococcus aureus* isolates with decreased susceptibility to vancomycin and teicoplanin: isolation and purification of a constitutively produced protein associated with decreased susceptibility. *J. Infect. Dis.* 166:1066–72
20. Domanski TL, de Jonge BL, Bayles KW. 1997. Transcription analysis of the *Staphylococcus aureus* gene encoding

penicillin-binding protein 4. *J. Bacteriol.* 179:2651–57

21. Fairbrother RW, Williams BL. 1956. Two new antibiotics: antibacterial activity of novobiocin and vancomycin. *Lancet*:1177–78
22. Finan JE, Archer GL, Pucci MJ, Climo MW. 2001. Role of penicillin-binding protein 4 in expression of vancomycin resistance among clinical isolates of oxacillin-resistant *Staphylococcus aureus. Antimicrob. Agents Chemother.* 45:3070–75
23. Fridkin SK, Hageman JC, McDougal L, Mohammed J, Kellum ME, et al. 2001. Nationwide epidemiologic study of *Staphylococcus aureus* with reduced susceptibility to vancomycin. *41st Intersci. Confer. Antimicrobial Agents and Chemotherapy*. Chicago: ASM
24. Garrod LP, Waterworth PM. 1956. Behaviour in vitro of some new antistaphylococcal antibiotics. *Br. Med. J.*:61–65
25. Geisel R, Schmitz FJ, Thomas L, Berns G, Zetsche O, et al. 1999. Emergence of heterogeneous intermediate vancomycin resistance in *Staphylococcus aureus* isolates in the Dusseldorf area. *J. Antimicrob. Chemother.* 43:846–48
26. Hanaki H, Inaba Y, Sasaki K, Hiramatsu K. 1998. A novel method of detecting *Staphylococcus aureus* heterogeneously resistant to vancomycin (hetero-VRSA). *Jpn. J. Antibiot.* 51:521–30
27. Hanaki H, Labischinski H, Inaba Y, Kondo N, Murakami H, Hiramatsu K. 1998. Increase in glutamine-non-amidated muropeptides in the peptidoglycan of vancomycin-resistant *Staphylococcus aureus* strain Mu50. *J. Antimicrob. Chemother.* 42: 315–20
28. Henze UU, Berger-Bachi B. 1995. *Staphylococcus aureus* penicillin-binding protein 4 and intrinsic beta-lactam resistance. *Antimicrob. Agents Chemother.* 39:2415–22
29. Henze UU, Berger-Bachi B. 1996. Penicillin-binding protein 4 overproduction increases beta-lactam resistance in *Staphylococcus aureus. Antimicrob. Agents Chemother.* 40:2121–25
30. Heyer G, Saba S, Adamo R, Rush W, Soong G, et al. 2002. *Staphylococcus aureus* agr and sarA functions are required for invasive infection but not inflammatory responses in the lung. *Infect. Immun.* 70:127–33
31. Hiramatsu K, Aritaka N, Hanaki H, Kawasaki S, Hosoda Y, et al. 1997. Dissemination in Japanese hospitals of strains of *Staphylococcus aureus* heterogeneously resistant to vancomycin. *Lancet* 350:1670–73
32. Hiramatsu K, Hanaki H, Ino T, Yabuta K, Oguri T, Tenover FC. 1997. Methicillin-resistant *Staphylococcus aureus* clinical strain with reduced vancomycin susceptibility. *J. Antimicrob. Chemother.* 40:135–36
33. Howe RA, Wootton M, Bennett PM, MacGowan AP, Walsh TR. 1999. Interactions between methicillin and vancomycin in methicillin-resistant *Staphylococcus aureus* strains displaying different phenotypes of vancomycin susceptibility. *J. Clin. Microbiol.* 37:3068–71
34. Hubert SK, Mohammed JM, Fridkin SK, Gaynes RP, McGowan JE, Tenover FC. 1999. Glycopeptide-intermediate *Staphylococcus aureus*: evaluation of a novel screening method and results of a survey of selected U.S. hospitals. *J. Clin. Microbiol.* 37:3590–93
35. Ike Y, Arakawa Y, Ma X, Tatewaki K, Nagasawa M, et al. 2001. Nationwide survey shows that methicillin-resistant *Staphylococcus aureus* strains heterogeneously and intermediately resistant to vancomycin are not disseminated throughout Japanese hospitals. *J. Clin. Microbiol.* 39:4445–51
36. Imada A, Nozaki Y, Kawashima F, Yoneda M. 1977. Regulation of glucosamine utilization in *Staphylococcus aureus* and *Escherichia coli. J. Gen. Microbiol.* 100:329–37
37. Jiang P, Peliska JA, Ninfa AJ. 1998. The regulation of *Escherichia coli* glutamine synthetase revisited: role of

2-ketoglutarate in the regulation of glutamine synthetase adenylylation state. *Biochemistry* 37:12802–10

38. Jung SI, Kiem S, Lee NY, Kim YS, Oh WS, et al. 2002. One-point population analysis and effect of osmolarity on detection of hetero-vancomycin-resistant *Staphylococcus aureus. J. Clin. Microbiol.* 40:1493–95
39. Kaatz GW, Seo SM, Dorman NJ, Lerner SA. 1990. Emergence of teicoplanin resistance during therapy of *Staphylococcus aureus* endocarditis. *J. Infect. Dis.* 162:103–8
40. Kuroda M, Kuwahara-Arai K, Hiramatsu K. 2000. Identification of the up- and down-regulated genes in vancomycin-resistant *Staphylococcus aureus* strains Mu3 and Mu50 by cDNA differential hybridization method. *Biochem. Biophys. Res. Commun.* 269:485–90
41. Kuroda M, Ohta T, Uchiyama I, Baba T, Yuzawa H, et al. 2001. Whole genome sequencing of meticillin-resistant *Staphylococcus aureus. Lancet* 357:1225–40
42. Linnett PE, Strominger JL. 1974. Amidation and cross-linking of the enzymatically synthesized peptidoglycan of *Bacillus stearothermophilus. J. Biol. Chem.* 249: 2489–96
43. Lowy FD. 1998. *Staphylococcus aureus* infections. *N. Engl. J. Med.* 339:520–30
44. Marchese A, Balistreri G, Tonoli E, Debbia EA, Schito GC. 2000. Heterogeneous vancomycin resistance in methicillin-resistant *Staphylococcus aureus* strains isolated in a large Italian hospital. *J. Clin. Microbiol.* 38: 866–69
45. Marlowe EM, Cohen MD, Hindler JF, Ward KW, Bruckner DA. 2001. Practical strategies for detecting and confirming vancomycin-intermediate *Staphylococcus aureus*: a tertiary-care hospital laboratory's experience. *J. Clin. Microbiol.* 39:2637–39
46. Nakel M, Ghuysen JM, Kandler O. 1971. Wall peptidoglycan in *Aerococcus viridans* strains 201 Evans and ATCC 11563 and in *Gaffkya homari* strain ATCC 10400. *Biochemistry* 10:2170–75
47. Navarre WW, Schneewind O. 1999. Surface proteins of gram-positive bacteria and mechanisms of their targeting to the cell wall envelope. *Microbiol. Mol. Biol. Rev.* 63:174–229
48. Noble WC, Virani Z, Cree RG. 1992. Cotransfer of vancomycin and other resistance genes from *Enterococcus faecalis* NCTC 12201 to *Staphylococcus aureus. FEMS Microbiol. Lett.* 72:195–98
49. Oliveira GA, Dell'Aquila AM, Masiero RL, Levy CE, Gomes MS, et al. 2001. Isolation in Brazil of nosocomial *Staphylococcus aureus* with reduced susceptibility to vancomycin. *Infect. Control Hosp. Epidemiol.* 22:443–58
50. Pfeltz RF, Singh VK, Schmidt JL, Batten MA, Baranyk CS, et al. 2000. Characterization of passage-selected vancomycin-resistant *Staphylococcus aureus* strains of diverse parental backgrounds. *Antimicrob. Agents Chemother.* 44:294–303
51. Ploy MC, Grelaud C, Martin C, de Lumley L, Denis F. 1998. First clinical isolate of vancomycin-intermediate *Staphylococcus aureus* in a French hospital. *Lancet* 351:1212
52. Rotun SS, McMath V, Schoonmaker DJ, Maupin PS, Tenover FC, et al. 1999. *Staphylococcus aureus* with reduced susceptibility to vancomycin isolated from a patient with fatal bacteremia. *Emerg. Infect. Dis.* 5:147–49
53. Sakoulas G, Eliopoulos GM, Moellering RC, Wennersten C, Venkataraman L, et al. 2001. Characterisation of the accessory gene regulator (agr) locus in geographically diverse *Staphylcoccus aureus* with reduced susceptibility to vancomycin. *41st Intersci. Conf. Antimicrobial Agents and Chemotherapy*. Chicago: ASM
54. Sakoulas G, Eliopoulos GM, Moellering RC, Wennersten C, Venkataraman L, et al. 2002. Accessory gene regulator (agr) locus in geographically diverse *Staphylococcus* isolates with reduced susceptibility to vancomycin. *Antimicrob. Agents Chemother.* 46:1492–502
55. Schrader-Fischer G, Berger-Bachi B. 2001.

The AbcA transporter of *Staphylococcus aureus* affects cell autolysis. *Antimicrob. Agents Chemother.* 45:407–12
56. Sieradzki K, Pinho MG, Tomasz A. 1999. Inactivated pbp4 in highly glycopeptide-resistant laboratory mutants of *Staphylococcus aureus*. *J. Biol. Chem.* 274:18942–46
57. Sieradzki K, Roberts RB, Haber SW, Tomasz A. 1999. The development of vancomycin resistance in a patient with methicillin-resistant *Staphylococcus aureus* infection. *N. Engl. J. Med.* 340:517–23
58. Sieradzki K, Tomasz A. 1996. A highly vancomycin-resistant laboratory mutant of *Staphylococcus aureus*. *FEMS Microbiol. Lett.* 142:161–66
59. Sieradzki K, Tomasz A. 1997. Inhibition of cell wall turnover and autolysis by vancomycin in a highly vancomycin-resistant mutant of *Staphylococcus aureus*. *J. Bacteriol.* 179:2557–66
60. Sieradzki K, Tomasz A. 1998. Suppression of glycopeptide resistance in a highly teicoplanin-resistant mutant of *Staphylococcus aureus* by transposon inactivation of genes involved in cell wall synthesis. *Microb. Drug Resist.* 4:159–68
61. Sieradzki K, Tomasz A. 1999. Gradual alterations in cell wall structure and metabolism in vancomycin-resistant mutants of *Staphylococcus aureus*. *J. Bacteriol.* 181:7566–70
62. Smith TL, Pearson ML, Wilcox KR, Cruz C, Lancaster MV, et al. 1999. Emergence of vancomycin resistance in *Staphylococcus aureus*. Glycopeptide-intermediate *Staphylococcus aureus* working group. *N. Engl. J. Med.* 340:493–501
63. Tenover FC, Biddle JW, Lancaster MV. 2001. Increasing resistance to vancomycin and other glycopeptides in *Staphylococcus aureus*. *Emerg. Infect. Dis.* 7:327–32
64. Tenover FC, Lancaster MV, Hill BC, Steward CD, Stocker SA, et al. 1998. Characterization of staphylococci with reduced susceptibilities to vancomycin and other glycopeptides. *J. Clin. Microbiol.* 36:1020–27. Erratum. 1998. *J. Clin.Microbiol.* 36(7):2167
65. Trakulsomboon S, Danchaivijitr S, Rongrungruang Y, Dhiraputra C, Susaemgrat W, et al. 2001. First report of methicillin-resistant *Staphylococcus aureus* with reduced susceptibility to vancomycin in Thailand. *J. Clin. Microbiol.* 39:591–95
66. Walsh TR, Bolmstrom A, Qwarnstrom A, Ho P, Wootton M, et al. 2001. Evaluation of current methods for detection of staphylococci with reduced susceptibility to glycopeptides. *J. Clin. Microbiol.* 39:2439–44
67. Walsh TR, Wootton M, Howe RA, Bennett PM, MacGowan AP. 1999. Increased biofilm formation and adherence to artificial surfaces of hVISA [heterogeneous vancomycin (V) intermediate *Staphylococcus aureus*] on exposure to vancomcyin. *39th Intersci. Conf. Antimicrob. Agents Chemother.* San Fransisco: ASM
68. Ward PB, Johnson PD, Grabsch EA, Mayall BC, Grayson ML. 2001. Treatment failure due to methicillin-resistant *Staphylococcus aureus* (MRSA) with reduced susceptibility to vancomycin. *Med. J. Aust.* 175:480–83
69. Watanakunakorn C. 1990. In-vitro selection of resistance of *Staphylococcus aureus* to teicoplanin and vancomycin. *J. Antimicrob. Chemother.* 25:69–72
70. Wilcox MH, Fawley W. 2001. Extremely low prevalence of UK *Staphylococcus aureus* isolates with reduced susceptibility to vancomycin. *J. Antimicrob. Chemother.* 48:144–45
71. Williams I, Venables WA, Lloyd D, Paul F, Critchley I. 1997. The effects of adherence to silicone surfaces on antibiotic susceptibility in *Staphylococcus aureus*. *Microbiology* 143 (Pt. 7):2407–13
72. Wong SS, Ho PL, Woo PC, Yuen KY. 1999. Bacteremia caused by staphylococci with inducible vancomycin heteroresistance. *Clin. Infect. Dis.* 29:760–67
73. Woodford N, Warner M, Aucken HM. 2000. Vancomycin resistance among epidemic strains of methicillin-resistant

Staphylococcus aureus in England and Wales. *J. Antimicrob. Chemother.* 45:258–59

74. Wootton M, Howe RA, Hillman R, Walsh TR, Bennett PM, MacGowan AP. 2001. A modified population analysis profile (PAP) method to detect hetero-resistance to vancomycin in *Staphylococcus aureus* in a UK hospital. *J. Antimicrob. Chemother.* 47: 399–403
75. Wyke AW, Ward JB, Hayes MV, Curtis NA. 1981. A role in vivo for penicillin-binding protein-4 of *Staphylococcus aureus*. *Eur. J. Biochem.* 119:389–93

Annu. Rev. Microbiol. 2002. 56:677–702
doi: 10.1146/annurev.micro.56.012302.160757

First published online as a Review in Advance on July 15, 2002

POLIOVIRUS CELL ENTRY: Common Structural Themes in Viral Cell Entry Pathways

James M. Hogle
Department of Biological Chemistry and Molecular Pharmacology, Harvard Medical School, Boston, Massachusetts 02115; e-mail: hogle@hogles.med.harvard.edu

Key Words virus structure, conformational changes, spring-loaded trap, virus-receptor interactions, virus-membrane interactions

■ **Abstract** Structural studies of polio- and closely related viruses have provided a series of snapshots along their cell entry pathways. Based on the structures and related kinetic, biochemical, and genetic studies, we have proposed a model for the cell entry pathway for polio- and closely related viruses. In this model a maturation cleavage of a capsid protein precursor locks the virus in a metastable state, and the receptor acts like a transition-state catalyst to overcome an energy barrier and release the mature virion from the metastable state. This initiates a series of conformational changes that allow the virus to attach to membranes, form a pore, and finally release its RNA genome into the cytoplasm. This model has striking parallels with emerging models for the maturation and cell entry of more complex enveloped viruses such as influenza virus and HIV.

CONTENTS

INTRODUCTION

In many ways simple nonenveloped viruses such as poliovirus and closely related picornaviruses present ideal models for understanding how viruses enter cells and initiate infection. As a result of years of study, arising first from a focus on development of poliovirus vaccines and subsequently from their emergence as model systems, this group of viruses is exceptionally well characterized biochemically and genetically. Receptors for a number of these viruses have been identified, and the viruses are amenable to genetic and structural characterization. However, despite years of characterization the mechanism that allows poliovirus (or any nonenveloped virus) to cross a membrane and release its genome into the cell for replication remains poorly understood. Paradoxically, this process is better understood for enveloped viruses whose structural organization is more complex. Indeed, in many ways the entry process for enveloped viruses is simpler. On a strictly topological basis the nucleocapsid of an enveloped virus is already in the same compartment as the cytoplasm, and this topological equivalence can be realized physically provided that there is a mechanism for facilitating fusion of the viral envelope with a cell membrane. In contrast, the nucleocapsid (virion) of a nonenveloped virus is topologically outside the cell and remains there even if it is internalized in a cellular vesicle. At some point the entire virion, or at least the viral genome, must cross a membrane.

SOME COMMON THEMES IN VIRAL ENTRY

Although there are significant differences in cell entry pathways among viruses, several common themes are emerging from the study of a wide variety of viruses. These common themes are the consequence of a central problem faced by all viruses in the passage from cell-to-cell or from host-to-host. Thus, in order to withstand the rigors of the extracellular environment, the virion must be stable to a variety of insults, which may include extremes of pH, ionic strength, and temperature, and the presence of proteases. The virion must be inert and not adhere to surfaces or to other cells until it reaches its target tissue and encounters its specific receptor. However, once the virion reaches its target cell and encounters the appropriate trigger(s), it must be capable of undergoing structural changes that allow it to cross a membrane and deliver its genome to the appropriate compartment of the cell to initiate replication.

A Link Between Assembly and Entry

For many viruses the solution to this problem involves a link between the final stages of assembly and the initiation of cell entry—an attractive model because it emphasizes the role of the virion as an intermediate (linking assembly and entry) in a cycle. Thus, the final stage of assembly for many viruses involves proteolytic processing of a structural protein (generally a surface glycoprotein for enveloped viruses or a capsid protein of a noneveloped virus). By first folding the structure as a precursor and subsequently introducing a covalent alteration, it becomes possible for the lowest energy structure of the product to differ from the lowest energy structure of the precursor. However, once the precursor has folded, the lowest energy structure of the mature glycoprotein (for an enveloped virus) or the intact mature virus (for the nonenveloped viruses) may not be accessible owing to energy barriers that block the pathway between the two structural forms. The protein or virus particle is then said to be kinetically trapped in a metastable state. When the virus encounters the appropriate trigger (either receptor binding, acidification of an endosome, or both), the protein (or virion) is released from the metastable state and proceeds to a lower-energy state that exposes hydrophobic sequences that allow it to attach to membranes. In enveloped viruses the membrane attachment facilitates fusion of the viral envelope with the cell membrane. In nonenveloped viruses the hydrophobic sequences must either generate a pore or disrupt a cellular membrane to facilitate entry.

The Influenza Hemagglutinin as an Example

Perhaps the best-characterized model for this link between assembly and cell entry is the influenza A virus hemagglutinin [reviewed in (100, 117)]. In influenza A, the hemagglutinin glycoprotein spike provides the site for binding to its cellular receptor (sialic acid residues on glycoproteins and glycolipids on the cell surface) and serves as the fusion protein. The hemagglutinin is first synthesized as a precursor HA_0 and is cleaved late in assembly to form two chains, HA_1 and HA_2. The newly generated N terminus of HA_2 begins with a string of nonpolar amino acids, "the fusion peptide," that ultimately facilitates fusion of the viral envelope with the cell membrane. When the virus binds to its cellular receptor, it is taken up into endosomes. Upon acidification of the endosomes, the hemagglutinin undergoes a massive conformational change that results in the exposure of the fusion peptide, its insertion into the membrane of the endosome, the fusion of the viral envelope with the endosome, and the release of the nucleocapsid into the cytoplasm. Thus, for influenza the receptor plays a single role, namely to concentrate virus at the surface of susceptible cells, and the trigger that releases the hemagglutinin from its metastable state is acidification of the endosome. In other enveloped viruses the receptor also serves as the trigger that induces the conformational changes required for entry, in which case the virus envelope may fuse directly with the plasma membrane.

High-resolution structures of ectodomains of HA_0 (21), the mature HA (118), and of the fusogenic form of HA_2, which is produced by proteolytic removal of HA_1 from HA subsequent to acidification (18), have been solved in Don Wiley's laboratory at Harvard University. The hemagglutinin is a trimer that is held together largely by coil-coil interactions between long helices in the HA_2 subunit (Figure 1, see color insert). The globular head of the trimer comprises much of the HA_1 chain and contains the receptor-binding sites. In HA_0 and HA at neutral pH, the fusion peptide is located at the base of the molecule near the site of attachment to the viral envelope. The fusion peptide is contained in an exposed loop in HA_0 (Figure 1*a*), and upon cleavage to produce HA the newly freed fusion peptide is inserted between the helices and is not exposed (Figure 1*b*). In mature HA at neutral pH, the fusion peptide is located nearly 100 Å away from the receptor-binding site (Figure 1*b*). Acidification induces a massive conformational rearrangement that moves both the fusion peptide and the C terminus of the ectodomain of HA_2 (which in the intact protein is anchored in the viral envelope) to a point near the top (cell-proximal end) of the molecule (Figure 1*c*). At this point the fusion peptide and the cell membrane to which it is attached, and the C terminus of HA_2 and the viral envelope to which it is attached, are in close proximity, allowing fusion of the two membranes to proceed. The detailed mechanism of induction of fusion is still poorly understood.

The Model May be General

The HA of influenza A is the only protein where both the metastable and fusogenic forms of the glycoprotein have been characterized structurally. However, similar conformational changes have been proposed for the envelope glycoproteins of a number of other viruses based on structures of analogues of the fusogenic form (5, 20, 32, 61, 71, 72, 113, 114, 123). Using poliovirus as an example, we demonstrate that similar mechanisms in which receptor binding releases virus from a metastable state and exposes hydrophobic sequences may also occur in nonenveloped viruses.

THE LIFE CYCLE OF POLIO- AND RELATED VIRUSES

Poliovirus is a member of the picornavirus family, which includes a number of significant pathogens of humans (e.g., rhinoviruses, Coxsackieviruses, echoviruses, enteroviruses, and hepatitis A virus) and livestock (e.g., foot-and-mouth disease viruses). The viruses are small (~300 Å in diameter) and consist entirely of an icosahedral protein coat that encapsidates a single-stranded plus-sense RNA genome. Despite their simple organization, the viruses have a complex life cycle [reviewed in (92)].

Assembly

PROTEOLYTIC PROCESSING IS ASSOCIATED WITH VIRAL ASSEMBLY A brief summary of the poliovirus life cycle is shown in Figure 2 (see color insert). Because the process is cyclic we may start at any point. We choose viral assembly as our starting point to develop the parallel with the model described above. Upon release into the cytoplasm, the viral RNA is translated in a single open reading frame to produce a polyprotein that is processed cotranslationally by viral proteases to yield the viral proteins (57, 87). The polyprotein is myristoylated at its N terminus (22). An early cotranslational cleavage of the polyprotein by the viral 2A protease releases a precursor protein myristoyl-P1 from the N terminus of the polyprotein. This cleavage may occur in *cis*. The P1 protein contains all the capsid protein sequences. Subsequent cleavage of P1 by the viral 3CD protease produces the capsid proteins VP1 and VP3 and the immature capsid protein myristoyl-VP0. This cleavage is associated with the assembly of the proteins into a pentameric assembly intermediate, which spontaneously assembles into empty capsid containing 60 copies each of VP0, VP3, and VP1. The empty capsids and pentamers are apparently in equilibrium within the cell, and therefore it has not been possible to determine whether RNA is encapsidated by pentamers or by insertion into the preformed empty capsids. Regardless, the encapsidation appears to be tightly linked to RNA replication, as there is an absolute dependence of encapsidation on de novo synthesis of progeny RNA (77, 78). Encapsidation leads to the formation of a precursor, called the provirion, that contains the RNA and 60 copies each of VP0, VP3, and VP1. The processing of the immature protein myristoyl-VP0 to yield myristoyl-VP4 and VP2 is associated with encapsidation of the RNA. There is no known protease requirement for this cleavage, and it is thought to be autocatalytic, depending only on the capsid proteins themselves and perhaps the viral RNA. The cleavage of VP0 to form the virion is associated with a significant increase in the stability of the particle. The mechanism of release of the virus from the cell is unclear.

Receptor Binding and Cell Entry

RECEPTORS The next round of replication is initiated when the virus encounters its receptor. The receptors for a number of picornaviruses have been identified (11–14, 43, 51, 76, 98, 107). The poliovirus receptor (Pvr) is a CAM-like molecule with three extracellular Ig-like domains (76). There is no known requirement for a coreceptor for poliovirus entry. The cellular function for Pvr is not known, but two paralogs in humans, called Nectin-1 and Nectin-2, are homotrophic adhesion proteins that interact with the actin skeleton through a cytoplasmic protein called afadin (108). Mice lacking the mouse homolog of Nectin-2 show specific defects in spermatogenesis (17). It is not clear whether Pvr plays similar roles, as it lacks sequence motifs in its cytoplasmic domain that mediate interactions with afadin. Interestingly, Pvr, Nectin-1, and Nectin-2 are coreceptors for alpha herpesviruses (38, 99). Several lines of evidence suggest that the first (N-terminal) domain of Pvr is responsible for poliovirus binding and infection. (*a*) Cells expressing the first domain by itself or as a chimera with other Ig-like proteins are susceptible to infection with poliovirus (58, 80, 96, 97). This functionality is independent of the cytoplasmic domain or the nature of the transmembrane anchor, suggesting that intracellular signaling is not critical for Pvr's function as a receptor. Indeed, the ectodomain fused to a GPI anchor is a functional receptor (M. Chow, personal communication). (*b*) Mutations in the first Ig-like domain alter virus binding (2, 15, 79).

THE SITE OF CELL ENTRY IS A MYSTERY Although many textbooks state that picornaviruses enter cells by receptor-mediated endocytosis (70), the actual site and mechanism of cell entry remains unknown. Indeed, the characterization of the entry mechanism is complicated by the relatively high particle-to-pfu (plaque-forming-unit) ratio for polio and its relatives (ranging from 10^2 to 10^3). Thus, when virus or viral-derived particles are detected in a given compartment during entry it is not clear whether the particles are involved in productive or nonproductive events. As a result, classical biochemical and electron microscopic methods (which "count" particles rather than infectious units) have led to contradictory findings. Early experiments with monensin and lysomotrophic amines (which do measure infectious units) suggested that poliovirus entry requires acidification (70). However, subsequent studies with bafilomycin A demonstrated that poliovirus entry is independent of acidification of endosomes (88) and suggested that the contradictory results of the earlier studies were an artifact of inhibition of downstream events in infection, most likely RNA replication. Similar studies indicate that most enteroviruses and some (8), but not all (91), rhinoviruses enter via mechanisms that do not depend on acidification of endosomes.

Others and we have taken advantage of the recent development of dominant-negative mutants of dynamin (28) to further probe the route of entry of poliovirus and several closely related entero- and rhinoviruses. Dynamin is a large GTPase that plays a key role in "pinching off" coated vesicle to form coated pits during classical clathrin-mediated endocytosis. Recent data suggest that dynamin also is required for internalization via caveoli. DeTulleo & Kirchhausen (30) have shown that expression of a dominant-negative dynamin mutant abrogates the ability of rhinovirus 14 but not poliovirus to initiate infection (30). We have characterized the dependence of several additional enteroviruses including echovirus 1, echovirus 7, Coxsackie B3 (Nancy), Coxsackie B3 (RD), and Coxsackievirus B4 (E.S. Mittler, J.M. Bergelson & J.M. Hogle, manuscript in preparation). These studies showed that closely related viruses differ in their dependence on dynamin for cell entry. Thus, echovirus 1 and the RD strain Coxsackievirus B3 were shown to be dynamin dependent, whereas echovirus 7, the Nancy strain of Coxsackievirus B3 (which is the parent of the RD strain), and Coxsackievirus B4 were not. Dynamin dependence seemed to be correlated with the receptor used, but it was not correlated with the nature of the transmembrane anchor of the receptor or the presence of known clathrin-recruiting signals in the cytoplasmic domain of the receptor. Mittler et al. also investigated the effect of domain truncations of ICAM-1 on the dynamin dependence of rhinovirus 14 (E.S. Mittler, J.M. Bergelson & J.M. Hogle, manuscript in preparation). They showed that while infection mediated by full-length (five domain) ICAM-1 was dynamin dependent, infection mediated by mutant receptors with deletions of domain 5 (membrane proximal), domains 4 and 5 or domain 3 were not. The results suggest that dynamin dependence is not an intrinsic property of the virus itself. Thus, the enteroviruses and rhinoviruses are not obligatorily dependent on dynamin-requiring pathways such as endocytosis for cell entry and may be promiscuous in their choice of pathways.

STRUCTURAL ALTERATIONS ASSOCIATED WITH CELL ENTRY For polioviruses, rhinoviruses, and most related enteroviruses, binding cells expressing the receptor at physiological temperatures result in the induction of conformational changes in the virus to produce a particle called the A particle or 135S particle (29, 33). In the course of a typical experimental infection, a significant fraction of the A particle subsequently elutes from cells in what is thought to represent an abortive infection. However, the A particle is also the predominant cell-associated form of the virus early in infection (within the first 20–30 min) (36). At later times (post infection) the levels of A particles begin to decrease. The timing of the disappearance of the A particle is correlated with the timing of the appearance of a second altered form of the virus that has lost its RNA and now sediments at 80S. The trigger for conversion to the 80S form is not known, but it does not require receptor.

The native-to-A particle conversion can also be induced by solubilized forms of the receptor (40, 55) and by the soluble ectodomain of the receptor in the absence of cells (3, 111). The A particle has altered sedimentation behavior (sedimenting at 135S versus 160S for the native virion) and altered antigenicity. In contrast to the virion (which is stable to the proteases and quite soluble), the A particle is sensitive to proteases and is hydrophobic (partitioning into detergent micelles). The A particle has externalized myristoyl-VP4 and the N-terminal extension of the capsid protein VP1 (36), both of which are in the interior of the native virion. The receptor-induced conversion to the A particle is apparently irreversible. Interestingly, transient and reversible exposure of VP4 and the N-terminal extension of VP1 occur when the virus is at physiological temperatures (but not room temperature) in a process that has been termed "breathing" (64). The breathing provides striking evidence for the dynamic nature of the poliovirus structure and suggests that the virus is literally primed to undergo larger, concerted, and irreversible changes associated with receptor binding. Breathing also occurs in other nonenveloped viruses such as Flock House virus (16) and rhinoviruses (63).

NEWLY EXPOSED SEQUENCES IN THE A PARTICLE FACILITATE MEMBRANE ATTACHMENT The exposed amino-terminal extension of VP1 (which is predicted to form an amphipathic helix in all entero- and rhinoviruses) enables the A particle to attach to liposomes (36). Insertion of the N terminus of VP1 (and perhaps the myristoyl group of VP4) may facilitate cell entry either by disrupting a membrane or by forming a pore in a membrane (36). The A particle (and the virus when incubated at 37°C where breathing is efficient) forms channels in planar bilayers (109). This observation, together with unpublished data that show that the channels become pores in the presence of receptor (M. Chow, personal communication), would support a model in which the inserted sequences formed a pore in the membrane through which the RNA could be extruded into the cytoplasm.

IS THE A PARTICLE A CELL ENTRY INTERMEDIATE? Although there is still considerable controversy concerning the role of the two altered particles, it is generally thought that the A particle may be an intermediate in the cell entry pathway and that the 80S empty particle is the final protein product that accumulates after the RNA is released into the cytoplasm to initiate translation and replication. There are several lines of evidence that the A particle is indeed an intermediate: (*a*) The antiviral activity of compounds that bind to the capsid correlates well with their ability to inhibit the formation of the A particle in vitro (110). (*b*) The kinetics of appearance and disappearance of the A particle in the cell during experimental infections at 37°C are consistent with it playing a role in both entry and RNA release (36). (*c*) The ability of the A particle to attach to liposomes provides a compelling model for membrane attachment during entry. (*d*) The ability of the A particle to form channels in membranes (109) could provide a mechanism for RNA translocation. (*e*) The A particle is infectious in a receptor-independent fashion (27). Although the efficiency of infection with the A particle is nearly four orders of magnitude lower than virions, this inefficiency is largely attributable to the lack of a receptor to bring the particle to high concentration at the cell surface. Indeed, the efficiency can be enhanced significantly (to within an order of magnitude of virus) by preincubating the A particles with non-neutralizing antibodies and using these virus/antibody complexes to infect cells expressing the Fc receptor (53).

However, the role of the A particle as a productive intermediate in the cell entry pathway has recently been questioned based on observations that it does not accumulate in cells when cold-adapted viruses are grown at 25°C (31). Although this observation raises an important *caveat* concerning the role of the A particle, the failure to observe an intermediate in a steady-state process in cell entry does not necessarily imply that the intermediate does not exist. In fact, in a steady-state process, an intermediate is expected to accumulate to appreciable levels only if a rate-limiting step in the pathway occurs downstream of the putative intermediate. Recent experiments using a neutral red assay to follow the kinetics of RNA release at 37°C and at 25°C for infections initiated by virions and by A particles demonstrate that RNA release is rate-limiting at 37°C, but that A particle formation is rate-limiting at 25°C (53). Thus, the A particle would not be expected to accumulate in infections at low temperature, and it (or perhaps a similar particle) remains a viable candidate as a cell entry intermediate.

IN VITRO PRODUCTION OF ALTERED PARTICLES The native-to–A particle and A particle–to–80S particle conversion can also be induced in the absence of receptor by warming the particle in hypotonic buffers in the presence of millimolar levels of calcium ions (27, 115). In the absence of calcium the reaction proceeds directly to the 80S particles, suggesting that calcium is required to stabilize the A particle and that depletion of calcium at some stage during the normal entry process may serve as a trigger for RNA release. Like the receptor-mediated native-to–A particle conversion, the thermal-mediated conversion is inhibited by capsid-binding antiviral agents (110, 116).

The ability to recapitulate the conformational alterations by simply warming the virion provides a convenient means for production of large amounts of the altered particles in vitro from purified virus. It also provides a convenient assay system for studying the kinetics of the virion-to-A particle conversion as a function of temperature. These kinetic studies have led to several observations that support the general

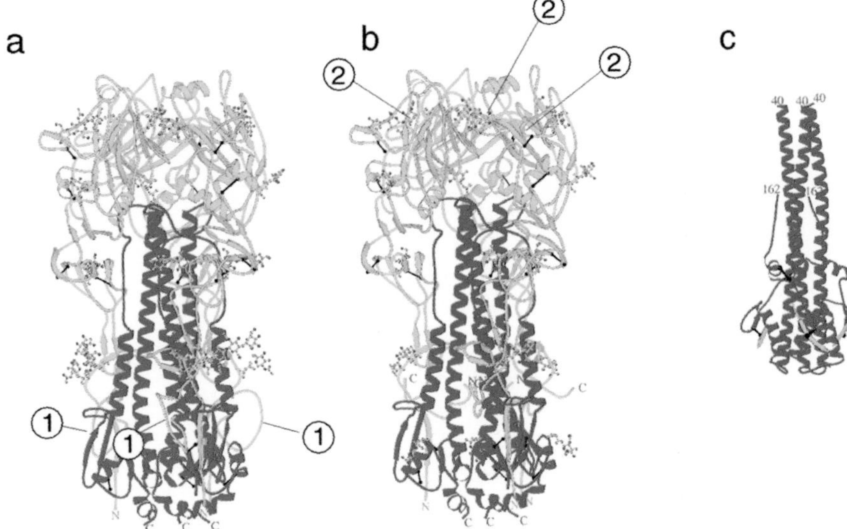

Figure 1 Structural changes on maturation and acidification of the influenza A virus hemagglutinin. The hemagglutinin is a homotrimer. (*a*) The monomer is initially synthesized as a single chain HA_0. (*b*) HA_0 is subsequently processed late in assembly to form two chains, HA_1 (*blue*) and HA_2 (*red*). (*c*) Upon acidification in early endosomes during entry, HA undergoes a significant conformational alteration to a fusogenic form. The newly generated amino terminus of HA_2 (*yellow*) facilitates fusion and is called the fusion peptide. In HA_0 the fusion peptide is in an exposed loop near the base of the molecule (indicated by the 1 in Figure 1*a*). (*b*) In the mature HA the fusion peptide is buried. The fusion peptide is located near the bottom of the molecule, which would be close to the viral membrane, and some 100 Å away from the receptor-binding site (indicated by 2 in Figure 1b). In the fusogenic form (*c*) the fusion peptide and the C terminus of HA_2 (which anchors the molecule in the viral membrane) and the fusion peptide (which is believed to insert into the cell membrane) are both near the top of the molecule. This would bring the viral membrane and cell membrane into close proximity. Figure reproduced from Skehel & Wiley (100), with permission.

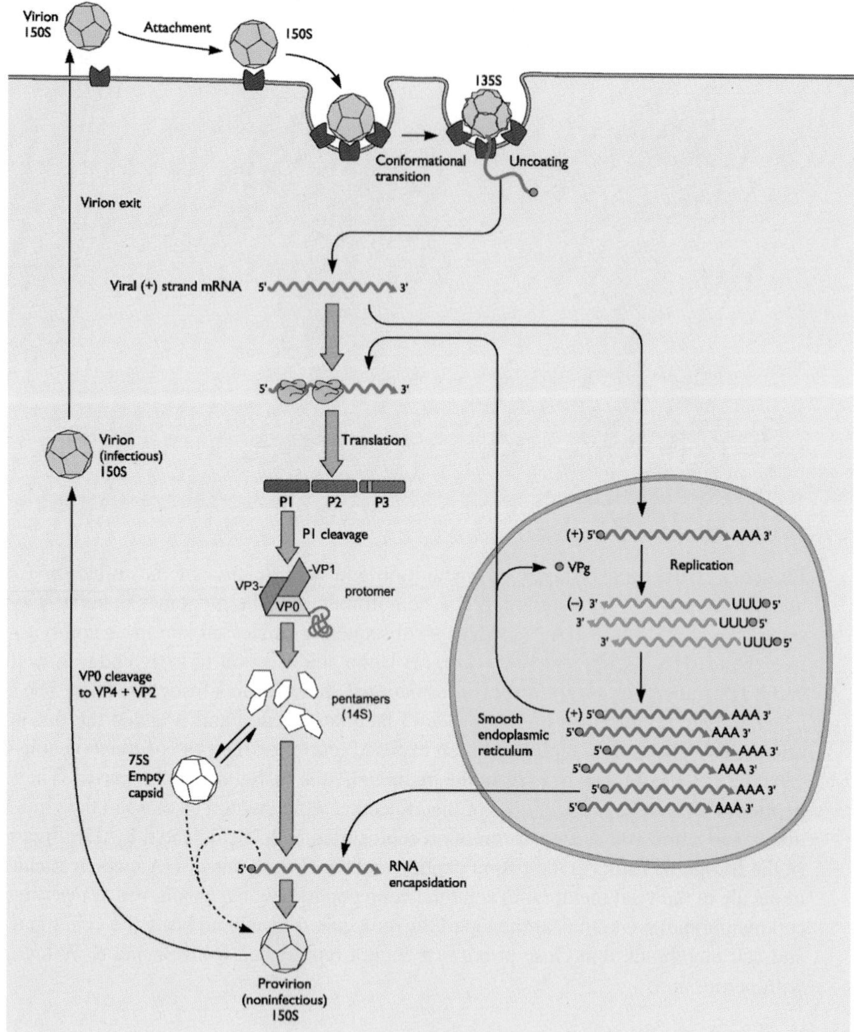

Figure 2 The life cycle of poliovirus and related picornaviruses. Infection is initiated by attachment to receptor, which induces conformational changes in the virus that facilitate translocation of the viral RNA into the cytoplasm where it is replicated to yield progeny RNAs and translated to yield viral proteins. Translation produces a long polyprotein that is processed by viral proteases. Assembly of the virus is linked to processing of the polyprotein and proceeds through a series of intermediates including a protomer, a pentamer, an empty capsid, a provirion, and ultimately the virus. Adapted from *Principles of Virology,* (S. J. Flint, V. R. Racaniello, L. W. Enquist, A. M. Skalka, & R. M. Krug) with permission.

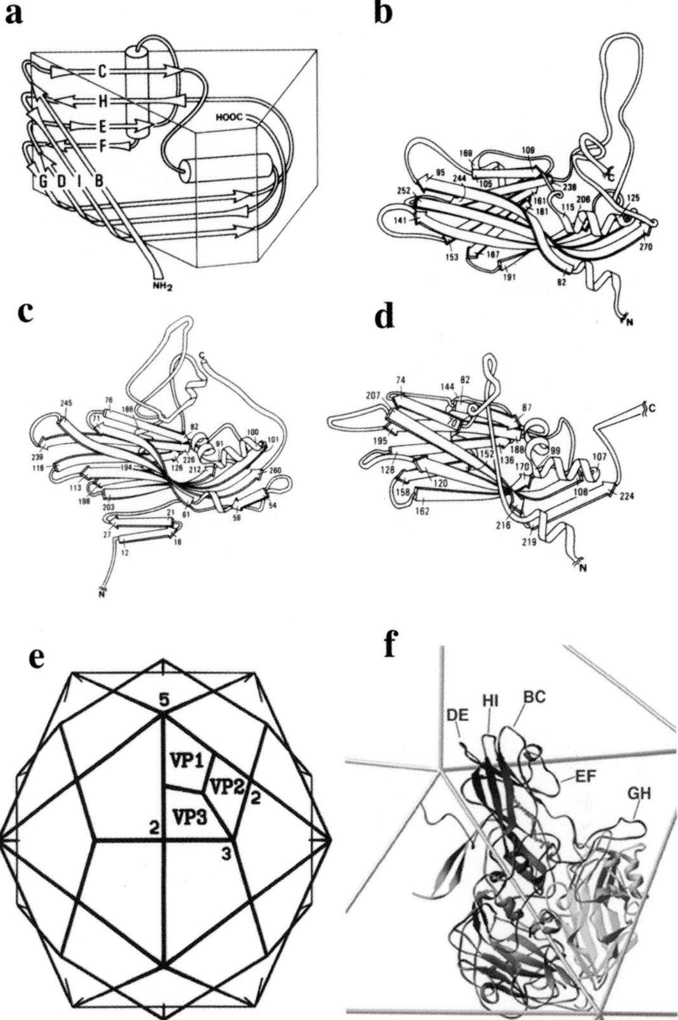

Figure 3 The structure of poliovirus. The virion is composed of 60 copies of 4 proteins, VP1, VP2, VP3, and VP4, arranged on an icosahedral surface. (*a*) A cartoon representation of the core structure eight-stranded beta-sandwich that is shared by VP1, VP2, and VP3 and the capsid proteins of a number of other icosahedral viruses. (*b–d*) Ribbon diagrams of VP1, VP2, and VP3 showing the common beta-sandwich core and the unique loops and terminal extensions of each of the subunits. The N-terminal extensions of VP1 and VP3 have been truncated for clarity. (*e*) An icosahedral framework showing the organization of VP1, VP2, and VP3 with respect to the symmetry axes of the particle. (*f*) A ribbon diagram representation of a single protomer arranged on a portion of the icosahedral framework, showing VP1 (*blue*), VP2 (*yellow*), VP3 (*red*), and VP4 (*green*). A fatty acid–like molecule, here modeled as a sphingosine (*magenta*), binds in the hydrophobic core of VP1.

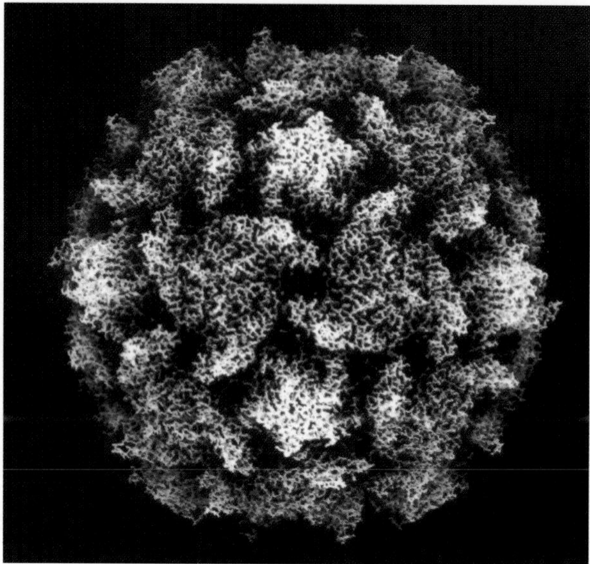

Figure 4 Radial depth cued view of the surface of poliovirus. The virus has been colored based on the distance of individual atoms from the center of the particle, with atoms closest to the center being dark and those farthest from the center being white. Note the star-shaped mesas at the fivefold axes, the three-bladed propellers at the threefold axes, and the deep canyon that separates the star-shaped mesas from the nearest blades of the propellers.

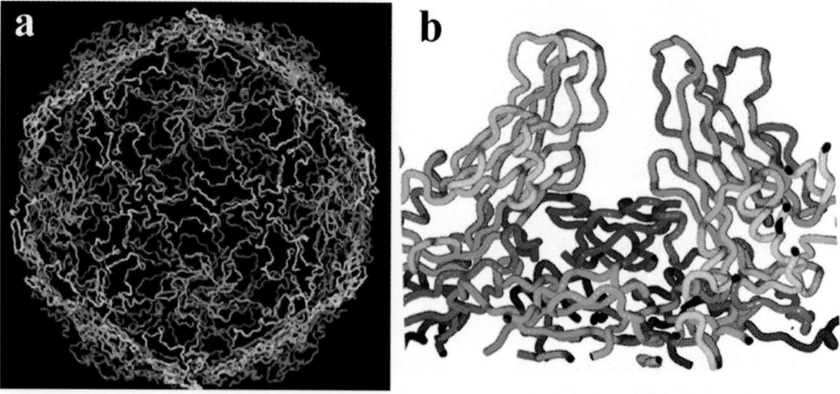

Figure 5 Network formed by the interaction of VP4 and the N-terminal extensions of VP1, VP2, and VP3 on the inner surface of the protein shell. (*a*) The network is viewed from the inside looking out. Only the VP4 (*green*) and N-terminal extensions of VP1 (*blue*), VP2 (*yellow*), and VP3 (*red*) are shown. Note the extensive interactions linking all portions of the structure. (*b*) A cutaway closeup of a plug formed by the N terminus of VP3 (*red*) and the N terminus of VP4 (*green*) that blocks otherwise open channels at the fivefold axes of the viral particle. In this view the inside of the virus is down and the outside is up. Two copies of VP1 flanking one such channel are shown in *blue*.

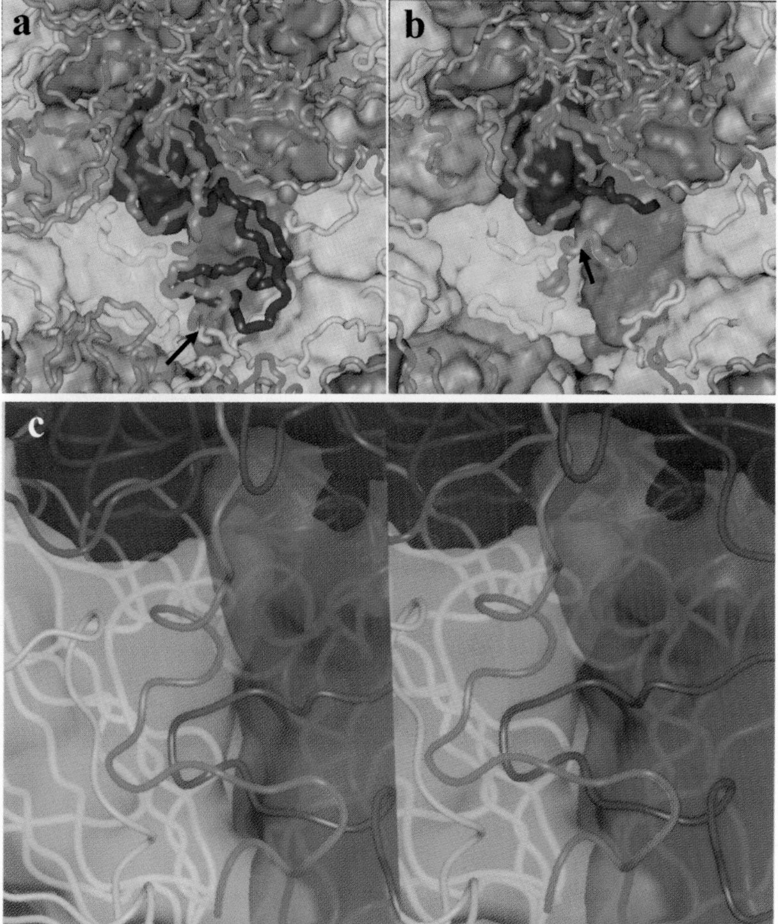

Figure 6 Changes on the inner network that occur upon cleavage of VP0. In these panels the view is from the inside of the particle. VP4 (*green*) and the N-terminal extensions of VP1 (*blue*), VP2 (*yellow*), and VP3 (*red*) are shown as tubes with the bodies of the subunits shown as either (*a*, *b*) opaque surfaces or (*c*) translucent surfaces. (*a*) The network in the mature virion. The N terminus of VP2 and the C terminus of VP4 (generated by the cleavage of VP0) are indicated by an *arrow*. (*b*) The network in the 75S particle prior to VP0 cleavage. The entire N-terminal extension of VP1 and the C terminus of VP4 are disordered, and the N-terminal extension of VP2 is either disordered or rearranged. Only the N-terminal extension of VP3 is intact. The peptide containing the scissile bond of VP0 (*green*) runs across the bottom center of the panel. The scissile bond is indicated by the *arrow*. Note that this peptide blocks access of the N-terminal extension of VP1 and the C terminus of VP4 from the site they occupy in the mature virion, and thus prevents the formation of the mature network. (*c*) A stereo view of the pocket in the inner surface where critical portions of the N terminus of VP1 and the C terminus of VP4 bind in the mature virus. Mutations in these residues suggest that the formation of this portion of the network (which extends otherwise tenuous interactions between VP2 and VP3 within a protomer) is important to viral stability.

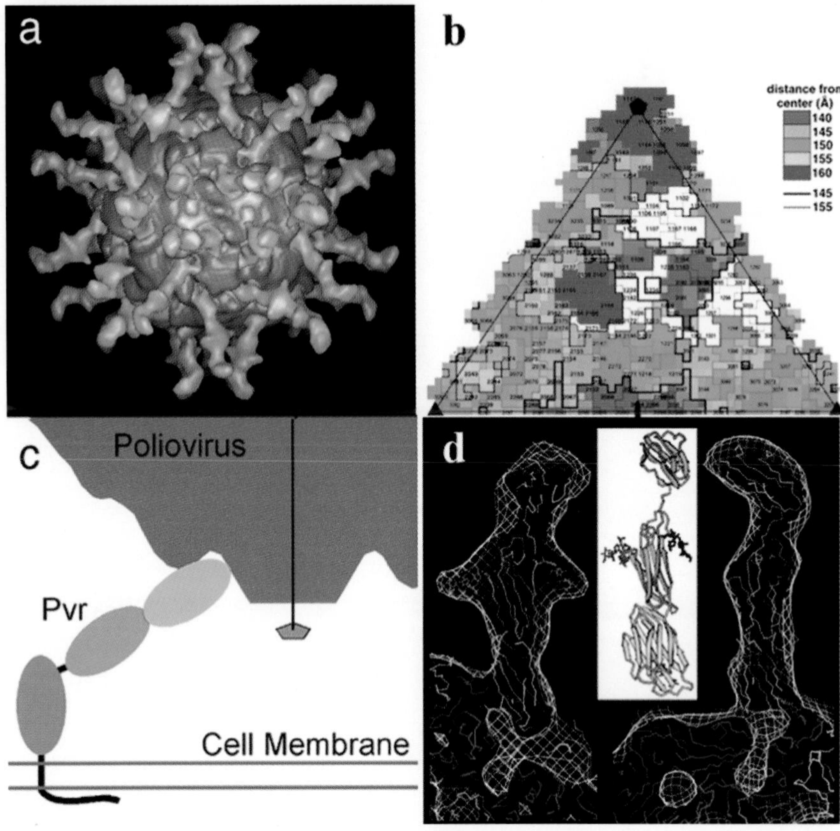

Figure 7 The virus-receptor complex. (*a*) Image reconstruction of the complex between poliovirus and its receptor Pvr. The portion of the reconstruction belonging to the virus is colored *red*, the remaining surface, which represents the receptor, is *blue*. There is compelling density for all three Ig-like domains of the receptor. Prominent arms on the density of the middle domain correspond to glycosylation sites. (*b*) A "roadmap" representation of the footprint of Pvr on the viral surface. A single icosahedral asymmetric unit is shown. The roadmap is color-coded by the distance of residues from the center of the particle with residues closest to the center in *blue* and those farthest from the center in *red*. The footprint is in *white*. (*c*) A cartoon showing how receptor binding would orient the virus with respect to the cell membrane. Note that the geometry of receptor binding observed in the complex would bring a particle fivefold axis (*line with pentagon*) in close proximity to the cell membrane. (*d*) A homology model for Pvr and the atomic model for the virus have been fit to the image reconstruction. Note the good fit to the density, including the fit of a model for glycosylation sites on the second domain to the prominent arms of the receptor reconstruction.

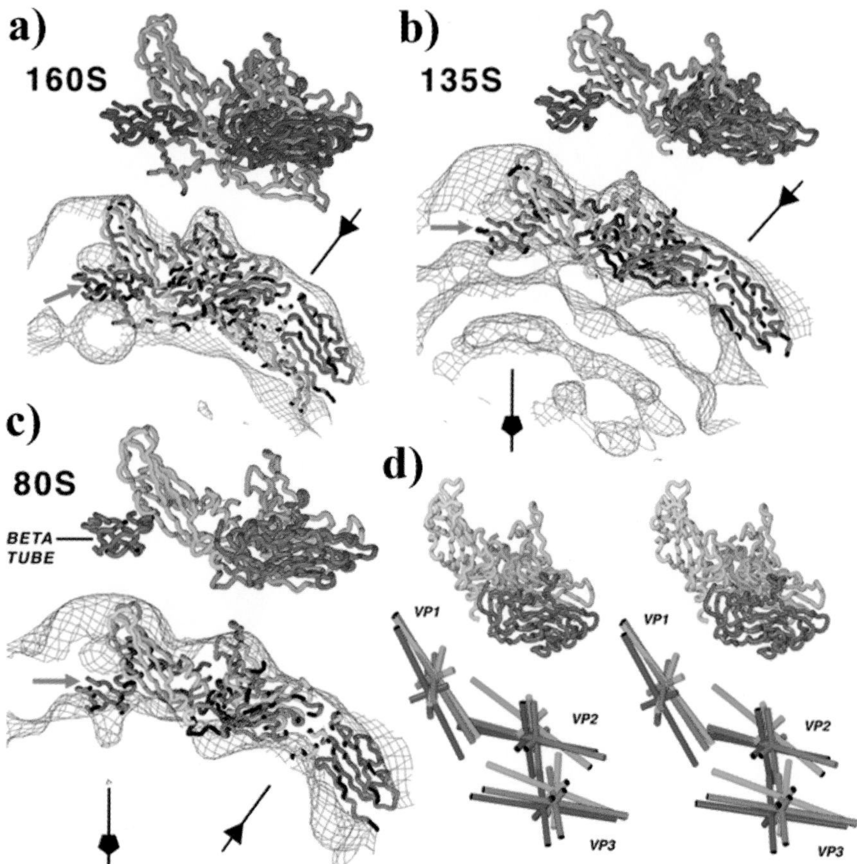

Figure 8 The structures of cell entry intermediates. (*a*) Models for the virus, (*b*) the A or 135S particle, and (*c*) the 80S empty particle have been fit to the reconstruction density, by treating the cores of VP1 (*blue*), VP2 (*yellow*), and VP3 (*red*) as rigid bodies. Note that all three reconstructions contain clear density for the plug formed by the VP3 beta-tube (*red*, indicated by *red arrow*). The particle fivefold axes (*line with pentagon*) and threefold axes (*line with triangle*) are indicated in each panel. (*d*) Stereo representations of simplified models for VP1, VP2, and VP3 in the virion (*purple*), the A particle (*green*), and the 80S particle (*magenta*), showing the movement of the proteins during the structural transitions. Umbrella-like motions of the subunits with VP1 pivoting about the fivefold axes and VP2 and VP3 pivoting about the threefold axes produce a flattened appearance of the model in the A or (*b*) 135S particle. This flattening results in a more angular appearance to the A particle than is readily apparent in reconstruction and in the original micrographs. Reproduced from (9) with permission.

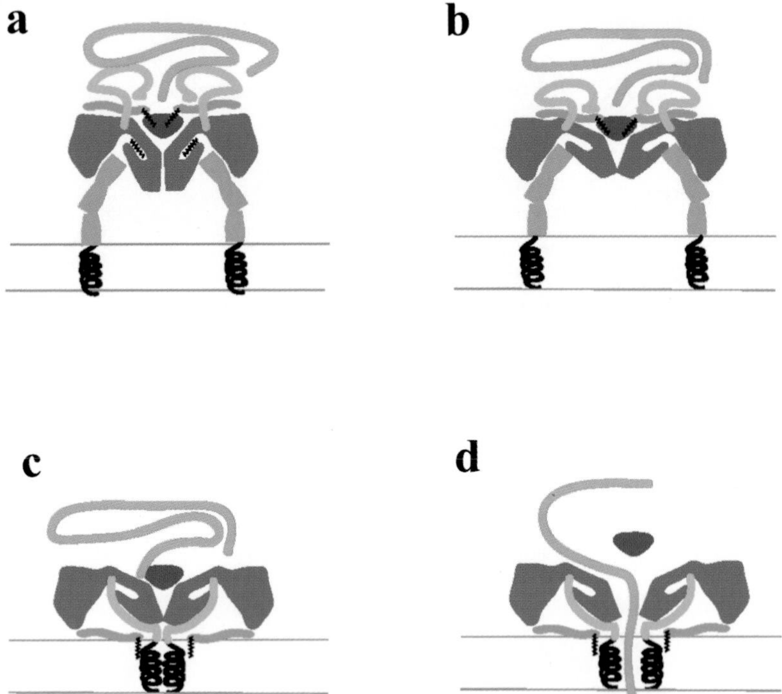

Figure 9 A cartoon representation of a working model for the cell entry mechanism of polio- and related viruses. (*a*) The virus attaches to the receptor on the membrane of a cell to form an initial complex. The three domains of the receptor are shown in *pink* with the transmembrane helices (*black*) anchored in the membrane. The body of the virus is in *blue*, with the pocket factor in *black*. The N-terminal extension of VP1 is *cyan*, the VP3 plug is *red*, VP4 is *green*, the myristate is *black*, and the viral RNA is *orange*. (*b*) At physiological temperatures the receptor induces a subtle conformational change that opens the receptor-binding site and presumably displaces the pocket factor to produce a tight-binding complex. (*c*) The virus undergoes conformational rearrangements that result in the insertion of the amphipathic helices of the N terminus of VP1 (*black spirals*) and the myristate group of VP4 into the membrane to form a channel. (*d*) A trigger (perhaps fluxes in Ca^{+2} concentration) results in the expansion of the particle, the expansion of the pore to form a channel, the temporary removal of the VP3 plug, and the release of the viral RNA into the cell.

model for cell entry discussed in the introduction and that provide additional insights into the role of the receptor in mediating the virion-to-A particle conversion.

THE VIRION IS KINETICALLY TRAPPED The dependence of the native-to–A particle conversion on elevated temperature suggests that there is a high activation energy barrier in the conversion pathway. This would be consistent with the model in which the virus is kinetically trapped in a metastable state. Analysis of the kinetics of both the thermal-induced and the receptor-induced virion-to–A particle conversion as a function of temperature supports this model (110). The rate of the thermal-induced conversion displays steep temperature dependence and is consistent with an activation energy barrier of 140 kcal/mole (110).

THE RECEPTOR FACILITATES CONVERSION BY LOWERING THE ACTIVATION BARRIER The dependence of the receptor-independent reaction on elevated temperature suggests that the receptor may facilitate the conversion at physiological temperature by lowering the activation barrier. Analysis of the kinetics of the receptor-mediated conversion as a function of temperature confirms that the receptor produces a significant enhancement of the rate, such that the rate becomes biologically relevant at physiological temperatures. The data also demonstrate that the receptor lowers the activation barrier for the reaction by nearly 50 kcal/mole. Thus, the receptor behaves like a classical transition-state catalyst (111).

THE CAPSID-BINDING ANTIVIRAL AGENTS STABILIZE VIRUS VIA ENTROPIC EFFECTS As a corollary to our original proposal that receptor would facilitate conversion by lowering the activation barrier, we had proposed that the capsid-binding antiviral agents that prevent receptor-mediated conversion at physiological temperatures would stabilize the virions by raising the activation barrier of the conversion. A model based on this prediction suggests that the antiviral agents act like a peg to rigidify the capsid. Surprisingly, kinetic studies of the thermal-mediated transition show that the capsid-binding antiviral agents have no effect on the activation barrier (an enthalpic term) and suggest that the drugs act via entropic stabilization of the virion (110). These experimental studies are consistent with computational studies of rhinovirus by Post and colleagues that show that binding the drugs increases rather than decreases the compressibility of the virus, and stabilizes the virus by providing a higher density of low-energy states to the virion (89, 90, 105).

THE RECEPTOR-CATALYZED REACTION PROCEEDS THROUGH AN ACTIVATED INTERMEDIATE Although the inhibition of receptor-independent conversion of virions to A particles by capsid-binding antivirals is mediated by strictly entropic effects, the inhibition of the receptor-mediated pathway includes both enthalpic and entropic contributions (111). These results can only be explained if the receptor-mediated pathway proceeds via an intermediate not present in the receptor-independent pathway. By analogy with classic enzyme kinetic models, we propose that this intermediate represents an activated virus-receptor complex in which the virus (and receptor?) undergoes conformational changes and switches from an initial complex to a "tight-binding" complex. This proposal is also consistent with the observation that soluble receptor has two affinities for the virus, with the low-affinity mode dominating at low temperature and the high-affinity mode becoming more prevalent as the temperature is increased (75). In classic models in enzymology, tight-binding complexes are "slow-binding" (the result of the enzyme closing down on the substrate), where a reduction in k_{on} is more than compensated for by a large reduction in k_{off}. In contrast, the tight-binding mode for the Pvr-poliovirus complex is characterized by a significant increase in k_{on}, which more than compensates for a small increase in k_{off} (75). This suggests that the tight-binding mode for the Pvr-poliovirus complex is the result of an opening of the receptor-binding site, making it more accessible to receptor binding.

SNAPSHOTS OF THE CELL ENTRY PATHWAY

In our studies of the poliovirus cell entry pathway we have attempted to obtain structural "snapshots" of stable intermediates and to couple the structural information with the results of genetic, biophysical, and biochemical observations to fill in the gaps. The structural snapshots begin with the virus structure, which has been determined at high resolution by X-ray crystallographic methods (52), and a high-resolution structure of the empty capsid assembly intermediate (which has not yet undergone the maturation cleavage of the immature capsid protein precursor VP0) (7). These snapshots also include structures of the virus-receptor complex (10, 46, 120), the A particle (9), and the 80S particle (9), which have been determined at approximately 22 Å by image reconstruction analysis of electron micrographs of samples in vitreous ice.

The Virion

The structures of the Mahoney strain of type 1 poliovirus (52) and of the closely related rhinovirus 14 (94) were first reported in 1985. Since then, structures of several other picornaviruses have been described (1, 34, 35, 41, 48, 56, 62, 65–67, 84, 85, 104, 112, 121, 122). The protein shell of the virion is composed of 60 copies of the 4 capsid proteins, VP1, VP2, VP3, and VP4, that are arranged on an icosahedral surface. The three large capsid proteins (VP1, VP2, and VP3) share a common fold (an eight-stranded beta-barrel) that is also seen in a number of other plant, insect, and animal viruses (Figure 3*a*, see color insert). The beta-barrel cores of the capsid proteins are decorated with unique loops connecting the beta-strands, and with unique C-terminal and rather long N-terminal extensions (Figure 3*b–d*, see color insert). The beta-barrel cores of the proteins make up the closed shell of the virus with the narrow end of VP1 packing around the fivefold axes and the narrow end of VP2 and VP3 alternating around the threefold axes (Figure 3*e*,*f*, see color insert).

THE OUTER SURFACE The shape of the structure formed by the cores alone is similar among all picornaviruses. The outer surface of the shell is decorated by the connecting loops and C-terminal extensions, which confer unique surface structure to each virus. The outer surface of the poliovirus and its closest relatives (rhinoviruses, Coxsackie A and B viruses, echoviruses, and enteroviruses) are dominated by star-shaped mesas at the fivefold axes and by a three-bladed propeller-like structure at the threefold axes (Figure 4, see color insert). These prominent surface features are punctuated by depressions surrounding the fivefold axes and crossing the twofold axes. In polioviruses, related enteroviruses, and rhinoviruses the depressions surrounding the star-shaped mesa at the fivefold axes are joined to form a moat-like canyon, which has since been the site of receptor attachment for major group rhinoviruses (86), polioviruses (10, 46, 120), and Coxsackie A (119) and B viruses (47).

A HYDROPHOBIC POCKET IN VP1 At the base of the canyon there is an opening into a hydrophobic pocket in the hydrophobic core of the capsid protein VP1. In poliovirus and other enteroviruses, and in most rhinoviruses whose structures are known, the hydrophobic pocket is occupied by a fatty acid–like ligand or "pocket factor" (34) (Figure 3*f*). The pocket factor has been variously modeled as sphingosine, palmitate, or other shorter-chain fatty acids and is probably a mixture of fatty acid–like molecules. Genetic data suggest that the pocket factor may function to regulate the thermal stability of the virion (34, 68). The hydrophobic pocket is also the binding site for the capsid-binding antiviral agents whose antiviral activity stems from their ability to prevent the conformational changes required for cell entry (42, 102).

INTERNAL NETWORK The N-terminal extensions of VP1, VP2, and VP3 together with VP4 decorate the inner surface of the protein shell, forming an elaborate network that contributes substantially to the protein-protein interactions stabilizing the protein shell (Figures 5*a* and 6*a,c*, see color insert). One particularly striking feature of this network is an interaction formed by five copies of the N terminus of VP3 as they intertwine around each of the fivefold axes to form a twisted parallel "beta-tube" (Figure 5*b*, see color insert). This beta-tube is cradled by five copies of the myristoyl moiety at the N terminus of VP4 and is flanked on its inner surface side by five copies of a short three-stranded beta-sheet consisting of two strands from the N terminus of VP4 and one strand that is believed to represent residues from the extreme N terminus of VP1. The beta-tube forms a plug on the inner surface of the shell that blocks a solvent-filled channel (five copies of VP1 define the walls of the channel) between the five copies of VP1 as they pack around each of the fivefold axes.

The Empty Capsid Assembly Intermediate

THE INTERNAL NETWORK IS DISRUPTED IN THE EMPTY CAPSID An intermediate consisting of 60 copies of VP0, VP3, and VP1 and sedimenting at 75S accumulates to low levels in cells infected with some (but not all) picornaviruses. When properly isolated this particle is antigenically indistinguishable from virus and can be completely dissociated to assembly-competent 14S pentamers by treatment at slightly alkaline pH (pH 8.3). The 75S particles are unstable and are readily converted to a faster sedimenting, nondissociable particle with altered antigenicity by exposure to room temperature or extremes of ionic strength (7). Levels of the 75S empty capsid can be increased significantly by adding millimolar levels of guanidine to cells 2–3 h post infection. The guanidine is a specific inhibitor of RNA replication. The structure of the 75S empty capsid intermediate was reported in 1994 (7). As expected the structure of the outer surface and the protein shell are virtually identical to the structure of the protein shell of the virion. However, the network formed by the VP4 and the N-terminal extensions of the other capsid proteins is nearly completely disrupted (Figure 5*a*,*b*). Indeed, with the exception of the N-terminal extension of VP3 (including the beta-tube) and the N terminus of VP4, the peptide segments contributing to the internal network are either rearranged or completely disordered in the empty capsids. The failure to form the network disrupts interactions between pentamers, accounting for the ability to dissociate the capsids into pentamers. The disruption of the network also eliminates extensive contacts between VP1, VP2, and VP3 within a protomer (the copies derived from proteolysis of the capsid precursor P1) and between protomers in the pentamer. The loss of these critical interactions may explain the decreased stability of the empty capsid with respect to the virus.

A MODEL FOR AUTOCATALYTIC CLEAVAGE OF VP0 One portion of the network that is ordered in the empty capsids but in a position that is substantially different from that observed in the virus is the region spanning the scissile bond of VP0 (corresponding to the C terminus of VP4 and the N terminus of VP2 in the virion). In the virion the newly generated chain termini are located close together near the threefold axes (Figure 6*a*), and in both poliovirus and rhinovirus 14 there is an interaction between the hydroxyl oxygen of serine 10 of VP2 and one of the carboxylate oxygens of the C terminus of VP4. This led to the proposal that the autocatalytic cleavage of VP0 proceeds via a serine protease-like mechanism with hydroxyl of serine 10 serving as the nucleophile (4). However, mutagenesis studies revealed that the substitution of an alanine at position 10 did not affect cleavage (45). In the empty capsids the peptide segment spanning the scissile bond is located nearly 25 Å away from the site occupied by the C terminus of VP4 and the N terminus of VP2 in the virion (Figure 6*b*, see color insert). The scissile bond itself is located immediately above a segment of VP2 that contains a stretch of residues, including His 195 of VP2, that is highly conserved in all picornavirus sequences, leading to the proposal that the imidazole side chain served as a base that activates a water that then serves as the nucleophile to initiate peptide bond cleavage. Subsequent mutagenesis studies revealed that mutations of His 195 prevent VP0 cleavage and result in the production of highly unstable provirions (50).

HOW DOES VP0 CLEAVAGE TRAP THE VIRUS IN A METASTABLE STATE? The residues spanning the scissile bond pass over the top of a depression or pocket in the inner surface of the protein shell that produces a thin spot in the protein shell immediately below the interface between VP2 and VP3 from a single protomer. In this virion, this depression is filled with a loop of the N-terminal extension of VP1 [residues 44–56], which is covered on the inner surface by residues from the C terminus of VP4 (Figure 6*c*, see color insert). Interactions of this loop of VP1 with the inner surface of VP2 and VP3 would be expected to contribute significantly to the stability of the capsid. Indeed, in the absence of the interactions involving this loop, the interactions between VP2 and VP3 from the protomer are restricted to a thin band of interactions at the outer edge of the interface joining their beta-barrel cores. Mutations in this loop of VP1, together with residues on the inner surface of the core of VP2 and residues from the C-terminal half of VP4 that contact it, cause a wide variety of phenotypes including alterations of thermal stability (34, 69), resistance to (and in some cases dependence on) capsid-binding drugs (83), mouse adaptation (25, 26, 81), and ability of the virus to grow in cells expressing mutant receptors (24). This suggests that these residues play an important role in regulating structural transitions of the virus and in cell entry. In the empty capsid structure the residues spanning the scissile bond block the N terminus of VP1 from accessing the site in the pocket that it occupies in the mature virion, suggesting that cleavage of VP0 is required for the completion of this portion of the network. This leads to a model in which the cleavage and subsequent rearrangement of the VP4 and the N-terminal extensions of VP1 and VP2 are responsible for locking the virus in the metastable state. The observation that some of the last segments of the network to be put in place (namely VP4 and the N terminus of VP1) are externalized reversibly when the virus "breathes" and irreversibly in receptor-mediated conformational rearrangements early in the entry process suggests that the cleavage and reorganization may also prime the virus for conformational changes required for cell entry. Indeed, analysis of thermal-induced conformational changes in the empty capsid demonstrates that exposure of sequences corresponding to VP4 cannot take place unless VP0 is cleaved (6). We have drawn a comparison of this process to the process of setting a mousetrap.

The Virus-Receptor Complex

THE CAM-LIKE RECEPTORS BIND IN THE CANYON The structures of picornavirus-receptor complexes have been solved at low resolution (15–25 Å) by image reconstruction analysis of cryo-electron micrographs. The complexes include complexes of rhinovirus 14 (59, 86), rhinovirus 16 (59), and Coxsackievirus A12 (119) with ICAM-1 (the receptor for the "major group" rhinoviruses), the complex of poliovirus 1 (Mahoney) with Pvr (10, 46, 120), the structure of the complex of Coxsackievirus B3 with CAR (Coxsackie adeno receptor) (47), and the structure of the complex of rhinovirus 2 with the VLDL receptor (the receptor for the "minor group" rhinoviruses) (49). The ectodomains of ICAM-1, Pvr, and CAR are

CAM-like molecules with 5, 3, and 2 Ig-like domains in their ectodomains respectively. All three CAM-like receptors (Pvr, ICAM-1, and CAR) bind in the canyon of their cognate viruses, but the footprints of the receptors on the virus surface and the geometry of approach of the receptor to the virus surface differ greatly. Only the N-terminal (membrane distal) domain of the receptor contacts the virus surface. The area of the footprint of the Pvr on the poliovirus surface is considerably greater than the footprint of ICAM-1 on rhinovirus, which may account in part for the observation that the affinity of soluble Pvr for poliovirus is higher than the affinity of soluble ICAM-1 for rhinoviruses. Because there is clear density for all three domains of Pvr in the reconstruction of the Pvr-poliovirus complex, it has been possible to model how the complex would orient the virus in the context of the membrane (Figure 7*c*, see color insert). If we assume that multiple receptors attach to the virus, the complex would result in the close approach of a particle fivefold axis to the membrane, which might facilitate insertion of multiple copies of VP4 and the N terminus of VP1 into the membrane upon conversion to the A particle.

The reconstructions have been used to develop pseudoatomic models for the virus-receptor complexes by fitting the model of the virus derived from high-resolution crystallographic studies and crystallographic (ICAM-1 and CAR) or homology (Pvr) models of the receptor to the reconstruction density (Figure 7*c*). There is no evidence for significant conformational changes in the virus upon receptor attachment for any of these viruses.

NOT ALL RECEPTORS BIND IN THE CANYON The structure of the rhinovirus 2-LDL receptor complex differs significantly from the complexes with the CAM-like receptors. The LDL receptor does not bind in the canyon. Instead, it binds to sites at the top of the star-shaped mesas at the fivefold axes, interacting primarily with residues from the BC and the HI loop of VP1 (49). These loops are components of one of the major antigenic sites of rhinoviruses. Even prior to structural studies, the canyon had been predicted to be the most probable site of receptor attachment for rhinoviruses and poliovirus. The original rationale for the "Canyon Hypothesis" was that the receptor-binding site of the virus needed to be inaccessible to the antigen-binding sites of antibodies to prevent selection of mutants incapable of binding receptor under the pressure of a vigorous immune response (94). The underlying logic of the hypothesis has been questioned on theoretical grounds based on the premise that RNA viruses sample all possible mutations, but receptor-binding mutants are nonviable and could therefore never be fixed regardless of the selective pressure. The logic had also been challenged by two earlier experimental observations: (a) Smith and colleagues showed that the binding site of an antibody to one of the primary antigenic sites of rhinovirus 14 overlaps extensively with the ICAM-1-binding site (101, 103). (b) Stuart and colleagues showed that both the RGD sequence that is the presumed binding site for the integrin receptor for FMDV (1) and a site that is used by some tissue culture adapted strains of FMDV as a binding site for cell surface heparin sulfate (37) are highly exposed on the surface of the virus in areas that have been identified as antigenic sites.

THE "CANYON HYPOTHESIS" REVISITED So why is it that the receptor-binding site for poliovirus and for major group rhinovirus is located in a deep invagination in the virus surface, whereas the receptor-binding sites for minor group rhinoviruses and for FMDV are located on highly exposed surfaces? The answer may rest in differences in the roles played by the receptors for these different viruses. Both rhinovirus 2 (91) and FMDV require acidification for productive cell entry, and the receptor for these viruses seems to play only one role, namely to increase the concentration of virus at the surface of the target cell. In contrast, in poliovirus, Coxsackie A and B viruses, and the major group rhinoviruses, the receptor must also trigger conformational changes that are necessary for subsequent events in entry. In order to induce the changes the receptor binding must provide sufficient energy to overcome the kinetic barrier that holds the virus in its native conformation. Thus, the receptor must donate some of its binding energy to facilitate structural rearrangements. To a first approximation the binding energy of a protein-protein interaction is proportional to the surface area that is buried in the interface between the proteins. Thus, one would anticipate that a virus whose receptor served the dual function of binding protein and "lever" to facilitate conformational change would have a larger area of contact with its receptor than a virus whose receptor served only as a binding protein. Locating the receptor-binding site in a deep invagination in the virus surface provides an efficient mechanism for increasing the surface area of the virus-receptor interaction and thus the binding energy available to do the work of initiating conformational changes required for subsequent steps in cell entry.

The A Particle and 80S Particle: Two Cell Entry Intermediates

STRUCTURES DERIVED FROM CRYO-ELECTRON MICROSCOPY Although we have obtained crystals of the A particle and the 80S empty particle, these crystals have not yet proved suitable for high-resolution studies. However, the structures have been solved at 22–23 Å resolution by cryo-electron microscopy (9). The three-dimensional reconstructions for the A particle and 80S particle differ significantly from either a low-resolution surface generated from the high-resolution model of the virion or a reconstruction of the virion at comparable resolution. Models for the coat protein subunits have been docked into the resulting reconstructions and refined using methods that are analogous to the methods used to refine high-resolution crystallographic structures (Figure 8, see color insert). Due to limited resolution, the modeling and refinement have assumed that the individual capsid proteins move as rigid bodies (a form of molecular tectonics). Despite this approximation, the models exhibit a remarkable fit to the reconstruction density, with the exception of several large loops (e.g., the GH loop of VP1), VP4, and the N-terminal extension of VP1, which are reorganized or externalized in the altered particles.

THE STRUCTURAL CHANGES IN THE A PARTICLE Comparison of the models suggests that the virion-to–A particle transition is characterized by shifts in all three major capsid proteins (VP1, VP2, and VP3). In the model for the A particle, the subunits of VP1 have undergone a movement similar to the opening of an umbrella, with the tips of the subunits at the fivefold axes serving as a pivot and the wide end of the subunit (which forms the north wall and part of the base of the canyon) moving radially outward (Figure 8*d*). This pivoting about the residues near the fivefold axes may provide an explanation for previous observations that viruses with mutations in residues in the BC, GH, and DE loops of VP1 often have phenotypes that include alterations in their ability to undergo thermal- and receptor-mediated conformational changes (24, 73, 74, 116). VP2 and VP3 undergo a similar movement, with the narrow ends of the subunits pivoting about the threefold axes and the wide ends pivoting outward radially. VP1, VP2, and VP3 also undergo significant tangential reorientation (in the plane of the virus surface), with the tangential shifts of VP2 being largest. Because of the umbrella-like motions, the base of the canyon is moved outward appreciably, giving the A particle a more angular appearance, especially when viewed down the twofold axes. In this view the models, the reconstructions, and even the micrographs take on a distinctly hexagonal appearance. This is in sharp contrast to the virion, which appears roughly spherical in all views. The movement of VP1 also results in opening gaps at the base of the canyon between fivefold-related copies of VP1. The separation is also apparent in the reconstructions (the density is noticeably thinner) but is not sufficient to result in the appearance of "holes" in the density at 22 Å resolution.

THE STRUCTURAL CHANGES ARE PARTIALLY REVERSED IN THE 80S PARTICLE In the models for the 80s particle the subunits have undergone a partial reversal of the structural rearrangements that characterized the virion-to–A particle transition (Figure 8*c*). Indeed the orientations of VP2 and VP3 in the 80S particle are similar to their orientations in the virion. In contrast the orientation of VP1 remains similar to that seen in the A particle.

WHERE DO VP4 AND THE N TERMINUS OF VP1 EXIT THE VIRION? Based on analogies to several other viruses, Rossmann and colleagues have proposed that during the N-to-A transition, VP4 and the N terminus of VP1 exit the virion via channels at the fivefold axes (44, 82, 95). The model is attractive because it would place five copies of the N terminus of VP1 (which is predicted to form an amphipathic helix) in a position where they could interact and insert into the membrane to form a channel. However, as discussed above, this channel is blocked at its base on the inside surface of the virion by a "plug" consisting of a beta-tube formed by five copies of the N terminus of VP3 interdigitating around the fivefold axis. In all three reconstructions (virion, A particle, and 80S particle) there is clear density for the VP3 plug occluding the channel (Figure 8*a–c*). More importantly, there is simply no room in the channel in either the A particle or the 80S particle reconstruction for five copies of the part N terminus of VP1, which must remain after the extreme N-terminal segment has been externalized (Figure 8*b,c*).

An alternative model for the egress of VP4 and the N terminus of VP1 has been proposed based on an analogy with the expansion of several structurally related plant viruses. These viruses undergo a significant (~10%) expansion when exposed to slightly basic pH in the presence of chelators of divalent cations (54). The structure of the expanded state of TBSV has been solved at 7 Å by X-ray crystallography (93), and the structure of the expanded state of CCMV has been solved at 25 Å by cryo-electron microscopy (106). In both structures, expansion is characterized by a coordinated rotation and outward movement of the subunits analogous to the movements of VP2 and VP3 along the threefold axes, and by a rotation and outward movement of the subunits analogous to the movement VP1 along the fivefold axes. These coordinated movements result in the opening of large holes at the interfaces between the capsid protein subunits, at positions structurally analogous to the base of the canyon in the picornaviruses. In the case of TBSV, the N-terminal domain of two of the capsid protein subunits exits through this pore upon expansion (93). The extrusion of the N-terminal domain is reversible if the particles are "annealed" by gradually lowering the pH, but the N terminus is trapped outside if the particle is "quenched" by rapid acidification (39).

Mutations in the interfaces of poliovirus that are analogous to the interfaces disrupted during expansion of the plant viruses have been shown to affect viral stability (34, 69) and to alter the sensitivity of the virus to neutralization by soluble receptor (23). Although the capsid protein movements are less pronounced in the virion-to–A particle transition of poliovirus than in the expansion of the plant viruses, the umbrella-like movements of VP1, VP2, and VP3 do result in significant gaps between the subunits at the base of the canyon in positions. These gaps are in positions that are entirely analogous to the positions of larger openings seen in the plant viruses. Inspection of the reconstruction density near the gaps (which are located right where the N-terminal extension of VP1 leaves the beta-barrel core and enters the interior of the virion) suggests that the N-terminal extensions of VP1 may exit through these gaps and proceed up the outer surfaces of the mesas at the fivefold axes. Again this model is attractive, as it would locate five copies of the presumed amphipathic helix at the N terminus of VP1 close together such that they could form a fivefold helical bundle once inserted into the membrane.

ARE THERE ADDITIONAL INTERMEDIATES? One of the more notable features of the reconstructions of the A particle and the 80S particle is the lack of openings in the viral surface that would be sufficiently large to allow the facile extrusion of VP4 and the N-terminal extension of VP1 in the N–to–A particle transition or the RNA in the A particle–to–80S particle transition. Indeed, the relatively small extent of the expansion that occurs during the N–to–A particle transition was a surprise. Given the rather large shift in sedimentation coefficient (135S for the A particle versus 160S for the virion), we had anticipated a more significant expansion (on the order of 10%–15%). We therefore postulate that there are additional, as yet undetected, intermediates in the pathway located between the virion and the A particle and the A particle and the 80S particle. In these intermediates we propose that the particles would be sufficiently expanded to create openings for the release of the VP4 and the N terminus of VP1 in the N–to–A particle transition and of RNA in the A particle–to–80S particle transition. The failure to detect these intermediates to date may indicate that they are transient. The additional intermediate in the N-to-A pathway would then be analogous to the expanded form of the plant virus, and the A particle itself would correspond to the quenched form of the expanded plant virus in which the N-terminal arm is trapped outside the particle. The transient and reversible externalization of the N terminus of VP1 that occurs when the virus breathes may indicate a reversible equilibrium between the virion and the new intermediate that is analogous to the expansion and slow acidification transition that returns the N-terminal arm to the inside of the particle in the plant viruses. Efforts to identify conditions that would allow these putative additional intermediates to be isolated and characterized are in progress.

HOW DOES THE RNA EXIT THE VIRION? The nature of the intermediate linking the A particle and the 80S particle is less clear. Indeed, there are significant questions that remain to be answered concerning how the viral RNA is released from the particle. Nonetheless, it is clear that some factor must dictate that the RNA be released from a unique site in an otherwise icosahedrally symmetric particle, lest the RNA and protein shell be tied in an inextricable knot. In the course of a natural infection, one factor that could influence which site is chosen for RNA release could be the presence of membrane near one surface of the virus. However, the presence of membrane is not a necessary factor because the RNA is efficiently released in the absence of membrane in the in vitro (thermal-mediated) conversion. Other factors that could regulate which site is used, among the many otherwise equivalent sites, could include steric factors (e.g., the proximity of a specific structure in the viral RNA, including perhaps the genome-linked protein VPg) or kinetic factors (e.g., a slow initiation, followed by rapid release at the first initiation site that is established). Because the structures observed to date are symmetric, they can tell us little about the site of release. Indeed, the only clues from the structures that may indicate the site of release are differences in the density of the RNA in the reconstructions of the virion and the A particle. The density in the interior region of the virion provides a view of the icosahedrally averaged structure of the linear genome. The appearance of the average RNA structure differs significantly in the virion and the A particle, with the RNA making a significantly closer approach to the fivefold axes in the A particle.

A WORKING MODEL FOR POLIOVIRUS CELL ENTRY

Based on a combination of the structural, genetic, and biochemical evidence available to date we propose a working model for the cell entry of poliovirus, related enteroviruses, and major group rhinoviruses (Figure 9, see color insert). The first step in the entry pathway is the formation of an initial binding complex with the receptor (Figure 9*a*). The formation would require no significant conformational alterations in either the virus or the receptor and would be the predominant form of the complex at low temperatures. We would propose that the published structures of the complexes represent this form of the complex. At more physiological temperatures the formation of this initial complex induces conformational changes in the virus (and perhaps the receptor) to form a tight-binding complex (Figure 9*b*).

Rossmann et al. (95) have proposed a model for the tight-binding complex in which the receptor is initially bound with most of the contacts between virus and receptor involving the "south wall" of the canyon. Subsequent structural changes in the virus cause the VP1 to move away from the fivefold axes, pinching the canyon such that the receptor makes bridging contacts with both the north and south walls of the canyon. In this model the pinching motion results in the opening of the channels at the fivefold axes. However, because the conformational alteration results in more extensive contacts with the receptor in a more "closed contact," this model would predict that the tight-binding complex be characterized by a slower on-rate and a much slower off-rate for the receptor. This prediction is contradicted by kinetic data for both poliovirus/Pvr and rhinovirus/ICAM-1 binding, which show that the tight-binding complex is characterized instead by a faster off-rate, and a much faster on-rate (19, 75, 120). The Rossmann model also predicts that VP4 and the N terminus of VP1 are released through the newly opened fivefold channel, which is inconsistent with the structure of the A particle.

We propose an alternative model in which the transition from the initial binding complex to the tight-binding complex is characterized by movements of VP1, VP2, and VP3 that mimic the umbrella-like movements of the virion–to–A particle transition (Figure 9*b*). Consistent with the kinetic data, this would open the receptor-binding site, providing for a faster association rate for binding additional receptors. This tight-binding state is detectable at room temperature, but it may be transient at physiological temperature. At physiological temperatures binding of multiple receptors destabilizes the particles, resulting in the release of the pocket factor as the particle begins to expand to form the first transient state. This expansion permits the externalization of VP1 and the N terminus of VP4. Once these peptide segments are externalized, the particle undergoes a partial reversal of the expansion to form the A particle. At some time in this process, five copies of the N terminus of VP1 insert into the membrane by a mechanism that may be facilitated by the myristoyl group at the N terminus of VP4 (Figure 9*c*). The membrane-associated portions of VP1 then associate to form a channel composed of five amphipathic helices, each with its hydrophobic surface facing the membrane and its hydrophilic surface facing the interior of the channel (Figure 9*c*). At some point later in the entry process, some as yet unknown trigger results in a second round of expansion and RNA release. We would suggest that in this expansion the wide ends of VP1, VP2, and VP3 may serve as the pivot points for an umbrella-like movement of VP1 away from the fivefold axes. These movements would be coupled with a rotation of the five amphipathic helices, and a movement of the internal plug formed by the N termini of VP3 to form a large channel through which RNA is released (Figure 9*d*). Because the A particle does not have an appreciable affinity for the receptor, we feel that it is unlikely that the receptor plays any specific role in triggering this expansion.

Although the trigger has not yet been identified, the observation that the 135S particle spontaneously converts to 80S particles in the absence of millimolar levels of calcium (115) suggests that changes in calcium concentration may play a role in triggering expansion and RNA release. After the RNA has been released, the VP3 plug is reinserted, the pore closes, and the particle shrinks to form the 80S structure.

ACKNOWLEDGMENTS

The author would like to acknowledge the contributions of the students and postdoctoral fellows whose works are cited in this review, the support of NIH (AI20566), and the collaborators who have contributed significantly to the ideas presented, including Dave Filman, Marie Chow, Jeff Bergelson, Vincent Racaniello, and Alasdair Steven. I would particularly like to acknowledge the contributions of Don Wiley, a teacher and colleague whose insights have shaped the way we all think about viruses and their interactions with cells.

The *Annual Review of Microbiology* is online at http://micro.annualreviews.org

LITERATURE CITED

1. Acharya R, Fry E, Stuart D, Fox G, Rowlands D, Brown F. 1989. The three-dimensional structure of foot-and-mouth disease virus at 2.9 Å resolution. *Nature* 337:709–16
2. Aoki J, Koike S, Ise I, Sato-Yoshida Y, Nomoto A. 1994. Amino acid residues on human poliovirus receptor involved in interaction with poliovirus. *J. Biol. Chem.* 269:8431–38
3. Arita M, Koike S, Aoki J, Horie H, Nomoto A. 1998. Interaction of poliovirus with its purified receptor and conformational alteration in the virion. *J. Virol.* 72:3578–86
4. Arnold E, Luo M, Vriend G, Rossmann MG, Palmenberg AC, et al. 1987. Implications of the picornavirus capsid structure for polyprotein processing. *Proc. Natl. Acad. Sci. USA* 84:21–25
5. Baker KA, Dutch RE, Lamb RA, Jardetzky TS. 1999. Structural basis for paramyxovirus-mediated membrane fusion. *Mol. Cell* 3:309–19
6. Basavappa R, Gomez-Yafal A, Hogle JM. 1998. The poliovirus empty capsid specifically recognizes the poliovirus receptor and undergoes some, but not all, of the transitions associated with cell entry. *J. Virol.* 72:7551–56
7. Basavappa R, Syed R, Flore O, Icenogle JP, Filman DJ, Hogle JM. 1994. Role and mechanism of the maturation cleavage of VP0 in poliovirus assembly: structure of the empty capsid assembly intermediate at 2.9 Å resolution. *Protein Sci.* 3:1651–69
8. Bayer N, Prchla E, Schwab M, Blaas D, Fuchs R. 1999. Human rhinovirus HRV14 uncoats from early endosomes in the presence of bafilomycin. *FEBS Lett.* 463:175–78
9. Belnap DM, Filman DJ, Trus BL, Cheng N, Booy FP, et al. 2000. Molecular tectonic model of virus structural transitions: the putative cell entry states of poliovirus. *J. Virol.* 74:1342–54
10. Belnap DM, McDermott BM Jr, Filman DJ, Cheng N, Trus BL, et al. 2000. Three-dimensional structure of poliovirus receptor bound to poliovirus. *Proc. Natl. Acad. Sci. USA* 97:73–78

11. Bergelson JM, Chan M, Solomon KR, St. John NF, Lin H, Finberg RW. 1994. Decay-accelerating factor (CD55), a glycosylphosphatidylinositol-anchored complement regulatory protein, is a receptor for several echoviruses. *Proc. Natl. Acad. Sci. USA* 91:6245–48
12. Bergelson JM, Cunningham JA, Droguett G, Kurt-Jones EA, Krithivas A, et al. 1997. Isolation of a common receptor for coxsackie B viruses and adenoviruses 2 and 5. *Science* 275:1320–23
13. Bergelson JM, Krithivas A, Celi L, Droguett G, Horwitz MS, et al. 1998. The murine CAR homolog is a receptor for coxsackie B viruses and adenoviruses. *J. Virol.* 72:415–19
14. Bergelson JM, Shepley MP, Chan BMC, Hemler ME, Finberg RW. 1995. Identification of the integrin VLA-2 as a receptor for echovirus 1. *Science* 255:1718–20
15. Bernhardt G, Harber J, Zibert A, DeCrombrugghe M, Wimmer E. 1994. The poliovirus receptor: identification of domains and amino acid residues critical for virus binding. *Virology* 203:344–56
16. Bothner B, Dong XF, Bibbs L, Johnson JE, Siuzdak G. 1998. Evidence of viral capsid dynamics using limited proteolysis and mass spectrometry. *J. Biol. Chem.* 273:673–76
17. Bouchard MJ, Dong Y, McDermott BM Jr, Lam DH, Brown KR, et al. 2000. Defects in nuclear and cytoskeletal morphology and mitochondrial localization in spermatozoa of mice lacking nectin-2, a component of cell-cell adherens junctions. *Mol. Cell Biol.* 20:2865–73
18. Bullough PA, Hughson FM, Skehel JJ, Wiley DC. 1994. Structure of influenza haemagglutinin at the pH of membrane fusion. *Nature* 371:37–43
19. Casasnovas JM, Springer TA. 1995. Kinetics and thermodynamics of virus binding to receptor. Studies with rhinovirus, intercellular adhesion molecule-1 (ICAM-1), and surface plasmon resonance. *J. Biol. Chem.* 270:13216–24
20. Chan DC, Fass D, Berger JM, Kim PS. 1997. Core structure of gp41 from the HIV envelope glycoprotein. *Cell* 89:263–73
21. Chen J, Lee KH, Steinhauer DA, Stevens DJ, Skehel JJ, Wiley DC. 1998. Structure of the hemagglutinin precursor cleavage site, a determinant of influenza pathogenicity and the origin of the labile conformation. *Cell* 95:409–17
22. Chow M, Newman JFE, Filman D, Hogle JM, Rowlands DJ, Brown F. 1987. Myristylation of picornavirus capsid protein VP4 and its structural significance. *Nature* 327:482–86
23. Colston E, Racaniello VR. 1994. Soluble receptor-resistant poliovirus mutants identify surface and internal capsid residues that control interaction with the cell receptor. *EMBO J.* 13:5855–62
24. Colston EM, Racaniello VR. 1995. Poliovirus variants selected on mutant receptor-expressing cells identify capsid residues that expand receptor recognition. *J. Virol.* 69:4823–29
25. Couderc T, Guédo N, Calvez V, Pelletier I, Hogle J, et al. 1994. Substitutions in the capsids of poliovirus mutants selected in human neuroblastoma cells confer on the Mahoney type 1 strain a phenotype neurovirulent in mice. *J. Virol.* 68:8386–91
26. Couderc T, Hogle J, Le Blay H, Horaud F, Blondel B. 1993. Molecular characterization of mouse-virulent poliovirus type 1 Mahoney mutants: involvement of residues of polypeptides VP1 and VP2 located on the inner surface of the capsid protein shell. *J. Virol.* 67:3808–17
27. Curry S, Chow M, Hogle JM. 1996. The poliovirus 135S particle is infectious. *J. Virol.* 70:7125–31
28. Damke H, Baba T, Warnock DE, Schmid SL. 1994. Induction of mutant dynamin specifically blocks endocytic coated vesicle formation. *J. Cell Biol.* 127:915–34
29. De Sena J, Mandel B. 1977. Studies on the in vitro uncoating of poliovirus. II. Characteristics of the membrane-modified particle. *Virology* 78:554–66

30. DeTulleo L, Kirchhausen T. 1998. The clathrin endocytic pathway in viral infection. *EMBO J.* 17:4585–93
31. Dove AW, Racaniello VR. 1997. Cold-adapted poliovirus mutants bypass a post-entry replication block. *J. Virol.* 71:4728–35
32. Fass D, Harrison SC, Kim PS. 1996. Retrovirus envelope domain at 1.7 angstrom resolution. *Nat. Struct. Biol.* 3:465–69
33. Fenwick ML, Cooper PD. 1962. Early interactions between poliovirus and ERK cells. Some observations on the nature and significance of the rejected particles. *Virology* 18:212–23
34. Filman DJ, Syed R, Chow M, Macadam AJ, Minor PD, Hogle JM. 1989. Structural factors that control conformational transitions and serotype specificity in type 3 poliovirus. *EMBO J.* 8:1567–79
35. Filman DJ, Wien MW, Cunningham JA, Bergelson JM, Hogle JM. 1998. Structure determination of echovirus 1. *Acta Crystallogr. D* 54:1261–72
36. Fricks CE, Hogle JM. 1990. Cell-induced conformational change of poliovirus: Externalization of the amino terminus of VP1 is responsible for liposome binding. *J. Virol.* 64:1934–45
37. Fry EE, Lea SM, Jackson T, Newman JW, Ellard FM, et al. 1999. The structure and function of a foot-and-mouth disease virus-oligosaccharide receptor complex. *EMBO J.* 18:543–54
38. Geraghty RJ, Krummenacher C, Cohen GH, Eisenberg RJ, Spear PG. 1998. Entry of alphaherpesviruses mediated by poliovirus receptor-related protein 1 and poliovirus receptor. *Science* 280:1618–20
39. Golden J, Harrison SC. 1982. Proteolytic dissection of turnip crinkle virus subunit in solution. *Biochemistry* 21:3862–66
40. Gomez Yafal A, Kaplan G, Racaniello VR, Hogle JM. 1993. Characterization of poliovirus conformational alteration mediated by soluble cell receptors. *Virology* 197:501–5
41. Grant RA, Filman DJ, Fujinami RS, Icenogle JP, Hogle JM. 1992. Three-dimensional structure of Theiler virus. *Proc. Natl. Acad. Sci. USA* 89:2061–65
42. Grant RA, Hiremath C, Filman DJ, Syed R, Andries K, Hogle JM. 1994. Structures of poliovirus complexes with antiviral drugs: implications for viral stability and drug design. *Curr. Biol.* 4:784–97
43. Greve JM, Davis G, Meyer AM, Forte CP, Yost SC, et al. 1989. The major human rhinovirus receptor is ICAM-1. *Cell* 56:839–47
44. Hadfield AT, Lee W-M, Zhao R, Oliveira MA, Minor I, et al. 1997. The refined structure of human rhinovirus 16 at 2.15 Å resolution: implications for the viral life cycle. *Structure* 5:427–41
45. Harber JJ, Bradley J, Anderson CW, Wimmer E. 1991. Catalysis of poliovirus VP0 maturation cleavage is not mediated by serine 10 of VP2. *J. Virol.* 65:326–34
46. He Y, Bowman VD, Mueller S, Bator CM, Bella J, et al. 2000. Interaction of the poliovirus receptor with poliovirus. *Proc. Natl. Acad. Sci. USA* 97:79–84
47. He Y, Chipman PR, Howitt J, Bator CM, Whitt MA, et al. 2001. Interaction of coxsackievirus B3 with the full length coxsackievirus-adenovirus receptor. *Nat. Struct. Biol.* 8:874–78
48. Hendry E, Hatanaka H, Fry E, Smyth M, Tate J, et al. 1999. The crystal structure of coxsackievirus A9: new insights into the uncoating mechanisms of enteroviruses. *Struct. Fold Des.* 7:1527–38
49. Hewat EA, Neumann E, Conway JF, Moser R, Ronacher B, et al. 2000. The cellular receptor to human rhinovirus 2 binds around the 5-fold axis and not in the canyon: a structural view. *EMBO J.* 19:6317–25
50. Hindiyeh M, Li QH, Basavappa R, Hogle JM, Chow M. 1999. Poliovirus mutants at histidine 195 of VP2 do not cleave VP0 into VP2 and VP4. *J. Virol.* 73:9072–79
51. Hofer F, Gruenberger M, Kowalski H, Machat H, Huettinger M, et al. 1994.

Members of the low density lipoprotein receptor family mediate cell entry of a minor-group common cold virus. *Proc. Natl. Acad. Sci. USA* 91:1839–42
52. Hogle JM, Chow M, Filman DJ. 1985. Three-dimensional structure of poliovirus at 2.9 Å resolution. *Science* 229:1358–65
53. Huang Y, Hogle JM, Chow M. 2000. Is the 135S poliovirus particle an intermediate during cell entry? *J. Virol.* 74:8757–61
54. Incardona NL, Kaesberg P. 1964. A pH-induced structural change in bromegrass mosaic virus. *Biophys. J.* 4:11–21
55. Kaplan G, Freistadt MS, Racaniello VR. 1990. Neutralization of poliovirus by cell receptors expressed in insect cells. *J. Virol.* 64:4697–702
56. Kim S, Smith TJ, Chapman MS, Rossmann MG, Pevear DC, et al. 1989. Crystal structure of human rhinovirus serotype 1A (HRV1A). *J. Mol. Biol.* 210:91–111
57. Kitamura N, Semler BL, Rothberg PG, Larsen GR, Adler CJ, et al. 1981. Primary structure, gene organization and polypeptide expression of poliovirus RNA. *Nature* 291:547–53
58. Koike S, Ise I, Nomoto A. 1991. Functional domains of the poliovirus receptor. *Proc. Natl. Acad. Sci. USA* 88:4104–8
59. Kolatkar PR, Bella J, Olson NH, Bator CM, Baker TS, Rossmann MG. 1999. Structural studies of two rhinovirus serotypes complexed with fragments of their cellular receptor. *EMBO J.* 18:6249–59
60. Deleted in proof
61. Lamb RA, Joshi SB, Dutch RE. 1999. The paramyxovirus fusion protein forms an extremely stable core trimer: structural parallels to influenza virus haemagglutinin and HIV-1 gp41. *Mol. Membr. Biol.* 16:11–19
62. Lentz KN, Smith AD, Geisler SC, Cox S, Buontempo P, et al. 1997. Structure of poliovirus type 2 Lansing complexed with antiviral agent SCH48973: comparison of the structural and biological properties of three poliovirus serotypes. *Structure* 5:961–78
63. Lewis JK, Bothner B, Smith TJ, Siuzdak G. 1998. Antiviral agent blocks breathing of the common cold virus. *Proc. Natl. Acad. Sci. USA* 95:6774–78
64. Li Q, Yafal AG, Lee YM-H, Hogle J, Chow M. 1994. Poliovirus neutralization by antibodies to internal epitopes of VP4 and VP1 results from reversible exposure of these sequences at physiological temperature. *J. Virol.* 68:3965–70
65. Luo M, He C, Toth KS, Zhang CX, Lipton HL. 1992. Three-dimensional structure of Theiler murine encephalomyelitis virus (BeAn strain). *Proc. Natl. Acad. Sci. USA* 89:2409–13
66. Luo M, Toth KS, Zhou L, Pritchard A, Lipton HL. 1996. The structure of a highly virulent Theiler's murine encephalomyelitis virus (GDVII) and implications for determinants of viral persistence. *Virology* 220:246–50
67. Luo M, Vriend G, Kamer G, Minor I, Arnold E, et al. 1987. The atomic structure of Mengo virus at 3.0 Å resolution. *Science* 235:182–91
68. Macadam AJ, Arnold C, Howlett J, John A, Marsden S, et al. 1989. Reversion of the attenuated and temperature-sensitive phenotypes of the Sabin type 3 strain of poliovirus in vaccinees. *Virology* 172:408–14
69. Macadam AJ, Ferguson M, Arnold C, Minor PD. 1991. An assembly defect as a result of an attenuating mutation in the capsid proteins of the poliovirus type 3 vaccine strain. *J. Virol.* 65:5225–31
70. Madshus IH, Olsnes S, Sandvig K. 1984. Requirements for entry of poliovirus into cells at low pH. *EMBO J.* 3:1945–50
71. Malashkevich VN, Chan DC, Chutkowski CT, Kim PS. 1998. Crystal structure of the simian immunodeficiency virus (SIV) gp41 core: Conserved helical interactions underlie the broad inhibitory activity of gp41 peptides. *Proc. Natl. Acad. Sci. USA* 95:9134–39

72. Malashkevich VN, Schneider BJ, McNally ML, Milhollen MA, Pang JX, Kim PS. 1999. Core structure of the envelope glycoprotein GP2 from Ebola virus at 1.9 Å resolution. *Proc. Natl. Acad. Sci. USA* 96:2662–67
73. Martin A, Benichou D, Couderc T, Hogle JM, Wychowski C, et al. 1991. Use of type 1/type 2 chimaeric polioviruses to study determinants of poliovirus type 1 neurovirulence in a mouse model. *Virology* 180:648–58
74. Martin A, Wychowski C, Couderc T, Crainic R, Hogle J, Girard M. 1988. Engineering a poliovirus type 2 antigenic site on a type 1 capsid results in a chimaeric virus which is neurovirulent for mice. *EMBO J.* 7:2839–47
75. McDermott BM, Rux AH, Eisenberg RJ, Cohen GH, Racaniello VR. 2000. Two distinct binding affinities of poliovirus for its cellular receptor. *J. Biol. Chem.* 275:23089–96
76. Mendelsohn CL, Wimmer E, Racaniello VR. 1989. Cellular receptor for poliovirus: molecular cloning, nucleotide sequence, and expression of a new member of the immunoglobulin superfamily. *Cell* 56:855–65
77. Molla A, Paul AV, Wimmer E. 1991. Cell-free, de novo synthesis of poliovirus. *Science* 254:1647–51
78. Molla A, Paul AV, Wimmer E. 1993. In vitro synthesis of poliovirus. *Dev. Biol. Stand.* 78:39–53
79. Morrison ME, He YJ, Wien MW, Hogle JM, Racaniello VR. 1994. Homolog-scanning mutagenesis reveals poliovirus receptor residues important for virus binding and replication. *J. Virol.* 68:2578–88
80. Morrison ME, Racaniello VR. 1992. Molecular cloning and expression of a murine homolog of the human poliovirus receptor gene. *J. Virol.* 66:2807–13
81. Moss EG, Racaniello VR. 1991. Host range determinants located on the interior of the poliovirus capsid. *EMBO J.* 10:1067–74
82. Mosser AG, Rueckert RR. 1993. WIN 51711-dependent mutants of poliovirus type 3: evidence that virions decay after release from cells unless drug is present. *J. Virol.* 67:1246–54
83. Mosser AG, Sgro JY, Rueckert RR. 1994. Distribution of drug resistance mutations in type 3 poliovirus identifies three regions involved in uncoating functions. *J. Virol.* 68:8193–201
84. Muckelbauer JK, Kremer M, Minor I, Diana G, Dutko FJ, et al. 1995. The structure of coxsackievirus B3 at 3.5 Å resolution. *Structure* 3:653–67
85. Oliveira MA, Zhao R, Lee WM, Kremer MJ, Minor I, et al. 1993. The structure of human rhinovirus 16. *Structure* 1:51–68
86. Olson NH, Kolatkar PR, Oliveira MA, Cheng RH, Greve JM, et al. 1993. Structure of a human rhinovirus complexed with its receptor molecule. *Proc. Natl. Acad. Sci. USA* 90:507–11
87. Pallansch MA, Kew O, Semler BL, Omilianowski DR, Anderson CW, et al. 1984. Protein processing map of poliovirus. *J. Virol.* 49:873–80
88. Perez L, Carrasco L. 1993. Entry of poliovirus into cells does not require a low-pH step. *J. Virol.* 67:4543–48
89. Phelps D, Post C. 1999. Molecular dynamics investigation of the effect of an antiviral compound on human rhinovirus. *Protein Sci.* 8:2281–89
90. Phelps DK, Speelman B, Post CB. 2000. Theoretical studies of viral capsid proteins. *Curr. Opin. Struct. Biol.* 10:170–73
91. Prchla E, Kuechler E, Blaas D, Fuchs R. 1994. Uncoating of human rhinovirus serotype 2 from late endosomes. *J. Virol.* 68:3713–23
92. Racaniello VR. 2001. *Picornaviridae*: the viruses and their replication. In *Fields Virology*, ed. D. Knipe, P. Howley, pp. 685–722. New York: Lippincott Williams & Wilkens
93. Robinson IK, Harrison SC. 1982. Structure of the expanded state of tomato bushy stunt virus. *Nature* 297:563–68

94. Rossmann MG, Arnold E, Erickson JW, Frankenberger EA, Griffith JP, et al. 1985. Structure of a human common cold virus and functional relationship to other picornaviruses. *Nature* 317:145–53
95. Rossmann MG, Bella J, Kolatkar PR, He Y, Wimmer E, et al. 2000. Cell recognition and entry by rhino- and enteroviruses. *Virology* 269:239–47
96. Selinka H-C, Zibert A, Wimmer E. 1991. Poliovirus can enter and infect mammalian cells by way of an intercellular adhesion molecule 1 pathway. *Proc. Natl. Acad. Sci. USA* 88:3598–602
97. Selinka H-C, Zibert A, Wimmer E. 1992. A chimeric poliovirus/CD4 receptor confers susceptibility to poliovirus on mouse cells. *J. Virol.* 66:2523–26
98. Shafren DR, Bates RC, Agrez MV, Herd RL, Burns GF, Barry RD. 1995. Coxsackieviruses B1, B3, and B5 use decay accelerating factor as a receptor for cell attachment. *J. Virol.* 69:3873–77
99. Shukla D, Rowe CL, Dong Y, Racaniello VR, Spear PG. 1999. The murine homolog (Mph) of human herpesvirus entry protein B (HveB) mediates entry of pseudorabies virus but not herpes simplex virus types 1 and 2. *J. Virol.* 73:4493–97
100. Skehel JJ, Wiley DC. 2000. Receptor binding and membrane fusion in virus entry: the influenza hemagglutinin. *Annu. Rev. Biochem.* 69:531–69
101. Smith TJ, Chase ES, Schmidt TJ, Olson NH, Baker TS. 1996. Neutralizing antibody to human rhinovirus 14 penetrates the receptor-binding canyon. *Nature* 383:350–54
102. Smith TJ, Kremer MJ, Luo M, Vriend G, Arnold E, et al. 1986. The site of attachment in human rhinovirus 14 for antiviral agents that inhibit uncoating. *Science* 233:1286–93
103. Smith TJ, Olson NH, Cheng RH, Liu H, Chase ES, et al. 1993. Structure of human rhinovirus complexed with Fab fragments from a neutralizing antibody. *J. Virol.* 67:1148–58
104. Smyth M, Tate J, Hoey E, Lyons C, Martin S, Stuart D. 1995. Implications for viral uncoating from the structure of bovine enterovirus. *Struct. Biol.* 2:224–31
105. Speelman B, Brooks BR, Post CB. 2001. Molecular dynamics simulations of human rhinovirus and an antiviral compound. *Biophys. J.* 80:121–29
106. Speir JA, Munshi S, Wang G, Baker TS, Johnson JE. 1995. Structures of the native and swollen forms of cowpea chlorotic mottle virus determined by X-ray crystallography and cryo-electron microscopy. *Structure* 3:63–78
107. Staunton DE, Merluzzi VJ, Rothlein R, Barton R, Marlin SD, Springer TA. 1989. A cell adhesion molecule, ICAM-1, is the major surface receptor for rhinoviruses. *Cell* 56:849–53
108. Takahashi K, Nakanishi H, Miyahara M, Mandai K, Satoh K, et al. 1999. Nectin/PRR: An immunoglobulin-like cell adhesion molecule recruited to cadherin-based adherens junctions through interaction with Afadin, a PDZ domain-containing protein. *J. Cell Biol.* 145:539–49
109. Tosteson MT, Chow M. 1997. Characterization of the ion channels formed by poliovirus in planar lipid membranes. *J. Virol.* 71:507–11
110. Tsang SK, Danthi P, Chow M, Hogle JM. 2000. Stabilization of poliovirus by capsid-binding antiviral drugs is due to entropic effects. *J. Mol. Biol.* 296:335–40
111. Tsang SK, McDermott BM, Racaniello VR, Hogle JM. 2001. A kinetic analysis of the effect of poliovirus receptor on viral uncoating: the receptor as a catalyst. *J. Virol.* 75:4984–89
112. Verdaguer N, Blaas D, Fita I. 2000. Structure of human rhinovirus serotype 2 (HRV2). *J. Mol. Biol.* 300:1179–94
113. Weissenhorn W, Calder LJ, Wharton SA, Skehel JJ, Wiley DC. 1998. The central structural feature of the membrane fusion protein subunit from the Ebola virus glycoprotein is a long triple-stranded

coiled coil. *Proc. Natl. Acad. Sci. USA* 95:6032–36

114. Weissenhorn W, Dessen A, Harrison SC, Skehel JJ, Wiley DC. 1997. Atomic structure of the ectodomain from HIV-1 gp41. *Nature* 387:426–30
115. Wetz K, Kucinski T. 1991. Influence of different ionic and pH environments on structural alterations of poliovirus and their possible relation to virus uncoating. *J. Gen. Virol.* 72:2541–44
116. Wien MW, Curry S, Filman DJ, Hogle JM. 1997. Structural studies of poliovirus mutants that overcome receptor defects. *Nat. Struct. Biol.* 4:666–74
117. Wiley DC, Skehel JJ. 1987. The structure and function of the hemagglutinin membrane glycoprotein of influenza virus. *Annu. Rev. Biochem.* 56:365–94
118. Wilson IA, Skehel JJ, Wiley DC. 1981. Structure of the haemagglutinin membrane glycoprotein of influenza virus at 3 Å resolution. *Nature* 289:366–73
119. Xiao C, Bator CM, Bowman VD, Rieder E, He Y, et al. 2001. Interaction of coxsackievirus A21 with its cellular receptor, ICAM-1. *J. Virol.* 75:2444–51
120. Xing L, Tjarnlund K, Lindqvist B, Kaplan GG, Feigelstock D, et al. 2000. Distinct cellular receptor interactions in poliovirus and rhinoviruses. *EMBO J.* 19:1207–16
121. Yeates TO, Jacobson DH, Martin A, Wychowski C, Girard M, et al. 1991. Three-dimensional structure of a mouse-adapted type 2/type 1 poliovirus chimera. *EMBO J.* 10:2331–41
122. Zhao R, Pevear DC, Kremer MJ, Giranda VL, Kofron JA, et al. 1996. Human rhinovirus 3 at 3.0 Å resolution. *Structure* 4:1205–20
123. Zhao X, Singh M, Malashkevich VN, Kim PS. 2000. Structural characterization of the human respiratory syncytial virus fusion protein core. *Proc. Natl. Acad. Sci. USA* 97:14172–77

Annu. Rev. Microbiol. 2002. 56:703–41
doi: 10.1146/annurev.micro.56.013002.100603

First published online as a Review in Advance on July 15, 2002

PRIONS AS PROTEIN-BASED GENETIC ELEMENTS

Susan M. Uptain[1] and Susan Lindquist[2]
[1]*Howard Hughes Medical Institute, Department of Molecular Genetics and Cell Biology, University of Chicago, Chicago, Illinois 60637; e-mail: s-uptain@uchicago.edu*
[2]*Whitehead Institute for Biomedical Research, Department of Biology, Massachusetts Institute of Biology, Nine Cambridge Center, Cambridge, Massachusetts 02142; e-mail: lindquist_admin@wi.mit.edu*

Key Words Sup35, Ure2, HET-s, Rnq1, amyloid

■ **Abstract** Fungal prions are fascinating protein-based genetic elements. They alter cellular phenotypes through self-perpetuating changes in protein conformation and are cytoplasmically partitioned from mother cell to daughter. The four prions of *Saccharomyces cerevisiae* and *Podospora anserina* affect diverse biological processes: translational termination, nitrogen regulation, inducibility of other prions, and heterokaryon incompatibility. They share many attributes, including unusual genetic behaviors, that establish criteria to identify new prions. Indeed, other fungal traits that baffled microbiologists meet some of these criteria and might be caused by prions. Recent research has provided notable insight about how prions are induced and propagated and their many biological roles. The ability to become a prion appears to be evolutionarily conserved in two cases. [*PSI*$^+$] provides a mechanism for genetic variation and phenotypic diversity in response to changing environments. All available evidence suggests that prions epigenetically modulate a wide variety of fundamental biological processes, and many await discovery.

CONTENTS

0066-4227/02/1013-0703$14.00

INTRODUCTION

Prion proteins are unique because they adopt at least two distinct conformational states, one of which, the prion form, can stimulate the nonprion conformation to convert into the prion form. The term "prion" first described the unusual proteinaceous infectious agent that causes devastating neurodegenerative diseases of mammals called transmissible spongiform encephalopathies (TSEs) (128). TSEs include mad-cow disease of cattle, scrapie of sheep and goats, as well as Creutzfeldt-Jakob disease and Kuru of humans (129). Most transmissible diseases are caused by nucleic acid–based agents; however, TSEs are probably caused by the aberrant folding of a cellular protein (PrP^C) into an infectious form (PrP^{Sc}) (129). The function of PrP^C is unknown. Although the protein-only nature of the infectious agent has not been unequivocally demonstrated, protein conformational change is closely linked to transmissibility.

The revolutionary prion concept has been extended to explain three unusual genetic elements of the yeast *Saccharomyces cerevisiae* and one of the filamentous fungus *Podospora anserina* (35, 147, 176). The term prion is no longer confined to the infectious agent of TSEs, but applies to any protein that can switch to a self-sustaining conformation. The agents of the four fungal prions are four endogenous cellular proteins, which participate in diverse biological processes and are apparently nonhomologous to each other or to the mammalian prion protein. The conformational switch to the prion state alters the protein's function and the cell's phenotype. The altered phenotypes are propagated from generation to generation as the protein in the prion state is transferred from mother to daughter cell, continuing the cycle of conformational conversion. Thus, yeast prions act as heritable protein-based genetic elements that cause biologically important phenotypic changes without any underlying nucleic acid change.

The yeast prion *[PSI⁺]* is caused by a conformationally altered form of Sup35 (122, 123, 176), one of two proteins that comprise the translational release factor (152). Some mutant Sup35 proteins cause ribosomes to read through stop codons at an appreciable frequency (71, 72). Such mutants suppress nonsense-codon mutations in other genes, hence their name. [PSI^+] strains also display a nonsense-suppression phenotype (38) because translational termination becomes impaired when Sup35 adopts the prion conformation (Figure 1; for nomenclature, see Figure 3). Unlike the recessive phenotype of *sup35* mutations, [PSI^+] is

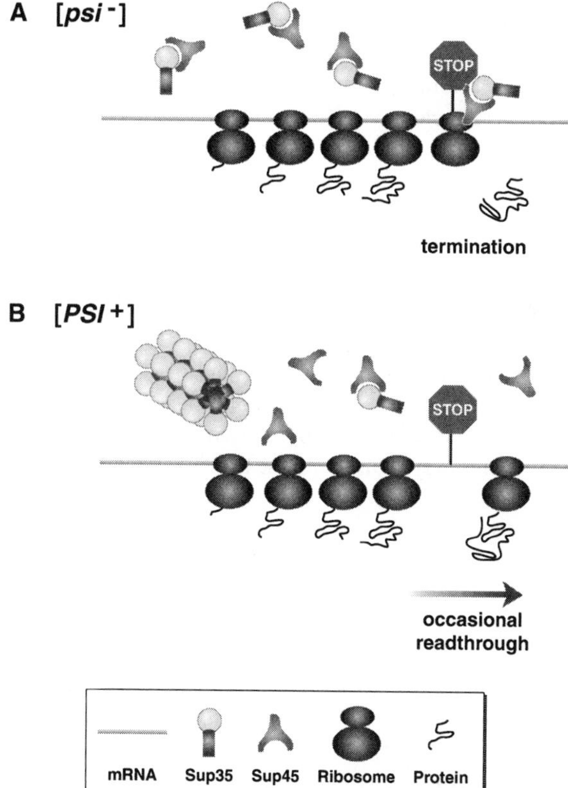

Figure 1 The effect of [PSI^+] on Sup35 and translational termination. (*A*) A complex of Sup35 (see legend at bottom) and Sup45 binds ribosomes at stop codons and mediates translational termination. Sup35 is composed of two regions, a prion-determining domain (PrD, *rectangle*) and a termination domain (Sup35C, *sphere*). In nonprion [psi^-] strains, translational termination occurs efficiently at stop codons at the ends of open reading frames, and the completed protein is released from the ribosome. (*B*) In [PSI^+] cells, most Sup35 proteins adopt the prion conformation and self-assemble into an aggregated, possibly amyloid structure (depicted as *large cylinder*). This conformational change impairs Sup35's ability to participate in translational termination and consequently, stop codons are read through occasionally, producing proteins with a C-terminal extension.

dominant, characterized by non-Mendelian inheritance, and efficiently transmitted from mother cell to daughter through the cytoplasm (38, 40). (The brackets in the name [*PSI*$^+$] denote the latter two characteristics, and the uppercase and italicized *PSI* denotes a dominant trait).

The prion [*URE3*] affects nitrogen catabolite repression. Normally, yeast growing on rich nitrogen sources such as ammonium repress production of proteins needed to metabolize poor nitrogen sources such as ureidosuccinate (158). Recessive mutants termed *ure* were isolated that utilize poor nitrogen sources in the presence of good sources (95, 96). The Ure2 protein regulates nitrogen catabolism by binding to and interfering with the transcriptional activator Gln3 (10). One dominant variant, called [*URE3*], showed non-Mendelian inheritance and was cytoplasmically transferable (2, 96). Conversion of Ure2 into its prion form renders Gln3 constitutively active, thereby eliminating nitrogen catabolite repression and enabling yeast to utilize ureidosuccinate in the presence of ammonium (176) (Figure 2).

Another yeast prion was identified by a directed computer search of yeast genome databases based upon its similarity to regions of Sup35 and Ure2. Rnq1, so named because its sequence is rich in asparagine (Asn, N) and glutamine (Gln, Q), is the protein determinant of the prion [*RNQ*$^+$] (147). Genetic, cell biological, and biochemical analyses proved [*RNQ*$^+$] is a prion, despite being caused by a nonessential protein of unknown function (45, 47, 48, 147). Moreover, [*RNQ*$^+$] is coincident with an epigenetic factor previously identified as [*PIN*$^+$] that affects [*PSI*$^+$] induction, providing the first evidence that one prion affects the appearance of another (45, 118).

The [Het-s] prion of *P. anserina* mediates heterokaryon incompatibility. Because mycelia of this fungus can fuse, resulting in cytoplasmic mixing and occasional exchange of nuclei, this organism has a mechanism to prevent heterokaryon formation between incompatible strains (134). Heteroallelism between any of as many as nine *het* loci triggers a lytic reaction that leads to cell death via an uncharacterized mechanism (134). On solid media, accumulation of these dead cells leads to a distinct, easily visible abnormal contact line, termed a barrage. There are two alleles at the *het-s* locus called *het-s* and *het-S*. The *het-s* allele encodes a protein with two different conformations: HET-s behaves as a prion and HET-s* does not (35). The *het-S* allele encodes the HET-S protein that is incompatible with [Het-s] prion strains but is compatible with neutral [Het-s*] strains (for nomenclature, see Figure 3).

Here, we review the literature on [*PSI*$^+$], [*URE3*], [*RNQ*$^+$], and [Het-s], focusing on their common characteristics, the mechanisms by which they are induced and propagated, their biological significance, and their possible roles in evolution. A comprehensive analysis of the many intricacies of these fascinating elements is beyond the scope of this review. Fortunately, many extensive reviews of individual prions are available (23, 129, 141, 143, 178) and we urge interested readers to consult them. Because many other unidentified prions probably exist (113, 143, 147), we seek to aid the reader in ascertaining whether proteins with unusual conformational

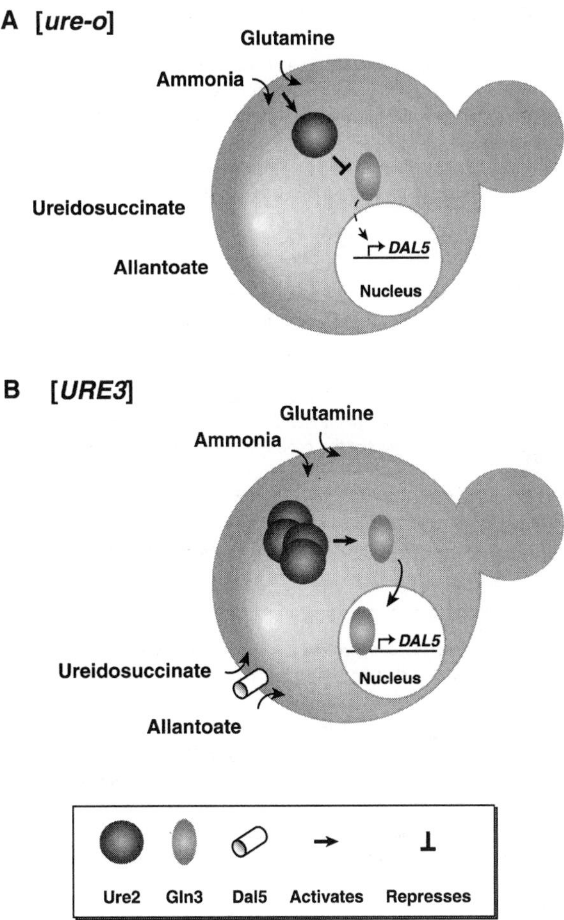

Figure 2 The effect of [*URE3*] on Ure2 and ureidosuccinate uptake. (*A*) In nonprion [*ure-o*] cells, uptake of poor nitrogen sources such as ureidosuccinate and allantoate is repressed in the presence of good nitrogen sources such as glutamine and ammonia. The availability of good nitrogen sources is relayed through Ure2 (see legend at bottom), which blocks the action of the transcription factor, Gln3. Without transcriptional activation, the allantoate transporter, Dal5, is not produced. (*B*) In [*URE3*] cells, conversion of Ure2 into its prion conformation interferes with its ability to repress Gln3. Thus, even in the presence of preferable nitrogen sources, Gln3 activates the transcription of *DAL5*. Ureidosuccinate is a structural mimic of allantoate and enters the yeast cell via Dal5.

[*PSI*$^+$] NOMENCLATURE

Alleles	
SUP35	Wild-type gene encoding a subunit of the translation termination factor
sup35	Mutant that impairs translational termination
Protein Products	
Sup35	Full length protein that can exist in nonprion and prion states
Sup35N	N-terminal fragment required for [*PSI*$^+$], includes the QN-rich region and imperfect repeats
Sup35NM	N-terminal fragment consisting of Sup35N and the highly charged M region
Phenotypes	
[*PSI*$^+$]	Prion phenotype causes nonsense suppression of all three types of stop codons
[*psi*$^-$]	Nonprion phenotype, wild-type translational termination

[*Het-s*] NOMENCLATURE

Alleles	
het-S	This allele does not encode a prion and is incompatible with the *het-s* allele
het-s	This allele encodes a protein that can exist in a prion and nonprion form
het-s^o	*het-s* locus disrupted by gene replacement
Protein Products	
HET-S	Protein product of the *het-S* allele, incompatible with HET-s
HET-s	Protein product of the *het-s* allele in the prion state, incompatible with HET-S
HET-s*	Protein product of the *het-s* allele in the nonprion state, compatible with HET-S or HET-s
Phenotypes	
[Het-S]	Phenotype of *het-S* allele, is incompatible with [Het-s] strains
[Het-s]	Prion phenotype of the *het-s* allele that is incompatible with [Het-S]
[Het-s*]	Neutral, nonprion phenotype of the *het-s* allele that is compatible with [Het-S] and [Het-s]

[*URE3*] NOMENCLATURE

Alleles	
URE2	Wild-type gene involved in nitrogen catabolite repression
ure2	Mutant allele that is impaired in nitrogen catabolite repression
Protein Products	
Ure2	Full length protein that exists in nonprion and prion states
Ure2N	N-terminal fragment containing the asparagine and glutamine-rich PrD of Ure2
Ure2C	C-terminal fragment that functions in nitrogen catabolite repression
Phenotypes	
[*URE3*]	Prion phenotype, faulty nitrogen catabolite repression
[*ure-o*]	Nonprion phenotype, wild-type nitrogen catabolite repression

[*RNQ*$^+$] NOMENCLATURE

Alleles	
RNQ1	Wild-type gene, function unknown
rnq1	Mutant with no obvious phenotype
Protein Products	
Rnq1	Full length protein that exists in nonprion and prion states
Rnq1N	N-terminal fragment, unknown function
Rnq1C	C-terminal fragment of Rnq1 containing the asparagine and glutamine-rich PrD of Rnq1
Phenotypes	
[*RNQ*$^+$]	Prion phenotype, function unknown
[*rnq*$^-$]	Nonprion phenotype, function unknown

Figure 3 Summary of fungal prion nomenclature. This is a brief guide to the nomenclature and terminology of fungal prions and their associated alleles and phenotypes. The yeast prion nomenclature is in accordance with the accepted rules for yeast nomenclature (142).

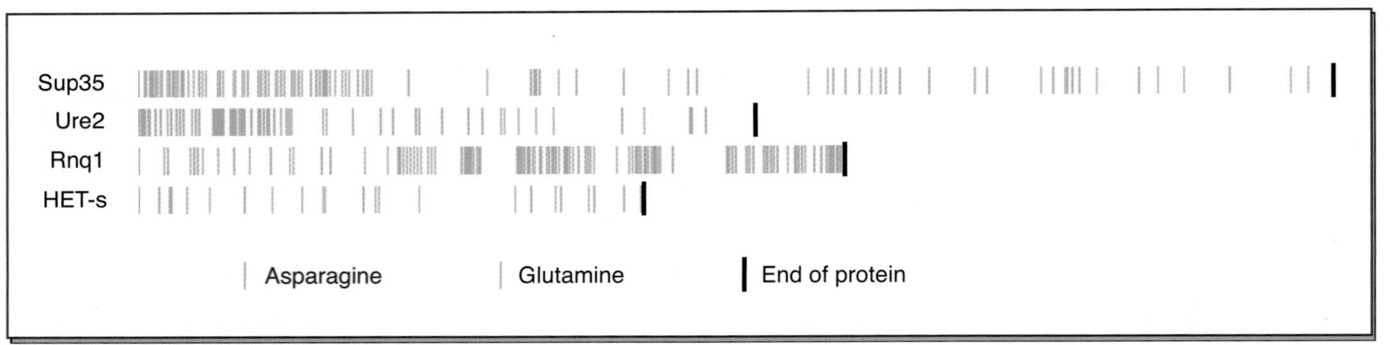

Figure 5 Comparison of the Gln- and Asn-rich regions of *S. cerevisiae* and *P. anserina* prions. Scale representation of the amino acid sequences of the four fungal prions. A *green bar* represents one Asn residue and a *purple bar* represents one Gln residue. All other amino acids are denoted as *white spaces*. The *black bar* designates the C-terminal end of each protein. GenBank accession numbers are Sup35 (NP_010457.1), Ure2 (NP_014170), Rnq1 (NP_009902), and HET-s (S16556).

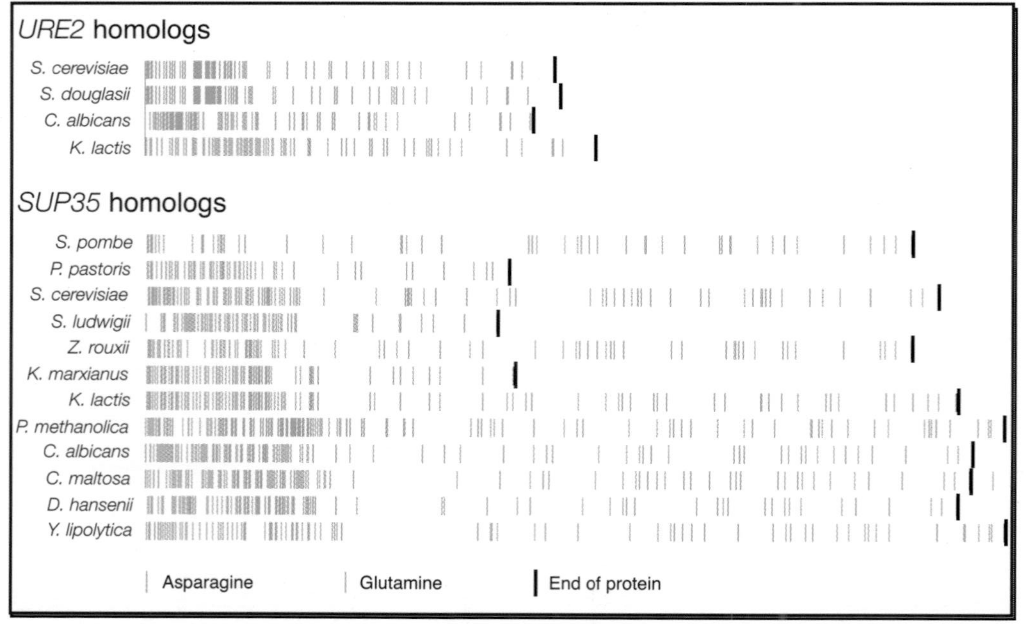

Figure 6 Comparison of the Gln- and Asn-rich regions of Ure2 and Sup35 yeast homologs. Scale representation of the amino acid sequences of the yeast homologs of Ure2 and Sup35. A *green bar* represents one Asn residue and a *purple bar* represents one Gln residue. All other amino acids are denoted as *white spaces*. A *black bar* designates the last residue in the sequence. The sequences for *Pichia pastoris*, *Saccharomycodes ludwigii*, and *Kluyveromyces marxianius* only include the regions homologous to the *S. cerevisiae* Sup35NM region. GenBank accession numbers for Ure2 are NP_014170 (*S. cerevisiae*), AAK51641 (*Saccharomyces douglasii*), AAK51643 (*Candida albicans*), and AAK51642 (*Kluyveromyces lactis*). GenBank accession numbers for Sup35 are BAA33530.1 (*Schizosaccharomyces pombe*), AAF14005.1 (*Pichia pastoris*), NP_010457.1 (*S. cerevisiae*), AAF14006.1 (*Saccharomycodes ludwigii*), BAB12684.1 (*Zygosaccharomyces rouxii*), AAF14004.1 (*Kluyveromyces marxianus*), BAB12680.1 (*Kluyveromyces lactis*), S12921 (*Pichia methanolica*), AAB82541.1 (*Candida albicans*), BAB12681.1 (*Candida maltosa*), BAB12682.1 (*Debaryomyces hansenii*), and BAB12683.1 (*Yarrowia lipolytica*).

properties or genetic traits with unusual epigenetic properties are prion-like. We hope that highlighting the distinctiveness and variety of prion behavior will fuel discovery of novel prions.

CHARACTERISTICS OF FUNGAL PRIONS

The four fungal prions, particularly [*PSI*$^+$] and [*URE3*], share many characteristics (Figure 4) that are widely cited as criteria to identify novel prions. Indeed, these criteria led to the identification of other prion elements (133, 147) and undoubtedly will lead to the discovery of more. There are some notable differences among [*PSI*$^+$], [*URE3*], [*RNQ*$^+$], and [Het-s], however, making it likely that prions differing in key aspects will be found. Ultimately, to be a prion we suggest that a protein must fulfill only one requirement: It must adopt at least two stable states, one of which is self-sustaining.

Fungal Prions are Metastable

Prions occur spontaneously in laboratory fungal strains at a low frequency. [*PSI*$^+$] and [*URE3*] arise in 1 per 10^5 to 10^7 cells, depending on the genetic background (2, 38, 105, 108, 176). The frequency at which [Het-s*] converts to [Het-s] is estimated at less than 1 per 10^7 (8). Incubating cells at low temperatures increases the induction rate of [*PSI*$^+$] and [*URE3*] somewhat (23, 26, 47). Overexpressing the gene encoding a prion increases the induction rate several orders of magnitude (22, 24, 35, 50, 110, 147, 176), presumably because this increases the chance that some proteins will spontaneously adopt the prion conformation. Importantly, [*PSI*$^+$] and [*URE3*] are maintained even if the increased expression of the prion determinant is transient (24, 176). Thus, [*PSI*$^+$] and [*URE3*] are self-perpetuating, even after the levels of Sup35 and Ure2 return to normal.

Once established, most prions propagate faithfully through mitosis and meiosis and are rarely spontaneously lost, although there are some exceptions (9, 35, 40, 176). Stability differences can be attributed to genetic background (96, 108) and the specific prion (50). In general, [*URE3*] is less meiotically stable than [*PSI*$^+$] (38, 96). Some growth conditions efficiently cure fungal prions. Perhaps the most widely used is growth on 1–5 mM guanidinium hydrochloride (GuHCl), which eliminates [*PSI*$^+$], [*URE3*], and [*RNQ*$^+$] (2, 108, 165, 176) but not [Het-s] (35). Prions can usually reappear in previously cured strains.

Altering the expression of cellular proteins that affect protein folding can also cure prions. All three yeast prions require Hsp104, a molecular chaperone that resolves misfolded protein aggregates (70, 121), and are irreversibly cured when *HSP104* is deleted (26, 114, 147). [*PSI*$^+$] requires a specific amount of Hsp104, as it can also be cured by *HSP104* overexpression (26). Because no other yeast prion is cured by Hsp104 overproduction (45, 48, 114, 148), the mechanisms by which overexpression and deletion of *HSP104* cure [*PSI*$^+$] might be distinct.

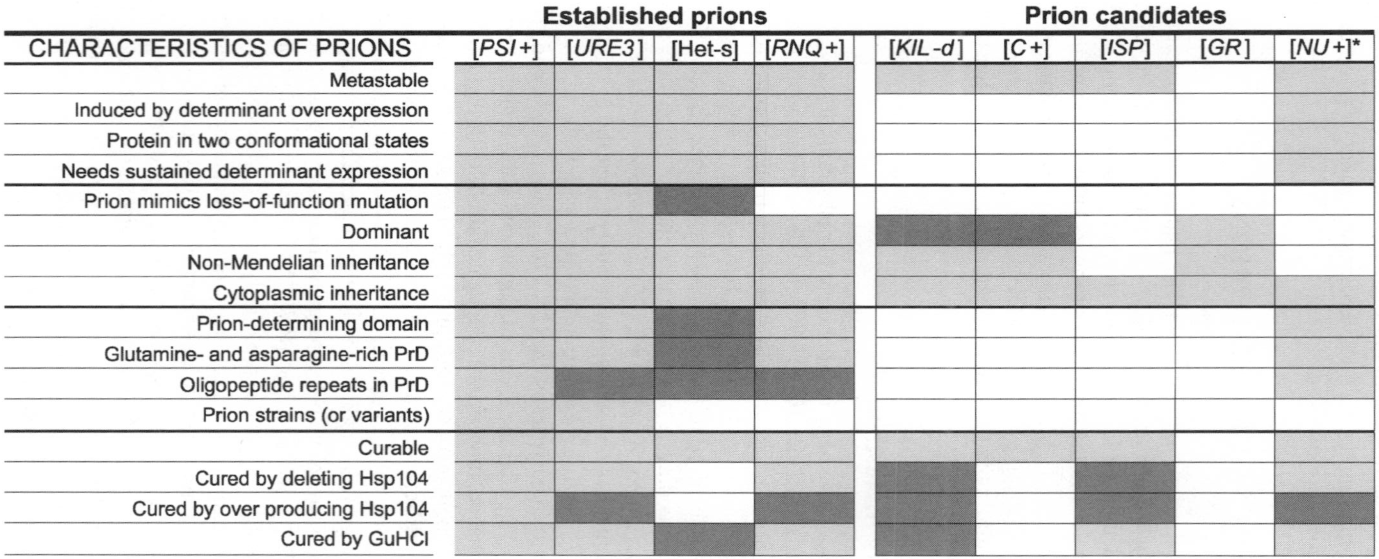

CHARACTERISTICS OF PRIONS	Established prions				Prion candidates				
	[PSI+]	*[URE3]*	[Het-s]	*[RNQ+]*	*[KIL-d]*	*[C+]*	*[ISP]*	*[GR]*	*[NU+]**
Metastable									
Induced by determinant overexpression									
Protein in two conformational states									
Needs sustained determinant expression									
Prion mimics loss-of-function mutation									
Dominant									
Non-Mendelian inheritance									
Cytoplasmic inheritance									
Prion-determining domain									
Glutamine- and asparagine-rich PrD									
Oligopeptide repeats in PrD									
Prion strains (or variants)									
Curable									
Cured by deleting Hsp104									
Cured by over producing Hsp104									
Cured by GuHCl									

Figure 4 Comparison of the characteristics of established prions to those of candidate prions. Gray indicates that a prion or trait displays a characteristic and black indicates that it does not. White denotes no information available. *based upon the behavior of New1 PrD fused to Sup35C. See text for references.

Capacity to Convert to a Self-Propagating Conformational State

Strong support for the prion model derives, in part, from biochemical characterization of prion proteins. When crude yeast lysates are fractionated by differential sedimentation, Sup35 and Rnq1 partition differently, depending on whether the cells contain a prion. Sup35 and Rnq1 proteins are mostly soluble in [*psi*$^-$] and [*rnq*$^-$] lysates, but they are insoluble in [*PSI*$^+$] and [*RNQ*$^+$] lysates (122, 123, 147). When [*PSI*$^+$] lysates are fractionated using size-exclusion chromatography, the prion form of Sup35 is associated with high-molecular-weight fractions (123). The Sup35 prion conformation is somewhat more resistant to proteinase K digestion than the nonprion form (122, 123). Together, these studies indicate that Sup35 and Rnq1 are aggregated in the prion state but not in the nonprion state, which provides a satisfying explanation for the translational termination defect of [*PSI*$^+$] strains.

To visualize the prion conformation converting newly made protein to the prion state in living cells, Rnq1 or certain prion-determining fragments of Sup35 were fused to green fluorescent protein (GFP) and expressed in yeast. In cells without either prion, these GFP fusion proteins distribute evenly throughout the cytoplasm (122, 147). In [*PSI*$^+$] or [*RNQ*$^+$] cells, such fusion proteins quickly coalesce into discrete, cytoplasmic foci that do not appear to localize to any cellular structure (122, 147). The particulate nature of the fusions in [*PSI*$^+$] cells was confirmed by differential sedimentation (122). For [*PSI*$^+$], double labeling demonstrated the capture of newly made Sup35-GFP fusions at the sites of pre-existing Sup35 aggregates tagged with hemagglutinin (104).

Similar approaches indicated that the prion forms of Ure2 and HET-s are aggregated in prion-containing cells; however, the nature of this conformational difference is not as well understood as with Sup35 and Rnq1. In some genetic backgrounds, the prion form of Ure2 is more resistant to proteolysis than the nonprion form (111, 137, 162). Consistent with these findings, differential sedimentation and size-exclusion chromatography of yeast lysates indicate that Ure2 (57, 137) is insoluble in [*URE3*] cells but soluble in nonprion [*ure-o*] cells. Moreover, ectopically expressed Ure2-GFP fusions coalesce into fluorescent foci only in [*URE3*] (57, 137). In other yeast genetic backgrounds, however, such Ure2-GFP fusions do not coalesce detectably (62, 63), and there is no apparent difference in how Ure2 partitions upon fractionation of [*URE3*] or [*ure-o*] lysates by differential sedimentation (63). One plausible explanation is that the extent of Ure2 aggregation depends on the genetic background.

For [Het-s], detection of prion aggregates apparently depends on the expression level of HET-s and the assay used. In crude cell lysates, the prion conformation, HET-s, is more resistant to proteinase K digestion than its nonprion form (35), which suggests the prion form is conformationally different and might be insoluble. By differential sedimentation or size-exclusion chromatography of [Het-s] lysates, however, HET-s aggregates are detectable only in [Het-s] cells that highly overexpress the prion protein (37). Similarly, HET-s-GFP fusions coalesced into fluorescent foci specifically in [Het-s] prion strains only when those fusions were highly expressed (37). At normal levels, fluorescence is diffusely distributed in the cytoplasm or is localized to vacuoles. Possibly, only a small fraction of the HET-s protein is aggregated in vivo. Alternatively, the prion conformation may be in small aggregates or preferentially degraded, thereby precluding its detection. Indeed, the steady-state level of HET-s is more than threefold lower in [Het-s] prion strains than in [Het-s*] neutral strains (37).

The conformational switch of all four fungal prions between soluble and insoluble states can be reconstituted in vitro under physiological conditions (53, 69, 146, 156). The insoluble conformation consists of highly ordered amyloid fibers (53, 69, 81, 156), remarkably similar to those associated with Alzheimer's and Huntington's diseases (17). The fiber's width varies depending on the prion and solution conditions (range: 4–22 nm) but is proportional to the molecular weight of the monomer (5, 53, 69, 81, 138, 156, 162). Fibers can be short or tens of microns long (53, 69, 162). Like other amyloids, prion fibers are rich in β-sheet structure (53, 69, 156) and exhibit a cross-β-pleated-sheet pattern by X-ray diffraction (5, 140).

Biochemical characterization of prion amyloid formation provided remarkable insight into the probable molecular mechanism by which prions arise and propagate in vivo. Just as prions arise spontaneously, but infrequently, in cells (2, 8, 38, 105, 108, 176), each prion protein forms fibers spontaneously, but slowly, in vitro (53, 69, 146, 156). Mutations that increase or decrease the rate of prion induction in vivo correspondingly affect fiber formation in vitro (44, 69, 105). Because most fungal prions are mitotically stable, some process must facilitate efficient conversion of nascent prion proteins into the prion conformation. Remarkably, small amounts of pre-existing amyloid can promote rapid conversion of unpolymerized prion proteins into fibers in vitro (53, 69, 81, 146, 156). That this reflects prion propagation in vivo is evidenced by the fact that lysates from [*PSI*$^+$] cells, but not [*psi*$^-$] cells, accelerate Sup35 fiber formation (69, 124, 168). Thus, amyloid formation in vitro serves as a model for how prions act as heritable elements in vivo.

Prions Require Sustained Expression of Their Determinant Genes, Yet Their Phenotypes Often Mimic Loss-of-Function Mutations

Although the fungal prions act as genetic elements, their maintenance depends upon continual expression of the genes encoding them (176). Poor or interrupted expression leads to the eventual loss of the prion (99, 147, 176). It is not surprising, therefore, that *SUP35*, *URE2*, and *RNQ1* are all expressed at nearly constant levels throughout the cell cycle (43, 150).

The conformational change to the prion state often decreases the protein's activity. Aggregation may sequester a prion from its substrate or impede its proper localization within the cell. Thus, the phenotypes of prions mimic loss-of-function mutations in their prion determinants (176). As with *ure2* mutants, nitrogen catabolite repression is impaired in [*URE3*] strains (54, 96, 176). Similarly, the nonsense-suppression phenotype of [PSI^+] reflects a partial loss of Sup35 activity (122, 123). However, not all prion phenotypes mimic loss-of-function mutations. A strain carrying a prion can be phenotypically indistinguishable from one lacking the prion, particularly when no obvious phenotype is associated with its determinant. Such is the case with [RNQ^+] and [rnq^-] strains (147). The portion of a protein that facilitates prion conversion is often distinct from the portion that is enzymatically active; thus, conversion to the prion conformation may not significantly affect an enzyme's activity, particularly if its substrate is diffusable (99, 111). In one case, the phenotype associated with cells in which the determinant is disrupted is opposite that of a prion. *P. anserina* $het\text{-}s^0$ strains, in which the *het-s* locus was inactivated by gene replacement, are neutral in compatibility assays (166), unlike [Het-s] prion strains, which are incompatible with [Het-S] strains.

Dominant, Non-Mendelian Segregation and Cytoplasmic Transmission

Unusual genetic properties are a hallmark of fungal prions. [*URE3*], [PSI^+], and [RNQ^+] are dominant traits that display non-Mendelian inheritance (38, 45, 48, 96). In matings between haploid [$PRION^+$] and [$prion^-$] yeast cells, the resulting diploid cells are [$PRION^+$], and when the diploid is sporulated all four spores are usually [$PRION^+$], although not always (2, 96, 100, 176). The peculiar inheritance of [PSI^+] and [*URE3*] led yeast geneticists to conclude that these elements were transmitted cytoplasmically. This was confirmed for [PSI^+] and [*URE3*] and later for [RNQ^+] by using strains that have a nuclear mutation that allows the cytoplasm of the two cells to mix upon mating but prohibits fusion of their nuclei (cytoduction) (2, 31, 40, 147). Subsequent experiments demonstrated that [PSI^+] and [*URE3*] were not caused by other cytoplasmically transferred genetic determinants, such as mitochondrial DNA, viruses, or plasmids (68, 96, 97, 164, 185).

The dominant, non-Mendelian, and cytoplasmic character of these unusual genetic elements is consistent with a self-perpetuating protein-based mechanism of inheritance. The prion conformation is self-propagating, unlike most protein aggregates. During mating, haploid yeast fuse, mixing cytoplasmic contents and allowing the prion conformation to convert proteins from the [$prion^-$] cytoplasm into the prion conformation. Upon cell division, the prion conformation is passed from the mother cell to the daughter cell through the cytoplasm.

The [Het-s] prion of *P. anserina* can also be transferred cytoplasmically (8, 35). Anastomosis is a fungal process, analogous to yeast cytoduction, by which hyphae fuse to create a network. This causes their cytoplasms to mix. When a [Het-s] strain undergoes anastomosis with a neutral [Het-s*] strain, the resulting mycelium contains the prion [Het-s] (8, 35). Cytoplasmic transmission also affects [Het-s] propagation during mating and meiosis. Because the female parent contributes most of the cytoplasm, nearly all meiotic progeny of a cross between [Het-s] and [Het-s*] strains have the phenotype of the female parent (8, 134).

Prion-Determining Region (PrD)

Sup35 contains three regions that are distinguished by their function and amino acid composition (78, 93, 94, 157, 180). The amino-terminal region, termed Sup35N (amino acids 1 to 123), is unusually rich in Gln (28%) and Asn (16%) residues, but has few aliphatic amino acids (e.g., alanine and valine, 6%) (Figure 5, see color insert). By comparison, the average protein has just 9% Gln and Asn (133) and 29% aliphatic residues (112). Sup35N also contains several imperfect oligopeptide PQGGYQQ_YN repeats that are similar in character to PHGGGWGQ repeats present in the mammalian prion protein, PrP (85, 105, 120). These repeats are the only immediately obvious similarity between Sup35 and PrP protein sequences. The middle region, Sup35M (amino acids 124 to 253), is highly charged, unlike the rest of the protein. Forty-one percent of the residues in this region are lysine, glutamic acid, or aspartic acid. Sup35NM is not required for viability (160); however, Sup35C (amino acids 254 to 685) is essential (91, 160). Sup35C is homologous to the translation elongation factor EF-1α, contains four putative GTP-binding sites, and is sufficient for translational termination (78, 94, 152, 180).

Ure2 and Rnq1 are also Gln- and Asn-rich proteins. Each can be subdivided into two functionally distinct regions. Ure2N [amino acids 1 to 93 (126, 162)] contains more Asn (36%) than Gln residues (11%) (Figure 5), and only 13% of its residues are aliphatic. The remainder of the protein, Ure2C (amino acids 94 to 354), encodes the nitrogen regulatory activity of Ure2 (33) and is homologous to bacterial glutathione S-transferases (GST), although it lacks GST enzymatic activity (13, 29, 33, 167). The Rnq1C (amino acids 153 to 405) has many Asn (16%) and Gln (27%) residues and contains few aliphatic residues (8%) (147). Rnq1N (amino acids 1 to 152) is not similar to any other protein and its function is unknown (147). Neither Ure2 nor Rnq1 are essential for viability (33, 147) and neither contains oligopeptide repeats similar to those of Sup35 or PrP.

Remarkably, distinct regions of Sup35, Ure2, and Rnq1, which are similar in character but not identical in sequence, enable these protein to act as prions and are called prion-determining regions (PrDs). Mutational analysis delimited the regions of Sup35 and Ure2 necessary for [PSI^+] and [*URE3*]. When N-terminal segments are deleted, [PSI^+] and [*URE3*] are irreversibly cured (111, 159). Moreover, those N-terminal fragments of Ure2 and Sup35 are sufficient to support prion induction and propagation (110, 159). Although Sup35M is not required for [PSI^+], it appears to play a role in [PSI^+] stability and Sup35 solubility (69, 104, 160). That Ure2C is dispensable for prion propagation was demonstrated by passaging the prion through strains lacking this region (110).

sup35 and *ure2* mutants that affect prion induction and propagation cluster to their N-terminal regions (44, 50, 52, 62, 105, 109, 111, 120). Most reduce the Asn and Gln content of these proteins or the number of oligopeptide repeats in Sup35, indicating that these two characteristics are important for these regions to function as PrDs. The absolute ratio of Asn to Gln is not crucial because [PSI^+] can be induced by overproducing a fusion of a mutant Sup35N region to GFP, in which a polyGln stretch replaced an Asn- and Gln-rich stretch (44).

Because prions are induced more frequently when their determinants are overexpressed, several groups measured the efficiency with which [PSI^+] and [*URE3*] arose when various prion protein subfragments were overproduced. Elevated expression of any fragment containing Sup35N or Ure2N increases the rate of prion induction (24, 50, 83, 109–111, 176). These fragments induce prion formation much more efficiently than full-length versions expressed at similar levels (50, 83, 109, 111). Together with the mutational analysis, these studies showed that the N-terminal regions of Sup35 and Ure2 are required for [PSI^+] and [*URE3*].

Because [RNQ^+] strains are phenotypically indistinguishable from [rnq^-] strains, the region of *RNQ1* that constitutes its PrD was ascertained using a novel approach. Sondheimer & Lindquist (147) reasoned that if Rnq1C, which is rich in Asn and Gln, was the PrD, it would functionally substitute for Sup35's endogenous PrD. Indeed, they found that a [PSI^+]-like state is induced in strains expressing Rnq1C fused to SupMC.

Other studies further demonstrated that the PrDs of yeast prions are modular and transferable. Fusions of the PrDs of Sup35, Rnq1, and Ure2 to GFP can join preexisting prion aggregates in vivo (57, 122, 147). In [PSI^+], [RNQ^+], or [*URE3*] cells, expression of these fusions leads to discrete fluorescent foci; whereas, in [psi^-], [rnq^-], and [*ure-o*] cells, the fluorescence is evenly distributed throughout the cytoplasm. In another case, the PrD of Sup35 was fused to a completely unrelated protein to create a novel chimeric prion. The mammalian glucocorticoid receptor (GR), a hormone-regulated transcriptional activator, was fused to the PrD of Sup35 and expressed in a strain bearing a GR-regulated *lacZ* reporter. This fusion protein can exist in two stable but interchangeable functional states, only one of which activated transcription of the reporter (99). Importantly, the fusion protein exhibited many of the unusual genetic characteristics of known yeast prions, and it converted to the nonfunctional state on its own in a mutant [psi^-] strain in which the endogenous Sup35 PrD was deleted.

The 289-amino-acid HET-s protein is the only identified fungal prion not Asn or Gln rich (166) and without a well-defined PrD. To identify regions important for conversion to the [Het-s] prion state and for heterokaryon incompatibility, a series of *het-s* deletion mutants were ectopically expressed in a *het-s^0* strain (36). An N-terminal fragment (amino acids 1 to 112) is sufficient for incompatibility and for conversion to [Het-s], suggesting that it contains the PrD. Consistent with this interpretation, point mutations in this region affect prion propagation (36). Further analysis revealed that a C-terminal fragment (amino acids 86 to 289) also mediates incompatibility and conversion. Although this suggests that the 27-amino-acid region common to both fragments is the minimal PrD, a fragment lacking only that region also supports incompatibility and conversion. Thus, no specific PrD was identified. Interestingly, an even shorter N-terminal fragment (amino acids

1 to 25) propagates the [Het-s] phenotype (36) but cannot mediate heterokaryon incompatibility.

PRION CANDIDATES

Yeast and filamentous fungi have other traits with unusual genetic properties that resemble those of prions (Figure 4). Although they are not yet characterized enough to judge whether they are caused by prions, all can be transmitted cytoplasmically. Most display non-Mendelian inheritance and are metastable and curable. However, some of their characteristics, such as not being cured by GuHCl treatment, are unlike those of known prions. If any are prions, those unique characteristics will broaden the criteria for identifying prions and aid the search for novel prions.

[*KIL-d*]

The [*KIL-d*] element epigenetically regulates viral gene expression in haploid *S. cerevisiae* cells (154, 155, 175). The killer virus is composed of two double-stranded RNAs, called L-A and M, that are transmitted cytoplasmically between yeast. L-A encodes for the virus's replication machinery and is required to maintain its satellite, M, which encodes a secreted toxin that kills surrounding uninfected yeast and a pre-toxin that provides the infected cell resistance to the toxin it produces. Yeast infected with L-A and M are phenotypically described as killer (K+) and resistant to toxin (R+). Uninfected yeast and those infected with only L-A are phenotypically nonkiller (K−) and susceptible to toxin (R−) [reviewed in (177)].

Like the fungal prions, [*KIL-d*] is a cytoplasmically transmitted trait (155). It causes haploid yeast infected with killer virus (both L-A and M) to display variegated, defective killer phenotypes (K*R*), such as being defective in killing or resistance or both (e.g., K−R+, K+R−, K−R−) and losing the M satellite at a higher rate. The [*KIL-d*] element does not map to L-A or M (155), mitochondrial DNA (155), or the yeast 2-μm plasmid (154). Like known prions, [*KIL-d*] is metastable. It is lost at a frequency of 10^{-4} to 10^{-5} and arises spontaneously at a rate of 10^{-3} (154, 175).

Some characteristics of [*KIL-d*] are unlike those of established fungal prions. [*KIL-d*] is not cured by GuHCl treatment (155), is recessive in the diploid (175), and is not dependent on the molecular chaperone, Hsp104. [*KIL-d*] persists in strains where the chromosomal copy of *HSP104* is deleted or in strains overexpressing *HSP104* (154).

Peculiarly, a cell only manifests the [*KIL-d*] variegated killer phenotype after it undergoes meiosis (175). [*KIL-d*] is phenotypically cryptic in diploids and in haploids that have not passed through meiosis. Mating a [*KIL-d*] strain harboring M to a nonkiller wild-type strain results in a diploid strain with the killer and resistance phenotypes (K+R+). After the diploids sporulate, however, their haploid meiotic progeny exhibit a range of pleiotropic, defective phenotypes that are indicative of [*KIL-d*]. Backcrossing any of the defective progeny (K*R*) generates diploid cells that are killer and resistant (K+R+). The meiotic progeny of this diploid again exhibit various degrees of killer and resistance defects (175). [*KIL-d*] can be transmitted by cytoduction, but is cryptic in the haploid cytoductants. Diploids formed by mating these cytoductants to a wild-type uninfected strain are also killer and resistant (K+R+). The meiotic progeny, however, display the range of killer and resistance phenotypes that are characteristic of [*KIL-d*] (155). Thus, [*KIL-d*] is transferred through the cytoplasm but its phenotypic expression requires cells first undergo meiosis.

The molecular determinant of [*KIL-d*] is unknown, precluding any direct biochemical assessment of whether it can adopt the prion conformation. One hypothesis is that [*KIL-d*] is a prion-like element whose phenotypic expression is "reset" by meiosis and "healed" by nuclear fusion (155). If so, the molecular determinant of [*KIL-d*] would alter expression of a nuclear chromosomal target, which epigenetically regulates M expression. Thus, the prion conformation underlying [*KIL-d*] would only access its chromosomal target during meiosis. Variation in the extent of this interaction would cause the variegated phenotypes typical of [*KIL-d*] meiotic progeny. Finally, because all [*KIL-d*] strains were isolated after ethyl methanesulfonate mutagenesis of a killer-infected yeast strain, the determinant of [*KIL-d*] may be a mutant more likely to spontaneously adopt the prion conformation, but effectively convert the wild-type protein into the prion. Some *sup35* and *ure2* mutants similarly affect [PSI^+] (105) and [*URE3*] (62).

[C^+]

Crippled growth, [C^+], is a cytoplasmically transferable trait of *P. anserina* that impairs mycelial growth, hence its name (144). It also causes abnormal hyphal morphology, pigment accumulation, shortened longevity, and reduced female fertility (144). Like known prions, [C^+] is metastable. Wild-type strains become [C^+] once they enter stationary phase. However, strains with certain ribosomal protein mutations that augment accurate translational termination become [C^+] during vegetative growth (144). Agents that stress cells, including high and low temperatures, high osmotic pressure, nutrient deprivation, and ultraviolet irradiation, cure [C^+] (144) and some also cure known prions (9, 35, 137, 145, 165). Some drugs that impair translational termination also cure [C^+] from vegetatively growing cells, suggesting that termination read through produces a factor that impairs [C^+] propagation (144). [C^+] can be efficiently transmitted during mitosis; however, it is inefficiently propagated through meiosis (144), possibly because its molecular determinant, which is unknown, is not expressed during meiosis. There is no evidence of mitochondrial DNA mutations or virus-like elements in [C^+] strains that could explain the unusual genetic behavior of this element (144).

[ISP^+]

[ISP^+], inversion of suppressor phenotype, is another prion-like trait involved in regulating translation termination in *S. cerevisiae* (169). This non-Mendelian element abrogates the nonsense-suppression phenotype of certain recessive *sup35* mutations. GuHCl treatment cures [ISP^+] strains but the element reappears spontaneously at a high frequency in those cured strains. [*ISP*$^+$] propagation does not depend on Hsp104, however. [*ISP*$^+$] is not induced by expressing the mutant *sup35* allele and it can be propagated in strains lacking the PrD of Sup35. Moreover, Sup35 is not aggregated in [*ISP*$^+$] cells. Thus, [*ISP*$^+$] interacts with some *sup35* mutants but is not a form of [*PSI*$^+$].

[*GR*]

[*GR*] is a cytoplasmically transferable trait of *S. cerevisiae* that provides resistance to glucosamine, a glucose analog that inhibits growth on nonfermentable carbon sources (6, 60). [*GR*] is a dominant, non-Mendelian element that is not caused by mitochondrial DNA, killer virus, [*PSI*$^+$], or [*URE3*]. It is not cured by heat or cycloheximide treatment, which eliminate the killer virus (87).

GRR1 is a plausible candidate for the determinant of [*GR*] (146). Grr1 is a component of the ubiquitin proteolysis machinery and appears to regulate the glucose signaling pathway and cell cycle progression (98). Grr1 contains several runs of polyAsn residues at its N and C termini. *grr1* deletion strains exhibit pleiotropic phenotypes related to loss of degradation of key regulatory proteins, some of which, including glucosamine resistance, are also exhibited by [*GR*] strains (65, 146, 183).

[*PIN*$^+$]

[*PIN*$^+$] is non-Mendelian trait of *S. cerevisiae* that is required for efficient [*PSI*$^+$] induction when Sup35 is overproduced (47, 48). Because all yeast prions are efficiently induced when their determinant genes are overexpressed, Derkatch et al. (45) searched for genes that converted cells from [*pin*$^-$] to [*PIN*$^+$] when overexpressed. Twelve such proteins were identified (45), two of which were the prion forms of Ure2 and Rnq1 (45). Thus, having one prion in a cell increases the frequency that another will appear. The other proteins participate in diverse cellular processes. Swi1 and Cyc8 regulate transcription, Yck1 and Ste18 play roles in signal transduction, Nup116 is important for nuclear transport, and Lsm4 is involved in mRNA processing (1, 45). *NEW1*, *PIN2*, *PIN3*, and *PIN4* are uncharacterized genes. All encode Gln- or Asn-rich proteins with obvious similarity to the PrDs of Sup35, Rnq1, and Ure2. Whether any form stable prions is unclear; however, the Asn-rich region of New1 is part of the determinant of [*NU*$^+$], another prion-like element (133).

[*NU*$^+$]

Because prion domains are modular and transferable, replacing Sup35's PrD with other proteins and assaying for [*PSI*$^+$]-like behavior is a powerful method to identify new prion candidates (147). [*NU*$^+$], whose putative prion determinant is encoded by an N-terminal fragment of the *NEW1* gene fused to Sup35C, was identified in this manner (133). The New1-Sup35C fusion can adopt two biochemically and functionally different states (133). New1 was analyzed because it is Gln and Asn rich and has several oligopeptide repeats reminiscent of those of Sup35 and PrP (133). [*NU*$^+$] can be transferred by cytoduction (118), cured by GuHCl treatment (118), and cannot be isolated in cells lacking *HSP104* (118). However, Hsp104 overproduction has no effect (119).

Although New1-Sup35C behaves like a prion, it is unclear whether the full-length wild-type New1 ever exists as a stable prion. [*NU*$^+$] cannot be supported by endogenous New1, but requires expression of New1-Sup35C (118). *NEW1* is a determinant of [*PIN*$^+$], however, suggesting that it can convert to a prion-like state (45, 118).

PRION STRAINS

One of the most controversial aspects of mammalian prions is the existence of distinct TSE strains, which differ in their disease latency periods, brain pathologies, neuropathological manifestations, and distribution of PrP^{Sc} in brain tissues (14, 16). Some clinically distinct human prion diseases are caused by mutations in the prion protein gene, *Prnp* (129). However, prion strains cannot be due to any mutations in *Prnp* or elsewhere in the mouse genome because they are passaged in the same inbred mice background. Opponents of the protein-only hypothesis argue that a viral TSE agent best explains prion strains (15, 51, 139); proponents posit that PrP^{Sc} conformational differences cause strains (20, 129, 139). Such conformational variation is present because different positions of cleavage occur upon protease treatment of infectious material from different strains [reviewed in (20, 30)]. Whether this is the underlying basis of mammalian prion strains remains unclear.

[*PSI*$^+$] and [*URE3*] can also exist as different strains (38, 50, 137), which are sometimes called "variants" to avoid confusion with the many yeast genetic backgrounds (strains) in which prions are passaged. [*PSI*$^+$] variants were discovered first and are the best characterized. *SUP35* overproduction in [*psi*$^-$] cells induces new [*PSI*$^+$] strains that exhibit a wide range of translational termination defects, despite having arisen in genetically identical cells (50). When any of these [*PSI*$^+$] variants are cured and *SUP35* is overexpressed, each [*psi*$^-$] derivative can give rise to [*PSI*$^+$] cells with the same range of termination defects. Thus, mutations in *sup35* or any other yeast gene do not cause [*PSI*$^+$] variants. They are epigenetic.

Most [*PSI*$^+$] variants are distinguished by their nonsense-suppression phenotypes and are called strong, moderate, or weak depending on their severity (50). Usually, weak variants do not spontaneously become strong or vice-versa; however, one such example was described (82). Compared to strong [*PSI*$^+$] variants, weak variants are less stable and are more easily cured by GuHCl treatment or by *HSP104* overexpression (50). Strong [*PSI*$^+$] is dominant to weak [*PSI*$^+$] in genetic crosses (49). The extent of Sup35NM-GFP aggregation differs among [*PSI*$^+$] variants. Many more strong [*PSI*$^+$] cells have punctate fluorescent foci than weak [*PSI*$^+$] cells (49). A few [*PSI*$^+$] variants can only be differentiated by their translational termination phenotypes when certain *sup35* mutations are expressed (49, 80).

[PSI^+] variants can also be distinguished by reproducible differences in the amount of soluble Sup35 in vivo and by the efficiencies with which they mediate conversion to the prion state in vitro (168). Strong [PSI^+] cells have much less soluble Sup35 than weak [PSI^+] cells, yet the total amount of Sup35 is the same (82, 168, 188). In cell-free conversion assays, the prion proteins from some [PSI^+] variants convert Sup35 to the prion state much more efficiently than others, differing by as much as 20-fold (168). Importantly, those that convert most efficiently originate from [PSI^+] variants with the most severe termination defects and the least soluble Sup35 in vivo (168).

These results led Uptain et al. (168) to propose a model that explains the phenotypic characteristics of [PSI^+] variants in molecular terms. The severity of a variant's translational defect is determined by the amount of soluble, functional Sup35 in the cells. The less soluble Sup35, the more pronounced the termination defect. The amount of soluble Sup35 is determined by how well the prion conformation of Sup35 captures and converts soluble Sup35 to the insoluble prion state. In strong [PSI^+] variants, conversion is efficient and consequently the amount of soluble Sup35 is low. In weak variants, conversion is less efficient, more soluble Sup35 is present, and the variants exhibit milder termination defects. Although prion strains were first discovered in mammals, the different physical states of PrP in vitro have not been correlated with the phenotypes associated with prion strains, primarily because the cellular function of PrP is unknown. Thus, this model for [PSI^+] variants provides the first coherent molecular explanation for the phenotypes of prion strains.

Whereas [PSI^+] variants have been observed for decades (38, 50, 100), [*URE3*] variants were described only recently (137). Three types of [*URE3*] variants were identified when Ure2 or its PrD was overproduced in [*ure-o*] cells (137). A novel reporter system involving the adenine biosynthesis pathway was developed to detect them (137). On rich media, Ade2-deficient cells are red because they accumulate a red metabolic intermediate, whereas wild-type cells are white. When *ADE2* expression is controlled by the *DAL5* promoter, [*ure-o*] colonies are red because Ure2 keeps Gln3 in the cytoplasm, the *DAL5* promoter is not activated, and *ADE2* is not expressed. Most [*URE3*] strains are pink or white. Type A [*URE3*] variants are pink and more difficult to cure than type B variants, which are redder. Type C variants are red and unlike the other two types of variants, they have soluble Ure2 and cannot be cured by GuHCl treatment or transferred by cytoduction (137). Although the molecular basis of these variants is unknown, no differences in the prion conformation of Ure2 were detected by proteinase K digestion (137). One attractive possibility is that differences in prion conversion efficiency similar to that of Sup35 may underlie them (137, 168).

What might lead to the differences in conversion efficiencies in yeast prion variants? Prion proteins may adopt conformations with intrinsically different abilities to convert the nonprion form into the prion form. Indeed, purified Sup35NM can form amyloid fibers with different structural states in vitro (69). Similarly, a Sup35NM chimera derived from two yeast species forms fibers with distinct physical properties in vitro related to different [*PSI*$^+$] variants in vivo (28). In the experiments of Uptain et al. (168), no conformational differences were observed between fibers nucleated by Sup35 from different wild-type [*PSI*$^+$] variants; however, in a second round of seeding these fibers did not retain their initial characteristic differences in conversion efficiency. Thus, if different Sup35 conformations cause [*PSI*$^+$] variants, they are not retained in the cell-free system. Perhaps interactions between the prions and other cellular factors, which might be stochastic in origin but self-perpetuating in nature, maintain variants in vivo. Alternatively, if prion aggregates of stronger variants were smaller and more numerous than those of weaker variants, they might convert soluble Sup35 more effectively and be propagated more faithfully in vivo, but not be sustainable in vitro.

YEAST PRIONS AS PROTEIN-BASED HERITABLE ELEMENTS

The ability of fungal prions to act as protein-based elements of inheritance depends upon two key processes: *induction*, whereby a protein spontaneously converts to the prion state, and *propagation*, which depends upon efficient, continual conversion of newly synthesized prion protein and partitioning of the prion conformation to daughter cells before cytokinesis. Since Wickner's (176) remarkable proposal that [*URE3*] and [*PSI*$^+$] are prions, an explosion of studies have provided notable insight into the mechanism of induction and propagation. Indeed, today much more is known about induction and propagation of yeast prions than mammalian prions (19).

Induction, Propagation, and Curing

Any model for prion induction must account for several general observations. The rate of spontaneous induction of [*PSI*$^+$], [*URE3*], [*RNQ*$^+$], and [Het-s] is slow, typically 10^{-5} to 10^{-7} (2, 8, 38, 105, 108, 176). However, induction frequency increases dramatically when the prion determinant or fragments containing the PrD are overproduced, depending on the level of overproduction and amino acid context of the PrD (22, 24, 35, 50, 109, 111, 147, 176). High-copy expression plasmids induce [*PSI*$^+$] more efficiently than low-copy plasmids (50). At the same copy number, plasmids expressing a certain region of Sup35N (amino acids 1 to 114) induce [*PSI*$^+$] more efficiently than other fragments (50, 80, 83).

Induction seems to involve a maturation period, during which nascent prions are particularly unstable. [*PSI*$^+$] and [*RNQ*$^+$] are unstable when newly induced and segregate [*prion*$^-$] and stable [*PRION*$^+$] clones (45, 47). Stabilizing the prion state may be more difficult for some prion-like proteins than others, and some may never complete this maturation period (45, 118). Perhaps a certain number of conversion events must occur to cross a threshold of stabilization. Alternatively, some proteins may adopt the prion conformation more readily than others.

Tremendous insight into how prions are induced and propagated came from studying the self-perpetuating conversion of prions in vitro. Conformational conversion of prion proteins into amyloid fibers proceeds by a cooperative process, which can be subdivided into a lag phase, during which nuclei form, and a templated assembly phase, which is characterized by rapid conversion of soluble proteins as it interacts with nuclei (53, 69, 81, 138, 140, 147, 162). The lag phase can be quite long and is reminiscent of the slow rate of spontaneous prion induction in vivo. It can be shortened, or even eliminated, by adding small amounts of preformed converted protein at the onset, mimicking how pre-existing prions convert newly synthesized protein in vivo.

The processes that convert Sup35 and Ure2 into amyloid in vitro and produce [*PSI*$^+$] and [*URE3*] in vivo are certainly closely related. Addition of [*PSI*$^+$], but not [*psi*$^-$], lysates greatly accelerates conversion of Sup35NM (69, 124, 168). Mutations that increase or decrease the rate of prion induction correspondingly alter the rate of amyloid formation in vitro (44, 105). The structural characteristics of Sup35 appear to be optimized to allow conversion (135). The optimal temperature for Sup35NM assembly into amyloid is 25–30°C, coinciding with the optimal growth temperature of yeast (135). Finally, [*PSI*$^+$] can be induced, albeit inefficiently, by introducing purified, recombinant Sup35NM into the yeast cytoplasm using a liposome transformation method (149). Although the conformational state of the Sup35NM used in the protein transformation experiments could not be determined, it also accelerated Sup35NM conversion in vitro.

Although prion proteins form amyloid fibers in vitro, it is not clear if the mature prion conformation in vivo is amyloid fibers. Amyloid-like filaments were detected by thin-section electron microscopy and immunogold-labeling techniques in [*URE3*] cells that overproduce Ure2, but not in [*URE3*] cells with wild-type Ure2 levels or in [*ure-o*] cells (151). Attempts to detect Sup35 filaments in [*PSI*$^+$] cells expressing wild-type levels of Sup35 were inconclusive (A. Kowal & S. Lindquist, unpublished data). Fluorescent, ribbon-like structures are observed in [*PSI*$^+$] (187) and [*URE3*] cells (137) when prion-GFP fusions are overproduced. Thus, the prions can promote the formation of amyloid-like filaments in yeast when highly overproduced, but perhaps not under wild-type conditions. Perhaps, one or more of the structural intermediates identified during formation of prion amyloids in vitro might constitute the prion state in vivo (67, 140, 161).

Two subregions of Sup35N crucial for conversion were identified through mutations. Mutant proteins containing fewer Gln and Asn residues than wild type between amino acids 8 and 24 do not propagate [*PSI*$^+$] (44). Deletion of an oligopeptide repeat or even certain point mutations within one interfere with [*PSI*$^+$] propagation (52, 105, 120). Some mutant proteins convert too slowly to propagate the prion state (44, 105, 120). Others can dominantly interfere with incorporation of wild-type protein into the prion aggregate, and [*PSI*$^+$] is cured when those mutants are coexpressed with wild-type Sup35 (44, 52, 184).

Although no equivalent *ure2* mutants are known, [*URE3*] propagation can be efficiently impeded by highly expressing Ure2N-, Ure2C-, or Ure2-GFP fusions (57). This is a particularly perplexing phenomenon because overexpressing full-length *URE2* efficiently induces [*URE3*] (176). Adding to the complexity, those same fragments or GFP fusions do not cure when overexpressed at low levels (57). Possibly, the overproduced proteins interfere with [*URE3*] propagation by interacting with wild-type Ure2 prion aggregates. Alternatively, a factor that cures the prion, such as a molecular chaperone, may be induced, or a limiting factor necessary for [*URE3*] propagation may be consumed.

One mystery is how prions are partitioned from one cell to another. Clearly, this process is efficient. Most yeast prions are mitotically stable and [*prion*$^-$] cells rarely appear spontaneously (38, 96). It is widely assumed that the partitioning of prion particles occurs stochastically as a portion of the yeast cytoplasm containing many prion particles is passed from mother cell to daughter. However, prion aggregates might segregate by an active mechanism, perhaps analogous to organelle partitioning (18, 170). Interestingly, [*PSI*$^+$] strains lacking Sla1, a protein that interacts with Sup35 and is involved in cortical actin polymerization, are more easily cured than wild-type [*PSI*$^+$] strains (3). Sla1 may help partition prions to daughter cells, but Sup35NM-GFP fusions do not colocalize with actin patches in [*PSI*$^+$] cells (4).

A remarkable yet poorly understood feature of fungal prions is their different rates of spontaneous loss. [*PSI*$^+$] is usually more stable than [*URE3*] during meiosis (38, 96). Even the stability of prions formed from the same determinant can vary. Notably, weak [*PSI*$^+$] variants become [*psi*$^-$] more often than isogenic strong [*PSI*$^+$] variants (50). For example, most strong [*PSI*$^+$] elements segregate to 100% of meiotic progeny (38); however, one form of [*PSI*$^+$], initially called [*ETA*$^+$], segregates to only 70%–85% (100, 188). This instability may relate to the intrinsically lower prion conversion rates in weak [*PSI*$^+$] variants (168). A few *sup35* mutants, including one lacking Sup35M, reduce the mitotic stability of [*PSI*$^+$] when they replace the wild-type *SUP35* allele (12, 25, 88, 104). Usually, prions are equally stable during meiosis and mitosis, although in rare instances these can be uncoupled (74).

FACTORS THAT MODULATE PRION INDUCTION FREQUENCY Certain mutations in prion determinants or in other genes increase the rate of spontaneous prion induction. The most effective *sup35* mutant tested contains two additional oligopeptide repeats. [*PSI*$^+$] arises 5000-fold more frequently in this mutant background (105). Compared to wild-type Sup35, this mutant protein is less structured and forms amyloid fibers more readily in vitro (105, 135). A Sup35 mutant lacking some of the repeats was more structured than wild-type and underwent conversion much less efficiently (105, 135). This and other work suggest that conformational flexibility increases prion induction frequency. Several *ure2* mutants increase the rate of [*URE3*] induction, one (*h2*) as much as 1000-fold over the wild type (62). Unexpectedly, most of the 10 amino acid substitutions in h2 that affect induction map outside the PrD. Perhaps these mutations destabilize normal Ure2 interactions with itself or with other proteins, allowing the PrD to exert a stronger influence on conversion.

[*PIN*$^+$] is a non-Mendelian element required to induce [*PSI*$^+$] efficiently but is not caused by *SUP35* and is independent of [*PSI*$^+$] (47, 48). Overproducing Sup35 readily converts [*PIN*$^+$] cells, but not [*pin*$^-$] cells, to [*PSI*$^+$] (48). [*PIN*$^+$] is unnecessary for [*PSI*$^+$] propagation and does not affect the translational defect of [*PSI*$^+$] cells (47). [*PIN*$^+$] is a dominant, cytoplasmically transferable element (45, 48) that occurs spontaneously at a low frequency and is cured by treating cells with GuHCl or deleting *HSP104* (47).

Remarkably, [*PIN*$^+$] is not caused by just one prion protein; rather, it is a susceptibility state in which the presence of any one of several different prions makes it more likely that a second prion will form (45, 118). One hypothesis is that one protein in the prion state can, with low efficiency, nucleate conversion of another to the prion state through a direct interaction (45, 118). Preliminary work with New1 and Sup35 failed to detect heterotypic interactions or cross-nucleation (118). Alternatively, a prion may titrate a factor that ordinarily impedes another prion protein's conversion. Because of their impact on [*PSI*$^+$], molecular chaperones are obvious candidates for such factors. Whether some prions can arise spontaneously without any other prion remains unclear. [*RNQ*$^+$] induction also apparently requires a [*PIN*$^+$] factor, such as [*PSI*$^+$] or [*URE3*]. Punctate fluorescent foci appear in [*PSI*$^+$][*rnq*$^-$] or [*URE3*][*rnq*$^-$] strains when Rnq1-GFP fusions are overexpressed; however, fluorescence is diffuse in [*psi*$^-$][*rnq*$^-$] or [*ure-o*][*rnq*$^-$] strains (45).

Unlike [*PIN*$^+$] determinants, some yeast proteins affect induction of specific prions. One such example is Sup45, which interacts with Sup35 to mediate translational termination (61, 66, 125, 152). Sup45 does not affect [*PSI*$^+$] propagation; however, *SUP45* overexpression inhibits [*PSI*$^+$] induction when *SUP35* is also overexpressed (46). By binding to Sup35, Sup45 may stabilize the nonprion state and thereby interfere with conversion (46). If so, this interaction need not affect prion propagation if conversion occurs before Sup35 associates with Sup45 or if subunit exchange is dynamic in Sup35-Sup45 complexes.

Sla1 is a Gln-rich protein involved in cortical actin assembly in yeast. Two-hybrid analysis identified an interaction between the Gln-rich regions of Sla1 and Sup35N that depends on *HSP104* expression (3). Sup35 and N-terminal fragments induce [*PSI*$^+$] less efficiently in *sla1* deletion strains than in wild-type strains (3). [*PSI*$^+$] propagation, [*PSI*$^+$]-mediated translational read through, and the amount of soluble Sup35 are unaffected in *sla1* deletion strains (3). Sla1 may play some role in translation because *sla1* deletion strains are sensitive to translational inhibitors (3). If Sla1 interacts with some aspect or component of translation that also interacts with Sup35, *SLA1* overexpression may titrate factors that normally limit Sup35 conversion. Alternatively, Sla1 may promote prion induction by interacting with Sup35.

As noted earlier, Hsp104 expression is required for all well-characterized yeast prions. Other chaperones can affect induction of [*PSI*$^+$], and perhaps induction of other prions. Two Hsp70 members, *SSB1* and *SSB2*, associate with the ribosome and nascent polypeptide chains and may function in cotranslational protein folding (11, 116, 127). Intriguingly, [*PSI*$^+$] spontaneously arises 10 times more often in *ssb1, ssb2* deletion strains than in wild-type strains (27). Perhaps conformational conversion to the prion state occurs cotranslationally, and *SSB1* and *SSB2* chaperone Sup35 folding into the translationally active nonprion form.

At least one signal transduction pathway can modulate [*URE3*] induction. Mks1 negatively regulates *URE2* in the nitrogen catabolite repression cascade (58). In response to environmental cues, Mks1 acts indirectly to positively regulate *DAL5*, the permease that imports ureidosuccinate (58). Mks1 is unnecessary for [*URE3*] propagation because the prion can be transferred to a [*ure-o*] *mks1* deletion strain by cytoduction (59). However, the efficiency of [*URE3*] induction is higher in strains overexpressing *MKS1* and is several orders of magnitude lower in *mks1* deletion strains. Mks1 does not alter the steady-state levels of Ure2 (58), and the mechanism by which it modulates [*URE3*] induction is yet to be discovered.

AGENTS THAT CAN CURE FUNGAL PRIONS A great variety of chemical, environmental, and protein-based agents cure yeast prions. These agents differ in curing efficiency, action, and specificity. In principle, curing agents may impair conformational conversion, inhibit partitioning to the daughter cell, or both. [*PSI*$^+$] and [Het-s] strains are readily cured if grown on media containing various solutes at high osmotic strength (9, 35, 145, 165). In contrast, [*URE3*] is selectively cured by switching from minimal media with ureidosuccinate to rich media with other nitrogen sources (137). The observation that many agents that cure [*PSI*$^+$], such as 10% dimethylsulfoxide, 10% methanol, as well as high concentrations of potassium chloride, ethylene glycol, sodium glutamate, and glycerol, are not mutagenic was one of the first indications that [*PSI*$^+$] is not caused by a nucleic acid–based agent (39, 108, 145, 165). Although other curing agents, such as ethyl methanesulfonate and ultraviolet light, are known mutagens, they probably cure [*PSI*$^+$] by affecting *HSP104* expression rather than by mutating *SUP35* (26, 103).

GuHCl is the most frequently used prion-curing agent because it effectively cures [*PSI*$^+$] (165), [*RNQ*$^+$] (45, 48), and [*URE3*] (176). At molar concentrations GuHCl denatures proteins, but protein denaturation is unlikely to explain curing because millimolar concentrations are effective. GuHCl cures [*PSI*$^+$] only in proliferating cells and appears to block [*PSI*$^+$] proliferation without affecting pre-existing prion aggregates (56). There is a lag phase of several generations before cells are cured, possibly because the numerous pre-existing prion particles are diluted with each cell division (56). It is unclear whether GuHCl cures by inactivating Hsp104 or by some other mechanism. Certainly, at low concentrations, GuHCl lowers the ATPase activity of Hsp104 in vitro (70). GuHCl also inhibits cells from acquiring thermotolerance and from refolding thermally denatured luciferase, two processes that require active Hsp104 (64, 75). However, the kinetics by which GuHCl and *hsp104* deletion cure [*PSI*$^+$] may be different, which would suggest that these agents cure by different mechanisms (171).

The toxin latrunculin A (Lat-A) disrupts the actin cytoskeleton by sequestering monomeric actin (34), and cures some forms of [*PSI*$^+$] but not [*RNQ*$^+$] (4,45). Because [*PSI*$^+$] can be propagated in *sla1* deletion strains (3), it is unlikely that Lat-A cures [*PSI*$^+$] by affecting Sla1. Moreover, the kinetics by which Lat-A disrupts the actin cytoskeleton and cures [*PSI*$^+$] are quite different, suggesting that different mechanisms of action underlie them (4). Because yeast cells do not divide in the presence of Lat-A, this drug may cure [*PSI*$^+$] by disrupting prion aggregates or by interfering with conversion (4).

The Role of Hsp104 in Prion Induction and Propagation

The molecular chaperone Hsp104 resolves thermally denatured proteins (70, 121). The *HSP104* gene was isolated in a genetic screen for factors that cure [*PSI*$^+$] when overproduced (26). When *HSP104* is highly expressed, most forms of [*PSI*$^+$] are efficiently cured (26). Remarkably, even transient high-level expression suffices (26). Moderate overexpression, however, does not cure efficiently, but it partially alleviates [*PSI*$^+$]-mediated nonsense suppression. Deletion of *HSP104* also cures (26). Thus, [*PSI*$^+$] propagation requires an intermediate level of Hsp104. [*URE3*] and [*RNQ*$^+$] are also cured when *HSP104* is deleted (45, 48, 114, 147). The dependence of these three prions on this protein-remodeling factor is strong evidence that prions are caused by protein-based agents.

Two models might explain why [*PSI*$^+$] requires an intermediate level of Hsp104. One postulates that Sup35 conversion proceeds through an unstable, oligomeric intermediate state promoted by Hsp104 (26, 103, 122, 140). In [*PSI*$^+$] cells, the intermediate state of Sup35 converts to the prion state when it associates with prion conformers. The intermediate state reverts to the nonprion state in [*psi*$^-$] cells and it is unlikely to form without Hsp104. Transient overexpression of *SUP35* efficiently induces [*PSI*$^+$] de novo because of the increased propensity of Hsp104 and Sup35 to interact. [*PSI*$^+$] propagation does not require Hsp104 to disaggregate Sup35 prion aggregates, although some disaggregation may occur when *HSP104* is highly overexpressed.

The other model posits Hsp104 simply disaggregates Sup35 prion aggregates, which creates many smaller aggregates that are more likely to be partitioned to the daughter cell (90, 123). Without Hsp104, fewer and larger aggregates accumulate, reducing the probability that enough prion proteins will partition to the daughter. High Hsp104 levels cure [*PSI*$^+$] cells simply because all prion aggregates are resolved. Unlike the first model, Hsp104 is not needed to convert Sup35 to the prion state.

Both models predict that Hsp104 interacts with Sup35. Stable complexes of Hsp104 with Sup35, or any other yeast prion, have not been detected; but there is indirect evidence for a transient interaction (136). Sup35 inhibits the ATPase activity of Hsp104 in vitro, and the circular dichroism spectrum of a mixture of

Hsp104 and Sup35 differs from that predicted by adding these proteins' individual spectra (136). Whether this interaction is necessary to convert Sup35 or to disrupt prion aggregates is less clear. Indeed, it may do either, depending on the presence of other chaperones. Consistent with the first model, Sup35 amyloid formation proceeds through an oligomeric state that forms without Hsp104 in vitro (69, 140). When Sup35-GFP fusions are expressed at low levels in [*psi*$^-$] *hsp104* deletion strains, no punctate fluorescent foci are visible. Expressing the fusion at higher levels overcomes the apparent requirement for Hsp104 (69). Finally, a greater proportion of Sup35 is soluble in [*PSI*$^+$] cells that moderately overexpress *HSP104* (123). This may indicate that Hsp104 disaggregates Sup35 or that Sup35 molecules are dispersed among more chaperone complexes, thereby lowering the number of oligomeric intermediates formed.

One apparent difficulty with the disaggregation model is that high *HSP104* expression does not cure [*URE3*] and [*RNQ*$^+$] (45, 48, 114). On the other hand, two observations support the disaggregation model. First, Sup35NM-GFP foci become larger and fewer, and [*PSI*$^+$] is rapidly cured when Hsp104 levels are reduced (171). Second, in a *sup35* mutant strain in which three oligopeptide repeats are disrupted, [*PSI*$^+$] is unstable and a greater proportion of Sup35 is insoluble than in wild-type [*PSI*$^+$] strains (12). One possible explanation is that large insoluble aggregates are inefficiently partitioned during cell division. Consistent with the disaggregation model, moderately overproducing Hsp104 bolstered the stability of [*PSI*$^+$] in the mutant strain.

The Role of Other Molecular Chaperones in Prion Induction and Propagation

[*PSI*$^+$] is cured inefficiently or not at all by many conditions that dramatically raise the level of Hsp104, including heat shock, or sporulation (131, 132, 145, 165). This was initially surprising because high *HSP104* overexpression cures [*PSI*$^+$] efficiently (26). Perhaps curing requires cellular division, which does not occur during heat shock, or other factors that are induced that counteract the effect of higher Hsp104 levels.

One such candidate is the heat-inducible molecular chaperone Hsp70. Yeast have four functionally redundant and nearly identical cytosolic Hsp70 proteins called Ssa1-4 (11, 41, 42). Ssa1 is constitutively expressed, but its level increases two- to threefold after heat stress (173, 174). Hsp104 overproduction cures [*PSI*$^+$] less effectively, and less Sup35 becomes soluble when Ssa1 is also overproduced (117). In contrast, overproduction of Ssb1 or Ssb2, two Hsp70 relatives that are not heat inducible (106, 172), enhance the efficiency with which Hsp104 overproduction cures [*PSI*$^+$] (27).

Although the role of chaperones in prion propagation is less well characterized for [*URE3*] than [*PSI*$^+$], there are intriguing differences. *HSP104* deletion cures [*URE3*] and [*PSI*$^+$], but *HSP104* overexpression cures only [*PSI*$^+$] (26, 114). Overproducing Ydj1, a cytosolic Hsp40, slowly cures [*URE3*] (114), but not [*PSI*$^+$] (89).

Because Hsp40s can specify the substrates of Hsp70-Hsp40 complexes (76), they may direct which prions and chaperones interact.

Despite the importance of chaperones in prion propagation, only one stable interaction between chaperones and prions has been identified. The cytosolic Hsp40 Sis1 is essential for viability and required for translational initiation (107, 186). The glycine- and phenylalanine-rich (G/F) region of Sis1 partly determines the chaperone's specificity (182). Rnq1 only binds stably to Sis1 when it is in the prion state, and [*RNQ*$^+$] cannot be propagated in a *sis1* mutant lacking the G/F region (148).

EVOLUTIONARY AND BIOLOGICAL SIGNIFICANCE OF PRIONS

Why do prions exist? At first glance, yeast prions do not appear beneficial. Indeed, [*PSI*$^+$] cells have a translation defect, and [*URE3*] cells grow slower than wild-type cells and wastefully utilize poor nitrogen sources (38, 96). Because yeast rapidly lose even mildly harmful markers, the ability to become a prion would not be conserved if prions are simply deleterious. However, the Sup35 and Ure2 homologs of many yeast species contain PrD-like regions positioned N-terminally to their functional domains (33, 92, 115, 133) (Figure 6, see color insert).

To determine if Sup35NM might be adaptive, the extent of nucleotide polymorphism within many laboratory, commercial, and clinical isolates of *S. cerevisiae* in Sup35 was compared to the extent of sequence divergence of a related species, *S. paradoxus* (73). The amino acid sequences of Sup35N, M, and C are constrained to varying extents, presumably by purifying selection against mutations that change a coding sequence. Sup35C is under strong constraint, presumably because of its essential role in translational termination; Sup35NM also appears to be under selection, albeit more weakly.

What might be the function of the N-terminal PrD-like regions of Sup35 and Ure2 homologs? Certainly, the PrDs of *S. cerevisiae* are not essential for translational termination or for nitrogen catabolite repression (33, 160), although nitrogen regulation of Gln3 is somewhat diminished when the PrD of Ure2 is deleted (86). Sup35NM and the ability to form [*PSI*$^+$] is essential in yeast strains containing nonsense mutations in vital genes (84, 102). But most strains do not contain such mutations. One possibility is that the putative PrDs have functions other than being able to act as prions. Under some growth conditions, Sup35NM deletion strains have phenotypes distinct from those of [*PSI*$^+$] or [*psi*$^-$] strains (163). Sup35N may link translation to the cytoskeleton via its interaction with Sla1, or it may play a role in glucose metabolism by interacting with Reg1 and Eno2 (3). A more interesting possibility is that a putative PrD is conserved to produce a prion, and thereby epigenetically modify the function of the protein domain to which it is attached.

One line of support for this hypothesis is the observation that all yeast Sup35 homologs tested can convert to the [*PSI*$^+$] state. *Kluyveromyces lactis* Sup35 was shown to have this capacity in *K. lactis* (115). Because most other yeast species lack a facile phenotypic assay for [*PSI*$^+$], the ability of diverse *SUP35* proteins to adopt a [*PSI*$^+$]-like state was tested in *S. cerevisiae* by expressing fusions of putative PrDs from many species to *S. cerevisiae* Sup35NM or Sup35C (25, 88, 115, 133). Most could not convert the endogenous *S. cerevisiae* Sup35 to the prion state. Remarkably, all could induce and propagate a prion-like state if the corresponding heterologous *SUP35* homolog was also expressed (25, 88, 115, 133).

Strikingly, the amino acid sequences of homologus PrDs are divergent, precluding precise sequence alignment. Moreover, the ratio of Asn to Gln residues varies tremendously, even between homologous PrDs. [Compare the *S. cerevisiae* Ure2 PrD with its *K. lactis* homolog and the *Saccharomycodes ludwigii* Sup35 PrD to that of *Pichia methanolica* (Figure 6)]. Thus, these regions have acquired many changes in sequence, yet they retain both the unusual amino acid composition characteristic of the *S. cerevisiae* PrDs and the capacity to form prions. We suggest that these observations support the hypothesis that the capacity of these domains to function as prions is evolutionarily conserved. Moreover, this capacity is derived from their unusual amino acid composition, not from their absolute sequences.

The potential evolutionary significance of [*URE3*] is not obvious. However, some *ure2*, and presumably [*URE3*] strains, grow better than wild-type strains in the presence of high concentrations of Na^+, Li^+, and Mn^{2+} (181), and reach a higher biomass than wild-type strains when grown in the presence of grape juices (130). Natural populations may also benefit from being able to constitutively use a wide array of good and poor nitrogen sources, as is the case in *ure2* and [*URE3*] strains.

One interesting hypothesis for an adaptive value for [*PSI*$^+$] was that the translational read through it causes would produce a constitutive heat shock response and make cells constitutively thermotolerant (55). Although this was true in some strains, it was not true in most strains tested [(55, 163); J. Taulein, Y. Chernoff & S. Lindquist, unpublished data]. A more provocative suggestion is that [*PSI*$^+$] provides a mechanism for genetic variation and phenotypic diversity in the face of changing environments. On rich media, the growth characteristics of isogenic sets of [*PSI*$^+$] and [*psi*$^-$] strains are usually indistinguishable. However, differences in growth or survival often occur when such sets are grown under other conditions, including a variety of carbon and nitrogen sources, and in the presence of potentially toxic salts, metals, and inhibitors of diverse cellular processes including DNA replication, signal transduction, protein glycosylation, and microtubule dynamics (163). In an extensive study using more than 150 different conditions, [*PSI*$^+$] exerted a substantial effect in at least one strain background in nearly half of the conditions tested, and [*PSI*$^+$] strains grew or survived better than [*psi*$^-$] strains in about 20 of these tests (163). Remarkably, each genetic background displayed unique and diverse constellations of phenotypes in response to different environmental conditions (163).

How might this extraordinary phenotypic diversity arise? Almost certainly, it is related to the translational defect caused by [*PSI*$^+$] (163). Some stop-codon read through events might append extra amino acids to proteins, and others might activate cryptic genes or pseudo-genes that had accumulated mutations while inactive. Because [*PSI*$^+$] is spontaneously induced or cured at a low frequency in a natural population of any substantial size, some cells will be present in a [*PSI*$^+$] state and others in a [*psi*$^-$] state. The different phenotypes of these cells would increase the chance that the genome will survive when environmental conditions change. Finally, because the level of Hsp104 affects [*PSI*$^+$] propagation and is increased by environmental stress, [*PSI*$^+$] provides an intriguing and plausible mechanism for yeast to adapt to different environmental niches in response to environmental change.

If prions are beneficial, how may novel prions be created under natural circumstances? One explanation is that prion domains are modular and transferable. Sup35NM can be transferred to an unrelated transcriptional activator to create a novel prion that regulates transcriptional initiation rather than translational termination (99). This chimeric prion protein can exist in two stable, but interchangeable, functional states, which are independent of the prion conformation of Sup35. Thus, new prions may arise stochastically through recombination that appends sequences encoding PrDs to other genes in the genome.

HOW WIDESPREAD ARE PRIONS?

How many other prion proteins are there? To address this fascinating question, an algorithm was used to search the proteomes of 31 organisms for proteins containing at least 30 Gln or Asn residues within an 80-amino-acid region (113). Proteins fitting this criterion were nearly absent from the 28 archeal, thermophilic, and mesophilic bacterial proteomes examined. In contrast, they constituted a surprisingly large fraction of eukaryotic proteins, as much as several percent of the total. *S. cerevisiae* had 107 candidates, or about 1.69% of the total proteins. This estimate is consistent with another, independently obtained with a more stringent algorithm (>50 consecutive residues containing at least 45% Q/N and 60% polar residues) (L. Li, M. Long & S. Lindquist, unpublished data). The prion candidates are involved in diverse biological processes and include transcription and translation factors, nucleoporins, DNA- and RNA-binding proteins, and proteins involved in vesicular trafficking. Although it is unclear how many of these proteins are prions, eight were independently identified as determinants of [*PIN*$^+$] when overexpressed (45). Moreover, [*RNQ*$^+$] and [*NU*$^+$] were found using similar database searches (133, 146). Notably, some prions that are not rich in Asn and Gln, such as HET-s and PrP, are missed by such searches, suggesting that these searches are still underestimating the number of possible prion proteins.

The results of directed searches and the mere fact that prions occur in yeast, filamentous fungi, and mammals strongly suggest that prions are widely dispersed throughout the living world and underlie a wide variety of biological processes. Prions may epigenetically modulate chromatin structure and function (101), mediate homologous chromosome pairing (153), comprise developmental switches (101), and participate in organelle inheritance.

Although the self-perpetuating mechanism of the fungal prions is closely related to amyloid formation, some as yet unrecognized prions might be propagated through entirely different mechanisms. Indeed, there are other novel self-assembling protein structures that are unrelated to amyloids (32, 77, 79). Similarly, there may be mechanisms to partition prions faithfully other than stochastic division of cytoplasmic content. Associations with cellular structures, such as organelles or outer membranes, can mediate inheritance of proteins. Indeed, one example of multigenerational cortical inheritance is the protein Rax2, which orients bud polarity in yeast (21). Another is cortical inheritance of the pattern of cilia on the surface of *Paramecium* (7, 179).

Finally, more than one prion can exist in a cell, and the presence of a prion can increase the likelihood that another will appear (45, 118). Thus, different combinations of prions within a cell might produce a wide variety of heritable phenotypes, all without necessitating any underlying changes in the genome. Clearly, identifying prions and ascertaining their roles in biology will occupy scientists for years to come.

ACKNOWLEDGMENTS

The work cited from the author's laboratory was funded by the NIH and HHMI; S.M. Uptain was supported a postdoctoral fellowship from the American Cancer Society. We thank our many colleagues who sent us reprints and preprints and communicated results prior to publication. We also thank C. Queitsch, O. Hormann, M. Duennwald, T.A. Sangster, and L. Li for critical comments on the manuscript.

The *Annual Review of Microbiology* is online at http://micro.annualreviews.org

LITERATURE CITED

1. http://genome-www.stanford.edu/Saccharomyces/. Saccharomyces Genome Database
2. Aigle M, Lacroute F. 1975. Genetical aspects of [*URE3*], a non-mitochondrial, cytoplasmically inherited mutation in yeast. *Mol. Gen. Genet.* 136:327–35
3. Bailleul PA, Newnam GP, Steenbergen JN, Chernoff YO. 1999. Genetic study of interactions between the cytoskeletal assembly protein Sla1 and prion-forming domain of the release factor Sup35 (eRF3) in *Saccharomyces cerevisiae*. *Genetics* 153:81–94
4. Bailleul-Winslett PA, Newnam GP, Wegrzyn RD, Chernoff YO. 2000. An anti-prion effect of the anticytoskeletal drug latrunculin A in yeast. *Gene Exp.* 9:145–56
5. Balbirnie M, Grothe R, Eisenberg DS. 2001. An amyloid-forming peptide from the yeast prion Sup35 reveals a dehydrated beta-sheet structure for amyloid. *Proc. Natl. Acad. Sci. USA* 98:2375–80

6. Ball AJ, Wong DK, Elliott JJ. 1976. Glucosamine resistance in yeast. I. A preliminary genetic analysis. *Genetics* 84:311–17
7. Beisson J, Sonneborn TM. 1965. Cytoplasmic inheritance of the organization of the cell cortex in *Paramecium aurelia*. *Proc. Natl. Acad. Sci. USA* 53:275–82
8. Beisson-Schecroun J. 1962. Incompatibilite cellulaire et interactions nucleo-cytoplasmiques dans les phenomenes de 'barrage' chez le *Podospora anserina*. *Ann. Genet.* 4:3–50
9. Belcour L. 1975. Cytoplasmic mutations isolated from protoplasts of *Podospora anserina*. *Genet. Res.* 25:155–61
10. Blinder D, Coschigano PW, Magasanik B. 1996. Interaction of the GATA factor Gln3p with the nitrogen regulator Ure2p in *Saccharomyces cerevisiae*. *J. Bacteriol.* 178:4734–36
11. Boorstein WR, Ziegelhoffer T, Craig EA. 1994. Molecular evolution of the *HSP70* multigene family. *J. Mol. Evol.* 38:1–17
12. Borchsenius AS, Wegrzyn RD, Newnam GP, Inge-Vechtomov SG, Chernoff YO. 2001. Yeast prion protein derivative defective in aggregate shearing and production of new 'seeds.' *EMBO J.* 20:6683–91
13. Bousset L, Belrhali H, Janin J, Melki R, Morera S. 2001. Structure of the globular region of the prion protein Ure2 from the yeast *Saccharomyces cerevisiae*. *Structure* 9:39–46
14. Bruce ME. 1993. Scrapie strain variation and mutation. *Br. Med. Bull.* 49:822–38
15. Bruce ME, Dickinson AG. 1987. Biological evidence that scrapie agent has an independent genome. *J. Gen. Virol.* 68:79–89
16. Bruce ME, Fraser H. 1991. Scrapie strain variation and its implications. *Curr. Top. Microbiol. Immunol.* 172:125–38
17. Carrell RW, Gooptu B. 1998. Conformational changes and disease—serpins, prions and Alzheimer's. *Curr. Opin. Struct. Biol.* 8:799–809
18. Catlett NL, Weisman LS. 2000. Divide and multiply: organelle partitioning in yeast. *Curr. Opin. Cell Biol.* 12:509–16
19. Caughey B. 2000. Transmissible spongiform encephalopathies, amyloidoses and yeast prions: common threads? *Nat. Med.* 6:751–54
20. Caughey B, Raymond GJ, Callahan MA, Wong C, Baron GS, Xiong LW. 2001. Interactions and conversions of prion protein isoforms. *Adv. Protein Chem.* 57:139–69
21. Chen T, Hiroko T, Chaudhuri A, Inose F, Lord M, et al. 2000. Multigenerational cortical inheritance of the Rax2 protein in orienting polarity and division in yeast. *Science* 290:1975–78
22. Chernoff Y, Derkatch I, Dagkesamanskaya A, Tikhomironva V, Ter-Avanesyan M, Inge-Vechtomov S. 1988. Nonsense-suppression by amplification of translational protein factor gene. *Dokl. Akad. Nauk SSSR* 301:1227–29
23. Chernoff YO. 2001. Mutation processes at the protein level: Is Lamarck back? *Mutat. Res.* 488:39–64
24. Chernoff YO, Derkatch IL, Inge-Vechtomov SG. 1993. Multicopy *SUP35* gene induces *de-novo* appearance of *psi*-like factors in the yeast *Saccharomyces cerevisiae*. *Curr. Genet.* 24:268–70
25. Chernoff YO, Galkin AP, Lewitin E, Chernova TA, Newnam GP, Belenkiy SM. 2000. Evolutionary conservation of prion-forming abilities of the yeast Sup35 protein. *Mol. Microbiol.* 35:865–76
26. Chernoff YO, Lindquist SL, Ono B, Inge-Vechtomov SG, Liebman SW. 1995. Role of the chaperone protein Hsp104 in propagation of the yeast prion-like factor [*PSI*$^+$]. *Science* 268:880–84
27. Chernoff YO, Newnam GP, Kumar J, Allen K, Zink AD. 1999. Evidence for a protein mutator in yeast: role of the

Hsp70-related chaperone Ssb in formation, stability, and toxicity of the *[PSI]* prion. *Mol. Cell Biol.* 19:8103–12
28. Chien P, Weissman JS. 2001. Conformational diversity in a yeast prion dictates its seeding specificity. *Nature* 410:223–27
29. Choi JH, Lou W, Vancura A. 1998. A novel membrane-bound glutathione S-transferase functions in the stationary phase of the yeast *Saccharomyces cerevisiae. J. Biol. Chem.* 273:29915–22
30. Collinge J. 2001. Prion diseases of humans and animals: their causes and molecular basis. *Annu. Rev. Neurosci.* 24:519–50
31. Conde J, Fink GR. 1976. A mutant of *Saccharomyces cerevisiae* defective for nuclear fusion. *Proc. Natl. Acad. Sci. USA* 73:3651–55
32. Contegno F, Cioce M, Pelicci PG, Minucci S. 2002. Targeting protein inactivation through an oligomerization chain reaction. *Proc. Natl. Acad. Sci. USA* 12:12
33. Coschigano PW, Magasanik B. 1991. The *URE2* gene product of *Saccharomyces cerevisiae* plays an important role in the cellular response to the nitrogen source and has homology to glutathione S-transferases. *Mol. Cell Biol.* 11:822–32
34. Coue M, Brenner SL, Spector I, Korn ED. 1987. Inhibition of actin polymerization by latrunculin A. *FEBS Lett.* 213:316–18
35. Coustou V, Deleu C, Saupe S, Begueret J. 1997. The protein product of the *het-s* heterokaryon incompatibility gene of the fungus *Podospora anserina* behaves as a prion analog. *Proc. Natl. Acad. Sci. USA* 94:9773–78
36. Coustou V, Deleu C, Saupe SJ, Begueret J. 1999. Mutational analysis of the [Het-s] prion analog of *Podospora anserina*. A short N-terminal peptide allows prion propagation. *Genetics* 153:1629–40
37. Coustou-Linares V, Maddelein ML, Begueret J, Saupe S. 2001. *In vivo* aggregation of the HET-s prion protein of the fungus *Podospora anserina. Mol. Microbiol.* 42:1325–35
38. Cox B. 1965. *[PSI]*, a cytoplasmic suppressor of super-suppression in yeast. *Heredity* 20:505–21
39. Cox BS, Tuite MF, McLaughlin CS. 1988. The *psi* factor of yeast: a problem in inheritance. *Yeast* 4:159–78
40. Cox BS, Tuite MF, Mundy CJ. 1980. Reversion from suppression to nonsuppression in *SUQ5* [psi^+] strains of yeast: the classification of mutations. *Genetics* 95:589–609
41. Craig E. 1990. Regulation and function of the *HSP70* multigene family of *Saccharomyces cerevisiae*. In *Stress Proteins in Biology and Medicine*, ed. R Morimoto, A Tissieres, C Georgopoulos, pp. 301–21. Cold Spring Harbor, NY: Cold Spring Harbor Lab. Press
42. Craig E, Ziegelhoffer T, Nelson J, Laloraya S, Halladay J. 1995. Complex multigene family of functionally distinct Hsp70s of yeast. *Cold Spring Harbor Symp. Quant. Biol.* 60:441–99
43. http://cellcycle-www.stanford.edu/. Yeast Cell Cycle Analysis Project
44. DePace AH, Santoso A, Hillner P, Weissman JS. 1998. A critical role for amino-terminal glutamine/asparagine repeats in the formation and propagation of a yeast prion. *Cell* 93:1241–52
45. Derkatch IL, Bradley ME, Hong JY, Liebman SW. 2001. Prions affect the appearance of other prions: the story of [PIN^+]. *Cell* 106:171–82
46. Derkatch IL, Bradley ME, Liebman SW. 1998. Overexpression of the *SUP45* gene encoding a Sup35p-binding protein inhibits the induction of the *de novo* appearance of the [PSI^+] prion. *Proc. Natl. Acad. Sci. USA* 95:2400–5
47. Derkatch IL, Bradley ME, Masse SV, Zadorsky SP, Polozkov GV, et al. 2000. Dependence and independence of [PSI^+] and [PIN^+]: a two-prion system in yeast? *EMBO J.* 19:1942–52

48. Derkatch IL, Bradley ME, Zhou P, Chernoff YO, Liebman SW. 1997. Genetic and environmental factors affecting the *de novo* appearance of the [PSI^{+}] prion in *Saccharomyces cerevisiae*. *Genetics* 147:507–19
49. Derkatch IL, Bradley ME, Zhou P, Liebman SW. 1999. The *PNM2* mutation in the prion protein domain of *SUP35* has distinct effects on different variants of the [PSI^{+}] prion in yeast. *Curr. Genet.* 35: 59–67
50. Derkatch IL, Chernoff YO, Kushnirov VV, Inge-Vechtomov SG, Liebman SW. 1996. Genesis and variability of [*PSI*] prion factors in *Saccharomyces cerevisiae*. *Genetics* 144:1375–86
51. Dickinson AG, Outram GW. 1979. The scrapie replication-site hypothesis and its implications for pathogenesis. In *Slow Transmissible Diseases of the Nervous System*, ed. SB Prusiner, pp. 13–31. New York: Academic
52. Doel SM, McCready SJ, Nierras CR, Cox BS. 1994. The dominant *PNM2*-mutation which eliminates the [*PSI*] factor of *Saccharomyces cerevisiae* is the result of a missense mutation in the *SUP35* gene. *Genetics* 137:659–70
53. Dos Reis S, Coulary-Salin B, Forge V, Lascu I, Begueret J, Saupe SJ. 2002. The HET-s prion protein of the filamentous fungus *Podospora anserina* aggregates *in vitro* into amyloid-like fibrils. *J. Biol. Chem.* 277:5703–6
54. Drillien R, Aigle M, Lacroute F. 1973. Yeast mutants pleiotropically impaired in the regulation of the two glutamate dehydrogenases. *Biochem. Biophys. Res. Commun.* 53:367–72
55. Eaglestone SS, Cox BS, Tuite MF. 1999. Translation termination efficiency can be regulated in *Saccharomyces cerevisiae* by environmental stress through a prion-mediated mechanism. *EMBO J.* 18: 1974–81
56. Eaglestone SS, Ruddock LW, Cox BS, Tuite MF. 2000. Guanidine hydrochloride blocks a critical step in the propagation of the prion-like determinant [PSI^{+}] of *Saccharomyces cerevisiae*. *Proc. Natl. Acad. Sci. USA* 97:240–44
57. Edskes HK, Gray VT, Wickner RB. 1999. The [*URE3*] prion is an aggregated form of Ure2p that can be cured by overexpression of Ure2p fragments. *Proc. Natl. Acad. Sci. USA* 96:1498–503
58. Edskes HK, Hanover JA, Wickner RB. 1999. Mks1p is a regulator of nitrogen catabolism upstream of Ure2p in *Saccharomyces cerevisiae*. *Genetics* 153:585–94
59. Edskes HK, Wickner RB. 2000. A protein required for prion generation: [*URE3*] induction requires the Ras-regulated Mks1 protein. *Proc. Natl. Acad. Sci. USA* 97:6625–29
60. Elliot JJ, Ball AJ. 1975. A new mitochondrial mutation in *Saccharomyces cerevisiae*. *Biochem. Biophys. Res. Commun.* 64:277–81
61. Eurwilaichitr L, Graves FM, Stansfield I, Tuite MF. 1999. The C-terminus of eRF1 defines a functionally important domain for translation termination in *Saccharomyces cerevisiae*. *Mol. Microbiol.* 32:485–96
62. Fernandez-Bellot E, Guillemet E, Cullin C. 2000. The yeast prion [*URE3*] can be greatly induced by a functional mutated *URE2* allele. *EMBO J.* 19:3215–22
63. Fernandez-Bellot E, Guillemet E, Ness F, Baudin-Baillieu A, Ripaud L, et al. 2002. The [*URE3*] phenotype: evidence for a soluble prion in yeast. *EMBO Rep.* 3:76–81
64. Ferreira PC, Ness F, Edwards SR, Cox BS, Tuite MF. 2001. The elimination of the yeast [PSI^{+}] prion by guanidine hydrochloride is the result of Hsp104 inactivation. *Mol. Microbiol.* 40:1357–69
65. Flick JS, Johnston M. 1991. *GRR1* of *Saccharomyces cerevisiae* is required for glucose repression and encodes a protein

with leucine-rich repeats. *Mol. Cell Biol.* 11:5101–12

66. Frolova L, Le Goff X, Rasmussen HH, Cheperegin S, Drugeon G, et al. 1994. A highly conserved eukaryotic protein family possessing properties of polypeptide chain release factor. *Nature* 372:701–3
67. Galani D, Fersht AR, Perrett S. 2002. Folding of the yeast prion protein Ure2: kinetic evidence for folding and unfolding intermediates. *J. Mol. Biol.* 315:213–27
68. Garvik B, Haber JE. 1978. New cytoplasmic genetic element that controls 20S RNA synthesis during sporulation in yeast. *J. Bacteriol.* 134:261–69
69. Glover JR, Kowal AS, Schirmer EC, Patino MM, Liu JJ, Lindquist S. 1997. Self-seeded fibers formed by Sup35, the protein determinant of [PSI^+], a heritable prion-like factor of *S. cerevisiae*. *Cell* 89:811–19
70. Glover JR, Lindquist S. 1998. Hsp104, Hsp70, and Hsp40: a novel chaperone system that rescues previously aggregated proteins. *Cell* 94:73–82
71. Hawthorne DC, Leupold U. 1974. Suppressors in yeast. *Curr. Top. Microbiol. Immunol.* 64:1–47
72. Inge-Vechtomov SG, Andriavnova VM. 1970. Recessive supersuppressors in yeast. *Genetika* 6:103–15
73. Jensen MA, True HL, Chernoff YO, Lindquist S. 2001. Molecular population genetics and evolution of a prion-like protein in *Saccharomyces cerevisiae*. *Genetics* 159:527–35
74. Jung G, Jones G, Wegrzyn RD, Masison DC. 2000. A role for cytosolic Hsp70 in yeast [PSI^+] prion propagation and [PSI^+] as a cellular stress. *Genetics* 156:559–70
75. Jung G, Masison DC. 2001. Guanidine hydrochloride inhibits Hsp104 activity *in vivo*: a possible explanation for its effect in curing yeast prions. *Curr. Microbiol.* 43:7–10
76. Kelley WL. 1999. Molecular chaperones: how J domains turn on Hsp70s. *Curr. Biol.* 9:R305–8
77. Kentsis A, Gordon RE, Borden KL. 2002. Self-assembly properties of a model RING domain. *Proc. Natl. Acad. Sci. USA* 99:667–72
78. Kikuchi Y, Shimatake H, Kikuchi A. 1988. A yeast gene required for the G1-to-S transition encodes a protein containing an A-kinase target site and GTPase domain. *EMBO J.* 7:1175–82
79. Kim CA, Phillips ML, Kim W, Gingery M, Tran HH, et al. 2001. Polymerization of the SAM domain of TEL in leukemogenesis and transcriptional repression. *EMBO J.* 20:4173–82
80. King CY. 2001. Supporting the structural basis of prion strains: induction and identification of [*PSI*] variants. *J. Mol. Biol.* 307:1247–60
81. King CY, Tittmann P, Gross H, Gebert R, Aebi M, Wuthrich K. 1997. Prion-inducing domain 2-114 of yeast Sup35 protein transforms *in vitro* into amyloid-like filaments. *Proc. Natl. Acad. Sci. USA* 94:6618–22
82. Kochneva-Pervukhova NV, Chechenova MB, Valouev IA, Kushnirov VV, Smirnov VN, Ter-Avanesyan MD. 2001. [PSI^+] prion generation in yeast: characterization of the 'strain' difference. *Yeast* 18:489–97
83. Kochneva-Pervukhova NV, Poznyakovski AI, Smirnov VN, Ter-Avanesyan MD. 1998. C-terminal truncation of the Sup35 protein increases the frequency of *de novo* generation of a prion-based [PSI^+] determinant in *Saccharomyces cerevisiae*. *Curr. Genet.* 34:146–51
84. Kokoska RJ, Stefanovic L, DeMai J, Petes TD. 2000. Increased rates of genomic deletions generated by mutations in the yeast gene encoding DNA polymerase delta or by decreases in the cellular levels of DNA polymerase delta. *Mol. Cell Biol.* 20:7490–504

85. Kretzschmar HA, Stowring LE, Westaway D, Stubblebine WH, Prusiner SB, Dearmond SJ. 1986. Molecular cloning of a human prion protein cDNA. *DNA* 5:315–24
86. Kulkarni AA, Abul-Hamd AT, Rai R, El Berry H, Cooper TG. 2001. Gln3p nuclear localization and interaction with Ure2p in *Saccharomyces cerevisiae*. *J. Biol. Chem.* 276:32136–44
87. Kunz BA, Ball AJ. 1977. Glucosamine resistance in yeast. II. Cytoplasmic determinants conferring resistance. *Mol. Gen. Genet.* 153:169–77
88. Kushnirov VV, Kochneva-Pervukhova NV, Chechenova MB, Frolova NS, Ter-Avanesyan MD. 2000. Prion properties of the Sup35 protein of yeast *Pichia methanolica*. *EMBO J.* 19:324–31
89. Kushnirov VV, Kryndushkin DS, Boguta M, Smirnov VN, Ter-Avanesyan MD. 2000. Chaperones that cure yeast artificial [*PSI*$^+$] and their prion-specific effects. *Curr. Biol.* 10:1443–46
90. Kushnirov VV, Ter-Avanesyan MD. 1998. Structure and replication of yeast prions. *Cell* 94:13–16
91. Kushnirov VV, Ter-Avanesyan MD, Dagkesamanskaia AR, Chernoff YO, Inge-Vechtomov SG, Smirnov VN. 1990. Deletion analysis of the *SUP2* gene in *Saccharomyces cerevisiae*. *Mol. Biol.* 24:1037–41
92. Kushnirov VV, Ter-Avanesyan MD, Smirnov VN, Chernoff YO, Derkatch IL, et al. 1990. Comparative analysis of the structure of *SUP2* genes in *Pichia pinus* and *Saccharomyces cerevisiae*. *Mol. Biol.* 24:1024–36
93. Kushnirov VV, Ter-Avanesyan MD, Surguchov AP, Smirnov VN, Inge-Vechtomov SG. 1987. Localization of possible functional domains in *sup2* gene product of the yeast *Saccharomyces cerevisiae*. *FEBS Lett.* 215:257–60
94. Kushnirov VV, Ter -Avanesyan MD, Telckov MV, Surguchov AP, Smirnov VN, Inge-Vechtomov SG. 1988. Nucleotide sequence of the *SUP2* (*SUP35*) gene of *Saccharomyces cerevisiae*. *Gene* 66:45–54
95. Lacroute F. 1968. Regulation of pyrimidine biosynthesis in *Saccharomyces cerevisiae*. *J. Bacteriol.* 95:824–32
96. Lacroute F. 1971. Non-Mendelian mutation allowing ureidosuccinic acid uptake in yeast. *J. Bacteriol.* 206:519–22
97. Leibowitz MJ, Wickner RB. 1978. Pet18: a chromosomal gene required for cell growth and for the maintenance of mitochondrial DNA and the killer plasmid of yeast. *Mol. Gen. Genet.* 165:115–21
98. Li FN, Johnston M. 1997. Grr1 of *Saccharomyces cerevisiae* is connected to the ubiquitin proteolysis machinery through Skp1: coupling glucose sensing to gene expression and the cell cycle. *EMBO J.* 16:5629–38
99. Li L, Lindquist S. 2000. Creating a protein-based element of inheritance. *Science* 287:661–64
100. Liebman SW, All-Robyn JA. 1984. A non-Mendelian factor, [*eta*$^+$], causes lethality of yeast omnipotent-suppressor strains. *Curr. Genet.* 8:567–73
101. Lindquist S. 1997. Mad cows meet psichotic yeast: the expansion of the prion hypothesis. *Cell* 89:495–98
102. Lindquist S, Kim G. 1996. Heat-shock protein 104 expression is sufficient for thermotolerance in yeast. *Proc. Natl. Acad. Sci. USA* 93:5301–6
103. Lindquist S, Schirmer EC. 1999. The role of Hsp104 in stress tolerance and prion maintenance. In *Molecular Chaperones and Folding Catalysts. Regulation, Cellular Function and Mechanisms*, ed. B Bukau, pp. 347–80. Amsterdam: Harwood Acad.
104. Liu JJ. 2000. *Characterization of a protein responsible for the protein-based inheritance of [PSI*$^+$*] in yeast.* PhD thesis. Univ. Chicago. 151 pp.
105. Liu JJ, Lindquist S. 1999. Oligopeptide-repeat expansions modulate

'protein-only' inheritance in yeast. *Nature* 400:573–76

106. Lopez N, Halladay J, Walter W, Craig EA. 1999. *SSB*, encoding a ribosome-associated chaperone, is coordinately regulated with ribosomal protein genes. *J. Bacteriol.* 181:3136–43
107. Luke MM, Sutton A, Arndt KT. 1991. Characterization of *SIS1*, a *Saccharomyces cerevisiae* homologue of bacterial *dnaJ* proteins. *J. Cell Biol.* 114:623–38
108. Lund PM, Cox BS. 1981. Reversion analysis of [*psi*$^-$] mutations in *Saccharomyces cerevisiae. Genet. Res.* 37:173–82
109. Maddelein ML, Wickner RB. 1999. Two prion-inducing regions of Ure2p are nonoverlapping. *Mol. Cell Biol.* 19: 4516–24
110. Masison DC, Maddelein ML, Wickner RB. 1997. The prion model for [*URE3*] of yeast: spontaneous generation and requirements for propagation. *Proc. Natl. Acad. Sci. USA* 94:12503–8
111. Masison DC, Wickner RB. 1995. Prion-inducing domain of yeast Ure2p and protease resistance of Ure2p in prion-containing cells. *Science* 270:93–95
112. McCaldon P, Argos P. 1988. Oligopeptide biases in protein sequences and their use in predicting protein coding regions in nucleotide sequences. *Proteins* 4:99–122
113. Michelitsch MD, Weissman JS. 2000. A census of glutamine/asparagine-rich regions: implications for their conserved function and the prediction of novel prions. *Proc. Natl. Acad. Sci. USA* 97: 11910–15
114. Moriyama H, Edskes HK, Wickner RB. 2000. [*URE3*] prion propagation in *Saccharomyces cerevisiae*: requirement for chaperone Hsp104 and curing by overexpressed chaperone Ydj1p. *Mol. Cell Biol.* 20:8916–22
115. Nakayashiki T, Ebihara K, Bannai H, Nakamura Y. 2001. Yeast [*PSI*$^+$] "prions" that are crosstransmissible and susceptible beyond a species barrier through a quasi-prion state. *Mol. Cell.* 7:1121–30
116. Nelson RJ, Ziegelhoffer T, Nicolet C, Werner-Washburne M, Craig EA. 1992. The translation machinery and 70 kD heat shock protein cooperate in protein synthesis. *Cell* 71:97–105
117. Newnam GP, Wegrzyn RD, Lindquist SL, Chernoff YO. 1999. Antagonistic interactions between yeast chaperones Hsp104 and Hsp70 in prion curing. *Mol. Cell Biol.* 19:1325–33
118. Osherovich LZ, Weissman JS. 2001. Multiple Gln/Asn-rich prion domains confer susceptibility to induction of the yeast [*PSI*$^+$] prion. *Cell* 106:183–94
119. Osherovich LZ, Weissman JS. 2002. The utility of prions. *Dev. Cell* 2:143–51
120. Parham SN, Resende CG, Tuite MF. 2001. Oligopeptide repeats in the yeast protein Sup35p stabilize intermolecular prion interactions. *EMBO J.* 20:2111–19
121. Parsell DA, Kowal AS, Singer MA, Lindquist S. 1994. Protein disaggregation mediated by heat-shock protein Hsp104. *Nature* 372:475–78
122. Patino MM, Liu JJ, Glover JR, Lindquist S. 1996. Support for the prion hypothesis for inheritance of a phenotypic trait in yeast. *Science* 273:622–26
123. Paushkin SV, Kushnirov VV, Smirnov VN, Ter-Avanesyan MD. 1996. Propagation of the yeast prion-like [*PSI*$^+$] determinant is mediated by oligomerization of the *SUP35*-encoded polypeptide chain release factor. *EMBO J.* 15:3127–34
124. Paushkin SV, Kushnirov VV, Smirnov VN, Ter-Avanesyan MD. 1997. *In vitro* propagation of the prion-like state of yeast Sup35 protein. *Science* 277:381–83
125. Paushkin SV, Kushnirov VV, Smirnov VN, Ter-Avanesyan MD. 1997. Interaction between yeast Sup45p (eRF1) and Sup35p (eRF3) polypeptide chain

release factors: implications for prion-dependent regulation. *Mol. Cell Biol.* 17:2798–805
126. Perrett S, Freeman SJ, Butler PJ, Fersht AR. 1999. Equilibrium folding properties of the yeast prion protein determinant Ure2. *J. Mol. Biol.* 290:331–45
127. Pfund C, Lopez-Hoyo N, Ziegelhoffer T, Schilke BA, Lopez-Buesa P, et al. 1998. The molecular chaperone Ssb from *Saccharomyces cerevisiae* is a component of the ribosome-nascent chain complex. *EMBO J.* 17:3981–89
128. Prusiner SB. 1982. Novel proteinaceous infectious particles cause scrapie. *Science* 216:136–44
129. Prusiner SB. 1998. Prions. *Proc. Natl. Acad. Sci. USA* 95:13363–83
129a. Pruisner SB, ed. 1999. *Prion Biology and Diseases*. New York: Cold Spring Harbor Lab. Press. 794 pp.
130. Salmon JM, Barre P. 1998. Improvement of nitrogen assimilation and fermentation kinetics under enological conditions by derepression of alternative nitrogen-assimilatory pathways in an industrial *Saccharomyces cerevisiae* strain. *Appl. Environ. Microbiol.* 64:3831–37
131. Sanchez Y, Lindquist SL. 1990. *HSP104* required for induced thermotolerance. *Science* 248:1112–15
132. Sanchez Y, Taulien J, Borkovich KA, Lindquist S. 1992. Hsp104 is required for tolerance to many forms of stress. *EMBO J.* 11:2357–64
133. Santoso A, Chien P, Osherovich LZ, Weissman JS. 2000. Molecular basis of a yeast prion species barrier. *Cell* 100:277–88
134. Saupe SJ. 2000. Molecular genetics of heterokaryon incompatibility in filamentous ascomycetes. *Microbiol. Mol. Biol. Rev.* 64:489–502
135. Scheibel T, Lindquist SL. 2001. The role of conformational flexibility in prion propagation and maintenance for Sup35p. *Nat. Struct. Biol.* 8:958–62
136. Schirmer EC, Lindquist S. 1997. Interactions of the chaperone Hsp104 with yeast Sup35 and mammalian PrP. *Proc. Natl. Acad. Sci. USA* 94:13932–37
137. Schlumpberger M, Prusiner SB, Herskowitz I. 2001. Induction of distinct [*URE3*] yeast prion strains. *Mol. Cell Biol.* 21:7035–46
138. Schlumpberger M, Wille H, Baldwin MA, Butler DA, Herskowitz I, Prusiner SB. 2000. The prion domain of yeast Ure2p induces autocatalytic formation of amyloid fibers by a recombinant fusion protein. *Protein Sci.* 9:440–51
139. Scott M, DeArmond SJ, Prusiner SB, Ridley RM, Baker HF. 1999. Transgenic investigations of the species barrier and prion strains. See Ref. 129a, pp. 307–47
140. Serio TR, Cashikar AG, Kowal AS, Sawicki GJ, Moslehi JJ, et al. 2000. Nucleated conformational conversion and the replication of conformational information by a prion determinant. *Science* 289:1317–21
141. Serio TR, Lindquist SL. 1999. [PSI^+]: an epigenetic modulator of translation termination efficiency. *Annu. Rev. Cell Dev. Biol.* 15:661–703
142. Sherman F. 1991. Getting started with yeast. In *Guide to Yeast Genetics and Molecular Biology*, ed. C Guthrie, GR Fink, pp. 3–21. San Diego: Academic
143. Silar P, Daboussi MJ. 1999. Non-conventional infectious elements in filamentous fungi. *Trends Genet.* 15:141–45
144. Silar P, Haedens V, Rossignol M, Lalucque H. 1999. Propagation of a novel cytoplasmic, infectious and deleterious determinant is controlled by translational accuracy in *Podospora anserina*. *Genetics* 151:87–95
145. Singh A, Helms C, Sherman F. 1979. Mutation of the non-Mendelian suppressor, [PSI^+] in yeast by hypertonic media. *Proc. Natl. Acad. Sci. USA* 76:1952–56
146. Sondheimer N. 2000. *The identification of novel prion elements in Saccharomyces cerevisiae*. PhD thesis. Univ. Chicago. 138 pp.

147. Sondheimer N, Lindquist S. 2000. Rnq1: an epigenetic modifier of protein function in yeast. *Mol. Cell* 5:163–72
148. Sondheimer N, Lopez N, Craig EA, Lindquist S. 2001. The role of Sis1 in the maintenance of the [*RNQ*$^+$] prion. *EMBO J.* 20:2435–42
149. Sparrer HE, Santoso A, Szoka FC Jr, Weissman JS. 2000. Evidence for the prion hypothesis: induction of the yeast [*PSI*$^+$] factor by *in vitro*-converted Sup35 protein. *Science* 289:595–99
150. Spellman PT, Sherlock G, Zhang MQ, Iyer VR, Anders K, et al. 1998. Comprehensive identification of cell cycle-regulated genes of the yeast *Saccharomyces cerevisiae* by microarray hybridization. *Mol. Biol. Cell* 9:3273–97
151. Speransky VV, Taylor KL, Edskes HK, Wickner RB, Steven AC. 2001. Prion filament networks in [*URE3*] cells of *Saccharomyces cerevisiae. J. Cell Biol.* 153:1327–36
152. Stansfield I, Jones KM, Kushnirov VV, Dagkesamanskaya AR, Poznyakovski AI, et al. 1995. The products of the *SUP45*(eRF1) and *SUP35* genes interact to mediate translation termination in *Saccharomyces cerevisiae. EMBO J.* 14: 4365–73
153. Sybenga J. 1999. What makes homologous chromosomes find each other in meiosis? A review and an hypothesis. *Chromosoma* 108:209–19
154. Talloczy Z, Mazar R, Georgopoulos DE, Ramos F, Leibowitz MJ. 2000. The [*KIL-d*] element specifically regulates viral gene expression in yeast. *Genetics* 155:601–9
155. Talloczy Z, Menon S, Neigeborn L, Leibowitz MJ. 1998. The [*KIL-d*] cytoplasmic genetic element of yeast results in epigenetic regulation of viral M double-stranded RNA gene expression. *Genetics* 150:21–30
156. Taylor KL, Cheng N, Williams RW, Steven AC, Wickner RB. 1999. Prion domain initiation of amyloid formation *in vitro* from native Ure2p. *Science* 283:1339–43
157. Telkov M, Surguchev A. 1986. Characterization of transcripts of genes *sup1* and *sup2* of *Saccharomyces. Dokl. Akad. Nauk SSSR* 290:988–90
158. ter Schure EG, van Riel NA, Verrips CT. 2000. The role of ammonia metabolism in nitrogen catabolite repression in *Saccharomyces cerevisiae. FEMS Microbiol. Rev.* 24:67–83
159. Ter-Avanesyan MD, Dagkesamanskaya AR, Kushnirov VV, Smirnov VN. 1994. The *SUP35* omnipotent suppressor gene is involved in the maintenance of the non-Mendelian determinant [*PSI*$^+$] in the yeast *Saccharomyces cerevisiae. Genetics* 137:671–76
160. Ter-Avanesyan MD, Kushnirov VV, Dagkesamanskaya AR, Didichenko SA, Chernoff YO, et al. 1993. Deletion analysis of the *SUP35* gene of the yeast *Saccharomyces cerevisiae* reveals two non-overlapping functional regions in the encoded protein. *Mol. Microbiol.* 7:683–92
161. Thual C, Bousset L, Komar AA, Walter S, Buchner J, et al. 2001. Stability, folding, dimerization, and assembly properties of the yeast prion Ure2p. *Biochemistry* 40:1764–73
162. Thual C, Komar AA, Bousset L, Fernandez-Bellot E, Cullin C, Melki R. 1999. Structural characterization of *Saccharomyces cerevisiae* prion-like protein Ure2. *J. Biol. Chem.* 274:13666–74
163. True HL, Lindquist SL. 2000. A yeast prion provides a mechanism for genetic variation and phenotypic diversity. *Nature* 407:477–83
164. Tuite MF, Lund PM, Futcher AB, Dobson MJ, Cox BS, McLaughlin CS. 1982. Relationship of the [*psi*] factor with other plasmids of *Saccharomyces cerevisiae. Plasmid* 8:103–11
165. Tuite MF, Mundy CR, Cox BS. 1981. Agents that cause a high frequency of genetic change from [*PSI*$^+$] to [*psi*$^-$] in

Saccharomyces cerevisiae. *Genetics* 98: 691–711

166. Turcq B, Deleu C, Denayrolles M, Begueret J. 1991. Two allelic genes responsible for vegetative incompatibility in the fungus *Podospora anserina* are not essential for cell viability. *Mol. Gen. Genet.* 228:265–69
167. Umland TC, Taylor KL, Rhee S, Wickner RB, Davies DR. 2001. The crystal structure of the nitrogen regulation fragment of the yeast prion protein Ure2p. *Proc. Natl. Acad. Sci. USA* 98:1459–64
168. Uptain SM, Sawicki GJ, Caughey B, Lindquist S. 2001. Strains of [PSI^+] are distinguished by their efficiencies of prion-mediated conformational conversion. *EMBO J.* 20:6236–45
169. Volkov KV, Aksenova AY, Soom MJ, Osipov KV, Svitin AV, et al. 2002. Novel non-Mendelian determinant involved in the control of translation accuracy in *Saccharomyces cerevisiae*. *Genetics* 160:25–36
170. Warren G, Wickner W. 1996. Organelle inheritance. *Cell* 84:395–400
171. Wegrzyn RD, Bapat K, Newnam GP, Zink AD, Chernoff YO. 2001. Mechanism of prion loss after Hsp104 inactivation in yeast. *Mol. Cell Biol.* 21:4656–69
172. Werner-Washburne M, Becker J, Kosic-Smithers J, Craig EA. 1989. Yeast *HSP70* RNA levels vary in response to the physiological status of the cell. *J. Bacteriol.* 171:2680–88
173. Werner-Washburne M, Craig EA. 1989. Expression of members of the *Saccharomyces cerevisiae HSP70* multigene family. *Genome* 31:684–89
174. Werner-Washburne M, Stone DE, Craig EA. 1987. Complex interactions among members of an essential subfamily of *HSP70* genes in *Saccharomyces cerevisiae*. *Mol. Cell Biol.* 7:2568–77
175. Wickner RB. 1976. Mutants of the killer plasmid of *Saccharomyces cerevisiae* dependent on chromosomal diploidy for expression and maintenance. *Genetics* 82:273–85
176. Wickner RB. 1994. [*URE3*] as an altered Ure2 protein: evidence for a prion analog in *Saccharomyces cerevisiae*. *Science* 264:566–69
177. Wickner RB. 1996. Double-stranded RNA viruses of *Saccharomyces cerevisiae*. *Microbiol. Rev.* 60:250–65
178. Wickner RB, Chernoff YO. 1999. Prions of fungi: [*URE3*], [*PSI*], and [Het-s] discovered as heritable traits. See Ref. 129a, pp. 229–72
179. Wickner RB, Edskes HK, Maddelein ML, Taylor KL, Moriyama H. 1999. Prions of yeast and fungi. Proteins as genetic material. *J. Biol. Chem.* 274:555–58
180. Wilson PG, Culbertson MR. 1988. *SUF12* suppressor protein of yeast. A fusion protein related to the EF-1 family of elongation factors. *J. Mol. Biol.* 199: 559–73
181. Withee JL, Sen R, Cyert MS. 1998. Ion tolerance of *Saccharomyces cerevisiae* lacking the Ca2+/CaM- dependent phosphatase (calcineurin) is improved by mutations in *URE2* or *PMA1*. *Genetics* 149:865–78
182. Yan W, Craig EA. 1999. The glycine-phenylalanine-rich region determines the specificity of the yeast Hsp40 Sis1. *Mol. Cell Biol.* 19:7751–58
183. Yao B, Sollitti P, Zhang X, Marmur J. 1994. Shared control of maltose induction and catabolite repression of the *MAL* structural genes in *Saccharomyces*. *Mol. Gen. Genet.* 243:622–30
184. Young CS, Cox BS. 1971. Extrachromosomal elements in a super-suppression system of yeast. I. A nuclear gene controlling the inheritance of the extrachromosomal elements. *Heredity* 26:413–22
185. Young CS, Cox BS. 1972. Extrachromosomal elements in a super-suppression system of yeast. II. Relations with other extrachromosomal elements. *Heredity* 28:189–99

186. Zhong T, Arndt KT. 1993. The yeast *SIS1* protein, a *dnaJ* homolog, is required for the initiation of translation. *Cell* 73:1175–86
187. Zhou P, Derkatch IL, Liebman SW. 2001. The relationship between visible intracellular aggregates that appear after overexpression of Sup35 and the yeast prion-like elements [*PSI*$^+$] and [*PIN*$^+$]. *Mol. Microbiol.* 39:37–46
188. Zhou P, Derkatch IL, Uptain SM, Patino MM, Lindquist S, Liebman SW. 1999. The yeast non-Mendelian factor [*ETA*$^+$] is a variant of [*PSI*$^+$], a prion-like form of release factor eRF3. *EMBO J.* 18:1182–91

Annu. Rev. Microbiol. 2002. 56:743–68
doi: 10.1146/annurev.micro.56.012302.161038

First published online as a Review in Advance on June 4, 2002

MECHANISMS OF SOLVENT TOLERANCE IN GRAM-NEGATIVE BACTERIA

Juan L. Ramos, Estrella Duque, María-Trinidad Gallegos, Patricia Godoy, María Isabel Ramos-González, Antonia Rojas, Wilson Terán, and Ana Segura
Department of Plant Biochemistry and Molecular and Cellular Biology, Estación Experimental del Zaidín, Consejo Superior de Investigaciones Científicas, E-18008 Granada, Spain; e-mail: jlramos@eez.csic.es

Key Words *cis/trans* lipids, *cis*-to-*trans* isomerase, RND efflux pumps, solvent tolerance, *Pseudomonas*, transcriptional repressors

■ **Abstract** Organic solvents can be toxic to microorganisms, depending on the inherent toxicity of the solvent and the intrinsic tolerance of the bacterial species and strains. The toxicity of a given solvent correlates with the logarithm of its partition coefficient in *n*-octanol and water (log P_{ow}). Organic solvents with a log P_{ow} between 1.5 and 4.0 are extremely toxic for microorganisms and other living cells because they partition preferentially in the cytoplasmic membrane, disorganizing its structure and impairing vital functions. Several possible mechanisms leading to solvent-tolerance in gram-negative bacteria have been proposed: (*a*) adaptive alterations of the membrane fatty acids and phospholipid headgroup composition, (*b*) formation of vesicles loaded with toxic compounds, and (*c*) energy-dependent active efflux pumps belonging to the resistance-nodulation–cell division (RND) family, which export toxic organic solvents to the external medium. In these mechanisms, changes in the phospholipid profile and extrusion of the solvents seem to be shared by different strains. The most significant changes in phospholipids are an increase in the melting temperature of the membranes by rapid *cis*-to-*trans* isomerization of unsaturated fatty acids and modifications in the phospholipid headgroups. Toluene efflux pumps are involved in solvent tolerance in several gram-negative strains, e.g., *Escherichia coli*, *Pseudomonas putida*, and *Pseudomonas aeruginosa*. The AcrAB-TolC and AcrEF-TolC efflux pumps are important for *n*-hexane tolerance in *E. coli*. A number of *P. putida* strains have been isolated that tolerate toxic hydrocarbons such as toluene, styrene, and *p*-xylene. At least three efflux pumps (TtgABC, TtgDEF, and TtgGHI) are present in the most extensively characterized solvent-tolerant strain, *P. putida* DOT-T1E, and the number of efflux pumps has been found to correlate with the degree of solvent tolerance in different *P. putida* strains. The operation of these efflux pumps seems to be coupled to the proton motive force via the TonB system, although the intimate mechanism of energy transfer remains elusive. Specific and global regulators control the expression of the efflux pump operons of *E. coli* and *P. putida* at the transcriptional level.

CONTENTS

INTRODUCTION

Since the Industrial Revolution the production and use of chemicals have increased immensely. As a consequence, many kinds of products are synthesized, and many chemicals such as herbicides or insecticides are released into the environment. Other products such as organic solvents or fuels reach the biosphere through losses during production or storage, accidents, and solvent evaporation. There is now growing awareness concerning the possible toxic or even carcinogenic effects of these chemicals. Although the release of many of them is restricted by legislation, a number of pollutants have already reached the biosphere and need to be eliminated.

The use of biological treatments to remove toxic chemicals seems promising (67). However, chemical toxicity can hamper the use of microorganisms in the removal of pollutants from waste streams and dump sites. This is a serious problem with microbial bioremediation in reactors, biofilters, and soils (24, 76–79), and it is particularly critical when organic solvents are present at high concentrations.

Solvent toxicity correlates with its log P_{ow}, the logarithm of the partitioning coefficient of a solvent in a defined octanol-water mixture (log P_{ow}) (78). Solvents with a log P_{ow} below 4.0, e.g., benzene (log P_{ow} 2.13), toluene (log P_{ow} 2.69), octanol (log P_{ow} 2.92), xylenes (log P_{ow} 3.12–3.2), and styrene (log P_{ow} 2.95), are extremely toxic for microorganisms because they accumulate in the cytoplasmic membrane of bacteria and disrupt the cell membrane structure. The main function of the cell membrane of microorganisms is to form a permeability barrier that regulates the passage of solutes between the cell and the external environment (61), and it is of special importance for energy transduction in the cell (78). Organic solvents damage the cell membrane by impairing vital functions (loss of ions, metabolites, lipids, and proteins; dissipation of the pH gradient and electrical potential) or by inhibiting membrane protein functions. This damage is often followed by cell lysis and death (15a, 78).

Solvent toxicity depends not only on the inherent toxicity of the compound but also on the intrinsic tolerance of the bacterial species and strains. For example, certain strains of *Escherichia coli* are tolerant to cyclohexane (log P_{ow} 3.44), whereas others are sensitive (3). Most microorganisms are highly sensitive to aromatic solvents with a log P_{ow} between 2.0 and 3.3 (78); nevertheless, there exist several *Pseudomonas* species that grow in the presence of high concentrations of toxic organic solvents such as toluene, styrene, and *p*-xylene (15, 30, 41, 69, 89). These *Pseudomonas* spp. strains are not only resistant, in a two-phase system, to the aromatic compounds noted above, but they can often use these chemicals as carbon and energy sources (15, 69).

Regarding tolerance to aromatic hydrocarbons, a number of elements have been suggested to be involved in the response to these toxic chemicals: (*a*) metabolism of toxic hydrocarbons, which can contribute to their transformation into nontoxic compounds; (*b*) rigidification of the cell membrane via alteration of the phospholipid composition; (*c*) alterations in the cell surface that make the cells less permeable; (*d*) efflux of the toxic compound in an energy-dependent process; and (*e*) formation of vesicles that remove the solvent from the cell surface (Table 1).

TABLE 1 *Pseudomonas* solvent-tolerant strains and their characteristics

Strain	*cis*-to-*trans* isomerization	Efflux pumps	Vesicle formation	References
IH-2000	n.d.	n.d.	+	43
DOT-T1E	+	+	+	55, 68–70, 73
S12	+	+	n.d.	24, 25, 39, 40
MTB6	+	+	n.d.	28
GM73	−	+	n.d.	41
F1	+	+	n.d.	28

+ indicates that the character has been found in the strain.

− indicates that the character has not been found in the strain.

n.d., not determined.

Although the metabolism of toxic chemicals can help reduce their toxicity, two lines of evidence suggest that this mechanism is of minor importance in protecting cell viability. A number of microorganisms tolerant to a particular organic solvent cannot metabolize the toxic compound; e.g., *E. coli* K12 strains tolerant to 1% (vol/vol) *n*-hexane (log P_{ow}, 3.9) do not use (or biotransform) this compound at all (3). In addition, some *Pseudomonas* strains tolerant to supersaturating concentrations of toluene do not use toluene as a C-source (30).

Pseudomonas putida DOT-T1E is a toluene-tolerant strain that degrades toluene via the toluene-dioxygenase (Tod) pathway. Mutants unable to metabolize this compound have been generated and are as tolerant as the wild type to toluene (56). In addition, the toxicity of octane (log P_{ow} 4.9) in *Pseudomonas oleovorans* GP12, which bears the OCT plasmid for alkane metabolism, is the result of the metabolism of the alkane to the corresponding alcohol (log P_{ow} 2.92), which is more toxic than the alkane (13). It follows that the metabolism of a chemical and the development of tolerance to it are two unlinked events. In contrast, lipid membrane rigidification in response to toxic compounds and the active removal of the chemicals seem to be common elements in solvent tolerance in all *Pseudomonas* strains characterized so far (Table 1). Some solvent-tolerant strains form vesicles loaded with toxic compounds (42, 43). Although the importance of the role of these different phenomena in solvent tolerance is not the same, this article reviews the most critical features of each strategy known to be used by gram-negative bacteria to survive (or even to thrive) in the presence of compounds that are normally toxic for bacteria.

ALTERATION IN THE PHOSPHOLIPID COMPOSITION OF THE MEMBRANES DUE TO EXPOSURE TO ORGANIC SOLVENTS

The initial stages of the damage organic solvents cause when bacterial cells are exposed to them are binding and penetration into the lipid bilayer (78). As a consequence, membrane fluidity is affected and bacteria launch appropriate responses to diminish the disruptive effects. Membrane fluidity is re-adjusted primarily by altering the composition of the lipid bilayer through compensatory mechanisms that resemble some of those observed in response to physical and chemical stresses imposed by the environment (36, 71, 74). The following sections discuss the major responses that bacteria can use to re-adjust membrane composition in response to solvent exposure.

Changes in Fatty Acid Composition

There are two major mechanisms for changing ester-linked fatty acid composition, and thus membrane fluidity, in bacterial lipid bilayers: *cis*-to-*trans* isomerization of unsaturated fatty acids as a short-term response and changes in the saturated-to-unsaturated fatty acid ratio as a long-term response to solvent exposure. The ratio of long-chain to short-chain fatty acids can also be altered to regulate membrane fluidity. The steric behavior of *trans* fatty acids and saturated fatty acids is similar, as both possess a long extended conformation allowing denser packing of the membrane. In contrast, the *cis* configuration of the acyl-chain has a nonmovable 30° bend that causes steric hindrance and disturbs the highly ordered fatty acid package. The *cis* fatty acids have a lower phase transition temperature than the corresponding *trans* isomers (36). Analysis of phospholipid biosynthesis in a solvent-tolerant *P. putida* strain showed that the basal rate of biosynthesis was higher in this strain than in solvent-sensitive strains of the same species and that the rate of biosynthesis increased in response to exposure to sublethal concentrations of organic solvents (65). These findings strongly suggest that changes in the rate of phospholipid biosynthesis and alteration of phospholipid composition are specific responses of the cell to exposure to toxic solvents.

Cis-to-*Trans* Fatty Acid Isomerization

The *cis* acyl-chain configuration of membrane-located esterified unsaturated fatty acid is the isomer synthesized by the cells. However, *trans* isomers in cellular lipids have been found in a number of microorganisms as the result of a postsynthetic modification reaction catalyzed by a *cis*-to-*trans* isomerase (36).

Various *P. putida* strains respond to solvents and aromatic compounds by shifting their *cis*-to-*trans* ratio when exposed to these chemicals, i.e., phenol, 4-chlorophenol, toluene, xylenes, or alcohols of different chain lengths (24, 25, 37, 65, 66, 69, 70, 89). These *P. putida* strains contain mainly palmitoleic acid (C16:1, 9) and vaccenic acid (C18:1, 11) as *trans* isomers and are directly synthesized from the *cis* isomer within 1 min of exposure to the solvent with no shift in the position of the double bond (49, 70). Because organic solvents increase membrane fluidity, an increase in *trans* fatty acid content could counteract this alteration.

The *cis*-to-*trans* isomerase (Cti) of *P. putida* P8 (27), *P. putida* KT2440 and *P. putida* DOT-T1E (34), and *P. oleovorans* Gpo12 (64) have been cloned and sequenced and found to be 95% identical at the amino acid sequence level. Primer extension analysis revealed that in *P. putida* DOT-T1E the *cti* gene was expressed constitutively. However, Cti is active in promoting *cis*-to-*trans* isomerization only in the presence of toluene, but the molecular basis that governs this activation in vivo is unknown. Subcellular fractioning studies have located Cti in the periplasmic space, and in-frame fusion of the Cti protein with periplasmic alkaline phosphatase further supported this location (34, 64).

The Cti enzyme of *P. oleovorans* Gpo12, which has been purified to homogeneity (64), seems to be a monomer of about 70 kDa, and comparison of the N-terminal sequence of the mature protein with the deduced sequence from the *cti* gene revealed that the signal peptide of the original translated gene product was cleaved. The purified enzyme acted in vitro on free phospholipids as the substrate (the preferred substrates were *cis*-9-hexadecenoic acid, *cis*-11-octadecenoic, and *cis*-13-eicosenoic acid). However, the enzyme did not isomerize phospholipids in vitro. In contrast, when crude membranes from *E. coli* or *Pseudomonas* sp. were used as phospholipid sources, *cis*-to-*trans* isomerization was detectable only in the presence of organic solvents.

It is still not understood how cells sense the presence of organic solvents to consequently modify the ratio of *cis* to *trans* unsaturated fatty acids. If the hypothesis put forward by Witholt and colleagues is correct, one would expect that under solvent-induced stress conditions, phospholipid turnover would increase, the isomerase would reach the double bond of the unsaturated fatty chains, and the fatty acids would be incorporated into phospholipid headgroups. It is also worth noting that the *cis*-to-*trans* ratio of unsaturated fatty acids changes in response to other stress agents such as temperature, increased concentrations of salt (13, 26, 36), octane-to-octanol transformation (13), and exposure of cells to heavy metals (P. Godoy & J.L. Ramos, unpublished data). This suggests that *cis*-to-*trans* isomerization is a general defense mechanism rather than a specific strategy for solvent tolerance.

Junker & Ramos (34) generated a Cti-null mutant in *P. putida* DOT-T1E that was unable to isomerize *cis*-unsaturated fatty acids. The Cti DOT-T1E mutant strain showed the following phenotypic differences in comparison to the parental strain: (*a*) It displayed a longer lag phase in comparison with the parental strain when grown with toluene supplied in the gas phase; (*b*) the mutant showed a lower survival rate when shocked with 0.08% (vol/vol); and (*c*) the Cti-null mutant strain grew significantly more slowly at temperatures above 37°C than the wild-type strain. Thus, *cis*-to-*trans* isomerization of fatty acids in phospholipids improved the survival of this strain in response to toluene and high temperatures. These findings are convincing evidence that the solvent-resistant nature of *P. putida* DOT-T1E arises not only from the *cis*-to-*trans* isomerase but also from other mechanisms (see RND efflux pumps below).

Cis-to-*trans* isomerization of esterified fatty acids represents a short-term response to environmental stress in *P. putida*. This allows cells to adapt immediately to new environmental conditions under which denser membrane packaging is a selective advantage. This gains the cells time for the de novo biosynthesis of membrane components, a process that allows more precise adjustments to specific stresses.

Saturated-to-Unsaturated Fatty Acid Ratio

An increase in the saturated fatty acid content of phospholipids, a well-known response to temperature changes and solvent exposure (14, 23, 49, 82), had been observed in *E. coli* in the presence of long-chain alcohols and aromatic compounds (29). A slight increase in the saturated fatty acid content in *P. putida* strain Idaho was observed as late as 15 min after solvent exposure, with maximum level after 2 h of exposure to *o*-xylene (65); this mechanism of adaptation is considered a long-term response. Other *Pseudomonas* strains do not alter their unsaturated-to-saturated fatty acid ratio in response to organic solvents (70, 89).

Other Changes in Phospholipids

Changes in phospholipid headgroups influence membrane fluidity, although these changes are a less-well-studied phenomenon and few data are available (88). The major classes of phospholipid in *P. putida* solvent-tolerant and -sensitive strains are phosphatidylethanolamine (PE), phosphatidylglycerol (PG), and diphosphatidylglycerol or cardiolipin (CL). Changes in polar headgroups have been analyzed in three strains: *P. putida* S12, *P. putida* DOT-T1E, and *P. putida* Idaho. In the presence of toluene, changes in phospholipid headgroup composition were observed in *P. putida* strain S-12 in a chemostat culture. The level of PE decreased and that of CL and PG increased (87). Cardiolipin has a higher (by 10°C) transition temperature than PE, which decreases membrane fluidity and thus has a stabilizing effect (88). Similar results were reported for *P. putida* DOT-T1E. Ramos et al. (70) measured the incorporation of ^{32}P label in phospholipids upon toluene exposure and reported that almost 90% of the total label was incorporated in CL. This was expected because the total amount of CL increased after the strain was exposed to toluene.

In *P. putida* strain Idaho growing in a chemostat with succinate as the C-source, the total amount of membrane phospholipid fatty acids increased when the cells were exposed to *o*-xylene. This contrasted with the results obtained in equivalent chemostats run with the solvent-sensitive *P. putida* MW1200 strain; the different results suggest that the Idaho strain is better able than MW1200 to repair damaged membranes through efficient turnover and increased phospholipid biosynthesis (65). The fact that the rate of phospholipid biosynthesis was greater upon exposure to *o*-xylene than in the absence of the aromatic compound suggests that the increased rate is an inducible event. Recently, Segura et al. (A. Segura, A. Rojas, A. Delgado & J.L. Ramos, unpublished results) identified a toluene-sensitive mutant from *P. putida* DOT-T1E that is unable to incorporate $^{13}CH_3$-$^{13}COOH$ into fatty acids in the presence of toluene. The limited turnover in fatty acid metabolism in this mutant strain results in the formation of blebs on the cell surface, which leads to the loss of membrane integrity in the presence of solvents, and the concomitant increase in solvent sensitivity.

A detailed analysis of phospholipid headgroup turnover in the absence and in the presence of *o*-xylene in *P. putida* strain Idaho revealed that in the absence of the solvent the incorporation of ^{32}P was greatest in PG, followed by PE and CL. However, in the presence of 200 ppm *o*-xylene, incorporation was greatest for PE, followed closely by PG. An increase in PE to counteract the effects of solvents is not common in bacteria (88). Phosphatidylethanolamine has a higher melting point than PG, thus an increase in PE tends to stabilize the cell membrane. In their work with this strain, Pinkart & White (65) showed that the isomerization of *cis*-to-*trans* unsaturated fatty acids occurs mainly in PE. Therefore, different *Pseudomonas* strains seem to have developed different strategies for changing phospholipid polar headgroup composition to increase membrane rigidity and in this way overcome the damaging effects of solvents.

Alterations in Lipopolysaccharides

Lipopolysaccharides (LPS) of gram-negative bacteria are major components of the outer membrane and are considered a defense barrier. In several *P. putida* strains the addition of divalent cations (Mg^{2+} and Ca^{2+}) to a growth medium supplemented with organic solvents was found to improve survival (30, 69, 88). It was hypothesized that the divalent cations electrostatically linked adjacent polyanionic LPS molecules and reduced charge repulsion. The cations allowed denser packing of the anionic membrane molecules.

Changes in outer membrane LPS after exposure to solvents have been monitored in various solvent-tolerant bacteria (66, 88). In permeability assays with difloxacin, a hydrophobic antibiotic that has the same log P_{ow} as *o*-xylene (log P_{ow} 3.1), no difference in permeability to this agent was observed (66). Therefore, this study provided no conclusive evidence that LPS reduce solvent access to the cells.

The *wbpL* gene encodes the enzyme that initiates synthesis of the *O*-antigenic side chain of LPS. To determine whether LPS play a role in resistance to solvents, Junker et al. (35) generated, by site-directed mutagenesis, a DOT-T1E knock-out *wbpL* mutant. The mutant strain was as tolerant as the parental strain to organic solvents (toluene, octanol, *p*-xylene, propylbenzene, and heptane), and the authors concluded that the *O*-antigen side chain of LPS did not play a significant role in solvent tolerance.

EFFLUX PUMPS AND SOLVENT EXTRUSION

In the course of evolution, bacteria have been exposed to different toxic compounds such as natural toxins, endogenous metabolic end products, and antibiotics. To protect themselves, microorganisms have evolved different devices to detoxify and extrude these substances. One group of such devices comprises the multidrug resistance (MDR) efflux systems, which catalyze the active extrusion of many structurally and functionally unrelated compounds from the bacterial cytoplasm (or internal membrane) to the external medium. Some of the substrates of these MDR pumps are xenobiotics that do not resemble any of the known natural substrates that these cells may have encountered during evolution (50, 51). The data available to date indicate that it is the physical characteristics of the compounds (e.g., charge, hydrophobicity, or amphipathicity), the van der Waals interactions they establish with active sites and effector pockets, and the flexibility of these sites in the target proteins that determine the specificity of these multidrug efflux systems (59).

Multidrug efflux systems have been the subject of recent reviews, and four main families of MDR transporters have been identified in bacteria: (*a*) the major facilitator superfamily (MFS), (*b*) small multidrug resistance elements, (*c*) the ATP-binding cassette, and (*d*) pumps belonging to the resistance-nodulation-cell division (RND) family (63). All efflux pumps for organic solvents identified so far in gram-negative bacteria belong to this latter family (Table 2).

The distinguishing characteristic of gram-negative bacteria is the presence of an inner and an outer membrane separated by the periplasmic space (60). The RND

TABLE 2 Resistance-nodulation-division efflux pumps involved in solvent extrusion

Inner membrane transporter	Membrane fusion protein	Outer membrane protein	Bacterial species	References
AcrB (1049)	AcrA (397)	TolC (495)	*E. coli*	6, 19, 94
AcrF (1034)	AcrE (385)	TolC (495)	*E. coli*	44, 53
MexB (1046)	MexA (383)	OprM (485)	*P. aeruginosa*	48
TtgB (1050)	TtgA (384)	TtgC (484)	*P. putida*	68
TtgE (1048)	TtgD (382)	TtgF (480)	*P. putida*	55
TtgH (1049)	TtgG (391)	TtgI (470)	*P. putida*	73
SrpB (1049)	SrpA (382)	SrpC (470)	*P. putida*	39–41
ArpB (1050)	ArpA (371)	ArpC (484)	*P. putida*	20

Numbers in parentheses indicate the size in amino acids of the corresponding protein.

pumps of gram-negative bacteria export toxic substances across both membranes of the cell envelope in a single energy-coupled step (46). These efflux pumps are made of three components: a cytoplasmic membrane export system that acts as an energy-dependent extrusion pump, a membrane fusion protein (MFP), and an outer membrane factor (OMP) (46, 94).

Escherichia coli Solvent Tolerance Involves Mainly the AcrAB-TolC and the AcrEF-TolC Efflux Pumps

Escherichia coli strains are, in general, highly sensitive to organic solvents, and most strains only survive in the presence of solvents with a log $P_{ow} \geq 4$. Many studies on solvent tolerance in *E. coli* have used the JA300 strain. Aono et al. (3, 5) isolated spontaneous ampicillin- and chloramphenicol-resistant clones of JA300 and showed that the mutants exhibited improved tolerance to organic solvents such as cyclohexane (log P_{ow} 3.44), *n*-pentane (log P_{ow} 3.39), or *p*-xylene (log P_{ow} 3.15). These results indicated that some system that leads to resistance toward multiple antibiotics had become active in the organic solvent-tolerant mutants. These were the first reports to establish a connection between organic solvent tolerance and antibiotic resistance. Further characterization of these mutants revealed that the most important device in *E. coli* for the efflux of organic solvents is the AcrAB-TolC system. In this RND efflux pump the TolC protein is the OMP (6, 19), AcrA is the MFP, and AcrB is a translocase thought to be a proton transporter (52, 94).

Properties of the AcrAB-TolC Efflux Pump

The drug transporter elements of the RND efflux pumps are located in the inner membrane; however, they have not been exhaustively characterized at the molecular level, and part of our knowledge derives from computer analyses. These proteins appear to contain several transmembrane α-helices (usually 12) and 2 large domains projecting into the periplasm (75, 84). RND transporters appear to use the proton motive force as an energy source to export the substrates. This is supported by the fact that pump activity of the AcrAB system is inhibited in the presence of proton conductors such as CCCP and FCCP (93, 94). There are no details on the active site(s) of AcrB, which contrasts with the relatively detailed information available on translocases characterized in eukaryotic cells (10).

Membrane fusion proteins are lipoproteins anchored to the cytoplasmic membrane (53, 63). The best-characterized MFP in efflux pumps is AcrA. Because AcrA is highly similar to other MFPs identified in other efflux pumps (63), its characteristics should be well conserved in other MFPs. The first 24 amino acids of AcrA show features typical of a bacterial lipoprotein signal peptide. After cleavage of the signal peptide, the amino-terminal cysteine residue of the processed protein is acylated with fatty acids of the inner membrane (93). The in vitro characterization of purified AcrA protein revealed that monomers of this protein are highly asymmetric, with an axial ratio of 8. The chemical cross-linking experiments of Zgurskaya & Nikaido (95) showed that AcrA forms oligomers, probably trimers. The predicted length of AcrA is 17 nm, which is compatible with the hypothesis that it spans the periplasmic space and coordinates the concerted operation of inner and outer membrane components of the RND efflux complex (93).

The outer membrane channel TolC protein is the third component of the AcrAB efflux pump. Identification of TolC as the third component of this efflux pump is based on mutations in *tolC*-abolished AcrAB-dependent multidrug resistance in *E. coli* (19). The work of Koronakis et al. (46), which led to the crystallization of TolC (Figure 1, see color insert), has helped to clarify how the translocase element in the inner membrane of the cell is connected to the surrounding external medium during export and thus how the efflux of substrates by these pumps bypasses the intervening periplasmic space. These authors showed that three TolC protomers assemble to form a continuous, solvent-accessible tunnel more than 140 Å long that spans both the outer membrane and the periplasmic space. The periplasmic (proximal) end of the tunnel is sealed by sets of coiled helices, which may be untwisted by an allosteric mechanism mediated by protein-protein interactions to open the tunnel.

The allosteric protein that promotes untwisting was proposed by Koronakis et al. (46) to be the inner membrane translocase; this role is supported by experimental evidence (83) for interactions between TolC and the inner membrane translocase. Cross-linking studies have suggested that AcrA and AcrB interact with each other, but TolC may well be transiently associated with the AcrAB complex. There is growing evidence that the role of AcrA is to bring the inner and the outer membrane closer together.

Suppressors of *acrAB* Mutants Identified Other Efflux Pumps Involved in Solvent Tolerance

Kobayashi et al. (44) showed that the solvent sensitivity of an *acrAB* mutant could be suppressed. They isolated suppressor mutants with high solvent resistance, which produced high levels of AcrE and AcrF proteins as deduced from the N-terminal amino acid sequence of the hyperproduced proteins. The *acrEF* genes are expressed under laboratory conditions at low levels, if at all (54). However, in the suppressor mutants an IS*1* or IS2 insertion upstream of *acrE* leads to a high level of expression in this operon. The *tolC* gene product was found by these authors to be essential for suppression; therefore, they postulated that an efflux pump made of AcrEF and TolC is also functional for organic solvent efflux. In connection with these results are those of Jellen-Ritter & Kern (33), who showed that overexpression of *acrEF* leads to increased multidrug efflux and to the development of fluoroquinolone resistance in *E. coli*. Tsukagoshi & Aono (85) reported that the solvent hypersensitivity of an *acrAB* mutant of *E. coli* could also be partially suppressed by overexpression of *emrAB* or *yhiLIV*, which also encode efflux pumps belonging to the RND family.

Regulation of the *acrAB* Operon Involves Specific and Global Regulators

Sequence analysis has revealed three genes at the *acr* locus organized in two divergent operons: *acrR* and *acrAB* (53). Gene *acrR* encodes a transcriptional regulator with a helix-turn-helix motif at its N terminus that represses its own synthesis. AcrR is the primary modulator of the *acrAB* operon, belongs to the TetR family of transcriptional regulators, and its overexpression represses the transcription of *acrAB*.

Wang et al. (86) analyzed the genetic basis for fluoroquinolone resistance in 30 high-level fluoroquinolone-resistant *E. coli* clinical isolates from Beijing, China. Each of the strains showed resistance to a variety of antibiotics. The AcrA protein of the AcrAB multidrug efflux pump was overexpressed in 19 of the 30 strains of *E. coli* tested, and all 19 strains were tolerant to a mixture of 3:1 hexane:cyclohexane (log $P_{ow} = 3.4$). This fact unequivocally links AcrR function to the level of expression of the *acrAB* operon and to solvent tolerance.

Asako et al. (7) observed high levels of expression of the *acrAB* genes in a series of organic solvent-tolerant mutants. Some of the mutations were mapped in the *marR* (multiple antibiotic resistance) gene. The *marRAB* operon is negatively autoregulated by the MarR protein, which binds to the *marRAB* promoter (1). The MarA protein is a positive regulator that induces a set of genes (*mar* and *soxRS* regulons) that provide *E. coli* cells with resistance to a large number of antibiotics and superoxide-generating reagents. The substitution of serine for arginine at position 73 in MarR was responsible for the cyclohexane-tolerant phenotype (7, 58), probably because the mutation in MarR interferes with the proper orientation of the DNA-binding lobes of the dimeric protein, which then fails to repress the *mar* operon (2). Synthesis of the MarA protein thus allowed the bacteria to induce the efflux pump genes and to resist the action of multiple antibiotics and organic solvents.

The *sox*S and *rob* gene products can replace MarA in the activation of the *mar* operon. The presence of *sox*S and *rob* in high copy number plasmids increased the organic solvent tolerance in *E. coli* (57, 58). In contrast, overexpression of the *marR* wild-type gene decreased the levels of organic solvent tolerance. The different susceptibility to organic solvents indicates that the functions of the *marA* and *soxS* genes are opposite to those of the *rob* and *marR* genes. Deletion of the *acrAB* genes conferred increased susceptibility to multiple drugs and organic solvents in wild-type and Mar strains. Hence, the Mar phenotype is linked to overexpression of the *acrAB* locus, and certain Mar mutants are therefore resistant to organic solvents and antibiotics (62, 91).

It is also worth noting that transcription of *acrAB* also increased under general stress conditions (i.e., 4% ethanol, 0.5 M NaCl, and the stationary phase in Luria-Bertani medium) in both wild-type and *mar-sox* mutants of *E. coli* (54). These general stress conditions increased transcription of *acrAB* in the absence of a functional AcrR.

Ma et al. (52) performed a series of gel mobility shift assays and demonstrated the formation of both AcrR-dependent and AcrR-independent protein-DNA complexes at the *acr* promotor region, although the nature of the protein bound to the *acr* promotor is at present unknown. It seems, therefore, that responses to a number of stresses can be induced by exploiting global transcription regulators. AcrR apparently fine-tunes *acrAB* expression; hence, this protein has been proposed as a specific modulator to fine-tune the production of the AcrAB proteins (52).

Multiple Efflux Pumps in *Pseudomonas putida* Strains Tolerant to Highly Toxic Chemicals

A number of *P. putida* strains are tolerant to highly toxic solvents such as *p*-xylene (log P_{ow} 3.15), styrene (log P_{ow} 3.0), octanol (log P_{ow} 2.92), and toluene (log P_{ow} 2.69) (4, 15, 30, 32, 41, 69, 70, 89). Isken & de Bont (32) and Ramos et al. (70) measured the accumulation of [^{14}C]-labeled aromatic hydrocarbons by *P. putida* S-12 and DOT-T1E grown in the absence of (nonadapted) or in the presence (adapted) of toluene. They found that the amount of [^{14}C]-labeled aromatic hydrocarbons that accumulated in the adapted cells was one-half to one-fifth that in nonadapted bacteria. The addition of the respiratory chain inhibitor potassium cyanide or the proton conductor CCCP to adapted cells resulted in significant accumulation of the aromatic hydrocarbon. This finding was interpreted as evidence that these strains had energy-dependent exclusion systems able to decrease the level of the solvent in the membranes.

The isolation and characterization of transposon mutants of *P. putida* S12, DOT-T1E, and GM73 made it possible to identify several efflux pumps involved in solvent tolerance (39–41, 68). This was definitive evidence for the role of efflux pumps in solvent tolerance in these strains. The molecular characterization of the insertion site in the mutant identified the *srpABC* (solvent-resistant pump) genes of *P. putida* S12 and the *ttgABC* and *ttgB* (toluene tolerance genes) of *P. putida* DOT-T1E and GM73. The deduced amino acid sequences of the proteins encoded by these clusters are similar to those of the AcrAB-TolC efflux pump of *E. coli* and the MexAB-OprM multidrug efflux system of *P. aeruginosa* (from 58% to 77% identity depending on the proteins compared).

The *srpA* and *ttgA* genes encode the inner membrane-anchored lipoprotein that spans the periplasm and interacts with the inner membrane transporter (encoded in these systems by *srpB* and *ttgB*) and probably with the OMP encoded by *srpC* and *ttgC*. Moreover, Fukimori et al. (20) isolated a toluene-resistant mutant of the solvent-sensitive *P. putida* KT2442 in which an efflux pump almost identical to the TtgABC pump of *P. putida* DOT-T1E was expressed at high levels.

Ramos et al. (69) realized that toluene tolerance in the wild-type *P. putida* DOT-T1E strain was influenced by growth conditions. Preexposure of cells to low concentrations of this aromatic hydrocarbon led to survival of almost 100% of the cells after a sudden toluene shock; in contrast, only a fraction (10^{-4}) of the cells that had not been preexposed survived. When these assays were done with a knock-out mutant in *ttgB*, the strain did not withstand the sudden toluene shock at all, and only a small fraction (about 1 out of 10^5 cells) survived the shock if preexposed to low toluene concentrations. This led to the suggestion that the TtgABC pump contributes to the innate tolerance of *P. putida* DOT-T1E to solvents (68). Ramos et al. (68) then postulated the existence of other efflux pumps involved in toluene extrusion. A second inducible efflux pump, called TtgDEF, that expels toluene in the solvent-tolerant *P. putida* DOT-T1E was then identified as linked to the *tod* genes for toluene degradation (55). Linkage of the genes that encode a catabolic pathway with those involved in the efflux of the compound degraded by the strain was conserved in a wide range of *Pseudomonas* sp. strains that degraded toluene through the Tod pathway (28). It is tempting to suggest that the degradation of a compound with intrinsic toxicity coevolved with its exclusion to avoid accumulation of the toxic compound above a certain threshold.

After the TtgABC and TtgDEF pumps in *P. putida* DOT-T1E were inactivated, 10^{-6} cells still survived if they were preexposed to low concentrations of toluene, although none ($<10^{-8}$) survived without induction (55). Mosqueda & Ramos (55) suggested the presence of at least one other undiscovered pump; Rojas et al. (73) subsequently identified the *ttgGHI* genes, which are almost identical to the sequence that encodes the *srpABC* pump. To test whether this third efflux pump was involved in toluene tolerance in *P. putida* DOT-T1E, Rojas et al. (73) constructed the mutant strain DOT-T1E-PS34, which had a knock-out in each of the three efflux pumps. This strain was so sensitive to toluene that it could not grow in rich medium with toluene supplied in the gas phase. These authors suggested that either no other pump was involved in toluene tolerance or that the contribution of other pumps to toluene tolerance was minimal. Figure 2 shows parsimonious trees to illustrate the evolutionary relationship between the different components of solvent-extrusion RND pumps.

Rojas et al. (73) constructed single, double, and triple mutants for each of the Ttg pumps described above to assess their involvement in the extrusion of different solvents. They found that the TtgABC and the TtgGHI pumps were involved in toluene, styrene, xylenes, ethylbenzene, and propylbenzene efflux, whereas the

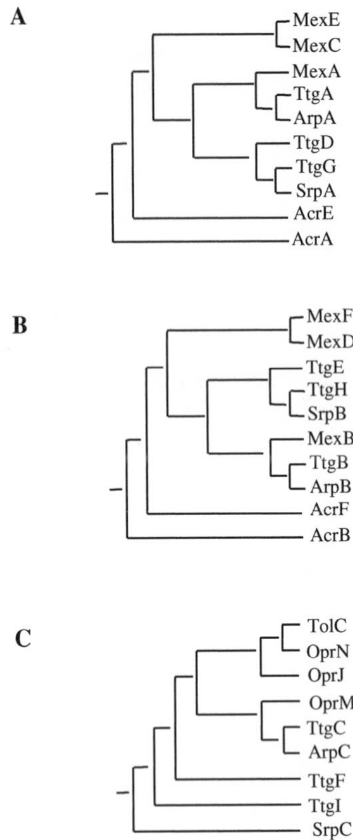

Figure 2 Parsimony trees showing the evolutionary relationship between RND components involved in solvent-extrusion. (*A*) Translocases, (*B*) MFPs, (*C*) OMPs.

TtgDEF efflux pump was involved in the efflux of styrene and toluene only. The work of Lee et al. (47) suggested that pumps that operate through a common mechanism have an additive effect in the extrusion of drugs. This may be the case with the RND solvent-extrusion pumps in *P. putida* DOT-T1E, as tolerance to propylbenzene and ethylbenzene in this strain was not compromised in mutants lacking TtgABC or TtgGHI but was compromised in double mutants lacking both these pumps.

Bugg et al. (12) showed that *Pseudomonas fluorescens* LP6a, which degrades a number of polycyclic aromatic hydrocarbons, exhibits a constitutively expressed energy-driven efflux pump that selectively removes phenanthrene, anthracene, and fluoranthene from the cell membrane. Therefore, efflux pumps are expected to be found in a wide range of toxic chemical–degrading microorganisms.

The Ttg and Srp efflux pumps are similar to the Mex antibiotic efflux pumps of *P. aeruginosa* (see Figure 2). In fact, TtgABC and TtgGHI also remove certain

Figure 1 Ribbon representation of the TtgC protein based on docking of the TtgC sequence on the TolC structure.

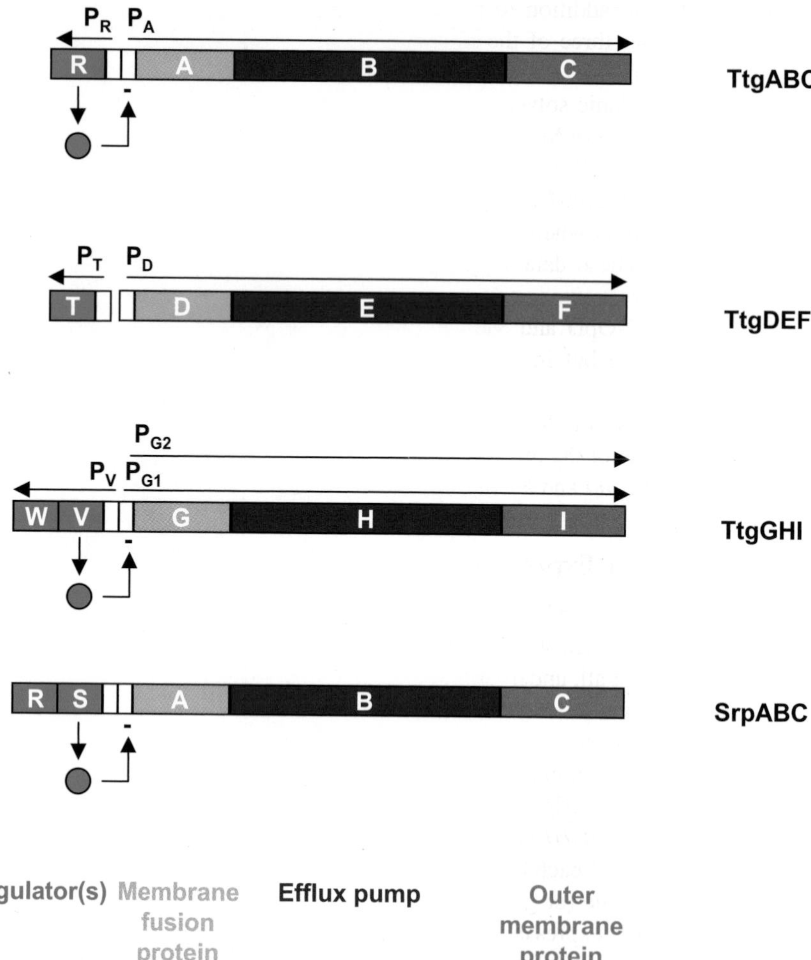

Figure 3 Organization of the efflux pump operons and their putative transcriptional regulators. Duque et al. (16) showed that TtgR negatively controls the expression from the *ttgABC* operon. The role of TtgT in the regulation of *ttgDEF* expression and that of TtgW in the regulation of *ttgGHI* expression remain unknown. A. Rojas, A. Segura & J. L. Ramos (unpublished data) have shown that TtgV negatively controls expression of the *ttgGHI* operon. Wery et al. (90) showed that SrpS is involved in the control of the *srpABC* gene cluster in *P. putida* S12, which is 99% identical to the TtgGHI/TtgVW system.

antibiotics in addition to aromatic hydrocarbons (48, 73). Li et al. (48) examined whether three of the multidrug efflux systems described in *P. aeruginosa* (MexAB-OprM, MexCD-OprJ, and MexEF-OprN) provided some level of tolerance to organic solvents. They showed that *P. aeruginosa* strains that express wild-type levels of MexAB-OprM, as well as mutants that hyperexpress this efflux pump (i.e., MexR mutants, in which the *mexR* gene product controls expression of the *mexAB-oprM* operon [17, 45, 81, 98]), tolerated *n*-hexane and *p*-xylene, whereas strains where these genes were partially or totally deleted showed no resistance. These data suggested that the MexAB-OprM multidrug efflux systems of *P. aeruginosa* can accommodate organic solvents as well as antibiotics. The MexCD-OprJ and MexEF-OprN efflux pumps may also contribute to solvent tolerance but in a less significant manner than the MexAB-OprM efflux pumps.

Overall it seems that the efflux of organic solvents mediated by an RND transport system is one of the main mechanisms that confers resistance to organic solvents in several *E. coli* and *Pseudomonas* sp. strains.

Regulation of the Expression of Efflux Pumps in *Pseudomonas*

The TtgABC pump is synthesized in large amounts in cells growing in the absence and in the presence of toluene (16). In contrast, the *ttgDEF* operon is expressed at a low level, if at all, under laboratory growth conditions, but expression increases in response to aromatic hydrocarbons in the culture medium (56). The *ttgGHI* operon is expressed from two overlapping promoters in the absence of toluene, but at a much higher level in the presence of toluene (73).

Reverse transcriptase-PCR assays have shown that for each of the *ttgABC*, *ttgDEF*, and *ttgGHI* clusters, the genes are part of a single operon (16, 56, 73). Transcription of each operon is controlled by one or more specific regulators encoded by a gene (or genes) adjacent to each operon (Figure 3, see color insert). TtgR controls expression of the *ttgABC* operon (16), TtgT seems to control the expression of the *ttgDEF* genes (W. Terán, J.L. Ramos & M.-T. Gallegos, unpublished data), and TtgV might be involved in the control of the *ttgGHI* operon. No role for TtgW in the transcription of *ttgGHI* has been demonstrated (A. Rojas, W. Terán, A. Segura & M.-T. Gallegos, unpublished data). TtgR and TtgW belong to the TetR family of transcriptional regulators, whereas TtgT and TtgV belong to the IclR family (A. Rojas, W. Terán, M.-T. Gallegos, J.L. Ramos & A. Segura, unpublished data). This may help identify their specific mode of action.

In a TtgR-mutant background, expression from the *ttgA* promoter increased about 20-fold, suggesting that TtgR repressed *ttgA* expression (16). In this mutant, background expression of the *ttgR* gene was also much higher than in the wild-type background, which indicates that TtgR also controls its own expression, although this higher level of expression increased still further in the presence of toluene and antibiotics. These results suggest the involvement of another regulator in TtgR expression. In the *ttgR*-null mutant background, expression from the *ttgD* and *ttgG* promoters followed the wild-type pattern of expression [(16); A. Rojas & W. Terán, unpublished data). The *ttgR*-null mutant, which overproduces the efflux pump proteins, was more resistant than the wild type to carbenicillin, chloramphenicol, nalidixic acid, and tetracycline, whereas the TtgB mutant was more sensitive than the wild type to these antibiotics. This provides further evidence of the involvement of the TtgABC pump in the efflux of these antibiotics and shows that an increase in expression of the *ttgABC* genes as a consequence of the lack of repression leads to increased tolerance to antibiotics.

The *ttgDEF* operon is expressed at low levels under laboratory growth conditions (56). This is probably because the *ttgT* gene is expressed at high levels. The TtgT protein is a member of the IclR family that inhibits expression from the promoter upstream from *ttgD* (W. Terán, J.L. Ramos & M.-T. Gallegos, unpublished data).

The way in which *ttgVW* gene products are involved in the control of expression of *ttgGHI* is unknown. TtgW has not yet been assigned a role in the regulation of *ttgGHI* because expression of the *ttgGHI* operon in a knock-out *ttgW* mutant followed the same pattern of expression as in the TtgW-proficient background (A. Rojas & A. Segura, unpublished data). No *ttgV* mutants are available at present for assays aimed at assigning a role to this protein in the control of the *ttgGHI* operon. In their work with *P. putida* S12, Kieboom et al. (40) showed that the *sprABC* efflux operon was inducible; later Wery et al. (90) reported the existence of two *srpSR* genes that were divergent with respect to the *srpABC* cluster and that were involved in the transcriptional control of the SrpABC operon (SrpSR and TtgVW are almost identical). They elegantly demonstrated that inactivation of *sprS* by IS*12* leads to a high level of expression of the *sprABC* genes and that solvent tolerance increases concomitantly (although the authors have not explored whether *sprABC* expression occurs from two promoters as is the case for *ttgGHI*).

Analyses of *P. putida* DOT-T1E-15, a mutant of DOT-T1E that exhibited increased sensitivity to sudden toluene shock regardless of whether it had been exposed previously to low concentrations of toluene (16), revealed that the level of expression of *ttgGHI* was negligible (W. Terán, J.L. Ramos & M.-T. Gallegos, unpublished data). In contrast, the *ttgR* gene was expressed at high levels in the DOT-T1E-15 mutant. This suggested that a global regulatory protein was involved in the solvent-tolerant response of this strain.

Duque et al. (16) showed that in the DOT-T1E-15 mutant, insertion of the mini-transposon prevented the expression of a gene homologous to *lrp*. This finding suggested that the Lrp-like protein was the global regulator. However, inactivation of the *lrp*-like gene by site-directed mutagenesis did not produce the phenotype observed in DOT-T1E-15, a result that rules out the possibility of the Lrp-like protein acting as the ultimate global regulator (E. Duque & J.L. Ramos, unpublished data). The mutation in DOT-T1E is now suspected to occur in a gene other than that inactivated by the transposon.

UNEXPECTED FUNCTIONS ASSOCIATED WITH SOLVENT TOLERANCE

Do Elements of Solvent Tolerance Parasitize the Flagellar Transport Machinery?

Segura et al. (76) isolated, after random mutagenesis with mini-Tn*5*, a mutant strain of *P. putida* DOT-T1E that was unable to tolerate toluene shocks and that exhibited a mini-Tn*5* insertion within the *fliP* gene. This gene is part of the *fliLMNOPQRflhBA* cluster, a set of genes that encodes for flagellar structure components. FliP is involved in the export of flagellar proteins, and in fact the *P. putida fliP* mutant was nonmotile. When the mutant allele was replaced with the wild-type one, the strain recovered the wild-type pattern of toluene tolerance and motility. This observation unequivocally assigned FliP a function in solvent resistance. A mutant of *P. putida* DOT-T1E with a knock-out in *flhB*, another component of the flagellar export apparatus, was also nonmotile and hypersensitive to toluene. A strain of *P. putida* S12 with a mutation in the *flgK* gene (which encodes the flagellar hook-associated HAP1 protein) was also toluene sensitive (38). However, in *P. putida* DOT-T1E a nonpolar mutation at the *fliL* gene, which encodes a cytoplasmic membrane protein associated with the flagellar basal body, yielded a nonmotile mutant that was toluene resistant. It is therefore likely that FliP, FlhB, and FlgK play a direct or an indirect role in toluene tolerance. The findings to date suggest that an intact flagellar export system is required for toluene tolerance. However, these results should not be interpreted to mean that motility and toluene tolerance are linked themselves, but rather that the flagellar transport system may be parasitized by elements involved in toluene tolerance.

This unusual case of parasitization of the flagellar system has precedents in *Yersinia enterocolitica* and *Xenorhadbus nematophilus*, where enzymes involved in virulence are exported to the outer medium via the flagellar system (21, 92). Mutants of *Yersinia* sp. or *Xenorhadbus* sp. exhibiting damage in the flagellar export apparatus were nonmotile and less virulent than the wild type.

Does the TonB System Transfer Energy to the RND Efflux Systems?

Several research groups have shown that the ionophore FCCP compromises the removal of drugs by RND efflux pumps, which suggests that the energy required for the process is provided by coupling with the proton motive force across the inner membrane (32, 70, 95). However, the molecular basis for this coupling is unknown. To shed light on this process, Godoy et al. (22) searched for mutants with increased susceptibility to different drugs, centering their work on genes that do not directly encode any component of the efflux machinery. A *P. putida* DOT-T1E mutant susceptible to 4-hydroxybenzoate (4HBA) and antibiotics as well as to toluene was isolated after mini-Tn*5* mutagenesis. The mutation affected the energy-transducing TonB system because the *exbD* gene of the *exbBDtonB* operon was knocked out. Because toluene and 4HBA accumulated in the cell membranes at higher levels than in strains with an intact TonB system (22, 72), it was concluded that the increased susceptibility to these drugs in the mutant strain was the result of the limited removal of these aromatic compounds by specific efflux pumps. It was mentioned above that preinduction of efflux pumps with low toluene concentrations in the wild-type strain led to a lower accumulation of ^{14}C-toluene in *P. putida* DOT-T1E cells (70); however, this did not occur in the mutant strain with a defective TonB system, indicating impaired operation of the TtgABC, TtgDEF, and TtgGHI efflux pumps.

The involvement of the TonB energy transduction system in the operation of efflux pumps has also been proposed in other bacterial strains. For instance, a *P. aeruginosa* strain with a mutation in the *tonB1* gene showed decreased efflux of antibiotics through RND efflux pumps. This finding again links sensitivity to drugs to a mutation in the TonB system (96). Several hypotheses might explain the participation of the TonB system in drug exclusion. (*a*) An imbalance in ExbD production in mutant strains may exert indirect effects on membrane functions, which may compromise the operation of the efflux pump systems. (*b*) An imbalance in ExbD may interfere with the incorporation of inner membrane components of the RND pumps in the cytoplasmic membrane, making the pumps nonfunctional. (*c*) Members of the RND efflux pump family may be activated through the TonB system, although neither the target element in the efflux component nor the mechanism of action has been elucidated.

Are Membrane Vesicles an Active Mechanism in Solvent Elimination?

Kobayashi et al. (43) recently reported that the solvent-tolerant *Pseudomonas* sp. IH-20000 produced membrane vesicles when toluene was added to the culture medium. They suggested that vesicle formation was a specific response of this strain to remove toluene. The vesicles consisted of phospholipids, LPS, and a small amount of proteins and contained 0.17–0.63 mol toluene/mol of lipid. In contrast, a solvent-sensitive mutant of this strain with a knocked-out *cyoC* gene (42) failed to produce vesicles, or produced incomplete ones, although the mutant also exhibited several pleiotropic defects: Some portions of outer membranes were lost and surface hydrophobicity increased. Membrane vesicles have been seen in cultures of *P. putida* DOT-T1E growing in the presence of high toluene concentrations (69); nonetheless, there is still no evidence in the latter strain that the function of membrane vesicles is to exclude solvents. It should be noted that certain pathogens release toxins to the outer medium through membrane vesicles (8), but evidence to determine whether this mechanism is specific or general for solvent exclusion awaits further study.

Alkylhydroperoxide Reductase

Ferrante et al. (18) isolated a mutant *E. coli* strain resistant to tetralin; the strain also tolerated cyclohexane, propylbenzene, and 1,2-dihydronaphthalene. A recombinant library revealed that the mutation was associated to the alkylhydroperoxide reductase operon *ahpCF*. The mutation was localized to a substitution of valine for glycine at codon 142 in the coding region of *ahpC*, which is the gene that codes the catalytic subunit of the enzyme. The *ahpC* mutant reduced tetralin hydroperoxide to 1,2,3,4,-tetrahydro-1-naphthol three times as quickly as the wild type. Moreover, the cloned mutant gene was sufficient to confer organic solvent resistance to *E. coli* wild-type cells. The authors concluded that the toxicity of tetralin and other hydrocarbons was caused by the formation of toxic hydroperoxides in the cell by a yet unidentified mechanism. The hydroperoxide hypothesis of organic solvent toxicity and resistance appears to be compatible with what is known about mutagen production and detoxification in animal cells (31).

CONCLUDING REMARKS AND FUTURE PROSPECTS

In gram-negative bacteria a variety of responses to counteract the effect of organic solvents can result in rigidification of the cell membrane, detoxification of toxic metabolites, and the extrusion of toxic chemicals. Not all mechanisms may operate in a single microorganism, and this variability may result in different levels of resistance to toxic chemicals (see Table 1).

Some *Pseudomonas* spp. strains responded rapidly to challenging agents via isomerization of the naturally synthesized *cis*-isomer of unsaturated lipids to the *trans*-isomer, a reaction catalyzed by a periplasmic *cis*-to-*trans* isomerase (Cti). Long-term exposure of *Pseudomonas* spp. and *E. coli* to nonlethal concentrations of solvents often leads to an increase in the total amount of phospholipids and to a higher proportion of saturated long-chain lipids. These changes are concomitant with alterations in the level of different phospholipid headgroups. The net result of this set of modifications is rigidification of the cell membrane. Research is needed to clarify the mechanism of action of Cti and how phospholipid headgroups and fatty acids are turned over when the cells are exposed to solvents. The element(s) that senses the presence of the solvent, the array of short- and long-term responses involved in the activation of *cis*-to-*trans* isomerase, and the transcription of genes that lead to altered phospholipid composition are potentially fruitful areas for future research.

Efflux pumps of the RND family are critical for solvent tolerance. Understanding the complexity of RND solvent exclusion pumps requires comprehension of their mechanisms of action and regulation. The molecular basis for the operation of these pumps is unknown, but all pumps exhibit a broad spectrum of potential substrates. Several efflux pumps for organic solvents have been identified in *Pseudomonas* spp. and in *E. coli*. These devices consist of an energy-dependent transporter bound to the inner membrane, an outer membrane protein that spans the periplasmic space and forms a tunnel [the three-dimensional structure has been solved for TolC (46)], and an inner membrane-anchored lipoprotein that expands through the periplasmic space and seems to agglutinate all pump elements. Knock-out mutants in the efflux pumps are significantly more sensitive to solvents than their parental strains, although there are different degrees of sensitivity to different solvents depending on the strain. These differences usually involve the differential expression of multiple pumps or the existence of yet unidentified elements involved in solvent tolerance. One of the main barriers toward the understanding of the molecular mechanisms of solvent exclusion is the lack of a high-resolution model for the inner membrane pump and the lipoprotein that expands through the periplasm.

How the efflux pumps function and whether they expel solvents that are in the membrane or in the cytoplasm are questions that need detailed in vitro analysis. Do the pumps have more than one active site? How are the pump domains organized, and how do they interact with the other efflux elements? These unexplored areas deserve attention. Advances in this area are likely to occur through the isolation of mutants with altered substrate specificity, i.e., mutants that maintain the ability to extrude certain drugs but that lose the ability to expel others. These mutants are available for multidrug pumps belonging to other families, i.e., QacA and MdfA (9, 11).

It is becoming clear that repressors govern genes that encode the efflux pumps, but several global regulators are also involved. How the signals are sensed and transmitted through a cascade of regulators, and how many regulators are actually involved in the cascade, remain to be discovered. Studies by Neyfakh and colleagues with the BmrR regulator, which controls the Bmr efflux pump, have revealed that this regulator interacts with multiple substrates because the inducer penetrates the hydrophobic core of the protein, where it participates in multiple van der Waals and stacking interactions with hydrophobic amino acids. In such a system, ligand specificity can be broadened by adjustments in the flexibility of the binding site (59, 97). Whether this is also the case for regulators that recognize solvents is worth exploring.

Solvent tolerance in bacteria is an area that deserves our utmost attention to elucidate the molecular basis of this intriguing property. It should one day be possible to exploit this ability in biotechnological applications such as the biotransformation of toxic and water-insoluble compounds into chemicals of added value, and the removal of pollutants from sites that have been heavily contaminated with toxic solvents and that remain invulnerable to colonization by microorganisms with biodegradative properties.

ACKNOWLEDGMENTS

Work in the authors' laboratory was supported by grants from the European Commission (QLK3-2000-CT-0170 and QLK3-2001-00435) and the Comisión

Interministerial de Ciencia y Tecnología (BIO 2000-0964 and 1FD97-1437). We thank M. Fandila and C. Lorente for secretarial assistance and K. Shashok for editing the manuscript.

The *Annual Review of Microbiology* is online at http://micro.annualreviews.org

LITERATURE CITED

1. Alekshun MN, Levy SB. 1997. Regulation of chromosomally mediated multiple antibiotic resistance: the *mar* regulon. *Antimicrob. Agents. Chemother.* 41:2067–75
2. Alekshun MN, Levy SB, Mealny TR, Seaton BA, Head JF. 2001. The crystal structure of MarR, a regulator of antibiotic resistance, at 2.3 Å resolution. *Nat. Struct. Biol.* 8:710–14
3. Aono R, Aibe K, Inoue A, Horikoshi K. 1991. Preparation of organic solvent-tolerant mutants from *Escherichia coli* K12. *Agric. Biol. Chem.* 55:1985–38
4. Aono R, Ito M, Inoue A, Horikoshi H. 1992. Isolation of a novel toluene-tolerant strain of *Pseudomonas aeruginosa*. *Biosci. Biotechnol. Biochem.* 1:145–46
5. Aono R, Kobayashi M, Nakajima H, Kobayashi H. 1995. A close correlation between improvement of organic solvent tolerance levels and alteration of resistance toward low levels of multiple antibiotics in *Escherichia coli*. *Biosci. Biotech. Biochem.* 59:213–18
6. Aono R, Tsukagoshi N, Yamamoto M. 1998. Involvement of outer membrane protein TolC, a possible member of the *mar-sox* regulon, in maintenance and improvement of organic solvent tolerance of *Escherichia coli* K-12. *J. Bacteriol.* 180:938–44
7. Asako H, Nakajima H, Kobayashi K, Kobayashi M, Aono R. 1997. Organic solvent tolerance and antibiotic resistance increased by overexpression of *marA* in *Escherichia coli*. *Appl. Environ. Microbiol.* 63:1428–33
8. Beveridge TJ. 1999. Structures of gram-negative cell walls and their derived membrane vesicles. *J. Bacteriol.* 181:4725–33
9. Bibi E, Adler J, Lewison O, Edgar R. 2001. MdfA, an interesting model protein for studying multidrug transport. *J. Mol. Microbiol. Biotechnol.* 3:171–77
10. Blackmore CG, McNaughton PA, van Veen HW. 2001. Multidrug transporters in prokaryotic and eukaryotic cells: physiological functions and transport mechanisms. *Mol. Membr. Biol.* 18:97–103
11. Brown MH, Skurray RA. 2001. Staphylococcal multidrug efflux protein QacA. *J. Mol. Microbiol. Biotechnol.* 3:163–70
12. Bugg T, Foght JM, Pickard MA, Gray MR. 2000. Uptake and active efflux of polycyclic aromatic hydrocarbons by *Pseudomonas fluorescens* LP6a. *Appl. Environ. Microbiol.* 66:5387–92
13. Chen Q, Janssen DB, Witholt B. 1995. Growth on octane alters the membrane lipid fatty acids of *Pseudomonas oleovorans* due to the induction of *alkB* and synthesis of octanol. *J. Bacteriol.* 177:6894–901
14. Cronan JE. 1968. Phospholipid alterations during growth of *Escherichia coli*. *J. Bacteriol.* 95:2054–61
15. Cruden DL, Wolfram JH, Rogers RD, Gibson DT. 1992. Physiological properties of a *Pseudomonas* strain which grows with *p*-xylene in a two-phase (organic-aqueous) medium. *Appl. Environ. Microbiol.* 58:2723–29

15a. de Smet MJ, Kingma J, Witholt B. 1978. The effect of toluene on the structure and permeability of the outer and cytoplasmic membranes of *Escherichia coli*. *Biochim. Biophys. Acta* 506:64–80

16. Duque E, Segura A, Mosqueda G, Ramos JL. 2001. Global and cognate regulators control the expression of the organic solvent efflux pumps TggABC and TtgDEF of *Pseudomonas putida*. *Mol. Microbiol.* 39:1100–6
17. Evans K, Adewage L, Poole K. 2001. MexR repressor of the *mexAB-oprM* multidrug efflux operon of *Pseudomonas aeruginosa*: identification of MexR binding sigres in the *mexA-mexR* intergenic region. *J. Bacteriol.* 183:807–12
18. Ferrante AA, Angliera J, Lewis K, Klibanov AM. 1995. Cloning of an organic solvent-resistance genes in *Escherichia coli*. The unexpected role of alkylhydroperoxide reductase. *Proc. Natl. Acad. Sci. USA* 92:7617–21
19. Fralick JA. 1996. Evidence that TolC is required for functioning of the Mar/AcrAB efflux pump of *Escherichia coli*. *J. Bacteriol.* 178:5803–5
20. Fukimori F, Hirayama H, Takami H, Inoue A, Horikoshi K. 1998. Isolation and transposon mutagenesis of a *Pseudomonas putida* KT2442 toluene-resistant variant: involvement of an efflux system in solvent resistance. *Extremophiles* 2:395–400
21. Giraudan A, Lanois A. 2000. *flhDC*, the flagellar master operon of *Xenorhadbus nematophilus*: requirement for motility, lypolysis, extracellular hemolysis, and full virulence in insects. *J. Bacteriol.* 182: 107–15
22. Godoy P, Ramos-González MI, Ramos JL. 2001. Involvement of the TonB system in tolerance to solvents and drugs in *Pseudomonas putida* DOT-T1E. *J. Bacteriol.* 183:5285–92
23. Hamamoto T, Takata N, Kudo T, Horikoshi K. 1994. Effect of temperature and growth phase on fatty acid composition of the psychrophilic *Vibrio* sp. strain no. 5710. *FEMS Microbiol. Lett.* 119:77–81
24. Heipieper HJ, de Bont JAM. 1994. Adaptation of *Pseudomonas putida* S12 to ethanol and toluene at the level of fatty acid composition of membranes. *Appl. Environ. Microbiol.* 60:4440–44
25. Heipieper HJ, Diefenbach R, Keweloh H. 1992. Conversion of *cis* unsaturated fatty acids to *trans*, a possible mechanism for the protection of phenol degrading *Pseudomonas putida* P8 from substrate toxicity. *Appl. Environ. Microbiol.* 58:1847–52
26. Heipieper H-J, Menlenbeld G, van Oirschot Q, de Bont JAM. 1996. Effect of environmental factors on the *cis/trans* ratio of unsaturated fatty acids in *Pseudomonas* S12. *Appl. Environ. Microbiol.* 62:2773–77
27. Holtwick R, Meinhardt F, Keweloh H. 1997. *Cis-trans* isomerization of unsaturated fatty acids: cloning and sequencing of the *cti* gene from *Pseudomonas putida* P8. *Appl. Environ. Microbiol.* 63:4292–97
28. Huertas MJ, Duque E, Molina L, Roselló-Mora R, Mosqueda G, et al. 2000. Tolerance to sudden organic solvent shocks by soil bacteria and characterization of *Pseudomonas putida* strains isolated from toluene polluted sites. *Environ. Sci. Technol.* 34:3395–400
29. Ingram LO. 1977. Changes in lipid composition of *Escherichia coli* resulting from growth with organic solvents and with food additives. *Appl. Environ. Microbiol.* 33:1233–36
30. Inoue A, Korikoshi K. 1989. A *Pseudomonas* thrives in high concentrations of toluene. *Nature* 338:264–66
31. Ishii T, Yamada M, Sato H, Matsue M, Taketani S, et al. 1993. Cloning and characterization of a 23-kDa stress-induced mouse peritoneal macrophage protein. *J. Biol. Chem.* 268:18633–36
32. Isken S, de Bont JAM. 1996. Active efflux of toluene in a solvent-resistant bacterium. *J. Bacteriol.* 178:6056–58
33. Jellen-Ritter AS, Kern WV. 2001. Enhanced expression of the multidrug efflux pumps AcrAB and AcrEF associated with insertion element transposition in *Escherichia coli* mutants selected with

a fluoroquinolone. *Antimicrob. Agents Chemother.* 45:1467–72

34. Junker F, Ramos JL. 1999. Involvement of the *cis-trans* isomerase CtiT1 in solvent resistance in *Pseudomonas putida* DOTT1. *J. Bacteriol.* 181:5693–700
35. Junker F, Rodríguez-Herva JJ, Duque E, Ramos-González MI, Llamas M, Ramos JL. 2001. A WbpL mutant of *Pseudomonas putida* DOT-T1E strain, which lacks the O-antigenic side chain of lipolysaccharides, is tolerant to organic solvent shocks. *Extremophiles* 5:93–99
36. Keweloh H, Heipieper HJ. 1996. *Trans* unsaturated fatty acids in bacteria. *Lipids* 31:129–37
37. Keweloh H, Weyrauch G, Rehm HJ. 1990. Phenol-induced membrane changes in free and immobilized *Escherichia coli*. *Appl. Microbiol. Biotechnol.* 33:66–71
38. Kieboom J, Bruinenberg R, Keizer-Gunnink I, de Bont JAM. 2001. Transposon mutations in the flagella biosynthetic pathway of the solvent-tolerant *Pseudomonas putida* S12 result in a decreased expression of solvent efflux genes. *FEMS Microbiol. Lett.* 198:117–22
39. Kieboom J, Dennis JJ, de Bont JAM, Zylstra GJ. 1998. Identification and molecular characterization of an efflux pump involved in *Pseudomonas putida* S12 solvent tolerance. *J. Biol. Chem.* 273:85–91
40. Kieboom J, Dennis JJ, Zylstra GJ, de Bont JAM. 1998. Active efflux of organic solvents by *Pseudomonas putida* S12 is induced by solvents. *J. Bacteriol.* 180:6769–72
41. Kim K, Lee S, Lee K, Lim D. 1998. Isolation and characterization of toluene-sensitive mutants from the toluene-resistant bacterium *Pseudomonas putida* GM73. *J. Bacteriol.* 180:3692–96
42. Kobayashi H, Takami H, Hirayama H, Kobata K, Usami R, Horikoshi K. 1999. Outer membrane changes in a toluene-sensitive mutant of toluene-tolerant *Pseudomonas putida* IH-2000. *J. Bacteriol.* 181:4493–98
43. Kobayashi H, Uematsu K, Hirayama H, Horikoshi K. 2000. Novel toluene elimination system in a toluene-tolerant microorganism. *J. Bacteriol.* 182:6451–55
44. Kobayashi K, Tsukagoshi N, Aono R. 2001. Suppression of hypersensitivity of *Escherichia coli acrB* mutant to organic solvents by integrational activation of the *acrEF* operon with the IS*1* or IS *2* element. *J. Bacteriol.* 183:2646–53
45. Köhler T, Epp SF, Kocjancic Curty L, Pechère J-C. 1999. Characterization of MexT, the regulator of the MexE-MexF-OprN multidrug efflux system of *Pseudomonas aeruginosa*. *J. Bacteriol.* 181: 6300–5
46. Koronakis V, Sharff A, Koronakis E, Luisi B, Hughes C. 2000. Crystal structure of the bacterial membrane protein TolC central to multidrug efflux and protein export. *Nature* 405:914–19
47. Lee A, Mao W, Warren MS, Mistry A, Hoshino K, et al. 2000. Interplay between efflux pumps may provide either additive of multiplicative effects on drug resistance. *J. Bacteriol.* 182:3142–50
48. Li X, Zhang L, Poole K. 1998. Role of the multidrug efflux systems of *Pseudomonas aeruginosa* in organic solvent tolerance. *J. Bacteriol.* 180:2987–91
49. Loffeld B, Keweloh H. 1996. *Cis/trans* isomerization of unsaturated fatty acids as possible control mechanism of membrane fluidity in *Pseudomonas putida* P8. *Lipids* 31:811–15
50. Lomovskaya O, Lee A, Hoshino K, Ishida H, Mistry A, et al. 1999. Use of a genetic approach to evaluate the consequences of inhibition of efflux pumps in *Pseudomonas aeruginosa*. *Antimicrob. Agents Chemother.* 43:1340–46
51. Lomovskaya O, Warren MS, Lee A, Galazzo J, Fronko R, et al. 2001. Identification and characterization of inhibitors of multidrug resistance efflux pumps in *Pseudomonas aeruginosa*: novel agents for combination therapy. *Antimicrob. Agents Chemother.* 45:105–16

52. Ma D, Alberti M, Lynch C, Nikaido H, Hearst JE. 1996. The local repressor AcrR plays a modulating role in regulation of *acrAB* genes of *Escherichia coli* by global stress signals. *Mol. Microbiol.* 19:101–12
53. Ma D, Cook DN, Alberti M, Pon NG, Nikaido H, Hearst JE. 1993. Molecular cloning and characterization of *acrA* and *acrE* genes of *Escherichia coli. J. Bacteriol.* 175:6299–313
54. Ma D, Cook DN, Alberti M, Pon NG, Nikaido H, Hearst JE. 1995. Genes *acrA* and *acrB* encode a stress-induced efflux system of *Escherichia coli. Mol. Microbiol.* 16:45–55
55. Mosqueda G, Ramos JL. 2000. A set of genes encoding a second toluene efflux system in *Pseudomonas putida* DOT-T1E is linked to the tod genes for toluene metabolism. *J. Bacteriol.* 181:937–43
56. Mosqueda G, Ramos-González MI, Ramos JL. 1999. Toluene metabolism by the solvent-tolerant *Pseudomonas putida* DOT-T1 strain, and its role in solvent impermeabilization. *Gene* 232:69–76
57. Nakajima H, Kobayashi K, Kobayashi M, Asako H, Aono R. 1995. Overexpression of the *robA* gene increases organic solvent tolerance and multiple antibiotic and heavy metal ion resistance in *Escherichia coli. Appl. Environ. Microbiol.* 61: 2302–7
58. Nakajima H, Kobayashi M, Negishi T, Aono R. 1995. *soxRS* gene increased the level of organic solvent tolerance in *Escherichia coli. Biosci. Biotech. Biochem.* 59:1323–25
59. Neyfakh AA. 2001. The ostensible paradox of multidrug recognition. *J. Mol. Microbiol. Biotechnol.* 3:151–54
60. Nikaido H. 1996. Multidrug efflux pumps of gram-negative bacteria. *J. Bacteriol.* 178:5853–59
61. Nikaido H. 1999. Microdermatology: cell surface in the interaction of microbes with the external world. *J. Bacteriol.* 181:4–8
62. Okusu H, Ma D, Nikaido H. 1996. AcrAB efflux pump plays a major role in the antibiotic resistance phenotype of *Escherichia coli.* Antibiotic-resistance (Mar) mutants. *J. Bacteriol.* 178:306–8
63. Paulsen IT, Chen J, Nelson KE, Seier MH Jr. 2001. Comparative genomics of microbial drug efflux systems. *J. Mol. Microbiol. Biotechnol.* 3:145–50
64. Pedrotta V, Witholt B. 1999. Isolation and characterization of the *cis-trans*-unsaturated fatty acid isomerase of *Pseudomonas oleovorans. J. Bacteriol.* 181: 3256–61
65. Pinkart HC, White DC. 1997. Phospholipid biosynthesis and solvent tolerance in *Pseudomonas putida* strains. *J. Bacteriol.* 179:4219–26
66. Pinkart HC, Wolfram JW, Rogers R, White DC. 1996. Cell envelope changes in solvent-tolerant and solvent-sensitive *Pseudomonas putida* strains following exposure to *o*-xylene. *Appl. Environ. Microbiol.* 62:1129–32
67. Ramos JL, Díaz E, Dowling D, de Lorenzo V, Molin S, et al. 1994. The behavior of bacteria designed for biodegradation. *Bio/Technology* 12:1349–58
68. Ramos JL, Duque E, Godoy P, Segura A. 1998. Efflux pumps involved in toluene tolerance in *Pseudomonas putida* DOT-T1E. *J. Bacteriol.* 180:3323–29
69. Ramos JL, Duque E, Huertas MJ, Haïdour A. 1995. Isolation and expansion of the catabolic potential of a *Pseudomonas putida* strain able to grow in the presence of high concentrations of aromatic hydrocarbons. *J. Bacteriol.* 177:3911–16
70. Ramos JL, Duque E, Rodríguez-Herva J-J, Godoy P, Haïdour A, et al. 1997. Mechanisms for solvent tolerance in bacteria. *J. Biol. Chem.* 272:3887–90
71. Ramos JL, Gallegos MT, Marqués S, Ramos-González MI, Espinosa-Urgel M, Segura A. 2001. Responses of gram-negative bacteria to certain environmental stresses. *Curr. Opin. Microbiol.* 4:166–71

72. Ramos-González MI, Godoy P, Alaminos M, Ben-Bassat A, Ramos JL. 2001. Physiological characterization of *Pseudomonas putida* DOT-T1E tolerance to *p*-hydroxybenzoate. *Appl. Environ. Microbiol.* 67:4338–41
73. Rojas A, Duque E, Mosqueda G, Golden G, Hurtado A, et al. 2001. Three efflux pumps are required to provide efficient tolerance to toluene in *Pseudomonas putida* DOT-T1E. *J. Bacteriol.* 183:3967–73
74. Russell N, Fukanaga N. 1990. A comparison of the thermal adaptation of membrane lipids in psychrophilic and thermophilic bacteria. *FEMS Microbiol. Rev.* 75:171–82
75. Saier MH Jr, Tam R, Reizer A, Reizer J. 1994. Two novel families of bacterial membrane proteins concerned with nodulation, cell division and transport. *Mol. Microbiol.* 11:841–47
76. Segura A, Duque E, Hurtado A, Ramos JL. 2001. Mutations in genes involved in the flagellar export apparatus of the solvent-tolerant *Pseudomonas putida* DOT-T1E strain impair motility and lead to hypersensibility to toluene shocks. *J. Bacteriol.* 183:4127–33
77. Segura A, Duque E, Mosqueda G, Ramos JL, Junker F. 1999. Multiple responses of gram-negative bacteria to organic solvents. *Environ. Microbiol.* 1:191–98
78. Sikkema J, de Bont JAM, Poolman B. 1995. Mechanisms of membrane toxicity of hydrocarbons. *Microbiol. Rev.* 59:201–22
79. Sikkema J, Poolman B, Konings WN, de Bont JAM. 1992. Effects of the membrane action of tetralin on the functional and structural properties of artificial and bacterial membranes. *J. Bacteriol.* 174:2986–92
80. Deleted in proof
81. Srikumar R, Paul CJ, Poole K. 2000. Influence of mutations in the mex*R* repressor gene on expression of the MexA-MexB-OprM multidrug efflux system of *Pseudomonas aeruginosa. J. Bacteriol.* 182:1410–14
82. Suutari M, Laakso S. 1994. Microbial fatty-acids and thermal adaptation. *Crit. Rev. Microbiol.* 20:129–37
83. Thanabalu T, Koronakis E, Hughes C, Koronakis V. 1998. Substrate-induced assembly of a contiguous channel for protein export from *E. coli* reversible bridging of an inner-membrane translocase to an outer membrane exit pore. *EMBO J.* 17:6487–96
84. Thanassi DG, Suh CSB, Nikaido H. 1995. Role of outer membrane barrier in efflux-mediated tetracycline resistance of *Escherichia coli. J. Bacteriol.* 177:998–1007
85. Tsukagoshi N, Aono R. 2000. Entry and release of solvents in *Escherichia coli* in an organic aqueous two-liquid phase system and substrate specificity of the AcrAB-TolC solvent-extruding pumps. *J. Bacteriol.* 182:4803–10
86. Wang H, Dzink-Fox JL, Chen M, Levy SB. 2001. Genetic characterization of highly fluoroquinolone-resistant clinical *Escherichia coli* strains from china: role of *acrR* mutations. *Antimicrob. Agents Chemother.* 45:1515–21
87. Weber FJ. 1994. *Toluene: biological waste-gas treatment, toxicity and microbial adaptation.* PhD thesis. Univ. Wageningen. The Netherlands.
88. Weber FJ, de Bont JAM. 1996. Adaptation mechanisms of microorganisms to the toxic effects of organic solvents on membranes. *Biochim. Biophys. Acta* 1286:225–45
89. Weber FJ, Isken S, de Bont JAM. 1994. *Cis/trans* isomerization of fatty acids as a defense mechanism of *Pseudomonas putida* strains to toxic concentrations of toluene. *Microbiology* 140:2013–17
90. Wery J, Hidayat B, Kieboom J, de Bont JAM. 2001. An insertion sequence prepares *Pseudomonas putida* S12 for severe solvent stress. *J. Biol. Chem.* 276:5700–6
91. White DG, Goldman JD, Demple B, Levy SD. 1997. Role of the *acrAB* locus in

organic solvent tolerance mediated by expression of *marA*, *soxS* or *robA* in *Escherichia coli*. *J. Bacteriol.* 179:6122–26

92. Young GM, Schmiel DH, Miller VL. 1999. A new pathway for the secretion of virulence factors by bacteria: The flagellar export apparatus functions as protein-secretion system. *Proc. Natl. Acad. Sci. USA* 96:6456–61
93. Zgurskaya HI, Nikaido H. 1999. AcrA is a highly asymmetric protein capable of spanning the periplasm. *J. Mol. Biol.* 285:409–20
94. Zgurskaya HI, Nikaido H. 1999. Bypassing the periplasm: reconstruction of the AcrAB multidrug efflux pump of *Escherichia coli*. *Proc. Natl. Acad. Sci. USA* 96:7190–95
95. Zgurskaya HI, Nikaido H. 2000. Cross-linked complex between oligomeric periplasmic lipoprotein AcrA and the inner-membrane-associated multidrug efflux pump AcrB from *Escherichia coli*. *J. Bacteriol.* 182:4264–67
96. Zhao Q, Li X-Z, Mistry A, Srikumar R, Zhang L, et al. 1998. Influence of the TonB energy-coupling protein on efflux-mediated multidrug resistance in *Pseudomonas aeruginosa*. *Antimicrob. Agents Chemother.* 42:2225–31
97. Zhleznova EE, Markham PN, Neyfakh AA, Brennan RG. 1999. Structural basis of multidrug recognition by BmrR, a transcription activator of a multidrug transporter. *Cell* 96:353–62
98. Ziha-Zarifi I, Llanes C, Köhler T, Pechère J-C, Plesiat P. 1999. *In vivo* emergence of multidrug-resistant mutants of *Pseudomonas aeruginosa* overexpressing the active efflux system MexA-MexB-OprM. *Antimicrob. Agents Chemother.* 43:287–91

Annu. Rev. Microbiol. 2002. 56:769–92
doi: 10.1146/annurev.micro.56.012302.160830

GROWING OLD: Metabolic Control and Yeast Aging

S. Michal Jazwinski
Department of Biochemistry and Molecular Biology, Louisiana State University Health Sciences Center, New Orleans, Louisiana 70112; e-mail: sjazwi@lsuhsc.edu

Key Words longevity, retrograde response, caloric restriction, age asymmetry, cell signaling

■ **Abstract** The metabolic characteristics of a yeast cell determine its life span. Depending on conditions, stress resistance can have either a salutary or a deleterious effect on longevity. Gene dysregulation increases with age, and countering it increases life span. These three determinants of yeast longevity may be interrelated, and they are joined by a potential fourth, genetic stability. These factors can also operate in phylogenetically diverse species. Adult longevity seems to borrow features from the genetic programs of dormancy to provide the metabolic and stress resistance resources necessary for extended survival. Both compensatory and preventive mechanisms determine life span, while epigenetic factors and the element of chance contribute to the role that genes and environment play in aging.

CONTENTS

WHAT IS AGING?

Definition

Biological aging is not easy to define, even though we know it when we see it. Based on the familiar features of human aging, the closest to a definition may be that biological aging is the progressive decline in the ability of the organism to resist stress, damage, and disease. Biological aging is characterized by the appearance of degenerative and neoplastic disorders.

At the demographic level, aging manifests itself as an exponential increase in mortality rate with the age of the cohort. This is coupled to an increase in the frailty of the aging population (117). Interestingly, mortality rate tends to plateau and even decreases at extreme ages (112). This has been explained variously, but because it can occur even in a homogeneous starting population, it has been ascribed to a process of winnowing out frail individuals, leaving the hardy ones behind. This phenomenon, often called demographic selection, requires the emergence of heterogeneity among the individuals in the population as it ages. This brings the problem of aging squarely to the level of the individual, which is as it should be because demographic changes do not have a life of their own but simply reflect biological events.

The issue then becomes recognition of the forces of aging as they occur in individuals. This is where problems arise. The manifestations of age-related decline or functional senescence, as it is called, are myriad, and they are not always easily separated from mere chronology, maturational events, and disease (27). In addition, functional senescence has a diverse penetrance among individuals. Thus, it is difficult to find a set of biomarkers of aging that are a better predictor of mortality than life span itself. The variability from individual to individual encountered in the progress of the aging-related changes in humans is readily seen in yeast as well.

Yeast Aging

The focus of this review is the yeast *Saccharomyces cerevisiae*. Although recently *Schizosaccharomyces pombe* has been shown to have a limited life span, this has not yet been exploited (4). Attention has been directed toward the finite capacity of individual yeast cells to divide, as a manifestation of an aging process (81). This limited replicative life span is quite independent of calendar time (83). More recently, survival of yeast cells in stationary phase (115) has been espoused as a model for chronological aging, measured by time rather than by cell divisions (72).

The replicative and chronological models of yeast aging seem disparate. The point has been taken that the replicative model may be useful for understanding the aging of dividing cells of higher eukaryotes, while the chronological model may be informative of events in post-mitotic cells (72). Such a stark distinction seems oversimplified (42). From the yeast spore, through the stationary phase cell, to the growing and rapidly dividing yeast cell, one can readily discern a metabolic continuum. There is a marked reduction in metabolic activity from exponential growth to stationary phase (61, 115). This reduction displays its zenith in the spore, which nevertheless has a measurable metabolic rate (80). Depending on the circumstances, the relative contributions of respiration and fermentation vary, indicating qualitative changes in metabolism. This presents difficulties in the choice of suitable indices of overall metabolic activity. It is clear, however, that it takes nondividing yeast much longer to carry out a given amount of metabolic activity than it does dividing yeast. Residence in stationary phase tends to shorten the subsequent replicative life span of yeast cells (2, 42), indicating the corresponding impact of metabolic activity on both chronologic and replicative aging and providing support for the metabolic continuum discussed here.

Nondividing cells are in general more stress resistant. Stationary phase yeasts are more resistant to heat and to oxidative stress (77), although they are more sensitive to ultraviolet radiation (88). This stress resistance may be a response to the nature of the metabolism of these cells, which are in a respiratory mode. There may also be a distinct fitness strategy at play here (40). Nondividing cells and spores must devote significant energy to survive in the face of stress. Dividing cells, on the other hand, deploy their metabolic resources to an increase in progeny, a few of which may survive in the face of stress by the force of their sheer numbers. Thus, a trade-off is at play.

The yeast system provides opportunities for examining the processes underlying demographic selection. Both dividing and nondividing yeast cells exhibit a plateau in mortality rate with age (46, 112). In the first case, it is clear that this is the result of the properties of the aging cohort. However, the dramatic and repeated fluctuations in viability of stationary cultures presented may reflect the lysis and regrowth of cells.

During their replicative life span, yeast cells undergo a variety of morphological and physiological changes (38, 45). Many of these alterations represent declining function. Thus, yeasts are subject to functional senescence. This is an important conclusion that cannot be extended to stationary phase yeasts at present. Without functional senescence, it is difficult to speak of a biological aging process.

Factors Contributing to Aging

Over 30 genes play a role in determining yeast replicative life span (45). A few genes have also been implicated in chronological aging (26), but there are likely to be many based on the genetics of survival in stationary phase. These genes encode a wide array of biochemical functions, suggesting that even in yeast aging is a complex process and that there is more than one cause of aging. Despite this complexity, there are only a handful of physiological principles underlying yeast aging (39). These include metabolic capacity, resistance to stress, gene dysregulation, and genetic stability.

The aging phenotype is elicited through the interaction of the longevity genes with the environment. It should already be clear how complicated this interaction can be from the discussion of the trade-off between growth and stress resistance. This interaction becomes even more complicated owing to the generation of stress by metabolic activity in the form of oxidative damage, among others.

In addition to the genes and the environment, chance or random events are important in aging. Of course, the environment introduces its own dose of uncertainty; however, here the focus is on epigenetic factors that are rooted in the fact that the organism constitutes a network of interacting structures, pathways, and processes. It has been proposed that change is the cause rather than the effect of aging, and this has been modeled mathematically (41, 46). Such change acts at various levels of biological organization, from the molecular to the cellular, and it constitutes a random epigenetic factor that operates according to deterministic rules [for a simple allegory, see (45)]. The mathematical model referenced here accurately predicts the epigenetic stratification of an aging yeast population, resulting from individual change, which leads to the plateau in mortality rate (46).

Deterministic rules can describe the impact of random events on aging. Is the opposite the case? Do stochastic events during aging contain information about the biological aging process? One such stochastic event is the change in the choice of bud site during the yeast replicative life span. The sum of these changes during the life span is termed the budding profile. Budding profiles are highly individual, as cells adopt bud sites in an ordered or random way, and they can be described mathematically (47). Significantly, these profiles accurately categorize individual longevity, long or short. The model predicts that a degree of randomness (random budding) is important for long life, and this expectation has been experimentally confirmed. Perhaps this can be interpreted as a need for some plasticity, so that the organism can more easily respond to different exigencies.

METABOLISM IS A DETERMINANT OF YEAST LIFE SPAN

The greater the metabolic capacity the longer the replicative life span of a yeast cell. This statement can be made a priori because the production of successive daughter cells requires expenditure of energy and production of biomass. However, two recent developments provide a strong foundation for the importance of metabolism in yeast aging. They are the implication of the retrograde response and caloric restriction in determining yeast replicative life span.

Retrograde Response

Induction of the retrograde response causes an increase in replicative life span (59) and entails an increase in stress resistance (59, 111). The retrograde response denotes changes in gene expression that occur in petite yeast cells (87). Petites are strains that lack fully functional mitochondria, and they can be spontaneous or induced. It is now clear from microarray studies that the retrograde response entails a wide array of changes in gene expression (25, 111). These changes in gene expression portend a broad metabolic remodeling of the cell in response to the metabolic duress associated with mitochondrial dysfunction (Figure 1). This includes the increased expression of genes that encode proteins involved in transport of small molecules, peroxisomal biogenesis and function, stress responses,

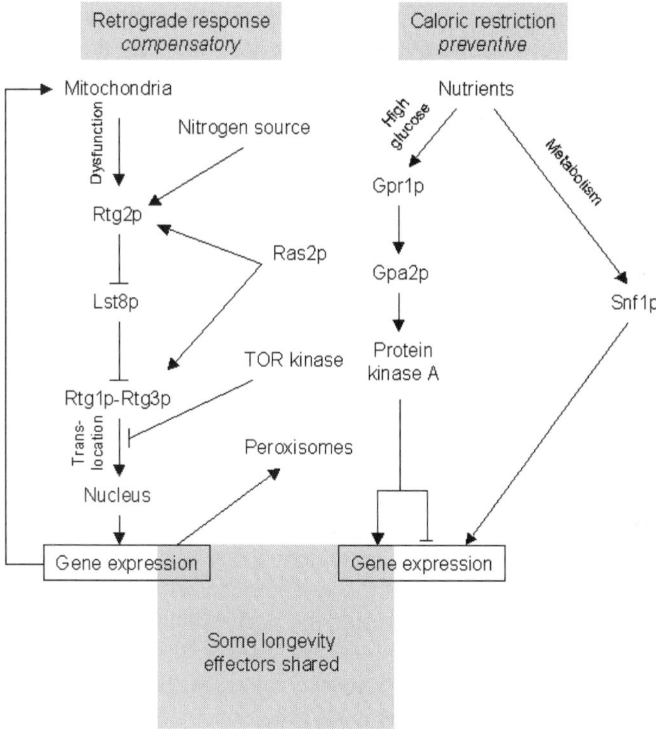

Figure 1 The retrograde response and caloric restriction operate through distinct pathways that share some replicative–life span effectors. The precise locus of action of Ras2p in the retrograde response is not known. ↓, indicates stimulation; ⊥, indicates inhibition. For details, see text.

and anaplerotic reactions. The latter involves the glyoxylate cycle and the generation and utilization of acetate, which would allow replenishment of Krebs cycle biosynthetic intermediates from acetate as a net source of carbon. Unlike the Krebs cycle, the glyoxylate cycle conserves carbon, facilitating more efficient utilization of resources.

The retrograde response depends on the action of three genes *RTG1*, *RTG2*, and *RTG3* (48, 68). Rtg1p and Rtg3p are the subunits of a heterodimeric, basic helix-loop-helix-leucine zipper transcription factor. Rtg2p is involved in the translocation of this transcription factor from the cytoplasm to the nucleus and in its activation (94). In cells under metabolic duress, the expression of Krebs cycle genes is under the control of the *RTG* genes, whose products replace the HAP transcription complex (70).

The retrograde response is repressed by glutamate, a rich nitrogen source (60, 70, 71). Rtg2p plays a special role in the response to the quality of the nitrogen source, acting upstream of the nitrogen catabolism regulation pathway (89). The Lst8p negatively regulates *RTG* gene function upstream through the expression of the Ssy1p-amino-acid sensor at the plasma membrane and downstream between Rtg2p and the Rtg1p-Rtg3p transcription factor (71). At least part of the effects of the retrograde response appears to be modulated by the target of rapamycin (TOR) kinase pathway (60). The retrograde response is potentiated by *RAS2* (59), yet another example of crosstalk between pathways that modulate metabolism. It should be stressed, however, that save for one limited study (101) the ramifications of the retrograde response at the transcriptional level have not been examined to determine their impact on cell biochemistry and physiology.

The enhancement of longevity by the retrograde response occurs in a variety of yeast strains, in which the conditions under which the retrograde response is induced vary. It has been shown to require *RTG2* and *RAS2* (*RTG1*, *RTG3*, and *LST8* have not yet been tested). The retrograde response is under glucose repression, and in some strains it appears to be constitutively activated, such that when glucose repression is relieved the response is induced and life span is extended. This may have potential consequences for the interpretation of experiments in which a reduction in glucose levels is found to increase life span.

The retrograde response does not appear to act as a simple on-off switch, but instead resembles more a rheostat that ratchets up the gene expression and associated life extension in response to the severity of the mitochondrial dysfunction (44). Mitochondrial dysfunction accumulates with age in mammals (97). It has recently been shown that this is also the case in yeast (61a). This loss of mitochondrial function with age may be due to the increase in oxidative stress with age (63). Thus, the retrograde response may have a function during normal aging to compensate for accumulating mitochondrial dysfunction, allowing cells to live as long as they do.

Caloric Restriction

For 70 years, caloric restriction, also called dietary or food restriction, has been known to extend mean and maximum life span and to postpone the manifestations of aging in rodents (78). Its effects are proportional to the extent to which dietary intake is curtailed, up to the point at which malnutrition sets in. The beneficial consequences of this regimen do not depend on the specific nutrient that is manipulated, prompting the notion that it is the calories consumed that are important. Caloric restriction in rodents reverses many of the senescent changes in gene expression in the various organs examined (32, 54, 64, 65). These changes do not necessarily correspond to those observed in older humans (114). One of the earliest events that occurs when caloric intake is lowered is a reduction in blood glucose levels (12), and it has been postulated that caloric restriction depends on a change in the way that glucose is metabolized (78).

A phenomenon resembling caloric restriction has recently been described in yeast (49). For simplicity, it will be called caloric restriction here, although there is no evidence that it is a reduction in the number of calories consumed that is relevant. Caloric restriction in yeast possesses all of the hallmarks of the mammalian phenomenon. Mean and maximum replicative life span are extended in proportion to the reduction in available nutrients in the growth medium, and this can be elicited by lowering either glucose or amino acid concentrations. Importantly, the increase in generation time, the interval between consecutive buddings, which normally occurs as yeasts age, is retarded. Thus, not only is life span increased but also aging is postponed.

Genetic evidence indicates that there is no overlap between the retrograde response and caloric restriction pathways (49). This conclusion was reached based on experiments in which the combined effects on longevity of nutrient manipulation and deletions of *RTG2* and *RTG3* were examined. This is significant because both retrograde response and caloric restriction represent metabolic mechanisms. Indeed, the same experiments suggest that, although the pathways are independent, there may be some interaction at the level of the longevity effectors that are under their control. Molecular analyses directed toward the identification of the caloric restriction effectors appear to rule out increased *CIT2* mRNA expression, which is a diagnostic of the retrograde response and of the glyoxylate pathway (49). However, it is important to keep in mind that posttranscriptional effects could be operating during caloric restriction.

Another study of caloric restriction in yeast has directly addressed signaling events (69). The negative impact on replicative life span of adenylate cyclase and its target protein kinase A was demonstrated, confirming the results of earlier studies (105). Ras2p stimulates adenylate cyclase in yeast, thus implicating the Ras-cAMP pathway. The life-extending effect of mutations in *CDC25*, which encodes a Ras2p-stimulatory factor, supports this association (69). However, *RAS2* functions in both cAMP-dependent and -independent pathways that affect longevity (105), and adenylate cyclase is the target of several signal transduction pathways (108). Epistasis studies were not carried out to resolve these issues. The connection between the Ras-cAMP pathway and caloric restriction was drawn from the presumed role of this pathway in glucose signaling (69). However, the involvement of *RAS2* in glucose signaling is questionable (108).

A more plausible candidate for the glucose signaling pathway in caloric restriction includes the glucose sensor Gpr1p and its target Gpa2p, based on the extended life span of mutants in the respective genes (69). Indeed, the life extension observed upon lowering glucose levels (69) may well operate through the Gpr1p/Gpa2p pathway. However, the combination of glucose reduction and manipulation of *GPR1* and/or *GPA2* was not explored in this study to confirm the association. It is possible that more than one glucose signaling pathway is involved in the caloric restriction effect, and it will be important to establish their relative contributions to determining life span. Because only the effect of a single, modest reduction in glucose levels was examined (69), it is possible that the observed effects on life span were the result of elimination of glucose repression. The indication that release from glucose repression may be the operative mechanism is suggested by the life extension observed by deletion of *HXK2* (69), which is known to be required for glucose repression (51). The retrograde response can elicit its life-extending effect on life span in some yeast strains simply through the release from glucose repression (59). It would be important to examine the effects of deletions of *RTG* genes on the life span effects observed, as in other studies of caloric restriction in yeast (49). Overall, the analysis of all the effects on life span of the glucose reduction and the various genetic manipulations (69) would require the use of a common genetic background to be congruent.

Despite the qualifications, the *GPR1/GPA2* pathway presents an attractive candidate for the glucose signaling pathway in yeast caloric restriction (Figure 1). This, however, still leaves open the identity of a putative amino acid signaling pathway, whose operation in caloric restriction is indicated by the life-extending effect of lowering amino acid concentrations in the growth medium (49). There is no direct data available to guide our speculation. However, the interaction of caloric restriction and the retrograde response at the level of some of the longevity effectors (49) raises interesting possibilities in the form of the TOR pathway. This pathway is also involved in nutrient responses, including nitrogen source, and it negatively regulates the nuclear accumulation of the Rtg1p-Rtg3p transcription factor (60). It is unclear at present whether caloric restriction in yeast exerts its effects through glucose and amino acid signaling and/or through changes in metabolism that pertain when the levels of these nutrients drop. In more general terms, caloric restriction may operate either through the positive actions of nutrient reduction or through the elimination of the negative effects of excess nutrients.

An attractive mechanism to explain the effects of caloric restriction on life span has been proposed recently (69). In this mechanism, the yeast cell would perceive its metabolic status through NAD levels and translate this to replicative life span through the *SIR2* NAD-dependent histone deacetylase. Clearly, deletion of *SIR2* shortens life span (36, 57). Deletion of *SIR2* prevents any life extension by mutation of *CDC25* (69). This latter effect has been taken as evidence that caloric restriction operates through *SIR2* (69). The connection with caloric restriction is tenuous because manipulation of *CDC25* was used as a surrogate for glucose reduction. *CDC25* operates upstream of *RAS2*, a known longevity gene in yeast (105). *RAS2* impacts life span through several pathways and processes [see (42) for review]. One of these pathways affects chromatin-dependent gene silencing (46), in which *SIR2* participates.

The role of *SIR2* as a master regulator of longevity, although attractive, requires additional justification. It has been proposed that caloric restriction by slowing metabolism enhances NAD levels. This, in turn, would increase the activity of the Sir2p histone deacetylase, preventing the inappropriate expression of genes and thus aging. Such a mechanism has its attractions (37, 39). However, it is not likely to operate as a normal cause of aging because NAD levels increase during the yeast replicative life span (1). Most importantly, deletion of *SIR2* does not suppress the increase in life span observed on reduction of glucose levels but only has an additive effect, indicating that caloric restriction and *SIR2* operate in separate pathways (50a).

Nutrient limitation in some yeast strains results in filamentous growth under certain conditions. Cells become elongated and they bud in a unipolar fashion providing for directional growth. In diploid strains this habit is termed pseudohyphal growth (30), and it is elicited by nitrogen starvation (75). In haploids it is known as invasive growth (75), and it is triggered by glucose deprivation (16). There are several similarities between filamentous growth and replicative life span. In each case, the number of cell divisions before growth ceases is at issue. Importantly, filamentous growth and replicative life span are potentiated by *RAS2*, acting in a cAMP-independent pathway (13, 30, 105), which prompts their juxtaposition (38). *RAS2* is involved in life-extension by the retrograde response (59). It will be interesting to see whether it modulates the caloric restriction effect on longevity. No evidence for filamentous growth was detected in the caloric restriction (49) or retrograde response (59) studies, indicating that these cellular responses are not equivalent to filamentous growth. Interestingly, the *GPR1/GPA2* pathway is required for filamentous growth (73).

The retrograde response appears to compensate for the mitochondrial dysfunction that accrues with aging (Figure 1). It has been proposed that caloric restriction affects many of the same cellular functions. However, unlike the retrograde response, caloric restriction may prevent damage and dysfunction, delaying the onset of aging and ameliorating its symptoms (49).

Snf1 Protein Kinase and Yeast Aging

Snf1p is required for the de-repression of glucose-repressed genes upon depletion of glucose in the growth medium, which allows the cell to utilize alternative carbon sources [for review see (51)]. The nature of the proximal signals that regulate Snf1p is not known. An attractive candidate is AMP or the AMP:ATP ratio. This is based on the similarity of the subunits of the Snf1 kinase (Snf1, Snf4, and Sip proteins) to the mammalian AMP-activated protein kinase, interpreted to be a fuel gauge. However, Snf1p is not directly activated by AMP.

The Sip2p, a component of the Snf1 protein kinase complex thought to be involved in targeting of this enzyme to its substrates, is required for yeast cells to have a life span of normal length (1). Elimination of Sip2p results in the hyperactivation of Snf1 protein kinase. It also reduces resistance to nutrient deprivation. The life span–shortening effect of deletion of *SIP2* was suppressed by deletion of *SNF4*, which encodes an activator of Snf1p. The results prompt the notion that inappropriate activation of certain pathways normally activated during glucose deprivation may be deleterious. There are at least two plausible candidates that may be activated by caloric restriction (Figure 1), one of which is the Snf1p pathway (1) and the other is the Gpr1p/Gpa2p pathway (69).

Energy Metabolism and Cellular Stress

Respiration brings with it oxidative stress, a consequence of leakage of electrons passing through the electron transport chain in the mitochondrion. As glucose levels drop in the growth medium, yeasts shift from fermentation to respiration to meet their energy needs (29). This is the situation yeasts encounter in stationary phase (115). Under these conditions, stress resistance is at a premium. Thus, it is not surprising that null mutations in *SOD1* and *SOD2*, encoding superoxide dismutases, shorten the chronological life span of yeasts in stationary phase (72). Given the metabolic continuum mentioned earlier, it is reassuring that the same pertains to replicative life span (3, 113).

The yeast chronological aging model system has been used to demonstrate the enhancement of survival by mutations in *CYR1*, which encodes adenylate cyclase (26). cAMP activates protein kinase A, which in turn blocks transcription of stress response genes by inhibiting the Msn2p-Msn4p transcription factor (77). The effect of the *cyr1* mutants on chronological aging appeared to be mediated through the Msn2p-Msn4p transcription factor because deletion of *MSN2* or *MSN4* reversed the effects of the mutation in *CYR1* (26). This supports the importance of stress resistance in chronological aging. Curiously, the deletion of *MSN2* and *MSN4* did not affect the survival of cells possessing an intact *CYR1* gene, indicating that in a normal wild-type cell the *MSN2/MSN4* pathway does not play a role in survival in stationary phase. These results might also indicate that the *cyr1* mutation simply results in a perturbation and that adenylate cyclase plays no role in normal aging.

The same study also showed that deletion of *SCH9*, encoding a protein kinase B homolog, improved survival in stationary phase (26). This effect was only partially suppressed by deletions of *MSN2* and *MSN4*, suggesting that not only stress responses are involved. The Sch9 protein kinase impinges on the protein kinase A signaling cascade downstream of the latter because some of the defects associated with the loss of protein kinase A activity can be suppressed by overexpression of *SCH9* (110), indicating it can partially substitute for protein kinase A. *SCH9*, like *RAS2*, modulates metabolic pathways, providing support for the role of metabolism in chronological aging. Also in common with the latter, the nature of the upstream signals that regulate Sch9p activity are not clear (108). Unlike Ras2p, knowledge of the downstream targets of Sch9p is scant. *SCH9*, in contrast to *RAS2*, is not involved in filamentous growth (73).

Longevity Effectors of Metabolic Pathways

Given the numerous changes in gene expression upon induction of the retrograde response, it would seem difficult to determine which are causal for life extension. The diagnostic gene for this response, *CIT2*, is not required (59). Lack of *CIT2* may be accommodated in various ways to allow extension of life span. Indeed, a response as complex and robust as this is likely to have a good deal of inherent flexibility, especially because it is a way in which the cell makes the best of a bad situation. It seems logical that the effect on life span should be multifactorial and combinatorial. Assignment of cause and effect in such a situation is indeed a complicated undertaking that may require more than simple deletion or overexpression studies.

The situation is not likely to be easier in the case of caloric restriction. If the many changes in gene expression that occur in mammalian systems are to be taken as a guide, caloric restriction in yeast will present a complex picture, which is yet to be examined. A similar situation pertains to genetic manipulations that affect global profiles of chromatin-dependent transcriptional silencing. The histone deacetylase genes *RPD3*, *HDA1*, and *SIR2* play a role in determining yeast replicative life span (36, 57). Deletions in these genes result in changes in expression of large suites of genes (8, 35, 116), which overlap partially. Combination of deletions in these genes with caloric restriction provided evidence for the existence of common longevity effectors for caloric restriction and the genes *RPD3* and *HDA1* (50a). The longevity effector genes appear to reside in pathways of energy metabolism. The outcome for *SIR2* was more complicated. The effects on life span of glucose reduction and deletion of this gene were purely additive. However, deletion of *SIR2* partially suppressed the life extension observed upon lowering amino acid concentrations, suggesting partial overlap between the effects of the two manipulations. In fact, *SIR2* deletion increases expression of many genes involved in the synthesis of amino acids. This could counter the reduction in amino acids available in the growth medium and thus prevent the life extension afforded by this nutritional treatment.

CELLULAR STRESS IS A DETERMINANT OF YEAST LIFE SPAN

The Janus Face of *RAS*

Yeasts are exposed to a variety of stresses in their environment. Thus, it is not surprising that resistance to stress affects the replicative life span. However, the effect of stress on life span depends on the nature of the stress and on its timing. The role of resistance to oxidative stress has already been discussed. Resistance to ultraviolet radiation increases with age, reaching a zenith in mid-life (53). This coincides with the peak in *RAS2* expression during the replicative life span (53, 105). *RAS2* is required for the response to ultraviolet radiation (24).

The impact of heat stress on yeast replicative life span is complicated (Figure 2). A single bout of sublethal heat stress has little or no effect on longevity. However, chronic episodes of stress markedly diminish it (95). This is exactly the type of situation that may normally pertain in nature. Deletion of *RAS2*, but not *RAS1*, exacerbates the curtailment of life span. This may seem surprising given all that has been said about the suppression of stress responses by *RAS2*. However, examination of the effects of sublethal heat stress on stress response and growth genes provides a satisfying explanation (95). Without *RAS2*, the induction of the former and the repression of the latter linger inordinately, preventing resumption of normal growth. Thus, it appears that it is just as important for the cell to rapidly recover from heat stress as it is to mount a response to it. Chronic stress response appears to be deleterious. This conclusion is supported by the rescue of longevity of yeasts exposed to repeated bouts of sublethal heat stress by overexpression of *RAS2* (95). Furthermore, this action of *RAS2* requires its adenylate

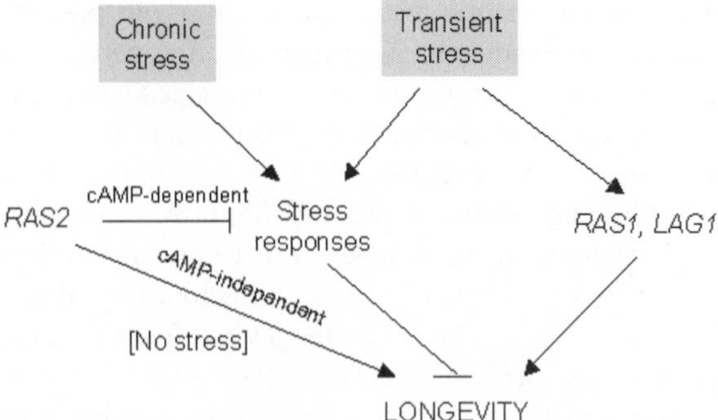

Figure 2 The effects of the *RAS* genes on replicative life span depend on the environmental conditions. *LAG1*, through its role in ceramide synthesis, is proposed to interact with *RAS1* in response to heat stress. For further details, see text.

cyclase-stimulatory activity (95), implying downregulation of stress response gene expression (77).

In contrast to the observations on heat-stressed cells, it is the cAMP-independent pathway that *RAS2* stimulates to increase replicative life span when it is overexpressed in unstressed cells (105). Indeed, deletion of *RAS2* shortens the life span of cells that are not under overt stress. On the other hand, Ras-cAMP pathway activity decreases survival of cells in stationary phase (26, 107). Even under growth conditions lacking stress, however, too much Ras activity shortens replicative life span (13, 105). The distinction is in the amount of Ras-cAMP pathway activity that displays the negative effect. These differences are not surprising. The yeast organism changes with age, and it is not the same in the presence and in the absence of stress. Thus, the effects of Ras2p on longevity would be expected to differ depending on environmental conditions and on whether growth and cell division are possible.

Contrary to the detrimental effects of chronic bouts of sublethal heat stress, transient sublethal heat shock early in the replicative life span results in an increase in longevity (96). This beneficial effect is due to a decrease in mortality rate that persists over several generations but is not permanent. This life extension requires active mitochondria and both *RAS1* and *RAS2*. The requirement for *RAS1* distinguishes this phenomenon from the effects on life span of chronic bouts of heat stress. Presumably, *RAS2* is necessary to downregulate stress response genes and to upregulate genes necessary for growth. However, the requirement for *RAS1* is perplexing because this gene is redundant to *RAS2*.

A clue to the role of *RAS1* in life extension by transient sublethal heat stress may be its role in glucose-stimulated inositol phospholipid turnover in yeast (52). Yeast sphingolipids contain inositol, and they are required for resistance to a variety of stresses, including heat (19). Transient sublethal heat stress may cause a Ras1p-dependent, persistent change in sphingolipid pools (40). Sphingolipids in yeast and in other eukaryotes can give rise to ceramide, which is a signaling molecule important in cell growth, proliferation, differentiation, apoptosis, and senescence (33, 79).

LAG1 at the Nexus of Growth Control and Stress Responses

The first yeast longevity gene cloned was *LAG1* (21). It was identified as a gene that is differentially expressed during the replicative life span. Subsequently, a deletion of the gene was found to extend the replicative life span. Homologs of this gene have been found in higher eukaryotes, including human (11, 50, 66, 86). However, the biochemical function of Lag1p has been elusive until recently. Lag1p has been found to improve the kinetics of transport of glycosylphosphatidylinositol (GPI)-anchored proteins from the endoplasmic reticulum to the Golgi (5). Evidence suggesting that the protein may be involved in sphingolipid metabolism has appeared (11), indicating that the effect on GPI-anchored protein transport may be secondary to this role. Sphingolipids are the source of these anchors in yeast (100).

Lag1p and its homolog Lac1p are essential components of ceramide synthase, which synthesizes ceramide from C26(acyl)-coenzyme A and dihydrosphingosine or phytosphingosine in a reaction that is sensitive to the antibiotic fumonisin B1 (31, 92). The presence of either Lag1p or Lac1p suffices. Absence of these proteins and of normal ceramide synthesis slows growth and renders cells sensitive to elevated temperatures in some genetic backgrounds while making them unviable in others. The involvement of the longevity gene *LAG1* in sphingolipid metabolism (31, 92) and in the response to heat stress [(92); J. Wang & S.M. Jazwinski, unpublished data] provides support for the putative role of sphingolipids in the life extension observed on transient sublethal heat stress.

In yeast, sphingolipid metabolites perform signaling functions in diverse phenomena (Figure 3), such as endocytosis, exocytosis, protein transport from the endoplasmic reticulum to the Golgi, G_1 cell cycle arrest, heat stress responses mediated by the Msn2p-Msn4p transcription factor, and heat stress–mediated degradation of plasma membrane permeases (14, 20, 76, 84, 119). Thus, *LAG1* is well positioned to play a central role in cellular homeostasis by modulating metabolism and stress responses, in similarity to the *RAS* genes. Indeed, there may be an interaction between *LAG1* and *RAS1* in their effects on heat stress responses through sphingolipid metabolites.

In mammalian cells, ceramide directly downregulates phosphatidylinositol-3 kinase (PI3K) activity (120) as well as protein kinase B (93), in addition to affecting other signal transduction pathways (79). Endocytosis in yeast is signaled by sphingolipid metabolites acting on Pkh1p and Pkh2p, which are 3-phosphoinositide-dependent kinases (28). This provides an intersection between sphingolipid metabolism and PI3K cell signaling in yeasts. On the other hand, G_1-arrest of yeast growth by ceramide involves a ceramide-activated protein phosphatase 2A (Figure 3), which has Sit4p as one of its three subunits (84). Sit4p

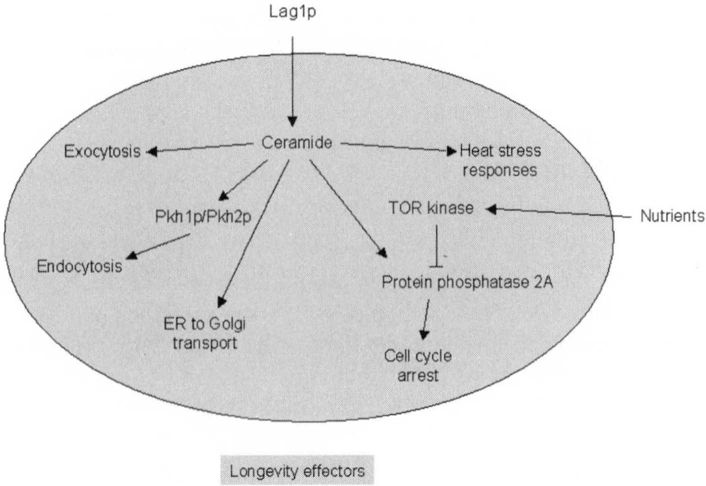

Figure 3 Lag1p modulates a variety of cellular processes. Lag1p (together with Lac1p) is an essential component of ceramide synthase and through the signaling molecule ceramide and other sphingolipid metabolites is proposed to impact the indicated processes, all of which are candidates for the longevity effectors of this protein. ER, endoplasmic reticulum. For additional information, see text.

is one of the mediators of the TOR kinase pathway in the cellular response to nutrients (7). Sphingolipid metabolites also affect nutrient availability directly by stimulating ubiquitin-dependent degradation of plasma membrane permeases during heat stress (14). Thus, it is possible that *LAG1* intersects with other metabolic mechanisms of yeast longevity, such as the retrograde response and caloric restriction. Through effects on nutrient availability, *LAG1* might also affect filamentous growth.

WHY ARE YEASTS BORN YOUNG?

Two fundamental questions must be answered to provide a full explanation of the limited replicative life span of yeasts. The first concerns the causes of aging of the mother cell, such that its life span is limited. This has been the subject of discussion thus far. The second question is directed to the immortality of the yeast strain or clone. It can be rephrased as a query into the mechanisms that guarantee the potential for a full replicative life span to daughter cells.

The issue of age asymmetry between mother and daughter has been addressed directly through the isolation of mutants in which it is disrupted (61a). Temperature-sensitive mutants that show clonal senescence at restrictive temperature were isolated and produced daughter cells of the replicative age of their mothers at the time of birth. One of these was a point mutant in *ATP2*, which encodes the β-subunit of mitochondrial ATP synthase, a protein that is highly conserved throughout phylogeny (10). This mutation did not affect the growth rate even on a nonfermentable carbon source, when oxidative phosphorylation is the source of the bulk of the energy. The *atp2* mutant, however, showed a decline in mitochondrial membrane potential, followed by the appearance of cells with a diminished content of mitochondria (61a). This was coupled to a change in mitochondrial morphology and distribution in the cell. Daughter cells received few or no mitochondria at restrictive temperature. This phenotype explains the clonal senescence of the mutant. Mutants in *EST1* also result in clonal senescence by inducing telomere shortening (74). However, this is not the result of a disruption of age asymmetry. Telomeres do not normally shorten during the yeast replicative life span (22).

Events similar to those that occur during induced clonal senescence in the *atp2* mutant also take place during the course of normal aging in yeast (61a). There is a progressive loss of mitochondrial membrane potential during the replicative life span of wild-type yeast. This could be due to oxidative stress, which increases with age in yeast (63). Furthermore, the older the mother cell is the more likely it is to generate daughter cells that have a low mitochondrial membrane potential (61a). Thus, daughter cells of *atp2*-mutant mothers would also be more likely to receive dysfunctional mitochondria from their mothers as these mothers aged, resulting in loss of age asymmetry and a curtailed life span. Dysfunctional mitochondria could readily constitute the "senescence factor," whose operation during yeast aging was deduced some time ago (23). A senescence factor has been invoked to explain the shortened life spans of daughter cells produced by old yeasts (38, 55).

The studies with the *atp2* mutant indicate the impact of mitochondria on the aging process. More generally, they suggest a requirement of proper partition of cell components and the "filtering" of damaged constituents from daughter cells in the maintenance of heritable age asymmetry. Examples of cellular asymmetry have been described in yeast, especially with respect to the mating-type switch. The filtering referred to here would benefit from a diffusion barrier between mother and daughter, based on plasma membrane compartmentalization as described recently (106). Filtering of damaged mitochondrial DNA, on the other hand, may occur through recombinational mechanisms (85).

It is virtually a given that the maintenance of age asymmetry and the filtering of damaged cell components require communication between subcellular compartments and organelles. It is likely that there are checkpoint controls that monitor these processes. Breakdown in this intracellular communication, or in other words the connectivity, could be an important factor in the biological aging process. The same principles may pertain to the perpetuation of stem cell lineages.

GENE DYSREGULATION

One cause of aging in yeast could be changes in the epigenetic inheritance of different regulatory states of chromatin (37), leading to unscheduled transcription and a state of gene dysregulation (39). The antecedents for such a process are the loss of transcriptional silencing of heterochromatic regions of the yeast genome during the replicative life span (58, 102). Such a loss of silencing may occur during senescence of human cells in culture (6). The loss of silencing of subtelomeric regions during aging is not uniform throughout the yeast genome (58), providing ample opportunity for chance to intervene in the aging process.

The association of silencing with aging received causal underpinnings with the implication of histone deacetylase genes in determining replicative life span. Deletion of the histone deacetylase genes *RPD3* or *HDA1* increases life span, whereas deletion of *SIR2* shortens it (57). These histone deacetylases regulate silencing at the three known heterochromatic regions in the yeast genome, including the silent mating-type locus, subtelomeric regions, and ribosomal DNA (57, 91, 104). However, the life span–extending effects noted correspond most closely to increased silencing at the ribosomal DNA locus (57). Sir2p is an NAD-dependent histone deacetylase (36, 62, 103), which has already been discussed in the context of metabolic mechanisms of aging. It has recently been found to extend the life span of the nematode *Caenorhabditis elegans* when provided as a transgene (109). The conclusion that the effects of Rpd3p and Hda1p on silencing are direct rather than through an effect on transcription of the Sir proteins has received experimental support recently (8).

It has been proposed that one of the ways in which the retrograde response and the histone deacetylase genes affect life span is through their influence on protein metabolism (43, 57). Protein synthesis declines with age in yeast, even though there is an increase in ribosomal RNA content (82). Perhaps there is an imbalance between production of ribosomal RNA and protein, which is at the root of the decrease in protein synthesis with age. Deletion of *RPD3* increases silencing of ribosomal DNA and would restore the balance. Induction of the retrograde response increases synthesis of ribosomal RNA (15), and it may do the same for ribosomal proteins (25). Thus, it would also balance ribosome production.

There are other genes that affect both transcriptional silencing and replicative life span in yeasts, including *ZDS1* and *ZDS2* (90) and *RAS2* (46). There are many genes whose transcription can be affected by these genes and the histone deacetylases discussed above. Just as in the case of the retrograde response and caloric restriction, it will not be easy to determine the factors proximal to life span determination caused by changes in silencing status of chromatin. Indeed, the individual changes may not be important in themselves. The issue may be the loss of homeostasis caused by gene dysregulation, which can be caused by many different combinations of gene expression changes and the secondary effects they may elicit. Such a situation could readily give rise to the individual differences in the aging phenotype. The opposing actions of Rpd3p and Hda1p on the one hand and Sir2p on the other would maintain the overall equilibrium between transcriptionally active and silenced chromatin, a balance that shifts in the direction of loss of silencing with replicative age.

It has been demonstrated that extrachromosomal ribosomal DNA circles (ERCs) accumulate during the yeast replicative life span and that artificial induction of

ERCs can cause cell death (99). ERCs arise from the tandem repeats of ribosomal DNA that are present in 100 to 200 copies in the genome. Changes in chromatin structure at the ribosomal DNA locus, such as would occur with loss of silencing, may facilitate ERC production during aging, a form of genetic instability. Deletion of *FOB1* eliminates ERCs and increases life span (18), which indicates a causal role for ERCs in yeast aging. This conclusion depends on the assumption that *FOB1* affects ERC production alone, which is not proven and not likely to be the case. This difficulty in arriving at causality in aging research is not an isolated one.

The significance of ERCs as a potential factor in human aging was suggested by the life span–shortening effect in yeast of deletion of the *SGS1* gene, which encodes a homolog of the human gene *WRN* (98). *WRN* encodes a DNA helicase, a mutation that results in a premature aging syndrome (118). In one set of studies, deletion of *SGS1* led to ERC appearance (99). However, in another, deletion of *SGS1* did not result in ERC accumulation, and the shortened life span of the *sgs1* mutant was rescued by human *BLM* and not by *WRN* (34). Interestingly, the retrograde response results in massive ERC accumulation (15), but at the same time it extends life span (59). Thus, ERCs are not sufficient to cause aging under all conditions; they are not necessary either (57).

YEAST AGING: A MICROCOSM OF AGING IN OTHER SPECIES

The study of aging in yeast points to four broad principles at play: metabolic capacity, stress resistance, gene dysregulation, and genetic stability. These four principles could already be discerned several years ago in species from yeast to human (39), and this convergence is even more obvious today (56). The extent of the similarity reaches the modulation of longevity by a common signal transduction pathway in *C. elegans* and in *Drosophila melanogaster* (Figure 4), certain components of which can also be seen at work in yeast and in mouse (56). A striking feature of the physiological responses whose operation is evident in life span determination across species is their essential function in dispersal forms, such as spores, and in the seasonal dormancy of diapause (39). The marshaling of metabolic resources and resistance to stress that this portends may be conducive to a long adult life span.

The significance of intracellular communication in determining life span in yeast has already been pointed out. In the retrograde response, this communication involves at least three organelles: mitochondria, peroxisomes, and the nucleus. The traces of several signal transduction pathways that impinge on longevity can be recognized in yeast. Certain key genes, such as *RAS2* or *LAG1*, modulate these pathways but do not appear to be integral components. This allows these genes to exert their effects on multiple pathways and processes and thus to function as homeostats. The connectivity of this elaborate circuitry determines life span.

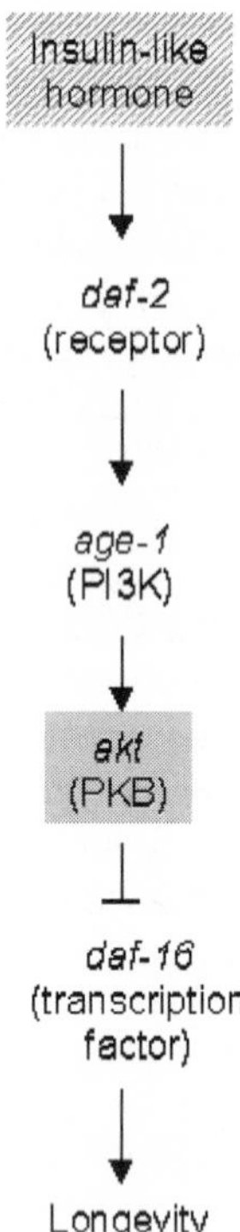

Figure 4 A signal transduction pathway for extended longevity in *C. elegans*. A similar pathway operates in the fruit fly. The *shaded* and *cross-hatched* components of the pathway have counterparts in yeast and in mouse, respectively. PKB, protein kinase B. For more information, see (56).

Changes in this connectivity, even subtle ones, can lead to new set points in the aging system and ultimately to loss of homeostasis (41, 46).

Intracellular communication is likely to play a role in aging in mammals. It is instructive that a mouse model of mitochondrial cardiomyopathy can survive several months and that its life span is determined by genetic background (67). Compensatory mechanisms are at play, and these may be akin to the retrograde response in yeast. The components of a retrograde response may also help determine extreme longevity in humans (9, 17). Thus, the yeast aging system serves as a facile means of generating hypotheses for testing in higher eukaryotes.

ACKNOWLEDGMENTS

Research in the author's laboratory is supported by grants from the National Institute on Aging of the National Institutes of Health (U.S.P.H.S.).

The *Annual Review of Microbiology* is online at http://micro.annualreviews.org

LITERATURE CITED

1. Ashrafi K, Lin SS, Manchester JK, Gordon JI. 2000. Sip2p and its partner snf1p kinase affect aging in *S. cerevisiae*. *Genes Dev.* 14:1872–85
2. Ashrafi K, Sinclair D, Gordon JI, Guarente L. 1999. Passage through stationary phase advances replicative aging in *Saccharomyces cerevisiae*. *Proc. Natl. Acad. Sci. USA* 96:9100–5
3. Barker MG, Brimage LJ, Smart KA. 1999. Effect of Cu, Zn superoxide dismutase disruption mutation on replicative senescence in *Saccharomyces cerevisiae*. *FEMS Microbiol. Lett.* 177:199–204
4. Barker MG, Walmsley RM. 1999. Replicative ageing in the fission yeast *Schizosaccharomyces pombe*. *Yeast* 15:1511–18
5. Barz WP, Walter P. 1999. Two endoplasmic reticulum (ER) membrane proteins that facilitate ER-to-Golgi transport of glycosylphosphatidylinositol-anchored proteins. *Mol. Biol. Cell* 10:1043–59
6. Baur JA, Zou Y, Shay JW, Wright WE. 2001. Telomere position effect in human cells. *Science* 292:2075–77
7. Beck T, Hall MN. 1999. The TOR signalling pathway controls nuclear localization of nutrient-regulated transcription factors. *Nature* 402:689–92
8. Bernstein BE, Tong JK, Schreiber SL. 2000. Genomewide studies of histone deacetylase function in yeast. *Proc. Natl. Acad. Sci. USA* 97:13708–13
9. Bonafè M, Barbi C, Olivieri F, Yashin A, Andreev KF, et al. 2002. An allele of *HRAS1* 3′ variable number of tandem repeats is a frailty allele: implication for an evolutionarily-conserved pathway involved in longevity. *Gene* 286:121–26
10. Boyer PD. 1997. The ATP synthase—a splendid molecular machine. *Annu. Rev. Biochem.* 66:717–49
11. Brandwagt BF, Mesbah LA, Takken FL, Laurent PL, Kneppers TJ, et al. 2000. A longevity assurance gene homolog of tomato mediates resistance to *Alternaria alternata* f. sp. *lycopersici* toxins and fumonisin B1. *Proc. Natl. Acad. Sci. USA* 97:4961–66
12. Cartee GD, Dean DJ. 1994. Glucose transport with brief dietary restriction: heterogenous responses in muscles. *Am. J. Physiol.* 266:E946–52
13. Chen JB, Sun J, Jazwinski SM. 1990. Prolongation of the yeast life span by the v-Ha-*RAS* oncogene. *Mol. Microbiol.* 4:2081–86
14. Chung N, Jenkins G, Hannun YA, Heitman J, Obeid LM. 2000. Sphingolipids signal heat stress-induced ubiquitin-dependent proteolysis. *J. Biol. Chem.* 275:17229–32
15. Conrad-Webb H, Butow RA. 1995. A polymerase switch in the synthesis of rRNA in *Saccharomyces cerevisiae*. *Mol. Cell Biol.* 15:2420–28
16. Cullen PJ, Sprague GF Jr. 2000. Glucose depletion causes haploid invasive growth in yeast. *Proc. Natl. Acad. Sci. USA* 97:13619–24
17. De Benedictis G, Carrieri G, Garasto S, Rose G, Varcasia O, et al. 2000. Does a retrograde response in human aging and longevity exist? *Exp. Gerontol.* 35:795–801
18. Defossez PA, Prusty R, Kaeberlein M, Lin SJ, Ferrigno P, et al. 1999. Elimination of replication block protein Fob1 extends the life span of yeast mother cells. *Mol. Cell* 3:447–55
19. Dickson RC. 1998. Sphingolipid functions in *Saccharomyces cerevisiae*: comparison to mammals. *Annu. Rev. Biochem.* 67:27–48
20. Dickson RC, Nagiec EE, Skrzypek M, Tillman P, Wells GB, Lester RL. 1997. Sphingolipids are potential heat stress

signals in *Saccharomyces. J. Biol. Chem.* 272:30196–200

21. D'mello NP, Childress AM, Franklin DS, Kale SP, Pinswasdi C, Jazwinski SM. 1994. Cloning and characterization of *LAG1*, a longevity-assurance gene in yeast. *J. Biol. Chem.* 269:15451–59
22. D'mello NP, Jazwinski SM. 1991. Telomere length constancy during aging of *Saccharomyces cerevisiae. J. Bacteriol.* 173:6709–13
23. Egilmez NK, Jazwinski SM. 1989. Evidence for the involvement of a cytoplasmic factor in the aging of the yeast *Saccharomyces cerevisiae. J. Bacteriol.* 171: 37–42
24. Engelberg D, Klein C, Martinetto H, Struhl K, Karin M. 1994. The UV response involving the Ras signaling pathway and AP-1 transcription factors is conserved between yeast and mammals. *Cell* 77:381–90
25. Epstein CB, Waddle JA, Hale WT, Dave V, Thornton J, et al. 2001. Genome-wide responses to mitochondrial dysfunction. *Mol. Biol. Cell* 12:297–308
26. Fabrizio P, Pozza F, Pletcher SD, Gendron CM, Longo VD. 2001. Regulation of longevity and stress resistance by Sch9 in yeast. *Science* 292:288–90
27. Finch CE. 1990. *Longevity, Senescence, and the Genome*. Chicago: Univ. Chicago Press
28. Friant S, Lombardi R, Schmelzle T, Hall MN, Riezman H. 2001. Sphingoid base signaling via Pkh kinases is required for endocytosis in yeast. *EMBO J.* 20:6783–92
29. Gancedo C, Serrano R. 1989. Energy-yielding metabolism. In *The Yeasts*, ed. AH Rose, JS Harrison, pp. 205–59. San Diego: Academic
30. Gimeno CJ, Ljungdahl PO, Styles CA, Fink GR. 1992. Unipolar cell divisions in the yeast *S. cerevisiae* lead to filamentous growth: regulation by starvation and *RAS*. *Cell* 68:1077–90
31. Guillas I, Kirchman PA, Chuard R, Pfefferli M, Jiang JC, et al. 2001. C26-CoA-dependent ceramide synthesis of *Saccharomyces cerevisiae* is operated by Lag1p and Lac1p. *EMBO J.* 20:2655–65
32. Han E, Hilsenbeck SG, Richardson A, Nelson JF. 2000. cDNA expression arrays reveal incomplete reversal of age-related changes in gene expression by calorie restriction. *Mech. Ageing Dev.* 115:157–74
33. Hannun YA, Luberto C. 2000. Ceramide in the eukaryotic stress response. *Trends Cell Biol.* 10:73–80
34. Heo SJ, Tatebayashi K, Ohsugi I, Shimamoto A, Furuichi Y, Ikeda H. 1999. Bloom's syndrome gene suppresses premature ageing caused by Sgs1 deficiency in yeast. *Genes Cells* 4:619–25
35. Hughes TR, Marton MJ, Jones AR, Roberts CJ, Stoughton R, et al. 2000. Functional discovery via a compendium of expression profiles. *Cell* 102:109–26
36. Imai S, Armstrong CM, Kaeberlein M, Guarente L. 2000. Transcriptional silencing and longevity protein Sir2 is an NAD-dependent histone deacetylase. *Nature* 403:795–800
37. Jazwinski SM. 1990. Aging and senescence of the budding yeast *Saccharomyces cerevisiae. Mol. Microbiol.* 4: 337–43
38. Jazwinski SM. 1993. The genetics of aging in the yeast *Saccharomyces cerevisiae. Genetica* 91:35–51
39. Jazwinski SM. 1996. Longevity, genes, and aging. *Science* 273:54–59
40. Jazwinski SM. 1999. Molecular mechanisms of yeast longevity. *Trends Microbiol.* 7:247–52
41. Jazwinski SM. 1999. Nonlinearity of the aging process revealed in studies with yeast. In *Molecular Biology of Aging*, ed. VA Bohr, BFC Clark, T Stevnsner, pp. 35–44. Copenhagen: Munksgaard
42. Jazwinski SM. 1999. The RAS genes: a homeostatic device in *Saccharomyces cerevisiae* longevity. *Neurobiol. Aging* 20:471–78

43. Jazwinski SM. 2000. Metabolic control and ageing. *Trends Genet.* 16:506–11
44. Jazwinski SM. 2000. Metabolic control and gene dysregulation in yeast aging. *Ann. NY Acad. Sci.* 908:21–30
45. Jazwinski SM. 2001. New clues to old yeast. *Mech. Ageing Dev.* 122:865–82
46. Jazwinski SM, Kim S, Lai CY, Benguria A. 1998. Epigenetic stratification: the role of individual change in the biological aging process. *Exp. Gerontol.* 33: 571–80
47. Jazwinski SM, Wawryn J. 2001. Profiles of random change during aging contain hidden information about longevity and the aging process. *J. Theor. Biol.* 213:599–608
48. Jia Y, Rothermel B, Thornton J, Butow RA. 1997. A basic helix-loop-helix-leucine zipper transcription complex in yeast functions in a signaling pathway from mitochondria to the nucleus. *Mol. Cell Biol.* 17:1110–17
49. Jiang JC, Jaruga E, Repnevskaya MV, Jazwinski SM. 2000. An intervention resembling caloric restriction prolongs life span and retards aging in yeast. *FASEB J.* 14:2135–37
50. Jiang JC, Kirchman PA, Zagulski M, Hunt J, Jazwinski SM. 1998. Homologs of the yeast longevity gene *LAG1* in *Caenorhabditis elegans* and human. *Genome Res.* 8: 1259–72
50a. Jiang JC, Wawryn J, Shantha Kumara HMC, Jazwinski SM. 2002. Distinct roles of processes modulated by histone deacetylases Rpd3p, Hda1p, and Sir2p in life extension by caloric restriction in yeast. *Exp. Gerontol.* In press
51. Johnston M. 1999. Feasting, fasting and fermenting. Glucose sensing in yeast and other cells. *Trends Genet.* 15:29–33
52. Kaibuchi K, Miyajima A, Arai K, Matsumoto K. 1986. Possible involvement of *RAS*-encoded proteins in glucose-induced inositolphospholipid turnover in *Saccharomyces cerevisiae*. *Proc. Natl. Acad. Sci. USA* 83:8172–76
53. Kale SP, Jazwinski SM. 1996. Differential response to UV stress and DNA damage during the yeast replicative life span. *Dev. Genet.* 18:154–60
54. Kayo T, Allison DB, Weindruch R, Prolla TA. 2001. Influences of aging and caloric restriction on the transcriptional profile of skeletal muscle from rhesus monkeys. *Proc. Natl. Acad. Sci. USA* 98:5093–98
55. Kennedy BK, Austriaco NR Jr, Guarente L. 1994. Daughter cells of *Saccharomyces cerevisiae* from old mothers display a reduced life span. *J. Cell Biol.* 127:1985–93
56. Kenyon C. 2001. A conserved regulatory system for aging. *Cell* 105:165–68
57. Kim S, Benguria A, Lai C-Y, Jazwinski SM. 1999. Modulation of life-span by histone deacetylase genes in *Saccharomyces cerevisiae*. *Mol. Biol. Cell* 10:3125–36
58. Kim S, Villeponteau B, Jazwinski SM. 1996. Effect of replicative age on transcriptional silencing near telomeres in *Saccharomyces cerevisiae*. *Biochem. Biophys. Res. Commun.* 219:370–76
59. Kirchman PA, Kim S, Lai C-Y, Jazwinski SM. 1999. Interorganelle signaling is a determinant of longevity in *Saccharomyces cerevisiae*. *Genetics* 152:179–90
60. Komeili A, Wedaman KP, O'Shea EK, Powers T. 2000. Mechanism of metabolic control. Target of rapamycin signaling links nitrogen quality to the activity of the Rtg1 and Rtg3 transcription factors. *J. Cell Biol.* 151:863–78
61. Lagunas R, Ruiz E. 1988. Balance of production and consumption of ATP in ammonium-starved *Saccharomyces cerevisiae*. *J. Gen. Microbiol.* 134(Pt. 9): 2507–11
61a. Lai C-Y, Jaruga E, Borghouts C, Jazwinski SM. 2002. A mutation in the *ATP2* gene abrogates the age symmetry between mother and daughter cells of the yeast *Saccharomyces cerevisiae*. *Genetics*. In press
62. Landry J, Sutton A, Tafrov ST, Heller RC, Stebbins J, et al. 2000. The silencing protein SIR2 and its homologs

are NAD-dependent protein deacetylases. *Proc. Natl. Acad. Sci. USA* 97:5807–11
63. Laun P, Pichova A, Madeo F, Fuchs J, Ellinger A, et al. 2001. Aged mother cells of *Saccharomyces cerevisiae* show markers of oxidative stress and apoptosis. *Mol. Microbiol.* 39:1166–73
64. Lee CK, Klopp RG, Weindruch R, Prolla TA. 1999. Gene expression profile of aging and its retardation by caloric restriction. *Science* 285:1390–93
65. Lee CK, Weindruch R, Prolla TA. 2000. Gene-expression profile of the ageing brain in mice. *Nat. Genet.* 25:294–97
66. Lee SJ. 1991. Expression of growth/differentiation factor 1 in the nervous system: conservation of a bicistronic structure. *Proc. Natl. Acad. Sci. USA* 88:4250–54
67. Li H, Wang J, Wilhelmsson H, Hansson A, Thoren P, et al. 2000. Genetic modification of survival in tissue-specific knockout mice with mitochondrial cardiomyopathy. *Proc. Natl. Acad. Sci. USA* 97:3467–72
68. Liao X, Butow RA. 1993. *RTG1* and *RTG2*: two yeast genes required for a novel path of communication from mitochondria to the nucleus. *Cell* 72:61–71
69. Lin SJ, Defossez PA, Guarente L. 2000. Requirement of NAD and *SIR2* for life-span extension by calorie restriction in *Saccharomyces cerevisiae*. *Science* 289: 2126–28
70. Liu Z, Butow RA. 1999. A transcriptional switch in the expression of yeast tricarboxylic acid cycle genes in response to a reduction or loss of respiratory function. *Mol. Cell Biol.* 19:6720–28
71. Liu Z, Sekito T, Epstein CB, Butow RA. 2001. *RTG*-dependent mitochondria to nucleus signaling is negatively regulated by the seven WD-repeat protein Lst8p. *EMBO J.* 20:7209–19
72. Longo VD. 1999. Mutations in signal transduction proteins increase stress resistance and longevity in yeast, nematodes, fruit flies, and mammalian neuronal cells. *Neurobiol. Aging* 20:479–86
73. Lorenz MC, Pan X, Harashima T, Cardenas ME, Xue Y, et al. 2000. The G protein-coupled receptor gpr1 is a nutrient sensor that regulates pseudohyphal differentiation in *Saccharomyces cerevisiae*. *Genetics* 154:609–22
74. Lundblad V, Szostak JW. 1989. A mutant with a defect in telomere elongation leads to senescence in yeast. *Cell* 57:633–43
75. Madhani HD, Fink GR. 1998. The control of filamentous differentiation and virulence in fungi. *Trends Cell Biol.* 8:348–53
76. Marash M, Gerst JE. 2001. t-SNARE dephosphorylation promotes SNARE assembly and exocytosis in yeast. *EMBO J.* 20:411–21
77. Marchler G, Schuller C, Adam G, Ruis H. 1993. A *Saccharomyces cerevisiae* UAS element controlled by protein kinase A activates transcription in response to a variety of stress conditions. *EMBO J.* 12:1997–2003
78. Masoro EJ. 1995. Dietary restriction. *Exp. Gerontol.* 30:291–98
79. Mathias S, Pena LA, Kolesnick RN. 1998. Signal transduction of stress via ceramide. *Biochem. J.* 335(Pt. 3):465–80
80. Miller JJ. 1989. Sporulation in *Saccharomyces cerevisiae*. In *The Yeasts*, ed. AH Rose, JS Harrison, pp. 489–550. San Diego: Academic
81. Mortimer RK, Johnston JR. 1959. Life span of individual yeast cells. *Nature* 183:1751–52
82. Motizuki M, Tsurugi K. 1992. The effect of aging on protein synthesis in the yeast *Saccharomyces cerevisiae*. *Mech. Ageing Dev.* 64:235–45
83. Müller I, Zimmermann M, Becker D, Flömer M. 1980. Calendar life span versus budding life span of *Saccharomyces cerevisiae*. *Mech. Ageing Dev.* 12:47–52
84. Nickels JT, Broach JR. 1996. A ceramide-activated protein phosphatase mediates ceramide-induced G1 arrest of *Saccharomyces cerevisiae*. *Genes Dev.* 10:382–94

85. Okamoto K, Perlman PS, Butow RA. 1998. The sorting of mitochondrial DNA and mitochondrial proteins in zygotes: preferential transmission of mitochondrial DNA to the medial bud. *J. Cell Biol.* 142:613–23
86. Pan H, Qin WX, Huo KK, Wan DF, Yu Y, et al. 2001. Cloning, mapping, and characterization of a human homologue of the yeast longevity assurance gene *LAG1*. *Genomics* 77:58–64
87. Parikh VS, Morgan MM, Scott R, Clements LS, Butow RA. 1987. The mitochondrial genotype can influence nuclear gene expression in yeast. *Science* 235:576–80
88. Parry JM, Davies PJ, Evans WE. 1976. The effects of "cell age" upon the lethal effects of physical and chemical mutagens in the yeast, *Saccharomyces cerevisiae*. *Mol. Gen. Genet.* 146:27–35
89. Pierce MM, Maddelein ML, Roberts BT, Wickner RB. 2001. A novel Rtg2p activity regulates nitrogen catabolism in yeast. *Proc. Natl. Acad. Sci. USA* 98:13213–18
90. Roy N, Runge KW. 2000. Two paralogs involved in transcriptional silencing that antagonistically control yeast life span. *Curr. Biol.* 10:111–14
91. Rundlett SE, Carmen AA, Kobayashi R, Bavykin S, Turner BM, Grunstein M. 1996. *HDA1* and *RPD3* are members of distinct yeast histone deacetylase complexes that regulate silencing and transcription. *Proc. Natl. Acad. Sci. USA* 93:14503–8
92. Schorling S, Vallee B, Barz WP, Riezman H, Oesterhelt D. 2001. Lag1p and Lac1p are essential for the Acyl-CoA-dependent ceramide synthase reaction in *Saccharomyces cerevisae*. *Mol. Biol. Cell* 12:3417–27
93. Schubert KM, Scheid MP, Duronio V. 2000. Ceramide inhibits protein kinase B/Akt by promoting dephosphorylation of serine 473. *J. Biol. Chem.* 275:13330–35
94. Sekito T, Thornton J, Butow RA. 2000. Mitochondria-to-nuclear signaling is regulated by the subcellular localization of the transcription factors Rtg1p and Rtg3p. *Mol. Biol. Cell* 11:2103–15
95. Shama S, Kirchman PA, Jiang JC, Jazwinski SM. 1998. Role of *RAS2* in recovery from chronic stress: effect on yeast life span. *Exp. Cell Res.* 245:368–78
96. Shama S, Lai C-Y, Antoniazzi JM, Jiang JC, Jazwinski SM. 1998. Heat stress-induced life span extension in yeast. *Exp. Cell Res.* 245:379–88
97. Shigenaga MK, Hagen TM, Ames BN. 1994. Oxidative damage and mitochondrial decay in aging. *Proc. Natl. Acad. Sci. USA* 91:10771–78
98. Sinclair DA, Mills K, Guarente L. 1997. Accelerated aging and nucleolar fragmentation in yeast *sgs1* mutants. *Science* 277:1313–16
99. Sinclair DA, Mills K, Guarente L. 1998. Molecular mechanisms of yeast aging. *Trends Biochem. Sci.* 23:131–34
100. Skrzypek M, Lester RL, Dickson RC. 1997. Suppressor gene analysis reveals an essential role for sphingolipids in transport of glycosylphosphatidylinositol-anchored proteins in *Saccharomyces cerevisiae*. *J. Bacteriol.* 179:1513–20
101. Small WC, Brodeur RD, Sandor A, Fedorova N, Li G, et al. 1995. Enzymatic and metabolic studies on retrograde regulation mutants of yeast. *Biochemistry* 34:5569–76
102. Smeal T, Claus J, Kennedy B, Cole F, Guarente L. 1996. Loss of transcriptional silencing causes sterility in old mother cells of *S. cerevisiae*. *Cell* 84:633–42
103. Smith JS, Brachmann CB, Celic I, Kenna MA, Muhammad S, et al. 2000. A phylogenetically conserved NAD^+-dependent protein deacetylase activity in the Sir2 protein family. *Proc. Natl. Acad. Sci. USA* 97:6658–63
104. Smith JS, Caputo E, Boeke JD. 1999. A genetic screen for ribosomal DNA silencing defects identifies multiple DNA replication and chromatin-modulating factors. *Mol. Cell Biol.* 19:3184–97

105. Sun J, Kale SP, Childress AM, Pinswasdi C, Jazwinski SM. 1994. Divergent roles of *RAS1* and *RAS2* in yeast longevity. *J. Biol. Chem.* 269:18638–45
106. Takizawa PA, DeRisi JL, Wilhelm JE, Vale RD. 2000. Plasma membrane compartmentalization in yeast by messenger RNA transport and a septin diffusion barrier. *Science* 290:341–44
107. Tatchell K. 1993. RAS genes in the budding yeast *Saccharomyces cerevisiae*. In *Signal Transduction: Prokaryotic and Simple Eukaryotic Systems*, ed. T Kurjan, pp. 147–88. San Diego: Academic
108. Thevelein JM, de Winde JH. 1999. Novel sensing mechanisms and targets for the cAMP-protein kinase A pathway in the yeast *Saccharomyces cerevisiae*. *Mol. Microbiol.* 33:904–18
109. Tissenbaum HA, Guarente L. 2001. Increased dosage of a sir-2 gene extends lifespan in *Caenorhabditis elegans*. *Nature* 410:227–30
110. Toda T, Cameron S, Sass P, Wigler M. 1988. SCH9, a gene of *Saccharomyces cerevisiae* that encodes a protein distinct from, but functionally and structurally related to, cAMP-dependent protein kinase catalytic subunits. *Genes Dev.* 2:517–27
111. Traven A, Wong JM, Xu D, Sopta M, Ingles CJ. 2000. Inter-organellar communication: altered nuclear gene expression profiles in a yeast mitochondrial DNA mutant. *J. Biol. Chem.* 276:4020–27
112. Vaupel JW, Carey JR, Christensen K, Johnson TE, Yashin AI, et al. 1998. Biodemographic trajectories of longevity. *Science* 280:855–60
113. Wawryn J, Krzepilko A, Myszka A, Bilinski T. 1999. Deficiency in superoxide dismutases shortens life span of yeast cells. *Acta Biochim. Pol.* 46:249–53
114. Welle S, Brooks A, Thornton CA. 2001. Senescence-related changes in gene expression in muscle: similarities and differences between mice and men. *Physiol. Genomics* 5:67–73
115. Werner-Washburne M, Braun E, Johnston GC, Singer RA. 1993. Stationary phase in the yeast *Saccharomyces cerevisiae*. *Microbiol. Rev.* 57:383–401
116. Wyrick JJ, Holstege FC, Jennings EG, Causton HC, Shore D, et al. 1999. Chromosomal landscape of nucleosome-dependent gene expression and silencing in yeast. *Nature* 402:418–21
117. Yashin AI, Begun AS, Boiko SI, Ukraintseva SV, Oeppen J. 2001. The new trends in survival improvement require a revision of traditional gerontological concepts. *Exp. Gerontol.* 37:157–67
118. Yu CE, Oshima J, Fu YH, Wijsman EM, Hisama F, et al. 1996. Positional cloning of the Werner's syndrome gene. *Science* 272:258–62
119. Zanolari B, Friant S, Funato K, Sutterlin C, Stevenson BJ, Riezman H. 2000. Sphingoid base synthesis requirement for endocytosis in *Saccharomyces cerevisiae*. *EMBO J.* 19:2824–33
120. Zundel W, Giaccia A. 1998. Inhibition of the anti-apoptotic PI(3)K/Akt/Bad pathway by stress. *Genes Dev.* 12:1941–46

Subject Index

F

H

I

N

S

Cumulative Indexes

CONTRIBUTING AUTHORS, VOLUMES 52–56

CHAPTER TITLES, VOLUMES 52–56

Prefatory Chapters

Animal Pathogens and Diseases

Applied Microbiology and Ecology

Chemotherapy and Chemotherapeutic Agents

Diversity and Systematics

Genetics and Physiology

Immunology

Morphology, Ultrastructure, and Differentiation

Plant-Bacterial Interactions

Virology